DATTORRO

CONVEX OPTIMIZATION & EUCLIDEAN DISTANCE GEOMETRY

Mεβoo

DATTORRO

CONVEX

OPTIMIZATION

&

EUCLIDEAN

DISTANCE

GEOMETRY

MEBOO

Convex Optimization
&
Euclidean Distance Geometry

Jon Dattorro

Meβoo Publishing

Meboo Publishing USA
PO Box 12
Palo Alto, California 94302

Dattorro, *Convex Optimization & Euclidean Distance Geometry,*
$\mathcal{M}\varepsilon\beta oo$, 2005, v2015.02.15.

ISBN 0976401304 (English)   ISBN 9780615193687 (International III)

This is version 2015.02.15: available in print, as conceived, in color.

cybersearch:

    I. convex optimization
   II. semidefinite program
  III. rank constraint
  IV. convex geometry
   V. distance matrix
  VI. convex cones
 VII. cardinality minimization

PROGRAMS BY MATLAB

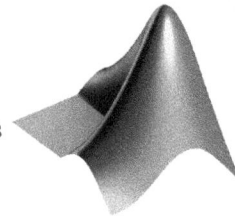

typesetting by *WinEdt*

with donations from **SIAM** and **AMS**.

This searchable electronic color pdfBook is click-navigable within the text by page, section, subsection, chapter, theorem, example, definition, cross reference, citation, equation, figure, table, and hyperlink. A pdfBook has no electronic copy protection and can be read and printed by most computers. The publisher hereby grants the right to reproduce this work in any format but limited to personal use.

for Jennie Columba

◊

Antonio
◊

◊
& Sze Wan

$$\mathbb{EDM} = \mathbb{S}_h \cap \left(\mathbb{S}_c^{\perp} - \mathbb{S}_+ \right)$$

Prelude

The constant demands of my department and university and the ever increasing work needed to obtain funding have stolen much of my precious thinking time, and I sometimes yearn for the halcyon days of Bell Labs.

−Steven Chu, Nobel laureate [83]

Convex Analysis is the calculus of inequalities while Convex Optimization is its application. Analysis is inherently the domain of a mathematician while Optimization belongs to the engineer. A convex optimization problem is conventionally regarded as minimization of a convex objective function subject to an artificial convex domain imposed upon it by the problem constraints. The constraints comprise equalities and inequalities of convex functions whose simultaneous solution set generally constitutes the imposed convex domain: called *feasible set*.

It is easy to minimize a convex function over any convex subset of its domain because any local minimum must be a global minimum. But it is difficult to find the maximum of a convex function over some convex domain because there can be many local maxima. Although this has practical application (Eternity II §4.6.0.0.15), it is not a convex problem. Tremendous benefit accrues when a mathematical problem can be transformed to an equivalent convex optimization problem, primarily because any locally optimal solution is then guaranteed globally optimal.[0.1]

Recognizing a problem as convex is an acquired skill; that being, to know when an objective function is convex and when constraints specify a convex feasible set. The challenge, which is indeed an art, is how to express difficult problems in a convex way: perhaps, problems previously believed nonconvex. Practitioners in the art of Convex Optimization engage themselves with discovery of which hard problems can be transformed into convex equivalents; because, once convex form of a problem is found, then a globally optimal solution is close at hand − the hard work is finished: Finding convex expression of a problem is itself, in a very real sense, its solution.

[0.1]Solving a nonlinear system, for example, by instead solving an equivalent convex optimization problem is therefore highly preferable and what motivates *geometric programming*; a form of convex optimization invented in 1960s [61] [81] that has driven great advances in the electronic circuit design industry. [34, §4.7] [254] [411] [414] [103] [186] [195] [196] [197] [198] [199] [270] [271] [317]

Yet, that skill acquired by understanding the geometry and application of Convex Optimization will remain more an art for some time to come; the reason being, there is generally no unique transformation of a given problem to its convex equivalent. This means, two researchers pondering the same problem are likely to formulate a convex equivalent differently; hence, one solution is likely different from the other although any convex combination of those two solutions remains optimal. Any presumption of only one right or correct solution becomes nebulous. Study of equivalence & sameness, uniqueness, and duality therefore pervade study of Optimization.

It can be difficult for the engineer to apply convex theory without an understanding of Analysis. These pages comprise my journal over a ten year period bridging gaps between engineer and mathematician; they constitute a translation, unification, and cohering of about four hundred papers, books, and reports from several different fields of mathematics and engineering. Beacons of historical accomplishment are cited throughout. Much of what is written here will not be found elsewhere. Care to detail, clarity, accuracy, consistency, and typography accompanies removal of ambiguity and verbosity out of respect for the reader. But the book is nonlinear in its presentation. Consequently there is much indexing, cross-referencing, and background material provided in the text, footnotes, and appendices so as to be more self-contained and to provide understanding of fundamental concepts.

−Jon Dattorro
Stanford, California
2013

Convex Optimization
&
Euclidean Distance Geometry

In my career, I found that the best people are the ones that really understand the content, and they're a pain in the butt to manage. But you put up with it because they're so great at the content. And that's what makes great products; it's not process, it's content.

—Steve Jobs, 1995

List of Figures

3 Geometry of convex functions   187

4   Semidefinite programming   239

List of Tables

CITATION: Dattorro, *Convex Optimization & Euclidean Distance Geometry*, $\mathcal{M}\varepsilon\beta oo$ Publishing USA, 2005, v2015.02.15.

Chapter 1

Overview

Convex Optimization
Euclidean Distance Geometry

People are so afraid of convex analysis.

−Claude Lemaréchal, 2003

In layman's terms, the mathematical science of Optimization is the study of how to make a good choice when confronted with conflicting requirements. The qualifier *convex* means: when an optimal solution is found, then it is guaranteed to be a best solution; there is no better choice.

Any convex optimization problem has geometric interpretation. If a given optimization problem can be transformed to a convex equivalent, then this interpretive benefit is acquired. That is a powerful attraction: the ability to visualize geometry of an optimization problem. Conversely, recent advances in geometry and in graph theory hold convex optimization within their proofs' core. [422] [327]

This book is about convex optimization, convex geometry (with particular attention to distance geometry), and nonconvex, combinatorial, and geometrical problems that can be relaxed or transformed into convex problems. A virtual flood of new applications follows by epiphany that many problems, presumed nonconvex, can be so transformed. [10] [11] [60] [95] [153] [155] [284] [306] [314] [370] [371] [419] [422] [34, §4.3, p.316-322]

Euclidean distance geometry is, fundamentally, a determination of point conformation (configuration, relative position or location) by inference from interpoint distance information. By *inference* we mean: *e.g.*, given only distance information, determine whether there corresponds a *realizable* conformation of points; a list of points in some dimension that attains the given interpoint distances. Each point may represent simply location or, abstractly, any entity expressible as a vector in finite-dimensional Euclidean space; *e.g.*, distance geometry of music [110].

21

CITATION: Dattorro, *Convex Optimization & Euclidean Distance Geometry*, Meßoo Publishing USA, 2005, v2015.02.15.

Figure 1: *Orion nebula.* (astrophotography by Massimo Robberto)

It is a common misconception to presume that some desired point conformation cannot be recovered in absence of complete interpoint distance information. We might, for example, want to realize a constellation given only interstellar distance (or, equivalently, parsecs from our Sun and relative angular measurement; the Sun as vertex to two distant stars); called *stellar cartography,* an application evoked by Figure 1. At first it may seem $O(N^2)$ data is required, yet there are many circumstances where this can be reduced to $O(N)$.

If we agree that a set of points may have a shape (three points can form a triangle and its interior, for example, four points a tetrahedron), then we can ascribe *shape* of a set of points to their convex hull. It should be apparent: from distance, these shapes can be determined only to within a *rigid transformation* (rotation, reflection, translation).

Absolute position information is generally lost, given only distance information, but we can determine the smallest possible dimension in which an unknown list of points can exist; that attribute is their *affine dimension* (a triangle in any ambient space has affine dimension 2, for example). In circumstances where stationary reference points are also provided, it becomes possible to determine absolute position or location; *e.g.,* Figure 2.

Geometric problems involving distance between points can sometimes be reduced to convex optimization problems. Mathematics of this combined study of geometry and optimization is rich and deep. Its application has already proven invaluable discerning organic *molecular conformation* by measuring interatomic distance along covalent bonds; *e.g.,* Figure 3. [90] [360] [143] [46] Many disciplines have already benefitted and simplified consequent to this theory; *e.g.,* distance-based *pattern recognition* (Figure 4), *localization*

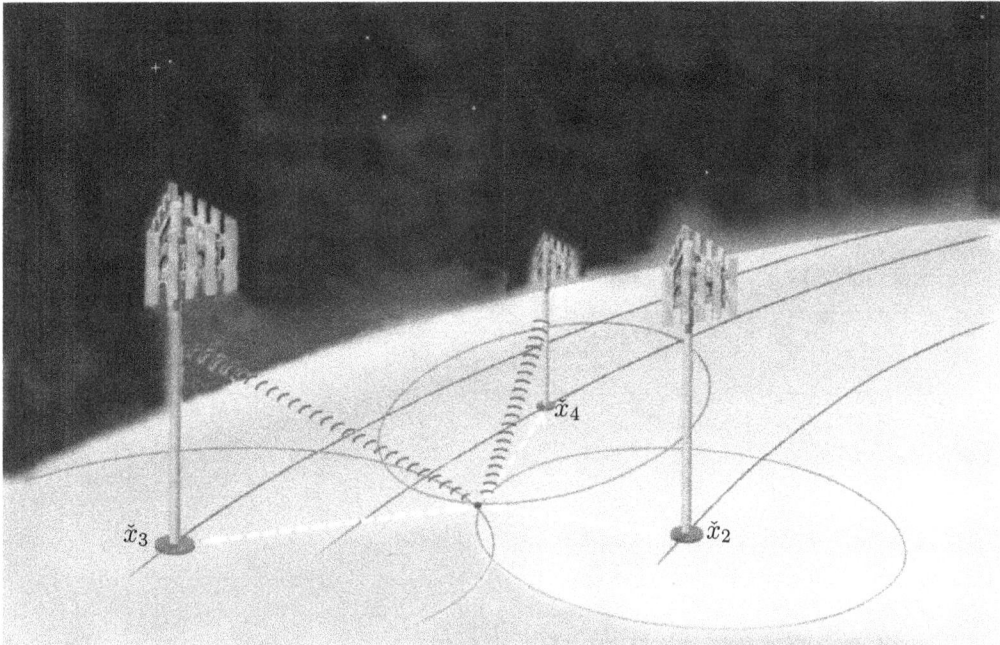

Figure 2: Application of trilateration (§5.4.2.2.8) is localization (determining position) of a radio signal source in 2 dimensions; more commonly known by radio engineers as the process "triangulation". In this scenario, anchors $\check{x}_2$, $\check{x}_3$, $\check{x}_4$ are illustrated as fixed antennae. [215] The radio signal source (a sensor • x_1) anywhere in affine hull of three antenna bases can be uniquely localized by measuring distance to each (dashed white arrowed line segments). Ambiguity of lone distance measurement to sensor is represented by circle about each antenna. Trilateration is expressible as a semidefinite program; hence, a convex optimization problem. [328]

in wireless sensor networks by measurement of intersensor distance along channels of communication, [47] [417] [45] *wireless location* of a radio-signal source such as a cell phone by multiple measurements of signal strength, the *global positioning system* (GPS), and *multidimensional scaling* which is a numerical representation of qualitative data by finding a low-dimensional scale.

Euclidean distance geometry together with convex optimization have also found application in *artificial intelligence*:

- to *machine learning* by discerning naturally occurring manifolds in:

  - Euclidean bodies (Figure 5, §6.7.0.0.1),

  - Fourier spectra of kindred utterances [218],

  - and photographic image sequences [400],

- to *robotics*; *e.g.*, automatic control of vehicles maneuvering in formation. (Figure 8)

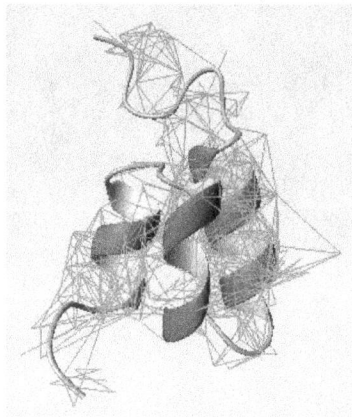

Figure 3: [194] [120] Distance data collected via nuclear magnetic resonance (NMR) helped render this 3-dimensional depiction of a protein molecule. *At the beginning of the 1980s, Kurt Wüthrich [Nobel laureate] developed an idea about how NMR could be extended to cover biological molecules such as proteins. He invented a systematic method of pairing each NMR signal with the right hydrogen nucleus (proton) in the macromolecule. The method is called* sequential assignment *and is today a cornerstone of all NMR structural investigations. He also showed how it was subsequently possible to determine pairwise distances between a large number of hydrogen nuclei and use this information with a mathematical method based on distance-geometry to calculate a three-dimensional structure for the molecule.* −[289] [406] [189]

by chapter

We study pervasive convex Euclidean bodies; their many manifestations and representations. In particular, we make convex polyhedra, cones, and dual cones visceral through illustration in **chapter 2 Convex geometry** where geometric relation of polyhedral cones to nonorthogonal bases (biorthogonal expansion) is examined. It is shown that coordinates are unique in any conic system whose basis cardinality equals or exceeds spatial dimension; for high cardinality, a new definition of *conic coordinate* is provided in Theorem 2.13.12.0.1. The conic analogue to linear independence, called *conic independence,* is introduced as a tool for study, analysis, and manipulation of cones; a natural extension and next logical step in progression: linear, affine, conic. We explain conversion between halfspace- and vertex-description of a convex cone, we motivate the dual cone and provide formulae for finding it, and we show how first-order optimality conditions or alternative systems of linear inequality or *linear matrix inequality* can be explained by *dual generalized inequalities* with respect to convex cones. Arcane theorems of alternative generalized inequality are, in fact, simply derived from cone *membership relations*; generalizations of algebraic *Farkas' lemma* translated to geometry of convex cones.

Any convex optimization problem can be visualized geometrically. Desire to visualize in high dimension [Sagan, *Cosmos − The Edge of Forever,* 22:55′] is deeply embedded in the mathematical psyche. [1] Chapter 2 provides tools to make visualization easier, and

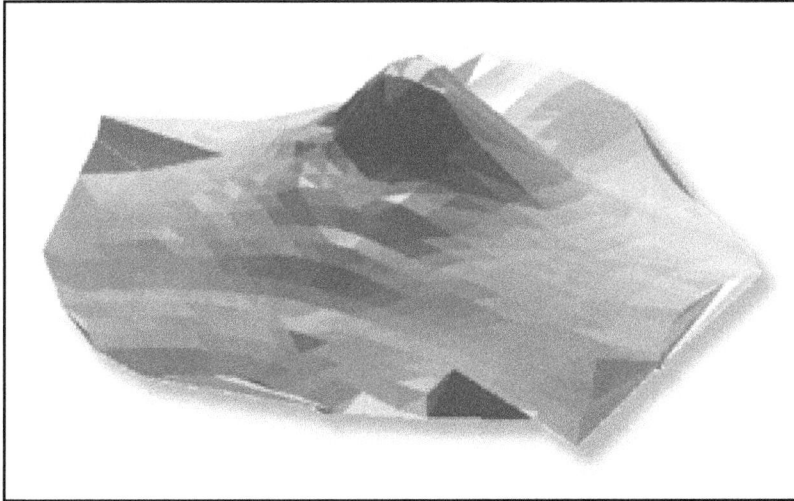

Figure 4: This coarsely discretized triangulated algorithmically flattened human face (made by Kimmel & the Bronsteins [232]) represents a stage in machine recognition of human identity; called *face recognition*. Distance geometry is applied to determine discriminating-features.

we teach how to visualize in high dimension. The concepts of face, extreme point, and extreme direction of a convex Euclidean body are explained here; crucial to understanding convex optimization. How to find the smallest face of any pointed closed convex cone containing convex set $\mathcal{C}$, for example, is divulged; later shown to have practical application to presolving convex programs. The convex cone of positive semidefinite matrices, in particular, is studied in depth:

- We interpret, for example, inverse image of the positive semidefinite cone under affine transformation. (Example 2.9.1.0.2)

- Subsets of the positive semidefinite cone, discriminated by rank exceeding some lower bound, are convex. In other words, high-rank subsets of the positive semidefinite cone boundary united with its interior are convex. (Theorem 2.9.2.9.3) There is a closed form for projection on those convex subsets.

- The positive semidefinite cone is a circular cone in low dimension, while *Geršgorin discs* specify inscription of a polyhedral cone into that positive semidefinite cone. (Figure 49)

Chapter 3 Geometry of convex functions observes Fenchel's analogy between convex sets and functions: We explain, for example, how the real affine function relates to convex functions as the hyperplane relates to convex sets. Partly a toolbox of practical useful convex functions and a cookbook for optimization problems, methods are drawn from the appendices about matrix calculus for determining convexity and discerning geometry.

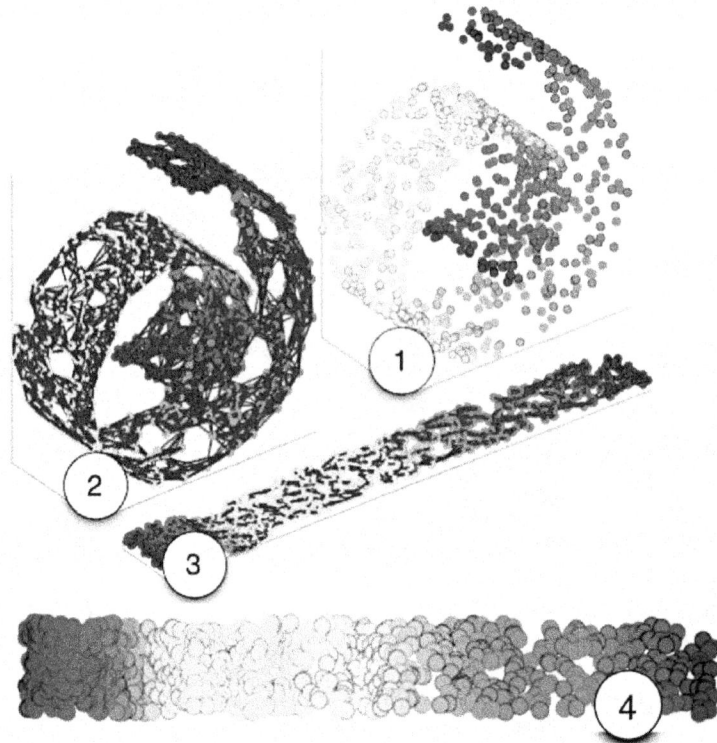

Figure 5: *Swiss roll* from Weinberger & Saul [400]. *The problem of manifold learning, illustrated for $N = 800$ data points sampled from a "Swiss roll"* ①. *A discretized manifold is revealed by connecting each data point and its $k=6$ nearest neighbors* ②. *An unsupervised learning algorithm unfolds the Swiss roll while preserving the local geometry of nearby data points* ③. *Finally, the data points are projected onto the two dimensional subspace that maximizes their variance, yielding a faithful embedding of the original manifold* ④.

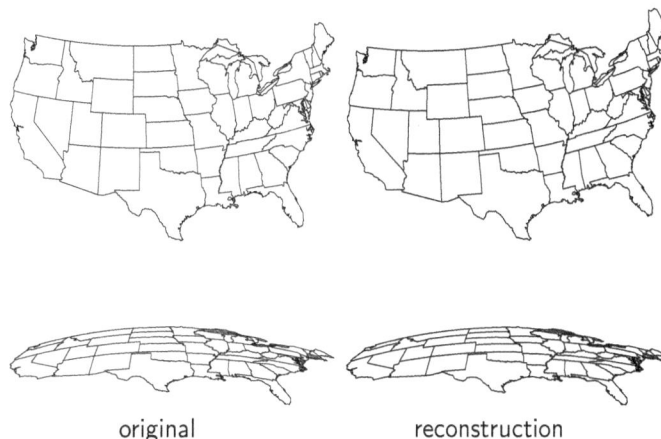

original   reconstruction

Figure 6: About five thousand points along borders constituting United States were used to create an exhaustive matrix of interpoint distance for each and every pair of points in an ordered set (a *list*); called *Euclidean distance matrix*. From that noiseless distance information, it is easy to reconstruct map exactly via Schoenberg criterion (987). (§5.13.1.0.1, *confer* Figure 145) Map reconstruction is exact (to within a rigid transformation) given any number of interpoint distances; the greater the number of distances, the greater the detail (as it is for all conventional map preparation).

Chapter 4. **Semidefinite programming** *has recently emerged to prominence because it admits a new class of problem previously unsolvable by convex optimization techniques, and because it theoretically subsumes other convex techniques: linear programming and quadratic programming and second-order cone programming.* (p.242) Semidefinite programming is reviewed with particular attention to optimality conditions for prototypical primal and dual problems, their interplay, and a perturbation method for rank reduction of optimal solutions (extant but not well known). *Positive definite Farkas' lemma* is derived, and we also show how to determine if a feasible set belongs exclusively to a positive semidefinite cone boundary. An arguably good three-dimensional polyhedral analogue to the positive semidefinite cone of 3×3 symmetric matrices is introduced: a new tool for visualizing coexistence of low- and high-rank optimal solutions in six isomorphic dimensions and a mnemonic aid for understanding semidefinite programs. We find a minimal cardinality Boolean solution to an instance of $Ax = b$:

$$
\begin{array}{ll}
\underset{x}{\text{minimize}} & \|x\|_0 \\
\text{subject to} & Ax = b \\
& x_i \in \{0, 1\}, \quad i = 1 \ldots n
\end{array}
\tag{710}
$$

The *sensor-network localization* problem is solved in any dimension in this chapter. We introduce a method of *convex iteration* for constraining rank in the form $\text{rank}\, G \leq \rho$ and cardinality in the form $\text{card}\, x \leq k$. Cardinality minimization is applied to a discrete

(a)   (b)

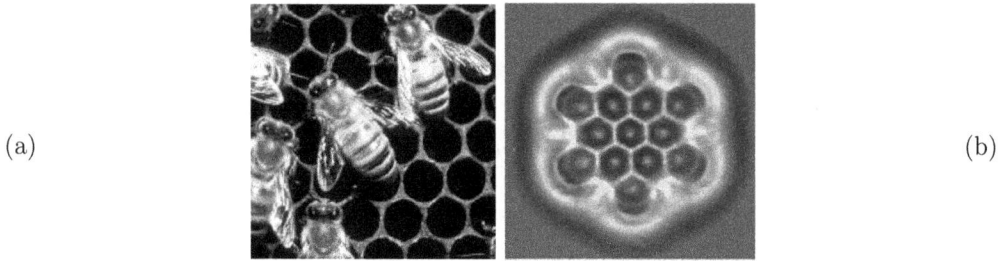

Figure 7: **(a)** These bees construct a honeycomb by solving a convex optimization problem (§5.4.2.2.6). The most dense packing of identical spheres about a central sphere in 2 dimensions is 6. Sphere centers describe a regular lattice. **(b)** *A hexabenzocoronene molecule (diameter: 1.4nm) imaged by noncontact atomic force microscopy using a microscope tip terminated with a single carbon monoxide molecule. The carbon-carbon bonds in the imaged molecule appear with different contrast and apparent lengths. Based on these disparities, the bond orders and lengths of the individual bonds can be distinguished.* (Image by Leo Gross)

image-gradient of the Shepp-Logan phantom, from Magnetic Resonance Imaging (MRI) in the field of medical imaging, for which we find a new lower bound of 1.9% cardinality. We show how to handle polynomial constraints, and how to transform a rank-constrained problem to a rank-1 problem.

The EDM is studied in **chapter 5 Euclidean Distance Matrix**; its properties and relationship to both positive semidefinite and Gram matrices. We relate the EDM to the four classical properties of Euclidean metric; thereby, observing existence of an infinity of properties of the Euclidean metric beyond triangle inequality. We proceed by deriving the fifth Euclidean metric property and then explain why furthering this endeavor is inefficient because the ensuing criteria (while describing polyhedra in angle or area, volume, content, and so on *ad infinitum*) grow linearly in complexity and number with problem size.

Reconstruction methods are explained and applied to a map of the United States; *e.g.*, Figure 6. We also experimentally test a conjecture of Borg & Groenen by reconstructing a distorted but recognizable isotonic map of the USA using only ordinal (comparative) distance data: Figure 145e-f. We demonstrate an elegant method for including dihedral (or *torsion*) angle constraints into a molecular conformation problem. We explain why *trilateration* (a.k.a *localization*) is a convex optimization problem. We show how to recover relative position given incomplete interpoint distance information, and how to pose EDM problems or transform geometrical problems to convex optimizations; *e.g.*, *kissing number* of packed spheres about a central sphere (solved in $\mathbb{R}^3$ by Isaac Newton).

The set of all Euclidean distance matrices forms a pointed closed convex cone called the *EDM cone*: $\mathbb{EDM}^N$. We offer a new proof of Schoenberg's seminal characterization of EDMs:

$$D \in \mathbb{EDM}^N \quad \Leftrightarrow \quad \begin{cases} -V_\mathcal{N}^\mathrm{T} D V_\mathcal{N} \succeq 0 \\ D \in \mathbb{S}_h^N \end{cases} \qquad (987)$$

Our proof relies on fundamental geometry; assuming, any EDM must correspond to a

Figure 8: Nanocopter swarm. Robotic vehicles in concert can move larger objects or localize a plume of gas, liquid, or radio waves. [142]

list of points contained in some polyhedron (possibly at its vertices) and *vice versa.* It is known, but not obvious, this *Schoenberg criterion* implies nonnegativity of the EDM entries; proved herein.

We characterize the eigenvalue spectrum of an EDM, and then devise a polyhedral spectral cone for determining membership of a given matrix (in Cayley-Menger form) to the convex cone of Euclidean distance matrices; *id est*, a matrix is an EDM if and only if its nonincreasingly ordered vector of eigenvalues belongs to a polyhedral spectral cone for $\mathbb{EDM}^N$;

$$D \in \mathbb{EDM}^N \quad \Leftrightarrow \quad \left\{ \begin{array}{l} \lambda\left(\left[\begin{array}{cc} 0 & \mathbf{1}^{\mathrm{T}} \\ \mathbf{1} & -D \end{array} \right] \right) \in \left[\begin{array}{c} \mathbb{R}_+^N \\ \mathbb{R}_- \end{array} \right] \cap \partial\mathcal{H} \\ D \in \mathbb{S}_h^N \end{array} \right. \quad (1205)$$

We will see: spectral cones are not unique.

In **chapter 6 Cone of distance matrices** we explain a geometric relationship between the cone of Euclidean distance matrices, two positive semidefinite cones, and the elliptope. We illustrate geometric requirements, in particular, for projection of a given matrix on a positive semidefinite cone that establish its membership to the EDM cone. The faces of the EDM cone are described, but still open is the question whether all its faces are exposed as they are for the positive semidefinite cone.

The *Schoenberg criterion*

$$D \in \mathbb{EDM}^N \;\Leftrightarrow\; \begin{cases} -V_{\mathcal{N}}^{\mathrm{T}} D V_{\mathcal{N}} \in \mathbb{S}_+^{N-1} \\ \qquad D \in \mathbb{S}_h^N \end{cases} \qquad (987)$$

for identifying a Euclidean distance matrix is revealed to be a discretized *membership relation* (*dual generalized inequalities,* a new Farkas'-like lemma) between the EDM cone and its ordinary dual: $\mathbb{EDM}^{N^*}$. A matrix criterion for membership to the dual EDM cone is derived that is simpler than the Schoenberg criterion:

$$D^* \in \mathbb{EDM}^{N^*} \;\Leftrightarrow\; \delta(D^*\mathbf{1}) - D^* \succeq 0 \qquad (1355)$$

There is a concise equality, relating the convex cone of Euclidean distance matrices to the positive semidefinite cone, apparently overlooked in the literature; an equality between two large convex Euclidean bodies:

$$\mathbb{EDM}^N = \mathbb{S}_h^N \cap \left(\mathbb{S}_c^{N\perp} - \mathbb{S}_+^N \right) \qquad (1349)$$

Seemingly innocuous problems in terms of point position $x_i \in \mathbb{R}^n$ like

$$\underset{\{x_i\}}{\text{minimize}} \sum_{i,j \in \mathcal{I}} \left(\|x_i - x_j\| - h_{ij} \right)^2 \qquad (1389)$$

$$\underset{\{x_i\}}{\text{minimize}} \sum_{i,j \in \mathcal{I}} \left(\|x_i - x_j\|^2 - h_{ij} \right)^2 \qquad (1390)$$

are difficult to solve. So, in **chapter 7 Proximity problems**, we instead explore methods of solution to a few fundamental and prevalent Euclidean distance matrix proximity problems; the problem of finding that distance matrix closest, in some sense, to a given matrix $H = [h_{ij}]$:

$$
\begin{array}{ll}
\underset{D}{\text{minimize}} & \|-V(D-H)V\|_{\mathrm{F}}^2 \\
\text{subject to} & \operatorname{rank} VDV \leq \rho \\
& D \in \mathbb{EDM}^N
\end{array}
\qquad
\begin{array}{ll}
\underset{\sqrt[\circ]{D}}{\text{minimize}} & \|\sqrt[\circ]{D} - H\|_{\mathrm{F}}^2 \\
\text{subject to} & \operatorname{rank} VDV \leq \rho \\
& \sqrt[\circ]{D} \in \sqrt{\mathbb{EDM}^N}
\end{array}
$$

$$
\begin{array}{ll}
\underset{D}{\text{minimize}} & \|D-H\|_{\mathrm{F}}^2 \\
\text{subject to} & \operatorname{rank} VDV \leq \rho \\
& D \in \mathbb{EDM}^N
\end{array}
\qquad
\begin{array}{ll}
\underset{\sqrt[\circ]{D}}{\text{minimize}} & \|-V(\sqrt[\circ]{D} - H)V\|_{\mathrm{F}}^2 \\
\text{subject to} & \operatorname{rank} VDV \leq \rho \\
& \sqrt[\circ]{D} \in \sqrt{\mathbb{EDM}^N}
\end{array}
\qquad (1391)
$$

We apply a convex iteration method for constraining rank. Known heuristics for rank minimization are also explained. We offer new geometrical proof, of a famous discovery by Eckart & Young in 1936 [133], with particular regard to Euclidean projection of a point on that generally nonconvex subset of the positive semidefinite cone boundary comprising all semidefinite matrices having rank not exceeding a prescribed bound ρ. We explain how this problem is transformed to a convex optimization for any rank ρ.

appendices

We presume a reader already comfortable with elementary vector operations; [13, §3] formally known as *analytic geometry*. [408] Sundry toolboxes are provided, in the form of appendices, so as to be more self-contained:

- linear algebra (**appendix A** is primarily concerned with proper statements of semidefiniteness for square matrices),

- simple matrices (dyad, doublet, elementary, Householder, Schoenberg, orthogonal, *etcetera*, in **appendix B**),

- collection of known analytical solutions to some important optimization problems (**appendix C**),

- matrix calculus remains somewhat unsystematized when compared to ordinary calculus (**appendix D** concerns matrix-valued functions, matrix differentiation and directional derivatives, Taylor series, and tables of first- and second-order gradients and matrix derivatives),

- an elaborate exposition offering insight into orthogonal and nonorthogonal projection on convex sets (the connection between projection and positive semidefiniteness, for example, or between projection and a linear objective function in **appendix E**),

- MATLAB code on *Wikimization* [390] to discriminate EDMs, to determine conic independence, to reduce or constrain rank of an optimal solution to a semidefinite program, compressed sensing (compressive sampling) for digital image and audio signal processing, and two distinct methods of reconstructing a map of the United States: one given only distance data, the other given only comparative distance.

Figure 9: Three-dimensional reconstruction of David from distance data.

Figure 10: *Digital Michelangelo Project*, Stanford University. Measuring distance to David by laser rangefinder. Spatial resolution is 0.29mm.

Chapter 2

Convex geometry

Convexity has an immensely rich structure and numerous applications. On the other hand, almost every "convex" idea can be explained by a two-dimensional picture.

−Alexander Barvinok [26, p.vii]

We study convex geometry because it is the easiest of geometries. For that reason, much of a practitioner's energy is expended seeking invertible transformation of problematic sets to convex ones.

As convex geometry and linear algebra are inextricably bonded by *asymmetry* (linear inequality), we provide much background material on linear algebra (especially in the appendices) although a reader is assumed comfortable with [338] [340] [208] or any other intermediate-level text. The essential references to convex analysis are [205] [315]. The reader is referred to [337] [26] [399] [42] [62] [312] [367] for a comprehensive treatment of convexity. There is relatively less published pertaining to convex matrix-valued functions. [221] [209, §6.6] [302]

2.1 Convex set

A set $\mathcal{C}$ is convex iff for all $Y, Z \in \mathcal{C}$ and $0 \leq \mu \leq 1$

$$\mu Y + (1 - \mu)Z \in \mathcal{C} \tag{1}$$

Under that defining condition on μ, the linear sum in (1) is called a *convex combination* of Y and Z. If Y and Z are points in real finite-dimensional Euclidean *vector space* [233] [408] $\mathbb{R}^n$ or $\mathbb{R}^{m \times n}$ (matrices), then (1) represents the closed line segment joining them. Line segments are thereby convex sets; $\mathcal{C}$ is convex iff the line segment connecting any two points in $\mathcal{C}$ is itself in $\mathcal{C}$. Apparent from this definition: a convex set is a connected set. [264, §3.4, §3.5] [42, p.2] A convex set can, but does not necessarily, contain the *origin* $\mathbf{0}$.

CITATION: Dattorro, *Convex Optimization & Euclidean Distance Geometry*, $\mathcal{M}\varepsilon\beta oo$ Publishing USA, 2005, v2015.02.15.

An *ellipsoid* centered at $x = a$ (Figure **13** p.39), given matrix $C \in \mathbb{R}^{m \times n}$

$$\{x \in \mathbb{R}^n \mid \|C(x-a)\|^2 = (x-a)^{\mathrm{T}} C^{\mathrm{T}} C(x-a) \leq 1\} \tag{2}$$

is a good icon for a convex set.[2.1]

2.1.1   subspace

A nonempty subset $\mathcal{R}$ of real Euclidean vector space $\mathbb{R}^n$ is called a *subspace* (§2.5) if every vector[2.2] of the form $\alpha x + \beta y$, for $\alpha, \beta \in \mathbb{R}$, is in $\mathcal{R}$ whenever vectors x and y are. [256, §2.3] A subspace is a convex set containing the origin, by definition. [315, p.4] Any subspace is therefore open in the sense that it contains no boundary, but closed in the sense [264, §2]

$$\mathcal{R} + \mathcal{R} = \mathcal{R} \tag{3}$$

It is not difficult to show

$$\mathcal{R} = -\mathcal{R} \tag{4}$$

as is true for any subspace $\mathcal{R}$, because $x \in \mathcal{R} \Leftrightarrow -x \in \mathcal{R}$. Given any $x \in \mathcal{R}$

$$\mathcal{R} = x + \mathcal{R} \tag{5}$$

The intersection of an arbitrary collection of subspaces remains a subspace. Any subspace not constituting the entire *ambient vector space* $\mathbb{R}^n$ is a *proper subspace*; *e.g.*,[2.3] any line (of infinite extent) through the origin in two-dimensional Euclidean space $\mathbb{R}^2$. The vector space $\mathbb{R}^n$ is itself a conventional subspace, inclusively, [233, §2.1] although not proper.

2.1.2   linear independence

Arbitrary given vectors in Euclidean space $\{\Gamma_i \in \mathbb{R}^n, \ i = 1 \ldots N\}$ are *linearly independent* (l.i.) if and only if, for all $\zeta \in \mathbb{R}^N$ ($\zeta_i \in \mathbb{R}$)

$$\Gamma_1 \zeta_1 + \cdots + \Gamma_{N-1} \zeta_{N-1} - \Gamma_N \zeta_N = \mathbf{0} \tag{6}$$

has only the *trivial solution* $\zeta = \mathbf{0}$; in other words, iff no vector from the given set can be expressed as a linear combination of those remaining.

Geometrically, two nontrivial vector subspaces are linearly independent iff they intersect only at the origin.

2.1.2.1   preservation of linear independence

(*confer* §2.4.2.4, §2.10.1) Linear transformation preserves linear dependence. [233, p.86] Conversely, linear independence can be preserved under linear transformation. Given $Y = [\, y_1 \ \ y_2 \ \cdots \ y_N \,] \in \mathbb{R}^{N \times N}$, consider the mapping

$$T(\Gamma) : \mathbb{R}^{n \times N} \to \mathbb{R}^{n \times N} \triangleq \Gamma Y \tag{7}$$

[2.1] This particular definition is slablike (Figure 11) in $\mathbb{R}^n$ when C has nontrivial nullspace.
[2.2] A *vector* is assumed, throughout, to be a column vector.
[2.3] We substitute abbreviation *e.g.* in place of the Latin *exempli gratia*.

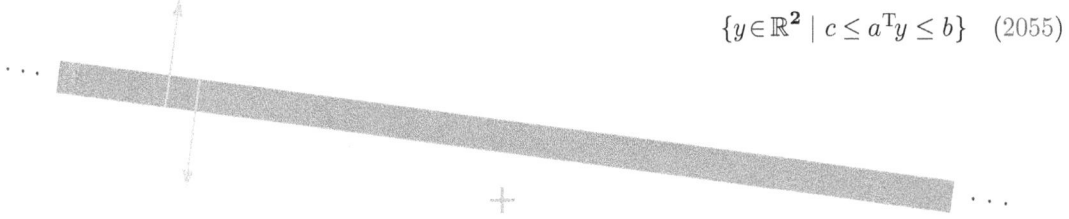

$$\{y \in \mathbb{R}^2 \mid c \le a^{\mathrm{T}} y \le b\} \quad (2055)$$

Figure 11: A *slab* is a convex Euclidean body infinite in extent but not affine. Illustrated in $\mathbb{R}^2$, it may be constructed by intersecting two opposing halfspaces whose bounding hyperplanes are parallel but not coincident. Because number of halfspaces used in its construction is finite, slab is a *polyhedron* (§2.12). (Cartesian axes + and vector inward-normal, to each halfspace-boundary, are drawn for reference.)

whose domain is the set of all matrices $\Gamma \in \mathbb{R}^{n \times N}$ holding a linearly independent set columnar. Linear independence of $\{\Gamma y_i \in \mathbb{R}^n, \ i = 1 \dots N\}$ demands, by definition, there exist no nontrivial solution $\zeta \in \mathbb{R}^N$ to

$$\Gamma y_1 \zeta_i + \dots + \Gamma y_{N-1} \zeta_{N-1} - \Gamma y_N \zeta_N = 0 \qquad (8)$$

By factoring out Γ, we see that triviality is ensured by linear independence of $\{y_i \in \mathbb{R}^N\}$.

2.1.3 Orthant:

name given to a closed convex set that is the higher-dimensional generalization of *quadrant* from the classical Cartesian partition of $\mathbb{R}^2$; a *Cartesian cone*. The most common is the nonnegative orthant $\mathbb{R}^n_+$ or $\mathbb{R}^{n \times n}_+$ (analogue to quadrant I) to which membership denotes nonnegative vector- or matrix-entries respectively; *e.g.*,

$$\mathbb{R}^n_+ \triangleq \{x \in \mathbb{R}^n \mid x_i \ge 0 \ \forall i\} \qquad (9)$$

The nonpositive orthant $\mathbb{R}^n_-$ or $\mathbb{R}^{n \times n}_-$ (analogue to quadrant III) denotes negative and 0 entries. Orthant convexity[2.4] is easily verified by definition (1).

2.1.4 affine set

A nonempty *affine set* (from the word *affinity*) is any subset of $\mathbb{R}^n$ that is a translation of some subspace. Any affine set is convex and open so contains no boundary: *e.g.*, empty set $\emptyset$, point, line, plane, *hyperplane* (§2.4.2), subspace, *etcetera*. The intersection of an arbitrary collection of affine sets remains affine.

[2.4]All orthants are selfdual simplicial cones. (§2.13.5.1, §2.12.3.1.1)

2.1.4.0.1 Definition.   *Affine subset.*
We analogize *affine subset* to subspace,[2.5] defining it to be any nonempty affine set of vectors; an affine subset of $\mathbb{R}^n$. $\triangle$

For some *parallel*[2.6] subspace $\mathcal{R}$ and any point $x \in \mathcal{A}$

$$\mathcal{A} \text{ is affine} \quad\Leftrightarrow\quad \mathcal{A} = x + \mathcal{R} \qquad\qquad (10)$$
$$= \{y \mid y - x \in \mathcal{R}\}$$

Affine hull of a set $\mathcal{C} \subseteq \mathbb{R}^n$ (§2.3.1) is the smallest affine set containing it.

2.1.5 dimension

Dimension of an arbitrary set $\mathcal{S}$ is Euclidean dimension of its affine hull; [399, p.14]

$$\dim \mathcal{S} \triangleq \dim \operatorname{aff} \mathcal{S} = \dim \operatorname{aff}(\mathcal{S} - s) , \quad s \in \mathcal{S} \qquad (11)$$

the same as dimension of the subspace parallel to that affine set $\operatorname{aff} \mathcal{S}$ when nonempty. Hence dimension (of a set) is synonymous with *affine dimension*. [205, A.2.1]

2.1.6 empty set *versus* empty interior

Emptiness $\emptyset$ of a set is handled differently than *interior* in the classical literature. It is common for a nonempty convex set to have empty interior; *e.g.*, paper in the real world:

- An ordinary flat sheet of paper is a nonempty convex set having empty interior in $\mathbb{R}^3$ but nonempty interior relative to its affine hull.

2.1.6.1 relative interior

Although it is always possible to pass to a smaller ambient Euclidean space where a nonempty set acquires an interior [26, §II.2.3], we prefer the qualifier *relative* which is the conventional fix to this ambiguous terminology.[2.7] So we distinguish *interior* from *relative interior* throughout: [337] [399] [367]

- Classical interior $\operatorname{int} \mathcal{C}$ is defined as a union of points: x is an interior point of $\mathcal{C} \subseteq \mathbb{R}^n$ if there exists an open ball of dimension n and nonzero radius centered at x that is contained in $\mathcal{C}$.

- Relative interior $\operatorname{rel int} \mathcal{C}$ of a convex set $\mathcal{C} \subseteq \mathbb{R}^n$ is interior relative to its affine hull.[2.8]

[2.5]The popular term *affine subspace* is an oxymoron.
[2.6]Two affine sets are *parallel* when one is a translation of the other. [315, p.4]
[2.7]Superfluous mingling of terms as in *relatively nonempty set* would be an unfortunate consequence. From the opposite perspective, some authors use the term *full* or *full-dimensional* to describe a set having nonempty interior.
[2.8]Likewise for *relative boundary* (§2.1.7.2), although *relative closure* is superfluous. [205, §A.2.1]

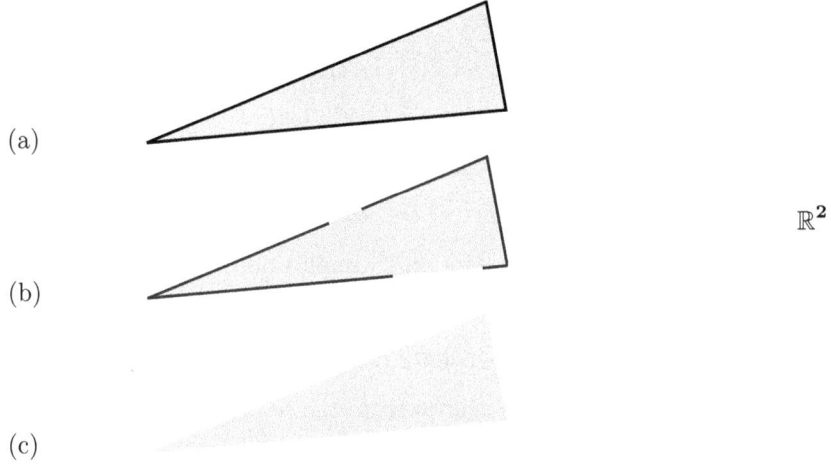

(a)

$\mathbb{R}^2$

(b)

(c)

Figure 12: **(a)** Closed convex set.   **(b)** Neither open, closed, or convex. Yet PSD cone can remain convex in absence of certain boundary components (§2.9.2.9.3). Nonnegative orthant with origin excluded (§2.6) and positive orthant with origin adjoined [315, p.49] are convex.   **(c)** Open convex set.

Thus defined, it is common (though confusing) for $\operatorname{int}\mathcal{C}$ the interior of $\mathcal{C}$ to be empty while its relative interior is not: this happens whenever dimension of its affine hull is less than dimension of the ambient space $(\dim\operatorname{aff}\mathcal{C}<n$; *e.g.*, were $\mathcal{C}$ paper**)** or in the exception when $\mathcal{C}$ is a single point; [264, §2.2.1]

$$\operatorname{rel\,int}\{x\} \triangleq \operatorname{aff}\{x\} = \{x\} \,, \qquad \operatorname{int}\{x\} = \emptyset \,, \qquad x \in \mathbb{R}^n \tag{12}$$

In any case, *closure* of the relative interior of a convex set $\mathcal{C}$ always yields closure of the set itself;

$$\overline{\operatorname{rel\,int}\mathcal{C}} = \overline{\mathcal{C}} \tag{13}$$

Closure is invariant to translation. If $\mathcal{C}$ is convex then $\operatorname{rel\,int}\mathcal{C}$ and $\overline{\mathcal{C}}$ are convex. [205, p.24] If $\mathcal{C}$ has nonempty interior, then

$$\operatorname{rel\,int}\mathcal{C} = \operatorname{int}\mathcal{C} \tag{14}$$

Given the intersection of convex set $\mathcal{C}$ with affine set $\mathcal{A}$

$$\operatorname{rel\,int}(\mathcal{C}\cap\mathcal{A}) = \operatorname{rel\,int}(\mathcal{C})\cap\mathcal{A} \quad \Leftarrow \quad \operatorname{rel\,int}(\mathcal{C})\cap\mathcal{A} \neq \emptyset \tag{15}$$

Because an affine set $\mathcal{A}$ is open

$$\operatorname{rel\,int}\mathcal{A} = \mathcal{A} \tag{16}$$

2.1.7   classical boundary

(*confer* §2.1.7.2) *Boundary* of a set $\mathcal{C}$ is the closure of $\mathcal{C}$ less its interior;

$$\partial\mathcal{C} = \overline{\mathcal{C}} \setminus \operatorname{int}\mathcal{C} \tag{17}$$

[56, §1.1] which follows from the fact

$$\overline{\operatorname{int}\mathcal{C}} = \overline{\mathcal{C}} \quad \Leftrightarrow \quad \partial\operatorname{int}\mathcal{C} = \partial\mathcal{C} \tag{18}$$

and presumption of nonempty interior.[2.9] Implications are:

- $\operatorname{int}\mathcal{C} = \overline{\mathcal{C}} \setminus \partial\mathcal{C}$

- a bounded open set has *boundary* defined but not contained in the set

- interior of an open set is equivalent to the set itself;

from which an open set is defined: [264, p.109]

$$\mathcal{C} \text{ is } open \quad \Leftrightarrow \quad \operatorname{int}\mathcal{C} = \mathcal{C} \tag{19}$$
$$\mathcal{C} \text{ is } closed \quad \Leftrightarrow \quad \overline{\operatorname{int}\mathcal{C}} = \mathcal{C} \tag{20}$$

The set illustrated in Figure 12b is not open because it is not equivalent to its interior, for example, it is not closed because it does not contain its boundary, and it is not convex because it does not contain all convex combinations of its boundary points.

2.1.7.1   Line intersection with boundary

A line can intersect the boundary of a convex set in any dimension at a point demarcating the line's entry to the set interior. On one side of that entry-point along the line is the exterior of the set, on the other side is the set interior. In other words,

- starting from any point of a convex set, a move toward the interior is an immediate entry into the interior. [26, §II.2]

When a line intersects the interior of a convex body in any dimension, the boundary appears to the line to be as thin as a point. This is intuitively plausible because, for example, a line intersects the boundary of the ellipsoids in Figure 13 at a point in $\mathbb{R}$, $\mathbb{R}^2$, and $\mathbb{R}^3$. Such thinness is a remarkable fact when pondering visualization of convex *polyhedra* (§2.12, §5.14.3) in four Euclidean dimensions, for example, having boundaries constructed from other three-dimensional convex polyhedra called *faces*.

We formally define *face* in (§2.6). For now, we observe the boundary of a convex body to be entirely constituted by all its faces of dimension lower than the body itself. Any face of a convex set is convex. For example: The ellipsoids in Figure 13 have boundaries composed only of zero-dimensional faces. The two-dimensional slab in Figure 11 is an unbounded *polyhedron* having one-dimensional faces making its boundary. The three-dimensional bounded polyhedron in Figure 20 has zero-, one-, and two-dimensional polygonal faces constituting its boundary.

[2.9]Otherwise, for $x \in \mathbb{R}^n$ as in (12), [264, §2.1-§2.3]

$$\overline{\operatorname{int}\{x\}} = \overline{\emptyset} = \emptyset$$

the empty set is both open and closed.

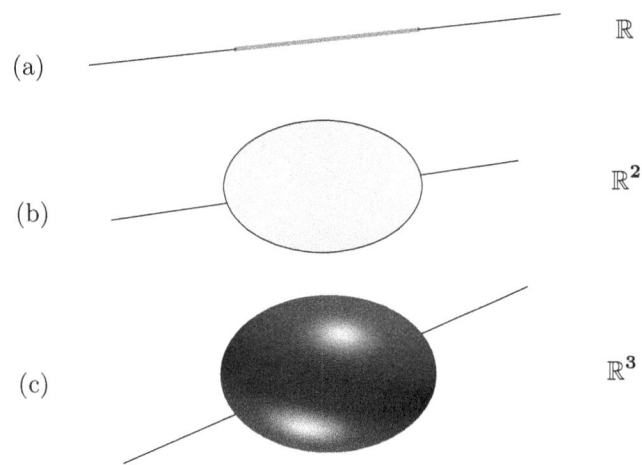

Figure 13: **(a)** Ellipsoid in $\mathbb{R}$ is a line segment whose boundary comprises two points. Intersection of line with ellipsoid in $\mathbb{R}$, **(b)** in $\mathbb{R}^2$, **(c)** in $\mathbb{R}^3$. Each ellipsoid illustrated has entire boundary constituted by zero-dimensional faces; in fact, by *vertices* (§2.6.1.0.1). Intersection of line with boundary is a point at entry to interior. These same facts hold in higher dimension.

2.1.7.1.1   Example.   *Intersection of line with boundary in $\mathbb{R}^6$.*
The convex cone of positive semidefinite matrices $\mathbb{S}_+^3$ (§2.9), in the ambient subspace of symmetric matrices $\mathbb{S}^3$ (§2.2.2.0.1), is a six-dimensional Euclidean body in *isometrically isomorphic* $\mathbb{R}^6$ (§2.2.1). Boundary of the positive semidefinite cone, in this dimension, comprises faces having only the dimensions 0, 1, and 3; *id est*, $\{\rho(\rho+1)/2, \ \rho=0,1,2\}$.

Unique minimum-distance projection PX (§E.9) of any point $X \in \mathbb{S}^3$ on that cone $\mathbb{S}_+^3$ is known in closed form (§7.1.2). Given, for example, $\lambda \in \text{int }\mathbb{R}_+^3$ and *diagonalization* (§A.5.1) of exterior point

$$X = Q\Lambda Q^\mathrm{T} \in \mathbb{S}^3, \quad \Lambda \triangleq \begin{bmatrix} \lambda_1 & & \mathbf{0} \\ & \lambda_2 & \\ \mathbf{0} & & -\lambda_3 \end{bmatrix} \tag{21}$$

where $Q \in \mathbb{R}^{3\times3}$ is an *orthogonal matrix,* then the projection on $\mathbb{S}_+^3$ in $\mathbb{R}^6$ is

$$PX = Q \begin{bmatrix} \lambda_1 & & \mathbf{0} \\ & \lambda_2 & \\ \mathbf{0} & & 0 \end{bmatrix} Q^\mathrm{T} \in \mathbb{S}_+^3 \tag{22}$$

This positive semidefinite matrix PX nearest X thus has rank 2, found by discarding all negative eigenvalues in Λ. The line connecting these two points is $\{X + (PX - X)t \mid t \in \mathbb{R}\}$ where $t=0 \Leftrightarrow X$ and $t=1 \Leftrightarrow PX$. Because this line intersects the boundary of the *positive semidefinite cone* $\mathbb{S}_+^3$ at point PX and passes through its interior (by assumption), then the matrix corresponding to an infinitesimally positive perturbation of t there should reside interior to the cone (rank 3). Indeed, for ε an arbitrarily small positive constant,

$$X + (PX-X)t|_{t=1+\varepsilon} = Q(\Lambda + (P\Lambda - \Lambda)(1+\varepsilon))Q^\mathrm{T} = Q \begin{bmatrix} \lambda_1 & & \mathbf{0} \\ & \lambda_2 & \\ \mathbf{0} & & \varepsilon\lambda_3 \end{bmatrix} Q^\mathrm{T} \in \text{int }\mathbb{S}_+^3 \tag{23}$$

$\square$

2.1.7.1.2   Example.   *Tangential line intersection with boundary.*
A higher-dimensional boundary $\partial\mathcal{C}$ of a convex Euclidean body $\mathcal{C}$ is simply a dimensionally larger set through which a line can pass when it does not intersect the body's interior. Still, for example, a line existing in five or more dimensions may pass *tangentially* (intersecting no point interior to $\mathcal{C}$ [225, §15.3]) through a single point relatively interior to a three-dimensional face on $\partial\mathcal{C}$. Let's understand why by inductive reasoning.

Figure 14a shows a vertical line-segment whose boundary comprises its two endpoints. For a line to pass through the boundary tangentially (intersecting no point relatively interior to the line-segment), it must exist in an ambient space of at least two dimensions. Otherwise, the line is confined to the same one-dimensional space as the line-segment and must pass along the segment to reach the end points.

Figure 14b illustrates a two-dimensional ellipsoid whose boundary is constituted entirely by zero-dimensional faces. Again, a line must exist in at least two dimensions to tangentially pass through any single arbitrarily chosen point on the boundary (without intersecting the ellipsoid interior).

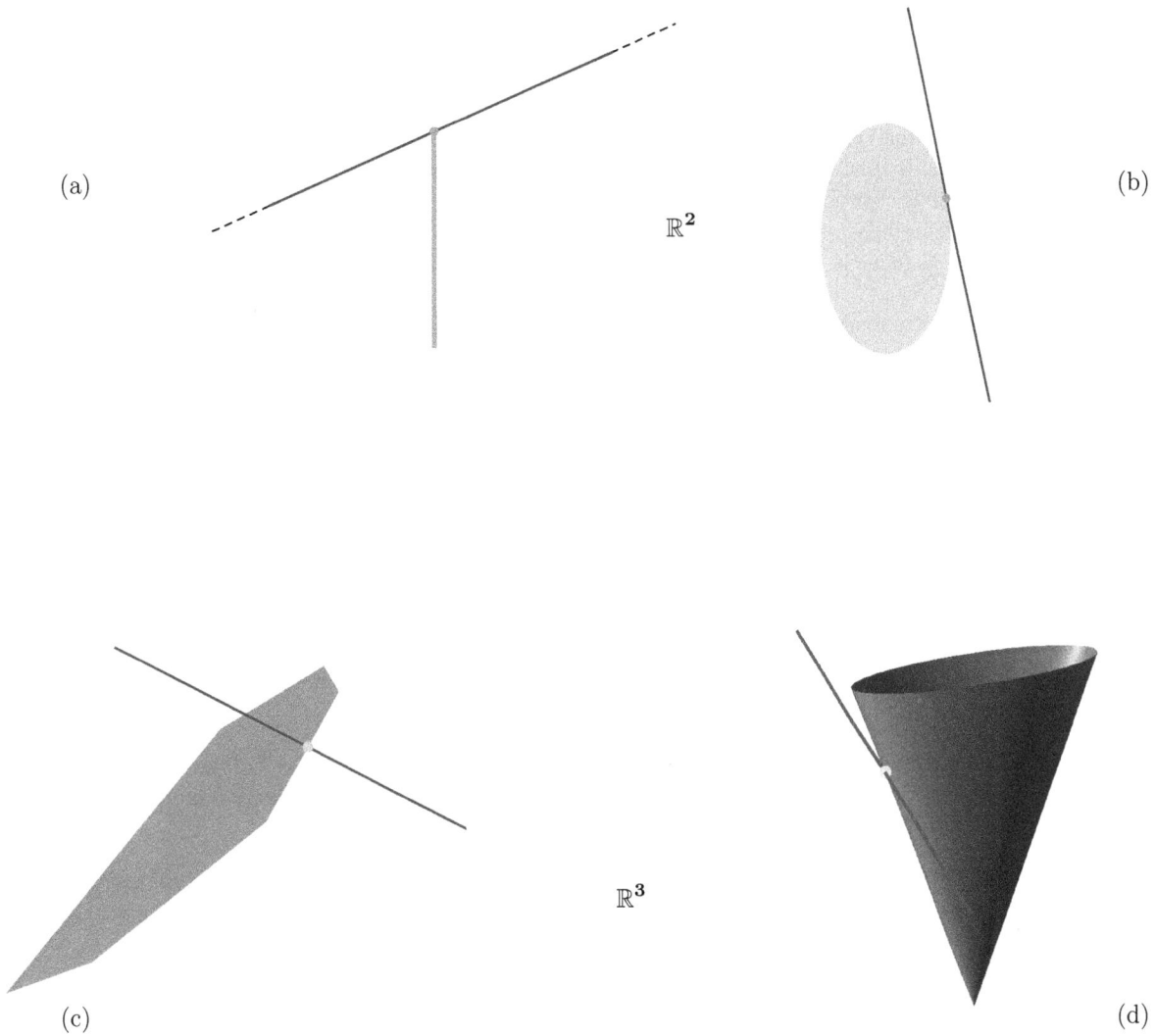

Figure 14: Line tangential: (a) (b) to relative interior of a zero-dimensional face in $\mathbb{R}^2$, (c) (d) to relative interior of a one-dimensional face in $\mathbb{R}^3$.

Now let's move to an ambient space of three dimensions. Figure 14c shows a polygon rotated into three dimensions. For a line to pass through its zero-dimensional boundary (one of its *vertices*) tangentially, it must exist in at least the two dimensions of the polygon. But for a line to pass tangentially through a single arbitrarily chosen point in the relative interior of a one-dimensional face on the boundary as illustrated, it must exist in at least three dimensions.

Figure 14d illustrates a solid circular cone (drawn truncated) whose one-dimensional faces are halflines emanating from its pointed end (*vertex*). This cone's boundary is constituted solely by those one-dimensional halflines. A line may pass through the boundary tangentially, striking only one arbitrarily chosen point relatively interior to a one-dimensional face, if it exists in at least the three-dimensional ambient space of the cone.

From these few examples, way deduce a general rule (without proof):

- A line may pass tangentially through a single arbitrarily chosen point relatively interior to a k-dimensional face on the boundary of a convex Euclidean body if the line exists in dimension at least equal to $k+2$.

Now the interesting part, with regard to Figure **20** showing a bounded polyhedron in $\mathbb{R}^3$; call it $\mathcal{P}$: A line existing in at least four dimensions is required in order to pass tangentially (without hitting $\operatorname{int}\mathcal{P}$) through a single arbitrary point in the relative interior of any two-dimensional polygonal face on the boundary of polyhedron $\mathcal{P}$. Now imagine that polyhedron $\mathcal{P}$ is itself a three-dimensional face of some other polyhedron in $\mathbb{R}^4$. To pass a line tangentially through polyhedron $\mathcal{P}$ itself, striking only one point from its relative interior $\operatorname{rel int}\mathcal{P}$ as claimed, requires a line existing in at least five dimensions.[2.10]

It is not too difficult to deduce:

- A line may pass through a single arbitrarily chosen point interior to a k-dimensional convex Euclidean body (hitting no other interior point) if that line exists in dimension at least equal to $k+1$.

In layman's terms, this means: a being capable of navigating four spatial dimensions (one Euclidean dimension beyond our physical reality) could see inside three-dimensional objects.   □

2.1.7.2 Relative boundary

The classical definition of *boundary* of a set $\mathcal{C}$ presumes nonempty interior:

$$\partial\mathcal{C} = \overline{\mathcal{C}} \setminus \operatorname{int}\mathcal{C} \qquad (17)$$

More suitable to study of convex sets is the *relative boundary*; defined [205, §A.2.1.2]

$$\operatorname{rel}\partial\mathcal{C} \triangleq \overline{\mathcal{C}} \setminus \operatorname{rel int}\mathcal{C} \qquad (24)$$

boundary relative to affine hull of $\mathcal{C}$.

[2.10]This rule can help determine whether there exists unique solution to a convex optimization problem whose *feasible set* is an intersecting line; *e.g.*, the *trilateration* problem (§5.4.2.2.8).

In the exception when $\mathcal{C}$ is a single point $\{x\}$, (12)

$$\operatorname{rel}\partial\{x\} = \overline{\{x\}}\backslash\{x\} = \emptyset, \qquad x \in \mathbb{R}^n \qquad (25)$$

A bounded convex polyhedron (§2.3.2, §2.12.0.0.1) in subspace $\mathbb{R}$, for example, has boundary constructed from two points, in $\mathbb{R}^2$ from at least three line segments, in $\mathbb{R}^3$ from convex polygons, while a convex *polychoron* (a bounded polyhedron in $\mathbb{R}^4$ [401]) has boundary constructed from three-dimensional convex polyhedra. A halfspace is partially bounded by a hyperplane; its interior therefore excludes that hyperplane. An affine set has no relative boundary.

2.1.8 intersection, sum, difference, product

2.1.8.0.1 Theorem. *Intersection.* [315, §2, thm.6.5]
Intersection of an arbitrary collection of convex sets $\{\mathcal{C}_i\}$ is convex. For a finite collection of N sets, a necessarily nonempty intersection of relative interior $\bigcap_{i=1}^{N} \operatorname{rel int} \mathcal{C}_i = \operatorname{rel int} \bigcap_{i=1}^{N} \mathcal{C}_i$ equals relative interior of intersection. And for a possibly infinite collection, $\bigcap \overline{\mathcal{C}_i} = \overline{\bigcap \mathcal{C}_i}$. $\diamond$

In converse this theorem is implicitly false insofar as a convex set can be formed by the intersection of sets that are not. Unions of convex sets are generally not convex. [205, p.22]

Vector sum of two convex sets $\mathcal{C}_1$ and $\mathcal{C}_2$ is convex [205, p.24]

$$\mathcal{C}_1 + \mathcal{C}_2 = \{x + y \mid x \in \mathcal{C}_1, \, y \in \mathcal{C}_2\} \qquad (26)$$

but not necessarily closed unless at least one set is closed and bounded.

By additive inverse, we can similarly define *vector difference* of two convex sets

$$\mathcal{C}_1 - \mathcal{C}_2 = \{x - y \mid x \in \mathcal{C}_1, \, y \in \mathcal{C}_2\} \qquad (27)$$

which is convex. Applying this definition to nonempty convex set $\mathcal{C}_1$, its self-difference $\mathcal{C}_1 - \mathcal{C}_1$ is generally nonempty, nontrivial, and convex; *e.g.*, for any *convex cone* $\mathcal{K}$, (§2.7.2) the set $\mathcal{K} - \mathcal{K}$ constitutes its affine hull. [315, p.15]

Cartesian product of convex sets

$$\mathcal{C}_1 \times \mathcal{C}_2 = \left\{ \begin{bmatrix} x \\ y \end{bmatrix} \mid x \in \mathcal{C}_1, \, y \in \mathcal{C}_2 \right\} = \begin{bmatrix} \mathcal{C}_1 \\ \mathcal{C}_2 \end{bmatrix} \qquad (28)$$

remains convex. The converse also holds; *id est*, a Cartesian product is convex iff each set is. [205, p.23]

Convex results are also obtained for scaling $\kappa\mathcal{C}$ of a convex set $\mathcal{C}$, rotation/reflection $Q\mathcal{C}$, or translation $\mathcal{C} + \alpha$; each similarly defined.

Given any operator T and convex set $\mathcal{C}$, we are prone to write $T(\mathcal{C})$ meaning

$$T(\mathcal{C}) \triangleq \{T(x) \mid x \in \mathcal{C}\} \qquad (29)$$

Given linear operator T, it therefore follows from (26),

$$\begin{aligned} T(\mathcal{C}_1 + \mathcal{C}_2) &= \{T(x+y) \mid x \in \mathcal{C}_1, \, y \in \mathcal{C}_2\} \\ &= \{T(x) + T(y) \mid x \in \mathcal{C}_1, \, y \in \mathcal{C}_2\} \\ &= T(\mathcal{C}_1) + T(\mathcal{C}_2) \end{aligned} \qquad (30)$$

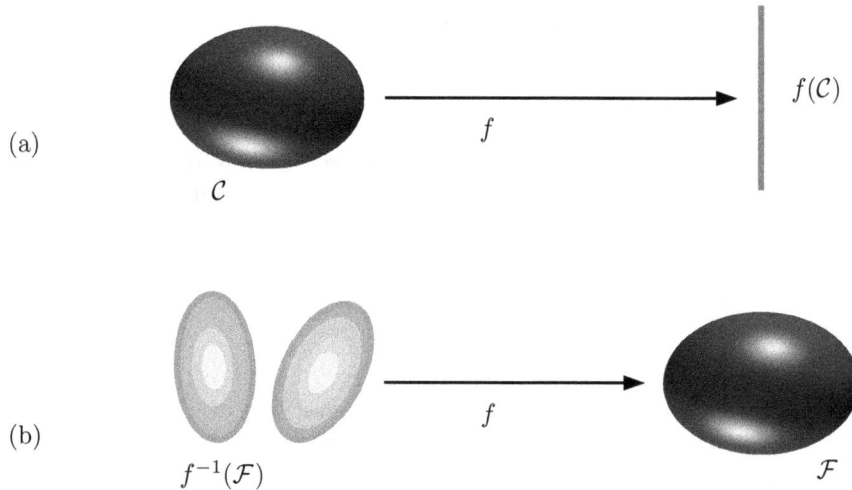

Figure 15: **(a)** Image of convex set in domain of any convex function f is convex, but there is no converse. **(b)** Inverse image under convex function f.

2.1.9 inverse image

While *epigraph* (§3.5) of a convex function must be convex, it generally holds that inverse image (Figure 15) of a convex function is not. The most prominent examples to the contrary are affine functions (§3.4):

2.1.9.0.1 Theorem. *Inverse image.*   [315, §3]
Let f be a mapping from $\mathbb{R}^{p \times k}$ to $\mathbb{R}^{m \times n}$.

- The image of a convex set $\mathcal{C}$ under any affine function

$$f(\mathcal{C}) = \{f(X) \mid X \in \mathcal{C}\} \subseteq \mathbb{R}^{m \times n} \tag{31}$$

  is convex.

- Inverse image of a convex set $\mathcal{F}$,

$$f^{-1}(\mathcal{F}) = \{X \mid f(X) \in \mathcal{F}\} \subseteq \mathbb{R}^{p \times k} \tag{32}$$

  a single- or many-valued mapping, under any affine function f is convex.   ◇

In particular, any affine transformation of an affine set remains affine. [315, p.8] Ellipsoids are invariant to any [*sic*] affine transformation.

$$\mathbb{R}^n \qquad\qquad A^\dagger A x = x_{\mathrm{p}} \qquad\qquad \mathbb{R}^m$$

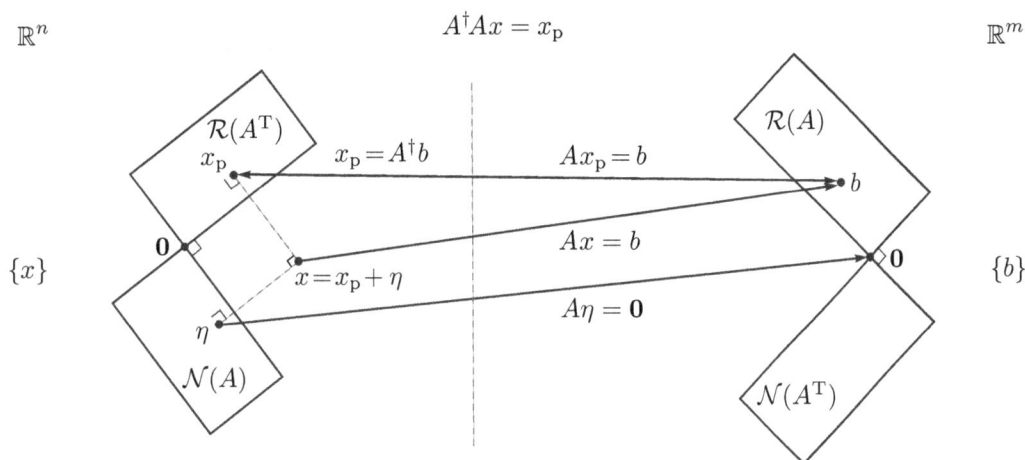

Figure 16: (*confer* Figure **172**) Action of linear map represented by $A \in \mathbb{R}^{m\times n}$: Component of vector x in nullspace $\mathcal{N}(A)$ maps to origin while component in rowspace $\mathcal{R}(A^{\mathrm{T}})$ maps to range $\mathcal{R}(A)$. For any $A \in \mathbb{R}^{m\times n}$, $AA^\dagger Ax = b$ (§E) and inverse image of $b \in \mathcal{R}(A)$ is a nonempty affine set: $x_{\mathrm{p}} + \mathcal{N}(A)$.

Although not precluded, this *inverse image theorem* does not require a uniquely invertible mapping f. Figure **16**, for example, mechanizes inverse image under a general linear map. Example 2.9.1.0.2 and Example 3.5.0.0.2 offer further applications.

Each converse of this two-part theorem is generally false; *id est*, given f affine, a convex image $f(\mathcal{C})$ does not imply that set $\mathcal{C}$ is convex, and neither does a convex inverse image $f^{-1}(\mathcal{F})$ imply set $\mathcal{F}$ is convex. A counterexample, invalidating a converse, is easy to visualize when the affine function is an orthogonal *projector* [338] [256]:

2.1.9.0.2 Corollary. *Projection on subspace.*[2.11]   (2027) [315, §3]
Orthogonal projection of a convex set on a subspace or nonempty affine set is another convex set.   ◇

Again, the converse is false. Shadows, for example, are umbral projections that can be convex when the body providing the shade is not.

2.2 Vectorized-matrix inner product

Euclidean space $\mathbb{R}^n$ comes equipped with a vector inner-product

$$\langle y, z \rangle \triangleq y^{\mathrm{T}} z = \|y\|\|z\| \cos\psi \tag{33}$$

[2.11] For hyperplane representations see §2.4.2. For projection of convex sets on hyperplanes see [399, §6.6]. A nonempty affine set is called an *affine subset* (§2.1.4.0.1). Orthogonal projection of points on affine subsets is reviewed in §E.4.

where ψ (996) represents angle (in radians) between vectors y and z. We prefer those angle brackets to connote a geometric rather than algebraic perspective; *e.g.*, vector y might represent a hyperplane normal (§2.4.2). Two vectors are *orthogonal* (*perpendicular*) to one another if and only if their inner product vanishes (iff ψ is an odd multiple of $\frac{\pi}{2}$);

$$y \perp z \;\Leftrightarrow\; \langle y, z \rangle = 0 \tag{34}$$

When orthogonal vectors each have unit *norm*, then they are *orthonormal*. A vector inner-product defines Euclidean norm (vector 2-norm, §A.7.1)

$$\|y\|_2 = \|y\| \triangleq \sqrt{y^{\mathrm{T}}y} \,, \qquad \|y\| = 0 \;\Leftrightarrow\; y = \mathbf{0} \tag{35}$$

For linear operator A, its *adjoint* A^{T} is a linear operator defined by [233, §3.10]

$$\langle y, A^{\mathrm{T}}z \rangle \triangleq \langle Ay, z \rangle \tag{36}$$

For linear operation on a vector, represented by real matrix A, the adjoint operator A^{T} is its *transposition*. This operator is *self-adjoint* when $A = A^{\mathrm{T}}$.

Vector inner-product for matrices is calculated just as it is for vectors; by first transforming a matrix in $\mathbb{R}^{p \times k}$ to a vector in $\mathbb{R}^{pk}$ by concatenating its columns in the *natural order*. For lack of a better term, we shall call that linear *bijective* (one-to-one and onto [233, App.A1.2]) transformation *vectorization*. For example, the vectorization of $Y = [\,y_1 \; y_2 \cdots y_k\,] \in \mathbb{R}^{p \times k}$ [172] [334] is

$$\mathrm{vec}\, Y \triangleq \begin{bmatrix} y_1 \\ y_2 \\ \vdots \\ y_k \end{bmatrix} \in \mathbb{R}^{pk} \tag{37}$$

Then the vectorized-matrix inner-product is trace of matrix inner-product; for $Z \in \mathbb{R}^{p \times k}$, [62, §2.6.1] [205, §0.3.1] [410, §8] [374, §2.2]

$$\langle Y, Z \rangle \triangleq \mathrm{tr}(Y^{\mathrm{T}}Z) \;=\; \mathrm{vec}(Y)^{\mathrm{T}} \mathrm{vec}\, Z \tag{38}$$

where (§A.1.1)

$$\mathrm{tr}(Y^{\mathrm{T}}Z) = \mathrm{tr}(ZY^{\mathrm{T}}) = \mathrm{tr}(YZ^{\mathrm{T}}) = \mathrm{tr}(Z^{\mathrm{T}}Y) = \mathbf{1}^{\mathrm{T}}(Y \circ Z)\mathbf{1} \tag{39}$$

and where $\circ$ denotes the *Hadamard product*[2.12] of matrices [164, §1.1.4]. The adjoint A^{T} operation on a matrix can therefore be defined in like manner:

$$\langle Y, A^{\mathrm{T}}Z \rangle \triangleq \langle AY, Z \rangle \tag{40}$$

Take any element $\mathcal{C}_1$ from a matrix-valued set in $\mathbb{R}^{p \times k}$, for example, and consider any particular dimensionally compatible real vectors v and w. Then vector inner-product of $\mathcal{C}_1$ with vw^{T} is

$$\langle vw^{\mathrm{T}}, \mathcal{C}_1 \rangle \;=\; \langle v, \mathcal{C}_1 w \rangle \;=\; v^{\mathrm{T}}\mathcal{C}_1 w \;=\; \mathrm{tr}(wv^{\mathrm{T}}\mathcal{C}_1) \;=\; \mathbf{1}^{\mathrm{T}}\big((vw^{\mathrm{T}}) \circ \mathcal{C}_1\big)\mathbf{1} \tag{41}$$

[2.12]Hadamard product is a simple entrywise product of corresponding entries from two matrices of like size; *id est*, not necessarily square. A commutative operation, the Hadamard product can be extracted from within a Kronecker product. [208, p.475]

$\mathbb{R}^2$ 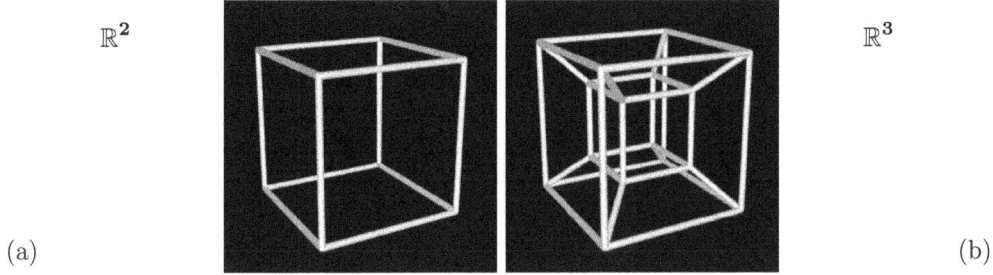 $\mathbb{R}^3$

(a) (b)

Figure 17: **(a)** Cube in $\mathbb{R}^3$ projected on paper-plane $\mathbb{R}^2$. Subspace projection operator is not an isomorphism because new adjacencies are introduced. **(b)** *Tesseract* is a projection of hypercube in $\mathbb{R}^4$ on $\mathbb{R}^3$.

Further, linear bijective vectorization is *distributive* with respect to Hadamard product; *id est,*

$$\operatorname{vec}(Y \circ Z) = \operatorname{vec}(Y) \circ \operatorname{vec}(Z) \tag{42}$$

2.2.0.0.1 Example. *Application of inverse image theorem.*
Suppose set $\mathcal{C} \subseteq \mathbb{R}^{p \times k}$ were convex. Then for any particular vectors $v \in \mathbb{R}^p$ and $w \in \mathbb{R}^k$, the set of vector inner-products

$$\mathcal{Y} \triangleq v^{\mathrm{T}} \mathcal{C} w = \langle vw^{\mathrm{T}}, \mathcal{C} \rangle \subseteq \mathbb{R} \tag{43}$$

is convex. It is easy to show directly that convex combination of elements from $\mathcal{Y}$ remains an element of $\mathcal{Y}$.[2.13] Instead given convex set $\mathcal{Y}$, $\mathcal{C}$ must be convex consequent to *inverse image theorem* 2.1.9.0.1.

More generally, vw^{T} in (43) may be replaced with any particular matrix $Z \in \mathbb{R}^{p \times k}$ while convexity of set $\langle Z, \mathcal{C} \rangle \subseteq \mathbb{R}$ persists. Further, by replacing v and w with any particular respective matrices U and W of dimension compatible with all elements of convex set $\mathcal{C}$, then set $U^{\mathrm{T}} \mathcal{C} W$ is convex by the *inverse image theorem* because it is a linear mapping of $\mathcal{C}$. □

2.2.1 Frobenius'

2.2.1.0.1 Definition. *Isomorphic.*
An *isomorphism* of a vector space is a transformation equivalent to a linear bijective mapping. Image and inverse image under the transformation operator are then called *isomorphic vector spaces.* △

[2.13]To verify that, take any two elements $\mathcal{C}_1$ and $\mathcal{C}_2$ from the convex matrix-valued set $\mathcal{C}$, and then form the vector inner-products (43) that are two elements of $\mathcal{Y}$ by definition. Now make a convex combination of those inner products; *videlicet,* for $0 \le \mu \le 1$

$$\mu \langle vw^{\mathrm{T}}, \mathcal{C}_1 \rangle + (1 - \mu) \langle vw^{\mathrm{T}}, \mathcal{C}_2 \rangle = \langle vw^{\mathrm{T}}, \mu \mathcal{C}_1 + (1 - \mu) \mathcal{C}_2 \rangle$$

The two sides are equivalent by linearity of inner product. The right-hand side remains a vector inner-product of vw^{T} with an element $\mu \mathcal{C}_1 + (1 - \mu) \mathcal{C}_2$ from the convex set $\mathcal{C}$; hence, it belongs to $\mathcal{Y}$. Since that holds true for any two elements from $\mathcal{Y}$, then it must be a convex set. ◆

Isomorphic vector spaces are characterized by preservation of *adjacency*; *id est*, if v and w are points connected by a line segment in one vector space, then their images will be connected by a line segment in the other. Two Euclidean bodies may be considered isomorphic if there exists an isomorphism, of their vector spaces, under which the bodies correspond. [376, §I.1] Projection (§E) is not an isomorphism, Figure 17 for example; hence, perfect reconstruction (inverse projection) is generally impossible without additional information.

When $Z = Y \in \mathbb{R}^{p \times k}$ in (38), *Frobenius' norm* is resultant from vector inner-product; (*confer* (1773))

$$\|Y\|_{\mathrm{F}}^2 = \|\operatorname{vec} Y\|_2^2 = \langle Y, Y \rangle = \operatorname{tr}(Y^{\mathrm{T}} Y)$$
$$= \sum_{i,j} Y_{ij}^2 = \sum_i \lambda(Y^{\mathrm{T}} Y)_i = \sum_i \sigma(Y)_i^2 \tag{44}$$

where $\lambda(Y^{\mathrm{T}} Y)_i$ is the ith eigenvalue of $Y^{\mathrm{T}} Y$, and $\sigma(Y)_i$ the ith singular value of Y. Were Y a *normal matrix* (§A.5.1), then $\sigma(Y) = |\lambda(Y)|$ [421, §8.1] thus

$$\|Y\|_{\mathrm{F}}^2 = \sum_i \lambda(Y)_i^2 = \|\lambda(Y)\|_2^2 = \langle \lambda(Y), \lambda(Y) \rangle = \langle Y, Y \rangle \tag{45}$$

The converse $(45) \Rightarrow$ normal matrix Y also holds. [208, §2.5.4]

Frobenius' norm is the Euclidean norm of vectorized matrices. Because the metrics are equivalent, for $X \in \mathbb{R}^{p \times k}$

$$\|\operatorname{vec} X - \operatorname{vec} Y\|_2 = \|X - Y\|_{\mathrm{F}} \tag{46}$$

and because vectorization (37) is a linear bijective map, then vector space $\mathbb{R}^{p \times k}$ is isometrically isomorphic with vector space $\mathbb{R}^{pk}$ in the Euclidean sense and vec is an *isometric isomorphism* of $\mathbb{R}^{p \times k}$. Because of this Euclidean structure, all the known results from convex analysis in Euclidean space $\mathbb{R}^n$ carry over directly to the space of real matrices $\mathbb{R}^{p \times k}$.

2.2.1.1   Injective linear operators

Injective mapping (transformation) means one-to-one mapping; synonymous with *uniquely invertible* linear mapping on Euclidean space.

- Linear injective mappings are fully characterized by lack of nontrivial nullspace.

2.2.1.1.1   Definition.   *Isometric isomorphism.*
An isometric isomorphism of a vector space having a *metric* defined on it is a linear bijective mapping T that preserves distance; *id est*, for all $x, y \in \operatorname{dom} T$

$$\|Tx - Ty\| = \|x - y\| \tag{47}$$

Then isometric isomorphism T is called a *bijective isometry*. $\triangle$

Unitary linear operator $Q : \mathbb{R}^k \to \mathbb{R}^k$, represented by orthogonal matrix $Q \in \mathbb{R}^{k \times k}$ (§B.5.2), is an isometric isomorphism; *e.g.*, discrete Fourier transform via (881). Suppose $T(X) = U X Q$ is a bijective isometry where U is a dimensionally compatible *orthonormal*

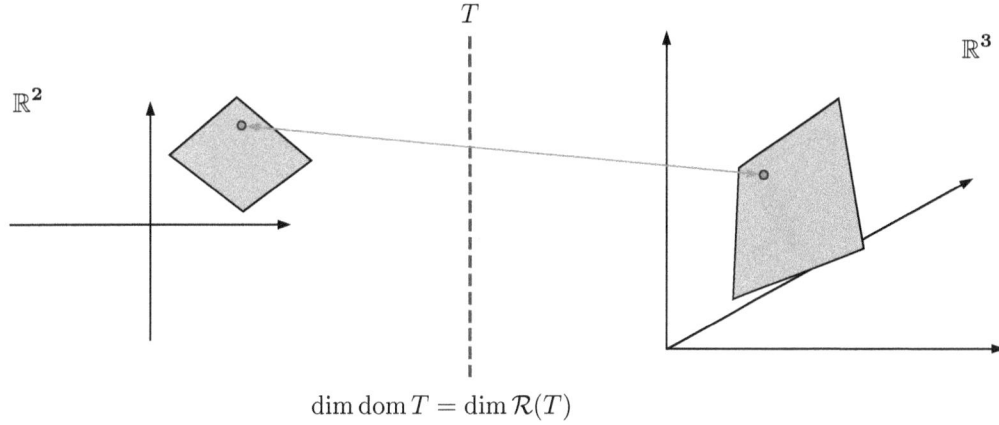

$$\dim \operatorname{dom} T = \dim \mathcal{R}(T)$$

Figure 18: Linear injective mapping $Tx = Ax : \mathbb{R}^2 \to \mathbb{R}^3$ of Euclidean body remains two-dimensional under mapping represented by skinny full-rank matrix $A \in \mathbb{R}^{3 \times 2}$; two bodies are isomorphic by Definition 2.2.1.0.1.

matrix.[2.14] Then we also say Frobenius' norm is *orthogonally invariant*; meaning, for $X, Y \in \mathbb{R}^{p \times k}$

$$\|U(X-Y)Q\|_{\mathrm{F}} = \|X-Y\|_{\mathrm{F}} \tag{48}$$

Yet isometric operator $T : \mathbb{R}^2 \to \mathbb{R}^3$, represented by $A = \begin{bmatrix} 1 & 0 \\ 0 & 1 \\ 0 & 0 \end{bmatrix}$ on $\mathbb{R}^2$, is injective but not a *surjective* map to $\mathbb{R}^3$. [233, §1.6, §2.6] This operator T can therefore be a bijective isometry only with respect to its range.

Any linear injective transformation on Euclidean space is uniquely invertible on its range. In fact, any linear injective transformation has a range whose dimension equals that of its domain. In other words, for any invertible linear transformation T [*ibidem*]

$$\dim \operatorname{dom}(T) = \dim \mathcal{R}(T) \tag{49}$$

e.g., T represented by skinny-or-square full-rank matrices. (Figure 18) An important consequence of this fact is:

- Affine dimension, of any n-dimensional Euclidean body in domain of operator T, is invariant to linear injective transformation.

2.2.1.1.2 Example. *Noninjective linear operators.*
Mappings in Euclidean space created by noninjective linear operators can be characterized in terms of an orthogonal projector (§E). Consider noninjective linear operator

---

[2.14] any matrix U whose columns are orthonormal with respect to each other ($U^{\mathrm{T}}U = I$); these include the orthogonal matrices.

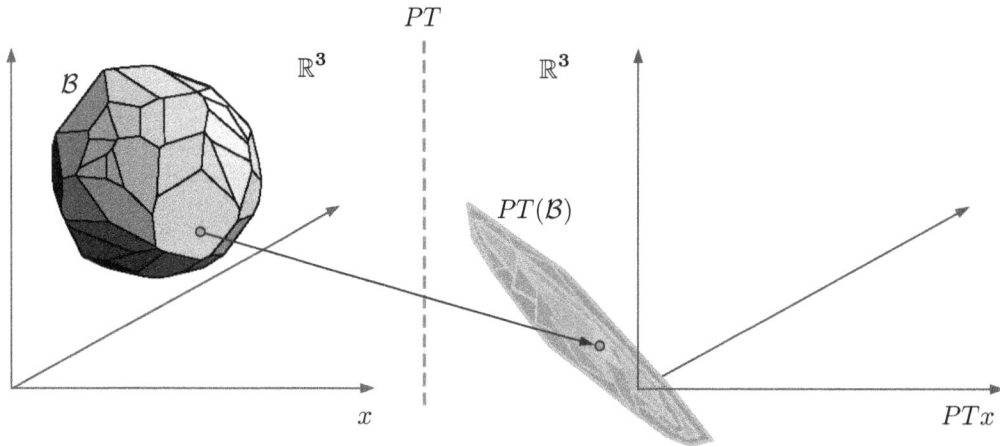

Figure 19: Linear noninjective mapping $PTx = A^{\dagger}Ax : \mathbb{R}^3 \to \mathbb{R}^3$ of three-dimensional Euclidean body $\mathcal{B}$ has affine dimension 2 under projection on rowspace of fat full-rank matrix $A \in \mathbb{R}^{2 \times 3}$. Set of coefficients of orthogonal projection $T\mathcal{B} = \{Ax \mid x \in \mathcal{B}\}$ is isomorphic with projection $P(T\mathcal{B})$ [sic].

$Tx = Ax : \mathbb{R}^n \to \mathbb{R}^m$ represented by fat matrix $A \in \mathbb{R}^{m \times n}$ $(m < n)$. What can be said about the nature of this m-dimensional mapping?

Concurrently, consider injective linear operator $Py = A^{\dagger}y : \mathbb{R}^m \to \mathbb{R}^n$ where $\mathcal{R}(A^{\dagger}) = \mathcal{R}(A^{\mathrm{T}})$. $P(Ax) = PTx$ achieves projection of vector x on the row space $\mathcal{R}(A^{\mathrm{T}})$. (§E.3.1) This means vector Ax can be succinctly interpreted as coefficients of orthogonal projection.

Pseudoinverse matrix $A^{\dagger}$ is skinny and full-rank, so operator Py is a linear *bijection* with respect to its range $\mathcal{R}(A^{\dagger})$. By Definition 2.2.1.0.1, image $P(T\mathcal{B})$ of projection $PT(\mathcal{B})$ on $\mathcal{R}(A^{\mathrm{T}})$ in $\mathbb{R}^n$ must therefore be isomorphic with the set of projection coefficients $T\mathcal{B} = \{Ax \mid x \in \mathcal{B}\}$ in $\mathbb{R}^m$ and have the same affine dimension by (49). To illustrate, we present a three-dimensional Euclidean body $\mathcal{B}$ in Figure 19 where any point x in the nullspace $\mathcal{N}(A)$ maps to the origin.   □

2.2.2 Symmetric matrices

2.2.2.0.1 Definition. *Symmetric matrix subspace.*
Define a subspace of $\mathbb{R}^{M \times M}$: the convex set of all symmetric $M \times M$ matrices;

$$\mathbb{S}^M \triangleq \left\{ A \in \mathbb{R}^{M \times M} \mid A = A^{\mathrm{T}} \right\} \subseteq \mathbb{R}^{M \times M} \tag{50}$$

This subspace comprising symmetric matrices $\mathbb{S}^M$ is isomorphic with the vector space $\mathbb{R}^{M(M+1)/2}$ whose dimension is the number of free variables in a symmetric $M \times M$ matrix. The *orthogonal complement* [338] [256] of $\mathbb{S}^M$ is

$$\mathbb{S}^{M\perp} \triangleq \left\{ A \in \mathbb{R}^{M \times M} \mid A = -A^{\mathrm{T}} \right\} \subset \mathbb{R}^{M \times M} \tag{51}$$

the subspace of *antisymmetric* matrices in $\mathbb{R}^{M \times M}$; *id est,*

$$\mathbb{S}^M \oplus \mathbb{S}^{M\perp} = \mathbb{R}^{M \times M} \tag{52}$$

where unique vector sum $\oplus$ is defined on page 666. $\triangle$

All antisymmetric matrices are *hollow* by definition (have **0** main diagonal). Any square matrix $A \in \mathbb{R}^{M \times M}$ can be written as a sum of its symmetric and antisymmetric parts: respectively,

$$A = \frac{1}{2}(A + A^{\mathrm{T}}) + \frac{1}{2}(A - A^{\mathrm{T}}) \tag{53}$$

The symmetric part is orthogonal in $\mathbb{R}^{M^2}$ to the antisymmetric part; *videlicet,*

$$\operatorname{tr}\big((A + A^{\mathrm{T}})(A - A^{\mathrm{T}})\big) = 0 \tag{54}$$

In the ambient space of real matrices, the antisymmetric matrix subspace can be described

$$\mathbb{S}^{M\perp} = \left\{ \frac{1}{2}(A - A^{\mathrm{T}}) \mid A \in \mathbb{R}^{M \times M} \right\} \subset \mathbb{R}^{M \times M} \tag{55}$$

because any matrix in $\mathbb{S}^M$ is orthogonal to any matrix in $\mathbb{S}^{M\perp}$. Further confined to the ambient subspace of symmetric matrices, because of antisymmetry, $\mathbb{S}^{M\perp}$ would become trivial.

2.2.2.1 Isomorphism of symmetric matrix subspace

When a matrix is symmetric in $\mathbb{S}^M$, we may still employ the vectorization transformation (37) to $\mathbb{R}^{M^2}$; vec, an isometric isomorphism. We might instead choose to realize in the lower-dimensional subspace $\mathbb{R}^{M(M+1)/2}$ by ignoring redundant entries (below the main diagonal) during transformation. Such a realization would remain isomorphic but not isometric. Lack of isometry is a spatial distortion due now to disparity in metric between $\mathbb{R}^{M^2}$ and $\mathbb{R}^{M(M+1)/2}$. To realize isometrically in $\mathbb{R}^{M(M+1)/2}$, we must make a correction: For $Y = [Y_{ij}] \in \mathbb{S}^M$ we take symmetric vectorization [221, §2.2.1]

$$\operatorname{svec} Y \triangleq \begin{bmatrix} Y_{11} \\ \sqrt{2}Y_{12} \\ Y_{22} \\ \sqrt{2}Y_{13} \\ \sqrt{2}Y_{23} \\ Y_{33} \\ \vdots \\ Y_{MM} \end{bmatrix} \in \mathbb{R}^{M(M+1)/2} \tag{56}$$

where all entries off the main diagonal have been scaled. Now for $Z \in \mathbb{S}^M$

$$\langle Y, Z \rangle \triangleq \operatorname{tr}(Y^{\mathrm{T}}Z) = \operatorname{vec}(Y)^{\mathrm{T}}\operatorname{vec} Z = \mathbf{1}^{\mathrm{T}}(Y \circ Z)\mathbf{1} = \operatorname{svec}(Y)^{\mathrm{T}}\operatorname{svec} Z \tag{57}$$

Then because the metrics become equivalent, for $X \in \mathbb{S}^M$

$$\| \operatorname{svec} X - \operatorname{svec} Y \|_2 = \| X - Y \|_{\mathrm{F}} \tag{58}$$

and because symmetric vectorization (56) is a linear bijective mapping, then svec is an isometric isomorphism of the symmetric matrix subspace. In other words, $\mathbb{S}^M$ is isometrically isomorphic with $\mathbb{R}^{M(M+1)/2}$ in the Euclidean sense under transformation svec.

The set of all symmetric matrices $\mathbb{S}^M$ forms a proper subspace in $\mathbb{R}^{M \times M}$, so for it there exists a standard orthonormal *basis* in isometrically isomorphic $\mathbb{R}^{M(M+1)/2}$

$$\{E_{ij} \in \mathbb{S}^M\} = \left\{ \begin{array}{ll} e_i e_i^{\mathrm{T}}, & i = j = 1 \ldots M \\ \frac{1}{\sqrt{2}} \left(e_i e_j^{\mathrm{T}} + e_j e_i^{\mathrm{T}} \right), & 1 \leq i < j \leq M \end{array} \right\} \tag{59}$$

where $M(M+1)/2$ *standard basis matrices* E_{ij} are formed from the *standard basis vectors*

$$e_i = \left[\left\{ \begin{array}{l} 1, \ i = j \\ 0, \ i \neq j \end{array} \right., \ j = 1 \ldots M \right] \in \mathbb{R}^M \tag{60}$$

Thus we have a basic *orthogonal expansion* for $Y \in \mathbb{S}^M$

$$Y = \sum_{j=1}^{M} \sum_{i=1}^{j} \langle E_{ij}, Y \rangle E_{ij} \tag{61}$$

whose unique coefficients

$$\langle E_{ij}, Y \rangle = \left\{ \begin{array}{ll} Y_{ii}, & i = 1 \ldots M \\ Y_{ij}\sqrt{2}, & 1 \leq i < j \leq M \end{array} \right. \tag{62}$$

correspond to entries of the symmetric vectorization (56).

2.2.3 Symmetric hollow subspace

2.2.3.0.1 Definition. *Hollow subspaces.*   [361]
Define a subspace of $\mathbb{R}^{M \times M}$: the convex set of all (real) symmetric $M \times M$ matrices having $\mathbf{0}$ main diagonal;

$$\mathbb{R}_h^{M \times M} \triangleq \left\{ A \in \mathbb{R}^{M \times M} \mid A = A^{\mathrm{T}}, \ \delta(A) = \mathbf{0} \right\} \subset \mathbb{R}^{M \times M} \tag{63}$$

where the main diagonal of $A \in \mathbb{R}^{M \times M}$ is denoted (§A.1)

$$\delta(A) \in \mathbb{R}^M \tag{1496}$$

Operating on a vector, linear operator δ naturally returns a diagonal matrix; $\delta(\delta(A))$ is a diagonal matrix. Operating recursively on a vector $\Lambda \in \mathbb{R}^N$ or diagonal matrix $\Lambda \in \mathbb{S}^N$, operator $\delta(\delta(\Lambda))$ returns Λ itself;

$$\delta^2(\Lambda) \equiv \delta(\delta(\Lambda)) = \Lambda \tag{1498}$$

The subspace $\mathbb{R}_h^{M \times M}$ (63) comprising (real) symmetric hollow matrices is isomorphic with subspace $\mathbb{R}^{M(M-1)/2}$; its orthogonal complement is

$$\mathbb{R}_h^{M \times M \perp} \triangleq \left\{ A \in \mathbb{R}^{M \times M} \mid A = -A^{\mathrm{T}} + 2\delta^2(A) \right\} \subseteq \mathbb{R}^{M \times M} \tag{64}$$

the subspace of *antisymmetric antihollow* matrices in $\mathbb{R}^{M \times M}$; *id est,*

$$\mathbb{R}_h^{M \times M} \oplus \mathbb{R}_h^{M \times M \perp} = \mathbb{R}^{M \times M} \tag{65}$$

Yet defined instead as a proper subspace of ambient $\mathbb{S}^M$

$$\mathbb{S}_h^M \triangleq \left\{ A \in \mathbb{S}^M \mid \delta(A) = \mathbf{0} \right\} \equiv \mathbb{R}_h^{M \times M} \subset \mathbb{S}^M \tag{66}$$

the orthogonal complement $\mathbb{S}_h^{M \perp}$ of *symmetric hollow subspace* $\mathbb{S}_h^M$,

$$\mathbb{S}_h^{M \perp} \triangleq \left\{ A \in \mathbb{S}^M \mid A = \delta^2(A) \right\} \subseteq \mathbb{S}^M \tag{67}$$

called *symmetric antihollow subspace,* is simply the subspace of diagonal matrices; *id est,*

$$\mathbb{S}_h^M \oplus \mathbb{S}_h^{M \perp} = \mathbb{S}^M \tag{68}$$

$$\triangle$$

Any matrix $A \in \mathbb{R}^{M \times M}$ can be written as a sum of its symmetric hollow and antisymmetric antihollow parts: respectively,

$$A = \left(\frac{1}{2}(A + A^{\mathrm{T}}) - \delta^2(A) \right) + \left(\frac{1}{2}(A - A^{\mathrm{T}}) + \delta^2(A) \right) \tag{69}$$

The symmetric hollow part is orthogonal to the antisymmetric antihollow part in $\mathbb{R}^{M^2}$; *videlicet,*

$$\mathrm{tr}\left(\left(\frac{1}{2}(A + A^{\mathrm{T}}) - \delta^2(A) \right) \left(\frac{1}{2}(A - A^{\mathrm{T}}) + \delta^2(A) \right) \right) = 0 \tag{70}$$

because any matrix in subspace $\mathbb{R}_h^{M \times M}$ is orthogonal to any matrix in the antisymmetric antihollow subspace

$$\mathbb{R}_h^{M \times M \perp} = \left\{ \frac{1}{2}(A - A^{\mathrm{T}}) + \delta^2(A) \mid A \in \mathbb{R}^{M \times M} \right\} \subseteq \mathbb{R}^{M \times M} \tag{71}$$

of the ambient space of real matrices; which reduces to the diagonal matrices in the ambient space of symmetric matrices

$$\mathbb{S}_h^{M \perp} = \left\{ \delta^2(A) \mid A \in \mathbb{S}^M \right\} = \left\{ \delta(u) \mid u \in \mathbb{R}^M \right\} \subseteq \mathbb{S}^M \tag{72}$$

In anticipation of their utility with Euclidean distance matrices (EDMs) in §5, for symmetric hollow matrices we introduce the linear bijective vectorization dvec that is the

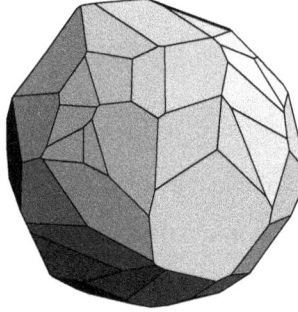

Figure 20: Convex hull of a random list of points in $\mathbb{R}^3$.   Some points from that generating list reside interior to this *convex polyhedron* (§2.12). [401, *Convex Polyhedron*] (Avis-Fukuda-Mizukoshi)

natural analogue to symmetric matrix vectorization svec (56): for $Y = [Y_{ij}] \in \mathbb{S}_h^M$

$$
\text{dvec}\, Y \;\triangleq\; \sqrt{2}
\begin{bmatrix}
Y_{12} \\
Y_{13} \\
Y_{23} \\
Y_{14} \\
Y_{24} \\
Y_{34} \\
\vdots \\
Y_{M-1,M}
\end{bmatrix}
\in \mathbb{R}^{M(M-1)/2}
\tag{73}
$$

Like svec, dvec is an isometric isomorphism on the symmetric hollow subspace. For $X \in \mathbb{S}_h^M$

$$
\| \text{dvec}\, X - \text{dvec}\, Y \|_2 = \| X - Y \|_{\text{F}}
\tag{74}
$$

The set of all symmetric hollow matrices $\mathbb{S}_h^M$ forms a proper subspace in $\mathbb{R}^{M \times M}$, so for it there must be a standard orthonormal basis in isometrically isomorphic $\mathbb{R}^{M(M-1)/2}$

$$
\{E_{ij} \in \mathbb{S}_h^M\} \;=\; \left\{ \frac{1}{\sqrt{2}}\left(e_i e_j^{\text{T}} + e_j e_i^{\text{T}}\right), \quad 1 \le i < j \le M \right\}
\tag{75}
$$

where $M(M-1)/2$ standard basis matrices E_{ij} are formed from the standard basis vectors $e_i \in \mathbb{R}^M$.

The *symmetric hollow majorization corollary* A.1.2.0.2 characterizes eigenvalues of symmetric hollow matrices.

2.3 Hulls

We focus on the affine, convex, and conic hulls: convex sets that may be regarded as kinds of Euclidean container or vessel united with its interior.

2.3.1   Affine hull, affine dimension

Affine dimension of any set in $\mathbb{R}^n$ is the dimension of the smallest affine set (empty set, point, line, plane, hyperplane (§2.4.2), translated subspace, $\mathbb{R}^n$) that contains it. For nonempty sets, affine dimension is the same as dimension of the subspace parallel to that affine set. [315, §1] [205, §A.2.1]

Ascribe the points in a list $\{x_\ell \in \mathbb{R}^n, \ \ell=1\dots N\}$ to the columns of matrix X:

$$X = [\, x_1 \ \cdots \ x_N \,] \in \mathbb{R}^{n\times N} \tag{76}$$

In particular, we define *affine dimension* r of the N-point list X as dimension of the smallest affine set in Euclidean space $\mathbb{R}^n$ that contains X;

$$r \triangleq \dim \operatorname{aff} X \tag{77}$$

Affine dimension r is a lower bound sometimes called *embedding dimension*. [361] [191] That affine set $\mathcal{A}$ in which those points are embedded is unique and called the *affine hull* [337, §2.1];

$$\begin{aligned}\mathcal{A} &\triangleq \operatorname{aff}\{x_\ell \in \mathbb{R}^n, \ \ell=1\dots N\} &= \operatorname{aff} X\\ &= x_1 + \mathcal{R}\{x_\ell - x_1, \ \ell=2\dots N\} &= \{Xa \mid a^{\mathrm{T}}\mathbf{1} = 1\} \subseteq \mathbb{R}^n\end{aligned} \tag{78}$$

for which we call list X a set of *generators.* Hull $\mathcal{A}$ is parallel to subspace

$$\mathcal{R}\{x_\ell - x_1, \ \ell=2\dots N\} = \mathcal{R}(X - x_1\mathbf{1}^{\mathrm{T}}) \subseteq \mathbb{R}^n \tag{79}$$

where

$$\mathcal{R}(A) = \{Ax \mid \forall x\} \tag{142}$$

Given some arbitrary set $\mathcal{C}$ and any $x \in \mathcal{C}$

$$\operatorname{aff}\mathcal{C} = x + \operatorname{aff}(\mathcal{C} - x) \tag{80}$$

where $\operatorname{aff}(\mathcal{C}-x)$ is a subspace.

$$\operatorname{aff}\emptyset \triangleq \emptyset \tag{81}$$

The affine hull of a point x is that point itself;

$$\operatorname{aff}\{x\} = \{x\} \tag{82}$$

Affine hull of two distinct points is the unique line through them. (Figure 21) The affine hull of three noncollinear points in any dimension is that unique plane containing the points, and so on. The subspace of symmetric matrices $\mathbb{S}^m$ is the affine hull of the cone of positive semidefinite matrices; (§2.9)

$$\operatorname{aff}\mathbb{S}^m_+ = \mathbb{S}^m \tag{83}$$

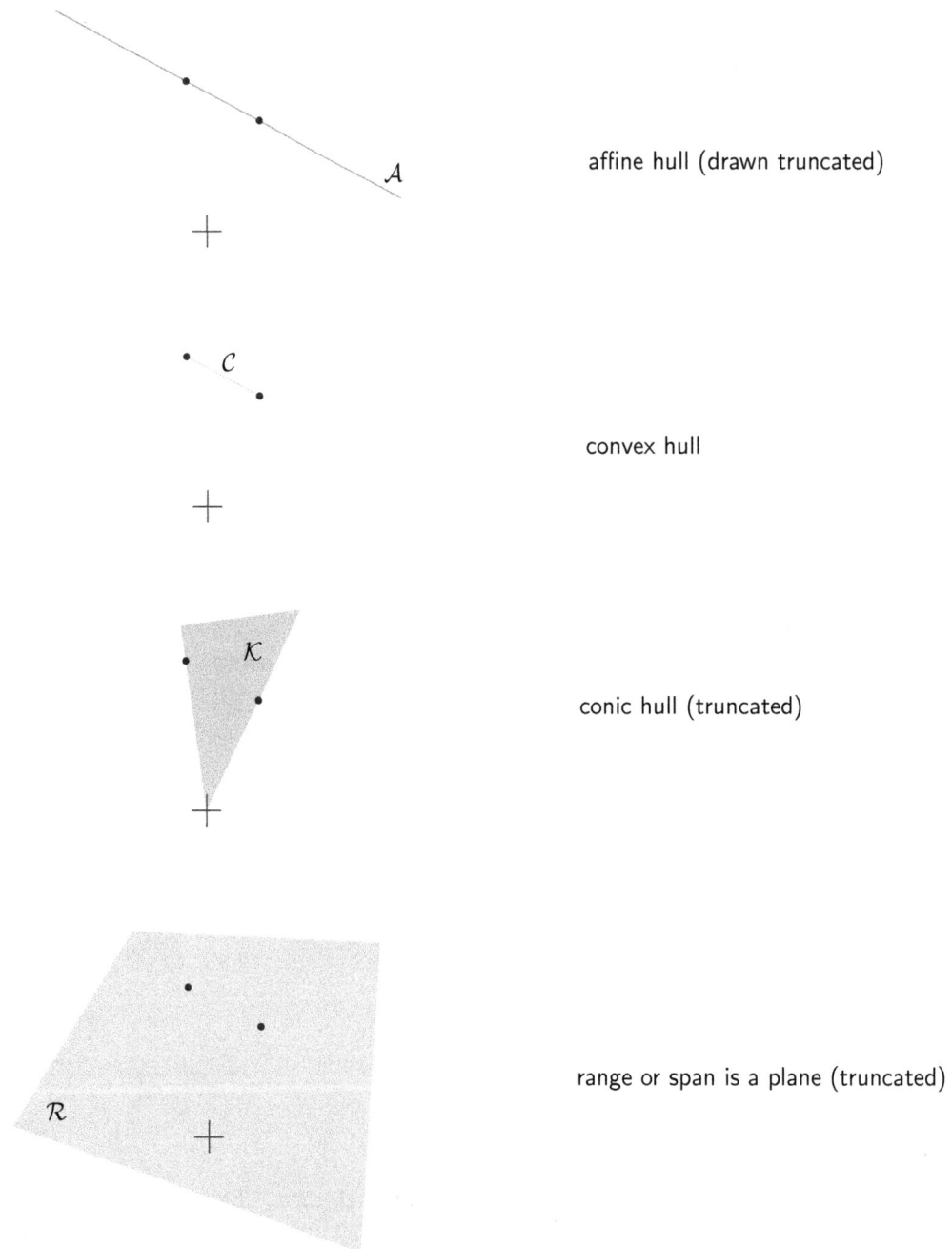

Figure 21: Given two points in Euclidean vector space of any dimension, their various hulls are illustrated. Each hull is a subset of range; generally, $\mathcal{A}, \mathcal{C}, \mathcal{K} \subseteq \mathcal{R} \ni \mathbf{0}$. (Cartesian axes drawn for reference.)

2.3.1.0.1 Example. *Affine hull of rank-1 correlation matrices.* [224]
The set of all $m \times m$ *rank*-1 *correlation matrices* is defined by all the binary vectors y in $\mathbb{R}^m$ (*confer* §5.9.1.0.1)

$$\{yy^{\mathrm{T}} \in \mathbb{S}_+^m \mid \delta(yy^{\mathrm{T}}) = \mathbf{1}\} \tag{84}$$

Affine hull of the rank-1 correlation matrices is equal to the set of normalized symmetric matrices; *id est*,

$$\mathrm{aff}\{yy^{\mathrm{T}} \in \mathbb{S}_+^m \mid \delta(yy^{\mathrm{T}}) = \mathbf{1}\} = \{A \in \mathbb{S}^m \mid \delta(A) = \mathbf{1}\} \tag{85}$$

$\square$

2.3.1.0.2 Exercise. *Affine hull of correlation matrices.*
Prove (85) via definition of affine hull. Find the convex hull instead. ▼

2.3.1.1 Partial order induced by $\mathbb{R}_+^N$ and $\mathbb{S}_+^M$

Notation $a \succeq 0$ means vector a belongs to nonnegative orthant $\mathbb{R}_+^N$ while $a \succ 0$ means vector a belongs to the nonnegative orthant's interior $\mathrm{int}\,\mathbb{R}_+^N$. $a \succeq b$ denotes comparison of vector a to vector b on $\mathbb{R}^N$ with respect to the nonnegative orthant; *id est*, $a \succeq b$ means $a - b$ belongs to the nonnegative orthant but neither a or b is necessarily nonnegative. With particular respect to the nonnegative orthant, $a \succeq b \Leftrightarrow a_i \geq b_i \; \forall i$ (369).

More generally, $a \succeq_\mathcal{K} b$ denotes comparison with respect to pointed closed convex cone $\mathcal{K}$, whereas comparison with respect to the cone's interior is denoted $a \succ_\mathcal{K} b$. But equivalence with entrywise comparison does not generally hold, and neither a or b necessarily belongs to $\mathcal{K}$. (§2.7.2.2)

The symbol $\geq$ is reserved for scalar comparison on the real line $\mathbb{R}$ with respect to the nonnegative real line $\mathbb{R}_+$ as in $a^{\mathrm{T}}y \geq b$. Comparison of matrices with respect to the positive semidefinite cone $\mathbb{S}_+^M$, like $I \succeq A \succeq 0$ in Example 2.3.2.0.1, is explained in §2.9.0.1.

2.3.2 Convex hull

The *convex hull* [205, §A.1.4] [315] of any bounded[2.15] list or set of N points $X \in \mathbb{R}^{n \times N}$ forms a unique bounded convex *polyhedron* (*confer* §2.12.0.0.1) whose vertices constitute some subset of that list;

$$\mathcal{P} \triangleq \mathrm{conv}\{x_\ell, \ \ell = 1 \ldots N\} = \mathrm{conv}\,X = \{Xa \mid a^{\mathrm{T}}\mathbf{1} = 1, \ a \succeq 0\} \subseteq \mathbb{R}^n \tag{86}$$

Union of relative interior and relative boundary (§2.1.7.2) of the polyhedron comprise its convex hull $\mathcal{P}$, the smallest closed convex set that contains the list X; *e.g.*, Figure **20**. Given $\mathcal{P}$, the *generating list* $\{x_\ell\}$ is not unique. But because every bounded polyhedron

---

[2.15]An arbitrary set $\mathcal{C}$ in $\mathbb{R}^n$ is *bounded* iff it can be contained in a Euclidean ball having finite radius. [116, §2.2] (*confer* §5.7.3.0.1) The smallest ball containing $\mathcal{C}$ has radius $\underset{x}{\inf}\,\underset{y \in \mathcal{C}}{\sup}\|x - y\|$ and center $x^\star$ whose determination is a convex problem because $\underset{y \in \mathcal{C}}{\sup}\|x - y\|$ is a convex function of x; but the supremum may be difficult to ascertain.

is the convex hull of its vertices, [337, §2.12.2] the vertices of $\mathcal{P}$ comprise a *minimal set of generators*.

Given some arbitrary set $\mathcal{C} \subseteq \mathbb{R}^n$, its convex hull $\operatorname{conv}\mathcal{C}$ is equivalent to the smallest convex set containing it. (*confer* §2.4.1.1.1) The convex hull is a subset of the affine hull;

$$\operatorname{conv}\mathcal{C} \ \subseteq \ \operatorname{aff}\mathcal{C} \ = \ \operatorname{aff}\overline{\mathcal{C}} \ = \ \overline{\operatorname{aff}\mathcal{C}} \ = \ \operatorname{aff}\operatorname{conv}\mathcal{C} \tag{87}$$

Any closed bounded convex set $\mathcal{C}$ is equal to the convex hull of its boundary;

$$\mathcal{C} = \operatorname{conv}\partial\mathcal{C} \tag{88}$$

$$\operatorname{conv}\emptyset \ \triangleq \ \emptyset \tag{89}$$

2.3.2.0.1 Example. *Hull of* rank-k *projection matrices.* [147] [295] [11, §4.1] [300, §3] [245, §2.4] [246] Convex hull of the set comprising outer product of orthonormal matrices has equivalent expression: for $1 \le k \le N$ (§2.9.0.1)

$$\operatorname{conv}\Big\{UU^{\mathrm{T}} \mid U \in \mathbb{R}^{N\times k},\ U^{\mathrm{T}}U=I\Big\} = \Big\{A \in \mathbb{S}^N \mid I \succeq A \succeq 0,\ \langle I,\,A\rangle = k\Big\} \subset \mathbb{S}_+^N \tag{90}$$

This important convex body we call *Fantope* (after mathematician Ky Fan). In case $k=1$, there is slight simplification: ((1702), Example 2.9.2.7.1)

$$\operatorname{conv}\Big\{UU^{\mathrm{T}} \mid U \in \mathbb{R}^N,\ U^{\mathrm{T}}U=1\Big\} = \Big\{A \in \mathbb{S}^N \mid A \succeq 0,\ \langle I,\,A\rangle = 1\Big\} \tag{91}$$

In case $k=N$, the Fantope is Identity matrix I. More generally, the set

$$\Big\{UU^{\mathrm{T}} \mid U \in \mathbb{R}^{N\times k},\ U^{\mathrm{T}}U=I\Big\} \tag{92}$$

comprises the extreme points (§2.6.0.0.1) of its convex hull. By (1544), each and every extreme point UU^{T} has only k nonzero eigenvalues λ and they all equal 1; *id est*, $\lambda(UU^{\mathrm{T}})_{1:k} = \lambda(U^{\mathrm{T}}U) = \mathbf{1}$. So Frobenius' norm of each and every extreme point equals the same constant

$$\|UU^{\mathrm{T}}\|_{\mathrm{F}}^2 = k \tag{93}$$

Each extreme point simultaneously lies on the boundary of the positive semidefinite cone (when $k<N$, §2.9) and on the boundary of a *hypersphere* of dimension $k(N-\frac{k}{2}+\frac{1}{2})$ and radius $\sqrt{k(1-\frac{k}{N})}$ centered at $\frac{k}{N}I$ along the ray (base $\mathbf{0}$) through the Identity matrix I in isomorphic vector space $\mathbb{R}^{N(N+1)/2}$ (§2.2.2.1).

Figure **22** illustrates extreme points (92) comprising the boundary of a Fantope, the boundary of a *disc* corresponding to $k=1$, $N=2$; but that circumscription does not hold in higher dimension. (§2.9.2.8) $\square$

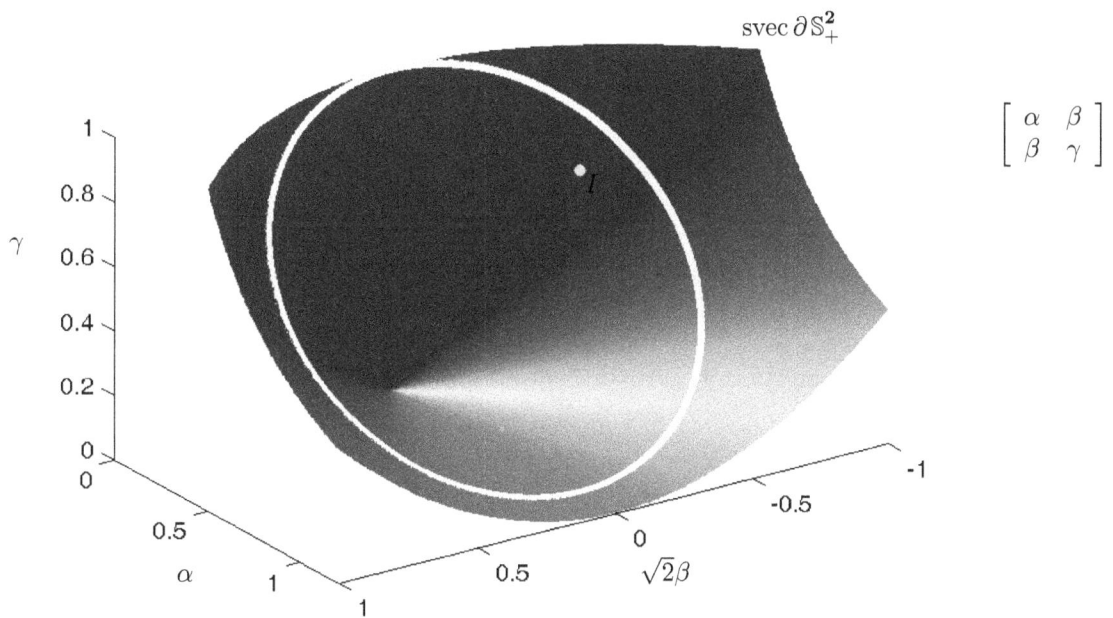

Figure 22: Two Fantopes. Circle (radius $1/\sqrt{2}$), shown here on boundary of positive semidefinite cone $\mathbb{S}_+^2$ in isometrically isomorphic $\mathbb{R}^3$ from Figure 44, comprises boundary of a Fantope (90) in this dimension ($k = 1$, $N = 2$). Lone point illustrated is Identity matrix I, interior to PSD cone, and is that Fantope corresponding to $k = 2$, $N = 2$. (View is from inside PSD cone looking toward origin.)

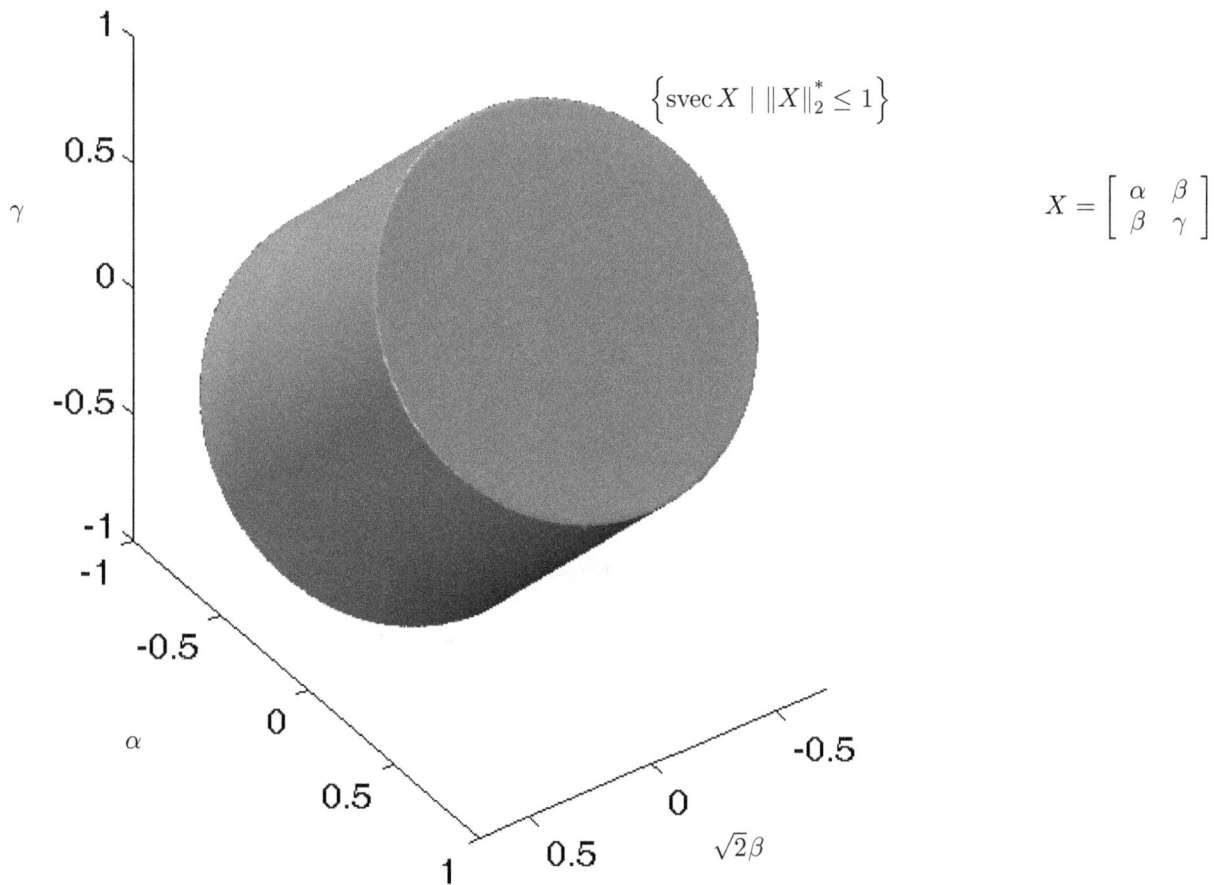

$$\left\{ \mathrm{svec}\, X \mid \|X\|_2^* \le 1 \right\}$$

$$X = \begin{bmatrix} \alpha & \beta \\ \beta & \gamma \end{bmatrix}$$

Figure 23: *Nuclear norm* is a sum of singular values; $\|X\|_2^* \triangleq \sum_i \sigma(X)_i$. Nuclear norm ball, in the subspace of 2×2 symmetric matrices, is a truncated cylinder in isometrically isomorphic $\mathbb{R}^3$.

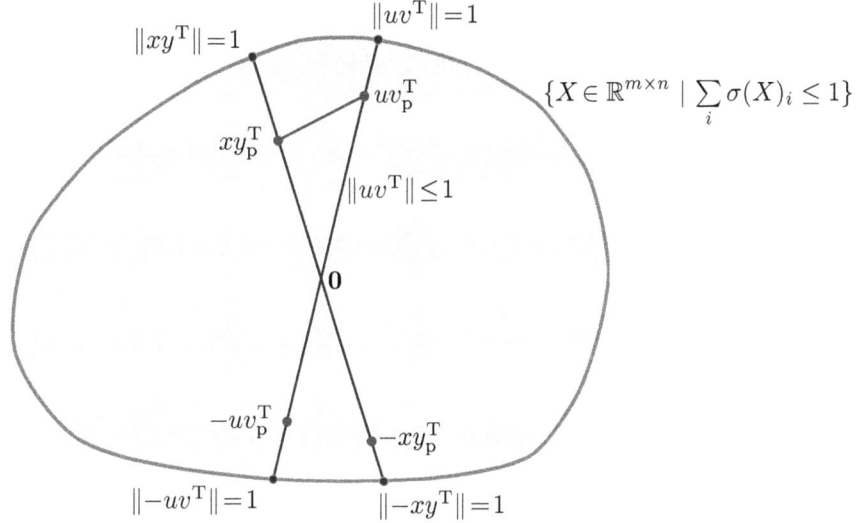

Figure 24: $uv_{\mathrm{p}}^{\mathrm{T}}$ is a convex combination of normalized dyads $\|\pm uv^{\mathrm{T}}\|=1$; similarly for $xy_{\mathrm{p}}^{\mathrm{T}}$. Any point in line segment joining $xy_{\mathrm{p}}^{\mathrm{T}}$ to $uv_{\mathrm{p}}^{\mathrm{T}}$ is expressible as a convex combination of two to four points indicated on boundary.

2.3.2.0.2 Example. *Nuclear norm ball: convex hull of rank-1 matrices.*
From (91), in Example 2.3.2.0.1, we learn that the convex hull of normalized symmetric rank-1 matrices is a slice of the positive semidefinite cone. In §2.9.2.7 we find the convex hull of all symmetric rank-1 matrices to be the entire positive semidefinite cone.

In the present example we abandon symmetry; instead posing, what is the convex hull of bounded nonsymmetric rank-1 matrices:

$$\mathrm{conv}\{uv^{\mathrm{T}} \mid \|uv^{\mathrm{T}}\| \leq 1,\ u\in\mathbb{R}^{m},\ v\in\mathbb{R}^{n}\} = \{X\in\mathbb{R}^{m\times n} \mid \sum_{i}\sigma(X)_{i} \leq 1\} \qquad (94)$$

where $\sigma(X)$ is a vector of singular values. (Since $\|uv^{\mathrm{T}}\|=\|u\|\|v\|$ (1693), norm of each vector constituting a *dyad* uv^{T} (§B.1) in the hull is effectively bounded above by 1.)

Proof. ($\Leftarrow$) Suppose $\sum\sigma(X)_{i} \leq 1$. As in §A.6, define a singular value decomposition: $X = U\Sigma V^{\mathrm{T}}$ where $U = [u_{1} \ldots u_{\min\{m,n\}}] \in \mathbb{R}^{m\times\min\{m,n\}}$, $V = [v_{1} \ldots v_{\min\{m,n\}}] \in \mathbb{R}^{n\times\min\{m,n\}}$, and whose sum of singular values is $\sum\sigma(X)_{i} = \mathrm{tr}\,\Sigma = \kappa \leq 1$. Then we may write $X = \sum\frac{\sigma_{i}}{\kappa}\sqrt{\kappa}u_{i}\sqrt{\kappa}v_{i}^{\mathrm{T}}$ which is a convex combination of dyads each of whose norm does not exceed 1. (Srebro)

($\Rightarrow$) Now suppose we are given a convex combination of dyads $X = \sum\alpha_{i}u_{i}v_{i}^{\mathrm{T}}$ such that $\sum\alpha_{i}=1$, $\alpha_{i}\geq 0\ \forall i$, and $\|u_{i}v_{i}^{\mathrm{T}}\|\leq 1\ \forall i$. Then by triangle inequality for singular values [209, cor.3.4.3] $\sum\sigma(X)_{i} \leq \sum\sigma(\alpha_{i}u_{i}v_{i}^{\mathrm{T}}) = \sum\alpha_{i}\|u_{i}v_{i}^{\mathrm{T}}\| \leq \sum\alpha_{i}$. $\blacklozenge$

Given any particular dyad $uv_{\mathrm{p}}^{\mathrm{T}}$ in the convex hull, because its polar $-uv_{\mathrm{p}}^{\mathrm{T}}$ and every convex combination of the two belong to that hull, then the unique line containing those two points $\pm uv_{\mathrm{p}}^{\mathrm{T}}$ (their affine combination (78)) must intersect the hull's boundary at the normalized dyads $\{\pm uv^{\mathrm{T}} \mid \|uv^{\mathrm{T}}\|=1\}$. Any point formed by convex combination of dyads in the hull must therefore be expressible as a convex combination of dyads on the boundary: Figure 24,

$$\mathrm{conv}\{uv^{\mathrm{T}} \mid \|uv^{\mathrm{T}}\| \leq 1, \ u\in\mathbb{R}^m, \ v\in\mathbb{R}^n\} \equiv \mathrm{conv}\{uv^{\mathrm{T}} \mid \|uv^{\mathrm{T}}\| = 1, \ u\in\mathbb{R}^m, \ v\in\mathbb{R}^n\} \quad (95)$$

id est, dyads may be normalized and the hull's boundary contains them;

$$\begin{aligned}\partial\{X\in\mathbb{R}^{m\times n} \mid \sum_i \sigma(X)_i \leq 1\} &= \{X\in\mathbb{R}^{m\times n} \mid \sum_i \sigma(X)_i = 1\}\\ &\supseteq \{uv^{\mathrm{T}} \mid \|uv^{\mathrm{T}}\| = 1, \ u\in\mathbb{R}^m, \ v\in\mathbb{R}^n\}\end{aligned} \quad (96)$$

Normalized dyads constitute the set of extreme points (§2.6.0.0.1) of this nuclear norm ball (*confer* Figure 23) which is, therefore, their convex hull.   □

2.3.2.0.3 Exercise. *Convex hull of outer product.*
Describe the interior of a Fantope.
Find the convex hull of nonorthogonal projection matrices (§E.1.1):

$$\{UV^{\mathrm{T}} \mid U\in\mathbb{R}^{N\times k}, \ V\in\mathbb{R}^{N\times k}, \ V^{\mathrm{T}}U=I\} \quad (97)$$

Find the convex hull of nonsymmetric matrices bounded under some norm:

$$\{UV^{\mathrm{T}} \mid U\in\mathbb{R}^{m\times k}, \ V\in\mathbb{R}^{n\times k}, \ \|UV^{\mathrm{T}}\| \leq 1\} \quad (98)$$

▼

2.3.2.0.4 Example. *Permutation polyhedron.*   [207] [325] [267]
A *permutation matrix* Ξ is formed by interchanging rows and columns of Identity matrix I. Since Ξ is square and $\Xi^{\mathrm{T}}\Xi = I$, the set of all permutation matrices is a *proper subset* of the nonconvex *manifold* of orthogonal matrices (§B.5). In fact, the only orthogonal matrices having all nonnegative entries are permutations of the Identity:

$$\Xi^{-1} = \Xi^{\mathrm{T}}, \qquad \Xi \geq \mathbf{0} \quad (99)$$

And the only positive semidefinite permutation matrix is the Identity. [340, §6.5 prob.20]

Regarding the permutation matrices as a set of points in Euclidean space, its convex hull is a bounded polyhedron (§2.12) described (Birkhoff, 1946)

$$\mathrm{conv}\{\Xi = \Pi_i(I\in\mathbb{S}^n)\in\mathbb{R}^{n\times n}, \ i=1\dots n!\} = \{X\in\mathbb{R}^{n\times n} \mid X^{\mathrm{T}}\mathbf{1}=\mathbf{1}, \ X\mathbf{1}=\mathbf{1}, \ X\geq\mathbf{0}\} \quad (100)$$

where Π_i is a linear operator here representing the ith permutation. This polyhedral hull, whose $n!$ vertices are the permutation matrices, is also known as the set of *doubly stochastic matrices*. The permutation matrices are the minimal cardinality (fewest nonzero entries) doubly stochastic matrices. The only orthogonal matrices belonging to this polyhedron are the permutation matrices.

It is remarkable that $n!$ permutation matrices can be described as the extreme points (§2.6.0.0.1) of a bounded polyhedron, of affine dimension $(n-1)^2$, that is itself described by $2n$ equalities ($2n-1$ linearly independent equality constraints in n^2 nonnegative variables). By *Carathéodory's theorem*, conversely, any doubly stochastic matrix can be described as a convex combination of at most $(n-1)^2+1$ permutation matrices. [208, §8.7] [56, thm.1.2.5] This polyhedron, then, can be a device for *relaxing* an integer, combinatorial, or Boolean optimization problem.[2.16] [67] [291, §3.1] □

2.3.2.0.5 Example. *Convex hull of orthonormal matrices.* [27, §1.2]
Consider rank-k matrices $U \in \mathbb{R}^{n \times k}$ such that $U^{\mathrm{T}}U = I$. These are the orthonormal matrices; a closed bounded submanifold, of all orthogonal matrices, having dimension $nk - \frac{1}{2}k(k+1)$ [53]. Their convex hull is expressed, for $1 \leq k \leq n$

$$\begin{aligned} \mathrm{conv}\{U \in \mathbb{R}^{n \times k} \mid U^{\mathrm{T}}U = I\} &= \{X \in \mathbb{R}^{n \times k} \mid \|X\|_2 \leq 1\} \\ &= \{X \in \mathbb{R}^{n \times k} \mid \|X^{\mathrm{T}}a\| \leq \|a\| \ \forall a \in \mathbb{R}^n\} \end{aligned} \quad (101)$$

By Schur complement (§A.4), the *spectral norm* $\|X\|_2$ constraining largest singular value $\sigma(X)_1$ can be expressed as a semidefinite constraint

$$\|X\|_2 \leq 1 \quad \Leftrightarrow \quad \begin{bmatrix} I & X \\ X^{\mathrm{T}} & I \end{bmatrix} \succeq 0 \quad (102)$$

because of equivalence $X^{\mathrm{T}}X \preceq I \Leftrightarrow \sigma(X) \preceq \mathbf{1}$ with singular values. (1647) (1531) (1532)

When $k = n$, matrices U are orthogonal and the convex hull is called the *spectral norm ball* which is the set of all *contractions*. [209, p.158] [336, p.313] The orthogonal matrices then constitute the extreme points (§2.6.0.0.1) of this hull. Hull intersection with the nonnegative orthant $\mathbb{R}_+^{n \times n}$ contains the permutation polyhedron (100). □

2.3.3 Conic hull

In terms of a finite-length point list (or set) arranged columnar in $X \in \mathbb{R}^{n \times N}$ (76), its conic hull is expressed

$$\mathcal{K} \triangleq \mathrm{cone}\{x_\ell, \ell = 1 \ldots N\} = \mathrm{cone}\, X = \{Xa \mid a \succeq 0\} \subseteq \mathbb{R}^n \quad (103)$$

id est, every nonnegative combination of points from the list. Conic hull of any finite-length list forms a *polyhedral cone* [205, §A.4.3] (§2.12.1.0.1; *e.g.*, Figure 51a); the smallest closed convex cone (§2.7.2) that contains the list.

By convention, the aberration [337, §2.1]

$$\mathrm{cone}\, \emptyset \triangleq \{\mathbf{0}\} \quad (104)$$

Given some arbitrary set $\mathcal{C}$, it is apparent

$$\mathrm{conv}\, \mathcal{C} \subseteq \mathrm{cone}\, \mathcal{C} \quad (105)$$

[2.16] *Relaxation* replaces an *objective function* with its *convex envelope* or expands a feasible set to one that is convex. Dantzig first showed in 1951 that, by this device, the so-called *assignment problem* can be formulated as a linear program. [324] [26, §II.5]

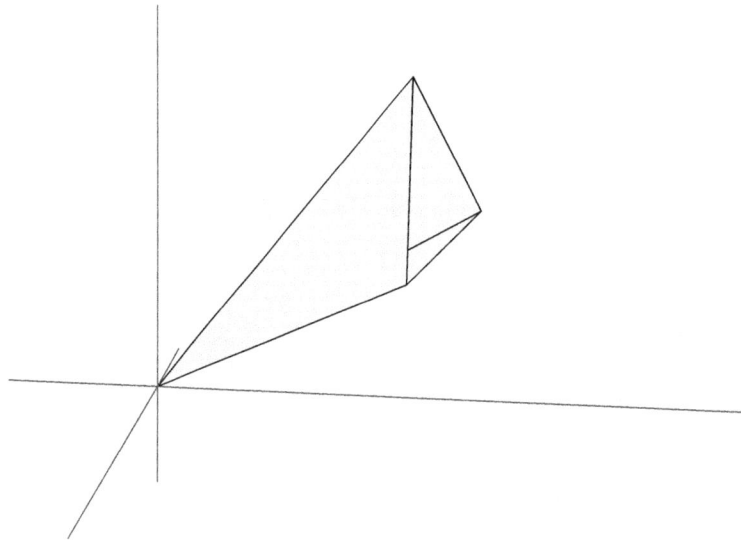

Figure 25: A simplicial cone (§2.12.3.1.1) in $\mathbb{R}^3$ whose boundary is drawn truncated; constructed using $A \in \mathbb{R}^{3 \times 3}$ and $C = \mathbf{0}$ in (286). By the most fundamental definition of a cone (§2.7.1), entire boundary can be constructed from an aggregate of rays emanating exclusively from the origin. Each of three extreme directions corresponds to an edge (§2.6.0.0.3); they are conically, affinely, and linearly independent for this cone. Because this set is polyhedral, exposed directions are in one-to-one correspondence with extreme directions; there are only three. Its extreme directions give rise to what is called a *vertex-description* of this polyhedral cone; simply, the conic hull of extreme directions. Obviously this cone can also be constructed by intersection of three halfspaces; hence the equivalent *halfspace-description*.

2.3.4 Vertex-description

The conditions in (78), (86), and (103) respectively define an *affine combination, convex combination,* and *conic combination* of elements from the set or list. Whenever a Euclidean body can be described as some hull or span of a set of points, then that representation is loosely called a *vertex-description* and those points are called *generators.*

2.4 Halfspace, Hyperplane

A two-dimensional affine subset is called a *plane.* An $(n-1)$-dimensional affine subset of $\mathbb{R}^n$ is called a *hyperplane.* [315] [205] Every hyperplane partially bounds a *halfspace* (which is convex, but not affine, and the only nonempty convex set in $\mathbb{R}^n$ whose complement is convex and nonempty).

2.4.1 Halfspaces $\mathcal{H}_+$ and $\mathcal{H}_-$

Euclidean space $\mathbb{R}^n$ is partitioned in two by any hyperplane $\partial\mathcal{H}$; *id est,* $\mathcal{H}_- + \mathcal{H}_+ = \mathbb{R}^n$. The resulting (closed convex) halfspaces, both partially bounded by $\partial\mathcal{H}$, may be described as an asymmetry

$$\mathcal{H}_- = \{y \mid a^\mathrm{T}y \leq b\} = \{y \mid a^\mathrm{T}(y - y_\mathrm{p}) \leq 0\} \subset \mathbb{R}^n \qquad (106)$$

$$\mathcal{H}_+ = \{y \mid a^\mathrm{T}y \geq b\} = \{y \mid a^\mathrm{T}(y - y_\mathrm{p}) \geq 0\} \subset \mathbb{R}^n \qquad (107)$$

where nonzero vector $a \in \mathbb{R}^n$ is an *outward-normal* to the hyperplane partially bounding $\mathcal{H}_-$ while an *inward-normal* with respect to $\mathcal{H}_+$. For any vector $y - y_\mathrm{p}$ that makes an obtuse angle with normal a, vector y will lie in the halfspace $\mathcal{H}_-$ on one side (shaded in Figure 26) of the hyperplane while acute angles denote y in $\mathcal{H}_+$ on the other side.

An equivalent more intuitive representation of a halfspace comes about when we consider all the points in $\mathbb{R}^n$ closer to point d than to point c or equidistant, in the Euclidean sense; from Figure 26,

$$\mathcal{H}_- = \{y \mid \|y - \mathrm{d}\| \leq \|y - \mathrm{c}\|\} \qquad (108)$$

This representation, in terms of proximity, is resolved with the more conventional representation of a halfspace (106) by squaring both sides of the inequality in (108);

$$\mathcal{H}_- = \left\{y \mid (\mathrm{c} - \mathrm{d})^\mathrm{T}y \leq \frac{\|\mathrm{c}\|^2 - \|\mathrm{d}\|^2}{2}\right\} = \left\{y \mid (\mathrm{c} - \mathrm{d})^\mathrm{T}\left(y - \frac{\mathrm{c} + \mathrm{d}}{2}\right) \leq 0\right\} \qquad (109)$$

2.4.1.1 PRINCIPLE 1: Halfspace-description of convex sets

The most fundamental principle in convex geometry follows from the *geometric Hahn-Banach theorem* [256, §5.12] [18, §1] [136, §I.1.2] which guarantees any closed convex set to be an intersection of halfspaces.

$$\mathcal{H}_+ = \{y \mid a^{\mathrm{T}}(y - y_{\mathrm{p}}) \geq 0\}$$

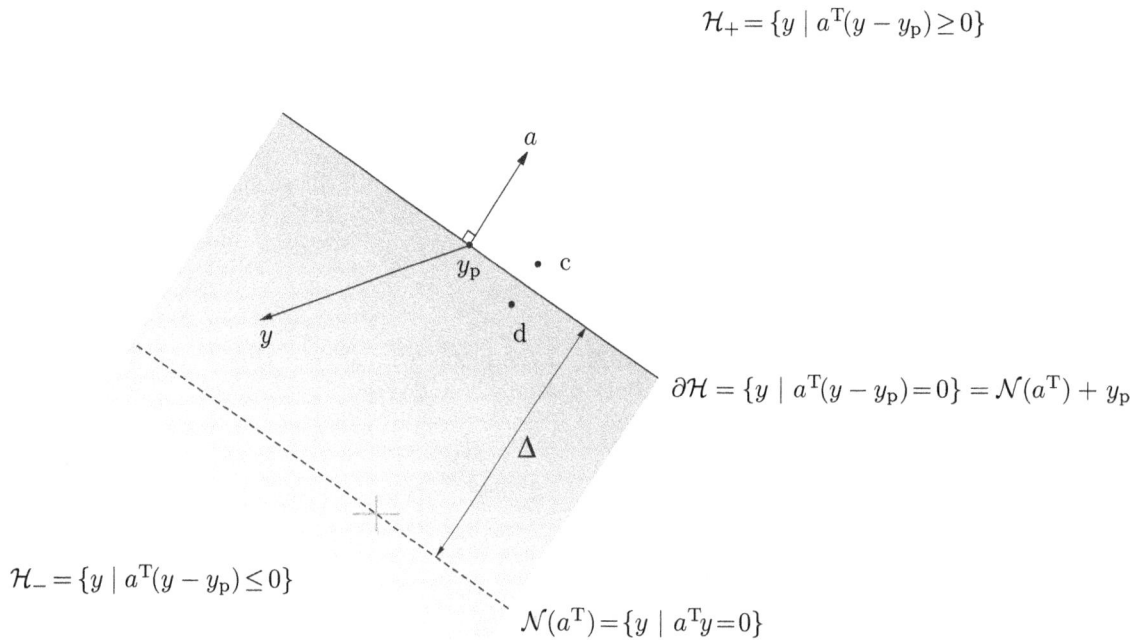

$$\partial\mathcal{H} = \{y \mid a^{\mathrm{T}}(y - y_{\mathrm{p}}) = 0\} = \mathcal{N}(a^{\mathrm{T}}) + y_{\mathrm{p}}$$

$$\mathcal{H}_- = \{y \mid a^{\mathrm{T}}(y - y_{\mathrm{p}}) \leq 0\}$$

$$\mathcal{N}(a^{\mathrm{T}}) = \{y \mid a^{\mathrm{T}}y = 0\}$$

Figure 26: Hyperplane illustrated $\partial\mathcal{H}$ is a line partially bounding halfspaces $\mathcal{H}_-$ and $\mathcal{H}_+$ in $\mathbb{R}^2$. Shaded is a rectangular piece of semiinfinite $\mathcal{H}_-$ with respect to which vector a is outward-normal to bounding hyperplane; vector a is inward-normal with respect to $\mathcal{H}_+$. Halfspace $\mathcal{H}_-$ contains nullspace $\mathcal{N}(a^{\mathrm{T}})$ (dashed line through origin) because $a^{\mathrm{T}}y_{\mathrm{p}} > 0$. Hyperplane, halfspace, and nullspace are each drawn truncated. Points c and d are equidistant from hyperplane, and vector c − d is normal to it. Δ is distance from origin to hyperplane.

2.4.1.1.1 Theorem. *Halfspaces.* [205, §A.4.2b] [42, §2.4]
A closed convex set in $\mathbb{R}^n$ is equivalent to the intersection of all halfspaces that contain it. $\diamond$

Intersection of multiple halfspaces in $\mathbb{R}^n$ may be represented using a matrix constant A

$$\bigcap_i \mathcal{H}_{i-} = \{y \mid A^\mathrm{T}y \preceq b\} = \{y \mid A^\mathrm{T}(y - y_\mathrm{p}) \preceq 0\} \tag{110}$$

$$\bigcap_i \mathcal{H}_{i+} = \{y \mid A^\mathrm{T}y \succeq b\} = \{y \mid A^\mathrm{T}(y - y_\mathrm{p}) \succeq 0\} \tag{111}$$

where b is now a vector, and the ith column of A is normal to a hyperplane $\partial\mathcal{H}_i$ partially bounding $\mathcal{H}_i$. By the *halfspaces theorem,* intersections like this can describe interesting convex Euclidean bodies such as polyhedra and cones, giving rise to the term *halfspace-description.*

2.4.2 Hyperplane $\partial\mathcal{H}$ representations

Every hyperplane $\partial\mathcal{H}$ is an affine set parallel to an $(n-1)$-dimensional subspace of $\mathbb{R}^n$; it is itself a subspace if and only if it contains the origin.

$$\dim \partial\mathcal{H} = n - 1 \tag{112}$$

so a hyperplane is a point in $\mathbb{R}$, a line in $\mathbb{R}^2$, a plane in $\mathbb{R}^3$, and so on. Every hyperplane can be described as the intersection of complementary halfspaces; [315, §19]

$$\partial\mathcal{H} = \mathcal{H}_- \cap \mathcal{H}_+ = \{y \mid a^\mathrm{T}y \leq b \,,\; a^\mathrm{T}y \geq b\} = \{y \mid a^\mathrm{T}y = b\} \tag{113}$$

a halfspace-description. Assuming *normal* $a \in \mathbb{R}^n$ to be nonzero, then any hyperplane in $\mathbb{R}^n$ can be described as the solution set to vector equation $a^\mathrm{T}y = b$ (illustrated in Figure **26** and Figure **27** for $\mathbb{R}^2$);

$$\partial\mathcal{H} \triangleq \{y \mid a^\mathrm{T}y = b\} = \{y \mid a^\mathrm{T}(y - y_\mathrm{p}) = 0\} = \{Z\xi + y_\mathrm{p} \mid \xi \in \mathbb{R}^{n-1}\} \subset \mathbb{R}^n \tag{114}$$

All solutions y constituting the hyperplane are *offset* from the nullspace of a^T by the same constant vector $y_\mathrm{p} \in \mathbb{R}^n$ that is any particular solution to $a^\mathrm{T}y = b$; *id est,*

$$y = Z\xi + y_\mathrm{p} \tag{115}$$

where the columns of $Z \in \mathbb{R}^{n \times n-1}$ constitute a basis for the nullspace $\mathcal{N}(a^\mathrm{T}) = \{x \in \mathbb{R}^n \mid a^\mathrm{T}x = 0\}$.[2.17]

Conversely, given any point y_p in $\mathbb{R}^n$, the unique hyperplane containing it having normal a is the affine set $\partial\mathcal{H}$ (114) where b equals $a^\mathrm{T}y_\mathrm{p}$ and where a basis for $\mathcal{N}(a^\mathrm{T})$ is arranged in Z columnar. Hyperplane dimension is apparent from dimension of Z; that hyperplane is parallel to the span of its columns.

[2.17]We will later find this expression for y in terms of nullspace of a^T (more generally, of matrix A^T (144)) to be a useful trick (a practical device) for eliminating affine equality constraints, much as we did here.

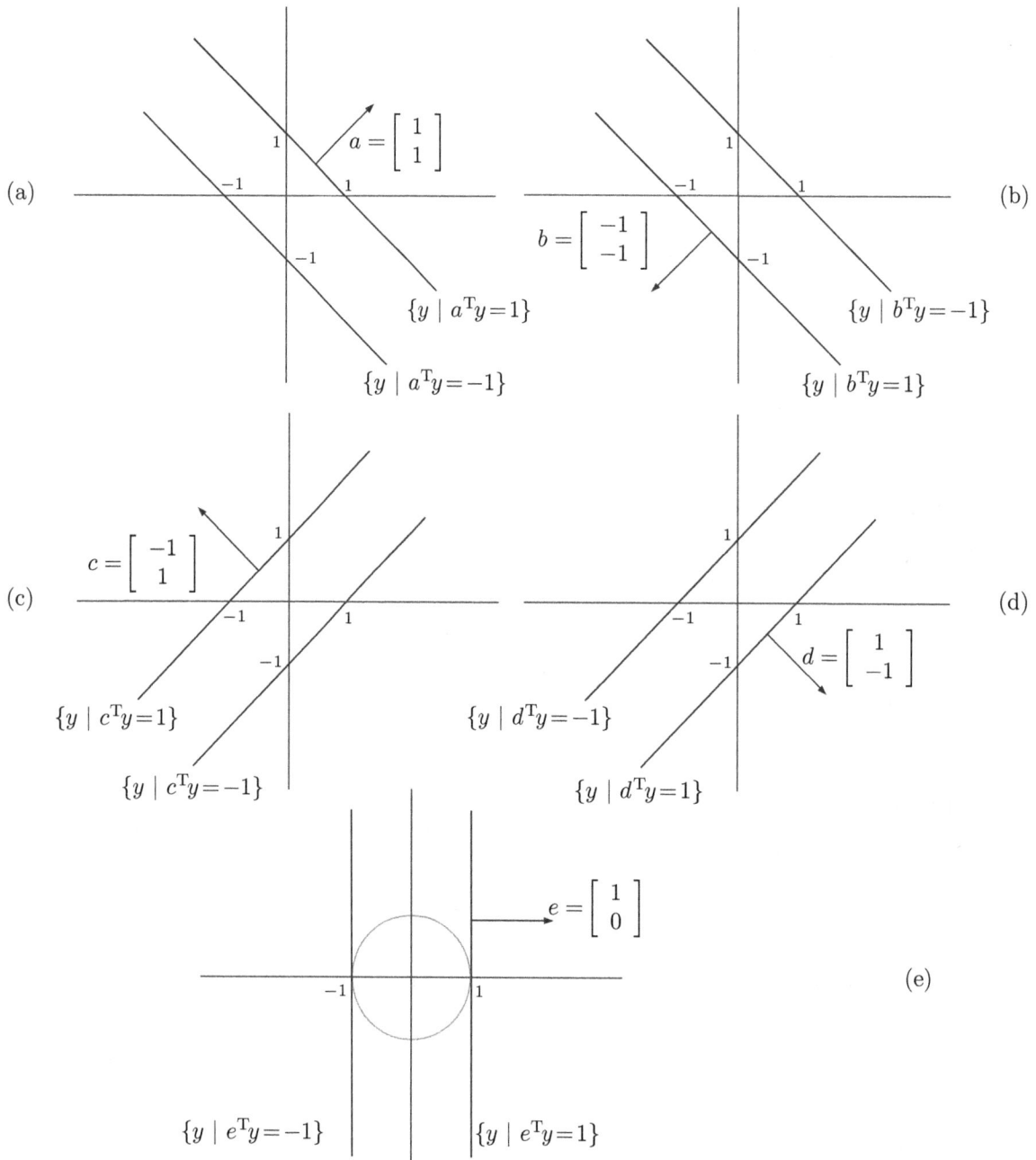

Figure 27: **(a)-(d)** Hyperplanes in $\mathbb{R}^2$ (truncated) redundantly emphasize: hyperplane movement opposite to its normal direction minimizes vector inner-product. This concept is exploited to attain analytical solution of linear programs by visual inspection; *e.g.*, §2.4.2.6.2, §2.5.1.2.2, §3.4.0.0.2, [62, exer.4.8-exer.4.20]. Each graph is also interpretable as contour plot of a real affine function of two variables as in Figure 74. **(e)** $|\beta|/\|\alpha\|$ from $\partial\mathcal{H} = \{x \mid \alpha^{\mathrm{T}}x = \beta\}$ represents radius of hypersphere about $\mathbf{0}$ supported by any hyperplane with same ratio |inner product|/norm.

2.4.2.0.1   Exercise.   *Hyperplane scaling.*

Given normal y, draw a hyperplane $\{x \in \mathbb{R}^2 \mid x^\mathrm{T}y = 1\}$. Suppose $z = \frac{1}{2}y$. On the same plot, draw the hyperplane $\{x \in \mathbb{R}^2 \mid x^\mathrm{T}z = 1\}$. Now suppose $z = 2y$, then draw the last hyperplane again with this new z. What is the apparent effect of scaling normal y?

$$\blacktriangledown$$

2.4.2.0.2   Example.   *Distance from origin to hyperplane.*

Given the (shortest) distance $\Delta \in \mathbb{R}_+$ from the origin to a hyperplane having normal vector a, we can find its representation $\partial\mathcal{H}$ by dropping a perpendicular. The point thus found is the orthogonal projection of the origin on $\partial\mathcal{H}$ (§E.5.0.0.5), equal to $a\Delta/\|a\|$ if the origin is known *a priori* to belong to halfspace $\mathcal{H}_-$ (Figure **26**), or equal to $-a\Delta/\|a\|$ if the origin belongs to halfspace $\mathcal{H}_+$; *id est*, when $\mathcal{H}_- \ni \mathbf{0}$

$$\partial\mathcal{H} \;=\; \{y \mid a^\mathrm{T}(y - a\Delta/\|a\|) = 0\} \;=\; \{y \mid a^\mathrm{T}y = \|a\|\Delta\} \tag{116}$$

or when $\mathcal{H}_+ \ni \mathbf{0}$

$$\partial\mathcal{H} \;=\; \{y \mid a^\mathrm{T}(y + a\Delta/\|a\|) = 0\} \;=\; \{y \mid a^\mathrm{T}y = -\|a\|\Delta\} \tag{117}$$

Knowledge of only distance Δ and normal a thus introduces ambiguity into the hyperplane representation. $\square$

2.4.2.1   Matrix variable

Any halfspace in $\mathbb{R}^{mn}$ may be represented using a matrix variable. For variable $Y \in \mathbb{R}^{m \times n}$, given constants $A \in \mathbb{R}^{m \times n}$ and $b = \langle A, Y_\mathrm{p} \rangle \in \mathbb{R}$

$$\mathcal{H}_- \;=\; \{Y \in \mathbb{R}^{mn} \mid \langle A, Y \rangle \leq b\} \;=\; \{Y \in \mathbb{R}^{mn} \mid \langle A, Y - Y_\mathrm{p} \rangle \leq 0\} \tag{118}$$
$$\mathcal{H}_+ \;=\; \{Y \in \mathbb{R}^{mn} \mid \langle A, Y \rangle \geq b\} \;=\; \{Y \in \mathbb{R}^{mn} \mid \langle A, Y - Y_\mathrm{p} \rangle \geq 0\} \tag{119}$$

Recall vector inner-product from §2.2: $\langle A, Y \rangle = \mathrm{tr}(A^\mathrm{T}Y) = \mathrm{vec}(A)^\mathrm{T}\mathrm{vec}(Y)$.

Hyperplanes in $\mathbb{R}^{mn}$ may, of course, also be represented using matrix variables.

$$\partial\mathcal{H} \;=\; \{Y \mid \langle A, Y \rangle = b\} \;=\; \{Y \mid \langle A, Y - Y_\mathrm{p} \rangle = 0\} \subset \mathbb{R}^{mn} \tag{120}$$

Vector a from Figure **26** is normal to the hyperplane illustrated. Likewise, nonzero vectorized matrix A is normal to hyperplane $\partial\mathcal{H}$;

$$A \perp \partial\mathcal{H} \ \text{ in } \mathbb{R}^{mn} \tag{121}$$

2.4.2.2   Vertex-description of hyperplane

Any hyperplane in $\mathbb{R}^n$ may be described as affine hull of a minimal set of points $\{x_\ell \in \mathbb{R}^n, \ \ell = 1 \ldots n\}$ arranged columnar in a matrix $X \in \mathbb{R}^{n \times n}$: (78)

$$
\begin{aligned}
\partial\mathcal{H} &= \mathrm{aff}\{x_\ell \in \mathbb{R}^n, \ \ell = 1 \ldots n\}, & \dim \mathrm{aff}\{x_\ell \ \forall\ell\} &= n-1 \\
&= \mathrm{aff}\,X, & \dim \mathrm{aff}\,X &= n-1 \\
&= x_1 + \mathcal{R}\{x_\ell - x_1, \ \ell = 2 \ldots n\}, & \dim \mathcal{R}\{x_\ell - x_1, \ \ell = 2 \ldots n\} &= n-1 \\
&= x_1 + \mathcal{R}(X - x_1\mathbf{1}^\mathrm{T}), & \dim \mathcal{R}(X - x_1\mathbf{1}^\mathrm{T}) &= n-1
\end{aligned}
\tag{122}
$$

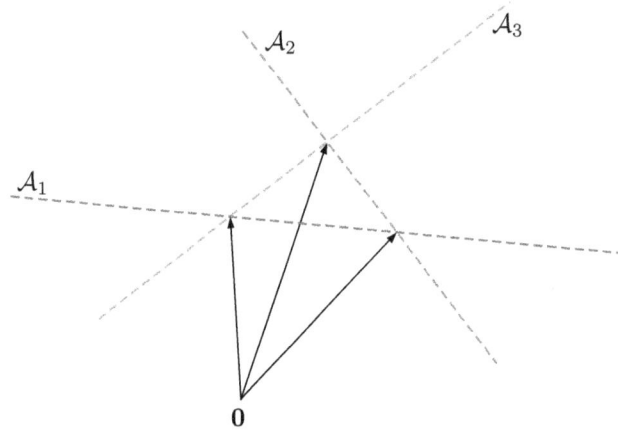

Figure 28: Of three points illustrated, any one particular point does not belong to affine hull $\mathcal{A}_i$ ($i \in 1,2,3$, each drawn truncated) of points remaining. Three corresponding vectors in $\mathbb{R}^2$ are, therefore, affinely independent (but neither linearly or conically independent).

where

$$\mathcal{R}(A) = \{Ax \mid \forall x\} \qquad (142)$$

2.4.2.3 Affine independence, minimal set

For any particular affine set, a *minimal set* of points constituting its vertex-description is an affinely independent generating set and *vice versa*.

Arbitrary given points $\{x_i \in \mathbb{R}^n,\ i=1\ldots N\}$ are *affinely independent* (a.i.) if and only if, over all $\zeta \in \mathbb{R}^N \ni \zeta^T \mathbf{1}=1$, $\zeta_k=0 \in \mathbb{R}$ (*confer* §2.1.2)

$$x_i\,\zeta_i + \cdots + x_j\,\zeta_j - x_k = \mathbf{0}, \qquad i \neq \cdots \neq j \neq k = 1\ldots N \qquad (123)$$

has no solution ζ; in words, iff no point from the given set can be expressed as an affine combination of those remaining. We deduce

$$\text{l.i.} \ \Rightarrow \ \text{a.i.} \qquad (124)$$

Consequently, $\{x_i,\ i=1\ldots N\}$ is an affinely independent set if and only if $\{x_i - x_1,\ i=2\ldots N\}$ is a linearly independent (l.i.) set. [211, §3] (Figure 28) This is equivalent to the property that the columns of $\begin{bmatrix} X \\ \mathbf{1}^T \end{bmatrix}$ (for $X \in \mathbb{R}^{n \times N}$ as in (76)) form a linearly independent set. [205, §A.1.3]

Two nontrivial affine subsets are affinely independent iff their intersection is empty $\{\emptyset\}$ or, analogously to subspaces, they intersect only at a point.

2.4.2.4 Preservation of affine independence

Independence in the linear (§2.1.2.1), affine, and conic (§2.10.1) senses can be preserved under linear transformation. Suppose a matrix $X \in \mathbb{R}^{n \times N}$ (76) holds an affinely independent set in its columns. Consider a transformation on the domain of such matrices

$$T(X) : \mathbb{R}^{n \times N} \to \mathbb{R}^{n \times N} \triangleq XY \qquad (125)$$

where fixed matrix $Y \triangleq [\, y_1 \ y_2 \ \cdots \ y_N \,] \in \mathbb{R}^{N \times N}$ represents linear operator T. Affine independence of $\{ Xy_i \in \mathbb{R}^n, \ i=1 \ldots N \}$ demands (by definition (123)) there exist no solution $\zeta \in \mathbb{R}^N \ni \zeta^{\mathrm{T}} \mathbf{1} = 1$, $\zeta_k = 0$, to

$$Xy_i \zeta_i + \cdots + Xy_j \zeta_j - Xy_k \zeta_k = \mathbf{0}, \qquad i \neq \cdots \neq j \neq k = 1 \ldots N \qquad (126)$$

By factoring out X, we see that is ensured by affine independence of $\{ y_i \in \mathbb{R}^N \}$ and by $\mathcal{R}(Y) \cap \mathcal{N}(X) = \mathbf{0}$ where

$$\mathcal{N}(A) = \{ x \mid Ax = \mathbf{0} \} \qquad (143)$$

2.4.2.5 Affine maps

Affine transformations preserve affine hulls. Given any affine mapping T of vector spaces and some arbitrary set $\mathcal{C}$ [315, p.8]

$$\mathrm{aff}(T\mathcal{C}) = T(\mathrm{aff}\,\mathcal{C}) \qquad (127)$$

2.4.2.6 PRINCIPLE 2: Supporting hyperplane

The second most fundamental principle of convex geometry also follows from the *geometric Hahn-Banach theorem* [256, §5.12] [18, §1] that guarantees existence of at least one hyperplane in $\mathbb{R}^n$ supporting a full-dimensional convex set[2.18] at each point on its boundary.

The partial boundary $\partial \mathcal{H}$ of a halfspace that contains arbitrary set $\mathcal{Y}$ is called a *supporting hyperplane* $\underline{\partial \mathcal{H}}$ to $\mathcal{Y}$ when the hyperplane contains at least one point of $\overline{\mathcal{Y}}$. [315, §11]

2.4.2.6.1 Definition. *Supporting hyperplane* $\underline{\partial \mathcal{H}}$.
Assuming set $\mathcal{Y}$ and some normal $a \neq \mathbf{0}$ reside in opposite halfspaces[2.19] (Figure 30a), then a hyperplane supporting $\mathcal{Y}$ at point $y_{\mathrm{p}} \in \partial \mathcal{Y}$ is described

$$\underline{\partial \mathcal{H}}_{-} = \{ y \mid a^{\mathrm{T}}(y - y_{\mathrm{p}}) = 0, \ \ y_{\mathrm{p}} \in \overline{\mathcal{Y}}, \ \ a^{\mathrm{T}}(z - y_{\mathrm{p}}) \leq 0 \ \ \forall z \in \overline{\mathcal{Y}} \} \qquad (128)$$

Given only normal a, the hyperplane supporting $\mathcal{Y}$ is equivalently described

$$\underline{\partial \mathcal{H}}_{-} = \{ y \mid a^{\mathrm{T}}y = \sup\{ a^{\mathrm{T}}z \mid z \in \mathcal{Y} \} \} \qquad (129)$$

[2.18]It is customary to speak of a hyperplane supporting set $\mathcal{C}$ but not containing $\mathcal{C}$; called *nontrivial support*. [315, p.100] Hyperplanes in support of lower-dimensional bodies are admitted.
[2.19]Normal a belongs to $\mathcal{H}_+$ by definition.

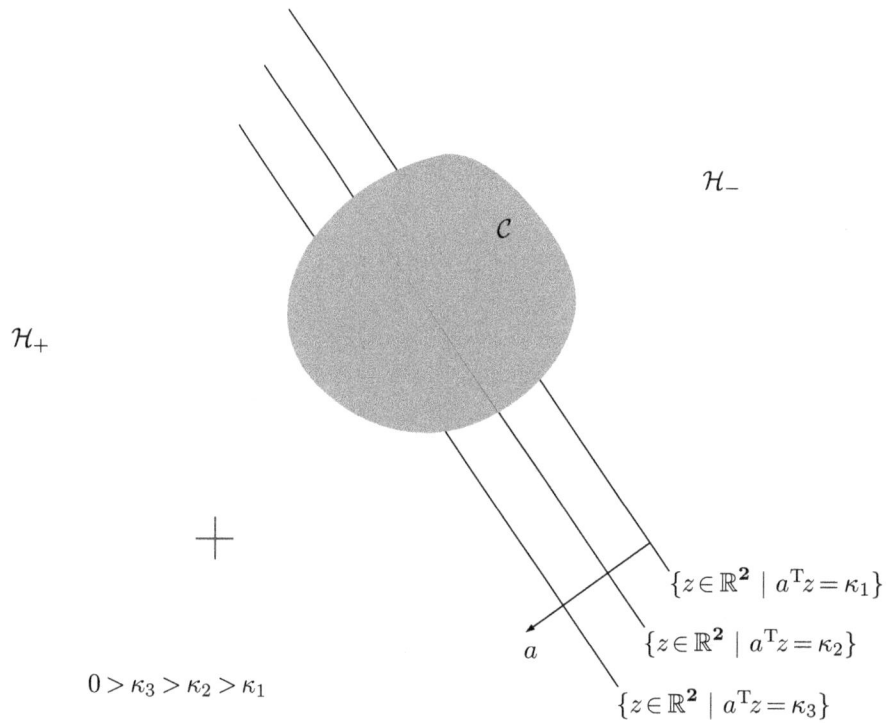

Figure 29: (*confer* Figure 74) Each linear contour, of equal inner product in vector z with normal a, represents i^{th} hyperplane in $\mathbb{R}^2$ parametrized by scalar κ_i. Inner product κ_i increases in direction of normal a. In convex set $\mathcal{C} \subset \mathbb{R}^2$, i^{th} line segment $\{z \in \mathcal{C} \mid a^{\text{T}}z = \kappa_i\}$ represents intersection with hyperplane. (Cartesian axes drawn for reference.)

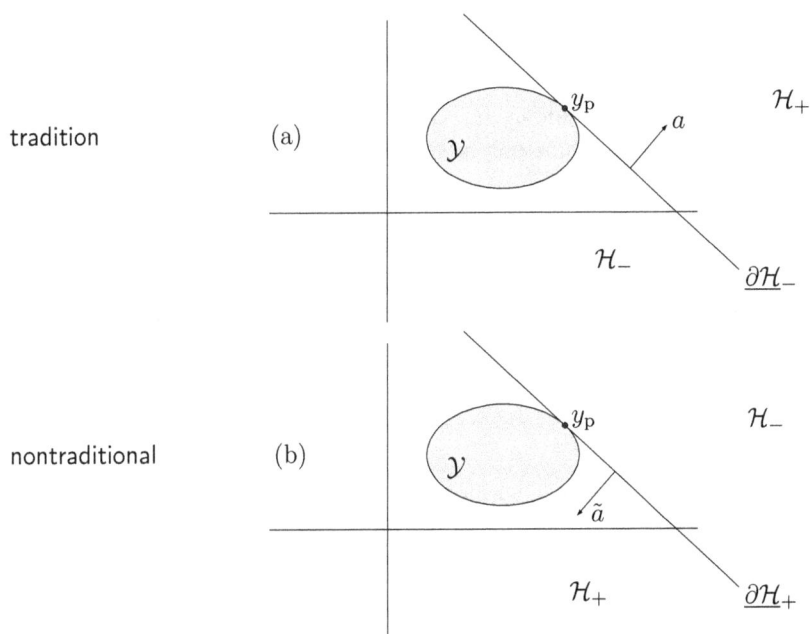

Figure 30: **(a)** Hyperplane $\underline{\partial\mathcal{H}_-}$ (128) supporting closed set $\mathcal{Y}\subset\mathbb{R}^2$. Vector a is inward-normal to hyperplane with respect to halfspace $\mathcal{H}_+$, but outward-normal with respect to set $\mathcal{Y}$. A supporting hyperplane can be considered the limit of an increasing sequence in the normal-direction like that in Figure **29**. **(b)** Hyperplane $\underline{\partial\mathcal{H}_+}$ nontraditionally supporting $\mathcal{Y}$. Vector $\tilde{a}$ is inward-normal to hyperplane now with respect to both halfspace $\mathcal{H}_+$ and set $\mathcal{Y}$. Tradition [205] [315] recognizes only positive normal polarity in support function $\sigma_{\mathcal{Y}}$ as in (129); *id est*, normal a, figure (a). But both interpretations of supporting hyperplane are useful.

where real function

$$\sigma_{\mathcal{Y}}(a) = \sup\{a^{\mathrm{T}}z \mid z \in \mathcal{Y}\} \qquad (554)$$

is called the *support function* for $\mathcal{Y}$.

Another equivalent but nontraditional representation[2.20] for a supporting hyperplane is obtained by reversing polarity of normal a; (1764)

$$
\begin{aligned}
\underline{\partial\mathcal{H}_+} &= \{y \mid \tilde{a}^{\mathrm{T}}(y - y_{\mathrm{p}}) = 0, \ \ y_{\mathrm{p}} \in \overline{\mathcal{Y}}, \ \ \tilde{a}^{\mathrm{T}}(z - y_{\mathrm{p}}) \geq 0 \ \ \forall z \in \overline{\mathcal{Y}}\} \\
&= \{y \mid \tilde{a}^{\mathrm{T}}y = -\inf\{\tilde{a}^{\mathrm{T}}z \mid z \in \mathcal{Y}\} = \sup\{-\tilde{a}^{\mathrm{T}}z \mid z \in \mathcal{Y}\}\}
\end{aligned}
\qquad (130)
$$

where normal $\tilde{a}$ and set $\mathcal{Y}$ both now reside in $\mathcal{H}_+$ (Figure 30b).

When a supporting hyperplane contains only a single point of $\overline{\mathcal{Y}}$, that hyperplane is termed *strictly supporting*.[2.21]   △

A full-dimensional set that has a supporting hyperplane at every point on its boundary, conversely, is convex. A convex set $\mathcal{C} \subset \mathbb{R}^n$, for example, can be expressed as the intersection of all halfspaces partially bounded by hyperplanes supporting it; *videlicet*, [256, p.135]

$$\overline{\mathcal{C}} = \bigcap_{a \in \mathbb{R}^n} \{y \mid a^{\mathrm{T}}y \leq \sigma_{\mathcal{C}}(a)\} \qquad (131)$$

by the *halfspaces theorem* (§2.4.1.1.1).

There is no geometric difference between supporting hyperplane $\underline{\partial\mathcal{H}_+}$ or $\underline{\partial\mathcal{H}_-}$ or $\underline{\partial\mathcal{H}}$ and[2.22] an ordinary hyperplane $\partial\mathcal{H}$ coincident with them.

2.4.2.6.2 Example. *Minimization over hypercube.*
Consider minimization of a linear function over a *hypercube*, given vector c

$$
\begin{aligned}
\underset{x}{\text{minimize}} \quad & c^{\mathrm{T}}x \\
\text{subject to} \quad & -\mathbf{1} \preceq x \preceq \mathbf{1}
\end{aligned}
\qquad (132)
$$

This convex optimization problem is called a *linear program*[2.23] because the *objective*[2.24] of minimization $c^{\mathrm{T}}x$ is a linear function of variable x and the constraints describe a *polyhedron* (intersection of a finite number of halfspaces and hyperplanes).

Any vector x satisfying the constraints is called a *feasible solution*. Applying graphical concepts from Figure 27, Figure 29, and Figure 30, $x^\star = -\operatorname{sgn}(c)$ is an *optimal solution*

[2.20] useful for constructing the *dual cone*; *e.g.*, Figure 56b. Tradition would instead have us construct the *polar cone*; which is, the negative dual cone.

[2.21] Rockafellar terms a strictly supporting hyperplane *tangent* to $\mathcal{Y}$ if it is unique there; [315, §18, p.169] a definition we do not adopt because our only criterion for tangency is intersection exclusively with a relative boundary. Hiriart-Urruty & Lemaréchal [205, p.44] (*confer* [315, p.100]) do not demand any tangency of a supporting hyperplane.

[2.22] If vector-normal polarity is unimportant, we may instead signify a supporting hyperplane by $\underline{\partial\mathcal{H}}$.

[2.23] The term *program* has its roots in economics. It was originally meant with regard to a plan or to efficient organization or systematization of some industrial process. [96, §2]

[2.24] The *objective* is the function that is argument to minimization or maximization.

to this minimization problem but is not necessarily unique. It generally holds for optimization problem solutions:

$$\text{optimal} \;\Rightarrow\; \text{feasible} \tag{133}$$

Because an optimal solution always exists at a hypercube vertex (§2.6.1.0.1) regardless of value of nonzero vector c in (132) [96, p.158] [15, p.2], mathematicians see this geometry as a means to relax a discrete problem (whose desired solution is integer or combinatorial, *confer* Example 4.2.3.1.1). [245, §3.1] [246] □

2.4.2.6.3 Exercise. *Unbounded below.*
Suppose instead we minimize over the unit hypersphere in Example 2.4.2.6.2; $\|x\| \leq 1$. What is an expression for optimal solution now? Is that program still linear?

Now suppose minimization of absolute value in (132). Are the following programs equivalent for some arbitrary real convex set $\mathcal{C}$? (*confer* (516))

$$\begin{array}{ll} \underset{x \in \mathbb{R}}{\text{minimize}} & |x| \\ \text{subject to} & -1 \leq x \leq 1 \\ & x \in \mathcal{C} \end{array} \quad \equiv \quad \begin{array}{ll} \underset{\alpha,\,\beta}{\text{minimize}} & \alpha + \beta \\ \text{subject to} & 1 \geq \beta \geq 0 \\ & 1 \geq \alpha \geq 0 \\ & \alpha - \beta \in \mathcal{C} \end{array} \tag{134}$$

Many optimization problems of interest and some methods of solution require nonnegative variables. The method illustrated below splits a variable into parts; $x = \alpha - \beta$ (extensible to vectors). Under what conditions on vector a and scalar b is an optimal solution $x^\star$ negative infinity?

$$\begin{array}{ll} \underset{\alpha \in \mathbb{R},\,\beta \in \mathbb{R}}{\text{minimize}} & \alpha - \beta \\ \text{subject to} & \beta \geq 0 \\ & \alpha \geq 0 \\ & a^{\mathrm{T}}\begin{bmatrix} \alpha \\ \beta \end{bmatrix} = b \end{array} \tag{135}$$

Minimization of the *objective function* entails maximization of β. ▼

2.4.2.7 PRINCIPLE 3: Separating hyperplane

The third most fundamental principle of convex geometry again follows from the *geometric Hahn-Banach theorem* [256, §5.12] [18, §1] [136, §I.1.2] that guarantees existence of a hyperplane separating two nonempty convex sets in $\mathbb{R}^n$ whose relative interiors are nonintersecting. *Separation* intuitively means each set belongs to a halfspace on an opposing side of the hyperplane. There are two cases of interest:

1) If the two sets intersect only at their relative boundaries (§2.1.7.2), then there exists a separating hyperplane $\partial\mathcal{H}$ containing the intersection but containing no points relatively interior to either set. If at least one of the two sets is open, conversely, then the existence of a separating hyperplane implies the two sets are nonintersecting. [62, §2.5.1]

2) A *strictly separating hyperplane* $\partial\mathcal{H}$ intersects the closure of neither set; its existence is guaranteed when intersection of the closures is empty and at least one set is bounded. [205, §A.4.1]

2.4.3 Angle between hyperspaces

Given halfspace-descriptions, *dihedral* angle between hyperplanes or halfspaces is defined as the angle between their defining normals. Given normals a and b respectively describing $\partial\mathcal{H}_a$ and $\partial\mathcal{H}_b$, for example

$$\measuredangle(\partial\mathcal{H}_a , \partial\mathcal{H}_b) \triangleq \arccos\left(\frac{\langle a , b\rangle}{\|a\| \, \|b\|}\right) \text{ radians} \tag{136}$$

2.5 Subspace representations

There are two common forms of expression for Euclidean subspaces, both coming from elementary linear algebra: *range form* $\mathcal{R}$ and *nullspace form* $\mathcal{N}$; a.k.a, vertex-description and halfspace-description respectively.

The fundamental vector subspaces associated with a matrix $A\in\mathbb{R}^{m\times n}$ [338, §3.1] are ordinarily related by orthogonal complement

$$\mathcal{R}(A^\mathrm{T}) \perp \mathcal{N}(A), \qquad \mathcal{N}(A^\mathrm{T}) \perp \mathcal{R}(A) \tag{137}$$

$$\mathcal{R}(A^\mathrm{T}) \oplus \mathcal{N}(A) = \mathbb{R}^n, \qquad \mathcal{N}(A^\mathrm{T}) \oplus \mathcal{R}(A) = \mathbb{R}^m \tag{138}$$

and of dimension:

$$\dim\mathcal{R}(A^\mathrm{T}) = \dim\mathcal{R}(A) = \operatorname{rank} A \leq \min\{m , n\} \tag{139}$$

with complementarity (a.k.a *conservation of dimension*)

$$\dim\mathcal{N}(A) = n - \operatorname{rank} A , \qquad \dim\mathcal{N}(A^\mathrm{T}) = m - \operatorname{rank} A \tag{140}$$

These equations (137)-(140) comprise the *fundamental theorem of linear algebra*. [338, p.95, p.138]

From these four fundamental subspaces, the rowspace and range identify one form of subspace description (range form or vertex-description (§2.3.4))

$$\mathcal{R}(A^\mathrm{T}) \triangleq \operatorname{span} A^\mathrm{T} = \{A^\mathrm{T}y \mid y\in\mathbb{R}^m\} = \{x\in\mathbb{R}^n \mid A^\mathrm{T}y=x , \;\; y\in\mathcal{R}(A)\} \tag{141}$$

$$\mathcal{R}(A) \triangleq \operatorname{span} A = \{Ax \mid x\in\mathbb{R}^n\} = \{y\in\mathbb{R}^m \mid Ax=y , \;\; x\in\mathcal{R}(A^\mathrm{T})\} \tag{142}$$

while the nullspaces identify the second common form (nullspace form or halfspace-description (113))

$$\mathcal{N}(A) \triangleq \{x\in\mathbb{R}^n \mid Ax=\mathbf{0}\} = \{x\in\mathbb{R}^n \mid x \perp \mathcal{R}(A^\mathrm{T})\} \tag{143}$$

$$\mathcal{N}(A^\mathrm{T}) \triangleq \{y\in\mathbb{R}^m \mid A^\mathrm{T}y=\mathbf{0}\} = \{y\in\mathbb{R}^m \mid y \perp \mathcal{R}(A)\} \tag{144}$$

Range forms (141) (142) are realized as the respective span of the column vectors in matrices A^T and A , whereas nullspace form (143) or (144) is the solution set to a linear equation similar to hyperplane definition (114). Yet because matrix A generally has multiple rows, halfspace-description $\mathcal{N}(A)$ is actually the intersection of as many hyperplanes through the origin; for (143), each row of A is normal to a hyperplane while each row of A^T is a normal for (144).

2.5.0.0.1 Exercise. *Subspace algebra.*
Given

$$\mathcal{R}(A) + \mathcal{N}(A^{\mathrm T}) = \mathcal{R}(B) + \mathcal{N}(B^{\mathrm T}) = \mathbb{R}^m \tag{145}$$

prove

$$\mathcal{R}(A) \supseteq \mathcal{N}(B^{\mathrm T}) \ \Leftrightarrow \ \mathcal{N}(A^{\mathrm T}) \subseteq \mathcal{R}(B) \tag{146}$$

$$\mathcal{R}(A) \supseteq \mathcal{R}(B) \ \Leftrightarrow \ \mathcal{N}(A^{\mathrm T}) \subseteq \mathcal{N}(B^{\mathrm T}) \tag{147}$$

e.g., Theorem A.3.1.0.6. ▼

2.5.1 Subspace or affine subset. . .

Any particular vector subspace $\mathcal{R}_{\mathrm p}$ can be described as nullspace $\mathcal{N}(A)$ of some matrix A or as range $\mathcal{R}(B)$ of some matrix B.

More generally, we have the choice of expressing an $n-m$-dimensional affine subset of $\mathbb{R}^n$ as the intersection of m hyperplanes, or as the offset span of $n-m$ vectors:

2.5.1.1 . . . as hyperplane intersection

Any affine subset $\mathcal{A}$ of dimension $n-m$ can be described as an intersection of m hyperplanes in $\mathbb{R}^n$; given *fat* ($m \le n$) *full-rank* (rank = $\min\{m,n\}$) matrix

$$A \triangleq \begin{bmatrix} a_1^{\mathrm T} \\ \vdots \\ a_m^{\mathrm T} \end{bmatrix} \in \mathbb{R}^{m \times n} \tag{148}$$

and vector $b \in \mathbb{R}^m$,

$$\mathcal{A} \triangleq \{x \in \mathbb{R}^n \mid Ax = b\} = \bigcap_{i=1}^m \{x \mid a_i^{\mathrm T}x = b_i\} \tag{149}$$

a halfspace-description. (113)

For example: The intersection of any two independent[2.25] hyperplanes in $\mathbb{R}^3$ is a line, whereas three independent hyperplanes intersect at a point. In $\mathbb{R}^4$, the intersection of two independent hyperplanes is a plane (Example 2.5.1.2.1), whereas three hyperplanes intersect at a line, four at a point, and so on. $\mathcal{A}$ describes a subspace whenever $b = \mathbf{0}$ in (149).

For $n > k$

$$\mathcal{A} \cap \mathbb{R}^k = \{x \in \mathbb{R}^n \mid Ax = b\} \cap \mathbb{R}^k = \bigcap_{i=1}^m \{x \in \mathbb{R}^k \mid a_i(1:k)^{\mathrm T}x = b_i\} \tag{150}$$

The result in §2.4.2.2 is extensible; *id est*, any affine subset $\mathcal{A}$ also has a vertex-description:

[2.25] Any number of hyperplanes are called *independent* when defining normals are linearly independent. This misuse departs from independence of two affine subsets that demands intersection only at a point or not at all. (§2.1.4.0.1)

2.5.1.2   . . . as span of nullspace basis

Alternatively, we may compute a basis for nullspace of matrix A (§E.3.1) and then
equivalently express affine subset $\mathcal{A}$ as its span plus an offset: Define

$$Z \triangleq \text{basis}\,\mathcal{N}(A) \in \mathbb{R}^{n \times n - \text{rank}\,A} \tag{151}$$

so $AZ = \mathbf{0}$. Then we have a vertex-description in Z,

$$\mathcal{A} = \{x \in \mathbb{R}^n \mid Ax = b\} = \left\{Z\xi + x_{\mathrm{p}} \mid \xi \in \mathbb{R}^{n - \text{rank}\,A}\right\} \subseteq \mathbb{R}^n \tag{152}$$

the offset span of $n - \text{rank}\,A$ column vectors, where x_{p} is any particular solution to
$Ax = b$; *e.g.*, $\mathcal{A}$ describes a subspace whenever $x_{\mathrm{p}} = \mathbf{0}$.

2.5.1.2.1   Example.   *Intersecting planes in 4-space.*
Two planes can intersect at a point in four-dimensional Euclidean vector space. It is easy
to visualize intersection of two planes in three dimensions; a line can be formed. In four
dimensions it is harder to visualize. So let's resort to the tools acquired.

   Suppose an intersection of two hyperplanes in four dimensions is specified by a fat
full-rank matrix $A_1 \in \mathbb{R}^{2 \times 4}$ ($m = 2$, $n = 4$) as in (149):

$$\mathcal{A}_1 \triangleq \left\{x \in \mathbb{R}^4 \ \middle| \ \begin{bmatrix} a_{11} & a_{12} & a_{13} & a_{14} \\ a_{21} & a_{22} & a_{23} & a_{24} \end{bmatrix} x = b_1 \right\} \tag{153}$$

The nullspace of A_1 is two dimensional (from Z in (152)), so $\mathcal{A}_1$ represents a plane in
four dimensions. Similarly define a second plane in terms of $A_2 \in \mathbb{R}^{2 \times 4}$:

$$\mathcal{A}_2 \triangleq \left\{x \in \mathbb{R}^4 \ \middle| \ \begin{bmatrix} a_{31} & a_{32} & a_{33} & a_{34} \\ a_{41} & a_{42} & a_{43} & a_{44} \end{bmatrix} x = b_2 \right\} \tag{154}$$

If the two planes are affinely independent and intersect, they intersect at a point because
$\begin{bmatrix} A_1 \\ A_2 \end{bmatrix}$ is invertible;

$$\mathcal{A}_1 \cap \mathcal{A}_2 = \left\{x \in \mathbb{R}^4 \ \middle| \ \begin{bmatrix} A_1 \\ A_2 \end{bmatrix} x = \begin{bmatrix} b_1 \\ b_2 \end{bmatrix} \right\} \tag{155}$$

$\square$

2.5.1.2.2   Exercise.   *Linear program.*
Minimize a hyperplane over affine set $\mathcal{A}$ in the nonnegative orthant

$$\begin{array}{ll} \underset{x}{\text{minimize}} & c^{\mathrm{T}}x \\ \text{subject to} & Ax = b \\ & x \succeq 0 \end{array} \tag{156}$$

where $\mathcal{A} = \{x \mid Ax = b\}$.   Two cases of interest are drawn in Figure **31**.   Graphically
illustrate and explain optimal solutions indicated in the caption. Why is $\alpha^\star$ negative in
both cases? Is there solution on the vertical axis? What causes objective unboundedness

(a)

$\mathcal{A} = \{x \mid Ax = b\}$

c

$\{x \mid c^{\mathrm{T}}x = \alpha\}$

(b)

$\mathcal{A} = \{x \mid Ax = b\}$

c

$\{x \mid c^{\mathrm{T}}x = \alpha\}$

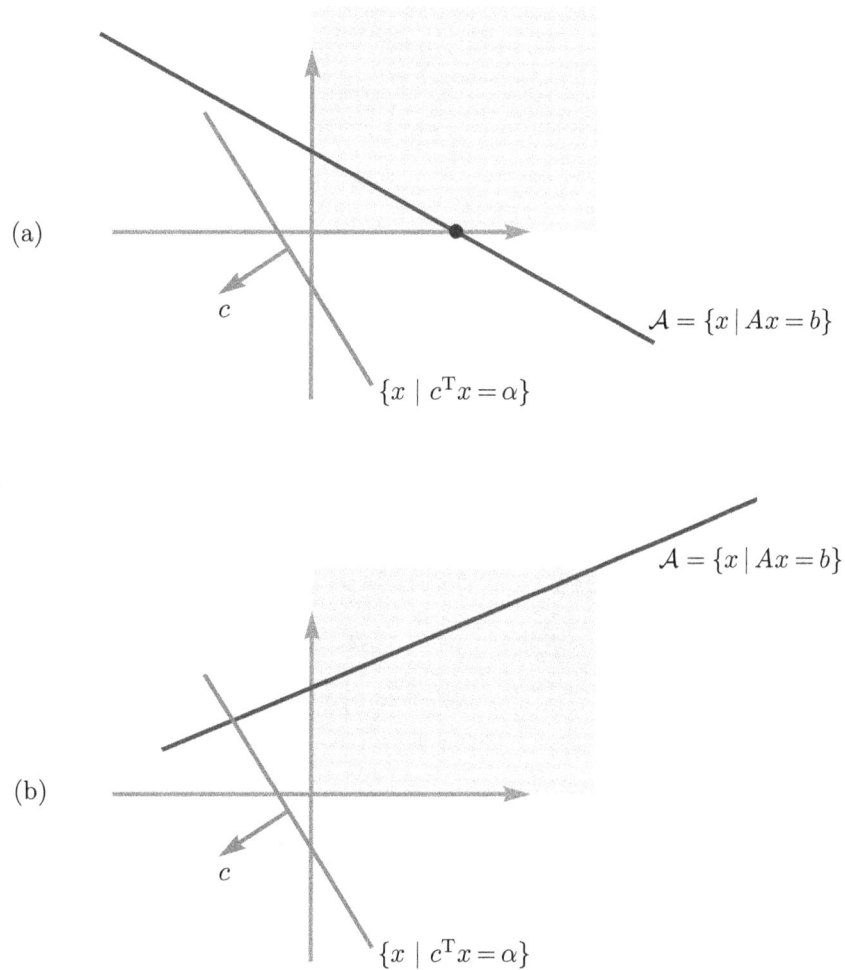

Figure 31: Minimizing hyperplane over affine set $\mathcal{A}$ in nonnegative orthant $\mathbb{R}_+^2$ whose extreme directions (§2.8.1) are the nonnegative Cartesian axes. Solutions are visually ascertainable: **(a)** Optimal solution is •. **(b)** Optimal objective $\alpha^\star = -\infty$.

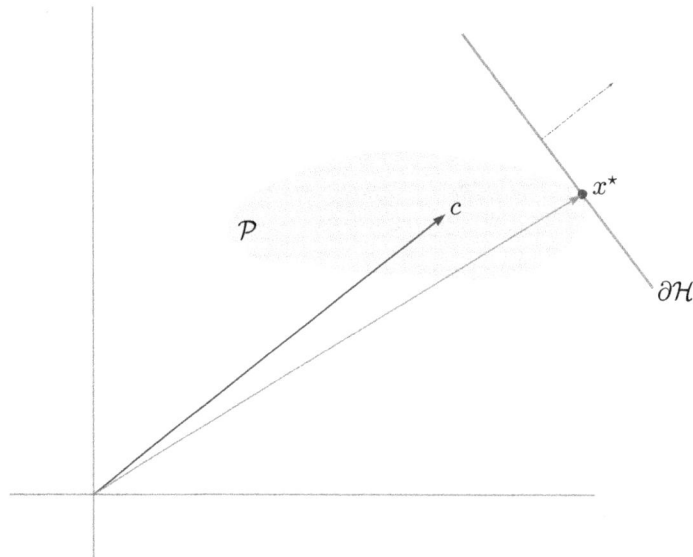

Figure 32: Maximizing hyperplane $\partial\mathcal{H}$, whose normal is vector $c\in\mathcal{P}$, over polyhedral set $\mathcal{P}$ in $\mathbb{R}^2$ is a linear program (157). Optimal solution $x^\star$ at •.

in the latter case (b)? Describe all vectors c that would yield finite optimal objective in (b).

Graphical solution to linear program

$$\begin{array}{ll} \underset{x}{\text{maximize}} & c^{\mathrm{T}}x \\ \text{subject to} & x \in \mathcal{P} \end{array} \tag{157}$$

is illustrated in Figure 32. Bounded set $\mathcal{P}$ is an intersection of many halfspaces. Why is optimal solution $x^\star$ not aligned with vector c as in Cauchy-Schwarz inequality (2152)?

▼

2.5.2   Intersection of subspaces

The intersection of nullspaces associated with two matrices $A\in\mathbb{R}^{m\times n}$ and $B\in\mathbb{R}^{k\times n}$ can be expressed most simply as

$$\mathcal{N}(A)\cap\mathcal{N}(B) = \mathcal{N}\left(\begin{bmatrix} A \\ B \end{bmatrix}\right) \triangleq \{x\in\mathbb{R}^n \mid \begin{bmatrix} A \\ B \end{bmatrix}x = \mathbf{0}\} \tag{158}$$

nullspace of their rowwise concatenation.

Suppose the columns of a matrix Z constitute a basis for $\mathcal{N}(A)$ while the columns of a matrix W constitute a basis for $\mathcal{N}(BZ)$. Then [164, §12.4.2]

$$\mathcal{N}(A)\cap\mathcal{N}(B) = \mathcal{R}(ZW) \tag{159}$$

If each basis is orthonormal, then the columns of ZW constitute an orthonormal basis for the intersection.

In the particular circumstance A and B are each positive semidefinite [21, §6], or in the circumstance A and B are two linearly independent dyads (§B.1.1), then

$$\mathcal{N}(A) \cap \mathcal{N}(B) = \mathcal{N}(A+B), \quad \begin{cases} A, B \in \mathbb{S}_+^M \\ \textbf{or} \\ A+B = u_1 v_1^{\mathrm{T}} + u_2 v_2^{\mathrm{T}} \quad \text{(l.i.)} \end{cases} \tag{160}$$

2.5.3 Visualization of matrix subspaces

Fundamental subspace relations, such as

$$\mathcal{R}(A^{\mathrm{T}}) \perp \mathcal{N}(A), \qquad \mathcal{N}(A^{\mathrm{T}}) \perp \mathcal{R}(A) \tag{137}$$

are partially defining. But to aid visualization of involved geometry, it sometimes helps to vectorize matrices. For any square matrix A, $s \in \mathcal{N}(A)$, and $w \in \mathcal{N}(A^{\mathrm{T}})$

$$\langle A, ss^{\mathrm{T}} \rangle = 0, \qquad \langle A, ww^{\mathrm{T}} \rangle = 0 \tag{161}$$

because $s^{\mathrm{T}} A s = w^{\mathrm{T}} A w = 0$. This innocuous observation becomes a sharp instrument for visualization of diagonalizable matrices (§A.5.1): for rank-ρ matrix $A \in \mathbb{R}^{M \times M}$

$$A = S \Lambda S^{-1} = [\, s_1 \cdots s_M \,] \, \Lambda \begin{bmatrix} w_1^{\mathrm{T}} \\ \vdots \\ w_M^{\mathrm{T}} \end{bmatrix} = \sum_{i=1}^{M} \lambda_i \, s_i \, w_i^{\mathrm{T}} \tag{1628}$$

where nullspace eigenvectors are real by Theorem A.5.0.0.1 and where (§B.1.1)

$$\begin{aligned} \mathcal{R}\{s_i \in \mathbb{R}^M \,|\, \lambda_i = 0\} &= \mathcal{R}\left(\sum_{i=\rho+1}^{M} s_i s_i^{\mathrm{T}} \right) = \mathcal{N}(A) \\ \mathcal{R}\{w_i \in \mathbb{R}^M \,|\, \lambda_i = 0\} &= \mathcal{R}\left(\sum_{i=\rho+1}^{M} w_i w_i^{\mathrm{T}} \right) = \mathcal{N}(A^{\mathrm{T}}) \end{aligned} \tag{162}$$

Define an unconventional basis among column vectors of each summation:

$$\begin{aligned} \text{basis}\,\mathcal{N}(A) &\subseteq \sum_{i=\rho+1}^{M} s_i s_i^{\mathrm{T}} \subseteq \mathcal{N}(A) \\ \text{basis}\,\mathcal{N}(A^{\mathrm{T}}) &\subseteq \sum_{i=\rho+1}^{M} w_i w_i^{\mathrm{T}} \subseteq \mathcal{N}(A^{\mathrm{T}}) \end{aligned} \tag{163}$$

We shall regard a *vectorized subspace* as vectorization of any $M \times M$ matrix whose columns comprise an *overcomplete* basis for that subspace; *e.g.*, §E.3.1

$$\begin{aligned} \text{vec basis}\,\mathcal{N}(A) &= \text{vec} \sum_{i=\rho+1}^{M} s_i s_i^{\mathrm{T}} \\ \text{vec basis}\,\mathcal{N}(A^{\mathrm{T}}) &= \text{vec} \sum_{i=\rho+1}^{M} w_i w_i^{\mathrm{T}} \end{aligned} \tag{164}$$

By this reckoning, vec basis $\mathcal{R}(A) = \text{vec}\,A$ but is not unique. Now, because

$$\left\langle A\,,\,\sum_{i=\rho+1}^{M} s_i s_i^{\mathrm{T}} \right\rangle = 0\,, \qquad \left\langle A\,,\,\sum_{i=\rho+1}^{M} w_i w_i^{\mathrm{T}} \right\rangle = 0 \qquad (165)$$

then vectorized matrix A is normal to a hyperplane (of dimension $M^2 - 1$) that contains both vectorized nullspaces (each of whose dimension is $M - \rho$);

$$\text{vec}\,A \perp \text{vec basis}\,\mathcal{N}(A)\,, \qquad \text{vec basis}\,\mathcal{N}(A^{\mathrm{T}}) \perp \text{vec}\,A \qquad (166)$$

These vectorized subspace orthogonality relations represent a departure (absent ^{T}) from fundamental subspace relations (137) stated at the outset.

2.6   Extreme, Exposed

2.6.0.0.1   Definition.   *Extreme point.*
An extreme point x_ε of a convex set $\mathcal{C}$ is a point, belonging to its closure $\overline{\mathcal{C}}$ [42, §3.3], that is not expressible as a convex combination of points in $\overline{\mathcal{C}}$ distinct from x_ε ; *id est*, for $x_\varepsilon \in \overline{\mathcal{C}}$ and all $x_1, x_2 \in \overline{\mathcal{C}} \setminus x_\varepsilon$

$$\mu\, x_1 + (1 - \mu)x_2 \neq x_\varepsilon\,, \qquad \mu \in [0, 1] \qquad (167)$$

$\triangle$

In other words, x_ε is an extreme point of $\mathcal{C}$ if and only if x_ε is not a point relatively interior to any line segment in $\overline{\mathcal{C}}$. [367, §2.10]

Borwein & Lewis offer: [56, §4.1.6] An extreme point of a convex set $\mathcal{C}$ is a point x_ε in $\overline{\mathcal{C}}$ whose *relative complement* $\overline{\mathcal{C}} \setminus x_\varepsilon$ is convex.

The set consisting of a single point $\mathcal{C} = \{x_\varepsilon\}$ is itself an extreme point.

2.6.0.0.2   Theorem.   *Extreme existence.*   [315, §18.5.3] [26, §II.3.5]
A nonempty closed convex set containing no lines has at least one extreme point. $\diamond$

2.6.0.0.3   Definition.   *Face, edge.* [205, §A.2.3]

- A *face* $\mathcal{F}$ of convex set $\mathcal{C}$ is a convex subset $\mathcal{F} \subseteq \overline{\mathcal{C}}$ such that every closed line segment $\overline{x_1 x_2}$ in $\overline{\mathcal{C}}$, having a relatively interior point ($x \in \text{rel int}\,\overline{x_1 x_2}$) in $\mathcal{F}$, has both endpoints in $\mathcal{F}$. The zero-dimensional faces of $\mathcal{C}$ constitute its extreme points. The empty set $\emptyset$ and $\overline{\mathcal{C}}$ itself are conventional faces of $\mathcal{C}$. [315, §18]

- All faces $\mathcal{F}$ are extreme sets by definition; *id est*, for $\mathcal{F} \subseteq \overline{\mathcal{C}}$ and all $x_1, x_2 \in \overline{\mathcal{C}} \setminus \mathcal{F}$

$$\mu\, x_1 + (1 - \mu)x_2 \notin \mathcal{F}\,, \qquad \mu \in [0, 1] \qquad (168)$$

- A one-dimensional face of a convex set is called an *edge*. $\triangle$

Dimension of a face is the penultimate number of affinely independent points (§2.4.2.3) belonging to it;

$$\dim \mathcal{F} = \sup_{\rho} \dim\{x_2 - x_1 \,,\, x_3 - x_1 \,,\, \ldots \,,\, x_\rho - x_1 \mid x_i \in \mathcal{F} \,,\, i = 1 \ldots \rho\} \qquad (169)$$

The point of intersection in $\overline{\mathcal{C}}$ with a strictly supporting hyperplane identifies an extreme point, but not *vice versa*. The nonempty intersection of any supporting hyperplane with $\overline{\mathcal{C}}$ identifies a face, in general, but not *vice versa*. To acquire a converse, the concept *exposed face* requires introduction:

2.6.1 Exposure

2.6.1.0.1 Definition. *Exposed face, exposed point, vertex, facet.*
[205, §A.2.3, A.2.4]

- $\mathcal{F}$ is an *exposed face* of an n-dimensional convex set $\mathcal{C}$ iff there is a supporting hyperplane $\underline{\partial \mathcal{H}}$ to $\overline{\mathcal{C}}$ such that

$$\mathcal{F} = \overline{\mathcal{C}} \cap \underline{\partial \mathcal{H}} \qquad (170)$$

  Only faces of dimension -1 through $n-1$ can be exposed by a hyperplane.

- An *exposed point,* the definition of *vertex,* is equivalent to a zero-dimensional exposed face; the point of intersection with a strictly supporting hyperplane.

- A *facet* is an $(n-1)$-dimensional exposed face of an n-dimensional convex set $\mathcal{C}$; facets exist in one-to-one correspondence with the $(n-1)$-dimensional faces.[2.26]

- $\overline{\{\text{exposed points}\}} = \{\text{extreme points}\}$
  $\{\text{exposed faces}\} \subseteq \{\text{faces}\}$ $\triangle$

2.6.1.1 Density of exposed points

For any closed convex set $\mathcal{C}$, its exposed points constitute a *dense* subset of its extreme points; [315, §18] [342] [337, §3.6, p.115] dense in the sense [401] that closure of that subset yields the set of extreme points.

For the convex set illustrated in Figure **33**, point B cannot be exposed because it relatively bounds both the facet $\overline{\text{AB}}$ and the closed quarter circle, each bounding the set. Since B is not relatively interior to any line segment in the set, then B is an extreme point by definition. Point B may be regarded as the limit of some sequence of exposed points beginning at vertex C.

[2.26]This coincidence occurs simply because the facet's dimension is the same as the dimension of the supporting hyperplane exposing it.

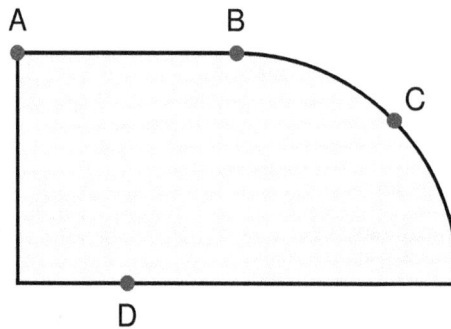

Figure 33: Closed convex set in $\mathbb{R}^2$. Point A is exposed hence extreme; a classical vertex. Point B is extreme but not an exposed point. Point C is exposed and extreme; zero-dimensional exposure makes it a vertex. Point D is neither an exposed or extreme point although it belongs to a one-dimensional exposed face. [205, §A.2.4] [337, §3.6] Closed face $\overline{AB}$ is exposed; a facet. The arc is not a conventional face, yet it is composed entirely of extreme points. Union of all rotations of this entire set about its vertical edge produces another convex set in three dimensions having no edges; but that convex set produced by rotation about horizontal edge containing D has edges.

2.6.1.2   Face transitivity and algebra

Faces of a convex set enjoy transitive relation. If $\mathcal{F}_1$ is a face (an extreme set) of $\mathcal{F}_2$ which in turn is a face of $\mathcal{F}_3$, then it is always true that $\mathcal{F}_1$ is a face of $\mathcal{F}_3$. (The parallel statement for exposed faces is false. [315, §18]) For example, any extreme point of $\mathcal{F}_2$ is an extreme point of $\mathcal{F}_3$; in this example, $\mathcal{F}_2$ could be a face exposed by a hyperplane supporting polyhedron $\mathcal{F}_3$. [229, def.115/6 p.358] Yet it is erroneous to presume that a face, of dimension 1 or more, consists entirely of extreme points. Nor is a face of dimension 2 or more entirely composed of edges, and so on.

For the polyhedron in $\mathbb{R}^3$ from Figure **20**, for example, the nonempty faces exposed by a hyperplane are the vertices, edges, and facets; there are no more. The zero-, one-, and two-dimensional faces are in one-to-one correspondence with the exposed faces in that example.

2.6.1.3   Smallest face

Define the smallest face $\mathcal{F}$, that contains some element G , of a convex set $\mathcal{C}$:

$$\mathcal{F}(\mathcal{C}\ni G) \tag{171}$$

videlicet, $\overline{\mathcal{C}} \supset \operatorname{rel int}\mathcal{F}(\mathcal{C}\ni G)\ni G$. An affine set has no faces except itself and the empty set. The smallest face, that contains G , of intersection of convex set $\mathcal{C}$ with an affine set $\mathcal{A}$ [245, §2.4] [246]

$$\mathcal{F}((\mathcal{C}\cap\mathcal{A})\ni G) = \mathcal{F}(\mathcal{C}\ni G)\cap\mathcal{A} \tag{172}$$

equals intersection of $\mathcal{A}$ with the smallest face, that contains G , of set $\mathcal{C}$.

2.6.1.4   Conventional boundary

(*confer* §2.1.7.2) Relative boundary

$$\operatorname{rel}\partial\mathcal{C} = \overline{\mathcal{C}} \setminus \operatorname{rel int}\mathcal{C} \tag{24}$$

is equivalent to:

2.6.1.4.1   Definition. *Conventional boundary of convex set.* [205, §C.3.1] The relative boundary $\partial\mathcal{C}$ of a nonempty convex set $\mathcal{C}$ is the union of all exposed faces of $\overline{\mathcal{C}}$.   △

Equivalence to (24) comes about because it is conventionally presumed that any supporting hyperplane, central to the definition of exposure, does not contain $\mathcal{C}$. [315, p.100] Any face $\mathcal{F}$ of convex set $\mathcal{C}$ (that is not $\mathcal{C}$ itself) belongs to $\operatorname{rel}\partial\mathcal{C}$. (§2.8.2.1)

2.7   Cones

In optimization, convex cones achieve prominence because they generalize subspaces. Most compelling is the projection analogy: Projection on a subspace can be ascertained from projection on its orthogonal complement (Figure **176**), whereas projection on a closed convex cone can be determined from projection instead on its *algebraic complement* (§2.13, Figure **177**, §E.9.2); called the *polar cone*.

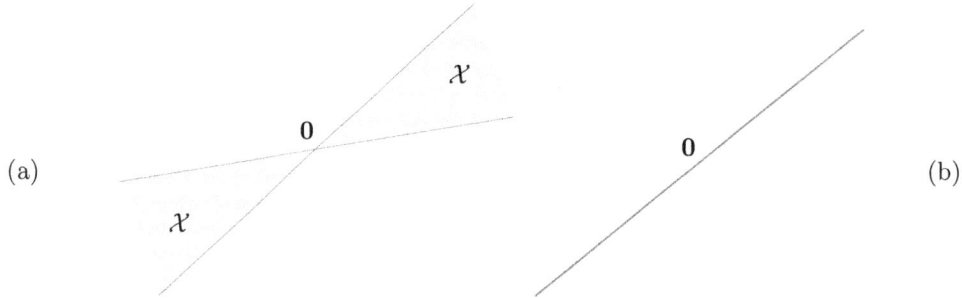

Figure 34: **(a)** Two-dimensional nonconvex cone drawn truncated. Boundary of this cone is itself a cone. Each polar half is itself a convex cone. **(b)** This convex cone (drawn truncated) is a line through the origin in any dimension. It has no relative boundary, while its relative interior comprises entire line.

2.7.0.0.1 Definition.   *Ray.*
The one-dimensional set

$$\{\zeta \Gamma + B \mid \zeta \geq 0, \ \Gamma \neq \mathbf{0}\} \subset \mathbb{R}^n \tag{173}$$

defines a *halfline* called a *ray* in nonzero *direction* $\Gamma \in \mathbb{R}^n$ having *base* $B \in \mathbb{R}^n$. When $B = \mathbf{0}$, a ray is the conic hull of direction Γ ; hence a convex cone. $\triangle$

Relative boundary of a single ray, base $\mathbf{0}$ in any dimension, is the origin because that is the union of all exposed faces not containing the entire set. Its relative interior is the ray itself excluding the origin.

2.7.1   Cone defined

A set $\mathcal{X}$ is called, simply, *cone* if and only if

$$\Gamma \in \mathcal{X} \ \Rightarrow \ \zeta \Gamma \in \overline{\mathcal{X}} \ \ \text{for all } \zeta \geq 0 \tag{174}$$

where $\overline{\mathcal{X}}$ denotes closure of cone $\mathcal{X}$. An example of such a cone is the union of two opposing quadrants; *e.g.*, $\mathcal{X} = \{x \in \mathbb{R}^2 \mid x_1 x_2 \geq 0\}$ which is not convex. [399, §2.5] Similar examples are shown in Figure **34** and Figure **38**.

All cones can be defined by an aggregate of rays emanating exclusively from the origin (but not all cones are convex). Hence all closed cones contain the origin $\mathbf{0}$ and are unbounded, excepting the simplest cone $\{\mathbf{0}\}$. The empty set $\emptyset$ is not a cone, but its conic hull is;

$$\text{cone}\, \emptyset \ = \ \{\mathbf{0}\} \tag{104}$$

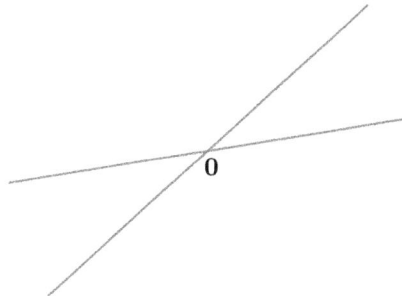

Figure 35: This nonconvex cone in $\mathbb{R}^2$ is a pair of lines through the origin. [256, §2.4] Because the lines are linearly independent, they are algebraic complements whose vector sum is $\mathbb{R}^2$ a convex cone.

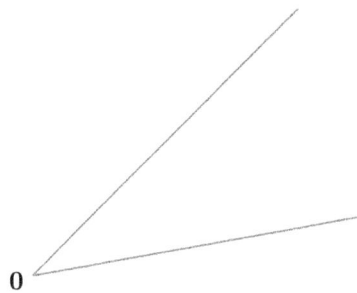

Figure 36: Boundary of a convex cone in $\mathbb{R}^2$ is a nonconvex cone; a pair of rays emanating from the origin.

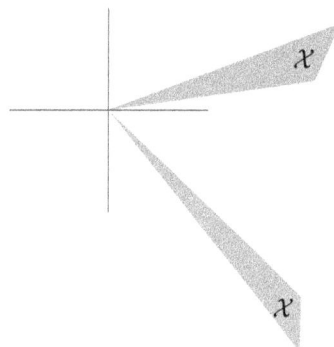

Figure 37: Union of two pointed closed convex cones is nonconvex cone $\mathcal{X}$.

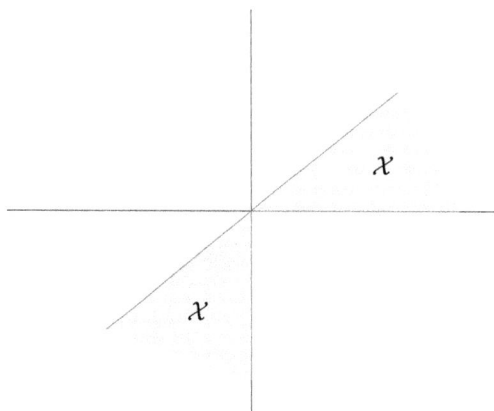

Figure 38: Truncated nonconvex cone $\mathcal{X} = \{x \in \mathbb{R}^2 \mid x_1 \geq x_2 \ , \ x_1 x_2 \geq 0\}$. Boundary is also a cone. [256, §2.4] Cartesian axes drawn for reference. Each half (about the origin) is itself a convex cone.

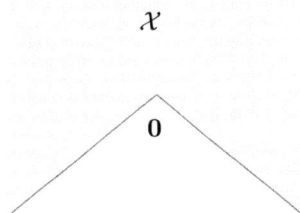

Figure 39: Nonconvex cone $\mathcal{X}$ drawn truncated in $\mathbb{R}^2$. Boundary is also a cone. [256, §2.4] Cone exterior is convex cone.

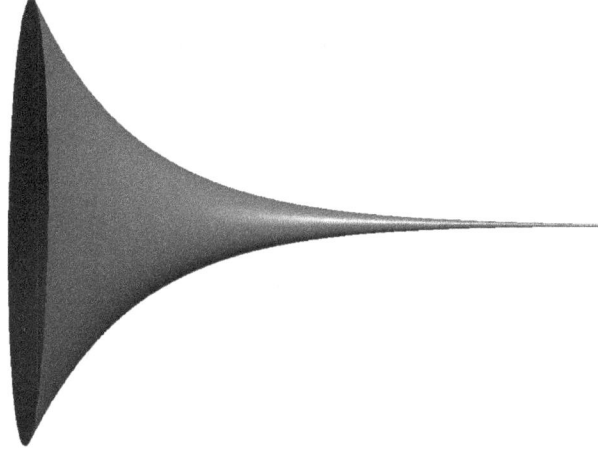

Figure 40: Not a cone; ironically, the three-dimensional *flared horn* (with or without its interior) resembling mathematical symbol ≻ denoting strict cone membership and partial order.

2.7.2   Convex cone

We call set $\mathcal{K}$ a *convex cone* iff

$$\Gamma_1, \Gamma_2 \in \mathcal{K} \;\Rightarrow\; \zeta\Gamma_1 + \xi\Gamma_2 \in \overline{\mathcal{K}} \;\; \text{for all } \zeta, \xi \geq 0 \qquad (175)$$

id est, if and only if any conic combination of elements from $\mathcal{K}$ belongs to its closure. Apparent from this definition, $\zeta\Gamma_1 \in \overline{\mathcal{K}}$ and $\xi\Gamma_2 \in \overline{\mathcal{K}}$ $\forall\, \zeta, \xi \geq 0$; meaning, $\mathcal{K}$ is a cone. Set $\mathcal{K}$ is convex since, for any particular $\zeta, \xi \geq 0$

$$\mu\zeta\Gamma_1 + (1 - \mu)\xi\Gamma_2 \in \overline{\mathcal{K}} \;\; \forall\, \mu \in [0, 1] \qquad (176)$$

because $\mu\zeta, (1 - \mu)\xi \geq 0$. Obviously,

$$\{\mathcal{X}\} \supset \{\mathcal{K}\} \qquad (177)$$

the set of all convex cones is a proper subset of all cones. The set of convex cones is a narrower but more familiar class of cone, any member of which can be equivalently described as the intersection of a possibly (but not necessarily) infinite number of hyperplanes (through the origin) and halfspaces whose bounding hyperplanes pass through the origin; a halfspace-description (§2.4). Convex cones need not be full-dimensional.

More familiar convex cones are *Lorentz cone* (*confer* Figure 47)[2.27]

$$\mathcal{K}_\ell = \left\{ \begin{bmatrix} x \\ t \end{bmatrix} \in \mathbb{R}^n \times \mathbb{R} \mid \|x\|_\ell \leq t \right\}, \qquad \ell = 2 \qquad (178)$$

[2.27] a.k.a: *second-order cone, quadratic cone, circular cone* (§2.9.2.8.1), unbounded *ice-cream cone* united with its interior.

and polyhedral cone (§2.12.1.0.1); *e.g.*, any orthant generated by Cartesian half-axes (§2.1.3). Esoteric examples of convex cones include the point at the origin, any line through the origin, any ray having the origin as base such as the nonnegative real line $\mathbb{R}_+$ in subspace $\mathbb{R}$, any halfspace partially bounded by a hyperplane through the origin, the positive semidefinite cone $\mathbb{S}_+^M$ (191), the cone of Euclidean distance matrices $\mathbb{EDM}^N$ (970) (§6), *completely positive* semidefinite matrices $\{CC^{\mathrm{T}} \,|\, C \geq 0\}$ [40, p.71], any subspace, and Euclidean vector space $\mathbb{R}^n$.

2.7.2.1   cone invariance

More Euclidean bodies are cones, it seems, than are not.[2.28] This class of convex body, the convex cone, is invariant to scaling, linear and single- or many-valued inverse linear transformation, vector summation, and Cartesian product, but is not invariant to translation. [315, p.22]

2.7.2.1.1   Theorem.   *Cone intersection (nonempty).*

- Intersection of an arbitrary collection of convex cones is a convex cone. [315, §2, §19]

- Intersection of an arbitrary collection of closed convex cones is a closed convex cone. [264, §2.3]

- Intersection of a finite number of polyhedral cones (Figure **51** p.125, §2.12.1.0.1) remains a polyhedral cone.   ◇

The property *pointedness* is associated with a convex cone; but,

- pointed cone ⇎ convex cone   (Figure **36**, Figure **37**)

2.7.2.1.2   Definition.   *Pointed convex cone.*   *(confer §2.12.2.2)*
A convex cone $\mathcal{K}$ is *pointed* iff it contains no line. Equivalently, $\mathcal{K}$ is not pointed iff there exists any nonzero direction $\Gamma \in \overline{\mathcal{K}}$ such that $-\Gamma \in \overline{\mathcal{K}}$. If the origin is an extreme point of $\overline{\mathcal{K}}$ or, equivalently, if

$$\overline{\mathcal{K}} \cap -\overline{\mathcal{K}} = \{\mathbf{0}\} \tag{179}$$

then $\mathcal{K}$ is pointed, and *vice versa*. [337, §2.10] A convex cone is pointed iff the origin is the smallest nonempty face of its closure.   △

Then a pointed closed convex cone, by principle of separating hyperplane (§2.4.2.7), has a strictly supporting hyperplane at the origin. The simplest and only bounded [399, p.75] convex cone $\mathcal{K} = \{\mathbf{0}\} \subseteq \mathbb{R}^n$ is pointed, by convention, but not full-dimensional. Its relative boundary is the empty set ∅ (25) while its relative interior is the point $\mathbf{0}$ itself (12). The pointed convex cone that is a halfline, emanating from the origin in $\mathbb{R}^n$, has relative boundary $\mathbf{0}$ while its relative interior is the halfline itself excluding $\mathbf{0}$. Pointed are any Lorentz cone, cone of Euclidean distance matrices $\mathbb{EDM}^N$ in symmetric hollow subspace $\mathbb{S}_h^N$, and positive semidefinite cone $\mathbb{S}_+^M$ in ambient $\mathbb{S}^M$.

[2.28]*confer* Figures: **25 34 35 36 39 38 40 42 44 51 55 58 60 61 63 64 65 66 67 149 162 187**

2.7.2.1.3 Theorem. *Pointed cones.* [56, §3.3.15, exer.20]

A closed convex cone $\mathcal{K} \subset \mathbb{R}^n$ is pointed if and only if there exists a normal α such that the set

$$\mathcal{C} \triangleq \{x \in \mathcal{K} \mid \langle x, \alpha \rangle = 1\} \tag{180}$$

is closed, bounded, and $\mathcal{K} = \mathrm{cone}\,\mathcal{C}$. Equivalently, $\mathcal{K}$ is pointed if and only if there exists a vector β normal to a hyperplane strictly supporting $\mathcal{K}$; *id est*, for some positive scalar ϵ

$$\langle x, \beta \rangle \geq \epsilon \|x\| \quad \forall x \in \mathcal{K} \tag{181}$$

$\diamond$

If closed convex cone $\mathcal{K}$ is not pointed, then it has no extreme point.[2.29] Yet a pointed closed convex cone has only one extreme point [42, §3.3]: the exposed point residing at the origin; its vertex. Pointedness is invariant to Cartesian product by (179). And from the *cone intersection theorem* it follows that an intersection of convex cones is pointed if at least one of the cones is; implying, each and every nonempty exposed face of a pointed closed convex cone is a pointed closed convex cone.

2.7.2.2 Pointed closed convex cone induces partial order

Relation $\preceq$ represents *partial order* on some set if that relation possesses[2.30]

reflexivity $(x \preceq x)$

antisymmetry $(x \preceq z,\ z \preceq x \Rightarrow x = z)$

transitivity $(x \preceq y,\ y \preceq z \Rightarrow x \preceq z)$, $\qquad\qquad (x \preceq y,\ y \prec z \Rightarrow x \prec z)$

A pointed closed convex cone $\mathcal{K}$ induces partial order on $\mathbb{R}^n$ or $\mathbb{R}^{m \times n}$, [21, §1] [331, p.7] essentially defined by vector or matrix inequality;

$$x \underset{\mathcal{K}}{\preceq} z \quad \Leftrightarrow \quad z - x \in \mathcal{K} \tag{182}$$

$$x \underset{\mathcal{K}}{\prec} z \quad \Leftrightarrow \quad z - x \in \mathrm{rel\,int}\,\mathcal{K} \tag{183}$$

Neither x or z is necessarily a member of $\mathcal{K}$ for these relations to hold. Only when $\mathcal{K}$ is a nonnegative orthant $\mathbb{R}^n_+$ do these inequalities reduce to ordinary entrywise comparison (§2.13.4.2.3) while partial order lingers. Inclusive of that special case, we ascribe nomenclature *generalized inequality* to comparison with respect to a pointed closed convex cone.

We say two points x and y are *comparable* when $x \preceq y$ or $y \preceq x$ with respect to pointed closed convex cone $\mathcal{K}$. Visceral mechanics of actually comparing points, when cone $\mathcal{K}$ is not an orthant, are well illustrated in the example of Figure 64 which relies on the equivalent membership-interpretation in definition (182) or (183).

[2.29] nor does it have extreme directions (§2.8.1).

[2.30] A set is *totally ordered* if it further obeys a comparability property of the relation: for each and every x and y from the set, $x \preceq y$ or $y \preceq x$; *e.g.*, one-dimensional real vector space $\mathbb{R}$ is the smallest unbounded totally ordered and connected set.

$\mathbb{R}^2$

$\mathcal{C}_1$

$x + \mathcal{K}$

(a)

y

$\mathcal{C}_2$

x

(b)

$y - \mathcal{K}$

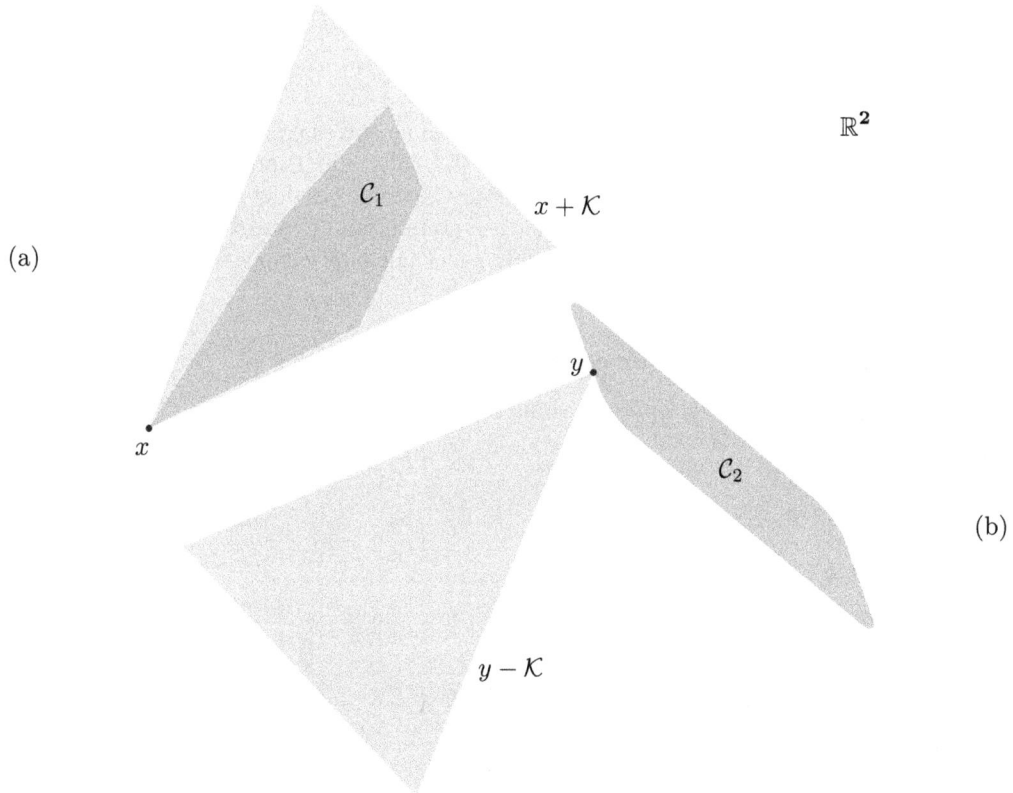

Figure 41: (*confer* Figure 70) (**a**) Point x is the minimum element of set $\mathcal{C}_1$ with respect to cone $\mathcal{K}$ because cone translated to $x \in \mathcal{C}_1$ contains entire set. (Cones drawn truncated.) (**b**) Point y is a minimal element of set $\mathcal{C}_2$ with respect to cone $\mathcal{K}$ because negative cone translated to $y \in \mathcal{C}_2$ contains only y. These concepts, minimum/minimal, become equivalent under a *total order*.

Comparable points and the *minimum element* of some vector- or matrix-valued partially ordered set are thus well defined, so nonincreasing sequences with respect to cone $\mathcal{K}$ can therefore converge in this sense: Point $x \in \mathcal{C}$ is the (unique) minimum element of set $\mathcal{C}$ with respect to cone $\mathcal{K}$ iff for each and every $z \in \mathcal{C}$ we have $x \preceq z$; equivalently, iff $\mathcal{C} \subseteq x + \mathcal{K}$.[2.31]

A closely related concept, *minimal element,* is useful for partially ordered sets having no minimum element: Point $x \in \mathcal{C}$ is a minimal element of set $\mathcal{C}$ with respect to pointed closed convex cone $\mathcal{K}$ if and only if $(x - \mathcal{K}) \cap \mathcal{C} = x$. (Figure 41) No uniqueness is implied here, although implicit is the assumption: $\dim \mathcal{K} \geq \dim \operatorname{aff} \mathcal{C}$. In words, a point that is a minimal element is smaller (with respect to $\mathcal{K}$) than any other point in the set to which it is comparable.

Further properties of partial order with respect to pointed closed convex cone $\mathcal{K}$ are not defining:

$$\text{homogeneity } (x \preceq y,\ \lambda \geq 0 \Rightarrow \lambda x \preceq \lambda z), \qquad\qquad (x \prec y,\ \lambda > 0 \Rightarrow \lambda x \prec \lambda z)$$

$$\text{additivity } (x \preceq z,\ u \preceq v \Rightarrow x + u \preceq z + v), \qquad\qquad (x \prec z,\ u \preceq v \Rightarrow x + u \prec z + v)$$

2.7.2.2.1 Definition. *Proper cone*: a cone that is

- pointed

- closed

- convex

- full-dimensional. △

A proper cone remains proper under injective linear transformation. [92, §5.1] Examples of proper cones are the positive semidefinite cone $\mathbb{S}_+^M$ in the ambient space of symmetric matrices (§2.9), the nonnegative real line $\mathbb{R}_+$ in vector space $\mathbb{R}$, or any orthant in $\mathbb{R}^n$, and the set of all coefficients of univariate degree-n polynomials nonnegative on interval $[0,1]$ [62, exmp.2.16] or univariate degree-$2n$ polynomials nonnegative over $\mathbb{R}$ [62, exer.2.37].

2.8 Cone boundary

Every hyperplane supporting a convex cone contains the origin. [205, §A.4.2] Because any supporting hyperplane to a convex cone must therefore itself be a cone, then from the *cone intersection theorem* (§2.7.2.1.1) it follows:

2.8.0.0.1 Lemma. *Cone faces.* [26, §II.8]
Each nonempty exposed face of a convex cone is a convex cone. ⋄

[2.31]Borwein & Lewis [56, §3.3 exer.21] ignore possibility of equality to $x + \mathcal{K}$ in this condition, and require a second condition: ... *and* $\mathcal{C} \subset y + \mathcal{K}$ *for some* y *in* $\mathbb{R}^n$ *implies* $x \in y + \mathcal{K}$.

2.8.0.0.2 Theorem. *Proper-cone boundary.*
Suppose a nonzero point Γ lies on the boundary $\partial\mathcal{K}$ of proper cone $\mathcal{K}$ in $\mathbb{R}^n$. Then it follows that the ray $\{\zeta\Gamma \mid \zeta \geq 0\}$ also belongs to $\partial\mathcal{K}$.   ⋄

Proof. By virtue of its propriety, a proper cone guarantees existence of a strictly supporting hyperplane at the origin. [315, cor.11.7.3][2.32] Hence the origin belongs to the boundary of $\mathcal{K}$ because it is the zero-dimensional exposed face. The origin belongs to the ray through Γ, and the ray belongs to $\mathcal{K}$ by definition (174). By the *cone faces lemma*, each and every nonempty exposed face must include the origin. Hence the closed line segment $\overline{0\Gamma}$ must lie in an exposed face of $\mathcal{K}$ because both endpoints do by Definition 2.6.1.4.1. That means there exists a supporting hyperplane $\partial\mathcal{H}$ to $\mathcal{K}$ containing $\overline{0\Gamma}$. So the ray through Γ belongs both to $\mathcal{K}$ and to $\partial\mathcal{H}$. $\partial\mathcal{H}$ must therefore expose a face of $\mathcal{K}$ that contains the ray; *id est,*

$$\{\zeta\Gamma \mid \zeta \geq 0\} \subseteq \mathcal{K} \cap \partial\mathcal{H} \subset \partial\mathcal{K} \tag{184}$$

◆

Proper cone $\{\mathbf{0}\}$ in $\mathbb{R}^0$ has no boundary (24) because (12)

$$\operatorname{rel\,int}\{\mathbf{0}\} = \{\mathbf{0}\} \tag{185}$$

The boundary of any proper cone in $\mathbb{R}$ is the origin.

The boundary of any convex cone whose dimension exceeds 1 can be constructed entirely from an aggregate of rays emanating exclusively from the origin.

2.8.1 Extreme direction

The property *extreme direction* arises naturally in connection with the pointed closed convex cone $\mathcal{K} \subset \mathbb{R}^n$, being analogous to extreme point. [315, §18, p.162][2.33] An extreme direction Γ_ε of pointed $\mathcal{K}$ is a vector corresponding to an edge that is a ray $\{\zeta\Gamma_\varepsilon \in \mathcal{K} \mid \zeta \geq 0\}$ emanating from the origin.[2.34] Nonzero direction Γ_ε in pointed $\mathcal{K}$ is extreme if and only if

$$\zeta_1\Gamma_1 + \zeta_2\Gamma_2 \neq \Gamma_\varepsilon \quad \forall\,\zeta_1,\zeta_2 \geq 0, \quad \forall\,\Gamma_1,\Gamma_2 \in \mathcal{K}\backslash\{\zeta\Gamma_\varepsilon \in \mathcal{K} \mid \zeta \geq 0\} \tag{186}$$

In words, an extreme direction in a pointed closed convex cone is the direction of a ray, called an *extreme ray,* that cannot be expressed as a conic combination of directions of any rays in the cone distinct from it.

An extreme ray is a one-dimensional face of $\mathcal{K}$. By (105), extreme direction Γ_ε is not a point relatively interior to any line segment in $\mathcal{K}\backslash\{\zeta\Gamma_\varepsilon \in \mathcal{K} \mid \zeta \geq 0\}$. Thus, by analogy, the corresponding extreme ray $\{\zeta\Gamma_\varepsilon \in \mathcal{K} \mid \zeta \geq 0\}$ is not a ray relatively interior to any *plane segment*[2.35] in $\mathcal{K}$.

[2.32]Rockafellar's corollary yields a supporting hyperplane at the origin to any convex cone in $\mathbb{R}^n$ not equal to $\mathbb{R}^n$.
[2.33]We diverge from Rockafellar's extreme direction: "extreme point at infinity".
[2.34]An edge (§2.6.0.0.3) of a convex cone is not necessarily a ray. A convex cone may contain an edge that is a line; *e.g.,* a wedge-shaped polyhedral cone ($\mathcal{K}^*$ in Figure 42).
[2.35]A planar fragment; in this context, a planar cone.

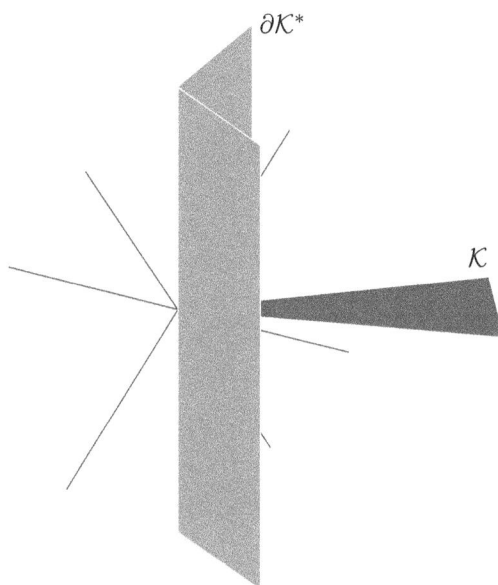

Figure 42: $\mathcal{K}$ is a pointed polyhedral cone not full-dimensional in $\mathbb{R}^3$ (drawn truncated in a plane parallel to the floor upon which you stand). Dual cone $\mathcal{K}^*$ is a *wedge* whose truncated boundary is illustrated (drawn perpendicular to the floor). In this particular instance, $\mathcal{K} \subset \operatorname{int} \mathcal{K}^*$ (excepting the origin). Cartesian coordinate axes drawn for reference.

2.8.1.1 extreme distinction, uniqueness

An extreme *direction* is unique, but its vector representation Γ_ε is not because any positive scaling of it produces another vector in the same (extreme) direction. Hence an extreme direction is *unique* to within a positive scaling. When we say extreme directions are *distinct,* we are referring to distinctness of rays containing them. Nonzero vectors of various length in the same extreme direction are therefore interpreted to be identical extreme directions.[2.36]

The extreme directions of the polyhedral cone in Figure **25** (p.64), for example, correspond to its three edges. For any pointed polyhedral cone, there is a one-to-one correspondence of one-dimensional faces with extreme directions.

The extreme directions of the positive semidefinite cone (§2.9) comprise the infinite set of all symmetric rank-one matrices. [21, §6] [201, §III] It is sometimes prudent to instead consider the less infinite but complete normalized set, for $M > 0$ (*confer*(235))

$$\{zz^{\mathrm{T}} \in \mathbb{S}^M \mid \|z\| = 1\} \tag{187}$$

The positive semidefinite cone in one dimension $M = 1$, $\mathbb{S}_+$ the nonnegative real line, has one extreme direction belonging to its relative interior; an idiosyncrasy of dimension 1.

Pointed closed convex cone $\mathcal{K} = \{\mathbf{0}\}$ has no extreme direction because extreme directions are nonzero by definition.

- If closed convex cone $\mathcal{K}$ is not pointed, then it has no extreme directions and no vertex. [21, §1]

Conversely, pointed closed convex cone $\mathcal{K}$ is equivalent to the convex hull of its vertex and all its extreme directions. [315, §18, p.167] That is the practical utility of extreme direction; to facilitate construction of polyhedral sets, apparent from the *extremes theorem*:

2.8.1.1.1 Theorem. (Klee) *Extremes.*   [337, §3.6] [315, §18, p.166]
(*confer* §2.3.2, §2.12.2.0.1) Any closed convex set containing no lines can be expressed as the convex hull of its extreme points and extreme rays.   ⋄

It follows that any element of a convex set containing no lines may be expressed as a linear combination of its extreme elements; *e.g.*, §2.9.2.7.1.

2.8.1.2 Generators

In the narrowest sense, *generators* for a convex set comprise any collection of points and directions whose convex hull constructs the set.

When the *extremes theorem* applies, the extreme points and directions are called generators of a convex set. An arbitrary collection of generators for a convex set includes its extreme elements as a subset; the set of extreme elements of a convex set is a minimal set of generators for that convex set. Any polyhedral set has a minimal set of generators whose *cardinality* is finite.

[2.36]Like vectors, an extreme direction can be identified with the Cartesian point at the vector's head with respect to the origin.

When the convex set under scrutiny is a closed convex cone, conic combination of generators during construction is implicit as shown in Example 2.8.1.2.1 and Example 2.10.2.0.1. So, a vertex at the origin (if it exists) becomes benign.

We can, of course, generate affine sets by taking the affine hull of any collection of points and directions. We broaden, thereby, the meaning of generator to be inclusive of all kinds of hulls.

Any hull of generators is loosely called a *vertex-description*. (§2.3.4) Hulls encompass subspaces, so any basis constitutes generators for a vertex-description; span basis $\mathcal{R}(A)$.

2.8.1.2.1 Example. *Application of extremes theorem.*
Given an extreme point at the origin and N extreme rays (§2.7.0.0.1) $\{\zeta\Gamma_i, \, i=1\dots N \mid \zeta \geq 0\}$, denoting the ith extreme direction by $\Gamma_i \in \mathbb{R}^n$, then their convex hull (86) is

$$
\begin{aligned}
\mathcal{P} &= \left\{ [\mathbf{0} \ \Gamma_1 \ \Gamma_2 \cdots \Gamma_N] \, a \, \zeta \mid a^{\text{T}}\mathbf{1} = 1, \ a \succeq 0, \ \zeta \geq 0 \right\} \\
&= \left\{ [\Gamma_1 \ \Gamma_2 \cdots \Gamma_N] \, a \, \zeta \mid a^{\text{T}}\mathbf{1} \leq 1, \ a \succeq 0, \ \zeta \geq 0 \right\} \\
&= \left\{ [\Gamma_1 \ \Gamma_2 \cdots \Gamma_N] \, b \quad \mid b \succeq 0 \right\} \subset \mathbb{R}^n
\end{aligned}
\tag{188}
$$

a closed convex set that is simply a conic hull like (103). □

2.8.2 Exposed direction

2.8.2.0.1 Definition. *Exposed point & direction of pointed convex cone.*
[315, §18] (*confer* §2.6.1.0.1)

- When a convex cone has a vertex, an exposed point, it resides at the origin; there can be only one.

- In the closure of a pointed convex cone, an *exposed direction* is the direction of a one-dimensional exposed face that is a ray emanating from the origin.

- {exposed directions} $\subseteq$ {extreme directions} △

For a proper cone in vector space $\mathbb{R}^n$ with $n \geq 2$, we can say more:

$$
\overline{\{\text{exposed directions}\}} = \{\text{extreme directions}\}
\tag{189}
$$

It follows from Lemma 2.8.0.0.1 for any pointed closed convex cone, there is one-to-one correspondence of one-dimensional exposed faces with exposed directions; *id est*, there is no one-dimensional exposed face that is not a ray base $\mathbf{0}$.

The pointed closed convex cone $\mathbb{EDM}^2$, for example, is a ray in isomorphic subspace $\mathbb{R}$ whose relative boundary (§2.6.1.4.1) is the origin. The conventionally exposed directions of $\mathbb{EDM}^2$ constitute the empty set $\emptyset \subset \{$extreme direction$\}$. This cone has one extreme direction belonging to its relative interior; an idiosyncrasy of dimension 1.

2.8.2.1   Connection between boundary and extremes

2.8.2.1.1   Theorem.   *Exposed.*   [315, §18.7]   (*confer* §2.8.1.1.1)
Any closed convex set $\mathcal{C}$ containing no lines (and whose dimension is at least 2) can be expressed as closure of the convex hull of its exposed points and exposed rays.   ⋄

From Theorem 2.8.1.1.1,

$$
\left.
\begin{aligned}
\operatorname{rel}\partial\mathcal{C} \; &= \; \overline{\mathcal{C}} \setminus \operatorname{rel}\operatorname{int}\mathcal{C} & (24) \\
&= \; \overline{\operatorname{conv}\{\text{exposed points and exposed rays}\}} \setminus \operatorname{rel}\operatorname{int}\mathcal{C} \\
&= \; \operatorname{conv}\{\text{extreme points and extreme rays}\} \setminus \operatorname{rel}\operatorname{int}\mathcal{C}
\end{aligned}
\right\} \qquad (190)
$$

Thus each and every extreme point of a convex set (that is not a point) resides on its relative boundary, while each and every extreme direction of a convex set (that is not a halfline and contains no line) resides on its relative boundary because extreme points and directions of such respective sets do not belong to relative interior by definition.

The relationship between extreme sets and the relative boundary actually goes deeper: Any face $\mathcal{F}$ of convex set $\mathcal{C}$ (that is not $\mathcal{C}$ itself) belongs to $\operatorname{rel}\partial\mathcal{C}$, so $\dim\mathcal{F} < \dim\mathcal{C}$. [315, §18.1.3]

2.8.2.2   Converse *caveat*

It is inconsequent to presume that each and every extreme point and direction is necessarily exposed, as might be erroneously inferred from the *conventional boundary definition* (§2.6.1.4.1); although it can correctly be inferred: each and every extreme point and direction belongs to some exposed face.

Arbitrary points residing on the relative boundary of a convex set are not necessarily exposed or extreme points. Similarly, the direction of an arbitrary ray, base $\mathbf{0}$, on the boundary of a convex cone is not necessarily an exposed or extreme direction. For the polyhedral cone illustrated in Figure 25, for example, there are three two-dimensional exposed faces constituting the entire boundary, each composed of an infinity of rays. Yet there are only three exposed directions.

Neither is an extreme direction on the boundary of a pointed convex cone necessarily an exposed direction. Lift the two-dimensional set in Figure 33, for example, into three dimensions such that no two points in the set are collinear with the origin. Then its conic hull can have an extreme direction B on the boundary that is not an exposed direction, illustrated in Figure 43.

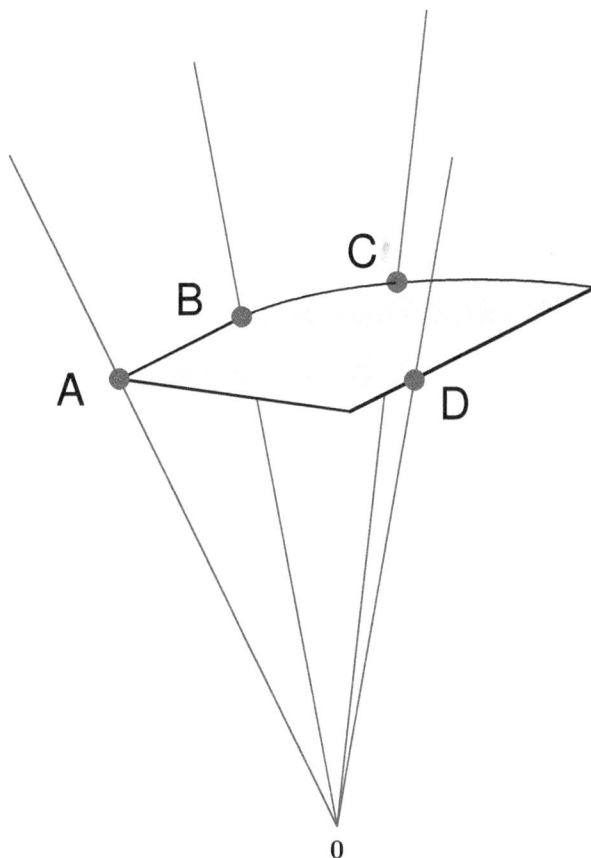

Figure 43: Properties of extreme points carry over to extreme directions. [315, §18] Four rays (drawn truncated) on boundary of conic hull of two-dimensional closed convex set from Figure **33** lifted to $\mathbb{R}^3$. Ray through point A is exposed hence extreme. Extreme direction B on cone boundary is not an exposed direction, although it belongs to the exposed face cone$\{A,B\}$. Extreme ray through C is exposed. Point D is neither an exposed or extreme direction although it belongs to a two-dimensional exposed face of the conic hull.

2.9 Positive semidefinite (PSD) cone

The cone of positive semidefinite matrices studied in this section is arguably the most important of all non-polyhedral cones whose facial structure we completely understand.

−Alexander Barvinok [26, p.78]

2.9.0.0.1 Definition. *Positive semidefinite cone.*
The set of all symmetric positive semidefinite matrices of particular dimension M is called the *positive semidefinite cone*:

$$
\begin{aligned}
\mathbb{S}_+^M \;\triangleq\; & \left\{ A \in \mathbb{S}^M \mid A \succeq 0 \right\} \\
= \; & \left\{ A \in \mathbb{S}^M \mid y^{\mathrm{T}} A y \geq 0 \;\; \forall \|y\| = 1 \right\} \\
= \; & \bigcap_{\|y\|=1} \left\{ A \in \mathbb{S}^M \mid \langle y y^{\mathrm{T}}, A \rangle \geq 0 \right\} \\
= \; & \{ A \in \mathbb{S}_+^M \mid \operatorname{rank} A \leq M \}
\end{aligned}
\tag{191}
$$

formed by the intersection of an infinite number of halfspaces (§2.4.1.1) in vectorized variable[2.37] A, each halfspace having partial boundary containing the origin in isomorphic $\mathbb{R}^{M(M+1)/2}$. It is a unique immutable proper cone in the ambient space of symmetric matrices $\mathbb{S}^M$.

The positive definite (full-rank) matrices comprise the cone interior

$$
\begin{aligned}
\operatorname{int} \mathbb{S}_+^M \;=\; & \left\{ A \in \mathbb{S}^M \mid A \succ 0 \right\} \\
= \; & \left\{ A \in \mathbb{S}^M \mid y^{\mathrm{T}} A y > 0 \;\; \forall \|y\| = 1 \right\} \\
= \; & \bigcap_{\|y\|=1} \left\{ A \in \mathbb{S}^M \mid \langle y y^{\mathrm{T}}, A \rangle > 0 \right\} \\
= \; & \{ A \in \mathbb{S}_+^M \mid \operatorname{rank} A = M \}
\end{aligned}
\tag{192}
$$

while all singular positive semidefinite matrices (having at least one 0 eigenvalue) reside on the cone boundary (Figure 44); (§A.7.5)

$$
\begin{aligned}
\partial \mathbb{S}_+^M \;=\; & \left\{ A \in \mathbb{S}^M \mid A \succeq 0, \; A \nsucc 0 \right\} \\
= \; & \left\{ A \in \mathbb{S}^M \mid \min\{\lambda(A)_i, \; i = 1 \ldots M\} = 0 \right\} \\
= \; & \left\{ A \in \mathbb{S}_+^M \mid \langle y y^{\mathrm{T}}, A \rangle = 0 \;\; \text{for some } \|y\| = 1 \right\} \\
= \; & \{ A \in \mathbb{S}_+^M \mid \operatorname{rank} A < M \}
\end{aligned}
\tag{193}
$$

where $\lambda(A) \in \mathbb{R}^M$ holds the eigenvalues of A. $\triangle$

The only symmetric positive semidefinite matrix in $\mathbb{S}_+^M$ having M 0-eigenvalues resides at the origin. (§A.7.3.0.1)

[2.37] infinite in number when $M > 1$. Because $y^{\mathrm{T}} A y = y^{\mathrm{T}} A^{\mathrm{T}} y$, matrix A is almost always assumed symmetric. (§A.2.1)

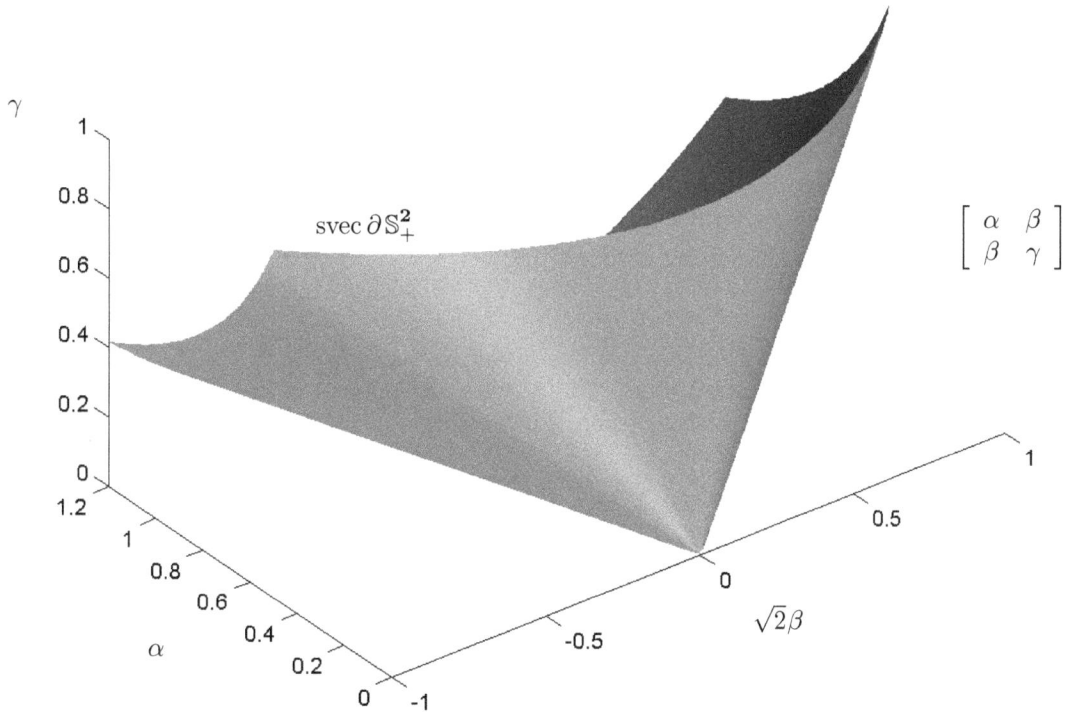

$$\begin{bmatrix} \alpha & \beta \\ \beta & \gamma \end{bmatrix}$$

Minimal set of generators are the extreme directions: $\mathrm{svec}\{yy^{\mathrm{T}} \mid y \in \mathbb{R}^{M}\}$

Figure 44: (d'Aspremont) Truncated boundary of PSD cone in $\mathbb{S}^{2}$ plotted in isometrically isomorphic $\mathbb{R}^{3}$ via svec (56); 0-contour of smallest eigenvalue (193). Lightest shading is closest, darkest shading is farthest and inside shell. Entire boundary can be constructed from an aggregate of rays (§2.7.0.0.1) emanating exclusively from origin: $\left\{ \kappa^{2} [\,z_{1}^{2} \quad \sqrt{2}z_{1}z_{2} \quad z_{2}^{2}\,]^{\mathrm{T}} \mid \kappa \in \mathbb{R}\,,\; z \in \mathbb{R}^{2} \right\}$. A circular cone in this dimension (§2.9.2.8), each and every ray on boundary corresponds to an extreme direction but such is not the case in any higher dimension (confer Figure 25). PSD cone geometry is not as simple in higher dimensions [26, §II.12] although PSD cone is selfdual (376) in ambient real space of symmetric matrices. [201, §II] PSD cone has no two-dimensional face in any dimension, its only extreme point residing at $\mathbf{0}$.

2.9.0.1   Membership

Observe notation $A \succeq 0$ denoting a positive semidefinite matrix;[2.38] meaning (*confer* §2.3.1.1), matrix A belongs to the positive semidefinite cone in the subspace of symmetric matrices whereas $A \succ 0$ denotes membership to that cone's interior. (§2.13.2) Notation $A \succ 0$, denoting a positive definite matrix, can be read: *symmetric matrix A exceeds the origin with respect to the positive semidefinite cone interior.* These notations further imply that coordinates [*sic*] for orthogonal expansion of a positive (semi)definite matrix must be its (nonnegative) positive eigenvalues (§2.13.7.1.1, §E.6.4.1.1) when expanded in its *eigenmatrices* (§A.5.0.3); *id est*, eigenvalues must be (nonnegative) positive.

Generalizing comparison on the real line, the notation $A \succeq B$ denotes comparison with respect to the positive semidefinite cone; (§A.3.1) *id est*, $A \succeq B \Leftrightarrow A - B \in \mathbb{S}_+^M$ but neither matrix A or B necessarily belongs to the positive semidefinite cone. Yet, (1567) $A \succeq B, \ B \succeq 0 \ \Rightarrow \ A \succeq 0$; *id est*, $A \in \mathbb{S}_+^M$. (*confer* Figure 64)

2.9.0.1.1   Example. *Equality constraints in semidefinite program* (679).
Employing properties of partial order (§2.7.2.2) for the pointed closed convex positive semidefinite cone, it is easy to show, given $A + S = C$

$$\begin{aligned} S \succeq 0 &\Leftrightarrow A \preceq C \\ S \succ 0 &\Leftrightarrow A \prec C \end{aligned} \tag{194}$$

$\square$

2.9.1   Positive semidefinite cone is convex

The set of all positive semidefinite matrices forms a convex cone in the ambient space of symmetric matrices because any pair satisfies definition (175); [208, §7.1] *videlicet*, for all $\zeta_1, \zeta_2 \geq 0$ and each and every $A_1, A_2 \in \mathbb{S}^M$

$$\zeta_1 A_1 + \zeta_2 A_2 \succeq 0 \ \Leftarrow \ A_1 \succeq 0, \ A_2 \succeq 0 \tag{195}$$

a fact easily verified by the definitive test for positive semidefiniteness of a symmetric matrix (§A):

$$A \succeq 0 \Leftrightarrow x^{\mathrm{T}} A \, x \geq 0 \ \text{ for each and every } \|x\| = 1 \tag{196}$$

id est, for $A_1, A_2 \succeq 0$ and each and every $\zeta_1, \zeta_2 \geq 0$

$$\zeta_1 \, x^{\mathrm{T}} A_1 \, x + \zeta_2 \, x^{\mathrm{T}} A_2 \, x \geq 0 \ \text{ for each and every normalized } x \in \mathbb{R}^M \tag{197}$$

The convex cone $\mathbb{S}_+^M$ is more easily visualized in the isomorphic vector space $\mathbb{R}^{M(M+1)/2}$ whose dimension is the number of free variables in a symmetric $M \times M$ matrix. When $M = 2$ the PSD cone is semiinfinite in expanse in $\mathbb{R}^3$, having boundary illustrated in Figure 44. When $M = 3$ the PSD cone is six-dimensional, and so on.

[2.38] the same as *nonnegative definite matrix.*

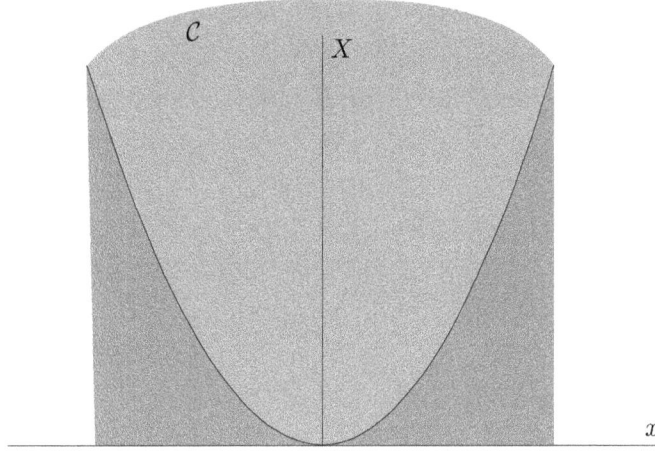

Figure 45: Convex set $\mathcal{C} = \{X \in \mathbb{S} \times x \in \mathbb{R} \mid X \succeq xx^{\mathrm{T}}\}$ drawn truncated.

2.9.1.0.1 Example. *Sets from maps of positive semidefinite cone.*
The set

$$\mathcal{C} = \{X \in \mathbb{S}^n \times x \in \mathbb{R}^n \mid X \succeq xx^{\mathrm{T}}\} \tag{198}$$

is convex because it has *Schur-form*; (§A.4)

$$X - xx^{\mathrm{T}} \succeq 0 \quad \Leftrightarrow \quad f(X,\, x) \triangleq \begin{bmatrix} X & x \\ x^{\mathrm{T}} & 1 \end{bmatrix} \succeq 0 \tag{199}$$

e.g., Figure 45. Set $\mathcal{C}$ is the inverse image (§2.1.9.0.1) of $\mathbb{S}_+^{n+1}$ under affine mapping f. The set $\{X \in \mathbb{S}^n \times x \in \mathbb{R}^n \mid X \preceq xx^{\mathrm{T}}\}$ is not convex, in contrast, having no Schur-form. Yet for fixed $x = x_{\mathrm{p}}$, the set

$$\{X \in \mathbb{S}^n \mid X \preceq x_{\mathrm{p}} x_{\mathrm{p}}^{\mathrm{T}}\} \tag{200}$$

is simply the negative semidefinite cone shifted to $x_{\mathrm{p}} x_{\mathrm{p}}^{\mathrm{T}}$. $\square$

2.9.1.0.2 Example. *Inverse image of positive semidefinite cone.*
Now consider finding the set of all matrices $X \in \mathbb{S}^N$ satisfying

$$AX + B \succeq 0 \tag{201}$$

given $A, B \in \mathbb{S}^N$. Define the set

$$\mathcal{X} \triangleq \{X \mid AX + B \succeq 0\} \subseteq \mathbb{S}^N \tag{202}$$

which is the inverse image of the positive semidefinite cone under affine transformation $g(X) \triangleq AX + B$. Set $\mathcal{X}$ must therefore be convex by Theorem 2.1.9.0.1.

Yet we would like a less amorphous characterization of this set, so instead we consider its vectorization (37) which is easier to visualize:

$$\operatorname{vec} g(X) \;=\; \operatorname{vec}(AX) + \operatorname{vec} B \;=\; (I \otimes A)\operatorname{vec} X + \operatorname{vec} B \qquad (203)$$

where

$$I \otimes A \triangleq Q \Lambda Q^{\mathrm{T}} \in \mathbb{S}^{N^2} \qquad (204)$$

is block-diagonal formed by *Kronecker product* (§A.1.1 *no.*31, §D.1.2.1). Assign

$$\begin{aligned} x &\triangleq \operatorname{vec} X \in \mathbb{R}^{N^2} \\ b &\triangleq \operatorname{vec} B \in \mathbb{R}^{N^2} \end{aligned} \qquad (205)$$

then make the equivalent problem: Find

$$\operatorname{vec} \mathcal{X} = \left\{ x \in \mathbb{R}^{N^2} \mid (I \otimes A)x + b \in \mathcal{K} \right\} \qquad (206)$$

where

$$\mathcal{K} \triangleq \operatorname{vec} \mathbb{S}_+^N \qquad (207)$$

is a proper cone isometrically isomorphic with the positive semidefinite cone in the subspace of symmetric matrices; the vectorization of every element of $\mathbb{S}_+^N$. Utilizing the diagonalization (204),

$$\begin{aligned} \operatorname{vec} \mathcal{X} &= \left\{ x \mid \Lambda Q^{\mathrm{T}} x \in Q^{\mathrm{T}}(\mathcal{K} - b) \right\} \\ &= \left\{ x \mid \Phi Q^{\mathrm{T}} x \in \Lambda^{\dagger} Q^{\mathrm{T}}(\mathcal{K} - b) \right\} \subseteq \mathbb{R}^{N^2} \end{aligned} \qquad (208)$$

where † denotes matrix *pseudoinverse* (§E) and

$$\Phi \triangleq \Lambda^{\dagger} \Lambda \qquad (209)$$

is a diagonal projection matrix whose entries are either 1 or 0 (§E.3). We have the complementary sum

$$\Phi Q^{\mathrm{T}} x + (I - \Phi) Q^{\mathrm{T}} x = Q^{\mathrm{T}} x \qquad (210)$$

So, adding $(I - \Phi)Q^{\mathrm{T}} x$ to both sides of the membership within (208) admits

$$\begin{aligned} \operatorname{vec} \mathcal{X} &= \left\{ x \in \mathbb{R}^{N^2} \mid Q^{\mathrm{T}} x \in \Lambda^{\dagger} Q^{\mathrm{T}}(\mathcal{K} - b) + (I - \Phi)Q^{\mathrm{T}} x \right\} \\ &= \left\{ x \mid Q^{\mathrm{T}} x \in \Phi\big(\Lambda^{\dagger} Q^{\mathrm{T}}(\mathcal{K} - b)\big) \oplus (I - \Phi)\mathbb{R}^{N^2} \right\} \\ &= \left\{ x \in Q\Lambda^{\dagger} Q^{\mathrm{T}}(\mathcal{K} - b) \oplus Q(I - \Phi)\mathbb{R}^{N^2} \right\} \\ &= (I \otimes A)^{\dagger}(\mathcal{K} - b) \oplus \mathcal{N}(I \otimes A) \end{aligned} \qquad (211)$$

where we used the facts: linear function $Q^{\mathrm{T}} x$ in x on $\mathbb{R}^{N^2}$ is a bijection, and $\Phi\Lambda^{\dagger} = \Lambda^{\dagger}$.

$$\operatorname{vec} \mathcal{X} = (I \otimes A)^{\dagger} \operatorname{vec}(\mathbb{S}_+^N - B) \oplus \mathcal{N}(I \otimes A) \qquad (212)$$

In words, set $\operatorname{vec} \mathcal{X}$ is the vector sum of the translated PSD cone (linearly mapped onto the rowspace of $I \otimes A$ (§E)) and the nullspace of $I \otimes A$ (synthesis of fact from §A.6.3 and §A.7.3.0.1). Should $I \otimes A$ have no nullspace, then $\operatorname{vec} \mathcal{X} = (I \otimes A)^{-1} \operatorname{vec}(\mathbb{S}_+^N - B)$ which is the expected result. $\square$

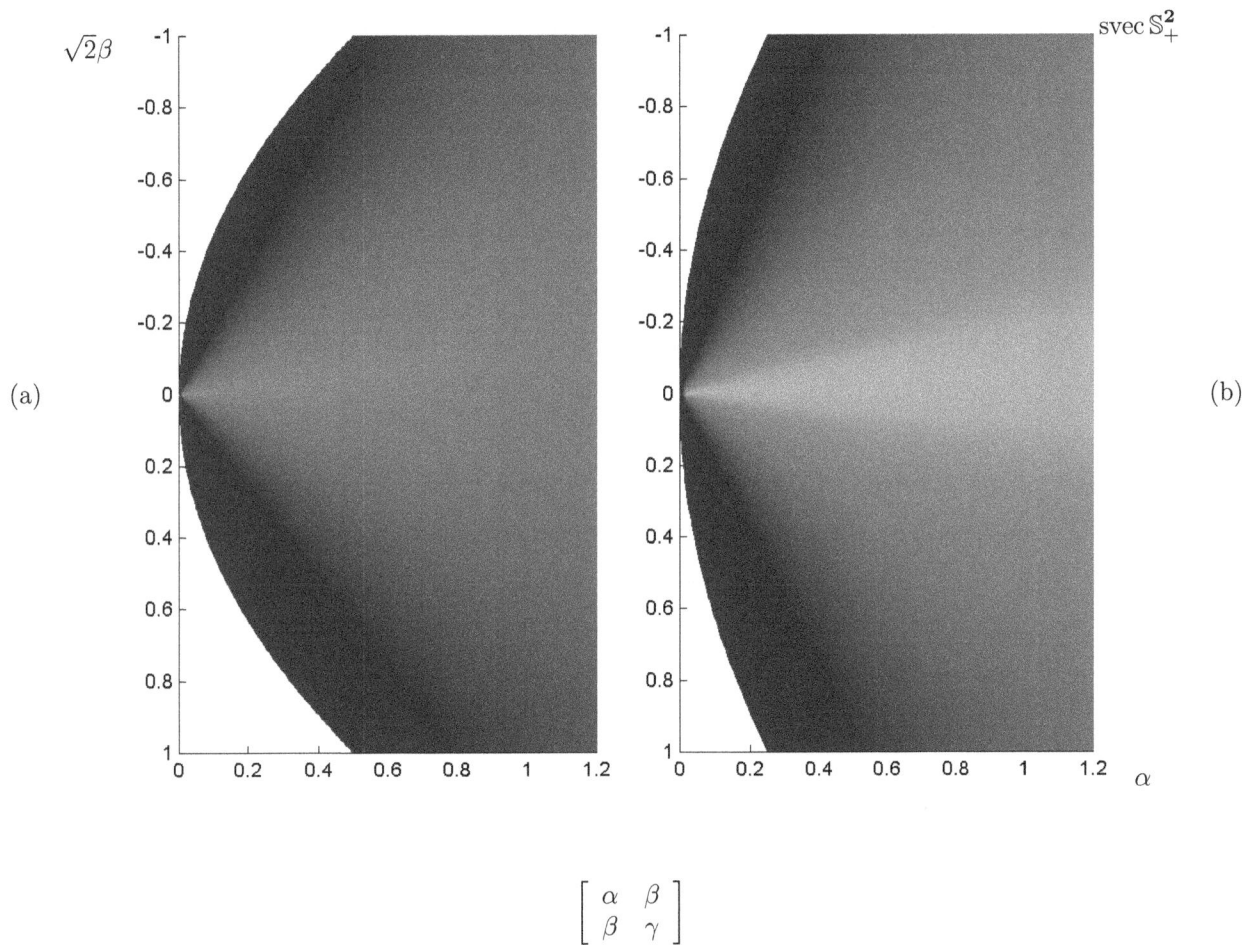

$$\begin{bmatrix} \alpha & \beta \\ \beta & \gamma \end{bmatrix}$$

Figure 46: **(a)** Projection of truncated PSD cone $\mathbb{S}^2_+$, truncated above $\gamma\!=\!1$, on $\alpha\beta$-plane in isometrically isomorphic $\mathbb{R}^3$. View is from above with respect to Figure 44. **(b)** Truncated above $\gamma\!=\!2$. From these plots we might infer, for example, line $\left\{[0 \quad 1/\sqrt{2} \quad \gamma]^{\mathrm{T}} \mid \gamma\!\in\!\mathbb{R}\right\}$ intercepts PSD cone at some large value of γ; in fact, $\gamma\!=\!\infty$.

2.9.2   Positive semidefinite cone boundary

For any symmetric positive semidefinite matrix A of rank ρ , there must exist a rank ρ matrix Y such that A be expressible as an outer product in Y ; [338, §6.3]

$$A = YY^{\mathrm{T}} \in \mathbb{S}_+^M \ , \quad \operatorname{rank} A = \operatorname{rank} Y = \rho , \quad Y \in \mathbb{R}^{M \times \rho} \tag{213}$$

Then the boundary of the positive semidefinite cone may be expressed

$$\partial\mathbb{S}_+^M = \left\{ A \in \mathbb{S}_+^M \mid \operatorname{rank} A < M \right\} = \left\{ YY^{\mathrm{T}} \mid Y \in \mathbb{R}^{M \times M-1} \right\} \tag{214}$$

Because the boundary of any convex body is obtained with closure of its relative interior (§2.1.7, §2.1.7.2), from (192) we must also have

$$\begin{aligned}
\mathbb{S}_+^M = \overline{\left\{ A \in \mathbb{S}_+^M \mid \operatorname{rank} A = M \right\}} &= \overline{\left\{ YY^{\mathrm{T}} \mid Y \in \mathbb{R}^{M \times M}, \ \operatorname{rank} Y = M \right\}} \\
&= \left\{ YY^{\mathrm{T}} \mid Y \in \mathbb{R}^{M \times M} \right\}
\end{aligned} \tag{215}$$

2.9.2.1   rank ρ subset of the positive semidefinite cone

For the same reason (closure), this applies more generally; for $0 \le \rho \le M$

$$\overline{\left\{ A \in \mathbb{S}_+^M \mid \operatorname{rank} A = \rho \right\}} = \left\{ A \in \mathbb{S}_+^M \mid \operatorname{rank} A \le \rho \right\} \tag{216}$$

For easy reference, we give such generally nonconvex sets a name: *rank ρ subset of a positive semidefinite cone.* For $\rho < M$ this subset, nonconvex for $M > 1$, resides on the positive semidefinite cone boundary.

2.9.2.1.1   Exercise.   *Closure and rank ρ subset.*
Prove equality in (216).   ▼

For example,

$$\partial\mathbb{S}_+^M = \overline{\left\{ A \in \mathbb{S}_+^M \mid \operatorname{rank} A = M-1 \right\}} = \left\{ A \in \mathbb{S}_+^M \mid \operatorname{rank} A \le M-1 \right\} \tag{217}$$

In $\mathbb{S}^2$, each and every ray on the boundary of the positive semidefinite cone in isomorphic $\mathbb{R}^3$ corresponds to a symmetric rank-1 matrix (Figure 44), but that does not hold in any higher dimension.

2.9.2.2   Subspace tangent to open rank ρ subset

When the positive semidefinite cone subset in (216) is left unclosed as in

$$\mathcal{M}(\rho) \triangleq \left\{ A \in \mathbb{S}_+^N \mid \operatorname{rank} A = \rho \right\} \tag{218}$$

then we can specify a subspace tangent to the positive semidefinite cone at a particular member of manifold $\mathcal{M}(\rho)$. Specifically, the subspace $\mathcal{R}_\mathcal{M}$ tangent to manifold $\mathcal{M}(\rho)$ at $B \in \mathcal{M}(\rho)$ [192, §5, prop.1.1]

$$\mathcal{R}_\mathcal{M}(B) \triangleq \{XB + BX^\mathrm{T} \mid X \in \mathbb{R}^{N \times N}\} \subseteq \mathbb{S}^N \tag{219}$$

has dimension

$$\dim \mathrm{svec}\, \mathcal{R}_\mathcal{M}(B) = \rho\left(N - \frac{\rho - 1}{2}\right) = \rho(N - \rho) + \frac{\rho(\rho + 1)}{2} \tag{220}$$

Tangent subspace $\mathcal{R}_\mathcal{M}$ contains no member of the positive semidefinite cone $\mathbb{S}_+^N$ whose rank exceeds ρ.

Subspace $\mathcal{R}_\mathcal{M}(B)$ is a hyperplane supporting $\mathbb{S}_+^N$ when $B \in \mathcal{M}(N-1)$. Another good example of tangent subspace is given in §E.7.2.0.2 by (2096); $\mathcal{R}_\mathcal{M}(\mathbf{1}\mathbf{1}^\mathrm{T}) = \mathbb{S}_c^{N\perp}$, orthogonal complement to the *geometric center subspace*. (Figure **159** p.468)

2.9.2.3 Faces of PSD cone, their dimension *versus* rank

Each and every face of the positive semidefinite cone, having dimension less than that of the cone, is exposed. [252, §6] [221, §2.3.4] Because each and every face of the positive semidefinite cone contains the origin (§2.8.0.0.1), each face belongs to a subspace of dimension the same as the face.

Define $\mathcal{F}(\mathbb{S}_+^M \ni A)$ (171) as the smallest face, that contains a given positive semidefinite matrix A, of positive semidefinite cone $\mathbb{S}_+^M$. Then matrix A, having ordered diagonalization $A = Q\Lambda Q^\mathrm{T} \in \mathbb{S}_+^M$ (§A.5.1), is relatively interior to[2.39] [26, §II.12] [116, §31.5.3] [245, §2.4] [246]

$$
\begin{aligned}
\mathcal{F}\left(\mathbb{S}_+^M \ni A\right) &= \{X \in \mathbb{S}_+^M \mid \mathcal{N}(X) \supseteq \mathcal{N}(A)\} \\
&= \{X \in \mathbb{S}_+^M \mid \langle Q(I - \Lambda\Lambda^\dagger)Q^\mathrm{T},\, X \rangle = 0\} \\
&= \{Q\Lambda\Lambda^\dagger \Psi \Lambda\Lambda^\dagger Q^\mathrm{T} \mid \Psi \in \mathbb{S}_+^M\} \\
&= Q\Lambda\Lambda^\dagger\, \mathbb{S}_+^M\, \Lambda\Lambda^\dagger Q^\mathrm{T} \\
&\simeq \mathbb{S}_+^{\mathrm{rank}\, A}
\end{aligned}
\tag{221}
$$

which is isomorphic with convex cone $\mathbb{S}_+^{\mathrm{rank}\, A}$; *e.g.*, $Q\mathbb{S}_+^M Q^\mathrm{T} = \mathbb{S}_+^M$. The larger the nullspace of A, the smaller the face. (140) Thus dimension of the smallest face that contains given matrix A is

$$\dim \mathcal{F}\left(\mathbb{S}_+^M \ni A\right) = \mathrm{rank}(A)(\mathrm{rank}(A) + 1)/2 \tag{222}$$

in isomorphic $\mathbb{R}^{M(M+1)/2}$, and each and every face of $\mathbb{S}_+^M$ is isomorphic with a positive semidefinite cone having dimension the same as the face. Observe: not all dimensions are represented, and the only zero-dimensional face is the origin. The positive semidefinite cone has no facets, for example.

[2.39] For $X \in \mathbb{S}_+^M$, $A = Q\Lambda Q^\mathrm{T} \in \mathbb{S}_+^M$, show $\mathcal{N}(X) \supseteq \mathcal{N}(A) \Leftrightarrow \langle Q(I - \Lambda\Lambda^\dagger)Q^\mathrm{T},\, X \rangle = 0$.
Given $\langle Q(I - \Lambda\Lambda^\dagger)Q^\mathrm{T},\, X \rangle = 0 \Leftrightarrow \mathcal{R}(X) \perp \mathcal{N}(A)$. (§A.7.4)
($\Rightarrow$) Assume $\mathcal{N}(X) \supseteq \mathcal{N}(A)$, then $\mathcal{R}(X) \perp \mathcal{N}(X) \supseteq \mathcal{N}(A)$.
($\Leftarrow$) Assume $\mathcal{R}(X) \perp \mathcal{N}(A)$, then $X\, Q(I - \Lambda\Lambda^\dagger)Q^\mathrm{T} = \mathbf{0} \Rightarrow \mathcal{N}(X) \supseteq \mathcal{N}(A)$. ◆

2.9.2.3.1   Table: Rank k *versus* dimension of $\mathbb{S}_+^3$ faces

	k	$\dim \mathcal{F}(\mathbb{S}_+^3 \ni \text{rank-}k \text{ matrix})$
	0	0
boundary	≤ 1	1
	≤ 2	3
interior	≤ 3	6

For the positive semidefinite cone $\mathbb{S}_+^2$ in isometrically isomorphic $\mathbb{R}^3$ depicted in Figure 44, rank-2 matrices belong to the interior of that face having dimension 3 (the entire closed cone), rank-1 matrices belong to relative interior of a face having dimension[2.40] 1, and the only rank-0 matrix is the point at the origin (the zero-dimensional face).

2.9.2.3.2   Exercise.   *Bijective isometry.*
Prove that the smallest face of positive semidefinite cone $\mathbb{S}_+^M$, containing a particular full-rank matrix A having ordered diagonalization $Q\Lambda Q^{\mathrm{T}}$, is the entire cone: *id est*, prove $Q\,\mathbb{S}_+^M Q^{\mathrm{T}} = \mathbb{S}_+^M$ from (221).   ▼

2.9.2.4   rank-k face of PSD cone

Any rank-$k < M$ positive semidefinite matrix A belongs to a face, of positive semidefinite cone $\mathbb{S}_+^M$, described by intersection with a hyperplane: for ordered diagonalization of $A = Q\Lambda Q^{\mathrm{T}} \in \mathbb{S}_+^M \ni \operatorname{rank}(A) = k < M$

$$
\begin{aligned}
\mathcal{F}\left(\mathbb{S}_+^M \ni A\right) &= \{X \in \mathbb{S}_+^M \mid \langle Q(I - \Lambda\Lambda^\dagger)Q^{\mathrm{T}},\, X \rangle = 0\} \\
&= \left\{ X \in \mathbb{S}_+^M \,\middle|\, \left\langle Q\left(I - \begin{bmatrix} I \in \mathbb{S}^k & \mathbf{0} \\ \mathbf{0}^{\mathrm{T}} & \mathbf{0} \end{bmatrix}\right)Q^{\mathrm{T}},\, X \right\rangle = 0 \right\} \\
&= \mathbb{S}_+^M \cap \partial\mathcal{H}_+ \\
&\simeq \mathbb{S}_+^k
\end{aligned}
\qquad (223)
$$

Faces are doubly indexed: continuously indexed by orthogonal matrix Q, and discretely indexed by rank k.   Each and every orthogonal matrix Q makes projectors $Q(:,k{+}1{:}M)Q(:,k{+}1{:}M)^{\mathrm{T}}$ indexed by k, in other words, each projector describing a normal[2.41] $\operatorname{svec}\!\left(Q(:,k{+}1{:}M)Q(:,k{+}1{:}M)^{\mathrm{T}}\right)$ to a supporting hyperplane $\partial\mathcal{H}_+$ (containing the origin) exposing a face (§2.11) of the positive semidefinite cone containing rank-k (and less) matrices.

2.9.2.4.1   Exercise.   *Simultaneously diagonalizable means commutative.*
Given diagonalization of rank-$k \leq M$ positive semidefinite matrix $A = Q\Lambda Q^{\mathrm{T}}$ and any particular $\Psi \succeq 0$, both in $\mathbb{S}^M$ from (221), show how $I - \Lambda\Lambda^\dagger$ and $\Lambda\Lambda^\dagger \Psi \Lambda\Lambda^\dagger$ share a complete set of eigenvectors.   ▼

[2.40] The boundary constitutes all the one-dimensional faces, in $\mathbb{R}^3$, which are rays emanating from the origin.
[2.41] Any vectorized nonzero matrix $\in \mathbb{S}_+^M$ is normal to a hyperplane supporting $\mathbb{S}_+^M$ (§2.13.1) because PSD cone is selfdual. Normal on boundary exposes nonzero face by (329) (330).

2.9.2.5 PSD cone face containing principal submatrix

A *principal submatrix* of a matrix $A \in \mathbb{R}^{M \times M}$ is formed by discarding any particular subset of its rows and columns having the same indices. There are $M!/(1!(M-1)!)$ principal 1×1 submatrices, $M!/(2!(M-2)!)$ principal 2×2 submatrices, and so on, totaling $2^M - 1$ principal submatrices including A itself. Principal submatrices of a symmetric matrix are symmetric. A given symmetric matrix has rank ρ iff it has a nonsingular principal $\rho \times \rho$ submatrix but none larger. [305, §5-10] By loading vector y in test $y^{\mathrm{T}}Ay$ (§A.2) with various binary patterns, it follows that any principal submatrix must be positive (semi)definite whenever A is (Theorem A.3.1.0.4). If positive semidefinite matrix $A \in \mathbb{S}_+^M$ has principal submatrix of dimension ρ with rank r, then $\operatorname{rank} A \leq M - \rho + r$ by (1617).

Because each and every principal submatrix of a positive semidefinite matrix in $\mathbb{S}^M$ is positive semidefinite, then each principal submatrix belongs to a certain face of positive semidefinite cone $\mathbb{S}_+^M$ by (222). Of special interest are full-rank positive semidefinite principal submatrices, for then description of smallest face becomes simpler. We can find the smallest face, that contains a particular complete full-rank principal submatrix of A, by embedding that submatrix in a $\mathbf{0}$ matrix of the same dimension as A: Were Φ a binary diagonal matrix

$$\Phi = \delta^2(\Phi) \in \mathbb{S}^M, \qquad \Phi_{ii} \in \{0,1\} \tag{224}$$

having diagonal entry 0 corresponding to a discarded row and column from $A \in \mathbb{S}_+^M$, then any principal submatrix[2.42] so embedded can be expressed $\Phi A \Phi$; *id est*, for an embedded principal submatrix $\Phi A \Phi \in \mathbb{S}_+^M \ni \operatorname{rank} \Phi A \Phi = \operatorname{rank} \Phi \leq \operatorname{rank} A$

$$\begin{aligned}
\mathcal{F}\left(\mathbb{S}_+^M \ni \Phi A \Phi\right) &= \{X \in \mathbb{S}_+^M \mid \mathcal{N}(X) \supseteq \mathcal{N}(\Phi A \Phi)\} \\
&= \{X \in \mathbb{S}_+^M \mid \langle I - \Phi, X \rangle = 0\} \\
&= \{\Phi \Psi \Phi \mid \Psi \in \mathbb{S}_+^M\} \\
&\simeq \mathbb{S}_+^{\operatorname{rank} \Phi}
\end{aligned} \tag{225}$$

Smallest face that contains an embedded principal submatrix, whose rank is not necessarily full, may be expressed like (221): For embedded principal submatrix $\Phi A \Phi \in \mathbb{S}_+^M \ni \operatorname{rank} \Phi A \Phi \leq \operatorname{rank} \Phi$, apply ordered diagonalization instead to

$$\hat{\Phi}^{\mathrm{T}} A \hat{\Phi} = U \Upsilon U^{\mathrm{T}} \in \mathbb{S}_+^{\operatorname{rank} \Phi} \tag{226}$$

where $U^{-1} = U^{\mathrm{T}}$ is an orthogonal matrix and $\Upsilon = \delta^2(\Upsilon)$ is diagonal. Then

$$\begin{aligned}
\mathcal{F}\left(\mathbb{S}_+^M \ni \Phi A \Phi\right) &= \{X \in \mathbb{S}_+^M \mid \mathcal{N}(X) \supseteq \mathcal{N}(\Phi A \Phi)\} \\
&= \{X \in \mathbb{S}_+^M \mid \langle \hat{\Phi} U(I - \Upsilon \Upsilon^\dagger) U^{\mathrm{T}} \hat{\Phi}^{\mathrm{T}} + I - \Phi, X \rangle = 0\} \\
&= \{\hat{\Phi} U \Upsilon \Upsilon^\dagger \Psi \Upsilon \Upsilon^\dagger U^{\mathrm{T}} \hat{\Phi}^{\mathrm{T}} \mid \Psi \in \mathbb{S}_+^{\operatorname{rank} \Phi}\} \\
&\simeq \mathbb{S}_+^{\operatorname{rank} \Phi A \Phi}
\end{aligned} \tag{227}$$

[2.42]To express a leading principal submatrix, for example, $\Phi = \begin{bmatrix} I & \mathbf{0} \\ \mathbf{0}^{\mathrm{T}} & \mathbf{0} \end{bmatrix}$.

where binary diagonal matrix Φ is partitioned into nonzero and zero columns by permutation $\Xi \in \mathbb{R}^{M \times M}$;

$$\Phi\Xi^{T} \triangleq [\,\hat{\Phi}\ \ \mathbf{0}\,] \in \mathbb{R}^{M \times M}, \quad \operatorname{rank}\hat{\Phi} = \operatorname{rank}\Phi, \quad \Phi = \hat{\Phi}\hat{\Phi}^{T} \in \mathbb{S}^{M}, \quad \hat{\Phi}^{T}\hat{\Phi} = I \qquad (228)$$

Any embedded principal submatrix may be expressed

$$\Phi A \Phi = \hat{\Phi}\hat{\Phi}^{T} A \,\hat{\Phi}\hat{\Phi}^{T} \in \mathbb{S}_{+}^{M} \qquad (229)$$

where $\hat{\Phi}^{T} A\, \hat{\Phi} \in \mathbb{S}_{+}^{\operatorname{rank}\Phi}$ extracts the principal submatrix whereas $\hat{\Phi}\hat{\Phi}^{T} A\, \hat{\Phi}\hat{\Phi}^{T}$ embeds it.

2.9.2.5.1 Example. *Smallest face containing disparate elements.*
Smallest face formula (221) can be altered to accommodate a union of points $\{A_i \in \mathbb{S}_{+}^{M}\}$:

$$\mathcal{F}\left(\mathbb{S}_{+}^{M} \supset \bigcup_{i} A_i\right) = \left\{ X \in \mathbb{S}_{+}^{M} \ \Big|\ \mathcal{N}(X) \supseteq \bigcap_{i} \mathcal{N}(A_i) \right\} \qquad (230)$$

To see that, imagine two vectorized matrices A_1 and A_2 on diametrically opposed sides of the positive semidefinite cone $\mathbb{S}_{+}^{2}$ boundary pictured in Figure 44. Regard $\operatorname{svec} A_1$ as normal to a hyperplane in $\mathbb{R}^{3}$ containing a vectorized basis for its nullspace: svec basis $\mathcal{N}(A_1)$ (§2.5.3). Similarly, there is a second hyperplane containing svec basis $\mathcal{N}(A_2)$ having normal $\operatorname{svec} A_2$. While each hyperplane is two-dimensional, each nullspace has only one affine dimension because A_1 and A_2 are rank-1. Because our interest is only that part of the nullspace in the positive semidefinite cone, then by

$$\langle X, A_i \rangle = 0 \ \Leftrightarrow\ X A_i = A_i X = \mathbf{0}, \qquad X, A_i \in \mathbb{S}_{+}^{M} \qquad (1673)$$

we may ignore the fact that vectorized nullspace svec basis $\mathcal{N}(A_i)$ is a proper subspace of the hyperplane. We may think instead in terms of whole hyperplanes because equivalence (1673) says that the positive semidefinite cone effectively filters that subset of the hyperplane, whose normal is A_i , constituting $\mathcal{N}(A_i)$.

  And so hyperplane intersection makes a line intersecting the positive semidefinite cone $\mathbb{S}_{+}^{2}$ but only at the origin. In this hypothetical example, smallest face containing those two matrices therefore comprises the entire cone because every positive semidefinite matrix has nullspace containing $\mathbf{0}$. The smaller the intersection, the larger the smallest face. □

2.9.2.5.2 Exercise. *Disparate elements.*
Prove that (230) holds for an arbitrary set $\{A_i \in \mathbb{S}_{+}^{M}\ \forall i \in \mathcal{I}\}$. One way is by showing $\bigcap \mathcal{N}(A_i) \cap \mathbb{S}_{+}^{M} = \operatorname{conv}\left(\{A_i\}\right)^{\perp} \cap \mathbb{S}_{+}^{M}$; with perpendicularity $\perp$ as in (372).[2.43]   ▼

2.9.2.6 face of all PSD matrices having same principal submatrix

Now we ask what is the smallest face of the positive semidefinite cone containing all matrices having a complete principal submatrix in common; in other words, that face containing all PSD matrices (of any rank) with particular entries fixed − the smallest

---
[2.43]Hint: (1673) (2004).

face containing all PSD matrices whose fixed entries correspond to some given embedded principal submatrix $\Phi A \Phi$. To maintain generality,[2.44] we move an extracted principal submatrix $\hat{\Phi}^{\mathrm{T}} A \hat{\Phi} \in \mathbb{S}_+^{\mathrm{rank}\,\Phi}$ into leading position via permutation Ξ from (228): for $A \in \mathbb{S}_+^M$

$$\Xi A \Xi^{\mathrm{T}} \triangleq \begin{bmatrix} \hat{\Phi}^{\mathrm{T}} A \hat{\Phi} & B \\ B^{\mathrm{T}} & C \end{bmatrix} \in \mathbb{S}_+^M \tag{231}$$

By properties of partitioned PSD matrices in §A.4.0.1,

$$\mathrm{basis}\,\mathcal{N}\left(\begin{bmatrix} \hat{\Phi}^{\mathrm{T}} A \hat{\Phi} & B \\ B^{\mathrm{T}} & C \end{bmatrix} \right) \supseteq \begin{bmatrix} \mathbf{0} \\ I - CC^{\dagger} \end{bmatrix} \tag{232}$$

Hence $\mathcal{N}(\Xi X \Xi^{\mathrm{T}}) \supseteq \mathcal{N}(\Xi A \Xi^{\mathrm{T}}) \not\supseteq \mathrm{span}\begin{bmatrix} \mathbf{0} \\ I \end{bmatrix}$ in a smallest face $\mathcal{F}$ formula[2.45] because all PSD matrices, given fixed principal submatrix, are admitted: Define a set of all PSD matrices

$$\mathcal{S} \triangleq \left\{ A = \Xi^{\mathrm{T}} \begin{bmatrix} \hat{\Phi}^{\mathrm{T}} A \hat{\Phi} & B \\ B^{\mathrm{T}} & C \end{bmatrix} \Xi \succeq 0 \,\middle|\, B \in \mathbb{R}^{\mathrm{rank}\,\Phi \times M - \mathrm{rank}\,\Phi},\, C \in \mathbb{S}_+^{M - \mathrm{rank}\,\Phi} \right\} \tag{233}$$

having fixed embedded principal submatrix $\Phi A \Phi = \Xi^{\mathrm{T}} \begin{bmatrix} \hat{\Phi}^{\mathrm{T}} A \hat{\Phi} & \mathbf{0} \\ \mathbf{0}^{\mathrm{T}} & \mathbf{0} \end{bmatrix} \Xi$. So

$$\begin{aligned} \mathcal{F}\left(\mathbb{S}_+^M \supseteq \mathcal{S} \right) &= \left\{ X \in \mathbb{S}_+^M \mid \mathcal{N}(X) \supseteq \mathcal{N}(\mathcal{S}) \right\} \\ &= \left\{ X \in \mathbb{S}_+^M \mid \langle \hat{\Phi} U(I - \Upsilon\Upsilon^{\dagger}) U^{\mathrm{T}} \hat{\Phi}^{\mathrm{T}}, X \rangle = 0 \right\} \\ &= \left\{ \Xi^{\mathrm{T}} \begin{bmatrix} U\Upsilon\Upsilon^{\dagger} & \mathbf{0} \\ \mathbf{0}^{\mathrm{T}} & I \end{bmatrix} \Psi \begin{bmatrix} \Upsilon\Upsilon^{\dagger}U^{\mathrm{T}} & \mathbf{0} \\ \mathbf{0}^{\mathrm{T}} & I \end{bmatrix} \Xi \,\middle|\, \Psi \in \mathbb{S}_+^M \right\} \\ &\simeq \mathbb{S}_+^{M - \mathrm{rank}\,\Phi + \mathrm{rank}\,\Phi A \Phi} \end{aligned} \tag{234}$$

$\Xi = I$ whenever $\Phi A \Phi$ denotes a leading principal submatrix. Smallest face of the positive semidefinite cone, containing all matrices having the same full-rank principal submatrix ($\Upsilon\Upsilon^{\dagger} = I$, $\Upsilon \succeq 0$), is the entire cone (Exercise 2.9.2.3.2).

2.9.2.7 Extreme directions of positive semidefinite cone

Because the positive semidefinite cone is pointed (§2.7.2.1.2), there is a one-to-one correspondence of one-dimensional faces with extreme directions in any dimension M; *id est*, because of the *cone faces lemma* (§2.8.0.0.1) and direct correspondence of exposed faces to faces of $\mathbb{S}_+^M$, it follows: there is no one-dimensional face of the positive semidefinite cone that is not a ray emanating from the origin.

Symmetric dyads constitute the set of all extreme directions: For $M > 1$

$$\{ yy^{\mathrm{T}} \in \mathbb{S}^M \mid y \in \mathbb{R}^M \} \subset \partial \mathbb{S}_+^M \tag{235}$$

[2.44] to fix any principal submatrix; not only leading principal submatrices.
[2.45] meaning, more pertinently, $I - \Phi$ is dropped from (227).

this superset of extreme directions (infinite in number, *confer*(187)) for the positive semidefinite cone is, generally, a subset of the boundary. By *extremes theorem* 2.8.1.1.1, the convex hull of extreme rays and origin is the positive semidefinite cone: (§2.8.1.2.1)

$$\text{conv}\{yy^{\mathrm{T}} \in \mathbb{S}^M \mid y \in \mathbb{R}^M\} = \left\{ \sum_{i=1}^{\infty} b_i\, z_i z_i^{\mathrm{T}} \mid z_i \in \mathbb{R}^M,\ b \succeq 0 \right\} = \mathbb{S}_+^M \qquad (236)$$

For two-dimensional matrices ($M = 2$, Figure 44)

$$\{yy^{\mathrm{T}} \in \mathbb{S}^{\mathbf{2}} \mid y \in \mathbb{R}^{\mathbf{2}}\} = \partial \mathbb{S}_+^{\mathbf{2}} \qquad (237)$$

while for one-dimensional matrices, in exception, ($M = 1$, §2.7)

$$\{yy \in \mathbb{S} \mid y \neq \mathbf{0}\} = \text{int}\,\mathbb{S}_+ \qquad (238)$$

Each and every extreme direction yy^{T} makes the same angle with the Identity matrix in isomorphic $\mathbb{R}^{M(M+1)/2}$, dependent only on dimension; *videlicet*,[2.46]

$$\sphericalangle(yy^{\mathrm{T}},\, I) = \arccos \frac{\langle yy^{\mathrm{T}},\, I \rangle}{\|yy^{\mathrm{T}}\|_{\mathrm{F}}\, \|I\|_{\mathrm{F}}} = \arccos\left(\frac{1}{\sqrt{M}}\right) \quad \forall\, y \in \mathbb{R}^M \qquad (239)$$

This means the positive semidefinite cone broadens in higher dimension.

2.9.2.7.1 Example. *Positive semidefinite matrix from extreme directions.*
Diagonalizability (§A.5) of symmetric matrices yields the following results:
    Any positive semidefinite matrix (1531) in $\mathbb{S}^M$ can be written in the form

$$A = \sum_{i=1}^{M} \lambda_i\, z_i z_i^{\mathrm{T}} = \hat{A}\hat{A}^{\mathrm{T}} = \sum_i \hat{a}_i \hat{a}_i^{\mathrm{T}} \succeq 0, \qquad \lambda \succeq 0 \qquad (240)$$

a conic combination of linearly independent extreme directions ($\hat{a}_i \hat{a}_i^{\mathrm{T}}$ or $z_i z_i^{\mathrm{T}}$ where $\|z_i\| = 1$), where λ is a vector of eigenvalues.
    If we limit consideration to all symmetric positive semidefinite matrices bounded via unity trace

$$\mathcal{C} \triangleq \{A \succeq 0 \mid \text{tr}\, A = 1\} \qquad (91)$$

then any matrix A from that set may be expressed as a convex combination of linearly independent extreme directions;

$$A = \sum_{i=1}^{M} \lambda_i\, z_i z_i^{\mathrm{T}} \in \mathcal{C}\,, \qquad \mathbf{1}^{\mathrm{T}}\lambda = 1, \quad \lambda \succeq 0 \qquad (241)$$

Implications are:

1. set $\mathcal{C}$ is convex (an intersection of PSD cone with hyperplane),

2. because the set of eigenvalues corresponding to a given square matrix A is unique (§A.5.0.1), no single eigenvalue can exceed 1; *id est*, $I \succeq A$

3. and the converse holds: set $\mathcal{C}$ is an instance of Fantope (91).   □

[2.46] Analogy with respect to the *EDM cone* is considered in [191, p.162] where it is found: angle is not constant. Extreme directions of the EDM cone can be found in §6.4.3.2. The cone's axis is $-E = \mathbf{11}^{\mathrm{T}} - I$ (1175).

2.9.2.7.2 Exercise. *Extreme directions of positive semidefinite cone.*
Prove, directly from definition (186), that symmetric dyads (235) constitute the set of all
extreme directions of the positive semidefinite cone. ▼

2.9.2.8 Positive semidefinite cone is generally not circular

Extreme angle equation (239) suggests that the positive semidefinite cone might be
invariant to rotation about its axis of revolution; *id est,* a circular cone. We investigate
this now:

2.9.2.8.1 Definition. *Circular cone:*[2.47]
a pointed closed convex cone having hyperspherical sections orthogonal to its *axis of
revolution* about which the cone is invariant to rotation. △

A *conic section* is the intersection of a cone with any hyperplane. In three dimensions,
an intersecting plane perpendicular to a circular cone's axis of revolution produces a
section bounded by a circle. (Figure 47) A prominent example of a circular cone in convex
analysis is Lorentz cone (178). We also find that the positive semidefinite cone and cone
of Euclidean distance matrices are circular cones, but only in low dimension.

The positive semidefinite cone has axis of revolution that is the ray (base **0**) through
the Identity matrix I. Consider a set of normalized extreme directions of the positive
semidefinite cone: for some arbitrary positive constant $a \in \mathbb{R}_+$

$$\{yy^{\mathrm{T}} \in \mathbb{S}^M \mid \|y\| = \sqrt{a}\} \subset \partial \mathbb{S}_+^M \qquad (242)$$

The distance from each extreme direction to the axis of revolution is radius

$$\mathrm{R} \triangleq \inf_c \|yy^{\mathrm{T}} - cI\|_{\mathrm{F}} = a\sqrt{1 - \frac{1}{M}} \qquad (243)$$

which is the distance from yy^{T} to $\frac{a}{M}I$; the length of vector $yy^{\mathrm{T}} - \frac{a}{M}I$.

Because distance R (in a particular dimension) from the axis of revolution to each and
every normalized extreme direction is identical, the extreme directions lie on the boundary
of a hypersphere in isometrically isomorphic $\mathbb{R}^{M(M+1)/2}$. From Example 2.9.2.7.1, the
convex hull (excluding vertex at the origin) of the normalized extreme directions is a
conic section

$$\mathcal{C} \triangleq \mathrm{conv}\{yy^{\mathrm{T}} \mid y \in \mathbb{R}^M, \ y^{\mathrm{T}}y = a\} = \mathbb{S}_+^M \cap \{A \in \mathbb{S}^M \mid \langle I, A \rangle = a\} \qquad (244)$$

orthogonal to Identity matrix I;

$$\left\langle \mathcal{C} - \frac{a}{M}I, \ I \right\rangle = \mathrm{tr}(\mathcal{C} - \frac{a}{M}I) = 0 \qquad (245)$$

[2.47] A circular cone is assumed convex throughout, although not so by other authors. We also assume a
right circular cone.

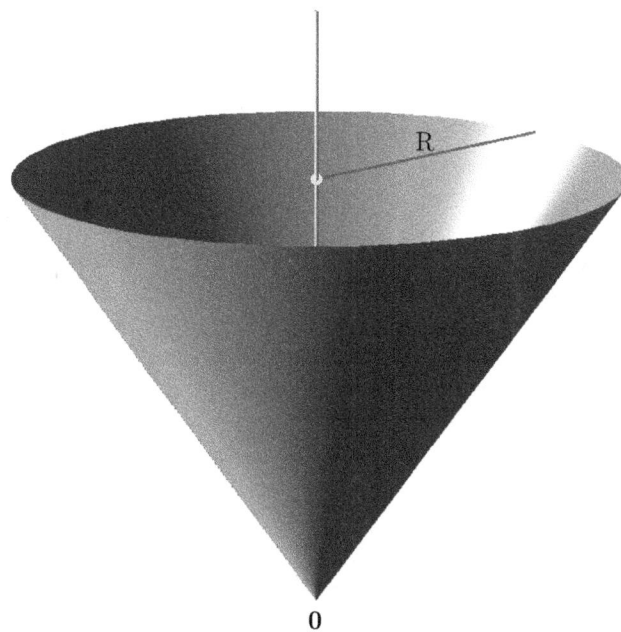

Figure 47: This solid circular cone in $\mathbb{R}^3$ continues upward infinitely. Axis of revolution is illustrated as vertical line through origin. R represents radius: distance measured from an extreme direction to axis of revolution. Were this a Lorentz cone, any plane slice containing axis of revolution would make a right angle.

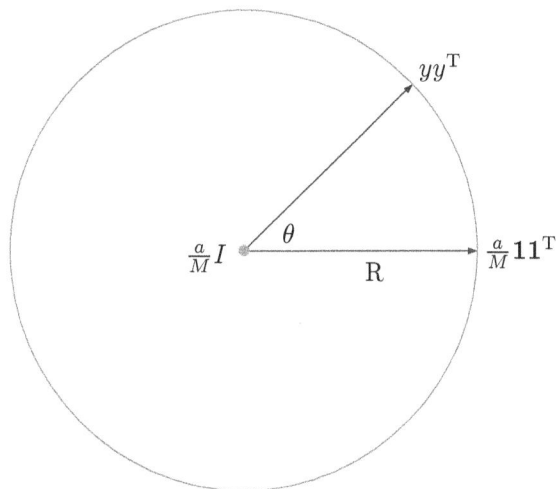

Figure 48: Illustrated is a section, perpendicular to axis of revolution, of circular cone from Figure 47. Radius R is distance from any extreme direction to axis at $\frac{a}{M}I$. Vector $\frac{a}{M}\mathbf{1}\mathbf{1}^{\mathrm{T}}$ is an arbitrary reference by which to measure angle θ.

Proof. Although the positive semidefinite cone possesses some characteristics of a circular cone, we can show it is not by demonstrating shortage of extreme directions; *id est*, some extreme directions corresponding to each and every angle of rotation about the axis of revolution are nonexistent: Referring to Figure 48, [408, §1-7]

$$\cos\theta = \frac{\left\langle \frac{a}{M}\mathbf{1}\mathbf{1}^{\mathrm{T}} - \frac{a}{M}I \ , \ yy^{\mathrm{T}} - \frac{a}{M}I \right\rangle}{a^2(1 - \frac{1}{M})} \tag{246}$$

Solving for vector y we get

$$a(1 + (M-1)\cos\theta) = (\mathbf{1}^{\mathrm{T}}y)^2 \tag{247}$$

which does not have real solution $\forall\, 0 \le \theta \le 2\pi$ in every matrix dimension M.   ♦

From the foregoing proof we can conclude that the positive semidefinite cone might be circular but only in matrix dimensions 1 and 2. Because of a shortage of extreme directions, conic section (244) cannot be hyperspherical by the *extremes theorem* (§2.8.1.1.1, Figure 43).

2.9.2.8.2 Exercise. *Circular semidefinite cone.*
Prove the PSD cone to be circular in matrix dimensions 1 and 2 while it is a rotation of Lorentz cone (178) in matrix dimension 2 .[2.48]   ▼

[2.48]Hint: Given cone $\left\{ \begin{bmatrix} \alpha & \beta/\sqrt{2} \\ \beta/\sqrt{2} & \gamma \end{bmatrix} \mid \sqrt{\alpha^2 + \beta^2} \le \gamma \right\}$, show $\frac{1}{\sqrt{2}}\begin{bmatrix} \gamma + \alpha & \beta \\ \beta & \gamma - \alpha \end{bmatrix}$ is a vector rotation that is positive semidefinite under the same inequality.

$$p = \begin{bmatrix} 1/2 \\ 1 \end{bmatrix}$$

$$p_1 A_{11} \geq \pm p_2 A_{12}$$
$$p_2 A_{22} \geq \pm p_1 A_{12}$$

$$p = \begin{bmatrix} 1 \\ 1 \end{bmatrix}$$

$$p = \begin{bmatrix} 2 \\ 1 \end{bmatrix}$$

A_{11}

$\sqrt{2}A_{12}$

A_{22}

svec $\partial \mathbb{S}^2_+$

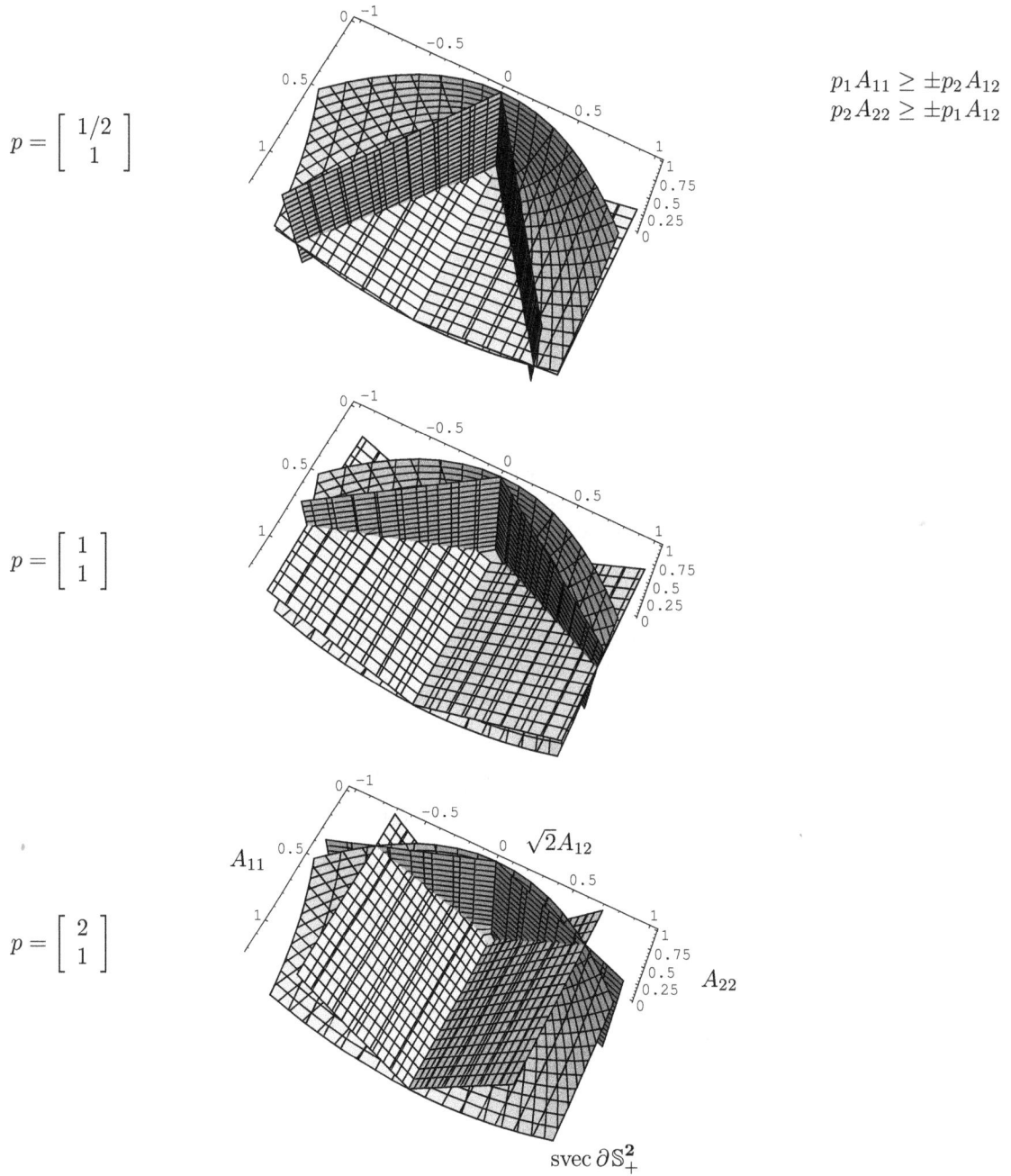

Figure 49: Proper polyhedral cone $\mathcal{K}$, created by intersection of halfspaces, inscribes PSD cone in isometrically isomorphic $\mathbb{R}^3$ as predicted by *Geršgorin discs theorem* for $A = [A_{ij}] \in \mathbb{S}^2$. Hyperplanes supporting $\mathcal{K}$ intersect along boundary of PSD cone. Four extreme directions of $\mathcal{K}$ coincide with extreme directions of PSD cone.

2.9.2.8.3 Example. *PSD cone inscription in three dimensions.*

Theorem. *Geršgorin discs.* [208, §6.1] [373] [261, p.140]
For $p \in \mathbb{R}_+^m$ given $A = [A_{ij}] \in \mathbb{S}^m$, then all eigenvalues of A belong to the union of m closed intervals on the real line;

$$\lambda(A) \in \bigcup_{i=1}^{m} \left\{ \xi \in \mathbb{R} \ \middle| \ |\xi - A_{ii}| \le \varrho_i \triangleq \frac{1}{p_i} \sum_{\substack{j=1 \\ j \ne i}}^{m} p_j |A_{ij}| \right\} = \bigcup_{i=1}^{m} [A_{ii} - \varrho_i \ , \ A_{ii} + \varrho_i] \qquad (248)$$

Furthermore, if a union of k of these m [intervals] forms a connected region that is disjoint from all the remaining $n-k$ [intervals], then there are precisely k eigenvalues of A in this region. ◇

To apply the theorem to determine positive semidefiniteness of symmetric matrix A, we observe that for each i we must have

$$A_{ii} \ge \varrho_i \qquad (249)$$

Suppose

$$m = 2 \qquad (250)$$

so $A \in \mathbb{S}^2$. Vectorizing A as in (56), svec A belongs to isometrically isomorphic $\mathbb{R}^3$. Then we have $m 2^{m-1} = 4$ inequalities, in the matrix entries A_{ij} with Geršgorin parameters $p = [p_i] \in \mathbb{R}_+^2$,

$$\begin{aligned} p_1 A_{11} &\ge \pm p_2 A_{12} \\ p_2 A_{22} &\ge \pm p_1 A_{12} \end{aligned} \qquad (251)$$

which describe an intersection of four halfspaces in $\mathbb{R}^{m(m+1)/2}$. That intersection creates the proper polyhedral cone $\mathcal{K}$ (§2.12.1) whose construction is illustrated in Figure 49. Drawn truncated is the boundary of the positive semidefinite cone svec $\mathbb{S}_+^2$ and the bounding hyperplanes supporting $\mathcal{K}$.

Created by means of Geršgorin discs, $\mathcal{K}$ always belongs to the positive semidefinite cone for any nonnegative value of $p \in \mathbb{R}_+^m$. Hence any point in $\mathcal{K}$ corresponds to some positive semidefinite matrix A. Only the extreme directions of $\mathcal{K}$ intersect the positive semidefinite cone boundary in this dimension; the four extreme directions of $\mathcal{K}$ are extreme directions of the positive semidefinite cone. As p_1/p_2 increases in value from 0, two extreme directions of $\mathcal{K}$ sweep the entire boundary of this positive semidefinite cone. Because the entire positive semidefinite cone can be swept by $\mathcal{K}$, the system of linear inequalities

$$Y^{\mathrm{T}} \mathrm{svec}\, A \triangleq \begin{bmatrix} p_1 & \pm p_2/\sqrt{2} & 0 \\ 0 & \pm p_1/\sqrt{2} & p_2 \end{bmatrix} \mathrm{svec}\, A \succeq 0 \qquad (252)$$

(when made dynamic) can replace a semidefinite constraint $A \succeq 0$; *id est*, for

$$\mathcal{K} = \{ z \mid Y^{\mathrm{T}} z \succeq 0 \} \subset \mathrm{svec}\, \mathbb{S}_+^m \qquad (253)$$

given p where $Y \in \mathbb{R}^{m(m+1)/2 \times m 2^{m-1}}$

$$\mathrm{svec}\, A \in \mathcal{K} \ \Rightarrow \ A \in \mathbb{S}_+^m \qquad (254)$$

but

$$\exists p \ \ni \ Y^{\mathrm{T}}\mathrm{svec}\, A \succeq 0 \ \Leftrightarrow \ A \succeq 0 \qquad (255)$$

In other words, *diagonal dominance* [208, p.349, §7.2.3]

$$A_{ii} \geq \sum_{\substack{j=1 \\ j \neq i}}^{m} |A_{ij}| \,, \quad \forall\, i = 1 \ldots m \qquad (256)$$

is generally only a sufficient condition for membership to the PSD cone. But by dynamic weighting p in this dimension, diagonal dominance was made necessary and sufficient.

$\square$

In higher dimension $(m > 2)$, boundary of the positive semidefinite cone is no longer constituted completely by its extreme directions (symmetric rank-one matrices); its geometry becomes intricate. How all the extreme directions can be swept by an inscribed polyhedral cone,[2.49] similarly to the foregoing example, remains an open question.

2.9.2.8.4   Exercise.   *Dual inscription.*
Find dual proper polyhedral cone $\mathcal{K}^*$ from Figure **49**.   ▼

2.9.2.9   Boundary constituents of the positive semidefinite cone

2.9.2.9.1   Lemma.   *Sum of positive semidefinite matrices.*   *(confer* (1547)*)*
For $A, B \in \mathbb{S}_+^M$

$$\mathrm{rank}(A + B) = \mathrm{rank}(\mu A + (1 - \mu)B) \qquad (257)$$

over open interval $(0, 1)$ of μ. $\diamond$

   Proof.   Any positive semidefinite matrix belonging to the PSD cone has an eigenvalue decomposition that is a positively scaled sum of linearly independent symmetric dyads. By the *linearly independent dyads definition* in §B.1.1.0.1, rank of the sum $A + B$ is equivalent to the number of linearly independent dyads constituting it. Linear independence is insensitive to further positive scaling by μ. The assumption of positive semidefiniteness prevents annihilation of any dyad from the sum $A + B$.   ◆

2.9.2.9.2   Example.   *Rank function quasiconcavity.*   *(confer* §3.8*)*
For $A, B \in \mathbb{R}^{m \times n}$ [208, §0.4]

$$\mathrm{rank}\, A + \mathrm{rank}\, B \ \geq \ \mathrm{rank}(A + B) \qquad (258)$$

that follows from the fact [338, §3.6]

$$\dim \mathcal{R}(A) + \dim \mathcal{R}(B) \ = \ \dim \mathcal{R}(A + B) + \dim\big(\mathcal{R}(A) \cap \mathcal{R}(B)\big) \qquad (259)$$

For $A, B \in \mathbb{S}_+^M$

$$\mathrm{rank}\, A + \mathrm{rank}\, B \ \geq \ \mathrm{rank}(A + B) \ \geq \ \min\{\mathrm{rank}\, A, \, \mathrm{rank}\, B\} \qquad (1547)$$

[2.49]It is not necessary to sweep the entire boundary in higher dimension.

that follows from the fact

$$\mathcal{N}(A + B) = \mathcal{N}(A) \cap \mathcal{N}(B) , \qquad A, B \in \mathbb{S}_+^M \qquad (160)$$

Rank is a *quasiconcave* function on $\mathbb{S}_+^M$ because the right-hand inequality in (1547) has the concave form (645); *videlicet*, Lemma 2.9.2.9.1. $\qquad\square$

From this example we see, unlike convex functions, *quasiconvex* functions are not necessarily continuous. (§3.8) We also glean:

2.9.2.9.3 Theorem. *Convex subsets of positive semidefinite cone.*
Subsets of the positive semidefinite cone $\mathbb{S}_+^M$, for $0 \leq \rho \leq M$

$$\mathbb{S}_+^M(\rho) \triangleq \{X \in \mathbb{S}_+^M \mid \operatorname{rank} X \geq \rho\} \qquad (260)$$

are pointed convex cones, but not closed unless $\rho = 0$; *id est*, $\mathbb{S}_+^M(0) = \mathbb{S}_+^M$. $\qquad\diamond$

Proof. Given ρ, a subset $\mathbb{S}_+^M(\rho)$ is convex if and only if convex combination of any two members has rank at least ρ. That is confirmed by applying identity (257) from Lemma 2.9.2.9.1 to (1547); *id est*, for $A, B \in \mathbb{S}_+^M(\rho)$ on closed interval $[0, 1]$ of μ

$$\operatorname{rank}(\mu A + (1 - \mu)B) \ \geq \ \min\{\operatorname{rank} A, \operatorname{rank} B\} \qquad (261)$$

It can similarly be shown, almost identically to proof of the lemma, any conic combination of A, B in subset $\mathbb{S}_+^M(\rho)$ remains a member; *id est*, $\forall \zeta, \xi \geq 0$

$$\operatorname{rank}(\zeta A + \xi B) \ \geq \ \min\{\operatorname{rank}(\zeta A), \operatorname{rank}(\xi B)\} \qquad (262)$$

Therefore, $\mathbb{S}_+^M(\rho)$ is a convex cone. $\qquad\blacklozenge$

Another proof of convexity can be made by projection arguments:

2.9.2.10 Projection on $\mathbb{S}_+^M(\rho)$

Because these cones $\mathbb{S}_+^M(\rho)$ indexed by ρ (260) are convex, projection on them is straightforward. Given a symmetric matrix H having diagonalization $H \triangleq Q\Lambda Q^{\mathrm{T}} \in \mathbb{S}^M$ (§A.5.1) with eigenvalues Λ arranged in nonincreasing order, then its *Euclidean projection* (minimum-distance projection) on $\mathbb{S}_+^M(\rho)$

$$P_{\mathbb{S}_+^M(\rho)} H = Q \Upsilon^\star Q^{\mathrm{T}} \qquad (263)$$

corresponds to a map of its eigenvalues:

$$\Upsilon_{ii}^\star = \begin{cases} \max\{\epsilon, \Lambda_{ii}\}, & i = 1 \ldots \rho \\ \max\{0, \Lambda_{ii}\}, & i = \rho + 1 \ldots M \end{cases} \qquad (264)$$

where ϵ is positive but arbitrarily close to 0.

2.9.2.10.1   Exercise.   *Projection on open convex cones.*
Prove (264) using Theorem E.9.2.0.1.   ▼

Because each $H \in \mathbb{S}^M$ has unique projection on $\mathbb{S}^M_+(\rho)$ (despite possibility of repeated eigenvalues in Λ), we may conclude it is a convex set by the *Bunt-Motzkin theorem* (§E.9.0.0.1).

Compare (264) to the well-known result regarding Euclidean projection on a rank ρ subset of the positive semidefinite cone (§2.9.2.1)

$$\mathbb{S}^M_+ \setminus \mathbb{S}^M_+(\rho+1) \;=\; \{X \in \mathbb{S}^M_+ \mid \operatorname{rank} X \le \rho\} \qquad (216)$$

$$P_{\mathbb{S}^M_+ \setminus \mathbb{S}^M_+(\rho+1)} H = Q\,\Upsilon^\star Q^{\mathrm{T}} \qquad (265)$$

As proved in §7.1.4, this projection of H corresponds to the eigenvalue map

$$\Upsilon^\star_{ii} = \begin{cases} \max\{0\,,\,\Lambda_{ii}\}\,, & i=1\dots\rho \\ 0\,, & i=\rho+1\dots M \end{cases} \qquad (1422)$$

Together these two results (264) and (1422) mean: A higher-rank solution to projection on the positive semidefinite cone lies arbitrarily close to any given lower-rank projection, but not *vice versa*. Were the number of nonnegative eigenvalues in Λ known *a priori* not to exceed ρ, then these two different projections would produce identical results in the limit $\epsilon \to 0$.

2.9.2.11   Uniting constituents

Interior of the PSD cone $\operatorname{int}\mathbb{S}^M_+$ is convex by Theorem 2.9.2.9.3, for example, because all positive semidefinite matrices having rank M constitute the cone interior.

All positive semidefinite matrices of rank less than M constitute the cone boundary; an amalgam of positive semidefinite matrices of different rank. Thus each nonconvex subset of positive semidefinite matrices, for $0 < \rho < M$

$$\{Y \in \mathbb{S}^M_+ \mid \operatorname{rank} Y = \rho\} \qquad (266)$$

having rank ρ successively 1 lower than M, appends a nonconvex constituent to the cone boundary; but only in their union is the boundary complete: (*confer* §2.9.2)

$$\partial\mathbb{S}^M_+ \;=\; \bigcup_{\rho=0}^{M-1} \{Y \in \mathbb{S}^M_+ \mid \operatorname{rank} Y = \rho\} \qquad (267)$$

The composite sequence, the cone interior in union with each successive constituent, remains convex at each step; *id est*, for $0 \le k \le M$

$$\bigcup_{\rho=k}^{M} \{Y \in \mathbb{S}^M_+ \mid \operatorname{rank} Y = \rho\} \qquad (268)$$

is convex for each k by Theorem 2.9.2.9.3.

2.9.2.12 Peeling constituents

Proceeding the other way: To peel constituents off the complete positive semidefinite cone boundary, one starts by removing the origin; the only rank-0 positive semidefinite matrix. What remains is convex. Next, the extreme directions are removed because they constitute all the rank-1 positive semidefinite matrices. What remains is again convex, and so on. Proceeding in this manner eventually removes the entire boundary leaving, at last, the convex interior of the PSD cone; all the positive definite matrices.

2.9.2.12.1 Exercise. *Difference $A - B$.*
What about a difference of matrices A, B belonging to the PSD cone? Show:

- Difference of any two points on the boundary belongs to the boundary or exterior.

- Difference $A - B$, where A belongs to the boundary while B is interior, belongs to the exterior.   ▼

2.9.3 Barvinok's proposition

Barvinok posits existence and quantifies an upper bound on rank of a positive semidefinite matrix belonging to the intersection of the PSD cone with an affine subset:

2.9.3.0.1 Proposition. *Affine intersection with PSD cone.*   [26, §II.13] [24, §2.2]
Consider finding a matrix $X \in \mathbb{S}^N$ satisfying

$$ X \succeq 0 \, , \qquad \langle A_j \, , \, X \rangle = b_j \, , \quad j = 1 \ldots m \tag{269} $$

given nonzero linearly independent (vectorized) $A_j \in \mathbb{S}^N$ and real b_j . Define the affine subset

$$ \mathcal{A} \triangleq \{ X \mid \langle A_j \, , \, X \rangle = b_j \, , \; j = 1 \ldots m \} \subseteq \mathbb{S}^N \tag{270} $$

If the intersection $\mathcal{A} \cap \mathbb{S}_+^N$ is nonempty given a number m of equalities, then there exists a matrix $X \in \mathcal{A} \cap \mathbb{S}_+^N$ such that

$$ \operatorname{rank} X \, (\operatorname{rank} X + 1)/2 \leq m \tag{271} $$

whence the upper bound[2.50]

$$ \operatorname{rank} X \leq \left\lfloor \frac{\sqrt{8m+1} - 1}{2} \right\rfloor \tag{272} $$

Given desired rank instead, equivalently,

$$ m < (\operatorname{rank} X + 1)(\operatorname{rank} X + 2)/2 \tag{273} $$

[2.50] §4.1.2.2 contains an intuitive explanation. This bound is itself limited above, of course, by N ; a *tight* limit corresponding to an interior point of $\mathbb{S}_+^N$.

An extreme point of $\mathcal{A} \cap \mathbb{S}_+^N$ satisfies (272) and (273). (*confer* §4.1.2.2) A matrix $X \triangleq R^{\mathrm{T}}R$ is an extreme point if and only if the smallest face, that contains X, of $\mathcal{A} \cap \mathbb{S}_+^N$ has dimension 0; [245, §2.4] [246] *id est*, iff (171)

$$\dim \mathcal{F}\Big((\mathcal{A} \cap \mathbb{S}_+^N) \ni X\Big)$$
$$= \operatorname{rank}(X)(\operatorname{rank}(X)+1)/2 - \operatorname{rank}\big[\, \operatorname{svec} R A_1 R^{\mathrm{T}} \ \operatorname{svec} R A_2 R^{\mathrm{T}} \cdots \ \operatorname{svec} R A_m R^{\mathrm{T}} \,\big] \tag{274}$$

equals 0 in isomorphic $\mathbb{R}^{N(N+1)/2}$.

Now the intersection $\mathcal{A} \cap \mathbb{S}_+^N$ is assumed bounded: Assume a given nonzero upper bound ρ on rank, a number of equalities

$$m = (\rho+1)(\rho+2)/2 \tag{275}$$

and matrix dimension $N \geq \rho + 2 \geq 3$. If the intersection is nonempty and bounded, then there exists a matrix $X \in \mathcal{A} \cap \mathbb{S}_+^N$ such that

$$\operatorname{rank} X \leq \rho \tag{276}$$

This represents a tightening of the upper bound; a reduction by exactly 1 of the bound provided by (272) given the same specified number m (275) of equalities; *id est*,

$$\operatorname{rank} X \leq \frac{\sqrt{8m+1}-1}{2} - 1 \tag{277}$$

◇

2.10   Conic independence (c.i.)

In contrast to extreme direction, the property *conically independent direction* is more generally applicable; inclusive of all closed convex cones (not only pointed closed convex cones). Arbitrary given directions $\{\Gamma_i \in \mathbb{R}^n, \ i=1\ldots N\}$ comprise a *conically independent set* if and only if (*confer* §2.1.2, §2.4.2.3)

$$\Gamma_i \zeta_i + \cdots + \Gamma_j \zeta_j - \Gamma_\ell = \mathbf{0}, \qquad i \neq \cdots \neq j \neq \ell = 1 \ldots N \tag{278}$$

has no solution $\zeta \in \mathbb{R}_+^N$ ($\zeta_i \in \mathbb{R}_+$); in words, iff no direction from the given set can be expressed as a conic combination of those remaining; *e.g.*, Figure 50 [380, *conic independence* test (278) MATLAB]. Arranging any set of generators for a particular closed convex cone in a matrix columnar,

$$X \triangleq [\,\Gamma_1 \ \Gamma_2 \ \cdots \ \Gamma_N\,] \in \mathbb{R}^{n \times N} \tag{279}$$

then this test of conic independence (278) may be expressed as a set of linear *feasibility problems*: for $\ell = 1 \ldots N$

$$\begin{aligned}
\text{find} \quad & \zeta \in \mathbb{R}^N \\
\text{subject to} \quad & X\zeta = \Gamma_\ell \\
& \zeta \succeq 0 \\
& \zeta_\ell = 0
\end{aligned} \tag{280}$$

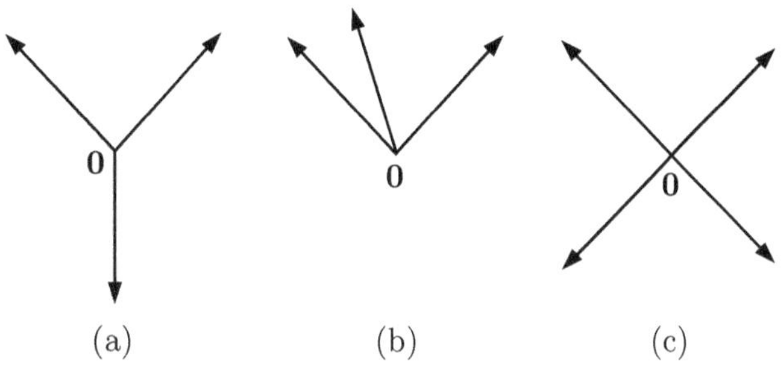

Figure 50: Vectors in $\mathbb{R}^2$: (a) affinely and conically independent, (b) affinely independent but not conically independent, (c) conically independent but not affinely independent. None of the examples exhibits linear independence. (In general, a.i. $\not\Leftrightarrow$ c.i.)

If feasible for any particular ℓ, then the set is not conically independent.

To find all conically independent directions from a set via (280), generator Γ_ℓ must be removed from the set once it is found (feasible) conically dependent on remaining generators in X. So, to continue testing remaining generators when Γ_ℓ is found to be dependent, Γ_ℓ must be discarded from matrix X before proceeding. A generator Γ_ℓ that is instead found conically independent of remaining generators in X, on the other hand, is conically independent of any subset of remaining generators. A c.i. set thus found is not necessarily unique.

It is evident that linear independence (l.i.) of N directions implies their conic independence;

- l.i. $\Rightarrow$ c.i.

which suggests, number of l.i. generators in the columns of X cannot exceed number of c.i. generators. Denoting by $\mathbf{k}$ the number of conically independent generators contained in X, we have the most fundamental rank inequality for convex cones

$$\dim\text{aff}\,\mathcal{K} = \dim\text{aff}[\,\mathbf{0}\ \ X\,] = \text{rank}\,X \leq \mathbf{k} \leq N \tag{281}$$

Whereas N directions in n dimensions can no longer be linearly independent once N exceeds n, conic independence remains possible:

2.10.0.0.1 Table: Maximum number of c.i. directions

dimension n	sup $\mathbf{k}$ (pointed)	sup $\mathbf{k}$ (not pointed)
0	0	0
1	1	2
2	2	4
3	∞	∞
$\vdots$	$\vdots$	$\vdots$

Assuming veracity of this table, there is an apparent vastness between two and three dimensions. The finite numbers of conically independent directions indicate:

- Convex cones in dimensions $0, 1,$ and 2 must be polyhedral. (§2.12.1)

Conic independence is certainly one convex idea that cannot be completely explained by a two-dimensional picture as Barvinok suggests [26, p.vii].

From this table it is also evident that dimension of Euclidean space cannot exceed the number of conically independent directions possible;

- $n \leq \sup \mathbf{k}$

2.10.1 Preservation of conic independence

Independence in the linear (§2.1.2.1), affine (§2.4.2.4), and conic senses can be preserved under linear transformation. Suppose a matrix $X \in \mathbb{R}^{n \times N}$ (279) holds a conically independent set columnar. Consider a transformation on the domain of such matrices

$$T(X) : \mathbb{R}^{n \times N} \to \mathbb{R}^{n \times N} \triangleq XY \qquad (282)$$

where fixed matrix $Y \triangleq [\, y_1 \ \ y_2 \ \cdots \ y_N \,] \in \mathbb{R}^{N \times N}$ represents linear operator T. Conic independence of $\{Xy_i \in \mathbb{R}^n, \ i = 1 \ldots N\}$ demands, by definition (278),

$$Xy_i \, \zeta_i + \cdots + Xy_j \, \zeta_j - Xy_\ell = \mathbf{0}, \qquad i \neq \cdots \neq j \neq \ell = 1 \ldots N \qquad (283)$$

have no solution $\zeta \in \mathbb{R}_+^N$. That is ensured by conic independence of $\{y_i \in \mathbb{R}^N\}$ and by $\mathcal{R}(Y) \cap \mathcal{N}(X) = \mathbf{0}$; seen by factoring out X.

2.10.1.1 linear maps of cones

[21, §7] If $\mathcal{K}$ is a convex cone in Euclidean space $\mathcal{R}$ and T is any linear mapping from $\mathcal{R}$ to Euclidean space $\mathcal{M}$, then $T(\mathcal{K})$ is a convex cone in $\mathcal{M}$ and $x \preceq y$ with respect to $\mathcal{K}$ implies $T(x) \preceq T(y)$ with respect to $T(\mathcal{K})$. If $\mathcal{K}$ is full-dimensional in $\mathcal{R}$, then so is $T(\mathcal{K})$ in $\mathcal{M}$.

If T is a linear bijection, then $x \preceq y \ \Leftrightarrow \ T(x) \preceq T(y)$. If $\mathcal{K}$ is pointed, then so is $T(\mathcal{K})$. And if $\mathcal{K}$ is closed, so is $T(\mathcal{K})$. If $\mathcal{F}$ is a face of $\mathcal{K}$, then $T(\mathcal{F})$ is a face of $T(\mathcal{K})$.

Linear bijection is only a sufficient condition for pointedness and closedness; convex polyhedra (§2.12) are invariant to any linear or inverse linear transformation [26, §I.9] [315, p.44, thm.19.3].

2.10.2 Pointed closed convex $\mathcal{K}$ & conic independence

The following bullets can be derived from definitions (186) and (278) in conjunction with the *extremes theorem* (§2.8.1.1.1):

The set of all extreme directions from a pointed closed convex cone $\mathcal{K} \subset \mathbb{R}^n$ is not necessarily a linearly independent set, yet it must be a conically independent set; (compare Figure **25** on page 64 with Figure **51**a)

- {extreme directions} $\Rightarrow$ {c.i.}

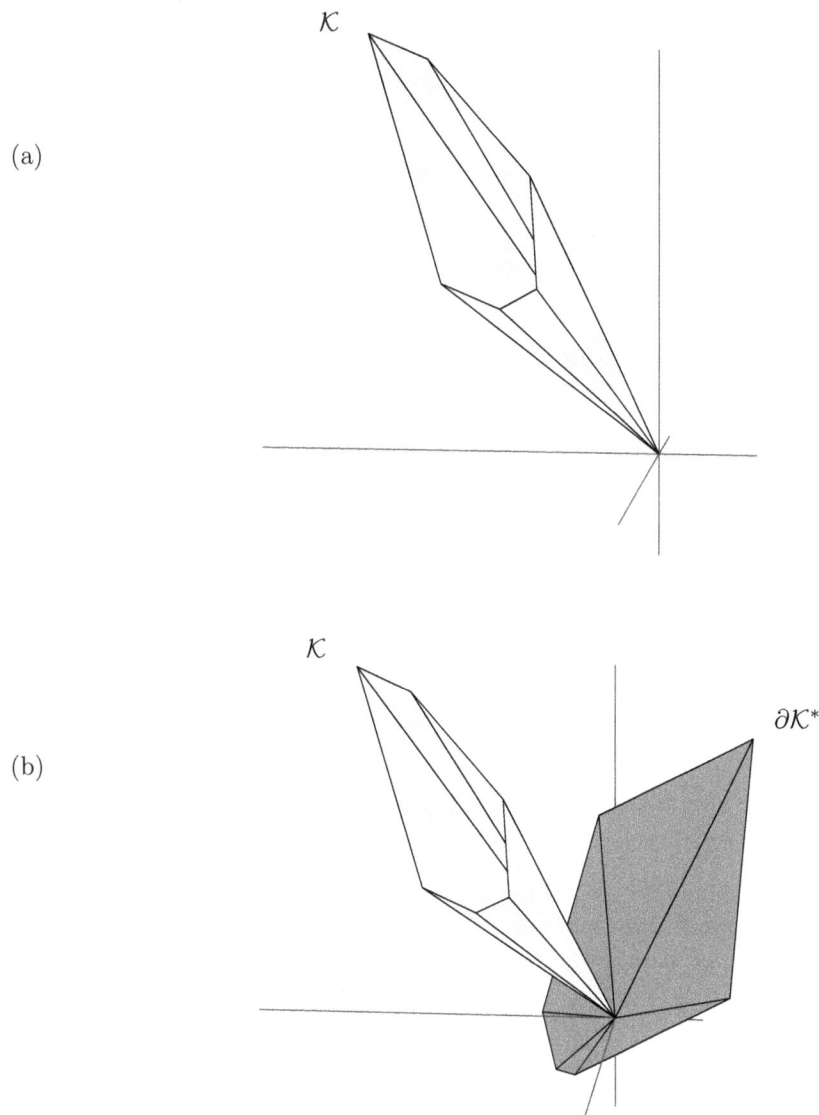

Figure 51: **(a)** A pointed polyhedral cone (drawn truncated) in $\mathbb{R}^3$ having six facets. The extreme directions, corresponding to six edges emanating from the origin, are generators for this cone; not linearly independent but they must be conically independent. **(b)** The boundary of dual cone $\mathcal{K}^*$ (drawn truncated) is now added to the drawing of same $\mathcal{K}$. $\mathcal{K}^*$ is polyhedral, proper, and has the same number of extreme directions as $\mathcal{K}$ has facets.

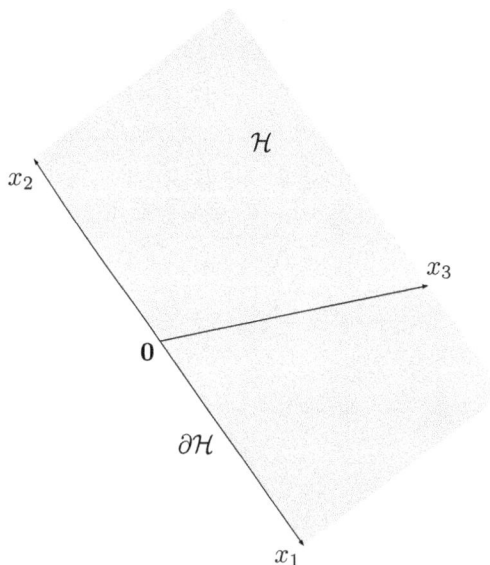

Figure 52: Minimal set of generators $X = [\, x_1 \; x_2 \; x_3 \,] \in \mathbb{R}^{2 \times 3}$ (not extreme directions) for halfspace about origin; affinely and conically independent.

When a conically independent set of directions from pointed closed convex cone $\mathcal{K}$ is known to comprise generators, conversely, then all directions from that set must be extreme directions of the cone;

- {extreme directions} $\Leftrightarrow$ {c.i. generators of pointed closed convex $\mathcal{K}$}

Barker & Carlson [21, §1] call the extreme directions a *minimal generating set* for a pointed closed convex cone. A minimal set of generators is therefore a conically independent set of generators, and *vice versa*,[2.51] for a pointed closed convex cone.

An arbitrary collection of n or fewer distinct extreme directions from pointed closed convex cone $\mathcal{K} \subset \mathbb{R}^n$ is not necessarily a linearly independent set; *e.g.*, dual extreme directions (482) from Example 2.13.11.0.3.

- {$\leq n$ extreme directions in $\mathbb{R}^n$} $\not\Rightarrow$ {l.i.}

Linear dependence of few extreme directions is another convex idea that cannot be explained by a two-dimensional picture as Barvinok suggests [26, p.vii]; indeed, it only first comes to light in four dimensions! But there is a converse: [337, §2.10.9]

- {extreme directions} $\Leftarrow$ {l.i. generators of closed convex $\mathcal{K}$}

[2.51]This converse does not hold for nonpointed closed convex cones as Table 2.10.0.0.1 implies; *e.g.*, ponder four conically independent generators for a plane ($n = 2$, Figure 50).

2.10.2.0.1 Example. *Vertex-description of halfspace $\mathcal{H}$ about origin.*
From $n+1$ points in $\mathbb{R}^n$ we can make a vertex-description of a convex cone that is a halfspace $\mathcal{H}$, where $\{x_\ell \in \mathbb{R}^n, \ \ell = 1 \ldots n\}$ constitutes a minimal set of generators for a hyperplane $\partial \mathcal{H}$ through the origin. An example is illustrated in Figure **52**. By demanding the augmented set $\{x_\ell \in \mathbb{R}^n, \ \ell = 1 \ldots n+1\}$ be affinely independent (we want vector x_{n+1} not parallel to $\partial \mathcal{H}$), then

$$
\begin{aligned}
\mathcal{H} &= \bigcup_{\zeta \geq 0} (\zeta \, x_{n+1} + \partial \mathcal{H}) \\
&= \{\zeta \, x_{n+1} + \operatorname{cone}\{x_\ell \in \mathbb{R}^n, \ \ell = 1 \ldots n\} \mid \zeta \geq 0\} \\
&= \operatorname{cone}\{x_\ell \in \mathbb{R}^n, \ \ell = 1 \ldots n+1\}
\end{aligned}
\tag{284}
$$

a union of parallel hyperplanes. Cardinality is one step beyond dimension of the ambient space, but $\{x_\ell \ \forall \ell\}$ is a minimal set of generators for this convex cone $\mathcal{H}$ which has no extreme elements. $\qquad\square$

2.10.2.0.2 Exercise. *Enumerating conically independent directions.*
Do Example 2.10.2.0.1 in $\mathbb{R}$ and $\mathbb{R}^3$ by drawing two figures corresponding to Figure **52** and enumerating $n+1$ conically independent generators for each. Describe a nonpointed polyhedral cone in three dimensions having more than 8 conically independent generators. (*confer* Table **2.10.0.0.1**) $\qquad\blacktriangledown$

2.10.3 Utility of conic independence

Perhaps the most useful application of conic independence is determination of the intersection of closed convex cones from their halfspace-descriptions, or representation of the sum of closed convex cones from their vertex-descriptions.

$\bigcap \mathcal{K}_i$ A halfspace-description for the intersection of any number of closed convex cones $\mathcal{K}_i$ can be acquired by pruning normals; specifically, only the conically independent normals from the aggregate of all the halfspace-descriptions need be retained.

$\sum \mathcal{K}_i$ Generators for the sum of any number of closed convex cones $\mathcal{K}_i$ can be determined by retaining only the conically independent generators from the aggregate of all the vertex-descriptions.

Such conically independent sets are not necessarily unique or minimal.

2.11 When extreme means exposed

For any convex full-dimensional polyhedral set in $\mathbb{R}^n$, distinction between the terms *extreme* and *exposed* vanishes [337, §2.4] [116, §2.2] for faces of all dimensions except n ; their meanings become equivalent as we saw in Figure **20** (discussed in §2.6.1.2). In other words, each and every face of any polyhedral set (except the set itself) can be exposed by a hyperplane, and *vice versa*; *e.g.*, Figure **25**.

    Lewis [252, §6] [221, §2.3.4] claims nonempty extreme proper subsets and the exposed subsets coincide for $\mathbb{S}_+^n$; *id est*, each and every face of the positive semidefinite cone, whose dimension is less than dimension of the cone, is exposed. A more general discussion of cones having this property can be found in [348]; *e.g.*, Lorentz cone (178) [20, §II.A].

2.12   Convex polyhedra

Every polyhedron, such as the convex hull (86) of a bounded list X, can be expressed as the solution set of a finite system of linear equalities and inequalities, and *vice versa.* [116, §2.2]

2.12.0.0.1   Definition.   *Convex polyhedra, halfspace-description.*
A convex polyhedron is the intersection of a finite number of halfspaces and hyperplanes;

$$\mathcal{P} = \{y \mid Ay \succeq b,\, Cy = d\} \subseteq \mathbb{R}^n \tag{285}$$

where coefficients A and C generally denote matrices. Each row of C is a vector normal to a hyperplane, while each row of A is a vector inward-normal to a hyperplane partially bounding a halfspace.   △

By the *halfspaces theorem* in §2.4.1.1.1, a polyhedron thus described is a closed convex set possibly not full-dimensional; *e.g.*, Figure **20**. Convex polyhedra[2.52] are finite-dimensional comprising all affine sets (§2.3.1), polyhedral cones, line segments, rays, halfspaces, convex polygons, *solids* [229, def.104/6 p.343], polychora, *polytopes,*[2.53] *etcetera.*

It follows from definition (285) by exposure that each face of a convex polyhedron is a convex polyhedron.

Projection of any polyhedron on a subspace remains a polyhedron. More generally, image and inverse image of a convex polyhedron under any linear transformation remains a convex polyhedron; [26, §I.9] [315, thm.19.3] the foremost consequence being, invariance of polyhedral set closedness.

When b and d in (285) are $\mathbf{0}$, the resultant is a polyhedral cone. The set of all polyhedral cones is a subset of convex cones:

2.12.1   Polyhedral cone

From our study of cones, we see: the number of intersecting hyperplanes and halfspaces constituting a convex cone is possibly but not necessarily infinite. When the number is finite, the convex cone is termed *polyhedral.* That is the primary distinguishing feature between the set of all convex cones and polyhedra; all polyhedra, including polyhedral cones, are *finitely generated* [315, §19]. (Figure **53**) We distinguish polyhedral cones in the set of all convex cones for this reason, although all convex cones of dimension 2 or less are polyhedral.

[2.52]We consider only convex polyhedra throughout, but acknowledge the existence of concave polyhedra. [401, *Kepler-Poinsot Solid*]
[2.53]Some authors distinguish bounded polyhedra via the designation *polytope.* [116, §2.2]

2.12.1.0.1 Definition. *Polyhedral cone, halfspace-description.*[2.54]
(*confer*(103)) A polyhedral cone is the intersection of a finite number of halfspaces and hyperplanes about the origin;

$$
\begin{aligned}
\mathcal{K} \;=\;& \{y \mid Ay \succeq 0,\, Cy = \mathbf{0}\} \;\subseteq\; \mathbb{R}^n && \text{(a)}\\
\;=\;& \{y \mid Ay \succeq 0,\, Cy \succeq 0,\, Cy \preceq 0\} && \text{(b)}\\
\;=\;& \left\{ y \;\bigg|\; \begin{bmatrix} A \\ C \\ -C \end{bmatrix} y \succeq 0 \right\} && \text{(c)}
\end{aligned}
\qquad (286)
$$

where coefficients A and C generally denote matrices of finite dimension. Each row of C is a vector normal to a hyperplane containing the origin, while each row of A is a vector inward-normal to a hyperplane containing the origin and partially bounding a halfspace.

$\triangle$

A polyhedral cone thus defined is closed, convex (§2.4.1.1), has only a finite number of generators (§2.8.1.2), and can be not full-dimensional. (Minkowski) Conversely, any finitely generated convex cone is polyhedral. (Weyl) [337, §2.8] [315, thm.19.1]

2.12.1.0.2 Exercise. *Unbounded convex polyhedra.*
Illustrate an unbounded polyhedron that is not a cone or its translation. ▼

From the definition it follows that any single hyperplane through the origin, or any halfspace partially bounded by a hyperplane through the origin is a polyhedral cone. The most familiar example of polyhedral cone is any quadrant (or orthant, §2.1.3) generated by Cartesian half-axes. Esoteric examples of polyhedral cone include the point at the origin, any line through the origin, any ray having the origin as base such as the nonnegative real line $\mathbb{R}_+$ in subspace $\mathbb{R}$, polyhedral flavor (proper) Lorentz cone (316), any subspace, and $\mathbb{R}^n$. More polyhedral cones are illustrated in Figure 51 and Figure 25.

2.12.2 Vertices of convex polyhedra

By definition, a vertex (§2.6.1.0.1) always lies on the relative boundary of a convex polyhedron. [229, def.115/6 p.358] In Figure 20, each vertex of the polyhedron is located at an intersection of three or more facets, and every edge belongs to precisely two facets [26, §VI.1 p.252]. In Figure 25, the only vertex of that polyhedral cone lies at the origin.

The set of all polyhedral cones is clearly a subset of convex polyhedra and a subset of convex cones (Figure 53). Not all convex polyhedra are bounded; evidently, neither can they all be described by the convex hull of a bounded set of points as defined in (86). Hence a universal vertex-description of polyhedra in terms of that same finite-length list X (76):

[2.54]Rockafellar [315, §19] proposes affine sets be handled via complementary pairs of affine inequalities; *e.g.*, $Cy \succeq d$ and $Cy \preceq d$ which, when taken together, can present severe difficulties to some *interior-point methods* of numerical solution.

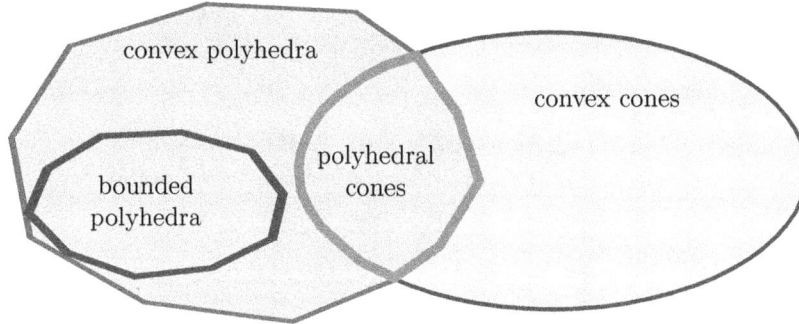

Figure 53: Polyhedral cones are finitely generated, unbounded, and convex.

2.12.2.0.1 Definition.   *Convex polyhedra, vertex-description.*
(*confer* §2.8.1.1.1) Denote the truncated a-vector

$$a_{i:\ell} = \begin{bmatrix} a_i \\ \vdots \\ a_\ell \end{bmatrix} \tag{287}$$

By discriminating a suitable finite-length *generating list* (or set) arranged columnar in $X \in \mathbb{R}^{n \times N}$, then any particular polyhedron may be described

$$\mathcal{P} = \left\{ Xa \mid a_{1:k}^{\mathrm{T}} \mathbf{1} = 1, \ a_{m:N} \succeq 0, \ \{1 \ldots k\} \cup \{m \ldots N\} = \{1 \ldots N\} \right\} \tag{288}$$

where $0 \le k \le N$ and $1 \le m \le N+1$. Setting $k=0$ removes the affine equality condition. Setting $m=N+1$ removes the inequality. $\triangle$

Coefficient indices in (288) may or may not be overlapping, but all the coefficients are assumed constrained. From (78), (86), and (103), we summarize how the coefficient conditions may be applied;

$$\left. \begin{array}{ccc} \text{affine sets} & \longrightarrow & a_{1:k}^{\mathrm{T}} \mathbf{1} = 1 \\ \text{polyhedral cones} & \longrightarrow & a_{m:N} \succeq 0 \end{array} \right\} \longleftarrow \text{convex hull } (m \le k) \tag{289}$$

It is always possible to describe a convex hull in the region of overlapping indices because, for $1 \le m \le k \le N$

$$\left\{ a_{m:k} \mid a_{m:k}^{\mathrm{T}} \mathbf{1} = 1, \ a_{m:k} \succeq 0 \right\} \subseteq \left\{ a_{m:k} \mid a_{1:k}^{\mathrm{T}} \mathbf{1} = 1, \ a_{m:N} \succeq 0 \right\} \tag{290}$$

Members of a generating list are not necessarily vertices of the corresponding polyhedron; certainly true for (86) and (288), some subset of list members reside in the polyhedron's relative interior. Conversely, when boundedness (86) applies, convex hull of the vertices is a polyhedron identical to convex hull of the generating list.

2.12.2.1 Vertex-description of polyhedral cone

Given closed convex cone $\mathcal{K}$ in a subspace of $\mathbb{R}^n$ having any set of generators for it arranged in a matrix $X \in \mathbb{R}^{n \times N}$ as in (279), then that cone is described setting $m=1$ and $k=0$ in vertex-description (288):

$$\mathcal{K} \;=\; \operatorname{cone} X \;=\; \{Xa \mid a \succeq 0\} \subseteq \mathbb{R}^n \qquad (103)$$

a conic hull of N generators.

2.12.2.2 Pointedness

(§2.7.2.1.2) [337, §2.10] Assuming all generators constituting the columns of $X \in \mathbb{R}^{n \times N}$ are nonzero, polyhedral cone $\mathcal{K}$ is pointed if and only if there is no nonzero $a \succeq 0$ that solves $Xa = \mathbf{0}$; *id est*, iff

$$
\begin{array}{rl}
\text{find} & a \\
\text{subject to} & Xa = \mathbf{0} \\
& \mathbf{1}^{\mathrm{T}}a = 1 \\
& a \succeq 0
\end{array}
\qquad (291)
$$

is infeasible or iff $\mathcal{N}(X) \cap \mathbb{R}_+^N = \mathbf{0}$.[2.55] Otherwise, the cone will contain at least one line and there can be no vertex; *id est*, the cone cannot otherwise be pointed. Any subspace, Euclidean vector space $\mathbb{R}^n$, or any halfspace are examples of nonpointed polyhedral cone; hence, no vertex. This null-pointedness criterion $Xa = \mathbf{0}$ means that a pointed polyhedral cone is invariant to linear injective transformation.

Examples of pointed polyhedral cone $\mathcal{K}$ include: the origin, any $\mathbf{0}$-based ray in a subspace, any two-dimensional V-shaped cone in a subspace, any orthant in $\mathbb{R}^n$ or $\mathbb{R}^{m \times n}$; *e.g.*, nonnegative real line $\mathbb{R}_+$ in vector space $\mathbb{R}$.

2.12.3 Unit simplex

A peculiar subset of the nonnegative orthant with halfspace-description

$$\mathcal{S} \;\triangleq\; \{s \mid s \succeq 0, \; \mathbf{1}^{\mathrm{T}}s \leq 1\} \subseteq \mathbb{R}_+^n \qquad (292)$$

is a unique bounded convex full-dimensional polyhedron called *unit simplex* (Figure 54) having $n+1$ facets, $n+1$ vertices, and dimension

$$\dim \mathcal{S} \;=\; n \qquad (293)$$

The origin supplies one vertex while heads of the *standard basis* [208] [338] $\{e_i, \; i=1\ldots n\}$ in $\mathbb{R}^n$ constitute those remaining;[2.56] thus its vertex-description:

$$
\begin{aligned}
\mathcal{S} \;&=\; \operatorname{conv}\{\mathbf{0}, \{e_i, \; i=1\ldots n\}\} \\
&=\; \{[\,\mathbf{0} \;\; e_1 \;\; e_2 \;\cdots\; e_n\,]\,a \mid a^{\mathrm{T}}\mathbf{1} = 1, \; a \succeq 0\}
\end{aligned}
\qquad (294)
$$

[2.55] If $\operatorname{rank} X = n$, then the dual cone $\mathcal{K}^*$ (§2.13.1) is pointed. (309)

[2.56] In $\mathbb{R}^0$ the unit simplex is the point at the origin, in $\mathbb{R}$ the unit simplex is the line segment $[0,1]$, in $\mathbb{R}^2$ it is a triangle and its relative interior, in $\mathbb{R}^3$ it is the convex hull of a tetrahedron (Figure 54), in $\mathbb{R}^4$ it is the convex hull of a *pentatope* [401], and so on.

$$\mathcal{S} = \{s \mid s \succeq 0, \ \mathbf{1}^{\mathrm{T}}s \leq 1\}$$

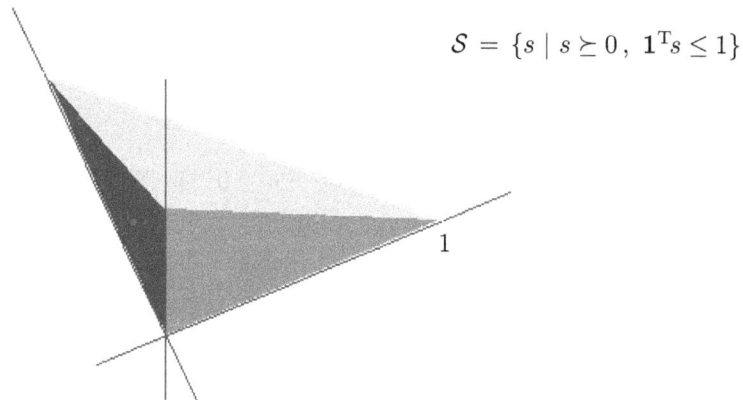

Figure 54: Unit simplex $\mathcal{S}$ in $\mathbb{R}^3$ is a unique solid tetrahedron but not *regular*.

2.12.3.1 Simplex

The unit simplex comes from a class of general polyhedra called *simplex*, having vertex-description: [96] [315] [399] [116]

$$\mathrm{conv}\{x_\ell \in \mathbb{R}^n\} \mid \ell = 1 \dots k+1, \quad \dim \mathrm{aff}\{x_\ell\} = k, \quad n \geq k \qquad (295)$$

So defined, a simplex is a closed bounded convex set possibly not full-dimensional. Examples of simplices, by increasing affine dimension, are: a point, any line segment, any triangle and its relative interior, a general tetrahedron, any five-vertex polychoron, and so on.

2.12.3.1.1 Definition. *Simplicial cone.*
A proper polyhedral (§2.7.2.2.1) cone $\mathcal{K}$ in $\mathbb{R}^n$ is called *simplicial* iff $\mathcal{K}$ has exactly n extreme directions; [20, §II.A] equivalently, iff proper $\mathcal{K}$ has exactly n linearly independent generators contained in any given set of generators. $\qquad \triangle$

- simplicial cone $\Rightarrow$ proper polyhedral cone

There are an infinite variety of simplicial cones in $\mathbb{R}^n$; *e.g.*, Figure 25, Figure 55, Figure 65. Any orthant is simplicial, as is any rotation thereof.

2.12.4 Converting between descriptions

Conversion between halfspace- (285) (286) and vertex-description (86) (288) is nontrivial, in general, [15] [116, §2.2] but the conversion is easy for simplices. [62, §2.2.4] Nonetheless, we tacitly assume the two descriptions to be equivalent. [315, §19 thm.19.1] We explore conversions in §2.13.4, §2.13.9, and §2.13.11:

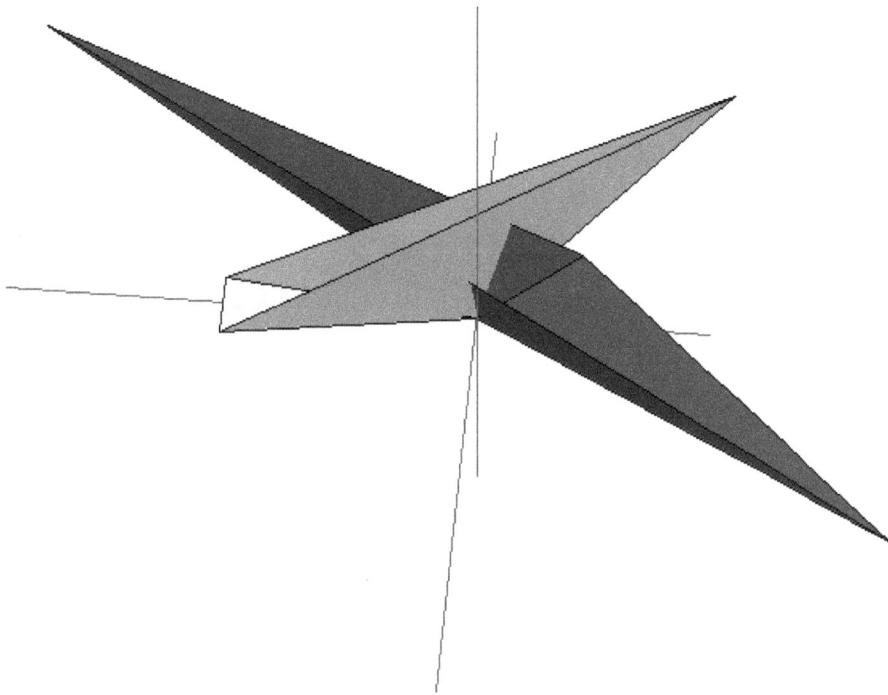

Figure 55: Two views of a simplicial cone and its dual in $\mathbb{R}^3$ (second view on next page). Semiinfinite boundary of each cone is truncated for illustration. Each cone has three facets (*confer* §2.13.11.0.3). Cartesian axes are drawn for reference.

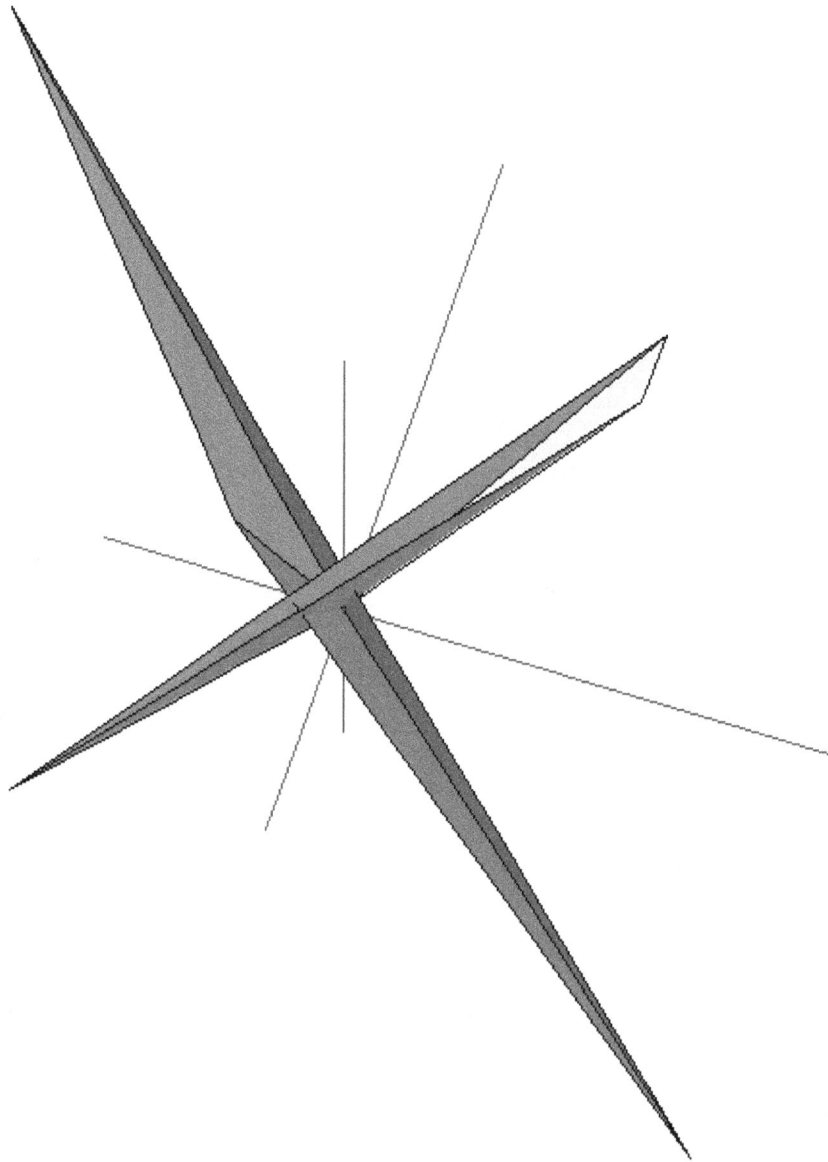

2.13 Dual cone & generalized inequality & biorthogonal expansion

These three concepts, dual cone, generalized inequality, and biorthogonal expansion, are inextricably melded; meaning, it is difficult to completely discuss one without mentioning the others. The dual cone is critical in tests for convergence by contemporary primal/dual methods for numerical solution of conic problems. [419] [284, §4.5] For unique minimum-distance projection on a closed convex cone $\mathcal{K}$, the negative dual cone $-\mathcal{K}^*$ plays the role that orthogonal complement plays for subspace projection.[2.57] (§E.9.2, Figure 177) Indeed, $-\mathcal{K}^*$ is the algebraic complement in $\mathbb{R}^n$;

$$\mathcal{K} \boxplus -\mathcal{K}^* = \mathbb{R}^n \qquad (2120)$$

where $\boxplus$ denotes unique orthogonal vector sum.

One way to think of a pointed closed convex cone is as a new kind of coordinate system whose basis is generally nonorthogonal; a conic system, very much like the familiar Cartesian system whose analogous cone is the first quadrant (the nonnegative orthant). Generalized inequality $\succeq_{\mathcal{K}}$ is a formalized means to determine membership to any pointed closed convex cone $\mathcal{K}$ (§2.7.2.2) whereas *biorthogonal expansion* is, fundamentally, an expression of coordinates in a pointed conic system whose axes are linearly independent but not necessarily orthogonal. When cone $\mathcal{K}$ is the nonnegative orthant, then these three concepts come into alignment with the Cartesian prototype: biorthogonal expansion becomes orthogonal expansion, the dual cone becomes identical to the orthant, and generalized inequality obeys a total order entrywise.

2.13.1 Dual cone

For any set $\mathcal{K}$ (convex or not), the *dual cone* [112, §4.2]

$$\mathcal{K}^* \triangleq \{y \in \mathbb{R}^n \mid \langle y, x \rangle \geq 0 \text{ for all } x \in \mathcal{K}\} \qquad (296)$$

is a unique cone[2.58] that is always closed and convex because it is an intersection of halfspaces (§2.4.1.1.1). Each halfspace has inward-normal x, belonging to $\mathcal{K}$, and boundary containing the origin; *e.g.*, Figure 56a.

When cone $\mathcal{K}$ is convex, there is a second and equivalent construction: Dual cone $\mathcal{K}^*$ is the union of each and every vector y inward-normal to a hyperplane supporting $\mathcal{K}$ (§2.4.2.6.1); *e.g.*, Figure 56b. When $\mathcal{K}$ is represented by a halfspace-description such as (286), for example, where

$$A \triangleq \begin{bmatrix} a_1^\mathrm{T} \\ \vdots \\ a_m^\mathrm{T} \end{bmatrix} \in \mathbb{R}^{m \times n}, \qquad C \triangleq \begin{bmatrix} c_1^\mathrm{T} \\ \vdots \\ c_p^\mathrm{T} \end{bmatrix} \in \mathbb{R}^{p \times n} \qquad (297)$$

[2.57]Namely, projection on a subspace is ascertainable from projection on its orthogonal complement (Figure 176).
[2.58]The dual cone is the negative polar cone defined by many authors; $\mathcal{K}^* = -\mathcal{K}^\circ$. [205, §A.3.2] [315, §14] [41] [26] [337, §2.7]

then the dual cone can be represented as the conic hull

$$\mathcal{K}^* = \text{cone}\{a_1, \ldots, a_m, \pm c_1, \ldots, \pm c_p\} \tag{298}$$

a vertex-description, because each and every conic combination of normals from the halfspace-description of $\mathcal{K}$ yields another inward-normal to a hyperplane supporting $\mathcal{K}$.

$\mathcal{K}^*$ can also be constructed pointwise using projection theory from §E.9.2: for $P_\mathcal{K} x$ the Euclidean projection of point x on closed convex cone $\mathcal{K}$

$$-\mathcal{K}^* = \{x - P_\mathcal{K} x \mid x \in \mathbb{R}^n\} = \{x \in \mathbb{R}^n \mid P_\mathcal{K} x = \mathbf{0}\} \tag{2121}$$

2.13.1.0.1 Exercise. *Manual dual cone construction.*
Perhaps the most instructive graphical method of dual cone construction is cut-and-try. Find the dual of each polyhedral cone from Figure **57** by using dual cone equation (296).
▼

2.13.1.0.2 Exercise. *Dual cone definitions.*
What is $\{x \in \mathbb{R}^n \mid x^\mathrm{T} z \geq 0 \ \forall z \in \mathbb{R}^n\}$?
What is $\{x \in \mathbb{R}^n \mid x^\mathrm{T} z \geq 1 \ \forall z \in \mathbb{R}^n\}$?
What is $\{x \in \mathbb{R}^n \mid x^\mathrm{T} z \geq 1 \ \forall z \in \mathbb{R}^n_+\}$?
▼

As defined, dual cone $\mathcal{K}^*$ exists even when the affine hull of the original cone is a proper subspace; *id est*, even when the original cone is not full-dimensional.[2.59]

To further motivate our understanding of the dual cone, consider the ease with which convergence can be ascertained in the following optimization problem (301p):

2.13.1.0.3 Example. *Dual problem.*   (*confer* §4.1)
Duality is a powerful and widely employed tool in applied mathematics for a number of reasons. First, the dual program is always convex even if the primal is not. Second, the number of variables in the dual is equal to the number of constraints in the primal which is often less than the number of variables in the primal program. Third, the maximum value achieved by the dual problem is often equal to the minimum of the primal. [307, §2.1.3]
When not equal, the dual always provides a bound on the primal optimal objective. For convex problems, the dual variables provide necessary and sufficient optimality conditions:

Essentially, *Lagrange duality theory* concerns representation of a given optimization problem as half of a *minimax problem*. [315, §36] [62, §5.4] Given any real function $f(x, z)$

$$\underset{x}{\text{minimize}}\,\underset{z}{\text{maximize}}\, f(x, z) \geq \underset{z}{\text{maximize}}\,\underset{x}{\text{minimize}}\, f(x, z) \tag{299}$$

always holds. When

$$\underset{x}{\text{minimize}}\,\underset{z}{\text{maximize}}\, f(x, z) = \underset{z}{\text{maximize}}\,\underset{x}{\text{minimize}}\, f(x, z) \tag{300}$$

[2.59]Rockafellar formulates dimension of $\mathcal{K}$ and $\mathcal{K}^*$. [315, §14.6.1] His monumental work *Convex Analysis* has not one figure or illustration. See [26, §II.16] for a good illustration of Rockafellar's *recession cone* [42].

(a)

(b)

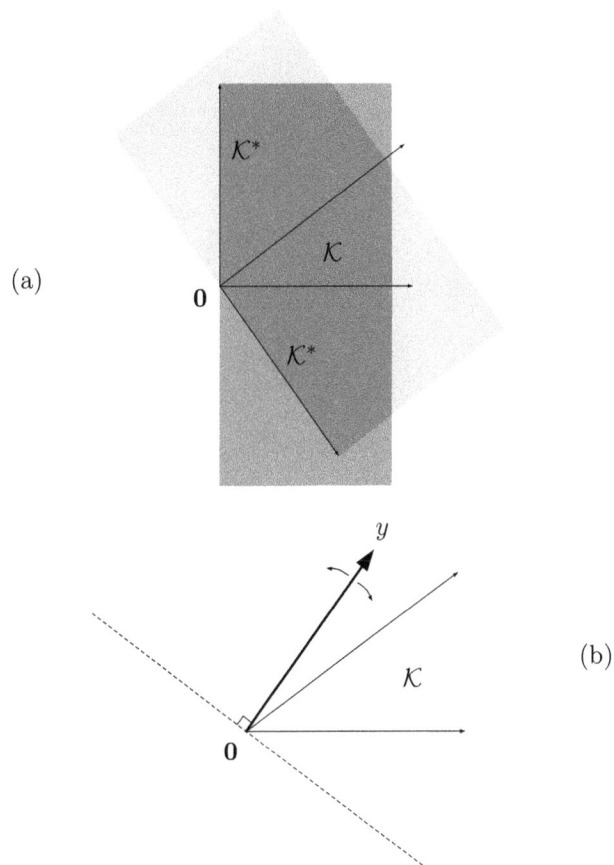

Figure 56: Two equivalent constructions of dual cone $\mathcal{K}^*$ in $\mathbb{R}^2$: (a) Showing construction by intersection of halfspaces about $\mathbf{0}$ (drawn truncated). Only those two halfspaces whose bounding hyperplanes have inward-normal corresponding to an extreme direction of this pointed closed convex cone $\mathcal{K} \subset \mathbb{R}^2$ need be drawn; by (368). (b) Suggesting construction by union of inward-normals y to each and every hyperplane $\partial\mathcal{H}_+$ supporting $\mathcal{K}$. This interpretation is valid when $\mathcal{K}$ is convex because existence of a supporting hyperplane is then guaranteed (§2.4.2.6).

$$\mathbb{R}^2 \qquad\qquad\qquad \mathbb{R}^3$$

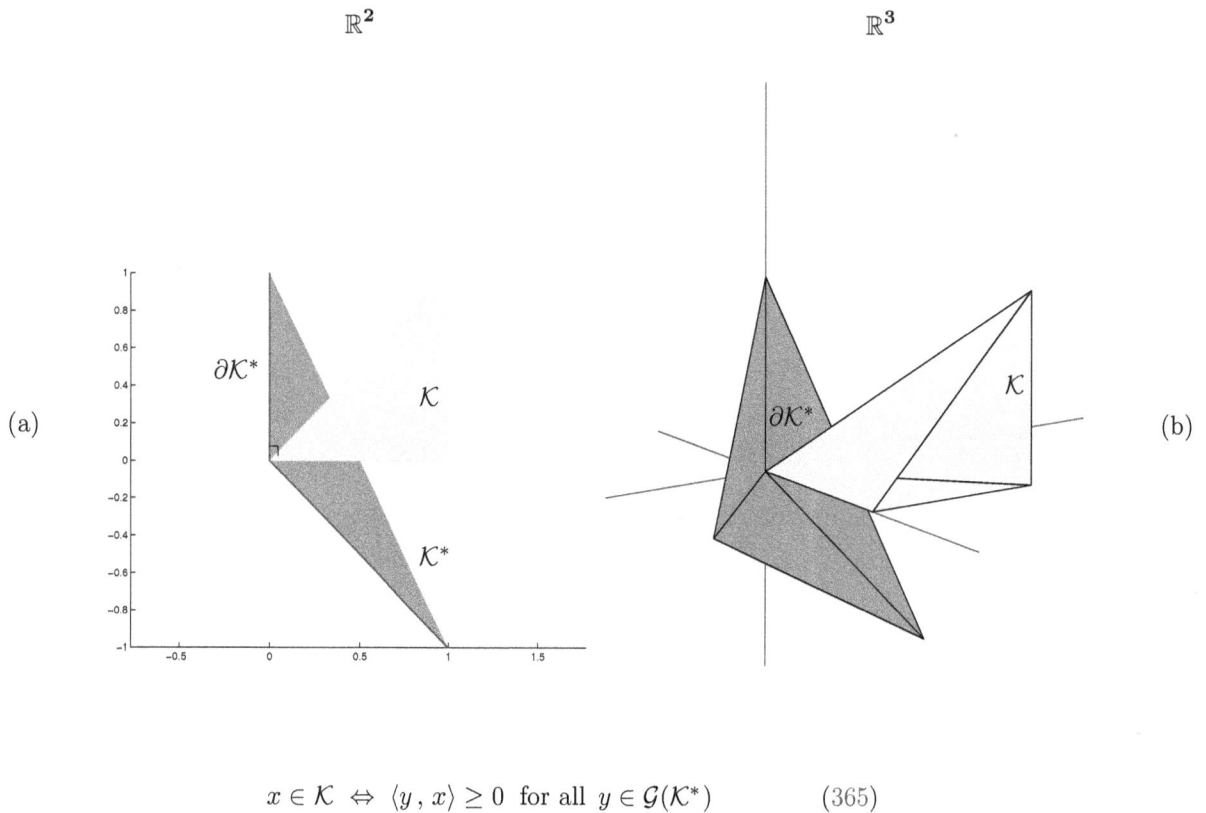

(a)   (b)

$$x \in \mathcal{K} \;\Leftrightarrow\; \langle y \,,\, x \rangle \geq 0 \;\text{ for all } y \in \mathcal{G}(\mathcal{K}^*) \qquad\qquad (365)$$

Figure 57: Dual cone construction by right angle. Each extreme direction of a proper polyhedral cone is orthogonal to a facet of its dual cone, and *vice versa,* in any dimension. (§2.13.6.1) **(a)** This characteristic guides graphical construction of dual cone in two dimensions: It suggests finding dual-cone boundary ∂ by making right angles with extreme directions of polyhedral cone. The construction is then pruned so that each dual boundary vector does not exceed $\pi/2$ radians in angle with each and every vector from polyhedral cone. Were dual cone in $\mathbb{R}^2$ to narrow, Figure 58 would be reached in limit. **(b)** Same polyhedral cone and its dual continued into three dimensions. (*confer* Figure 65)

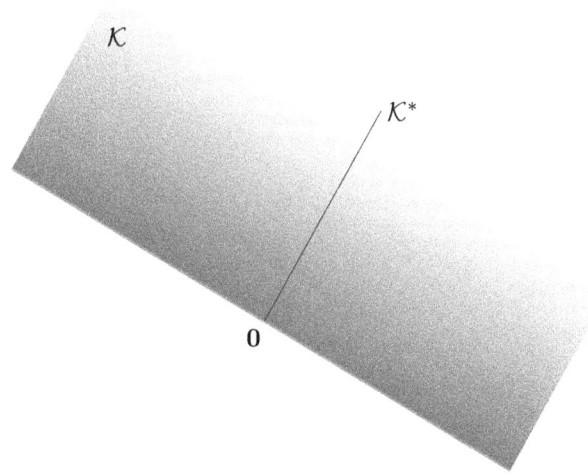

Figure 58: Polyhedral cone $\mathcal{K}$ is a halfspace about origin in $\mathbb{R}^2$. Dual cone $\mathcal{K}^*$ is a ray base $\mathbf{0}$, hence not full-dimensional in $\mathbb{R}^2$; so $\mathcal{K}$ cannot be pointed, hence has no extreme directions. (Both convex cones appear truncated.)

we have *strong duality* and then a *saddle value* [156] exists. (Figure **59**) [312, p.3] Consider primal conic problem (p) (over cone $\mathcal{K}$) and its corresponding *dual problem* (d): [301, §3.3.1] [245, §2.1] [246] given vectors α, β and matrix constant C

$$
\begin{array}{llll}
& \underset{x}{\text{minimize}} & \alpha^{\mathrm{T}}x & \qquad \underset{y,z}{\text{maximize}} \quad \beta^{\mathrm{T}}z \\
\text{(p)} & \text{subject to} & x \in \mathcal{K} & \qquad \text{subject to} \quad y \in \mathcal{K}^* \qquad \text{(d)} \\
& & Cx = \beta & \qquad \qquad \qquad \quad C^{\mathrm{T}}z + y = \alpha
\end{array}
\qquad (301)
$$

Observe: the dual problem is also conic, and its objective function value never exceeds that of the primal;

$$
\begin{aligned}
\alpha^{\mathrm{T}}x &\geq \beta^{\mathrm{T}}z \\
x^{\mathrm{T}}(C^{\mathrm{T}}z + y) &\geq (Cx)^{\mathrm{T}}z \\
x^{\mathrm{T}}y &\geq 0
\end{aligned}
\qquad (302)
$$

which holds by definition (296). Under the sufficient condition that (301p) is a *convex problem*[2.60] satisfying *Slater's condition* (p.249), then equality

$$
x^{\star\mathrm{T}}y^{\star} = 0 \qquad (303)
$$

is achieved; which is necessary and sufficient for optimality (§2.13.10.1.5); each problem (p) and (d) attains the same optimal value of its objective and each problem is called a *strong dual* to the other because the *duality gap* (optimal primal−dual objective difference)

[2.60]In this context, problems (p) and (d) are convex if $\mathcal{K}$ is a convex cone.

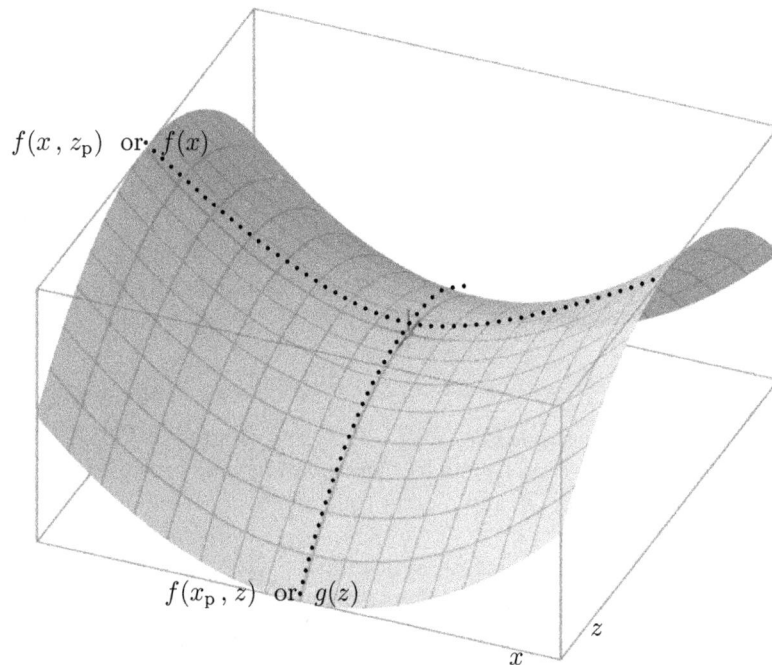

Figure 59: (Drawing by Lucas V. Barbosa) This serves as mnemonic icon for primal and dual problems, although objective functions from conic problems (301p) (301d) are linear. When problems are *strong duals*, duality gap is 0; meaning, functions $f(x)$, $g(z)$ (dotted) *kiss* at saddle value as depicted at center. Otherwise, dual functions never meet $(f(x) > g(z))$ by (299).

becomes 0. Then (p) and (d) are together equivalent to the minimax problem

$$
\begin{array}{ll}
\underset{x,y,z}{\text{minimize}} & \alpha^{\mathrm T}x - \beta^{\mathrm T}z \\
\text{subject to} & x \in \mathcal K , \quad y \in \mathcal K^* \\
& Cx = \beta , \quad C^{\mathrm T}z + y = \alpha
\end{array}
\qquad \text{(p)}-\text{(d)} \qquad (304)
$$

whose optimal objective always has the saddle value 0 (regardless of the particular convex cone $\mathcal K$ and other problem parameters). [368, §3.2] Thus determination of convergence for either primal or dual problem is facilitated.

Were convex cone $\mathcal K$ polyhedral (§2.12.1), then problems (p) and (d) would be linear programs. Selfdual nonnegative orthant $\mathcal K$ yields the primal prototypical linear program and its dual. Were $\mathcal K$ a positive semidefinite cone, then problem (p) has the form of prototypical *semidefinite program* (679) with (d) its dual. It is sometimes possible to solve a primal problem by way of its dual; advantageous *when the dual problem is easier to solve than the primal problem, for example, because it can be solved analytically, or has some special structure that can be exploited.* [62, §5.5.5] (§4.2.3.1)   □

2.13.1.1 Key properties of dual cone

- For any cone, $(-\mathcal K)^* = -\mathcal K^*$

- For any cones $\mathcal K_1$ and $\mathcal K_2$, $\mathcal K_1 \subseteq \mathcal K_2 \Rightarrow \mathcal K_1^* \supseteq \mathcal K_2^*$ [337, §2.7]

- (Cartesian product) For closed convex cones $\mathcal K_1$ and $\mathcal K_2$, their Cartesian product $\mathcal K = \mathcal K_1 \times \mathcal K_2$ is a closed convex cone, and

$$
\mathcal K^* = \mathcal K_1^* \times \mathcal K_2^* \qquad (305)
$$

  where each dual is determined with respect to a cone's ambient space.

- (conjugation) [315, §14] [112, §4.5] [337, p.52] When $\mathcal K$ is any convex cone, dual of the dual cone equals closure of the original cone;

$$
\mathcal K^{**} = \overline{\mathcal K} \qquad (306)
$$

  is the intersection of all halfspaces about the origin that contain $\mathcal K$. Because $\mathcal K^{***} = \mathcal K^*$ always holds,

$$
\mathcal K^* = (\overline{\mathcal K})^* \qquad (307)
$$

  When convex cone $\mathcal K$ is closed, then dual of the dual cone is the original cone; $\mathcal K^{**} = \mathcal K \Leftrightarrow \mathcal K$ is a closed convex cone: [337, p.53, p.95]

$$
\mathcal K = \{x \in \mathbb R^n \mid \langle y , x \rangle \geq 0 \; \forall y \in \mathcal K^*\} \qquad (308)
$$

- If any cone $\mathcal K$ is full-dimensional, then $\mathcal K^*$ is pointed;

$$
\mathcal K \text{ full-dimensional} \Rightarrow \mathcal K^* \text{ pointed} \qquad (309)
$$

If the closure of any convex cone $\mathcal{K}$ is pointed, conversely, then $\mathcal{K}^*$ is full-dimensional;

$$\overline{\mathcal{K}} \text{ pointed } \Rightarrow \mathcal{K}^* \text{ full-dimensional} \tag{310}$$

Given that a cone $\mathcal{K} \subset \mathbb{R}^n$ is closed and convex, $\mathcal{K}$ is pointed if and only if $\mathcal{K}^* - \mathcal{K}^* = \mathbb{R}^n$; *id est*, iff $\mathcal{K}^*$ is full-dimensional. [56, §3.3 exer.20]

- (vector sum) [315, thm.3.8] For convex cones $\mathcal{K}_1$ and $\mathcal{K}_2$

$$\mathcal{K}_1 + \mathcal{K}_2 = \text{conv}(\mathcal{K}_1 \cup \mathcal{K}_2) \tag{311}$$

is a convex cone.

- (dual vector-sum) [315, §16.4.2] [112, §4.6] For convex cones $\mathcal{K}_1$ and $\mathcal{K}_2$

$$\mathcal{K}_1^* \cap \mathcal{K}_2^* = (\mathcal{K}_1 + \mathcal{K}_2)^* = (\mathcal{K}_1 \cup \mathcal{K}_2)^* \tag{312}$$

- (closure of vector sum of duals)[2.61] For closed convex cones $\mathcal{K}_1$ and $\mathcal{K}_2$

$$(\mathcal{K}_1 \cap \mathcal{K}_2)^* = \overline{\mathcal{K}_1^* + \mathcal{K}_2^*} = \overline{\text{conv}(\mathcal{K}_1^* \cup \mathcal{K}_2^*)} \tag{313}$$

[337, p.96] where operation closure becomes superfluous under the sufficient condition $\mathcal{K}_1 \cap \text{int}\, \mathcal{K}_2 \neq \emptyset$ [56, §3.3 exer.16, §4.1 exer.7].

- (Krein-Rutman) Given closed convex cones $\mathcal{K}_1 \subseteq \mathbb{R}^m$ and $\mathcal{K}_2 \subseteq \mathbb{R}^n$ and any linear map $A : \mathbb{R}^n \to \mathbb{R}^m$, then provided $\text{int}\, \mathcal{K}_1 \cap A\mathcal{K}_2 \neq \emptyset$ [56, §3.3.13, *confer* §4.1 exer.9]

$$(A^{-1}\mathcal{K}_1 \cap \mathcal{K}_2)^* = A^{\mathrm{T}}\mathcal{K}_1^* + \mathcal{K}_2^* \tag{314}$$

where dual of cone $\mathcal{K}_1$ is with respect to its ambient space $\mathbb{R}^m$ and dual of cone $\mathcal{K}_2$ is with respect to $\mathbb{R}^n$, where $A^{-1}\mathcal{K}_1$ denotes inverse image (§2.1.9.0.1) of $\mathcal{K}_1$ under mapping A, and where A^{T} denotes adjoint operator. The particularly important case $\mathcal{K}_2 = \mathbb{R}^n$ is easy to show: for $A^{\mathrm{T}}A = I$

$$\begin{aligned}
(A^{\mathrm{T}}\mathcal{K}_1)^* &\triangleq \{y \in \mathbb{R}^n \mid x^{\mathrm{T}}y \geq 0 \;\; \forall x \in A^{\mathrm{T}}\mathcal{K}_1\} \\
&= \{y \in \mathbb{R}^n \mid (A^{\mathrm{T}}z)^{\mathrm{T}}y \geq 0 \;\; \forall z \in \mathcal{K}_1\} \\
&= \{A^{\mathrm{T}}w \mid z^{\mathrm{T}}w \geq 0 \;\; \forall z \in \mathcal{K}_1\} \\
&= A^{\mathrm{T}}\mathcal{K}_1^*
\end{aligned} \tag{315}$$

- $\mathcal{K}$ is proper if and only if $\mathcal{K}^*$ is proper.

- $\mathcal{K}$ is polyhedral if and only if $\mathcal{K}^*$ is polyhedral. [337, §2.8]

- $\mathcal{K}$ is simplicial if and only if $\mathcal{K}^*$ is simplicial. (§2.13.9.2) A simplicial cone and its dual are proper polyhedral cones (Figure **65**, Figure **55**), but not the converse.

- $\mathcal{K} \boxplus -\mathcal{K}^* = \mathbb{R}^n \Leftrightarrow \mathcal{K}$ is closed and convex. (2120)

- Any direction in a proper cone $\mathcal{K}$ is normal to a hyperplane separating $\mathcal{K}$ from $-\mathcal{K}^*$.

[2.61]These parallel analogous results for subspaces $\mathcal{R}_1, \mathcal{R}_2 \subseteq \mathbb{R}^n$; [112, §4.6]

$$(\mathcal{R}_1 + \mathcal{R}_2)^\perp = \mathcal{R}_1^\perp \cap \mathcal{R}_2^\perp$$

$$(\mathcal{R}_1 \cap \mathcal{R}_2)^\perp = \overline{\mathcal{R}_1^\perp + \mathcal{R}_2^\perp}$$

$\mathcal{R}^{\perp\perp} = \mathcal{R}$ for any subspace $\mathcal{R}$.

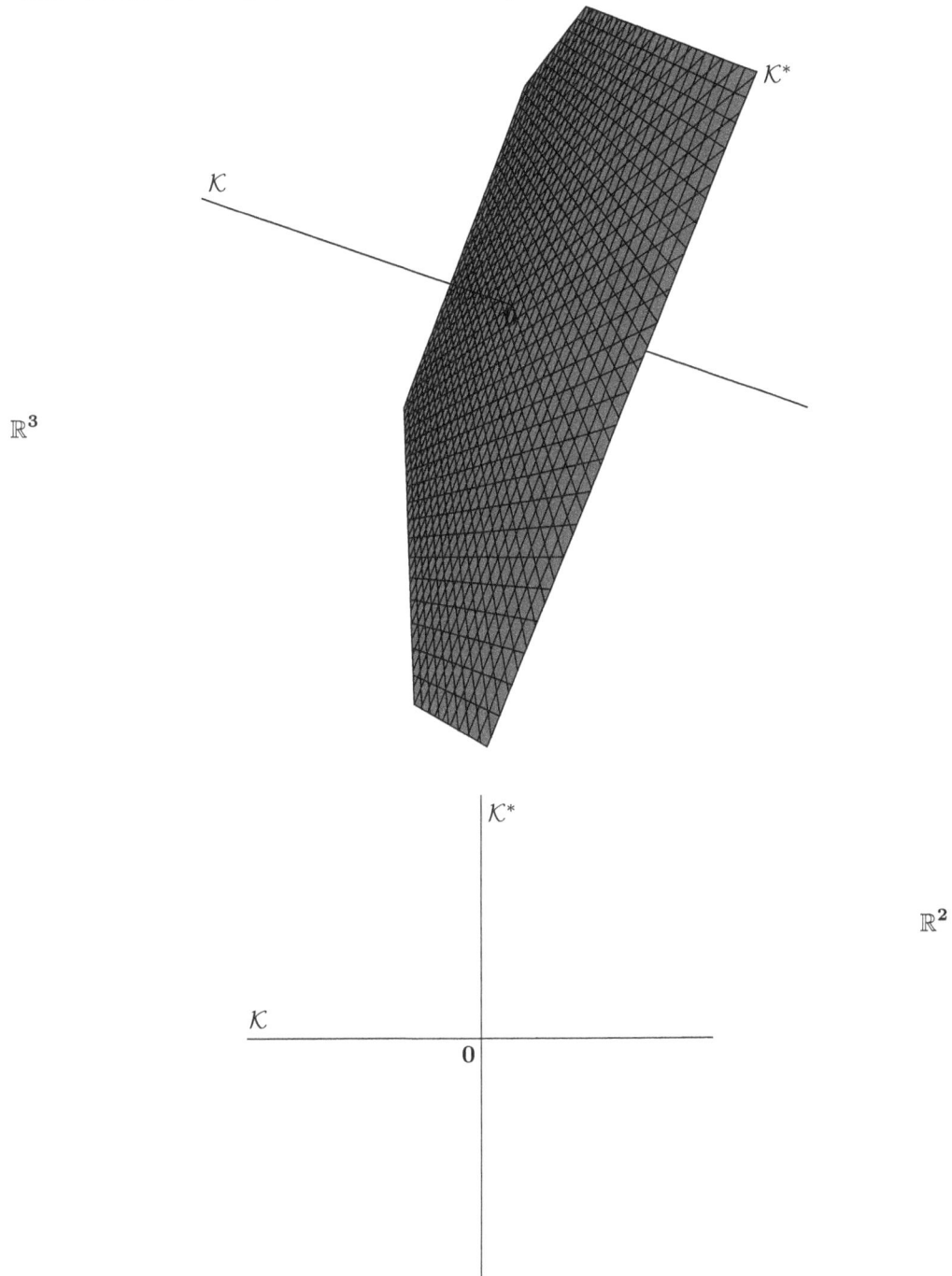

Figure 60: When convex cone $\mathcal{K}$ is any one Cartesian axis, its dual cone is the convex hull of all axes remaining; its orthogonal complement. In $\mathbb{R}^3$, dual cone $\mathcal{K}^*$ (drawn tiled and truncated) is a hyperplane through origin; its normal belongs to line $\mathcal{K}$. In $\mathbb{R}^2$, dual cone $\mathcal{K}^*$ is a line through origin while convex cone $\mathcal{K}$ is that line orthogonal.

2.13.1.2   Examples of dual cone

When $\mathcal{K}$ is $\mathbb{R}^n$, $\mathcal{K}^*$ is the point at the origin, and *vice versa*.

When $\mathcal{K}$ is a subspace, $\mathcal{K}^*$ is its orthogonal complement, and *vice versa*. (§E.9.2.1, Figure **60**)

When cone $\mathcal{K}$ is a halfspace in $\mathbb{R}^n$ with $n>0$ (Figure **58** for example), the dual cone $\mathcal{K}^*$ is a ray (base **0**) belonging to that halfspace but orthogonal to its bounding hyperplane (that contains the origin), and *vice versa*.

When convex cone $\mathcal{K}$ is a closed halfplane in $\mathbb{R}^3$ (Figure **61**), it is neither pointed or full-dimensional; hence, the dual cone $\mathcal{K}^*$ can be neither full-dimensional or pointed.

When $\mathcal{K}$ is any particular orthant in $\mathbb{R}^n$, the dual cone is identical; *id est*, $\mathcal{K}=\mathcal{K}^*$.

When $\mathcal{K}$ is any quadrant in subspace $\mathbb{R}^2$, $\mathcal{K}^*$ is a *wedge*-shaped polyhedral cone in $\mathbb{R}^3$; *e.g.*, for $\mathcal{K}$ equal to quadrant I, $\mathcal{K}^*=\begin{bmatrix} \mathbb{R}^2_+ \\ \mathbb{R} \end{bmatrix}$.

When $\mathcal{K}$ is a polyhedral flavor Lorentz cone (*confer*(178))

$$\mathcal{K}_\ell = \left\{ \begin{bmatrix} x \\ t \end{bmatrix} \in \mathbb{R}^n \times \mathbb{R} \mid \|x\|_\ell \leq t \right\}, \qquad \ell \in \{1,\,\infty\} \tag{316}$$

its dual is the proper cone [62, exmp.2.25]

$$\mathcal{K}_q = \mathcal{K}_\ell^* = \left\{ \begin{bmatrix} x \\ t \end{bmatrix} \in \mathbb{R}^n \times \mathbb{R} \mid \|x\|_q \leq t \right\}, \qquad \ell \in \{1,\,2,\,\infty\} \tag{317}$$

where $\|x\|_\ell^* = \|x\|_q$ is that norm dual to $\|x\|_\ell$ determined via solution to $1/\ell + 1/q = 1$. Figure **63** illustrates $\mathcal{K}=\mathcal{K}_1$ and $\mathcal{K}^*=\mathcal{K}_1^*=\mathcal{K}_\infty$ in $\mathbb{R}^2 \times \mathbb{R}$.

2.13.2   Abstractions of *Farkas' lemma*

2.13.2.0.1   Corollary. *Generalized inequality and membership relation.*
[205, §A.4.2] Let $\mathcal{K}$ be any closed convex cone and $\mathcal{K}^*$ its dual, and let x and y belong to a vector space $\mathbb{R}^n$. Then

$$y \in \mathcal{K}^* \Leftrightarrow \langle y,\,x \rangle \geq 0 \text{ for all } x \in \mathcal{K} \tag{318}$$

which is, merely, a statement of fact by definition of dual cone (296). By closure we have conjugation: [315, thm.14.1]

$$x \in \mathcal{K} \Leftrightarrow \langle y,\,x \rangle \geq 0 \text{ for all } y \in \mathcal{K}^* \tag{319}$$

which may be regarded as a simple translation of *Farkas' lemma* [137] as in [315, §22] to the language of convex cones, and a generalization of the well-known Cartesian cone fact

$$x \succeq 0 \Leftrightarrow \langle y,\,x \rangle \geq 0 \text{ for all } y \succeq 0 \tag{320}$$

for which implicitly $\mathcal{K} = \mathcal{K}^* = \mathbb{R}^n_+$ the nonnegative orthant.

Membership relation (319) is often written instead as *dual generalized inequalities*, when $\mathcal{K}$ and $\mathcal{K}^*$ are pointed closed convex cones,

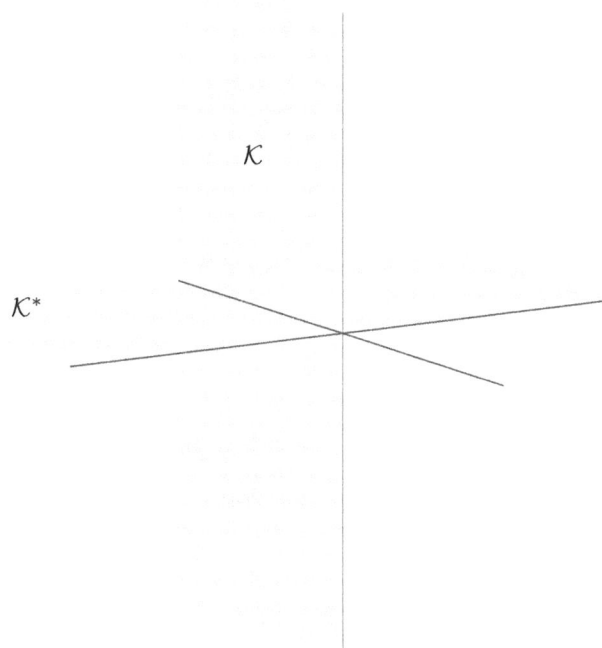

Figure 61: $\mathcal{K}$ and $\mathcal{K}^*$ are halfplanes in $\mathbb{R}^3$; *blades*. Both semiinfinite convex cones appear truncated. Each cone is like $\mathcal{K}$ from Figure 58 but embedded in a two-dimensional subspace of $\mathbb{R}^3$. Cartesian coordinate axes drawn for reference.

$$x \underset{\mathcal{K}}{\succeq} 0 \;\Leftrightarrow\; \langle y\,,x \rangle \geq 0 \;\text{ for all }\; y \underset{\mathcal{K}^*}{\succeq} 0 \qquad\qquad (321)$$

meaning, coordinates for biorthogonal expansion of x (§2.13.7.1.2, §2.13.8) [374] must be nonnegative when x belongs to $\mathcal{K}$. Conjugating,

$$y \underset{\mathcal{K}^*}{\succeq} 0 \;\Leftrightarrow\; \langle y\,,x \rangle \geq 0 \;\text{ for all }\; x \underset{\mathcal{K}}{\succeq} 0 \qquad\qquad (322)$$

$\diamond$

When pointed closed convex cone $\mathcal{K}$ is not polyhedral, coordinate axes for biorthogonal expansion asserted by the corollary are taken from extreme directions of $\mathcal{K}$; expansion is assured by *Carathéodory's theorem* (§E.6.4.1.1).

We presume, throughout, the obvious:

$$\begin{aligned} x \in \mathcal{K} \;&\Leftrightarrow\; \langle y\,,x \rangle \geq 0 \;\text{ for all }\; y \in \mathcal{K}^* \qquad (319) \\ &\Leftrightarrow \\ x \in \mathcal{K} \;&\Leftrightarrow\; \langle y\,,x \rangle \geq 0 \;\text{ for all }\; y \in \mathcal{K}^*,\; \|y\|=1 \end{aligned} \qquad (323)$$

2.13.2.0.2 Exercise. *Dual generalized inequalities.*
Test Corollary 2.13.2.0.1 and (323) graphically on the two-dimensional polyhedral cone and its dual in Figure 57.   ▼

(*confer* §2.7.2.2) When pointed closed convex cone $\mathcal{K}$ is implicit from context:

$$\begin{aligned} x \succeq 0 \;&\Leftrightarrow\; x \in \mathcal{K} \\ x \succ 0 \;&\Leftrightarrow\; x \in \operatorname{rel int} \mathcal{K} \end{aligned} \qquad (324)$$

Strict inequality $x \succ 0$ means coordinates for biorthogonal expansion of x must be positive when x belongs to $\operatorname{rel int}\mathcal{K}$. Strict membership relations are useful; *e.g.*, for any proper cone[2.62] $\mathcal{K}$ and its dual $\mathcal{K}^*$

$$x \in \operatorname{int}\mathcal{K} \;\Leftrightarrow\; \langle y\,,x \rangle > 0 \;\text{ for all }\; y \in \mathcal{K}^*,\; y \neq \mathbf{0} \qquad (325)$$

$$x \in \mathcal{K},\; x \neq \mathbf{0} \;\Leftrightarrow\; \langle y\,,x \rangle > 0 \;\text{ for all }\; y \in \operatorname{int}\mathcal{K}^* \qquad (326)$$

Conjugating, we get the dual relations:

$$y \in \operatorname{int}\mathcal{K}^* \;\Leftrightarrow\; \langle y\,,x \rangle > 0 \;\text{ for all }\; x \in \mathcal{K},\; x \neq \mathbf{0} \qquad (327)$$

$$y \in \mathcal{K}^*,\; y \neq \mathbf{0} \;\Leftrightarrow\; \langle y\,,x \rangle > 0 \;\text{ for all }\; x \in \operatorname{int}\mathcal{K} \qquad (328)$$

Boundary-membership relations for proper cones are also useful:

$$x \in \partial\mathcal{K} \;\Leftrightarrow\; \exists\, y \neq \mathbf{0} \ni \langle y\,,x \rangle = 0,\; y \in \mathcal{K}^*,\; x \in \mathcal{K} \qquad (329)$$

$$y \in \partial\mathcal{K}^* \;\Leftrightarrow\; \exists\, x \neq \mathbf{0} \ni \langle y\,,x \rangle = 0,\; x \in \mathcal{K},\; y \in \mathcal{K}^* \qquad (330)$$

which are consistent; *e.g.*, $x \in \partial\mathcal{K} \;\Leftrightarrow\; x \notin \operatorname{int}\mathcal{K}$ **and** $x \in \mathcal{K}$.

[2.62] An open cone $\mathcal{K}$ is admitted to (325) and (328) by (19).

2.13.2.0.3 Example. *Linear inequality.* [344, §4]

(*confer* §2.13.5.1.1) Consider a given matrix A and closed convex cone $\mathcal{K}$. By membership relation we have

$$
\begin{aligned}
Ay \in \mathcal{K}^* &\Leftrightarrow x^{\mathrm{T}} A\, y \geq 0 \quad \forall x \in \mathcal{K} \\
&\Leftrightarrow y^{\mathrm{T}} z \geq 0 \quad \forall z \in \{A^{\mathrm{T}} x \mid x \in \mathcal{K}\} \\
&\Leftrightarrow y \in \{A^{\mathrm{T}} x \mid x \in \mathcal{K}\}^*
\end{aligned}
\tag{331}
$$

This implies

$$
\{y \mid Ay \in \mathcal{K}^*\} = \{A^{\mathrm{T}} x \mid x \in \mathcal{K}\}^*
\tag{332}
$$

When $\mathcal{K}$ is the *selfdual* nonnegative orthant (§2.13.5.1), for example, then

$$
\{y \mid Ay \succeq 0\} = \{A^{\mathrm{T}} x \mid x \succeq 0\}^*
\tag{333}
$$

and the dual relation

$$
\{y \mid Ay \succeq 0\}^* = \{A^{\mathrm{T}} x \mid x \succeq 0\}
\tag{334}
$$

comes by a theorem of Weyl (p.129) that yields closedness for any vertex-description of a polyhedral cone. $\qquad\square$

2.13.2.1 Null certificate, Theorem of the alternative

If in particular $x_{\mathrm{p}} \notin \mathcal{K}$ a closed convex cone, then construction in Figure 56b suggests there exists a supporting hyperplane (having inward-normal belonging to dual cone $\mathcal{K}^*$) separating x_{p} from $\mathcal{K}$; indeed, (319)

$$
x_{\mathrm{p}} \notin \mathcal{K} \Leftrightarrow \exists\, y \in \mathcal{K}^* \ni \langle y, x_{\mathrm{p}} \rangle < 0
\tag{335}
$$

Existence of any one such y is a certificate of null membership. From a different perspective,

$$
\begin{aligned}
&x_{\mathrm{p}} \in \mathcal{K} \\
&\text{or in the alternative} \\
&\exists\, y \in \mathcal{K}^* \ni \langle y, x_{\mathrm{p}} \rangle < 0
\end{aligned}
\tag{336}
$$

By *alternative* is meant: these two systems are incompatible; one system is feasible while the other is not.

2.13.2.1.1 Example. *Theorem of the alternative for linear inequality.*

Myriad alternative systems of linear inequality can be explained in terms of pointed closed convex cones and their duals.

Beginning from the simplest Cartesian dual generalized inequalities (320) (with respect to nonnegative orthant $\mathbb{R}^m_+$),

$$
y \succeq 0 \Leftrightarrow x^{\mathrm{T}} y \geq 0 \ \text{ for all } x \succeq 0
\tag{337}
$$

Given $A \in \mathbb{R}^{n \times m}$, we make vector substitution $y \leftarrow A^{\mathrm{T}} y$

$$
A^{\mathrm{T}} y \succeq 0 \Leftrightarrow x^{\mathrm{T}} A^{\mathrm{T}} y \geq 0 \ \text{ for all } x \succeq 0
\tag{338}
$$

Introducing a new vector by calculating $b \triangleq Ax$ we get

$$A^{\mathrm{T}}y \succeq 0 \quad \Leftrightarrow \quad b^{\mathrm{T}}y \geq 0, \quad b = Ax \ \text{ for all } \ x \succeq 0 \tag{339}$$

By complementing sense of the scalar inequality:

$$A^{\mathrm{T}}y \succeq 0$$

$$\text{or in the alternative} \tag{340}$$

$$b^{\mathrm{T}}y < 0, \quad \exists \, b = Ax, \quad x \succeq 0$$

If one system has a solution, then the other does not; define a convex cone $\mathcal{K} = \{y \mid A^{\mathrm{T}}y \succeq 0\}$, then $y \in \mathcal{K}$ or in the alternative $y \notin \mathcal{K}$.

Scalar inequality $b^{\mathrm{T}}y < 0$ is movable to the other side of alternative (340), but that requires some explanation: From results in Example 2.13.2.0.3, the dual cone is $\mathcal{K}^* = \{Ax \mid x \succeq 0\}$. From (319) we have

$$\begin{aligned} y \in \mathcal{K} \quad &\Leftrightarrow \quad b^{\mathrm{T}}y \geq 0 \ \text{ for all } \ b \in \mathcal{K}^* \\ A^{\mathrm{T}}y \succeq 0 \quad &\Leftrightarrow \quad b^{\mathrm{T}}y \geq 0 \ \text{ for all } \ b \in \{Ax \mid x \succeq 0\} \end{aligned} \tag{341}$$

Given some b vector and $y \in \mathcal{K}$, then $b^{\mathrm{T}}y < 0$ can only mean $b \notin \mathcal{K}^*$. An alternative system is therefore simply $b \in \mathcal{K}^*$: [205, p.59] (Farkas/Tucker)

$$A^{\mathrm{T}}y \succeq 0, \quad b^{\mathrm{T}}y < 0$$

$$\text{or in the alternative} \tag{342}$$

$$b = Ax, \quad x \succeq 0$$

Geometrically this means: either vector b belongs to convex cone $\mathcal{K}^*$ or it does not. When $b \notin \mathcal{K}^*$, then there is a vector $y \in \mathcal{K}$ normal to a hyperplane separating point b from cone $\mathcal{K}^*$.

For another example, from membership relation (318) with affine transformation of dual variable we may write: Given $A \in \mathbb{R}^{n \times m}$ and $b \in \mathbb{R}^n$

$$b - Ay \in \mathcal{K}^* \quad \Leftrightarrow \quad x^{\mathrm{T}}(b - Ay) \geq 0 \qquad \forall \, x \in \mathcal{K} \tag{343}$$

$$A^{\mathrm{T}}x = \mathbf{0}, \quad b - Ay \in \mathcal{K}^* \quad \Rightarrow \quad x^{\mathrm{T}}b \geq 0 \qquad \qquad \forall \, x \in \mathcal{K} \tag{344}$$

From membership relation (343), conversely, suppose we allow any $y \in \mathbb{R}^m$. Then because $-x^{\mathrm{T}}Ay$ is unbounded below, $x^{\mathrm{T}}(b - Ay) \geq 0$ implies $A^{\mathrm{T}}x = \mathbf{0}$: for $y \in \mathbb{R}^m$

$$A^{\mathrm{T}}x = \mathbf{0}, \quad b - Ay \in \mathcal{K}^* \quad \Leftarrow \quad x^{\mathrm{T}}(b - Ay) \geq 0 \qquad \forall \, x \in \mathcal{K} \tag{345}$$

In toto,

$$b - Ay \in \mathcal{K}^* \quad \Leftrightarrow \quad x^{\mathrm{T}}b \geq 0, \ A^{\mathrm{T}}x = \mathbf{0} \ \ \forall \, x \in \mathcal{K} \tag{346}$$

Vector x belongs to cone $\mathcal{K}$ but is also constrained to lie in a subspace of $\mathbb{R}^n$ specified by an intersection of hyperplanes through the origin $\{x \in \mathbb{R}^n \mid A^{\mathrm{T}}x = \mathbf{0}\}$. From this, alternative systems of generalized inequality with respect to pointed closed convex cones $\mathcal{K}$ and $\mathcal{K}^*$

$$Ay \underset{\mathcal{K}^*}{\preceq} b$$

or in the alternative

$$x^{\mathrm{T}}b < 0, \quad A^{\mathrm{T}}x = \mathbf{0}, \quad x \underset{\mathcal{K}}{\succeq} 0 \tag{347}$$

derived from (346) simply by taking the complementary sense of the inequality in $x^{\mathrm{T}}b$. These two systems are alternatives; if one system has a solution, then the other does not.[2.63] [315, p.201]

By invoking a strict membership relation between proper cones (325), we can construct a more exotic alternative strengthened by demand for an interior point;

$$b - Ay \underset{\mathcal{K}^*}{\succ} 0 \quad \Leftrightarrow \quad x^{\mathrm{T}}b > 0, \quad A^{\mathrm{T}}x = \mathbf{0} \quad \forall x \underset{\mathcal{K}}{\succeq} 0, \quad x \neq \mathbf{0} \tag{348}$$

From this, alternative systems of generalized inequality [62, pages:50,54,262]

$$Ay \underset{\mathcal{K}^*}{\prec} b$$

or in the alternative

$$x^{\mathrm{T}}b \leq 0, \quad A^{\mathrm{T}}x = \mathbf{0}, \quad x \underset{\mathcal{K}}{\succeq} 0, \quad x \neq \mathbf{0} \tag{349}$$

derived from (348) by taking complementary sense of the inequality in $x^{\mathrm{T}}b$.

And from this, alternative systems with respect to the nonnegative orthant attributed to Gordan in 1873: [165] [56, §2.2] substituting $A \leftarrow A^{\mathrm{T}}$ and setting $b = \mathbf{0}$

$$A^{\mathrm{T}}y \prec 0$$

or in the alternative

$$Ax = \mathbf{0}, \quad x \succeq 0, \quad \|x\|_1 = 1 \tag{350}$$

Ben-Israel collects related results from Farkas, Motzkin, Gordan, and Stiemke in *Motzkin transposition theorem.* [33] □

[2.63]If solutions at $\pm\infty$ are disallowed, then the alternative systems become instead *mutually exclusive* with respect to nonpolyhedral cones. Simultaneous infeasibility of the two systems is not precluded by mutual exclusivity; called a *weak alternative*. Ye provides an example illustrating simultaneous infeasibility with respect to the positive semidefinite cone: $x \in \mathbb{S}^2$, $y \in \mathbb{R}$, $A = \begin{bmatrix} 1 & 0 \\ 0 & 0 \end{bmatrix}$, and $b = \begin{bmatrix} 0 & 1 \\ 1 & 0 \end{bmatrix}$ where $x^{\mathrm{T}}b$ means $\langle x, b \rangle$. A better strategy than disallowing solutions at $\pm\infty$ is to demand an interior point as in (349) or Lemma 4.2.1.1.2. Then question of simultaneous infeasibility is moot.

2.13.3 Optimality condition

(*confer* §2.13.10.1) The general first-order necessary and sufficient condition for optimality of solution $x^\star$ to a minimization problem ((301p) for example) with real differentiable convex objective function $f(x) : \mathbb{R}^n \to \mathbb{R}$ is [314, §3]

$$\nabla f(x^\star)^{\mathrm{T}}(x - x^\star) \geq 0 \quad \forall\, x \in \mathcal{C}, \;\; x^\star \in \mathcal{C} \tag{351}$$

where $\mathcal{C}$ is a convex *feasible set*,[2.64] and where $\nabla f(x^\star)$ is the *gradient* (§3.6) of f with respect to x evaluated at $x^\star$. In words, negative gradient is normal to a hyperplane supporting the feasible set at a point of optimality. (Figure 68)

Direct solution to *variation inequality* (351), instead of the corresponding minimization, spawned from *calculus of variations*. [256, p.178] [136, p.37] One solution method solves an equivalent fixed point-of-projection problem

$$x = P_{\mathcal{C}}(x - \nabla f(x)) \tag{352}$$

that follows from a necessary and sufficient condition for projection on convex set $\mathcal{C}$ (Theorem E.9.1.0.2)

$$P(x^\star - \nabla f(x^\star)) \in \mathcal{C}, \qquad \langle x^\star - \nabla f(x^\star) - x^\star, \; x - x^\star\rangle \leq 0 \quad \forall\, x \in \mathcal{C} \tag{2106}$$

Proof of equivalence [377, *Complementarity problem*] is provided by Németh. Given minimum-distance projection problem

$$\begin{array}{ll} \underset{x}{\text{minimize}} & \frac{1}{2}\|x - y\|^2 \\ \text{subject to} & x \in \mathcal{C} \end{array} \tag{353}$$

on convex feasible set $\mathcal{C}$ for example, the equivalent fixed point problem

$$x = P_{\mathcal{C}}(x - \nabla f(x)) = P_{\mathcal{C}}(y) \tag{354}$$

is solved in one step.

In the unconstrained case ($\mathcal{C} = \mathbb{R}^n$), optimality condition (351) reduces to the classical rule (p.214)

$$\nabla f(x^\star) = \mathbf{0}, \quad x^\star \in \operatorname{dom} f \tag{355}$$

which can be inferred from the following application:

2.13.3.0.1 Example. *Optimality for equality constrained problem.*
Given a real differentiable convex function $f(x) : \mathbb{R}^n \to \mathbb{R}$ defined on domain $\mathbb{R}^n$, a fat full-rank matrix $C \in \mathbb{R}^{p \times n}$, and vector $d \in \mathbb{R}^p$, the convex optimization problem

$$\begin{array}{ll} \underset{x}{\text{minimize}} & f(x) \\ \text{subject to} & Cx = d \end{array} \tag{356}$$

[2.64] presumably nonempty set of all variable values satisfying all given problem constraints; the set of *feasible solutions*.

is characterized by the well-known necessary and sufficient optimality condition [62, §4.2.3]

$$\nabla f(x^\star) + C^{\mathrm{T}}\nu = \mathbf{0} \tag{357}$$

where $\nu \in \mathbb{R}^p$ is the eminent *Lagrange multiplier.* [313] [256, p.188] [235] In other words, solution $x^\star$ is optimal if and only if $\nabla f(x^\star)$ belongs to $\mathcal{R}(C^{\mathrm{T}})$.

Via membership relation, we now derive condition (357) from the general first-order condition for optimality (351): For problem (356)

$$\mathcal{C} \triangleq \{x \in \mathbb{R}^n \mid Cx = d\} = \{Z\xi + x_{\mathrm{p}} \mid \xi \in \mathbb{R}^{n-\operatorname{rank}C}\} \tag{358}$$

is the feasible set where $Z \in \mathbb{R}^{n \times n - \operatorname{rank}C}$ holds basis $\mathcal{N}(C)$ columnar, and x_{p} is any particular solution to $Cx = d$. Since $x^\star \in \mathcal{C}$, we arbitrarily choose $x_{\mathrm{p}} = x^\star$ which yields an equivalent optimality condition

$$\nabla f(x^\star)^{\mathrm{T}} Z\xi \geq 0 \quad \forall \xi \in \mathbb{R}^{n-\operatorname{rank}C} \tag{359}$$

when substituted into (351). But this is simply half of a membership relation where the cone dual to $\mathbb{R}^{n-\operatorname{rank}C}$ is the origin in $\mathbb{R}^{n-\operatorname{rank}C}$. We must therefore have

$$Z^{\mathrm{T}}\nabla f(x^\star) = \mathbf{0} \;\Leftrightarrow\; \nabla f(x^\star)^{\mathrm{T}} Z\xi \geq 0 \quad \forall \xi \in \mathbb{R}^{n-\operatorname{rank}C} \tag{360}$$

meaning, $\nabla f(x^\star)$ must be orthogonal to $\mathcal{N}(C)$. These conditions

$$Z^{\mathrm{T}}\nabla f(x^\star) = \mathbf{0}, \qquad Cx^\star = d \tag{361}$$

are necessary and sufficient for optimality. $\qquad\qquad\qquad\qquad\qquad\qquad \square$

2.13.4 Discretization of membership relation

2.13.4.1 Dual halfspace-description

Halfspace-description of dual cone $\mathcal{K}^*$ is equally simple as vertex-description

$$\mathcal{K} = \operatorname{cone}(X) = \{Xa \mid a \succeq 0\} \subseteq \mathbb{R}^n \qquad (103)$$

for corresponding closed convex cone $\mathcal{K}$: By definition (296), for $X \in \mathbb{R}^{n \times N}$ as in (279), (*confer* (286))

$$\begin{aligned}
\mathcal{K}^* &= \{y \in \mathbb{R}^n \mid z^{\mathrm{T}}y \geq 0 \text{ for all } z \in \mathcal{K}\} \\
&= \{y \in \mathbb{R}^n \mid z^{\mathrm{T}}y \geq 0 \text{ for all } z = Xa, \; a \succeq 0\} \\
&= \{y \in \mathbb{R}^n \mid a^{\mathrm{T}}X^{\mathrm{T}}y \geq 0, \; a \succeq 0\} \\
&= \{y \in \mathbb{R}^n \mid X^{\mathrm{T}}y \succeq 0\}
\end{aligned} \tag{362}$$

that follows from the *generalized inequality and membership corollary* (320). The semi-infinity of tests specified by all $z \in \mathcal{K}$ has been reduced to a set of generators for $\mathcal{K}$ constituting the columns of X; *id est*, the test has been discretized.

Whenever cone $\mathcal{K}$ is known to be closed and convex, the conjugate statement must also hold; *id est*, given any set of generators for dual cone $\mathcal{K}^*$ arranged columnar in Y, then the consequent vertex-description of dual cone connotes a halfspace-description for $\mathcal{K}$: [337, §2.8]

$$\mathcal{K}^* = \{Ya \mid a \succeq 0\} \;\Leftrightarrow\; \mathcal{K}^{**} = \mathcal{K} = \{z \mid Y^{\mathrm{T}}z \succeq 0\} \tag{363}$$

2.13.4.2   First dual-cone formula

From these two results (362) and (363) we deduce a general principle:

- From any [*sic*] given vertex-description (103) of closed convex cone $\mathcal{K}$, a halfspace-description of the dual cone $\mathcal{K}^*$ is immediate by matrix transposition (362); conversely, from any given halfspace-description (286) of $\mathcal{K}$, a dual vertex-description is immediate (363). [315, p.122]

Various other converses are just a little trickier. (§2.13.9, §2.13.11)

We deduce further: For any polyhedral cone $\mathcal{K}$, the dual cone $\mathcal{K}^*$ is also polyhedral and $\mathcal{K}^{**} = \mathcal{K}$. [337, p.56]

The *generalized inequality and membership corollary* is discretized in the following theorem inspired by (362) and (363):

2.13.4.2.1   Theorem.   *Discretized membership.*   (*confer* §2.13.2.0.1)[2.65]
Given any set of generators, (§2.8.1.2) denoted by $\mathcal{G}(\mathcal{K})$ for closed convex cone $\mathcal{K} \subseteq \mathbb{R}^n$, and any set of generators denoted $\mathcal{G}(\mathcal{K}^*)$ for its dual such that

$$\mathcal{K} = \text{cone}\,\mathcal{G}(\mathcal{K})\ ,\qquad \mathcal{K}^* = \text{cone}\,\mathcal{G}(\mathcal{K}^*) \tag{364}$$

then discretization of the *generalized inequality and membership corollary* (p.144) is necessary and sufficient for certifying cone membership: for x and y in vector space $\mathbb{R}^n$

$$x \in \mathcal{K} \ \Leftrightarrow\ \langle \gamma^*, x \rangle \geq 0 \ \text{ for all } \ \gamma^* \in \mathcal{G}(\mathcal{K}^*) \tag{365}$$

$$y \in \mathcal{K}^* \ \Leftrightarrow\ \langle \gamma, y \rangle \geq 0 \ \text{ for all } \ \gamma \in \mathcal{G}(\mathcal{K}) \tag{366}$$

$$\diamond$$

Proof. $\mathcal{K}^* = \{\mathcal{G}(\mathcal{K}^*)a \mid a \succeq 0\}$. $y \in \mathcal{K}^* \Leftrightarrow y = \mathcal{G}(\mathcal{K}^*)a$ for   some $a \succeq 0$.
$x \in \mathcal{K} \ \Leftrightarrow\ \langle y, x \rangle \geq 0 \ \forall y \in \mathcal{K}^* \Leftrightarrow \langle \mathcal{G}(\mathcal{K}^*)a, x \rangle \geq 0 \ \forall a \succeq 0$ (319). $a \triangleq \sum_i \alpha_i e_i$ where e_i is the
i^{th} member of a standard basis of possibly infinite cardinality. $\langle \mathcal{G}(\mathcal{K}^*)a, x \rangle \geq 0 \ \forall a \succeq 0$
$\Leftrightarrow \sum_i \alpha_i \langle \mathcal{G}(\mathcal{K}^*)e_i, x \rangle \geq 0 \ \forall \alpha_i \geq 0 \ \Leftrightarrow \ \langle \mathcal{G}(\mathcal{K}^*)e_i, x \rangle \geq 0 \ \forall i$.   Conjugate relation (366) is
similarly derived.   ♦

2.13.4.2.2   Exercise.   *Discretized dual generalized inequalities.*
Test Theorem 2.13.4.2.1 on Figure 57a using extreme directions there as generators.
                                                                                                                    ▼

From the *discretized membership theorem* we may further deduce a more surgical description of closed convex cone that prescribes only a finite number of halfspaces for its construction when polyhedral: (Figure **56**a)

$$\mathcal{K} = \{ x \in \mathbb{R}^n \mid \langle \gamma^*, x \rangle \geq 0 \text{ for all } \gamma^* \in \mathcal{G}(\mathcal{K}^*) \} \tag{367}$$

$$\mathcal{K}^* = \{ y \in \mathbb{R}^n \mid \langle \gamma, y \rangle \geq 0 \text{ for all } \gamma \in \mathcal{G}(\mathcal{K}) \} \tag{368}$$

[2.65]Stated in [21, §1] without proof for pointed closed convex case.

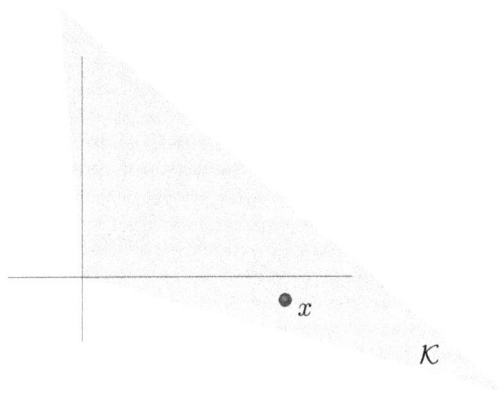

Figure 62: $x \succeq 0$ with respect to $\mathcal{K}$ but not with respect to nonnegative orthant $\mathbb{R}^2_+$ (pointed convex cone $\mathcal{K}$ drawn truncated).

2.13.4.2.3 Exercise. *Partial order induced by orthant.*
When comparison is with respect to the nonnegative orthant $\mathcal{K} = \mathbb{R}^n_+$, then from the *discretized membership theorem* it directly follows:

$$x \preceq z \ \Leftrightarrow \ x_i \leq z_i \ \forall i \tag{369}$$

Generate simple counterexamples demonstrating that this equivalence with entrywise inequality holds only when the underlying cone inducing partial order is the nonnegative orthant; *e.g.*, explain Figure 62.   ▼

2.13.4.2.4 Example. *Boundary membership to proper polyhedral cone.*
For a polyhedral cone, test (329) of boundary membership can be formulated as a linear program. Say proper polyhedral cone $\mathcal{K}$ is specified completely by generators that are arranged columnar in

$$X = [\Gamma_1 \ \cdots \ \Gamma_N] \in \mathbb{R}^{n \times N} \tag{279}$$

id est, $\mathcal{K} = \{Xa \mid a \succeq 0\}$. Then membership relation

$$c \in \partial\mathcal{K} \ \Leftrightarrow \ \exists \, y \neq \mathbf{0} \ni \langle y, c \rangle = 0, \ y \in \mathcal{K}^*, \ c \in \mathcal{K} \tag{329}$$

may be expressed[2.66]

$$\begin{array}{ll} \underset{a,\,y}{\text{find}} & y \neq \mathbf{0} \\ \text{subject to} & c^{\mathrm{T}} y = 0 \\ & X^{\mathrm{T}} y \succeq 0 \\ & Xa = c \\ & a \succeq 0 \end{array} \tag{370}$$

This linear feasibility problem has a solution iff $c \in \partial\mathcal{K}$.   □

[2.66]Nonzero $y \in \mathbb{R}^n$ may be realized in many ways; *e.g.*, via constraint $\mathbf{1}^{\mathrm{T}} y = 1$ if $c \neq \mathbf{1}$.

2.13.4.3   smallest face of pointed closed convex cone

Given nonempty convex subset $\mathcal{C}$ of a convex set $\mathcal{K}$, the *smallest face* of $\mathcal{K}$ containing $\mathcal{C}$ is equivalent to intersection of all faces of $\mathcal{K}$ that contain $\mathcal{C}$. [315, p.164] By (308), membership relation (329) means that each and every point on boundary $\partial\mathcal{K}$ of proper cone $\mathcal{K}$ belongs to a hyperplane supporting $\mathcal{K}$ whose normal y belongs to dual cone $\mathcal{K}^*$. It follows that the smallest face $\mathcal{F}$, containing $\mathcal{C} \subset \partial\mathcal{K} \subset \mathbb{R}^n$ on boundary of proper cone $\mathcal{K}$, is the intersection of all hyperplanes containing $\mathcal{C}$ whose normals are in $\mathcal{K}^*$;

$$\mathcal{F}(\mathcal{K} \supset \mathcal{C}) = \{x \in \mathcal{K} \mid x \perp \mathcal{K}^* \cap \mathcal{C}^{\perp}\} \tag{371}$$

where

$$\mathcal{C}^{\perp} \triangleq \{y \in \mathbb{R}^n \mid \langle z, y \rangle = 0 \ \ \forall z \in \mathcal{C}\} \tag{372}$$

When $\mathcal{C} \cap \operatorname{int}\mathcal{K} \neq \emptyset$ then $\mathcal{F}(\mathcal{K} \supset \mathcal{C}) = \mathcal{K}$.

2.13.4.3.1   Example.   *Finding smallest face of cone.*
Suppose polyhedral cone $\mathcal{K}$ is completely specified by generators arranged columnar in

$$X = [\Gamma_1 \ \cdots \ \Gamma_N] \in \mathbb{R}^{n \times N} \tag{279}$$

To find its smallest face $\mathcal{F}(\mathcal{K} \ni c)$ containing a given point $c \in \mathcal{K}$, by the *discretized membership theorem* 2.13.4.2.1, it is necessary and sufficient to find generators for the smallest face. We may do so one generator at a time:[2.67] Consider generator Γ_i. If there exists a vector $z \in \mathcal{K}^*$ orthogonal to c but not to Γ_i, then Γ_i cannot belong to the smallest face of $\mathcal{K}$ containing c. Such a vector z can be realized by a linear feasibility problem:

$$\begin{array}{ll} \text{find} & z \in \mathbb{R}^n \\ \text{subject to} & c^{\mathrm{T}}z = 0 \\ & X^{\mathrm{T}}z \succeq 0 \\ & \Gamma_i^{\mathrm{T}}z = 1 \end{array} \tag{373}$$

If there exists a solution z for which $\Gamma_i^{\mathrm{T}}z = 1$, then

$$\Gamma_i \not\perp \mathcal{K}^* \cap c^{\perp} = \{z \in \mathbb{R}^n \mid X^{\mathrm{T}}z \succeq 0, \ c^{\mathrm{T}}z = 0\} \tag{374}$$

so $\Gamma_i \notin \mathcal{F}(\mathcal{K} \ni c)$; solution z is a certificate of null membership. If this problem is infeasible for generator $\Gamma_i \in \mathcal{K}$, conversely, then $\Gamma_i \in \mathcal{F}(\mathcal{K} \ni c)$ by (371) and (362) because $\Gamma_i \perp \mathcal{K}^* \cap c^{\perp}$; in that case, Γ_i is a generator of $\mathcal{F}(\mathcal{K} \ni c)$.

Since the constant in constraint $\Gamma_i^{\mathrm{T}}z = 1$ is arbitrary positively, then by *theorem of the alternative* there is correspondence between (373) and (347) admitting the alternative linear problem: for a given point $c \in \mathcal{K}$

$$\begin{array}{ll} \underset{a \in \mathbb{R}^N, \ \mu \in \mathbb{R}}{\text{find}} & a, \mu \\ \text{subject to} & \mu c - \Gamma_i = Xa \\ & a \succeq 0 \end{array} \tag{375}$$

Now if this problem is feasible (bounded) for generator $\Gamma_i \in \mathcal{K}$, then (373) is infeasible and $\Gamma_i \in \mathcal{F}(\mathcal{K} \ni c)$ is a generator of the smallest face that contains c.   □

[2.67] When finding a smallest face, generators of $\mathcal{K}$ in matrix X may not be diminished in number (by discarding columns) until all generators of the smallest face have been found.

2.13.4.3.2 Exercise. *Finding smallest face of pointed closed convex cone.*
Show that formula (371) and algorithms (373) and (375) apply more broadly; *id est*, a
full-dimensional cone $\mathcal{K}$ is an unnecessary condition.[2.68]   ▼

2.13.4.3.3 Exercise. *Smallest face of positive semidefinite cone.*
Derive (221) from (371).   ▼

2.13.5 Dual PSD cone and generalized inequality

The *dual positive semidefinite cone* $\mathcal{K}^*$ is confined to $\mathbb{S}^M$ by convention;

$$\mathbb{S}_+^{M\,*} \triangleq \{Y \in \mathbb{S}^M \mid \langle Y\,,\, X\rangle \geq 0 \text{ for all } X \in \mathbb{S}_+^M\} = \mathbb{S}_+^M \tag{376}$$

The positive semidefinite cone is selfdual in the ambient space of symmetric matrices
[62, exmp.2.24] [39] [201, §II]; $\mathcal{K} = \mathcal{K}^*$.

Dual generalized inequalities with respect to the positive semidefinite cone in the
ambient space of symmetric matrices can therefore be simply stated: (Fejér)

$$X \succeq 0 \quad \Leftrightarrow \quad \operatorname{tr}(Y^{\mathrm{T}}X) \geq 0 \text{ for all } Y \succeq 0 \tag{377}$$

Membership to this cone can be determined in the isometrically isomorphic Euclidean
space $\mathbb{R}^{M^2}$ via (38). (§2.2.1) By the two interpretations in §2.13.1, positive semidefinite
matrix Y can be interpreted as inward-normal to a hyperplane supporting the positive
semidefinite cone.

The fundamental statement of positive semidefiniteness, $y^{\mathrm{T}}Xy \geq 0 \; \forall y$ (§A.3.0.0.1),
evokes a particular instance of these dual generalized inequalities (377):

$$X \succeq 0 \quad \Leftrightarrow \quad \langle yy^{\mathrm{T}},\, X\rangle \geq 0 \;\; \forall yy^{\mathrm{T}}(\succeq 0) \qquad (1523)$$

Discretization (§2.13.4.2.1) allows replacement of positive semidefinite matrices Y with
this minimal set of generators comprising the extreme directions of the positive semidefinite
cone (§2.9.2.7).

2.13.5.1 selfdual cones

From (131) (a consequence of the *halfspaces theorem*, §2.4.1.1.1), where the only finite
value of the support function for a convex cone is 0 [205, §C.2.3.1], or from discretized
definition (368) of the dual cone we get a rather self-evident characterization of selfdual
cones:

$$\mathcal{K} = \mathcal{K}^* \quad \Leftrightarrow \quad \mathcal{K} = \bigcap_{\gamma \in \mathcal{G}(\mathcal{K})} \{y \mid \gamma^{\mathrm{T}}y \geq 0\} \tag{378}$$

In words: Cone $\mathcal{K}$ is *selfdual* iff its own extreme directions are inward-normals to a
(minimal) set of hyperplanes bounding halfspaces whose intersection constructs it. This
means each extreme direction of $\mathcal{K}$ is normal to a hyperplane exposing one of its own
faces; a necessary but insufficient condition for selfdualness (Figure 63, for example).

[2.68]Hint: A hyperplane, with normal in $\mathcal{K}^*$, containing cone $\mathcal{K}$ is admissible.

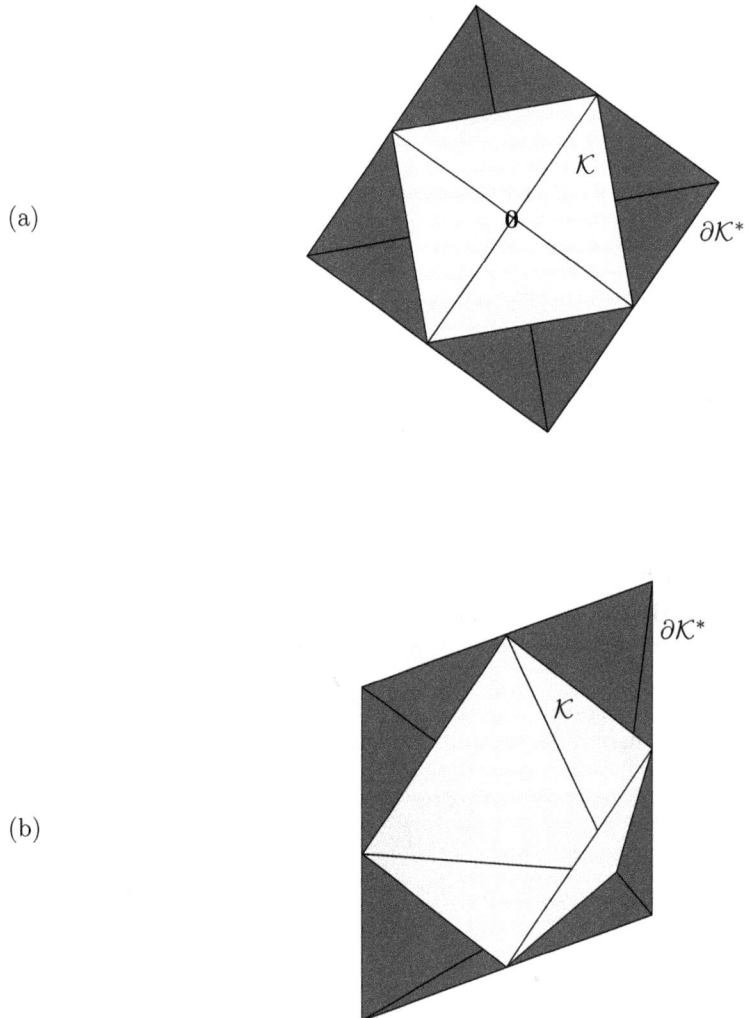

Figure 63: Two (truncated) views of a polyhedral cone $\mathcal{K}$ and its dual in $\mathbb{R}^3$. Each of four extreme directions from $\mathcal{K}$ belongs to a face of dual cone $\mathcal{K}^*$. Cone $\mathcal{K}$, shrouded by its dual, is symmetrical about its axis of revolution. Each pair of diametrically opposed extreme directions from $\mathcal{K}$ makes a right angle. An orthant (or any rotation thereof; a simplicial cone) is not the only selfdual polyhedral cone in three or more dimensions; [20, §2.A.21] e.g., consider an *equilateral* having five extreme directions. In fact, every selfdual polyhedral cone in $\mathbb{R}^3$ has an odd number of extreme directions. [22, thm.3]

Selfdual cones are necessarily full-dimensional. [30, §I] Their most prominent representatives are the orthants (Cartesian cones), the positive semidefinite cone $\mathbb{S}_+^M$ in the ambient space of symmetric matrices (376), and Lorentz cone (178) [20, §II.A] [62, exmp.2.25]. In three dimensions, a plane containing the axis of revolution of a selfdual cone (and the origin) will produce a *slice* whose boundary makes a right angle.

2.13.5.1.1 Example. *Linear matrix inequality.* (*confer* §2.13.2.0.3)
Consider a peculiar vertex-description for a convex cone $\mathcal{K}$ defined over a positive semidefinite cone (instead of a nonnegative orthant as in definition (103)): for $X \in \mathbb{S}_+^n$ given $A_j \in \mathbb{S}^n$, $j = 1 \ldots m$

$$
\begin{aligned}
\mathcal{K} &= \left\{ \begin{bmatrix} \langle A_1, X \rangle \\ \vdots \\ \langle A_m, X \rangle \end{bmatrix} \mid X \succeq 0 \right\} \subseteq \mathbb{R}^m \\
&= \left\{ \begin{bmatrix} \operatorname{svec}(A_1)^{\mathrm{T}} \\ \vdots \\ \operatorname{svec}(A_m)^{\mathrm{T}} \end{bmatrix} \operatorname{svec} X \mid X \succeq 0 \right\} \\
&\triangleq \{ A \operatorname{svec} X \mid X \succeq 0 \}
\end{aligned}
\tag{379}
$$

where $A \in \mathbb{R}^{m \times n(n+1)/2}$, and where symmetric vectorization svec is defined in (56). Cone $\mathcal{K}$ is indeed convex because, by (175)

$$
A \operatorname{svec} X_{\mathrm{p}_1}, \ A \operatorname{svec} X_{\mathrm{p}_2} \in \mathcal{K} \ \Rightarrow \ A(\zeta \operatorname{svec} X_{\mathrm{p}_1} + \xi \operatorname{svec} X_{\mathrm{p}_2}) \in \mathcal{K} \ \text{for all} \ \zeta, \xi \geq 0 \tag{380}
$$

since a nonnegatively weighted sum of positive semidefinite matrices must be positive semidefinite. (§A.3.1.0.2) Although matrix A is finite-dimensional, $\mathcal{K}$ is generally not a polyhedral cone (unless $m = 1$ or 2) simply because $X \in \mathbb{S}_+^n$.

Theorem. *Inverse image closedness.* [205, prop.A.2.1.12]
[315, thm.6.7] Given affine operator $g : \mathbb{R}^m \to \mathbb{R}^p$, convex set $\mathcal{D} \subseteq \mathbb{R}^m$, and convex set $\mathcal{C} \subseteq \mathbb{R}^p \ni g^{-1}(\operatorname{rel int} \mathcal{C}) \neq \emptyset$, then

$$
\operatorname{rel int} g(\mathcal{D}) = g(\operatorname{rel int} \mathcal{D}), \quad \operatorname{rel int} g^{-1}\mathcal{C} = g^{-1}(\operatorname{rel int} \mathcal{C}), \quad \overline{g^{-1}\mathcal{C}} = g^{-1}\overline{\mathcal{C}} \tag{381}
$$

$\diamond$

By this theorem, relative interior of $\mathcal{K}$ may always be expressed

$$
\operatorname{rel int} \mathcal{K} = \{ A \operatorname{svec} X \mid X \succ 0 \} \tag{382}
$$

Because $\dim(\operatorname{aff} \mathcal{K}) = \dim(A \operatorname{svec} \mathbb{S}^n)$ (127) then, provided the vectorized A_j matrices are linearly independent,

$$
\operatorname{rel int} \mathcal{K} = \operatorname{int} \mathcal{K} \tag{14}
$$

meaning, cone $\mathcal{K}$ is full-dimensional $\Rightarrow$ dual cone $\mathcal{K}^*$ is pointed by (309). Convex cone $\mathcal{K}$ can be closed, by this corollary:

Corollary. *Cone closedness invariance.*   [57, §3] [52, §3]
Given linear operator $A : \mathbb{R}^p \to \mathbb{R}^m$ and closed convex cone $\mathcal{X} \subseteq \mathbb{R}^p$, convex cone $\mathcal{K} = A(\mathcal{X})$ is closed $\left(\overline{A(\mathcal{X})} = A(\mathcal{X})\right)$ if and only if

$$\mathcal{N}(A) \cap \mathcal{X} = \{\mathbf{0}\} \quad \text{or} \quad \mathcal{N}(A) \cap \mathcal{X} \nsubseteq \operatorname{rel} \partial \mathcal{X} \tag{383}$$

Otherwise, $\overline{\mathcal{K}} = \overline{A(\mathcal{X})} \supseteq A(\overline{\mathcal{X}}) \supseteq A(\mathcal{X})$. [315, thm.6.6]   ◇

If matrix A has no nontrivial nullspace, then $A \operatorname{svec} X$ is an isomorphism in X between cone $\mathbb{S}_+^n$ and range $\mathcal{R}(A)$ of matrix A; (§2.2.1.0.1, §2.10.1.1) sufficient for convex cone $\mathcal{K}$ to be closed and have relative boundary

$$\operatorname{rel} \partial \mathcal{K} = \{A \operatorname{svec} X \mid X \succeq 0, \ X \neq 0\} \tag{384}$$

Now consider the (closed convex) dual cone:

$$\begin{aligned}
\mathcal{K}^* &= \{y \mid \langle z, y \rangle \geq 0 \text{ for all } z \in \mathcal{K}\} \subseteq \mathbb{R}^m \\
&= \{y \mid \langle z, y \rangle \geq 0 \text{ for all } z = A \operatorname{svec} X, \ X \succeq 0\} \\
&= \{y \mid \langle A \operatorname{svec} X, y \rangle \geq 0 \text{ for all } X \succeq 0\} \\
&= \{y \mid \langle \operatorname{svec} X, A^{\mathrm{T}} y \rangle \geq 0 \text{ for all } X \succeq 0\} \\
&= \{y \mid \operatorname{svec}^{-1}(A^{\mathrm{T}} y) \succeq 0\}
\end{aligned} \tag{385}$$

that follows from (377) and leads to an equally peculiar halfspace-description

$$\mathcal{K}^* = \{y \in \mathbb{R}^m \mid \sum_{j=1}^m y_j A_j \succeq 0\} \tag{386}$$

The summation inequality with respect to positive semidefinite cone $\mathbb{S}_+^n$ is known as *linear matrix inequality*. [60] [155] [268] [371] Although we already know that the dual cone is convex (§2.13.1), *inverse image theorem* 2.1.9.0.1 certifies convexity of $\mathcal{K}^*$ which is the inverse image of positive semidefinite cone $\mathbb{S}_+^n$ under linear transformation $g(y) \triangleq \sum y_j A_j$. And although we already know that the dual cone is closed, it is certified by (381). By the *inverse image closedness theorem,* dual cone relative interior may always be expressed

$$\operatorname{rel int} \mathcal{K}^* = \{y \in \mathbb{R}^m \mid \sum_{j=1}^m y_j A_j \succ 0\} \tag{387}$$

Function $g(y)$ on $\mathbb{R}^m$ is an isomorphism when the vectorized A_j matrices are linearly independent; hence, uniquely invertible. Inverse image $\mathcal{K}^*$ must therefore have dimension equal to $\dim(\mathcal{R}(A^{\mathrm{T}}) \cap \operatorname{svec} \mathbb{S}_+^n)$ (49) and relative boundary

$$\operatorname{rel} \partial \mathcal{K}^* = \{y \in \mathbb{R}^m \mid \sum_{j=1}^m y_j A_j \succeq 0, \ \sum_{j=1}^m y_j A_j \neq 0\} \tag{388}$$

When this dimension equals m, then dual cone $\mathcal{K}^*$ is full-dimensional

$$\operatorname{rel int} \mathcal{K}^* = \operatorname{int} \mathcal{K}^* \tag{14}$$

which implies: closure of convex cone $\mathcal{K}$ is pointed (309).   □

2.13.6 Dual of pointed polyhedral cone

In a subspace of $\mathbb{R}^n$, now we consider a pointed polyhedral cone $\mathcal{K}$ given in terms of its extreme directions Γ_i arranged columnar in

$$X = [\,\Gamma_1 \ \ \Gamma_2 \ \cdots \ \Gamma_N\,] \in \mathbb{R}^{n \times N} \qquad (279)$$

The *extremes theorem* (§2.8.1.1.1) provides the vertex-description of a pointed polyhedral cone in terms of its finite number of extreme directions and its lone vertex at the origin:

2.13.6.0.1 Definition. *Pointed polyhedral cone, vertex-description.*
Given pointed polyhedral cone $\mathcal{K}$ in a subspace of $\mathbb{R}^n$, denoting its ith extreme direction by $\Gamma_i \in \mathbb{R}^n$ arranged in a matrix X as in (279), then that cone may be described: (86) (*confer* (188) (292))

$$
\begin{aligned}
\mathcal{K} &= \big\{ [\,\mathbf{0} \ \ X\,]\, a\,\zeta \ \mid \ a^{\mathrm{T}} \mathbf{1} = 1, \ a \succeq 0, \ \zeta \ge 0 \big\} \\
&= \quad \big\{ X a \,\zeta \ \mid \ a^{\mathrm{T}} \mathbf{1} \le 1, \ a \succeq 0, \ \zeta \ge 0 \big\} \\
&= \quad \big\{ X b \quad \mid \ b \succeq 0 \big\} \subseteq \mathbb{R}^n
\end{aligned} \qquad (389)
$$

that is simply a conic hull (like (103)) of a finite number N of directions. Relative interior may always be expressed

$$\operatorname{rel\,int} \mathcal{K} = \{ X b \mid b \succ 0 \} \subset \mathbb{R}^n \qquad (390)$$

but identifying the cone's relative boundary in this manner

$$\operatorname{rel} \partial \mathcal{K} = \{ X b \mid b \succeq 0, \ b \neq 0 \} \qquad (391)$$

holds only when matrix X represents a bijection onto its range; in other words, some coefficients meeting lower bound zero ($b \in \partial \mathbb{R}_+^N$) do not necessarily provide membership to relative boundary of cone $\mathcal{K}$. $\triangle$

Whenever cone $\mathcal{K}$ is pointed, closed, and convex (not only polyhedral), then dual cone $\mathcal{K}^*$ has a halfspace-description in terms of the extreme directions Γ_i of $\mathcal{K}$:

$$\mathcal{K}^* = \big\{ y \mid \gamma^{\mathrm{T}} y \ge 0 \ \text{ for all } \ \gamma \in \{\Gamma_i \,, \ i = 1 \ldots N\} \subseteq \operatorname{rel} \partial \mathcal{K} \big\} \qquad (392)$$

because when $\{\Gamma_i\}$ constitutes any set of generators for $\mathcal{K}$, the discretization result in §2.13.4.1 allows relaxation of the requirement $\forall x \in \mathcal{K}$ in (296) to $\forall \gamma \in \{\Gamma_i\}$ directly.[2.69] That dual cone so defined is unique, identical to (296), polyhedral whenever the number of generators N is finite

$$\mathcal{K}^* = \big\{ y \mid X^{\mathrm{T}} y \succeq 0 \big\} \subseteq \mathbb{R}^n \qquad (362)$$

and is full-dimensional because $\mathcal{K}$ is assumed pointed. But $\mathcal{K}^*$ is not necessarily pointed unless $\mathcal{K}$ is full-dimensional (§2.13.1.1).

[2.69]The extreme directions of $\mathcal{K}$ constitute a minimal set of generators. Formulae and conversions to vertex-description of the dual cone are in §2.13.9 and §2.13.11.

2.13.6.1   Facet normal & extreme direction

We see from (362) that the conically independent generators of cone $\mathcal{K}$ (namely, the extreme directions of pointed closed convex cone $\mathcal{K}$ constituting the N columns of X) each define an inward-normal to a hyperplane supporting dual cone $\mathcal{K}^*$ (§2.4.2.6.1) and exposing a dual facet when N is finite. Were $\mathcal{K}^*$ pointed and finitely generated, then by closure the conjugate statement would also hold; *id est*, the extreme directions of pointed $\mathcal{K}^*$ each define an inward-normal to a hyperplane supporting $\mathcal{K}$ and exposing a facet when N is finite. Examine Figure **57** or Figure **63**, for example.

We may conclude, the extreme directions of proper polyhedral $\mathcal{K}$ are respectively orthogonal to the facets of $\mathcal{K}^*$; likewise, the extreme directions of proper polyhedral $\mathcal{K}^*$ are respectively orthogonal to the facets of $\mathcal{K}$.

2.13.7   Biorthogonal expansion by example

2.13.7.0.1   Example.   *Relationship to dual polyhedral cone.*
Simplicial cone $\mathcal{K}$ illustrated in Figure **64** induces a partial order on $\mathbb{R}^2$. All points greater than x with respect to $\mathcal{K}$, for example, are contained in the translated cone $x + \mathcal{K}$. The extreme directions Γ_1 and Γ_2 of $\mathcal{K}$ do not make an orthogonal set; neither do extreme directions Γ_3 and Γ_4 of dual cone $\mathcal{K}^*$; rather, we have the *biorthogonality condition* [374]

$$\begin{aligned} \Gamma_4^T\Gamma_1 = \Gamma_3^T\Gamma_2 = 0 \\ \Gamma_3^T\Gamma_1 \neq 0, \quad \Gamma_4^T\Gamma_2 \neq 0 \end{aligned} \tag{393}$$

Biorthogonal expansion of $x \in \mathcal{K}$ is then

$$x = \Gamma_1 \frac{\Gamma_3^T x}{\Gamma_3^T\Gamma_1} + \Gamma_2 \frac{\Gamma_4^T x}{\Gamma_4^T\Gamma_2} \tag{394}$$

where $\Gamma_3^T x/(\Gamma_3^T\Gamma_1)$ is the nonnegative coefficient of nonorthogonal projection (§E.6.1) of x on Γ_1 in the direction orthogonal to Γ_3 (y in Figure **174** p.619), and where $\Gamma_4^T x/(\Gamma_4^T\Gamma_2)$ is the nonnegative coefficient of nonorthogonal projection of x on Γ_2 in the direction orthogonal to Γ_4 (z in Figure **174**); they are coordinates in this nonorthogonal system. Those coefficients must be nonnegative $x \succeq_{\mathcal{K}} 0$ because $x \in \mathcal{K}$ (324) and $\mathcal{K}$ is simplicial.

If we ascribe the extreme directions of $\mathcal{K}$ to the columns of a matrix

$$X \triangleq [\,\Gamma_1 \;\; \Gamma_2\,] \tag{395}$$

then we find that the pseudoinverse transpose matrix

$$X^{\dagger T} = \left[\, \Gamma_3 \frac{1}{\Gamma_3^T\Gamma_1} \quad \Gamma_4 \frac{1}{\Gamma_4^T\Gamma_2} \,\right] \tag{396}$$

holds the extreme directions of the dual cone. Therefore

$$x = XX^\dagger x \tag{402}$$

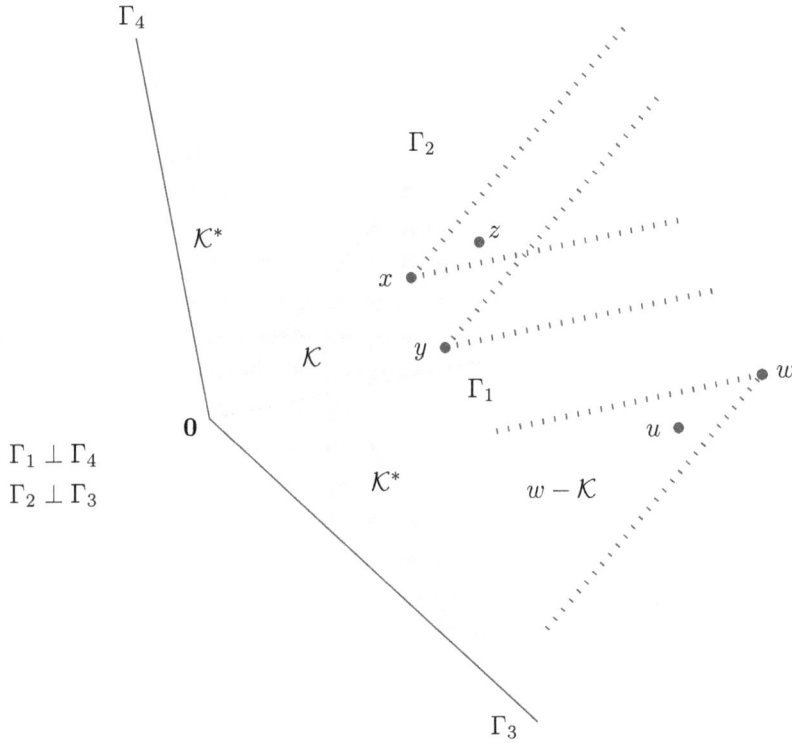

Figure 64: (*confer* Figure 174)   Simplicial cone $\mathcal{K}$ in $\mathbb{R}^2$ and its dual $\mathcal{K}^*$ drawn truncated. Conically independent generators Γ_1 and Γ_2 constitute extreme directions of $\mathcal{K}$ while Γ_3 and Γ_4 constitute extreme directions of $\mathcal{K}^*$. Dotted ray-pairs bound translated cones $\mathcal{K}$. Point x is comparable to point z (and *vice versa*) but not to y; $z \succeq_{\mathcal{K}} x \Leftrightarrow z - x \in \mathcal{K} \Leftrightarrow z - x \succeq_{\mathcal{K}} 0$ iff $\exists$ nonnegative coordinates for biorthogonal expansion of $z - x$. Point y is not comparable to z because z does not belong to $y \pm \mathcal{K}$. Translating a negated cone is quite helpful for visualization: $u \preceq_{\mathcal{K}} w \Leftrightarrow u \in w - \mathcal{K} \Leftrightarrow u - w \preceq_{\mathcal{K}} 0$. Points need not belong to $\mathcal{K}$ to be comparable; *e.g.*, all points less than w (with respect to $\mathcal{K}$) belong to $w - \mathcal{K}$.

is biorthogonal expansion (394) (§E.0.1), and biorthogonality condition (393) can be expressed succinctly (§E.1.1)[2.70]

$$X^{\dagger}X = I \qquad (403)$$

Expansion $w = XX^{\dagger}w$, for any particular $w \in \mathbb{R}^n$ more generally, is unique w.r.t X if and only if the extreme directions of $\mathcal{K}$ populating the columns of $X \in \mathbb{R}^{n \times N}$ are linearly independent; *id est*, iff X has no nullspace.   □

2.13.7.0.2   Exercise. *Visual comparison of real sums.*
Given $y \preceq x$ with respect to the nonnegative orthant, draw a figure showing a negated shifted orthant (like the cone in Figure 64) illustrating why $\mathbf{1}^{\mathrm{T}}y \leq \mathbf{1}^{\mathrm{T}}x$ for y and x anywhere in $\mathbb{R}^{\mathbf{2}}$. Incorporate two hyperplanes into your drawing, one containing y and another containing x with reference to Figure 27. Does this result hold in higher dimension?   ▼

2.13.7.1   Pointed cones and biorthogonality

Biorthogonality condition $X^{\dagger}X = I$ from Example 2.13.7.0.1 means Γ_1 and Γ_2 are linearly independent generators of $\mathcal{K}$ (§B.1.1.1); generators because every $x \in \mathcal{K}$ is their conic combination. From §2.10.2 we know that means Γ_1 and Γ_2 must be extreme directions of $\mathcal{K}$.

A biorthogonal expansion is necessarily associated with a pointed closed convex cone; pointed, otherwise there can be no extreme directions (§2.8.1). We will address biorthogonal expansion with respect to a pointed polyhedral cone not full-dimensional in §2.13.8.

2.13.7.1.1   Example. *Expansions implied by diagonalization.*
(*confer* §6.4.3.2.1) When matrix $X \in \mathbb{R}^{M \times M}$ is *diagonalizable* (§A.5),

$$X = S \Lambda S^{-1} = [\, s_1 \cdots s_M \,] \, \Lambda \begin{bmatrix} w_1^{\mathrm{T}} \\ \vdots \\ w_M^{\mathrm{T}} \end{bmatrix} = \sum_{i=1}^{M} \lambda_i \, s_i w_i^{\mathrm{T}} \qquad (1628)$$

coordinates for biorthogonal expansion are its eigenvalues λ_i (contained in diagonal matrix Λ) when expanded in S;

$$X = SS^{-1}X = [\, s_1 \cdots s_M \,] \begin{bmatrix} w_1^{\mathrm{T}}X \\ \vdots \\ w_M^{\mathrm{T}}X \end{bmatrix} = \sum_{i=1}^{M} \lambda_i \, s_i w_i^{\mathrm{T}} \qquad (397)$$

Coordinate values depend upon geometric relationship of X to its linearly independent eigenmatrices $s_i w_i^{\mathrm{T}}$. (§A.5.0.3, §B.1.1)

[2.70]Possibly confusing is the fact that formula $XX^{\dagger}x$ is simultaneously: the orthogonal projection of x on $\mathcal{R}(X)$ (2003), and a sum of nonorthogonal projections of $x \in \mathcal{R}(X)$ on the range of each and every column of full-rank X skinny-or-square (§E.5.0.0.2).

- Eigenmatrices $s_i w_i^{\mathrm{T}}$ are linearly independent dyads constituted by right and left eigenvectors of diagonalizable X and are generators of some pointed polyhedral cone $\mathcal{K}$ in a subspace of $\mathbb{R}^{M \times M}$.

When S is real and X belongs to that polyhedral cone $\mathcal{K}$, for example, then coordinates of expansion (the eigenvalues λ_i) must be nonnegative.

When $X = Q\Lambda Q^{\mathrm{T}}$ is symmetric, coordinates for biorthogonal expansion are its eigenvalues when expanded in Q; *id est*, for $X \in \mathbb{S}^M$

$$X = QQ^{\mathrm{T}}X = \sum_{i=1}^{M} q_i\, q_i^{\mathrm{T}}X = \sum_{i=1}^{M} \lambda_i\, q_i q_i^{\mathrm{T}} \in \mathbb{S}^M \tag{398}$$

becomes an orthogonal expansion with *orthonormality condition* $Q^{\mathrm{T}}Q = I$ where λ_i is the ith eigenvalue of X, q_i is the corresponding i^{th} eigenvector arranged columnar in orthogonal matrix

$$Q = [\, q_1 \;\; q_2 \;\; \cdots \;\; q_M \,] \in \mathbb{R}^{M \times M} \tag{399}$$

and where eigenmatrix $q_i q_i^{\mathrm{T}}$ is an extreme direction of some pointed polyhedral cone $\mathcal{K} \subset \mathbb{S}^M$ and an extreme direction of the positive semidefinite cone $\mathbb{S}_+^M$.

- Orthogonal expansion is a special case of biorthogonal expansion of $X \in \operatorname{aff}\mathcal{K}$ occurring when polyhedral cone $\mathcal{K}$ is any rotation about the origin of an orthant belonging to a subspace.

Similarly, when $X = Q\Lambda Q^{\mathrm{T}}$ belongs to the positive semidefinite cone in the subspace of symmetric matrices, coordinates for orthogonal expansion must be its nonnegative eigenvalues (1531) when expanded in Q; *id est*, for $X \in \mathbb{S}_+^M$

$$X = QQ^{\mathrm{T}}X = \sum_{i=1}^{M} q_i\, q_i^{\mathrm{T}}X = \sum_{i=1}^{M} \lambda_i\, q_i q_i^{\mathrm{T}} \in \mathbb{S}_+^M \tag{400}$$

where $\lambda_i \geq 0$ is the ith eigenvalue of X. This means matrix X simultaneously belongs to the positive semidefinite cone and to the pointed polyhedral cone $\mathcal{K}$ formed by the conic hull of its eigenmatrices. $\qquad\square$

2.13.7.1.2 Example. *Expansion respecting nonpositive orthant.*
Suppose $x \in \mathcal{K}$ any orthant in $\mathbb{R}^n$.[2.71] Then coordinates for biorthogonal expansion of x must be nonnegative; in fact, absolute value of the Cartesian coordinates.

Suppose, in particular, x belongs to the nonpositive orthant $\mathcal{K} = \mathbb{R}_-^n$. Then biorthogonal expansion becomes orthogonal expansion

$$x = XX^{\mathrm{T}}x = \sum_{i=1}^{n} -e_i(-e_i^{\mathrm{T}}x) = \sum_{i=1}^{n} -e_i|e_i^{\mathrm{T}}x| \in \mathbb{R}_-^n \tag{401}$$

and the coordinates of expansion are nonnegative. For this orthant $\mathcal{K}$ we have orthonormality condition $X^{\mathrm{T}}X = I$ where $X = -I$, $e_i \in \mathbb{R}^n$ is a standard basis vector, and $-e_i$ is an extreme direction (§2.8.1) of $\mathcal{K}$.

[2.71]An orthant is simplicial and selfdual.

Of course, this expansion $x = X X^{\mathrm{T}} x$ applies more broadly to domain $\mathbb{R}^n$, but then the coordinates each belong to all of $\mathbb{R}$. $\square$

2.13.8 Biorthogonal expansion, derivation

Biorthogonal expansion is a means for determining coordinates in a pointed conic coordinate system characterized by a nonorthogonal basis. Study of nonorthogonal bases invokes pointed polyhedral cones and their duals; extreme directions of a cone $\mathcal{K}$ are assumed to constitute the *basis* while those of the dual cone $\mathcal{K}^*$ determine coordinates.

Unique biorthogonal expansion with respect to $\mathcal{K}$ relies upon existence of its linearly independent extreme directions: Polyhedral cone $\mathcal{K}$ must be pointed; then it possesses extreme directions. Those extreme directions must be linearly independent to uniquely represent any point in their span.

We consider nonempty pointed polyhedral cone $\mathcal{K}$ possibly not full-dimensional; *id est*, we consider a basis spanning a subspace. Then we need only observe that section of dual cone $\mathcal{K}^*$ in the affine hull of $\mathcal{K}$ because, by *expansion* of x, membership $x \in \operatorname{aff} \mathcal{K}$ is implicit and because any breach of the ordinary dual cone into ambient space becomes irrelevant (§2.13.9.3). *Biorthogonal expansion*

$$x = X X^{\dagger} x \in \operatorname{aff} \mathcal{K} = \operatorname{aff} \operatorname{cone}(X) \tag{402}$$

is expressed in the extreme directions $\{\Gamma_i\}$ of $\mathcal{K}$ arranged columnar in

$$X = [\,\Gamma_1 \ \ \Gamma_2 \ \cdots \ \Gamma_N\,] \in \mathbb{R}^{n \times N} \tag{279}$$

under assumption of *biorthogonality*

$$X^{\dagger} X = I \tag{403}$$

where † denotes matrix pseudoinverse (§E). We therefore seek, in this section, a vertex-description for $\mathcal{K}^* \cap \operatorname{aff} \mathcal{K}$ in terms of linearly independent dual generators $\{\Gamma_i^*\} \subset \operatorname{aff} \mathcal{K}$ in the same finite quantity[2.72] as the extreme directions $\{\Gamma_i\}$ of

$$\mathcal{K} = \operatorname{cone}(X) = \{X a \mid a \succeq 0\} \subseteq \mathbb{R}^n \tag{103}$$

We assume the quantity of extreme directions N does not exceed the dimension n of ambient vector space because, otherwise, expansion w.r.t $\mathcal{K}$ could not be unique; *id est*, assume N linearly independent extreme directions hence $N \leq n$ (X *skinny*[2.73]-or-square full-rank). In other words, fat full-rank matrix X is prohibited by uniqueness because of existence of an infinity of right inverses;

- polyhedral cones whose extreme directions number in excess of the ambient space dimension are precluded in biorthogonal expansion.

2.13.8.1 $x \in \mathcal{K}$

Suppose x belongs to $\mathcal{K} \subseteq \mathbb{R}^n$. Then $x = X a$ for some $a \succeq 0$. Coordinate vector a is unique only when $\{\Gamma_i\}$ is a linearly independent set.[2.74] Vector $a \in \mathbb{R}^N$ can take the

---

[2.72] When $\mathcal{K}$ is contained in a proper subspace of $\mathbb{R}^n$, the ordinary dual cone $\mathcal{K}^*$ will have more generators in any minimal set than $\mathcal{K}$ has extreme directions.
[2.73] "Skinny" meaning thin; more rows than columns.
[2.74] Conic independence alone (§2.10) is insufficient to guarantee uniqueness.

form $a = Bx$ if $\mathcal{R}(B) = \mathbb{R}^N$. Then we require $Xa = XBx = x$ and $Bx = BXa = a$. The pseudoinverse $B = X^\dagger \in \mathbb{R}^{N \times n}$ (§E) is suitable when X is skinny-or-square and full-rank. In that case rank $X = N$, and for all $c \succeq 0$ and $i = 1 \dots N$

$$a \succeq 0 \;\Leftrightarrow\; X^\dagger Xa \succeq 0 \;\Leftrightarrow\; a^\mathrm{T} X^\mathrm{T} X^{\dagger\mathrm{T}} c \geq 0 \;\Leftrightarrow\; \Gamma_i^\mathrm{T} X^{\dagger\mathrm{T}} c \geq 0 \qquad (404)$$

The penultimate inequality follows from the *generalized inequality and membership corollary*, while the last inequality is a consequence of that corollary's discretization (§2.13.4.2.1).[2.75] From (404) and (392) we deduce

$$\mathcal{K}^* \cap \operatorname{aff} \mathcal{K} = \operatorname{cone}(X^{\dagger\mathrm{T}}) = \{X^{\dagger\mathrm{T}} c \mid c \succeq 0\} \subseteq \mathbb{R}^n \qquad (405)$$

is the vertex-description for that section of $\mathcal{K}^*$ in the affine hull of $\mathcal{K}$ because $\mathcal{R}(X^{\dagger\mathrm{T}}) = \mathcal{R}(X)$ by definition of the pseudoinverse. From (309), we know $\mathcal{K}^* \cap \operatorname{aff} \mathcal{K}$ must be pointed if $\operatorname{rel int} \mathcal{K}$ is logically assumed nonempty with respect to $\operatorname{aff} \mathcal{K}$.

Conversely, suppose full-rank skinny-or-square matrix $(N \leq n)$

$$X^{\dagger\mathrm{T}} \triangleq \left[\Gamma_1^* \; \Gamma_2^* \; \cdots \; \Gamma_N^* \right] \in \mathbb{R}^{n \times N} \qquad (406)$$

comprises the extreme directions $\{\Gamma_i^*\} \subset \operatorname{aff} \mathcal{K}$ of the dual-cone intersection with the affine hull of $\mathcal{K}$.[2.76] From the *discretized membership theorem* and (313) we get a partial dual to (392); *id est*, assuming $x \in \operatorname{aff} \operatorname{cone} X$

$$x \in \mathcal{K} \;\Leftrightarrow\; \gamma^{*\mathrm{T}} x \geq 0 \text{ for all } \gamma^* \in \left\{ \Gamma_i^*, \; i = 1 \dots N \right\} \subset \partial \mathcal{K}^* \cap \operatorname{aff} \mathcal{K} \qquad (407)$$

$$\Leftrightarrow\; X^\dagger x \succeq 0 \qquad (408)$$

that leads to a partial halfspace-description,

$$\mathcal{K} = \left\{ x \in \operatorname{aff} \operatorname{cone} X \mid X^\dagger x \succeq 0 \right\} \qquad (409)$$

For $\gamma^* = X^{\dagger\mathrm{T}} e_i$, any $x = Xa$, and for all i we have $e_i^\mathrm{T} X^\dagger Xa = e_i^\mathrm{T} a \geq 0$ only when $a \succeq 0$. Hence $x \in \mathcal{K}$.

When X is full-rank, then unique biorthogonal expansion of $x \in \mathcal{K}$ becomes (402)

$$x = XX^\dagger x = \sum_{i=1}^N \Gamma_i \Gamma_i^{*\mathrm{T}} x \qquad (410)$$

2.75

$$a \succeq 0 \;\Leftrightarrow\; a^\mathrm{T} X^\mathrm{T} X^{\dagger\mathrm{T}} c \geq 0 \quad \forall (c \succeq 0 \;\Leftrightarrow\; a^\mathrm{T} X^\mathrm{T} X^{\dagger\mathrm{T}} c \geq 0 \quad \forall a \succeq 0)$$
$$\forall (c \succeq 0 \;\Leftrightarrow\; \Gamma_i^\mathrm{T} X^{\dagger\mathrm{T}} c \geq 0 \qquad \forall i) \qquad \blacklozenge$$

Intuitively, any nonnegative vector a is a conic combination of the standard basis $\{e_i \in \mathbb{R}^N\}$; $a \succeq 0 \Leftrightarrow a_i e_i \succeq 0$ for all i. The last inequality in (404) is a consequence of the fact that $x = Xa$ may be any extreme direction of $\mathcal{K}$, in which case a is a standard basis vector; $a = e_i \succeq 0$. Theoretically, because $c \succeq 0$ defines a pointed polyhedral cone (in fact, the nonnegative orthant in $\mathbb{R}^N$), we can take (404) one step further by discretizing c:

$$a \succeq 0 \;\Leftrightarrow\; \Gamma_i^\mathrm{T} \Gamma_j^* \geq 0 \text{ for } i, j = 1 \dots N \;\Leftrightarrow\; X^\dagger X \geq \mathbf{0}$$

In words, $X^\dagger X$ must be a matrix whose entries are each nonnegative.
2.76 When closed convex cone $\mathcal{K}$ is not full-dimensional, $\mathcal{K}^*$ has no extreme directions.

whose *coordinates* $a = \Gamma_i^{*\mathrm{T}} x$ must be nonnegative because $\mathcal{K}$ is assumed pointed, closed, and convex. Whenever X is full-rank, so is its pseudoinverse $X^\dagger$. (§E) In the present case, the columns of $X^{\dagger\mathrm{T}}$ are linearly independent and generators of the dual cone $\mathcal{K}^* \cap \mathrm{aff}\,\mathcal{K}$; hence, the columns constitute its extreme directions. (§2.10.2) That section of the dual cone is itself a polyhedral cone (by (286) or the *cone intersection theorem,* §2.7.2.1.1) having the same number of extreme directions as $\mathcal{K}$.

2.13.8.2 $x \in \mathrm{aff}\,\mathcal{K}$

The extreme directions of $\mathcal{K}$ and $\mathcal{K}^* \cap \mathrm{aff}\,\mathcal{K}$ have a distinct relationship; because $X^\dagger X = I$, then for $i,j = 1 \ldots N$, $\Gamma_i^{\mathrm{T}}\Gamma_i^* = 1$, while for $i \neq j$, $\Gamma_i^{\mathrm{T}}\Gamma_j^* = 0$. Yet neither set of extreme directions, $\{\Gamma_i\}$ nor $\{\Gamma_i^*\}$, is necessarily orthogonal. This is a biorthogonality condition, precisely, [374, §2.2.4] [208] implying each set of extreme directions is linearly independent. (§B.1.1.1)

Biorthogonal expansion therefore applies more broadly; meaning, for any $x \in \mathrm{aff}\,\mathcal{K}$, vector x can be uniquely expressed $x = Xb$ where $b \in \mathbb{R}^N$ because $\mathrm{aff}\,\mathcal{K}$ contains the origin. Thus, for any such $x \in \mathcal{R}(X)$ (*confer* §E.1.1), biorthogonal expansion (410) becomes $x = XX^\dagger Xb = Xb$.

2.13.9   Formulae finding dual cone

2.13.9.1   Pointed $\mathcal{K}$, dual, X skinny-or-square full-rank

We wish to derive expressions for a convex cone and its ordinary dual under the general assumptions: pointed polyhedral $\mathcal{K}$ denoted by its linearly independent extreme directions arranged columnar in matrix X such that

$$\mathrm{rank}(X \in \mathbb{R}^{n \times N}) = N \triangleq \dim \mathrm{aff}\,\mathcal{K} \leq n \tag{411}$$

The vertex-description is given:

$$\mathcal{K} = \{Xa \mid a \succeq 0\} \subseteq \mathbb{R}^n \tag{412}$$

from which a halfspace-description for the dual cone follows directly:

$$\mathcal{K}^* = \{y \in \mathbb{R}^n \mid X^{\mathrm{T}} y \succeq 0\} \tag{413}$$

By defining a matrix

$$X^\perp \triangleq \mathrm{basis}\,\mathcal{N}(X^{\mathrm{T}}) \tag{414}$$

(a columnar basis for the orthogonal complement of $\mathcal{R}(X)$), we can say

$$\mathrm{aff\,cone}\,X = \mathrm{aff}\,\mathcal{K} = \{x \mid X^{\perp\mathrm{T}} x = \mathbf{0}\} \tag{415}$$

meaning $\mathcal{K}$ lies in a subspace, perhaps $\mathbb{R}^n$. Thus a halfspace-description

$$\mathcal{K} = \{x \in \mathbb{R}^n \mid X^\dagger x \succeq 0, \ X^{\perp\mathrm{T}} x = \mathbf{0}\} \tag{416}$$

and a vertex-description[2.77] from (313)

$$\mathcal{K}^* = \{\, [\,X^{\dagger\mathrm{T}}\ \ X^\perp\ \ -X^\perp\,]b \mid b \succeq 0\,\} \subseteq \mathbb{R}^n \tag{417}$$

These results are summarized for a pointed polyhedral cone, having linearly independent generators, and its ordinary dual:

Cone Table 1	$\mathcal{K}$	$\mathcal{K}^*$
vertex-description	X	$X^{\dagger\mathrm{T}},\ \pm X^\perp$
halfspace-description	$X^\dagger,\ X^{\perp\mathrm{T}}$	X^T

2.13.9.2 Simplicial case

When a convex cone is simplicial (§2.12.3), Cone Table **1** simplifies because then aff cone $X = \mathbb{R}^n$: For square X and assuming simplicial $\mathcal{K}$ such that

$$\operatorname{rank}(X \in \mathbb{R}^{n\times N}) = N \triangleq \dim\operatorname{aff}\mathcal{K} = n \tag{418}$$

we have

Cone Table S	$\mathcal{K}$	$\mathcal{K}^*$
vertex-description	X	$X^{\dagger\mathrm{T}}$
halfspace-description	$X^\dagger$	X^T

For example, vertex-description (417) simplifies to

$$\mathcal{K}^* = \{X^{\dagger\mathrm{T}}b \mid b \succeq 0\} \subset \mathbb{R}^n \tag{419}$$

Now, because $\dim\mathcal{R}(X) = \dim\mathcal{R}(X^{\dagger\mathrm{T}})$, (§E) the dual cone $\mathcal{K}^*$ is simplicial whenever $\mathcal{K}$ is.

2.13.9.3 Cone membership relations in a subspace

It is obvious by definition (296) of ordinary dual cone $\mathcal{K}^*$, in ambient vector space $\mathcal{R}$, that its determination instead in subspace $\mathcal{S} \subseteq \mathcal{R}$ is identical to its intersection with $\mathcal{S}$; *id est*, assuming closed convex cone $\mathcal{K} \subseteq \mathcal{S}$ and $\mathcal{K}^* \subseteq \mathcal{R}$

$$(\mathcal{K}^* \text{ were ambient } \mathcal{S}) \equiv (\mathcal{K}^* \text{ in ambient } \mathcal{R}) \cap \mathcal{S} \tag{420}$$

because

$$\{y \in \mathcal{S} \mid \langle y, x\rangle \geq 0 \text{ for all } x \in \mathcal{K}\} = \{y \in \mathcal{R} \mid \langle y, x\rangle \geq 0 \text{ for all } x \in \mathcal{K}\} \cap \mathcal{S} \tag{421}$$

From this, a constrained membership relation for the ordinary dual cone $\mathcal{K}^* \subseteq \mathcal{R}$, assuming $x, y \in \mathcal{S}$ and closed convex cone $\mathcal{K} \subseteq \mathcal{S}$

$$y \in \mathcal{K}^* \cap \mathcal{S} \ \Leftrightarrow \ \langle y, x\rangle \geq 0 \text{ for all } x \in \mathcal{K} \tag{422}$$

[2.77]These descriptions are not unique. A vertex-description of the dual cone, for example, might use four conically independent generators for a plane (§2.10.0.0.1, Figure 50) when only three would suffice.

168 CHAPTER 2. CONVEX GEOMETRY

By closure in subspace $\mathcal{S}$ we have conjugation (§2.13.1.1):

$$x \in \mathcal{K} \;\Leftrightarrow\; \langle y, x \rangle \geq 0 \text{ for all } y \in \mathcal{K}^* \cap \mathcal{S} \tag{423}$$

This means membership determination in subspace $\mathcal{S}$ requires knowledge of dual cone only in $\mathcal{S}$. For sake of completeness, for proper cone $\mathcal{K}$ with respect to subspace $\mathcal{S}$ (*confer* (325))

$$x \in \operatorname{int}\mathcal{K} \;\Leftrightarrow\; \langle y, x \rangle > 0 \text{ for all } y \in \mathcal{K}^* \cap \mathcal{S}, \; y \neq 0 \tag{424}$$

$$x \in \mathcal{K}, \; x \neq 0 \;\Leftrightarrow\; \langle y, x \rangle > 0 \text{ for all } y \in \operatorname{int}\mathcal{K}^* \cap \mathcal{S} \tag{425}$$

(By closure, we also have the conjugate relations.) Yet when $\mathcal{S}$ equals aff $\mathcal{K}$ for $\mathcal{K}$ a closed convex cone

$$x \in \operatorname{rel int}\mathcal{K} \;\Leftrightarrow\; \langle y, x \rangle > 0 \text{ for all } y \in \mathcal{K}^* \cap \operatorname{aff}\mathcal{K}, \; y \neq 0 \tag{426}$$

$$x \in \mathcal{K}, \; x \neq 0 \;\Leftrightarrow\; \langle y, x \rangle > 0 \text{ for all } y \in \operatorname{rel int}(\mathcal{K}^* \cap \operatorname{aff}\mathcal{K}) \tag{427}$$

2.13.9.4 Subspace $\mathcal{S} = \operatorname{aff}\mathcal{K}$

Assume now a subspace $\mathcal{S}$ that is the affine hull of cone $\mathcal{K}$: Consider again a pointed polyhedral cone $\mathcal{K}$ denoted by its extreme directions arranged columnar in matrix X such that

$$\operatorname{rank}(X \in \mathbb{R}^{n \times N}) = N \triangleq \dim \operatorname{aff}\mathcal{K} \leq n \tag{411}$$

We want expressions for the convex cone and its dual in subspace $\mathcal{S} = \operatorname{aff}\mathcal{K}$:

Cone Table A	$\mathcal{K}$	$\mathcal{K}^* \cap \operatorname{aff}\mathcal{K}$
vertex-description	X	$X^{\dagger\mathrm{T}}$
halfspace-description	$X^\dagger, X^{\perp\mathrm{T}}$	$X^\mathrm{T}, X^{\perp\mathrm{T}}$

When $\dim \operatorname{aff}\mathcal{K} = n$, this table reduces to Cone Table **S**. These descriptions facilitate work in a proper subspace. The subspace of symmetric matrices $\mathbb{S}^N$, for example, often serves as ambient space.[2.78]

2.13.9.4.1 Exercise. *Conically independent columns and rows.*
We suspect the number of conically independent columns (rows) of X to be the same for $X^{\dagger\mathrm{T}}$, where $\dagger$ denotes matrix pseudoinverse (§E). Prove whether it holds that the columns (rows) of X are c.i. $\Leftrightarrow$ the columns (rows) of $X^{\dagger\mathrm{T}}$ are c.i. ▼

[2.78]The dual cone of positive semidefinite matrices $\mathbb{S}_+^{N*} = \mathbb{S}_+^N$ remains in $\mathbb{S}^N$ by convention, whereas the ordinary dual cone would venture into $\mathbb{R}^{N \times N}$.

(a)

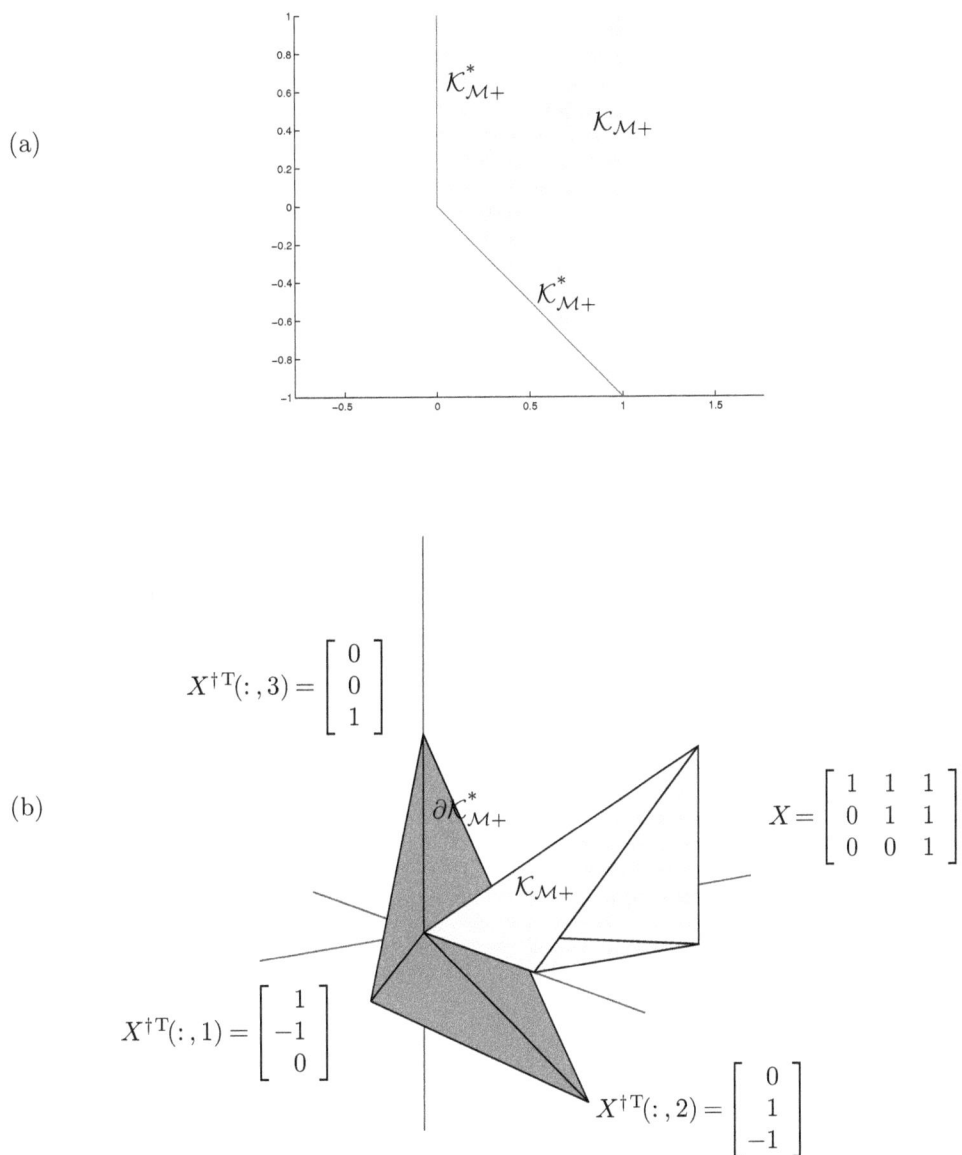

(b)

$$X^{\dagger \mathrm{T}}(:,3) = \begin{bmatrix} 0 \\ 0 \\ 1 \end{bmatrix}$$

$$X = \begin{bmatrix} 1 & 1 & 1 \\ 0 & 1 & 1 \\ 0 & 0 & 1 \end{bmatrix}$$

$$X^{\dagger \mathrm{T}}(:,1) = \begin{bmatrix} 1 \\ -1 \\ 0 \end{bmatrix}$$

$$X^{\dagger \mathrm{T}}(:,2) = \begin{bmatrix} 0 \\ 1 \\ -1 \end{bmatrix}$$

Figure 65: Simplicial cones. (a) Monotone nonnegative cone $\mathcal{K}_{\mathcal{M}+}$ and its dual $\mathcal{K}^*_{\mathcal{M}+}$ (drawn truncated) in $\mathbb{R}^2$. (b) Monotone nonnegative cone and boundary of its dual (both drawn truncated) in $\mathbb{R}^3$. Extreme directions of $\mathcal{K}^*_{\mathcal{M}+}$ are indicated.

2.13.9.4.2   Example.   *Monotone nonnegative cone.*
[62, exer.2.33] [362, §2] Simplicial cone (§2.12.3.1.1) $\mathcal{K}_{\mathcal{M}+}$ is the cone of all nonnegative vectors having their entries sorted in nonincreasing order:

$$
\begin{aligned}
\mathcal{K}_{\mathcal{M}+} &\triangleq \{x \mid x_1 \geq x_2 \geq \cdots \geq x_n \geq 0\} \subseteq \mathbb{R}_+^n \\
&= \{x \mid (e_i - e_{i+1})^{\mathrm{T}}x \geq 0,\ i = 1 \ldots n-1,\ e_n^{\mathrm{T}}x \geq 0\} \\
&= \{x \mid X^{\dagger}x \succeq 0\}
\end{aligned}
\tag{428}
$$

a halfspace-description where e_i is the ith standard basis vector, and where[2.79]

$$
X^{\dagger\mathrm{T}} \triangleq [\, e_1 - e_2 \quad e_2 - e_3 \quad \cdots \quad e_n \,] \in \mathbb{R}^{n \times n}
\tag{429}
$$

For any vectors x and y, simple algebra demands

$$
x^{\mathrm{T}}y = \sum_{i=1}^{n} x_i y_i = (x_1 - x_2)y_1 + (x_2 - x_3)(y_1 + y_2) + (x_3 - x_4)(y_1 + y_2 + y_3) + \cdots
\tag{430}
$$
$$
+ (x_{n-1} - x_n)(y_1 + \cdots + y_{n-1}) + x_n(y_1 + \cdots + y_n)
$$

Because $x_i - x_{i+1} \geq 0\ \forall i$ by assumption whenever $x \in \mathcal{K}_{\mathcal{M}+}$, we can employ dual generalized inequalities (322) with respect to the selfdual nonnegative orthant $\mathbb{R}_+^n$ to find the halfspace-description of dual monotone nonnegative cone $\mathcal{K}_{\mathcal{M}+}^*$. We can say $x^{\mathrm{T}}y \geq 0$ for all $X^{\dagger}x \succeq 0$ [*sic*] if and only if

$$
y_1 \geq 0, \quad y_1 + y_2 \geq 0, \quad \ldots, \quad y_1 + y_2 + \cdots + y_n \geq 0
\tag{431}
$$

id est,

$$
x^{\mathrm{T}}y \geq 0 \ \forall X^{\dagger}x \succeq 0 \ \Leftrightarrow \ X^{\mathrm{T}}y \succeq 0
\tag{432}
$$

where

$$
X = [\, e_1 \quad e_1 + e_2 \quad e_1 + e_2 + e_3 \quad \cdots \quad \mathbf{1} \,] \in \mathbb{R}^{n \times n}
\tag{433}
$$

Because $X^{\dagger}x \succeq 0$ connotes membership of x to pointed $\mathcal{K}_{\mathcal{M}+}$, then by (296) the dual cone we seek comprises all y for which (432) holds; thus its halfspace-description

$$
\mathcal{K}_{\mathcal{M}+}^* = \{y \underset{\mathcal{K}_{\mathcal{M}+}^*}{\succeq} 0\} = \{y \mid \textstyle\sum_{i=1}^{k} y_i \geq 0,\ k = 1 \ldots n\} = \{y \mid X^{\mathrm{T}}y \succeq 0\} \subset \mathbb{R}^n
\tag{434}
$$

The monotone nonnegative cone and its dual are simplicial, illustrated for two Euclidean spaces in Figure 65.

From §2.13.6.1, the extreme directions of proper $\mathcal{K}_{\mathcal{M}+}$ are respectively orthogonal to the facets of $\mathcal{K}_{\mathcal{M}+}^*$. Because $\mathcal{K}_{\mathcal{M}+}^*$ is simplicial, the inward-normals to its facets constitute the linearly independent rows of X^{T} by (434). Hence the vertex-description for $\mathcal{K}_{\mathcal{M}+}$ employs the columns of X in agreement with Cone Table **S** because $X^{\dagger} = X^{-1}$. Likewise, the extreme directions of proper $\mathcal{K}_{\mathcal{M}+}^*$ are respectively orthogonal to the facets of $\mathcal{K}_{\mathcal{M}+}$ whose inward-normals are contained in the rows of $X^{\dagger}$ by (428). So the vertex-description for $\mathcal{K}_{\mathcal{M}+}^*$ employs the columns of $X^{\dagger\mathrm{T}}$.   □

[2.79]With $X^{\dagger}$ in hand, we might concisely scribe the remaining vertex- and halfspace-descriptions from the tables for $\mathcal{K}_{\mathcal{M}+}$ and its dual. Instead we use dual generalized inequalities in their derivation.

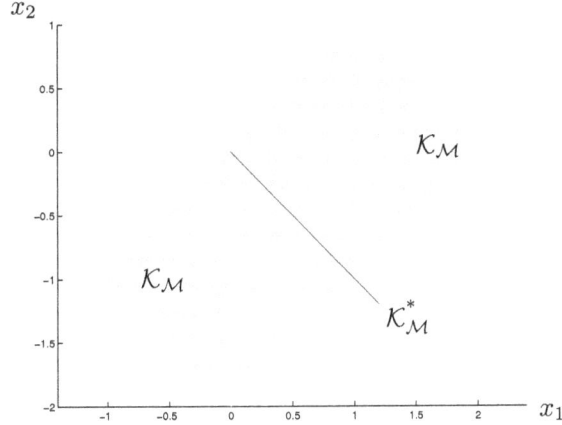

Figure 66: Monotone cone $\mathcal{K_M}$ and its dual $\mathcal{K}_\mathcal{M}^*$ (drawn truncated) in $\mathbb{R}^2$.

2.13.9.4.3   Example. *Monotone cone.*
(Figure **66**, Figure **67**) Full-dimensional but not pointed, the monotone cone is polyhedral and defined by the halfspace-description

$$\mathcal{K_M} \triangleq \{x \in \mathbb{R}^n \mid x_1 \geq x_2 \geq \cdots \geq x_n\} = \{x \in \mathbb{R}^n \mid X^{*\mathrm{T}}x \succeq 0\} \qquad (435)$$

Its dual is therefore pointed but not full-dimensional;

$$\mathcal{K}_\mathcal{M}^* = \{ X^* b \triangleq [\, e_1 - e_2 \;\; e_2 - e_3 \; \cdots \; e_{n-1} - e_n \,]\, b \mid b \succeq 0 \} \subset \mathbb{R}^n \qquad (436)$$

the dual cone vertex-description where the columns of X^* comprise its extreme directions. Because dual monotone cone $\mathcal{K}_\mathcal{M}^*$ is pointed and satisfies

$$\mathrm{rank}(X^* \in \mathbb{R}^{n \times N}) = N \triangleq \dim \mathrm{aff}\, \mathcal{K}^* \leq n \qquad (437)$$

where $N = n-1$, and because $\mathcal{K_M}$ is closed and convex, we may adapt Cone Table **1** (p.167) as follows:

Cone Table 1*	$\mathcal{K}^*$	$\mathcal{K}^{**} = \mathcal{K}$
vertex-description	X^*	$X^{*\dagger\mathrm{T}},\ \pm X^{*\perp}$
halfspace-description	$X^{*\dagger},\ X^{*\perp\mathrm{T}}$	$X^{*\mathrm{T}}$

The vertex-description for $\mathcal{K_M}$ is therefore

$$\mathcal{K_M} = \{[\, X^{*\dagger\mathrm{T}} \;\; X^{*\perp} \;\; -X^{*\perp} \,]\, a \mid a \succeq 0 \} \subset \mathbb{R}^n \qquad (438)$$

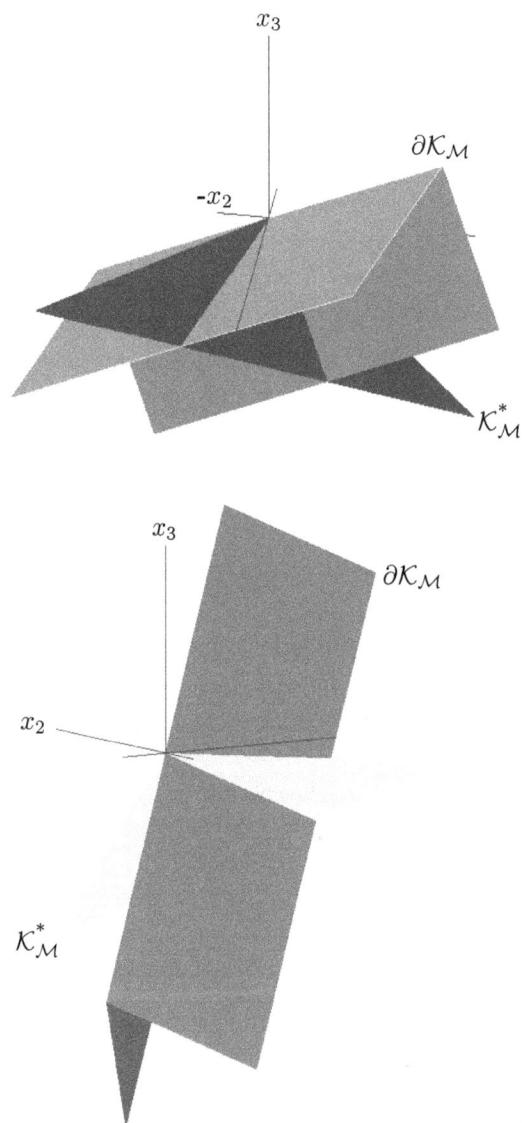

Figure 67: Two views of monotone cone $\mathcal{K}_\mathcal{M}$ and its dual $\mathcal{K}_\mathcal{M}^*$ (drawn truncated) in $\mathbb{R}^3$. Monotone cone is not pointed. Dual monotone cone is not full-dimensional. Cartesian coordinate axes are drawn for reference.

where $X^{*\perp} = \mathbf{1}$ and

$$X^{*\dagger} = \frac{1}{n} \begin{bmatrix} n-1 & -1 & -1 & \cdots & -1 & -1 & -1 \\ n-2 & n-2 & -2 & \ddots & \cdots & -2 & -2 \\ \vdots & n-3 & n-3 & \ddots & -(n-4) & \vdots & -3 \\ 3 & \vdots & n-4 & \ddots & -(n-3) & -(n-3) & \vdots \\ 2 & 2 & \cdots & \ddots & 2 & -(n-2) & -(n-2) \\ 1 & 1 & 1 & \cdots & 1 & 1 & -(n-1) \end{bmatrix} \in \mathbb{R}^{n-1 \times n} \quad (439)$$

while

$$\mathcal{K}_{\mathcal{M}}^* = \{y \in \mathbb{R}^n \mid X^{*\dagger}y \succeq 0, \ X^{*\perp T}y = \mathbf{0}\} \quad (440)$$

is the dual monotone cone halfspace-description.   □

2.13.9.4.4 Exercise. *Inside the monotone cones.*
Mathematically describe the respective interior of the monotone nonnegative cone and monotone cone. In three dimensions, also describe the relative interior of each face. ▼

2.13.9.5 More pointed cone descriptions with equality condition

Consider pointed polyhedral cone $\mathcal{K}$ having a linearly independent set of generators and whose subspace membership is explicit; *id est*, we are given the ordinary halfspace-description

$$\mathcal{K} = \{x \mid Ax \succeq 0, \ Cx = \mathbf{0}\} \subseteq \mathbb{R}^n \quad (286a)$$

where $A \in \mathbb{R}^{m \times n}$ and $C \in \mathbb{R}^{p \times n}$. This can be equivalently written in terms of nullspace of C and vector ξ:

$$\mathcal{K} = \{Z\xi \in \mathbb{R}^n \mid AZ\xi \succeq 0\} \quad (441)$$

where $\mathcal{R}(Z \in \mathbb{R}^{n \times n - \text{rank} C}) \triangleq \mathcal{N}(C)$. Assuming (411) is satisfied

$$\text{rank} X \triangleq \text{rank}\Big((AZ)^\dagger \in \mathbb{R}^{n - \text{rank} C \times m}\Big) = m - \ell = \dim \text{aff} \mathcal{K} \leq n - \text{rank} C \quad (442)$$

where ℓ is the number of conically dependent rows in AZ which must be removed to make $\hat{A}Z$ before the Cone Tables become applicable.[2.80] Then results collected there admit assignment $\hat{X} \triangleq (\hat{A}Z)^\dagger \in \mathbb{R}^{n - \text{rank} C \times m - \ell}$, where $\hat{A} \in \mathbb{R}^{m - \ell \times n}$, followed with linear transformation by Z. So we get the vertex-description, for full-rank $(\hat{A}Z)^\dagger$ skinny-or-square,

$$\mathcal{K} = \{Z(\hat{A}Z)^\dagger b \mid b \succeq 0\} \quad (443)$$

From this and (362) we get a halfspace-description of the dual cone

$$\mathcal{K}^* = \{y \in \mathbb{R}^n \mid (Z^T\hat{A}^T)^\dagger Z^T y \succeq 0\} \quad (444)$$

[2.80]When the conically dependent rows (§2.10) are removed, the rows remaining must be linearly independent for the Cone Tables (p.19) to apply.

From this and Cone Table **1** (p.167) we get a vertex-description, (1965)

$$\mathcal{K}^* = \{[\, Z^{\dagger\mathrm{T}}(\hat{A}Z)^{\mathrm{T}} \quad C^{\mathrm{T}} \quad -C^{\mathrm{T}} \,]c \mid c \succeq 0\} \tag{445}$$

Yet because

$$\mathcal{K} = \{x \mid Ax \succeq 0\} \cap \{x \mid Cx = \mathbf{0}\} \tag{446}$$

then, by (313), we get an equivalent vertex-description for the dual cone

$$
\begin{aligned}
\mathcal{K}^* &= \overline{\{x \mid Ax \succeq 0\}^* + \{x \mid Cx = \mathbf{0}\}^*} \\
&= \{[\, A^{\mathrm{T}} \quad C^{\mathrm{T}} \quad -C^{\mathrm{T}} \,]b \mid b \succeq 0\}
\end{aligned}
\tag{447}
$$

from which the conically dependent columns may, of course, be removed.

2.13.10   Dual cone-translate

(§E.10.3.2.1) First-order optimality condition (351) inspires a dual-cone variant: For any set $\mathcal{K}$, the negative dual of its translation by any $a \in \mathbb{R}^n$ is

$$
\begin{aligned}
-(\mathcal{K} - a)^* &= \{y \in \mathbb{R}^n \mid \langle y\,,\, x - a\rangle \le 0 \text{ for all } x \in \mathcal{K}\} \triangleq \mathcal{K}^\perp(a) \\
&= \{y \in \mathbb{R}^n \mid \langle y\,,\, x\rangle \le 0 \text{ for all } x \in \mathcal{K} - a\}
\end{aligned}
\tag{448}
$$

a closed convex cone called *normal cone* to $\mathcal{K}$ at point a. From this, a new membership relation like (319):

$$y \in -(\mathcal{K} - a)^* \;\Leftrightarrow\; \langle y\,,\, x - a\rangle \le 0 \text{ for all } x \in \mathcal{K} \tag{449}$$

and by closure the conjugate, for closed convex cone $\mathcal{K}$

$$x \in \mathcal{K} \;\Leftrightarrow\; \langle y\,,\, x - a\rangle \le 0 \text{ for all } y \in -(\mathcal{K} - a)^* \tag{450}$$

2.13.10.1   first-order optimality condition - restatement

(*confer* §2.13.3) The general first-order necessary and sufficient condition for optimality of solution $x^\star$ to a minimization problem with real differentiable convex objective function $f(x) : \mathbb{R}^n \to \mathbb{R}$ over convex feasible set $\mathcal{C}$ is [314, §3]

$$-\nabla f(x^\star) \in -(\mathcal{C} - x^\star)^*, \qquad x^\star \in \mathcal{C} \tag{451}$$

id est, the negative gradient (§3.6) belongs to the normal cone to $\mathcal{C}$ at $x^\star$ as in Figure **68**.

2.13.10.1.1   Example. *Normal cone to orthant.*
Consider proper cone $\mathcal{K} = \mathbb{R}^n_+$, the selfdual nonnegative orthant in $\mathbb{R}^n$. The normal cone to $\mathbb{R}^n_+$ at $a \in \mathcal{K}$ is (2188)

$$\mathcal{K}^\perp_{\mathbb{R}^n_+}(a \in \mathbb{R}^n_+) = -(\mathbb{R}^n_+ - a)^* = -\mathbb{R}^n_+ \cap a^\perp, \quad a \in \mathbb{R}^n_+ \tag{452}$$

where $-\mathbb{R}^n_+ = -\mathcal{K}^*$ is the algebraic complement of $\mathbb{R}^n_+$, and $a^\perp$ is the orthogonal complement to range of vector a. This means: When point a is interior to $\mathbb{R}^n_+$, the normal cone is the origin. If n_p represents number of nonzero entries in vector $a \in \partial\mathbb{R}^n_+$, then $\dim(-\mathbb{R}^n_+ \cap a^\perp) = n - n_\mathrm{p}$ and there is a complementary relationship between the nonzero entries in vector a and the nonzero entries in any vector $x \in -\mathbb{R}^n_+ \cap a^\perp$. □

$$\alpha \geq \beta \geq \gamma$$

$$-\nabla f(x^\star)$$

$$\{z \mid f(z) = \alpha\}$$

$$\{y \mid \nabla f(x^\star)^{\mathrm{T}}(y - x^\star) = 0, \; f(x^\star) = \gamma\}$$

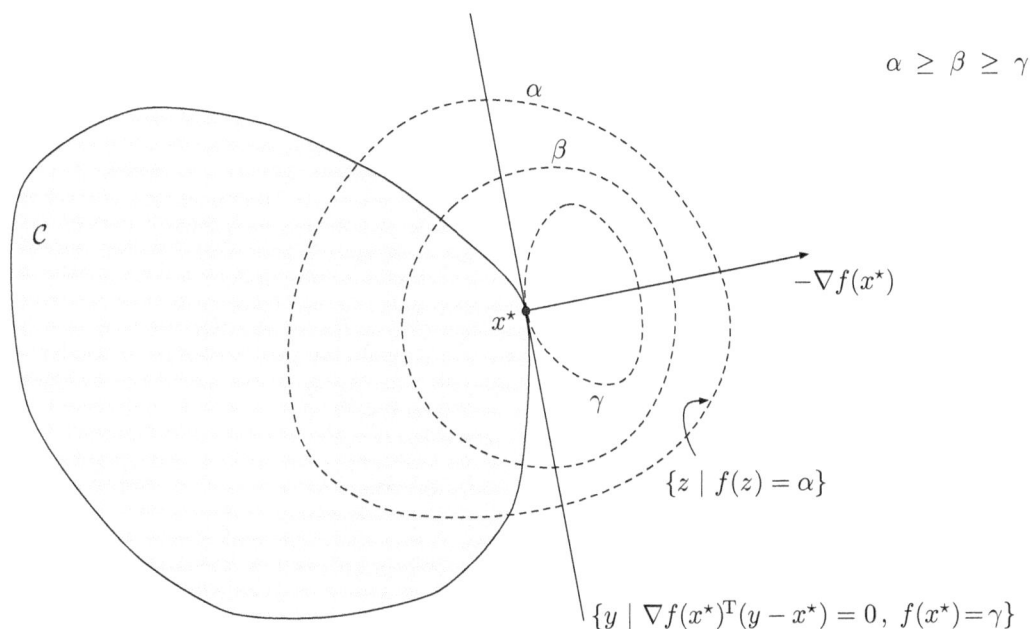

Figure 68: (*confer* Figure 79) Shown is a plausible contour plot in $\mathbb{R}^{\mathbf{2}}$ of some arbitrary differentiable convex real function $f(x)$ at selected levels α, β, and γ; *id est*, contours of equal level f (*level sets*) drawn dashed in function's domain. From results in §3.6.2 (p.220), gradient $\nabla f(x^\star)$ is normal to γ-*sublevel set* $\mathcal{L}_\gamma f$ (560) by Definition E.9.1.0.1. From §2.13.10.1, function is minimized over convex set $\mathcal{C}$ at point $x^\star$ iff negative gradient $-\nabla f(x^\star)$ belongs to normal cone to $\mathcal{C}$ there. In circumstance depicted, normal cone is a ray whose direction is coincident with negative gradient. So, gradient is normal to a hyperplane supporting both $\mathcal{C}$ and the γ-sublevel set.

2.13.10.1.2   Example.   *Optimality conditions for conic problem.*
Consider a convex optimization problem having real differentiable convex objective function $f(x) : \mathbb{R}^n \to \mathbb{R}$ defined on domain $\mathbb{R}^n$

$$
\begin{array}{ll}
\underset{x}{\text{minimize}} & f(x) \\
\text{subject to} & x \in \mathcal{K}
\end{array}
\tag{453}
$$

Let's first suppose that the feasible set is a pointed polyhedral cone $\mathcal{K}$ possessing a linearly independent set of generators and whose subspace membership is made explicit by fat full-rank matrix $C \in \mathbb{R}^{p \times n}$; *id est*, we are given the halfspace-description, for $A \in \mathbb{R}^{m \times n}$

$$
\mathcal{K} = \{x \mid Ax \succeq 0, \; Cx = \mathbf{0}\} \subseteq \mathbb{R}^n
\tag{286a}
$$

(We'll generalize to any convex cone $\mathcal{K}$ shortly.) Vertex-description of this cone, assuming $(\hat{A}Z)^\dagger$ skinny-or-square full-rank, is

$$
\mathcal{K} = \{Z(\hat{A}Z)^\dagger b \mid b \succeq 0\}
\tag{443}
$$

where $\hat{A} \in \mathbb{R}^{m-\ell \times n}$, ℓ is the number of conically dependent rows in AZ (§2.10) which must be removed, and $Z \in \mathbb{R}^{n \times n - \text{rank}\, C}$ holds basis $\mathcal{N}(C)$ columnar.
   From optimality condition (351),

$$
\nabla f(x^\star)^{\mathrm{T}} \big(Z(\hat{A}Z)^\dagger b - x^\star\big) \geq 0 \quad \forall b \succeq 0
\tag{454}
$$

$$
-\nabla f(x^\star)^{\mathrm{T}} Z(\hat{A}Z)^\dagger (b - b^\star) \leq 0 \quad \forall b \succeq 0
\tag{455}
$$

because

$$
x^\star \triangleq Z(\hat{A}Z)^\dagger b^\star \in \mathcal{K}
\tag{456}
$$

From membership relation (449) and Example 2.13.10.1.1

$$
\begin{array}{c}
\langle -(Z^{\mathrm{T}}\hat{A}^{\mathrm{T}})^\dagger Z^{\mathrm{T}} \nabla f(x^\star), \, b - b^\star \rangle \leq 0 \text{ for all } b \in \mathbb{R}^{m-\ell}_+ \\
\Leftrightarrow \\
-(Z^{\mathrm{T}}\hat{A}^{\mathrm{T}})^\dagger Z^{\mathrm{T}} \nabla f(x^\star) \in -\mathbb{R}^{m-\ell}_+ \cap b^{\star\perp}
\end{array}
\tag{457}
$$

Then equivalent necessary and sufficient conditions for optimality of conic problem (453) with feasible set $\mathcal{K}$ are: (*confer* (361))

$$
\underset{\mathbb{R}^{m-\ell}_+}{(Z^{\mathrm{T}}\hat{A}^{\mathrm{T}})^\dagger Z^{\mathrm{T}} \nabla f(x^\star) \succeq 0}, \qquad \underset{\mathbb{R}^{m-\ell}_+}{b^\star \succeq 0}, \qquad \nabla f(x^\star)^{\mathrm{T}} Z(\hat{A}Z)^\dagger b^\star = 0
\tag{458}
$$

expressible, by (444),

$$
\nabla f(x^\star) \in \mathcal{K}^*, \qquad x^\star \in \mathcal{K}, \qquad \nabla f(x^\star)^{\mathrm{T}} x^\star = 0
\tag{459}
$$

   This result (459) actually applies more generally to any convex cone $\mathcal{K}$ comprising the feasible set: Necessary and sufficient optimality conditions are in terms of objective gradient

$$
-\nabla f(x^\star) \in -(\mathcal{K} - x^\star)^*, \qquad x^\star \in \mathcal{K}
\tag{451}
$$

whose membership to normal cone, assuming only cone $\mathcal{K}$ convexity,

$$-(\mathcal{K} - x^\star)^* = \mathcal{K}_{\mathcal{K}}^\perp(x^\star \in \mathcal{K}) = -\mathcal{K}^* \cap x^{\star\perp} \qquad (2188)$$

equivalently expresses conditions (459).

When $\mathcal{K} = \mathbb{R}_+^n$, in particular, then $C = \mathbf{0}$, $A = Z = I \in \mathbb{S}^n$; *id est,*

$$\begin{array}{ll} \underset{x}{\text{minimize}} & f(x) \\ \text{subject to} & x \underset{\mathbb{R}_+^n}{\succeq} 0 \end{array} \qquad (460)$$

Necessary and sufficient optimality conditions become (*confer* [62, §4.2.3])

$$\nabla f(x^\star) \underset{\mathbb{R}_+^n}{\succeq} 0, \qquad x^\star \underset{\mathbb{R}_+^n}{\succeq} 0, \qquad \nabla f(x^\star)^{\mathrm{T}} x^\star = 0 \qquad (461)$$

equivalent to condition $(329)^{2.81}$ (under nonzero gradient) for membership to the nonnegative orthant boundary $\partial\mathbb{R}_+^n$.   □

2.13.10.1.3 Example. *Complementarity problem.*   [216]
A complementarity problem in nonlinear function f is nonconvex

$$\begin{array}{ll} \text{find} & z \in \mathcal{K} \\ \text{subject to} & f(z) \in \mathcal{K}^* \\ & \langle z\,,\,f(z)\rangle = 0 \end{array} \qquad (462)$$

yet bears strong resemblance to (459) and to (2124) Moreau's decomposition theorem on page 638 for projection P on mutually polar cones $\mathcal{K}$ and $-\mathcal{K}^*$. Identify a sum of mutually orthogonal projections $x \triangleq z - f(z)$; in Moreau's terms, $z = P_{\mathcal{K}}x$ and $-f(z) = P_{-\mathcal{K}^*}x$. Then $f(z) \in \mathcal{K}^*$ (§E.9.2.2 *no.*4) and z is a solution to the complementarity problem iff it is a fixed point of

$$z = P_{\mathcal{K}}x = P_{\mathcal{K}}(z - f(z)) \qquad (463)$$

Given that a solution exists, existence of a fixed point would be guaranteed by theory of *contraction.* [233, p.300] But because only *nonexpansivity* (Theorem E.9.3.0.1) is achievable by a projector, uniqueness cannot be assured. [209, p.155] Elegant proofs of equivalence between complementarity problem (462) and fixed point problem (463) are provided by Németh [387, *Fixed point problems*].   □

2.13.10.1.4 Example. *Linear complementarity problem.* [89] [278] [319]
Given matrix $B \in \mathbb{R}^{n \times n}$ and vector $q \in \mathbb{R}^n$, a prototypical complementarity problem on the nonnegative orthant $\mathcal{K} = \mathbb{R}_+^n$ is linear in $w = f(z)$:

$$\begin{array}{ll} \text{find} & z \succeq 0 \\ \text{subject to} & w \succeq 0 \\ & w^{\mathrm{T}}z = 0 \\ & w = q + Bz \end{array} \qquad (464)$$

$^{2.81}$ and equivalent to well-known Karush-Kuhn-Tucker (KKT) optimality conditions [62, §5.5.3] because the dual variable becomes gradient $\nabla f(x)$.

This problem is not convex when both vectors w and z are variable.[2.82] Notwithstanding, this linear complementarity problem can be solved by identifying $w \leftarrow \nabla f(z) = q + Bz$ then substituting that gradient into (462)

$$\begin{aligned} \text{find} \quad & z \in \mathcal{K} \\ \text{subject to} \quad & \nabla f(z) \in \mathcal{K}^* \\ & \langle z, \nabla f(z)\rangle = 0 \end{aligned} \tag{465}$$

which is simply a restatement of optimality conditions (459) for conic problem (453). Suitable $f(z)$ is the quadratic objective from convex problem

$$\begin{aligned} \underset{z}{\text{minimize}} \quad & \tfrac{1}{2}z^{\mathrm{T}}Bz + q^{\mathrm{T}}z \\ \text{subject to} \quad & z \succeq 0 \end{aligned} \tag{466}$$

which means $B \in \mathbb{S}_+^n$ should be (symmetric) positive semidefinite for solution of (464) by this method. Then (464) has solution iff (466) does. $\square$

2.13.10.1.5 Exercise. *Optimality for equality constrained conic problem.*
Consider a conic optimization problem like (453) having real differentiable convex objective function $f(x) : \mathbb{R}^n \to \mathbb{R}$

$$\begin{aligned} \underset{x}{\text{minimize}} \quad & f(x) \\ \text{subject to} \quad & Cx = d \\ & x \in \mathcal{K} \end{aligned} \tag{467}$$

minimized over convex cone $\mathcal{K}$ but, this time, constrained to affine set $\mathcal{A} = \{x \mid Cx = d\}$. Show, by means of first-order optimality condition (351) or (451), that necessary and sufficient optimality conditions are: (*confer* (459))

$$\begin{aligned} x^\star &\in \mathcal{K} \\ Cx^\star &= d \\ \nabla f(x^\star) + C^{\mathrm{T}}\nu^\star &\in \mathcal{K}^* \\ \langle \nabla f(x^\star) + C^{\mathrm{T}}\nu^\star, x^\star\rangle &= 0 \end{aligned} \tag{468}$$

where $\nu^\star$ is any vector[2.83] satisfying these conditions. ▼

[2.82]But if one of them is fixed, then the problem becomes convex with a very simple geometric interpretation: Define the affine subset

$$\mathcal{A} \triangleq \{y \in \mathbb{R}^n \mid By = w - q\}$$

For $w^{\mathrm{T}}z$ to vanish, there must be a complementary relationship between the nonzero entries of vectors w and z; *id est*, $w_i z_i = 0 \ \forall i$. Given $w \succeq 0$, then z belongs to the convex set of solutions:

$$z \in -\mathcal{K}_{\mathbb{R}_+^n}^\perp(w \in \mathbb{R}_+^n) \cap \mathcal{A} = \mathbb{R}_+^n \cap w^\perp \cap \mathcal{A}$$

where $\mathcal{K}_{\mathbb{R}_+^n}^\perp(w)$ is the normal cone to $\mathbb{R}_+^n$ at w (452). If this intersection is nonempty, then the problem is solvable.
[2.83] an optimal dual variable, these optimality conditions are equivalent to the KKT conditions [62, §5.5.3].

2.13.11 Proper nonsimplicial $\mathcal{K}$, dual, X fat full-rank

Since conically dependent columns can always be removed from X to construct $\mathcal{K}$ or to determine $\mathcal{K}^*$ [380], then assume we are given a set of N conically independent generators (§2.10) of an arbitrary proper polyhedral cone $\mathcal{K}$ in $\mathbb{R}^n$ arranged columnar in $X \in \mathbb{R}^{n \times N}$ such that $N > n$ (fat) and rank $X = n$. Having found formula (419) to determine the dual of a simplicial cone, the easiest way to find a vertex-description of proper dual cone $\mathcal{K}^*$ is to first decompose $\mathcal{K}$ into simplicial parts $\mathcal{K}_i$ so that $\mathcal{K} = \bigcup \mathcal{K}_i$.[2.84] Each component simplicial cone in $\mathcal{K}$ corresponds to some subset of n linearly independent columns from X. The key idea, here, is how the extreme directions of the simplicial parts must remain extreme directions of $\mathcal{K}$. Finding the dual of $\mathcal{K}$ amounts to finding the dual of each simplicial part:

2.13.11.0.1 Theorem. *Dual cone intersection.* [337, §2.7]
Suppose proper cone $\mathcal{K} \subset \mathbb{R}^n$ equals the union of M simplicial cones $\mathcal{K}_i$ whose extreme directions all coincide with those of $\mathcal{K}$. Then proper dual cone $\mathcal{K}^*$ is the intersection of M dual simplicial cones $\mathcal{K}_i^*$; *id est,*

$$\mathcal{K} = \bigcup_{i=1}^{M} \mathcal{K}_i \quad \Rightarrow \quad \mathcal{K}^* = \bigcap_{i=1}^{M} \mathcal{K}_i^* \tag{469}$$

$\diamond$

Proof. For $X_i \in \mathbb{R}^{n \times n}$, a complete matrix of linearly independent extreme directions (p.126) arranged columnar, corresponding simplicial $\mathcal{K}_i$ (§2.12.3.1.1) has vertex-description

$$\mathcal{K}_i = \{X_i\, c \mid c \succeq 0\} \tag{470}$$

Now suppose,

$$\mathcal{K} = \bigcup_{i=1}^{M} \mathcal{K}_i = \bigcup_{i=1}^{M} \{X_i\, c \mid c \succeq 0\} \tag{471}$$

The union of all $\mathcal{K}_i$ can be equivalently expressed

$$\mathcal{K} = \left\{ [\, X_1\ X_2\ \cdots\ X_M\,] \begin{bmatrix} a \\ b \\ \vdots \\ c \end{bmatrix} \mid a,\, b \ldots c \succeq 0 \right\} \tag{472}$$

Because extreme directions of the simplices $\mathcal{K}_i$ are extreme directions of $\mathcal{K}$ by assumption,

---

[2.84]That proposition presupposes, of course, that we know how to perform simplicial decomposition efficiently; also called "triangulation". [311] [179, §3.1] [180, §3.1] Existence of multiple simplicial parts means expansion of $x \in \mathcal{K}$, like (410), can no longer be unique because number N of extreme directions in $\mathcal{K}$ exceeds dimension n of the space.

then

$$\mathcal{K} \;=\; \{\,[\,X_1\ X_2\ \cdots\ X_M\,]\,d \mid d \succeq 0\,\} \tag{473}$$

by the *extremes theorem* (§2.8.1.1.1). Defining $X \triangleq [\,X_1\ X_2\ \cdots\ X_M\,]$ (with any redundant [*sic*] columns optionally removed from X), then $\mathcal{K}^*$ can be expressed ((362), Cone Table **S** p.167)

$$\mathcal{K}^* \;=\; \{y \mid X^{\mathrm{T}}y \succeq 0\} \;=\; \bigcap_{i=1}^{M} \{y \mid X_i^{\mathrm{T}}y \succeq 0\} \;=\; \bigcap_{i=1}^{M} \mathcal{K}_i^* \tag{474}$$

$\blacklozenge$

To find the extreme directions of the dual cone, first we observe that some facets of each simplicial part $\mathcal{K}_i$ are common to facets of $\mathcal{K}$ by assumption, and the union of all those common facets comprises the set of all facets of $\mathcal{K}$ by design. For any particular proper polyhedral cone $\mathcal{K}$, the extreme directions of dual cone $\mathcal{K}^*$ are respectively orthogonal to the facets of $\mathcal{K}$. (§2.13.6.1) Then the extreme directions of the dual cone can be found among inward-normals to facets of the component simplicial cones $\mathcal{K}_i$; those normals are extreme directions of the dual simplicial cones $\mathcal{K}_i^*$. From the theorem and Cone Table **S** (p.167),

$$\mathcal{K}^* \;=\; \bigcap_{i=1}^{M} \mathcal{K}_i^* \;=\; \bigcap_{i=1}^{M} \{X_i^{\dagger\mathrm{T}}c \mid c \succeq 0\} \tag{475}$$

The set of extreme directions $\{\Gamma_i^*\}$ for proper dual cone $\mathcal{K}^*$ is therefore constituted by those conically independent generators, from the columns of all the dual simplicial matrices $\{X_i^{\dagger\mathrm{T}}\}$, that do not violate discrete definition (362) of $\mathcal{K}^*$;

$$\left\{\Gamma_1^*,\, \Gamma_2^* \dots \Gamma_N^*\right\} \;=\; \text{c.i.}\left\{X_i^{\dagger\mathrm{T}}(:,j),\ i=1\dots M,\ j=1\dots n \mid X_i^{\dagger}(j,:)\Gamma_\ell \ge 0,\ \ell=1\dots N\right\} \tag{476}$$

where c.i. denotes selection of only the conically independent vectors from the argument set, argument $(:,j)$ denotes the jth column while $(j,:)$ denotes the jth row, and $\{\Gamma_\ell\}$ constitutes the extreme directions of $\mathcal{K}$. Figure 51b (p.125) shows a cone and its dual found via this algorithm.

2.13.11.0.2 Example. *Dual of $\mathcal{K}$ nonsimplicial in subspace* aff $\mathcal{K}$.
Given conically independent generators for pointed closed convex cone $\mathcal{K}$ in $\mathbb{R}^4$ arranged columnar in

$$X = [\,\Gamma_1\ \Gamma_2\ \Gamma_3\ \Gamma_4\,] = \begin{bmatrix} 1 & 1 & 0 & 0 \\ -1 & 0 & 1 & 0 \\ 0 & -1 & 0 & 1 \\ 0 & 0 & -1 & -1 \end{bmatrix} \tag{477}$$

having $\dim \operatorname{aff} \mathcal{K} = \operatorname{rank} X = 3$, (281) then performing the most inefficient simplicial decomposition in $\operatorname{aff} \mathcal{K}$ we find

$$X_1 = \begin{bmatrix} 1 & 1 & 0 \\ -1 & 0 & 1 \\ 0 & -1 & 0 \\ 0 & 0 & -1 \end{bmatrix}, \quad X_2 = \begin{bmatrix} 1 & 1 & 0 \\ -1 & 0 & 0 \\ 0 & -1 & 1 \\ 0 & 0 & -1 \end{bmatrix}$$

$$X_3 = \begin{bmatrix} 1 & 0 & 0 \\ -1 & 1 & 0 \\ 0 & 0 & 1 \\ 0 & -1 & -1 \end{bmatrix}, \quad X_4 = \begin{bmatrix} 1 & 0 & 0 \\ 0 & 1 & 0 \\ -1 & 0 & 1 \\ 0 & -1 & -1 \end{bmatrix} \tag{478}$$

The corresponding dual simplicial cones in $\operatorname{aff} \mathcal{K}$ have generators respectively columnar in

$$4X_1^{\dagger \mathrm{T}} = \begin{bmatrix} 2 & 1 & 1 \\ -2 & 1 & 1 \\ 2 & -3 & 1 \\ -2 & 1 & -3 \end{bmatrix}, \quad 4X_2^{\dagger \mathrm{T}} = \begin{bmatrix} 1 & 2 & 1 \\ -3 & 2 & 1 \\ 1 & -2 & 1 \\ 1 & -2 & -3 \end{bmatrix}$$

$$4X_3^{\dagger \mathrm{T}} = \begin{bmatrix} 3 & 2 & -1 \\ -1 & 2 & -1 \\ -1 & -2 & 3 \\ -1 & -2 & -1 \end{bmatrix}, \quad 4X_4^{\dagger \mathrm{T}} = \begin{bmatrix} 3 & -1 & 2 \\ -1 & 3 & -2 \\ -1 & -1 & 2 \\ -1 & -1 & -2 \end{bmatrix} \tag{479}$$

Applying algorithm (476) we get

$$\begin{bmatrix} \Gamma_1^* & \Gamma_2^* & \Gamma_3^* & \Gamma_4^* \end{bmatrix} = \frac{1}{4} \begin{bmatrix} 1 & 2 & 3 & 2 \\ 1 & 2 & -1 & -2 \\ 1 & -2 & -1 & 2 \\ -3 & -2 & -1 & -2 \end{bmatrix} \tag{480}$$

whose rank is 3, and is the known result;[2.85] a conically independent set of generators for that pointed section of the dual cone $\mathcal{K}^*$ in $\operatorname{aff} \mathcal{K}$; *id est*, $\mathcal{K}^* \cap \operatorname{aff} \mathcal{K}$. □

2.13.11.0.3 Example. *Dual of proper polyhedral $\mathcal{K}$ in $\mathbb{R}^4$.*
Given conically independent generators for a full-dimensional pointed closed convex cone $\mathcal{K}$

$$X = \begin{bmatrix} \Gamma_1 & \Gamma_2 & \Gamma_3 & \Gamma_4 & \Gamma_5 \end{bmatrix} = \begin{bmatrix} 1 & 1 & 0 & 1 & 0 \\ -1 & 0 & 1 & 0 & 1 \\ 0 & -1 & 0 & 1 & 0 \\ 0 & 0 & -1 & -1 & 0 \end{bmatrix} \tag{481}$$

[2.85]These calculations proceed so as to be consistent with [118, §6]; as if the ambient vector space were proper subspace $\operatorname{aff} \mathcal{K}$ whose dimension is 3. In that ambient space, $\mathcal{K}$ may be regarded as a proper cone. Yet that author (from the citation) erroneously states dimension of the ordinary dual cone to be 3; it is, in fact, 4.

we count $5!/((5-4)!\,4!)=5$ component simplices.[2.86] Applying algorithm (476), we find
the six extreme directions of dual cone $\mathcal{K}^*$ (with $\Gamma_2 = \Gamma_5^*$)

$$X^* = \begin{bmatrix} \Gamma_1^* & \Gamma_2^* & \Gamma_3^* & \Gamma_4^* & \Gamma_5^* & \Gamma_6^* \end{bmatrix} = \begin{bmatrix} 1 & 0 & 0 & 1 & 1 & 1 \\ 1 & 0 & 0 & 1 & 0 & 0 \\ 1 & 0 & -1 & 0 & -1 & 1 \\ 1 & -1 & -1 & 1 & 0 & 0 \end{bmatrix} \qquad (482)$$

which means, (§2.13.6.1) this proper polyhedral $\mathcal{K} = \text{cone}(X)$ has six (three-dimensional)
facets generated $\mathcal{G}$ by its {extreme directions}:

$$\mathcal{G}\left\{\begin{matrix}\mathcal{F}_1\\\mathcal{F}_2\\\mathcal{F}_3\\\mathcal{F}_4\\\mathcal{F}_5\\\mathcal{F}_6\end{matrix}\right\} = \left\{\begin{matrix}\Gamma_1 & \Gamma_2 & \Gamma_3 & & & \\ \Gamma_1 & \Gamma_2 & & & \Gamma_5 & \\ \Gamma_1 & & & \Gamma_4 & \Gamma_5 & \\ \Gamma_1 & & \Gamma_3 & \Gamma_4 & & \\ & & \Gamma_3 & \Gamma_4 & \Gamma_5 & \\ & \Gamma_2 & \Gamma_3 & & \Gamma_5 & \end{matrix}\right\} \qquad (483)$$

whereas dual proper polyhedral cone $\mathcal{K}^*$ has only five:

$$\mathcal{G}\left\{\begin{matrix}\mathcal{F}_1^*\\\mathcal{F}_2^*\\\mathcal{F}_3^*\\\mathcal{F}_4^*\\\mathcal{F}_5^*\end{matrix}\right\} = \left\{\begin{matrix}\Gamma_1^* & \Gamma_2^* & \Gamma_3^* & \Gamma_4^* & & \\ \Gamma_1^* & \Gamma_2^* & & & & \Gamma_6^* \\ \Gamma_1^* & & & \Gamma_4^* & \Gamma_5^* & \Gamma_6^* \\ & & \Gamma_3^* & \Gamma_4^* & \Gamma_5^* & \\ & \Gamma_2^* & \Gamma_3^* & & \Gamma_5^* & \Gamma_6^* \end{matrix}\right\} \qquad (484)$$

Six two-dimensional cones, having generators respectively $\{\Gamma_1^*\ \Gamma_3^*\}\ \{\Gamma_2^*\ \Gamma_4^*\}\ \{\Gamma_1^*\ \Gamma_5^*\}$
$\{\Gamma_4^*\ \Gamma_6^*\}\ \{\Gamma_2^*\ \Gamma_5^*\}\ \{\Gamma_3^*\ \Gamma_6^*\}$, are relatively interior to dual facets; so cannot be
two-dimensional faces of $\mathcal{K}^*$ (by Definition 2.6.0.0.3).

   We can check this result (482) by reversing the process; we find $6!/((6-4)!\,4!)-3=12$
component simplices in the dual cone.[2.87] Applying algorithm (476) to those simplices
returns a conically independent set of generators for $\mathcal{K}$ equivalent to (481).   □

2.13.11.0.4   Exercise.   *Reaching proper polyhedral cone interior.*
Name two extreme directions Γ_i of cone $\mathcal{K}$ from Example 2.13.11.0.3 whose convex hull
passes through that cone's interior. Explain why. Are there two such extreme directions
of dual cone $\mathcal{K}^*$?   ▼

2.13.12   coordinates in proper nonsimplicial system

A natural question pertains to whether a theory of unique coordinates, like biorthogonal
expansion w.r.t pointed closed convex $\mathcal{K}$, is extensible to proper cones whose extreme
directions number in excess of ambient spatial dimensionality.

[2.86]There are no linearly dependent combinations of three or four extreme directions in the primal cone.
[2.87]Three combinations of four dual extreme directions are linearly dependent; they belong to the dual
facets. But there are no linearly dependent combinations of three dual extreme directions.

2.13.12.0.1 Theorem. *Conic coordinates.*

With respect to vector v in some finite-dimensional Euclidean space $\mathbb{R}^n$, define a coordinate $t_v^\star$ of point x in full-dimensional pointed closed convex cone $\mathcal{K}$

$$t_v^\star(x) \triangleq \sup\{t \in \mathbb{R} \mid x - tv \in \mathcal{K}\} \qquad (485)$$

Given points x and y in cone $\mathcal{K}$, if $t_v^\star(x) = t_v^\star(y)$ for each and every extreme direction v of $\mathcal{K}$ then $x = y$. $\diamond$

Conic coordinate definition (485) acquires its heritage from conditions (375) for generator membership to a smallest face. Coordinate $t_v^\star(c) = 0$, for example, corresponds to unbounded μ in (375); indicating, extreme direction v cannot belong to the smallest face of cone $\mathcal{K}$ that contains c.

2.13.12.0.2 Proof. Vector $x - t^\star v$ must belong to the cone boundary $\partial\mathcal{K}$ by definition (485). So there must exist a nonzero vector λ that is inward-normal to a hyperplane supporting cone $\mathcal{K}$ and containing $x - t^\star v$; *id est*, by boundary membership relation for full-dimensional pointed closed convex cones (§2.13.2)

$$x - t^\star v \in \partial\mathcal{K} \;\Leftrightarrow\; \exists\, \lambda \neq \mathbf{0} \ni \langle\lambda, x - t^\star v\rangle = 0, \;\; \lambda \in \mathcal{K}^*, \;\; x - t^\star v \in \mathcal{K} \quad (329)$$

where

$$\mathcal{K}^* = \{w \in \mathbb{R}^n \mid \langle v, w\rangle \geq 0 \text{ for all } v \in \mathcal{G}(\mathcal{K})\} \qquad (368)$$

is the full-dimensional pointed closed convex dual cone. The set $\mathcal{G}(\mathcal{K})$, of possibly infinite cardinality N, comprises generators for cone $\mathcal{K}$; *e.g.*, its extreme directions which constitute a minimal generating set. If $x - t^\star v$ is nonzero, any such vector λ must belong to the dual cone boundary by conjugate boundary membership relation

$$\lambda \in \partial\mathcal{K}^* \;\Leftrightarrow\; \exists\, x - t^\star v \neq \mathbf{0} \ni \langle\lambda, x - t^\star v\rangle = 0, \;\; x - t^\star v \in \mathcal{K}, \;\; \lambda \in \mathcal{K}^* \quad (330)$$

where

$$\mathcal{K} = \{z \in \mathbb{R}^n \mid \langle\lambda, z\rangle \geq 0 \text{ for all } \lambda \in \mathcal{G}(\mathcal{K}^*)\} \qquad (367)$$

This description of $\mathcal{K}$ means: cone $\mathcal{K}$ is an intersection of halfspaces whose inward-normals are generators of the dual cone. Each and every face of cone $\mathcal{K}$ (except the cone itself) belongs to a hyperplane supporting $\mathcal{K}$. Each and every vector $x - t^\star v$ on the cone boundary must therefore be orthogonal to an extreme direction constituting generators $\mathcal{G}(\mathcal{K}^*)$ of the dual cone.

To the ith extreme direction $v = \Gamma_i \in \mathbb{R}^n$ of cone $\mathcal{K}$, ascribe a coordinate $t_i^\star(x) \in \mathbb{R}$ of x from definition (485). On domain $\mathcal{K}$, the mapping

$$t^\star(x) = \begin{bmatrix} t_1^\star(x) \\ \vdots \\ t_N^\star(x) \end{bmatrix} : \mathbb{R}^n \to \mathbb{R}^N \qquad (486)$$

has no nontrivial nullspace. Because $x - t^\star v$ must belong to $\partial\mathcal{K}$ by definition, the mapping $t^\star(x)$ is equivalent to a convex problem (separable in index i) whose objective (by (329)) is tightly bounded below by 0:

$$t^\star(x) \equiv \underset{t \in \mathbb{R}^N}{\arg\text{minimize}} \quad \sum_{i=1}^N \Gamma_{j(i)}^{*\mathrm{T}}(x - t_i \Gamma_i) \qquad (487)$$
$$\text{subject to} \quad x - t_i \Gamma_i \in \mathcal{K}\,, \qquad i = 1 \dots N$$

where index $j \in \mathcal{I}$ is dependent on i and where (by (367)) $\lambda = \Gamma_j^* \in \mathbb{R}^n$ is an extreme direction of dual cone $\mathcal{K}^*$ that is normal to a hyperplane supporting $\mathcal{K}$ and containing $x - t_i^\star \Gamma_i$. Because extreme-direction cardinality N for cone $\mathcal{K}$ is not necessarily the same as for dual cone $\mathcal{K}^*$, index j must be judiciously selected from a set $\mathcal{I}$.

To prove injectivity when extreme-direction cardinality $N > n$ exceeds spatial dimension, we need only show mapping $t^\star(x)$ to be invertible; [132, thm.9.2.3] *id est*, x is recoverable given $t^\star(x)$:

$$x = \underset{\tilde{x} \in \mathbb{R}^n}{\arg\text{minimize}} \quad \sum_{i=1}^N \Gamma_{j(i)}^{*\mathrm{T}}(\tilde{x} - t_i^\star \Gamma_i) \qquad (488)$$
$$\text{subject to} \quad \tilde{x} - t_i^\star \Gamma_i \in \mathcal{K}\,, \qquad i = 1 \dots N$$

The feasible set of this nonseparable convex problem is an intersection of translated full-dimensional pointed closed convex cones $\bigcap_i \mathcal{K} + t_i^\star \Gamma_i$. The objective function's linear part describes movement in normal-direction $-\Gamma_j^*$ for each of N hyperplanes. The optimal point of hyperplane intersection is the unique solution x when $\{\Gamma_j^*\}$ comprises n linearly independent normals that come from the dual cone and make the objective vanish. Because the dual cone $\mathcal{K}^*$ is full-dimensional, pointed, closed, and convex by assumption, there exist N extreme directions $\{\Gamma_j^*\}$ from $\mathcal{K}^* \subset \mathbb{R}^n$ that span $\mathbb{R}^n$. So we need simply choose N spanning dual extreme directions that make the optimal objective vanish. Because such dual extreme directions preexist by (329), $t^\star(x)$ is invertible.

Otherwise, in the case $N \leq n$, $t^\star(x)$ holds coordinates for biorthogonal expansion. Reconstruction of x is therefore unique. ♦

2.13.12.1   reconstruction from conic coordinates

The foregoing proof of the *conic coordinates theorem* is not constructive; it establishes existence of dual extreme directions $\{\Gamma_j^*\}$ that will reconstruct a point x from its coordinates $t^\star(x)$ via (488), but does not prescribe the index set $\mathcal{I}$. There are at least two computational methods for specifying $\{\Gamma_{j(i)}^*\}$: one is combinatorial but sure to succeed, the other is a geometric method that searches for a minimum of a nonconvex function. We describe the latter:

Convex problem (P)

$$\text{(P)} \quad \begin{array}{ll} \underset{t \in \mathbb{R}}{\text{maximize}} & t \\ \text{subject to} & x - tv \in \mathcal{K} \end{array} \qquad \begin{array}{ll} \underset{\lambda \in \mathbb{R}^n}{\text{minimize}} & \lambda^\mathrm{T} x \\ \text{subject to} & \lambda^\mathrm{T} v = 1 \\ & \lambda \in \mathcal{K}^* \end{array} \text{(D)} \qquad (489)$$

is equivalent to definition (485) whereas convex problem (D) is its dual;[2.88] meaning, primal and dual optimal objectives are equal $t^\star = \lambda^{\star\mathrm{T}} x$ assuming Slater's condition (p.249)

[2.88] Form a Lagrangian associated with primal problem (P):

is satisfied. Under this assumption of strong duality, $\lambda^{\star\mathrm{T}}(x - t^\star v) = t^\star(1 - \lambda^{\star\mathrm{T}}v) = 0$; which implies, the primal problem is equivalent to

$$\begin{array}{ll} \underset{t\in\mathbb{R}}{\text{minimize}} & \lambda^{\star\mathrm{T}}(x - tv) \\ \text{subject to} & x - tv \in \mathcal{K} \end{array} \quad \text{(p)} \tag{490}$$

while the dual problem is equivalent to

$$\begin{array}{ll} \underset{\lambda\in\mathbb{R}^n}{\text{minimize}} & \lambda^{\mathrm{T}}(x - t^\star v) \\ \text{subject to} & \lambda^{\mathrm{T}}v = 1 \\ & \lambda \in \mathcal{K}^* \end{array} \quad \text{(d)} \tag{491}$$

Instead given coordinates $t^\star(x)$ and a description of cone $\mathcal{K}$, we propose inversion by alternating solution of respective primal and dual problems

$$\begin{array}{ll} \underset{x\in\mathbb{R}^n}{\text{minimize}} & \sum_{i=1}^{N} \Gamma_i^{*\mathrm{T}}(x - t_i^\star\Gamma_i) \\ \text{subject to} & x - t_i^\star\Gamma_i \in \mathcal{K}, \qquad i=1\dots N \end{array} \tag{492}$$

$$\begin{array}{ll} \underset{\Gamma_i^*\in\mathbb{R}^n}{\text{minimize}} & \sum_{i=1}^{N} \Gamma_i^{*\mathrm{T}}(x^\star - t_i^\star\Gamma_i) \\ \text{subject to} & \Gamma_i^{*\mathrm{T}}\Gamma_i = 1, \qquad i=1\dots N \\ & \Gamma_i^* \in \mathcal{K}^*, \qquad i=1\dots N \end{array} \tag{493}$$

where dual extreme directions Γ_i^* are initialized arbitrarily and ultimately ascertained by the alternation. Convex problems (492) and (493) are iterated until convergence which is guaranteed by virtue of a monotonically nonincreasing real sequence of objective values. Convergence can be fast. The mapping $t^\star(x)$ is uniquely inverted when the necessarily nonnegative objective vanishes; *id est*, when $\Gamma_i^{*\mathrm{T}}(x^\star - t_i^\star\Gamma_i) = 0 \ \forall i$. Here, a zero objective can occur only at the true solution x. But this global optimality condition cannot be guaranteed by the alternation because the common objective function, when regarded in both primal x and dual Γ_i^* variables simultaneously, is generally neither quasiconvex or monotonic. (§3.8.0.0.3)

Conversely, a nonzero objective at convergence is a certificate that inversion was not performed properly. A nonzero objective indicates that a global minimum of a multimodal objective function could not be found by this alternation. That is a flaw in this particular iterative algorithm for inversion; not in theory.[2.89] A numerical remedy is to reinitialize the Γ_i^* to different values.

$$\mathcal{L}(t,\lambda) = t + \lambda^{\mathrm{T}}(x - tv) = \lambda^{\mathrm{T}}x + t(1 - \lambda^{\mathrm{T}}v), \qquad \lambda \underset{\mathcal{K}^*}{\succeq} 0$$

$$\sup_t \mathcal{L}(t,\lambda) = \lambda^{\mathrm{T}}x, \qquad 1 - \lambda^{\mathrm{T}}v = 0$$

Dual variable (Lagrange multiplier [256, p.216]) λ generally has a nonnegative sense $\succeq$ for primal maximization with any cone membership constraint, whereas λ would have a nonpositive sense $\preceq$ were the primal instead a minimization problem with a cone membership constraint.

[2.89] The Proof 2.13.12.0.2, that suitable dual extreme directions $\{\Gamma_j^*\}$ always exist, means that a global optimization algorithm would always find the zero objective of alternation (492) (493); hence, the unique inversion x. But such an algorithm can be combinatorial.

Chapter 3

Geometry of convex functions

The link between convex sets and convex functions is via the epigraph: A function is convex if and only if its epigraph is a convex set.

−Werner Fenchel

We limit our treatment of *multidimensional functions*[3.1] to finite-dimensional Euclidean space.   Then an icon for a one-dimensional (real) *convex function* is bowl-shaped (Figure 78), whereas the *concave* icon is the inverted bowl; respectively characterized by a unique global minimum and maximum whose existence is assumed. Because of this simple relationship, usage of the term *convexity* is often implicitly inclusive of *concavity*. Despite iconic imagery, the reader is reminded that the set of all convex, concave, quasiconvex, and quasiconcave functions contains the *monotonic functions* [209] [221, §2.3.5].

3.1   Convex function

3.1.1   real and vector-valued function

Vector-valued function

$$f(X) : \mathbb{R}^{p \times k} \to \mathbb{R}^M = \begin{bmatrix} f_1(X) \\ \vdots \\ f_M(X) \end{bmatrix} \tag{494}$$

assigns each X in its domain dom f (a subset of ambient vector space $\mathbb{R}^{p \times k}$) to a specific element [264, p.3] of its range (a subset of $\mathbb{R}^M$).   Function $f(X)$ is *linear* in X on its domain if and only if, for each and every $Y, Z \in \text{dom } f$ and $\alpha, \beta \in \mathbb{R}$

$$f(\alpha Y + \beta Z) = \alpha f(Y) + \beta f(Z) \tag{495}$$

[3.1] vector- or matrix-valued functions including the real functions. Appendix D, with its tables of first- and second-order gradients, is the practical adjunct to this chapter.

187
CITATION: Dattorro, *Convex Optimization & Euclidean Distance Geometry*, $\mathcal{M}\varepsilon\beta oo$ Publishing USA, 2005, v2015.02.15.

A vector-valued function $f(X) : \mathbb{R}^{p \times k} \to \mathbb{R}^M$ is *convex* in X if and only if $\operatorname{dom} f$ is a convex set and, for each and every $Y, Z \in \operatorname{dom} f$ and $0 \le \mu \le 1$

$$f(\mu Y + (1 - \mu)Z) \underset{\mathbb{R}^M_+}{\preceq} \mu f(Y) + (1 - \mu)f(Z) \qquad (496)$$

As defined, continuity is implied but not differentiability (nor *smoothness*).[3.2] Apparently some, but not all, nonlinear functions are convex. Reversing sense of the inequality flips this definition to concavity. Linear (and affine §3.4)[3.3] functions attain equality in this definition. Linear functions are therefore simultaneously convex and concave.

Vector-valued functions are most often compared (182) as in (496) with respect to the M-dimensional selfdual nonnegative orthant $\mathbb{R}^M_+$, a proper cone.[3.4] In this case, the test prescribed by (496) is simply a comparison on $\mathbb{R}$ of each *entry* f_i of a vector-valued function f. (§2.13.4.2.3) The vector-valued function case is therefore a straightforward generalization of conventional convexity theory for a real function. This conclusion follows from theory of dual generalized inequalities (§2.13.2.0.1) which asserts

$$f \text{ convex w.r.t } \mathbb{R}^M_+ \iff w^{\mathrm{T}}f \text{ convex } \forall w \in \mathcal{G}(\mathbb{R}^{M^*}_+) \qquad (497)$$

shown by substitution of the defining inequality (496). Discretization allows relaxation (§2.13.4.2.1) of a semiinfinite number of conditions $\{w \in \mathbb{R}^{M^*}_+\}$ to:

$$\{w \in \mathcal{G}(\mathbb{R}^{M^*}_+)\} \equiv \{e_i \in \mathbb{R}^M, \ i = 1 \dots M\} \qquad (498)$$

(the standard basis for $\mathbb{R}^M$ and a minimal set of generators (§2.8.1.2) for $\mathbb{R}^M_+$) from which the stated conclusion follows; *id est*, the test for convexity of a vector-valued function is a comparison on $\mathbb{R}$ of each entry.

3.1.2   strict convexity

When $f(X)$ instead satisfies, for each and every distinct Y and Z in $\operatorname{dom} f$ and all $0 < \mu < 1$ on an open interval

$$f(\mu Y + (1 - \mu)Z) \underset{\mathbb{R}^M_+}{\prec} \mu f(Y) + (1 - \mu)f(Z) \qquad (499)$$

then we shall say f is a *strictly convex function*.

Similarly to (497)

$$f \text{ strictly convex w.r.t } \mathbb{R}^M_+ \iff w^{\mathrm{T}}f \text{ strictly convex } \forall w \in \mathcal{G}(\mathbb{R}^{M^*}_+) \qquad (500)$$

discretization allows relaxation of the semiinfinite number of conditions $\{w \in \mathbb{R}^{M^*}_+, \ w \neq \mathbf{0}\}$ (325) to a finite number (498). More tests for strict convexity are given in §3.6.1.0.2, §3.6.4, and §3.7.3.0.2.

[3.2]Figure 69b illustrates a nondifferentiable convex function.   Differentiability is certainly not a requirement for optimization of convex functions by numerical methods; *e.g.*, [249].

[3.3]While linear functions are not invariant to translation (offset), convex functions are.

[3.4]Definition of convexity can be broadened to other (not necessarily proper) cones. Referred to in the literature as $\mathcal{K}$-*convexity*, [302] $\mathbb{R}^{M^*}_+$ (497) generalizes to $\mathcal{K}^*$.

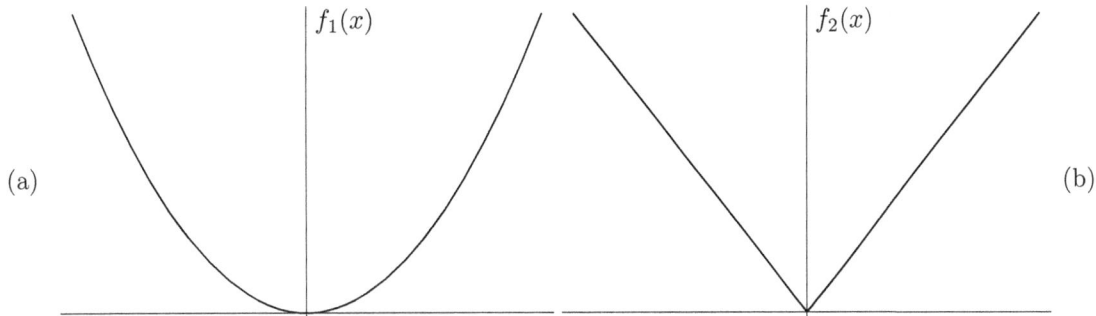

(a)   (b)

Figure 69: Convex real functions here have a unique minimizer $x^\star$. For $x \in \mathbb{R}$, $f_1(x) = x^2 = \|x\|_2^2$ is strictly convex whereas nondifferentiable function $f_2(x) = \sqrt{x^2} = |x| = \|x\|_2$ is convex but not strictly. Strict convexity of a real function is only a sufficient condition for minimizer uniqueness.

3.1.2.1 local/global minimum, uniqueness of solution

A local minimum of any convex real function is also its global minimum. In fact, any convex real function $f(X)$ has one minimum value over any convex subset of its domain. [310, p.123] Yet solution to some convex optimization problem is, in general, not unique; *id est*, given minimization of a convex real function over some convex feasible set $\mathcal{C}$

$$\begin{array}{ll} \underset{X}{\text{minimize}} & f(X) \\ \text{subject to} & X \in \mathcal{C} \end{array} \qquad (501)$$

any *optimal solution* $X^\star$ comes from a convex set of optimal solutions

$$X^\star \in \{X \mid f(X) = \inf_{Y \in \mathcal{C}} f(Y)\} \subseteq \mathcal{C} \qquad (502)$$

But a strictly convex real function has a unique minimizer $X^\star$; *id est*, for the optimal solution set in (502) to be a single point, it is sufficient (Figure 69) that $f(X)$ be a strictly convex real[3.5] function and set $\mathcal{C}$ convex. But strict convexity is not necessary for minimizer uniqueness: existence of any strictly supporting hyperplane to a convex function epigraph (p.187, §3.5) at its minimum over $\mathcal{C}$ is necessary and sufficient for uniqueness.

Quadratic real functions $x^{\mathrm{T}}Ax + b^{\mathrm{T}}x + c$ are convex in x iff $A \succeq 0$. (§3.6.4.0.1) Quadratics characterized by positive definite matrix $A \succ 0$ are strictly convex and *vice versa*. The vector 2-norm square $\|x\|^2$ (Euclidean norm square) and Frobenius' norm square $\|X\|_{\mathrm{F}}^2$, for example, are strictly convex functions of their respective argument (each absolute norm is convex but not strictly). Figure 69a illustrates a strictly convex real function.

[3.5]It is more customary to consider only a real function for the objective of a convex optimization problem because vector- or matrix-valued functions can introduce ambiguity into the optimal objective value. (§2.7.2.2, §3.1.2.2) Study of multidimensional objective functions is called *multicriteria-* [331] or *multiobjective-* or *vector-optimization*.

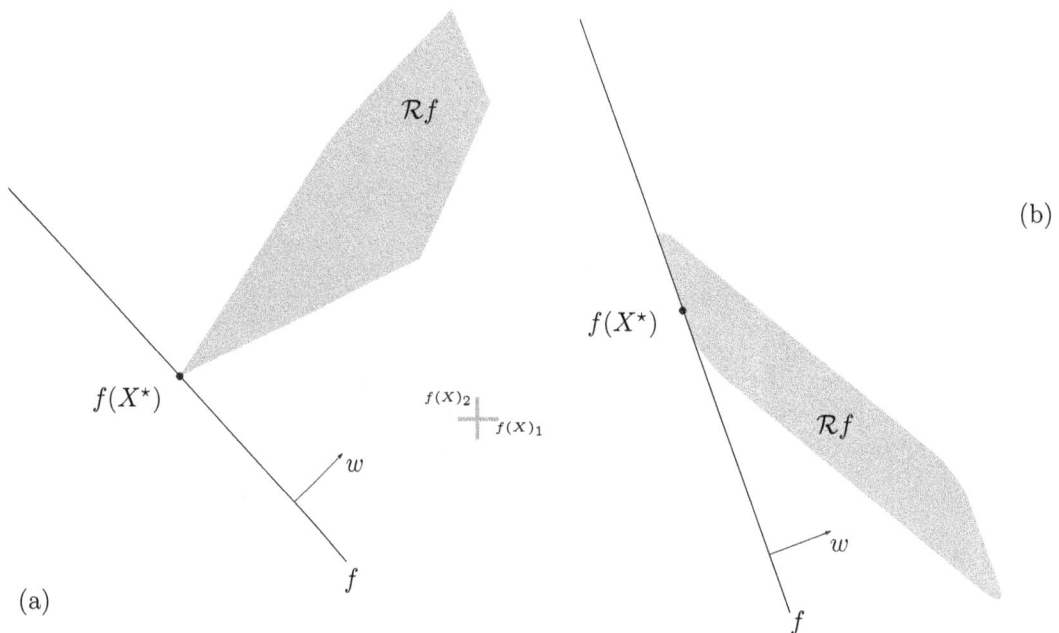

Figure 70: (*confer* Figure 41) Function range is convex for a convex problem.
(a) Point $f(X^\star)$ is the unique minimum element of function range $\mathcal{R}f$.
(b) Point $f(X^\star)$ is a minimal element of depicted range.
(Cartesian axes drawn for reference.)

3.1.2.2 minimum/minimal element, dual cone characterization

$f(X^\star)$ is the *minimum element* of its range if and only if, for each and every $w \in \text{int } \mathbb{R}_+^{M^*}$, it is the unique minimizer of $w^\mathrm{T} f$. (Figure 70) [62, §2.6.3]

If $f(X^\star)$ is a *minimal element* of its range, then there exists a nonzero $w \in \mathbb{R}_+^{M^*}$ such that $f(X^\star)$ minimizes $w^\mathrm{T} f$. If $f(X^\star)$ minimizes $w^\mathrm{T} f$ for some $w \in \text{int } \mathbb{R}_+^{M^*}$, conversely, then $f(X^\star)$ is a minimal element of its range.

3.1.2.2.1 Exercise. *Cone of convex functions.*
Prove that relation (497) implies: The set of all convex vector-valued functions forms a convex cone in some space. Indeed, any nonnegatively weighted sum of convex functions remains convex. So trivial function $f = \mathbf{0}$ is convex. Relatively interior to each face of this cone are the strictly convex functions of corresponding dimension.[3.6] How do convex real functions fit into this cone? Where are the affine functions? ▼

3.1.2.2.2 Example. *Conic origins of Lagrangian.*
The cone of convex functions, implied by membership relation (497), provides foundation for what is known as a *Lagrangian function.* [258, p.398] [288] Consider a conic optimization problem, for proper cone $\mathcal{K}$ and affine subset $\mathcal{A}$

$$\begin{array}{ll} \underset{x}{\text{minimize}} & f(x) \\ \text{subject to} & g(x) \succeq_\mathcal{K} 0 \\ & h(x) \in \mathcal{A} \end{array} \qquad (503)$$

A Cartesian product of convex functions remains convex, so we may write

$$\begin{bmatrix} f \\ g \\ h \end{bmatrix} \text{ convex w.r.t } \begin{bmatrix} \mathbb{R}_+^M \\ \mathcal{K} \\ \mathcal{A} \end{bmatrix} \Leftrightarrow [w^\mathrm{T} \lambda^\mathrm{T} \nu^\mathrm{T}] \begin{bmatrix} f \\ g \\ h \end{bmatrix} \text{ convex } \forall \begin{bmatrix} w \\ \lambda \\ \nu \end{bmatrix} \in \begin{bmatrix} \mathbb{R}_+^{M^*} \\ \mathcal{K}^* \\ \mathcal{A}^\perp \end{bmatrix} \quad (504)$$

where $\mathcal{A}^\perp$ is a normal cone to $\mathcal{A}$. A normal cone to an affine subset is the orthogonal complement of its parallel subspace (§E.10.3.2.1).

Membership relation (504) holds because of equality for h in convexity criterion (496) and because normal-cone membership relation (450), given point $a \in \mathcal{A}$, becomes

$$h \in \mathcal{A} \Leftrightarrow \langle \nu, h - a \rangle = 0 \text{ for all } \nu \in \mathcal{A}^\perp \qquad (505)$$

In other words: all affine functions are convex (with respect to any given proper cone), all convex functions are translation invariant, whereas any affine function must satisfy (505).

A real Lagrangian arises from the scalar term in (504); *id est,*

$$\mathfrak{L} \triangleq [w^\mathrm{T} \lambda^\mathrm{T} \nu^\mathrm{T}] \begin{bmatrix} f \\ g \\ h \end{bmatrix} = w^\mathrm{T} f + \lambda^\mathrm{T} g + \nu^\mathrm{T} h \qquad (506)$$

□

[3.6]Strict case excludes cone's point at origin and zero weighting.

3.2   Practical norm functions, absolute value

To some mathematicians, *"all norms on $\mathbb{R}^n$ are equivalent"* [164, p.53]; meaning, ratios of different norms are bounded above and below by finite predeterminable constants. But to statisticians and engineers, all norms are certainly not created equal; as evidenced by the *compressed sensing* (*sparsity*) revolution, begun in 2004, whose focus is predominantly the argument of a 1-norm minimization.

A *norm* on $\mathbb{R}^n$ is a convex function $f : \mathbb{R}^n \to \mathbb{R}$ satisfying: for $x, y \in \mathbb{R}^n$, $\alpha \in \mathbb{R}$ [233, p.59] [164, p.52]

1. $f(x) \geq 0$ $(f(x) = 0 \Leftrightarrow x = 0)$ (nonnegativity)

2. $f(x + y) \leq f(x) + f(y)$ [3.7](triangle inequality)

3. $f(\alpha x) = |\alpha| f(x)$ (nonnegative homogeneity)

Convexity follows by properties 2 and 3. Most useful are 1-, 2-, and ∞-norm:

$$\|x\|_1 \;=\; \begin{array}{c}\text{minimize}\\ {\scriptstyle t \in \mathbb{R}^n}\end{array} \quad \mathbf{1}^{\mathrm{T}} t \tag{507}$$
$$\text{subject to} \quad -t \preceq x \preceq t$$

where $|x| = t^\star$ (entrywise absolute value equals optimal t).[3.8]

$$\|x\|_2 \;=\; \begin{array}{c}\text{minimize}\\ {\scriptstyle t \in \mathbb{R}}\end{array} \quad t$$
$$\text{subject to} \quad \begin{bmatrix} tI & x \\ x^{\mathrm{T}} & t \end{bmatrix}_{\mathbb{S}^{n+1}_+} \succeq 0 \tag{508}$$

where $\|x\|_2 = \|x\| \triangleq \sqrt{x^{\mathrm{T}} x} = t^\star$.

$$\|x\|_\infty \;=\; \begin{array}{c}\text{minimize}\\ {\scriptstyle t \in \mathbb{R}}\end{array} \quad t \tag{509}$$
$$\text{subject to} \quad -t\mathbf{1} \preceq x \preceq t\mathbf{1}$$

where $\max\{|x_i|,\; i = 1 \ldots n\} = t^\star$ because $\|x\|_\infty = \max\{|x_i|\} \leq t \Leftrightarrow |x| \preceq t\mathbf{1}$.   Absolute value $|x|$ inequality, in this sense, describes a norm ball.

$$\|x\|_1 \;=\; \begin{array}{c}\text{minimize}\\ {\scriptstyle \alpha \in \mathbb{R}^n,\ \beta \in \mathbb{R}^n}\end{array} \quad \mathbf{1}^{\mathrm{T}}(\alpha + \beta)$$
$$\text{subject to} \quad \alpha, \beta \succeq 0 \tag{510}$$
$$x = \alpha - \beta$$

where $|x| = \alpha^\star + \beta^\star$ because of complementarity $\alpha^{\star \mathrm{T}} \beta^\star = 0$ at optimality. (507) (509) (510) represent linear programs, (508) is a semidefinite program.

Over some convex set $\mathcal{C}$, given vector constant y or matrix constant Y,

$$\arg \inf_{x \in \mathcal{C}} \|x - y\| = \arg \inf_{x \in \mathcal{C}} \|x - y\|^2 \tag{511}$$

[3.7] $\|x + y\| \leq \|x\| + \|y\|$ for any norm, with equality iff $x = \kappa y$ where $\kappa \geq 0$.
[3.8] Vector $\mathbf{1}$ may be replaced with any positive [*sic*] vector to get absolute value, theoretically, although $\mathbf{1}$ provides the 1-norm.

$$\arg\inf_{X\in\mathcal{C}}\|X-Y\| = \arg\inf_{X\in\mathcal{C}}\|X-Y\|^2 \tag{512}$$

are unconstrained convex problems for any convex norm and any affine transformation of variable. (But equality would not hold for, instead, a sum of norms; *e.g.*, §5.4.2.2.4.) Optimal solution is norm dependent: [62, p.297]

$$\begin{array}{ll}\underset{x\in\mathbb{R}^n}{\text{minimize}} & \|x\|_1 \\ \text{subject to} & x\in\mathcal{C}\end{array} \equiv \begin{array}{ll}\underset{x\in\mathbb{R}^n,\,t\in\mathbb{R}^n}{\text{minimize}} & \mathbf{1}^{\mathrm{T}}t \\ \text{subject to} & -t\preceq x\preceq t \\ & x\in\mathcal{C}\end{array} \tag{513}$$

$$\begin{array}{ll}\underset{x\in\mathbb{R}^n}{\text{minimize}} & \|x\|_2 \\ \text{subject to} & x\in\mathcal{C}\end{array} \equiv \begin{array}{ll}\underset{x\in\mathbb{R}^n,\,t\in\mathbb{R}}{\text{minimize}} & t \\ \text{subject to} & \begin{bmatrix} tI & x \\ x^{\mathrm{T}} & t \end{bmatrix}\underset{\mathbb{S}_+^{n+1}}{\succeq} 0 \\ & x\in\mathcal{C}\end{array} \tag{514}$$

$$\begin{array}{ll}\underset{x\in\mathbb{R}^n}{\text{minimize}} & \|x\|_\infty \\ \text{subject to} & x\in\mathcal{C}\end{array} \equiv \begin{array}{ll}\underset{x\in\mathbb{R}^n,\,t\in\mathbb{R}}{\text{minimize}} & t \\ \text{subject to} & -t\mathbf{1}\preceq x\preceq t\mathbf{1} \\ & x\in\mathcal{C}\end{array} \tag{515}$$

In $\mathbb{R}^n$ these norms represent: $\|x\|_1$ is length measured along a grid (*taxicab distance*), $\|x\|_2$ is Euclidean length, $\|x\|_\infty$ is maximum |coordinate|.

$$\begin{array}{ll}\underset{x\in\mathbb{R}^n}{\text{minimize}} & \|x\|_1 \\ \text{subject to} & x\in\mathcal{C}\end{array} \equiv \begin{array}{ll}\underset{\alpha\in\mathbb{R}^n,\,\beta\in\mathbb{R}^n}{\text{minimize}} & \mathbf{1}^{\mathrm{T}}(\alpha+\beta) \\ \text{subject to} & \alpha,\beta\succeq 0 \\ & x=\alpha-\beta \\ & x\in\mathcal{C}\end{array} \tag{516}$$

These foregoing problems (507)-(516) are convex whenever set $\mathcal{C}$ is. Their equivalence transformations make objectives smooth.

3.2.0.0.1 Example. *Projecting the origin, on an affine subset, in 1-norm.*
In (1977) we interpret *least norm* solution to linear system $Ax=b$ as orthogonal projection of the origin $\mathbf{0}$ on affine subset $\mathcal{A}=\{x\in\mathbb{R}^n\,|\,Ax=b\}$ where $A\in\mathbb{R}^{m\times n}$ is fat full-rank. Suppose, instead of the Euclidean metric, we use taxicab distance to do projection. Then the least 1-norm problem is stated, for $b\in\mathcal{R}(A)$

$$\begin{array}{ll}\underset{x}{\text{minimize}} & \|x\|_1 \\ \text{subject to} & Ax=b\end{array} \tag{517}$$

`a.k.a` *compressed sensing problem.* Optimal solution can be interpreted as an *oblique projection* of the origin on $\mathcal{A}$ simply because the Euclidean metric is not employed. This problem statement sometimes returns optimal $x^\star$ having minimal cardinality; which can be explained intuitively with reference to Figure **71**: [19]

Projection of the origin, in 1-norm, on affine subset $\mathcal{A}$ is equivalent to maximization (in this case) of the 1-norm ball $\mathcal{B}_1$ until it kisses $\mathcal{A}$; rather, a kissing point in $\mathcal{A}$ achieves

$$\mathcal{A} = \{x \in \mathbb{R}^3 \mid Ax = b\}$$

$\mathbb{R}^3$

$$\mathcal{B}_1 = \{x \in \mathbb{R}^3 \mid \|x\|_1 \le 1\}$$

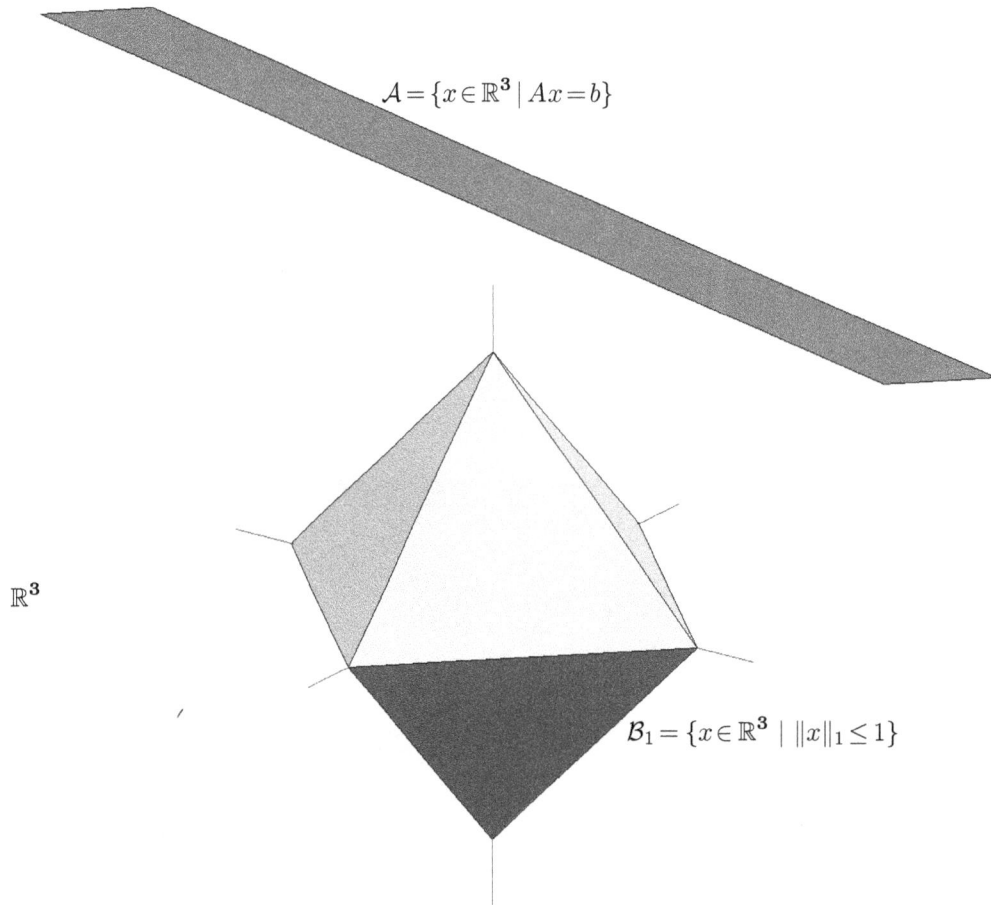

Figure 71: 1-norm ball $\mathcal{B}_1$ is convex hull of all cardinality-1 vectors of unit norm (its vertices). Ball boundary contains all points equidistant from origin in 1-norm. Cartesian axes drawn for reference. Plane $\mathcal{A}$ is overhead (drawn truncated). If 1-norm ball is expanded until it *kisses* $\mathcal{A}$ (intersects ball only at boundary), then distance (in 1-norm) from origin to $\mathcal{A}$ is achieved. Euclidean ball would be spherical in this dimension. Only were $\mathcal{A}$ parallel to two axes could there be a minimal cardinality least Euclidean norm solution. Yet 1-norm ball offers infinitely many, but not all, $\mathcal{A}$-orientations resulting in a minimal cardinality solution. (1-norm ball is an octahedron in this dimension while ∞-norm ball is a cube.)

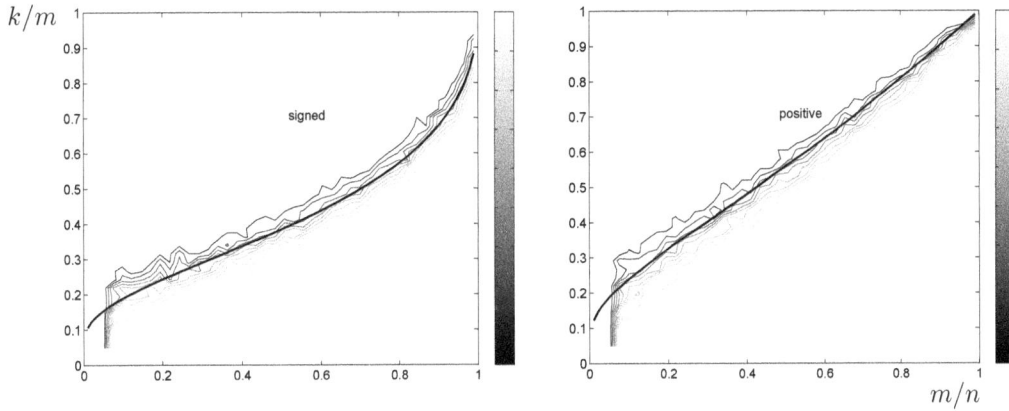

$$(517) \quad \begin{array}{ll} \underset{x}{\text{minimize}} & \|x\|_1 \\ \text{subject to} & Ax = b \end{array} \qquad\qquad \begin{array}{ll} \underset{x}{\text{minimize}} & \|x\|_1 \\ \text{subject to} & Ax = b \\ & x \succeq 0 \end{array} \quad (522)$$

Figure 72: Exact recovery transition: Respectively signed [121] [123] or positive [128] [126] [127] solutions x to $Ax = b$ with sparsity k below thick curve are recoverable. For Gaussian random matrix $A \in \mathbb{R}^{m \times n}$, thick curve demarcates phase transition in ability to find sparsest solution x by linear programming. These results were empirically reproduced in [38].

the distance in 1-norm from the origin to $\mathcal{A}$. For the example illustrated ($m = 1$, $n = 3$), it appears that a vertex of the ball will be first to touch $\mathcal{A}$. 1-norm ball vertices in $\mathbb{R}^3$ represent nontrivial points of minimal cardinality 1, whereas edges represent cardinality 2, while relative interiors of facets represent maximal cardinality 3. By reorienting affine subset $\mathcal{A}$ so it were parallel to an edge or facet, it becomes evident as we expand or contract the ball that a kissing point is not necessarily unique.[3.9]

The 1-norm ball in $\mathbb{R}^n$ has 2^n facets and $2n$ vertices.[3.10] For $n > 0$

$$\mathcal{B}_1 = \{x \in \mathbb{R}^n \mid \|x\|_1 \leq 1\} = \text{conv}\{\|x \in \mathbb{R}^n\| = 1 \mid \text{card } x = 1\} = \text{conv}\{\pm e_i \in \mathbb{R}^n, \ i = 1 \ldots n\} \tag{518}$$

is a vertex-description of the unit 1-norm ball. Maximization of the 1-norm ball, until it kisses $\mathcal{A}$, is equivalent to minimization of the 1-norm ball until it no longer intersects $\mathcal{A}$. Then projection of the origin on affine subset $\mathcal{A}$ is

$$\begin{array}{ll} \underset{x \in \mathbb{R}^n}{\text{minimize}} & \|x\|_1 \\ \text{subject to} & Ax = b \end{array} \quad \equiv \quad \begin{array}{ll} \underset{c \in \mathbb{R}, \ x \in \mathbb{R}^n}{\text{minimize}} & c \\ \text{subject to} & x \in c\mathcal{B}_1 \\ & Ax = b \end{array} \tag{519}$$

[3.9] This is unlike the case for the Euclidean ball (1977) where minimum-distance projection on a convex set is unique (§E.9); all kissable faces of the Euclidean ball are single points (vertices).
[3.10] The ∞-norm ball in $\mathbb{R}^n$ has $2n$ facets and 2^n vertices.

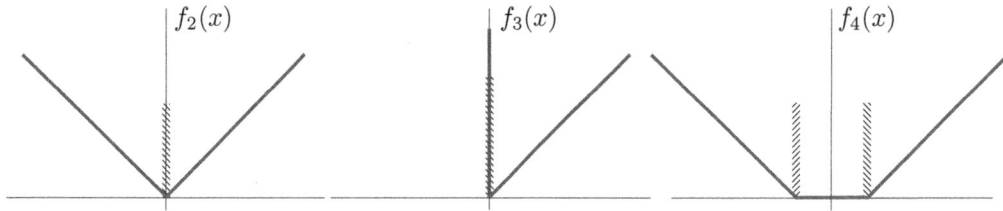

Figure 73: Under 1-norm $f_2(x)$, histogram (hatched) of residual amplitudes $Ax-b$ exhibits predominant accumulation of zero-residuals. Nonnegatively constrained 1-norm $f_3(x)$ from (522) accumulates more zero-residuals than $f_2(x)$. Under norm $f_4(x)$ (not discussed), histogram would exhibit predominant accumulation of (nonzero) residuals at gradient discontinuities.

where

$$cB_1 = \{[\,I \in \mathbb{R}^{n \times n} \quad -I \in \mathbb{R}^{n \times n}\,]a \mid a^{\mathrm{T}}\mathbf{1} = c, \ a \succeq 0\} \tag{520}$$

which is the convex hull of 1-norm ball vertices. Then (519) is equivalent to

$$
\begin{array}{ll}
\underset{c \in \mathbb{R}, \ x \in \mathbb{R}^n, \ a \in \mathbb{R}^{2n}}{\text{minimize}} \quad c \\
\text{subject to} \quad
\begin{aligned}
x &= [\,I \quad -I\,]a \\
a^{\mathrm{T}}\mathbf{1} &= c \\
a &\succeq 0 \\
Ax &= b
\end{aligned}
\end{array}
\quad \equiv \quad
\begin{array}{ll}
\underset{a \in \mathbb{R}^{2n}}{\text{minimize}} \quad \|a\|_1 \\
\text{subject to} \quad
\begin{aligned}
[\,A \quad -A\,]a &= b \\
a &\succeq 0
\end{aligned}
\end{array}
\tag{521}
$$

where $x^\star = [\,I \quad -I\,]a^\star$. (confer(516)) Significance of this result:

- (confer p.346) Any vector 1-norm minimization problem may have its variable replaced with a nonnegative variable of the same optimal cardinality but twice the length.

All other things being equal, nonnegative variables are easier to solve for sparse solutions. (Figure **72**, Figure **73**, Figure **104**) The compressed sensing problem (517) becomes easier to interpret; e.g., for $A \in \mathbb{R}^{m \times n}$

$$
\begin{array}{ll}
\underset{x}{\text{minimize}} \quad \|x\|_1 \\
\text{subject to} \quad
\begin{aligned}
Ax &= b \\
x &\succeq 0
\end{aligned}
\end{array}
\quad \equiv \quad
\begin{array}{ll}
\underset{x}{\text{minimize}} \quad \mathbf{1}^{\mathrm{T}}x \\
\text{subject to} \quad
\begin{aligned}
Ax &= b \\
x &\succeq 0
\end{aligned}
\end{array}
\tag{522}
$$

movement of a hyperplane (Figure **27**, Figure **31**) over a bounded polyhedron always has a *vertex solution* [96, p.158]. Or vector b might lie on the relative boundary of a pointed polyhedral cone $\mathcal{K} = \{Ax \mid x \succeq 0\}$. In the latter case, we find practical application of the smallest face $\mathcal{F}$ containing b from §2.13.4.3 to remove all columns of matrix A not belonging to $\mathcal{F}$; because those columns correspond to 0-entries in vector x (and *vice versa*).   □

3.2.0.0.2 Exercise. *Combinatorial optimization.*
A device commonly employed to relax combinatorial problems is to arrange desirable solutions at vertices of bounded polyhedra; *e.g.*, the permutation matrices of dimension n, which are factorial in number, are the extreme points of a polyhedron in the nonnegative orthant described by an intersection of $2n$ hyperplanes (§2.3.2.0.4). Minimizing a linear objective function over a bounded polyhedron is a convex problem (a linear program) that always has an optimal solution residing at a vertex.

What about minimizing other functions? Given some *nonsingular* matrix A, geometrically describe three circumstances under which there are likely to exist vertex solutions to

$$\begin{array}{ll} \underset{x\in\mathbb{R}^n}{\text{minimize}} & \|Ax\|_1 \\ \text{subject to} & x\in\mathcal{P} \end{array} \tag{523}$$

optimized over some bounded polyhedron $\mathcal{P}$.[3.11] ▼

3.2.1 k smallest entries

Sum of the k smallest entries of $x\in\mathbb{R}^n$ is the optimal objective value from: for $1\le k\le n$

$$\sum_{i=n-k+1}^{n}\pi(x)_i = \begin{array}{ll} \underset{y\in\mathbb{R}^n}{\text{minimize}} & x^{\mathrm{T}}y \\ \text{subject to} & 0\preceq y\preceq\mathbf{1} \\ & \mathbf{1}^{\mathrm{T}}y=k \end{array} \equiv \sum_{i=n-k+1}^{n}\pi(x)_i = \begin{array}{ll} \underset{z\in\mathbb{R}^n,\,t\in\mathbb{R}}{\text{maximize}} & k\,t+\mathbf{1}^{\mathrm{T}}z \\ \text{subject to} & x\succeq t\,\mathbf{1}+z \\ & z\preceq 0 \end{array} \tag{524}$$

which are dual linear programs, where $\pi(x)_1 = \max\{x_i,\ i=1\ldots n\}$ where π is a nonlinear permutation-operator sorting its vector argument into nonincreasing order. Finding k smallest entries of an n-length vector x is expressible as an infimum of $n!/(k!(n-k)!)$ linear functions of x. The sum $\sum\pi(x)_i$ is therefore a concave function of x; in fact, monotonic (§3.6.1.0.1) in this instance.

3.2.2 k largest entries

Sum of the k largest entries of $x\in\mathbb{R}^n$ is the optimal objective value from: [62, exer.5.19]

$$\sum_{i=1}^{k}\pi(x)_i = \begin{array}{ll} \underset{y\in\mathbb{R}^n}{\text{maximize}} & x^{\mathrm{T}}y \\ \text{subject to} & 0\preceq y\preceq\mathbf{1} \\ & \mathbf{1}^{\mathrm{T}}y=k \end{array} \equiv \sum_{i=1}^{k}\pi(x)_i = \begin{array}{ll} \underset{z\in\mathbb{R}^n,\,t\in\mathbb{R}}{\text{minimize}} & k\,t+\mathbf{1}^{\mathrm{T}}z \\ \text{subject to} & x\preceq t\,\mathbf{1}+z \\ & z\succeq 0 \end{array} \tag{525}$$

which are dual linear programs. Finding k largest entries of an n-length vector x is expressible as a supremum of $n!/(k!(n-k)!)$ linear functions of x. (Figure 75) The summation is therefore a convex function (and monotonic in this instance, §3.6.1.0.1).

[3.11]Hint: Suppose, for example, $\mathcal{P}$ belongs to an orthant and A were orthogonal. Begin with $A=I$ and apply level sets of the objective, as in Figure 68 and Figure 71. Or rewrite the problem as a linear program like (513) and (515) but in a composite variable $\begin{bmatrix} x \\ t \end{bmatrix}\leftarrow y$.

3.2.2.1 k-largest norm

Let $\Pi\,x$ be a permutation of entries x_i such that their absolute value becomes arranged in nonincreasing order: $|\Pi\,x|_1 \geq |\Pi\,x|_2 \geq \cdots \geq |\Pi\,x|_n$. Sum of the k largest entries of $|x| \in \mathbb{R}^n$ is a norm, by properties of vector norm (§3.2), and is the optimal objective value of a linear program:

$$
\|x\|_{\substack{n \\ k}} \;\triangleq\; \sum_{i=1}^{k} |\Pi\,x|_i = \sum_{i=1}^{k} \pi(|x|)_i \quad =\quad
\begin{array}{ll}
\underset{z \in \mathbb{R}^n,\; t \in \mathbb{R}}{\text{minimize}} & k\,t + \mathbf{1}^{\mathrm{T}} z \\[4pt]
\text{subject to} & -t\,\mathbf{1} - z \preceq x \preceq t\,\mathbf{1} + z \\[2pt]
& z \succeq 0
\end{array}
$$

$$
=\; \sup_{i \in \mathcal{I}} \left\{ a_i^{\mathrm{T}} x \;\middle|\; \begin{array}{l} a_{ij} \in \{-1,0,1\} \\ \operatorname{card} a_i = k \end{array} \right\} \quad =\quad
\begin{array}{ll}
\underset{y_1,\,y_2 \in \mathbb{R}^n}{\text{maximize}} & (y_1 - y_2)^{\mathrm{T}} x \\[4pt]
\text{subject to} & 0 \preceq y_1 \preceq \mathbf{1} \\[2pt]
& 0 \preceq y_2 \preceq \mathbf{1} \\[2pt]
& (y_1 + y_2)^{\mathrm{T}} \mathbf{1} = k
\end{array}
\tag{526}
$$

where the norm subscript derives from a binomial coefficient $\dbinom{n}{k}$, and

$$
\begin{aligned}
\|x\|_{\substack{n \\ n}} &= \|x\|_1 \\
\|x\|_{\substack{n \\ 1}} &= \|x\|_\infty \\
\|x\|_{\substack{n \\ k}} &= \|\pi(|x|)_{1:k}\|_1
\end{aligned}
\tag{527}
$$

Sum of k largest absolute entries of an n-length vector x is expressible as a supremum of $2^k n!/(k!(n-k)!)$ linear functions of x; (Figure 75) hence, this norm is convex in x. [62, exer.6.3e]

$$
\begin{array}{ll}
\underset{x \in \mathbb{R}^n}{\text{minimize}} & \|x\|_{\substack{n \\ k}} \\[4pt]
\text{subject to} & x \in \mathcal{C}
\end{array}
\quad \equiv \quad
\begin{array}{ll}
\underset{z \in \mathbb{R}^n,\; t \in \mathbb{R},\; x \in \mathbb{R}^n}{\text{minimize}} & k\,t + \mathbf{1}^{\mathrm{T}} z \\[4pt]
\text{subject to} & -t\,\mathbf{1} - z \preceq x \preceq t\,\mathbf{1} + z \\[2pt]
& z \succeq 0 \\[2pt]
& x \in \mathcal{C}
\end{array}
\tag{528}
$$

3.2.2.1.1 Exercise. *Polyhedral epigraph of k-largest norm.*
Make those $\operatorname{card}\mathcal{I} = 2^k n!/(k!(n-k)!)$ linear functions explicit for $\|x\|_{\substack{2 \\ 2}}$ and $\|x\|_{\substack{2 \\ 1}}$ on $\mathbb{R}^2$ and $\|x\|_{\substack{3 \\ 2}}$ on $\mathbb{R}^3$. Plot $\|x\|_{\substack{2 \\ 2}}$ and $\|x\|_{\substack{2 \\ 1}}$ in three dimensions.   ▼

3.2.2.1.2 Example. *Compressed sensing problem.*
Conventionally posed as convex problem (517), we showed: the compressed sensing problem can always be posed equivalently with a nonnegative variable as in convex statement (522). The 1-norm predominantly appears in the literature because it is convex, it inherently minimizes cardinality under some technical conditions, [69] and because the desirable 0-norm is intractable.

Assuming a cardinality-k solution exists, the compressed sensing problem may be written as a difference of two convex functions: for $A \in \mathbb{R}^{m \times n}$

$$
\begin{array}{ll}
\underset{x \in \mathbb{R}^n}{\text{minimize}} & \|x\|_1 - \|x\|_{\frac{n}{k}} \\
\text{subject to} & Ax = b \\
& x \succeq 0
\end{array}
\quad \equiv \quad
\begin{array}{ll}
\text{find} & x \in \mathbb{R}^n \\
\text{subject to} & Ax = b \\
& x \succeq 0 \\
& \|x\|_0 \leq k
\end{array}
\tag{529}
$$

which is a nonconvex statement, a minimization of $n-k$ smallest entries of variable vector x, minimization of a concave function on $\mathbb{R}_+^n$ (§3.2.1) [315, §32]; but a statement of compressed sensing more precise than (522) because of its equivalence to 0-norm. $\|x\|_{\frac{n}{k}}$ is the convex k-largest norm of x (monotonic on $\mathbb{R}_+^n$) while $\|x\|_0$ (quasiconcave on $\mathbb{R}_+^n$) expresses its cardinality. Global optimality occurs at a zero objective of minimization; *id est*, when the smallest $n-k$ entries of variable vector x are zeroed. Under nonnegativity constraint, this compressed sensing problem (529) becomes the same as

$$
\begin{array}{ll}
\underset{z(x),\, x \in \mathbb{R}^n}{\text{minimize}} & (\mathbf{1} - z)^{\mathrm{T}} x \\
\text{subject to} & Ax = b \\
& x \succeq 0
\end{array}
\tag{530}
$$

where

$$
\left.
\begin{aligned}
\mathbf{1} &= \nabla \|x\|_1 = \nabla \mathbf{1}^{\mathrm{T}} x \\
z &= \nabla \|x\|_{\frac{n}{k}} = \nabla z^{\mathrm{T}} x
\end{aligned}
\right\}, \qquad x \succeq 0
\tag{531}
$$

where gradient of k-largest norm is an optimal solution to a convex problem:

$$
\left.
\begin{array}{rl}
\|x\|_{\frac{n}{k}} = & \underset{y \in \mathbb{R}^n}{\text{maximize}} \quad y^{\mathrm{T}} x \\
& \text{subject to} \quad 0 \preceq y \preceq \mathbf{1} \\
& \qquad\qquad\; y^{\mathrm{T}} \mathbf{1} = k \\[2ex]
\nabla \|x\|_{\frac{n}{k}} = & \underset{y \in \mathbb{R}^n}{\arg\max} \quad y^{\mathrm{T}} x \\
& \text{subject to} \quad 0 \preceq y \preceq \mathbf{1} \\
& \qquad\qquad\; y^{\mathrm{T}} \mathbf{1} = k
\end{array}
\right\}, \qquad x \succeq 0
\tag{532}
$$

$\square$

3.2.2.1.3 Exercise. *k-largest norm gradient.*
Prove (531). Is $\nabla \|x\|_{\frac{n}{k}}$ unique? Find $\nabla \|x\|_1$ and $\nabla \|x\|_{\frac{n}{k}}$ on $\mathbb{R}^n$.[3.12]   ▼

3.2.3 clipping

Zeroing negative vector entries under 1-norm is accomplished:

$$
\|x_+\|_1 = \begin{array}{ll}
\underset{t \in \mathbb{R}^n}{\text{minimize}} & \mathbf{1}^{\mathrm{T}} t \\
\text{subject to} & x \preceq t \\
& 0 \preceq t
\end{array}
\tag{533}
$$

[3.12] Hint: §D.2.1.

where, for $x = [x_i,\ i=1\ldots n] \in \mathbb{R}^n$

$$x_+ \triangleq t^\star = \left[\begin{array}{ll} x_i\,, & x_i \geq 0 \\ 0\,, & x_i < 0 \end{array}\right\}\,, \quad i=1\ldots n\right] = \frac{1}{2}(x + |x|) \tag{534}$$

(clipping)

$$\begin{array}{ll} \underset{x \in \mathbb{R}^n}{\text{minimize}} & \|x_+\|_1 \\ \text{subject to} & x \in \mathcal{C} \end{array} \quad \equiv \quad \begin{array}{ll} \underset{x \in \mathbb{R}^n,\ t \in \mathbb{R}^n}{\text{minimize}} & \mathbf{1}^{\mathrm{T}} t \\ \text{subject to} & x \preceq t \\ & 0 \preceq t \\ & x \in \mathcal{C} \end{array} \tag{535}$$

3.3　Inverted functions and roots

A given function f is convex iff $-f$ is concave. Both functions are loosely referred to as *convex* since $-f$ is simply f inverted about the abscissa axis, and minimization of f is equivalent to maximization of $-f$.

A given positive function f is convex iff $1/f$ is concave; f inverted about ordinate 1 is concave. Minimization of f is maximization of $1/f$.

We wish to implement objectives of the form x^{-1}. Suppose we have a 2×2 matrix

$$T \triangleq \left[\begin{array}{cc} x & z \\ z & y \end{array}\right] \in \mathbb{R}^{\mathbf{2}} \tag{536}$$

which is positive semidefinite by (1595) when

$$T \succeq 0 \quad \Leftrightarrow \quad x > 0 \ \textbf{and} \ xy \geq z^2 \tag{537}$$

A polynomial constraint such as this is therefore called a *conic constraint*.[3.13] This means we may formulate convex problems, having inverted variables, as semidefinite programs in Schur-form (§A.4); *e.g.*,

$$\begin{array}{ll} \underset{x \in \mathbb{R}}{\text{minimize}} & x^{-1} \\ \text{subject to} & x > 0 \\ & x \in \mathcal{C} \end{array} \quad \equiv \quad \begin{array}{ll} \underset{x,\,y \in \mathbb{R}}{\text{minimize}} & y \\ \text{subject to} & \left[\begin{array}{cc} x & 1 \\ 1 & y \end{array}\right] \succeq 0 \\ & x \in \mathcal{C} \end{array} \tag{538}$$

rather

$$x > 0\,, \quad y \geq \frac{1}{x} \quad \Leftrightarrow \quad \left[\begin{array}{cc} x & 1 \\ 1 & y \end{array}\right] \succeq 0 \tag{539}$$

(inverted) For vector $x = [x_i,\ i=1\ldots n] \in \mathbb{R}^n$

$$\begin{array}{ll} \underset{x \in \mathbb{R}^n}{\text{minimize}} & \sum_{i=1}^{n} x_i^{-1} \\ \text{subject to} & x \succ 0 \\ & x \in \mathcal{C} \end{array} \quad \equiv \quad \begin{array}{ll} \underset{x \in \mathbb{R}^n,\ y \in \mathbb{R}}{\text{minimize}} & y \\ \text{subject to} & \left[\begin{array}{cc} x_i & \sqrt{n} \\ \sqrt{n} & y \end{array}\right] \succeq 0\,, \quad i=1\ldots n \\ & x \in \mathcal{C} \end{array} \tag{540}$$

[3.13] In this dimension, the convex cone formed from the set of all values $\{x,y,z\}$ satisfying constraint (537) is called a *rotated quadratic* or circular cone or positive semidefinite cone.

rather

$$x \succ 0, \quad y \geq \mathrm{tr}\big(\delta(x)^{-1}\big) \quad \Leftrightarrow \quad \left[\begin{array}{cc} x_i & \sqrt{n} \\ \sqrt{n} & y \end{array}\right] \succeq 0, \quad i=1\ldots n \qquad (541)$$

3.3.1 fractional power

[153] To implement an objective of the form x^α for positive α, we quantize α and work instead with that approximation. Choose nonnegative integer q for adequate quantization of α like so:

$$\alpha \triangleq \frac{k}{2^q}, \qquad k \in \{0,\, 1,\, 2 \ldots 2^q - 1\} \qquad (542)$$

Any k from that set may be written $k = \sum_{i=1}^{q} b_i\, 2^{i-1}$ where $b_i \in \{0,1\}$. Define vector $y = [y_i\,,\ i=0\ldots q] \in \mathbb{R}^{q+1}$ with $y_0 = 1$:

3.3.1.1 positive

Then we have the equivalent semidefinite program for maximizing a concave function x^α, for quantized $0 \leq \alpha < 1$

$$\begin{array}{ll} \underset{x \in \mathbb{R}}{\text{maximize}} & x^\alpha \\ \text{subject to} & x > 0 \\ & x \in \mathcal{C} \end{array} \quad \equiv \quad \begin{array}{ll} \underset{x \in \mathbb{R},\ y \in \mathbb{R}^{q+1}}{\text{maximize}} & y_q \\ \text{subject to} & \left[\begin{array}{cc} y_{i-1} & y_i \\ y_i & x^{b_i} \end{array}\right] \succeq 0, \quad i=1\ldots q \\ & x \in \mathcal{C} \end{array} \qquad (543)$$

where nonnegativity of y_q is enforced by maximization; *id est*,

$$x > 0, \quad y_q \leq x^\alpha \quad \Leftrightarrow \quad \left[\begin{array}{cc} y_{i-1} & y_i \\ y_i & x^{b_i} \end{array}\right] \succeq 0, \quad i=1\ldots q \qquad (544)$$

3.3.1.1.1 Example. *Square root.*
$\alpha = \frac{1}{2}$. Choose $q=1$ and $k=1=2^0$.

$$\begin{array}{ll} \underset{x \in \mathbb{R}}{\text{maximize}} & \sqrt{x} \\ \text{subject to} & x > 0 \\ & x \in \mathcal{C} \end{array} \quad \equiv \quad \begin{array}{ll} \underset{x \in \mathbb{R},\ y \in \mathbb{R}^2}{\text{maximize}} & y_1 \\ \text{subject to} & \left[\begin{array}{cc} y_0 = 1 & y_1 \\ y_1 & x \end{array}\right] \succeq 0 \\ & x \in \mathcal{C} \end{array} \qquad (545)$$

where

$$x > 0, \quad y_1 \leq \sqrt{x} \quad \Leftrightarrow \quad \left[\begin{array}{cc} 1 & y_1 \\ y_1 & x \end{array}\right] \succeq 0 \qquad (546)$$

$\square$

3.3.1.2   negative

It is also desirable to implement an objective of the form $x^{-\alpha}$ for positive α. The technique is nearly the same as before: for quantized $0 \le \alpha < 1$

$$
\begin{array}{ll}
\underset{x\in\mathbb{R}}{\text{minimize}} & x^{-\alpha} \\
\text{subject to} & x > 0 \\
& x \in \mathcal{C}
\end{array}
\quad \equiv \quad
\begin{array}{ll}
\underset{x,\,z\in\mathbb{R}\,,\,y\in\mathbb{R}^{q+1}}{\text{minimize}} & z \\
\text{subject to} & \begin{bmatrix} y_{i-1} & y_i \\ y_i & x^{b_i} \end{bmatrix} \succeq 0\,, \quad i=1\ldots q \\
& \begin{bmatrix} z & 1 \\ 1 & y_q \end{bmatrix} \succeq 0 \\
& x \in \mathcal{C}
\end{array}
\tag{547}
$$

rather

$$
x > 0\,, \quad z \ge x^{-\alpha} \quad \Leftrightarrow \quad
\begin{array}{l}
\begin{bmatrix} y_{i-1} & y_i \\ y_i & x^{b_i} \end{bmatrix} \succeq 0\,, \quad i=1\ldots q \\
\begin{bmatrix} z & 1 \\ 1 & y_q \end{bmatrix} \succeq 0
\end{array}
\tag{548}
$$

3.3.1.3   positive inverted

Now define vector $t = [t_i\,,\ i=0\ldots q] \in \mathbb{R}^{q+1}$ with $t_0 = 1$. To implement an objective $x^{1/\alpha}$ for quantized $0 \le \alpha < 1$ as in (542)

$$
\begin{array}{ll}
\underset{x\in\mathbb{R}}{\text{minimize}} & x^{1/\alpha} \\
\text{subject to} & x > 0 \\
& x \in \mathcal{C}
\end{array}
\quad \equiv \quad
\begin{array}{ll}
\underset{x,\,y\in\mathbb{R}\,,\,t\in\mathbb{R}^{q+1}}{\text{minimize}} & y \\
\text{subject to} & \begin{bmatrix} t_{i-1} & t_i \\ t_i & y^{b_i} \end{bmatrix} \succeq 0\,, \quad i=1\ldots q \\
& x = t_q > 0 \\
& x \in \mathcal{C}
\end{array}
\tag{549}
$$

rather

$$
x > 0\,, \quad y \ge x^{1/\alpha} \quad \Leftrightarrow \quad
\begin{array}{l}
\begin{bmatrix} t_{i-1} & t_i \\ t_i & y^{b_i} \end{bmatrix} \succeq 0\,, \quad i=1\ldots q \\
x = t_q > 0
\end{array}
\tag{550}
$$

3.4   Affine function

A function $f(X)$ is *affine* when it is continuous and has the dimensionally extensible form (*confer* §2.9.1.0.2)

$$
f(X) = AX + B \tag{551}
$$

All affine functions are simultaneously convex and concave. Both $-AX+B$ and $AX+B$, for example, are convex functions of X. The linear functions constitute a proper subset of affine functions; *e.g.*, when $B = \mathbf{0}$, function $f(X)$ is linear.

Unlike other convex functions, affine function convexity holds with respect to any dimensionally compatible proper cone substituted into convexity definition (496). All affine functions satisfy a membership relation, for some normal cone, like (505). Affine multidimensional functions are more easily recognized by existence of no multiplicative multivariate terms and no polynomial terms of degree higher than 1 ; *id est*, entries of the function are characterized by only linear combinations of argument entries plus constants.

$A^\mathrm{T}XA + B^\mathrm{T}B$ is affine in X, for example. Trace is an affine function; actually, a real linear function expressible as inner product $f(X) = \langle A, X \rangle$ with matrix A being the Identity. The real affine function in Figure 74 illustrates hyperplanes, in its domain, constituting contours of equal function-value (*level sets* (556)).

3.4.0.0.1 Example. *Engineering control.* [416, §2.2][3.14]
For $X \in \mathbb{S}^M$ and matrices A, B, Q, R of any compatible dimensions, for example, the expression XAX is not affine in X whereas

$$g(X) = \begin{bmatrix} R & B^\mathrm{T}X \\ XB & Q + A^\mathrm{T}X + XA \end{bmatrix} \tag{552}$$

is an affine multidimensional function. Such a function is typical in engineering control. [60] [155] □

(*confer* Figure 16) Any single- or many-valued inverse of an affine function is affine.

3.4.0.0.2 Example. *Linear objective.*
Consider minimization of a real affine function $f(z) = a^\mathrm{T}z + b$ over the convex feasible set $\mathcal{C}$ in its domain $\mathbb{R}^2$ illustrated in Figure 74. Since scalar b is fixed, the problem posed is the same as the convex optimization

$$\begin{array}{ll} \underset{z}{\text{minimize}} & a^\mathrm{T}z \\ \text{subject to} & z \in \mathcal{C} \end{array} \tag{553}$$

whose objective of minimization is a real linear function. Were convex set $\mathcal{C}$ polyhedral (§2.12), then this problem would be called a linear program. Were convex set $\mathcal{C}$ an intersection with a positive semidefinite cone, then this problem would be called a semidefinite program.

There are two distinct ways to visualize this problem: one in the objective function's domain $\mathbb{R}^2$, the other including the ambient space of the objective function's range as in $\begin{bmatrix} \mathbb{R}^2 \\ \mathbb{R} \end{bmatrix}$. Both visualizations are illustrated in Figure 74. Visualization in the function domain is easier because of lower dimension and because

• level sets (556) of any affine function are affine. (§2.1.9)

In this circumstance, the level sets are parallel hyperplanes with respect to $\mathbb{R}^2$. One solves optimization problem (553) graphically by finding that hyperplane intersecting feasible set $\mathcal{C}$ furthest right (in the direction of negative gradient $-a$ (§3.6)). □

[3.14]The interpretation from this citation of $\{X \in \mathbb{S}^M \mid g(X) \succeq 0\}$ as "an intersection between a linear subspace and the cone of positive semidefinite matrices" is incorrect. (See §2.9.1.0.2 for a similar example.) The conditions they state under which strong duality holds for semidefinite programming are conservative. (*confer* §4.2.3.0.1)

CHAPTER 3. GEOMETRY OF CONVEX FUNCTIONS

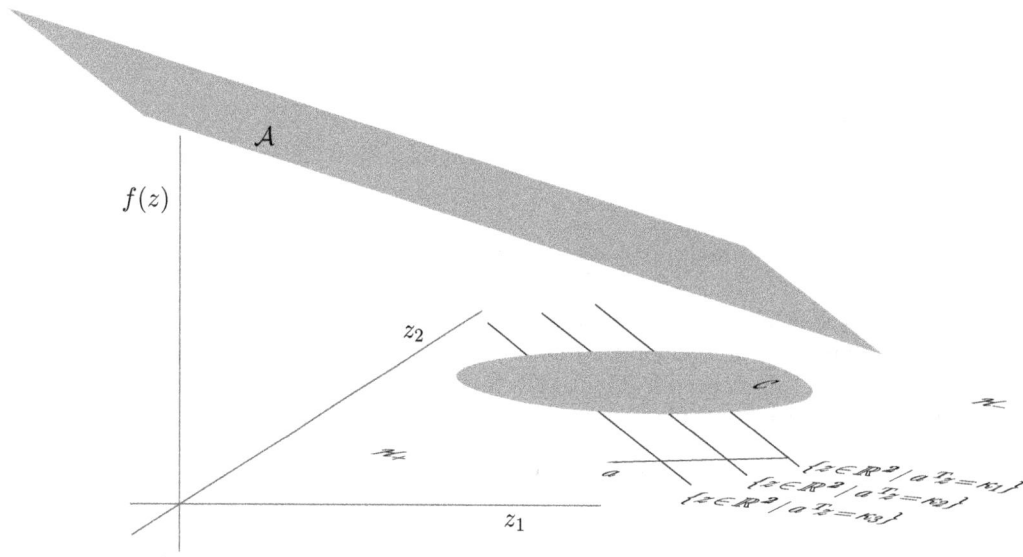

Figure 74: Three hyperplanes intersecting convex set $\mathcal{C} \subset \mathbb{R}^2$ from Figure 29. Cartesian axes in $\mathbb{R}^3$: Plotted is affine subset $\mathcal{A} = f(\mathbb{R}^2) \subset \mathbb{R}^2 \times \mathbb{R}$; a plane with third dimension. We say sequence of hyperplanes, w.r.t domain $\mathbb{R}^2$ of affine function $f(z) = a^{\mathrm{T}}z + b : \mathbb{R}^2 \to \mathbb{R}$, is increasing in normal direction (Figure 27) because affine function increases in direction of gradient a (§3.6.0.0.3). Minimization of $a^{\mathrm{T}}z + b$ over $\mathcal{C}$ is equivalent to minimization of $a^{\mathrm{T}}z$.

$\{a^{\mathrm{T}}z_1 + b_1 \mid a \in \mathbb{R}\}$

$\{a^{\mathrm{T}}z_2 + b_2 \mid a \in \mathbb{R}\}$

$\sup_i a_{\mathrm{p}}^{\mathrm{T}}z_i + b_i$

$\{a^{\mathrm{T}}z_3 + b_3 \mid a \in \mathbb{R}\}$

a

$\{a^{\mathrm{T}}z_4 + b_4 \mid a \in \mathbb{R}\}$

$\{a^{\mathrm{T}}z_5 + b_5 \mid a \in \mathbb{R}\}$

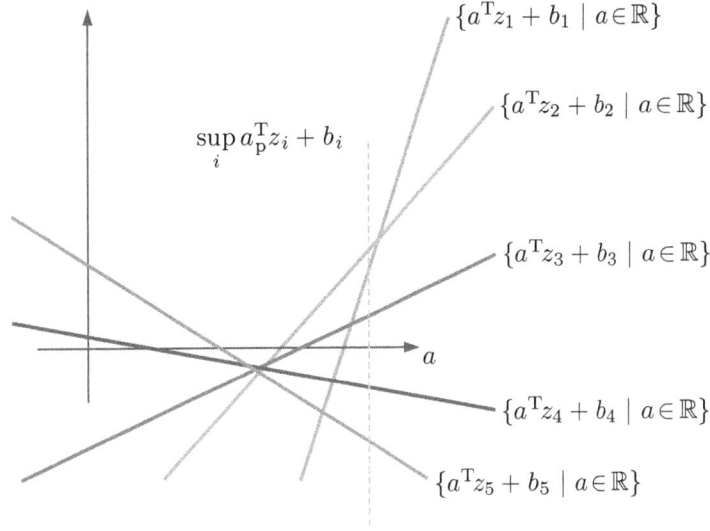

Figure 75: Pointwise supremum of any convex functions remains convex; by epigraph intersection. Supremum of affine functions in variable a evaluated at argument a_{p} is illustrated. Topmost affine function per a is supremum.

When a differentiable convex objective function f is nonlinear, the negative gradient $-\nabla f$ is a viable search direction (replacing $-a$ in (553)). (§2.13.10.1, Figure 68) [156] Then the nonlinear objective function can be replaced with a dynamic linear objective; linear as in (553).

3.4.0.0.3 Example. *Support function.*   [205, §C.2.1-§C.2.3.1]
For arbitrary set $\mathcal{Y} \subseteq \mathbb{R}^n$, its *support function* $\sigma_{\mathcal{Y}}(a) : \mathbb{R}^n \to \mathbb{R}$ is defined

$$\sigma_{\mathcal{Y}}(a) \triangleq \sup_{z \in \mathcal{Y}} a^{\mathrm{T}}z \tag{554}$$

a positively homogeneous function of direction a whose range contains $\pm\infty$. [256, p.135] For each $z \in \mathcal{Y}$, $a^{\mathrm{T}}z$ is a linear function of vector a. Because $\sigma_{\mathcal{Y}}(a)$ is a pointwise supremum of linear functions, it is convex in a (Figure 75). Application of the support function is illustrated in Figure 30a for one particular normal a. Given nonempty closed bounded convex sets $\mathcal{Y}$ and $\mathcal{Z}$ in $\mathbb{R}^n$ and nonnegative scalars β and γ [399, p.234]

$$\sigma_{\beta\mathcal{Y}+\gamma\mathcal{Z}}(a) = \beta\sigma_{\mathcal{Y}}(a) + \gamma\sigma_{\mathcal{Z}}(a) \tag{555}$$

$\square$

3.4.0.0.4 Exercise. *Level sets.*
Given a function f and constant κ, its level sets are defined

$$\mathcal{L}_\kappa^\kappa f \triangleq \{z \mid f(z) = \kappa\} \tag{556}$$

Give two distinct examples of convex function, that are not affine, having convex level sets.   ▼

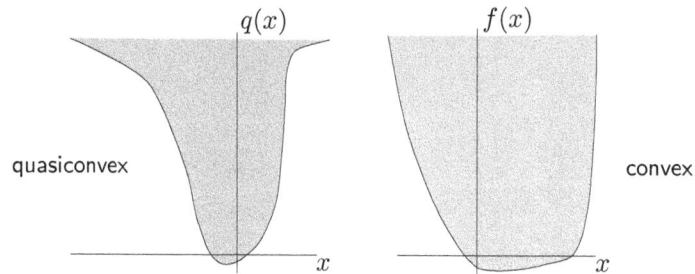

Figure 76: Quasiconvex function q epigraph is not necessarily convex, but convex function f epigraph is convex in any dimension. Sublevel sets are necessarily convex for either function, but sufficient only for quasiconvexity.

3.4.0.0.5   Exercise.   *Epigraph intersection.*   (*confer* Figure 75)
Draw three hyperplanes in $\mathbb{R}^3$ representing $\max(0, x) \triangleq \sup\{0, x_i \mid x \in \mathbb{R}^n\}$ in $\mathbb{R}^2 \times \mathbb{R}$ to see why maximum of nonnegative vector entries is a convex function of variable x. What is the normal to each hyperplane? (hint: p.218) Why is $\max(x)$ convex?   ▼

3.5   Epigraph, Sublevel set

It is well established that a continuous real function is convex if and only if its epigraph makes a convex set; [205] [315] [367] [399] [256] epigraph is the connection between convex sets and convex functions (p.187). Piecewise-continuous convex functions are admitted, thereby, and all invariant properties of convex sets carry over directly to convex functions. Generalization of *epigraph* to a vector-valued function $f(X) : \mathbb{R}^{p \times k} \to \mathbb{R}^M$ is straightforward: [302]

$$\operatorname{epi} f \triangleq \{ (X, t) \in \mathbb{R}^{p \times k} \times \mathbb{R}^M \mid X \in \operatorname{dom} f, \ f(X) \underset{\mathbb{R}_+^M}{\preceq} t \} \tag{557}$$

id est,

$$f \text{ convex } \Leftrightarrow \operatorname{epi} f \text{ convex} \tag{558}$$

Necessity is proven: [62, exer.3.60] Given any $(X, u), (Y, v) \in \operatorname{epi} f$, we must show for all $\mu \in [0, 1]$ that $\mu(X, u) + (1 - \mu)(Y, v) \in \operatorname{epi} f$; *id est*, we must show

$$f(\mu X + (1 - \mu)Y) \underset{\mathbb{R}_+^M}{\preceq} \mu\, u + (1 - \mu)v \tag{559}$$

Yet this holds by definition because $f(\mu X + (1 - \mu)Y) \preceq \mu f(X) + (1 - \mu)f(Y)$. The converse also holds.   ◆

3.5.0.0.1 Exercise. *Epigraph sufficiency.*
Prove that converse: Given any $(X, u), (Y, v) \in \operatorname{epi} f$, if for all $\mu \in [0, 1]$
$\mu(X, u) + (1 - \mu)(Y, v) \in \operatorname{epi} f$ holds, then f must be convex. ▼

Sublevel sets of a convex real function are convex. Likewise, corresponding to each and every $\nu \in \mathbb{R}^M$

$$\mathcal{L}_\nu f \triangleq \{X \in \operatorname{dom} f \mid f(X) \underset{\mathbb{R}_+^M}{\preceq} \nu\} \subseteq \mathbb{R}^{p \times k} \tag{560}$$

sublevel sets of a convex vector-valued function are convex. As for convex real functions, the converse does not hold. (Figure 76)

To prove necessity of convex sublevel sets: For any $X, Y \in \mathcal{L}_\nu f$ we must show for each and every $\mu \in [0, 1]$ that $\mu X + (1 - \mu)Y \in \mathcal{L}_\nu f$. By definition,

$$f(\mu X + (1 - \mu)Y) \underset{\mathbb{R}_+^M}{\preceq} \mu f(X) + (1 - \mu)f(Y) \underset{\mathbb{R}_+^M}{\preceq} \nu \tag{561}$$

◆

When an epigraph (557) is artificially bounded above, $t \preceq \nu$, then the corresponding sublevel set can be regarded as an orthogonal projection of epigraph on the function domain.

Sense of the inequality is reversed in (557), for concave functions, and we use instead the nomenclature *hypograph*. Sense of the inequality in (560) is reversed, similarly, with each convex set then called *superlevel set*.

3.5.0.0.2 Example. *Matrix pseudofractional function.*
Consider a real function of two variables

$$f(A, x) : \mathbb{S}^n \times \mathbb{R}^n \to \mathbb{R} = x^{\mathrm{T}} A^\dagger x \tag{562}$$

on $\operatorname{dom} f = \mathbb{S}_+^n \times \mathcal{R}(A)$. This function is convex simultaneously in both variables when variable matrix A belongs to the entire positive semidefinite cone $\mathbb{S}_+^n$ and variable vector x is confined to range $\mathcal{R}(A)$ of matrix A.

To explain this, we need only demonstrate that the function epigraph is convex. Consider Schur-form (1592) from §A.4: for $t \in \mathbb{R}$

$$\begin{aligned} G(A, z, t) = \begin{bmatrix} A & z \\ z^{\mathrm{T}} & t \end{bmatrix} \succeq 0 \\ \Leftrightarrow \\ z^{\mathrm{T}}(I - AA^\dagger) = 0 \\ t - z^{\mathrm{T}} A^\dagger z \geq 0 \\ A \succeq 0 \end{aligned} \tag{563}$$

Inverse image of the positive semidefinite cone $\mathbb{S}_+^{n+1}$ under affine mapping $G(A, z, t)$ is convex by Theorem 2.1.9.0.1. Of the equivalent conditions for positive semidefiniteness of

G, the first is an equality demanding vector z belong to $\mathcal{R}(A)$. Function $f(A,z) = z^{\mathrm{T}}A^{\dagger}z$ is convex on convex domain $\mathbb{S}_+^n \times \mathcal{R}(A)$ because the Cartesian product constituting its epigraph

$$\operatorname{epi} f(A,z) = \left\{(A,z,t) \mid A \succeq 0,\ z \in \mathcal{R}(A),\ z^{\mathrm{T}}A^{\dagger}z \leq t\right\} = G^{-1}\left(\mathbb{S}_+^{n+1}\right) \qquad (564)$$

is convex.   □

3.5.0.0.3   Exercise.   *Matrix product function.*
Continue Example 3.5.0.0.2 by introducing vector variable x and making the substitution $z \leftarrow Ax$. Because of matrix symmetry (§E), for all $x \in \mathbb{R}^n$

$$f(A,z(x)) \;=\; z^{\mathrm{T}}A^{\dagger}z \;=\; x^{\mathrm{T}}A^{\mathrm{T}}A^{\dagger}Ax \;=\; x^{\mathrm{T}}Ax \;=\; f(A,x) \qquad (565)$$

whose epigraph is

$$\operatorname{epi} f(A,x) \;=\; \left\{(A,x,t) \mid A \succeq 0,\ x^{\mathrm{T}}Ax \leq t\right\} \qquad (566)$$

Provide two simple explanations why $f(A,x) = x^{\mathrm{T}}Ax$ is not a function convex simultaneously in positive semidefinite matrix A and vector x on $\operatorname{dom} f = \mathbb{S}_+^n \times \mathbb{R}^n$.
                                                                               ▼

3.5.0.0.4   Example.   *Matrix fractional function.*   (*confer* §3.7.2.0.1)
Continuing Example 3.5.0.0.2, now consider a real function of two variables on $\operatorname{dom} f = \mathbb{S}_+^n \times \mathbb{R}^n$ for small positive constant ϵ (*confer* (1962))

$$f(A,x) = \epsilon x^{\mathrm{T}}(A + \epsilon I)^{-1}x \qquad (567)$$

where the inverse always exists by (1532). This function is convex simultaneously in both variables over the entire positive semidefinite cone $\mathbb{S}_+^n$ and all $x \in \mathbb{R}^n$: Consider Schur-form (1595) from §A.4: for $t \in \mathbb{R}$

$$\begin{aligned}
G(A,z,t) &= \begin{bmatrix} A + \epsilon I & z \\ z^{\mathrm{T}} & \epsilon^{-1}t \end{bmatrix} \succeq 0 \\
&\Leftrightarrow \\
t - \epsilon z^{\mathrm{T}}&(A + \epsilon I)^{-1}z \geq 0 \\
A + \epsilon I &\succ 0
\end{aligned} \qquad (568)$$

Inverse image of the positive semidefinite cone $\mathbb{S}_+^{n+1}$ under affine mapping $G(A,z,t)$ is convex by Theorem 2.1.9.0.1. Function $f(A,z)$ is convex on $\mathbb{S}_+^n \times \mathbb{R}^n$ because its epigraph is that inverse image:

$$\operatorname{epi} f(A,z) = \left\{(A,z,t) \mid A + \epsilon I \succ 0,\ \epsilon z^{\mathrm{T}}(A+\epsilon I)^{-1}z \leq t\right\} = G^{-1}\left(\mathbb{S}_+^{n+1}\right) \qquad (569)$$

□

3.5.1 matrix fractional projector

Consider nonlinear function f having orthogonal projector W as argument:

$$f(W, x) = \epsilon\, x^{\mathrm{T}}(W + \epsilon\, I)^{-1}x \tag{570}$$

Projection matrix W has property $W^{\dagger} = W^{\mathrm{T}} = W \succeq 0$ (2009). Any orthogonal projector can be decomposed into an outer product of orthonormal matrices $W = UU^{\mathrm{T}}$ where $U^{\mathrm{T}}U = I$ as explained in §E.3.2. From (1962) for any $\epsilon > 0$ and idempotent symmetric W, $\epsilon(W + \epsilon\, I)^{-1} = I - (1 + \epsilon)^{-1}W$ from which

$$f(W, x) = \epsilon\, x^{\mathrm{T}}(W + \epsilon\, I)^{-1}x = x^{\mathrm{T}}\big(I - (1 + \epsilon)^{-1}W\big)x \tag{571}$$

Therefore

$$\lim_{\epsilon \to 0^+} f(W, x) = \lim_{\epsilon \to 0^+} \epsilon\, x^{\mathrm{T}}(W + \epsilon\, I)^{-1}x = x^{\mathrm{T}}(I - W)x \tag{572}$$

where $I - W$ is also an orthogonal projector (§E.2).

We learned from Example 3.5.0.0.4 that $f(W, x) = \epsilon\, x^{\mathrm{T}}(W + \epsilon\, I)^{-1}x$ is convex simultaneously in both variables over all $x \in \mathbb{R}^n$ when $W \in \mathbb{S}^n_+$ is confined to the entire positive semidefinite cone (including its boundary). It is now our goal to incorporate f into an optimization problem such that an optimal solution returned always comprises a projection matrix W. The set of orthogonal projection matrices is a nonconvex subset of the positive semidefinite cone. So f cannot be convex on the projection matrices, and its equivalent (for idempotent W)

$$f(W, x) = x^{\mathrm{T}}\big(I - (1 + \epsilon)^{-1}W\big)x \tag{573}$$

cannot be convex simultaneously in both variables on either the positive semidefinite or symmetric projection matrices.

Suppose we allow $\operatorname{dom} f$ to constitute the entire positive semidefinite cone but constrain W to a Fantope (90); *e.g.,* for convex set $\mathcal{C}$ and $0 < k < n$ as in

$$\begin{array}{ll} \underset{x \in \mathbb{R}^n,\, W \in \mathbb{S}^n}{\text{minimize}} & \epsilon\, x^{\mathrm{T}}(W + \epsilon\, I)^{-1}x \\ \text{subject to} & 0 \preceq W \preceq I \\ & \operatorname{tr} W = k \\ & x \in \mathcal{C} \end{array} \tag{574}$$

Although this is a convex problem, there is no guarantee that optimal W is a projection matrix because only extreme points of a Fantope are orthogonal projection matrices UU^{T}.

Let's try partitioning the problem into two convex parts (one for x and one for W), substitute equivalence (571), and then iterate solution of convex problem

$$\begin{array}{ll} \underset{x \in \mathbb{R}^n}{\text{minimize}} & x^{\mathrm{T}}(I - (1 + \epsilon)^{-1}W)x \\ \text{subject to} & x \in \mathcal{C} \end{array} \tag{575}$$

with convex problem

$$\text{(a)} \quad \begin{array}{ll} \underset{W \in \mathbb{S}^n}{\text{minimize}} & x^{\star \mathrm{T}}(I - (1 + \epsilon)^{-1}W)x^{\star} \\ \text{subject to} & 0 \preceq W \preceq I \\ & \operatorname{tr} W = k \end{array} \quad \equiv \quad \begin{array}{ll} \underset{W \in \mathbb{S}^n}{\text{maximize}} & x^{\star \mathrm{T}}Wx^{\star} \\ \text{subject to} & 0 \preceq W \preceq I \\ & \operatorname{tr} W = k \end{array} \tag{576}$$

until convergence, where $x^\star$ represents an optimal solution of (575) from any particular iteration. The idea is to optimally solve for the partitioned variables which are later combined to solve the original problem (574). What makes this approach sound is that the constraints are separable, the partitioned feasible sets are not interdependent, and the fact that the original problem (though nonlinear) is convex simultaneously in both variables.[3.15]

But partitioning alone does not guarantee a projector as solution. To make orthogonal projector W a certainty, we must invoke a known analytical optimal solution to problem (576): Diagonalize optimal solution from problem (575) $x^\star x^{\star\mathrm{T}} \triangleq Q\Lambda Q^{\mathrm{T}}$ (§A.5.1) and set $U^\star = Q(:, 1\!:\!k) \in \mathbb{R}^{n \times k}$ per (1791c);

$$W = U^\star U^{\star\mathrm{T}} = \frac{x^\star x^{\star\mathrm{T}}}{\|x^\star\|^2} + Q(:, 2\!:\!k)Q(:, 2\!:\!k)^{\mathrm{T}} \qquad (577)$$

Then optimal solution $(x^\star, U^\star)$ to problem (574) is found, for small ϵ , by iterating solution to problem (575) with optimal (projector) solution (577) to convex problem (576).

Proof.   Optimal vector $x^\star$ is orthogonal to the last $n-1$ columns of orthogonal matrix Q, so

$$f^\star_{(575)} = \|x^\star\|^2 (1 - (1+\epsilon)^{-1}) \qquad (578)$$

after each iteration. Convergence of $f^\star_{(575)}$ is proven with the observation that iteration (575) (576a) is a nonincreasing sequence that is bounded below by 0. Any bounded monotonic sequence in $\mathbb{R}$ is convergent. [264, §1.2] [42, §1.1] Expression (577) for optimal projector W holds at each iteration, therefore $\|x^\star\|^2 (1 - (1+\epsilon)^{-1})$ must also represent the optimal objective value $f^\star_{(575)}$ at convergence.

Because the objective $f_{(574)}$ from problem (574) is also bounded below by 0 on the same domain, this convergent optimal objective value $f^\star_{(575)}$ (for positive ϵ arbitrarily close to 0) is necessarily optimal for (574); *id est*,

$$f^\star_{(575)} \geq f^\star_{(574)} \geq 0 \qquad (579)$$

by (1770), and

$$\lim_{\epsilon \to 0^+} f^\star_{(575)} = 0 \qquad (580)$$

Since optimal $(x^\star, U^\star)$ from problem (575) is feasible to problem (574), and because their objectives are equivalent for projectors by (571), then converged $(x^\star, U^\star)$ must also be optimal to (574) in the limit. Because problem (574) is convex, this represents a globally optimal solution.   ◆

[3.15] A convex problem has convex feasible set, and the objective *surface* has one and only one global minimum. [310, p.123]

3.5.2 semidefinite program via Schur

Schur complement (1592) can be used to convert a projection problem to an optimization problem in *epigraph form*. Suppose, for example, we are presented with the constrained projection problem studied by Hayden & Wells in [190] (who provide analytical solution): Given $A \in \mathbb{R}^{M \times M}$ and some full-rank matrix $S \in \mathbb{R}^{M \times L}$ with $L < M$

$$\begin{aligned} \underset{X \in \mathbb{S}^M}{\text{minimize}} \quad & \|A - X\|_{\text{F}}^2 \\ \text{subject to} \quad & S^{\text{T}} X S \succeq 0 \end{aligned} \tag{581}$$

Variable X is constrained to be positive semidefinite, but only on a subspace determined by S. First we write the epigraph form:

$$\begin{aligned} \underset{X \in \mathbb{S}^M,\ t \in \mathbb{R}}{\text{minimize}} \quad & t \\ \text{subject to} \quad & \|A - X\|_{\text{F}}^2 \le t \\ & S^{\text{T}} X S \succeq 0 \end{aligned} \tag{582}$$

Next we use Schur complement [284, §6.4.3] [254] and matrix vectorization (§2.2):

$$\begin{aligned} \underset{X \in \mathbb{S}^M,\ t \in \mathbb{R}}{\text{minimize}} \quad & t \\ \text{subject to} \quad & \begin{bmatrix} tI & \text{vec}(A - X) \\ \text{vec}(A - X)^{\text{T}} & 1 \end{bmatrix} \succeq 0 \\ & S^{\text{T}} X S \succeq 0 \end{aligned} \tag{583}$$

This semidefinite program (§4) is an epigraph form in disguise, equivalent to (581); it demonstrates how a quadratic objective or constraint can be converted to a semidefinite constraint.

Were problem (581) instead equivalently expressed without the square

$$\begin{aligned} \underset{X \in \mathbb{S}^M}{\text{minimize}} \quad & \|A - X\|_{\text{F}} \\ \text{subject to} \quad & S^{\text{T}} X S \succeq 0 \end{aligned} \tag{584}$$

then we get a subtle variation:

$$\begin{aligned} \underset{X \in \mathbb{S}^M,\ t \in \mathbb{R}}{\text{minimize}} \quad & t \\ \text{subject to} \quad & \|A - X\|_{\text{F}} \le t \\ & S^{\text{T}} X S \succeq 0 \end{aligned} \tag{585}$$

that leads to an equivalent for (584) (and for (581) by (512))

$$\begin{aligned} \underset{X \in \mathbb{S}^M,\ t \in \mathbb{R}}{\text{minimize}} \quad & t \\ \text{subject to} \quad & \begin{bmatrix} tI & \text{vec}(A - X) \\ \text{vec}(A - X)^{\text{T}} & t \end{bmatrix} \succeq 0 \\ & S^{\text{T}} X S \succeq 0 \end{aligned} \tag{586}$$

3.5.2.0.1   Example.   *Schur anomaly.*
Consider a problem abstract in the convex constraint, given symmetric matrix A

$$\begin{array}{ll} \underset{X \in \mathbb{S}^M}{\text{minimize}} & \|X\|_{\text{F}}^2 - \|A - X\|_{\text{F}}^2 \\ \text{subject to} & X \in \mathcal{C} \end{array} \tag{587}$$

the minimization of a difference of two quadratic functions each convex in matrix X. Observe equality

$$\|X\|_{\text{F}}^2 - \|A - X\|_{\text{F}}^2 = 2\operatorname{tr}(XA) - \|A\|_{\text{F}}^2 \tag{588}$$

So problem (587) is equivalent to the convex optimization

$$\begin{array}{ll} \underset{X \in \mathbb{S}^M}{\text{minimize}} & \operatorname{tr}(XA) \\ \text{subject to} & X \in \mathcal{C} \end{array} \tag{589}$$

But this problem (587) does not have Schur-form

$$\begin{array}{ll} \underset{X \in \mathbb{S}^M, \, \alpha \, , \, t}{\text{minimize}} & t - \alpha \\ \text{subject to} & X \in \mathcal{C} \\ & \|X\|_{\text{F}}^2 \leq t \\ & \|A - X\|_{\text{F}}^2 \geq \alpha \end{array} \tag{590}$$

because the constraint in α is nonconvex. (§2.9.1.0.1)   □

Matrix 2-norm (*spectral norm*) coincides with largest singular value.

$$\|X\|_2 \triangleq \sup_{\|a\|=1} \|Xa\|_2 = \sigma(X)_1 = \sqrt{\lambda(X^{\text{T}}X)_1} = \begin{array}{l} \underset{t \in \mathbb{R}}{\text{minimize}} \quad t \\ \text{subject to} \quad \begin{bmatrix} tI & X \\ X^{\text{T}} & tI \end{bmatrix} \succeq 0 \end{array} \tag{591}$$

This supremum of a family of convex functions in X must be convex because it constitutes an intersection of epigraphs of convex functions.

3.6   Gradient

Gradient ∇f of any differentiable multidimensional function f (formally defined in §D.1) maps each entry f_i to a space having the same dimension as the ambient space of its domain. Notation ∇f is shorthand for gradient $\nabla_x f(x)$ of f with respect to x. $\nabla f(y)$ can mean $\nabla_y f(y)$ or gradient $\nabla_x f(y)$ of $f(x)$ with respect to x evaluated at y; a distinction that should become clear from context.

Gradient of a differentiable real function $f(x): \mathbb{R}^K \to \mathbb{R}$, with respect to its vector argument, is uniquely defined

$$\nabla f(x) = \begin{bmatrix} \frac{\partial f(x)}{\partial x_1} \\ \frac{\partial f(x)}{\partial x_2} \\ \vdots \\ \frac{\partial f(x)}{\partial x_K} \end{bmatrix} \in \mathbb{R}^K \qquad (1851)$$

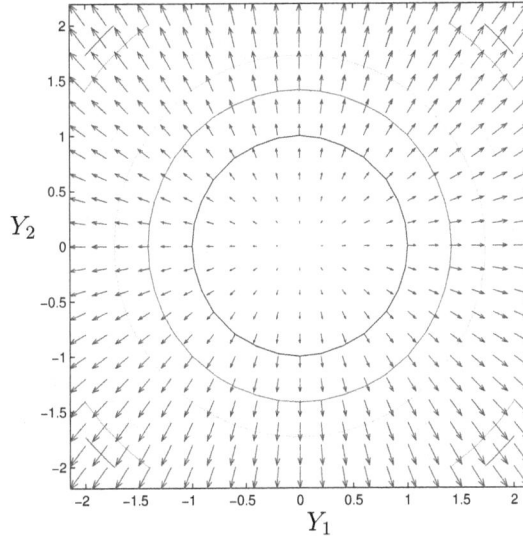

Figure 77: Gradient in $\mathbb{R}^2$ evaluated on grid over some open disc in domain of convex quadratic bowl $f(Y)=Y^{\mathrm{T}}Y : \mathbb{R}^2 \to \mathbb{R}$ illustrated in Figure 78. Circular contours are level sets; each defined by a constant function-value.

while the second-order gradient of the twice differentiable real function, with respect to its vector argument, is traditionally called the *Hessian* ;[3.16]

$$\nabla^2 f(x) = \begin{bmatrix} \frac{\partial^2 f(x)}{\partial x_1^2} & \frac{\partial^2 f(x)}{\partial x_1 \partial x_2} & \cdots & \frac{\partial^2 f(x)}{\partial x_1 \partial x_K} \\ \frac{\partial^2 f(x)}{\partial x_2 \partial x_1} & \frac{\partial^2 f(x)}{\partial x_2^2} & \cdots & \frac{\partial^2 f(x)}{\partial x_2 \partial x_K} \\ \vdots & \vdots & \ddots & \vdots \\ \frac{\partial^2 f(x)}{\partial x_K \partial x_1} & \frac{\partial^2 f(x)}{\partial x_K \partial x_2} & \cdots & \frac{\partial^2 f(x)}{\partial x_K^2} \end{bmatrix} \in \mathbb{S}^K \qquad (1852)$$

The gradient can be interpreted as a vector pointing in the direction of greatest change, [225, §15.6] or polar to direction of *steepest descent*.[3.17] [403] The gradient can also be interpreted as that vector normal to a sublevel set; *e.g.*, Figure 79, Figure 68.

For the quadratic bowl in Figure 78, the gradient maps to $\mathbb{R}^2$; illustrated in Figure 77. For a one-dimensional function of real variable, for example, the gradient evaluated at any point in the function domain is just the *slope* (or *derivative*) of that function there. (*confer* §D.1.4.1)

- For any differentiable multidimensional function, zero gradient $\nabla f = 0$ is a condition necessary for its unconstrained minimization [156, §3.2]:

[3.16] *Jacobian* is the Hessian transpose, so commonly confused in matrix calculus.
[3.17] *Newton's direction* $-\nabla^2 f(x)^{-1}\nabla f(x)$ is better for optimization of nonlinear functions well approximated locally by a quadratic function. [156, p.105]

3.6.0.0.1   Example. *Projection on rank-1 subset.*
For $A \in \mathbb{S}^N$ having eigenvalues $\lambda(A) = [\lambda_i] \in \mathbb{R}^N$, consider the unconstrained nonconvex optimization that is a projection on the rank-1 subset (§2.9.2.1) of positive semidefinite cone $\mathbb{S}_+^N$: Defining $\lambda_1 \triangleq \max_i \{\lambda(A)_i\}$ and corresponding eigenvector v_1

$$\underset{x}{\text{minimize}} \, \|xx^\mathrm{T} - A\|_\mathrm{F}^2 \;=\; \underset{x}{\text{minimize}} \, \text{tr}(xx^\mathrm{T}(x^\mathrm{T}x) - 2Axx^\mathrm{T} + A^\mathrm{T}A)$$

$$= \left\{ \begin{array}{ll} \|\lambda(A)\|^2 \, , & \lambda_1 \leq 0 \\ \|\lambda(A)\|^2 - \lambda_1^2 \, , & \lambda_1 > 0 \end{array} \right. \tag{1786}$$

$$\arg \underset{x}{\text{minimize}} \, \|xx^\mathrm{T} - A\|_\mathrm{F}^2 \;=\; \left\{ \begin{array}{ll} \mathbf{0} \, , & \lambda_1 \leq 0 \\ v_1 \sqrt{\lambda_1} \, , & \lambda_1 > 0 \end{array} \right. \tag{1787}$$

From (1881) and §D.2.1, the gradient of $\|xx^\mathrm{T} - A\|_\mathrm{F}^2$ is

$$\nabla_x \big((x^\mathrm{T}x)^2 - 2x^\mathrm{T}A\,x \big) = 4(x^\mathrm{T}x)x - 4Ax \tag{592}$$

Setting the gradient to $\mathbf{0}$

$$Ax = x(x^\mathrm{T}x) \tag{593}$$

is necessary for optimal solution. Replace vector x with a normalized eigenvector v_i of $A \in \mathbb{S}^N$, corresponding to a positive eigenvalue λ_i, scaled by square root of that eigenvalue. Then (593) is satisfied

$$x \leftarrow v_i \sqrt{\lambda_i} \;\Rightarrow\; Av_i = v_i \lambda_i \tag{594}$$

$xx^\mathrm{T} = \lambda_i v_i v_i^\mathrm{T}$ is a rank-1 matrix on the positive semidefinite cone boundary, and the minimum is achieved (§7.1.2) when $\lambda_i = \lambda_1$ is the largest positive eigenvalue of A. If A has no positive eigenvalue, then $x = \mathbf{0}$ yields the minimum. $\square$

Differentiability is a prerequisite neither to convexity or to numerical solution of a convex optimization problem. The gradient provides a necessary and sufficient condition (351) (451) for optimality in the constrained case, nevertheless, as it does in the unconstrained case:

- For any differentiable multidimensional *convex* function, zero gradient $\nabla f = \mathbf{0}$ is a necessary and sufficient condition for its unconstrained minimization [62, §5.5.3]:

3.6.0.0.2   Example. *Pseudoinverse.*
The pseudoinverse matrix is the unique solution to an unconstrained convex optimization problem [164, §5.5.4]: given $A \in \mathbb{R}^{m \times n}$

$$\underset{X \in \mathbb{R}^{n \times m}}{\text{minimize}} \|XA - I\|_\mathrm{F}^2 \tag{595}$$

where

$$\|XA - I\|_\mathrm{F}^2 = \text{tr}\big(A^\mathrm{T}X^\mathrm{T}XA - XA - A^\mathrm{T}X^\mathrm{T} + I \big) \tag{596}$$

whose gradient (§D.2.3)

$$\nabla_X \|XA - I\|_\mathrm{F}^2 = 2\big(XAA^\mathrm{T} - A^\mathrm{T} \big) = \mathbf{0} \tag{597}$$

vanishes when

$$XAA^{\mathrm{T}} = A^{\mathrm{T}} \tag{598}$$

When A is fat full-rank, then AA^{T} is invertible, $X^\star = A^{\mathrm{T}}(AA^{\mathrm{T}})^{-1}$ is the pseudoinverse $A^\dagger$, and $AA^\dagger = I$. Otherwise, we can make AA^{T} invertible by adding a positively scaled Identity, for any $A \in \mathbb{R}^{m \times n}$

$$X = A^{\mathrm{T}}(AA^{\mathrm{T}} + t\,I)^{-1} \tag{599}$$

Invertibility is guaranteed for any finite positive value of t by (1532). Then matrix X becomes the pseudoinverse $X \to A^\dagger \triangleq X^\star$ in the limit $t \to 0^+$. Minimizing instead $\|AX - I\|_{\mathrm{F}}^2$ yields the second flavor in (1961). $\qquad\square$

3.6.0.0.3 Example. *Hyperplane, line, described by affine function.*
Consider the real affine function of vector variable, (*confer* Figure 74)

$$f(x): \mathbb{R}^p \to \mathbb{R} \;=\; a^{\mathrm{T}}x + b \tag{600}$$

whose domain is $\mathbb{R}^p$ and whose gradient $\nabla f(x) = a$ is a constant vector (independent of x). This function describes the real line $\mathbb{R}$ (its range), and it describes a *nonvertical* [205, §B.1.2] hyperplane $\partial\mathcal{H}$ in the space $\mathbb{R}^p \times \mathbb{R}$ for any particular vector a (*confer* §2.4.2);

$$\partial\mathcal{H} = \left\{ \begin{bmatrix} x \\ a^{\mathrm{T}}x + b \end{bmatrix} \;\middle|\; x \in \mathbb{R}^p \right\} \subset \mathbb{R}^p \times \mathbb{R} \tag{601}$$

having nonzero normal

$$\eta = \begin{bmatrix} a \\ -1 \end{bmatrix} \in \mathbb{R}^p \times \mathbb{R} \tag{602}$$

This equivalence to a hyperplane holds only for real functions.[3.18] Epigraph of real affine function $f(x)$ is therefore a halfspace in $\begin{bmatrix} \mathbb{R}^p \\ \mathbb{R} \end{bmatrix}$, so we have:

The real affine function is to convex functions

as

the hyperplane is to convex sets.

[3.18]To prove that, consider a vector-valued affine function

$$f(x): \mathbb{R}^p \to \mathbb{R}^M \;=\; Ax + b$$

having gradient $\nabla f(x) = A^{\mathrm{T}} \in \mathbb{R}^{p \times M}$: The affine set

$$\left\{ \begin{bmatrix} x \\ Ax + b \end{bmatrix} \;\middle|\; x \in \mathbb{R}^p \right\} \subset \mathbb{R}^p \times \mathbb{R}^M$$

is perpendicular to

$$\eta \triangleq \begin{bmatrix} \nabla f(x) \\ -I \end{bmatrix} \in \mathbb{R}^{p \times M} \times \mathbb{R}^{M \times M}$$

because

$$\eta^{\mathrm{T}}\left(\begin{bmatrix} x \\ Ax + b \end{bmatrix} - \begin{bmatrix} 0 \\ b \end{bmatrix} \right) = 0 \quad \forall x \in \mathbb{R}^p$$

Yet η is a vector (in $\mathbb{R}^p \times \mathbb{R}^M$) only when $M = 1$. $\qquad\blacklozenge$

Similarly, the matrix-valued affine function of real variable x, for any particular matrix $A \in \mathbb{R}^{M \times N}$,

$$h(x) : \mathbb{R} \to \mathbb{R}^{M \times N} = Ax + B \qquad (603)$$

describes a line in $\mathbb{R}^{M \times N}$ in direction A

$$\{Ax + B \mid x \in \mathbb{R}\} \subseteq \mathbb{R}^{M \times N} \qquad (604)$$

and describes a line in $\mathbb{R} \times \mathbb{R}^{M \times N}$

$$\left\{ \begin{bmatrix} x \\ Ax + B \end{bmatrix} \mid x \in \mathbb{R} \right\} \subset \mathbb{R} \times \mathbb{R}^{M \times N} \qquad (605)$$

whose slope with respect to x is A.   □

3.6.1   monotonic function

A real function of real argument is called *monotonic* when it is exclusively nonincreasing or nondecreasing over the whole of its domain. A real differentiable function of real argument is monotonic when its first derivative (not necessarily continuous) maintains sign over the function domain.

3.6.1.0.1   Definition.   *Monotonicity.*
In pointed closed convex cone $\mathcal{K}$, multidimensional function $f(X)$ is

$$\begin{array}{lll} \text{nondecreasing monotonic when} & Y \succeq_{\mathcal{K}} X \;\Rightarrow\; f(Y) \succeq f(X) \\ \text{nonincreasing monotonic when} & Y \succeq_{\mathcal{K}} X \;\Rightarrow\; f(Y) \preceq f(X) \end{array} \qquad (606)$$

$\forall\, X, Y \in \operatorname{dom} f$. Multidimensional function $f(X)$ is

$$\begin{array}{lll} \text{increasing monotonic when} & Y \succ_{\mathcal{K}} X \;\Rightarrow\; f(Y) \succ f(X) \\ \text{decreasing monotonic when} & Y \succ_{\mathcal{K}} X \;\Rightarrow\; f(Y) \prec f(X) \end{array} \qquad (607)$$

These latter inequalities define *strict monotonicity* when they hold over all $X, Y \in \operatorname{dom} f$.
$\triangle$

For monotonicity of vector-valued functions, f compared with respect to the nonnegative orthant, it is necessary and sufficient for each entry f_i to be monotonic in the same sense.

Any affine function is monotonic. In $\mathcal{K} = \mathbb{S}_+^M$, for example, $\operatorname{tr}(Z^{\mathrm{T}} X)$ is a nondecreasing monotonic function of matrix $X \in \mathbb{S}^M$ when constant matrix Z is positive semidefinite; which follows from a result (377) of Fejér.

A convex function can be characterized by another kind of nondecreasing monotonicity of its gradient:

3.6.1.0.2 Theorem. *Gradient monotonicity.* [205, §B.4.1.4]
[56, §3.1 exer.20] Given real differentiable function $f(X) : \mathbb{R}^{p \times k} \to \mathbb{R}$ with matrix argument on open convex domain, the condition

$$\langle \nabla f(Y) - \nabla f(X) , Y - X \rangle \geq 0 \text{ for each and every } X, Y \in \operatorname{dom} f \qquad (608)$$

is necessary and sufficient for convexity of f. Strict inequality and *caveat* $Y \neq X$ constitute necessary and sufficient conditions for strict convexity. ◇

3.6.1.0.3 Example. *Composition of functions.* [62, §3.2.4] [205, §B.2.1]
Monotonic functions play a vital role determining convexity of functions constructed by transformation. Given functions $g : \mathbb{R}^k \to \mathbb{R}$ and $h : \mathbb{R}^n \to \mathbb{R}^k$, their composition $f = g(h) : \mathbb{R}^n \to \mathbb{R}$ defined by

$$f(x) = g(h(x)) , \qquad \operatorname{dom} f = \{ x \in \operatorname{dom} h \mid h(x) \in \operatorname{dom} g \} \qquad (609)$$

is convex if

- g is convex nondecreasing monotonic **and** h is convex

- g is convex nonincreasing monotonic **and** h is concave

and composite function f is concave if

- g is concave nondecreasing monotonic **and** h is concave

- g is concave nonincreasing monotonic **and** h is convex

where ∞ $(-\infty)$ is assigned to convex (concave) g when evaluated outside its domain. For differentiable functions, these rules are consequent to (1882).

- Convexity (concavity) of any g is preserved when h is affine. □

If f and g are nonnegative convex real functions, then $\left(f(x)^k + g(x)^k \right)^{1/k}$ is also convex for integer $k \geq 1$. [255, p.44] A squared norm is convex having the same minimum because a squaring operation is convex nondecreasing monotonic on the nonnegative real line.

3.6.1.0.4 Exercise. *Order of composition.*
Real function $f = x^{-2}$ is convex on $\mathbb{R}_+$ but not predicted so by results in Example 3.6.1.0.3 when $g = h(x)^{-1}$ and $h = x^2$. Explain this anomaly. ▼

The following result for product of real functions is extensible to inner product of multidimensional functions on real domain:

3.6.1.0.5 Exercise. *Product and ratio of convex functions.*   [62, exer.3.32]
In general the product or ratio of two convex functions is not convex. [236] *However, there are some results that apply to functions on* $\mathbb{R}$ [real domain]. *Prove the following.*[3.19]

  (a) *If f and g are convex, both nondecreasing (or nonincreasing), and positive functions on an interval, then fg is convex.*

  (b) *If f, g are concave, positive, with one nondecreasing and the other nonincreasing, then fg is concave.*

  (c) *If f is convex, nondecreasing, and positive, and g is concave, nonincreasing, and positive, then f/g is convex.*   ▼

3.6.2   first-order convexity condition, real function

Discretization of $w \succeq 0$ in (497) invites refocus to the real-valued function:

3.6.2.0.1   Theorem. *Necessary and sufficient convexity condition.*
[42, §1.2] [62, §3.1.3] [136, §I.5.2] [314, §3] [337, §4.2] [419, §1.2.3]
For real differentiable function $f(X) : \mathbb{R}^{p \times k} \to \mathbb{R}$ with matrix argument on open convex domain, the condition (*confer* §D.1.7)

$$f(Y) \geq f(X) + \langle \nabla f(X) , Y - X \rangle \quad \text{for each and every } X, Y \in \operatorname{dom} f \qquad (610)$$

is necessary and sufficient for convexity of f. *Caveat* $Y \neq X$ and strict inequality again constitute necessary and sufficient conditions for strict convexity. [205, §B.4.1.1]   ◇

When $f(X) : \mathbb{R}^p \to \mathbb{R}$ is a real differentiable convex function with vector argument on open convex domain, there is simplification of the first-order condition (610); for each and every $X, Y \in \operatorname{dom} f$

$$f(Y) \geq f(X) + \nabla f(X)^{\mathrm{T}}(Y - X) \qquad (611)$$

From this we can find $\underline{\partial \mathcal{H}}_-$ a unique [399, p.220-229] nonvertical [205, §B.1.2] hyperplane (§2.4), expressed in terms of function gradient, supporting $\operatorname{epi} f$ at $\begin{bmatrix} X \\ f(X) \end{bmatrix}$: *videlicet,* defining $f(Y \notin \operatorname{dom} f) \triangleq \infty$ [62, §3.1.7]

$$\begin{bmatrix} Y \\ t \end{bmatrix} \in \operatorname{epi} f \;\Leftrightarrow\; t \geq f(Y) \;\Rightarrow\; \begin{bmatrix} \nabla f(X)^{\mathrm{T}} & -1 \end{bmatrix}\left(\begin{bmatrix} Y \\ t \end{bmatrix} - \begin{bmatrix} X \\ f(X) \end{bmatrix} \right) \leq 0 \qquad (612)$$

This means, for each and every point X in the domain of a convex real function $f(X)$, there exists a hyperplane $\underline{\partial \mathcal{H}}_-$ in $\mathbb{R}^p \times \mathbb{R}$ having normal $\begin{bmatrix} \nabla f(X) \\ -1 \end{bmatrix}$ supporting the function epigraph at $\begin{bmatrix} X \\ f(X) \end{bmatrix} \in \underline{\partial \mathcal{H}}_-$

$$\underline{\partial \mathcal{H}}_- = \left\{ \begin{bmatrix} Y \\ t \end{bmatrix} \in \begin{bmatrix} \mathbb{R}^p \\ \mathbb{R} \end{bmatrix} \;\middle|\; \begin{bmatrix} \nabla f(X)^{\mathrm{T}} & -1 \end{bmatrix}\left(\begin{bmatrix} Y \\ t \end{bmatrix} - \begin{bmatrix} X \\ f(X) \end{bmatrix} \right) = 0 \right\} \qquad (613)$$

[3.19]Hint: Prove §3.6.1.0.5a by verifying Jensen's inequality ((496) at $\mu = \frac{1}{2}$).

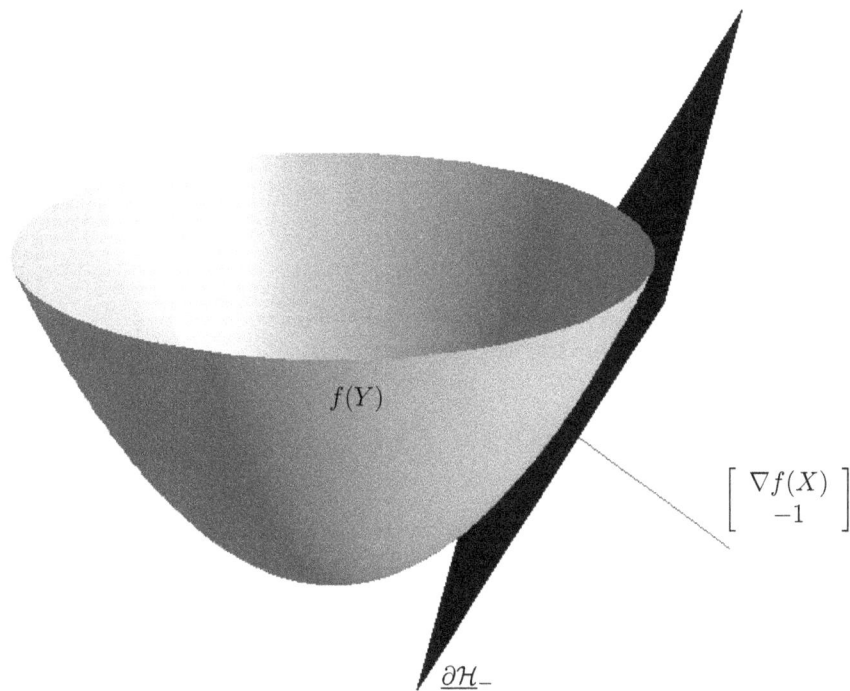

Figure 78: When a real function f is differentiable at each point in its open domain, there is an intuitive geometric interpretation of function convexity in terms of its gradient ∇f and its epigraph: Drawn is a convex quadratic bowl in $\mathbb{R}^2 \times \mathbb{R}$ (*confer* Figure 171 p.586); $f(Y) = Y^{\mathrm{T}}Y : \mathbb{R}^2 \to \mathbb{R}$ *versus* Y on some open disc in $\mathbb{R}^2$. Unique strictly supporting hyperplane $\underline{\partial \mathcal{H}_-} \subset \mathbb{R}^2 \times \mathbb{R}$ (only partially drawn) and its normal vector $[\,\nabla f(X)^{\mathrm{T}} \quad -1\,]^{\mathrm{T}}$ at the particular point of support $[\,X^{\mathrm{T}} \quad f(X)\,]^{\mathrm{T}}$ are illustrated. The interpretation: At each and every coordinate Y, there is a unique hyperplane containing $[\,Y^{\mathrm{T}} \quad f(Y)\,]^{\mathrm{T}}$ and supporting the epigraph of convex differentiable f.

Such a hyperplane is strictly supporting whenever a function is strictly convex. One such supporting hyperplane (*confer* Figure **30**a) is illustrated in Figure **78** for a convex quadratic.

From (611) we deduce, for each and every $X, Y \in \mathrm{dom}\, f$ in the domain,

$$\nabla f(X)^{\mathrm{T}}(Y - X) \geq 0 \;\Rightarrow\; f(Y) \geq f(X) \tag{614}$$

meaning, the gradient at X identifies a supporting hyperplane there in $\mathbb{R}^p$

$$\{ Y \in \mathbb{R}^p \mid \nabla f(X)^{\mathrm{T}}(Y - X) = 0 \} \tag{615}$$

to the convex sublevel sets of convex function f (*confer* (560))

$$\mathcal{L}_{f(X)} f \triangleq \{ Z \in \mathrm{dom}\, f \mid f(Z) \leq f(X) \} \subseteq \mathbb{R}^p \tag{616}$$

illustrated for an arbitrary convex real function in Figure **79** and Figure **68**. That supporting hyperplane is unique for twice differentiable f. [225, p.501]

3.6.3 first-order convexity condition, vector-valued f

Now consider the first-order necessary and sufficient condition for convexity of a vector-valued function: Differentiable function $f(X) : \mathbb{R}^{p \times k} \to \mathbb{R}^M$ is convex if and only if $\mathrm{dom}\, f$ is open, convex, and for each and every $X, Y \in \mathrm{dom}\, f$

$$f(Y) \;\underset{\mathbb{R}_+^M}{\succeq}\; f(X) + \overset{\to Y - X}{df(X)} \;\triangleq\; f(X) + \frac{d}{dt}\bigg|_{t=0} f(X + t\,(Y - X)) \tag{617}$$

where $\overset{\to Y - X}{df(X)}$ is the *directional derivative*[3.20] [225] [339] of f at X in direction $Y - X$. This, of course, follows from the real-valued function case: by dual generalized inequalities (§2.13.2.0.1),

$$f(Y) - f(X) - \overset{\to Y - X}{df(X)} \underset{\mathbb{R}_+^M}{\succeq} 0 \;\Leftrightarrow\; \left\langle f(Y) - f(X) - \overset{\to Y - X}{df(X)} ,\, w \right\rangle \geq 0 \;\; \forall w \underset{\mathbb{R}_+^{M*}}{\succeq} 0 \tag{618}$$

where

$$\overset{\to Y - X}{df(X)} = \begin{bmatrix} \mathrm{tr}\big(\nabla f_1(X)^{\mathrm{T}}(Y - X)\big) \\ \mathrm{tr}\big(\nabla f_2(X)^{\mathrm{T}}(Y - X)\big) \\ \vdots \\ \mathrm{tr}\big(\nabla f_M(X)^{\mathrm{T}}(Y - X)\big) \end{bmatrix} \in \mathbb{R}^M \tag{619}$$

Necessary and sufficient discretization (497) allows relaxation of the semiinfinite number of conditions $\{w \succeq 0\}$ instead to $\{w \in \{e_i ,\, i = 1 \ldots M\}\}$ the extreme directions of the selfdual nonnegative orthant. Each extreme direction picks out a real entry f_i and $\overset{\to Y - X}{df(X)}_i$ from the vector-valued function and its directional derivative, then Theorem 3.6.2.0.1 applies.

The vector-valued function case (617) is therefore a straightforward application of the first-order convexity condition for real functions to each entry of the vector-valued function.

[3.20]We extend the traditional definition of directional derivative in §D.1.4 so that direction may be indicated by a vector or a matrix, thereby broadening scope of the Taylor series (§D.1.7). The right-hand side of inequality (617) is the first-order Taylor series expansion of f about X.

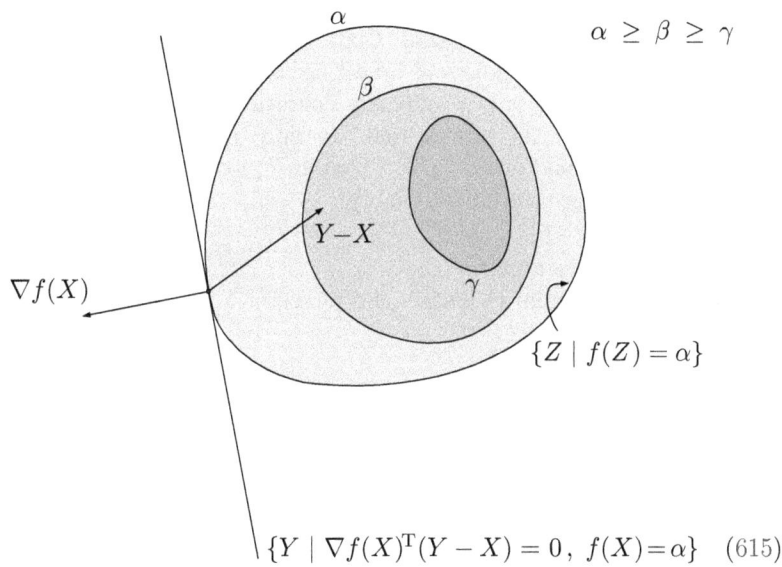

$$\{Y \mid \nabla f(X)^{\mathrm{T}}(Y - X) = 0, \; f(X) = \alpha\} \quad (615)$$

Figure 79: (*confer* Figure 68) Shown is a plausible contour plot in $\mathbb{R}^2$ of some arbitrary real differentiable convex function $f(Z)$ at selected levels α, β, and γ; contours of equal level f (level sets) drawn in the function's domain. A convex function has convex sublevel sets $\mathcal{L}_{f(X)}f$ (616). [315, §4.6] The sublevel set whose boundary is the level set at α, for instance, comprises all shaded regions. For any particular convex function, the family comprising all its sublevel sets is nested. [205, p.75] Were sublevel sets not convex, we may certainly conclude the corresponding function is neither convex. Contour plots of real affine functions are illustrated in Figure 27 and Figure 74.

3.6.4   second-order convexity condition

Again, by discretization (497), we are obliged only to consider each individual entry f_i of a vector-valued function f; *id est*, the real functions $\{f_i\}$.

For $f(X):\mathbb{R}^p \!\to\! \mathbb{R}^M$, a twice differentiable vector-valued function with vector argument on open convex domain,

$$\underset{\mathbb{S}^p_+}{\nabla^2 f_i(X) \,\succeq\, 0} \quad \forall\, X \in \operatorname{dom} f\,, \quad i=1\ldots M \tag{620}$$

is a necessary and sufficient condition for convexity of f. Obviously, when $M\!=\!1$, this convexity condition also serves for a real function. Condition (620) demands nonnegative *curvature,* intuitively, hence precluding points of inflection as in Figure **80** (p.228).

Strict inequality in (620) provides only a sufficient condition for strict convexity, but that is nothing new; *videlicet,* strictly convex real function $f_i(x)\!=\!x^4$ does not have positive second derivative at each and every $x\in\mathbb{R}$. Quadratic forms constitute a notable exception where the strict-case converse holds reliably.

3.6.4.0.1   Example.   *Convex quadratic.*
Real quadratic multivariate polynomial in matrix A and vector b

$$x^{\mathrm{T}}\!A\,x + 2b^{\mathrm{T}}x + c \tag{621}$$

is convex if and only if $A\succeq 0$. Proof follows by observing second order gradient: (§D.2.1)

$$\nabla_x^2\big(x^{\mathrm{T}}\!A\,x + 2b^{\mathrm{T}}x + c\big) = A + A^{\mathrm{T}} \tag{622}$$

Because $x^{\mathrm{T}}(A+A^{\mathrm{T}})x = 2x^{\mathrm{T}}\!A\,x$, matrix A can be assumed symmetric.   □

3.6.4.0.2   Exercise.   *Real fractional function.*
(*confer* §3.3, §3.5.0.0.4, §3.8.2.0.1)   Prove that real function $f(x,y)=y/x$ is not convex on the first quadrant. Also exhibit this in a plot of the function. (f is *quasilinear* (p.237) on $\{x>0\}$ and nonmonotonic; even on the first quadrant.)   ▼

3.6.4.0.3   Exercise.   *Stress function.*
Define $|x-y| \triangleq \sqrt{(x-y)^2}$ and

$$X = [\,x_1 \;\cdots\; x_N\,] \in \mathbb{R}^{1\times N} \tag{76}$$

Given symmetric nonnegative data $[h_{ij}] \in \mathbb{S}^N \cap \mathbb{R}_+^{N\times N}$, consider function

$$f(\operatorname{vec} X) = \sum_{i=1}^{N-1}\sum_{j=i+1}^{N} \big(|x_i-x_j| - h_{ij}\big)^2 \in \mathbb{R} \tag{1389}$$

Find a gradient and Hessian for f. Then explain why f is not a convex function; *id est,* why doesn't second-order condition (620) apply to the constant positive semidefinite Hessian matrix you found. For $N\!=\!6$ and h_{ij} data from (1462), apply *line theorem* 3.7.3.0.1 to plot f along some arbitrary lines through its domain.   ▼

3.6.4.1   second-order ⇒ first-order condition

For a twice-differentiable real function $f_i(X) : \mathbb{R}^p \to \mathbb{R}$ having open domain, a consequence of the *mean value theorem* from calculus allows compression of its complete Taylor series expansion about $X \in \operatorname{dom} f_i$ (§D.1.7) to three terms: On some open interval of $\|Y\|_2$, so that each and every line segment $[X, Y]$ belongs to $\operatorname{dom} f_i$, there exists an $\alpha \in [0, 1]$ such that [419, §1.2.3] [42, §1.1.4]

$$f_i(Y) = f_i(X) + \nabla f_i(X)^{\mathrm{T}}(Y - X) + \frac{1}{2}(Y - X)^{\mathrm{T}}\nabla^2 f_i(\alpha X + (1 - \alpha)Y)(Y - X) \quad (623)$$

The first-order condition for convexity (611) follows directly from this and the second-order condition (620).

3.7   Convex matrix-valued function

We need different tools for matrix argument: We are primarily interested in continuous matrix-valued functions $g(X)$. We choose symmetric $g(X) \in \mathbb{S}^M$ because matrix-valued functions are most often compared (624) with respect to the positive semidefinite cone $\mathbb{S}^M_+$ in the ambient space of symmetric matrices.[3.21]

3.7.0.0.1   Definition.   *Convex matrix-valued function:*
1) *Matrix-definition.*
A function $g(X) : \mathbb{R}^{p \times k} \to \mathbb{S}^M$ is convex in X iff $\operatorname{dom} g$ is a convex set and, for each and every $Y, Z \in \operatorname{dom} g$ and all $0 \le \mu \le 1$ [221, §2.3.7]

$$g(\mu Y + (1 - \mu)Z) \underset{\mathbb{S}^M_+}{\preceq} \mu g(Y) + (1 - \mu)g(Z) \quad (624)$$

Reversing sense of the inequality flips this definition to concavity. Strict convexity is defined less a stroke of the pen in (624) similarly to (499).
2) *Scalar-definition.*
It follows that $g(X) : \mathbb{R}^{p \times k} \to \mathbb{S}^M$ is convex in X iff $w^{\mathrm{T}} g(X)w : \mathbb{R}^{p \times k} \to \mathbb{R}$ is convex in X for each and every $\|w\| = 1$; shown by substituting the defining inequality (624). By dual generalized inequalities we have the equivalent but more broad criterion, (§2.13.5)

$$g \text{ convex w.r.t } \mathbb{S}^M_+ \Leftrightarrow \langle W, g \rangle \text{ convex } \text{ for each and every } W \underset{\mathbb{S}^{M*}_+}{\succeq} 0 \quad (625)$$

Strict convexity on both sides requires *caveat* $W \neq \mathbf{0}$. Because the set of all extreme directions for the selfdual positive semidefinite cone (§2.9.2.7) comprises a minimal set of generators for that cone, discretization (§2.13.4.2.1) allows replacement of matrix W with symmetric dyad ww^{T} as proposed. △

[3.21] Function symmetry is not a necessary requirement for convexity; indeed, for $A \in \mathbb{R}^{m \times p}$ and $B \in \mathbb{R}^{m \times k}$, $g(X) = AX + B$ is a convex (affine) function in X on domain $\mathbb{R}^{p \times k}$ with respect to the nonnegative orthant $\mathbb{R}^{m \times k}_+$. Symmetric convex functions share the same benefits as symmetric matrices. Horn & Johnson [208, §7.7] liken symmetric matrices to real numbers, and (symmetric) positive definite matrices to positive real numbers.

3.7.0.0.2   Example.   *Taxicab distance matrix.*
Consider an n-dimensional vector space $\mathbf{R}^n$ with metric induced by the 1-norm. Then distance between points x_1 and x_2 is the norm of their difference: $\|x_1 - x_2\|_1$. Given a list of points arranged columnar in a matrix

$$X = [\,x_1 \;\cdots\; x_N\,] \in \mathbf{R}^{n \times N} \qquad (76)$$

then we could define a taxicab distance matrix

$$
\begin{aligned}
\mathbf{D}_1(X) \;&\triangleq\; (I \otimes \mathbf{1}_n^{\mathrm{T}})\,|\,\mathrm{vec}(X)\mathbf{1}^{\mathrm{T}} - \mathbf{1} \otimes X\,| \;\in\; \mathbf{S}_h^N \cap \mathbf{R}_+^{N \times N} \\[2mm]
&= \begin{bmatrix}
0 & \|x_1 - x_2\|_1 & \|x_1 - x_3\|_1 & \cdots & \|x_1 - x_N\|_1 \\
\|x_1 - x_2\|_1 & 0 & \|x_2 - x_3\|_1 & \cdots & \|x_2 - x_N\|_1 \\
\|x_1 - x_3\|_1 & \|x_2 - x_3\|_1 & 0 & & \|x_3 - x_N\|_1 \\
\vdots & \vdots & & \ddots & \vdots \\
\|x_1 - x_N\|_1 & \|x_2 - x_N\|_1 & \|x_3 - x_N\|_1 & \cdots & 0
\end{bmatrix}
\end{aligned}
\qquad (626)
$$

where $\mathbf{1}_n$ is a vector of ones having $\dim \mathbf{1}_n = n$ and where $\otimes$ represents Kronecker product. This matrix-valued function is convex with respect to the nonnegative orthant since, for each and every $Y, Z \in \mathbf{R}^{n \times N}$ and all $0 \le \mu \le 1$

$$\mathbf{D}_1(\mu Y + (1 - \mu)Z) \;\underset{\mathbf{R}_+^{N \times N}}{\preceq}\; \mu\,\mathbf{D}_1(Y) + (1 - \mu)\mathbf{D}_1(Z) \qquad (627)$$

$\square$

3.7.0.0.3   Exercise.   *1-norm distance matrix.*
The 1-norm is called *taxicab distance* because to go from one point to another in a city by car, road distance is a sum of grid lengths. Prove (627).   ▼

3.7.1   first-order convexity condition, matrix-valued f

From the *scalar-definition* (§3.7.0.0.1) of a convex matrix-valued function, for differentiable function g and for each and every real vector w of unit norm $\|w\| = 1$, we have

$$w^{\mathrm{T}} g(Y) w \ge w^{\mathrm{T}} g(X) w + w^{\mathrm{T}} \overset{\rightarrow Y-X}{dg(X)}\, w \qquad (628)$$

that follows immediately from the first-order condition (610) for convexity of a real function because

$$w^{\mathrm{T}} \overset{\rightarrow Y-X}{dg(X)}\, w = \left\langle \nabla_X\, w^{\mathrm{T}} g(X) w \,,\, Y - X \right\rangle \qquad (629)$$

where $\overset{\rightarrow Y-X}{dg(X)}$ is the directional derivative (§D.1.4) of function g at X in direction $Y - X$. By discretized dual generalized inequalities, (§2.13.5)

$$g(Y) - g(X) - \overset{\rightarrow Y-X}{dg(X)} \underset{\mathbb{S}_+^M}{\succeq} 0 \;\Leftrightarrow\; \left\langle g(Y) - g(X) - \overset{\rightarrow Y-X}{dg(X)} \,,\, ww^{\mathrm{T}} \right\rangle \ge 0 \;\; \forall ww^{\mathrm{T}}(\underset{\mathbb{S}_+^{M*}}{\succeq} 0)$$

$$(630)$$

For each and every $X, Y \in \mathrm{dom}\, g$ *(confer* (617)*)*

$$g(Y) \underset{\mathbb{S}_+^M}{\succeq} g(X) + \overset{\rightarrow Y - X}{dg(X)} \tag{631}$$

must therefore be necessary and sufficient for convexity of a matrix-valued function of matrix variable on open convex domain.

3.7.2 epigraph of matrix-valued function, sublevel sets

We generalize *epigraph* to a continuous matrix-valued function [34, p.155] $g(X) : \mathbb{R}^{p \times k} \to \mathbb{S}^M$:

$$\text{epi}\, g \triangleq \{ (X, T) \in \mathbb{R}^{p \times k} \times \mathbb{S}^M \mid X \in \text{dom}\, g, \ g(X) \underset{\mathbb{S}_+^M}{\preceq} T \} \tag{632}$$

from which it follows

$$g \text{ convex} \Leftrightarrow \text{epi}\, g \text{ convex} \tag{633}$$

Proof of necessity is similar to that in §3.5 on page 206.

Sublevel sets of a convex matrix-valued function corresponding to each and every $S \in \mathbb{S}^M$ (*confer* (560))

$$\mathcal{L}_S g \triangleq \{ X \in \text{dom}\, g \mid g(X) \underset{\mathbb{S}_+^M}{\preceq} S \} \subseteq \mathbb{R}^{p \times k} \tag{634}$$

are convex. There is no converse.

3.7.2.0.1 Example. *Matrix fractional function* redux. [34, p.155]
Generalizing Example 3.5.0.0.4 consider a matrix-valued function of two variables on $\text{dom}\, g = \mathbb{S}_+^N \times \mathbb{R}^{n \times N}$ for small positive constant ϵ (*confer* (1962))

$$g(A, X) = \epsilon X (A + \epsilon I)^{-1} X^{\mathrm{T}} \tag{635}$$

where the inverse always exists by (1532). This function is convex simultaneously in both variables over the entire positive semidefinite cone $\mathbb{S}_+^N$ and all $X \in \mathbb{R}^{n \times N}$: Consider Schur-form (1595) from §A.4: for $T \in \mathbb{S}^n$

$$\begin{aligned} G(A, X, T) = \begin{bmatrix} A + \epsilon I & X^{\mathrm{T}} \\ X & \epsilon^{-1} T \end{bmatrix} \succeq 0 \\ \Leftrightarrow \\ T - \epsilon X (A + \epsilon I)^{-1} X^{\mathrm{T}} \succeq 0 \\ A + \epsilon I \succ 0 \end{aligned} \tag{636}$$

By Theorem 2.1.9.0.1, inverse image of the positive semidefinite cone $\mathbb{S}_+^{N+n}$ under affine mapping $G(A, X, T)$ is convex. Function $g(A, X)$ is convex on $\mathbb{S}_+^N \times \mathbb{R}^{n \times N}$ because its epigraph is that inverse image:

$$\text{epi}\, g(A, X) = \{ (A, X, T) \mid A + \epsilon I \succ 0, \ \epsilon X (A + \epsilon I)^{-1} X^{\mathrm{T}} \preceq T \} = G^{-1}\left(\mathbb{S}_+^{N+n} \right) \tag{637}$$

$\square$

3.7.3   second-order convexity condition, matrix-valued f

The following *line theorem* is a potent tool for establishing convexity of a multidimensional function. To understand it, what is meant by *line* must first be solidified. Given a function $g(X) : \mathbb{R}^{p \times k} \to \mathbb{S}^M$ and particular $X, Y \in \mathbb{R}^{p \times k}$ not necessarily in that function's domain, then we say a line $\{X + tY \mid t \in \mathbb{R}\}$ (infinite in extent) passes through $\operatorname{dom} g$ when $X + tY \in \operatorname{dom} g$ over some interval of $t \in \mathbb{R}$.

3.7.3.0.1   Theorem.   *Line theorem.*   (*confer* [62, §3.1.1])
Multidimensional function $f(X) : \mathbb{R}^{p \times k} \to \mathbb{R}^M$ or $g(X) : \mathbb{R}^{p \times k} \to \mathbb{S}^M$ is convex in X if and only if it remains convex on the intersection of any line with its domain.   ◇

Now we assume a twice differentiable function.

3.7.3.0.2   Definition.   *Differentiable convex matrix-valued function.*
Matrix-valued function $g(X) : \mathbb{R}^{p \times k} \to \mathbb{S}^M$ is convex in X iff $\operatorname{dom} g$ is an open convex set, and its second derivative $g''(X + tY) : \mathbb{R} \to \mathbb{S}^M$ is positive semidefinite on each point of intersection along every line $\{X + tY \mid t \in \mathbb{R}\}$ that intersects $\operatorname{dom} g$; *id est*, iff for each and every $X, Y \in \mathbb{R}^{p \times k}$ such that $X + tY \in \operatorname{dom} g$ over some open interval of $t \in \mathbb{R}$

$$\frac{d^2}{dt^2} g(X + tY) \underset{\mathbb{S}^M_+}{\succeq} 0 \qquad\qquad (638)$$

Similarly, if

$$\frac{d^2}{dt^2} g(X + tY) \underset{\mathbb{S}^M_+}{\succ} 0 \qquad\qquad (639)$$

then g is strictly convex; the converse is generally false. [62, §3.1.4][3.22]   △

3.7.3.0.3   Example.   *Matrix inverse.*   (*confer* §3.3.1)
The matrix-valued function X^μ is convex on $\operatorname{int} \mathbb{S}^M_+$ for $-1 \leq \mu \leq 0$ or $1 \leq \mu \leq 2$ and concave for $0 \leq \mu \leq 1$. [62, §3.6.2] In particular, the function $g(X) = X^{-1}$ is convex on $\operatorname{int} \mathbb{S}^M_+$. For each and every $Y \in \mathbb{S}^M$ (§D.2.1, §A.3.1.0.5)

$$\frac{d^2}{dt^2} g(X + tY) = 2(X + tY)^{-1} Y (X + tY)^{-1} Y (X + tY)^{-1} \underset{\mathbb{S}^M_+}{\succeq} 0 \qquad (640)$$

on some open interval of $t \in \mathbb{R}$ such that $X + tY \succ 0$. Hence, $g(X)$ is convex in X. This result is extensible;[3.23] $\operatorname{tr} X^{-1}$ is convex on that same domain. [208, §7.6 prob.2] [56, §3.1 exer.25]   □

[3.22] The strict-case converse is reliably true for quadratic forms.
[3.23] $d/dt \operatorname{tr} g(X + tY) = \operatorname{tr} d/dt \, g(X + tY)$. [209, p.491]

3.7.3.0.4 Example. *Matrix squared.*

Iconic real function $f(x) = x^2$ is strictly convex on $\mathbb{R}$. The matrix-valued function $g(X) = X^2$ is convex on the domain of symmetric matrices; for $X, Y \in \mathbb{S}^M$ and any open interval of $t \in \mathbb{R}$ (§D.2.1)

$$\frac{d^2}{dt^2} g(X + tY) = \frac{d^2}{dt^2}(X + tY)^2 = 2Y^2 \tag{641}$$

which is positive semidefinite when Y is symmetric because then $Y^2 = Y^{\mathrm{T}}Y$ (1538).[3.24]

A more appropriate matrix-valued counterpart for f is $g(X) = X^{\mathrm{T}}X$ which is a convex function on domain $\{X \in \mathbb{R}^{m \times n}\}$, and strictly convex whenever X is skinny-or-square full-rank. This matrix-valued function can be generalized to $g(X) = X^{\mathrm{T}}AX$ which is convex whenever matrix A is positive semidefinite (p.596), and strictly convex when A is positive definite and X is skinny-or-square full-rank (Corollary A.3.1.0.5). $\qquad \square$

3.7.3.0.5 Exercise. *Squared maps.*

Give seven examples of distinct polyhedra $\mathcal{P}$ for which the set

$$\{X^{\mathrm{T}}X \mid X \in \mathcal{P}\} \subseteq \mathbb{S}_+^n \tag{642}$$

were convex. Is this set convex, in general, for any polyhedron $\mathcal{P}$? (*confer* (1314)(1321)) Is the epigraph of function $g(X) = X^{\mathrm{T}}X$ convex for any polyhedral domain? $\qquad \blacktriangledown$

3.7.3.0.6 Exercise. *Squared inverse.* $\qquad$ (*confer* §3.7.2.0.1)

For positive scalar a, real function $f(x) = ax^{-2}$ is convex on the nonnegative real line. Given positive definite matrix constant A, prove via *line theorem* that $g(X) = \mathrm{tr}((X^{\mathrm{T}}A^{-1}X)^{-1})$ is generally not convex unless $X \succ 0$.[3.25] From this result, show how it follows via Definition 3.7.0.0.1-2 that $h(X) = (X^{\mathrm{T}}A^{-1}X)^{-1}$ is generally neither convex. $\qquad \blacktriangledown$

3.7.3.0.7 Example. *Matrix exponential.*

The matrix-valued function $g(X) = e^X : \mathbb{S}^M \to \mathbb{S}^M$ is convex on the subspace of *circulant* [174] symmetric matrices. Applying the *line theorem*, for all $t \in \mathbb{R}$ and circulant $X, Y \in \mathbb{S}^M$, from Table **D.2.7** we have

$$\frac{d^2}{dt^2} e^{X + tY} = Y e^{X + tY} Y \underset{\mathbb{S}_+^M}{\succeq} 0, \qquad (XY)^{\mathrm{T}} = XY \tag{643}$$

because all circulant matrices are *commutative* and, for symmetric matrices, $XY = YX \Leftrightarrow (XY)^{\mathrm{T}} = XY$ (1555). Given symmetric argument, the matrix exponential always resides interior to the cone of positive semidefinite matrices in the symmetric matrix subspace; $e^A \succ 0 \ \forall A \in \mathbb{S}^M$ (1959). Then for any matrix Y of compatible dimension, $Y^{\mathrm{T}}e^A Y$ is positive semidefinite. (§A.3.1.0.5)

[3.24] By (1556) in §A.3.1, changing the domain instead to all symmetric and nonsymmetric positive semidefinite matrices, for example, will not produce a convex function.
[3.25] Hint: §D.2.3.

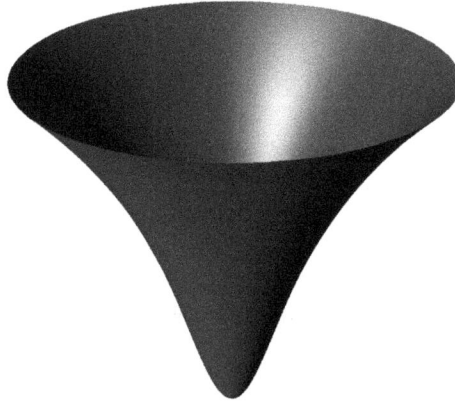

Figure 80: Iconic unimodal differentiable quasiconvex function of two variables graphed in $\mathbb{R}^2 \times \mathbb{R}$ on some open disc in $\mathbb{R}^2$. Note reversal of curvature in direction of gradient.

The subspace of circulant symmetric matrices contains all diagonal matrices. The matrix exponential of any diagonal matrix e^Λ exponentiates each individual entry on the main diagonal. [257, §5.3] So, changing the function domain to the subspace of real diagonal matrices reduces the matrix exponential to a vector-valued function in an isometrically isomorphic subspace $\mathbb{R}^M$; known convex (§3.1.1) from the real-valued function case [62, §3.1.5].   □

There are more methods for determining function convexity; [42] [62] [136] one can be more efficient than another depending on the function in hand.

3.7.3.0.8   Exercise.   *log det function.*
Matrix determinant is neither a convex or concave function, in general, but its inverse is convex when domain is restricted to interior of a positive semidefinite cone. [34, p.149] Show by three different methods: On interior of the positive semidefinite cone, $\log \det X = -\log \det X^{-1}$ is concave.   ▼

3.8   Quasiconvex

Quasiconvex functions [205] [337] [399] [250, §2] are valuable, pragmatically, because they are *unimodal* (by definition when nonmonotonic); a global minimum is guaranteed to exist over any convex set in the function domain; *e.g.*, Figure 80. Optimal solution to quasiconvex problems is by method of bisection (**a.k.a** *binary search*). [62, §4.2.5]

3.8.0.0.1 Definition. *Quasiconvex function.*   (*confer* (496))
$f(X) : \mathbb{R}^{p \times k} \to \mathbb{R}$ is a quasiconvex function of matrix X iff $\operatorname{dom} f$ is a convex set and for each and every $Y, Z \in \operatorname{dom} f$ and $0 \le \mu \le 1$

$$f(\mu Y + (1 - \mu)Z) \le \max\{f(Y), f(Z)\} \tag{644}$$

A quasiconcave function is similarly defined:

$$f(\mu Y + (1 - \mu)Z) \ge \min\{f(Y), f(Z)\} \tag{645}$$

Caveat $Y \ne Z$ and strict inequality on an open interval $0 < \mu < 1$ constitute necessary and sufficient conditions for strict quasiconvexity.   △

Unlike convex functions, quasiconvex functions are not necessarily continuous; *e.g.*, quasiconcave $\operatorname{rank}(X)$ on $\mathbb{S}_+^M$ (§2.9.2.9.2) and $\operatorname{card}(x)$ on $\mathbb{R}_+^M$. Although insufficient for convex functions, convexity of each and every sublevel set serves as a definition of quasiconvexity:

3.8.0.0.2 Definition. *Quasiconvex multidimensional function.*
Scalar-, vector-, or matrix-valued function $g(X) : \mathbb{R}^{p \times k} \to \mathbb{S}^M$ is a quasiconvex function of matrix X iff $\operatorname{dom} g$ is a convex set and its sublevel set

$$\mathcal{L}_S g = \{X \in \operatorname{dom} g \mid g(X) \preceq S\} \subseteq \mathbb{R}^{p \times k} \qquad (634)$$

(corresponding to each and every $S \in \mathbb{S}^M$) is convex. Vectors are compared with respect to the nonnegative orthant $\mathbb{R}_+^M$ while matrices are with respect to the positive semidefinite cone $\mathbb{S}_+^M$.

Likewise, convexity of each and every *superlevel set*

$$\mathcal{L}^S g = \{X \in \operatorname{dom} g \mid g(X) \succeq S\} \subseteq \mathbb{R}^{p \times k} \tag{646}$$

is necessary and sufficient for quasiconcavity.   △

3.8.0.0.3 Exercise. *Nonconvexity of matrix product.*
Consider real function f on a positive definite domain

$$f(X) = \operatorname{tr}(X_1 X_2), \qquad X \triangleq \begin{bmatrix} X_1 \\ X_2 \end{bmatrix} \in \operatorname{dom} f \triangleq \begin{bmatrix} \operatorname{rel int} \mathbb{S}_+^N \\ \operatorname{rel int} \mathbb{S}_+^N \end{bmatrix} \tag{647}$$

with superlevel sets

$$\mathcal{L}^s f = \{X \in \operatorname{dom} f \mid \langle X_1, X_2 \rangle \ge s\} \tag{648}$$

Prove that $f(X)$ is not quasiconcave except when $N = 1$, nor is it quasiconvex unless $X_1 = X_2$.   ▼

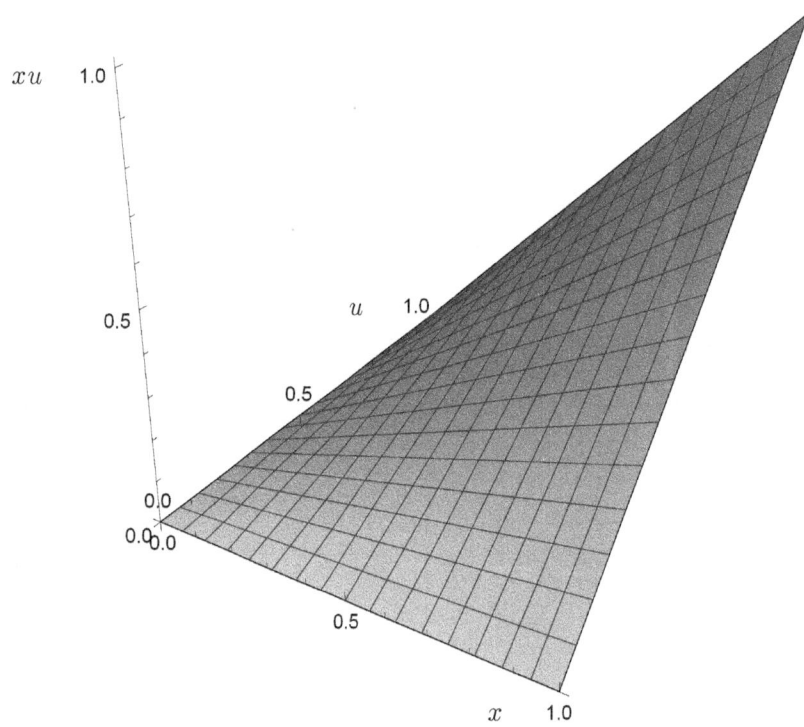

Figure 81: Strictly monotonic quasiconcave real product function xu is bowed (not affine) on the nonnegative orthants.

3.8.1 bilinear

Bilinear (inner product) function[3.26] $x^{\mathrm{T}}u$ of vectors x and u is quasiconcave and strictly monotonic on the nonnegative orthants $\mathbb{R}_+^\eta \times \mathbb{R}_+^\eta$ only when dimension η equals 1. (Figure 81)

When variable $x \leftarrow \beta$ has dimension 1 but u is a vector of arbitrary dimension η, real function $f(\beta, u) = \beta \mathbf{1}^{\mathrm{T}}u = \beta\operatorname{tr}\delta(u)$ is quasiconcave strictly monotonic on the nonnegative orthants $\mathbb{R}_+ \times \mathbb{R}_+^\eta$. $f(\beta, u)$ is quasiconcave strictly monotonic on $\mathbb{R}_+ \times \mathbb{R}^\eta$ when $\mathbf{1}^{\mathrm{T}}u \geq 0$.

3.8.1.0.1 Proof.   Domain of function

$$f(\beta, u) : \mathbb{R}_+ \times \mathbb{R}^\eta \to \mathbb{R}_+ = \beta\mathbf{1}^{\mathrm{T}}u , \quad \mathbf{1}^{\mathrm{T}}u \geq 0 \tag{649}$$

is a nonpointed polyhedral cone in $\mathbb{R}^{\eta+1}$, its range a halfline $\mathbb{R}_+$.

(quasiconcavity)   Because this function spans an orthant, its 0-superlevel set is the deepest superlevel set and identical to its domain. Higher superlevel sets of the function, given some fixed nonzero scalar $\zeta \geq 0$

$$
\begin{aligned}
\{\beta, u \mid f(\beta, u) \geq \zeta, \; \beta > 0, \; \mathbf{1}^{\mathrm{T}}u \geq 0\} &= \{\beta, u \mid \beta\mathbf{1}^{\mathrm{T}}u \geq \zeta, \; \beta > 0, \; \mathbf{1}^{\mathrm{T}}u \geq 0\} \\
&= \{\beta, u \mid \mathbf{1}^{\mathrm{T}}u \geq \tfrac{\zeta}{\beta}, \; \beta > 0\}
\end{aligned}
\tag{650}
$$

are not polyhedral but they are convex because (§A.4)

$$\mathbf{1}^{\mathrm{T}}u \geq \frac{\zeta}{\beta}, \quad \beta > 0 \quad \Leftrightarrow \quad \begin{bmatrix} \beta & \sqrt{\zeta} \\ \sqrt{\zeta} & \operatorname{tr}\delta(u) \end{bmatrix} \succeq 0 \tag{651}$$

and because inverse image of a positive semidefinite cone under affine transformation is convex by Theorem 2.1.9.0.1. Convex superlevel sets are necessary and sufficient for quasiconcavity by Definition 3.8.0.0.2.

(monotonicity)   By Definition 3.6.1.0.1,

$$f \text{ is increasing monotonic when } \begin{bmatrix} \beta \\ u \end{bmatrix} \succ \begin{bmatrix} \tau \\ z \end{bmatrix} \Rightarrow f(\beta, u) \succ f(\tau, z) \tag{652}$$

for all β, u, τ, z in the domain. Assuming $\beta > \tau \geq 0$ and $u \succ z$, it follows that $\mathbf{1}^{\mathrm{T}}u > \mathbf{1}^{\mathrm{T}}z$. (Exercise 2.13.7.0.2) Therefore $\beta\mathbf{1}^{\mathrm{T}}u > \tau\mathbf{1}^{\mathrm{T}}z$ and so $f(\beta, u) = \beta\mathbf{1}^{\mathrm{T}}u$ is strictly monotonic.   ◆

3.8.1.0.2 Example.   *Analog filter design.*
Traditional filter design is a determination of Laplace transfer function coefficients to meet specified tolerances in magnitude and phase over given frequency bands. The problem posed by this example is to find a stable minimum phase filter $H(\jmath\omega)$, of given order η, closest to specified samples of frequency-domain magnitude

$$\{g_i \mid \omega_i \in \Omega\} \tag{653}$$

[3.26]Convex envelope of bilinear functions is well known. [3]

We consider a recursive filter whose real causal impulse response has Laplace transform

$$H(s) = \frac{1 + b_1 s + b_2 s^2 + \ldots + b_\eta s^\eta}{1 + a_1 s + a_2 s^2 + \ldots + a_\eta s^\eta} \qquad (654)$$

whose poles and zeros lie in the left halfplane ($\mathrm{re}\, s = \sigma < 0$) and whose gain is unity at DC (at $s = 0$). This transfer function is a ratio of polynomials in complex variable s defined

$$s \triangleq \sigma + \jmath\omega \qquad (655)$$

To reduce passive component sensitivity, physical implementation is sometimes facilitated by factoring transform (654) into cascaded (or parallel) second order sections which are needed to realize complex poles and zeros. Magnitude square of a second order transfer function evaluated along the $\jmath\omega$ axis (the Fourier domain) is

$$\frac{1 + v_1\omega^2 + v_2\omega^4}{1 + u_1\omega^2 + u_2\omega^4} = \frac{1 + (b_1^2 - 2b_2)\omega^2 + b_2^2\omega^4}{1 + (a_1^2 - 2a_2)\omega^2 + a_2^2\omega^4} \qquad (656)$$

Real filter coefficient vectors b, a, v, u are independent of radian frequency $\omega = 2\pi f$. Coefficients b, a translate directly to passive component values. Stability requires coefficients to obey $0 < a_1$ and $0 < a_2 \leq a_1^2/4$, while minimum phase demands $0 < b_1$ and $0 < b_2 \leq b_1^2/4$. These two requirements imply nonnegative magnitude square filter coefficients v, u.

A cascade implementation of second order sections can realize a high order unity gain filter. Magnitude square of ηth order transfer function H, evaluated along the $\jmath\omega$ axis, is

$$|H(\jmath\omega)|^2 = H(\jmath\omega)H(-\jmath\omega) \triangleq \frac{V(\omega)}{U(\omega)} = \frac{1 + v_1\omega^2 + v_2\omega^4 + \ldots + v_\eta\omega^{2\eta}}{1 + u_1\omega^2 + u_2\omega^4 + \ldots + u_\eta\omega^{2\eta}} \qquad (657)$$

A cascade of two second order sections, $\eta = 4$ for example, has form

$$\frac{1 + \ddot{v}_1\omega^2 + \ddot{v}_2\omega^4}{1 + \ddot{u}_1\omega^2 + \ddot{u}_2\omega^4} \frac{1 + \ddot{v}_3\omega^2 + \ddot{v}_4\omega^4}{1 + \ddot{u}_3\omega^2 + \ddot{u}_4\omega^4} = \frac{1 + v_1\omega^2 + v_2\omega^4 + v_3\omega^6 + v_4\omega^8}{1 + u_1\omega^2 + u_2\omega^4 + u_3\omega^6 + u_4\omega^8} \qquad (658)$$

where (*confer* §4.4.1.2.5)

$$\begin{aligned}
v_1 &= \ddot{v}_1 + \ddot{v}_3 , & u_1 &= \ddot{u}_1 + \ddot{u}_3 \\
v_2 &= \ddot{v}_2 + \ddot{v}_4 + \ddot{v}_1\ddot{v}_3 , & u_2 &= \ddot{u}_2 + \ddot{u}_4 + \ddot{u}_1\ddot{u}_3 \\
v_3 &= \ddot{v}_2\ddot{v}_3 + \ddot{v}_1\ddot{v}_4 , & u_3 &= \ddot{u}_2\ddot{u}_3 + \ddot{u}_1\ddot{u}_4 \\
v_4 &= \ddot{v}_2\ddot{v}_4 , & u_4 &= \ddot{u}_2\ddot{u}_4
\end{aligned} \qquad (659)$$

Odd η is implemented by cascading one first order section.

To guise filter design as an optimization problem, magnitude square filter coefficients

$$v \triangleq \begin{bmatrix} v_1 \\ v_2 \\ \vdots \\ v_\eta \end{bmatrix} \in \mathbb{R}^\eta , \qquad u \triangleq \begin{bmatrix} u_1 \\ u_2 \\ \vdots \\ u_\eta \end{bmatrix} \in \mathbb{R}^\eta \qquad (660)$$

become the variables. Given a set of frequencies Ω at which finite $g(\omega)$ is sampled, our notation reflects this role reversal:

$$\frac{V_i}{U_i} = \frac{V_i(v)}{U_i(u)} \triangleq \frac{1 + v_1\omega_i^2 + v_2\omega_i^4 + \ldots + v_\eta\omega_i^{2\eta}}{1 + u_1\omega_i^2 + u_2\omega_i^4 + \ldots + u_\eta\omega_i^{2\eta}} \,, \qquad \omega_i \in \Omega \qquad (661)$$

A better filter design may be obtained if stability and minimum phase requirements are ignored by the optimization.[3.27] Then coefficients v, u need not be nonnegatively constrained. A *least peak deviation filter* design problem may be expressed

$$\begin{array}{ll} \underset{v,\,u,\,\beta}{\text{minimize}} & \beta \\ \text{subject to} & \dfrac{1}{\beta} \leq \dfrac{U_i}{V_i} g_i^2 \leq \beta \,, \qquad \omega_i \in \Omega \\ & V_i \geq 0 \,, \quad U_i \geq 0 \,, \qquad \omega_i \in \Omega \end{array} \qquad (662)$$

which is a nonconvex optimization problem because of a ratio in variables $v \,\&\, u$. Eliminating the ratio:

$$\begin{array}{ll} \underset{v,\,u,\,\beta}{\text{minimize}} & \beta \\ \text{subject to} & U_i\, g_i^2 \leq \beta V_i \,, \qquad \omega_i \in \Omega \\ & V_i\, g_i^{-2} \leq \beta U_i \,, \qquad \omega_i \in \Omega \\ & V_i \geq 0 \,, \quad U_i \geq 0 \,, \qquad \omega_i \in \Omega \end{array} \qquad (663)$$

This equivalent problem is likewise nonconvex because of new products in variables $\beta \,\&\, v$ and $\beta \,\&\, u$. The feasible set in variables v, u, β is also nonconvex.

But problem (663) can be solved by applying the fact that each product $\beta V_i(v)$, $\beta U_i(u)$ increases monotonically. We may then rely on convexity of the feasible set described by linear constraints for fixed β. One strategy is to choose a sufficiently large positive β so that (663) is initially feasible; say $\beta = \beta_o$. Then halve β until (663) becomes infeasible; say $\beta = \beta_\infty$. At infeasibility, backtrack to a feasible β. Toward a perfect fit, in the sense that all g_i are collocated, optimal β approaches 1. This proposed iteration represents a binary search (**a.k.a** *bisection*) for minimum β, say $\beta = \beta^\star$, that is assumed to lie in the interval $[\beta_\infty \,, \beta_o]$:

$$\begin{array}{l} \beta = \beta_o \\ \beta_\infty = 0 \\ \texttt{for } k = 1, 2, \ldots \texttt{until convergence } \{ \\ \quad \texttt{if (663) feasible } \{ \\ \qquad \beta_o = \beta_{k-1} \\ \quad \texttt{else} \\ \qquad \beta_\infty = \beta_{k-1} \\ \quad \} \\ \quad \beta = \beta_k = (\beta_o + \beta_\infty)/2 \\ \} \end{array} \qquad (664)$$

[3.27] Poles and zeros of $H(\jmath\omega)$ possess conjugate symmetry because of its real coefficients. Poles and zeros of $H(\jmath\omega)H(-\jmath\omega)$ possess both conjugate and real symmetry. Stability and minimum phase may be imposed postoptimization by picking only those poles and zeros, respectively from $H(\jmath\omega)H(-\jmath\omega)$, that reside in the left half s-plane.

Global convergence to any desired tolerance is certain, in absence of numerical error, characterized by an ever narrowing gap between β_∞ and β_o. At convergence of (664), the nonnegative $\beta^\star$ that minimizes (662) is found.

Proof. Problem (663) may be equivalently written

$$
\begin{array}{ll}
\underset{v,\,u,\,\beta,\,t,\,r}{\text{minimize}} & \beta \\
\text{subject to} & U_i\,g_i^2 \;\leq\; r\,, \qquad\qquad \omega_i \in \Omega \\
& r \leq \beta\,V_i\,, \qquad\qquad\quad \omega_i \in \Omega \\
& V_i\,g_i^{-2} \leq t\,, \qquad\qquad \omega_i \in \Omega \\
& t \leq \beta\,U_i\,, \qquad\qquad\quad \omega_i \in \Omega \\
& V_i \geq 0\,, \quad U_i \geq 0\,, \qquad \omega_i \in \Omega
\end{array}
\tag{665}
$$

where $\beta\,,t\,,r \in \mathbb{R}_+$ are implicitly nonnegative. The feasible set is not convex because t is variable in $t \leq \beta\,U_i$; this implicit union of superlevel sets is nonconvex. The same holds for $r \leq \beta\,V_i$. By Schur complement (§A.4), strictly monotonic quasiconcave functions $\beta\,V_i$ and $\beta\,U_i$ (§3.8.1.0.1) may be decomposed

$$
\begin{array}{ll}
\beta\,V_i \geq r\,, \quad \beta > 0 \quad \Leftrightarrow \quad \begin{bmatrix} V_i & \sqrt{r} \\ \sqrt{r} & \beta \end{bmatrix} \succeq 0 \\[2em]
\beta\,U_i \geq t\,, \quad \beta > 0 \quad \Leftrightarrow \quad \begin{bmatrix} \beta & \sqrt{t} \\ \sqrt{t} & U_i \end{bmatrix} \succeq 0
\end{array}
\tag{666}
$$

By Theorem A.3.1.0.4, problem (665) is thereby equivalent to

$$
\begin{array}{ll}
\underset{v,\,u,\,\beta,\,t,\,r}{\text{minimize}} & \begin{bmatrix} 0 & 1 & 0 \end{bmatrix} \begin{bmatrix} V(v) & \sqrt{r} & 0 \\ \sqrt{r} & \beta & \sqrt{t} \\ 0 & \sqrt{t} & U(u) \end{bmatrix} \begin{bmatrix} 0 \\ 1 \\ 0 \end{bmatrix} \\[3em]
\text{subject to} & U_i\,g_i^2 \;\leq\; r\,, \qquad\qquad \omega_i \in \Omega \\
& V_i\,g_i^{-2} \leq t\,, \qquad\qquad \omega_i \in \Omega \\
& \begin{bmatrix} V_i & \sqrt{r} & 0 \\ \sqrt{r} & \beta & \sqrt{t} \\ 0 & \sqrt{t} & U_i \end{bmatrix} \succeq 0\,, \quad \omega_i \in \Omega
\end{array}
\tag{667}
$$

This means minimizing β is like simultaneously minimizing functions $\beta\,V(v)$ and $\beta\,U(u)$ over a Cartesian subspace that is the intersection of their domains; namely, over a line containing the β axis in an increasing direction. Minimization on any line retains quasiconcavity [117], while minimization on any line in an increasing direction maintains strict monotonicity. But local minima cannot be precluded because minimizing a strictly monotonic function over a more general convex feasible set does not imply increasing direction (§3.6.1.0.1, Figure 81) and because, to begin with, the feasible set in variables $v\,,u\,,\beta\,,t\,,r$ is nonconvex due to $\sqrt{r}$ and $\sqrt{t}$.

One recourse is to perform bisection (664) on β to find its globally optimal value $\beta^\star$ in problem (663):

$$
\begin{array}{ll}
\text{find} & v\,,u \\
\text{subject to} & U_i\,g_i^2 \;\leq\; \beta\,V_i\,, \qquad \omega_i \in \Omega \\
& V_i\,g_i^{-2} \leq \beta\,U_i\,, \qquad \omega_i \in \Omega \\
& V_i \geq 0\,, \quad U_i \geq 0\,, \qquad \omega_i \in \Omega
\end{array}
\tag{668}
$$

At each iteration, the feasible set in variables v, u is convex because all the inequalities are linear when β is fixed. A necessary and sufficient condition for bisection to find $\beta^\star$ requires feasibility for all $\beta \geq \beta^\star$ and infeasibility for all $0 \leq \beta < \beta^\star$. β is bounded below by 0 but not bounded above (nor need it be). Feasibility is guaranteed by existence of a pointed polyhedral cone $\mathcal{K}$ described by the linear inequalities as an intersection of halfspaces about the origin in $\mathbb{R}^2$; namely

$$
\mathcal{K} = \left\{ \bigcap_i \left\{ \begin{bmatrix} V_i \\ U_i \end{bmatrix} \in \mathbb{R}^2 \right\} \middle| \begin{array}{c} [\beta \; -g_i^2] \begin{bmatrix} V_i \\ U_i \end{bmatrix} \geq 0 \\ [-g_i^{-2} \; \beta] \begin{bmatrix} V_i \\ U_i \end{bmatrix} \geq 0 \\ \begin{bmatrix} V_i \\ U_i \end{bmatrix} \succeq 0 \end{array} \right\} \subseteq \mathbb{R}_+^2 \tag{669}
$$

Cone $\mathcal{K}$, the feasible set in dependent variables V_i and U_i, ceases to exist for any $\beta < \beta^\star$. This fact becomes evident by considering reach of the hyperplane normals which are linear in β: $[\beta \; -g_i^2]^{\mathrm{T}}$ is confined to the fourth quadrant, $[-g_i^{-2} \; \beta]^{\mathrm{T}}$ is confined to the second. As $\beta \to \infty$, the extreme directions of $\mathcal{K}$ approach the Cartesian axes in the first quadrant. As $\beta \to 0$, extreme directions collapse inward so as to narrow the cone then empty the feasible set. When $\beta = \beta^\star \geq 1$, $\mathcal{K}$ becomes a ray emanating from the origin in the first quadrant. Hyperplane-normal linearity in β and lack of an objective function here obviate the local-minima obstacle. ♦

High powers of radian frequency, in magnitude square transfer function $|H(\jmath\omega)|^2$ (657), demand excessive precision from floating-point numerics. Even some second order filter design optimization problems can cause the best linear program solvers to fail because double precision (64 bits, 52-bit mantissa) is inadequate. (Saunders provides a quadruple precision solver: 128-bit wordlength, 112-bit mantissa.) High powers of frequency are ameliorated simply by scaling specification (653):

$$
\Omega' \triangleq \frac{1}{c}\Omega \tag{670}
$$

Scaled frequency-domain magnitude specifications become

$$
\{ g_i' \mid \omega_i' \in \Omega' \} = \{ g_i \mid \omega_i \in \Omega \} \tag{671}
$$

where

$$
\omega' \triangleq \frac{\omega}{c} \tag{672}
$$

and $c > 1$ is a constant. A filter is designed as before except magnitude square transfer function (657) is replaced with

$$
|H(\jmath\omega')|^2 \triangleq \frac{V'(\omega')}{U'(\omega')} = \frac{1 + v_1'\omega'^2 + v_2'\omega'^4 + \ldots + v_\eta'\omega'^{2\eta}}{1 + u_1'\omega'^2 + u_2'\omega'^4 + \ldots + u_\eta'\omega'^{2\eta}} \tag{673}
$$

Optimization problem (663) is solved for primed magnitude square filter coefficients v', u':

$$
\frac{V_i'(v')}{U_i'(u')} \triangleq \frac{1 + v_1'\omega_i'^2 + v_2'\omega_i'^4 + \ldots + v_\eta'\omega_i'^{2\eta}}{1 + u_1'\omega_i'^2 + u_2'\omega_i'^4 + \ldots + u_\eta'\omega_i'^{2\eta}}, \qquad \omega_i' \in \Omega' \tag{674}
$$

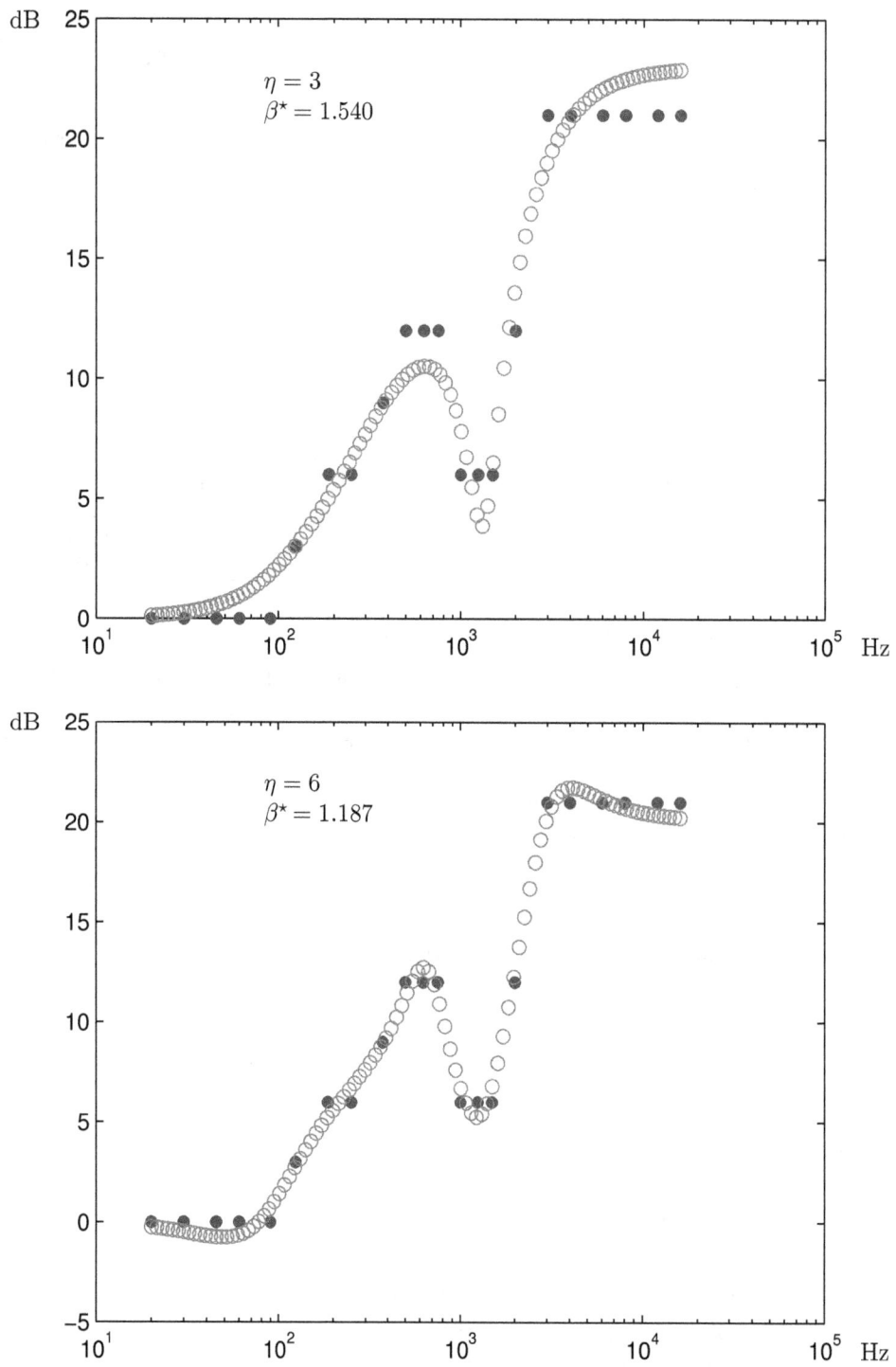

Figure 82: Arbitrary magnitude analog filter design specifications represented by blue dots. Red circles represent fit.

Optimal unprimed coefficients are scaled replicas of optimal primed:

$$
v^\star \triangleq \begin{bmatrix} \dfrac{v_1'^\star}{c^2} \\[2mm] \dfrac{v_2'^\star}{c^4} \\[1mm] \vdots \\[1mm] \dfrac{v_\eta'^\star}{c^{2\eta}} \end{bmatrix}, \qquad u^\star \triangleq \begin{bmatrix} \dfrac{u_1'^\star}{c^2} \\[2mm] \dfrac{u_2'^\star}{c^4} \\[1mm] \vdots \\[1mm] \dfrac{u_\eta'^\star}{c^{2\eta}} \end{bmatrix} \tag{675}
$$

The originally desired optimal fit is achieved as primed frequency expands:

$$
|H(\jmath\omega)^\star|^2 = \frac{1 + v_1^\star(\omega'c)^2 + v_2^\star(\omega'c)^4 + \ldots + v_\eta^\star(\omega'c)^{2\eta}}{1 + u_1^\star(\omega'c)^2 + u_2^\star(\omega'c)^4 + \ldots + u_\eta^\star(\omega'c)^{2\eta}} \tag{676}
$$

A benefit, of this design methodology, is an unexpectedly low order η required to meet a given filter specification (653) to within reasonable tolerance $\beta^\star$; *e.g.*, Figure **82**. □

3.8.2 quasilinear

When a function is simultaneously quasiconvex and quasiconcave, it is called *quasilinear*. Quasilinear functions are completely determined by convex level sets. Multidimensional quasilinear functions are not necessarily monotonic; *e.g.*, Exercise 3.6.4.0.2.

One-dimensional function $f(x) = x^3$ and vector-valued signum function $\mathrm{sgn}(x)$ are quasilinear, for example.

3.8.2.0.1 Exercise. *Quasiconcave product function.* *(confer §3.6.4.0.2)*
Show that vector-valued function $\beta u : \mathbb{R}_+ \times \mathbb{R}_+^\eta \to \mathbb{R}_+^\eta$ is quasiconcave and strictly monotonic but not quasilinear. ▼

3.9 Salient properties
of convex and quasiconvex functions

1. • A convex function is assumed continuous but not necessarily differentiable on the relative interior of its domain. [315, §10]

   • A quasiconvex function is not necessarily a continuous function.

2. convex epigraph $\Leftrightarrow$ convexity $\Rightarrow$ quasiconvexity $\Leftrightarrow$ convex sublevel sets.
   convex hypograph $\Leftrightarrow$ concavity $\Rightarrow$ quasiconcavity $\Leftrightarrow$ convex superlevel.
   quasilinearity $\Leftrightarrow$ convex level sets.

3. log-convex $\Rightarrow$ convex $\Rightarrow$ quasiconvex. $^{3.28}$
   concave $\Rightarrow$ quasiconcave $\Leftarrow$ log-concave $\Leftarrow$ positive concave.

$^{3.28}$*Log-convex* means: logarithm of function f is convex on dom f.

4. *Line Theorem* 3.7.3.0.1 translates identically to quasiconvexity (quasiconcavity). [117]

5.   • g convex $\Leftrightarrow$ $-g$ concave.
        g quasiconvex $\Leftrightarrow$ $-g$ quasiconcave.
        g log-convex $\Leftrightarrow$ $1/g$ log-concave.

     • (*translation, homogeneity*) Function convexity, concavity, quasiconvexity, and quasiconcavity are invariant to offset and nonnegative scaling.

6. (*affine transformation of argument*) Composition $g(h(X))$ of convex (concave) function g with any affine function $h : \mathbb{R}^{m \times n} \to \mathbb{R}^{p \times k}$ remains convex (concave) in $X \in \mathbb{R}^{m \times n}$, where $h(\mathbb{R}^{m \times n}) \cap \mathrm{dom}\, g \neq \emptyset$. [205, §B.2.1] Likewise for the quasiconvex (quasiconcave) functions g.

7.   •   – Nonnegatively weighted sum of convex (concave) functions remains convex (concave). (§3.1.2.2.1)
         – Nonnegatively weighted nonzero sum of strictly convex (concave) functions remains strictly convex (concave).
         – Nonnegatively weighted maximum (minimum) of convex[3.29] (concave) functions remains convex (concave).
         – Pointwise supremum (infimum) of convex (concave) functions remains convex (concave). (Figure **75**) [315, §5]

     •   – Sum of quasiconvex functions not necessarily quasiconvex.
         – Nonnegatively weighted maximum (minimum) of quasiconvex (quasiconcave) functions remains quasiconvex (quasiconcave).
         – Pointwise supremum (infimum) of quasiconvex (quasiconcave) functions remains quasiconvex (quasiconcave).

[3.29]Supremum and maximum of convex functions are proven convex by intersection of epigraphs.

Chapter 4

Semidefinite programming

Prior to 1984, linear and nonlinear programming,[4.1] one a subset of the other, had evolved for the most part along unconnected paths, without even a common terminology. (The use of "programming" to mean "optimization" serves as a persistent reminder of these differences.)

−Forsgren, Gill, & Wright (2002) [149]

Given some practical application of convex analysis, it may at first seem puzzling why a search for its solution ends abruptly with a formalized statement of the problem itself as a constrained optimization. The explanation is: typically we do not seek analytical solution because there are relatively few. (§3.5.2, §C) If a problem can be expressed in convex form, rather, then there exist computer programs providing efficient numerical global solution. [173] [412] [413] [411] [357] [343] The goal, then, becomes conversion of a given problem (perhaps a nonconvex or combinatorial problem statement) to an equivalent convex form or to an *alternation* of convex subproblems convergent to a solution of the original problem:

By the *fundamental theorem of Convex Optimization,* any locally optimal point (or solution) of a convex problem is globally optimal. [62, §4.2.2] [314, §1] Given convex real objective function g and convex feasible set $\mathcal{D} \subseteq \operatorname{dom} g$, which is the set of all variable values satisfying the problem constraints, we pose a generic convex optimization problem

$$
\begin{array}{ll}
\underset{X}{\text{minimize}} & g(X) \\
\text{subject to} & X \in \mathcal{D}
\end{array}
\tag{677}
$$

[4.1] nascence of polynomial-time *interior-point methods* of solution [372] [409].
Linear programming $\subset$ (convex $\cap$ nonlinear) programming.

CITATION: Dattorro, *Convex Optimization & Euclidean Distance Geometry,* Mεβoo Publishing USA, 2005, v2015.02.15.

where constraints are abstract here in membership of variable X to convex feasible set $\mathcal{D}$. Inequality constraint functions of a convex optimization problem are convex while equality constraint functions are conventionally affine, but not necessarily so. Affine equality constraint functions, as opposed to the superset of all convex equality constraint functions having convex level sets (§3.4.0.0.4), make convex optimization tractable.

Similarly, the problem

$$\begin{array}{ll} \underset{X}{\text{maximize}} & g(X) \\ \text{subject to} & X \in \mathcal{D} \end{array} \tag{678}$$

is called *convex* were g a real concave function and feasible set $\mathcal{D}$ convex. As conversion to convex form is not always possible, there is much ongoing research to determine which problem classes have convex expression or relaxation. [34] [60] [155] [284] [353] [153]

4.1 Conic problem

Still, we are surprised to see the relatively small number of submissions to semidefinite programming (SDP) solvers, as this is an area of significant current interest to the optimization community. We speculate that semidefinite programming is simply experiencing the fate of most new areas: Users have yet to understand how to pose their problems as semidefinite programs, and the lack of support for SDP solvers in popular modelling languages likely discourages submissions.

−*SIAM News*, 2002. [119, p.9]

(*confer* p.139) Consider a *conic problem* (p) and its dual (d): [301, §3.3.1] [245, §2.1] [246]

$$\text{(p)}\quad \begin{array}{ll} \underset{x}{\text{minimize}} & c^{\mathrm{T}}x \\ \text{subject to} & x \in \mathcal{K} \\ & Ax = b \end{array} \qquad \begin{array}{ll} \underset{y,s}{\text{maximize}} & b^{\mathrm{T}}y \\ \text{subject to} & s \in \mathcal{K}^* \\ & A^{\mathrm{T}}y + s = c \end{array} \quad \text{(d)} \tag{301}$$

where $\mathcal{K}$ is a closed convex cone, $\mathcal{K}^*$ is its dual, matrix A is fixed, and the remaining quantities are vectors.

When $\mathcal{K}$ is a polyhedral cone (§2.12.1), then each conic problem becomes a *linear program*; the selfdual nonnegative orthant providing the prototypical primal linear program and its dual. [96, §3-1][4.2] More generally, each optimization problem is convex when $\mathcal{K}$ is a closed convex cone. Solution to each convex problem is not necessarily unique; the optimal solution sets $\{x^\star\}$ and $\{y^\star, s^\star\}$ are convex and may comprise more than a single point.

[4.2] Dantzig explains reasoning behind a nonnegativity constraint: *... negative quantities of activities are not possible. ...a negative number of cases cannot be shipped.*

4.1.1   a semidefinite program

When $\mathcal{K}$ is the selfdual cone of positive semidefinite matrices $\mathbb{S}_+^n$ in the subspace of symmetric matrices $\mathbb{S}^n$, then each conic problem is called a *semidefinite program* (SDP); [284, §6.4] primal problem (P) having matrix variable $X \in \mathbb{S}^n$ while corresponding dual (D) has *slack variable* $S \in \mathbb{S}^n$ and vector variable $y = [y_i] \in \mathbb{R}^m$: [10] [11, §2] [419, §1.3.8]

$$
\begin{array}{cc}
\begin{array}{ll}
& \underset{X \in \mathbb{S}^n}{\text{minimize}} \quad \langle C\,,\, X \rangle \\
\text{(P)} & \text{subject to} \quad X \succeq 0 \\
& \qquad\qquad A \operatorname{svec} X = b
\end{array}
&
\begin{array}{ll}
\underset{y \in \mathbb{R}^m,\, S \in \mathbb{S}^n}{\text{maximize}} \quad \langle b\,,\, y \rangle & \\
\text{subject to} \quad S \succeq 0 & \text{(D)} \\
\qquad\qquad \operatorname{svec}^{-1}(A^{\mathrm{T}}y) + S = C &
\end{array}
\end{array}
\qquad (679)
$$

This is the *prototypical semidefinite program* and its dual, where matrix $C \in \mathbb{S}^n$ and vector $b \in \mathbb{R}^m$ are fixed as is

$$
A \triangleq \begin{bmatrix} \operatorname{svec}(A_1)^{\mathrm{T}} \\ \vdots \\ \operatorname{svec}(A_m)^{\mathrm{T}} \end{bmatrix} \in \mathbb{R}^{m \times n(n+1)/2}
\qquad (680)
$$

because $\{A_i \in \mathbb{S}^n,\ i = 1 \ldots m\}$ is given. Thus

$$
A \operatorname{svec} X = \begin{bmatrix} \langle A_1\,,\, X \rangle \\ \vdots \\ \langle A_m\,,\, X \rangle \end{bmatrix}
\qquad (681)
$$

$$
\operatorname{svec}^{-1}(A^{\mathrm{T}}y) = \sum_{i=1}^{m} y_i A_i
$$

The vector inner-product for matrices is defined in the Euclidean/Frobenius sense in the isomorphic vector space $\mathbb{R}^{n(n+1)/2}$; *id est,*

$$
\langle C\,,\, X \rangle \triangleq \operatorname{tr}(C^{\mathrm{T}}X) = \operatorname{svec}(C)^{\mathrm{T}} \operatorname{svec} X
\qquad (38)
$$

where $\operatorname{svec} X$ defined by (56) denotes symmetric vectorization.

> *In a national planning problem of some size, one may easily run into several hundred variables and perhaps a hundred or more degrees of freedom. ... It should always be remembered that any mathematical method and particularly methods in linear programming must be judged with reference to the type of computing machinery available. Our outlook may perhaps be changed when we get used to the super modern, high capacity electronic computor that will be available here from the middle of next year.*   −Ragnar Frisch [151]

The Simplex method of solution for linear programming, invented by Dantzig in 1947 [96], is now integral to modern technology. The same cannot yet be said for semidefinite programming whose roots trace back to systems of positive semidefinite linear inequalities studied by Bellman & Fan in 1963 [31] [104] who provided saddle convergence criteria. Interior-point methods for numerical solution of linear programs can be traced back to the logarithmic barrier of Frisch in 1954 and Fiacco & McCormick in 1968 [145]. Karmarkar's polynomial-time interior-point method sparked a log-barrier renaissance

$\mathcal{PC}$

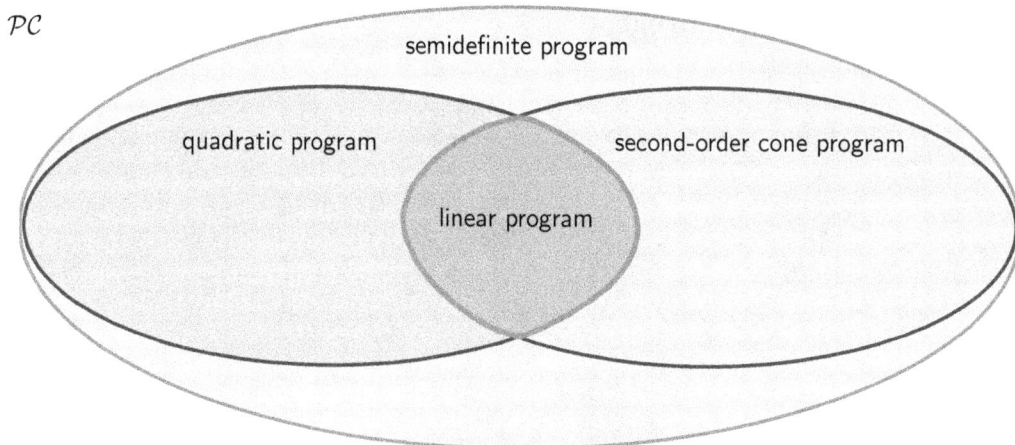

Figure 83: Venn diagram of programming hierarchy. Semidefinite program is a subset of convex program $\mathcal{PC}$. Semidefinite program subsumes other convex program classes excepting geometric program. Second-order cone program and quadratic program each subsume linear program. Nonconvex program $\backslash\mathcal{PC}$ comprises those for which convex equivalents have not yet been found.

in 1984, [281, §11] [409] [372] [284, p.3] but numerical performance of contemporary general-purpose semidefinite program solvers remains limited: Computational intensity for dense systems varies as $\mathrm{O}(m^2 n)$ (m constraints $\ll n$ variables) based on interior-point methods that produce solutions no more relatively accurate than 1E-8. There are no solvers capable of handling in excess of n=100,000 variables without significant, sometimes crippling, loss of precision or time.[4.3] [35] [283, p.258] [68, p.3]

Nevertheless, semidefinite programming has recently emerged to prominence because it admits a new class of problem previously unsolvable by convex optimization techniques, [60] and because it theoretically subsumes other convex techniques: (Figure 83) linear programming and *quadratic programming* and *second-order cone programming*.[4.4] Determination of the Riemann mapping function from complex analysis [294] [29, §8, §13], for example, can be posed as a semidefinite program.

4.1.2   Maximal complementarity

It has been shown [419, §2.5.3] that contemporary interior-point methods [410] [297] [284] [11] [62, §11] [149] (developed *circa* 1990 [155] for numerical solution of semidefinite

---

[4.3]*Heuristics are not ruled out by SIOPT; indeed I would suspect that most successful methods have (appropriately described) heuristics under the hood - my codes certainly do. ...Of course, there are still questions relating to high-accuracy and speed, but for many applications a few digits of accuracy suffices and overnight runs for non-real-time delivery is acceptable.* —Nicholas I. M. Gould, Stanford alumnus, SIOPT Editor in Chief

[4.4]SOCP came into being in the 1990s; it is not posable as a quadratic program. [254]

programs) can converge to a solution of *maximal complementarity*; [182, §5] [418] [259] [162] not a vertex solution but a solution of highest cardinality or rank among all optimal solutions.[4.5]

This phenomenon can be explained by recognizing that interior-point methods generally find solutions relatively interior to a feasible set by design.[4.6] [6, p.3] Log barriers are designed to fail numerically at the feasible set boundary. So low-rank solutions, all on the boundary, are rendered more difficult to find as numerical error becomes more prevalent there.

4.1.2.1 Reduced-rank solution

A simple rank reduction algorithm, for construction of a primal optimal solution $X^\star$ to (679P) satisfying an upper bound on rank governed by Proposition 2.9.3.0.1, is presented in §4.3. That proposition asserts existence of feasible solutions with an upper bound on their rank; [26, §II.13.1] specifically, it asserts an extreme point (§2.6.0.0.1) of *primal feasible set* $\mathcal{A} \cap \mathbb{S}_+^n$ satisfies upper bound

$$\operatorname{rank} X \leq \left\lfloor \frac{\sqrt{8m+1}-1}{2} \right\rfloor \qquad (272)$$

where, given $A \in \mathbb{R}^{m \times n(n+1)/2}$ and $b \in \mathbb{R}^m$,

$$\mathcal{A} \triangleq \{X \in \mathbb{S}^n \mid A \operatorname{svec} X = b\} \qquad (682)$$

is the affine subset from primal problem (679P).

4.1.2.2 Coexistence of low- and high-rank solutions; analogy

That low-rank and high-rank optimal solutions $\{X^\star\}$ of (679P) coexist may be grasped with the following analogy: We compare a proper polyhedral cone $\mathcal{S}_+^3$ in $\mathbb{R}^3$ (illustrated in Figure 84) to the positive semidefinite cone $\mathbb{S}_+^3$ in isometrically isomorphic $\mathbb{R}^6$, difficult to visualize. The analogy is good:

- int $\mathbb{S}_+^3$ is constituted by rank-3 matrices.
  int $\mathcal{S}_+^3$ has three dimensions.

- boundary $\partial\mathbb{S}_+^3$ contains rank-0, rank-1, and rank-2 matrices.
  boundary $\partial\mathcal{S}_+^3$ contains 0-, 1-, and 2-dimensional faces.

- the only rank-0 matrix resides in the vertex at the origin.

- Rank-1 matrices are in one-to-one correspondence with extreme directions of $\mathbb{S}_+^3$ and $\mathcal{S}_+^3$. The set of all rank-1 symmetric matrices in this dimension

$$\{G \in \mathbb{S}_+^3 \mid \operatorname{rank} G = 1\} \qquad (683)$$

  is not a connected set.

[4.5]This characteristic might be regarded as a disadvantage to interior-point methods of numerical solution, but this behavior is not certain and depends on solver implementation.
[4.6]Simplex methods, in contrast, find vertex solutions. [96, p.158] [15, p.2]

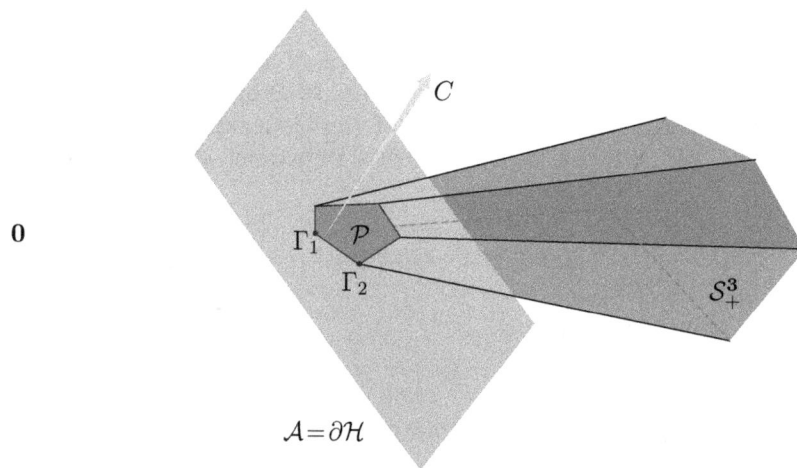

Figure 84: Visualizing positive semidefinite cone in high dimension: Proper polyhedral cone $\mathcal{S}_+^3 \subset \mathbb{R}^3$ representing positive semidefinite cone $\mathbb{S}_+^3 \subset \mathbb{S}^3$; analogizing its intersection with hyperplane $\mathbb{S}_+^3 \cap \partial\mathcal{H}$. Number of facets is arbitrary (analogy is not inspired by eigenvalue decomposition). The rank-0 positive semidefinite matrix corresponds to the origin in $\mathbb{R}^3$, rank-1 positive semidefinite matrices correspond to the edges of the polyhedral cone, rank-2 to the facet relative interiors, and rank-3 to the polyhedral cone interior. Vertices Γ_1 and Γ_2 are extreme points of polyhedron $\mathcal{P} = \partial\mathcal{H} \cap \mathcal{S}_+^3$, and extreme directions of $\mathcal{S}_+^3$. A given vector C is normal to another hyperplane (not illustrated but independent w.r.t $\partial\mathcal{H}$) containing line segment $\overline{\Gamma_1\Gamma_2}$ minimizing real linear function $\langle C, X \rangle$ on $\mathcal{P}$. (confer Figure 27, Figure 31)

- Rank of a sum of members $F+G$ in Lemma 2.9.2.9.1 and location of a difference $F-G$ in §2.9.2.12.1 similarly hold for $\mathbb{S}^3_+$ and $\mathcal{S}^3_+$.

- Euclidean distance from any particular rank-3 positive semidefinite matrix (in the cone interior) to the closest rank-2 positive semidefinite matrix (on the boundary) is generally less than the distance to the closest rank-1 positive semidefinite matrix. (§7.1.2)

- distance from any point in $\partial\mathbb{S}^3_+$ to $\text{int}\,\mathbb{S}^3_+$ is infinitesimal (§2.1.7.1.1).
  distance from any point in $\partial\mathcal{S}^3_+$ to $\text{int}\,\mathcal{S}^3_+$ is infinitesimal.

- faces of $\mathbb{S}^3_+$ correspond to faces of $\mathcal{S}^3_+$ (*confer* Table **2.9.2.3.1**):

	k	$\dim\mathcal{F}(\mathcal{S}^3_+)$	$\dim\mathcal{F}(\mathbb{S}^3_+)$	$\dim\mathcal{F}(\mathbb{S}^3_+ \ni \text{rank-}k \text{ matrix})$
	0	0	0	0
boundary	1	1	1	1
	2	2	3	3
interior	3	3	6	6

Integer k indexes k-dimensional faces $\mathcal{F}$ of $\mathcal{S}^3_+$. Positive semidefinite cone $\mathbb{S}^3_+$ has four kinds of faces, including cone itself ($k=3$, boundary + interior), whose dimensions in isometrically isomorphic $\mathbb{R}^6$ are listed under $\dim\mathcal{F}(\mathbb{S}^3_+)$. Smallest face $\mathcal{F}(\mathbb{S}^3_+ \ni \text{rank-}k \text{ matrix})$ that contains a rank-k positive semidefinite matrix has dimension $k(k+1)/2$ by (222).

- For $\mathcal{A}$ equal to intersection of m hyperplanes having linearly independent normals, and for $X \in \mathcal{S}^3_+ \cap \mathcal{A}$, we have $\text{rank}\,X \leq m$; the analogue to (272).

Proof. With reference to Figure **84**: Assume one ($m=1$) hyperplane $\mathcal{A} = \partial\mathcal{H}$ intersects the polyhedral cone. Every intersecting plane contains at least one matrix having rank less than or equal to 1; *id est*, from all $X \in \partial\mathcal{H} \cap \mathcal{S}^3_+$ there exists an X such that $\text{rank}\,X \leq 1$. Rank 1 is therefore an upper bound in this case.

Now visualize intersection of the polyhedral cone with two ($m=2$) hyperplanes having linearly independent normals. The hyperplane intersection $\mathcal{A}$ makes a line. Every intersecting line contains at least one matrix having rank less than or equal to 2, providing an upper bound. In other words, there exists a positive semidefinite matrix X belonging to any line intersecting the polyhedral cone such that $\text{rank}\,X \leq 2$.

In the case of three independent intersecting hyperplanes ($m=3$), the hyperplane intersection $\mathcal{A}$ makes a point that can reside anywhere in the polyhedral cone. The upper bound on a point in $\mathcal{S}^3_+$ is also the greatest upper bound: $\text{rank}\,X \leq 3$. ◆

4.1.2.2.1   Example. *Optimization over* $\mathcal{A} \cap \mathcal{S}_+^3$.
Consider minimization of the real linear function $\langle C\,,X\rangle$ over

$$\mathcal{P} \triangleq \mathcal{A} \cap \mathcal{S}_+^3 \tag{684}$$

a polyhedral feasible set;

$$f_0^\star \triangleq \underset{X}{\text{minimize}} \quad \langle C\,,X\rangle \\ \text{subject to} \quad X \in \mathcal{A} \cap \mathcal{S}_+^3 \tag{685}$$

As illustrated for particular vector C and hyperplane $\mathcal{A} = \partial\mathcal{H}$ in Figure 84, this linear function is minimized on any X belonging to the face of $\mathcal{P}$ containing extreme points $\{\Gamma_1\,,\Gamma_2\}$ and all the rank-2 matrices in between; *id est*, on any X belonging to the face of $\mathcal{P}$

$$\mathcal{F}(\mathcal{P}) = \{X \mid \langle C\,,X\rangle = f_0^\star\} \cap \mathcal{A} \cap \mathcal{S}_+^3 \tag{686}$$

exposed by the hyperplane $\{X \mid \langle C\,,X\rangle = f_0^\star\}$. In other words, the set of all optimal points $X^\star$ is a face of $\mathcal{P}$

$$\{X^\star\} = \mathcal{F}(\mathcal{P}) = \overline{\Gamma_1\Gamma_2} \tag{687}$$

comprising rank-1 and rank-2 positive semidefinite matrices. Rank 1 is the upper bound on existence in the feasible set $\mathcal{P}$ for this case $m = 1$ hyperplane constituting $\mathcal{A}$. The rank-1 matrices Γ_1 and Γ_2 in face $\mathcal{F}(\mathcal{P})$ are extreme points of that face and (by transitivity (§2.6.1.2)) extreme points of the intersection $\mathcal{P}$ as well. As predicted by analogy to Barvinok's Proposition 2.9.3.0.1, the upper bound on rank of X existent in the feasible set $\mathcal{P}$ is satisfied by an extreme point. The upper bound on rank of an optimal solution $X^\star$ existent in $\mathcal{F}(\mathcal{P})$ is thereby also satisfied by an extreme point of $\mathcal{P}$ precisely because $\{X^\star\}$ constitutes $\mathcal{F}(\mathcal{P})$;[4.7] in particular,

$$\{X^\star \in \mathcal{P} \mid \text{rank}\,X^\star \le 1\} = \{\Gamma_1\,,\Gamma_2\} \subseteq \mathcal{F}(\mathcal{P}) \tag{688}$$

As all linear functions on a polyhedron are minimized on a face, [96] [258] [280] [288] by analogy we so demonstrate coexistence of optimal solutions $X^\star$ of (679P) having assorted rank.   □

4.1.2.3   Previous work

Barvinok showed, [24, §2.2] when given a positive definite matrix C and an arbitrarily small neighborhood of C comprising positive definite matrices, there exists a matrix $\tilde{C}$ from that neighborhood such that optimal solution $X^\star$ to (679P) (substituting $\tilde{C}$) is an extreme point of $\mathcal{A} \cap \mathbb{S}_+^n$ and satisfies upper bound (272).[4.8] Given arbitrary positive definite C, this means nothing inherently guarantees that an optimal solution $X^\star$ to problem (679P) satisfies (272); certainly nothing given any symmetric matrix C, as the problem is posed. This can be proved by example:

[4.7] and every face contains a subset of the extreme points of $\mathcal{P}$ by the *extreme existence theorem* (§2.6.0.0.2). This means: because the affine subset $\mathcal{A}$ and hyperplane $\{X \mid \langle C\,,X\rangle = f_0^\star\}$ must intersect a whole face of $\mathcal{P}$, calculation of an upper bound on rank of $X^\star$ ignores counting the hyperplane when determining m in (272).

[4.8] Further, the set of all such $\tilde{C}$ in that neighborhood is open and dense.

4.1.2.3.1 Example. (Ye) *Maximal Complementarity.*

Assume dimension n to be an even positive number. Then the particular instance of problem (679P),

$$\begin{array}{ll} \underset{X \in \mathbb{S}^n}{\text{minimize}} & \left\langle \begin{bmatrix} I & \mathbf{0} \\ \mathbf{0} & 2I \end{bmatrix} ,\, X \right\rangle \\ \text{subject to} & X \succeq 0 \\ & \langle I ,\, X \rangle = n \end{array} \qquad (689)$$

has optimal solution

$$X^\star = \begin{bmatrix} 2I & \mathbf{0} \\ \mathbf{0} & \mathbf{0} \end{bmatrix} \in \mathbb{S}^n \qquad (690)$$

with an equal number of twos and zeros along the main diagonal. Indeed, optimal solution (690) is a terminal solution along the *central path* taken by the interior-point method as implemented in [419, §2.5.3]; it is also a solution of highest rank among all optimal solutions to (689). Clearly, rank of this primal optimal solution exceeds by far a rank-1 solution predicted by upper bound (272). □

4.1.2.4 Later developments

This rational example (689) indicates the need for a more generally applicable and simple algorithm to identify an optimal solution $X^\star$ satisfying Barvinok's Proposition 2.9.3.0.1. We will review such an algorithm in §4.3, but first we provide more background.

4.2 Framework

4.2.1 Feasible sets

Denote by $\mathcal{D}$ and $\mathcal{D}^*$ the convex sets of primal and dual points respectively satisfying the primal and dual constraints in (679), each assumed nonempty;

$$\begin{aligned} \mathcal{D} &= \left\{ X \in \mathbb{S}^n_+ \mid \begin{bmatrix} \langle A_1 ,\, X \rangle \\ \vdots \\ \langle A_m ,\, X \rangle \end{bmatrix} = b \right\} = \mathcal{A} \cap \mathbb{S}^n_+ \\ \mathcal{D}^* &= \left\{ S \in \mathbb{S}^n_+ ,\; y = [y_i] \in \mathbb{R}^m \mid \sum_{i=1}^m y_i A_i + S = C \right\} \end{aligned} \qquad (691)$$

These are the *primal feasible set* and *dual feasible set*. Geometrically, primal feasible $\mathcal{A} \cap \mathbb{S}^n_+$ represents an intersection of the positive semidefinite cone $\mathbb{S}^n_+$ with an affine subset $\mathcal{A}$ of the subspace of symmetric matrices $\mathbb{S}^n$ in isometrically isomorphic $\mathbb{R}^{n(n+1)/2}$. $\mathcal{A}$ has dimension $n(n+1)/2 - m$ when the vectorized A_i are linearly independent. Dual feasible set $\mathcal{D}^*$ is a Cartesian product of the positive semidefinite cone with its inverse image (§2.1.9.0.1) under affine transformation[4.9] $C - \sum y_i A_i$. Both feasible sets are convex, and

---

[4.9]Inequality $C - \sum y_i A_i \succeq 0$ follows directly from (679D) (§2.9.0.1.1) and is known as a *linear matrix inequality*. (§2.13.5.1.1) Because $\sum y_i A_i \preceq C$, matrix S is known as a *slack variable* (a term borrowed from linear programming [96]) since its inclusion raises this inequality to equality.

the objective functions are linear on a Euclidean vector space. Hence, (679P) and (679D) are convex optimization problems.

4.2.1.1 $\mathcal{A} \cap \mathbb{S}_+^n$ emptiness determination via Farkas' lemma

4.2.1.1.1   Lemma. *Semidefinite Farkas' lemma.*
Given set $\{A_i \in \mathbb{S}^n,\ i=1\ldots m\}$, vector $b=[b_i] \in \mathbb{R}^m$, and affine subset

$$(682) \quad \mathcal{A} = \{X \in \mathbb{S}^n \mid \langle A_i,X \rangle = b_i,\ i=1\ldots m\} \ni \{A \operatorname{svec} X \mid X \succeq 0\} \ (379) \text{ is closed,}$$

then primal feasible set $\mathcal{A} \cap \mathbb{S}_+^n$ is nonempty if and only if $y^{\mathrm{T}}b \geq 0$ holds for each and every vector $y=[y_i] \in \mathbb{R}^m$ such that $\sum\limits_{i=1}^m y_i A_i \succeq 0$.

Equivalently, primal feasible set $\mathcal{A} \cap \mathbb{S}_+^n$ is nonempty if and only if $y^{\mathrm{T}}b \geq 0$ holds for each and every vector $\|y\|=1$ such that $\sum\limits_{i=1}^m y_i A_i \succeq 0$.   ◇

Semidefinite Farkas' lemma provides necessary and sufficient conditions for a set of hyperplanes to have nonempty intersection $\mathcal{A} \cap \mathbb{S}_+^n$ with the positive semidefinite cone. Given

$$A = \begin{bmatrix} \operatorname{svec}(A_1)^{\mathrm{T}} \\ \vdots \\ \operatorname{svec}(A_m)^{\mathrm{T}} \end{bmatrix} \in \mathbb{R}^{m \times n(n+1)/2} \qquad (680)$$

semidefinite Farkas' lemma assumes that a convex cone

$$\mathcal{K} = \{A \operatorname{svec} X \mid X \succeq 0\} \qquad (379)$$

is closed per membership relation (319) from which the lemma springs: [238, §I] $\mathcal{K}$ closure is attained when matrix A satisfies the *cone closedness invariance corollary* (p.158). Given closed convex cone $\mathcal{K}$ and its dual from Example 2.13.5.1.1

$$\mathcal{K}^* = \{y \mid \sum_{j=1}^m y_j A_j \succeq 0\} \qquad (386)$$

then we can apply membership relation

$$b \in \mathcal{K} \ \Leftrightarrow\ \langle y, b \rangle \geq 0 \ \forall y \in \mathcal{K}^* \qquad (319)$$

to obtain the lemma

$$
\begin{aligned}
b \in \mathcal{K} \quad &\Leftrightarrow\quad \exists X \succeq 0 \ni A \operatorname{svec} X = b \quad &\Leftrightarrow\quad \mathcal{A} \cap \mathbb{S}_+^n \neq \emptyset \qquad &(692) \\
b \in \mathcal{K} \quad &\Leftrightarrow\quad \langle y,b \rangle \geq 0 \ \forall y \in \mathcal{K}^* \quad &\Leftrightarrow\quad \mathcal{A} \cap \mathbb{S}_+^n \neq \emptyset \qquad &(693)
\end{aligned}
$$

The final equivalence synopsizes *semidefinite Farkas' lemma.*

While the lemma is correct as stated, a positive definite version is required for semidefinite programming [419, §1.3.8] because existence of a feasible solution in the cone interior $\mathcal{A} \cap \operatorname{int} \mathbb{S}_+^n$ is required by *Slater's condition*[4.10] to achieve 0 duality gap (optimal primal–dual objective difference §4.2.3, Figure **59**). Geometrically, a positive definite lemma is required to insure that a point of intersection closest to the origin is not at infinity; *e.g.*, Figure **46**. Then given $A \in \mathbb{R}^{m \times n(n+1)/2}$ having rank m, we wish to detect existence of nonempty primal feasible set interior to the PSD cone;[4.11] (382)

$$b \in \operatorname{int} \mathcal{K} \quad \Leftrightarrow \quad \langle y, b \rangle > 0 \ \forall y \in \mathcal{K}^*, \ y \neq \mathbf{0} \quad \Leftrightarrow \quad \mathcal{A} \cap \operatorname{int} \mathbb{S}_+^n \neq \emptyset \tag{694}$$

Positive definite Farkas' lemma is made from proper cones, $\mathcal{K}$ (379) and $\mathcal{K}^*$ (386), and membership relation (325) for which $\mathcal{K}$ closedness is unnecessary:

4.2.1.1.2 Lemma. *Positive definite Farkas' lemma.*
Given l.i. set $\{A_i \in \mathbb{S}^n, \ i=1 \ldots m\}$ and vector $b=[b_i] \in \mathbb{R}^m$, make affine set

$$\mathcal{A} = \{X \in \mathbb{S}^n \,|\, \langle A_i, X \rangle = b_i, \ i=1 \ldots m\} \tag{682}$$

Primal feasible cone interior $\mathcal{A} \cap \operatorname{int} \mathbb{S}_+^n$ is nonempty if and only if $y^{\mathrm{T}} b > 0$ holds for each and every vector $y = [y_i] \neq \mathbf{0}$ such that $\sum_{i=1}^m y_i A_i \succeq 0$.

Equivalently, primal feasible cone interior $\mathcal{A} \cap \operatorname{int} \mathbb{S}_+^n$ is nonempty if and only if $y^{\mathrm{T}} b > 0$ holds for each and every vector $\|y\|=1 \ni \sum_{i=1}^m y_i A_i \succeq 0$. $\qquad\qquad \diamond$

4.2.1.1.3 Example. *"New" Farkas' lemma.*
Lasserre [238, §III] presented an example in 1995, originally offered by Ben-Israel in 1969 [32, p.378], to support closedness in *semidefinite Farkas' Lemma* 4.2.1.1.1:

$$A \triangleq \begin{bmatrix} \operatorname{svec}(A_1)^{\mathrm{T}} \\ \operatorname{svec}(A_2)^{\mathrm{T}} \end{bmatrix} = \begin{bmatrix} 0 & 1 & 0 \\ 0 & 0 & 1 \end{bmatrix}, \qquad b = \begin{bmatrix} 1 \\ 0 \end{bmatrix} \tag{695}$$

Intersection $\mathcal{A} \cap \mathbb{S}_+^n$ is practically empty because the solution set

$$\{X \succeq 0 \mid A \operatorname{svec} X = b\} = \left\{ \begin{bmatrix} \alpha & \frac{1}{\sqrt{2}} \\ \frac{1}{\sqrt{2}} & 0 \end{bmatrix} \succeq 0 \mid \alpha \in \mathbb{R} \right\} \tag{696}$$

is positive semidefinite only asymptotically $(\alpha \to \infty)$. Yet the dual system $\sum_{i=1}^m y_i A_i \succeq 0 \Rightarrow y^{\mathrm{T}} b \geq 0$ erroneously indicates nonempty intersection because $\mathcal{K}$ (379) violates a closedness condition of the lemma; *videlicet*, for $\|y\|=1$

$$y_1 \begin{bmatrix} 0 & \frac{1}{\sqrt{2}} \\ \frac{1}{\sqrt{2}} & 0 \end{bmatrix} + y_2 \begin{bmatrix} 0 & 0 \\ 0 & 1 \end{bmatrix} \succeq 0 \quad \Leftrightarrow \quad y = \begin{bmatrix} 0 \\ 1 \end{bmatrix} \quad \Rightarrow \quad y^{\mathrm{T}} b = 0 \tag{697}$$

[4.10]Slater's sufficient constraint qualification is satisfied whenever any primal or dual *strictly feasible solution* exists; *id est*, any point satisfying the respective affine constraints and relatively interior to the convex cone. [337, §6.6] [41, p.325] If the cone were polyhedral, then Slater's constraint qualification is satisfied when any feasible solution exists (relatively interior to the cone or on its relative boundary). [62, §5.2.3]
[4.11]Detection of $\mathcal{A} \cap \operatorname{int} \mathbb{S}_+^n \neq \emptyset$ by examining $\operatorname{int} \mathcal{K}$ instead is a trick need not be lost.

On the other hand, *positive definite Farkas' Lemma* 4.2.1.1.2 certifies that $\mathcal{A} \cap \operatorname{int} \mathbb{S}_+^n$ is empty; what we need to know for semidefinite programming.

Lasserre suggested addition of another condition to *semidefinite Farkas' lemma* (§4.2.1.1.1) to make a new lemma having no closedness condition. But *positive definite Farkas' lemma* (§4.2.1.1.2) is simpler and obviates the additional condition proposed.

$\square$

4.2.1.2   Theorem of the alternative for semidefinite programming

Because these Farkas' lemmas follow from membership relations, we may construct alternative systems from them. Applying the method of §2.13.2.1.1, then from *positive definite Farkas' lemma* we get

$$\mathcal{A} \cap \operatorname{int} \mathbb{S}_+^n \neq \emptyset$$
$$\text{or in the alternative}$$
$$y^{\mathrm{T}} b \leq 0, \quad \sum_{i=1}^{m} y_i A_i \succeq 0, \quad y \neq \mathbf{0} \tag{698}$$

Any single vector y satisfying the alternative certifies $\mathcal{A} \cap \operatorname{int} \mathbb{S}_+^n$ is empty. Such a vector can be found as a solution to another semidefinite program: for linearly independent (vectorized) set $\{A_i \in \mathbb{S}^n, \ i=1 \dots m\}$

$$\begin{aligned} \underset{y}{\text{minimize}} \quad & y^{\mathrm{T}} b \\ \text{subject to} \quad & \sum_{i=1}^{m} y_i A_i \succeq 0 \\ & \|y\|^2 \leq 1 \end{aligned} \tag{699}$$

If an optimal vector $y^\star \neq \mathbf{0}$ can be found such that $y^{\star \mathrm{T}} b \leq 0$, then primal feasible cone interior $\mathcal{A} \cap \operatorname{int} \mathbb{S}_+^n$ is empty.

4.2.1.3   Boundary-membership criterion

(*confer* (693)(694)) From boundary-membership relation (329), for proper cones $\mathcal{K}$ (379) and $\mathcal{K}^*$ (386) of linear matrix inequality,

$$b \in \partial \mathcal{K} \quad \Leftrightarrow \quad \exists \, y \neq \mathbf{0} \ni \langle y, b \rangle = 0, \ \ y \in \mathcal{K}^*, \ \ b \in \mathcal{K} \quad \Leftrightarrow \quad \partial \mathbb{S}_+^n \supset \mathcal{A} \cap \mathbb{S}_+^n \neq \emptyset \tag{700}$$

Whether vector $b \in \partial \mathcal{K}$ belongs to cone $\mathcal{K}$ boundary, that is a determination we can indeed make; one that is certainly expressible as a feasibility problem: Given linearly independent set[4.12] $\{A_i \in \mathbb{S}^n, \ i=1 \dots m\}$, for $b \in \mathcal{K}$ (692)

$$\begin{aligned} \text{find} \quad & y \neq \mathbf{0} \\ \text{subject to} \quad & y^{\mathrm{T}} b = 0 \\ & \sum_{i=1}^{m} y_i A_i \succeq 0 \end{aligned} \tag{701}$$

[4.12]From the results of Example 2.13.5.1.1, vector b on the boundary of $\mathcal{K}$ cannot be detected simply by looking for 0 eigenvalues in matrix X. We do not consider a skinny-or-square matrix A because then feasible set $\mathcal{A} \cap \mathbb{S}_+^n$ is at most a single point.

Any such nonzero solution y certifies that affine subset $\mathcal{A}$ (682) intersects the positive semidefinite cone $\mathbb{S}_+^n$ only on its boundary; in other words, nonempty feasible set $\mathcal{A} \cap \mathbb{S}_+^n$ belongs to the positive semidefinite cone boundary $\partial\mathbb{S}_+^n$.

4.2.2 Duals

The dual objective function from (679D) evaluated at any feasible solution represents a lower bound on the primal optimal objective value from (679P). We can see this by direct substitution: Assume the feasible sets $\mathcal{A} \cap \mathbb{S}_+^n$ and $\mathcal{D}^*$ are nonempty. Then it is always true:

$$
\begin{aligned}
\langle C \,,\, X \rangle &\geq \langle b \,,\, y \rangle \\
\left\langle \sum_i y_i A_i + S \,,\, X \right\rangle &\geq \left[\, \langle A_1 \,,\, X \rangle \,\cdots\, \langle A_m \,,\, X \rangle \,\right] y \\
\langle S \,,\, X \rangle &\geq 0
\end{aligned}
\qquad (702)
$$

The converse also follows because

$$
X \succeq 0 \,,\;\; S \succeq 0 \;\;\Rightarrow\;\; \langle S \,,\, X \rangle \geq 0 \qquad (1566)
$$

Optimal value of the dual objective thus represents the greatest lower bound on the primal. This fact is known as the *weak duality theorem* for semidefinite programming, [419, §1.3.8] and can be used to detect convergence in any primal/dual numerical method of solution.

4.2.2.1 Dual problem statement is not unique

Even subtle but equivalent restatements of a primal convex problem can lead to vastly different statements of a corresponding dual problem. This phenomenon is of interest because a particular instantiation of dual problem might be easier to solve numerically or it might take one of few forms for which analytical solution is known.

Here is a canonical restatement of prototypical dual semidefinite program (679D), for example, equivalent by (194):

$$
\text{(D)}\quad
\begin{array}{ll}
\underset{y \in \mathbb{R}^m,\, S \in \mathbb{S}^n}{\text{maximize}} & \langle b \,,\, y \rangle \\
\text{subject to} & S \succeq 0 \\
& \text{svec}^{-1}(A^{\mathrm{T}} y) + S = C
\end{array}
\;\;\equiv\;\;
\begin{array}{ll}
\underset{y \in \mathbb{R}^m}{\text{maximize}} & \langle b \,,\, y \rangle \\
\text{subject to} & \text{svec}^{-1}(A^{\mathrm{T}} y) \preceq C
\end{array}
\qquad (679\tilde{\text{D}})
$$

Dual feasible cone interior in $\operatorname{int}\mathbb{S}_+^n$ (691) (681) thereby corresponds with canonical dual ($\tilde{\text{D}}$) feasible interior

$$
\operatorname{rel\,int} \tilde{\mathcal{D}}^* \triangleq \left\{ y \in \mathbb{R}^m \mid \sum_{i=1}^m y_i A_i \prec C \right\} \qquad (703)
$$

4.2.2.1.1 Exercise. *Primal prototypical semidefinite program.*
Derive prototypical primal (679P) from its canonical dual (679$\tilde{\text{D}}$); *id est*, demonstrate that particular connectivity in Figure **85**. ▼

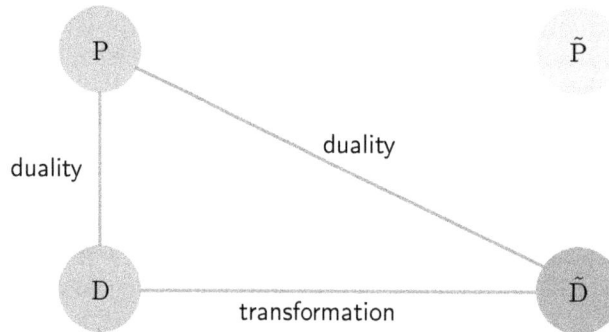

Figure 85: Connectivity indicates paths between particular primal and dual problems from Exercise 4.2.2.1.1. More generally, any path between primal problems P (and equivalent $\tilde{P}$) and dual D (and equivalent $\tilde{D}$) is possible: implying, any given path is not necessarily circuital; dual of a dual problem is not necessarily stated in precisely same manner as corresponding primal convex problem, in other words, although its solution set is equivalent to within some transformation.

4.2.3   Optimality conditions

When primal feasible cone interior $\mathcal{A} \cap \operatorname{int} \mathbb{S}_+^n$ exists in $\mathbb{S}^n$ or when canonical dual feasible interior $\operatorname{rel\,int} \tilde{\mathcal{D}}^*$ exists in $\mathbb{R}^m$, then these two problems (679P) (679D) become strong duals by Slater's sufficient condition (p.249). In other words, the primal optimal objective value becomes equal to the dual optimal objective value: there is no duality gap (Figure 59) and so determination of convergence is facilitated; *id est*, if $\exists\, X \in \mathcal{A} \cap \operatorname{int} \mathbb{S}_+^n$ or $\exists\, y \in \operatorname{rel\,int} \tilde{\mathcal{D}}^*$ then

$$
\begin{aligned}
\langle C ,\, X^\star \rangle &= \langle b ,\, y^\star \rangle \\
\left\langle \sum_i y_i^\star A_i + S^\star ,\, X^\star \right\rangle &= \left[\, \langle A_1 ,\, X^\star \rangle \,\cdots\, \langle A_m ,\, X^\star \rangle \,\right] y^\star \\
\langle S^\star ,\, X^\star \rangle &= 0
\end{aligned}
\tag{704}
$$

where $S^\star,\, y^\star$ denote a dual optimal solution.[4.13] We summarize this:

4.2.3.0.1   Corollary.   *Optimality and strong duality.*   [368, §3.1]
[419, §1.3.8]   For semidefinite programs (679P) and (679D), assume primal and dual feasible sets $\mathcal{A} \cap \mathbb{S}_+^n \subset \mathbb{S}^n$ and $\mathcal{D}^* \subset \mathbb{S}^n \times \mathbb{R}^m$ (691) are nonempty. Then

- $X^\star$ is optimal for (679P)

- $S^\star,\, y^\star$ are optimal for (679D)

- duality gap $\langle C, X^\star \rangle - \langle b,\, y^\star \rangle$ is 0

[4.13]Optimality condition $\langle S^\star ,\, X^\star \rangle = 0$ is called a *complementary slackness condition,* in keeping with linear programming tradition [96], that forbids dual inequalities in (679) to simultaneously hold strictly. [314, §4]

if and only if

i) $\exists\, X \in \mathcal{A} \cap \operatorname{int} \mathbb{S}_+^n$ **or** $\exists\, y \in \operatorname{rel\,int} \tilde{\mathcal{D}}^*$

and

ii) $\langle S^\star, X^\star \rangle = 0$ ◇

For symmetric positive semidefinite matrices, requirement **ii** is equivalent to the *complementarity* (§A.7.4)

$$\langle S^\star, X^\star \rangle = 0 \quad \Leftrightarrow \quad S^\star X^\star = X^\star S^\star = \mathbf{0} \tag{705}$$

Commutativity of diagonalizable matrices is a necessary and sufficient condition [208, §1.3.12] for these two optimal symmetric matrices to be simultaneously diagonalizable. Therefore

$$\operatorname{rank} X^\star + \operatorname{rank} S^\star \le n \tag{706}$$

Proof. To see that, the product of symmetric optimal matrices $X^\star, S^\star \in \mathbb{S}^n$ must itself be symmetric because of commutativity. (1555) The symmetric product has diagonalization [11, cor.2.11]

$$S^\star X^\star = X^\star S^\star = Q\,\Lambda_{S\star}\Lambda_{X\star}\,Q^{\mathrm{T}} = \mathbf{0} \quad \Leftrightarrow \quad \Lambda_{X\star}\Lambda_{S\star} = \mathbf{0} \tag{707}$$

where Q is an orthogonal matrix. The product of the nonnegative diagonal Λ matrices can be $\mathbf{0}$ if their main diagonal zeros are complementary or coincide. Due only to symmetry, $\operatorname{rank} X^\star = \operatorname{rank} \Lambda_{X\star}$ and $\operatorname{rank} S^\star = \operatorname{rank} \Lambda_{S\star}$ for these optimal primal and dual solutions. (1541) So, because of the complementarity, the total number of nonzero diagonal entries from both Λ cannot exceed n. ♦

When equality is attained in (706)

$$\operatorname{rank} X^\star + \operatorname{rank} S^\star = n \tag{708}$$

there are no coinciding main diagonal zeros in $\Lambda_{X\star}\Lambda_{S\star}$, and so we have what is called *strict complementarity*.[4.14] Logically it follows that a necessary and sufficient condition for strict complementarity of an optimal primal and dual solution is

$$X^\star + S^\star \succ 0 \tag{709}$$

4.2.3.1 solving primal problem via dual

The beauty of Corollary 4.2.3.0.1 is its conjugacy; *id est*, one can solve either the primal or dual problem in (679) and then find a solution to the other via the optimality conditions. When a dual optimal solution is known, for example, a primal optimal solution is any primal feasible solution in hyperplane $\{X \mid \langle S^\star, X \rangle = 0\}$.

[4.14] distinct from maximal complementarity (§4.1.2).

4.2.3.1.1   Example.   *Minimal cardinality Boolean.*   [95] [34, §4.3.4] [353]
(*confer* Example 4.5.1.5.1) Consider finding a *minimal cardinality* Boolean solution x to
the classic linear algebra problem $Ax = b$ given noiseless data $A \in \mathbb{R}^{m \times n}$ and $b \in \mathbb{R}^m$;

$$\begin{array}{ll} \underset{x}{\text{minimize}} & \|x\|_0 \\ \text{subject to} & Ax = b \\ & x_i \in \{0, 1\}, \quad i = 1 \ldots n \end{array} \qquad (710)$$

where $\|x\|_0$ denotes cardinality of vector x (a.k.a 0-*norm*; not a convex function).

A minimal cardinality solution answers the question: "Which fewest linear combination
of columns in A constructs vector b?" *Cardinality problems* have extraordinarily wide
appeal, arising in many fields of science and across many disciplines. [326] [220] [177] [176]
Yet designing an efficient algorithm to optimize cardinality has proved difficult. In this
example, we also constrain the variable to be Boolean. The Boolean constraint forces an
identical solution were the norm in problem (710) instead the 1-norm or 2-norm; *id est*,
the two problems

$$(710) \quad \begin{array}{ll} \underset{x}{\text{minimize}} & \|x\|_0 \\ \text{subject to} & Ax = b \\ & x_i \in \{0, 1\}, \quad i = 1 \ldots n \end{array} \quad = \quad \begin{array}{ll} \underset{x}{\text{minimize}} & \|x\|_1 \\ \text{subject to} & Ax = b \\ & x_i \in \{0, 1\}, \quad i = 1 \ldots n \end{array} \qquad (711)$$

are the same. The Boolean constraint makes the 1-norm problem nonconvex.

Given data

$$A = \begin{bmatrix} -1 & 1 & 8 & 1 & 1 & 0 \\ -3 & 2 & 8 & \frac{1}{2} & \frac{1}{3} & \frac{1}{2} - \frac{1}{3} \\ -9 & 4 & 8 & \frac{1}{4} & \frac{1}{9} & \frac{1}{4} - \frac{1}{9} \end{bmatrix}, \qquad b = \begin{bmatrix} 1 \\ \frac{1}{2} \\ \frac{1}{4} \end{bmatrix} \qquad (712)$$

the obvious and desired solution to the problem posed,

$$x^\star = e_4 \in \mathbb{R}^6 \qquad (713)$$

has norm $\|x^\star\|_2 = 1$ and minimal cardinality; the minimum number of nonzero entries in
vector x. The MATLAB backslash command x=A\b , for example, finds

$$x_{\text{M}} = \begin{bmatrix} \frac{2}{128} \\ 0 \\ \frac{5}{128} \\ 0 \\ \frac{90}{128} \\ 0 \end{bmatrix} \qquad (714)$$

having norm $\|x_{\text{M}}\|_2 = 0.7044$. Coincidentally, x_{M} is a 1-norm solution; *id est*, an optimal
solution to

$$\begin{array}{ll} \underset{x}{\text{minimize}} & \|x\|_1 \\ \text{subject to} & Ax = b \end{array} \qquad (517)$$

The pseudoinverse solution (rounded)

$$x_{\mathbf{P}} = A^{\dagger}b = \begin{bmatrix} -0.0456 \\ -0.1881 \\ 0.0623 \\ 0.2668 \\ 0.3770 \\ -0.1102 \end{bmatrix} \tag{715}$$

has least norm $\|x_{\mathbf{P}}\|_2 = 0.5165$; *id est*, the optimal solution to (§E.0.1.0.1)

$$\begin{array}{ll} \underset{x}{\text{minimize}} & \|x\|_2 \\ \text{subject to} & Ax = b \end{array} \tag{716}$$

Certainly none of the traditional methods provide $x^{\star} = e_4$ (713) because, and in general, for $Ax = b$

$$\left\|\arg\inf\|x\|_2\right\|_2 \leq \left\|\arg\inf\|x\|_1\right\|_2 \leq \left\|\arg\inf\|x\|_0\right\|_2 \tag{717}$$

We can reformulate this minimal cardinality Boolean problem (710) as a semidefinite program: First transform the variable

$$x \triangleq (\hat{x}+1)\tfrac{1}{2} \tag{718}$$

so $\hat{x}_i \in \{-1,1\}$; equivalently,

$$\begin{array}{ll} \underset{\hat{x}}{\text{minimize}} & \|(\hat{x}+1)\tfrac{1}{2}\|_0 \\ \text{subject to} & A(\hat{x}+1)\tfrac{1}{2} = b \\ & \delta(\hat{x}\hat{x}^{\mathrm{T}}) = 1 \end{array} \tag{719}$$

where δ is the main-diagonal linear operator (§A.1). By assigning (§B.1)

$$G = \begin{bmatrix} \hat{x} \\ 1 \end{bmatrix} \begin{bmatrix} \hat{x}^{\mathrm{T}} & 1 \end{bmatrix} = \begin{bmatrix} X & \hat{x} \\ \hat{x}^{\mathrm{T}} & 1 \end{bmatrix} \triangleq \begin{bmatrix} \hat{x}\hat{x}^{\mathrm{T}} & \hat{x} \\ \hat{x}^{\mathrm{T}} & 1 \end{bmatrix} \in \mathbb{S}^{n+1} \tag{720}$$

problem (719) becomes equivalent to: (Theorem A.3.1.0.7)

$$\begin{array}{ll} \underset{X \in \mathbb{S}^n,\ \hat{x} \in \mathbb{R}^n}{\text{minimize}} & \mathbf{1}^{\mathrm{T}}\hat{x} \\ \text{subject to} & A(\hat{x}+1)\tfrac{1}{2} = b \\ & G = \begin{bmatrix} X & \hat{x} \\ \hat{x}^{\mathrm{T}} & 1 \end{bmatrix} (\succeq 0) \\ & \delta(X) = 1 \\ & \operatorname{rank} G = 1 \end{array} \tag{721}$$

where solution is confined to rank-1 vertices of the *elliptope* in $\mathbb{S}^{n+1}$ (§5.9.1.0.1) by the rank constraint, the positive semidefiniteness, and the equality constraints $\delta(X) = 1$. The

rank constraint makes this problem nonconvex; by removing it[4.15] we get the semidefinite
program

$$
\begin{array}{ll}
\underset{X\in\mathbb{S}^n,\ \hat{x}\in\mathbb{R}^n}{\text{minimize}} & \mathbf{1}^{\mathrm{T}}\hat{x} \\
\text{subject to} & A(\hat{x}+\mathbf{1})\tfrac{1}{2}=b \\
& G=\begin{bmatrix} X & \hat{x} \\ \hat{x}^{\mathrm{T}} & 1 \end{bmatrix}\succeq 0 \\
& \delta(X)=\mathbf{1}
\end{array}
\tag{722}
$$

whose optimal solution $x^\star$ (718) is identical to that of minimal cardinality Boolean problem
(710) if and only if $\operatorname{rank} G^\star = 1$.

Hope[4.16] of acquiring a rank-1 solution is not ill-founded because 2^n elliptope vertices
have rank 1, and we are minimizing an affine function on a subset of the elliptope
(Figure 141) containing rank-1 vertices; *id est*, by assumption that the feasible set of
minimal cardinality Boolean problem (710) is nonempty, a desired solution resides on the
elliptope relative boundary at a rank-1 vertex.[4.17]

For that data given in (712), our semidefinite program solver sdpsol [412] [413]
(accurate in solution to approximately 1E-8)[4.18] finds optimal solution to (722)

$$
\operatorname{round}(G^\star)=\begin{bmatrix}
1 & 1 & 1 & -1 & 1 & 1 & -1 \\
1 & 1 & 1 & -1 & 1 & 1 & -1 \\
1 & 1 & 1 & -1 & 1 & 1 & -1 \\
-1 & -1 & -1 & 1 & -1 & -1 & 1 \\
1 & 1 & 1 & -1 & 1 & 1 & -1 \\
1 & 1 & 1 & -1 & 1 & 1 & -1 \\
-1 & -1 & -1 & 1 & -1 & -1 & 1
\end{bmatrix}
\tag{723}
$$

near a rank-1 vertex of the elliptope in $\mathbb{S}^{n+1}$ (Theorem 5.9.1.0.2); its sorted eigenvalues,

$$
\lambda(G^\star)=\begin{bmatrix}
6.99999977799099 \\
0.00000022687241 \\
0.00000002250296 \\
0.00000000262974 \\
-0.00000000999738 \\
-0.00000000999875 \\
-0.00000001000000
\end{bmatrix}
\tag{724}
$$

Negative eigenvalues are undoubtedly finite-precision effects.   Because the largest
eigenvalue predominates by many orders of magnitude, we can expect to find a good

---

[4.15]Relaxed problem (722) can also be derived via Lagrange duality; it is a dual of a dual program [*sic*] to
(721). [312] [62, §5, exer.5.39] [405, §IV] [154, §11.3.4] The relaxed problem must therefore be convex
having a larger feasible set; its optimal objective value represents a generally *loose* lower bound (1770) on
the optimal objective of problem (721).

[4.16]A more deterministic approach to constraining rank and cardinality is in §4.6.0.0.12.

[4.17]Confinement to the elliptope can be regarded as a kind of normalization akin to matrix A column
normalization suggested in [124] and explored in Example 4.2.3.1.2.

[4.18]A typically ignored limitation of interior-point solution methods is their relative accuracy of only about
1E-8 on a machine using 64-bit (*double precision*) floating-point arithmetic; *id est*, optimal solution $x^\star$
cannot be more accurate than square root of machine epsilon ($\epsilon=$2.2204E-16). Nonzero primal–dual
objective difference is not a good measure of solution accuracy.

approximation to a minimal cardinality Boolean solution by truncating all smaller eigenvalues. We find, indeed, the desired result (713)

$$x^\star = \text{round}\left(\begin{bmatrix} 0.00000000127947 \\ 0.00000000527369 \\ 0.00000000181001 \\ 0.99999997469044 \\ 0.00000001408950 \\ 0.00000000482903 \end{bmatrix}\right) = e_4 \qquad (725)$$

These numerical results are solver dependent; insofar, not all SDP solvers will return a rank-1 vertex solution. $\qquad\square$

4.2.3.1.2 Example. *Optimization over elliptope* versus *1-norm polyhedron for minimal cardinality Boolean Example 4.2.3.1.1.*
A minimal cardinality problem is typically formulated via, what is by now, the standard practice [124] [69, §3.2, §3.4] of column normalization applied to a 1-norm problem surrogate like (517). Suppose we define a diagonal matrix

$$\Lambda \triangleq \begin{bmatrix} \|A(:,1)\|_2 & & & \mathbf{0} \\ & \|A(:,2)\|_2 & & \\ & & \ddots & \\ \mathbf{0} & & & \|A(:,6)\|_2 \end{bmatrix} \in \mathbb{S}^6 \qquad (726)$$

used to normalize the columns (assumed nonzero) of given noiseless data matrix A. Then approximate the minimal cardinality Boolean problem

$$\begin{array}{ll} \underset{x}{\text{minimize}} & \|x\|_0 \\ \text{subject to} & Ax = b \\ & x_i \in \{0,1\}, \quad i=1\dots n \end{array} \qquad (710)$$

as

$$\begin{array}{ll} \underset{\tilde{y}}{\text{minimize}} & \|\tilde{y}\|_1 \\ \text{subject to} & A\Lambda^{-1}\tilde{y} = b \\ & \mathbf{1} \succeq \Lambda^{-1}\tilde{y} \succeq 0 \end{array} \qquad (727)$$

where optimal solution

$$y^\star = \text{round}(\Lambda^{-1}\tilde{y}^\star) \qquad (728)$$

The inequality in (727) relaxes Boolean constraint $y_i \in \{0,1\}$ from (710); bounding any solution $y^\star$ to a nonnegative unit hypercube whose vertices are binary numbers. Convex problem (727) is justified by the *convex envelope*

$$\text{cenv}\,\|x\|_0 \quad \text{on} \quad \{x \in \mathbb{R}^n \mid \|x\|_\infty \le \kappa\} = \frac{1}{\kappa}\|x\|_1 \qquad (1447)$$

Donoho concurs with this particular formulation, equivalently expressible as a linear program via (513).

Approximation (727) is therefore equivalent to minimization of an affine function (§3.2) on a bounded polyhedron, whereas semidefinite program

$$
\begin{array}{ll}
\underset{X \in \mathbb{S}^n,\ \hat{x} \in \mathbb{R}^n}{\text{minimize}} & \mathbf{1}^{\mathrm{T}}\hat{x} \\
\text{subject to} & A(\hat{x}+\mathbf{1})\tfrac{1}{2} = b \\
& G = \begin{bmatrix} X & \hat{x} \\ \hat{x}^{\mathrm{T}} & 1 \end{bmatrix} \succeq 0 \\
& \delta(X) = \mathbf{1}
\end{array}
\tag{722}
$$

minimizes an affine function on an intersection of the elliptope with hyperplanes. Although the same Boolean solution is obtained from this approximation (727) as compared with semidefinite program (722), when given that particular data from Example 4.2.3.1.1, Singer confides a counterexample: Instead, given data

$$
A = \begin{bmatrix} 1 & 0 & \frac{1}{\sqrt{2}} \\ 0 & 1 & \frac{1}{\sqrt{2}} \end{bmatrix}, \qquad b = \begin{bmatrix} 1 \\ 1 \end{bmatrix}
\tag{729}
$$

then solving approximation (727) yields

$$
y^\star = \text{round}\left(\begin{bmatrix} 1-\frac{1}{\sqrt{2}} \\ 1-\frac{1}{\sqrt{2}} \\ 1 \end{bmatrix}\right) = \begin{bmatrix} 0 \\ 0 \\ 1 \end{bmatrix}
\tag{730}
$$

(infeasible, with or without rounding, with respect to original problem (710)) whereas solving semidefinite program (722) produces

$$
\text{round}(G^\star) = \begin{bmatrix} 1 & 1 & -1 & 1 \\ 1 & 1 & -1 & 1 \\ -1 & -1 & 1 & -1 \\ 1 & 1 & -1 & 1 \end{bmatrix}
\tag{731}
$$

with sorted eigenvalues

$$
\lambda(G^\star) = \begin{bmatrix} 3.99999965057264 \\ 0.00000035942736 \\ -0.00000000000000 \\ -0.00000001000000 \end{bmatrix}
\tag{732}
$$

Truncating all but the largest eigenvalue, from (718) we obtain (*confer* $y^\star$)

$$
x^\star = \text{round}\left(\begin{bmatrix} 0.99999999625299 \\ 0.99999999625299 \\ 0.00000001434518 \end{bmatrix}\right) = \begin{bmatrix} 1 \\ 1 \\ 0 \end{bmatrix}
\tag{733}
$$

the desired minimal cardinality Boolean result.   □

4.2.3.1.3   Exercise.   *Minimal cardinality Boolean art.*
Assess general performance of standard-practice approximation (727) as compared with the proposed semidefinite program (722).   ▼

4.2.3.1.4 Exercise. *Conic independence.*
Matrix A from (712) is full-rank having three-dimensional nullspace. Find its four conically independent columns. (§2.10) To what part of proper cone $\mathcal{K} = \{Ax \mid x \succeq 0\}$ does vector b belong? ▼

4.2.3.1.5 Exercise. *Linear independence.*
Show why fat matrix A, from compressed sensing problem (517) or (522), may be regarded full-rank without loss of generality. In other words: Is a minimal cardinality solution invariant to linear dependence of rows? ▼

4.3 Rank reduction

> *...it is not clear generally how to predict* $\operatorname{rank} X^\star$ *or* $\operatorname{rank} S^\star$ *before solving the SDP problem.*
>
> −Farid Alizadeh (1995) [11, p.22]

The premise of rank reduction in semidefinite programming is: an optimal solution $X^\star$ found does not satisfy Barvinok's upper bound (272) on rank. The particular numerical algorithm solving a semidefinite program may have instead returned a high-rank optimal solution (§4.1.2; *e.g.*, (690)) when a lower-rank optimal solution was expected. Rank reduction is a means to adjust rank of an optimal solution, returned by a solver, until it satisfies Barvinok's upper bound.

4.3.1 Posit a perturbation of $X^\star$

Recall from §4.1.2.1, there is an extreme point of $\mathcal{A} \cap \mathbb{S}_+^n$ (682) satisfying upper bound (272) on rank. [24, §2.2] It is therefore sufficient to locate an extreme point of the intersection whose primal objective value (679P) is optimal:[4.19] [116, §31.5.3] [245, §2.4] [246] [7, §3] [299]

Consider again affine subset

$$\mathcal{A} = \{X \in \mathbb{S}^n \mid A \operatorname{svec} X = b\} \qquad (682)$$

where for $A_i \in \mathbb{S}^n$

$$A \triangleq \begin{bmatrix} \operatorname{svec}(A_1)^{\mathrm{T}} \\ \vdots \\ \operatorname{svec}(A_m)^{\mathrm{T}} \end{bmatrix} \in \mathbb{R}^{m \times n(n+1)/2} \qquad (680)$$

Given any optimal solution $X^\star$ to

$$\begin{array}{ll} \underset{X \in \mathbb{S}^n}{\text{minimize}} & \langle C\,,\, X \rangle \\ \text{subject to} & X \in \mathcal{A} \cap \mathbb{S}_+^n \end{array} \qquad (679\mathrm{P})$$

[4.19] *There is no known construction for Barvinok's tighter result* (277).

−Monique Laurent (2004)

whose rank does not satisfy upper bound (272), we posit existence of a set of perturbations

$$\{t_j B_j \mid t_j \in \mathbb{R} \ , \ B_j \in \mathbb{S}^n, \ j=1\ldots n\} \tag{734}$$

such that, for some $0 \leq i \leq n$ and scalars $\{t_j \ , \ j=1\ldots i\}$,

$$X^\star + \sum_{j=1}^{i} t_j B_j \tag{735}$$

becomes an extreme point of $\mathcal{A} \cap \mathbb{S}_+^n$ and remains an optimal solution of (679P). Membership of (735) to affine subset $\mathcal{A}$ is secured for the ith perturbation by demanding

$$\langle B_i \, , A_j \rangle = 0 \, , \quad j = 1 \ldots m \tag{736}$$

while membership to the positive semidefinite cone $\mathbb{S}_+^n$ is insured by small perturbation (745). Feasibility of (735) is insured in this manner, optimality is proved in §4.3.3.

The following simple algorithm has very low computational intensity and locates an optimal extreme point, assuming a nontrivial solution:

4.3.1.0.1 Procedure. *Rank reduction.*   [392]
`initialize:` $B_i = \mathbf{0} \ \forall i$
`for iteration i=1...n`
`{`

1. `compute` a nonzero perturbation matrix B_i of $\ X^\star + \sum_{j=1}^{i-1} t_j B_j$

2. `maximize` t_i
   `subject to` $X^\star + \sum_{j=1}^{i} t_j B_j \ \in \mathbb{S}_+^n$

`}`   ¶

A rank-reduced optimal solution is then

$$X^\star \leftarrow X^\star + \sum_{j=1}^{i} t_j B_j \tag{737}$$

4.3.2 Perturbation form

Perturbations of $X^\star$ are independent of constants $C \in \mathbb{S}^n$ and $b \in \mathbb{R}^m$ in primal and dual problems (679). Numerical accuracy of any rank-reduced result, found by perturbation of an initial optimal solution $X^\star$, is therefore quite dependent upon initial accuracy of $X^\star$.

4.3.2.0.1 Definition. *Matrix step function.*   (*confer* §A.6.5.0.1)
Define the signum-like quasiconcave real function $\psi : \mathbb{S}^n \to \mathbb{R}$

$$\psi(Z) \triangleq \left\{ \begin{array}{ll} 1, & Z \succeq 0 \\ -1, & \text{otherwise} \end{array} \right. \tag{738}$$

The value -1 is taken for indefinite or nonzero negative semidefinite argument.   △

Deza & Laurent [116, §31.5.3] prove: every perturbation matrix B_i , $i=1\ldots n$, is of the form

$$B_i = -\psi(Z_i)R_i Z_i R_i^{\mathrm{T}} \in \mathbb{S}^n \tag{739}$$

where

$$X^\star \triangleq R_1 R_1^{\mathrm{T}} , \qquad X^\star + \sum_{j=1}^{i-1} t_j B_j \triangleq R_i R_i^{\mathrm{T}} \in \mathbb{S}^n \tag{740}$$

where the t_j are scalars and $R_i \in \mathbb{R}^{n\times\rho}$ is full-rank and skinny where

$$\rho \triangleq \mathrm{rank}\left(X^\star + \sum_{j=1}^{i-1} t_j B_j\right) \tag{741}$$

and where matrix $Z_i \in \mathbb{S}^\rho$ is found at each iteration i by solving a very simple feasibility problem: [4.20]

$$\begin{array}{ll} \text{find} & Z_i \in \mathbb{S}^\rho \\ \text{subject to} & \langle Z_i \,,\, R_i^{\mathrm{T}} A_j R_i \rangle = 0 , \qquad j=1\ldots m \end{array} \tag{742}$$

Were there a sparsity pattern common to each member of the set $\{R_i^{\mathrm{T}} A_j R_i \in \mathbb{S}^\rho,\ j=1\ldots m\}$, then a good choice for Z_i has 1 in each entry corresponding to a 0 in the pattern; *id est*, a sparsity pattern complement. At iteration i

$$X^\star + \sum_{j=1}^{i-1} t_j B_j + t_i B_i = R_i(I - t_i \psi(Z_i)Z_i)R_i^{\mathrm{T}} \tag{743}$$

By fact (1531), therefore

$$X^\star + \sum_{j=1}^{i-1} t_j B_j + t_i B_i \succeq 0 \;\Leftrightarrow\; \mathbf{1} - t_i \psi(Z_i)\lambda(Z_i) \succeq 0 \tag{744}$$

where $\lambda(Z_i) \in \mathbb{R}^\rho$ denotes the eigenvalues of Z_i .

Maximization of each t_i in step 2 of the Procedure reduces rank of (743) and locates a new point on the boundary $\partial(\mathcal{A} \cap \mathbb{S}_+^n)$.[4.21] Maximization of t_i thereby has closed form;

$$(t_i^\star)^{-1} = \max\{\psi(Z_i)\lambda(Z_i)_j \,,\; j=1\ldots\rho\} \tag{745}$$

[4.20]A simple method of solution is closed-form projection of a random nonzero point on that proper subspace of isometrically isomorphic $\mathbb{R}^{\rho(\rho+1)/2}$ specified by the constraints. (§E.5.0.0.6) Such a solution is nontrivial assuming the specified intersection of hyperplanes is not the origin; guaranteed by $\rho(\rho+1)/2 > m$. Indeed, this geometric intuition about forming the perturbation is what bounds any solution's rank from below; m is fixed by the number of equality constraints in (679P) while rank ρ decreases with each iteration i . Otherwise, we might iterate indefinitely.

[4.21]This holds because rank of a positive semidefinite matrix in $\mathbb{S}^n$ is diminished below n by the number of its 0 eigenvalues (1541), and because a positive semidefinite matrix having one or more 0 eigenvalues corresponds to a point on the PSD cone boundary (193). Necessity and sufficiency are due to the facts: R_i can be completed to a nonsingular matrix (§A.3.1.0.5), and $I - t_i\psi(Z_i)Z_i$ can be padded with zeros while maintaining equivalence in (743).

When Z_i is indefinite, direction of perturbation (determined by $\psi(Z_i)$) is arbitrary. We may take an early exit from the Procedure were Z_i to become $\mathbf{0}$ or were ρ to become equal to 1 (assuming a nontrivial solution) or were

$$\text{rank}\left[\, \text{svec}\, R_i^{\mathrm{T}} A_1 R_i \quad \text{svec}\, R_i^{\mathrm{T}} A_2 R_i \quad \cdots \quad \text{svec}\, R_i^{\mathrm{T}} A_m R_i \,\right] = \rho(\rho+1)/2 \qquad (746)$$

which characterizes rank ρ of any [sic] extreme point in $\mathcal{A} \cap \mathbb{S}_+^n$. [245, §2.4] [246]

Proof.   Assuming the form of every perturbation matrix is indeed (739), then by (742)

$$\text{svec}\, Z_i \perp \left[\, \text{svec}(R_i^{\mathrm{T}} A_1 R_i) \quad \text{svec}(R_i^{\mathrm{T}} A_2 R_i) \quad \cdots \quad \text{svec}(R_i^{\mathrm{T}} A_m R_i) \,\right] \qquad (747)$$

By orthogonal complement we have

$$\begin{aligned}
&\text{rank}\left[\, \text{svec}(R_i^{\mathrm{T}} A_1 R_i) \quad \cdots \quad \text{svec}(R_i^{\mathrm{T}} A_m R_i) \,\right]^{\perp} \\
&\quad + \ \text{rank}\left[\, \text{svec}(R_i^{\mathrm{T}} A_1 R_i) \quad \cdots \quad \text{svec}(R_i^{\mathrm{T}} A_m R_i) \,\right] = \rho(\rho+1)/2
\end{aligned} \qquad (748)$$

When Z_i can only be $\mathbf{0}$, then the perturbation is null because an extreme point has been found; thus

$$\left[\, \text{svec}(R_i^{\mathrm{T}} A_1 R_i) \quad \cdots \quad \text{svec}(R_i^{\mathrm{T}} A_m R_i) \,\right]^{\perp} = \mathbf{0} \qquad (749)$$

from which the stated result (746) directly follows.   ◆

4.3.3   Optimality of perturbed $X^\star$

We show that the optimal objective value is unaltered by perturbation (739); id est,

$$\left\langle C \,,\, X^\star + \sum_{j=1}^{i} t_j B_j \right\rangle = \langle C \,,\, X^\star \rangle \qquad (750)$$

Proof.   From Corollary 4.2.3.0.1 we have the necessary and sufficient relationship between optimal primal and dual solutions under assumption of nonempty primal feasible cone interior $\mathcal{A} \cap \text{int}\,\mathbb{S}_+^n$:

$$S^\star X^\star = S^\star R_1 R_1^{\mathrm{T}} = X^\star S^\star = R_1 R_1^{\mathrm{T}} S^\star = \mathbf{0} \qquad (751)$$

This means $\mathcal{R}(R_1) \subseteq \mathcal{N}(S^\star)$ and $\mathcal{R}(S^\star) \subseteq \mathcal{N}(R_1^{\mathrm{T}})$. From (740) and (743) we get the sequence:

$$X^\star = R_1 R_1^{\mathrm{T}}$$
$$X^\star + t_1 B_1 = R_2 R_2^{\mathrm{T}} = R_1 (I - t_1 \psi(Z_1) Z_1) R_1^{\mathrm{T}}$$
$$X^\star + t_1 B_1 + t_2 B_2 = R_3 R_3^{\mathrm{T}} = R_2 (I - t_2 \psi(Z_2) Z_2) R_2^{\mathrm{T}} = R_1 (I - t_1 \psi(Z_1) Z_1)(I - t_2 \psi(Z_2) Z_2) R_1^{\mathrm{T}}$$
$$\vdots$$
$$X^\star + \sum_{j=1}^{i} t_j B_j = R_1 \left(\prod_{j=1}^{i} (I - t_j \psi(Z_j) Z_j) \right) R_1^{\mathrm{T}} \qquad (752)$$

Substituting $C = \text{svec}^{-1}(A^{\mathrm{T}}y^\star) + S^\star$ from (679),

$$
\begin{aligned}
\langle C\,,\, X^\star + \sum_{j=1}^{i} t_j B_j \rangle &= \left\langle \text{svec}^{-1}(A^{\mathrm{T}}y^\star) + S^\star\,,\, R_1\left(\prod_{j=1}^{i}(I - t_j \psi(Z_j)Z_j)\right)R_1^{\mathrm{T}} \right\rangle \\
&= \left\langle \sum_{k=1}^{m} y_k^\star A_k\,,\, X^\star + \sum_{j=1}^{i} t_j B_j \right\rangle \\
&= \left\langle \sum_{k=1}^{m} y_k^\star A_k + S^\star\,,\, X^\star \right\rangle = \langle C\,,\, X^\star \rangle
\end{aligned}
\tag{753}
$$

because $\langle B_i\,, A_j \rangle = 0 \;\; \forall\, i\,, j$ by design (736). ◆

4.3.3.0.1 Example. $A\,\delta(X) = b$.
This academic example demonstrates that a solution found by rank reduction can certainly have rank less than Barvinok's upper bound (272): Assume a given vector b belongs to the conic hull of columns of a given matrix A

$$
A = \begin{bmatrix} -1 & 1 & 8 & 1 & 1 \\ -3 & 2 & 8 & \frac{1}{2} & \frac{1}{3} \\ -9 & 4 & 8 & \frac{1}{4} & \frac{1}{9} \end{bmatrix} \in \mathbb{R}^{m \times n}\,, \qquad b = \begin{bmatrix} 1 \\ \frac{1}{2} \\ \frac{1}{4} \end{bmatrix} \in \mathbb{R}^{m}
\tag{754}
$$

Consider the convex optimization problem

$$
\begin{aligned}
&\underset{X \in \mathbb{S}^5}{\text{minimize}} && \text{tr}\, X \\
&\text{subject to} && X \succeq 0 \\
& && A\,\delta(X) = b
\end{aligned}
\tag{755}
$$

that minimizes the 1-norm of the main diagonal; *id est*, problem (755) is the same as

$$
\begin{aligned}
&\underset{X \in \mathbb{S}^5}{\text{minimize}} && \|\delta(X)\|_1 \\
&\text{subject to} && X \succeq 0 \\
& && A\,\delta(X) = b
\end{aligned}
\tag{756}
$$

that finds a solution to $A\,\delta(X) = b$. Rank-3 solution $X^\star = \delta(x_{\mathrm{M}})$ is optimal, where (*confer* (714))

$$
x_{\mathrm{M}} = \begin{bmatrix} \frac{2}{128} \\ 0 \\ \frac{5}{128} \\ 0 \\ \frac{90}{128} \end{bmatrix}
\tag{757}
$$

Yet upper bound (272) predicts existence of at most a

$$
\text{rank-}\left(\left\lfloor \frac{\sqrt{8m+1}-1}{2} \right\rfloor = 2\right)
\tag{758}
$$

feasible solution from $m = 3$ equality constraints. To find a lower rank ρ optimal solution to (755) (barring combinatorics), we invoke Procedure 4.3.1.0.1:

Initialize:
$\quad C = I$, $\rho = 3$, $A_j \triangleq \delta(A(j,:))$, $j = 1, 2, 3$, $X^\star = \delta(x_{\mathrm{M}})$, $m = 3$, $n = 5$.
{

Iteration i=1:

$$\text{Step 1:}\ R_1 = \begin{bmatrix} \sqrt{\frac{2}{128}} & 0 & 0 \\ 0 & 0 & 0 \\ 0 & \sqrt{\frac{5}{128}} & 0 \\ 0 & 0 & 0 \\ 0 & 0 & \sqrt{\frac{90}{128}} \end{bmatrix}.$$

$$\begin{aligned} &\text{find} \quad Z_1 \in \mathbb{S}^3 \\ &\text{subject to} \quad \langle Z_1, R_1^{\mathrm{T}} A_j R_1 \rangle = 0, \qquad j = 1, 2, 3 \end{aligned} \tag{759}$$

A nonzero randomly selected matrix Z_1, having $\mathbf{0}$ main diagonal, is a solution yielding nonzero perturbation matrix B_1. Choose arbitrarily

$$Z_1 = \mathbf{1}\mathbf{1}^{\mathrm{T}} - I \in \mathbb{S}^3 \tag{760}$$

then (rounding)

$$B_1 = \begin{bmatrix} 0 & 0 & 0.0247 & 0 & 0.1048 \\ 0 & 0 & 0 & 0 & 0 \\ 0.0247 & 0 & 0 & 0 & 0.1657 \\ 0 & 0 & 0 & 0 & 0 \\ 0.1048 & 0 & 0.1657 & 0 & 0 \end{bmatrix} \tag{761}$$

Step 2: $t_1^\star = 1$ because $\lambda(Z_1) = [\,-1 \quad -1 \quad 2\,]^{\mathrm{T}}$. So,

$$X^\star \leftarrow \delta(x_{\mathrm{M}}) + B_1 = \begin{bmatrix} \frac{2}{128} & 0 & 0.0247 & 0 & 0.1048 \\ 0 & 0 & 0 & 0 & 0 \\ 0.0247 & 0 & \frac{5}{128} & 0 & 0.1657 \\ 0 & 0 & 0 & 0 & 0 \\ 0.1048 & 0 & 0.1657 & 0 & \frac{90}{128} \end{bmatrix} \tag{762}$$

has rank $\rho \leftarrow 1$ and produces the same optimal objective value.

} $\square$

4.3.3.0.2   Exercise.   *Rank reduction of maximal complementarity.*
Apply rank reduction Procedure 4.3.1.0.1 to the *maximal complementarity example* (§4.1.2.3.1). Demonstrate a rank-1 solution; which can certainly be found (by Barvinok's Proposition 2.9.3.0.1) because there is only one equality constraint.   ▼

4.3.4 thoughts regarding rank reduction

Because rank reduction Procedure 4.3.1.0.1 is guaranteed only to produce another optimal solution conforming to Barvinok's upper bound (272), the Procedure will not necessarily produce solutions of arbitrarily low rank; but if they exist, the Procedure can. Arbitrariness of search direction when matrix Z_i becomes indefinite, mentioned on page 262, and the enormity of choices for Z_i (742) are liabilities for this algorithm.

4.3.4.1 inequality constraints

The question naturally arises: what to do when a semidefinite program (not in prototypical form (679))[4.22] has linear inequality constraints of the form

$$\alpha_i^{\mathrm{T}}\mathrm{svec}\, X \preceq \beta_i \,, \quad i = 1 \ldots k \tag{763}$$

where $\{\beta_i\}$ are given scalars and $\{\alpha_i\}$ are given vectors. One expedient way to handle this circumstance is to convert the inequality constraints to equality constraints by introducing a slack variable γ ; *id est,*

$$\alpha_i^{\mathrm{T}}\mathrm{svec}\, X + \gamma_i = \beta_i \,, \quad i = 1 \ldots k \,, \qquad \gamma \succeq 0 \tag{764}$$

thereby converting the problem to prototypical form.

Alternatively, we say the ith inequality constraint is *active* when it is met with equality; *id est,* when for particular i in (763), $\alpha_i^{\mathrm{T}}\mathrm{svec}\, X^\star = \beta_i$. An optimal high-rank solution $X^\star$ is, of course, feasible (satisfying all the constraints). But for the purpose of rank reduction, inactive inequality constraints are ignored while active inequality constraints are interpreted as equality constraints. In other words, we take the union of active inequality constraints (as equalities) with equality constraints $A\,\mathrm{svec}\, X = b$ to form a composite affine subset $\hat{\mathcal{A}}$ substituting for (682). Then we proceed with rank reduction of $X^\star$ as though the semidefinite program were in prototypical form (679P).

4.4 Rank-constrained semidefinite program

We generalize the trace heuristic (§7.2.2.1), for finding low-rank optimal solutions to SDPs of a more general form:

4.4.1 rank constraint by convex iteration

Consider a *semidefinite feasibility problem* of the form

$$\begin{aligned}
\underset{G \in \mathbb{S}^N}{\text{find}} \quad & G \\
\text{subject to} \quad & G \in \mathcal{C} \\
& G \succeq 0 \\
& \mathrm{rank}\, G \leq n
\end{aligned} \tag{765}$$

[4.22]Contemporary numerical packages for solving semidefinite programs can solve a range of problems wider than prototype (679). Generally, they do so by transforming a given problem into prototypical form by introducing new constraints and variables. [11] [413] We are momentarily considering a departure from the primal prototype that augments the constraint set with linear inequalities.

266 CHAPTER 4. SEMIDEFINITE PROGRAMMING

where $\mathcal{C}$ is a convex set presumed to contain positive semidefinite matrices of rank n or less; *id est*, $\mathcal{C}$ intersects the positive semidefinite cone boundary. We propose that this rank-constrained feasibility problem can be equivalently expressed as iteration of the convex problem sequence (766) and (1791a):

$$\begin{array}{ll} \underset{G \in \mathbb{S}^N}{\text{minimize}} & \langle G \, , W \rangle \\ \text{subject to} & G \in \mathcal{C} \\ & G \succeq 0 \end{array} \qquad (766)$$

where *direction vector*[4.23] W is an optimal solution to semidefinite program, for $0 \leq n \leq N-1$

$$\sum_{i=n+1}^{N} \lambda(G^\star)_i \;=\; \begin{array}{ll} \underset{W \in \mathbb{S}^N}{\text{minimize}} & \langle G^\star , W \rangle \\ \text{subject to} & 0 \preceq W \preceq I \\ & \text{tr}\, W = N - n \end{array} \qquad (1791\text{a})$$

whose feasible set is a Fantope (§2.3.2.0.1), and where $G^\star$ is an optimal solution to problem (766) given some iterate W. The idea is to iterate solution of (766) and (1791a) until convergence as defined in §4.4.1.2:[4.24] $\big(\text{confer}\,(802)\big)$

$$\sum_{i=n+1}^{N} \lambda(G^\star)_i = \langle G^\star , W^\star \rangle = \lambda(G^\star)^{\text{T}} \lambda(W^\star) \triangleq 0 \qquad (767)$$

defines *global convergence* of the iteration; a vanishing objective that is a certificate of global optimality but cannot be guaranteed. *Optimal direction vector* $W^\star$ is defined as any positive semidefinite matrix yielding optimal solution $G^\star$ of rank n or less to then convex equivalent (766) of feasibility problem (765):

$$(765) \qquad \begin{array}{ll} \underset{G \in \mathbb{S}^N}{\text{find}} & G \\ \text{subject to} & G \in \mathcal{C} \\ & G \succeq 0 \\ & \text{rank}\, G \leq n \end{array} \quad \equiv \quad \begin{array}{ll} \underset{G \in \mathbb{S}^N}{\text{minimize}} & \langle G \, , W^\star \rangle \\ \text{subject to} & G \in \mathcal{C} \\ & G \succeq 0 \end{array} \qquad (766)$$

id est, any direction vector for which the last $N-n$ nonincreasingly ordered eigenvalues λ of $G^\star$ are zero.

In any semidefinite feasibility problem, a solution of least rank must be an extreme point of the feasible set.[4.25] This means there exists a hyperplane supporting the feasible set at that extreme point. Then there must exist a linear objective function such that this least-rank feasible solution optimizes the resultant semidefinite program.

[4.23]Search *direction* W is a hyperplane-normal pointing opposite to direction of movement describing minimization of a real linear function $\langle G\,, W \rangle$ (p.68).

[4.24]Proposed iteration is neither *dual projection* (Figure **175**) or *alternating projection* (Figure **179**). Sum of eigenvalues follows from results of Ky Fan (page 567). Inner product of eigenvalues follows from (1673) and properties of commutative matrix products (page 524).

[4.25] which follows by *extremes theorem* 2.8.1.1.1, by rank of a sum of positive semidefinite matrices (1547) (257), and by definition of extreme point (167) for which no convex combination can produce it: If a least rank solution were expressible as a convex combination of feasible points, then there could exist feasible matrices of lesser rank.

We emphasize that convex problem (766) is not a relaxation of rank-constrained feasibility problem (765); at global convergence, convex iteration (766) (1791a) makes it instead an *equivalent problem.*

4.4.1.1 direction matrix interpretation

(*confer* §4.5.1.2) The feasible set of direction matrices in (1791a) is the convex hull of outer product of all rank-$(N-n)$ orthonormal matrices; *videlicet,*

$$\text{conv}\left\{UU^{\mathrm{T}} \mid U \in \mathbb{R}^{N \times N-n}, \ U^{\mathrm{T}}U = I\right\} = \left\{A \in \mathbb{S}^N \mid I \succeq A \succeq 0, \ \langle I, A \rangle = N - n\right\} \quad (90)$$

This set (92), argument to conv{ }, comprises the extreme points of this Fantope (90). An optimal solution W to (1791a), that is an extreme point, is known in closed form (p.567): Given ordered diagonalization $G^\star = Q \Lambda Q^{\mathrm{T}} \in \mathbb{S}_+^N$ (§A.5.1), then direction matrix $W = U^\star U^{\star\mathrm{T}}$ is optimal and extreme where $U^\star = Q(:, n+1:N) \in \mathbb{R}^{N \times N-n}$. Eigenvalue vector $\lambda(W)$ has 1 in each entry corresponding to the $N - n$ smallest entries of $\delta(\Lambda)$ and has 0 elsewhere. By (221) (223), polar direction $-W$ can be regarded as pointing toward the set of all rank-n (or less) positive semidefinite matrices whose nullspace contains that of $G^\star$. For that particular closed-form solution W, consequently, (*confer* (804))

$$\sum_{i=n+1}^{N} \lambda(G^\star)_i = \langle G^\star, W \rangle = \lambda(G^\star)^{\mathrm{T}} \lambda(W) \geq 0 \quad (768)$$

This is the connection to cardinality minimization of vectors;[4.26] *id est*, eigenvalue λ cardinality (rank) is analogous to vector x cardinality via (804): for positive semidefinite X,

$$\begin{array}{rclcl} \sum_i \lambda(X)_i = \operatorname{tr} X &=& \|X\|_2^* &\Leftrightarrow& \|x\|_1 \\ \sqrt{\sum_i \lambda(X)_i^2} = \sqrt{\operatorname{tr} X^2} &=& \|X\|_{\mathrm{F}} &\Leftrightarrow& \|x\|_2 \\ \max_i\{\lambda(X)_i\} &=& \|X\|_2 &\Leftrightarrow& \|x\|_\infty \end{array} \quad (769)$$

So that this method, for constraining rank, will not be misconstrued under closed-form solution W to (1791a): Define (*confer* (221))

$$\mathcal{S}_n \triangleq \{(I-W)G(I-W) \mid G \in \mathbb{S}^N\} = \{X \in \mathbb{S}^N \mid \mathcal{N}(X) \supseteq \mathcal{N}(G^\star)\} \quad (770)$$

as the symmetric subspace of rank$\leq n$ matrices whose nullspace contains $\mathcal{N}(G^\star)$. Then projection of $G^\star$ on $\mathcal{S}_n$ is $(I-W)G^\star(I-W)$. (§E.7) Direction of projection is $-WG^\star W$. (Figure 86) $\operatorname{tr}(WG^\star W)$ is a measure of proximity to $\mathcal{S}_n$ because its orthogonal complement is $\mathcal{S}_n^\perp = \{WGW \mid G \in \mathbb{S}^N\}$; the point being, convex iteration incorporating constrained $\operatorname{tr}(WGW) = \langle G, W \rangle$ minimization is not a projection method: certainly, not on these two subspaces.

[4.26] not trace minimization of a nonnegative diagonal matrix $\delta(x)$ as in [139, §1] [308, §2]. To make rank-constrained problem (765) resemble cardinality problem (529), we could make $\mathcal{C}$ an affine subset:

$$\begin{array}{ll} \text{find} & X \in \mathbb{S}^N \\ \text{subject to} & A\operatorname{svec} X = b \\ & X \succeq 0 \\ & \operatorname{rank} X \leq n \end{array}$$

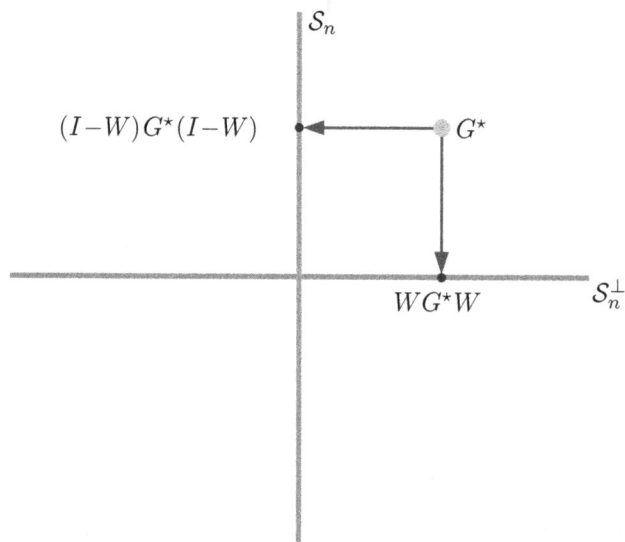

Figure 86: (*confer* Figure 176) Projection of $G^\star$ on subspace $\mathcal{S}_n$ of rank$\,\leq n$ matrices whose nullspace contains $\mathcal{N}(G^\star)$. This direction W is closed-form solution to (1791a).

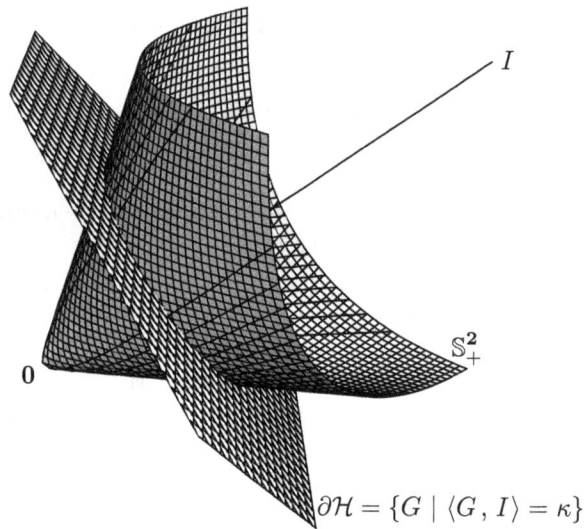

(*confer* Figure 103)

Figure 87: Trace heuristic can be interpreted as minimization of a hyperplane, with normal I, over positive semidefinite cone drawn here in isometrically isomorphic $\mathbb{R}^3$. Polar of direction vector $W = I$ points toward origin.

Closed-form solution W to problem (1791a), though efficient, comes with a *caveat*: there exist cases where this projection matrix solution W does not provide the shortest route to an optimal rank-n solution $G^\star$; *id est*, direction W is not unique. So we sometimes choose to solve (1791a) instead of employing a known closed-form solution.

When direction matrix $W = I$, as in the trace heuristic for example, then $-W$ points directly at the origin (the rank-0 PSD matrix, Figure 87). Vector inner-product of an optimization variable with direction matrix W is therefore a generalization of the trace heuristic (§7.2.2.1) for rank minimization; $-W$ is instead trained toward the boundary of the positive semidefinite cone.

4.4.1.2 convergence

We study convergence to ascertain conditions under which a direction matrix will reveal a feasible solution G, of rank n or less, to semidefinite program (766). Denote by $W^\star$ a particular optimal direction matrix from semidefinite program (1791a) such that (767) holds (feasible rank $G \leq n$ found). Then we define *global convergence* of the iteration (766) (1791a) to correspond with this vanishing vector inner-product (767) of optimal solutions.

Because this iterative technique for constraining rank is not a projection method, it can find a rank-n solution $G^\star$ ((767) will be satisfied) only if at least one exists in the feasible set of program (766).

4.4.1.2.1 Proof. Suppose $\langle G^\star, W \rangle = \tau$ is satisfied for some nonnegative constant τ after any particular iteration (766) (1791a) of the two minimization problems. Once a particular value of τ is achieved, it can never be exceeded by subsequent iterations because existence of feasible G and W having that vector inner-product τ has been established simultaneously in each problem. Because the infimum of vector inner-product of two positive semidefinite matrix variables is zero, the nonincreasing sequence of iterations is thus bounded below hence convergent because any bounded monotonic sequence in $\mathbb{R}$ is convergent. [264, §1.2] [42, §1.1] *Local convergence* to some nonnegative objective value τ is thereby established.   ♦

Local convergence, in this context, means convergence to a *fixed point* of possibly infeasible rank. Only local convergence can be established because objective $\langle G, W \rangle$, when instead regarded simultaneously in two variables (G, W), is generally multimodal. (§3.8.0.0.3)

Local convergence, convergence to $\tau \neq 0$ and definition of a *stall*, never implies nonexistence of a rank-n feasible solution to (766). A nonexistent rank-n feasible solution would mean certain failure to converge globally by definition (767) (convergence to $\tau \neq 0$) but, as proved, convex iteration always converges locally if not globally.

When a rank-n feasible solution to (766) exists, it remains an open problem to state conditions under which $\langle G^\star, W^\star \rangle = \tau = 0$ (767) is achieved by iterative solution of semidefinite programs (766) and (1791a). Then rank $G^\star \leq n$ and pair $(G^\star, W^\star)$ becomes a globally optimal fixed point of iteration. There can be no proof of global convergence because of the implicit high-dimensional multimodal manifold in variables (G, W). When stall occurs, direction vector W can be manipulated to steer out; *e.g.*, reversal of search direction as in Example 4.6.0.0.1, or reinitialization to a random rank-$(N-n)$ matrix in the same positive semidefinite cone face (§2.9.2.3) demanded by the current iterate: given ordered diagonalization $G^\star = Q\Lambda Q^T \in \mathbb{S}^N$, then $W = U^\star \Phi U^{\star T}$ where $U^\star = Q(:, n+1:N) \in \mathbb{R}^{N \times N-n}$ and where eigenvalue vector $\lambda(W)_{1:N-n} = \lambda(\Phi)$ has nonnegative uniformly distributed random entries in $(0,1]$ by selection of $\Phi \in \mathbb{S}_+^{N-n}$ while $\lambda(W)_{N-n+1:N} = \mathbf{0}$. Zero eigenvalues act as memory while randomness largely reduces likelihood of stall. When this direction works, rank and objective sequence $\langle G^\star, W \rangle$ with respect to iteration tend to be noisily monotonic.

4.4.1.2.2 Exercise. *Completely positive semidefinite matrix.*   [40]

Given rank-2 positive semidefinite matrix $G = \begin{bmatrix} 0.50 & 0.55 & 0.20 \\ 0.55 & 0.61 & 0.22 \\ 0.20 & 0.22 & 0.08 \end{bmatrix}$, find a positive factorization $G = X^T X$ (977) by solving

$$
\begin{aligned}
\underset{X \in \mathbb{R}^{2 \times 3}}{\text{find}} \quad & X \geq \mathbf{0} \\
\text{subject to} \quad & Z = \begin{bmatrix} I & X \\ X^T & G \end{bmatrix} \succeq 0 \\
& \operatorname{rank} Z \leq 2
\end{aligned}
\tag{771}
$$

via convex iteration.   ▼

4.4.1.2.3 Exercise. *Nonnegative matrix factorization.*

Given rank-2 nonnegative matrix $X = \begin{bmatrix} 17 & 28 & 42 \\ 16 & 47 & 51 \\ 17 & 82 & 72 \end{bmatrix}$, find a nonnegative factorization

$$
X = WH
\tag{772}
$$

by solving

$$
\begin{aligned}
\underset{A \in \mathbb{S}^3, B \in \mathbb{S}^3, W \in \mathbb{R}^{3 \times 2}, H \in \mathbb{R}^{2 \times 3}}{\text{find}} \quad & W, H \\
\text{subject to} \quad & Z = \begin{bmatrix} I & W^T & H \\ W & A & X \\ H^T & X^T & B \end{bmatrix} \succeq 0 \\
& W \geq \mathbf{0} \\
& H \geq \mathbf{0} \\
& \operatorname{rank} Z \leq 2
\end{aligned}
\tag{773}
$$

which follows from the fact, at optimality,

$$
Z^\star = \begin{bmatrix} I \\ W \\ H^T \end{bmatrix} [I \ \ W^T \ \ H]
\tag{774}
$$

Use the known closed-form solution for a direction vector Y to regulate rank by convex iteration; set $Z^\star = Q\Lambda Q^T \in \mathbb{S}^8$ to an ordered diagonalization and $U^\star = Q(:, 3:8) \in \mathbb{R}^{8 \times 6}$, then $Y = U^\star U^{\star T}$ (§4.4.1.1).

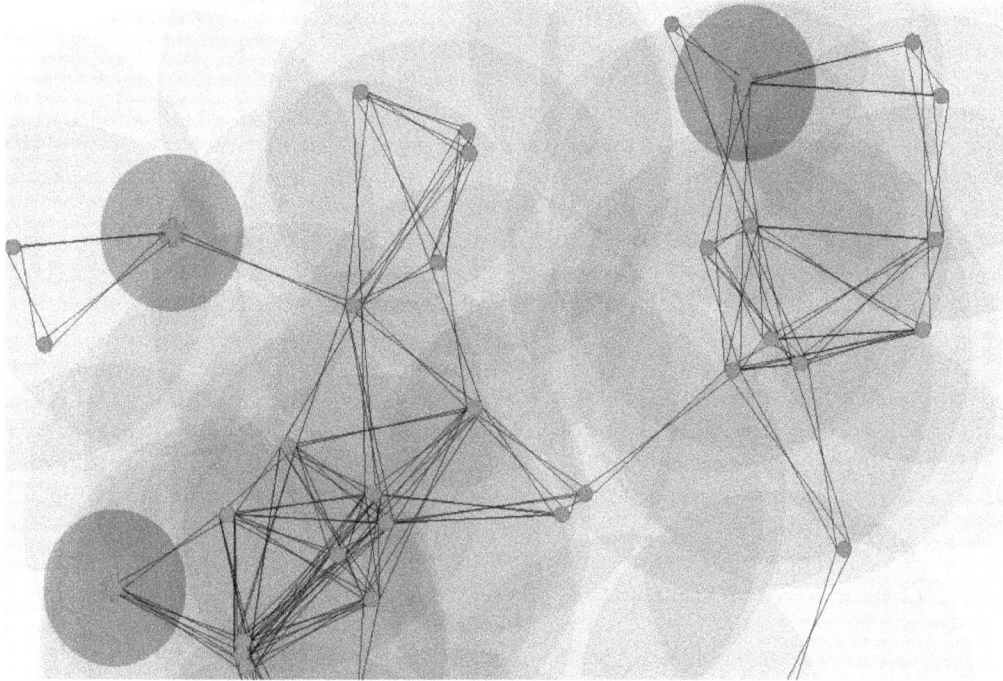

Figure 88: Sensor-network localization in $\mathbb{R}^2$, illustrating connectivity and circular radio-range per *sensor*. Smaller dark grey regions each hold an *anchor* at their center; known fixed sensor positions. Sensor/anchor distance is measurable with negligible uncertainty for sensor within those grey regions. (Graphic by Geoff Merrett.)

In summary, initialize Y then iterate numerical solution of (convex) semidefinite program

$$
\begin{array}{cl}
\underset{A\in\mathbb{S}^3,\, B\in\mathbb{S}^3,\, W\in\mathbb{R}^{3\times 2},\, H\in\mathbb{R}^{2\times 3}}{\text{minimize}} & \langle Z\,,\,Y\rangle \\[2ex]
\text{subject to} & Z = \begin{bmatrix} I & W^{\mathrm{T}} & H \\ W & A & X \\ H^{\mathrm{T}} & X^{\mathrm{T}} & B \end{bmatrix} \succeq 0 \\[2ex]
& W \geq \mathbf{0} \\
& H \geq \mathbf{0}
\end{array}
\tag{775}
$$

with $Y = U^{\star}U^{\star\mathrm{T}}$ until convergence (which is global and occurs in very few iterations for this instance). ▼

Now, an application to optimal regulation of affine dimension:

4.4.1.2.4 Example. *Sensor-Network Localization and Wireless Location.*
Heuristic solution to a sensor-network localization problem, proposed by Carter, Jin,

Saunders, & Ye in [73],[4.27] is limited to two Euclidean dimensions and applies semidefinite programming (SDP) to little subproblems. There, a large network is partitioned into smaller subnetworks (as small as one *sensor* − a mobile point, whereabouts unknown) and then semidefinite programming and heuristics called SPASELOC are applied to localize each and every partition by two-dimensional distance geometry. Their partitioning procedure is one-pass, yet termed *iterative*; a term applicable only insofar as adjoining partitions can share localized sensors and *anchors* (absolute sensor positions known *a priori*). But there is no iteration on the entire network, hence the term "iterative" is perhaps inappropriate. As partitions are selected based on "rule sets" (heuristics, not geographics), they also term the partitioning *adaptive.* But no adaptation of a partition actually occurs once it has been determined.

One can reasonably argue that semidefinite programming methods are unnecessary for localization of small partitions of large sensor networks. [286] [88] In the past, these nonlinear localization problems were solved algebraically and computed by least squares solution to hyperbolic equations; called *multilateration.*[4.28] [234] [272] Indeed, practical contemporary numerical methods for global positioning (GPS) by satellite do not rely on convex optimization. [298]

Modern distance geometry is inextricably melded with semidefinite programming. The beauty of semidefinite programming, as relates to localization, lies in convex expression of classical multilateration: So & Ye showed [328] that the problem of finding unique solution, to a noiseless nonlinear system describing the common point of intersection of hyperspheres in real Euclidean vector space, can be expressed as a semidefinite program via distance geometry.

But the need for SDP methods in Carter & Jin *et alii* is enigmatic for two more reasons: 1) guessing solution to a partition whose intersensor measurement data or connectivity is inadequate for localization by distance geometry, 2) reliance on complicated and extensive heuristics for partitioning a large network that could instead be efficiently solved whole by one semidefinite program [230, §3]. While partitions range in size between 2 and 10 sensors, 5 sensors optimally, heuristics provided are only for two spatial dimensions (no higher-dimensional heuristics are proposed). For these small numbers it remains unclarified as to precisely what advantage is gained over traditional least squares: it is difficult to determine what part of their noise performance is attributable to SDP and what part is attributable to their heuristic geometry.

Partitioning of large sensor networks is a compromise to rapid growth of SDP computational intensity with problem size. But when impact of noise on distance measurement is of most concern, one is averse to a partitioning scheme because noise-effects vary inversely with problem size. [54, §2.2] (§5.13.2) Since an individual partition's solution is not iterated in Carter & Jin and is interdependent with adjoining partitions, we expect errors to propagate from one partition to the next; the ultimate partition solved, expected to suffer most.

Heuristics often fail on real-world data because of unanticipated circumstances.

[4.27]The paper constitutes Jin's dissertation for University of Toronto [222] although her name appears as second author. Ye's authorship is honorary.
[4.28]Multilateration − literally, *having many sides*; shape of a geometric figure formed by nearly intersecting lines of position. In navigation systems, therefore: Obtaining a *fix* from multiple lines of position. Multilateration can be regarded as noisy trilateration.

When heuristics fail, generally they are repaired by adding more heuristics. Tenuous is any presumption, for example, that distance measurement errors have distribution characterized by circular contours of equal probability about an unknown sensor-location. (Figure 88) That presumption effectively appears within Carter & Jin's optimization problem statement as affine equality constraints relating unknowns to distance measurements that are corrupted by noise. Yet in most all urban environments, this measurement noise is more aptly characterized by ellipsoids of varying orientation and eccentricity as one recedes from a sensor. (Figure 137) Each unknown sensor must therefore instead be bound to its own particular range of distance, primarily determined by the terrain.[4.29] The nonconvex problem we must instead solve is:

$$\begin{array}{cl} \underset{i,\,j\,\in\,\mathcal{I}}{\text{find}} & \{x_i\,,\,x_j\} \\ \text{subject to} & \underline{d_{ij}} \leq \|x_i - x_j\|^2 \leq \overline{d_{ij}} \end{array} \qquad (776)$$

where x_i represents sensor location, and where $\underline{d_{ij}}$ and $\overline{d_{ij}}$ respectively represent lower and upper bounds on measured distance-square from i^{th} to j^{th} sensor (or from sensor to anchor). Figure 93 illustrates contours of equal sensor-location uncertainty. By establishing these individual upper and lower bounds, orientation and eccentricity can effectively be incorporated into the problem statement.

Generally speaking, there can be no unique solution to the sensor-network localization problem because there is no unique formulation; that is the art of Optimization. Any optimal solution obtained depends on whether or how a network is partitioned, whether distance data is complete, presence of noise, and how the problem is formulated. When a particular formulation is a convex optimization problem, then the set of all optimal solutions forms a convex set containing the actual or true localization. Measurement noise precludes equality constraints representing distance. The optimal solution set is consequently expanded; necessitated by introduction of distance inequalities admitting more and higher-rank solutions. Even were the optimal solution set a single point, it is not necessarily the true localization because there is little hope of exact localization by any algorithm once significant noise is introduced.

Carter & Jin gauge performance of their heuristics to the SDP formulation of author Biswas whom they regard as vanguard to the art. [14, §1] Biswas posed localization as an optimization problem minimizing a distance measure. [47] [45] Intuitively, minimization of any distance measure yields compacted solutions; (*confer* §6.7.0.0.1) precisely the anomaly motivating Carter & Jin. Their two-dimensional heuristics outperformed Biswas' localizations both in execution-time and proximity to the desired result. Perhaps, instead of heuristics, Biswas' approach to localization can be improved: [44] [46].

The sensor-network localization problem is considered difficult. [14, §2] Rank constraints in optimization are considered more difficult. Control of affine dimension in Carter & Jin is suboptimal because of implicit projection on $\mathbb{R}^2$. In what follows, we present the localization problem as a semidefinite program (equivalent to (776)) having an explicit rank constraint which controls affine dimension of an optimal solution. We show how to achieve that rank constraint only if the feasible set contains a matrix of desired rank. Our problem formulation is extensible to any spatial dimension.

[4.29]A distinct contour map corresponding to each anchor is required in practice.

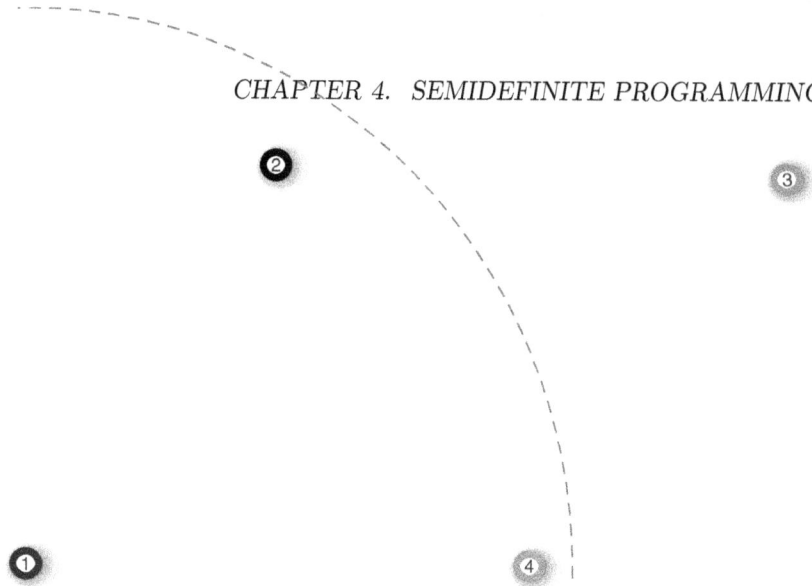

Figure 89: 2-lattice in $\mathbb{R}^2$, hand-drawn. Nodes 3 and 4 are anchors; remaining nodes are sensors. Radio range of sensor 1 indicated by arc.

proposed standardized test

Jin proposes an academic test in two-dimensional real Euclidean space $\mathbb{R}^2$ that we adopt. In essence, this test is a localization of sensors and anchors arranged in a regular triangular lattice. Lattice connectivity is solely determined by sensor radio range; a connectivity graph is assumed incomplete. In the interest of test standardization, we propose adoption of a few small examples: Figure 89 through Figure 92 and their particular connectivity represented by matrices (777) through (780) respectively.

$$
\begin{bmatrix}
0 & \bullet & ? & \bullet \\
\bullet & 0 & \bullet & \bullet \\
? & \bullet & 0 & \circ \\
\bullet & \bullet & \circ & 0
\end{bmatrix}
\tag{777}
$$

Matrix entries *dot* $\bullet$ indicate measurable distance between *nodes* while unknown distance is denoted by ? (*question mark*). Matrix entries *hollow dot* $\circ$ represent known distance between anchors (to high accuracy) while zero distance is denoted 0. Because measured distances are quite unreliable in practice, our solution to the localization problem substitutes a distinct range of possible distance for each measurable distance; equality constraints exist only for anchors.

Anchors are chosen so as to increase difficulty for algorithms dependent on existence of sensors in their convex hull. The challenge is to find a solution in two dimensions close to the true sensor positions given incomplete noisy intersensor distance information.

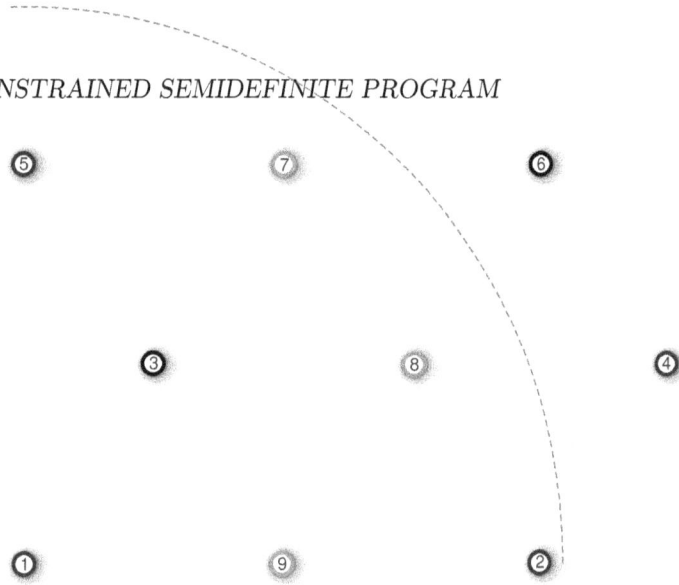

Figure 90: 3-lattice in $\mathbb{R}^2$, hand-drawn. Nodes 7, 8, and 9 are anchors; remaining nodes are sensors. Radio range of sensor 1 indicated by arc.

$$
\begin{bmatrix}
0 & \bullet & \bullet & ? & \bullet & ? & ? & \bullet & \bullet \\
\bullet & 0 & \bullet & \bullet & ? & \bullet & ? & \bullet & \bullet \\
\bullet & \bullet & 0 & \bullet & \bullet & \bullet & \bullet & \bullet & \bullet \\
? & \bullet & \bullet & 0 & ? & \bullet & \bullet & \bullet & \bullet \\
\bullet & ? & \bullet & ? & 0 & \bullet & \bullet & \bullet & \bullet \\
? & \bullet & \bullet & \bullet & \bullet & 0 & \bullet & \bullet & \bullet \\
? & ? & \bullet & \bullet & \bullet & \bullet & 0 & \circ & \circ \\
\bullet & \bullet & \bullet & \bullet & \bullet & \bullet & \circ & 0 & \circ \\
\bullet & \bullet & \bullet & \bullet & \bullet & \bullet & \circ & \circ & 0
\end{bmatrix}
\tag{778}
$$

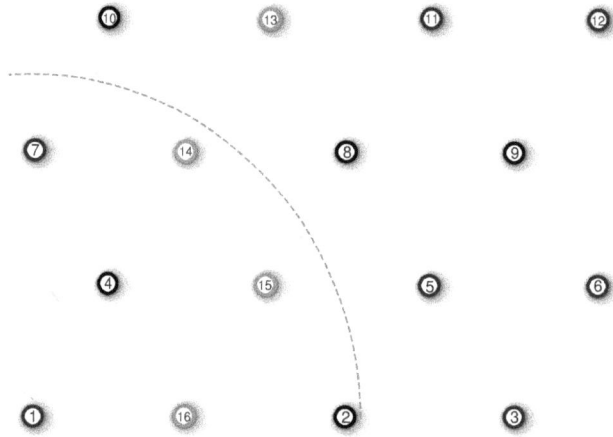

Figure 91: 4-lattice in $\mathbb{R}^2$, hand-drawn. Nodes 13, 14, 15, and 16 are anchors; remaining nodes are sensors. Radio range of sensor 1 indicated by arc.

$$
\begin{array}{ccccccccccccccccc}
0 & ? & ? & \bullet & ? & ? & \bullet & ? & ? & ? & ? & ? & ? & ? & \bullet & \bullet \\
? & 0 & \bullet & \bullet & \bullet & \bullet & ? & \bullet & ? & ? & ? & ? & ? & \bullet & \bullet & \bullet \\
? & \bullet & 0 & ? & \bullet & \bullet & ? & ? & \bullet & ? & ? & ? & ? & ? & \bullet & \bullet \\
\bullet & \bullet & ? & 0 & \bullet & ? & \bullet & \bullet & ? & \bullet & ? & ? & \bullet & \bullet & \bullet & \bullet \\
? & \bullet & \bullet & \bullet & 0 & \bullet & ? & \bullet & \bullet & ? & \bullet & \bullet & \bullet & \bullet & \bullet & \bullet \\
? & \bullet & \bullet & ? & \bullet & 0 & ? & \bullet & \bullet & ? & \bullet & \bullet & ? & ? & ? & ? \\
\bullet & ? & ? & \bullet & ? & ? & 0 & ? & ? & \bullet & ? & ? & \bullet & \bullet & \bullet & \bullet \\
? & \bullet & ? & \bullet & \bullet & \bullet & ? & 0 & \bullet & \bullet & \bullet & \bullet & \bullet & \bullet & \bullet & \bullet \\
? & ? & \bullet & ? & \bullet & \bullet & ? & \bullet & 0 & ? & \bullet & \bullet & \bullet & ? & \bullet & ? \\
? & ? & ? & \bullet & ? & ? & \bullet & \bullet & ? & 0 & \bullet & ? & \bullet & \bullet & \bullet & ? \\
? & ? & ? & ? & \bullet & \bullet & ? & \bullet & \bullet & \bullet & 0 & \bullet & \bullet & \bullet & \bullet & ? \\
? & ? & ? & ? & \bullet & \bullet & ? & \bullet & \bullet & ? & \bullet & 0 & ? & ? & ? & ? \\
? & ? & ? & \bullet & \bullet & ? & \bullet & \bullet & \bullet & \bullet & \bullet & ? & 0 & \circ & \circ & \circ \\
? & \bullet & ? & \bullet & \bullet & ? & \bullet & \bullet & ? & \bullet & \bullet & ? & \circ & 0 & \circ & \circ \\
\bullet & \bullet & \bullet & \bullet & \bullet & ? & \bullet & \bullet & \bullet & \bullet & \bullet & ? & \circ & \circ & 0 & \circ \\
\bullet & \bullet & \bullet & \bullet & \bullet & ? & \bullet & \bullet & ? & ? & ? & ? & \circ & \circ & \circ & 0
\end{array}
\qquad (779)
$$

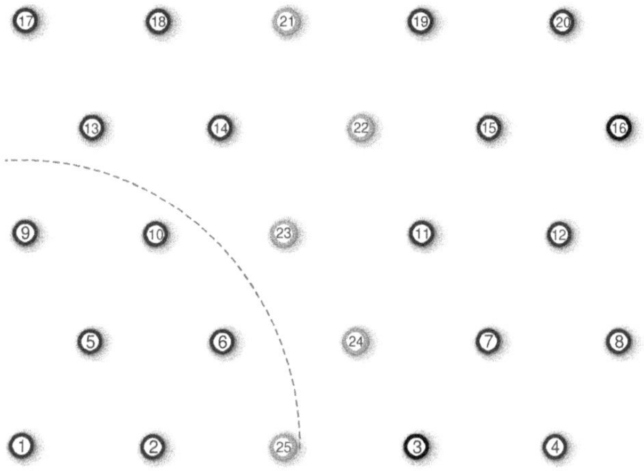

Figure 92: 5-lattice in $\mathbb{R}^2$. Nodes 21 through 25 are anchors.

$$
\begin{array}{ccccccccccccccccccccccccc}
0 & \bullet & ? & ? & \bullet & \bullet & ? & ? & \bullet & ? & ? & ? & ? & ? & ? & ? & ? & ? & ? & ? & ? & ? & ? & ? & ? \\
\bullet & 0 & ? & ? & \bullet & \bullet & ? & ? & ? & \bullet & ? & ? & ? & ? & ? & ? & ? & ? & ? & ? & ? & ? & ? & \bullet & \bullet \\
? & ? & 0 & \bullet & ? & \bullet & \bullet & \bullet & ? & ? & \bullet & \bullet & ? & ? & ? & ? & ? & ? & ? & ? & ? & ? & \bullet & \bullet & \bullet \\
? & ? & \bullet & 0 & ? & ? & \bullet & \bullet & ? & ? & ? & \bullet & ? & ? & ? & ? & ? & ? & ? & ? & ? & ? & ? & \bullet & ? \\
\bullet & \bullet & ? & ? & 0 & \bullet & ? & ? & \bullet & \bullet & ? & ? & \bullet & \bullet & ? & ? & \bullet & ? & ? & ? & ? & ? & \bullet & ? & \bullet \\
\bullet & \bullet & \bullet & ? & \bullet & 0 & \bullet & ? & \bullet & \bullet & \bullet & ? & ? & \bullet & ? & ? & ? & ? & ? & ? & ? & ? & \bullet & \bullet & \bullet \\
? & ? & \bullet & \bullet & ? & \bullet & 0 & \bullet & ? & ? & \bullet & \bullet & ? & ? & \bullet & \bullet & ? & ? & ? & ? & ? & ? & \bullet & \bullet & \bullet \\
? & ? & \bullet & \bullet & ? & ? & \bullet & 0 & ? & ? & \bullet & \bullet & ? & ? & \bullet & \bullet & ? & ? & ? & ? & ? & ? & ? & \bullet & ? \\
\bullet & ? & ? & ? & \bullet & \bullet & ? & ? & 0 & \bullet & ? & ? & \bullet & \bullet & ? & ? & \bullet & \bullet & ? & ? & ? & ? & ? & ? & ? \\
? & \bullet & ? & ? & \bullet & \bullet & ? & ? & \bullet & 0 & \bullet & ? & \bullet & \bullet & ? & ? & ? & \bullet & ? & ? & \bullet & \bullet & \bullet & \bullet & \bullet \\
? & ? & \bullet & ? & ? & \bullet & \bullet & \bullet & ? & \bullet & 0 & \bullet & ? & \bullet & \bullet & \bullet & ? & ? & \bullet & ? & ? & \bullet & \bullet & \bullet & \bullet \\
? & ? & \bullet & \bullet & ? & ? & \bullet & \bullet & ? & ? & \bullet & 0 & ? & ? & \bullet & \bullet & ? & ? & \bullet & \bullet & ? & \bullet & \bullet & \bullet & ? \\
? & ? & ? & ? & \bullet & ? & ? & ? & \bullet & \bullet & ? & ? & 0 & \bullet & ? & ? & \bullet & \bullet & ? & ? & \bullet & \bullet & ? & ? & ? \\
? & ? & ? & ? & \bullet & \bullet & ? & ? & \bullet & \bullet & \bullet & ? & \bullet & 0 & \bullet & ? & \bullet & \bullet & \bullet & ? & \bullet & \bullet & \bullet & \bullet & ? \\
? & ? & ? & ? & ? & ? & \bullet & \bullet & ? & ? & \bullet & \bullet & ? & \bullet & 0 & \bullet & ? & ? & \bullet & \bullet & \bullet & \bullet & \bullet & \bullet & ? \\
? & ? & ? & ? & ? & ? & \bullet & \bullet & ? & ? & \bullet & \bullet & ? & ? & \bullet & 0 & ? & ? & \bullet & \bullet & ? & \bullet & ? & ? & ? \\
? & ? & ? & ? & \bullet & ? & ? & ? & \bullet & ? & ? & ? & \bullet & \bullet & ? & ? & 0 & \bullet & ? & ? & \bullet & ? & ? & ? & ? \\
? & ? & ? & ? & ? & ? & ? & ? & \bullet & \bullet & ? & ? & \bullet & \bullet & ? & ? & \bullet & 0 & \bullet & ? & \bullet & \bullet & \bullet & ? & ? \\
? & ? & ? & ? & ? & ? & ? & ? & ? & ? & \bullet & \bullet & ? & \bullet & \bullet & \bullet & ? & \bullet & 0 & \bullet & \bullet & \bullet & \bullet & ? & ? \\
? & ? & ? & ? & ? & ? & ? & ? & ? & ? & ? & \bullet & ? & ? & \bullet & \bullet & ? & ? & \bullet & 0 & \bullet & \bullet & ? & ? & ? \\
? & ? & ? & ? & ? & ? & ? & ? & ? & \bullet & ? & ? & \bullet & \bullet & \bullet & ? & \bullet & \bullet & \bullet & \bullet & 0 & \circ & \circ & \circ & \circ \\
? & ? & ? & ? & ? & ? & ? & ? & ? & \bullet & \bullet & \bullet & \bullet & \bullet & \bullet & \bullet & ? & \bullet & \bullet & \bullet & \circ & 0 & \circ & \circ & \circ \\
? & ? & \bullet & ? & \bullet & \bullet & \bullet & ? & ? & \bullet & \bullet & \bullet & ? & \bullet & \bullet & ? & ? & \bullet & \bullet & ? & \circ & \circ & 0 & \circ & \circ \\
? & \bullet & \bullet & \bullet & ? & \bullet & \bullet & \bullet & ? & \bullet & \bullet & \bullet & ? & \bullet & \bullet & ? & ? & ? & ? & ? & \circ & \circ & \circ & 0 & \circ \\
? & \bullet & \bullet & ? & \bullet & \bullet & \bullet & ? & ? & \bullet & \bullet & ? & ? & ? & ? & ? & ? & ? & ? & ? & \circ & \circ & \circ & \circ & 0
\end{array}
\tag{780}
$$

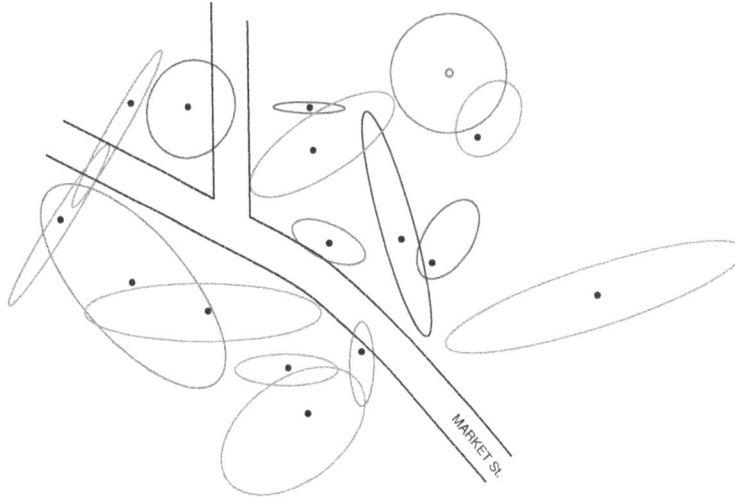

Figure 93: Location uncertainty ellipsoid in $\mathbb{R}^2$ for each of 15 sensors • within three city blocks in downtown San Francisco. Data by Polaris Wireless.

problem statement

Ascribe points in a list $\{x_\ell \in \mathbb{R}^n,\ \ell = 1 \ldots N\}$ to the columns of a matrix X;

$$X = [\, x_1 \ \cdots \ x_N \,] \in \mathbb{R}^{n \times N} \qquad (76)$$

where N is regarded as cardinality of list X. Positive semidefinite matrix $X^{\mathrm{T}}X$, formed from inner product of the list, is a *Gram matrix*; [256, §3.6]

$$G = X^{\mathrm{T}}X = \begin{bmatrix} \|x_1\|^2 & x_1^{\mathrm{T}}x_2 & x_1^{\mathrm{T}}x_3 & \cdots & x_1^{\mathrm{T}}x_N \\ x_2^{\mathrm{T}}x_1 & \|x_2\|^2 & x_2^{\mathrm{T}}x_3 & \cdots & x_2^{\mathrm{T}}x_N \\ x_3^{\mathrm{T}}x_1 & x_3^{\mathrm{T}}x_2 & \|x_3\|^2 & \ddots & x_3^{\mathrm{T}}x_N \\ \vdots & \vdots & \ddots & \ddots & \vdots \\ x_N^{\mathrm{T}}x_1 & x_N^{\mathrm{T}}x_2 & x_N^{\mathrm{T}}x_3 & \cdots & \|x_N\|^2 \end{bmatrix} \in \mathbb{S}_+^N \qquad (977)$$

where $\mathbb{S}_+^N$ is the convex cone of $N \times N$ positive semidefinite matrices in the symmetric matrix subspace $\mathbb{S}^N$.

Existence of noise precludes measured distance from the input data. We instead assign measured distance to a range estimate specified by individual upper and lower bounds: $\overline{d_{ij}}$ is an upper bound on distance-square from i^{th} to j^{th} sensor, while $\underline{d_{ij}}$ is a lower bound. These bounds become the input data. Each measurement range is presumed different from the others because of measurement uncertainty; *e.g.*, Figure 93.

Our mathematical treatment of anchors and sensors is not dichotomized.[4.30] A sensor position that is known *a priori* to high accuracy (with absolute certainty) $\check{x}_i$ is called an *anchor*. Then the sensor-network localization problem (776) can be expressed equivalently: Given a number of anchors m, and $\mathcal{I}$ a set of indices (corresponding to all measurable distances •), for $0 < n < N$

$$
\begin{aligned}
&\underset{G \in \mathbb{S}^N, \ X \in \mathbb{R}^{n \times N}}{\text{find}} \quad X \\
&\text{subject to} \quad
\begin{aligned}
d_{ij} \leq \langle G, (e_i - e_j)(e_i - e_j)^{\mathrm{T}} \rangle &\leq \overline{d_{ij}} \qquad \forall (i,j) \in \mathcal{I} \\
\langle G, e_i e_i^{\mathrm{T}} \rangle &= \|\check{x}_i\|^2, \quad i = N - m + 1 \ldots N \\
\langle G, (e_i e_j^{\mathrm{T}} + e_j e_i^{\mathrm{T}})/2 \rangle &= \check{x}_i^{\mathrm{T}} \check{x}_j, \quad i < j, \quad \forall i,j \in \{N - m + 1 \ldots N\} \\
X(:, N - m + 1 : N) &= [\check{x}_{N-m+1} \ \cdots \ \check{x}_N] \\
Z = \begin{bmatrix} I & X \\ X^{\mathrm{T}} & G \end{bmatrix} &\succeq 0 \\
\operatorname{rank} Z &= n
\end{aligned}
\end{aligned}
\tag{781}
$$

where e_i is the ith member of the standard basis for $\mathbb{R}^N$. Distance-square

$$
d_{ij} = \|x_i - x_j\|_2^2 = \langle x_i - x_j, \ x_i - x_j \rangle \tag{964}
$$

is related to Gram matrix entries $G \triangleq [g_{ij}]$ by vector inner-product

$$
\begin{aligned}
d_{ij} &= g_{ii} + g_{jj} - 2g_{ij} \\
&= \langle G, (e_i - e_j)(e_i - e_j)^{\mathrm{T}} \rangle = \operatorname{tr}(G^{\mathrm{T}}(e_i - e_j)(e_i - e_j)^{\mathrm{T}})
\end{aligned}
\tag{979}
$$

hence the scalar inequalities. Each linear equality constraint in $G \in \mathbb{S}^N$ represents a hyperplane in isometrically isomorphic Euclidean vector space $\mathbb{R}^{N(N+1)/2}$, while each linear inequality pair represents a convex Euclidean body known as *slab*.[4.31] By Schur complement (§A.4), any solution (G, X) provides comparison with respect to the positive semidefinite cone

$$
G \succeq X^{\mathrm{T}} X \tag{1017}
$$

which is a convex relaxation of the desired equality constraint

$$
\begin{bmatrix} I & X \\ X^{\mathrm{T}} & G \end{bmatrix} = \begin{bmatrix} I \\ X^{\mathrm{T}} \end{bmatrix} [I \ \ X] \tag{1018}
$$

The rank constraint insures this equality holds, by Theorem A.4.0.1.3, thus restricting solution to $\mathbb{R}^n$. Assuming full-rank solution (list) X

$$
\operatorname{rank} Z = \operatorname{rank} G = \operatorname{rank} X \tag{782}
$$

[4.30] *Wireless location* problem thus stated identically; difference being: fewer sensors.
[4.31] an intersection of two parallel but opposing halfspaces (Figure 11). In terms of position X, this distance slab can be thought of as a thick *hypershell* instead of a hypersphere boundary.

convex equivalent problem statement

Problem statement (781) is nonconvex because of the rank constraint. We do not eliminate or ignore the rank constraint; rather, we find a convex way to enforce it: for $0 < n < N$

$$
\begin{array}{ll}
\underset{G \in \mathbb{S}^N, \, X \in \mathbb{R}^{n \times N}}{\text{minimize}} & \langle Z \,, W \rangle \\[1ex]
\text{subject to} & \underline{d_{ij}} \leq \langle G \,, (e_i - e_j)(e_i - e_j)^{\mathrm{T}} \rangle \leq \overline{d_{ij}} \qquad \forall (i,j) \in \mathcal{I} \\
& \langle G \,, e_i e_i^{\mathrm{T}} \rangle \qquad\qquad\quad = \|\check{x}_i\|^2 \,, \quad i = N - m + 1 \ldots N \\
& \langle G \,, (e_i e_j^{\mathrm{T}} + e_j e_i^{\mathrm{T}})/2 \rangle \quad\; = \check{x}_i^{\mathrm{T}} \check{x}_j \,, \quad i < j \,, \quad \forall i, j \in \{N - m + 1 \ldots N\} \\
& X(:, N - m + 1{:}N) \qquad\qquad = [\,\check{x}_{N-m+1} \; \cdots \; \check{x}_N\,] \\
& Z = \begin{bmatrix} I & X \\ X^{\mathrm{T}} & G \end{bmatrix} \qquad\qquad\qquad \succeq 0
\end{array}
\qquad (783)
$$

Convex function $\operatorname{tr} Z$ is a well-known heuristic whose sole purpose is to represent convex envelope of $\operatorname{rank} Z$. (§7.2.2.1) In this convex optimization problem (783), a semidefinite program, we substitute a vector inner-product objective function for trace;

$$
\operatorname{tr} Z = \langle Z \,, I \rangle \leftarrow \langle Z \,, W \rangle
\qquad (784)
$$

a generalization of the trace heuristic for minimizing convex envelope of rank, where $W \in \mathbb{S}_+^{N+n}$ is constant with respect to (783). Matrix W is normal to a hyperplane in $\mathbb{S}^{N+n}$ minimized over a convex feasible set specified by the constraints in (783). Matrix W is chosen so $-W$ points in direction of rank-n feasible solutions G. For properly chosen W, problem (783) becomes an equivalent to (781). Thus the purpose of vector inner-product objective (784) is to locate a rank-n feasible Gram matrix assumed existent on the boundary of positive semidefinite cone $\mathbb{S}_+^N$, as explained beginning in §4.4.1; how to choose direction vector W is explained there and in what follows:

direction matrix W

Denote by $Z^\star$ an optimal composite matrix from semidefinite program (783). Then for $Z^\star \in \mathbb{S}^{N+n}$ whose eigenvalues $\lambda(Z^\star) \in \mathbb{R}^{N+n}$ are arranged in nonincreasing order, (Ky Fan)

$$
\sum_{i=n+1}^{N+n} \lambda(Z^\star)_i = \begin{array}{ll} \underset{W \in \mathbb{S}^{N+n}}{\text{minimize}} & \langle Z^\star, W \rangle \\[1ex] \text{subject to} & 0 \preceq W \preceq I \\ & \operatorname{tr} W = N \end{array}
\qquad (1791\text{a})
$$

which has an optimal solution that is known in closed form (p.567, §4.4.1.1). This eigenvalue sum is zero when $Z^\star$ has rank n or less.

Foreknowledge of optimal $Z^\star$, to make possible this search for W, implies iteration; *id est*, semidefinite program (783) is solved for $Z^\star$ initializing $W = I$ or $W = \mathbf{0}$. Once found, $Z^\star$ becomes constant in semidefinite program (1791a) where a new normal direction W is found as its optimal solution. Then this cycle (783) (1791a) iterates until convergence. When $\operatorname{rank} Z^\star = n$, solution via this convex iteration solves sensor-network localization problem (776) and its equivalent (781).

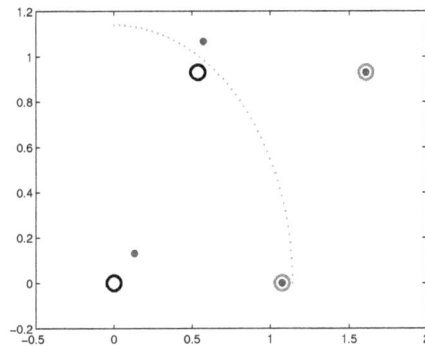

Figure 94: Typical solution for 2-lattice in Figure **89** with noise factor $\eta = 0.1$. Two red rightmost nodes are anchors; two remaining nodes are sensors. Radio range of sensor 1 indicated by arc; radius $= 1.14$. Actual sensor indicated by target $\bigcirc$ while its localization is indicated by bullet $\bullet$. Rank-2 solution found in 1 iteration (783) (1791a) subject to reflection error.

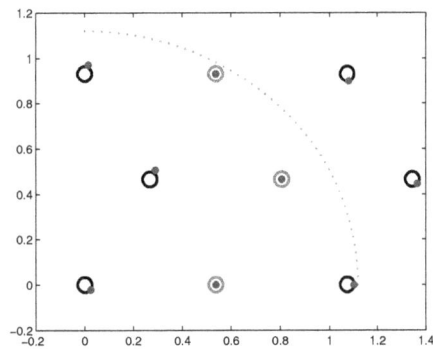

Figure 95: Typical solution for 3-lattice in Figure **90** with noise factor $\eta = 0.1$. Three red vertical middle nodes are anchors; remaining nodes are sensors. Radio range of sensor 1 indicated by arc; radius $= 1.12$. Actual sensor indicated by target $\bigcirc$ while its localization is indicated by bullet $\bullet$. Rank-2 solution found in 2 iterations (783) (1791a).

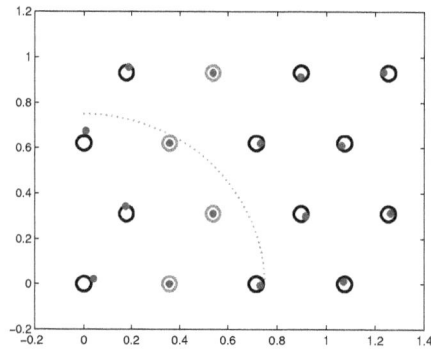

Figure 96: Typical solution for 4-lattice in Figure 91 with noise factor $\eta = 0.1$. Four red vertical middle-left nodes are anchors; remaining nodes are sensors. Radio range of sensor 1 indicated by arc; radius $= 0.75$. Actual sensor indicated by target $\bigcirc$ while its localization is indicated by bullet $\bullet$. Rank-2 solution found in 7 iterations (783) (1791a).

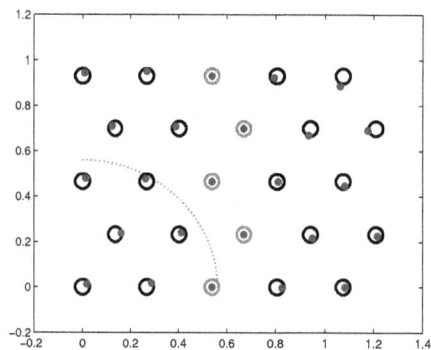

Figure 97: Typical solution for 5-lattice in Figure 92 with noise factor $\eta = 0.1$. Five red vertical middle nodes are anchors; remaining nodes are sensors. Radio range of sensor 1 indicated by arc; radius $= 0.56$. Actual sensor indicated by target $\bigcirc$ while its localization is indicated by bullet $\bullet$. Rank-2 solution found in 3 iterations (783) (1791a).

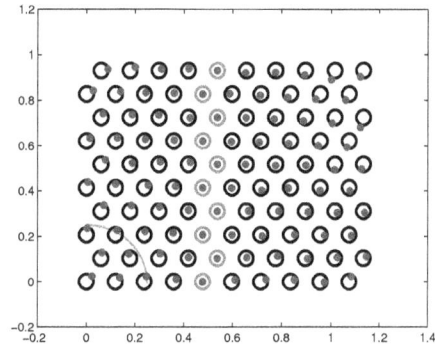

Figure 98: Typical solution for 10-lattice with noise factor $\eta = 0.1$ compares better than Carter & Jin [73, fig.4.2]. Ten red vertical middle nodes are anchors; the rest are sensors. Radio range of sensor 1 indicated by arc; radius $= 0.25$. Actual sensor indicated by target $\bigcirc$ while its localization is indicated by bullet $\bullet$. Rank-2 solution found in 5 iterations (783) (1791a).

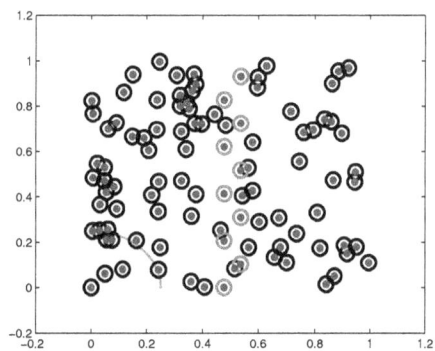

Figure 99: Typical localization of 100 randomized noiseless sensors $(\eta = 0)$ is exact despite incomplete EDM. Ten red vertical middle nodes are anchors; remaining nodes are sensors. Radio range of sensor at origin indicated by arc; radius $= 0.25$. Actual sensor indicated by target $\bigcirc$ while its localization is indicated by bullet $\bullet$. Rank-2 solution found in 3 iterations (783) (1791a).

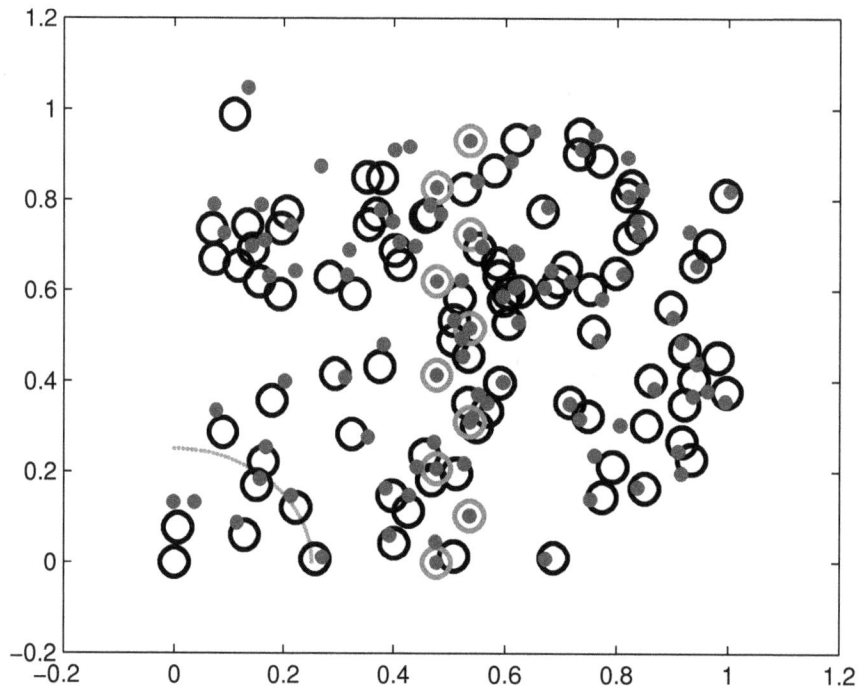

Figure 100: Typical solution for 100 randomized sensors with noise factor $\eta = 0.1$; worst measured average sensor error ≈ 0.0044 compares better than Carter & Jin's 0.0154 computed in 0.71s [73, p.19]. Ten red vertical middle nodes are anchors; same as before. Remaining nodes are sensors. Interior anchor placement makes localization difficult. Radio range of sensor at origin indicated by arc; radius $= 0.25$. Actual sensor indicated by target $\bigcirc$ while its localization is indicated by bullet $\bullet$. After 1 iteration rank $G = 92$, after 2 iterations rank $G = 4$.   Rank-2 solution found in 3 iterations (783) (1791a). (Regular lattice in Figure 98 is actually harder to solve, requiring more iterations.) Runtime for SDPT3 [357] under cvx [173] is a few minutes on 2009 vintage laptop Core 2 Duo CPU (Intel T6400@2GHz, 800MHz FSB).

numerical solution

In all examples to follow, number of anchors

$$m = \sqrt{N} \tag{785}$$

equals square root of cardinality N of list X. Indices set $\mathcal{I}$ identifying all measurable distances ● is ascertained from connectivity matrix (777), (778), (779), or (780). We solve iteration (783) (1791a) in dimension $n = 2$ for each respective example illustrated in Figure **89** through Figure **92**.

In presence of negligible noise, true position is reliably localized for every standardized example; noteworthy insofar as each example represents an incomplete graph. This implies that the set of all optimal solutions having least rank must be small.

To make the examples interesting and consistent with previous work, we randomize each range of distance-square that bounds $\langle G , (e_i - e_j)(e_i - e_j)^{\mathrm{T}} \rangle$ in (783); *id est*, for each and every $(i,j) \in \mathcal{I}$

$$\begin{aligned} \overline{d_{ij}} &= d_{ij}(1 + \sqrt{3}\,\eta\,\chi_l)^2 \\ \underline{d_{ij}} &= d_{ij}(1 - \sqrt{3}\,\eta\,\chi_{l+1})^2 \end{aligned} \tag{786}$$

where $\eta = 0.1$ is a constant noise factor, χ_l is the lth sample of a noise process realization uniformly distributed in the interval $(0,1)$ like `rand(1)` from MATLAB, and d_{ij} is actual distance-square from i^{th} to j^{th} sensor. Because of distinct function calls to `rand()`, each range of distance-square $[\underline{d_{ij}}, \overline{d_{ij}}]$ is not necessarily centered on actual distance-square d_{ij}. Unit stochastic variance is provided by factor $\sqrt{3}$.

Figure **94** through Figure **97** each illustrate one realization of numerical solution to the standardized lattice problems posed by Figure **89** through Figure **92** respectively. Exact localization, by any method, is impossible because of measurement noise. Certainly, by inspection of their published graphical data, our results are better than those of Carter & Jin. (Figure **98**, **99**, **100**) Obviously our solutions do not suffer from those compaction-type errors (clustering of localized sensors) exhibited by Biswas' graphical results for the same noise factor η.

localization example conclusion

Solution to this sensor-network localization problem became apparent by understanding geometry of optimization. Trace of a matrix, to a student of linear algebra, is perhaps a sum of eigenvalues. But to us, trace represents the normal I to some hyperplane in Euclidean vector space. (Figure **87**)

Our solutions are globally optimal, requiring: **1)** no centralized-gradient postprocessing heuristic refinement as in [44] because there is effectively no relaxation of (781) at global optimality, **2)** no implicit postprojection on rank-2 positive semidefinite matrices induced by nonzero $G - X^{\mathrm{T}}X$ denoting suboptimality as occurs in [45] [46] [47] [73] [222] [230]; indeed, $G^\star = X^{\star\mathrm{T}}X^\star$ by convex iteration.

Numerical solution to noisy problems, containing sensor variables well in excess of 100, becomes difficult via the holistic semidefinite program we proposed. When problem size is within reach of contemporary general-purpose semidefinite program solvers, then the convex iteration we presented inherently overcomes limitations of Carter & Jin with respect to both noise performance and ability to localize in any desired affine dimension.

The legacy of Carter, Jin, Saunders, & Ye [73] is a sobering demonstration of the need for more efficient methods for solution of semidefinite programs, while that of So & Ye [328] forever bonds distance geometry to semidefinite programming. Elegance of our semidefinite problem statement (783), for constraining affine dimension of sensor-network localization, should provide some *impetus* to focus more research on computational intensity of general-purpose semidefinite program solvers. An approach different from interior-point methods is required; higher speed and greater accuracy from a simplex-like solver is what is needed.   □

4.4.1.2.5   Example. *Nonnegative spectral factorization. (confer §3.8.1.0.2)*
Having found optimal real coefficient vectors $v^\star, u^\star$ for a sixteenth order magnitude square transfer function, evaluated along the $\jmath\omega$ axis (p.231),

$$|H(\jmath\omega)|^2 = H(\jmath\omega)H(-\jmath\omega) = \frac{1 + v_1^\star\omega^2 + v_2^\star\omega^4 + \ldots + v_8^\star\omega^{16}}{1 + u_1^\star\omega^2 + u_2^\star\omega^4 + \ldots + u_8^\star\omega^{16}} \qquad (657)$$

we wish to find real coefficients b, a for corresponding Fourier transform

$$H(\jmath\omega) = \frac{1 + b_1\jmath\omega + b_2(\jmath\omega)^2 + \ldots + b_8(\jmath\omega)^8}{1 + a_1\jmath\omega + a_2(\jmath\omega)^2 + \ldots + a_8(\jmath\omega)^8} \qquad (654)$$

These coefficients $b, a, v^\star, u^\star$ are related through simultaneous nonlinear algebraic equations:

$$
\begin{aligned}
& v_1^\star = b_1^2 - 2b_2 \,, && u_1^\star = a_1^2 - 2a_2 \\
& v_2^\star = b_2^2 - 2b_1b_3 + 2b_4 \,, && u_2^\star = a_2^2 - 2a_1a_3 + 2a_4 \\
& v_3^\star = b_3^2 - 2b_2b_4 + 2b_1b_5 - 2b_6 \,, && u_3^\star = a_3^2 - 2a_2a_4 + 2a_1a_5 - 2a_6 \\
& v_4^\star = b_4^2 - 2b_3b_5 + 2b_2b_6 - 2b_1b_7 + 2b_8 \,, && u_4^\star = a_4^2 - 2a_3a_5 + 2a_2a_6 - 2a_1a_7 + 2a_8 \\
& v_5^\star = b_5^2 - 2b_4b_6 + 2b_3b_7 - 2b_2b_8 \,, && u_5^\star = a_5^2 - 2a_4a_6 + 2a_3a_7 - 2a_2a_8 \\
& v_6^\star = b_6^2 - 2b_5b_7 + 2b_4b_8 \,, && u_6^\star = a_6^2 - 2a_5a_7 + 2a_4a_8 \\
& v_7^\star = b_7^2 - 2b_6b_8 \,, && u_7^\star = a_7^2 - 2a_6a_8 \\
& v_8^\star = b_8^2 \,, && u_8^\star = a_8^2 \qquad (787)
\end{aligned}
$$

Define a rank-one matrix

$$G(b) \triangleq \begin{bmatrix} 1 \\ b \end{bmatrix} \begin{bmatrix} 1 & b^{\mathrm{T}} \end{bmatrix} = \begin{bmatrix}
1 & b_1 & b_2 & b_3 & b_4 & b_5 & b_6 & b_7 & b_8 \\
b_1 & b_1^2 & b_1b_2 & b_1b_3 & b_1b_4 & b_1b_5 & b_1b_6 & b_1b_7 & b_1b_8 \\
b_2 & b_1b_2 & b_2^2 & b_2b_3 & b_2b_4 & b_2b_5 & b_2b_6 & b_2b_7 & b_2b_8 \\
b_3 & b_1b_3 & b_2b_3 & b_3^2 & b_3b_4 & b_3b_5 & b_3b_6 & b_3b_7 & b_3b_8 \\
b_4 & b_1b_4 & b_2b_4 & b_3b_4 & b_4^2 & b_4b_5 & b_4b_6 & b_4b_7 & b_4b_8 \\
b_5 & b_1b_5 & b_2b_5 & b_3b_5 & b_4b_5 & b_5^2 & b_5b_6 & b_5b_7 & b_5b_8 \\
b_6 & b_1b_6 & b_2b_6 & b_3b_6 & b_4b_6 & b_5b_6 & b_6^2 & b_6b_7 & b_6b_8 \\
b_7 & b_1b_7 & b_2b_7 & b_3b_7 & b_4b_7 & b_5b_7 & b_6b_7 & b_7^2 & b_7b_8 \\
b_8 & b_1b_8 & b_2b_8 & b_3b_8 & b_4b_8 & b_5b_8 & b_6b_8 & b_7b_8 & b_8^2
\end{bmatrix} \in \mathbb{S}^9 \qquad (788)$$

(Matrix $G(a)$ is similarly defined.)   Observe that $v^\star$ in (787) is formed by summing antidiagonals of $G(b)$ whose entries alternate sign. A particular sum is specified by a

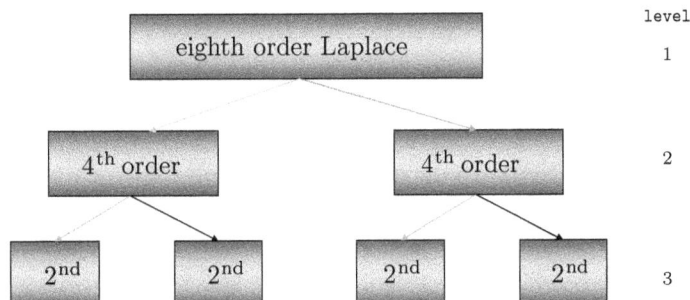

Figure 101: Nonnegative spectral factorization, high order bisection strategy. $\eta = $ eighth order Laplace transform corresponds to $2\eta = $ sixteenth order magnitude square transfer function. Because numerator v and denominator u are factored separately, number of factorizations $ = 2(\log_2(\eta) - 1)$. In the text, double dots $\ddot{v}, \ddot{u}$ connote first bifurcation (`level` 2). Triple dots $\dddot{v}, \dddot{u}$ connote second bifurcations (`level` 3). Factors per level $ = 2^{\text{level}-1}$.

predetermined symmetric constant matrix A_i (*confer* (57)) from a set $\{A_i \in \mathbb{S}^9, \; i = 1 \ldots 8\}$. With

$$A = \begin{bmatrix} \text{svec}(A_1)^{\mathrm{T}} \\ \vdots \\ \text{svec}(A_8)^{\mathrm{T}} \end{bmatrix} \in \mathbb{R}^{8 \times 9(9+1)/2} \qquad (680)$$

as previously defined in §4.1.1, all the sums (787) may be stated as two linear equalities $A \, \text{svec} \, G(b) = v^\star$ and $A \, \text{svec} \, G(a) = u^\star$. Then the problem of finding coefficients b may be stated as a feasibility problem[4.32]

$$\begin{array}{rl} \underset{G \in \mathbb{S}^9}{\text{find}} & b \in \mathbb{R}^8 \\ \text{subject to} & A \, \text{svec} \, G = v^\star \\ & \begin{bmatrix} 1 \\ b \end{bmatrix} = G(:,1) \\ & b \succeq 0 \\ & (G \succeq 0) \\ & \text{rank} \, G = 1 \end{array} \qquad (789)$$

The rank-one constraint is handled by convex iteration, as explained in §4.4.1. Positive semidefiniteness is parenthetical here because, for rank-one matrices, symmetry is necessary and sufficient (§A.3.1.0.7). $\qquad \square$

[4.32] separately from the similar optimization problem to find vector a. Stability requires $a \succeq 0$ with additional constraints on a. Minimum phase requires $b \succeq 0$ plus more constraints on b that are missing from problem statement (789). Both stability and minimum phase may be enforced, subsequent to spectral factorization, by negating positive real parts of poles and zeros respectively in order to move them into the left half (Laplace) s-plane with no impact to $|H(\jmath\omega)|$.

4.4.1.2.6   Example.   *Nonnegative spectral factorization* II.

The purpose of spectral factorization, in electronics, is to facilitate high order filter implementation in the form of passive and active circuitry. Cascades of second order (Laplace) sections are preferred because component sensitivity becomes manageable and because needed complex poles and zeros cannot be obtained from a first order section.

Nonnegative spectral factorization on a magnitude square transfer function, evaluated along the $\jmath\omega$ axis, was performed in Example 4.4.1.2.5 to recover its corresponding Fourier transform.[4.33] In this example, we nonnegatively decompose a high order magnitude square transfer function into a product of successively lower order magnitude square transfer functions. Once fourth order magnitude square functions are found, then corresponding second order Laplace transfer function coefficients are ascertained from (656) and then passive component values can be determined from those coefficients.

Our strategy, for an eighth order Laplace transfer function, is illustrated in Figure 101. We begin at the tree's `level` 2 factorization. Nonnegative decomposition of a 16th order magnitude square transfer function into two 8th order functions

$$\frac{1 + v_1^\star\omega^2 + v_2^\star\omega^4 + \ldots + v_8^\star\omega^{16}}{1 + u_1^\star\omega^2 + u_2^\star\omega^4 + \ldots + u_8^\star\omega^{16}} = \frac{1 + \ddot{v}_1\omega^2 + \ddot{v}_2\omega^4 + \ddot{v}_3\omega^6 + \ddot{v}_4\omega^8}{1 + \ddot{u}_1\omega^2 + \ddot{u}_2\omega^4 + \ddot{u}_3\omega^6 + \ddot{u}_4\omega^8}\; \frac{1 + \ddot{v}_5\omega^2 + \ddot{v}_6\omega^4 + \ddot{v}_7\omega^6 + \ddot{v}_8\omega^8}{1 + \ddot{u}_5\omega^2 + \ddot{u}_6\omega^4 + \ddot{u}_7\omega^6 + \ddot{u}_8\omega^8}$$
$$(790)$$

implies these simultaneous algebraic identifications with known real coefficient vectors $v^\star, u^\star$:

$$
\begin{aligned}
v_1^\star &= \ddot{v}_1 + \ddot{v}_5 , & u_1^\star &= \ddot{u}_1 + \ddot{u}_5 \\
v_2^\star &= \ddot{v}_2 + \ddot{v}_6 + \ddot{v}_1\ddot{v}_5 , & u_2^\star &= \ddot{u}_2 + \ddot{u}_6 + \ddot{u}_1\ddot{u}_5 \\
v_3^\star &= \ddot{v}_3 + \ddot{v}_7 + \ddot{v}_1\ddot{v}_6 + \ddot{v}_2\ddot{v}_5 , & u_3^\star &= \ddot{u}_3 + \ddot{u}_7 + \ddot{u}_1\ddot{u}_6 + \ddot{u}_2\ddot{u}_5 \\
v_4^\star &= \ddot{v}_4 + \ddot{v}_8 + \ddot{v}_1\ddot{v}_7 + \ddot{v}_2\ddot{v}_6 + \ddot{v}_3\ddot{v}_5 , & u_4^\star &= \ddot{u}_4 + \ddot{u}_8 + \ddot{u}_1\ddot{u}_7 + \ddot{u}_2\ddot{u}_6 + \ddot{u}_3\ddot{u}_5 \\
v_5^\star &= \ddot{v}_4\ddot{v}_5 + \ddot{v}_3\ddot{v}_6 + \ddot{v}_2\ddot{v}_7 + \ddot{v}_1\ddot{v}_8 , & u_5^\star &= \ddot{u}_4\ddot{u}_5 + \ddot{u}_3\ddot{u}_6 + \ddot{u}_2\ddot{u}_7 + \ddot{u}_1\ddot{u}_8 \\
v_6^\star &= \ddot{v}_4\ddot{v}_6 + \ddot{v}_3\ddot{v}_7 + \ddot{v}_2\ddot{v}_8 , & u_6^\star &= \ddot{u}_4\ddot{u}_6 + \ddot{u}_3\ddot{u}_7 + \ddot{u}_2\ddot{u}_8 \\
v_7^\star &= \ddot{v}_4\ddot{v}_7 + \ddot{v}_3\ddot{v}_8 , & u_7^\star &= \ddot{u}_4\ddot{u}_7 + \ddot{u}_3\ddot{u}_8 \\
v_8^\star &= \ddot{v}_4\ddot{v}_8 , & u_8^\star &= \ddot{u}_4\ddot{u}_8
\end{aligned}
$$
$$(791)$$

Now define a rank-one matrix for the numerator

$$
G(\ddot{v}) \triangleq \begin{bmatrix} 1 \\ \ddot{v} \end{bmatrix} \begin{bmatrix} 1 & \ddot{v}^{\mathrm{T}} \end{bmatrix} = \begin{bmatrix}
1 & \ddot{v}_1 & \ddot{v}_2 & \ddot{v}_3 & \ddot{v}_4 & \ddot{v}_5 & \ddot{v}_6 & \ddot{v}_7 & \ddot{v}_8 \\
\ddot{v}_1 & \ddot{v}_1^2 & \ddot{v}_1\ddot{v}_2 & \ddot{v}_1\ddot{v}_3 & \ddot{v}_1\ddot{v}_4 & \ddot{v}_1\ddot{v}_5 & \ddot{v}_1\ddot{v}_6 & \ddot{v}_1\ddot{v}_7 & \ddot{v}_1\ddot{v}_8 \\
\ddot{v}_2 & \ddot{v}_1\ddot{v}_2 & \ddot{v}_2^2 & \ddot{v}_2\ddot{v}_3 & \ddot{v}_2\ddot{v}_4 & \ddot{v}_2\ddot{v}_5 & \ddot{v}_2\ddot{v}_6 & \ddot{v}_2\ddot{v}_7 & \ddot{v}_2\ddot{v}_8 \\
\ddot{v}_3 & \ddot{v}_1\ddot{v}_3 & \ddot{v}_2\ddot{v}_3 & \ddot{v}_3^2 & \ddot{v}_3\ddot{v}_4 & \ddot{v}_3\ddot{v}_5 & \ddot{v}_3\ddot{v}_6 & \ddot{v}_3\ddot{v}_7 & \ddot{v}_3\ddot{v}_8 \\
\ddot{v}_4 & \ddot{v}_1\ddot{v}_4 & \ddot{v}_2\ddot{v}_4 & \ddot{v}_3\ddot{v}_4 & \ddot{v}_4^2 & \ddot{v}_4\ddot{v}_5 & \ddot{v}_4\ddot{v}_6 & \ddot{v}_4\ddot{v}_7 & \ddot{v}_4\ddot{v}_8 \\
\ddot{v}_5 & \ddot{v}_1\ddot{v}_5 & \ddot{v}_2\ddot{v}_5 & \ddot{v}_3\ddot{v}_5 & \ddot{v}_4\ddot{v}_5 & \ddot{v}_5^2 & \ddot{v}_5\ddot{v}_6 & \ddot{v}_5\ddot{v}_7 & \ddot{v}_5\ddot{v}_8 \\
\ddot{v}_6 & \ddot{v}_1\ddot{v}_6 & \ddot{v}_2\ddot{v}_6 & \ddot{v}_3\ddot{v}_6 & \ddot{v}_4\ddot{v}_6 & \ddot{v}_5\ddot{v}_6 & \ddot{v}_6^2 & \ddot{v}_6\ddot{v}_7 & \ddot{v}_6\ddot{v}_8 \\
\ddot{v}_7 & \ddot{v}_1\ddot{v}_7 & \ddot{v}_2\ddot{v}_7 & \ddot{v}_3\ddot{v}_7 & \ddot{v}_4\ddot{v}_7 & \ddot{v}_5\ddot{v}_7 & \ddot{v}_6\ddot{v}_7 & \ddot{v}_7^2 & \ddot{v}_7\ddot{v}_8 \\
\ddot{v}_8 & \ddot{v}_1\ddot{v}_8 & \ddot{v}_2\ddot{v}_8 & \ddot{v}_3\ddot{v}_8 & \ddot{v}_4\ddot{v}_8 & \ddot{v}_5\ddot{v}_8 & \ddot{v}_6\ddot{v}_8 & \ddot{v}_7\ddot{v}_8 & \ddot{v}_8^2
\end{bmatrix} \in \mathbb{S}^9 \quad (792)
$$

(Matrix $G(\ddot{u})$ is defined similarly for the denominator.) Terms in (791) are picked out of $G(\ddot{v})$ by a predetermined symmetric constant matrix $\ddot{A}_i$ (*confer* (57)) from a set

---

[4.33]When there are no poles on the $\jmath\omega$ axis, a Laplace transform can be recovered from a Fourier transform by substitution $\jmath\omega \leftarrow s$.

$\{\ddot{A}_i \in \mathbb{S}^\mathbf{9}, \ i=1\ldots 8\}$. Populating rows of

$$A = \begin{bmatrix} \mathrm{svec}(\ddot{A}_1)^\mathrm{T} \\ \vdots \\ \mathrm{svec}(\ddot{A}_8)^\mathrm{T} \end{bmatrix} \in \mathbb{R}^{\mathbf{8}\times\mathbf{9}(\mathbf{9}+\mathbf{1})/\mathbf{2}} \qquad (680)$$

with vectorized $\ddot{A}_i$ (as in §4.1.1), sums (791) are succinctly represented by two linear equalities $A\,\mathrm{svec}\,G(\ddot{v}) = v^\star$ and $A\,\mathrm{svec}\,G(\ddot{u}) = u^\star$. Then this spectral factorization in $\ddot{v}$ may be posed as a feasibility problem

$$\begin{array}{cl} \underset{G\in\mathbb{S}^\mathbf{9}}{\text{find}} & \ddot{v} \in \mathbb{R}^\mathbf{8} \\ \text{subject to} & A\,\mathrm{svec}\,G = v^\star \\ & \begin{bmatrix} 1 \\ \ddot{v} \end{bmatrix} = G(:,1) \\ & \ddot{v} \succeq 0 \\ & (G \succeq 0) \\ & \mathrm{rank}\,G = 1 \end{array} \qquad (793)$$

Having found two 8th order square spectral factors in nonnegative $\ddot{v}^\star$ from (793), two pairs of 4th order `level` 3 factors remain to be found:

$$\frac{1 + \ddot{v}_1^\star\omega^2 + \ddot{v}_2^\star\omega^4 + \ddot{v}_3^\star\omega^6 + \ddot{v}_4^\star\omega^8}{1 + \ddot{u}_1^\star\omega^2 + \ddot{u}_2^\star\omega^4 + \ddot{u}_3^\star\omega^6 + \ddot{u}_4^\star\omega^8} = \frac{1 + \ddot{v}_1\omega^2 + \ddot{v}_2\omega^4}{1 + \ddot{u}_1\omega^2 + \ddot{u}_2\omega^4}\frac{1 + \ddot{v}_3\omega^2 + \ddot{v}_4\omega^4}{1 + \ddot{u}_3\omega^2 + \ddot{u}_4\omega^4} \qquad (794)$$

$$\frac{1 + \ddot{v}_5^\star\omega^2 + \ddot{v}_6^\star\omega^4 + \ddot{v}_7^\star\omega^6 + \ddot{v}_8^\star\omega^8}{1 + \ddot{u}_5^\star\omega^2 + \ddot{u}_6^\star\omega^4 + \ddot{u}_7^\star\omega^6 + \ddot{u}_8^\star\omega^8} = \frac{1 + \ddot{v}_5\omega^2 + \ddot{v}_6\omega^4}{1 + \ddot{u}_5\omega^2 + \ddot{u}_6\omega^4}\frac{1 + \ddot{v}_7\omega^2 + \ddot{v}_8\omega^4}{1 + \ddot{u}_7\omega^2 + \ddot{u}_8\omega^4} \qquad (795)$$

$$\begin{array}{ll} \ddot{v}_1^\star = \ddot{v}_1 + \ddot{v}_3 \ , & \ddot{u}_1^\star = \ddot{u}_1 + \ddot{u}_3 \\ \ddot{v}_2^\star = \ddot{v}_2 + \ddot{v}_4 + \ddot{v}_1\ddot{v}_3 \ , & \ddot{u}_2^\star = \ddot{u}_2 + \ddot{u}_4 + \ddot{u}_1\ddot{u}_3 \\ \ddot{v}_3^\star = \ddot{v}_1\ddot{v}_4 + \ddot{v}_2\ddot{v}_3 \ , & \ddot{u}_3^\star = \ddot{u}_1\ddot{u}_4 + \ddot{u}_2\ddot{u}_3 \\ \ddot{v}_4^\star = \ddot{v}_2\ddot{v}_4 \ , & \ddot{u}_4^\star = \ddot{u}_2\ddot{u}_4 \end{array} \qquad (796)$$

$$\begin{array}{ll} \ddot{v}_5^\star = \ddot{v}_5 + \ddot{v}_7 \ , & \ddot{u}_5^\star = \ddot{u}_5 + \ddot{u}_7 \\ \ddot{v}_6^\star = \ddot{v}_6 + \ddot{v}_8 + \ddot{v}_5\ddot{v}_7 \ , & \ddot{u}_6^\star = \ddot{u}_6 + \ddot{u}_8 + \ddot{u}_5\ddot{u}_7 \\ \ddot{v}_7^\star = \ddot{v}_5\ddot{v}_8 + \ddot{v}_6\ddot{v}_7 \ , & \ddot{u}_7^\star = \ddot{u}_5\ddot{u}_8 + \ddot{u}_6\ddot{u}_7 \\ \ddot{v}_8^\star = \ddot{v}_6\ddot{v}_8 \ , & \ddot{u}_8^\star = \ddot{u}_6\ddot{u}_8 \end{array} \qquad (797)$$

$$G(\ddot{v}) \triangleq \begin{bmatrix} 1 \\ \ddot{v} \end{bmatrix} \begin{bmatrix} 1 & \ddot{v}^\mathrm{T} \end{bmatrix} = \begin{bmatrix} 1 & \ddot{v}_1 & \ddot{v}_2 & \ddot{v}_3 & \ddot{v}_4 & \ddot{v}_5 & \ddot{v}_6 & \ddot{v}_7 & \ddot{v}_8 \\ \ddot{v}_1 & \ddot{v}_1^2 & \ddot{v}_1\ddot{v}_2 & \ddot{v}_1\ddot{v}_3 & \ddot{v}_1\ddot{v}_4 & \ddot{v}_1\ddot{v}_5 & \ddot{v}_1\ddot{v}_6 & \ddot{v}_1\ddot{v}_7 & \ddot{v}_1\ddot{v}_8 \\ \ddot{v}_2 & \ddot{v}_1\ddot{v}_2 & \ddot{v}_2^2 & \ddot{v}_2\ddot{v}_3 & \ddot{v}_2\ddot{v}_4 & \ddot{v}_2\ddot{v}_5 & \ddot{v}_2\ddot{v}_6 & \ddot{v}_2\ddot{v}_7 & \ddot{v}_2\ddot{v}_8 \\ \ddot{v}_3 & \ddot{v}_1\ddot{v}_3 & \ddot{v}_2\ddot{v}_3 & \ddot{v}_3^2 & \ddot{v}_3\ddot{v}_4 & \ddot{v}_3\ddot{v}_5 & \ddot{v}_3\ddot{v}_6 & \ddot{v}_3\ddot{v}_7 & \ddot{v}_3\ddot{v}_8 \\ \ddot{v}_4 & \ddot{v}_1\ddot{v}_4 & \ddot{v}_2\ddot{v}_4 & \ddot{v}_3\ddot{v}_4 & \ddot{v}_4^2 & \ddot{v}_4\ddot{v}_5 & \ddot{v}_4\ddot{v}_6 & \ddot{v}_4\ddot{v}_7 & \ddot{v}_4\ddot{v}_8 \\ \ddot{v}_5 & \ddot{v}_1\ddot{v}_5 & \ddot{v}_2\ddot{v}_5 & \ddot{v}_3\ddot{v}_5 & \ddot{v}_4\ddot{v}_5 & \ddot{v}_5^2 & \ddot{v}_5\ddot{v}_6 & \ddot{v}_5\ddot{v}_7 & \ddot{v}_5\ddot{v}_8 \\ \ddot{v}_6 & \ddot{v}_1\ddot{v}_6 & \ddot{v}_2\ddot{v}_6 & \ddot{v}_3\ddot{v}_6 & \ddot{v}_4\ddot{v}_6 & \ddot{v}_5\ddot{v}_6 & \ddot{v}_6^2 & \ddot{v}_6\ddot{v}_7 & \ddot{v}_6\ddot{v}_8 \\ \ddot{v}_7 & \ddot{v}_1\ddot{v}_7 & \ddot{v}_2\ddot{v}_7 & \ddot{v}_3\ddot{v}_7 & \ddot{v}_4\ddot{v}_7 & \ddot{v}_5\ddot{v}_7 & \ddot{v}_6\ddot{v}_7 & \ddot{v}_7^2 & \ddot{v}_7\ddot{v}_8 \\ \ddot{v}_8 & \ddot{v}_1\ddot{v}_8 & \ddot{v}_2\ddot{v}_8 & \ddot{v}_3\ddot{v}_8 & \ddot{v}_4\ddot{v}_8 & \ddot{v}_5\ddot{v}_8 & \ddot{v}_6\ddot{v}_8 & \ddot{v}_7\ddot{v}_8 & \ddot{v}_8^2 \end{bmatrix} \in \mathbb{S}^\mathbf{9} \qquad (798)$$

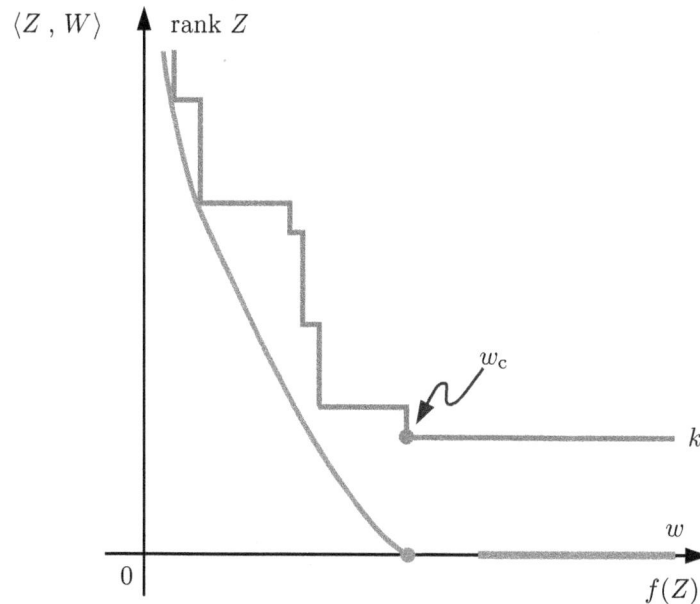

Figure 102: Regularization curve, parametrized by weight w for real convex objective f minimization (800) with rank constraint to k by convex iteration.

Setting

$$A = \begin{bmatrix} \mathrm{svec}(\ddot{A}_1)^{\mathrm{T}} \\ \vdots \\ \mathrm{svec}(\ddot{A}_8)^{\mathrm{T}} \end{bmatrix} \in \mathbb{R}^{\mathbf{8} \times \mathbf{9(9+1)/2}} \qquad (680)$$

then all `level 3` (Figure 101) nonnegative spectral factorization coefficients $\ddot{v}$ are found at once by solving

$$
\begin{aligned}
\underset{G \in \mathbb{S}^{\mathbf{9}}}{\mathrm{find}} \quad & \ddot{v} \in \mathbb{R}^{\mathbf{8}} \\
\text{subject to} \quad & A\,\mathrm{svec}\,G = \ddot{v}^{\star} \\
& \begin{bmatrix} 1 \\ \ddot{v} \end{bmatrix} = G(:,1) \\
& \ddot{v} \succeq 0 \\
& (G \succeq 0) \\
& \mathrm{rank}\,G = 1
\end{aligned}
\qquad (799)
$$

The feasibility problem to find $\ddot{u}$ is similar. All second order Laplace transfer function coefficients can be found via (656).   □

4.4.1.3   regularization

We test the convex iteration technique, for constraining rank, over a wide range of problems beyond localization of randomized positions (Figure 100); *e.g.*, *stress* (§7.2.2.7.1), *ball packing* (§5.4.2.2.6), and cardinality (§4.6). We have had some success introducing the direction matrix inner-product (784) as a *regularization* term[4.34]

$$\begin{array}{ll} \underset{Z\in\mathbb{S}^N}{\text{minimize}} & f(Z) + w\langle Z\,,\,W\rangle \\ \text{subject to} & Z \in \mathcal{C} \\ & Z \succeq 0 \end{array} \qquad (800)$$

$$\begin{array}{ll} \underset{W\in\mathbb{S}^N}{\text{minimize}} & f(Z^\star) + w\langle Z^\star, W\rangle \\ \text{subject to} & 0 \preceq W \preceq I \\ & \operatorname{tr} W = N - n \end{array} \qquad (801)$$

whose purpose is to constrain rank, affine dimension, or cardinality:

The abstraction, that is Figure 102, is a synopsis; a broad generalization of accumulated empirical evidence: There exists a critical (smallest) weight w_c • for which a minimal-rank constraint is met. Graphical discontinuity can subsequently exist when there is a range of greater w providing required rank k but not necessarily increasing a minimization objective function f; *e.g.*, §4.6.0.0.2. Positive scalar w is well chosen by cut-and-try.

4.5 Constraining cardinality

The convex iteration technique for constraining rank can be applied to cardinality problems. There are parallels in its development analogous to how prototypical semidefinite program (679) resembles linear program (301) on page 240 [410]:

4.5.1 nonnegative variable

Our goal has been to reliably constrain rank in a semidefinite program. There is a direct analogy to linear programming that is simpler to present but, perhaps, more difficult to solve. In Optimization, that analogy is known as the *cardinality problem*.

Consider a feasibility problem $Ax = b$, but with an upper bound k on cardinality $\|x\|_0$ of a nonnegative solution x: for $A\in\mathbb{R}^{m\times n}$ and vector $b\in\mathcal{R}(A)$

$$\begin{array}{ll} \text{find} & x \in \mathbb{R}^n \\ \text{subject to} & Ax = b \\ & x \succeq 0 \\ & \|x\|_0 \leq k \end{array} \qquad (529)$$

where $\|x\|_0 \leq k$ means[4.35] vector x has at most k nonzero entries; such a vector is presumed existent in the feasible set. Nonnegativity constraint $x \succeq 0$ is analogous to positive semidefiniteness; the notation means vector x belongs to the nonnegative orthant $\mathbb{R}^n_+$. Cardinality is quasiconcave on $\mathbb{R}^n_+$ just as rank is quasiconcave on $\mathbb{S}^n_+$. [62, §3.4.2]

[4.34] called *multiobjective-* or *vector optimization*. Proof of convergence for this convex iteration is identical to that in §4.4.1.2.1 because f is a convex real function, hence bounded below, and $f(Z^\star)$ is constant in (801).

[4.35] Although it is a metric (§5.2), cardinality $\|x\|_0$ cannot be a norm (§3.2) because it is not positively homogeneous.

4.5.1.1   direction vector

We propose that cardinality-constrained feasibility problem (529) can be equivalently expressed as iteration of a sequence of two convex problems: for $0 \leq k \leq n-1$

$$\begin{array}{ll} \underset{x \in \mathbb{R}^n}{\text{minimize}} & \langle x, y \rangle \\ \text{subject to} & Ax = b \\ & x \succeq 0 \end{array} \qquad (156)$$

$$\sum_{i=k+1}^{n} \pi(x^\star)_i = \begin{array}{ll} \underset{y \in \mathbb{R}^n}{\text{minimize}} & \langle x^\star, y \rangle \\ \text{subject to} & 0 \preceq y \preceq \mathbf{1} \\ & y^{\mathrm{T}}\mathbf{1} = n - k \end{array} \qquad (524)$$

where π is the (nonincreasing) presorting function. This sequence is iterated until $x^{\star\mathrm{T}}y^\star$ vanishes; *id est*, until desired cardinality is achieved. But this *global convergence* cannot always be guaranteed.[4.36]

Problem (524) is analogous to the rank constraint problem; (p.266)

$$\sum_{i=k+1}^{N} \lambda(G^\star)_i = \begin{array}{ll} \underset{W \in \mathbb{S}^N}{\text{minimize}} & \langle G^\star, W \rangle \\ \text{subject to} & 0 \preceq W \preceq I \\ & \operatorname{tr} W = N - k \end{array} \qquad (1791a)$$

Linear program (524) sums the $n-k$ smallest entries from vector x. In context of problem (529), we want $n-k$ entries of x to sum to zero; *id est*, we want a globally optimal objective $x^{\star\mathrm{T}}y^\star$ to vanish: more generally, (*confer* (767))

$$\sum_{i=k+1}^{n} \pi(|x^\star|)_i = \langle |x^\star|, y^\star \rangle = |x^\star|^{\mathrm{T}} y^\star \triangleq 0 \qquad (802)$$

defines *global convergence* for the iteration. Then $n-k$ entries of $x^\star$ are themselves zero whenever their absolute sum is, and cardinality of $x^\star \in \mathbb{R}^n$ is at most k. *Optimal direction vector* $y^\star$ is defined as any nonnegative vector for which

$$(529) \quad \begin{array}{ll} \text{find} & x \in \mathbb{R}^n \\ \text{subject to} & Ax = b \\ & x \succeq 0 \\ & \|x\|_0 \leq k \end{array} \quad \equiv \quad \begin{array}{ll} \underset{x \in \mathbb{R}^n}{\text{minimize}} & \langle x, y^\star \rangle \\ \text{subject to} & Ax = b \\ & x \succeq 0 \end{array} \quad (156)$$

Existence of such a $y^\star$, whose nonzero entries are complementary to those of $x^\star$, is obvious assuming existence of a cardinality-k solution $x^\star$.

[4.36]When it succeeds, a sequence may be regarded as a *homotopy* to minimal 0-norm.

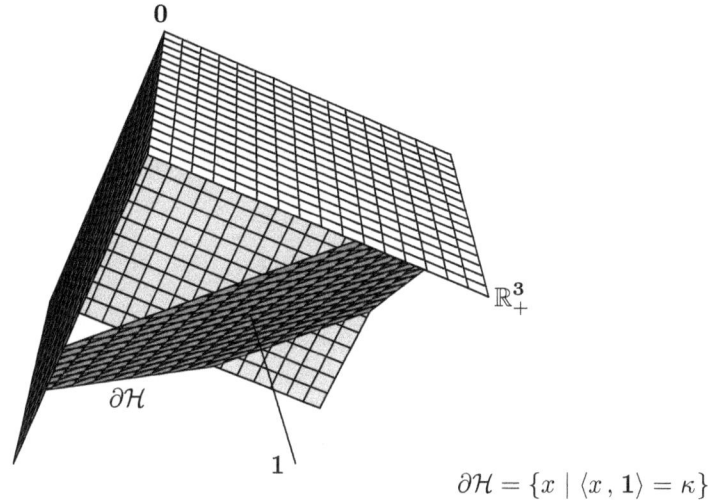

$$\partial \mathcal{H} = \{x \mid \langle x,\mathbf{1} \rangle = \kappa\}$$

(*confer* Figure **87**)

Figure 103: 1-norm heuristic for cardinality minimization can be interpreted as minimization of a hyperplane, $\partial\mathcal{H}$ with normal $\mathbf{1}$, over nonnegative orthant drawn here in $\mathbb{R}^3$. Polar of direction vector $y = \mathbf{1}$ points toward origin.

4.5.1.2 direction vector interpretation

(*confer* §4.4.1.1) Vector y may be interpreted as a negative search direction; it points opposite to direction of movement of hyperplane $\{x \mid \langle x,y \rangle = \tau\}$ in a minimization of real linear function $\langle x,y \rangle$ over the feasible set in linear program (156). (p.68) Direction vector y is not unique. The feasible set of direction vectors in (524) is the convex hull of all cardinality-$(n-k)$ one-vectors; *videlicet*,

$$\text{conv}\{u \in \mathbb{R}^n \mid \text{card } u = n-k,\ u_i \in \{0,1\}\} = \{a \in \mathbb{R}^n \mid \mathbf{1} \succeq a \succeq 0,\ \langle \mathbf{1},a \rangle = n-k\} \quad (803)$$

This set, argument to conv{ }, comprises the extreme points of set (803) which is a *nonnegative hypercube slice*. An optimal solution y to (524), that is an extreme point of its feasible set, is known in closed form: it has 1 in each entry corresponding to the $n-k$ smallest entries of $x^\star$ and has 0 elsewhere. That particular polar direction $-y$ can be interpreted[4.37] by Proposition 7.1.3.0.3 as pointing toward the nonnegative orthant in the *Cartesian subspace,* whose basis is a subset of the Cartesian axes, containing all cardinality k (or less) vectors having the same ordering as $x^\star$. Consequently, for that closed-form solution, (*confer* (768))

$$\sum_{i=k+1}^n \pi(|x^\star|)_i = \langle |x^\star|,y \rangle = |x^\star|^{\mathrm{T}} y \geq 0 \quad (804)$$

[4.37] Convex iteration (156) (524) is not a projection method because there is no thresholding or discard of variable-vector x entries. An optimal direction vector y must always reside on the feasible set boundary in (524) page 292; *id est*, it is ill-advised to attempt simultaneous optimization of variables x and y.

When $y = \mathbf{1}$, as in 1-norm minimization for example, then polar direction $-y$ points directly at the origin (the cardinality-0 nonnegative vector) as in Figure **103**. We sometimes solve (524) instead of employing a known closed form because a direction vector is not unique. Setting direction vector y instead in accordance with an iterative inverse weighting scheme, called *reweighting* [166], was described for the 1-norm by Huo [213, §4.11.3] in 1999.

4.5.1.3   convergence can mean stalling

Convex iteration (156) (524) always converges to a *locally optimal solution,* a fixed point of possibly infeasible cardinality, by virtue of a monotonically nonincreasing real objective sequence. [264, §1.2] [42, §1.1] There can be no proof of global convergence, defined by (802). Constraining cardinality, solution to problem (529), can often be achieved but simple examples can be contrived that *stall* at a fixed point of infeasible cardinality; at a positive objective value $\langle x^\star, y \rangle = \tau > 0$. Direction vector y is then manipulated, as countermeasure, to steer out of local minima; *e.g.,* complete randomization as in Example 4.5.1.5.1, or reinitialization to a random cardinality-$(n-k)$ vector in the same nonnegative orthant face demanded by the current iterate: y has nonnegative uniformly distributed random entries in $(0,1]$ corresponding to the $n-k$ smallest entries of $x^\star$ and has 0 elsewhere. Zero entries behave like memory or state while randomness greatly diminishes likelihood of a stall. When this particular heuristic is successful, cardinality and objective sequence $\langle x^\star, y \rangle$ *versus* iteration are characterized by noisy monotonicity.

4.5.1.4   algebraic derivation of direction vector for convex iteration

In §3.2.2.1.2, the compressed sensing problem was precisely represented as a nonconvex difference of convex functions bounded below by 0

$$\begin{array}{ll} \text{find} & x \in \mathbb{R}^n \\ \text{subject to} & Ax = b \\ & x \succeq 0 \\ & \|x\|_0 \leq k \end{array} \quad \equiv \quad \begin{array}{ll} \underset{x \in \mathbb{R}^n}{\text{minimize}} & \|x\|_1 - \|x\|_k^n \\ \text{subject to} & Ax = b \\ & x \succeq 0 \end{array} \quad (529)$$

where convex k-largest norm $\|x\|_k^n$ is monotonic on $\mathbb{R}_+^n$. There we showed how (529) is equivalently stated in terms of gradients

$$\begin{array}{ll} \underset{x \in \mathbb{R}^n}{\text{minimize}} & \left\langle x, \nabla\|x\|_1 - \nabla\|x\|_k^n \right\rangle \\ \text{subject to} & Ax = b \\ & x \succeq 0 \end{array} \quad (805)$$

because

$$\|x\|_1 = x^{\mathrm{T}}\nabla\|x\|_1, \qquad \|x\|_k^n = x^{\mathrm{T}}\nabla\|x\|_k^n, \qquad x \succeq 0 \quad (806)$$

The objective function from (805) is a directional derivative (at x in direction x, §D.1.6, *confer* §D.1.4.1.1) of the objective function from (529) while the direction vector of convex iteration

$$y = \nabla\|x\|_1 - \nabla\|x\|_k^n \quad (807)$$

is an objective gradient where $\nabla \|x\|_1 = \nabla \mathbf{1}^{\mathrm{T}} x = \mathbf{1}$ under nonnegativity and

$$
\left.
\begin{array}{rl}
\nabla \|x\|_{\substack{n \\ k}} \;=\; \nabla z^{\mathrm{T}} x \;=\; & \underset{z \in \mathbb{R}^n}{\arg \text{maximize}} \quad z^{\mathrm{T}} x \\[4pt]
& \text{subject to} \quad 0 \preceq z \preceq \mathbf{1} \\[4pt]
& \qquad\qquad\;\; z^{\mathrm{T}} \mathbf{1} = k
\end{array}
\right\}, \qquad x \succeq 0 \qquad (532)
$$

is not unique. Substituting $\mathbf{1} - z \leftarrow z$ the direction vector becomes

$$
\begin{array}{rlcl}
y \;=\; \mathbf{1} - & \underset{z \in \mathbb{R}^n}{\arg \text{maximize}} \quad z^{\mathrm{T}} x & \leftarrow & \underset{z \in \mathbb{R}^n}{\arg \text{minimize}} \quad z^{\mathrm{T}} x \\[4pt]
& \text{subject to} \quad 0 \preceq z \preceq \mathbf{1} & & \text{subject to} \quad 0 \preceq z \preceq \mathbf{1} \\[4pt]
& \qquad\qquad\;\; z^{\mathrm{T}} \mathbf{1} = k & & \qquad\qquad\;\; z^{\mathrm{T}} \mathbf{1} = n - k
\end{array} \qquad (524)
$$

4.5.1.5 optimality conditions for compressed sensing

Now we see how global optimality conditions can be stated without reference to a dual problem: From conditions (468) for optimality of (529), it is necessary [62, §5.5.3] that

$$
\begin{array}{rl}
x^{\star} \succeq 0 & \quad (1) \\[4pt]
A x^{\star} = b & \quad (2) \\[4pt]
\nabla \|x^{\star}\|_1 - \nabla \|x^{\star}\|_{\substack{n \\ k}} + A^{\mathrm{T}} \nu^{\star} \succeq 0 & \quad (3) \\[4pt]
\langle \nabla \|x^{\star}\|_1 - \nabla \|x^{\star}\|_{\substack{n \\ k}} + A^{\mathrm{T}} \nu^{\star}, \, x^{\star} \rangle = 0 & \quad (4\ell)
\end{array} \qquad (808)
$$

These conditions must hold at any optimal solution (locally or globally). By (806), the fourth condition is identical to

$$
\|x^{\star}\|_1 - \|x^{\star}\|_{\substack{n \\ k}} + \nu^{\star \mathrm{T}} A x^{\star} = 0 \qquad (4\ell) \qquad (809)
$$

Because a 1-norm

$$
\|x\|_1 = \|x\|_{\substack{n \\ k}} + \|\pi(|x|)_{k+1:n}\|_1 \qquad (810)
$$

is separable into k largest and $n - k$ smallest absolute entries,

$$
\|\pi(|x|)_{k+1:n}\|_1 = 0 \;\Leftrightarrow\; \|x\|_0 \le k \qquad (4g) \qquad (811)
$$

is a necessary condition for global optimality. By assumption, matrix A is fat and $b \ne \mathbf{0} \;\Rightarrow\; A x^{\star} \ne \mathbf{0}$. This means $\nu^{\star} \in \mathcal{N}(A^{\mathrm{T}}) \subset \mathbb{R}^m$, and $\nu^{\star} = \mathbf{0}$ when A is full-rank. By definition, $\nabla \|x\|_1 \succeq \nabla \|x\|_{\substack{n \\ k}}$ always holds. Assuming existence of a cardinality-k solution, then only three of the four conditions are necessary and sufficient for global optimality of (529):

$$
\begin{array}{rl}
x^{\star} \succeq 0 & \quad (1) \\[4pt]
A x^{\star} = b & \quad (2) \\[4pt]
\|x^{\star}\|_1 - \|x^{\star}\|_{\substack{n \\ k}} = 0 & \quad (4g)
\end{array} \qquad (812)
$$

meaning, global optimality of a feasible solution to (529) is identified by a zero objective.

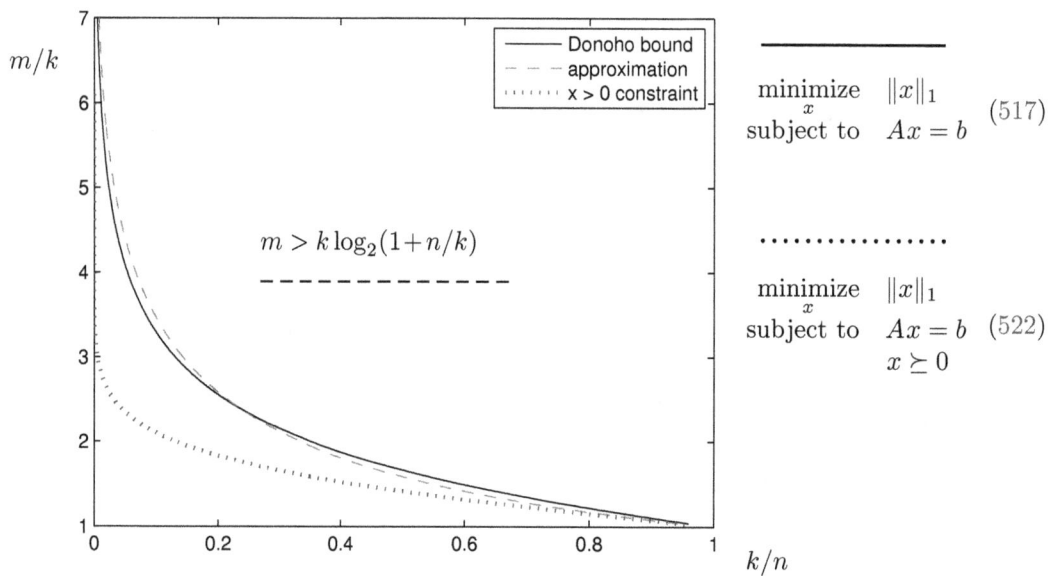

Figure 104: For Gaussian random matrix $A \in \mathbb{R}^{m \times n}$, graph illustrates Donoho/Tanner least lower bound on number of measurements m below which recovery of k-sparse n-length signal x by linear programming fails with overwhelming probability. Hard problems are below curve, but not the reverse; *id est*, failure above depends on proximity. Inequality demarcates *approximation* (dashed curve) empirically observed in [23]. Problems having nonnegativity constraint (dotted) are easier to solve. [126] [127]

4.5.1.5.1 Example. *Sparsest solution to $Ax = b$.* [71] [122]
(*confer* Example 4.5.2.0.4) Problem (712) has sparsest solution not easily recoverable by least 1-norm; *id est*, not by compressed sensing because of proximity to a theoretical lower bound on number of *measurements* m depicted in Figure 104: for $A \in \mathbb{R}^{m \times n}$

- Given data from Example 4.2.3.1.1, for $m = 3$, $n = 6$, $k = 1$

$$A = \begin{bmatrix} -1 & 1 & 8 & 1 & 1 & 0 \\ -3 & 2 & 8 & \frac{1}{2} & \frac{1}{3} & \frac{1}{2} - \frac{1}{3} \\ -9 & 4 & 8 & \frac{1}{4} & \frac{1}{9} & \frac{1}{4} - \frac{1}{9} \end{bmatrix}, \qquad b = \begin{bmatrix} 1 \\ \frac{1}{2} \\ \frac{1}{4} \end{bmatrix} \qquad (712)$$

the sparsest solution to classical linear equation $Ax = b$ is $x = e_4 \in \mathbb{R}^6$ (*confer* (725)).

Although the sparsest solution is recoverable by inspection, we discern it instead by convex iteration; namely, by iterating problem sequence (156) (524) on page 292. From the numerical data given, cardinality $\|x\|_0 = 1$ is expected. Iteration continues until $x^{\mathrm{T}}y$ vanishes (to within some numerical precision); *id est*, until desired cardinality is achieved. But this comes not without a stall.

Stalling, whose occurrence is sensitive to initial conditions of convex iteration, is a consequence of finding a local minimum of a multimodal objective $\langle x, y \rangle$ when regarded as simultaneously variable in x and y. (§3.8.0.0.3) Stalls are simply detected as fixed points x of infeasible cardinality, sometimes remedied by reinitializing direction vector y to a random positive state.

Bolstered by success in breaking out of a stall, we then apply convex iteration to 22,000 randomized problems:

- Given random data for $m = 3$, $n = 6$, $k = 1$, in MATLAB notation

  A $=$ randn(3, 6),   index $=$ round(5 $*$ rand(1)) $+$ 1,   b $=$ rand(1) $*$ A(:, index)   (813)

  the sparsest solution $x \propto e_{\mathrm{index}}$ is a scaled standard basis vector.

Without convex iteration or a nonnegativity constraint $x \succeq 0$, rate of failure for this minimal cardinality problem $Ax = b$ by 1-norm minimization of x is 22%. That failure rate drops to 6% with a nonnegativity constraint. If we then engage convex iteration, detect stalls, and randomly reinitialize the direction vector, failure rate drops to 0% but the amount of computation is approximately doubled.   □

Stalling is not an inevitable behavior. For some problem types (beyond mere $Ax = b$), convex iteration succeeds nearly all the time. Here is a cardinality problem, with noise, whose statement is just a bit more intricate but easy to solve in a few convex iterations:

4.5.1.5.2 Example. *Signal dropout.* [125, §6.2]
Signal dropout is an old problem; well studied from both an industrial and academic perspective. Essentially *dropout* means momentary loss or gap in a signal, while passing through some channel, caused by some man-made or natural phenomenon. The signal lost is assumed completely destroyed somehow. What remains within the time-gap is system or idle channel noise. The signal could be voice over Internet protocol (VoIP), for

example, audio data from a *compact disc* (CD) or video data from a digital video disc (DVD), a television transmission over cable or the airwaves, or a typically ravaged cell phone communication, *etcetera*.

Here we consider signal dropout in a discrete-time signal corrupted by additive white noise assumed uncorrelated to the signal. The linear channel is assumed to introduce no filtering. We create a discretized windowed signal for this example by positively combining k randomly chosen vectors from a *discrete cosine transform* (DCT) basis denoted $\Psi \in \mathbb{R}^{n \times n}$. Frequency increases, in the Fourier sense, from DC toward Nyquist as column index of basis Ψ increases. Otherwise, details of the basis are unimportant except for its orthogonality $\Psi^{\mathrm{T}} = \Psi^{-1}$. Transmitted signal is denoted

$$s = \Psi z \in \mathbb{R}^n \tag{814}$$

whose upper bound on DCT basis coefficient cardinality card$\,z \leq k$ is assumed known;[4.38] hence a critical assumption: transmitted signal s is sparsely supported ($k < n$) on the DCT basis. It is further assumed that nonzero signal coefficients in vector z place each chosen basis vector above the noise floor.

We also assume that the gap's beginning and ending in time are precisely localized to within a sample; *id est*, index ℓ locates the last sample prior to the gap's onset, while index $n-\ell+1$ locates the first sample subsequent to the gap: for rectangularly windowed received signal g possessing a time-gap loss and additive noise $\eta \in \mathbb{R}^n$

$$g = \begin{bmatrix} s_{1:\ell} & + & \eta_{1:\ell} \\ & & \eta_{\ell+1:n-\ell} \\ s_{n-\ell+1:n} & + & \eta_{n-\ell+1:n} \end{bmatrix} \in \mathbb{R}^n \tag{815}$$

The window is thereby centered on the gap and short enough so that the DCT spectrum of signal s can be assumed static over the window's duration n. Signal to noise ratio within this window is defined

$$\mathrm{SNR} \triangleq 20 \log \frac{\left\| \begin{bmatrix} s_{1:\ell} \\ s_{n-\ell+1:n} \end{bmatrix} \right\|}{\|\eta\|} \tag{816}$$

In absence of noise, knowing the signal DCT basis and having a good estimate of basis coefficient cardinality makes perfectly reconstructing gap-loss easy: it amounts to solving a linear system of equations and requires little or no optimization; with caveat, number of equations exceeds cardinality of signal representation (roughly $\ell \geq k$) with respect to DCT basis.

But addition of a significant amount of noise η increases level of difficulty dramatically; a 1-norm based method of reducing cardinality, for example, almost always returns DCT basis coefficients numbering in excess of minimal cardinality. We speculate that is because signal cardinality 2ℓ becomes the predominant cardinality. DCT basis coefficient cardinality is an explicit constraint to the optimization problem we shall pose: In presence of noise, constraints equating reconstructed signal f to received signal g are not possible.

[4.38]This simplifies exposition, although it may be an unrealistic assumption in many applications.

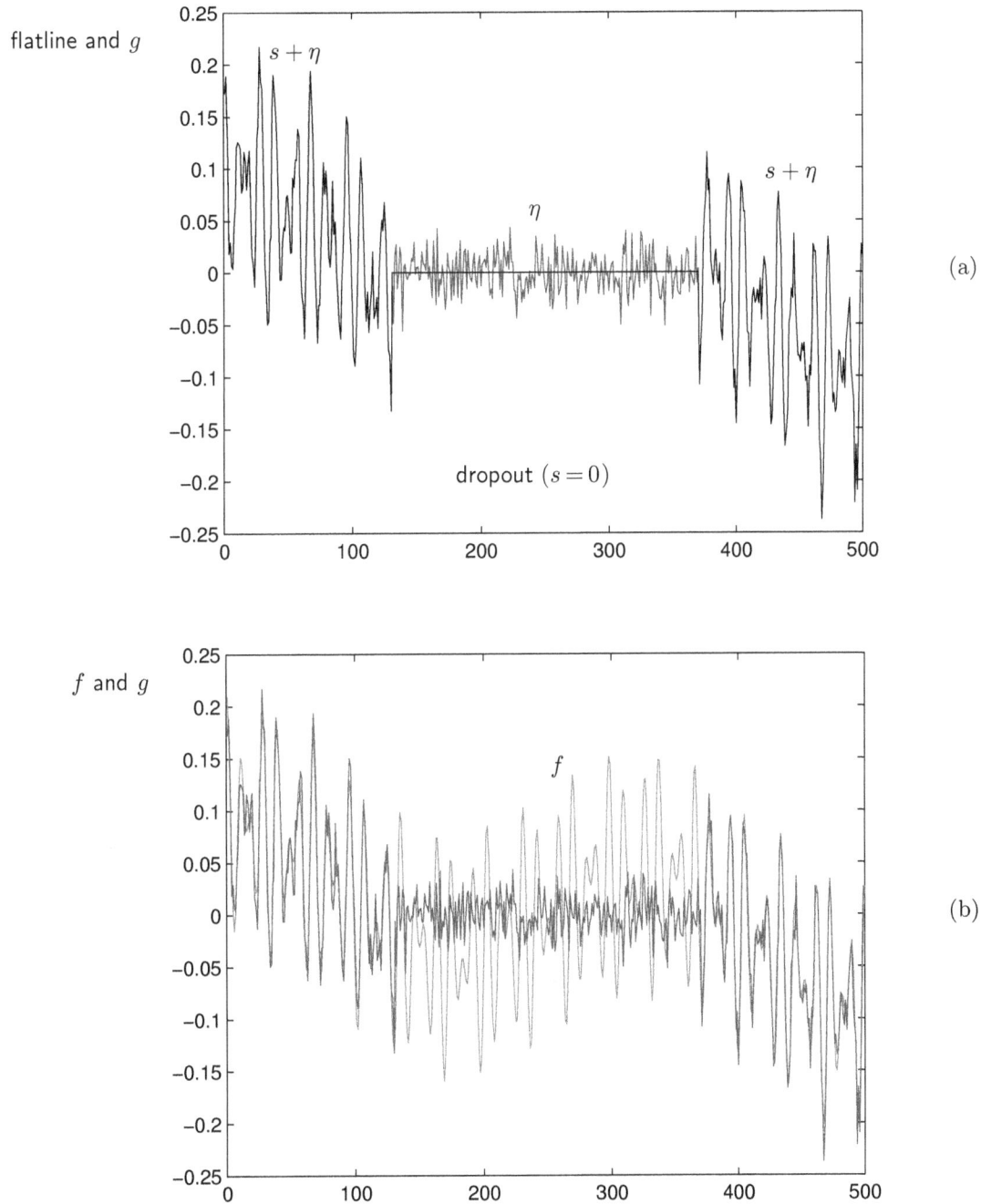

Figure 105: (a) Signal dropout in signal s corrupted by noise η (SNR $=$ 10dB, $g = s + \eta$). Flatline indicates duration of signal dropout. (b) Reconstructed signal f (red) overlaid with corrupted signal g.

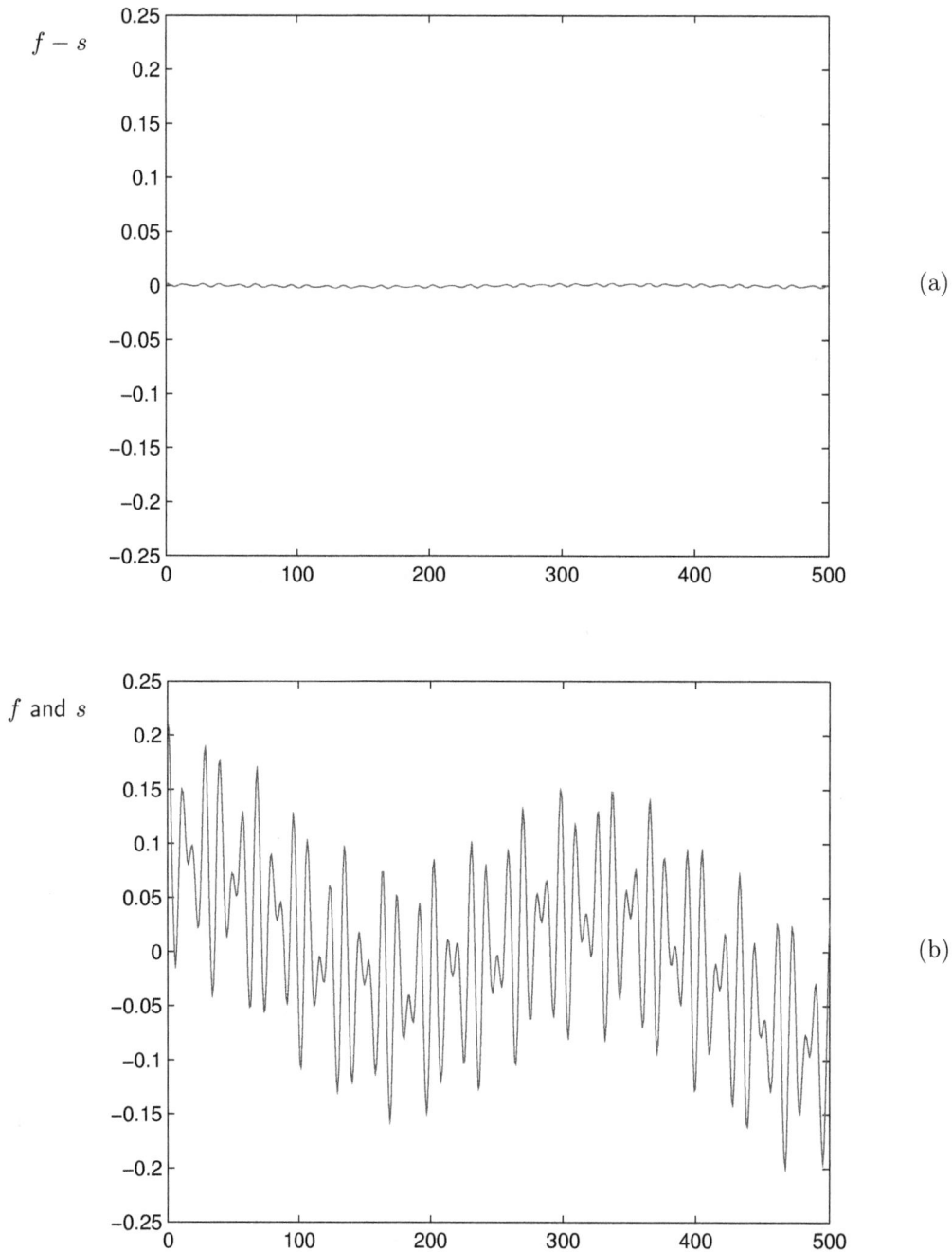

Figure 106: (a) Error signal power (reconstruction f less original noiseless signal s) is 36dB below s. (b) Original signal s overlaid with reconstruction f (red) from signal g having dropout plus noise.

We can instead formulate the dropout recovery problem as a best approximation:

$$\begin{array}{ll} \underset{x \in \mathbb{R}^n}{\text{minimize}} & \left\| \begin{bmatrix} f_{1:\ell} - g_{1:\ell} \\ f_{n-\ell+1:n} - g_{n-\ell+1:n} \end{bmatrix} \right\| \\ \text{subject to} & f = \Psi x \\ & x \succeq 0 \\ & \text{card}\, x \leq k \end{array} \qquad (817)$$

We propose solving this nonconvex problem (817) by moving the cardinality constraint to the objective as a regularization term as explained in §4.5; *id est*, by iteration of two convex problems until convergence:

$$\begin{array}{ll} \underset{x \in \mathbb{R}^n}{\text{minimize}} & \langle x \,,\, y \rangle + \left\| \begin{bmatrix} f_{1:\ell} - g_{1:\ell} \\ f_{n-\ell+1:n} - g_{n-\ell+1:n} \end{bmatrix} \right\| \\ \text{subject to} & f = \Psi x \\ & x \succeq 0 \end{array} \qquad (818)$$

and

$$\begin{array}{ll} \underset{y \in \mathbb{R}^n}{\text{minimize}} & \langle x^\star \,,\, y \rangle \\ \text{subject to} & 0 \preceq y \preceq \mathbf{1} \\ & y^{\mathrm{T}} \mathbf{1} = n - k \end{array} \qquad (524)$$

Signal cardinality 2ℓ is implicit to the problem statement. When number of samples in the dropout region exceeds half the window size, then that deficient cardinality of signal remaining becomes a source of degradation to reconstruction in presence of noise. Thus, by observation, we divine a reconstruction rule for this signal dropout problem to attain good noise suppression: ℓ must exceed a maximum of cardinality bounds; $2\ell \geq \max\{2k\,,\, n/2\}$.

Figure **105** and Figure **106** show one realization of this dropout problem. Original signal s is created by adding four ($k\!=\!4$) randomly selected DCT basis vectors, from Ψ ($n\!=\!500$ in this example), whose amplitudes are randomly selected from a uniform distribution above the noise floor; in the interval $[10^{-10/20},\, 1]$. Then a 240-sample dropout is realized ($\ell\!=\!130$) and Gaussian noise η added to make corrupted signal g (from which a best approximation f will be made) having 10dB signal to noise ratio (816). The time gap contains much noise, as apparent from Figure **105**a. But in only a few iterations (818) (524), original signal s is recovered with relative error power 36dB down; illustrated in Figure **106**. Correct cardinality is also recovered (card $x =$ card z) along with the basis vector indices used to make original signal s. Approximation error is due to DCT basis coefficient estimate error. When this experiment is repeated 1000 times on noisy signals averaging 10dB SNR, the correct cardinality and indices are recovered 99% of the time with average relative error power 30dB down. Without noise, we get perfect reconstruction in one iteration. [394, Matlab code] $\qquad\qquad\square$

4.5.1.6 Compressed sensing geometry with a nonnegative variable

It is well known that cardinality problem (529), on page 199, is easier to solve by linear programming when variable x is nonnegatively constrained than when not; *e.g.*, Figure **72**, Figure **104**. We postulate a simple geometrical explanation:

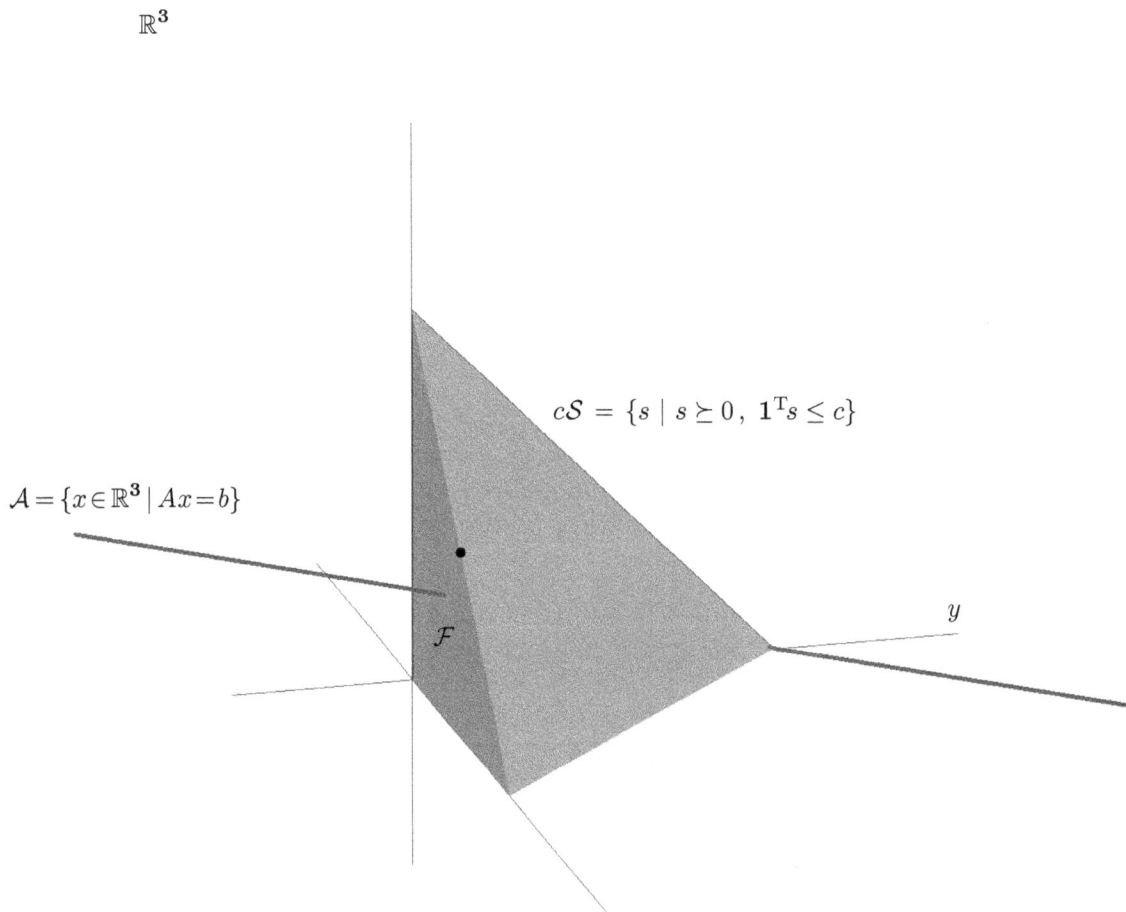

Figure 107: Simplex $\mathcal{S}$ is convex hull of origin and all cardinality-1 nonnegative vectors of unit norm (its vertices). Line $\mathcal{A}$, intersecting two-dimensional (cardinality-2) face $\mathcal{F}$ of nonnegative simplex $c\mathcal{S}$, emerges from $c\mathcal{S}$ at a cardinality-1 vertex. $\mathcal{S}$ equals nonnegative orthant $\mathbb{R}^3_+ \cap$ 1-norm ball $\mathcal{B}_1$ (Figure 71). Kissing point achieved when $\bullet$ (on edge) meets $\mathcal{A}$ as simplex contracts (as scalar c diminishes) under optimization (522).

Figure **71** illustrates 1-norm ball $\mathcal{B}_1$ in $\mathbb{R}^3$ and affine subset $\mathcal{A}$ defined $\{x \in \mathbb{R}^3 \mid Ax = b\}$. Prototypical compressed sensing problem, for $A \in \mathbb{R}^{m \times n}$

$$
\begin{array}{ll}
\underset{x}{\text{minimize}} & \|x\|_1 \\
\text{subject to} & Ax = b
\end{array}
\tag{517}
$$

is solved when the 1-norm ball $\mathcal{B}_1$ kisses the affine subset.

If variable x is constrained to the nonnegative orthant

$$
\begin{array}{ll}
\underset{x \in \mathbb{R}^n}{\text{minimize}} & \|x\|_1 \\
\text{subject to} & Ax = b \\
& x \succeq 0
\end{array}
\equiv
\begin{array}{ll}
\underset{x \in \mathbb{R}^n}{\text{minimize}} & \mathbf{1}^{\mathrm{T}} x \\
\text{subject to} & Ax = b \\
& x \succeq 0
\end{array}
\equiv
\begin{array}{ll}
\underset{c \in \mathbb{R},\, x \in \mathbb{R}^n}{\text{minimize}} & c \\
\text{subject to} & Ax = b \\
& x \in c\mathcal{S}
\end{array}
\tag{522}
$$

then 1-norm ball $\mathcal{B}_1$ becomes nonnegative simplex $\mathcal{S}$ in Figure **107** where

$$
c\mathcal{S} = \{[\,I \in \mathbb{R}^{n \times n} \quad \mathbf{0} \in \mathbb{R}^n\,]a \mid a^{\mathrm{T}}\mathbf{1} = c,\ a \succeq 0\} = \{s \mid s \succeq 0,\ \mathbf{1}^{\mathrm{T}}s \le c\}
\tag{819}
$$

Nonnegative simplex $\mathcal{S}$ is the convex hull of its vertices. All $n+1$ vertices of $\mathcal{S}$ are constituted by standard basis vectors and the origin. In other words, all its nonzero extreme points are cardinality-1.

Affine subset $\mathcal{A}$ kisses nonnegative simplex $c^\star \mathcal{S}$ at optimality of (522). A kissing point is achieved at $x^\star$ for optimal $c^\star$ as $\mathcal{B}_1$ or $\mathcal{S}$ contracts. Whereas 1-norm ball $\mathcal{B}_1$ has only six vertices in $\mathbb{R}^3$ corresponding to cardinality-1 solutions, simplex $\mathcal{S}$ has three edges (along the Cartesian axes) containing an infinity of cardinality-1 solutions. And whereas $\mathcal{B}_1$ has twelve edges containing cardinality-2 solutions, $\mathcal{S}$ has three (out of total four) facets constituting cardinality-2 solutions. In other words, likelihood of a low-cardinality solution is higher by kissing nonnegative simplex $\mathcal{S}$ (522) than by kissing 1-norm ball $\mathcal{B}_1$ (517) because facial dimension (corresponding to given cardinality) is higher in $\mathcal{S}$.

Empirically, this observation also holds in other Euclidean dimensions (Figure **72**, Figure **104**).

4.5.1.7 cardinality-1 compressed sensing problem always solvable

In the special case of cardinality-1 feasible solution to nonnegative compressed sensing problem (522), there is a geometrical interpretation that leads to an algorithm.

Figure **107** illustrates a cardinality-1 feasible solution to problem (522) in $\mathbb{R}^3$; a vertex solution. But *first-octant* $\mathcal{S}$ of 1-norm ball $\mathcal{B}_1$ does not kiss line $\mathcal{A}$; which would be an optimality condition. How can we perform optimization and make $\mathcal{A}$ intersect $\mathcal{S}$ at a vertex? Assuming that nonnegative cardinality-1 solutions exist in the feasible set, it so happens:

4.5.1.7.1 Algorithm. *Deprecation.*

Columns of *measurement matrix* A, corresponding to high cardinality solution of (522)[4.39] found by Simplex method [96], may be *deprecated* and the problem solved

---

[4.39]Because signed compressed sensing problem (517) can be equivalently expressed in a nonnegative variable, as we learned in Example 3.2.0.0.1 (p.196), and because a cardinality-1 constraint in (517) transforms to a cardinality-1 constraint in its nonnegative equivalent (521), then this cardinality-1 recursive reconstruction algorithm continues to hold for a signed variable as in (517).

again with those columns missing. Such columns are recursively removed from A until a
cardinality-1 solution is found.									¶

This algorithm intimates that either a solution to problem (522) is cardinality-1 or
column indices of A, corresponding to a higher cardinality solution, do not intersect that
index corresponding to a cardinality-1 feasible solution.

When problem (522) is first solved, in the example of Figure 107, solution is
cardinality-2 at a kissing point on that edge of simplex $c\mathcal{S}$ indicated by •. Imagining
that the corresponding cardinality-2 face $\mathcal{F}$ has collapsed, as a result of zeroing those two
extreme points whose convex hull constructs that same edge • of $\mathcal{F}$, then the simplex
collapses to a line segment along the y axis. When that line segment kisses $\mathcal{A}$, then the
cardinality-1 vertex solution illustrated has been found.[4.40]

4.5.1.7.2 Proof (pending). *Deprecation algorithm* 4.5.1.7.1.

We require proof that a cardinality-1 feasible solution to (522) cannot exist within
a higher cardinality optimal solution found by Simplex method; for only then can
corresponding columns of A be eliminated without precluding cardinality-1 at optimality
of the deprecated problem. Crucial is the Simplex method of solution because then an
optimal solution is guaranteed to reside at a vertex of the feasible set. [96, p.158] [15, p.2]
									■

Although it is more efficient (compared with our algorithm) to search over individual
columns of matrix A for a cardinality-1 solution known *a priori* to exist, tables are turned
when cardinality exceeds 1:

4.5.2   cardinality-k geometric presolver

This idea of deprecating columns has foundation in convex cone theory. (§2.13.4.3)
Removing columns (and rows)[4.41] from $A\in\mathbb{R}^{m\times n}$ in a linear program like (522) (§3.2)
is known in the industry as *presolving*;[4.42] the elimination of redundant constraints and
identically zero variables prior to numerical solution. We offer a different and geometric
presolver:

Two interpretations of the constraints from problem (522) are realized in Figure 108.
Assuming that a cardinality-k solution exists and matrix A describes a pointed polyhedral
cone $\mathcal{K}=\{Ax\mid x\succeq0\}$, as in Figure 108b, columns are removed from A if they do not
belong to the smallest face $\mathcal{F}$ of $\mathcal{K}$ containing vector b; those columns correspond to
0-entries in variable vector x (and *vice versa*). Generators of that smallest face always hold
a minimal cardinality solution, in other words, because a generator outside the smallest
face (having positive coefficient) would violate the assumption that b belongs to that face.

[4.40]A similar argument holds for any orientation of line $\mathcal{A}$ and cardinality-1 point of emergence from
simplex $c\mathcal{S}$. This cardinality-1 reconstruction algorithm also holds more generally when affine subset $\mathcal{A}$
has any higher dimension $n-m$.
[4.41]Rows of matrix A are removed based upon linear dependence. Assuming $b\in\mathcal{R}(A)$, corresponding
entries of vector b may also be removed without loss of generality.
[4.42]The conic technique proposed here can exceed performance of the best industrial presolvers in terms
of number of columns removed, but not in execution time. This geometric presolver becomes attractive
when a linear or *integer program* is not solvable by other means; perhaps because of sheer dimension.

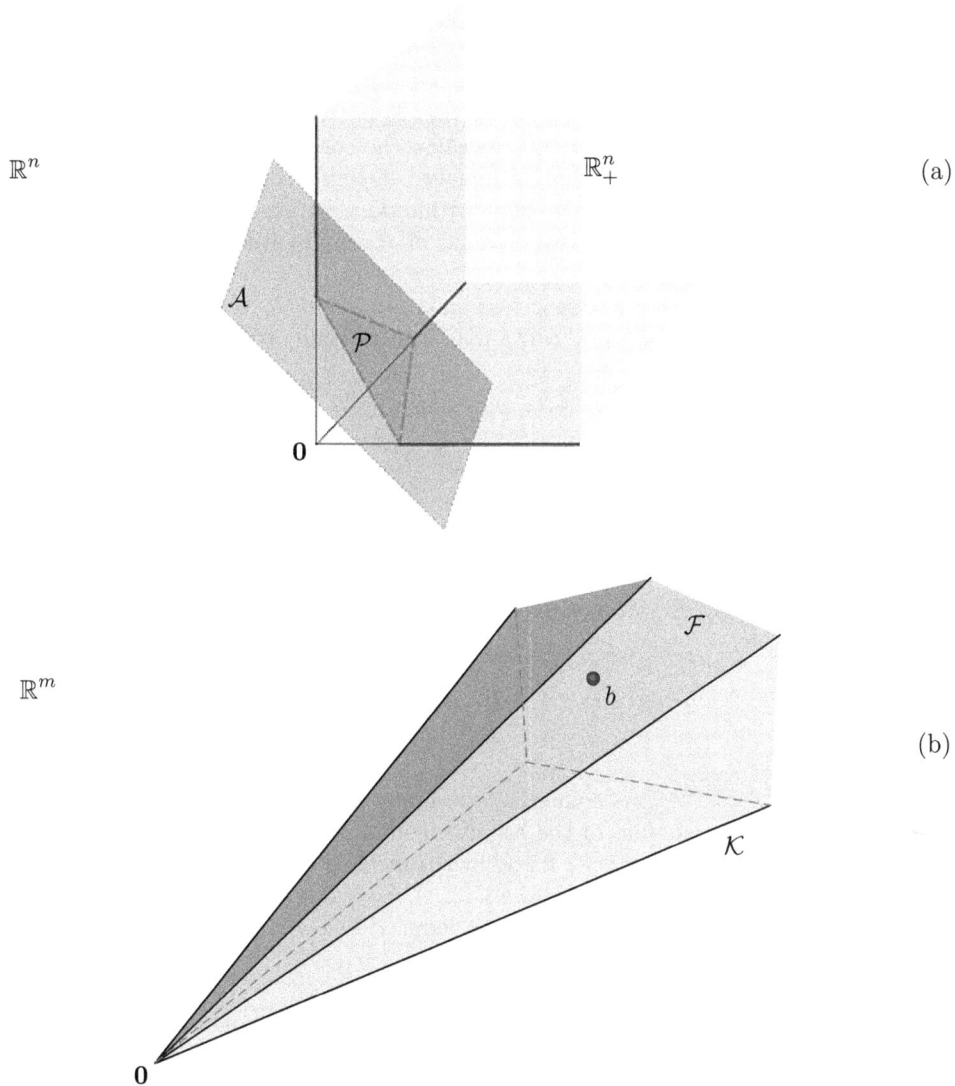

Figure 108: Constraint interpretations: **(a)** Halfspace-description of feasible set in problem (522) is a polyhedron $\mathcal{P}$ formed by intersection of nonnegative orthant $\mathbb{R}^n_+$ with hyperplanes $\mathcal{A}$ prescribed by equality constraint. (Drawing by Pedro Sánchez) **(b)** Vertex-description of constraints in problem (522): point b belongs to polyhedral cone $\mathcal{K} = \{Ax \mid x \succeq 0\}$. Number of extreme directions in $\mathcal{K}$ may exceed dimensionality of ambient space.

Benefit accrues when vector b does not belong to relative interior of $\mathcal{K}$; there would be no columns to remove were $b \in \operatorname{rel} \operatorname{int} \mathcal{K}$ since the smallest face becomes cone $\mathcal{K}$ itself (Example 4.5.2.0.4). Were b an extreme direction, at the other end of the spectrum, then the smallest face is an edge that is a ray containing b; this geometrically describes a cardinality-1 case where all columns, save one, would be removed from A.

When vector b resides in a face $\mathcal{F}$ of $\mathcal{K}$ that is not cone $\mathcal{K}$ itself, benefit is realized as a reduction in computational intensity because the consequent equivalent problem has smaller dimension. Number of columns removed depends completely on geometry of a given problem; particularly, location of b within $\mathcal{K}$. In the example of Figure 108b, interpreted literally in $\mathbb{R}^3$, all but two columns of A are discarded by our presolver when b belongs to facet $\mathcal{F}$.

There are always exactly n linear feasibility problems to solve in order to discern generators of the smallest face of $\mathcal{K}$ containing b; the topic of §2.13.4.3.[4.43]

4.5.2.0.3   Exercise.   *Minimal cardinality generators.*
Prove that generators of the smallest face $\mathcal{F}$ of $\mathcal{K} = \{Ax \mid x \succeq 0\}$ containing vector b always hold a minimal cardinality solution to $Ax = b$.   ▼

4.5.2.0.4   Example.   *Presolving for cardinality-2 solution to $Ax = b$.*
(*confer* Example 4.5.1.5.1)   Again   taking   data   from   Example 4.2.3.1.1,   for $m=3$, $n=6$, $k=2$ ($A \in \mathbb{R}^{m \times n}$, desired cardinality of x is k)

$$A = \begin{bmatrix} -1 & 1 & 8 & 1 & 1 & 0 \\ -3 & 2 & 8 & \frac{1}{2} & \frac{1}{3} & \frac{1}{2}-\frac{1}{3} \\ -9 & 4 & 8 & \frac{1}{4} & \frac{1}{9} & \frac{1}{4}-\frac{1}{9} \end{bmatrix}, \qquad b = \begin{bmatrix} 1 \\ \frac{1}{2} \\ \frac{1}{4} \end{bmatrix} \qquad (712)$$

proper cone $\mathcal{K} = \{Ax \mid x \succeq 0\}$ is pointed as proven by method of §2.12.2.2. A cardinality-2 solution is known to exist; sum of the last two columns of matrix A. Generators of the smallest face that contains vector b, found by the method in Example 2.13.4.3.1, comprise the entire A matrix because $b \in \operatorname{int} \mathcal{K}$ (§2.13.4.2.4). So geometry of this particular problem does not permit number of generators to be reduced below n by discerning the smallest face.[4.44]   □

There is wondrous bonus to presolving when a constraint matrix is sparse. After columns are removed by theory of convex cones (finding the smallest face), some remaining rows may become $\mathbf{0}^{\mathrm{T}}$, identical to other rows, or nonnegative. When nonnegative rows appear in an equality constraint to $\mathbf{0}$, all nonnegative variables corresponding to nonnegative entries in those rows must vanish (§A.7.1); meaning, more columns may be removed. Once rows and columns have been removed from a constraint matrix, even more rows and columns may be removed by repeating the presolver procedure.

[4.43]Comparison of computational intensity for this conic presolver to a brute force search would pit combinatorial complexity, a binomial coefficient $\propto \binom{n}{k}$, against polynomial complexity; n linear feasibility problems plus numerical solution of the presolved problem.
[4.44]But a canonical set of conically independent generators of $\mathcal{K}$ comprise only the first two and last two columns of A.

4.5.3　constraining cardinality of signed variable

Now consider a feasibility problem equivalent to the classical problem from linear algebra $Ax = b$, but with an upper bound k on cardinality $\|x\|_0$: for vector $b \in \mathcal{R}(A)$

$$
\begin{array}{ll}
\text{find} & x \in \mathbb{R}^n \\
\text{subject to} & Ax = b \\
& \|x\|_0 \leq k
\end{array}
\tag{820}
$$

where $\|x\|_0 \leq k$ means vector x has at most k nonzero entries; such a vector is presumed existent in the feasible set. Convex iteration (§4.5.1) utilizes a nonnegative variable; so absolute value $|x|$ is needed here. We propose that nonconvex problem (820) can be equivalently written as a sequence of convex problems that move the cardinality constraint to the objective:

$$
\begin{array}{ll}
\underset{x \in \mathbb{R}^n}{\text{minimize}} & \langle |x|, y \rangle \\
\text{subject to} & Ax = b
\end{array}
\equiv
\begin{array}{ll}
\underset{x \in \mathbb{R}^n, \, t \in \mathbb{R}^n}{\text{minimize}} & \langle t, y + \varepsilon \mathbf{1} \rangle \\
\text{subject to} & Ax = b \\
& -t \preceq x \preceq t
\end{array}
\tag{821}
$$

$$
\begin{array}{ll}
\underset{y \in \mathbb{R}^n}{\text{minimize}} & \langle t^\star, y + \varepsilon \mathbf{1} \rangle \\
\text{subject to} & 0 \preceq y \preceq \mathbf{1} \\
& y^{\mathrm{T}} \mathbf{1} = n - k
\end{array}
\tag{524}
$$

where ε is a relatively small positive constant. This sequence is iterated until a direction vector y is found that makes $|x^\star|^{\mathrm{T}} y^\star$ vanish. The term $\langle t, \varepsilon \mathbf{1} \rangle$ in (821) is necessary to determine absolute value $|x^\star| = t^\star$ (§3.2) because vector y can have zero-valued entries. By initializing y to $(1-\varepsilon)\mathbf{1}$, the first iteration of problem (821) is a 1-norm problem (513); *id est*,

$$
\begin{array}{ll}
\underset{x \in \mathbb{R}^n, \, t \in \mathbb{R}^n}{\text{minimize}} & \langle t, \mathbf{1} \rangle \\
\text{subject to} & Ax = b \\
& -t \preceq x \preceq t
\end{array}
\equiv
\begin{array}{ll}
\underset{x \in \mathbb{R}^n}{\text{minimize}} & \|x\|_1 \\
\text{subject to} & Ax = b
\end{array}
\tag{517}
$$

Subsequent iterations of problem (821) engaging cardinality term $\langle t, y \rangle$ can be interpreted as corrections to this 1-norm problem leading to a 0-norm solution; vector y can be interpreted as a direction of search.

4.5.3.1　local convergence

As before (§4.5.1.3), convex iteration (821) (524) always converges to a locally optimal solution; a fixed point of possibly infeasible cardinality.

4.5.3.2　simple variations on a signed variable

Several useful equivalents to linear programs (821) (524) are easily devised, but their geometrical interpretation is not as apparent: *e.g.*, equivalent in the limit $\varepsilon \to 0^+$

$$
\begin{array}{ll}
\underset{x \in \mathbb{R}^n, \, t \in \mathbb{R}^n}{\text{minimize}} & \langle t, y \rangle \\
\text{subject to} & Ax = b \\
& -t \preceq x \preceq t
\end{array}
\tag{822}
$$

$$\begin{array}{ll}\underset{y\in\mathbb{R}^n}{\text{minimize}} & \langle |x^\star| , y\rangle \\ \text{subject to} & 0 \preceq y \preceq \mathbf{1} \\ & y^{\mathrm{T}}\mathbf{1} = n - k \end{array} \qquad (524)$$

We get another equivalent to linear programs (821) (524), in the limit, by interpreting problem (517) as infimum to a vertex-description of the 1-norm ball (Figure 71, Example 3.2.0.0.1, *confer* (516)):

$$\begin{array}{ll}\underset{x\in\mathbb{R}^n}{\text{minimize}} & \|x\|_1 \\ \text{subject to} & Ax = b \end{array} \equiv \begin{array}{ll}\underset{a\in\mathbb{R}^{2n}}{\text{minimize}} & \langle a , y\rangle \\ \text{subject to} & [A \;\; -A]a = b \\ & a \succeq 0 \end{array} \qquad (823)$$

$$\begin{array}{ll}\underset{y\in\mathbb{R}^{2n}}{\text{minimize}} & \langle a^\star , y\rangle \\ \text{subject to} & 0 \preceq y \preceq \mathbf{1} \\ & y^{\mathrm{T}}\mathbf{1} = 2n - k \end{array} \qquad (524)$$

where $x^\star = [I \;\; -I]a^\star$; from which it may be construed that any vector 1-norm minimization problem has equivalent expression in a nonnegative variable.

4.6 Cardinality and rank constraint examples

4.6.0.0.1   Example.   *Projection on ellipsoid boundary.*   [53]
[152, §5.1] [253, §2] Consider classical linear equation $Ax = b$ but with constraint on norm of solution x, given matrices C, fat A, and vector $b\in\mathcal{R}(A)$

$$\begin{array}{ll}\text{find} & x \in \mathbb{R}^N \\ \text{subject to} & Ax = b \\ & \|Cx\| = 1 \end{array} \qquad (824)$$

The set $\{x \mid \|Cx\| = 1\}$ (2) describes an ellipsoid boundary (Figure 13).   This is a nonconvex problem because solution is constrained to that boundary. Assign

$$G = \begin{bmatrix} Cx \\ 1 \end{bmatrix} \begin{bmatrix} x^{\mathrm{T}}C^{\mathrm{T}} & 1 \end{bmatrix} = \begin{bmatrix} X & Cx \\ x^{\mathrm{T}}C^{\mathrm{T}} & 1 \end{bmatrix} \triangleq \begin{bmatrix} Cxx^{\mathrm{T}}C^{\mathrm{T}} & Cx \\ x^{\mathrm{T}}C^{\mathrm{T}} & 1 \end{bmatrix} \in \mathbb{S}^{N+1} \qquad (825)$$

Any rank-1 solution must have this form. (§B.1.0.2) Ellipsoidally constrained feasibility problem (824) is equivalent to:

$$\begin{array}{ll}\underset{X\in\mathbb{S}^N}{\text{find}} & x \in \mathbb{R}^N \\ \text{subject to} & Ax = b \\ & G = \begin{bmatrix} X & Cx \\ x^{\mathrm{T}}C^{\mathrm{T}} & 1 \end{bmatrix} (\succeq 0) \\ & \text{rank}\, G = 1 \\ & \text{tr}\, X = 1 \end{array} \qquad (826)$$

This is transformed to an equivalent convex problem by moving the rank constraint to the objective: We iterate solution of

$$
\begin{aligned}
\underset{X \in \mathbb{S}^N,\ x \in \mathbb{R}^N}{\text{minimize}} \quad & \langle G,\, Y \rangle \\
\text{subject to} \quad & Ax = b \\
& G = \begin{bmatrix} X & Cx \\ x^{\mathrm{T}} C^{\mathrm{T}} & 1 \end{bmatrix} \succeq 0 \\
& \operatorname{tr} X = 1
\end{aligned}
\tag{827}
$$

with

$$
\begin{aligned}
\underset{Y \in \mathbb{S}^{N+1}}{\text{minimize}} \quad & \langle G^{\star},\, Y \rangle \\
\text{subject to} \quad & 0 \preceq Y \preceq I \\
& \operatorname{tr} Y = N
\end{aligned}
\tag{828}
$$

until convergence. Direction matrix $Y \in \mathbb{S}^{N+1}$, initially $\mathbf{0}$, regulates rank. (1791a) Singular value decomposition $G^{\star} = U \Sigma Q^{\mathrm{T}} \in \mathbb{S}_+^{N+1}$ (§A.6) provides a new direction matrix $Y = U(:, 2\!:\!N\!+\!1) U(:, 2\!:\!N\!+\!1)^{\mathrm{T}}$ that optimally solves (828) at each iteration. An optimal solution to (824) is thereby found in a few iterations, making convex problem (827) its equivalent.

It remains possible for the iteration to stall; were a rank-1 G matrix not found. In that case, the current search direction is momentarily reversed with an added randomized element:

$$
\texttt{Y = -U(:, 2:N+1) * (U(:, 2:N+1)' + randn(N,1) * U(:,1)')}
\tag{829}
$$

in MATLAB notation. This heuristic is quite effective for problem (824) which is exceptionally easy to solve by convex iteration.

When $b \notin \mathcal{R}(A)$ then problem (824) must be restated as a projection:

$$
\begin{aligned}
\underset{x \in \mathbb{R}^N}{\text{minimize}} \quad & \| Ax - b \| \\
\text{subject to} \quad & \| Cx \| = 1
\end{aligned}
\tag{830}
$$

This is a projection of point b on an ellipsoid boundary because any affine transformation of an ellipsoid remains an ellipsoid. Problem (827) in turn becomes

$$
\begin{aligned}
\underset{X \in \mathbb{S}^N,\ x \in \mathbb{R}^N}{\text{minimize}} \quad & \langle G,\, Y \rangle + \| Ax - b \| \\
\text{subject to} \quad & G = \begin{bmatrix} X & Cx \\ x^{\mathrm{T}} C^{\mathrm{T}} & 1 \end{bmatrix} \succeq 0 \\
& \operatorname{tr} X = 1
\end{aligned}
\tag{831}
$$

We iterate this with calculation (828) of direction matrix Y as before until a rank-1 G matrix is found. $\qquad\qquad\square$

4.6.0.0.2 Example. *Orthonormal Procrustes.* [53]
Example 4.6.0.0.1 is extensible. An orthonormal matrix $Q \in \mathbb{R}^{n \times p}$ is characterized $Q^{\mathrm{T}} Q = I$. Consider the particular case $Q = [\, x \ \ y \,] \in \mathbb{R}^{n \times 2}$ as variable to a Procrustes problem (§C.3): given $A \in \mathbb{R}^{m \times n}$ and $B \in \mathbb{R}^{m \times 2}$

$$
\begin{aligned}
\underset{Q \in \mathbb{R}^{n \times 2}}{\text{minimize}} \quad & \| AQ - B \|_{\mathrm{F}} \\
\text{subject to} \quad & Q^{\mathrm{T}} Q = I
\end{aligned}
\tag{832}
$$

which is nonconvex. By vectorizing matrix Q we can make the assignment:

$$G = \begin{bmatrix} x \\ y \\ 1 \end{bmatrix} \begin{bmatrix} x^{\mathrm{T}} & y^{\mathrm{T}} & 1 \end{bmatrix} = \begin{bmatrix} X & Z & x \\ Z^{\mathrm{T}} & Y & y \\ x^{\mathrm{T}} & y^{\mathrm{T}} & 1 \end{bmatrix} \triangleq \begin{bmatrix} xx^{\mathrm{T}} & xy^{\mathrm{T}} & x \\ yx^{\mathrm{T}} & yy^{\mathrm{T}} & y \\ x^{\mathrm{T}} & y^{\mathrm{T}} & 1 \end{bmatrix} \in \mathbb{S}^{2n+1} \qquad (833)$$

Now orthonormal Procrustes problem (832) can be equivalently restated:

$$\begin{array}{ll} \underset{X,\,Y\in\mathbb{S},\,Z,\,x,\,y}{\text{minimize}} & \|A[x \ \ y] - B\|_{\mathrm{F}} \\[2mm] \text{subject to} & G = \begin{bmatrix} X & Z & x \\ Z^{\mathrm{T}} & Y & y \\ x^{\mathrm{T}} & y^{\mathrm{T}} & 1 \end{bmatrix} (\succeq 0) \\[4mm] & \operatorname{rank} G = 1 \\ & \operatorname{tr} X = 1 \\ & \operatorname{tr} Y = 1 \\ & \operatorname{tr} Z = 0 \end{array} \qquad (834)$$

To solve this, we form the convex problem sequence:

$$\begin{array}{ll} \underset{X,\,Y,\,Z,\,x,\,y}{\text{minimize}} & \|A[x \ \ y] - B\|_{\mathrm{F}} + \langle G,\,W \rangle \\[2mm] \text{subject to} & G = \begin{bmatrix} X & Z & x \\ Z^{\mathrm{T}} & Y & y \\ x^{\mathrm{T}} & y^{\mathrm{T}} & 1 \end{bmatrix} \succeq 0 \\[4mm] & \operatorname{tr} X = 1 \\ & \operatorname{tr} Y = 1 \\ & \operatorname{tr} Z = 0 \end{array} \qquad (835)$$

and

$$\begin{array}{ll} \underset{W\in\mathbb{S}^{2n+1}}{\text{minimize}} & \langle G^{\star},\,W \rangle \\[2mm] \text{subject to} & 0 \preceq W \preceq I \\ & \operatorname{tr} W = 2n \end{array} \qquad (836)$$

which has an optimal solution W that is known in closed form (p.567). These two problems are iterated until convergence and a rank-1 G matrix is found. A good initial value for direction matrix W is $\mathbf{0}$. Optimal $Q^{\star}$ equals $[x^{\star} \ \ y^{\star}]$.

Numerically, this Procrustes problem is easy to solve; a solution seems always to be found in one or few iterations. This problem formulation is extensible, of course, to orthogonal (square) matrices Q. $\qquad\qquad\qquad\qquad\qquad\qquad\qquad\qquad\qquad\qquad\qquad\qquad$ $\square$

4.6.0.0.3 Example. *Combinatorial Procrustes problem.*
In case $A,B\in\mathbb{R}^{n}$, when vector $A = \Xi B$ is known to be a permutation of vector B, solution to orthogonal Procrustes problem

$$\begin{array}{ll} \underset{X\in\mathbb{R}^{n\times n}}{\text{minimize}} & \|A - XB\|_{\mathrm{F}} \\[2mm] \text{subject to} & X^{\mathrm{T}} = X^{-1} \end{array} \qquad (1803)$$

is not necessarily a permutation matrix Ξ even though an optimal objective value of 0 is found by the known analytical solution (§C.3). The simplest method of solution finds permutation matrix $X^\star = \Xi$ simply by sorting vector B with respect to A.

Instead of sorting, we design two different convex problems each of whose optimal solution is a permutation matrix: one design is based on rank constraint, the other on cardinality. Because permutation matrices are sparse by definition, we depart from a traditional Procrustes problem by instead demanding a vector 1-norm which is known to produce solutions more sparse than Frobenius' norm.

There are two principal facts exploited by the first convex iteration design (§4.4.1) we propose. Permutation matrices Ξ constitute:

1) the set of all nonnegative orthogonal matrices,

2) all points extreme to the polyhedron (100) of doubly stochastic matrices.

That means:

1) norm of each row and column is 1 ,[4.45]

$$\|\Xi(:\,,i)\| = 1\,, \quad \|\Xi(i\,,:)\| = 1\,, \qquad i=1\ldots n \tag{837}$$

2) sum of each nonnegative row and column is 1, (§2.3.2.0.4)

$$\Xi^{\mathrm{T}}\mathbf{1}=\mathbf{1}\,, \quad \Xi\mathbf{1}=\mathbf{1}\,, \quad \Xi\geq\mathbf{0} \tag{838}$$

solution via rank constraint

The idea is to individually constrain each column of variable matrix X to have unity norm. Matrix X must also belong to that polyhedron, (100) in the nonnegative orthant, implied by constraints (838); so each row-sum and column-sum of X must also be unity. It is this combination of nonnegativity, sum, and sum square constraints that extracts the permutation matrices: (Figure 109) given nonzero vectors $A\,, B$

$$
\begin{aligned}
\underset{X\in\mathbb{R}^{n\times n},\,G_i\in\mathbb{S}^{n+1}}{\text{minimize}} \quad & \|A - XB\|_1 \,+\, w\sum_{i=1}^{n}\langle G_i\,,\,W_i\rangle \\
\text{subject to} \quad & \left.\begin{aligned} G_i &= \begin{bmatrix} G_i(1\!:\!n\,,1\!:\!n) & X(:\,,i) \\ X(:\,,i)^{\mathrm{T}} & 1 \end{bmatrix} \succeq 0 \\ \operatorname{tr} G_i &= 2 \end{aligned}\right\}\,, \quad i=1\ldots n \\
& X^{\mathrm{T}}\mathbf{1}=\mathbf{1} \\
& X\mathbf{1}=\mathbf{1} \\
& X\geq\mathbf{0}
\end{aligned}
\tag{839}
$$

[4.45]This fact would be superfluous were the objective of minimization linear, because the permutation matrices reside at the extreme points of a polyhedron (100) implied by (838). But as posed, only either rows or columns need be constrained to unit norm because matrix orthogonality implies transpose orthogonality. (§B.5.2) Absence of vanishing inner product constraints that help define orthogonality, like $\operatorname{tr} Z = 0$ from Example 4.6.0.0.2, is a consequence of nonnegativity; *id est*, the only orthogonal matrices having exclusively nonnegative entries are permutations of the Identity.

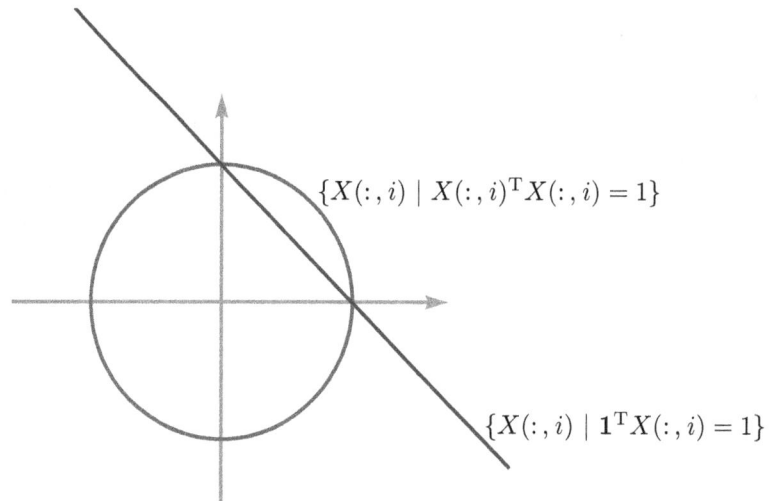

Figure 109: Permutation matrix i^{th} column-sum and column-norm constraint, abstract in two dimensions, when rank-1 constraint is satisfied. Optimal solutions reside at intersection of hyperplane with unit circle.

where $w \approx 10$ positively weights the rank regularization term. Optimal solutions $G_i^\star$ are key to finding direction matrices W_i for the next iteration of semidefinite programs (839) (840):

$$
\left.
\begin{array}{ll}
\underset{W_i \in \mathbb{S}^{n+1}}{\text{minimize}} & \langle G_i^\star \,,\, W_i \rangle \\[1ex]
\text{subject to} & 0 \preceq W_i \preceq I \\[1ex]
& \operatorname{tr} W_i = n
\end{array}
\right\} , \qquad i = 1 \ldots n
\tag{840}
$$

Direction matrices thus found lead toward rank-1 matrices $G_i^\star$ on subsequent iterations. Constraint on trace of $G_i^\star$ normalizes the ith column of $X^\star$ to unity because (*confer* p.375)

$$
G_i^\star = \begin{bmatrix} X^\star(:,i) \\ 1 \end{bmatrix} \begin{bmatrix} X^\star(:,i)^{\text{T}} & 1 \end{bmatrix}
\tag{841}
$$

at convergence. Binary-valued $X^\star$ column entries result from the further sum constraint $X\mathbf{1} = \mathbf{1}$. Columnar orthogonality is a consequence of the further transpose-sum constraint $X^{\text{T}}\mathbf{1} = \mathbf{1}$ in conjunction with nonnegativity constraint $X \geq \mathbf{0}$; but we leave proof of orthogonality an exercise. The optimal objective value is 0 for both semidefinite programs when vectors A and B are related by permutation. In any case, optimal solution $X^\star$ becomes a permutation matrix Ξ.

Because there are n direction matrices W_i to find, it can be advantageous to invoke a known closed-form solution for each from page 567. What makes this combinatorial problem more tractable are relatively small semidefinite constraints in (839). (*confer* (835)) When a permutation A of vector B exists, number of iterations can be as small as 1. But this combinatorial Procrustes problem can be made even more challenging when vector A has repeated entries.

solution via cardinality constraint

Now the idea is to force solution at a vertex of permutation polyhedron (100) by finding a solution of desired sparsity. Because permutation matrix X is n-sparse by assumption, this combinatorial Procrustes problem may instead be formulated as a compressed sensing problem with convex iteration on cardinality of vectorized X (§4.5.1): given nonzero vectors A, B

$$
\begin{aligned}
&\underset{X \in \mathbb{R}^{n \times n}}{\text{minimize}} && \|A - XB\|_1 + w\langle X, Y \rangle \\
&\text{subject to} && X^{\mathrm{T}}\mathbf{1} = \mathbf{1} \\
& && X\mathbf{1} = \mathbf{1} \\
& && X \geq \mathbf{0}
\end{aligned}
\tag{842}
$$

where direction vector Y is an optimal solution to

$$
\begin{aligned}
&\underset{Y \in \mathbb{R}^{n \times n}}{\text{minimize}} && \langle X^\star, Y \rangle \\
&\text{subject to} && \mathbf{0} \leq Y \leq \mathbf{1} \\
& && \mathbf{1}^{\mathrm{T}} Y \mathbf{1} = n^2 - n
\end{aligned}
\tag{524}
$$

each a linear program. In this circumstance, use of closed-form solution for direction vector Y is discouraged. When vector A is a permutation of B, both linear programs have objectives that converge to 0. When vectors A and B are permutations and no entries of A are repeated, optimal solution $X^\star$ can be found as soon as the first iteration.

In any case, $X^\star = \Xi$ is a permutation matrix. □

4.6.0.0.4 Exercise. *Combinatorial Procrustes constraints.*
Assume that the objective of semidefinite program (839) is 0 at optimality. Prove that the constraints in program (839) are necessary and sufficient to produce a permutation matrix as optimal solution. Alternatively and equivalently, prove those constraints necessary and sufficient to optimally produce a nonnegative orthogonal matrix. ▼

4.6.0.0.5 Example. *Tractable polynomial constraint.*
The set of all coefficients for which a multivariate polynomial were convex is generally difficult to determine. But the ability to handle rank constraints makes any nonconvex polynomial constraint transformable to a convex constraint. *All optimization problems having polynomial objective and polynomial constraints can be reformulated as a semidefinite program with a rank-1 constraint.* [293] Suppose we require

$$
3 + 2x - xy \leq 0
\tag{843}
$$

Identify

$$
G = \begin{bmatrix} x \\ y \\ 1 \end{bmatrix} \begin{bmatrix} x & y & 1 \end{bmatrix} = \begin{bmatrix} x^2 & xy & x \\ xy & y^2 & y \\ x & y & 1 \end{bmatrix} \in \mathbb{S}^3
\tag{844}
$$

Then nonconvex polynomial constraint (843) is equivalent to constraint set

$$
\begin{aligned}
&\operatorname{tr}(GA) \leq 0 \\
&G_{33} = 1 \\
&(G \succeq 0) \\
&\operatorname{rank} G = 1
\end{aligned}
\tag{845}
$$

with direct correspondence to sense of trace inequality where G is assumed symmetric (§B.1.0.2) and

$$A = \begin{bmatrix} 0 & -\frac{1}{2} & 1 \\ -\frac{1}{2} & 0 & 0 \\ 1 & 0 & 3 \end{bmatrix} \in \mathbb{S}^3 \qquad (846)$$

Then the method of convex iteration from §4.4.1 is applied to implement the rank constraint.   □

4.6.0.0.6   Exercise.   *Binary Pythagorean theorem.*
Technique of Example 4.6.0.0.5 is extensible to any quadratic constraint; *e.g.*, $x^{\mathrm{T}}Ax + 2b^{\mathrm{T}}x + c \leq 0$, $x^{\mathrm{T}}Ax + 2b^{\mathrm{T}}x + c \geq 0$, and $x^{\mathrm{T}}Ax + 2b^{\mathrm{T}}x + c = 0$.   Write a rank-constrained semidefinite program to solve (Figure 109)

$$\begin{cases} x + y = 1 \\ x^2 + y^2 = 1 \end{cases} \qquad (847)$$

whose feasible set is not connected. Implement this system in cvx [173] by convex iteration.
▼

4.6.0.0.7   Example.   *High order polynomials.*
Consider nonconvex problem from Canadian Mathematical Olympiad 1999:

$$\begin{array}{ll} \underset{x,\,y,\,z\in\mathbb{R}}{\text{find}} & x\,,\,y\,,\,z \\ \text{subject to} & x^2y + y^2z + z^2x = \frac{2^2}{3^3} \\ & x + y + z = 1 \\ & x\,,\,y\,,\,z \geq 0 \end{array} \qquad (848)$$

We wish to solve for, what is known to be, a tight upper bound $\frac{2^2}{3^3}$ on the constrained polynomial $x^2y + y^2z + z^2x$ by transformation to a rank-constrained semidefinite program. First identify

$$G = \begin{bmatrix} x \\ y \\ z \\ 1 \end{bmatrix}\begin{bmatrix} x & y & z & 1 \end{bmatrix} = \begin{bmatrix} x^2 & xy & zx & x \\ xy & y^2 & yz & y \\ zx & yz & z^2 & z \\ x & y & z & 1 \end{bmatrix} \in \mathbb{S}^4 \qquad (849)$$

$$X = \begin{bmatrix} x^2 \\ y^2 \\ z^2 \\ x \\ y \\ z \\ 1 \end{bmatrix}\begin{bmatrix} x^2 & y^2 & z^2 & x & y & z & 1 \end{bmatrix} = \begin{bmatrix} x^4 & x^2y^2 & z^2x^2 & x^3 & x^2y & zx^2 & x^2 \\ x^2y^2 & y^4 & y^2z^2 & xy^2 & y^3 & y^2z & y^2 \\ z^2x^2 & y^2z^2 & z^4 & z^2x & yz^2 & z^3 & z^2 \\ x^3 & xy^2 & z^2x & x^2 & xy & zx & x \\ x^2y & y^3 & yz^2 & xy & y^2 & yz & y \\ zx^2 & y^2z & z^3 & zx & yz & z^2 & z \\ x^2 & y^2 & z^2 & x & y & z & 1 \end{bmatrix} \in \mathbb{S}^7 \qquad (850)$$

then apply convex iteration (§4.4.1) to implement rank constraints:

$$\begin{aligned}
&\underset{A,\,C\in\mathbb{S},\,b}{\text{find}} && b \\
&\text{subject to} && \operatorname{tr}(XE) = \tfrac{2^2}{3^3} \\
& && G = \begin{bmatrix} A & b \\ b^{\mathrm{T}} & 1 \end{bmatrix} (\succeq 0) \\
& && X = \begin{bmatrix} C & \begin{bmatrix} \delta(A) \\ b \end{bmatrix} \\ \begin{bmatrix} \delta(A)^{\mathrm{T}} & b^{\mathrm{T}} \end{bmatrix} & 1 \end{bmatrix} (\succeq 0) \\
& && \mathbf{1}^{\mathrm{T}}b = 1 \\
& && b \succeq 0 \\
& && \operatorname{rank} G = 1 \\
& && \operatorname{rank} X = 1
\end{aligned} \tag{851}$$

where

$$E = \begin{bmatrix}
0 & 0 & 0 & 0 & 1 & 0 & 0 \\
0 & 0 & 0 & 0 & 0 & 1 & 0 \\
0 & 0 & 0 & 1 & 0 & 0 & 0 \\
0 & 0 & 1 & 0 & 0 & 0 & 0 \\
1 & 0 & 0 & 0 & 0 & 0 & 0 \\
0 & 1 & 0 & 0 & 0 & 0 & 0 \\
0 & 0 & 0 & 0 & 0 & 0 & 0
\end{bmatrix} \frac{1}{2} \in \mathbb{S}^7 \tag{852}$$

[388, MATLAB code]. Positive semidefiniteness is optional only when rank-1 constraints are explicit by Theorem A.3.1.0.7. Optimal solution $(x\,,\,y\,,\,z) = (0\,,\,\tfrac{2}{3}\,,\,\tfrac{1}{3})$ to problem (848) is not unique. $\qquad\square$

4.6.0.0.8 Exercise. *Motzkin polynomial.*
Prove $xy^2 + x^2y - 3xy + 1$ to be nonnegative on the nonnegative orthant. $\qquad\blacktriangledown$

4.6.0.0.9 Example. *Boolean vector satisfying $Ax \preceq b$.* (*confer* §4.2.3.1.1)
Now we consider solution to a discrete problem whose only known analytical method of solution is combinatorial in complexity: given $A \in \mathbb{R}^{M\times N}$ and $b \in \mathbb{R}^M$

$$\begin{aligned}
&\text{find} && x \in \mathbb{R}^N \\
&\text{subject to} && Ax \preceq b \\
& && \delta(xx^{\mathrm{T}}) = \mathbf{1}
\end{aligned} \tag{853}$$

This nonconvex problem demands a Boolean solution $[\,x_i = \pm 1,\; i = 1\ldots N\,]$.

Assign a rank-1 matrix of variables; symmetric variable matrix X and solution vector x:

$$G = \begin{bmatrix} x \\ 1 \end{bmatrix} \begin{bmatrix} x^{\mathrm{T}} & 1 \end{bmatrix} = \begin{bmatrix} X & x \\ x^{\mathrm{T}} & 1 \end{bmatrix} \triangleq \begin{bmatrix} xx^{\mathrm{T}} & x \\ x^{\mathrm{T}} & 1 \end{bmatrix} \in \mathbb{S}^{N+1} \tag{854}$$

Then design an equivalent semidefinite feasibility problem to find a Boolean solution to

$Ax \preceq b$:

$$
\begin{array}{ll}
\underset{X \in \mathbb{S}^N}{\text{find}} & x \in \mathbb{R}^N \\
\text{subject to} & Ax \preceq b \\
& G = \begin{bmatrix} X & x \\ x^{\mathrm{T}} & 1 \end{bmatrix} (\succeq 0) \\
& \text{rank}\, G = 1 \\
& \delta(X) = \mathbf{1}
\end{array}
\tag{855}
$$

where $x_i^\star \in \{-1, 1\}$, $i = 1 \ldots N$. The two variables X and x are made dependent via their assignment to rank-1 matrix G. By (1703), an optimal rank-1 matrix $G^\star$ must take the form (854).

As before, we regularize the rank constraint by introducing a direction matrix Y into the objective:

$$
\begin{array}{ll}
\underset{X \in \mathbb{S}^N, \; x \in \mathbb{R}^N}{\text{minimize}} & \langle G, Y \rangle \\
\text{subject to} & Ax \preceq b \\
& G = \begin{bmatrix} X & x \\ x^{\mathrm{T}} & 1 \end{bmatrix} \succeq 0 \\
& \delta(X) = \mathbf{1}
\end{array}
\tag{856}
$$

Solution of this semidefinite program is iterated with calculation of the direction matrix Y from semidefinite program (828). At convergence, in the sense (767), convex problem (856) becomes equivalent to nonconvex Boolean problem (853).

Direction matrix Y can be an orthogonal projector having closed-form expression, by (1791a), although convex iteration is not a projection method. (§4.4.1.1) Given randomized data A and b for a large problem, we find that stalling becomes likely (convergence of the iteration to a positive objective $\langle G^\star, Y \rangle$). To overcome this behavior, we introduce a heuristic into the implementation on *Wικιmization* [378] that momentarily reverses direction of search (like (829)) upon stall detection. We find that rate of convergence can be sped significantly by detecting stalls early. □

4.6.0.0.10 Example. *Variable-vector normalization.*
Suppose, within some convex optimization problem, we want vector variables $x, y \in \mathbb{R}^N$ constrained by a nonconvex equality:

$$
x \|y\| = y
\tag{857}
$$

id est, $\|x\| = 1$ and x points in the same direction as $y \neq \mathbf{0}$; *e.g.*,

$$
\begin{array}{ll}
\underset{x, y}{\text{minimize}} & f(x, y) \\
\text{subject to} & (x, y) \in \mathcal{C} \\
& x \|y\| = y
\end{array}
\tag{858}
$$

where f is some convex function and $\mathcal{C}$ is some convex set. We can realize the nonconvex equality by constraining rank and adding a regularization term to the objective. Make the assignment:

$$
G = \begin{bmatrix} x \\ y \\ 1 \end{bmatrix} \begin{bmatrix} x^T & y^T & 1 \end{bmatrix} = \begin{bmatrix} X & Z & x \\ Z & Y & y \\ x^T & y^T & 1 \end{bmatrix} \triangleq \begin{bmatrix} xx^T & xy^T & x \\ yx^T & yy^T & y \\ x^T & y^T & 1 \end{bmatrix} \in \mathbb{S}^{2N+1} \tag{859}
$$

where $X, Y \in \mathbb{S}^N$, also $Z \in \mathbb{S}^N$ [*sic*]. Any rank-1 solution must take the form of (859). (§B.1) The problem statement equivalent to (858) is then written

$$
\begin{array}{cl}
\underset{X, Y \in \mathbb{S}, Z, x, y}{\text{minimize}} & f(x, y) + \|X - Y\|_F \\
\text{subject to} & (x, y) \in \mathcal{C} \\
& G = \begin{bmatrix} X & Z & x \\ Z & Y & y \\ x^T & y^T & 1 \end{bmatrix} (\succeq 0) \\
& \text{rank}\, G = 1 \\
& \text{tr}(X) = 1 \\
& \delta(Z) \succeq 0
\end{array} \tag{860}
$$

The trace constraint on X normalizes vector x while the diagonal constraint on Z maintains sign between respective entries of x and y. Regularization term $\|X - Y\|_F$ then makes x equal to y to within a real scalar; (§C.2.0.0.2) in this case, a positive scalar. To make this program solvable by convex iteration, as explained in Example 4.4.1.2.4 and other previous examples, we move the rank constraint to the objective

$$
\begin{array}{cl}
\underset{X, Y, Z, x, y}{\text{minimize}} & f(x, y) + \|X - Y\|_F + \langle G, W \rangle \\
\text{subject to} & (x, y) \in \mathcal{C} \\
& G = \begin{bmatrix} X & Z & x \\ Z & Y & y \\ x^T & y^T & 1 \end{bmatrix} \succeq 0 \\
& \text{tr}(X) = 1 \\
& \delta(Z) \succeq 0
\end{array} \tag{861}
$$

by introducing a direction matrix W found from (1791a):

$$
\begin{array}{cl}
\underset{W \in \mathbb{S}^{2N+1}}{\text{minimize}} & \langle G^\star, W \rangle \\
\text{subject to} & 0 \preceq W \preceq I \\
& \text{tr}\, W = 2N
\end{array} \tag{862}
$$

This semidefinite program has an optimal solution that is known in closed form. Iteration (861) (862) terminates when $\text{rank}\, G = 1$ and linear regularization $\langle G, W \rangle$ vanishes to within some numerical tolerance in (861); typically, in two iterations. If function f competes too much with the regularization, positively weighting each regularization term will become required. At convergence, problem (861) becomes a convex equivalent to the original nonconvex problem (858). $\qquad \square$

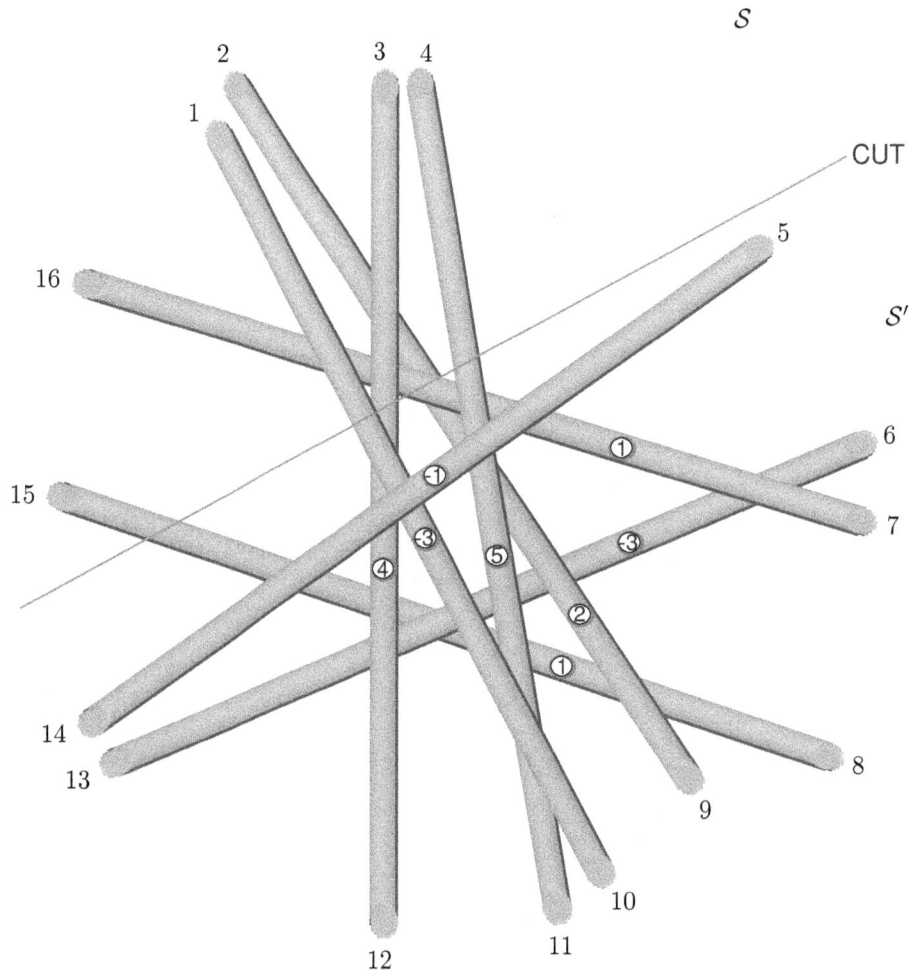

Figure 110: A CUT partitions nodes $\{i=1\ldots16\}$ of this graph into $\mathcal{S}$ and $\mathcal{S}'$. Linear arcs have circled weights. The problem is to find a cut maximizing total weight of all arcs linking partitions made by the cut.

4.6.0.0.11 Example. FAST MAX CUT. [116]

> *Let Γ be an n-node graph, and let the arcs (i,j) of the graph be associated with [ ] weights a_{ij}. The problem is to find a cut of the largest possible weight, i.e., to partition the set of nodes into two parts $\mathcal{S}, \mathcal{S}'$ in such a way that the total weight of all arcs linking $\mathcal{S}$ and $\mathcal{S}'$ (i.e., with one incident node in $\mathcal{S}$ and the other one in $\mathcal{S}'$ [Figure 110]) is as large as possible. [34, §4.3.3]*

Literature on the MAX CUT problem is vast because this problem has elegant primal and dual formulation, its solution is very difficult, and there exist many commercial applications; e.g., semiconductor design [129], quantum computing [415].

Our purpose here is to demonstrate how iteration of two simple convex problems can quickly converge to an optimal solution of the MAX CUT problem with a 98% success rate, on average.[4.46] MAX CUT is stated:

$$\underset{x}{\text{maximize}} \quad \sum_{1 \le i < j \le n} a_{ij}(1 - x_i x_j)\tfrac{1}{2} \\ \text{subject to} \quad \delta(xx^{\mathrm{T}}) = \mathbf{1} \tag{863}$$

where $[a_{ij}]$ are real arc weights, and binary vector $x = [x_i] \in \mathbb{R}^n$ corresponds to the n nodes; specifically,

$$\begin{aligned} \text{node } i \in \mathcal{S} &\Leftrightarrow x_i = 1 \\ \text{node } i \in \mathcal{S}' &\Leftrightarrow x_i = -1 \end{aligned} \tag{864}$$

If nodes i and j have the same binary value x_i and x_j, then they belong to the same partition and contribute nothing to the cut. Arc (i,j) traverses the cut, otherwise, adding its weight a_{ij} to the cut.

MAX CUT statement (863) is the same as, for $A = [a_{ij}] \in \mathbb{S}^n$

$$\underset{x}{\text{maximize}} \quad \tfrac{1}{4}\langle \mathbf{11}^{\mathrm{T}} - xx^{\mathrm{T}}, A \rangle \\ \text{subject to} \quad \delta(xx^{\mathrm{T}}) = \mathbf{1} \tag{865}$$

Because of Boolean assumption $\delta(xx^{\mathrm{T}}) = \mathbf{1}$

$$\langle \mathbf{11}^{\mathrm{T}} - xx^{\mathrm{T}}, A \rangle = \langle xx^{\mathrm{T}}, \delta(A\mathbf{1}) - A \rangle \tag{866}$$

so problem (865) is the same as

$$\underset{x}{\text{maximize}} \quad \tfrac{1}{4}\langle xx^{\mathrm{T}}, \delta(A\mathbf{1}) - A \rangle \\ \text{subject to} \quad \delta(xx^{\mathrm{T}}) = \mathbf{1} \tag{867}$$

This MAX CUT problem is combinatorial (nonconvex).

Because an estimate of upper bound to MAX CUT is needed to ascertain convergence when vector x has large dimension, we digress to derive the dual problem: Directly from (867), its Lagrangian is [62, §5.1.5] (1499)

$$\begin{aligned} \mathcal{L}(x, \nu) &= \tfrac{1}{4}\langle xx^{\mathrm{T}}, \delta(A\mathbf{1}) - A \rangle + \langle \nu, \delta(xx^{\mathrm{T}}) - \mathbf{1} \rangle \\ &= \tfrac{1}{4}\langle xx^{\mathrm{T}}, \delta(A\mathbf{1}) - A \rangle + \langle \delta(\nu), xx^{\mathrm{T}} \rangle - \langle \nu, \mathbf{1} \rangle \\ &= \tfrac{1}{4}\langle xx^{\mathrm{T}}, \delta(A\mathbf{1} + 4\nu) - A \rangle - \langle \nu, \mathbf{1} \rangle \end{aligned} \tag{868}$$

[4.46]We term our solution to MAX CUT *fast* because we sacrifice a little accuracy to achieve speed; *id est*, only about two or three convex iterations, achieved by heavily weighting a rank regularization term.

where quadratic $x^{\mathrm{T}}(\delta(A\mathbf{1}+4\nu)-A)x$ has supremum 0 if $\delta(A\mathbf{1}+4\nu)-A$ is negative semidefinite, and has supremum ∞ otherwise. The finite supremum of dual function

$$g(\nu) \;=\; \sup_{x} \mathfrak{L}(x,\nu) \;=\; \begin{cases} -\langle \nu\,,\,\mathbf{1}\rangle\,, & A-\delta(A\mathbf{1}+4\nu) \succeq 0 \\ \infty & \text{otherwise} \end{cases} \qquad (869)$$

is chosen to be the objective of minimization to dual (convex) problem

$$\begin{array}{ll} \underset{\nu}{\text{minimize}} & -\nu^{\mathrm{T}}\mathbf{1} \\ \text{subject to} & A-\delta(A\mathbf{1}+4\nu) \succeq 0 \end{array} \qquad (870)$$

whose optimal value provides a least upper bound to MAX CUT, but is not tight ($\frac{1}{4}\langle xx^{\mathrm{T}},\,\delta(A\mathbf{1})-A\rangle < g(\nu)$, duality gap is nonzero). [161] In fact, we find that the bound's variance with problem instance is too large to be useful for this problem; thus ending our digression.[4.47]

To transform MAX CUT to its convex equivalent, first define

$$X = xx^{\mathrm{T}} \in \mathbb{S}^{n} \qquad (875)$$

then MAX CUT (867) becomes

$$\begin{array}{ll} \underset{X\in\mathbb{S}^{n}}{\text{maximize}} & \tfrac{1}{4}\langle X,\,\delta(A\mathbf{1})-A\rangle \\ \text{subject to} & \delta(X)=\mathbf{1} \\ & (X \succeq 0) \\ & \text{rank}\,X = 1 \end{array} \qquad (871)$$

whose rank constraint can be regularized as in

$$\begin{array}{ll} \underset{X\in\mathbb{S}^{n}}{\text{maximize}} & \tfrac{1}{4}\langle X,\,\delta(A\mathbf{1})-A\rangle - w\langle X,\,W\rangle \\ \text{subject to} & \delta(X)=\mathbf{1} \\ & X \succeq 0 \end{array} \qquad (872)$$

where $w\approx 1000$ is a nonnegative fixed weight, and W is a direction matrix determined from

$$\sum_{i=2}^{n} \lambda(X^{\star})_{i} \;=\; \begin{array}{ll} \underset{W\in\mathbb{S}^{n}}{\text{minimize}} & \langle X^{\star},\,W\rangle \\ \text{subject to} & 0 \preceq W \preceq I \\ & \text{tr}\,W = n-1 \end{array} \qquad (1791\text{a})$$

which has an optimal solution that is known in closed form. These two problems (872) and (1791a) are iterated until convergence as defined on page 266.

Because convex problem statement (872) is so elegant, it is numerically solvable for large binary vectors within reasonable time.[4.48] To test our convex iterative method, we

---

[4.47] Taking the dual of dual problem (870) would provide (871) but without the rank constraint. [154] Dual of a dual of even a convex primal problem is not necessarily the same primal problem; although, optimal solution of one can be obtained from the other.
[4.48] We solved for a length-250 binary vector in only a few minutes and convex iterations on a 2006 vintage laptop Core 2 CPU (Intel T7400@2.16GHz, 666MHz FSB).

compare an optimal convex result to an actual solution of the MAX CUT problem found by performing a brute force combinatorial search of $(867)^{4.49}$ for a tight upper bound. Search-time limits binary vector lengths to 24 bits (about five days CPU time). 98% accuracy, actually obtained, is independent of binary vector length (12, 13, 20, 24) when averaged over more than 231 problem instances including planar, randomized, and toroidal graphs.[4.50] When failure occurred, large and small errors were manifest. That same 98% average accuracy is presumed maintained when binary vector length is further increased. A MATLAB program is provided on *Wıкımization* [383]. □

4.6.0.0.12 Example. *Cardinality/rank problem.*
d'Aspremont, El Ghaoui, Jordan, & Lanckriet [97] propose approximating a positive semidefinite matrix $A \in \mathbb{S}_+^N$ by a rank-one matrix having constraint on cardinality c : for $0 < c < N$

$$\begin{array}{ll} \underset{z}{\text{minimize}} & \|A - zz^{\mathrm{T}}\|_{\mathrm{F}} \\ \text{subject to} & \text{card}\, z \leq c \end{array} \tag{873}$$

which, they explain, is a hard problem equivalent to

$$\begin{array}{ll} \underset{x}{\text{maximize}} & x^{\mathrm{T}}A\,x \\ \text{subject to} & \|x\| = 1 \\ & \text{card}\, x \leq c \end{array} \tag{874}$$

where $z \triangleq \sqrt{\lambda}\, x$ and where optimal solution $x^\star$ is a *principal eigenvector* (1785) (§A.5) of A and $\lambda = x^{\star \mathrm{T}}A\, x^\star$ is the *principal eigenvalue* [164, p.331] when c is true cardinality of that eigenvector. This is *principal component analysis* with a cardinality constraint which controls solution sparsity. Define the matrix variable

$$X \triangleq xx^{\mathrm{T}} \in \mathbb{S}^N \tag{875}$$

whose desired rank is 1, and whose desired diagonal cardinality

$$\text{card}\,\delta(X) \equiv \text{card}\, x \tag{876}$$

is equivalent to cardinality c of vector x. Then we can transform cardinality problem (874) to an equivalent problem in new variable X :[4.51]

$$\begin{array}{ll} \underset{X \in \mathbb{S}^N}{\text{maximize}} & \langle X ,\, A \rangle \\ \text{subject to} & \langle X ,\, I \rangle = 1 \\ & (X \succeq 0) \\ & \text{rank}\, X = 1 \\ & \text{card}\,\delta(X) \leq c \end{array} \tag{877}$$

[4.49] more computationally intensive than the proposed convex iteration by many orders of magnitude. Solving MAX CUT by searching over all binary vectors of length 100, for example, would occupy a contemporary supercomputer for a million years.

[4.50] Existence of a polynomial-time approximation to MAX CUT with accuracy provably better than 94.11% would refute NP-hardness; which Håstad believes to be highly unlikely. [187, thm.8.2] [188]

[4.51] A semidefiniteness constraint $X \succeq 0$ is not required, theoretically, because positive semidefiniteness of a rank-1 matrix is enforced by symmetry. (Theorem A.3.1.0.7)

CHAPTER 4. SEMIDEFINITE PROGRAMMING

We transform problem (877) to an equivalent convex problem by introducing two direction matrices into regularization terms: W to achieve desired cardinality $\operatorname{card} \delta(X)$, and Y to find an approximating rank-one matrix X :

$$
\begin{array}{ll}
\underset{X \in \mathbb{S}^N}{\operatorname{maximize}} & \langle X ,\, A - w_1 Y \rangle - w_2 \langle \delta(X) ,\, \delta(W) \rangle \\
\text{subject to} & \langle X ,\, I \rangle = 1 \\
& X \succeq 0
\end{array}
\tag{878}
$$

where w_1 and w_2 are positive scalars respectively weighting $\operatorname{tr}(XY)$ and $\delta(X)^{\mathrm{T}} \delta(W)$ just enough to insure that they vanish to within some numerical precision, where direction matrix Y is an optimal solution to semidefinite program

$$
\begin{array}{ll}
\underset{Y \in \mathbb{S}^N}{\operatorname{minimize}} & \langle X^{\star} ,\, Y \rangle \\
\text{subject to} & 0 \preceq Y \preceq I \\
& \operatorname{tr} Y = N - 1
\end{array}
\tag{879}
$$

and where diagonal direction matrix $W \in \mathbb{S}^N$ optimally solves linear program

$$
\begin{array}{ll}
\underset{W = \delta^2(W)}{\operatorname{minimize}} & \langle \delta(X^{\star}) ,\, \delta(W) \rangle \\
\text{subject to} & 0 \preceq \delta(W) \preceq \mathbf{1} \\
& \operatorname{tr} W = N - c
\end{array}
\tag{880}
$$

Both direction matrix programs are derived from (1791a) whose analytical solution is known but is not necessarily unique. We emphasize (*confer* p.266): because this iteration (878) (879) (880) (initial $Y, W = \mathbf{0}$) is not a projection method (§4.4.1.1), success relies on existence of matrices in the feasible set of (878) having desired rank and diagonal cardinality. In particular, the feasible set of convex problem (878) is a Fantope (91) whose extreme points constitute the set of all normalized rank-one matrices; among those are found rank-one matrices of any desired diagonal cardinality.

Convex problem (878) is neither a relaxation of cardinality problem (874); instead, problem (878) becomes a convex equivalent to (874) at global convergence of iteration (878) (879) (880). Because the feasible set of convex problem (878) contains all normalized (§B.1) symmetric rank-one matrices of every nonzero diagonal cardinality, a constraint too low or high in cardinality c will not prevent solution. An optimal rank-one solution $X^{\star}$, whose diagonal cardinality is equal to cardinality of a principal eigenvector of matrix A, will produce the least residual Frobenius norm (to within machine noise processes) in the original problem statement (873). $\qquad \square$

4.6.0.0.13 Example. *Compressive sampling of a phantom.*
In summer of 2004, Candès, Romberg, & Tao [71] and Donoho [122] released papers on perfect signal reconstruction from samples that stand in violation of Shannon's classical sampling theorem. These defiant signals are assumed sparse inherently or under some sparsifying affine transformation. Essentially, they proposed *sparse sampling theorems* asserting average sample rate independent of signal bandwidth and less than Shannon's rate.

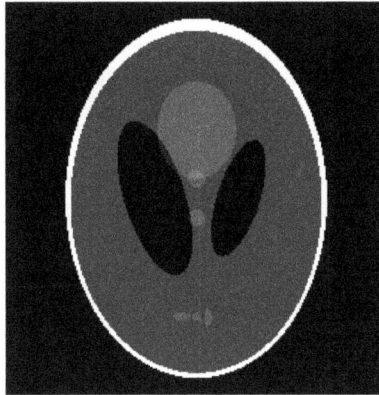

phantom(256)

Figure 111: Shepp-Logan phantom from MATLAB *image processing toolbox*.

MINIMUM SAMPLING RATE:

- OF Ω-BANDLIMITED SIGNAL: 2Ω (Shannon)

- OF k-SPARSE LENGTH-n SIGNAL: $k\log_2(1+n/k)$ (Candès/Donoho)

(Figure 104).   Certainly, much was already known about nonuniform or random sampling [36] [214] and about subsampling or *multirate systems* [93] [366]. Vetterli, Marziliano, & Blu [375] had congealed a theory of noiseless signal reconstruction, in May 2001, from samples that violate the Shannon rate. [393, *Sampling Sparsity*] They anticipated the sparsifying transform by recognizing: it is the *innovation* (onset) of functions constituting a (not necessarily bandlimited) signal that determines minimum sampling rate for perfect reconstruction.   Average onset (sparsity), Vetterli *et alii* call, the *rate of innovation*. Vector inner-products that Candès/Donoho call *samples* or *measurements*, Vetterli calls *projections*. From those projections Vetterli demonstrates reconstruction (by digital signal processing and "root finding") of a Dirac comb, the very same prototypical signal from which Candès probabilistically derives minimum sampling rate [*Compressive Sampling and Frontiers in Signal Processing*, University of Minnesota, June 6, 2007]. Combining their terminology, we paraphrase a sparse sampling theorem:

- Minimum sampling rate, asserted by Candès/Donoho, is proportional to Vetterli's rate of innovation (a.k.a: *information rate, degrees of freedom* [*ibidem* June 5, 2007]).

What distinguishes these researchers are their methods of reconstruction.

Properties of the 1-norm were also well understood by June 2004 finding applications in *deconvolution* of linear systems [85], constrained *linear regression* (*Lasso*) [356] [326], and *basis pursuit* [79] [219]. But never before had there been a formalized and rigorous sense that perfect reconstruction were possible by convex optimization of 1-norm when information lost in a subsampling process became nonrecoverable by classical methods.

Donoho named this discovery *compressed sensing* to describe a nonadaptive perfect reconstruction method by means of linear programming. By the time Candès' and Donoho's landmark papers were finally published by IEEE in 2006, compressed sensing was old news that had spawned intense research which still persists; notably, from prominent members of the *wavelet* community.

Reconstruction of the *Shepp-Logan phantom* (Figure 111), from a severely aliased image (Figure 113) obtained by Magnetic Resonance Imaging (MRI), was the *impetus* driving Candès' quest for a sparse sampling theorem. He realized that line segments appearing in the aliased image were regions of high *total variation*. There is great motivation, in the medical community, to apply compressed sensing to MRI because it translates to reduced scan-time which brings great technological and physiological benefits. MRI is now about 35 years old, beginning in 1973 with Nobel laureate Paul Lauterbur from Stony Brook USA. There has been much progress in MRI and compressed sensing since 2004, but there have also been indications of 1-norm abandonment (indigenous to reconstruction by compressed sensing) in favor of criteria closer to 0-norm because of a correspondingly smaller number of measurements required to accurately reconstruct a sparse signal:[4.52]

5481 complex samples (22 radial lines, $\approx$256 complex samples per) were required in June 2004 to reconstruct a noiseless 256×256-pixel Shepp-Logan phantom by 1-norm minimization of an image-gradient integral estimate called *total variation*; *id est*, 8.4% subsampling of 65536 data. [71, §1.1] [70, §3.2] It was soon discovered that reconstruction of the Shepp-Logan phantom were possible with only 2521 complex samples (10 radial lines, Figure 112); 3.8% subsampled data input to a (nonconvex) $\frac{1}{2}$-norm total-variation minimization. [77, §IIIA] The closer to 0-norm, the fewer the samples required for perfect reconstruction.

Passage of a few years witnessed an algorithmic speedup and dramatic reduction in minimum number of samples required for perfect reconstruction of the noiseless Shepp-Logan phantom. But minimization of total variation is ideally suited to recovery of any piecewise-constant image, like a phantom, because gradient of such images is highly sparse by design.

There is no inherent characteristic of real-life MRI images that would make reasonable an expectation of sparse gradient. Sparsification of a discrete image-gradient tends to preserve edges. Then minimization of total variation seeks an image having fewest edges. There is no deeper theoretical foundation than that. When applied to human brain scan or angiogram, with as much as 20% of 256×256 Fourier samples, we have observed[4.53] a 30dB image/reconstruction-error ratio[4.54] barrier that seems impenetrable by the total-variation objective. Total-variation minimization has met with moderate success, in retrospect, only because some medical images are moderately piecewise-constant signals. One simply

---

[4.52]Efficient techniques continually emerge urging 1-norm criteria abandonment; [82] [365] [364, §IID] *e.g.*, five techniques for compressed sensing are compared in [37] demonstrating that 1-norm performance limits for cardinality minimization can be reliably exceeded.

[4.53]Experiments with real-life images were performed by Christine Law at Lucas Center for Imaging, Stanford University.

[4.54]Noise considered here is due only to the reconstruction process itself; *id est*, noise in excess of that produced by the best reconstruction of an image from a complete set of samples in the sense of Shannon. At less than 30dB image/error, artifacts generally remain visible to the naked eye. We estimate about 50dB is required to eliminate noticeable distortion in a visual A/B comparison.

hopes a reconstruction, that is in some sense equal to a known subset of samples and whose gradient is most sparse, is that unique image we seek.[4.55]

The total-variation objective, operating on an image, is expressible as norm of a linear transformation (899). It is natural to ask whether there exist other sparsifying transforms that might break the real-life 30dB barrier (any sampling pattern @20% 256×256 data) in MRI. There has been much research into application of wavelets, discrete cosine transform (DCT), randomized orthogonal bases, splines, *etcetera*, but with suspiciously little focus on objective measures like image/error or illustration of difference images; the predominant basis of comparison instead being subjectively visual (Duensing & Huang, ISMRM Toronto 2008).[4.56] Despite choice of transform, there seems yet to have been a breakthrough of the 30dB barrier. Application of compressed sensing to MRI, therefore, remains fertile in 2008 for continued research.

Lagrangian form of compressed sensing in imaging

We now repeat Candès' image reconstruction experiment from 2004 which led to discovery of sparse sampling theorems. [71, §1.2] But we achieve perfect reconstruction with an algorithm based on vanishing gradient of a compressed sensing problem's Lagrangian, which is computationally efficient. Our *contraction method* (p.330) is fast also because matrix multiplications are replaced by fast Fourier transforms and number of constraints is cut in half by sampling symmetrically. Convex iteration for cardinality minimization (§4.5) is incorporated which allows perfect reconstruction of a phantom at 4.1% subsampling rate; 50% Candès' rate. By making neighboring-pixel selection adaptive, convex iteration reduces discrete image-gradient sparsity of the Shepp-Logan phantom to 1.9%; 33% lower than previously reported.

We demonstrate application of discrete image-gradient sparsification to the $n \times n = 256 \times 256$ Shepp-Logan phantom, simulating idealized acquisition of MRI data by radial sampling in the Fourier domain (Figure 112).[4.57] Define a Nyquist-centric *discrete Fourier transform* (DFT) matrix

$$
F \triangleq \begin{bmatrix}
1 & 1 & 1 & 1 & \cdots & 1 \\
1 & e^{-j2\pi/n} & e^{-j4\pi/n} & e^{-j6\pi/n} & \cdots & e^{-j(n-1)2\pi/n} \\
1 & e^{-j4\pi/n} & e^{-j8\pi/n} & e^{-j12\pi/n} & \cdots & e^{-j(n-1)4\pi/n} \\
1 & e^{-j6\pi/n} & e^{-j12\pi/n} & e^{-j18\pi/n} & \cdots & e^{-j(n-1)6\pi/n} \\
\vdots & \vdots & \vdots & \vdots & \ddots & \vdots \\
1 & e^{-j(n-1)2\pi/n} & e^{-j(n-1)4\pi/n} & e^{-j(n-1)6\pi/n} & \cdots & e^{-j(n-1)^2 2\pi/n}
\end{bmatrix} \frac{1}{\sqrt{n}} \in \mathbb{C}^{n \times n} \tag{881}
$$

[4.55] In vascular radiology, diagnoses are almost exclusively based on morphology of vessels and, in particular, presence of stenoses. There is a compelling argument for total-variation reconstruction of magnetic resonance angiogram because it helps isolate structures of particular interest.

[4.56] *I have never calculated the PSNR of these reconstructed images* [of Barbara].

—Jean-Luc Starck

The sparsity of the image is the percentage of transform coefficients sufficient for diagnostic-quality reconstruction. Of course the term "diagnostic quality" is subjective. ...I have yet to see an "objective" measure of image quality. Difference images, in my experience, definitely do not tell the whole story. Often I would show people some of my results and get mixed responses, but when I add artificial Gaussian noise to an image, often people say that it looks better. —Michael Lustig

[4.57] k-*space* is conventional acquisition terminology indicating domain of the continuous raw data provided by an MRI machine. An image is reconstructed by inverse discrete Fourier transform of that data interpolated on a Cartesian grid in two dimensions.

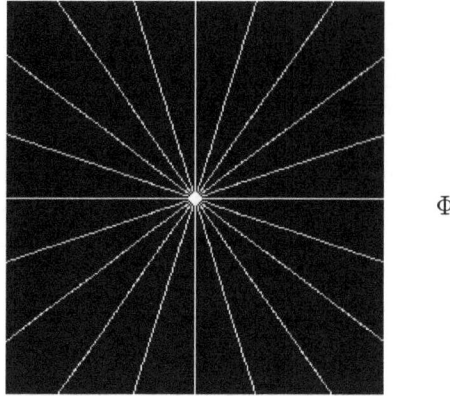

Figure 112: MRI radial sampling pattern, in DC-centric Fourier domain, representing 4.1% (10 lines) subsampled data. Only half of these complex samples, in any halfspace about the origin in theory, need be acquired for a real image because of conjugate symmetry. Due to MRI machine imperfections, samples are generally taken over full extent of each radial line segment. MRI acquisition time is proportional to number of lines.

a symmetric (nonHermitian) unitary matrix characterized

$$
\begin{aligned}
F &= F^{\mathrm{T}} \\
F^{-1} &= F^{\mathrm{H}}
\end{aligned}
$$
(882)

Denoting an unknown image $\mathcal{U} \in \mathbb{R}^{n \times n}$, its two-dimensional discrete Fourier transform $\mathfrak{F}$ is

$$
\mathfrak{F}(\mathcal{U}) \triangleq F\mathcal{U}F
$$
(883)

hence the inverse discrete transform

$$
\mathcal{U} = F^{\mathrm{H}}\mathfrak{F}(\mathcal{U})F^{\mathrm{H}}
$$
(884)

From §A.1.1 *no.*31 we have a vectorized two-dimensional DFT via Kronecker product $\otimes$

$$
\operatorname{vec}\mathfrak{F}(\mathcal{U}) \triangleq (F \otimes F)\operatorname{vec}\mathcal{U}
$$
(885)

and from (884) its inverse [172, p.24]

$$
\operatorname{vec}\mathcal{U} = (F^{\mathrm{H}} \otimes F^{\mathrm{H}})(F \otimes F)\operatorname{vec}\mathcal{U} = (F^{\mathrm{H}}F \otimes F^{\mathrm{H}}F)\operatorname{vec}\mathcal{U}
$$
(886)

Idealized radial sampling in the Fourier domain can be simulated by Hadamard product $\circ$ with a binary mask $\Phi \in \mathbb{R}^{n \times n}$ whose nonzero entries could, for example, correspond with the radial line segments in Figure 112. To make the mask Nyquist-centric, like DFT matrix F, define a circulant [174] symmetric permutation matrix[4.58]

$$
\Theta \triangleq \begin{bmatrix} \mathbf{0} & I \\ I & \mathbf{0} \end{bmatrix} \in \mathbb{S}^n
$$
(887)

[4.58]MATLAB `fftshift()`

Then given subsampled Fourier domain (MRI 𝕜-space) measurements in incomplete $K \in \mathbb{C}^{n \times n}$, we might constrain $\mathfrak{F}(\mathcal{U})$ thus:

$$\Theta \Phi \Theta \circ FUF = K \tag{888}$$

and in vector form, (42) (1879)

$$\delta(\text{vec}\,\Theta\Phi\Theta)(F \otimes F)\,\text{vec}\,\mathcal{U} = \text{vec}\,K \tag{889}$$

Because measurements K are complex, there are actually twice the number of equality constraints as there are measurements.

We can cut that number of constraints in half via vertical and horizontal mask Φ symmetry which forces the imaginary inverse transform to $\mathbf{0}$: The inverse subsampled transform in matrix form is

$$F^{\mathrm{H}}(\Theta\Phi\Theta \circ FUF)F^{\mathrm{H}} = F^{\mathrm{H}}KF^{\mathrm{H}} \tag{890}$$

and in vector form

$$(F^{\mathrm{H}} \otimes F^{\mathrm{H}})\delta(\text{vec}\,\Theta\Phi\Theta)(F \otimes F)\,\text{vec}\,\mathcal{U} = (F^{\mathrm{H}} \otimes F^{\mathrm{H}})\,\text{vec}\,K \tag{891}$$

later abbreviated

$$P\,\text{vec}\,\mathcal{U} = f \tag{892}$$

where

$$P \triangleq (F^{\mathrm{H}} \otimes F^{\mathrm{H}})\delta(\text{vec}\,\Theta\Phi\Theta)(F \otimes F) \in \mathbb{C}^{n^2 \times n^2} \tag{893}$$

Because of idempotence $P = P^2$, P is a projection matrix. Because of its Hermitian symmetry [172, p.24]

$$P = (F^{\mathrm{H}} \otimes F^{\mathrm{H}})\delta(\text{vec}\,\Theta\Phi\Theta)(F \otimes F) = (F \otimes F)^{\mathrm{H}}\delta(\text{vec}\,\Theta\Phi\Theta)(F^{\mathrm{H}} \otimes F^{\mathrm{H}})^{\mathrm{H}} = P^{\mathrm{H}} \tag{894}$$

P is an orthogonal projector.[4.59] $P\,\text{vec}\,\mathcal{U}$ is real when P is real; *id est*, when for positive even integer n

$$\Phi = \begin{bmatrix} \Phi_{11} & \Phi(1,2\!:\!n)\Xi \\ \Xi\Phi(2\!:\!n,1) & \Xi\Phi(2\!:\!n,2\!:\!n)\Xi \end{bmatrix} \in \mathbb{R}^{n \times n} \tag{895}$$

where $\Xi \in \mathbb{S}^{n-1}$ is the order-reversing permutation matrix (1819). In words, this necessary and sufficient condition on Φ (for a real inverse subsampled transform [292, p.53]) demands vertical symmetry about row $\frac{n}{2}+1$ and horizontal symmetry[4.60] about column $\frac{n}{2}+1$.

Define

$$\Delta \triangleq \begin{bmatrix} 1 & 0 & & & & & \mathbf{0} \\ -1 & 1 & 0 & & & & \\ & -1 & 1 & \ddots & & & \\ & & \ddots & \ddots & \ddots & & \\ & & & \ddots & 1 & 0 \\ \mathbf{0}^{\mathrm{T}} & & & & -1 & 1 \end{bmatrix} \in \mathbb{R}^{n \times n} \tag{896}$$

[4.59] (893) is a diagonalization of matrix P whose binary eigenvalues are $\delta(\text{vec}\,\Theta\Phi\Theta)$ while the corresponding eigenvectors constitute the columns of unitary matrix $F^{\mathrm{H}} \otimes F^{\mathrm{H}}$.
[4.60] This condition on Φ applies to both DC- and Nyquist-centric DFT matrices.

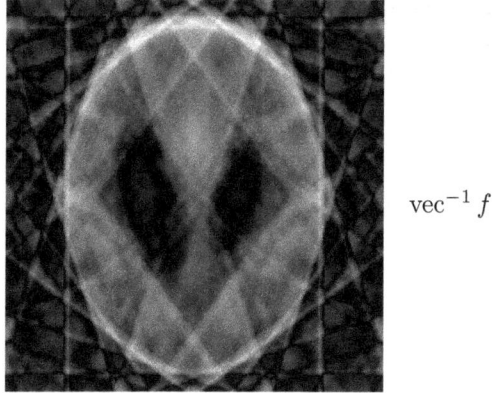

$$\mathrm{vec}^{-1} f$$

Figure 113: Aliasing of Shepp-Logan phantom in Figure 111 resulting from $\Bbbk$-space subsampling pattern in Figure 112. This image is real because binary mask Φ is vertically and horizontally symmetric. It is remarkable that the phantom can be reconstructed, by convex iteration, given only $\mathcal{U}_0 = \mathrm{vec}^{-1} f$.

Express an image-gradient estimate

$$\nabla \mathcal{U} \triangleq \begin{bmatrix} \mathcal{U} \Delta \\ \mathcal{U} \Delta^{\mathrm{T}} \\ \Delta \mathcal{U} \\ \Delta^{\mathrm{T}} \mathcal{U} \end{bmatrix} \in \mathbb{R}^{4n \times n} \tag{897}$$

that is a simple first-order difference of neighboring pixels (Figure 114) to the right, left, above, and below.[4.61] By §A.1.1 *no.*31, its vectorization: for $\Psi_i \in \mathbb{R}^{n^2 \times n^2}$

$$\mathrm{vec}\, \nabla \mathcal{U} = \begin{bmatrix} \Delta^{\mathrm{T}} \otimes I \\ \Delta \otimes I \\ I \otimes \Delta \\ I \otimes \Delta^{\mathrm{T}} \end{bmatrix} \mathrm{vec}\, \mathcal{U} \triangleq \begin{bmatrix} \Psi_1 \\ \Psi_1^{\mathrm{T}} \\ \Psi_2 \\ \Psi_2^{\mathrm{T}} \end{bmatrix} \mathrm{vec}\, \mathcal{U} \triangleq \Psi\, \mathrm{vec}\, \mathcal{U} \in \mathbb{R}^{4n^2} \tag{898}$$

where $\Psi \in \mathbb{R}^{4n^2 \times n^2}$. A total-variation minimization for reconstructing MRI image $\mathcal{U}$, that is known suboptimal [213] [72], may be concisely posed

$$\begin{array}{ll} \underset{\mathcal{U}}{\mathrm{minimize}} & \|\Psi\, \mathrm{vec}\, \mathcal{U}\|_1 \\ \mathrm{subject\ to} & P\, \mathrm{vec}\, \mathcal{U} = f \end{array} \tag{899}$$

where

$$f = (F^{\mathrm{H}} \otimes F^{\mathrm{H}})\, \mathrm{vec}\, K \in \mathbb{C}^{n^2} \tag{900}$$

[4.61]There is significant improvement in reconstruction quality by augmentation of a nominally two-point discrete image-gradient estimate to four points per pixel by inclusion of two polar directions. Improvement is due to centering; symmetry of discrete differences about a central pixel. We find small improvement on real-life images, $\approx$1dB empirically, by further augmentation with diagonally adjacent pixel differences.

is the known inverse subsampled Fourier data (a vectorized aliased image, Figure 113), and where a norm of discrete image-gradient $\nabla\mathcal{U}$ is equivalently expressed as norm of a linear transformation $\Psi\,\mathrm{vec}\,\mathcal{U}$.

Although this simple problem statement (899) is equivalent to a linear program (§3.2), its numerical solution is beyond the capability of even the most highly regarded of contemporary commercial solvers.[4.62] Our only recourse is to recast the problem in Lagrangian form (§3.1.2.2.2) and write customized code to solve it:

$$
\begin{array}{c}
\underset{\mathcal{U}}{\text{minimize}} \quad \langle\,|\Psi\,\mathrm{vec}\,\mathcal{U}|\,,\,y\rangle \\
\text{subject to} \quad P\,\mathrm{vec}\,\mathcal{U} = f
\end{array} \qquad \text{(a)}
$$

$$
\equiv
$$

$$
\underset{\mathcal{U}}{\text{minimize}} \ \langle\,|\Psi\,\mathrm{vec}\,\mathcal{U}|\,,\,y\rangle + \tfrac{1}{2}\lambda\|P\,\mathrm{vec}\,\mathcal{U} - f\|_2^2 \qquad \text{(b)}
$$

(901)

where multiobjective parameter $\lambda\in\mathbb{R}_+$ is quite large ($\lambda\approx 1\mathrm{E}8$) so as to enforce the equality constraint: $P\,\mathrm{vec}\,\mathcal{U}-f=\mathbf{0} \Leftrightarrow \|P\,\mathrm{vec}\,\mathcal{U}-f\|_2^2=0$ (§A.7.1). We introduce a direction vector $y\in\mathbb{R}_+^{4n^2}$ as part of a convex iteration (§4.5.3) to overcome that known suboptimal minimization of discrete image-gradient cardinality: *id est*, there exists a vector $y^\star$ with entries $y_i^\star\in\{0,1\}$ such that

$$
\begin{array}{c}
\underset{\mathcal{U}}{\text{minimize}} \quad \|\Psi\,\mathrm{vec}\,\mathcal{U}\|_0 \\
\text{subject to} \quad P\,\mathrm{vec}\,\mathcal{U} = f
\end{array}
\equiv
\underset{\mathcal{U}}{\text{minimize}} \ \langle\,|\Psi\,\mathrm{vec}\,\mathcal{U}|\,,\,y^\star\rangle + \tfrac{1}{2}\lambda\|P\,\mathrm{vec}\,\mathcal{U} - f\|_2^2 \qquad \text{(902)}
$$

Existence of such a $y^\star$, complementary to an optimal vector $\Psi\,\mathrm{vec}\,\mathcal{U}^\star$, is obvious by definition of global optimality $\langle\,|\Psi\,\mathrm{vec}\,\mathcal{U}^\star|\,,\,y^\star\rangle=0$ (802) under which a cardinality-c optimal objective $\|\Psi\,\mathrm{vec}\,\mathcal{U}^\star\|_0$ is assumed to exist.

Because (901b) is an unconstrained convex problem, a zero objective function gradient is necessary and sufficient for optimality (§2.13.3); *id est*, (§D.2.1)

$$
\Psi^{\mathrm{T}}\delta(y)\,\mathrm{sgn}(\Psi\,\mathrm{vec}\,\mathcal{U}) + \lambda P^{\mathrm{H}}(P\,\mathrm{vec}\,\mathcal{U} - f) = \mathbf{0} \qquad (903)
$$

Because of P idempotence and Hermitian symmetry and $\mathrm{sgn}(x)=x/|x|$, this is equivalent to

$$
\lim_{\epsilon\to 0}\ \big(\Psi^{\mathrm{T}}\delta(y)\delta(|\Psi\,\mathrm{vec}\,\mathcal{U}| + \epsilon\mathbf{1})^{-1}\Psi + \lambda P\big)\,\mathrm{vec}\,\mathcal{U} = \lambda P f \qquad (904)
$$

where small positive constant $\epsilon\in\mathbb{R}_+$ has been introduced for invertibility. Speaking more analytically, introduction of ϵ serves to uniquely define the objective's gradient everywhere in the function domain; *id est*, it transforms absolute value in (901b) from a function differentiable almost everywhere into a differentiable function. An example of such a transformation in one dimension is illustrated in Figure 115. When small enough for practical purposes[4.63] ($\epsilon\approx 1\mathrm{E}\text{-}3$), we may ignore the limiting operation. Then the

---

[4.62] for images as small as 128×128 pixels. Obstacle to numerical solution is not a computer resource: *e.g.*, execution time, memory. The obstacle is, in fact, inadequate numerical precision. Even when all dependent equality constraints are manually removed, the best commercial solvers fail simply because computer numerics become nonsense; *id est*, numerical errors enter significant digits and the algorithm exits prematurely, loops indefinitely, or produces an infeasible solution.

[4.63] We are looking for at least 50dB image/error ratio from only 4.1% subsampled data (10 radial lines in k-space). With this setting of ϵ, we actually attain in excess of 100dB from a simple MATLAB program in about a minute on a 2006 vintage laptop Core 2 CPU (Intel T7400@2.16GHz, 666MHz FSB). By trading execution time and treating discrete image-gradient cardinality as a known quantity for this phantom, over 160dB is achievable.

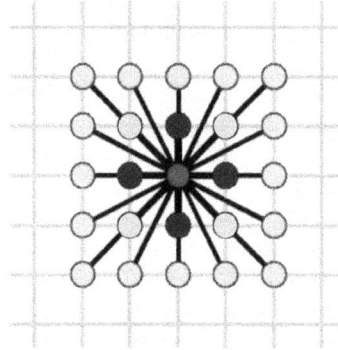

Figure 114: Neighboring-pixel stencil [365] for image-gradient estimation on Cartesian grid. Implementation selects adaptively from darkest four • about central. Continuous image-gradient from two pixels holds only in a limit. For discrete differences, better practical estimates are obtained when centered.

mapping, for $0 \preceq y \preceq \mathbf{1}$

$$\operatorname{vec}\mathcal{U}_{t+1} = \left(\Psi^{\mathrm{T}}\delta(y)\delta(|\Psi\operatorname{vec}\mathcal{U}_t| + \epsilon\mathbf{1})^{-1}\Psi + \lambda P\right)^{-1}\lambda P f \qquad (905)$$

is a *contraction* in $\mathcal{U}_t$ that can be solved recursively in t for its unique *fixed point*; *id est*, until $\mathcal{U}_{t+1} \to \mathcal{U}_t$. [233, p.300] [209, p.155] Calculating this inversion directly is not possible for large matrices on contemporary computers because of numerical precision, so instead we apply the *conjugate gradient* method of solution to

$$\left(\Psi^{\mathrm{T}}\delta(y)\delta(|\Psi\operatorname{vec}\mathcal{U}_t| + \epsilon\mathbf{1})^{-1}\Psi + \lambda P\right)\operatorname{vec}\mathcal{U}_{t+1} = \lambda P f \qquad (906)$$

which is linear in $\mathcal{U}_{t+1}$ at each recursion in the MATLAB program [379].[4.64]

Observe that P (893), in the equality constraint from problem (901a), is not a fat matrix.[4.65] Although number of Fourier samples taken is equal to the number of nonzero entries in binary mask Φ, matrix P is square but never actually formed during computation. Rather, a two-dimensional fast Fourier transform of $\mathcal{U}$ is computed followed by masking with $\Theta\Phi\Theta$ and then an inverse fast Fourier transform. This technique significantly reduces memory requirements and, together with contraction method of solution, is the principal reason for relatively fast computation.

[4.64]Conjugate gradient method requires positive definiteness. [156, §4.8.3.2]
[4.65]Fat is typical of compressed sensing problems; *e.g.*, [70] [77].

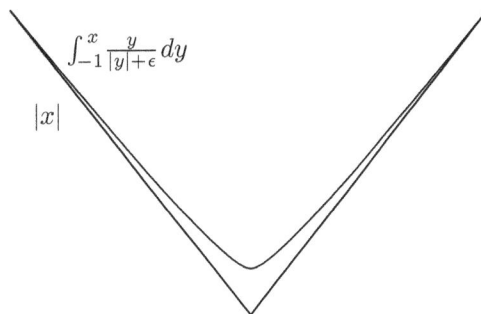

Figure 115: Real absolute value function $f_2(x) = |x|$ on $x \in [-1, 1]$ from Figure **69**b superimposed upon integral of its derivative at $\epsilon = 0.05$ which smooths objective function.

convex iteration

By *convex iteration* we mean alternation of solution to (901a) and (907) until convergence. Direction vector y is initialized to $\mathbf{1}$ until the first fixed point is found; which means, the contraction recursion begins calculating a (1-norm) solution $\mathcal{U}^\star$ to (899) via problem (901b). Once $\mathcal{U}^\star$ is found, vector y is updated according to an estimate of discrete image-gradient cardinality c: Sum of the $4n^2 - c$ smallest entries of $|\Psi \, \text{vec} \, \mathcal{U}^\star| \in \mathbb{R}^{4n^2}$ is the optimal objective value from a linear program, for $0 \leq c \leq 4n^2 - 1$ (524)

$$\sum_{i=c+1}^{4n^2} \pi(|\Psi \, \text{vec} \, \mathcal{U}^\star|)_i \;\; = \;\; \begin{array}{l} \underset{y \in \mathbb{R}^{4n^2}}{\text{minimize}} \quad \langle |\Psi \, \text{vec} \, \mathcal{U}^\star| \, , \, y \rangle \\ \text{subject to} \quad 0 \preceq y \preceq \mathbf{1} \\ \phantom{\text{subject to}} \quad y^{\mathrm{T}} \mathbf{1} = 4n^2 - c \end{array} \qquad (907)$$

where π is the nonlinear permutation-operator sorting its vector argument into nonincreasing order. *An optimal solution y to (907), that is an extreme point of its feasible set, is known in closed form: it has 1 in each entry corresponding to the $4n^2 - c$ smallest entries of $|\Psi \, \text{vec} \, \mathcal{U}^\star|$ and has 0 elsewhere* (page 293). Updated image $\mathcal{U}^\star$ is assigned to $\mathcal{U}_t$, the contraction is recomputed solving (901b), direction vector y is updated again, and so on until convergence which is guaranteed by virtue of a monotonically nonincreasing real sequence of objective values in (901a) and (907).

There are two features that distinguish problem formulation (901b) and our particular implementation of it [379, MATLAB code]:

1) An image-gradient estimate may engage any combination of four adjacent pixels. In other words, the algorithm is not locked into a four-point gradient estimate (Figure **114**); number of points constituting an estimate is directly determined by direction vector y .[4.66] Indeed, we find only $c = 5092$ zero entries in $y^\star$ for the Shepp-Logan phantom; meaning, discrete image-gradient sparsity is actually closer to 1.9% than the 3% reported elsewhere; *e.g.,* [364, §IIB].

[4.66]This adaptive gradient was not contrived. It is an artifact of the convex iteration method for minimal cardinality solution; in this case, cardinality minimization of a discrete image-gradient.

2) Numerical precision of the fixed point of contraction (905) ($\approx$1E-2 for perfect reconstruction @$-$103dB error) is a parameter to the implementation; meaning, direction vector y is updated after contraction begins but prior to its culmination. Impact of this idiosyncrasy tends toward simultaneous optimization in variables $\mathcal{U}$ and y while insuring y settles on a boundary point of its feasible set (nonnegative hypercube slice) in (907) at every iteration; for only a boundary point[4.67] can yield the sum of smallest entries in $|\Psi \operatorname{vec} \mathcal{U}^\star|$.

Perfect reconstruction of the Shepp-Logan phantom (at 103dB image/error) is achieved in a MATLAB minute with 4.1% subsampled data (2671 complex samples); well below an 11% least lower bound predicted by the sparse sampling theorem. Because reconstruction approaches optimal solution to a 0-norm problem, minimum number of Fourier-domain samples is bounded below by cardinality of discrete image-gradient at 1.9%.   □

4.6.0.0.14 Exercise. *Contraction operator.*
Determine conditions on λ and ϵ under which (905) is a contraction and $\Psi^{\mathrm{T}}\delta(y)\delta(|\Psi \operatorname{vec}\mathcal{U}_t| + \epsilon\mathbf{1})^{-1}\Psi + \lambda P$ from (906) is positive definite.   ▼

4.6.0.0.15 Example. *Eternity* II.
A tessellation puzzle game, playable by children, commenced world-wide in July 2007; introduced in London by Christopher Walter Monckton, 3rd Viscount Monckton of Brenchley. Called Eternity II, its name derives from an estimate of time that would pass while trying all allowable tilings of puzzle pieces before obtaining a complete solution. By the end of 2008, a complete solution had not yet been found although a $10,000 USD prize was awarded for a high score 467 (out of $480 = 2\sqrt{M}(\sqrt{M}-1)$) obtained by heuristic methods.[4.68] No prize was awarded for 2009 and 2010. Game-rules state that a $2M prize would be awarded to the first person who completely solves the puzzle before December 31, 2010, but the prize remains unclaimed after the deadline.

The full game comprises $M = 256$ square pieces and a 16×16 gridded board (Figure 117) whose complete tessellation is considered *NP-hard*.[4.69] [351] [109] A player may tile, retile, and rotate pieces, indexed 1 through 256, in any order face-up on the square board. Pieces are immutable in the sense that each is characterized by 4 colors (and their uniquely associated symbols), one at each edge, which are not necessarily the same per piece or from piece to piece; *id est*, different pieces may or may not have some edge-colors in common. There are $L = 22$ distinct edge-colors plus a solid grey. The object of the game is to completely tile the board with pieces whose touching edges have identical color. The boundary of the board must be colored grey.

[4.67]Simultaneous optimization of these two variables $\mathcal{U}$ and y should never be a pinnacle of aspiration; for then, optimal y might not attain a boundary point.

[4.68]That score means all but a few of the 256 pieces had been placed successfully (including the mandatory piece). Although distance between 467 to 480 is relatively small, there is apparently vast distance to a solution because no complete solution was found in 2009.

[4.69]Even so, combinatorial-intensity brute-force backtracking methods can solve similar puzzles in minutes given $M = 196$ pieces on a 14×14 test board; as demonstrated by Yannick Kirschhoffer. There is a steep rise in level of difficulty going to a 15×15 board.

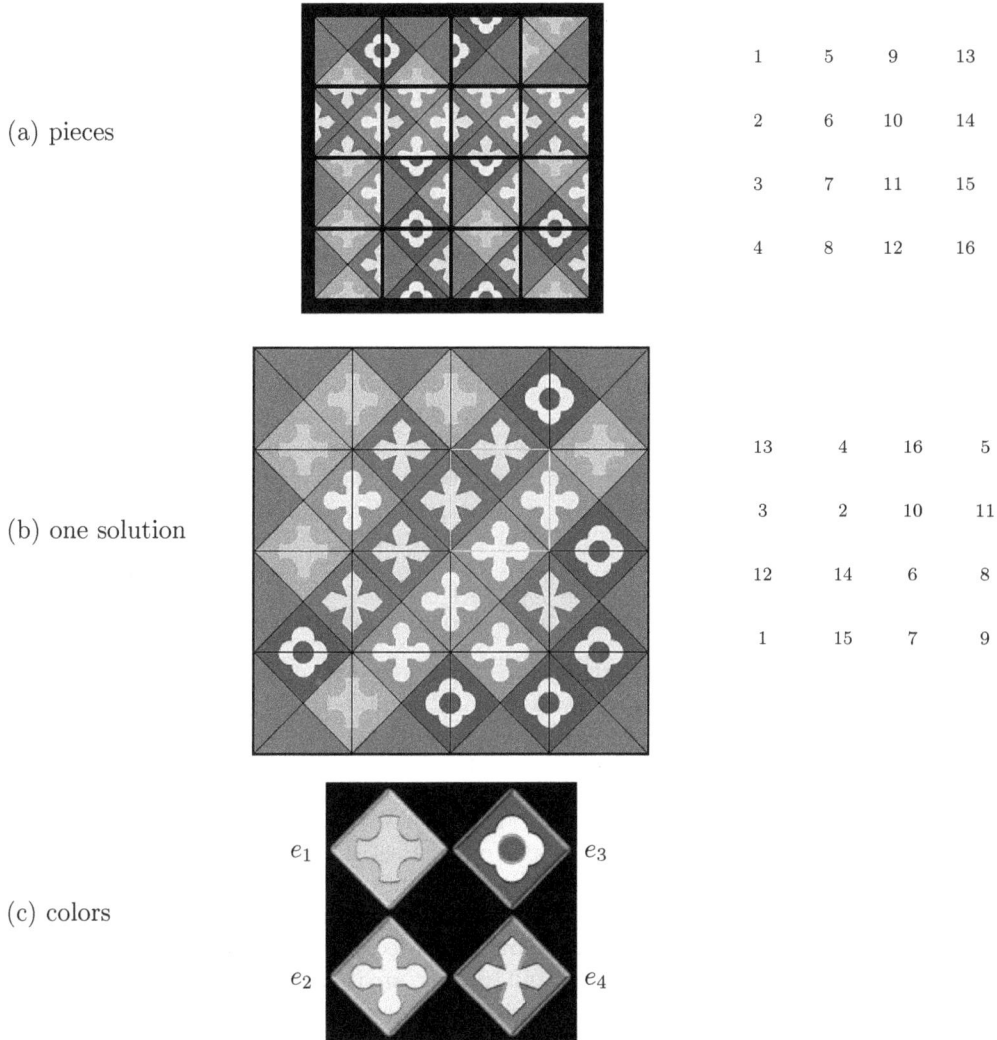

(a) pieces

1	5	9	13
2	6	10	14
3	7	11	15
4	8	12	16

(b) one solution

13	4	16	5
3	2	10	11
12	14	6	8
1	15	7	9

(c) colors

Figure 116: *Eternity* II is a board game in the puzzle genre. **(a)** Shown are all of the 16 puzzle pieces (indexed as in the tableau alongside) from a scaled-down computerized demonstration-version on the TOMY website. Puzzle pieces are square and triangularly partitioned into four colors (with associated symbols). Pieces may be moved, removed, and rotated at random on a 4×4 board. **(b)** Illustrated is one complete solution to this puzzle whose solution is not unique. The piece, whose border is lightly outlined, was placed last in this realization. There is no mandatory piece placement as for the full game, except the grey board-boundary. Solution time for a human is typically on the order of a minute. **(c)** This puzzle has four colors, indexed 1 through 4; grey corresponds to **0**.

full-game rules

1) Any puzzle piece may be rotated face-up in quadrature and placed or replaced on the square board.

2) Only one piece may occupy any particular cell on the board.

3) All adjacent pieces must match in color (and symbol) at their touching edges.

4) Solid grey edges must appear all along the board's boundary.

5) One mandatory piece (numbered 139 in the full game) must have a predetermined orientation in a predetermined cell on the board.

6) The board must be tiled completely (*covered*).

A scaled-down demonstration version of the game is illustrated in Figure 116. Differences between the full game (Figure 117) and scaled-down game are the number of edge-colors L (22 *versus* 4, ignoring solid grey), number of pieces M (256 *versus* 16), and a single mandatory piece placement interior to the board in the full game. The scaled-down game has four distinct edge-colors, plus a solid grey, whose coding is illustrated in Figure 116c.

- $L = 22$ distinct edge-colors and number of puzzle pieces $M = 256$ and board-dimension $\sqrt{M} \times \sqrt{M} = 16 \times 16$ for the full game.

- There are $L = 4$ distinct edge-colors and $M = 16$ pieces and dimension $\sqrt{M} \times \sqrt{M} = 4 \times 4$ for the scaled-down demonstration board.

Euclidean distance intractability

If each square puzzle piece were characterized by four points in quadrature, one point representing board coordinates and color per edge, then Euclidean distance geometry would be suitable for solving this puzzle. Since all interpoint distances per piece are known, this game may be regarded as a Euclidean distance matrix completion problem[4.70] in $\mathbb{EDM}^{4M}$. Because distance information provides for reconstruction of point position to within an isometry (§5.5), piece translation and rotation are isometric transformations that abide by rules of the game.[4.71] Convex constraints can be devised to prevent puzzle-piece reflection and to quantize rotation such that piece-edges stay aligned with the board boundary. (§5.5.2.0.1)

But manipulating such a large EDM is too numerically difficult for contemporary general-purpose semidefinite program solvers which incorporate interior-point methods; indeed, they are hard-pressed to find a solution for variable matrices of dimension as small as 100. Our challenge, therefore, is to express this game's rules as constraints in a convex and numerically tractable way so as to find one solution from a googol of possible combinations.[4.72]

[4.70](§6.7) Were edge-points ordered sequentially with piece number, then this EDM would have a block-diagonal structure of known entries.

[4.71]Translation occurs when a piece moves on the board in Figure 117, rotation occurs when colors are aligned with a neighboring piece.

[4.72]There exists at least one solution; their exact number is unknown although Monckton insists they

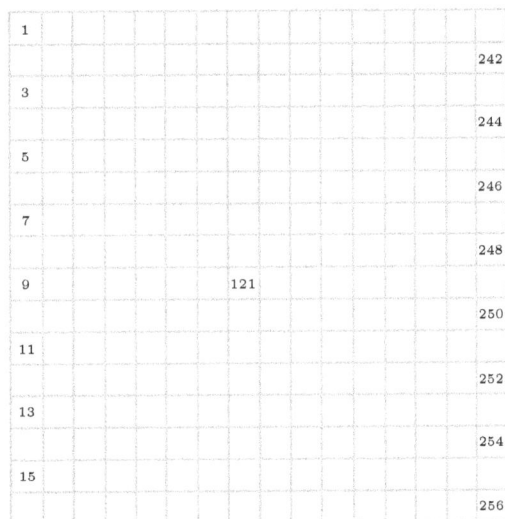

Figure 117: *Eternity* II full-game board (16×16, not actual size). Grid facilitates piece placement within unit-square cell; one piece per cell.

permutation polyhedron strategy

To each puzzle piece, from a set of M pieces $\{P_i \,,\ i=1\ldots M\}$, assign an index i representing a unique piece-number. Each square piece is characterized by four colors, in quadrature, corresponding to its four edges. Each color $p_{ij} \in \mathbb{R}^L$ is represented by $e_\ell \in \mathbb{R}^L$ an L-dimensional standard basis vector or $\mathbf{0}$ if grey. These four edge-colors are represented in a $4 \times L$-dimensional matrix; one matrix per piece

$$P_i \triangleq [\,p_{i1} \ \ p_{i2} \ \ p_{i3} \ \ p_{i4}\,]^{\mathrm{T}} \in \mathbb{R}^{4 \times L}, \qquad i=1\ldots M \tag{908}$$

In other words, each distinct nongrey color is assigned a unique corresponding index $\ell \in \{1\ldots L\}$ identifying a standard basis vector $e_\ell \in \mathbb{R}^L$ (Figure 116c) that becomes a vector $p_{ij} \in \{e_1 \ldots e_L\,,\ \mathbf{0}\} \subset \mathbb{R}^L$ constituting matrix P_i representing a particular piece. Rows $\{p_{ij}^{\mathrm{T}}\,,\ j=1\ldots 4\}$ of P_i are ordered counterclockwise as in Figure 118. Color data is given in Figure 119 for the demonstration board. Then matrix P_i describes the ith piece, excepting its orientation and location on the board.

Our strategy to solve Eternity II is to first vectorize the board, with respect to the whole pieces, and then relax a very hard combinatorial problem: All pieces are initially placed, as in Figure 119, in order of their given index. Then the vectorized game-board

---

number in thousands. Ignoring boundary constraints and the single mandatory piece placement in the full game, a loose upper bound on number of combinations is $M!\,4^M = 256!\,4^{256}$. That number gets loosened: $150638!/(256!(150638-256)!)$ after presolving Eternity II (929).

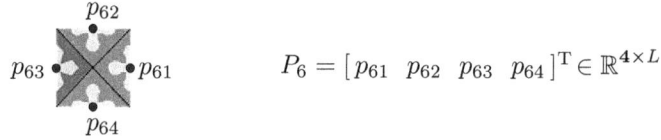

$$P_6 = [\,p_{61}\ \ p_{62}\ \ p_{63}\ \ p_{64}\,]^{\mathrm{T}} \in \mathbb{R}^{4 \times L}$$

Figure 118: Demo-game piece P_6 illustrating edge-color $\bullet$ $p_{6j} \in \mathbb{R}^L$ counterclockwise ordering in j beginning from right.

has initial state, as in Figure 119, represented within a matrix

$$P \triangleq \begin{bmatrix} P_1 \\ \vdots \\ P_M \end{bmatrix} \in \mathbb{R}^{4M \times L} \tag{909}$$

Moving pieces about the square board all at once corresponds to permuting pieces P_i on the vectorized board represented by matrix P, while rotating the ith piece is equivalent to circularly shifting row indices of P_i (rowwise permutation). This permutation problem, as stated, is doubly combinatorial ($M!\,4^M$ combinations) because we must find a permutation of pieces ($M!$)

$$\Xi \in \mathbb{R}^{M \times M} \tag{910}$$

and a rotation $\Pi_i \in \mathbb{R}^{4 \times 4}$ of each individual piece (4^M) that solve the puzzle;

$$(\Xi \otimes I_4)\Pi P = (\Xi \otimes I_4)\begin{bmatrix} \Pi_1 P_1 \\ \vdots \\ \Pi_M P_M \end{bmatrix} \in \mathbb{R}^{4M \times L} \tag{911}$$

where

$$\Pi_i \in \{\pi_1\,,\ \pi_2\,,\ \pi_3\,,\ \pi_4\} \triangleq \left\{ \begin{bmatrix} 1 & 0 & 0 & 0 \\ 0 & 1 & 0 & 0 \\ 0 & 0 & 1 & 0 \\ 0 & 0 & 0 & 1 \end{bmatrix}, \begin{bmatrix} 0 & 1 & 0 & 0 \\ 0 & 0 & 1 & 0 \\ 0 & 0 & 0 & 1 \\ 1 & 0 & 0 & 0 \end{bmatrix}, \begin{bmatrix} 0 & 0 & 1 & 0 \\ 0 & 0 & 0 & 1 \\ 1 & 0 & 0 & 0 \\ 0 & 1 & 0 & 0 \end{bmatrix}, \begin{bmatrix} 0 & 0 & 0 & 1 \\ 1 & 0 & 0 & 0 \\ 0 & 1 & 0 & 0 \\ 0 & 0 & 1 & 0 \end{bmatrix} \right\} \tag{912}$$

$$\Pi \triangleq \begin{bmatrix} \Pi_1 & & \mathbf{0} \\ & \ddots & \\ \mathbf{0} & & \Pi_M \end{bmatrix} \in \mathbb{R}^{4M \times 4M} \tag{913}$$

Initial game-board state P (909) corresponds to $\Xi = I$ and $\Pi_i = \pi_1 = I\ \forall i$. Circulant [174] permutation matrices $\{\pi_1\,,\ \pi_2\,,\ \pi_3\,,\ \pi_4\} \subset \mathbb{R}^{4 \times 4}$ correspond to clockwise piece-rotations $\{0°,\ 90°,\ 180°,\ 270°\}$.

Rules of the game dictate that adjacent pieces on the square board have colors that match at their touching edges as in Figure 116b.[4.73] A complete match is therefore equivalent to demanding that a constraint, comprising numeric color differences between $2\sqrt{M}(\sqrt{M}-1)$ touching edges, vanish. Because the vectorized board layout is fixed and

---

[4.73]Piece adjacencies on the square board map linearly to the vectorized board, of course.

P_1 $[\,e_3 \quad \mathbf{0} \quad \mathbf{0} \quad e_1\,]^{\mathrm{T}}$

P_2 $[\,e_2 \quad e_4 \quad e_4 \quad e_4\,]^{\mathrm{T}}$

P_3 $[\,e_2 \quad e_1 \quad \mathbf{0} \quad e_1\,]^{\mathrm{T}}$

P_4 $[\,e_4 \quad e_1 \quad \mathbf{0} \quad e_1\,]^{\mathrm{T}}$

P_5 $[\,\mathbf{0} \quad \mathbf{0} \quad e_3 \quad e_1\,]^{\mathrm{T}}$

P_6 $[\,e_2 \quad e_2 \quad e_4 \quad e_2\,]^{\mathrm{T}}$

P_7 $[\,e_2 \quad e_3 \quad \mathbf{0} \quad e_3\,]^{\mathrm{T}}$

P_8 $[\,e_4 \quad e_3 \quad \mathbf{0} \quad e_3\,]^{\mathrm{T}}$

P_9 $[\,\mathbf{0} \quad e_3 \quad e_3 \quad \mathbf{0}\,]^{\mathrm{T}}$

P_{10} $[\,e_2 \quad e_2 \quad e_4 \quad e_4\,]^{\mathrm{T}}$

P_{11} $[\,e_2 \quad e_3 \quad \mathbf{0} \quad e_1\,]^{\mathrm{T}}$

P_{12} $[\,e_4 \quad e_1 \quad \mathbf{0} \quad e_3\,]^{\mathrm{T}}$

P_{13} $[\,\mathbf{0} \quad e_1 \quad e_1 \quad \mathbf{0}\,]^{\mathrm{T}}$

P_{14} $[\,e_2 \quad e_2 \quad e_4 \quad e_4\,]^{\mathrm{T}}$

P_{15} $[\,e_2 \quad e_1 \quad \mathbf{0} \quad e_3\,]^{\mathrm{T}}$

P_{16} $[\,e_4 \quad e_3 \quad \mathbf{0} \quad e_1\,]^{\mathrm{T}}$

Figure 119: Vectorized demo-game board has $M=16$ matrices in $\mathbb{R}^{4\times L}$ describing initial state of game pieces; 4 colors per puzzle-piece (Figure 118), $L=4$ colors total in game (Figure 116c). So color difference measurement remains unweighted, standard basis vectors in $\mathbb{R}^L$ represent color.

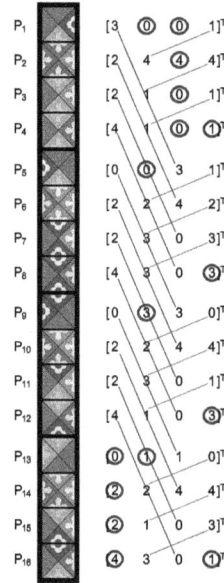

Figure 120: ◯ indicate boundary β, line segments indicate differences Δ. Entries are indices ℓ to standard basis vectors $e_\ell \in \mathbb{R}^L$ from Figure 119.

its cells are loaded or reloaded with pieces during play, locations of adjacent edges in $\mathbb{R}^{4M \times L}$ are known *a priori*. We need simply form differences between colors from adjacent edges of pieces loaded into those known locations (Figure 120). Each difference may be represented by a constant vector from a set $\{\Delta_i \in \mathbb{R}^{4M}, \ i = 1 \ldots 2\sqrt{M}(\sqrt{M}-1)\}$. Defining sparse constant fat matrix

$$\Delta \triangleq \begin{bmatrix} \Delta_1^{\mathrm{T}} \\ \vdots \\ \Delta_{2\sqrt{M}(\sqrt{M}-1)}^{\mathrm{T}} \end{bmatrix} \in \mathbb{R}^{2\sqrt{M}(\sqrt{M}-1) \times 4M} \tag{914}$$

whose entries belong to $\{-1, 0, 1\}$, then the desired constraint is

$$\Delta(\Xi \otimes I_4)\Pi P = \mathbf{0} \in \mathbb{R}^{2\sqrt{M}(\sqrt{M}-1) \times L} \tag{915}$$

Boundary of the square board must be colored grey. This means there are $4\sqrt{M}$ boundary locations in $\mathbb{R}^{4M \times L}$ that must have value $\mathbf{0}^{\mathrm{T}}$. These can all be lumped into one equality constraint

$$\beta^{\mathrm{T}}(\Xi \otimes I_4)\Pi P \mathbf{1} = 0 \tag{916}$$

where $\beta \in \mathbb{R}^{4M}$ is a sparse constant vector having entries in $\{0, 1\}$ complementary to the known $4\sqrt{M}$ zeros (Figure 120).

By combining variables:

$$\Phi \triangleq (\Xi \otimes I_4)\Pi \in \mathbb{R}^{4M \times 4M} \tag{917}$$

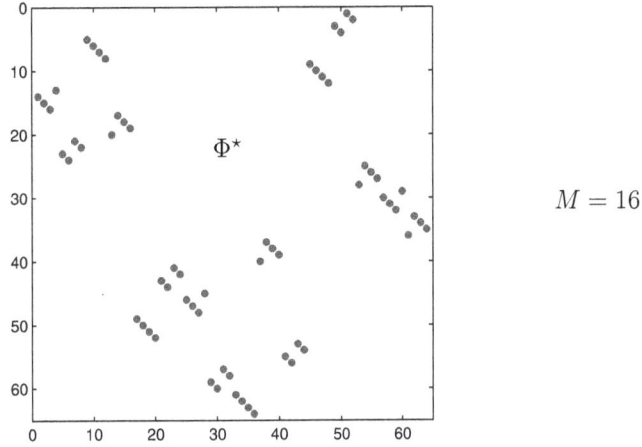

$M = 16$

Figure 121: Sparsity pattern for composite permutation matrix $\Phi^\star \in \mathbb{R}^{4M \times 4M}$ representing solution from Figure 116b. Each four-point cluster represents a circulant permutation matrix from (912). Any $M = 16$-piece solution may be verified on the TOMY website.

this square matrix becomes a structured permutation matrix replacing the product of permutation matrices. Partition the composite variable Φ into blocks

$$\Phi \triangleq \begin{bmatrix} \phi_{11} & \cdots & \phi_{1M} \\ \vdots & \ddots & \vdots \\ \phi_{M1} & \cdots & \phi_{MM} \end{bmatrix} \in \mathbb{R}^{4M \times 4M} \tag{918}$$

where $\Phi_{ij}^\star \in \{0, 1\}$ because (912)

$$\phi_{ij}^\star \in \{\pi_1, \pi_2, \pi_3, \pi_4, \mathbf{0}\} \subset \mathbb{R}^{4 \times 4} \tag{919}$$

An optimal composite permutation matrix $\Phi^\star$ is represented pictorially in Figure 121. Now we ask what are necessary conditions on $\Phi^\star$ at optimality:

- $4M$-sparsity and nonnegativity.

- Each column has one 1. Each row has one 1.

- Entries along each and every diagonal of each and every 4×4 block $\phi_{ij}^\star$ are equal.

- Corner pair of 2×2 submatrices on antidiagonal of each and every 4×4 block $\phi_{ij}^\star$ are equal.

We want an objective function whose global optimum, if attained, certifies that the puzzle has been solved. Then, in terms of this Φ partitioning (918), the Eternity II problem is a minimization of cardinality

$$
\begin{aligned}
\underset{\Phi \in \mathbb{R}^{4M \times 4M}}{\text{minimize}} \quad & \| \operatorname{vec} \Phi \|_0 \\
\text{subject to} \quad & \Delta \Phi P = \mathbf{0} \\
& \beta^{\mathrm{T}} \Phi P \mathbf{1} = 0 \\
& \Phi \mathbf{1} = \mathbf{1} \\
& \Phi^{\mathrm{T}} \mathbf{1} = \mathbf{1} \\
& (I \otimes R_{\mathrm{d}}) \Phi (I \otimes R_{\mathrm{d}}^{\mathrm{T}}) = (I \otimes S_{\mathrm{d}}) \Phi (I \otimes S_{\mathrm{d}}^{\mathrm{T}}) \\
& (I \otimes R_{\phi}) \Phi (I \otimes S_{\phi}^{\mathrm{T}}) = (I \otimes S_{\phi}) \Phi (I \otimes R_{\phi}^{\mathrm{T}}) \\
& (e_{121} \otimes I_4)^{\mathrm{T}} \Phi (e_{139} \otimes I_4) = \pi_3 \\
& \Phi \geq \mathbf{0}
\end{aligned}
\tag{920}
$$

which is convex in the constraints where e_{121}, $e_{139} \in \mathbb{R}^M$ are members of the standard basis representing mandatory piece placement in the full game,[4.74] and where

$$
R_{\mathrm{d}} \triangleq \begin{bmatrix} 1 & 0 & & \\ & 1 & 0 & \\ & & 1 & 0 \end{bmatrix} \in \mathbb{R}^{\mathbf{3 \times 4}}, \quad S_{\mathrm{d}} \triangleq \begin{bmatrix} 0 & 1 & & \\ & 0 & 1 & \\ & & 0 & 1 \end{bmatrix} \in \mathbb{R}^{\mathbf{3 \times 4}} \tag{921}
$$

$$
R_{\phi} \triangleq \begin{bmatrix} 1 & 0 & 0 & 0 \\ 0 & 1 & 0 & 0 \end{bmatrix} \in \mathbb{R}^{\mathbf{2 \times 4}}, \quad S_{\phi} \triangleq \begin{bmatrix} 0 & 0 & 1 & 0 \\ 0 & 0 & 0 & 1 \end{bmatrix} \in \mathbb{R}^{\mathbf{2 \times 4}} \tag{922}
$$

These matrices R and S enforce circulance.[4.75] Mandatory piece placement in the full game requires the equality constraint in π_3. Constraints $\Phi\mathbf{1}=\mathbf{1}$ and $\Phi^{\mathrm{T}}\mathbf{1}=\mathbf{1}$ confine nonnegative Φ to the permutation polyhedron (100) in $\mathbb{R}^{4M \times 4M}$. The feasible set of problem (920) is an intersection of the permutation polyhedron with a large number of hyperplanes. Any vertex in the permutation polyhedron, which is the convex hull of permutation matrices, has minimal cardinality. (§2.3.2.0.4) The intersection contains a vertex of the permutation polyhedron because a solution $\Phi^{\star}$ cannot otherwise be a permutation matrix; such a solution is known to exist, so it must also be a vertex of the intersection.[4.76]

In the vectorized variable, problem (920) is equivalent to

$$
\begin{aligned}
\underset{\Phi \in \mathbb{R}^{\mathbf{4M \times 4M}}}{\text{minimize}} \quad & \| \operatorname{vec} \Phi \|_0 \\
\text{subject to} \quad & (P^{\mathrm{T}} \otimes \Delta) \operatorname{vec} \Phi = \mathbf{0} \\
& (P\mathbf{1} \otimes \beta)^{\mathrm{T}} \operatorname{vec} \Phi = 0 \\
& (\mathbf{1}_{4M}^{\mathrm{T}} \otimes I_{4M}) \operatorname{vec} \Phi = \mathbf{1}_{4M} \\
& (I_{4M} \otimes \mathbf{1}_{4M}^{\mathrm{T}}) \operatorname{vec} \Phi = \mathbf{1}_{4M} \\
& (I \otimes R_{\mathrm{d}} \otimes I \otimes R_{\mathrm{d}} - I \otimes S_{\mathrm{d}} \otimes I \otimes S_{\mathrm{d}}) \operatorname{vec} \Phi = \mathbf{0} \\
& (I \otimes S_{\phi} \otimes I \otimes R_{\phi} - I \otimes R_{\phi} \otimes I \otimes S_{\phi}) \operatorname{vec} \Phi = \mathbf{0} \\
& (e_{139} \otimes I_4 \otimes e_{121} \otimes I_4)^{\mathrm{T}} \operatorname{vec} \Phi = \operatorname{vec} \pi_3 \\
& \Phi \geq \mathbf{0}
\end{aligned}
\tag{923}
$$

[4.74] meaning, that piece numbered 139 by the game designer must be placed in cell 121 on the vectorized board (Figure 117) with orientation π_3 (p.336).

[4.75] Since $\mathbf{0}$ is the trivial circulant matrix, application is democratic over all blocks ϕ_{ij}.

[4.76] *Vertex* means *zero-dimensional exposed face* (§2.6.1.0.1); intersection with a strictly supporting hyperplane. There can be no further intersection with a feasible affine subset that would enlarge that face; *id est*, a vertex of the permutation polyhedron persists in the feasible set.

This problem is abbreviated:

$$
\begin{array}{ll}
\underset{\Phi \in \mathbb{R}^{4M \times 4M}}{\text{minimize}} & \| \operatorname{vec} \Phi \|_0 \\
\text{subject to} & E \operatorname{vec} \Phi = \tau \\
& \Phi \geq 0
\end{array}
\tag{924}
$$

where $E \in \mathbb{R}^{2L\sqrt{M}(\sqrt{M}-1)+8M+13M^2+17 \times 16M^2}$ is sparse and optimal objective value is $4M$; dimension of E is $864{,}593 \times 1{,}048{,}576$. A compressed sensing paradigm [71] is not inherent here. To solve this by linear programming, a direction vector is introduced for cardinality minimization as in §4.5. It is critical, in this case, to add enough random noise to the direction vector so as to insure a vertex solution [96, p.158]. For the demonstration game, in fact, choosing a direction vector randomly will find an optimal solution in only a few iterations.[4.77] But for the full game, numerical errors prevent solution of (924); number of equality constraints 864,593 is too large.[4.78] So again, we reformulate the problem:

canonical Eternity II
Because each block ϕ_{ij} of Φ (918) is optimally circulant having only four degrees of freedom (919), we may take as variable every fourth column of Φ:

$$
\tilde{\Phi} \triangleq [\, \Phi(:,1) \ \ \Phi(:,5) \ \ \Phi(:,9) \ \cdots \ \Phi(:,4M-3)\,] \in \mathbb{R}^{4M \times M}
\tag{925}
$$

where $\tilde{\Phi}_{ij} \in \{0,1\}$. Then, for $e_i \in \mathbb{R}^4$

$$
\Phi = (\tilde{\Phi} \otimes e_1^{\mathrm{T}}) + (I \otimes \pi_4)(\tilde{\Phi} \otimes e_2^{\mathrm{T}}) + (I \otimes \pi_3)(\tilde{\Phi} \otimes e_3^{\mathrm{T}}) + (I \otimes \pi_2)(\tilde{\Phi} \otimes e_4^{\mathrm{T}}) \in \mathbb{R}^{4M \times 4M}
\tag{926}
$$

From §A.1.1 *no*.31 and *no*.40

$$
\begin{aligned}
\operatorname{vec} \Phi &= \left(I \otimes e_1 \otimes I_{4M} + I \otimes e_2 \otimes I \otimes \pi_4 + I \otimes e_3 \otimes I \otimes \pi_3 + I \otimes e_4 \otimes I \otimes \pi_2\right) \operatorname{vec} \tilde{\Phi} \\
&\triangleq Y \operatorname{vec} \tilde{\Phi} \in \mathbb{R}^{16M^2}
\end{aligned}
\tag{927}
$$

where $Y \in \mathbb{R}^{16M^2 \times 4M^2}$. Permutation polyhedron (100) now demands that each consecutive quadruple of adjacent rows of $\tilde{\Phi}$ sum to 1: $(I \otimes 1_4^{\mathrm{T}})\tilde{\Phi}1 = 1$. Constraints in R and S (which are most numerous) may be dropped because circulance of ϕ_{ij} is built into Φ-reconstruction formula (926). Eternity II (923) is thus equivalently transformed

$$
\begin{array}{ll}
\underset{\tilde{\Phi} \in \mathbb{R}^{4M \times M}}{\text{minimize}} & \| \operatorname{vec} \tilde{\Phi} \|_0 \\
\text{subject to} & (P^{\mathrm{T}} \otimes \Delta) Y \operatorname{vec} \tilde{\Phi} = 0 \\
& (P1 \otimes \beta)^{\mathrm{T}} Y \operatorname{vec} \tilde{\Phi} = 0 \\
& (1^{\mathrm{T}} \otimes I \otimes 1_4^{\mathrm{T}}) \operatorname{vec} \tilde{\Phi} = 1 \\
& (I \otimes 1_{4M}^{\mathrm{T}}) \operatorname{vec} \tilde{\Phi} = 1 \\
& (e_{139} \otimes e_1 \otimes e_{121} \otimes I_4)^{\mathrm{T}} Y \operatorname{vec} \tilde{\Phi} = \pi_3 e_1 \\
& \tilde{\Phi} \geq 0
\end{array}
\tag{928}
$$

[4.77] This can only mean: there are many optimal solutions. A simplex-method solver is required for numerical solution; interior-point methods will not work. A randomized direction vector also works for Clue Puzzles provided by the toy maker: similar 6×6 and 6×12 puzzles whose solution each provide a clue to solution of the full game. Even better is a nonnegative uniformly distributed randomized direction vector having $4M$ entries (M entries, in case (929)), corresponding to the largest entries of $\Phi^\star$, zeroed.
[4.78] Saunders' program `lusol` can reduce that number to 797,508 constraints by eliminating linearly dependent rows of matrix E, but that reduction is not enough to overcome numerical issues with the best contemporary linear program solvers.

whose optimal objective value is M with $\tilde{\Phi}^\star$-entries in $\{0, 1\}$ and where $e_1 \in \mathbb{R}^4$ (§A.1.1 *no.*39) and $e_{121}, e_{139} \in \mathbb{R}^M$. Number of equality constraints in abbreviation of reformulation (928)

$$
\begin{array}{ll}
\underset{\tilde{\Phi} \in \mathbb{R}^{4M \times M}}{\text{minimize}} & \| \operatorname{vec} \tilde{\Phi} \|_0 \\
\text{subject to} & \tilde{E} \operatorname{vec} \tilde{\Phi} = \tilde{\tau} \\
& \tilde{\Phi} \geq 0
\end{array}
\tag{929}
$$

is now 11,077 (an order of magnitude fewer constraints than (924)) where sparse $\tilde{E} \in \mathbb{R}^{2L\sqrt{M}(\sqrt{M}-1)+2M+5 \times 4M^2}$ replaces the E matrix from (924). Number of columns in matrix $\tilde{E}$ has been reduced from a million to 262,144; dimension of $\tilde{E}$ goes to $11,077 \times 262,144$. But this dimension remains out of reach of most highly regarded contemporary academic and commercial linear program solvers because of numerical failure; especially disappointing insofar as sparsity of $\tilde{E}$ is high with only 0.07% nonzero entries $\in \{-1, 0, 1, 2\}$; element $\{2\}$ arising only in the β constraint which is soon to disappear after presolving.

Variable $\operatorname{vec} \tilde{\Phi}$ itself is too large in dimension. Notice that the constraint in β, which zeroes the board at its edges, has all positive coefficients. The zero sum means that all $\operatorname{vec} \tilde{\Phi}$ entries, corresponding to nonzero entries in row vector $(P\mathbf{1} \otimes \beta)^{\mathrm{T}} Y$, must be zero. For the full game, this means we may immediately eliminate 57,840 variables from 262,144. After zero-row and dependent row removal, dimension of $\tilde{E}$ goes to $10,054 \times 204,304$ with entries in $\{-1, 0, 1\}$.

polyhedral cone theory

Eternity II problem (929) constraints are interpretable in the language of convex cones: The columns of matrix $\tilde{E}$ constitute a set of generators for a pointed polyhedral cone $\mathcal{K} = \{\tilde{E} \operatorname{vec} \tilde{\Phi} \mid \tilde{\Phi} \geq 0\}$. (§2.12.2.2) Even more intriguing is the observation: vector $\tilde{\tau}$ resides on that polyhedral cone's boundary. (§2.13.4.2.4)

We may apply techniques from §2.13.4.3 to prune generators not belonging to the smallest face of that cone, to which $\tilde{\tau}$ belongs, because generators of that smallest face must hold a minimal cardinality solution. Matrix dimension is thereby reduced:[4.79] The ith column $\tilde{E}(:,i)$ of matrix $\tilde{E}$ belongs to the smallest face $\mathcal{F}$ of $\mathcal{K}$ that contains $\tilde{\tau}$ if and only if

$$
\begin{array}{ll}
\underset{\tilde{\Phi} \in \mathbb{R}^{4M \times M}, \; \mu \in \mathbb{R}}{\text{find}} & \tilde{\Phi}, \mu \\
\text{subject to} & \mu \tilde{\tau} - \tilde{E}(:,i) = \tilde{E} \operatorname{vec} \tilde{\Phi} \\
& \operatorname{vec} \tilde{\Phi} \succeq 0
\end{array}
\tag{375}
$$

is feasible. By a transformation of Saunders, this linear feasibility problem is the same as

---

[4.79]Column elimination can be quite dramatic but is dependent upon problem geometry. By method of convex cones, we discard 53,666 more columns via Saunders' pdco; a total of 111,506 columns removed from 262,144 leaving all remaining column entries unaltered. Following dependent row removal via lusol, dimension of $\tilde{E}$ becomes $7,362 \times 150,638$; call that A. Any process of discarding rows and columns prior to optimization is *presolving*.

$$\begin{aligned}
\underset{\tilde{\Phi}\in\mathbb{R}^{4M\times M},\ \mu\in\mathbb{R}}{\text{find}}\quad & \tilde{\Phi},\mu \\
\text{subject to}\quad & \tilde{E}\operatorname{vec}\tilde{\Phi}=\mu\tilde{\tau} \\
& \operatorname{vec}\tilde{\Phi}\succeq 0 \\
& (\operatorname{vec}\tilde{\Phi})_i\geq 1
\end{aligned} \qquad (930)$$

A minimal cardinality solution to Eternity II (929) implicitly constrains $\tilde{\Phi}^\star$ to be binary. So this test (930) of membership to $\mathcal{F}(\mathcal{K}\ni\tilde{\tau})$ may be tightened to a test of $(\operatorname{vec}\tilde{\Phi})_i=1$; *id est*, for $i=1\ldots 4M^2$ distinct feasibility problems

$$\begin{aligned}
\underset{\tilde{\Phi}\in\mathbb{R}^{4M\times M}}{\text{find}}\quad & \tilde{\Phi} \\
\text{subject to}\quad & \tilde{E}\operatorname{vec}\tilde{\Phi}=\tilde{\tau} \\
& \operatorname{vec}\tilde{\Phi}\succeq 0 \\
& (\operatorname{vec}\tilde{\Phi})_i=1
\end{aligned} \qquad (931)$$

whose feasible set is a proper subset of that in (930). Real variable μ can be set to 1 because if it must not be, then feasible $(\operatorname{vec}\tilde{\Phi})_i=1$ could not be feasible to Eternity II (929). If infeasible here in (931), then the only choice remaining for $(\operatorname{vec}\tilde{\Phi})_i$ is 0; meaning, column $\tilde{E}(:,i)$ may be discarded after all columns have been tested. This tightened problem (931) therefore tells us two things when feasible: $\tilde{E}(:,i)$ belongs to the smallest face of $\mathcal{K}$ that contains $\tilde{\tau}$, and $(\operatorname{vec}\tilde{\Phi})_i$ constitutes a nonzero vertex-coordinate of permutation polyhedron (100). After presolving via this conic pruning method (with subsequent zero-row and dependent row removal), dimension of $\tilde{E}$ goes to $7{,}362\times 150{,}638$.

generators of smallest face are conically independent
Designate $A\in\mathbb{R}^{7362\times 150638}\triangleq\mathbb{R}^{m\times n}$ as matrix $\tilde{E}$ after having discarded all generators not in the smallest face $\mathcal{F}$ of cone $\mathcal{K}$ that contains $\tilde{\tau}$. The Eternity II problem (929) becomes

$$\begin{aligned}
\underset{x\in\mathbb{R}^n}{\text{minimize}}\quad & \|x\|_0 \\
\text{subject to}\quad & Ax=b \\
& x\succeq 0
\end{aligned} \qquad (932)$$

To further prune all generators relatively interior to that smallest face, we may subsequently test for conic dependence as described in §2.10 (280): for $i=1\ldots 150{,}638$

$$\begin{aligned}
\text{find}\quad & x \\
\text{subject to}\quad & Ax=A(:,i) \\
& x\succeq 0 \\
& x_i=0
\end{aligned} \qquad (933)$$

where x is $\operatorname{vec}\tilde{\Phi}$ corresponding to columns of $\tilde{E}$ not previously discarded by (931).[4.80] If feasible, then column $A(:,i)$ is a conically dependent generator of the smallest face and must be discarded from matrix A before proceeding with test of remaining columns. It turns out, for Eternity II: generators of the smallest face, previously found via (931),

---

[4.80] Discarded entries in $\operatorname{vec}\tilde{\Phi}$ are optimally 0.

comprise a minimal set; *id est*, (933) is never feasible and so no column of A can be discarded (A remains unaltered).[4.81]

affinity for maximization
Designate vector $b \in \mathbb{R}^m$ to be $\tilde{\tau}$ after discarding all entries corresponding to dependent rows in $\tilde{E}$; *id est*, b is $\tilde{\tau}$ subsequent to presolving. Then Eternity II resembles Figure **31**a (not (b)) because variable x is implicitly bounded above by design; $\mathbf{1} \succeq x$ by confinement of Φ in (920) to the permutation polyhedron (100), for $i = 1 \ldots 150{,}638$

$$1 = \underset{x}{\text{maximize}} \quad x_i$$
$$\text{subject to} \quad Ax = b \tag{934}$$
$$x \succeq 0$$

Unity is always attainable, by (931). By (925) this means (§4.5.1.4)

$$M = \underset{y(x),\,x}{\text{maximize}} \quad (\mathbf{1}-y)^{\mathrm{T}}x \qquad \underset{x}{\text{maximize}} \quad \|x\|_{\underset{M}{n}}$$
$$\text{subject to} \quad Ax = b \quad \equiv \quad \text{subject to} \quad Ax = b \tag{935}$$
$$x \succeq 0 \qquad\qquad\qquad x \succeq 0$$

where

$$y = \mathbf{1} - \nabla\|x\|_{\underset{M}{n}} \tag{807}$$

is a direction vector from the technique of convex iteration in §4.5.1.1 and $\|x\|_{\underset{M}{n}}$ is a k-largest norm (§3.2.2.1, $k = M$). When upper bound M in (935) is met, solution x will be optimal for Eternity II because it must then be a Boolean vector with minimal cardinality M.

Maximization of convex function $\|x\|_{\underset{M}{n}}$ (monotonic on $\mathbb{R}^n_+$) is not a convex problem, though the constraints are convex. [315, §32] This problem formulation is unusual, nevertheless, insofar as its geometrical visualization is quite clear. We therefore choose to work with a complementary direction vector z, in what follows, in predilection for a mental picture of convex function maximization.

direction vector is optimal solution at global convergence
Instead of solving (935), which is difficult, we propose iterating a convex problem sequence: for $\mathbf{1} - y \leftarrow z$

$$\underset{x \in \mathbb{R}^n}{\text{maximize}} \quad z^{\mathrm{T}}x$$
$$\text{subject to} \quad Ax = b \tag{936}$$
$$x \succeq 0$$

$$\underset{z \in \mathbb{R}^n}{\text{maximize}} \quad z^{\mathrm{T}}x^\star$$
$$\text{subject to} \quad 0 \preceq z \preceq \mathbf{1} \tag{525}$$
$$z^{\mathrm{T}}\mathbf{1} = M$$

[4.81] One cannot help but notice a binary selection of variable by tests (931) and (933): Geometrical test (931) (smallest face) checks feasibility of vector entry 1 while geometrical test (933) (conic independence) checks feasibility of 0. Changing 1 to 0 in (931) is always feasible for Eternity II.

Variable x is implicitly bounded above at unity by design of A. A globally optimal complementary direction vector $z^\star$ will always exactly match an optimal solution $x^\star$ for convex iteration of any problem formulated as maximization of a Boolean variable; here we have

$$z^{\star\mathrm{T}}x^\star \triangleq M \tag{937}$$

rumination

- Adding a few more clue pieces makes the problem harder in the sense that solution space is diminished; the target gets smaller.

- Because $z^\star = x^\star$, Eternity II can instead be formulated equivalently as a convex-quadratic maximization:

$$\begin{aligned}
&\underset{x\in\mathbb{R}^n}{\text{maximize}} && x^{\mathrm{T}}x \\
&\text{subject to} && Ax = b \\
& && x \succeq 0
\end{aligned} \tag{938}$$

a nonconvex problem but requiring no convex iteration. If it were possible to form a nullspace basis Z, for A of about equal sparsity,[4.82] such that

$$x = Z\xi + x_\mathrm{p} \tag{115}$$

then the equivalent problem formulation

$$\begin{aligned}
&\underset{\xi}{\text{maximize}} && (Z\xi + x_\mathrm{p})^{\mathrm{T}}(Z\xi + x_\mathrm{p}) \\
&\text{subject to} && Z\xi + x_\mathrm{p} \succeq 0
\end{aligned} \tag{939}$$

might invoke optimality conditions as obtained in [203, thm.8][4.83]. □

4.7 Constraining rank of indefinite matrices

Example 4.7.0.0.1, which follows, demonstrates that convex iteration is more generally applicable: to indefinite or nonsquare matrices $X\in\mathbb{R}^{m\times n}$; not only to positive semidefinite matrices. Indeed,

$$\begin{aligned}
&\underset{X\in\mathbb{R}^{m\times n}}{\text{find}} && X \\
&\text{subject to} && X \in \mathcal{C} \\
& && \operatorname{rank}X \leq k
\end{aligned}
\quad\equiv\quad
\begin{aligned}
&\underset{X,Y,Z}{\text{find}} && X \\
&\text{subject to} && X \in \mathcal{C} \\
& && G = \begin{bmatrix} Y & X \\ X^{\mathrm{T}} & Z \end{bmatrix} \\
& && \operatorname{rank}G \leq k
\end{aligned} \tag{940}$$

[4.82] Define *sparsity* as ratio of number of nonzero entries to matrix-dimension product. *For matrices, the average number of nonzeros per row or column is easier to understand and likely to be small for typical LP problems, independent of the dimensions.* −Michael Saunders

[4.83] *... the assumptions in Theorem 8 ask for the Q_i being positive definite (see the top of the page of Theorem 8). I must confess that I do not remember why.* −Jean-Baptiste Hiriart-Urruty

Proof. $\operatorname{rank} G \leq k \Rightarrow \operatorname{rank} X \leq k$ because X is the projection of composite matrix G on subspace $\mathbb{R}^{m \times n}$. For symmetric Y and Z, any rank-k positive semidefinite composite matrix G can be factored into rank-k terms R; $G = R^{\mathrm{T}} R$ where $R \triangleq [B \ \ C]$ and $\operatorname{rank} B, \operatorname{rank} C \leq \operatorname{rank} R$ and $B \in \mathbb{R}^{k \times m}$ and $C \in \mathbb{R}^{k \times n}$. Because Y and Z and $X = B^{\mathrm{T}} C$ are variable, (1546) $\operatorname{rank} X \leq \operatorname{rank} B, \operatorname{rank} C \leq \operatorname{rank} R = \operatorname{rank} G$ is tight. $\qquad\blacklozenge$

So there must exist an optimal direction vector $W^\star$ such that

$$
\begin{array}{ll}
\underset{X,Y,Z}{\text{find}} & X \\
\text{subject to} & X \in \mathcal{C} \\
& G = \begin{bmatrix} Y & X \\ X^{\mathrm{T}} & Z \end{bmatrix} \\
& \operatorname{rank} G \leq k
\end{array}
\quad \equiv \quad
\begin{array}{ll}
\underset{X,Y,Z}{\text{minimize}} & \langle G, W^\star \rangle \\
\text{subject to} & X \in \mathcal{C} \\
& G = \begin{bmatrix} Y & X \\ X^{\mathrm{T}} & Z \end{bmatrix} \succeq 0
\end{array}
\tag{941}
$$

Were $W^\star = I$, by (1777) the optimal resulting trace objective would be equivalent to the minimization of nuclear norm of X over $\mathcal{C}$. This means:

- (*confer* p.196) Any nuclear norm minimization problem may have its variable replaced with a composite semidefinite variable of the same optimal rank but doubly dimensioned.

Then Figure **87** becomes an accurate geometrical description of a consequent composite semidefinite problem objective. But there are better direction vectors than Identity I which occurs only under special conditions:

4.7.0.0.1 Example. *Compressed sensing, compressive sampling.*   [308]
As our modern technology-driven civilization acquires and exploits ever-increasing amounts of data, everyone now knows that most of the data we acquire can be thrown away with almost no perceptual loss — witness the broad success of lossy compression formats for sounds, images, and specialized technical data. The phenomenon of ubiquitous compressibility raises very natural questions: Why go to so much effort to acquire all the data when most of what we get will be thrown away? Can't we just directly measure the part that won't end up being thrown away?   —David Donoho [122]

Lossy data compression techniques like JPEG are popular, but it is also well known that compression artifacts become quite perceptible with signal postprocessing that goes beyond mere playback of a compressed signal. [226] [251] Spatial or audio frequencies presumed masked by a simultaneity are not encoded, for example, so rendered imperceptible even with significant postfiltering (of the compressed signal) that is meant to reveal them; *id est*, desirable artifacts are forever lost, so highly compressed data is not amenable to analysis and postprocessing: *e.g.*, sound effects [99] [100] [101] or image enhancement (*Photoshop*).[4.84]   Further, there can be no universally acceptable unique metric of perception for gauging exactly how much data can be tossed. For these reasons, there will always be need for raw (noncompressed) data.

[4.84] As simple a process as upward scaling of signal amplitude or image size will always introduce noise; even to a noncompressed signal. But scaling-noise is particularly noticeable in a JPEG-compressed image; *e.g.*, text or any sharp edge.

Figure 122: Massachusetts Institute of Technology (MIT) logo, including its white boundary, may be interpreted as a rank-5 matrix. (Stanford University logo rank is much higher;) This constitutes *Scene Y* observed by the one-pixel camera in Figure 123 for Example 4.7.0.0.1.

In this example we throw out only so much information as to leave perfect reconstruction within reach. Specifically, the MIT logo in Figure 122 is perfectly reconstructed from 700 time-sequential samples $\{y_i\}$ acquired by the one-pixel camera illustrated in Figure 123. The MIT-logo image in this example impinges a 46×81 array micromirror device. This mirror array is modulated by a pseudonoise source that independently positions all the individual mirrors. A single photodiode (one pixel) integrates incident light from all mirrors. After stabilizing the mirrors to a fixed but pseudorandom pattern, light so collected is then digitized into one sample y_i by analog-to-digital (A/D) conversion. This sampling process is repeated with the micromirror array modulated to a new pseudorandom pattern.

The most important questions are: How many samples do we need for perfect reconstruction? Does that number of samples represent compression of the original data?

We claim that perfect reconstruction of the MIT logo can be reliably achieved with as few as 700 samples $y = [y_i] \in \mathbb{R}^{700}$ from this one-pixel camera. That number represents only 19% of information obtainable from 3726 micromirrors.[4.85] (Figure 124)

Our approach to reconstruction is to look for low-rank solution to an *underdetermined* system:

$$\begin{array}{ll} \underset{X \in \mathbb{R}^{46 \times 81}}{\text{find}} & X \\ \text{subject to} & A \operatorname{vec} X = y \\ & \operatorname{rank} X \leq 5 \end{array} \qquad (942)$$

where $\operatorname{vec} X$ is the vectorized (37) unknown image matrix. Each row of fat matrix A is one realization of a pseudorandom pattern applied to the micromirrors. Since these patterns are deterministic (known), then the ith sample y_i equals $A(i,:) \operatorname{vec} Y$; *id est*, $y = A \operatorname{vec} Y$. *Perfect reconstruction* here means optimal solution $X^\star$ equals scene $Y \in \mathbb{R}^{46 \times 81}$ to within machine precision.

[4.85]That number (700 samples) is difficult to achieve, as reported in [308, §6]. If a minimal basis for the MIT logo were instead constructed, only five rows or columns worth of data (from a 46×81 matrix) are linearly independent. This means a lower bound on achievable compression is about $5 \times 46 = 230$ samples plus 81 samples column encoding; which corresponds to 8% of the original information. (Figure 124)

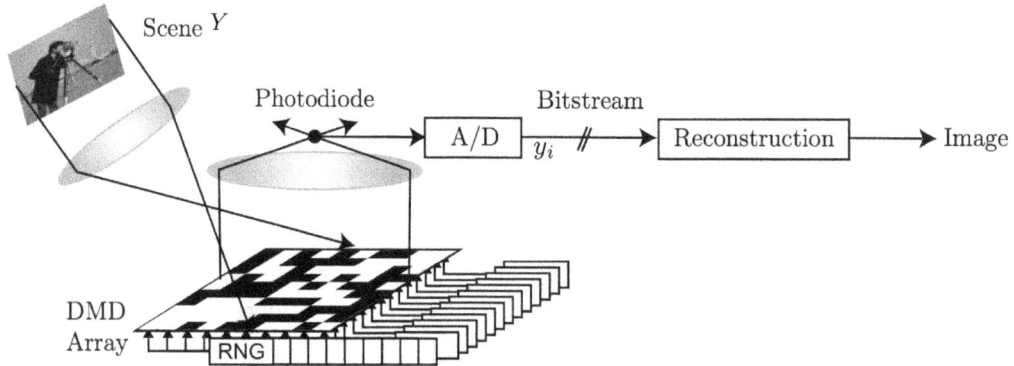

Figure 123: One-pixel camera. [352] [398] *Compressive imaging camera block diagram. Incident lightfield (corresponding to the desired image Y) is reflected off a digital micromirror device (DMD) array whose mirror orientations are modulated in the pseudorandom pattern supplied by the random number generators (RNG). Each different mirror pattern produces a voltage at the single photodiode that corresponds to one measurement y_i.*

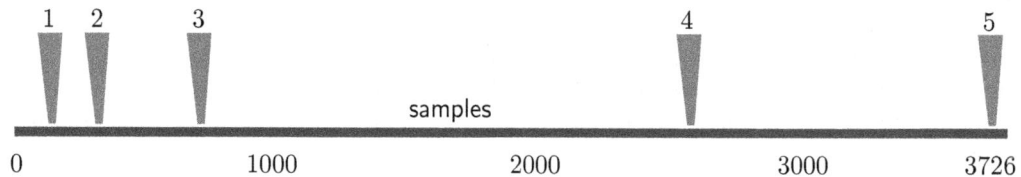

Figure 124: Estimates of compression for various encoding methods:
1) linear interpolation (140 samples),
2) minimal columnar basis (311 samples),
3) convex iteration (700 samples) can achieve lower bound predicted by compressed sensing (670 samples, $n = 46 \times 81$, $k = 140$, Figure 104) whereas nuclear norm minimization alone does not [308, §6],
4) JPEG @100% *quality* (2588 samples),
5) no compression (3726 samples).

Because variable matrix X is generally not square or positive semidefinite, we constrain its rank by rewriting the problem equivalently

$$
\begin{array}{cc}
\underset{W_1 \in \mathbb{R}^{46 \times 46},\, W_2 \in \mathbb{R}^{81 \times 81},\, X \in \mathbb{R}^{46 \times 81}}{\text{find}} & X \\
\text{subject to} & A \operatorname{vec} X = y \\
& \operatorname{rank} \begin{bmatrix} W_1 & X \\ X^{\mathrm{T}} & W_2 \end{bmatrix} \leq 5
\end{array} \tag{943}
$$

This rank constraint on the composite (block) matrix insures $\operatorname{rank} X \leq 5$ for any choice of dimensionally compatible matrices W_1 and W_2. But to solve this problem by convex iteration, we alternate solution of semidefinite program

$$
\begin{array}{cc}
\underset{W_1 \in \mathbb{S}^{46},\, W_2 \in \mathbb{S}^{81},\, X \in \mathbb{R}^{46 \times 81}}{\text{minimize}} & \operatorname{tr}\left(\begin{bmatrix} W_1 & X \\ X^{\mathrm{T}} & W_2 \end{bmatrix} Z \right) \\
\text{subject to} & A \operatorname{vec} X = y \\
& \begin{bmatrix} W_1 & X \\ X^{\mathrm{T}} & W_2 \end{bmatrix} \succeq 0
\end{array} \tag{944}
$$

with semidefinite program

$$
\begin{array}{cc}
\underset{Z \in \mathbb{S}^{46+81}}{\text{minimize}} & \operatorname{tr}\left(\begin{bmatrix} W_1 & X \\ X^{\mathrm{T}} & W_2 \end{bmatrix}^{\star} Z \right) \\
\text{subject to} & 0 \preceq Z \preceq I \\
& \operatorname{tr} Z = 46 + 81 - 5
\end{array} \tag{945}
$$

(which has an optimal solution that is known in closed form, p.567) until a rank-5 composite matrix is found.

With 1000 samples $\{y_i\}$, convergence occurs in two iterations; 700 samples require more than ten iterations but reconstruction remains perfect. Iterating more admits taking of fewer samples. Reconstruction is independent of pseudorandom sequence parameters; *e.g.*, binary sequences succeed with the same efficiency as Gaussian or uniformly distributed sequences. □

4.7.1 rank-constraint midsummary

We find that this *direction matrix* idea works well and quite independently of desired rank or affine dimension. This idea of direction matrix is good principally because of its simplicity: When confronted with a problem otherwise convex if not for a rank or cardinality constraint, then that constraint becomes a linear regularization term in the objective.

There exists a common thread through all these Examples; that being, convex iteration with a direction matrix as normal to a linear regularization (a generalization of the well-known trace heuristic). But each problem type (per Example) possesses its own idiosyncrasies that slightly modify how a rank-constrained optimal solution is actually obtained: The *ball packing* problem in Chapter 5.4.2.2.6, for example, requires a problem sequence in a progressively larger number of balls to find a good initial value for the direction matrix, whereas many of the examples in the present chapter require an initial value of **0**. Finding a Boolean solution in Example 4.6.0.0.9 requires a procedure to detect stalls, while other problems have no such requirement. The combinatorial Procrustes problem in Example 4.6.0.0.3 allows use of a known closed-form solution for direction vector when solved via rank constraint, but not when solved via cardinality constraint. Some problems require a careful weighting of the regularization term, whereas other problems do not, and so on. It would be nice if there were a universally applicable method for constraining rank; one that is less susceptible to quirks of a particular problem type.

Poor initialization of the direction matrix from the regularization can lead to an erroneous result. We speculate one reason to be a simple dearth of optimal solutions of desired rank or cardinality;[4.86] an unfortunate choice of initial search direction leading astray. Ease of solution by convex iteration occurs when optimal solutions abound. With this speculation in mind, we now propose a further generalization of convex iteration for constraining rank that attempts to ameliorate quirks and unify problem types:

4.8 Convex Iteration rank-1

We now develop a general method for constraining rank that first decomposes a given problem via standard diagonalization of matrices (§A.5). This method is motivated by observation (§4.4.1.1) that an optimal direction matrix can be simultaneously diagonalizable with an optimal variable matrix. This suggests minimization of an objective function directly in terms of eigenvalues. A second motivating observation is that variable orthogonal matrices seem easily found by convex iteration; *e.g.*, Procrustes Example 4.6.0.0.2.

4.8.1 rank-1 transformation

It turns out that this general method always requires solution to a rank-1 constrained problem regardless of desired rank ρ from the original problem. To demonstrate, we pose a semidefinite feasibility problem

$$
\begin{aligned}
\text{find} \quad & X \in \mathbb{S}^n \\
\text{subject to} \quad & A\,\text{svec}\,X = b \\
& X \succeq 0 \\
& \text{rank}\,X \leq \rho
\end{aligned}
\tag{946}
$$

given an upper bound $0 < \rho < n$ on rank, a vector $b \in \mathbb{R}^m$, and typically fat full-rank

$$
A = \begin{bmatrix} \text{svec}(A_1)^{\mathrm{T}} \\ \vdots \\ \text{svec}(A_m)^{\mathrm{T}} \end{bmatrix} \in \mathbb{R}^{m \times n(n+1)/2}
\tag{680}
$$

[4.86]In Convex Optimization, an optimal solution generally comes from a convex set of optimal solutions; (§3.1.2.1) that set can be large.

where $A_i \in \mathbb{S}^n$, $i = 1 \ldots m$. So, for symmetric matrix vectorization svec as defined in (56),

$$A \operatorname{svec} X = \begin{bmatrix} \operatorname{tr}(A_1 X) \\ \vdots \\ \operatorname{tr}(A_m X) \end{bmatrix} \qquad (681)$$

This program (946) is a statement of the classical problem of finding a matrix X of maximum rank ρ in the intersection of the positive semidefinite cone with a given number m of hyperplanes in the subspace of symmetric matrices $\mathbb{S}^n$. [26, §II.13] [24, §2.2] Such a matrix is presumed to exist.

To begin transformation of (946), express the nonincreasingly ordered diagonalization (§A.5.1) of positive semidefinite variable matrix

$$X \triangleq Q\Lambda Q^{\mathrm{T}} = \sum_{i=1}^{n} \lambda_i Q_{ii} \in \mathbb{S}^n \qquad (947)$$

which is a sum of rank-1 orthogonal projection matrices Q_{ii} weighted by eigenvalues λ_i where $Q_{ij} \triangleq q_i q_j^{\mathrm{T}} \in \mathbb{R}^{n \times n}$, $Q = [q_1 \cdots q_n] \in \mathbb{R}^{n \times n}$, $Q^{\mathrm{T}} = Q^{-1}$, $\Lambda_{ii} = \lambda_i \in \mathbb{R}$, and

$$\Lambda = \begin{bmatrix} \lambda_1 & & & \mathbf{0} \\ & \lambda_2 & & \\ & & \ddots & \\ \mathbf{0}^{\mathrm{T}} & & & \lambda_n \end{bmatrix} \in \mathbb{S}^n \qquad (948)$$

where $\lambda_1 \geq \lambda_2 \geq \cdots \geq \lambda_n \geq 0$. Recall the fact:

$$\Lambda \succeq 0 \quad \Leftrightarrow \quad X \succeq 0 \qquad (1531)$$

From orthogonal matrix Q in ordered diagonalization (947) of variable X, take a matrix

$$U \triangleq [u_1 \cdots u_\rho] \triangleq Q(:, 1:\rho)\sqrt{\Lambda(1:\rho, 1:\rho)} = \left[\sqrt{\lambda_1}\, q_1 \cdots \sqrt{\lambda_\rho}\, q_\rho\right] \in \mathbb{R}^{n \times \rho} \qquad (949)$$

Then U has orthogonal but unnormalized columns;

$$X = UU^{\mathrm{T}} = \sum_{i=1}^{\rho} u_i u_i^{\mathrm{T}} \triangleq \sum_{i=1}^{\rho} U_{ii} = \sum_{i=1}^{\rho} \lambda_i q_i q_i^{\mathrm{T}} \in \mathbb{S}^n \qquad (950)$$

Make an assignment

$$Z = \begin{bmatrix} u_1 \\ \vdots \\ u_\rho \end{bmatrix} [u_1^{\mathrm{T}} \cdots u_\rho^{\mathrm{T}}] \quad \in \mathbb{S}^{n\rho}$$

$$= \begin{bmatrix} U_{11} & \cdots & U_{1\rho} \\ \vdots & \ddots & \vdots \\ U_{1\rho}^{\mathrm{T}} & \cdots & U_{\rho\rho} \end{bmatrix} \triangleq \begin{bmatrix} u_1 u_1^{\mathrm{T}} & \cdots & u_1 u_\rho^{\mathrm{T}} \\ \vdots & \ddots & \vdots \\ u_\rho u_1^{\mathrm{T}} & \cdots & u_\rho u_\rho^{\mathrm{T}} \end{bmatrix} \qquad (951)$$

Then transformation of (946) to its rank-1 equivalent is:

$$
\begin{aligned}
&\underset{U_{ii}\in\mathbb{S}^n,\ U_{ij}\in\mathbb{R}^{n\times n}}{\text{find}} && X=\sum_{i=1}^{\rho}U_{ii}\\
&\text{subject to} && Z=\begin{bmatrix} U_{11} & \cdots & U_{1\rho}\\ \vdots & \ddots & \vdots\\ U_{1\rho}^{\mathrm{T}} & \cdots & U_{\rho\rho}\end{bmatrix}(\succeq 0)\\
& && A\operatorname{svec}\sum_{i=1}^{\rho}U_{ii}=b\\
& && \operatorname{tr}U_{ij}=0 \qquad\quad i<j=2\ldots\rho\\
& && \operatorname{rank}Z=1
\end{aligned}
\tag{952}
$$

Symmetry is necessary and sufficient for positive semidefiniteness of a rank-1 matrix. (§A.3.1.0.7) Matrix X is positive semidefinite whenever Z is. (§A.3.1.0.4, §A.3.1.0.2) This new problem always enforces a rank-1 constraint on matrix Z; *id est*, regardless of upper bound on rank ρ of variable matrix X, this equivalent problem always poses a rank-1 constraint. We propose solving (952) by iteration of convex problem

$$
\begin{aligned}
&\underset{U_{ii}\in\mathbb{S}^n,\ U_{ij}\in\mathbb{R}^{n\times n}}{\text{minimize}} && \operatorname{tr}(ZW)\\
&\text{subject to} && Z=\begin{bmatrix} U_{11} & \cdots & U_{1\rho}\\ \vdots & \ddots & \vdots\\ U_{1\rho}^{\mathrm{T}} & \cdots & U_{\rho\rho}\end{bmatrix}\succeq 0\\
& && A\operatorname{svec}\sum_{i=1}^{\rho}U_{ii}=b\\
& && \operatorname{tr}U_{ij}=0 \qquad\quad i<j=2\ldots\rho
\end{aligned}
\tag{953}
$$

with convex problem

$$
\begin{aligned}
&\underset{W\in\mathbb{S}^{n\rho}}{\text{minimize}} && \operatorname{tr}(Z^{\star}W)\\
&\text{subject to} && 0\preceq W\preceq I\\
& && \operatorname{tr}W=n\rho-1
\end{aligned}
\tag{954}
$$

the latter providing direction of search W for a rank-1 matrix Z in (953). These convex problems (953) (954) are iterated until a rank-1 Z matrix is found (until the objective of (953) vanishes). Initial value of direction matrix W is the Identity. For subsequent iterations, an optimal solution to (954) has closed form (p.567).

Because of the nonconvex nature of a rank-constrained problem, there can be no proof of global convergence of this convex iteration. But this iteration always converges to a local minimum because the sequence of objective values is monotonic and nonincreasing; any monotonically nonincreasing real sequence converges. [264, §1.2] [42, §1.1] A rank ρ matrix X solving the original problem (946) is found when the objective in (953) converges to 0; a certificate of global optimality for the convex iteration. In practice, incidence of global convergence is quite high (99.99% [381]); failures being mostly attributable to numerical accuracy. Upper bound ρ on rank of positive semidefinite matrix X is assured by rank-1 optimal matrix Z.

4.8.1.0.1 Example. *Singular value decomposition by convex iteration.* [163]
This diagonal decomposition technique (transformation to a rank-1 problem) is extensible to other problem types; *e.g.*, [231, §III]. Rank-1 transformation makes singular value decomposition (SVD, §A.6) possible by convex iteration because orthogonality constraints may then be introduced. We learn that any uniqueness properties, the SVD of rank-ρ matrix $X \triangleq USV^\mathrm{T} \in \mathbb{R}^{m\times n}$ might enjoy, stem from demand for singular vector orthonormality.[4.87]
Assignment $Z \in \mathbb{S}_+^{2m\rho+n\rho+\rho+1}$ is key to finding the SVD of X by convex optimization:

$$
\begin{aligned}
&\underset{H,\,J}{\text{find}} \quad U,\,\delta(S),\,V \\
&\text{subject to} \quad Z = \begin{bmatrix} 1 & \mathrm{vec}(H)^\mathrm{T} & \mathrm{vec}(U)^\mathrm{T} & \delta(S)^\mathrm{T} & \mathrm{vec}(V)^\mathrm{T} \\ \mathrm{vec}\,H & & & & \\ \mathrm{vec}\,U & & J & & \\ \delta(S) & & & & \\ \mathrm{vec}\,V & & & & \end{bmatrix} \succeq 0 \\
&\qquad\qquad \delta(S) \succeq 0 \\
&\qquad\qquad H = US \\
&\qquad\qquad X = HV^\mathrm{T} \\
&\qquad\qquad HU^\mathrm{T} \text{ symmetry} \\
&\qquad\qquad U^\mathrm{T}H \text{ perpendicularity} \\
&\qquad\qquad \mathrm{tr}\big(H(:,i)\,H(:,i)^\mathrm{T}\big) = S(i,i)^2 \qquad i=1\ldots\rho \\
&\qquad\qquad \mathrm{tr}\big(H(:,i)\,U(:,i)^\mathrm{T}\big) = S(i,i) \qquad i=1\ldots\rho \\
&\qquad\qquad H \text{ orthogonality} \\
&\qquad\qquad U,\,V \text{ orthonormality} \\
&\qquad\qquad \mathrm{rank}\,Z = 1
\end{aligned}
\tag{955}
$$

where variable matrix $J \in \mathbb{S}_+^{2m\rho+n\rho+\rho}$ is a large partition of Z, where given rank-ρ matrix $X \in \mathbb{R}^{m\times n}$ is subject to SVD in unknown orthonormal matrices $U \in \mathbb{R}^{m\times\rho}$ and $V \in \mathbb{R}^{n\times\rho}$ and unknown diagonal matrix of singular values $S \in \mathbb{R}^{\rho\times\rho}$, and where introduction of variable $H \triangleq US \in \mathbb{R}^{m\times\rho}$ makes identification of input $X = HV^\mathrm{T}$ possible within partition J. Orthogonality constraints on columns of H, within J, and orthonormality constraints on columns of U and V are critical; *videlicet*, $h \perp v \Leftrightarrow \mathrm{tr}(hv^\mathrm{T})=0$; $v^\mathrm{T}v=1 \Leftrightarrow \mathrm{tr}(vv^\mathrm{T})=1$.

Symmetric matrix Z is positive semidefinite rank-1 at optimality, regardless of ρ. That rank constraint is the only nonconvex constraint in (955); the only constraint that cannot be directly implemented in a convex manner per partition J. But the rank constraint is handled well by convex iteration. Matlab implementation of SVD by convex iteration is intricate although incidence of global convergence is 99.99% [395], barring numerical error.

□

[4.87]Otherwise, there exist many similarly structured tripartite nonorthogonal matrix decompositions; in place of ρ nonzero singular values, diagonal matrix S would instead hold exactly ρ *coordinates;* orthonormal columns in U and V would become merely linearly independent.

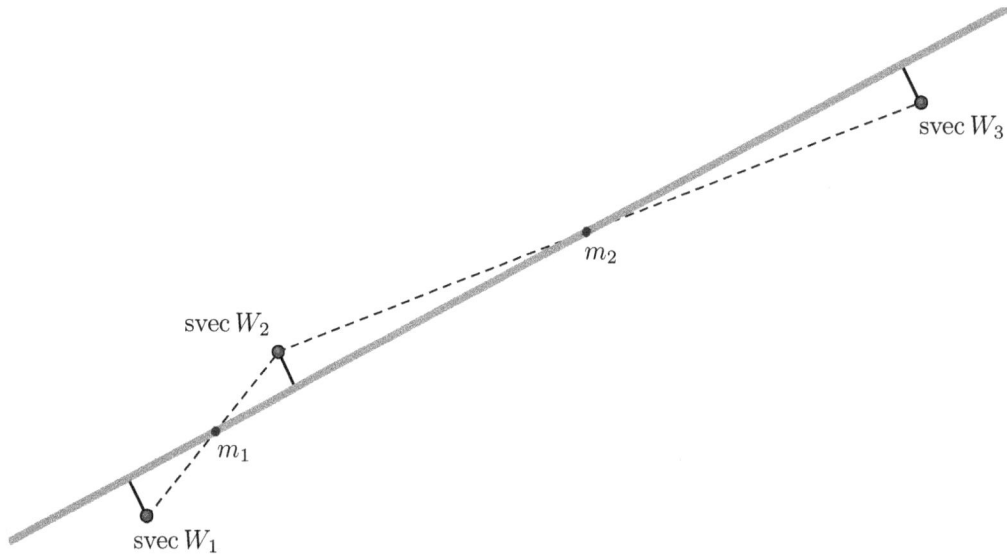

Figure 125: W_1, W_2, and W_3 represent the last three direction vectors in a sequence. m_1 represents the midpoint between direction vectors W_1 and W_2; m_2 is the midpoint of W_2 and W_3. Straight line passes through midpoints.

4.8.2   convex iteration accelerant

Convex iteration can be made to converge faster; sometimes, by orders of magnitude. The idea here is to determine whether the last three direction vectors are close to their fit to a straight line. When three direction vectors are close to a straight line, then the last direction vector may be replaced with its extrapolation along that line.

To reduce computation time, the fitted line is not a best fit. Instead, the midpoint between each pair of iteration-adjacent direction vectors is calculated (Figure 125). A straight line is uniquely defined by two midpoints in any dimension. The distance of each direction vector to the line is calculated, then those three distances summed. When a sum is small, three direction vectors are deemed close to the line determined by them.

What is meant above by *close* and *small* depends on the particular problem type at hand. For the parameters and normalized random data chosen for two Matlab realizations [381] [395] on *Wıκımization* (corresponding to problems (952) and (955)), *small* is numerically defined to be 1 or less in the statement `if straight < 1` whose purpose is to determine straightness of the last three direction vectors of convex iteration. The smaller the value of the normalized sum called `straight`, the closer the last three direction vectors are to a straight line. Variable `straight` is bounded below by 0 which indicates three direction vectors precisely on the line going through them.

If linear extrapolation goes too far, then the objective of convex iteration will increase or a solver may fail numerically. In either case, one must forget the last iteration and back up the linear extrapolation until the objective decreases. These techniques are illustrated by the Matlab programs.

Chapter 5

Euclidean Distance Matrix

These results [(987)] were obtained by Schoenberg (1935), a surprisingly late date for such a fundamental property of Euclidean geometry.

−John Clifford Gower [168, §3]

By itself, distance information between many points in Euclidean space is lacking. We might want to know more; such as, relative or absolute position or dimension of some hull. A question naturally arising in some fields (*e.g.*, geodesy, economics, genetics, psychology, biochemistry, engineering) [106] asks what facts can be deduced given only distance information. What can we know about the underlying points that the distance information purports to describe? We also ask what it means when given distance information is incomplete; or suppose the distance information is not reliable, available, or specified only by certain tolerances (affine inequalities). These questions motivate a study of interpoint distance, well represented in any spatial dimension by a simple matrix from linear algebra.[5.1] In what follows, we will answer some of these questions via Euclidean distance matrices.

[5.1] *e.g.*, $\sqrt[\circ]{D} \in \mathbb{R}^{N \times N}$, a classical two-dimensional matrix representation of absolute interpoint distance because its entries (in ordered rows and columns) can be written neatly on a piece of paper. Matrix D will be reserved throughout to hold distance-square.

CITATION: Dattorro, *Convex Optimization & Euclidean Distance Geometry*, $\mathcal{M}\varepsilon\beta oo$ Publishing USA, 2005, v2015.02.15.

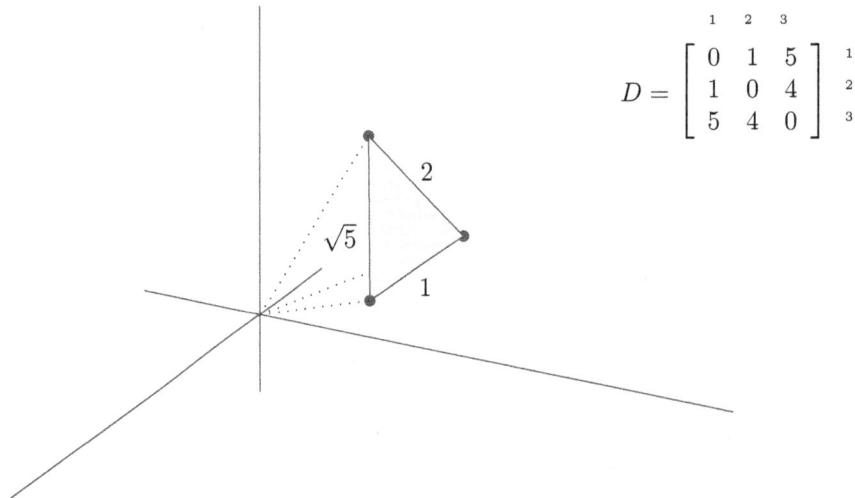

$$D = \begin{bmatrix} 0 & 1 & 5 \\ 1 & 0 & 4 \\ 5 & 4 & 0 \end{bmatrix} \begin{smallmatrix} 1 \\ 2 \\ 3 \end{smallmatrix}$$

Figure 126: Convex hull of three points ($N\!=\!3$) is shaded in $\mathbb{R}^{\mathbf{3}}$ ($n\!=\!3$). Dotted lines are imagined vectors to points whose affine dimension is 2.

5.1   EDM

Euclidean space $\mathbb{R}^n$ is a finite-dimensional real vector space having an inner product defined on it, inducing a *metric*. [233, §3.1] A Euclidean distance matrix, an EDM in $\mathbb{R}_+^{N \times N}$, is an exhaustive table of distance-square d_{ij} between points taken by pair from a list of N points $\{x_\ell \,,\;\ell\!=\!1\ldots N\}$ in $\mathbb{R}^n$; the squared metric, the measure of distance-square:

$$d_{ij} \;=\; \|x_i - x_j\|_2^2 \;\triangleq\; \langle x_i - x_j \,,\, x_i - x_j \rangle \tag{956}$$

Each point is labelled ordinally, hence the row or column index of an EDM, i or $j\!=\!1\ldots N$, individually addresses all the points in the list.

Consider the following example of an EDM for the case $N\!=\!3$:

$$D \;=\; [d_{ij}] \;=\; \begin{bmatrix} d_{11} & d_{12} & d_{13} \\ d_{21} & d_{22} & d_{23} \\ d_{31} & d_{32} & d_{33} \end{bmatrix} \;=\; \begin{bmatrix} 0 & d_{12} & d_{13} \\ d_{12} & 0 & d_{23} \\ d_{13} & d_{23} & 0 \end{bmatrix} \;=\; \begin{bmatrix} 0 & 1 & 5 \\ 1 & 0 & 4 \\ 5 & 4 & 0 \end{bmatrix} \tag{957}$$

Matrix D has N^2 entries but only $N(N\!-\!1)/2$ pieces of information. In Figure **126** are shown three points in $\mathbb{R}^{\mathbf{3}}$ that can be arranged in a list to correspond to D in (957). But such a list is not unique because any rotation, reflection, or translation (§5.5) of those points would produce the same EDM D.

5.2 First metric properties

For $i,j = 1 \ldots N$, absolute distance between points x_i and x_j must satisfy the defining requirements imposed upon any *metric space*: [233, §1.1] [264, §1.7] namely, for Euclidean metric $\sqrt{d_{ij}}$ (§5.4) in $\mathbb{R}^n$

1. $\sqrt{d_{ij}} \geq 0 , \quad i \neq j$ nonnegativity

2. $\sqrt{d_{ij}} = 0 \iff x_i = x_j$ self-distance

3. $\sqrt{d_{ij}} = \sqrt{d_{ji}}$ symmetry

4. $\sqrt{d_{ij}} \leq \sqrt{d_{ik}} + \sqrt{d_{kj}} , \quad i \neq j \neq k$ triangle inequality

Then all entries of an EDM must be in concord with these Euclidean metric properties: specifically, each entry must be nonnegative,[5.2] the main diagonal must be $\mathbf{0}$,[5.3] and an EDM must be symmetric. The fourth property provides upper and lower bounds for each entry. Property 4 is true more generally when there are no restrictions on indices i,j,k , but furnishes no new information.

5.3 $\exists$ fifth Euclidean metric property

The four properties of the Euclidean metric provide information insufficient to certify that a bounded convex polyhedron more complicated than a triangle has a Euclidean realization. [168, §2] Yet any list of points or the vertices of any bounded convex polyhedron must conform to the properties.

5.3.0.0.1 Example. *Triangle.*
Consider the EDM in (957), but missing one of its entries:

$$D = \begin{bmatrix} 0 & 1 & d_{13} \\ 1 & 0 & 4 \\ d_{31} & 4 & 0 \end{bmatrix} \tag{958}$$

Can we determine unknown entries of D by applying the metric properties? Property 1 demands $\sqrt{d_{13}}, \sqrt{d_{31}} \geq 0$, property 2 requires the main diagonal be $\mathbf{0}$, while property 3 makes $\sqrt{d_{31}} = \sqrt{d_{13}}$. The fourth property tells us

$$1 \leq \sqrt{d_{13}} \leq 3 \tag{959}$$

Indeed, described over that closed interval $[1,3]$ is a family of triangular polyhedra whose angle at vertex x_2 varies from 0 to π radians. So, yes we can determine the unknown entries of D , but they are not unique; nor should they be from the information given for this example. $\square$

[5.2] Implicit from the terminology, $\sqrt{d_{ij}} \geq 0 \iff d_{ij} \geq 0$ is always assumed.
[5.3] What we call self-distance, Marsden calls *nondegeneracy.* [264, §1.6] Kreyszig calls these first metric properties *axioms of the metric*; [233, p.4] Blumenthal refers to them as *postulates.* [51, p.15]

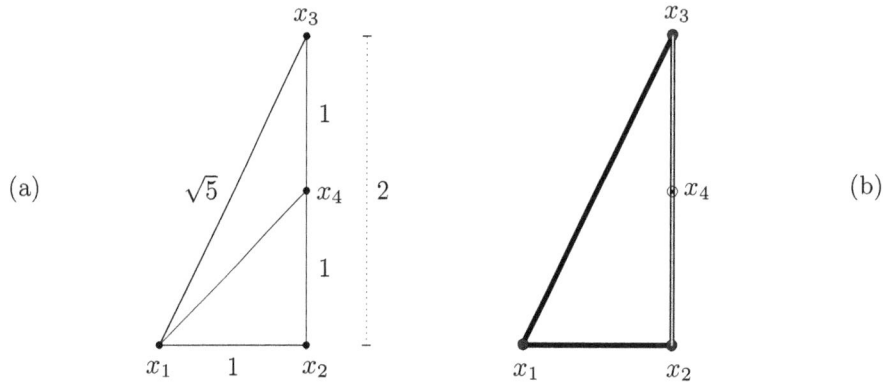

Figure 127: **(a)** Complete dimensionless *EDM graph*. **(b)** Emphasizing obscured segments $\overline{x_2 x_4}$, $\overline{x_4 x_3}$, and $\overline{x_2 x_3}$, now only five $(2N-3)$ absolute distances are specified. EDM so represented is incomplete, missing d_{14} as in (960), yet the isometric reconstruction (§5.4.2.2.10) is unique as proved in §5.9.3.0.1 and §5.14.4.1.1. First four properties of Euclidean metric are not a recipe for reconstruction of this polyhedron.

5.3.0.0.2 Example. *Small completion problem*, I.
Now consider the polyhedron in Figure **127**b formed from an unknown list $\{x_1, x_2, x_3, x_4\}$. The corresponding EDM less one critical piece of information, d_{14}, is given by

$$D = \begin{bmatrix} 0 & 1 & 5 & d_{14} \\ 1 & 0 & 4 & 1 \\ 5 & 4 & 0 & 1 \\ d_{14} & 1 & 1 & 0 \end{bmatrix} \tag{960}$$

From metric property 4 we may write a few inequalities for the two triangles common to d_{14}; we find

$$\sqrt{5}-1 \;\leq\; \sqrt{d_{14}} \;\leq\; 2 \tag{961}$$

We cannot further narrow those bounds on $\sqrt{d_{14}}$ using only the four metric properties (§5.8.3.1.1). Yet there is only one possible choice for $\sqrt{d_{14}}$ because points x_2, x_3, x_4 must be collinear. All other values of $\sqrt{d_{14}}$ in the interval $[\sqrt{5}-1,\, 2]$ specify impossible distances in any dimension; *id est,* in this particular example the triangle inequality does not yield an interval for $\sqrt{d_{14}}$ over which a family of convex polyhedra can be reconstructed.   □

We will return to this simple Example 5.3.0.0.2 to illustrate more elegant methods of solution in §5.8.3.1.1, §5.9.3.0.1, and §5.14.4.1.1. Until then, we can deduce some general principles from the foregoing examples:

- Unknown d_{ij} of an EDM are not necessarily uniquely determinable.

- The triangle inequality does not produce necessarily tight bounds.[5.4]

- Four Euclidean metric properties are insufficient for reconstruction.

5.3.1   Lookahead

There must exist at least one requirement more than the four properties of the Euclidean metric that makes them altogether necessary and sufficient to certify realizability of bounded convex polyhedra. Indeed, there are infinitely many more; there are precisely $N+1$ necessary and sufficient Euclidean metric requirements for N points constituting a generating list (§2.3.2). Here is the fifth requirement:

5.3.1.0.1   Fifth Euclidean metric property.   *Relative-angle inequality.*
(*confer* §5.14.2.1.1) Augmenting the four fundamental properties of the Euclidean metric in $\mathbb{R}^n$, for all $i, j, \ell \neq k \in \{1 \ldots N\}$, $i < j < \ell$, and for $N \geq 4$ distinct points $\{x_k\}$, the inequalities

$$\cos(\theta_{ik\ell} + \theta_{\ell kj}) \;\leq\; \cos\theta_{ikj} \;\leq\; \cos(\theta_{ik\ell} - \theta_{\ell kj})$$
$$0 \leq \theta_{ik\ell}, \theta_{\ell kj}, \theta_{ikj} \leq \pi \tag{962}$$

where $\theta_{ikj} = \theta_{jki}$ represents angle between vectors at vertex x_k (1034) (Figure **128**), must be satisfied at each point x_k regardless of affine dimension.   ◇

We will explore this in §5.14. One of our early goals is to determine matrix criteria that subsume all the Euclidean metric properties and any further requirements. Looking ahead, we will find (1313) (987) (991)

$$\left.\begin{array}{r} -z^{\mathrm{T}}Dz \geq 0 \\ \mathbf{1}^{\mathrm{T}}z = 0 \\ (\forall \|z\| = 1) \\ D \in \mathbb{S}_h^N \end{array}\right\} \;\Leftrightarrow\; D \in \mathbb{EDM}^N \tag{963}$$

where the convex cone of Euclidean distance matrices $\mathbb{EDM}^N \subseteq \mathbb{S}_h^N$ belongs to the subspace of symmetric hollow[5.5] matrices (§2.2.3.0.1). (Numerical test `isedm(D)` provided on *Wιкιmization* [397].) Having found equivalent matrix criteria, we will see there is a bridge from bounded convex polyhedra to EDMs in §5.9 .[5.6]

Now we develop some invaluable concepts, moving toward a link of the Euclidean metric properties to matrix criteria.

[5.4]The term *tight* with reference to an inequality means equality is achievable.
[5.5]**0** main diagonal.
[5.6]From an EDM, a generating list (§2.3.2, §2.12.2) for a polyhedron can be found (§5.12) correct to within a rotation, reflection, and translation (§5.5).

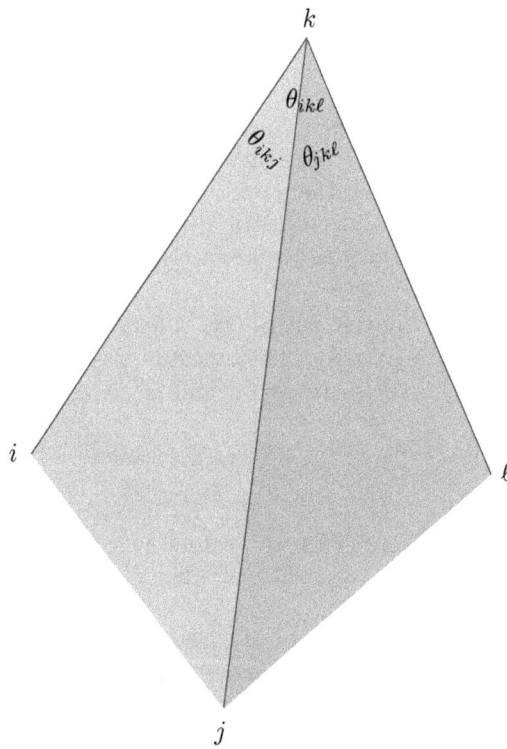

Figure 128: Fifth Euclidean metric property nomenclature. Each angle θ is made by a vector pair at vertex k while i, j, k, ℓ index four points at the vertices of a generally irregular tetrahedron. The fifth property is necessary for realization of four or more points; a reckoning by three angles in any dimension. Together with the first four Euclidean metric properties, this fifth property is necessary and sufficient for realization of four points.

5.4 EDM definition

Ascribe points in a list $\{x_\ell \in \mathbb{R}^n, \ \ell = 1 \ldots N\}$ to the columns of a matrix

$$X = [\, x_1 \ \cdots \ x_N \,] \in \mathbb{R}^{n \times N} \qquad (76)$$

where N is regarded as *cardinality* of list X. When matrix $D = [d_{ij}]$ is an EDM, its entries must be related to those points constituting the list by the Euclidean distance-square: for $i, j = 1 \ldots N$ (§A.1.1 *no.*33)

$$\begin{aligned}
d_{ij} &= \|x_i - x_j\|^2 = (x_i - x_j)^{\mathrm{T}}(x_i - x_j) = \|x_i\|^2 + \|x_j\|^2 - 2x_i^{\mathrm{T}}x_j \\
&= [\, x_i^{\mathrm{T}} \ \ x_j^{\mathrm{T}} \,] \begin{bmatrix} I & -I \\ -I & I \end{bmatrix} \begin{bmatrix} x_i \\ x_j \end{bmatrix} \\
&= \mathrm{vec}(X)^{\mathrm{T}}(\Phi_{ij} \otimes I)\,\mathrm{vec}\,X = \langle \Phi_{ij} \,,\, X^{\mathrm{T}}X \rangle
\end{aligned} \qquad (964)$$

where

$$\mathrm{vec}\,X = \begin{bmatrix} x_1 \\ x_2 \\ \vdots \\ x_N \end{bmatrix} \in \mathbb{R}^{nN} \qquad (965)$$

and where $\otimes$ signifies Kronecker product (§D.1.2.1). $\Phi_{ij} \otimes I$ is positive semidefinite (1563) having $I \in \mathbb{S}^n$ in its ii^{th} and jj^{th} block of entries while $-I \in \mathbb{S}^n$ fills its ij^{th} and ji^{th} block; *id est*,

$$\begin{aligned}
\Phi_{ij} &\triangleq \delta\big((e_i e_j^{\mathrm{T}} + e_j e_i^{\mathrm{T}})\mathbf{1}\big) - (e_i e_j^{\mathrm{T}} + e_j e_i^{\mathrm{T}}) \in \mathbb{S}_+^N \\
&= e_i e_i^{\mathrm{T}} + e_j e_j^{\mathrm{T}} - e_i e_j^{\mathrm{T}} - e_j e_i^{\mathrm{T}} \\
&= (e_i - e_j)(e_i - e_j)^{\mathrm{T}}
\end{aligned} \qquad (966)$$

where $\{e_i \in \mathbb{R}^N, \ i = 1 \ldots N\}$ is the set of standard basis vectors. Thus each entry d_{ij} is a convex quadratic function (§A.4.0.0.2) of $\mathrm{vec}\,X$ (37). [315, §6]

The collection of all Euclidean distance matrices $\mathbb{EDM}^N$ is a convex subset of $\mathbb{R}_+^{N \times N}$ called the *EDM cone* (§6, Figure **163** p.481);

$$\mathbf{0} \in \mathbb{EDM}^N \subseteq \mathbb{S}_h^N \cap \mathbb{R}_+^{N \times N} \subset \mathbb{S}^N \qquad (967)$$

An EDM D must be expressible as a function of some list X; *id est*, it must have the form

$$\begin{aligned}
\mathbf{D}(X) &\triangleq \delta(X^{\mathrm{T}}X)\mathbf{1}^{\mathrm{T}} + \mathbf{1}\,\delta(X^{\mathrm{T}}X)^{\mathrm{T}} - 2X^{\mathrm{T}}X \ \in \mathbb{EDM}^N && (968) \\
&= [\mathrm{vec}(X)^{\mathrm{T}}(\Phi_{ij} \otimes I)\,\mathrm{vec}\,X \,, \ \ i, j = 1 \ldots N] && (969)
\end{aligned}$$

Function $\mathbf{D}(X)$ will make an EDM given any $X \in \mathbb{R}^{n \times N}$, conversely, but $\mathbf{D}(X)$ is not a convex function of X (§5.4.1). Now the EDM cone may be described:

$$\mathbb{EDM}^N = \Big\{ \mathbf{D}(X) \mid X \in \mathbb{R}^{N-1 \times N} \Big\} \qquad (970)$$

Expression $\mathbf{D}(X)$ is a matrix definition of EDM and so conforms to the Euclidean metric properties:

Nonnegativity of EDM entries (property 1, §5.2) is obvious from the distance-square definition (964), so holds for any D expressible in the form $\mathbf{D}(X)$ in (968).

When we say D is an EDM, reading from (968), it implicitly means the main diagonal must be $\mathbf{0}$ (property 2, self-distance) and D must be symmetric (property 3); $\delta(D) = \mathbf{0}$ and $D^{\mathrm{T}} = D$ or, equivalently, $D \in \mathbb{S}_h^N$ are necessary matrix criteria.

5.4.0.1   homogeneity

Function $\mathbf{D}(X)$ is homogeneous in the sense, for $\zeta \in \mathbb{R}$

$$\sqrt[\circ]{\mathbf{D}(\zeta X)} = |\zeta| \sqrt[\circ]{\mathbf{D}(X)} \tag{971}$$

where the positive square root is entrywise ($\circ$).

Any nonnegatively scaled EDM remains an EDM; *id est*, the matrix class EDM is invariant to nonnegative scaling ($\alpha \mathbf{D}(X)$ for $\alpha \geq 0$) because all EDMs of dimension N constitute a convex cone $\mathbb{EDM}^N$ (§6, Figure 155).

5.4.1 $-V_{\mathcal{N}}^{\mathrm{T}} \mathbf{D}(X) V_{\mathcal{N}}$ convexity

We saw that EDM entries $d_{ij}\left(\begin{bmatrix} x_i \\ x_j \end{bmatrix}\right)$ are convex quadratic functions. Yet $-\mathbf{D}(X)$ (968) is not a quasiconvex function of matrix $X \in \mathbb{R}^{n \times N}$ because the second directional derivative (§3.8)

$$-\frac{d^2}{dt^2}\bigg|_{t=0} \mathbf{D}(X + tY) = 2\big(-\delta(Y^{\mathrm{T}}Y)\mathbf{1}^{\mathrm{T}} - \mathbf{1}\delta(Y^{\mathrm{T}}Y)^{\mathrm{T}} + 2Y^{\mathrm{T}}Y\big) \tag{972}$$

is indefinite for any $Y \in \mathbb{R}^{n \times N}$ since its main diagonal is $\mathbf{0}$. [164, §4.2.8] [208, §7.1 prob.2] Hence $-\mathbf{D}(X)$ can neither be convex in X.

The outcome is different when instead we consider

$$-V_{\mathcal{N}}^{\mathrm{T}} \mathbf{D}(X) V_{\mathcal{N}} = 2 V_{\mathcal{N}}^{\mathrm{T}} X^{\mathrm{T}} X V_{\mathcal{N}} \tag{973}$$

where we introduce the full-rank skinny *Schoenberg auxiliary matrix* (§B.4.2)

$$V_{\mathcal{N}} \triangleq \frac{1}{\sqrt{2}} \begin{bmatrix} -1 & -1 & \cdots & -1 \\ 1 & & & 0 \\ & 1 & & \\ & & \ddots & \\ 0 & & & 1 \end{bmatrix} = \frac{1}{\sqrt{2}} \begin{bmatrix} -\mathbf{1}^{\mathrm{T}} \\ I \end{bmatrix} \in \mathbb{R}^{N \times N-1} \tag{974}$$

$(\mathcal{N}(V_{\mathcal{N}}) = \mathbf{0})$ having range

$$\mathcal{R}(V_{\mathcal{N}}) = \mathcal{N}(\mathbf{1}^{\mathrm{T}}) , \quad V_{\mathcal{N}}^{\mathrm{T}} \mathbf{1} = \mathbf{0} \tag{975}$$

Matrix-valued function (973) meets the criterion for convexity in §3.7.3.0.2 over its domain that is all of $\mathbb{R}^{n \times N}$; *videlicet,* for any $Y \in \mathbb{R}^{n \times N}$

$$-\frac{d^2}{dt^2} V_{\mathcal{N}}^{\mathrm{T}} \mathbf{D}(X + tY) V_{\mathcal{N}} = 4 V_{\mathcal{N}}^{\mathrm{T}} Y^{\mathrm{T}} Y V_{\mathcal{N}} \succeq 0 \qquad (976)$$

Quadratic matrix-valued function $-V_{\mathcal{N}}^{\mathrm{T}} \mathbf{D}(X) V_{\mathcal{N}}$ is therefore convex in X achieving its minimum, with respect to a positive semidefinite cone (§2.7.2.2), at $X = \mathbf{0}$. When the penultimate number of points exceeds the dimension of the space $n < N - 1$, strict convexity of the quadratic (973) becomes impossible because (976) could not then be positive definite.

5.4.2 Gram-form EDM definition

Positive semidefinite matrix $X^{\mathrm{T}}X$ in (968), formed from inner product of list X, is known as a *Gram matrix*; [256, §3.6]

$$
\begin{aligned}
G &\triangleq X^{\mathrm{T}}X = \begin{bmatrix} x_1^{\mathrm{T}} \\ \vdots \\ x_N^{\mathrm{T}} \end{bmatrix} \begin{bmatrix} x_1 \cdots x_N \end{bmatrix} = \begin{bmatrix} \|x_1\|^2 & x_1^{\mathrm{T}}x_2 & x_1^{\mathrm{T}}x_3 & \cdots & x_1^{\mathrm{T}}x_N \\ x_2^{\mathrm{T}}x_1 & \|x_2\|^2 & x_2^{\mathrm{T}}x_3 & \cdots & x_2^{\mathrm{T}}x_N \\ x_3^{\mathrm{T}}x_1 & x_3^{\mathrm{T}}x_2 & \|x_3\|^2 & \ddots & x_3^{\mathrm{T}}x_N \\ \vdots & \vdots & \ddots & \ddots & \vdots \\ x_N^{\mathrm{T}}x_1 & x_N^{\mathrm{T}}x_2 & x_N^{\mathrm{T}}x_3 & \cdots & \|x_N\|^2 \end{bmatrix} \in \mathbb{S}_+^N \\[1em]
&= \delta\left(\begin{bmatrix} \|x_1\| \\ \|x_2\| \\ \vdots \\ \|x_N\| \end{bmatrix}\right) \begin{bmatrix} 1 & \cos\psi_{12} & \cos\psi_{13} & \cdots & \cos\psi_{1N} \\ \cos\psi_{12} & 1 & \cos\psi_{23} & \cdots & \cos\psi_{2N} \\ \cos\psi_{13} & \cos\psi_{23} & 1 & \ddots & \cos\psi_{3N} \\ \vdots & \vdots & \ddots & \ddots & \vdots \\ \cos\psi_{1N} & \cos\psi_{2N} & \cos\psi_{3N} & \cdots & 1 \end{bmatrix} \delta\left(\begin{bmatrix} \|x_1\| \\ \|x_2\| \\ \vdots \\ \|x_N\| \end{bmatrix}\right) \\[1em]
&\triangleq \sqrt{\delta^2(G)} \, \Psi \sqrt{\delta^2(G)}
\end{aligned}
\qquad (977)
$$

where ψ_{ij} (996) is angle between vectors x_i and x_j, and where δ^2 denotes a diagonal matrix in this case. Positive semidefiniteness of *interpoint angle matrix* Ψ implies positive semidefiniteness of Gram matrix G;

$$G \succeq 0 \Leftarrow \Psi \succeq 0 \qquad (978)$$

When $\delta^2(G)$ is nonsingular, then $G \succeq 0 \Leftrightarrow \Psi \succeq 0$. (§A.3.1.0.5)

Distance-square d_{ij} (964) is related to Gram matrix entries $G^{\mathrm{T}} = G \triangleq [g_{ij}]$

$$
\begin{aligned}
d_{ij} &= g_{ii} + g_{jj} - 2g_{ij} \\
&= \langle \Phi_{ij}, G \rangle
\end{aligned}
\qquad (979)
$$

where Φ_{ij} is defined in (966). Hence the linear EDM definition

$$
\left.\begin{aligned}
\mathbf{D}(G) &\triangleq \delta(G)\mathbf{1}^{\mathrm{T}} + \mathbf{1}\delta(G)^{\mathrm{T}} - 2G \in \mathbb{EDM}^N \\
&= [\langle \Phi_{ij}, G \rangle, \; i,j = 1 \dots N]
\end{aligned}\right\} \quad \Leftarrow \quad G \succeq 0 \qquad (980)
$$

The EDM cone may be described, (*confer* (1069)(1075))

$$\mathbb{EDM}^N = \left\{ \mathbf{D}(G) \mid G \in \mathbb{S}_+^N \right\} \tag{981}$$

5.4.2.1   First point at origin

Assume the first point x_1 in an unknown list X resides at the origin;

$$Xe_1 = \mathbf{0} \;\Leftrightarrow\; Ge_1 = \mathbf{0} \tag{982}$$

Consider the symmetric translation $(I - \mathbf{1}e_1^{\mathrm{T}})\mathbf{D}(G)(I - e_1\mathbf{1}^{\mathrm{T}})$ that shifts the first row and column of $\mathbf{D}(G)$ to the origin; setting Gram-form EDM operator $\mathbf{D}(G) = D$ for convenience,

$$-\left(D - (De_1\mathbf{1}^{\mathrm{T}} + \mathbf{1}e_1^{\mathrm{T}}D) + \mathbf{1}e_1^{\mathrm{T}}De_1\mathbf{1}^{\mathrm{T}}\right)\tfrac{1}{2} = G - (Ge_1\mathbf{1}^{\mathrm{T}} + \mathbf{1}e_1^{\mathrm{T}}G) + \mathbf{1}e_1^{\mathrm{T}}Ge_1\mathbf{1}^{\mathrm{T}} \tag{983}$$

where

$$e_1 \triangleq \begin{bmatrix} 1 \\ 0 \\ \vdots \\ 0 \end{bmatrix} \tag{984}$$

is the first vector from the standard basis. Then it follows for $D \in \mathbb{S}_h^N$

$$\begin{aligned}
G &= -\left(D - (De_1\mathbf{1}^{\mathrm{T}} + \mathbf{1}e_1^{\mathrm{T}}D)\right)\tfrac{1}{2}, \qquad x_1 = \mathbf{0} \\
&= -\begin{bmatrix} \mathbf{0} & \sqrt{2}V_{\mathcal{N}} \end{bmatrix}^{\mathrm{T}} D \begin{bmatrix} \mathbf{0} & \sqrt{2}V_{\mathcal{N}} \end{bmatrix}\tfrac{1}{2} \\
&= \begin{bmatrix} 0 & \mathbf{0}^{\mathrm{T}} \\ \mathbf{0} & -V_{\mathcal{N}}^{\mathrm{T}}DV_{\mathcal{N}} \end{bmatrix} \\
V_{\mathcal{N}}^{\mathrm{T}}GV_{\mathcal{N}} &= -V_{\mathcal{N}}^{\mathrm{T}}DV_{\mathcal{N}}\tfrac{1}{2} \qquad\qquad \forall X
\end{aligned} \tag{985}$$

where

$$I - e_1\mathbf{1}^{\mathrm{T}} = \begin{bmatrix} \mathbf{0} & \sqrt{2}V_{\mathcal{N}} \end{bmatrix} \tag{986}$$

is a projector nonorthogonally projecting (§E.1) on subspace

$$\begin{aligned}
\mathbb{S}_1^N &= \{G \in \mathbb{S}^N \mid Ge_1 = \mathbf{0}\} \\
&= \left\{ \begin{bmatrix} \mathbf{0} & \sqrt{2}V_{\mathcal{N}} \end{bmatrix}^{\mathrm{T}} Y \begin{bmatrix} \mathbf{0} & \sqrt{2}V_{\mathcal{N}} \end{bmatrix} \mid Y \in \mathbb{S}^N \right\}
\end{aligned} \tag{2098}$$

in the Euclidean sense. From (985) we get sufficiency of the first matrix criterion for an EDM proved by Schoenberg in 1935; [320][5.7]

$$D \in \mathbb{EDM}^N \;\Leftrightarrow\; \begin{cases} -V_{\mathcal{N}}^{\mathrm{T}}DV_{\mathcal{N}} \in \mathbb{S}_+^{N-1} \\ D \in \mathbb{S}_h^N \end{cases} \tag{987}$$

[5.7] From (975) we know $\mathcal{R}(V_{\mathcal{N}}) = \mathcal{N}(\mathbf{1}^{\mathrm{T}})$, so (987) is the same as (963). In fact, any matrix V in place of $V_{\mathcal{N}}$ will satisfy (987) whenever $\mathcal{R}(V) = \mathcal{R}(V_{\mathcal{N}}) = \mathcal{N}(\mathbf{1}^{\mathrm{T}})$. But $V_{\mathcal{N}}$ is the matrix implicit in Schoenberg's seminal exposition.

We provide a rigorous complete more geometric proof of this fundamental *Schoenberg criterion* in §5.9.1.0.4. [397, `isedm(D)`]

By substituting $G = \begin{bmatrix} 0 & \mathbf{0}^{\mathrm{T}} \\ \mathbf{0} & -V_{\mathcal{N}}^{\mathrm{T}} D V_{\mathcal{N}} \end{bmatrix}$ (985) into $\mathbf{D}(G)$ (980),

$$D = \begin{bmatrix} 0 \\ \delta\big(-V_{\mathcal{N}}^{\mathrm{T}} D V_{\mathcal{N}}\big) \end{bmatrix} \mathbf{1}^{\mathrm{T}} + \mathbf{1} \begin{bmatrix} 0 & \delta\big(-V_{\mathcal{N}}^{\mathrm{T}} D V_{\mathcal{N}}\big)^{\mathrm{T}} \end{bmatrix} - 2 \begin{bmatrix} 0 & \mathbf{0}^{\mathrm{T}} \\ \mathbf{0} & -V_{\mathcal{N}}^{\mathrm{T}} D V_{\mathcal{N}} \end{bmatrix} \quad (1089)$$

assuming $x_1 = \mathbf{0}$. Details of this bijection are provided in §5.6.2.

5.4.2.2   0 geometric center

Assume the *geometric center* (§5.5.1.0.1) of an unknown list X is the origin;

$$X\mathbf{1} = \mathbf{0} \;\Leftrightarrow\; G\mathbf{1} = \mathbf{0} \qquad (988)$$

Now consider the calculation $(I - \frac{1}{N}\mathbf{1}\mathbf{1}^{\mathrm{T}})\mathbf{D}(G)(I - \frac{1}{N}\mathbf{1}\mathbf{1}^{\mathrm{T}})$, a geometric centering or projection operation. (§E.7.2.0.2) Setting $\mathbf{D}(G) = D$ for convenience as in §5.4.2.1,

$$\begin{aligned} G &= -\big(D - \tfrac{1}{N}(D\mathbf{1}\mathbf{1}^{\mathrm{T}} + \mathbf{1}\mathbf{1}^{\mathrm{T}}D) + \tfrac{1}{N^2}\mathbf{1}\mathbf{1}^{\mathrm{T}}D\mathbf{1}\mathbf{1}^{\mathrm{T}}\big)\tfrac{1}{2}, \quad X\mathbf{1} = \mathbf{0} \\ &= -VDV\tfrac{1}{2} \\ VGV &= -VDV\tfrac{1}{2} \qquad\qquad\qquad\qquad\qquad\qquad\qquad \forall X \end{aligned} \qquad (989)$$

where more properties of the auxiliary (*geometric centering*, projection) matrix

$$V \triangleq I - \frac{1}{N}\mathbf{1}\mathbf{1}^{\mathrm{T}} \in \mathbb{S}^N \qquad (990)$$

are found in §B.4. VGV may be regarded as a *covariance matrix* of means $\mathbf{0}$. From (989) and the assumption $D \in \mathbb{S}_h^N$ we get sufficiency of the more popular form of Schoenberg criterion:

$$D \in \mathbb{EDM}^N \;\Leftrightarrow\; \begin{cases} -VDV \in \mathbb{S}_+^N \\ D \in \mathbb{S}_h^N \end{cases} \qquad (991)$$

Of particular utility when $D \in \mathbb{EDM}^N$ is the fact, (§B.4.2 *no.*20) (964)

$$\begin{aligned} \mathrm{tr}\big(-VDV\tfrac{1}{2}\big) &= \tfrac{1}{2N}\sum_{i,j} d_{ij} &&= \tfrac{1}{2N}\,\mathrm{vec}(X)^{\mathrm{T}}\bigg(\sum_{i,j}\Phi_{ij}\otimes I\bigg)\mathrm{vec}\,X \\ &= \mathrm{tr}(VGV), \quad G \succeq 0 \\ &= \mathrm{tr}\,G &&= \sum_{\ell=1}^N \|x_\ell\|^2 = \|X\|_{\mathrm{F}}^2, \qquad X\mathbf{1} = \mathbf{0} \end{aligned} \qquad (992)$$

where $\sum\Phi_{ij} \in \mathbb{S}_+^N$ (966), therefore convex in $\mathrm{vec}\,X$. We will find this trace useful as a heuristic to minimize affine dimension of an unknown list arranged columnar in X (§7.2.2), but it tends to facilitate reconstruction of a list configuration having least energy; *id est*, it compacts a reconstructed list by minimizing total norm-square of the vertices.

By substituting $G = -VDV\tfrac{1}{2}$ (989) into $\mathbf{D}(G)$ (980), assuming $X\mathbf{1} = \mathbf{0}$

$$D = \delta\big(-VDV\tfrac{1}{2}\big)\mathbf{1}^{\mathrm{T}} + \mathbf{1}\delta\big(-VDV\tfrac{1}{2}\big)^{\mathrm{T}} - 2\big(-VDV\tfrac{1}{2}\big) \qquad (1079)$$

Details of this bijection can be found in §5.6.1.1.

5.4.2.2.1   Example. *Hypersphere.*
These foregoing relationships allow combination of distance and Gram constraints in any optimization problem we might pose:

- Interpoint angle Ψ can be constrained by fixing all the individual point lengths $\sqrt[\circ]{\delta(G)}$; then

$$\Psi = -\sqrt{\delta^2(G)}^{-1} V D V \tfrac{1}{2} \sqrt{\delta^2(G)}^{-1}, \qquad X\mathbf{1} = \mathbf{0} \tag{993}$$

- (*confer* §5.9.1.0.3, (1178) (1322)) Constraining all main diagonal entries g_{ii} of a Gram matrix to 1, for example, is equivalent to the constraint that all points lie on a hypersphere of radius 1 centered at the origin.

$$D = 2(g_{11}\mathbf{1}\mathbf{1}^{\mathrm{T}} - G) \in \mathbb{EDM}^N \tag{994}$$

Requiring $\mathbf{0}$ geometric center then becomes equivalent to the constraint: $D\mathbf{1} = 2N\mathbf{1}$. [91, p.116] Any further constraint on that Gram matrix applies only to interpoint angle matrix $\Psi = G$.

Because any point list may be constrained to lie on a hypersphere boundary whose affine dimension exceeds that of the list, a Gram matrix may always be constrained to have equal positive values along its main diagonal. (Laura Klanfer 1933 [320, §3]) This observation renewed interest in the elliptope (§5.9.1.0.1).   □

5.4.2.2.2   Example. *List-member constraints via Gram matrix.*
Capitalizing on identity (989) relating Gram and EDM D matrices, a constraint set such as
$$\left.\begin{array}{rcl} \mathrm{tr}\big(-\tfrac{1}{2}VDVe_ie_i^{\mathrm{T}}\big) &=& \|x_i\|^2 \\ \mathrm{tr}\big(-\tfrac{1}{2}VDV(e_ie_j^{\mathrm{T}}+e_je_i^{\mathrm{T}})\tfrac{1}{2}\big) &=& x_i^{\mathrm{T}}x_j \\ \mathrm{tr}\big(-\tfrac{1}{2}VDVe_je_j^{\mathrm{T}}\big) &=& \|x_j\|^2 \end{array}\right\} \tag{995}$$
relates list member x_i to x_j to within an isometry through inner-product identity [408, §1-7]
$$\cos\psi_{ij} = \frac{x_i^{\mathrm{T}}x_j}{\|x_i\|\,\|x_j\|} \tag{996}$$
where ψ_{ij} is angle between the two vectors as in (977). For M list members, there total $M(M+1)/2$ such constraints. Angle constraints are incorporated in Example 5.4.2.2.5 and Example 5.4.2.2.13.   □

5.4.2.2.3   Example. *Gram matrix as optimization problem.*
Consider the academic problem of finding a Gram matrix (989) subject to constraints on each and every entry of the corresponding EDM:
$$\begin{array}{ll} \underset{D\in\mathbb{S}_h^N}{\mathrm{find}} & -VDV\tfrac{1}{2} \in \mathbb{S}^N \\ \mathrm{subject\ to} & \langle D, (e_ie_j^{\mathrm{T}}+e_je_i^{\mathrm{T}})\tfrac{1}{2}\rangle = \check{d}_{ij}, \quad i,j=1\ldots N, \quad i<j \\ & -VDV \succeq 0 \end{array} \tag{997}$$

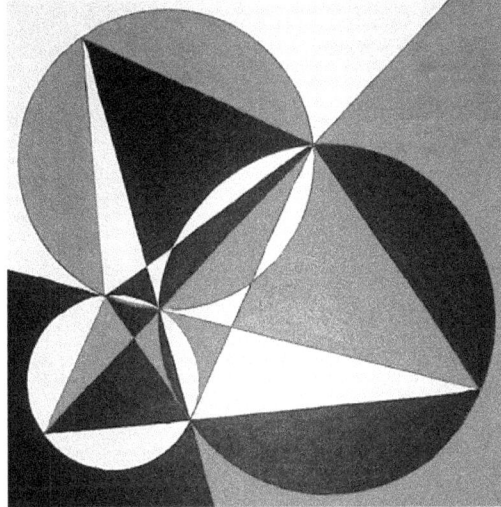

Figure 129: Rendering of *Fermat point* in acrylic on canvas by Suman Vaze. Three circles intersect at Fermat point of minimum total distance from three vertices of (and interior to) red/black/white triangle.

where the $\check{d}_{ij}$ are given nonnegative constants. EDM D can, of course, be replaced with the equivalent Gram-form (980). Requiring only the self-adjointness property (1499) of the main-diagonal linear operator δ we get, for $A\in\mathbb{S}^N$

$$\langle D\,,A\rangle = \langle \delta(G)\mathbf{1}^{\mathrm{T}}+\mathbf{1}\delta(G)^{\mathrm{T}}-2G\,,A\rangle = \langle G\,,\delta(A\mathbf{1})-A\rangle 2 \qquad (998)$$

Then the problem equivalent to (997) becomes, for $G\in\mathbb{S}^N_c \Leftrightarrow G\mathbf{1}=\mathbf{0}$

$$\begin{aligned}\underset{G\in\mathbb{S}^N_c}{\text{find}}\quad & G\in\mathbb{S}^N\\ \text{subject to}\quad & \Big\langle G\,,\delta\big((e_ie_j^{\mathrm{T}}+e_je_i^{\mathrm{T}})\mathbf{1}\big)-(e_ie_j^{\mathrm{T}}+e_je_i^{\mathrm{T}})\Big\rangle = \check{d}_{ij}\,,\quad i,j=1\ldots N,\quad i<j\\ & G\succeq 0\end{aligned} \qquad (999)$$

Barvinok's Proposition 2.9.3.0.1 predicts existence for either formulation (997) or (999) such that implicit equality constraints induced by subspace membership are ignored

$$\operatorname{rank}G\,,\ \operatorname{rank}VDV \le \left\lfloor\frac{\sqrt{8(N(N-1)/2)+1}-1}{2}\right\rfloor = N-1 \qquad (1000)$$

because, in each case, the Gram matrix is confined to a face of positive semidefinite cone $\mathbb{S}^N_+$ isomorphic with $\mathbb{S}^{N-1}_+$ (§6.6.1). (§E.7.2.0.2) This bound is tight (§5.7.1.1) and is the greatest upper bound.[5.8] $\qquad\square$

[5.8] $-VDV|_{N\leftarrow 1}=0$ (§B.4.1)

5.4.2.2.4   Example.   *First duality.*

Kuhn reports that the first dual optimization problem[5.9] to be recorded in the literature dates back to 1755. [385] Perhaps more intriguing is the fact: this earliest instance of duality is a two-dimensional Euclidean distance geometry problem known as *Fermat point* (Figure **129**) named after the French mathematician. Given N distinct points in the plane $\{x_i \in \mathbb{R}^2, \ i = 1 \ldots N\}$, the Fermat point y is an optimal solution to

$$\underset{y}{\text{minimize}} \ \sum_{i=1}^{N} \|y - x_i\| \tag{1001}$$

a convex minimization of total distance. The historically first dual problem formulation asks for the smallest equilateral triangle encompassing ($N = 3$) three points x_i. Another problem dual to (1001) (Kuhn 1967)

$$\begin{aligned} \underset{\{z_i\}}{\text{maximize}} \quad & \sum_{i=1}^{N} \langle z_i, x_i \rangle \\ \text{subject to} \quad & \sum_{i=1}^{N} z_i = \mathbf{0} \\ & \|z_i\| \le 1 \quad \forall i \end{aligned} \tag{1002}$$

has interpretation as minimization of work required to balance potential energy in an N-way tug-of-war between equally matched opponents situated at $\{x_i\}$. [402]

It is not so straightforward to write the Fermat point problem (1001) equivalently in terms of a Gram matrix from this section. Squaring instead

$$\underset{\alpha}{\text{minimize}} \sum_{i=1}^{N} \|\alpha - x_i\|^2 \ \equiv \ \begin{aligned} \underset{D \in \mathbb{S}^{N+1}}{\text{minimize}} \quad & \langle -V, D \rangle \\ \text{subject to} \quad & \langle D, e_i e_j^{\mathrm{T}} + e_j e_i^{\mathrm{T}} \rangle \tfrac{1}{2} = \check{d}_{ij} \quad \forall (i,j) \in \mathcal{I} \\ & -VDV \succeq 0 \end{aligned} \tag{1003}$$

yields an inequivalent convex geometric centering problem whose equality constraints comprise EDM D main-diagonal zeros and known distances-square.[5.10] Going the other way, a problem dual to total distance-square maximization (Example 6.7.0.0.1) is a penultimate minimum eigenvalue problem having application to *PageRank* calculation by search engines [237, §4]. [347]

Fermat function (1001) is empirically compared with (1003) in [62, §8.7.3], but for multiple unknowns in $\mathbb{R}^2$, where propensity of (1001) for producing zero distance between unknowns is revealed. An optimal solution to (1001) gravitates toward gradient discontinuities (§D.2.1), as in Figure **73**, whereas optimal solution to (1003) is less compact in the unknowns.[5.11] $\square$

[5.9]By *dual problem* is meant, in the strongest sense: the optimal objective achieved by a maximization problem, dual to a given minimization problem (related to each other by a Lagrangian function), is always equal to the optimal objective achieved by the minimization. (Figure 59 Example 2.13.1.0.3) A dual problem is always convex.

[5.10]$\alpha^\star$ is geometric center of points x_i (1053). For three points, $\mathcal{I} = \{1, 2, 3\}$; optimal affine dimension (§5.7) must be 2 because a third dimension can only increase total distance. Minimization of $\langle -V, D \rangle$ is a heuristic for rank minimization. (§7.2.2)

[5.11]Optimal solution to (1001) has mechanical interpretation in terms of interconnecting springs with constant force when distance is nonzero; otherwise, 0 force. Problem (1003) is interpreted instead using linear springs.

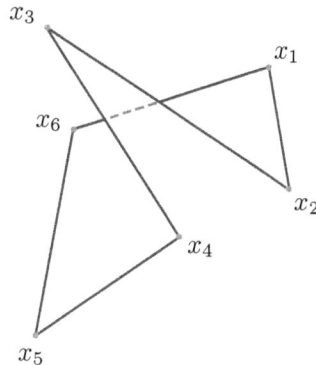

Figure 130: Arbitrary hexagon in $\mathbb{R}^3$ whose vertices are labelled clockwise.

5.4.2.2.5 Example. *Hexagon.*
Barvinok [25, §2.6] poses a problem in *geometric realizability* of an arbitrary hexagon
(Figure 130) having:

1. prescribed (one-dimensional) face-lengths l
2. prescribed angles φ between the three pairs of opposing faces
3. a constraint on the sum of norm-square of each and every vertex x;

ten affine equality constraints in all on a Gram matrix $G \in \mathbb{S}^6$ (989). Let's realize this as
a convex feasibility problem (with constraints written in the same order) also assuming $\mathbf{0}$
geometric center (988):

$$
\begin{aligned}
\underset{D \in \mathbb{S}_h^6}{\text{find}} \quad & -VDV\tfrac{1}{2} \in \mathbb{S}^6 \\
\text{subject to} \quad & \operatorname{tr}\!\big(D(e_i e_j^{\mathrm{T}} + e_j e_i^{\mathrm{T}})\tfrac{1}{2}\big) = l_{ij}^2 , \qquad j-1 = (i=1\ldots 6)\bmod 6 \\
& \operatorname{tr}\!\big(-\tfrac{1}{2}VDV(A_i + A_i^{\mathrm{T}})\tfrac{1}{2}\big) = \cos\varphi_i , \quad i = 1,2,3 \\
& \operatorname{tr}\!\big(-\tfrac{1}{2}VDV\big) = 1 \\
& -VDV \succeq 0
\end{aligned}
\tag{1004}
$$

where, for $A_i \in \mathbb{R}^{6\times 6}$ (996)

$$
\begin{aligned}
A_1 &= (e_1 - e_6)(e_3 - e_4)^{\mathrm{T}}/(l_{61}l_{34}) \\
A_2 &= (e_2 - e_1)(e_4 - e_5)^{\mathrm{T}}/(l_{12}l_{45}) \\
A_3 &= (e_3 - e_2)(e_5 - e_6)^{\mathrm{T}}/(l_{23}l_{56})
\end{aligned}
\tag{1005}
$$

and where the first constraint on length-square l_{ij}^2 can be equivalently written as a
constraint on the Gram matrix $-VDV\tfrac{1}{2}$ via (998). We show how to numerically solve
such a problem by *alternating projection* in §E.10.2.1.1. Barvinok's Proposition 2.9.3.0.1
asserts existence of a list, corresponding to Gram matrix G solving this feasibility problem,
whose affine dimension (§5.7.1.1) does not exceed 3 because the convex feasible set is
bounded by the third constraint $\operatorname{tr}(-\tfrac{1}{2}VDV) = 1$ (992).   □

Figure 131: Sphere-packing illustration from [401, *kissing number*]. Translucent balls illustrated all have the same diameter.

5.4.2.2.6   Example.   *Kissing number of sphere packing.*
Two nonoverlapping Euclidean balls are said to *kiss* if they touch.   An elementary geometrical problem can be posed: *Given hyperspheres, each having the same diameter* 1, *how many hyperspheres can simultaneously kiss one central hypersphere?* [422] Noncentral hyperspheres are allowed, but not required, to kiss.

As posed, the problem seeks the maximal number of spheres K kissing a central sphere in a particular dimension. The total number of spheres is $N = K + 1$. In one dimension, the answer to this kissing problem is 2. In two dimensions, 6. (Figure 7)

The question was presented, in three dimensions, to Isaac Newton by David Gregory in the context of celestial mechanics. And so was born a controversy between the two scholars on the campus of Trinity College Cambridge in 1694. Newton correctly identified the kissing number as 12 (Figure 131) while Gregory argued for 13. Their dispute was finally resolved in 1953 by Schütte & van der Waerden. [306] In 2003, Oleg Musin tightened the upper bound on kissing number K in four dimensions from 25 to $K = 24$ by refining a method of Philippe Delsarte from 1973. Delsarte's method provides an infinite number [16] of linear inequalities necessary for converting a rank-constrained semidefinite program[5.12] to a linear program.[5.13] [279]

There are no proofs known for kissing number in higher dimension excepting dimensions eight and twenty four. Interest persists [86] because sphere packing has found application to error correcting codes from the fields of communications and information theory; specifically to quantum computing. [94]

Translating this problem to an *EDM graph* realization (Figure 127, Figure 132) is suggested by Pfender & Ziegler. Imagine the centers of each sphere are connected by line segments. Then the distance between centers must obey simple criteria: Each sphere touching the central sphere has a line segment of length exactly 1 joining its center to

---

[5.12] whose feasible set belongs to that subset of an elliptope (§5.9.1.0.1) bounded above by some desired rank.
[5.13]Simplex-method solvers for linear programs produce numerically better results than contemporary log-barrier (interior-point method) solvers, for semidefinite programs, by about 7 orders of magnitude; they are far more predisposed to vertex solutions [96, p.158].

the central sphere's center. All spheres, excepting the central sphere, must have centers separated by a distance of at least 1.

From this perspective, the kissing problem can be posed as a semidefinite program. Assign index 1 to the central sphere assuming a total of N spheres:

$$
\begin{aligned}
\underset{D \in \mathbb{S}^N}{\text{minimize}} \quad & -\operatorname{tr}(W V_{\mathcal{N}}^{\mathrm{T}} D V_{\mathcal{N}}) \\
\text{subject to} \quad & D_{1j} = 1, & j = 2 \ldots N \\
& D_{ij} \geq 1, & 2 \leq i < j = 3 \ldots N \\
& D \in \mathbb{EDM}^N
\end{aligned}
\tag{1006}
$$

Then kissing number

$$
K = N_{\max} - 1
\tag{1007}
$$

is found from the maximal number N of spheres that solve this semidefinite program in a given affine dimension r whose realization is assured by 0 optimal objective. Matrix W is constant, in this program, determined by a method disclosed in §4.4.1. Matrix $W \in \mathbb{S}_+^{N-1}$ can be interpreted as direction of search through the positive semidefinite cone for a rank-r optimal solution $-V_{\mathcal{N}}^{\mathrm{T}} D^\star V_{\mathcal{N}} \in \mathbb{S}_+^{N-1}$: In one dimension, optimal direction matrix $W^\star$ has rank $= K - r = 2 - 1 = 1$;

$$
W^\star = \begin{bmatrix} 1 & 1 \\ 1 & 1 \end{bmatrix} \frac{1}{2}
\tag{1008}
$$

In two dimensions, optimal $W^\star$ has rank $= K - r = 6 - 2 = 4$;

$$
W^\star = \begin{bmatrix}
4 & 1 & 2 & -1 & -1 & 1 \\
1 & 4 & -1 & -1 & 2 & 1 \\
2 & -1 & 4 & 1 & 1 & -1 \\
-1 & -1 & 1 & 4 & 1 & 2 \\
-1 & 2 & 1 & 1 & 4 & -1 \\
1 & 1 & -1 & 2 & -1 & 4
\end{bmatrix} \frac{1}{6}
\tag{1009}
$$

In three dimensions, we leave it an exercise to find a rational optimal direction matrix $W^\star$ having rank $= K - r = 12 - 3 = 9$. Here is a full-rank rational optimal direction matrix:

$$
W^\star = \begin{bmatrix}
9 & 1 & -2 & -1 & 3 & -1 & -1 & 1 & 2 & 1 & -2 & 1 \\
1 & 9 & 3 & -1 & -1 & 1 & 1 & -2 & 1 & 2 & -1 & -1 \\
-2 & 3 & 9 & 1 & 2 & -1 & -1 & 2 & -1 & -1 & 1 & 2 \\
-1 & -1 & 1 & 9 & 1 & -1 & 1 & -1 & 3 & 2 & -1 & 1 \\
3 & -1 & 2 & 1 & 9 & 1 & 1 & -1 & -1 & -1 & 1 & -1 \\
-1 & 1 & -1 & -1 & 1 & 9 & 2 & -1 & 2 & -1 & 2 & 3 \\
-1 & 1 & -1 & 1 & 1 & 2 & 9 & 3 & -1 & 1 & -2 & -1 \\
1 & -2 & 2 & -1 & -1 & -1 & 3 & 9 & 2 & -1 & 1 & 1 \\
2 & 1 & -1 & 3 & -1 & 2 & -1 & 2 & 9 & -1 & 1 & -1 \\
1 & 2 & -1 & 2 & -1 & -1 & 1 & -1 & -1 & 9 & 3 & 1 \\
-2 & -1 & 1 & -1 & 1 & 2 & -2 & 1 & 1 & 3 & 9 & -1 \\
1 & -1 & 2 & 1 & -1 & 3 & -1 & 1 & -1 & 1 & -1 & 9
\end{bmatrix} \frac{1}{12}
\tag{1010}
$$

A four-dimensional solution also has rational optimal direction matrix $W^\star$ with rank $= K - r = 24 - 4 = 20$;

$$
W^\star = \frac{1}{24}
\begin{bmatrix}
20 & -2 & 2 & -2 & 0 & 0 & -2 & 2 & 2 & -2 & 2 & 0 & 2 & 4 & 2 & 2 & 0 & -2 & -2 & -2 & 2 & 0 & 0 & -2 \\
-2 & 20 & 2 & 0 & 2 & -2 & -2 & 0 & 2 & 0 & 2 & -2 & 0 & 2 & -2 & 4 & 2 & -2 & 2 & 0 & -2 & 2 & 2 & -2 \\
2 & 2 & 20 & 2 & 2 & 2 & 0 & 2 & 0 & -2 & 0 & 2 & -2 & -2 & 0 & -2 & -2 & 0 & 0 & 2 & -2 & -2 & -2 & 4 \\
-2 & 0 & 2 & 20 & -2 & 2 & -2 & 0 & -2 & 0 & 2 & -2 & 4 & 2 & 0 & -2 & 2 & -2 & 0 & 0 & 2 & 2 & 2 & -2 \\
0 & 2 & 2 & -2 & 20 & 0 & 2 & -2 & -2 & 2 & -2 & 0 & 2 & 0 & 2 & -2 & 0 & 2 & -2 & -2 & 2 & 4 & 0 & -2 \\
0 & -2 & 2 & 2 & 0 & 20 & 2 & -2 & 2 & 2 & -2 & 0 & -2 & 2 & 4 & -2 & 2 & -2 & 2 & 0 & 2 & 0 & 2 & -2 \\
-2 & -2 & 0 & -2 & 2 & 2 & 20 & 2 & 0 & -2 & 4 & -2 & 2 & 2 & 0 & 2 & 2 & 0 & 0 & 2 & -2 & -2 & 2 & 0 \\
2 & 0 & 2 & 0 & -2 & -2 & 2 & 20 & -2 & 4 & 2 & -2 & 0 & -2 & -2 & 0 & 2 & 2 & 2 & 0 & 0 & 2 & 2 & -2 \\
2 & 2 & 0 & -2 & -2 & 2 & 0 & -2 & 20 & 2 & 2 & 0 & 2 & 2 & 0 & -2 & -2 & -2 & 0 & 0 & -2 & -2 & 2 & 0 \\
-2 & 0 & -2 & 0 & 2 & 2 & -2 & 4 & 2 & 20 & 2 & 2 & 0 & 2 & 2 & 0 & -2 & -2 & -2 & 0 & 0 & -2 & -2 & 2 \\
2 & 2 & 0 & 2 & -2 & -2 & 4 & 2 & 2 & 2 & 20 & 2 & 0 & -2 & 2 & 0 & 2 & 2 & 2 & -2 & 0 & 4 & -2 & 2 \\
0 & -2 & 2 & -2 & 0 & 0 & -2 & -2 & 2 & 2 & 2 & 20 & 2 & 0 & -2 & 2 & 0 & 2 & 2 & 2 & -2 & 0 & 4 & -2 \\
2 & 0 & -2 & 4 & 2 & -2 & 2 & 0 & 2 & 0 & 0 & 2 & 20 & -2 & -2 & 0 & 2 & 4 & 2 & 2 & -2 & -2 & 2 & 0 \\
4 & 2 & -2 & 2 & 0 & 2 & 2 & -2 & 2 & 2 & -2 & 0 & -2 & 20 & 2 & -2 & 0 & 2 & -2 & 0 & 2 & -2 & -2 & 2 \\
2 & -2 & 0 & 0 & 2 & 4 & 0 & -2 & 0 & 2 & 2 & -2 & -2 & 2 & 20 & 2 & -2 & 2 & -2 & 0 & 0 & 2 & -2 & 0 \\
2 & 4 & -2 & -2 & -2 & -2 & 2 & 0 & -2 & 0 & 0 & 2 & 0 & -2 & 2 & 20 & 2 & -2 & 2 & -2 & 0 & 0 & 2 & 0 \\
0 & 2 & -2 & 2 & 0 & 2 & 2 & 2 & -2 & -2 & 2 & 0 & 2 & 0 & -2 & 2 & 20 & 2 & -2 & 2 & -2 & 0 & 0 & 2 \\
-2 & -2 & 0 & -2 & 2 & -2 & 0 & 2 & -2 & -2 & 2 & 2 & 4 & 2 & 2 & -2 & 2 & 20 & -2 & 2 & 2 & -2 & -2 & 0 \\
-2 & 2 & 0 & 0 & -2 & 2 & 0 & 2 & 0 & -2 & 2 & 2 & 2 & -2 & -2 & 2 & -2 & -2 & 20 & 4 & 2 & -2 & 2 & -2 \\
-2 & 0 & 2 & 0 & -2 & 0 & 2 & 0 & 0 & 0 & -2 & 2 & 2 & 0 & 0 & -2 & 2 & 2 & 4 & 20 & -2 & 2 & 2 & 2 \\
2 & -2 & -2 & 2 & 2 & 2 & -2 & 0 & -2 & 0 & 0 & 2 & -2 & 2 & 2 & 4 & 20 & -2 & 2 & 2 \\
0 & 2 & -2 & 2 & 4 & 0 & -2 & 2 & -2 & -2 & 4 & 0 & -2 & 2 & 2 & -2 & 2 & 2 & -2 & 20 & 0 & 2 \\
0 & 2 & -2 & 2 & 0 & 2 & 2 & 2 & 2 & -2 & -2 & 4 & 0 & -2 & -2 & -2 & 2 & 0 & 20 & 2 \\
-2 & -2 & 4 & -2 & -2 & -2 & 0 & -2 & 0 & 2 & 0 & -2 & 2 & 2 & 0 & 2 & 2 & 0 & 0 & -2 & 2 & 2 & 2 & 20
\end{bmatrix}
$$

(1011)

but these direction matrices are not unique and their precision not critical. Here is an optimal four-dimensional point list,[5.14] in MATLAB output format, reconstructed by a method in §5.12:

```
Columns 1 through 6
X =      0    -0.1983    -0.4584     0.1657     0.9399     0.7416
         0     0.6863     0.2936     0.6239    -0.2936     0.3927
         0    -0.4835     0.8146    -0.6448     0.0611    -0.4224
         0     0.5059     0.2004    -0.4093    -0.1632     0.3427
Columns 7 through 12
   -0.4815    -0.9399    -0.7416     0.1983     0.4584    -0.2832
         0     0.2936    -0.3927    -0.6863    -0.2936    -0.6863
   -0.8756    -0.0611     0.4224     0.4835    -0.8146    -0.3922
   -0.0372     0.1632    -0.3427    -0.5059    -0.2004    -0.5431
Columns 13 through 18
    0.2832    -0.2926    -0.6473     0.0943     0.3640    -0.3640
    0.6863     0.9176    -0.6239    -0.2313    -0.0624     0.0624
    0.3922     0.1698    -0.2309    -0.6533    -0.1613     0.1613
    0.5431    -0.2088     0.3721     0.7147    -0.9152     0.9152
Columns 19 through 25
   -0.0943     0.6473    -0.1657     0.2926    -0.5759     0.5759     0.4815
    0.2313     0.6239    -0.6239    -0.9176     0.2313    -0.2313          0
    0.6533     0.2309     0.6448    -0.1698    -0.2224     0.2224     0.8756
   -0.7147    -0.3721     0.4093     0.2088    -0.7520     0.7520     0.0372
```

[5.14] An optimal five-dimensional point list is known: *The answer was known at least 175 years ago. I believe Gauss knew it. Moreover, Korkine & Zolotarev proved in 1882 that D_5 is the densest lattice in five dimensions. So they proved that if a kissing arrangement in five dimensions can be extended to some lattice, then $k(5)=40$. Of course, the conjecture in the general case also is: $k(5)=40$. You would like to see coordinates? Easily. Let $A=\sqrt 2$. Then $p(1)=(A,A,0,0,0)$, $p(2)=(-A,A,0,0,0)$, $p(3)=(A,-A,0,0,0)$, ... $p(40)=(0,0,0,-A,-A)$; i.e., we are considering points with coordinates that have two A and three 0 with any choice of signs and any ordering of the coordinates; the same coordinates-expression in dimensions 3 and 4.*

The first miracle happens in dimension 6. There are better packings than D_6 (Conjecture: $k(6)=72$). It's a real miracle how dense the packing is in eight dimensions (E_8= Korkine & Zolotarev packing that was discovered in 1880s) and especially in dimension 24, that is the so-called Leech lattice.

Actually, people in coding theory have conjectures on the kissing numbers for dimensions up to 32 (or even greater?). However, sometimes they found better lower bounds. I know that Ericson & Zinoviev a few years ago discovered (by hand, no computer) in dimensions 13 and 14 better kissing arrangements than were known before. —Oleg Musin

The r nonzero optimal eigenvalues of $-V_\mathcal{N}^{\mathrm{T}}D^\star V_\mathcal{N}$ are equal; remaining eigenvalues are zero as per $-\operatorname{tr}(W^\star V_\mathcal{N}^{\mathrm{T}}D^\star V_\mathcal{N}) = 0$ (767). Numerical problems begin to arise with matrices of this size due to interior-point methods of solution to (1006). By eliminating some equality constraints from the kissing number problem, matrix size can be reduced: From §5.8.3 we have

$$-V_\mathcal{N}^{\mathrm{T}}DV_\mathcal{N} = \mathbf{1}\mathbf{1}^{\mathrm{T}} - [\,\mathbf{0}\ \ I\,] D \begin{bmatrix} \mathbf{0}^{\mathrm{T}} \\ I \end{bmatrix} \tfrac{1}{2} \tag{1012}$$

(which does not hold more generally) where Identity matrix $I \in \mathbb{S}^{N-1}$ has one less dimension than EDM D. By defining an EDM principal submatrix

$$\hat{D} \triangleq [\,\mathbf{0}\ \ I\,] D \begin{bmatrix} \mathbf{0}^{\mathrm{T}} \\ I \end{bmatrix} \in \mathbb{S}_h^{N-1} \tag{1013}$$

so that

$$-V_\mathcal{N}^{\mathrm{T}}DV_\mathcal{N} = \mathbf{1}\mathbf{1}^{\mathrm{T}} - \hat{D}\tfrac{1}{2} \tag{1014}$$

we get a convex problem equivalent to (1006)

$$\begin{array}{ll} \underset{\hat{D}\in\mathbb{S}^K}{\text{minimize}} & -\operatorname{tr}(W\hat{D}) \\ \text{subject to} & \hat{D}_{ij} \geq 1, \qquad 1 \leq i < j = 2\ldots K \\ & \mathbf{1}\mathbf{1}^{\mathrm{T}} - \hat{D}\tfrac{1}{2} \succeq 0 \\ & \delta(\hat{D}) = \mathbf{0} \end{array} \tag{1015}$$

Any feasible solution $\mathbf{1}\mathbf{1}^{\mathrm{T}} - \hat{D}\tfrac{1}{2}$ belongs to an elliptope (§5.9.1.0.1). $\qquad\square$

5.4.2.2.7 Exercise. *Rational optimal kissing direction matrix $W^\star$.*
Replace (1010) with a rational $W^\star$ having rank $= K - r = 12 - 3 = 9$, main diagonal 9, and common denominator 12. $\qquad\blacktriangledown$

This next example shows how finding the common point of intersection for three circles in a plane, a nonlinear problem, has convex expression.

5.4.2.2.8 Example. *Trilateration in wireless sensor network.* [167]
Given three known absolute point positions in $\mathbb{R}^2$ (three anchors $\check{x}_2, \check{x}_3, \check{x}_4$) and only one unknown point (one sensor x_1), the sensor's absolute position is determined from its noiseless measured distance-square $\check{d}_{i1}$ to each of three anchors (Figure 2, Figure 132a). This trilateration can be expressed as a convex optimization problem in terms of list $X \triangleq [\,x_1\ \check{x}_2\ \check{x}_3\ \check{x}_4\,] \in \mathbb{R}^{2\times4}$ and Gram matrix $G \in \mathbb{S}^4$ (977):

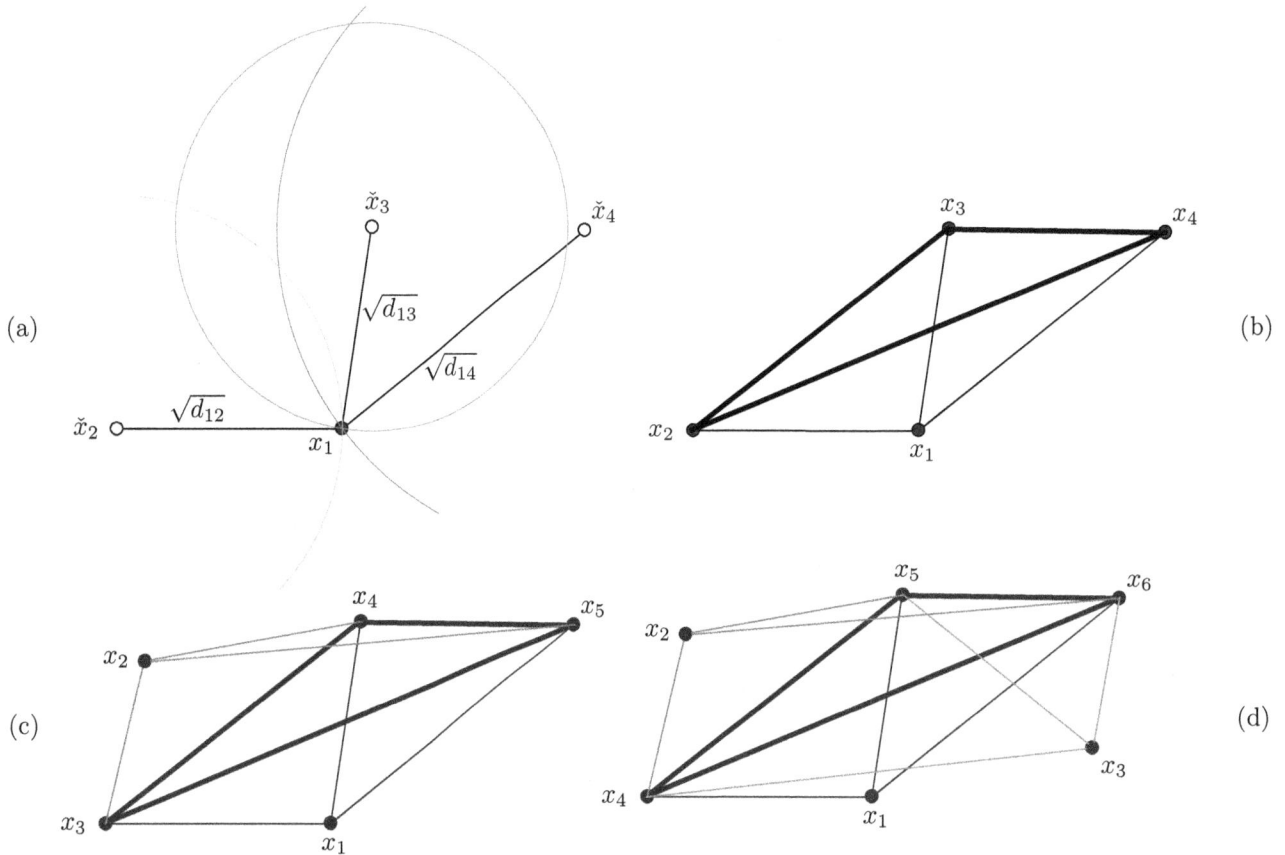

Figure 132: (a) Given three distances indicated with absolute point positions $\check{x}_2, \check{x}_3, \check{x}_4$ known and noncollinear, absolute position of x_1 in $\mathbb{R}^2$ can be precisely and uniquely determined by *trilateration*; solution to a system of nonlinear equations. Dimensionless EDM graphs (b) (c) (d) represent EDMs in various states of completion. Line segments represent known absolute distances and may cross without vertex at intersection. (b) Four-point list must always be embeddable in affine subset having dimension rank $V_\mathcal{N}^{\mathrm{T}} D V_\mathcal{N}$ not exceeding 3. To determine relative position of x_2, x_3, x_4, triangle inequality is necessary and sufficient (§5.14.1). Knowing all distance information, then (by injectivity of $\mathbf{D}$ (§5.6)) point position x_1 is uniquely determined to within an isometry in any dimension. (c) When fifth point is introduced, only distances to x_3, x_4, x_5 are required to determine relative position of x_2 in $\mathbb{R}^2$. Graph represents first instance of missing distance information; $\sqrt{d_{12}}$. (d) Three distances are absent ($\sqrt{d_{12}}$, $\sqrt{d_{13}}$, $\sqrt{d_{23}}$) from complete set of interpoint distances, yet unique isometric reconstruction (§5.4.2.2.10) of six points in $\mathbb{R}^2$ is certain.

$$\begin{array}{ll}
\underset{G\in\mathbb{S}^4,\ X\in\mathbb{R}^{2\times4}}{\text{minimize}} & \text{tr}\,G \\
\text{subject to} & \text{tr}(G\,\Phi_{i1}) & = \check{d}_{i1}\,, & i=2,3,4 \\
& \text{tr}\big(G\,e_i e_i^{\text{T}}\big) & = \|\check{x}_i\|^2\,, & i=2,3,4 \\
& \text{tr}(G(e_i e_j^{\text{T}}+e_j e_i^{\text{T}})/2) & = \check{x}_i^{\text{T}}\check{x}_j\,, & 2\le i<j=3,4 \\
& X(:,2\!:\!4) & = [\,\check{x}_2\ \check{x}_3\ \check{x}_4\,] \\
& \begin{bmatrix} I & X \\ X^{\text{T}} & G \end{bmatrix} & \succeq\ 0
\end{array} \qquad (1016)$$

where

$$\Phi_{ij} = (e_i - e_j)(e_i - e_j)^{\text{T}} \in \mathbb{S}_+^N \qquad (966)$$

and where the constraint on distance-square $\check{d}_{i1}$ is equivalently written as a constraint on the Gram matrix via (979). There are 9 linearly independent affine equality constraints on that Gram matrix while the sensor is constrained, only by dimensioning, to lie in $\mathbb{R}^2$. Although the objective $\text{tr}\,G$ of minimization[5.15] insures a solution on the boundary of positive semidefinite cone $\mathbb{S}_+^4$, for this problem, we claim that the set of feasible Gram matrices forms a line (§2.5.1.1) in isomorphic $\mathbb{R}^{10}$ tangent (§2.1.7.1.2) to the positive semidefinite cone boundary. (§5.4.2.2.9, *confer* §4.2.1.3)

By Schur complement (§A.4, §2.9.1.0.1), any feasible G and X provide

$$G \succeq X^{\text{T}}X \qquad (1017)$$

which is a convex relaxation of the desired (nonconvex) equality constraint

$$\begin{bmatrix} I & X \\ X^{\text{T}} & G \end{bmatrix} = \begin{bmatrix} I \\ X^{\text{T}} \end{bmatrix}\begin{bmatrix} I & X \end{bmatrix} \qquad (1018)$$

expected positive semidefinite rank-2 under noiseless conditions. But, by (1572), the relaxation admits

$$(3 \ge)\,\text{rank}\,G \ge \text{rank}\,X \qquad (1019)$$

(a third dimension corresponding to an intersection of three spheres, not circles, were there noise). If rank of an optimal solution equals 2,

$$\text{rank}\begin{bmatrix} I & X^\star \\ X^{\star\text{T}} & G^\star \end{bmatrix} = 2 \qquad (1020)$$

then $G^\star = X^{\star\text{T}}X^\star$ by Theorem A.4.0.1.3.

As posed, this *localization* problem does not require affinely independent (Figure 28, three noncollinear) anchors. Assuming the anchors exhibit no rotational or reflective symmetry in their affine hull (§5.5.2) and assuming the sensor x_1 lies in that affine hull, then sensor position solution $x_1^\star = X^\star(:,1)$ is unique under noiseless measurement. [328]
$\square$

[5.15]Trace $(\text{tr}\,G = \langle I,\,G\rangle)$ minimization is a heuristic for rank minimization. (§7.2.2.1) It may be interpreted as squashing G which is bounded below by $X^{\text{T}}X$ as in (1017); *id est*, $G-X^{\text{T}}X\succeq 0 \Rightarrow \text{tr}\,G \ge \text{tr}\,X^{\text{T}}X$ (1570). $\delta(G-X^{\text{T}}X)=\mathbf{0} \Leftrightarrow G=X^{\text{T}}X$ (§A.7.2) $\Rightarrow \text{tr}\,G = \text{tr}\,X^{\text{T}}X$ which is a condition necessary for equality.

This preceding transformation of trilateration to a semidefinite program works all the time ((1020) holds) despite relaxation (1017) because the optimal solution set is a unique point.

5.4.2.2.9 Proof (sketch).   Only the sensor location x_1 is unknown. The objective function together with the equality constraints make a linear system of equations in Gram matrix variable G

$$
\begin{aligned}
\operatorname{tr} G &= \|x_1\|^2 + \|\check{x}_2\|^2 + \|\check{x}_3\|^2 + \|\check{x}_4\|^2 \\
\operatorname{tr}(G\,\Phi_{i1}) &= \check{d}_{i1} , & i &= 2,3,4 \\
\operatorname{tr}\!\big(G\,e_i e_i^{\mathrm{T}}\big) &= \|\check{x}_i\|^2 , & i &= 2,3,4 \\
\operatorname{tr}(G(e_i e_j^{\mathrm{T}} + e_j e_i^{\mathrm{T}})/2) &= \check{x}_i^{\mathrm{T}} \check{x}_j , & 2 &\le i < j = 3,4
\end{aligned}
\tag{1021}
$$

which is invertible:

$$
\operatorname{svec} G =
\begin{bmatrix}
\operatorname{svec}(I)^{\mathrm{T}} \\
\operatorname{svec}(\Phi_{21})^{\mathrm{T}} \\
\operatorname{svec}(\Phi_{31})^{\mathrm{T}} \\
\operatorname{svec}(\Phi_{41})^{\mathrm{T}} \\
\operatorname{svec}(e_2 e_2^{\mathrm{T}})^{\mathrm{T}} \\
\operatorname{svec}(e_3 e_3^{\mathrm{T}})^{\mathrm{T}} \\
\operatorname{svec}(e_4 e_4^{\mathrm{T}})^{\mathrm{T}} \\
\operatorname{svec}\!\big((e_2 e_3^{\mathrm{T}} + e_3 e_2^{\mathrm{T}})/2\big)^{\mathrm{T}} \\
\operatorname{svec}\!\big((e_2 e_4^{\mathrm{T}} + e_4 e_2^{\mathrm{T}})/2\big)^{\mathrm{T}} \\
\operatorname{svec}\!\big((e_3 e_4^{\mathrm{T}} + e_4 e_3^{\mathrm{T}})/2\big)^{\mathrm{T}}
\end{bmatrix}^{-1}
\begin{bmatrix}
\|x_1\|^2 + \|\check{x}_2\|^2 + \|\check{x}_3\|^2 + \|\check{x}_4\|^2 \\
\check{d}_{21} \\
\check{d}_{31} \\
\check{d}_{41} \\
\|\check{x}_2\|^2 \\
\|\check{x}_3\|^2 \\
\|\check{x}_4\|^2 \\
\check{x}_2^{\mathrm{T}} \check{x}_3 \\
\check{x}_2^{\mathrm{T}} \check{x}_4 \\
\check{x}_3^{\mathrm{T}} \check{x}_4
\end{bmatrix}
\tag{1022}
$$

That line in the ambient space $\mathbb{S}^4$ of G, claimed on page 375, is traced by $\|x_1\|^2 \in \mathbb{R}$ on the right-hand side, as it turns out. One must show this line to be tangential (§2.1.7.1.2) to $\mathbb{S}_+^4$ in order to prove uniqueness. Tangency is possible for affine dimension 1 or 2 while its occurrence depends completely on the known measurement data.   ■

But as soon as significant noise is introduced or whenever distance data is incomplete, such problems can remain convex although the set of all optimal solutions generally becomes a convex set bigger than a single point (and still containing the noiseless solution).

5.4.2.2.10 Definition.   *Isometric reconstruction.*   (*confer* §5.5.3)
Isometric reconstruction from an EDM means building a list X correct to within a rotation, reflection, and translation; in other terms, reconstruction of relative position, correct to within an isometry, correct to within a rigid transformation.   △

How much distance information is needed to uniquely localize a sensor (to recover actual relative position)? The narrative in Figure **132** helps dispel any notion of distance data proliferation in *low affine dimension* $(r < N-2)$.[5.16] Huang, Liang,

---

[5.16]When affine dimension r reaches $N-2$, then all distances-square in the EDM must be known for unique isometric reconstruction in $\mathbb{R}^r$; going the other way, when $r=1$ then the condition that the dimensionless EDM graph be connected is necessary and sufficient. [193, §2.2]

and Pardalos [211, §4.2] claim O($2N$) distances is a least lower bound (independent of affine dimension r) for unique isometric reconstruction; achievable under certain noiseless conditions on graph connectivity and point position. Alfakih shows how to ascertain uniqueness over all affine dimensions via *Gale matrix.* [9] [4] [5] Figure **127**b (p.358, from *small completion problem* Example 5.3.0.0.2) is an example in $\mathbb{R}^2$ requiring only $2N-3 = 5$ known symmetric entries for unique isometric reconstruction, although the four-point example in Figure **132**b will not yield a unique reconstruction when any one of the distances is left unspecified.

The list represented by the particular dimensionless *EDM graph* in Figure **133**, having only $2N-3 = 9$ absolute distances specified, has only one realization in $\mathbb{R}^2$ but has more realizations in higher dimensions. Unique r-dimensional isometric reconstruction by semidefinite relaxation like (1017) occurs iff realization in $\mathbb{R}^r$ is unique and there exist no nontrivial higher-dimensional realizations. [328] For sake of reference, we provide the complete corresponding EDM:

$$D = \begin{bmatrix} 0 & 50641 & 56129 & 8245 & 18457 & 26645 \\ 50641 & 0 & 49300 & 25994 & 8810 & 20612 \\ 56129 & 49300 & 0 & 24202 & 31330 & 9160 \\ 8245 & 25994 & 24202 & 0 & 4680 & 5290 \\ 18457 & 8810 & 31330 & 4680 & 0 & 6658 \\ 26645 & 20612 & 9160 & 5290 & 6658 & 0 \end{bmatrix} \tag{1023}$$

We consider paucity of distance information in this next example which shows it is possible to recover exact relative position given incomplete noiseless distance information. An *ad hoc* method for recovery of the least-rank optimal solution under noiseless conditions is introduced:

5.4.2.2.11 Example. *Tandem trilateration in wireless sensor network.*
Given three known absolute point-positions in $\mathbb{R}^2$ (three anchors $\check{x}_3, \check{x}_4, \check{x}_5$), two unknown sensors $x_1, x_2 \in \mathbb{R}^2$ have absolute position determinable from their noiseless distances-square (as indicated in Figure **134**) assuming the anchors exhibit no rotational or reflective symmetry in their affine hull (§5.5.2). This example differs from Example 5.4.2.2.8 insofar as trilateration of each sensor is now in terms of one unknown position: the other sensor. We express this localization as a convex optimization problem (a semidefinite program, §4.1) in terms of list $X \triangleq [\,x_1\ x_2\ \check{x}_3\ \check{x}_4\ \check{x}_5\,] \in \mathbb{R}^{2\times5}$ and Gram matrix $G \in \mathbb{S}^5$ (977) via relaxation (1017):

$$\begin{aligned}
\underset{G\in\mathbb{S}^5,\ X\in\mathbb{R}^{2\times5}}{\text{minimize}} \quad & \operatorname{tr} G \\
\text{subject to} \quad & \operatorname{tr}(G\,\Phi_{i1}) = \check{d}_{i1}, & i = 2,4,5 \\
& \operatorname{tr}(G\,\Phi_{i2}) = \check{d}_{i2}, & i = 3,5 \\
& \operatorname{tr}\!\big(G\,e_i e_i^{\mathrm{T}}\big) = \|\check{x}_i\|^2, & i = 3,4,5 \\
& \operatorname{tr}(G(e_i e_j^{\mathrm{T}} + e_j e_i^{\mathrm{T}})/2) = \check{x}_i^{\mathrm{T}}\check{x}_j, & 3 \le i < j = 4,5 \\
& X(:,3{:}5) = [\,\check{x}_3\ \check{x}_4\ \check{x}_5\,] \\
& \begin{bmatrix} I & X \\ X^{\mathrm{T}} & G \end{bmatrix} \succeq 0
\end{aligned} \tag{1024}$$

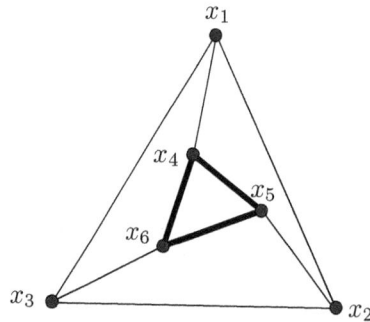

Figure 133: Incomplete EDM corresponding to this dimensionless EDM graph provides unique isometric reconstruction in $\mathbb{R}^2$. (drawn freehand; no symmetry intended, *confer*(1023))

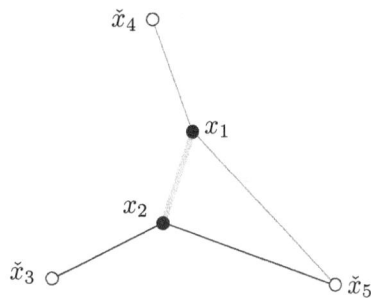

Figure 134: (Ye) Two sensors • and three anchors ○ in $\mathbb{R}^2$. Connecting line-segments denote known absolute distances. Incomplete EDM corresponding to this dimensionless EDM graph provides unique isometric reconstruction in $\mathbb{R}^2$.

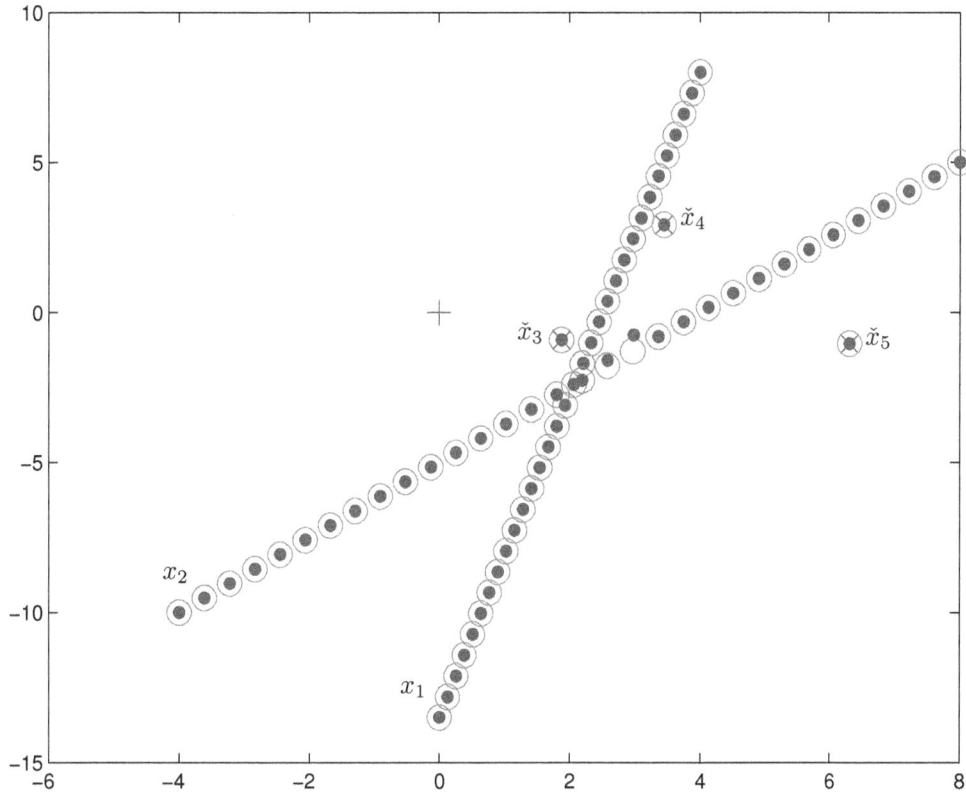

Figure 135: Given in red ○ are two discrete linear trajectories of sensors x_1 and x_2 in $\mathbb{R}^2$ localized by algorithm (1024) as indicated by blue bullets ●. Anchors $\check{x}_3$, $\check{x}_4$, $\check{x}_5$ corresponding to Figure 134 are indicated by ⊗. When targets ○ and bullets ● coincide under these noiseless conditions, localization is successful. On this run, two visible localization errors are due to rank-3 Gram optimal solutions. These errors can be corrected by choosing a different normal in objective of minimization.

where

$$\Phi_{ij} = (e_i - e_j)(e_i - e_j)^{\mathrm{T}} \in \mathbb{S}_+^N \qquad (966)$$

This problem realization is fragile because of the unknown distances between sensors and anchors. Yet there is no more information we may include beyond the 11 independent equality constraints on the Gram matrix (nonredundant constraints not antithetical) to reduce the feasible set.[5.17]

Exhibited in Figure 135 are two mistakes in solution $X^\star(:,1\!:\!2)$ due to a rank-3 optimal Gram matrix $G^\star$. The trace objective is a heuristic minimizing convex envelope of quasiconcave function[5.18] rank G. (§2.9.2.9.2, §7.2.2.1) A rank-2 optimal Gram matrix can be found and the errors corrected by choosing a different normal for the linear objective function, now implicitly the Identity matrix I; *id est,*

$$\mathrm{tr}\, G = \langle G\,,\,I\rangle \;\leftarrow\; \langle G\,,\,\delta(u)\rangle \qquad (1025)$$

where vector $u \in \mathbb{R}^5$ is randomly selected. A random search for a good normal $\delta(u)$ in only a few iterations is quite easy and effective because: the problem is small, an optimal solution is known *a priori* to exist in two dimensions, a good normal direction is not necessarily unique, and (we speculate) because the feasible affine-subset slices the positive semidefinite cone thinly in the Euclidean sense.[5.19]   □

We explore ramifications of noise and incomplete data throughout; their individual effect being to expand the optimal solution set, introducing more solutions and higher-rank solutions. Hence our focus shifts in §4.4 to discovery of a reliable means for diminishing the optimal solution set by introduction of a rank constraint.

Now we illustrate how a problem in distance geometry can be solved without equality constraints representing measured distance; instead, we have only upper and lower bounds on distances measured:

5.4.2.2.12   Example.   *Wireless location in a cellular telephone network.*
Utilizing measurements of distance, time of flight, angle of arrival, or signal power in the context of wireless telephony, *multilateration* is the process of localizing (determining absolute position of) a radio signal source • by inferring geometry relative to multiple fixed *base stations* ○ whose locations are known.

We consider localization of a cellular telephone by distance geometry, so we assume distance to any particular base station can be inferred from received signal power. On a large open flat expanse of terrain, signal-power measurement corresponds well with inverse distance. But it is not uncommon for measurement of signal power to suffer

---

[5.17]By virtue of their dimensioning, the sensors are already constrained to $\mathbb{R}^2$ the affine hull of the anchors.
[5.18]Projection on that nonconvex subset of all $N \times N$-dimensional positive semidefinite matrices, in an affine subset, whose rank does not exceed 2 is a problem considered difficult to solve. [362, §4]
[5.19]The log det rank-heuristic from §7.2.2.4 does not work here because it chooses the wrong normal. Rank reduction (§4.1.2.1) is unsuccessful here because Barvinok's upper bound (§2.9.3.0.1) on rank of $G^\star$ is 4.

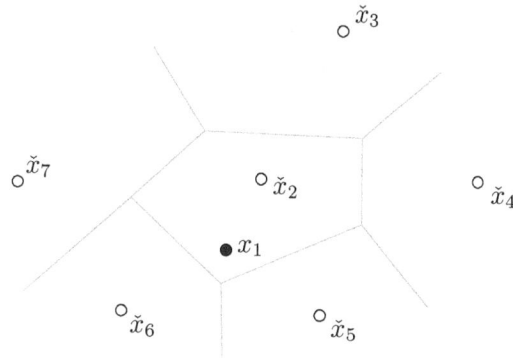

Figure 136: Regions of coverage by base stations ○ in a cellular telephone network. The term *cellular* arises from packing of regions best covered by neighboring base stations. Illustrated is a pentagonal *cell* best covered by base station $\check{x}_2$. Like a Voronoi diagram, cell geometry depends on base-station arrangement. In some US urban environments, it is not unusual to find base stations spaced approximately 1 mile apart. There can be as many as 20 base-station antennae capable of receiving signal from any given cell phone ● ; practically, that number is closer to 6.

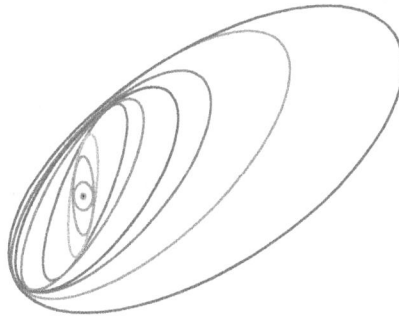

Figure 137: Some fitted contours of equal signal power in $\mathbb{R}^2$ transmitted from a commercial cellular telephone ● over about 1 mile suburban terrain outside San Francisco in 2005. Data by courtesy Polaris Wireless.

20 decibels in loss caused by factors such as *multipath* interference (signal reflections), mountainous terrain, man-made structures, turning one's head, or rolling the windows up in an automobile. Consequently, contours of equal signal power are no longer circular; their geometry is irregular and would more aptly be approximated by translated ellipsoids of graduated orientation and eccentricity as in Figure **137**.

Depicted in Figure **136** is one cell phone x_1 whose signal power is automatically and repeatedly measured by 6 base stations ○ nearby.[5.20] Those signal power measurements are transmitted from that cell phone to base station $\breve{x}_2$ who decides whether to transfer (*hand-off* or *hand-over*) responsibility for that call should the user roam outside its cell.[5.21]

Due to noise, at least one distance measurement more than the minimum number of measurements is required for reliable localization in practice; 3 measurements are minimum in two dimensions, 4 in three.[5.22] Existence of noise precludes measured distance from the input data. We instead assign measured distance to a range estimate specified by individual upper and lower bounds: $\overline{d_{i1}}$ is the upper bound on distance-square from the cell phone to i^{th} base station, while $\underline{d_{i1}}$ is the lower bound. These bounds become the input data. Each measurement range is presumed different from the others.

Then convex problem (1016) takes the form:

$$
\begin{array}{lll}
\underset{G\in\mathbb{S}^7,\ X\in\mathbb{R}^{2\times 7}}{\text{minimize}} & \operatorname{tr} G & \\[1ex]
\text{subject to} & \underline{d_{i1}} \leq \quad \operatorname{tr}(G\,\Phi_{i1}) \quad \leq \overline{d_{i1}}\,, & i=2\dots 7 \\[1ex]
& \operatorname{tr}\big(G\,e_i e_i^{\mathrm{T}}\big) = \|\breve{x}_i\|^2\,, & i=2\dots 7 \\[1ex]
& \operatorname{tr}(G(e_i e_j^{\mathrm{T}}+e_j e_i^{\mathrm{T}})/2) = \breve{x}_i^{\mathrm{T}}\breve{x}_j\,, & 2\leq i<j=3\dots 7 \\[1ex]
& X(:,2\!:\!7) = [\,\breve{x}_2\ \ \breve{x}_3\ \ \breve{x}_4\ \ \breve{x}_5\ \ \breve{x}_6\ \ \breve{x}_7\,] & \\[1ex]
& \begin{bmatrix} I & X \\ X^{\mathrm{T}} & G \end{bmatrix} \succeq 0 & \hspace{2em}(1026)
\end{array}
$$

where

$$
\Phi_{ij} = (e_i - e_j)(e_i - e_j)^{\mathrm{T}} \in \mathbb{S}_+^N \qquad (966)
$$

This semidefinite program realizes the wireless location problem illustrated in Figure **136**. Location $X^\star(:,1)$ is taken as solution, although measurement noise will often cause rank $G^\star$ to exceed 2. Randomized search for a rank-2 optimal solution is not so easy here as in Example 5.4.2.2.11. We introduce a method in §4.4 for enforcing the stronger rank-constraint (1020). To formulate this same problem in three dimensions, point list X is simply redimensioned in the semidefinite program.   □

[5.20] Cell phone signal power is typically encoded logarithmically with 1-decibel increment and 64-decibel dynamic range.

[5.21] Because distance to base station is quite difficult to infer from signal power measurements in an urban environment, localization of a particular cell phone ● by distance geometry would be far easier were the whole cellular system instead conceived so cell phone x_1 also transmits (to base station $\breve{x}_2$) its signal power as received by all other cell phones within range.

[5.22] In Example 4.4.1.2.4, we explore how this convex optimization algorithm fares in the face of measurement noise.

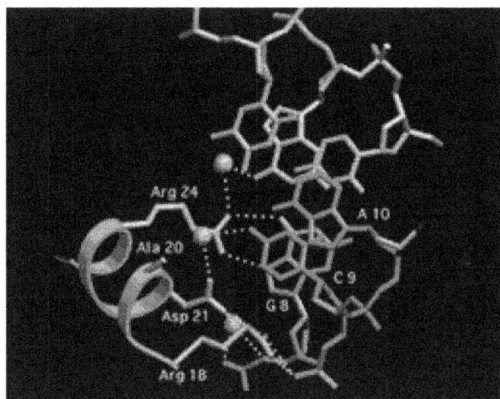

Figure 138: A depiction of molecular conformation. [120]

5.4.2.2.13 Example. (Biswas, Nigam, Ye) *Molecular Conformation.*
The subatomic measurement technique called *nuclear magnetic resonance spectroscopy*
(NMR) is employed to ascertain physical conformation of molecules; *e.g.*, Figure **3**,
Figure **138**. From this technique, distance, angle, and dihedral angle measurements can
be obtained. Dihedral angles arise consequent to a phenomenon where atom subsets are
physically constrained to Euclidean planes.

> *In the* rigid covalent geometry *approximation, the bond lengths and angles are
> treated as completely fixed, so that a given spatial structure can be described very
> compactly indeed by a list of torsion angles alone... These are the dihedral
> angles between the planes spanned by the two consecutive triples in a chain of
> four covalently bonded atoms.*
>
> −G. M. Crippen & T. F. Havel (1988) [90, §1.1]

Crippen & Havel recommend working exclusively with distance data because they consider
angle data to be mathematically cumbersome. The present example shows instead how
inclusion of dihedral angle data into a problem statement can be made elegant and convex.

As before, ascribe position information to the matrix

$$X = [\, x_1 \; \cdots \; x_N \,] \in \mathbb{R}^{3 \times N} \qquad (76)$$

and introduce a matrix $\aleph$ holding normals η to planes respecting dihedral angles φ :

$$\aleph \triangleq [\, \eta_1 \; \cdots \; \eta_M \,] \in \mathbb{R}^{3 \times M} \tag{1027}$$

As in the other examples, we preferentially work with Gram matrices G because of the
bridge they provide between other variables; we define

$$\begin{bmatrix} G_\aleph & Z \\ Z^T & G_X \end{bmatrix} \triangleq \begin{bmatrix} \aleph^T\aleph & \aleph^TX \\ X^T\aleph & X^TX \end{bmatrix} = \begin{bmatrix} \aleph^T \\ X^T \end{bmatrix} \begin{bmatrix} \aleph & X \end{bmatrix} \in \mathbb{R}^{N+M \times N+M} \qquad (1028)$$

whose rank is 3 by assumption. So our problem's variables are the two Gram matrices G_X and $G_\aleph$ and matrix $Z = \aleph^TX$ of cross products. Then measurements of distance-square d can be expressed as linear constraints on G_X as in (1026), dihedral angle φ measurements can be expressed as linear constraints on $G_\aleph$ by (996), and normal-vector η conditions can be expressed by vanishing linear constraints on cross-product matrix Z: Consider three points x labelled $1, 2, 3$ assumed to lie in the ℓth plane whose normal is η_ℓ. There might occur, for example, the independent constraints

$$\begin{array}{r} \eta_\ell^T(x_1 - x_2) = 0 \\ \eta_\ell^T(x_2 - x_3) = 0 \end{array} \qquad (1029)$$

which are expressible in terms of constant matrices $A_k \in \mathbb{R}^{M \times N}$;

$$\begin{array}{r} \langle Z, A_{\ell 12} \rangle = 0 \\ \langle Z, A_{\ell 23} \rangle = 0 \end{array} \qquad (1030)$$

Although normals η can be constrained exactly to unit length,

$$\delta(G_\aleph) = \mathbf{1} \qquad (1031)$$

NMR data is noisy; so measurements are given as $\overline{\text{upper}}$ and $\underline{\text{lower}}$ bounds. Given $\overline{\text{bounds}}$ on dihedral angles respecting $0 \leq \varphi_j \leq \pi$ and $\overline{\text{bounds}}$ on distances d_i and given constant matrices A_k (1030) and symmetric matrices Φ_i (966) and B_j per (996), then a molecular conformation problem can be expressed:

$$\begin{array}{rrccl} \underset{G_\aleph \in \mathbb{S}^M,\, G_X \in \mathbb{S}^N,\, Z \in \mathbb{R}^{M \times N}}{\text{find}} & G_X & & & \\[2mm] \text{subject to} & \underline{d_i} & \leq & \text{tr}(G_X \Phi_i) & \leq \overline{d_i} \qquad \forall i \in \mathcal{I}_1 \\[1mm] & \cos\underline{\varphi_j} & \leq & \text{tr}(G_\aleph B_j) & \leq \overline{\cos\varphi_j} \quad \forall j \in \mathcal{I}_2 \\[1mm] & \langle Z, A_k \rangle & & & = 0 \qquad \forall k \in \mathcal{I}_3 \\[1mm] & G_X \mathbf{1} & & & = \mathbf{0} \\[1mm] & \delta(G_\aleph) & & & = \mathbf{1} \\[1mm] & \begin{bmatrix} G_\aleph & Z \\ Z^T & G_X \end{bmatrix} & & & \succeq 0 \\[3mm] & \text{rank}\begin{bmatrix} G_\aleph & Z \\ Z^T & G_X \end{bmatrix} & & & = 3 \end{array} \qquad (1032)$$

where $G_X\mathbf{1} = \mathbf{0}$ provides a geometrically centered list X (§5.4.2.2). Ignoring the rank constraint would tend to force cross-product matrix Z to zero. What binds these three variables is the rank constraint; we show how to satisfy it in §4.4. $\square$

5.4.3 Inner-product form EDM definition

We might, for example, want to realize a constellation given only interstellar distance (or, equivalently, parsecs from our Sun and relative angular measurement; the Sun as vertex to two distant stars); called stellar cartography.   (p.22)

Equivalent to (964) is [408, §1-7] [338, §3.2]

$$
\begin{aligned}
d_{ij} &= d_{ik} + d_{kj} - 2\sqrt{d_{ik}d_{kj}}\cos\theta_{ikj}\\
&= \begin{bmatrix}\sqrt{d_{ik}} & \sqrt{d_{kj}}\end{bmatrix}
\begin{bmatrix} 1 & -e^{\imath\theta_{ikj}} \\ -e^{-\imath\theta_{ikj}} & 1 \end{bmatrix}
\begin{bmatrix}\sqrt{d_{ik}} \\ \sqrt{d_{kj}}\end{bmatrix}
\end{aligned}
\tag{1033}
$$

called *law of cosines* where $\imath \triangleq \sqrt{-1}$, i, j, k are positive integers, and θ_{ikj} is the angle at vertex x_k formed by vectors $x_i - x_k$ and $x_j - x_k$;

$$
\cos\theta_{ikj} = \frac{\frac{1}{2}(d_{ik} + d_{kj} - d_{ij})}{\sqrt{d_{ik}d_{kj}}} = \frac{(x_i - x_k)^{\mathrm{T}}(x_j - x_k)}{\|x_i - x_k\|\,\|x_j - x_k\|}
\tag{1034}
$$

where the numerator forms an inner product of vectors. Distance-square $d_{ij}\left(\begin{bmatrix}\sqrt{d_{ik}} \\ \sqrt{d_{kj}}\end{bmatrix}\right)$ is a convex quadratic function[5.23] on $\mathbb{R}^2_+$ whereas $d_{ij}(\theta_{ikj})$ is quasiconvex (§3.8) minimized over domain $\{-\pi \le \theta_{ikj} \le \pi\}$ by $\theta^\star_{ikj} = 0$, we get the *Pythagorean theorem* when $\theta_{ikj} = \pm\pi/2$, and $d_{ij}(\theta_{ikj})$ is maximized when $\theta^\star_{ikj} = \pm\pi$;

$$
\begin{aligned}
d_{ij} &= \left(\sqrt{d_{ik}} + \sqrt{d_{kj}}\right)^2, & \theta_{ikj} &= \pm\pi \\
d_{ij} &= d_{ik} + d_{kj}\,, & \theta_{ikj} &= \pm\tfrac{\pi}{2} \\
d_{ij} &= \left(\sqrt{d_{ik}} - \sqrt{d_{kj}}\right)^2, & \theta_{ikj} &= 0
\end{aligned}
\tag{1035}
$$

so

$$
\left|\sqrt{d_{ik}} - \sqrt{d_{kj}}\right| \le \sqrt{d_{ij}} \le \sqrt{d_{ik}} + \sqrt{d_{kj}}
\tag{1036}
$$

Hence the triangle inequality, Euclidean metric property 4, holds for any EDM D .

We may construct an inner-product form of the EDM definition for matrices by evaluating (1033) for $k=1$: By defining

$$
\Theta^{\mathrm{T}}\Theta \triangleq
\begin{bmatrix}
d_{12} & \sqrt{d_{12}d_{13}}\cos\theta_{213} & \sqrt{d_{12}d_{14}}\cos\theta_{214} & \cdots & \sqrt{d_{12}d_{1N}}\cos\theta_{21N} \\
\sqrt{d_{12}d_{13}}\cos\theta_{213} & d_{13} & \sqrt{d_{13}d_{14}}\cos\theta_{314} & \cdots & \sqrt{d_{13}d_{1N}}\cos\theta_{31N} \\
\sqrt{d_{12}d_{14}}\cos\theta_{214} & \sqrt{d_{13}d_{14}}\cos\theta_{314} & d_{14} & \ddots & \sqrt{d_{14}d_{1N}}\cos\theta_{41N} \\
\vdots & \vdots & \ddots & \ddots & \vdots \\
\sqrt{d_{12}d_{1N}}\cos\theta_{21N} & \sqrt{d_{13}d_{1N}}\cos\theta_{31N} & \sqrt{d_{14}d_{1N}}\cos\theta_{41N} & \cdots & d_{1N}
\end{bmatrix}
\in \mathbb{S}^{N-1}
\tag{1037}
$$

[5.23] $\begin{bmatrix} 1 & -e^{\imath\theta_{ikj}} \\ -e^{-\imath\theta_{ikj}} & 1 \end{bmatrix} \succeq 0$, having eigenvalues $\{0, 2\}$.

Minimum is attained for $\begin{bmatrix}\sqrt{d_{ik}} \\ \sqrt{d_{kj}}\end{bmatrix} = \begin{cases} \mu\mathbf{1}, & \mu \ge 0,\ \theta_{ikj} = 0 \\ \mathbf{0}, & -\pi \le \theta_{ikj} \le \pi,\ \theta_{ikj} \ne 0 \end{cases}$ · (§D.2.1, [62, exmp.4.5]) then any EDM may be expressed

$$
\mathbf{D}(\Theta) \triangleq \begin{bmatrix} 0 \\ \delta(\Theta^{\mathrm{T}}\Theta) \end{bmatrix} \mathbf{1}^{\mathrm{T}} + \mathbf{1} \begin{bmatrix} 0 & \delta(\Theta^{\mathrm{T}}\Theta)^{\mathrm{T}} \end{bmatrix} - 2 \begin{bmatrix} 0 & \mathbf{0}^{\mathrm{T}} \\ \mathbf{0} & \Theta^{\mathrm{T}}\Theta \end{bmatrix} \in \mathbb{EDM}^{N}
$$
$$
= \begin{bmatrix} 0 & \delta(\Theta^{\mathrm{T}}\Theta)^{\mathrm{T}} \\ \delta(\Theta^{\mathrm{T}}\Theta) & \delta(\Theta^{\mathrm{T}}\Theta)\mathbf{1}^{\mathrm{T}} + \mathbf{1}\delta(\Theta^{\mathrm{T}}\Theta)^{\mathrm{T}} - 2\Theta^{\mathrm{T}}\Theta \end{bmatrix}
$$
(1038)

$$
\mathbb{EDM}^{N} = \left\{ \mathbf{D}(\Theta) \mid \Theta \in \mathbb{R}^{N-1 \times N-1} \right\}
$$
(1039)

for which all Euclidean metric properties hold. Entries of $\Theta^{\mathrm{T}}\Theta$ result from vector inner-products as in (1034); *id est*,

$$
\Theta = \begin{bmatrix} x_2 - x_1 & x_3 - x_1 & \cdots & x_N - x_1 \end{bmatrix} = X\sqrt{2}V_{\mathcal{N}} \in \mathbb{R}^{n \times N-1}
$$
(1040)

Inner product $\Theta^{\mathrm{T}}\Theta$ is obviously related to a Gram matrix (977),

$$
G = \begin{bmatrix} 0 & \mathbf{0}^{\mathrm{T}} \\ \mathbf{0} & \Theta^{\mathrm{T}}\Theta \end{bmatrix}, \qquad x_1 = \mathbf{0}
$$
(1041)

For $D = \mathbf{D}(\Theta)$ and no condition on the list X (*confer* (985) (989))

$$
\Theta^{\mathrm{T}}\Theta = -V_{\mathcal{N}}^{\mathrm{T}}DV_{\mathcal{N}} \in \mathbb{R}^{N-1 \times N-1}
$$
(1042)

5.4.3.1   Relative-angle form

The inner-product form EDM definition is not a unique definition of Euclidean distance matrix; there are approximately five flavors distinguished by their argument to operator $\mathbf{D}$. Here is another one:

Like $\mathbf{D}(X)$ (968), $\mathbf{D}(\Theta)$ will make an EDM given any $\Theta \in \mathbb{R}^{n \times N-1}$, it is neither a convex function of Θ (§5.4.3.2), and it is homogeneous in the sense (971). Scrutinizing $\Theta^{\mathrm{T}}\Theta$ (1037) we find that because of the arbitrary choice $k=1$, distances therein are all with respect to point x_1. Similarly, relative angles in $\Theta^{\mathrm{T}}\Theta$ are between all vector pairs having vertex x_1. Yet picking arbitrary θ_{i1j} to fill $\Theta^{\mathrm{T}}\Theta$ will not necessarily make an EDM; inner product (1037) must be positive semidefinite.

$$
\Theta^{\mathrm{T}}\Theta = \sqrt{\delta(d)}\,\Omega\,\sqrt{\delta(d)} \triangleq
$$

$$
\begin{bmatrix} \sqrt{d_{12}} & & & \mathbf{0} \\ & \sqrt{d_{13}} & & \\ & & \ddots & \\ \mathbf{0} & & & \sqrt{d_{1N}} \end{bmatrix} \begin{bmatrix} 1 & \cos\theta_{213} & \cdots & \cos\theta_{21N} \\ \cos\theta_{213} & 1 & \ddots & \cos\theta_{31N} \\ \vdots & \ddots & \ddots & \vdots \\ \cos\theta_{21N} & \cos\theta_{31N} & \cdots & 1 \end{bmatrix} \begin{bmatrix} \sqrt{d_{12}} & & & \mathbf{0} \\ & \sqrt{d_{13}} & & \\ & & \ddots & \\ \mathbf{0} & & & \sqrt{d_{1N}} \end{bmatrix}
$$
(1043)

Expression $\mathbf{D}(\Theta)$ defines an EDM for any positive semidefinite *relative-angle matrix*

$$
\Omega = \begin{bmatrix} \cos\theta_{i1j}, & i,j = 2\ldots N \end{bmatrix} \in \mathbb{S}^{N-1}
$$
(1044)

and any nonnegative distance vector

$$d = [d_{1j} , \; j = 2 \ldots N] = \delta(\Theta^{\mathrm{T}}\Theta) \in \mathbb{R}^{N-1} \tag{1045}$$

because (§A.3.1.0.5)

$$\Omega \succeq 0 \;\Rightarrow\; \Theta^{\mathrm{T}}\Theta \succeq 0 \tag{1046}$$

Decomposition (1043) and the *relative-angle matrix inequality* $\Omega \succeq 0$ lead to a different expression of an inner-product form EDM definition (1038)

$$
\begin{aligned}
\mathbf{D}(\Omega,d) \;&\triangleq\; \begin{bmatrix} 0 \\ d \end{bmatrix} \mathbf{1}^{\mathrm{T}} + \mathbf{1} \begin{bmatrix} 0 & d^{\mathrm{T}} \end{bmatrix} - 2\sqrt{\delta\left(\begin{bmatrix} 0 \\ d \end{bmatrix}\right)} \begin{bmatrix} 0 & \mathbf{0}^{\mathrm{T}} \\ \mathbf{0} & \Omega \end{bmatrix} \sqrt{\delta\left(\begin{bmatrix} 0 \\ d \end{bmatrix}\right)} \\[2mm]
&= \begin{bmatrix} 0 & d^{\mathrm{T}} \\ d & d\mathbf{1}^{\mathrm{T}} + \mathbf{1}d^{\mathrm{T}} - 2\sqrt{\delta(d)}\,\Omega\,\sqrt{\delta(d)} \end{bmatrix} \in \mathbb{EDM}^{N}
\end{aligned}
\tag{1047}
$$

and another expression of the EDM cone:

$$\mathbb{EDM}^{N} = \left\{ \mathbf{D}(\Omega,d) \mid \Omega \succeq 0, \; \sqrt{\delta(d)} \succeq 0 \right\} \tag{1048}$$

In the particular circumstance $x_1 = \mathbf{0}$, we can relate interpoint angle matrix Ψ from the Gram decomposition in (977) to relative-angle matrix Ω in (1043). Thus,

$$\Psi \equiv \begin{bmatrix} 1 & \mathbf{0}^{\mathrm{T}} \\ \mathbf{0} & \Omega \end{bmatrix}, \qquad x_1 = \mathbf{0} \tag{1049}$$

5.4.3.2 Inner-product form $-V_{\mathcal{N}}^{\mathrm{T}}\mathbf{D}(\Theta)V_{\mathcal{N}}$ convexity

On page 385 we saw that each EDM entry d_{ij} is a convex quadratic function of $\begin{bmatrix} \sqrt{d_{ik}} \\ \sqrt{d_{kj}} \end{bmatrix}$ and a quasiconvex function of θ_{ikj}. Here the situation for inner-product form EDM operator $\mathbf{D}(\Theta)$ (1038) is identical to that in §5.4.1 for list-form $\mathbf{D}(X)$; $-\mathbf{D}(\Theta)$ is not a quasiconvex function of Θ by the same reasoning, and from (1042)

$$-V_{\mathcal{N}}^{\mathrm{T}}\mathbf{D}(\Theta)V_{\mathcal{N}} = \Theta^{\mathrm{T}}\Theta \tag{1050}$$

is a convex quadratic function of Θ on domain $\mathbb{R}^{n \times N-1}$ achieving its minimum at $\Theta = \mathbf{0}$.

5.4.3.3 Inner-product form, discussion

We deduce that knowledge of interpoint distance is equivalent to knowledge of distance and angle from the perspective of one point, x_1 in our chosen case. The total amount of information $N(N-1)/2$ in $\Theta^{\mathrm{T}}\Theta$ is unchanged[5.24] with respect to EDM D.

[5.24]The reason for amount $\mathrm{O}(N^2)$ information is because of the relative measurements. Use of a fixed reference in measurement of angles and distances would reduce required information but is antithetical. In the particular case $n = 2$, for example, ordering all points x_ℓ (in a length-N list) by increasing angle of vector $x_\ell - x_1$ with respect to $x_2 - x_1$, θ_{i1j} becomes equivalent to $\sum_{k=i}^{j-1} \theta_{k,1,k+1} \leq 2\pi$ and the amount of information is reduced to $2N-3$; rather, $\mathrm{O}(N)$.

5.5   Invariance

When D is an EDM, there exist an infinite number of corresponding N-point lists X (76) in Euclidean space. All those lists are related by isometric transformation: rotation, reflection, and translation (*offset* or *shift*).

5.5.1   Translation

Any translation common among all the points x_ℓ in a list will be cancelled in the formation of each d_{ij}. Proof follows directly from (964). Knowing that translation α in advance, we may remove it from the list constituting the columns of X by subtracting $\alpha \mathbf{1}^{\mathrm{T}}$. Then it stands to reason by list-form definition (968) of an EDM, for any translation $\alpha \in \mathbb{R}^n$

$$\mathbf{D}(X - \alpha \mathbf{1}^{\mathrm{T}}) = \mathbf{D}(X) \tag{1051}$$

In words, interpoint distances are unaffected by offset; EDM D is *translation invariant*. When $\alpha = x_1$ in particular,

$$[\,x_2 - x_1 \quad x_3 - x_1 \quad \cdots \quad x_N - x_1\,] = X\sqrt{2}V_{\mathcal{N}} \in \mathbb{R}^{n \times N-1} \tag{1040}$$

and so

$$\mathbf{D}(X - x_1 \mathbf{1}^{\mathrm{T}}) \;=\; \mathbf{D}(X - Xe_1 \mathbf{1}^{\mathrm{T}}) \;=\; \mathbf{D}\!\left(X\begin{bmatrix} \mathbf{0} & \sqrt{2}V_{\mathcal{N}} \end{bmatrix}\right) \;=\; \mathbf{D}(X) \tag{1052}$$

5.5.1.0.1   Example.   *Translating geometric center to origin.*
We might choose to shift the geometric center α_c of an N-point list $\{x_\ell\}$ (arranged columnar in X) to the origin; [361] [169]

$$\alpha = \alpha_c \triangleq Xb_c \triangleq X\mathbf{1}\tfrac{1}{N} \in \mathcal{P} \subseteq \mathcal{A} \tag{1053}$$

where $\mathcal{A}$ represents the list's affine hull. If we were to associate a point-mass m_ℓ with each of the points x_ℓ in the list, then their *center of mass* (or *gravity*) would be $\left(\sum x_\ell m_\ell\right)/\sum m_\ell$. The geometric center is the same as the center of mass under the assumption of uniform mass density across points. [225] The geometric center always lies in the convex hull $\mathcal{P}$ of the list; *id est*, $\alpha_c \in \mathcal{P}$ because $b_c^{\mathrm{T}}\mathbf{1} = 1$ and $b_c \succeq 0$.[5.25] Subtracting the geometric center from every list member,

$$X - \alpha_c \mathbf{1}^{\mathrm{T}} \;=\; X - \tfrac{1}{N}X\mathbf{1}\mathbf{1}^{\mathrm{T}} \;=\; X(I - \tfrac{1}{N}\mathbf{1}\mathbf{1}^{\mathrm{T}}) \;=\; XV \in \mathbb{R}^{n \times N} \tag{1054}$$

where V is the geometric centering matrix (990). So we have (*confer* (968))

$$\mathbf{D}(X) \;=\; \mathbf{D}(XV) \;=\; \delta(V^{\mathrm{T}}X^{\mathrm{T}}XV)\mathbf{1}^{\mathrm{T}} + \mathbf{1}\delta(V^{\mathrm{T}}X^{\mathrm{T}}XV)^{\mathrm{T}} - 2V^{\mathrm{T}}X^{\mathrm{T}}XV \in \mathbb{EDM}^N \tag{1055}$$

$$\square$$

[5.25]Any b from $\alpha = Xb$ chosen such that $b^{\mathrm{T}}\mathbf{1} = 1$, more generally, makes an auxiliary V-matrix. (§B.4.5)

5.5.1.1   Gram-form invariance

Following from (1055) and the linear Gram-form EDM operator (980):

$$\mathbf{D}(G) = \mathbf{D}(VGV) = \delta(VGV)\mathbf{1}^{\mathrm{T}} + \mathbf{1}\delta(VGV)^{\mathrm{T}} - 2VGV \in \mathbb{EDM}^N \qquad (1056)$$

The Gram-form consequently exhibits invariance to translation by a *doublet* (§B.2) $u\mathbf{1}^{\mathrm{T}} + \mathbf{1}u^{\mathrm{T}}$

$$\mathbf{D}(G) = \mathbf{D}(G - (u\mathbf{1}^{\mathrm{T}} + \mathbf{1}u^{\mathrm{T}})) \qquad (1057)$$

because, for any $u \in \mathbb{R}^N$, $\mathbf{D}(u\mathbf{1}^{\mathrm{T}} + \mathbf{1}u^{\mathrm{T}}) = \mathbf{0}$. The collection of all such doublets forms the nullspace (1073) to the operator; the *translation-invariant subspace* $\mathbb{S}_c^{N\perp}$ (2096) of the symmetric matrices $\mathbb{S}^N$. This means matrix G is not unique and can belong to an expanse more broad than a positive semidefinite cone; *id est*, $G \in \mathbb{S}_+^N - \mathbb{S}_c^{N\perp}$. So explains Gram matrix sufficiency in EDM definition (980).[5.26]

5.5.2   Rotation/Reflection

Rotation of the list $X \in \mathbb{R}^{n \times N}$ about some arbitrary point $\alpha \in \mathbb{R}^n$, or reflection through some affine subset containing α, can be accomplished via $Q(X - \alpha\mathbf{1}^{\mathrm{T}})$ where Q is an orthogonal matrix (§B.5).

We rightfully expect

$$\mathbf{D}\big(Q(X - \alpha\mathbf{1}^{\mathrm{T}})\big) = \mathbf{D}(QX - \beta\mathbf{1}^{\mathrm{T}}) = \mathbf{D}(QX) = \mathbf{D}(X) \qquad (1058)$$

Because list-form $\mathbf{D}(X)$ is translation invariant, we may safely ignore offset and consider only the impact of matrices that premultiply X. Interpoint distances are unaffected by rotation or reflection; we say, EDM D is *rotation/reflection invariant*. Proof follows from the fact, $Q^{\mathrm{T}} = Q^{-1} \Rightarrow X^{\mathrm{T}}Q^{\mathrm{T}}QX = X^{\mathrm{T}}X$. So (1058) follows directly from (968).

The class of premultiplying matrices for which interpoint distances are unaffected is a little more broad than orthogonal matrices. Looking at EDM definition (968), it appears that any matrix Q_{p} such that

$$X^{\mathrm{T}}Q_{\mathrm{p}}^{\mathrm{T}}Q_{\mathrm{p}}X = X^{\mathrm{T}}X \qquad (1059)$$

will have the property

$$\mathbf{D}(Q_{\mathrm{p}}X) = \mathbf{D}(X) \qquad (1060)$$

An example is skinny $Q_{\mathrm{p}} \in \mathbb{R}^{m \times n}$ $(m > n)$ having orthonormal columns; an orthonormal matrix.

5.5.2.0.1   Example.   *Reflection prevention and quadrature rotation.*
Consider the EDM graph in Figure 139b representing known distance between vertices (Figure 139a) of a tilted-square diamond in $\mathbb{R}^2$. Suppose some geometrical optimization problem were posed where isometric transformation is allowed excepting reflection, and where rotation must be quantized so that only *quadrature* rotations are allowed; only multiples of $\pi/2$.

[5.26]A constraint $G\mathbf{1} = \mathbf{0}$ would prevent excursion into the translation-invariant subspace (numerical unboundedness).

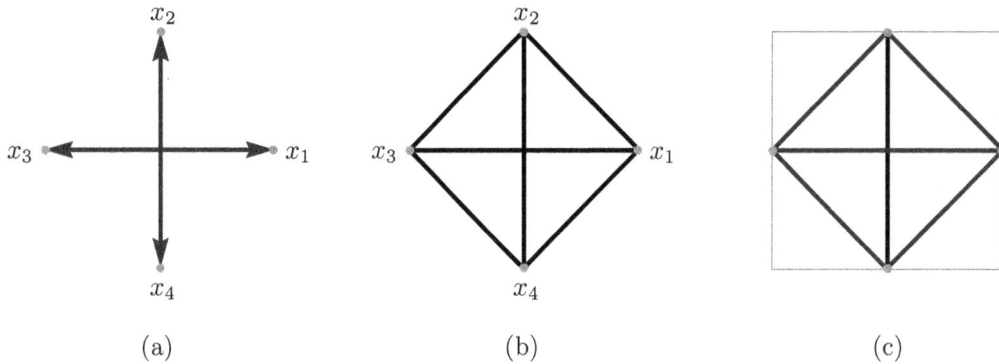

Figure 139: **(a)** Four points in quadrature in two dimensions about their geometric center. **(b)** Complete EDM graph of diamond-shaped vertices. **(c)** Quadrature rotation of a Euclidean body in $\mathbb{R}^2$ first requires a shroud that is the smallest Cartesian square containing it.

In two dimensions, a counterclockwise rotation of any vector about the origin by angle θ is prescribed by the orthogonal matrix

$$Q = \begin{bmatrix} \cos\theta & -\sin\theta \\ \sin\theta & \cos\theta \end{bmatrix} \qquad (1061)$$

whereas reflection of any point through a hyperplane containing the origin

$$\partial\mathcal{H} = \left\{ x \in \mathbb{R}^2 \,\middle|\, \begin{bmatrix} \cos\theta \\ \sin\theta \end{bmatrix}^{\mathrm{T}} x = 0 \right\} \qquad (1062)$$

is accomplished via multiplication with symmetric orthogonal matrix (§B.5.3)

$$R = \begin{bmatrix} \sin(\theta)^2 - \cos(\theta)^2 & -2\sin(\theta)\cos(\theta) \\ -2\sin(\theta)\cos(\theta) & \cos(\theta)^2 - \sin(\theta)^2 \end{bmatrix} \qquad (1063)$$

Rotation matrix Q is characterized by identical diagonal entries and by antidiagonal entries equal but opposite in sign, whereas reflection matrix R is characterized in the reverse sense.

Assign the diamond vertices $\{x_\ell \in \mathbb{R}^2, \ \ell = 1 \dots 4\}$ to columns of a matrix

$$X = [\, x_1 \ \ x_2 \ \ x_3 \ \ x_4 \,] \in \mathbb{R}^{2 \times 4} \qquad (76)$$

Our scheme to prevent reflection enforces a rotation matrix characteristic upon the coordinates of adjacent points themselves: First shift the geometric center of X to the origin; for geometric centering matrix $V \in \mathbb{S}^4$ (§5.5.1.0.1), define

$$Y \triangleq XV \in \mathbb{R}^{2 \times 4} \qquad (1064)$$

To maintain relative quadrature between points (Figure 139a) and to prevent reflection, it is sufficient that all interpoint distances be specified and that adjacencies $Y(:, 1\!:\!2)$,

$Y(:,2\!:\!3)$, and $Y(:,3\!:\!4)$ be proportional to $2\!\times\!2$ rotation matrices; any clockwise rotation would ascribe a reflection matrix characteristic. Counterclockwise rotation is thereby enforced by constraining equality among diagonal and antidiagonal entries as prescribed by (1061);

$$Y(:,1\!:\!3) = \begin{bmatrix} 0 & 1 \\ -1 & 0 \end{bmatrix} Y(:,2\!:\!4) \tag{1065}$$

Quadrature quantization of rotation can be regarded as a constraint on tilt of the smallest Cartesian square containing the diamond as in Figure **139**c. Our scheme to quantize rotation requires that all square vertices be described by vectors whose entries are nonnegative when the square is translated anywhere interior to the nonnegative orthant. We capture the four square vertices as columns of a product YC where

$$C = \begin{bmatrix} 1 & 0 & 0 & 1 \\ 1 & 1 & 0 & 0 \\ 0 & 1 & 1 & 0 \\ 0 & 0 & 1 & 1 \end{bmatrix} \tag{1066}$$

Then, assuming a unit-square shroud, the affine constraint

$$YC + \begin{bmatrix} 1/2 \\ 1/2 \end{bmatrix} \mathbf{1}^{\mathrm{T}} \geq \mathbf{0} \tag{1067}$$

quantizes rotation, as desired. $\qquad\qquad\square$

5.5.2.1 Inner-product form invariance

Likewise, $\mathbf{D}(\Theta)$ (1038) is rotation/reflection invariant;

$$\mathbf{D}(Q_{\mathrm{p}}\Theta) = \mathbf{D}(Q\Theta) = \mathbf{D}(\Theta) \tag{1068}$$

so (1059) and (1060) similarly apply.

5.5.3 Invariance conclusion

In the making of an EDM, absolute rotation, reflection, and translation information is lost. Given an EDM, reconstruction of point position (§5.12, the list X) can be guaranteed correct only in affine dimension r and relative position. Given a noiseless complete EDM, this isometric reconstruction is unique insofar as every realization of a corresponding list X is *congruent*:

5.6 Injectivity of D & unique reconstruction

Injectivity implies uniqueness of isometric reconstruction (§5.4.2.2.10); hence, we endeavor to demonstrate it.

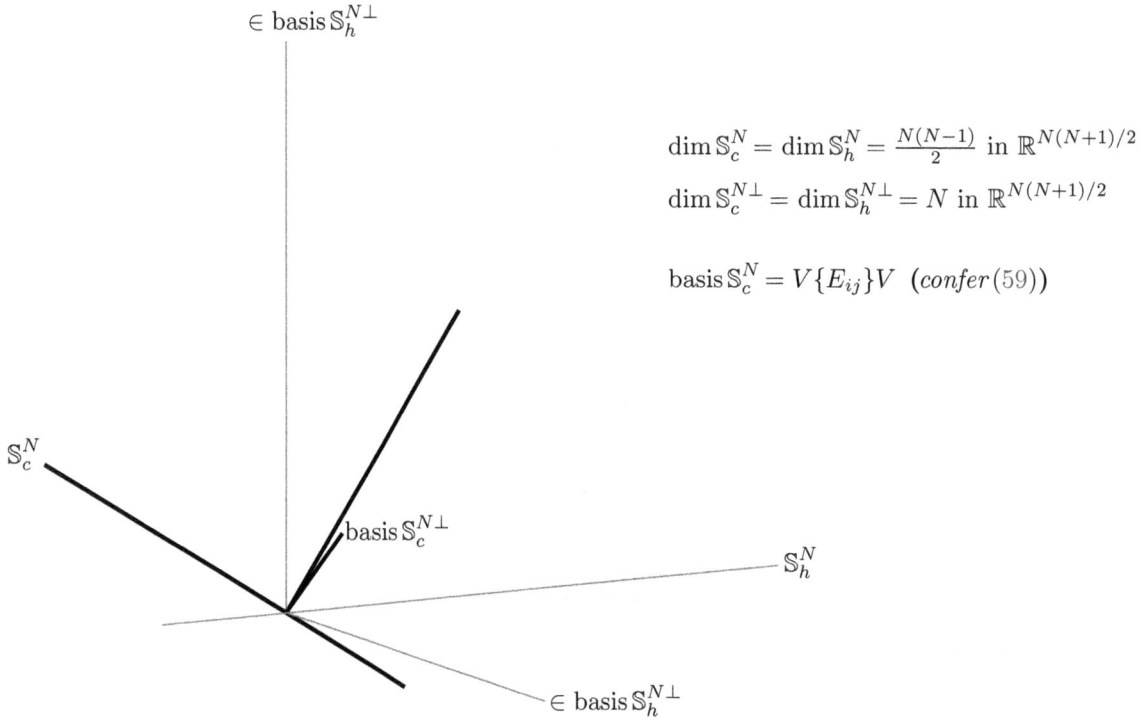

$$\dim \mathbb{S}_c^N = \dim \mathbb{S}_h^N = \frac{N(N-1)}{2} \text{ in } \mathbb{R}^{N(N+1)/2}$$

$$\dim \mathbb{S}_c^{N\perp} = \dim \mathbb{S}_h^{N\perp} = N \text{ in } \mathbb{R}^{N(N+1)/2}$$

$$\text{basis}\,\mathbb{S}_c^N = V\{E_{ij}\}V \quad (confer\,(59))$$

Figure 140: Orthogonal complements in $\mathbb{S}^N$ abstractly oriented in isometrically isomorphic $\mathbb{R}^{N(N+1)/2}$. Case $N=2$ accurately illustrated in $\mathbb{R}^3$. Orthogonal projection of basis for $\mathbb{S}_h^{N\perp}$ on $\mathbb{S}_c^{N\perp}$ yields another basis for $\mathbb{S}_c^{N\perp}$. (Basis vectors for $\mathbb{S}_c^{N\perp}$ are illustrated lying in a plane orthogonal to $\mathbb{S}_c^N$ in this dimension. Basis vectors for each $\perp$ space outnumber those for its respective orthogonal complement; such is not the case in higher dimension.)

EDM operators list-form $\mathbf{D}(X)$ (968), Gram-form $\mathbf{D}(G)$ (980), and inner-product form $\mathbf{D}(\Theta)$ (1038) are many-to-one surjections (§5.5) onto the same range; the EDM cone (§6): (confer (981) (1075))

$$\begin{aligned}
\text{EDM}^N &= \left\{ \mathbf{D}(X) : \mathbb{R}^{N-1\times N} \to \mathbb{S}_h^N \mid X \in \mathbb{R}^{N-1\times N} \right\} \\
&= \left\{ \mathbf{D}(G) : \mathbb{S}^N \to \mathbb{S}_h^N \mid G \in \mathbb{S}_+^N - \mathbb{S}_c^{N\perp} \right\} \\
&= \left\{ \mathbf{D}(\Theta) : \mathbb{R}^{N-1\times N-1} \to \mathbb{S}_h^N \mid \Theta \in \mathbb{R}^{N-1\times N-1} \right\}
\end{aligned} \qquad (1069)$$

where (§5.5.1.1)

$$\mathbb{S}_c^{N\perp} = \{u\mathbf{1}^T + \mathbf{1}u^T \mid u \in \mathbb{R}^N\} \subseteq \mathbb{S}^N \qquad (2096)$$

5.6.1   Gram-form bijectivity

Because linear Gram-form EDM operator

$$\mathbf{D}(G) \;=\; \delta(G)\mathbf{1}^{\mathrm{T}} + \mathbf{1}\delta(G)^{\mathrm{T}} - 2G \qquad (980)$$

has no nullspace [87, §A.1] on the geometric center subspace[5.27] (§E.7.2.0.2)

$$
\begin{aligned}
\mathbb{S}_c^N \;&\triangleq\; \{G \in \mathbb{S}^N \mid G\mathbf{1} = \mathbf{0}\} && (2094) \\
&= \{G \in \mathbb{S}^N \mid \mathcal{N}(G) \supseteq \mathbf{1}\} = \{G \in \mathbb{S}^N \mid \mathcal{R}(G) \subseteq \mathcal{N}(\mathbf{1}^{\mathrm{T}})\} \\
&= \{VYV \mid Y \in \mathbb{S}^N\} \subset \mathbb{S}^N && (2095) \\
&\equiv \{V_{\mathcal{N}} A V_{\mathcal{N}}^{\mathrm{T}} \mid A \in \mathbb{S}^{N-1}\}
\end{aligned}
\qquad (1070)
$$

then $\mathbf{D}(G)$ on that subspace is injective.

To prove injectivity of $\mathbf{D}(G)$ on $\mathbb{S}_c^N$: Any matrix $Y \in \mathbb{S}^N$ can be decomposed into orthogonal components in $\mathbb{S}^N$;

$$Y = VYV + (Y - VYV) \qquad (1071)$$

where $VYV \in \mathbb{S}_c^N$ and $Y - VYV \in \mathbb{S}_c^{N\perp}$ (2096). Because of translation invariance (§5.5.1.1) and linearity, $\mathbf{D}(Y - VYV) = \mathbf{0}$ hence $\mathcal{N}(\mathbf{D}) \supseteq \mathbb{S}_c^{N\perp}$. It remains only to show

$$\mathbf{D}(VYV) = \mathbf{0} \;\Leftrightarrow\; VYV = \mathbf{0} \qquad (1072)$$

$\left(\Leftrightarrow Y = u\mathbf{1}^{\mathrm{T}} + \mathbf{1}u^{\mathrm{T}} \text{ for some } u \in \mathbb{R}^N\right)$. $\mathbf{D}(VYV)$ will   vanish   whenever $2VYV = \delta(VYV)\mathbf{1}^{\mathrm{T}} + \mathbf{1}\delta(VYV)^{\mathrm{T}}$. But this implies $\mathcal{R}(\mathbf{1})$ (§B.2) were a subset of $\mathcal{R}(VYV)$, which is contradictory. Thus we have

$$\mathcal{N}(\mathbf{D}) \;=\; \{Y \mid \mathbf{D}(Y) = \mathbf{0}\} \;=\; \{Y \mid VYV = \mathbf{0}\} \;=\; \mathbb{S}_c^{N\perp} \qquad (1073)$$

♦

Since $G\mathbf{1} = \mathbf{0} \Leftrightarrow X\mathbf{1} = \mathbf{0}$ (988) simply means list X is geometrically centered at the origin, and because the Gram-form EDM operator $\mathbf{D}$ is translation invariant and $\mathcal{N}(\mathbf{D})$ is the translation-invariant subspace $\mathbb{S}_c^{N\perp}$, then EDM definition $\mathbf{D}(G)$ (1069) on[5.28] (*confer* §6.5.1, §6.6.1, §A.7.4.0.1)

$$\mathbb{S}_c^N \cap \mathbb{S}_+^N \;=\; \{VYV \succeq 0 \mid Y \in \mathbb{S}^N\} \;\equiv\; \{V_{\mathcal{N}} A V_{\mathcal{N}}^{\mathrm{T}} \mid A \in \mathbb{S}_+^{N-1}\} \subset \mathbb{S}^N \qquad (1074)$$

must be surjective onto $\mathbb{EDM}^N$; (*confer*(981))

$$\mathbb{EDM}^N = \left\{\mathbf{D}(G) \mid G \in \mathbb{S}_c^N \cap \mathbb{S}_+^N\right\} \qquad (1075)$$

[5.27]The equivalence $\equiv$ in (1070) follows from the fact: Given $B = VYV = V_{\mathcal{N}} A V_{\mathcal{N}}^{\mathrm{T}} \in \mathbb{S}_c^N$ with only matrix $A \in \mathbb{S}^{N-1}$ unknown, then $V_{\mathcal{N}}^{\dagger} B V_{\mathcal{N}}^{\dagger\mathrm{T}} = A$ or $V_{\mathcal{N}}^{\dagger} Y V_{\mathcal{N}}^{\dagger\mathrm{T}} = A$.

[5.28]Equivalence $\equiv$ in (1074) follows from the fact: Given $B = VYV = V_{\mathcal{N}} A V_{\mathcal{N}}^{\mathrm{T}} \in \mathbb{S}_+^N$ with only matrix A unknown, then $V_{\mathcal{N}}^{\dagger} B V_{\mathcal{N}}^{\dagger\mathrm{T}} = A$ and $A \in \mathbb{S}_+^{N-1}$ must be positive semidefinite by positive semidefiniteness of B and Corollary A.3.1.0.5.

5.6.1.1   Gram-form operator **D** inversion

Define the linear *geometric centering operator* **V** ; (*confer*(989))

$$\mathbf{V}(D) : \mathbb{S}^N \to \mathbb{S}^N \triangleq -VDV\tfrac{1}{2} \tag{1076}$$

[91, §4.3][5.29] This orthogonal projector **V** has no nullspace on

$$\mathbb{S}_h^N = \mathrm{aff}\,\mathbb{EDM}^N \qquad (1330)$$

because the projection of $-D/2$ on $\mathbb{S}_c^N$ (2094) can be **0** if and only if $D \in \mathbb{S}_c^{N\perp}$; but $\mathbb{S}_c^{N\perp} \cap \mathbb{S}_h^N = \mathbf{0}$ (Figure 140). Projector **V** on $\mathbb{S}_h^N$ is therefore injective hence uniquely invertible. Further, $-V\mathbb{S}_h^N V/2$ is equivalent to the geometric center subspace $\mathbb{S}_c^N$ in the ambient space of symmetric matrices; a surjection,

$$\mathbb{S}_c^N = \mathbf{V}(\mathbb{S}^N) = \mathbf{V}\big(\mathbb{S}_h^N \oplus \mathbb{S}_h^{N\perp}\big) = \mathbf{V}\big(\mathbb{S}_h^N\big) \tag{1077}$$

because (72)

$$\mathbf{V}\big(\mathbb{S}_h^N\big) \supseteq \mathbf{V}\big(\mathbb{S}_h^{N\perp}\big) = \mathbf{V}\big(\delta^2(\mathbb{S}^N)\big) \tag{1078}$$

Because $\mathbf{D}(G)$ on $\mathbb{S}_c^N$ is injective, and $\mathrm{aff}\,\mathbf{D}\big(\mathbf{V}(\mathbb{EDM}^N)\big) = \mathbf{D}\big(\mathbf{V}(\mathrm{aff}\,\mathbb{EDM}^N)\big)$ by property (127) of the affine hull, we find for $D \in \mathbb{S}_h^N$

$$\mathbf{D}(-VDV\tfrac{1}{2}) = \delta(-VDV\tfrac{1}{2})\mathbf{1}^{\mathrm T} + \mathbf{1}\delta(-VDV\tfrac{1}{2})^{\mathrm T} - 2(-VDV\tfrac{1}{2}) \tag{1079}$$

id est,

$$D = \mathbf{D}\big(\mathbf{V}(D)\big) \tag{1080}$$

$$-VDV = \mathbf{V}\big(\mathbf{D}(-VDV)\big) \tag{1081}$$

or

$$\mathbb{S}_h^N = \mathbf{D}\big(\mathbf{V}(\mathbb{S}_h^N)\big) \tag{1082}$$

$$-V\mathbb{S}_h^N V = \mathbf{V}\big(\mathbf{D}(-V\mathbb{S}_h^N V)\big) \tag{1083}$$

These operators **V** and **D** are mutual inverses.

The Gram-form $\mathbf{D}\big(\mathbb{S}_c^N\big)$ (980) is equivalent to $\mathbb{S}_h^N$;

$$\mathbf{D}\big(\mathbb{S}_c^N\big) = \mathbf{D}\big(\mathbf{V}(\mathbb{S}_h^N \oplus \mathbb{S}_h^{N\perp})\big) = \mathbb{S}_h^N + \mathbf{D}\big(\mathbf{V}(\mathbb{S}_h^{N\perp})\big) = \mathbb{S}_h^N \tag{1084}$$

because $\mathbb{S}_h^N \supseteq \mathbf{D}\big(\mathbf{V}(\mathbb{S}_h^{N\perp})\big)$. In summary, for the Gram-form we have the isomorphisms [92, §2] [91, p.76, p.107] [7, §2.1][5.30] [6, §2] [8, §18.2.1] [2, §2.1]

$$\mathbb{S}_h^N = \mathbf{D}(\mathbb{S}_c^N) \tag{1085}$$
$$\mathbb{S}_c^N = \mathbf{V}(\mathbb{S}_h^N) \tag{1086}$$

[5.29]Critchley cites Torgerson (1958) [358, ch.11, §2] for a history and derivation of (1076).
[5.30]In [7, p.6, line 20], delete sentence: *Since G is also ... not a singleton set.*
[7, p.10, line 11] $x_3 = 2$ (not 1).

and from bijectivity results in §5.6.1,

$$\mathbb{EDM}^N = \mathbf{D}(\mathbb{S}_c^N \cap \mathbb{S}_+^N) \tag{1087}$$

$$\mathbb{S}_c^N \cap \mathbb{S}_+^N = \mathbf{V}(\mathbb{EDM}^N) \tag{1088}$$

5.6.2 Inner-product form bijectivity

The Gram-form EDM operator $\mathbf{D}(G) = \delta(G)\mathbf{1}^T + \mathbf{1}\,\delta(G)^T - 2G$ (980) is an injective map, for example, on the domain that is the subspace of symmetric matrices having all zeros in the first row and column

$$
\begin{aligned}
\mathbb{S}_1^N &= \{G \in \mathbb{S}^N \mid Ge_1 = \mathbf{0}\} \\
&= \left\{ \begin{bmatrix} 0 & \mathbf{0}^T \\ \mathbf{0} & I \end{bmatrix} Y \begin{bmatrix} 0 & \mathbf{0}^T \\ \mathbf{0} & I \end{bmatrix} \mid Y \in \mathbb{S}^N \right\}
\end{aligned} \tag{2098}
$$

because it obviously has no nullspace there. Since $Ge_1 = \mathbf{0} \Leftrightarrow Xe_1 = \mathbf{0}$ (982) means the first point in the list X resides at the origin, then $\mathbf{D}(G)$ on $\mathbb{S}_1^N \cap \mathbb{S}_+^N$ must be surjective onto $\mathbb{EDM}^N$.

Substituting $\Theta^T\Theta \leftarrow -V_{\mathcal{N}}^T D V_{\mathcal{N}}$ (1050) into inner-product form EDM definition $\mathbf{D}(\Theta)$ (1038), it may be further decomposed:

$$\mathbf{D}(D) = \begin{bmatrix} 0 \\ \delta\!\left(-V_{\mathcal{N}}^T D V_{\mathcal{N}}\right) \end{bmatrix} \mathbf{1}^T + \mathbf{1}\begin{bmatrix} 0 & \delta\!\left(-V_{\mathcal{N}}^T D V_{\mathcal{N}}\right)^T \end{bmatrix} - 2\begin{bmatrix} 0 & \mathbf{0}^T \\ \mathbf{0} & -V_{\mathcal{N}}^T D V_{\mathcal{N}} \end{bmatrix} \tag{1089}$$

This linear operator $\mathbf{D}$ is another flavor of inner-product form and an injective map of the EDM cone onto itself. Yet when its domain is instead the entire symmetric hollow subspace $\mathbb{S}_h^N = \text{aff}\,\mathbb{EDM}^N$, $\mathbf{D}(D)$ becomes an injective map onto that same subspace. Proof follows directly from the fact: linear $\mathbf{D}$ has no nullspace [87, §A.1] on $\mathbb{S}_h^N = \text{aff}\,\mathbf{D}(\mathbb{EDM}^N) = \mathbf{D}(\text{aff}\,\mathbb{EDM}^N)$ (127).

5.6.2.1 Inversion of $\mathbf{D}\!\left(-V_{\mathcal{N}}^T D V_{\mathcal{N}}\right)$

Injectivity of $\mathbf{D}(D)$ suggests inversion of (*confer* (985))

$$\mathbf{V}_{\mathcal{N}}(D) : \mathbb{S}^N \to \mathbb{S}^{N-1} \triangleq -V_{\mathcal{N}}^T D V_{\mathcal{N}} \tag{1090}$$

a linear surjective[5.31] mapping onto $\mathbb{S}^{N-1}$ having nullspace[5.32] $\mathbb{S}_c^{N\perp}$;

$$\mathbf{V}_{\mathcal{N}}(\mathbb{S}_h^N) = \mathbb{S}^{N-1} \tag{1091}$$

[5.31]Surjectivity of $\mathbf{V}_{\mathcal{N}}(D)$ is demonstrated via the Gram-form EDM operator $\mathbf{D}(G)$: Since $\mathbb{S}_h^N = \mathbf{D}(\mathbb{S}_c^N)$ (1084), then for any $Y \in \mathbb{S}^{N-1}$, $-V_{\mathcal{N}}^T \mathbf{D}(V_{\mathcal{N}}^{\dagger T} Y V_{\mathcal{N}}^{\dagger}/2)V_{\mathcal{N}} = Y$.

[5.32]$\mathcal{N}(\mathbf{V}_{\mathcal{N}}) \supseteq \mathbb{S}_c^{N\perp}$ is apparent. There exists a linear mapping

$$T(\mathbf{V}_{\mathcal{N}}(D)) \triangleq V_{\mathcal{N}}^{\dagger T}\mathbf{V}_{\mathcal{N}}(D)V_{\mathcal{N}}^{\dagger} = -VDV\tfrac{1}{2} = \mathbf{V}(D)$$

such that

$$\mathcal{N}(T(\mathbf{V}_{\mathcal{N}})) = \mathcal{N}(\mathbf{V}) \supseteq \mathcal{N}(\mathbf{V}_{\mathcal{N}}) \supseteq \mathbb{S}_c^{N\perp} = \mathcal{N}(\mathbf{V})$$

where the equality $\mathbb{S}_c^{N\perp} = \mathcal{N}(\mathbf{V})$ is known (§E.7.2.0.2). ◆

injective on domain $\mathbb{S}_h^N$ because $\mathbb{S}_c^{N\perp} \cap \mathbb{S}_h^N = \mathbf{0}$.   Revising the argument of this inner-product form (1089), we get another flavor

$$\mathbf{D}(-V_{\mathcal{N}}^{\mathrm{T}} D V_{\mathcal{N}}) = \begin{bmatrix} 0 \\ \delta(-V_{\mathcal{N}}^{\mathrm{T}} D V_{\mathcal{N}}) \end{bmatrix} \mathbf{1}^{\mathrm{T}} + \mathbf{1} \begin{bmatrix} 0 & \delta(-V_{\mathcal{N}}^{\mathrm{T}} D V_{\mathcal{N}})^{\mathrm{T}} \end{bmatrix} - 2 \begin{bmatrix} 0 & \mathbf{0}^{\mathrm{T}} \\ \mathbf{0} & -V_{\mathcal{N}}^{\mathrm{T}} D V_{\mathcal{N}} \end{bmatrix}$$

(1092)

and we obtain mutual inversion of operators $\mathbf{V}_{\mathcal{N}}$ and $\mathbf{D}$, for $D \in \mathbb{S}_h^N$

$$D = \mathbf{D}\big(\mathbf{V}_{\mathcal{N}}(D)\big) \tag{1093}$$

$$-V_{\mathcal{N}}^{\mathrm{T}} D V_{\mathcal{N}} = \mathbf{V}_{\mathcal{N}}\big(\mathbf{D}(-V_{\mathcal{N}}^{\mathrm{T}} D V_{\mathcal{N}})\big) \tag{1094}$$

or

$$\mathbb{S}_h^N = \mathbf{D}\big(\mathbf{V}_{\mathcal{N}}(\mathbb{S}_h^N)\big) \tag{1095}$$

$$-V_{\mathcal{N}}^{\mathrm{T}} \mathbb{S}_h^N V_{\mathcal{N}} = \mathbf{V}_{\mathcal{N}}\big(\mathbf{D}(-V_{\mathcal{N}}^{\mathrm{T}} \mathbb{S}_h^N V_{\mathcal{N}})\big) \tag{1096}$$

Substituting $\Theta^{\mathrm{T}}\Theta \leftarrow \Phi$ into inner-product form EDM definition (1038), any EDM may be expressed by the new flavor

$$\mathbf{D}(\Phi) \triangleq \begin{bmatrix} 0 \\ \delta(\Phi) \end{bmatrix} \mathbf{1}^{\mathrm{T}} + \mathbf{1} \begin{bmatrix} 0 & \delta(\Phi)^{\mathrm{T}} \end{bmatrix} - 2 \begin{bmatrix} 0 & \mathbf{0}^{\mathrm{T}} \\ \mathbf{0} & \Phi \end{bmatrix} \in \mathbb{EDM}^N$$
$$\Leftrightarrow$$
$$\Phi \succeq 0$$

(1097)

where this $\mathbf{D}$ is a linear surjective operator onto $\mathbb{EDM}^N$ by definition, injective because it has no nullspace on domain $\mathbb{S}_+^{N-1}$. More broadly, aff $\mathbf{D}(\mathbb{S}_+^{N-1}) = \mathbf{D}(\text{aff}\,\mathbb{S}_+^{N-1})$ (127),

$$\mathbb{S}_h^N = \mathbf{D}(\mathbb{S}^{N-1})$$
$$\mathbb{S}^{N-1} = \mathbf{V}_{\mathcal{N}}(\mathbb{S}_h^N)$$

(1098)

demonstrably isomorphisms, and by bijectivity of this inner-product form:

$$\mathbb{EDM}^N = \mathbf{D}(\mathbb{S}_+^{N-1}) \tag{1099}$$

$$\mathbb{S}_+^{N-1} = \mathbf{V}_{\mathcal{N}}(\mathbb{EDM}^N) \tag{1100}$$

5.7 Embedding in affine hull

The affine hull $\mathcal{A}$ (78) of a point list $\{x_\ell\}$ (arranged columnar in $X \in \mathbb{R}^{n \times N}$ (76)) is identical to the affine hull of that polyhedron $\mathcal{P}$ (86) formed from all convex combinations of the x_ℓ; [62, §2] [315, §17]

$$\mathcal{A} = \text{aff}\,X = \text{aff}\,\mathcal{P} \tag{1101}$$

Comparing hull definitions (78) and (86), it becomes obvious that the x_ℓ and their convex hull $\mathcal{P}$ are embedded in their unique affine hull $\mathcal{A}$;

$$\mathcal{A} \supseteq \mathcal{P} \supseteq \{x_\ell\} \tag{1102}$$

Recall: *affine dimension* r is a lower bound on embedding, equal to dimension of the subspace parallel to that nonempty affine set $\mathcal{A}$ in which the points are embedded. (§2.3.1) We define dimension of the convex hull $\mathcal{P}$ to be the same as dimension r of the affine hull $\mathcal{A}$ [315, §2], but r is not necessarily equal to rank of X (1121).

For the particular example illustrated in Figure **126**, $\mathcal{P}$ is the triangle in union with its relative interior while its three vertices constitute the entire list X. Affine hull $\mathcal{A}$ is the unique plane that contains the triangle, so affine dimension $r = 2$ in that example while rank of X is 3. Were there only two points in Figure **126**, then the affine hull would instead be the unique line passing through them; r would become 1 while rank would then be 2.

5.7.1 Determining affine dimension

Knowledge of affine dimension r becomes important because we lose any absolute offset common to all the generating x_ℓ in $\mathbb{R}^n$ when reconstructing convex polyhedra given only distance information. (§5.5.1) To calculate r, we first remove any offset that serves to increase dimensionality of the subspace required to contain polyhedron $\mathcal{P}$; subtracting any $\alpha \in \mathcal{A}$ in the affine hull from every list member will work,

$$X - \alpha \mathbf{1}^{\mathrm{T}} \tag{1103}$$

translating $\mathcal{A}$ to the origin:[5.33]

$$\mathcal{A} - \alpha \;=\; \mathrm{aff}(X - \alpha \mathbf{1}^{\mathrm{T}}) \;=\; \mathrm{aff}(X) - \alpha \tag{1104}$$
$$\mathcal{P} - \alpha \;=\; \mathrm{conv}(X - \alpha \mathbf{1}^{\mathrm{T}}) \;=\; \mathrm{conv}(X) - \alpha \tag{1105}$$

Because (1101) and (1102) translate,

$$\mathbb{R}^n \;\supseteq\; \mathcal{A} - \alpha \;=\; \mathrm{aff}(X - \alpha \mathbf{1}^{\mathrm{T}}) \;=\; \mathrm{aff}(\mathcal{P} - \alpha) \;\supseteq\; \mathcal{P} - \alpha \;\supseteq\; \{x_\ell - \alpha\} \tag{1106}$$

where from the previous relations it is easily shown

$$\mathrm{aff}(\mathcal{P} - \alpha) = \mathrm{aff}(\mathcal{P}) - \alpha \tag{1107}$$

Translating $\mathcal{A}$ neither changes its dimension or the dimension of the embedded polyhedron $\mathcal{P}$; (77)

$$r \;\triangleq\; \dim \mathcal{A} \;=\; \dim(\mathcal{A} - \alpha) \;\triangleq\; \dim(\mathcal{P} - \alpha) \;=\; \dim \mathcal{P} \tag{1108}$$

For any $\alpha \in \mathbb{R}^n$, (1104)-(1108) remain true. [315, p.4, p.12] Yet when $\alpha \in \mathcal{A}$, the affine set $\mathcal{A} - \alpha$ becomes a unique subspace of $\mathbb{R}^n$ in which the $\{x_\ell - \alpha\}$ and their convex hull $\mathcal{P} - \alpha$ are embedded (1106), and whose dimension is more easily calculated.

5.7.1.0.1 Example. *Translating first list-member to origin.*
Subtracting the first member $\alpha \triangleq x_1$ from every list member will translate their affine hull $\mathcal{A}$ and their convex hull $\mathcal{P}$ and, in particular, $x_1 \in \mathcal{P} \subseteq \mathcal{A}$ to the origin in $\mathbb{R}^n$; *videlicet*,

$$X - x_1 \mathbf{1}^{\mathrm{T}} \;=\; X - X e_1 \mathbf{1}^{\mathrm{T}} \;=\; X(I - e_1 \mathbf{1}^{\mathrm{T}}) \;=\; X \begin{bmatrix} \mathbf{0} & \sqrt{2} V_{\mathcal{N}} \end{bmatrix} \in \mathbb{R}^{n \times N} \tag{1109}$$

[5.33]The manipulation of hull functions aff and conv follows from their definitions.

where $V_{\mathcal{N}}$ is defined in (974), and e_1 in (984). Applying (1106) to (1109),

$$\mathbb{R}^n \supseteq \mathcal{R}(XV_{\mathcal{N}}) = \mathcal{A} - x_1 = \mathrm{aff}(X - x_1\mathbf{1}^{\mathrm{T}}) = \mathrm{aff}(\mathcal{P} - x_1) \supseteq \mathcal{P} - x_1 \ni \mathbf{0} \quad (1110)$$

where $XV_{\mathcal{N}} \in \mathbb{R}^{n \times N-1}$. Hence

$$r = \dim \mathcal{R}(XV_{\mathcal{N}}) \qquad (1111)$$

$\square$

Since shifting the geometric center to the origin (§5.5.1.0.1) translates the affine hull to the origin as well, then it must also be true

$$r = \dim \mathcal{R}(XV) \qquad (1112)$$

For any matrix whose range is $\mathcal{R}(V) = \mathcal{N}(\mathbf{1}^{\mathrm{T}})$ we get the same result; *e.g.*,

$$r = \dim \mathcal{R}(XV_{\mathcal{N}}^{\dagger\mathrm{T}}) \qquad (1113)$$

because

$$\mathcal{R}(XV) = \{Xz \mid z \in \mathcal{N}(\mathbf{1}^{\mathrm{T}})\} \qquad (1114)$$

and $\mathcal{R}(V) = \mathcal{R}(V_{\mathcal{N}}) = \mathcal{R}(V_{\mathcal{N}}^{\dagger\mathrm{T}})$ (§E). These auxiliary matrices (§B.4.2) are more closely related;

$$V = V_{\mathcal{N}}V_{\mathcal{N}}^{\dagger} \qquad (1739)$$

5.7.1.1 Affine dimension r *versus* rank

Now, suppose D is an EDM as defined by

$$\mathbf{D}(X) = \delta(X^{\mathrm{T}}X)\mathbf{1}^{\mathrm{T}} + \mathbf{1}\delta(X^{\mathrm{T}}X)^{\mathrm{T}} - 2X^{\mathrm{T}}X \in \mathbb{EDM}^N \qquad (968)$$

and we premultiply by $-V_{\mathcal{N}}^{\mathrm{T}}$ and postmultiply by $V_{\mathcal{N}}$. Then because $V_{\mathcal{N}}^{\mathrm{T}}\mathbf{1} = \mathbf{0}$ (975), it is always true that

$$-V_{\mathcal{N}}^{\mathrm{T}}DV_{\mathcal{N}} = 2V_{\mathcal{N}}^{\mathrm{T}}X^{\mathrm{T}}XV_{\mathcal{N}} = 2V_{\mathcal{N}}^{\mathrm{T}}GV_{\mathcal{N}} \in \mathbb{S}^{N-1} \qquad (1115)$$

where G is a Gram matrix. Similarly pre- and postmultiplying by V (*confer* (989))

$$-VDV = 2VX^{\mathrm{T}}XV = 2VGV \in \mathbb{S}^N \qquad (1116)$$

always holds because $V\mathbf{1} = \mathbf{0}$ (1729). Likewise, multiplying inner-product form EDM definition (1038), it always holds:

$$-V_{\mathcal{N}}^{\mathrm{T}}DV_{\mathcal{N}} = \Theta^{\mathrm{T}}\Theta \in \mathbb{S}^{N-1} \qquad (1042)$$

For any matrix A, $\mathrm{rank}\, A^{\mathrm{T}}A = \mathrm{rank}\, A = \mathrm{rank}\, A^{\mathrm{T}}$. [208, §0.4][5.34] So, by (1114), affine dimension

$$\begin{aligned} r &= \mathrm{rank}\, XV = \mathrm{rank}\, XV_{\mathcal{N}} = \mathrm{rank}\, XV_{\mathcal{N}}^{\dagger\mathrm{T}} = \mathrm{rank}\, \Theta \\ &= \mathrm{rank}\, VDV = \mathrm{rank}\, VGV = \mathrm{rank}\, V_{\mathcal{N}}^{\mathrm{T}}DV_{\mathcal{N}} = \mathrm{rank}\, V_{\mathcal{N}}^{\mathrm{T}}GV_{\mathcal{N}} \end{aligned} \qquad (1117)$$

[5.34]For $A \in \mathbb{R}^{m \times n}$, $\mathcal{N}(A^{\mathrm{T}}A) = \mathcal{N}(A)$. [338, §3.3]

By conservation of dimension, (§A.7.3.0.1)

$$r + \dim \mathcal{N}(V_{\mathcal{N}}^{\mathrm{T}} D V_{\mathcal{N}}) = N-1 \tag{1118}$$

$$r + \dim \mathcal{N}(VDV) = N \tag{1119}$$

For $D \in \mathbb{EDM}^N$

$$-V_{\mathcal{N}}^{\mathrm{T}} D V_{\mathcal{N}} \succ 0 \;\Leftrightarrow\; r = N-1 \tag{1120}$$

but $-VDV \not\succ 0$. The general fact[5.35] $(confer\,(1000))$

$$r \le \min\{n, N-1\} \tag{1121}$$

is evident from (1109) but can be visualized in the example illustrated in Figure **126**. There we imagine a vector from the origin to each point in the list. Those three vectors are linearly independent in $\mathbb{R}^3$, but affine dimension r is 2 because the three points lie in a plane. When that plane is translated to the origin, it becomes the only subspace of dimension $r=2$ that can contain the translated triangular polyhedron.

5.7.2 *Précis*

We collect expressions for affine dimension r: for list $X \in \mathbb{R}^{n \times N}$ and Gram matrix $G \in \mathbb{S}_+^N$

$$
\begin{aligned}
r \;\triangleq\;& \dim(\mathcal{P}-\alpha) \;=\; \dim \mathcal{P} \;=\; \dim \operatorname{conv} X \\
=\,& \dim(\mathcal{A}-\alpha) \;=\; \dim \mathcal{A} \;=\; \dim \operatorname{aff} X \\
=\,& \operatorname{rank}(X - x_1 \mathbf{1}^{\mathrm{T}}) \;=\; \operatorname{rank}(X - \alpha_c \mathbf{1}^{\mathrm{T}}) \\
=\,& \operatorname{rank} \Theta \quad (1040) \\
=\,& \operatorname{rank} X V_{\mathcal{N}} \;=\; \operatorname{rank} XV \;=\; \operatorname{rank} X V_{\mathcal{N}}^{\dagger \mathrm{T}} \\
=\,& \operatorname{rank} X, \quad Xe_1 = \mathbf{0} \;\text{ or }\; X\mathbf{1} = \mathbf{0} \\
=\,& \operatorname{rank} V_{\mathcal{N}}^{\mathrm{T}} G V_{\mathcal{N}} \;=\; \operatorname{rank} VGV \;=\; \operatorname{rank} V_{\mathcal{N}}^{\dagger} G V_{\mathcal{N}} \\
=\,& \operatorname{rank} G, \quad Ge_1 = \mathbf{0} \;(985) \;\text{ or }\; G\mathbf{1}=\mathbf{0} \;(989) \\
=\,& \operatorname{rank} V_{\mathcal{N}}^{\mathrm{T}} D V_{\mathcal{N}} \;=\; \operatorname{rank} VDV \;=\; \operatorname{rank} V_{\mathcal{N}}^{\dagger} D V_{\mathcal{N}} \;=\; \operatorname{rank} V_{\mathcal{N}}(V_{\mathcal{N}}^{\mathrm{T}} D V_{\mathcal{N}})V_{\mathcal{N}}^{\mathrm{T}} \\
=\,& \operatorname{rank} \Lambda \quad (1208) \\
=\,& N-1 - \dim \mathcal{N}\!\left(\begin{bmatrix} 0 & \mathbf{1}^{\mathrm{T}} \\ \mathbf{1} & -D \end{bmatrix}\right) = \operatorname{rank} \begin{bmatrix} 0 & \mathbf{1}^{\mathrm{T}} \\ \mathbf{1} & -D \end{bmatrix} - 2 \quad (1129)
\end{aligned}
\;\Bigg\} \; D \in \mathbb{EDM}^N
\tag{1122}
$$

5.7.3 Eigenvalues of $-VDV$ *versus* $-V_{\mathcal{N}}^{\dagger} D V_{\mathcal{N}}$

Suppose for $D \in \mathbb{EDM}^N$ we are given eigenvectors $v_i \in \mathbb{R}^N$ of $-VDV$ and corresponding eigenvalues $\lambda \in \mathbb{R}^N$ so that

$$-VDV v_i = \lambda_i v_i, \quad i = 1 \dots N \tag{1123}$$

From these we can determine the eigenvectors and eigenvalues of $-V_{\mathcal{N}}^{\dagger} D V_{\mathcal{N}}$: Define

$$\nu_i \triangleq V_{\mathcal{N}}^{\dagger} v_i, \quad \lambda_i \ne 0 \tag{1124}$$

[5.35] $\operatorname{rank} X \le \min\{n, N\}$

Then we have:

$$-VDV_{\mathcal{N}}V_{\mathcal{N}}^{\dagger}\,v_i = \lambda_i\,v_i \tag{1125}$$

$$-V_{\mathcal{N}}^{\dagger}VDV_{\mathcal{N}}\,\nu_i = \lambda_i V_{\mathcal{N}}^{\dagger}\,v_i \tag{1126}$$

$$-V_{\mathcal{N}}^{\dagger}DV_{\mathcal{N}}\,\nu_i = \lambda_i\,\nu_i \tag{1127}$$

the eigenvectors of $-V_{\mathcal{N}}^{\dagger}DV_{\mathcal{N}}$ are given by (1124) while its corresponding nonzero eigenvalues are identical to those of $-VDV$ although $-V_{\mathcal{N}}^{\dagger}DV_{\mathcal{N}}$ is not necessarily positive semidefinite. In contrast, $-V_{\mathcal{N}}^{\mathrm{T}}DV_{\mathcal{N}}$ is positive semidefinite but its nonzero eigenvalues are generally different.

5.7.3.0.1   Theorem.   *EDM rank versus affine dimension* r.
[169, §3] [190, §3] [168, §3]   For $D\in\mathbb{EDM}^N$ (*confer*(1282))

1. $r = \mathrm{rank}(D) - 1 \;\Leftrightarrow\; \mathbf{1}^{\mathrm{T}}D^{\dagger}\mathbf{1} \neq 0$
    Points constituting a list X generating the polyhedron corresponding to D lie on the relative boundary of an r-dimensional *circumhypersphere* having

$$\begin{aligned}
\text{diameter} &= \sqrt{2}\left(\mathbf{1}^{\mathrm{T}}D^{\dagger}\mathbf{1}\right)^{-1/2} \\
\text{circumcenter} &= \frac{XD^{\dagger}\mathbf{1}}{\mathbf{1}^{\mathrm{T}}D^{\dagger}\mathbf{1}}
\end{aligned} \tag{1128}$$

2. $r = \mathrm{rank}(D) - 2 \;\Leftrightarrow\; \mathbf{1}^{\mathrm{T}}D^{\dagger}\mathbf{1} = 0$
    There can be no circumhypersphere whose relative boundary contains a generating list for the corresponding polyhedron.

3. In *Cayley-Menger form* [116, §6.2] [90, §3.3] [51, §40] (§5.11.2),

$$r = N - 1 - \dim\mathcal{N}\left(\begin{bmatrix} 0 & \mathbf{1}^{\mathrm{T}} \\ \mathbf{1} & -D \end{bmatrix}\right) = \mathrm{rank}\begin{bmatrix} 0 & \mathbf{1}^{\mathrm{T}} \\ \mathbf{1} & -D \end{bmatrix} - 2 \tag{1129}$$

    Circumhyperspheres exist for $r < \mathrm{rank}(D) - 2$. [355, §7] $\diamond$

For all practical purposes, (1121)

$$\max\{0\,,\,\mathrm{rank}(D) - 2\} \le r \le \min\{n\,,\,N-1\} \tag{1130}$$

5.8   Euclidean metric *versus* matrix criteria

5.8.1   Nonnegativity property 1

When $D = [d_{ij}]$ is an EDM (968), then it is apparent from (1115)

$$2V_{\mathcal{N}}^{\mathrm{T}}X^{\mathrm{T}}XV_{\mathcal{N}} \;=\; -V_{\mathcal{N}}^{\mathrm{T}}DV_{\mathcal{N}} \succeq 0 \tag{1131}$$

because for any matrix A, $A^TA \succeq 0$.[5.36] We claim nonnegativity of the d_{ij} is enforced primarily by the matrix inequality (1131); *id est*,

$$\left.\begin{array}{c} -V_{\mathcal{N}}^T D V_{\mathcal{N}} \succeq 0 \\ D \in \mathbb{S}_h^N \end{array}\right\} \Rightarrow d_{ij} \geq 0, \ i \neq j \qquad (1132)$$

(The matrix inequality to enforce strict positivity differs by a stroke of the pen. (1135))

We now support our claim: If any matrix $A \in \mathbb{R}^{m \times m}$ is positive semidefinite, then its main diagonal $\delta(A) \in \mathbb{R}^m$ must have all nonnegative entries. [164, §4.2] Given $D \in \mathbb{S}_h^N$

$$-V_{\mathcal{N}}^T D V_{\mathcal{N}} =$$

$$\begin{bmatrix} d_{12} & \frac{1}{2}(d_{12}+d_{13}-d_{23}) & \frac{1}{2}(d_{1,i+1}+d_{1,j+1}-d_{i+1,j+1}) & \cdots & \frac{1}{2}(d_{12}+d_{1N}-d_{2N}) \\ \frac{1}{2}(d_{12}+d_{13}-d_{23}) & d_{13} & \frac{1}{2}(d_{1,i+1}+d_{1,j+1}-d_{i+1,j+1}) & \cdots & \frac{1}{2}(d_{13}+d_{1N}-d_{3N}) \\ \frac{1}{2}(d_{1,j+1}+d_{1,i+1}-d_{j+1,i+1}) & \frac{1}{2}(d_{1,j+1}+d_{1,i+1}-d_{j+1,i+1}) & d_{1,i+1} & \ddots & \frac{1}{2}(d_{14}+d_{1N}-d_{4N}) \\ \vdots & \vdots & \ddots & \ddots & \vdots \\ \frac{1}{2}(d_{12}+d_{1N}-d_{2N}) & \frac{1}{2}(d_{13}+d_{1N}-d_{3N}) & \frac{1}{2}(d_{14}+d_{1N}-d_{4N}) & \cdots & d_{1N} \end{bmatrix}$$

$$= \tfrac{1}{2}(\mathbf{1}D_{1,2:N} + D_{2:N,1}\mathbf{1}^T - D_{2:N,2:N}) \in \mathbb{S}^{N-1} \qquad (1133)$$

where row,column indices $i,j \in \{1 \ldots N-1\}$. [320] It follows:

$$\left.\begin{array}{c} -V_{\mathcal{N}}^T D V_{\mathcal{N}} \succeq 0 \\ D \in \mathbb{S}_h^N \end{array}\right\} \Rightarrow \delta(-V_{\mathcal{N}}^T D V_{\mathcal{N}}) = \begin{bmatrix} d_{12} \\ d_{13} \\ \vdots \\ d_{1N} \end{bmatrix} \succeq 0 \qquad (1134)$$

Multiplication of $V_{\mathcal{N}}$ by any permutation matrix Ξ has null effect on its range and nullspace. In other words, any permutation of the rows or columns of $V_{\mathcal{N}}$ produces a basis for $\mathcal{N}(\mathbf{1}^T)$; *id est*, $\mathcal{R}(\Xi_r V_{\mathcal{N}}) = \mathcal{R}(V_{\mathcal{N}} \Xi_c) = \mathcal{R}(V_{\mathcal{N}}) = \mathcal{N}(\mathbf{1}^T)$. Hence, $-V_{\mathcal{N}}^T D V_{\mathcal{N}} \succeq 0$ $\Leftrightarrow -V_{\mathcal{N}}^T \Xi_r^T D \Xi_r V_{\mathcal{N}} \succeq 0$ $(\Leftrightarrow -\Xi_c^T V_{\mathcal{N}}^T D V_{\mathcal{N}} \Xi_c \succeq 0)$. Various permutation matrices[5.37] will sift the remaining d_{ij} similarly to (1134) thereby proving their nonnegativity. Hence $-V_{\mathcal{N}}^T D V_{\mathcal{N}} \succeq 0$ is a sufficient test for the first property (§5.2) of the Euclidean metric, nonnegativity. ♦

When affine dimension r equals 1, in particular, nonnegativity symmetry and hollowness become necessary and sufficient criteria satisfying matrix inequality (1131). (§6.5.0.0.1)

[5.36]For $A \in \mathbb{R}^{m \times n}$, $A^T A \succeq 0 \Leftrightarrow y^T A^T A y = \|Ay\|^2 \geq 0$ for all $\|y\| = 1$. When A is full-rank skinny-or-square, $A^T A \succ 0$.
[5.37]The rule of thumb is: If $\Xi_r(i,1) = 1$, then $\delta(-V_{\mathcal{N}}^T \Xi_r^T D \Xi_r V_{\mathcal{N}}) \in \mathbb{R}^{N-1}$ is some permutation of the ith row or column of D excepting the 0 entry from the main diagonal.

5.8.1.1 Strict positivity

Should we require the points in $\mathbb{R}^n$ to be distinct, then entries of D off the main diagonal must be strictly positive $\{d_{ij} > 0,\ i \neq j\}$ and only those entries along the main diagonal of D are 0. By similar argument, the strict matrix inequality is a sufficient test for strict positivity of Euclidean distance-square;

$$\left.\begin{array}{r} -V_{\mathcal{N}}^{\mathrm{T}} D V_{\mathcal{N}} \succ 0 \\ D \in \mathbb{S}_h^N \end{array}\right\} \ \Rightarrow\ d_{ij} > 0,\ \ i \neq j \tag{1135}$$

5.8.2 Triangle inequality property 4

In light of Kreyszig's observation [233, §1.1 prob.15] that properties 2 through 4 of the Euclidean metric (§5.2) together imply nonnegativity property 1,

$$2\sqrt{d_{jk}} = \sqrt{d_{jk}} + \sqrt{d_{kj}} \geq \sqrt{d_{jj}} = 0,\ \ \ j \neq k \tag{1136}$$

nonnegativity criterion (1132) suggests that matrix inequality $-V_{\mathcal{N}}^{\mathrm{T}} D V_{\mathcal{N}} \succeq 0$ might somehow take on the role of triangle inequality; *id est,*

$$\left.\begin{array}{r} \delta(D) = \mathbf{0} \\ D^{\mathrm{T}} = D \\ -V_{\mathcal{N}}^{\mathrm{T}} D V_{\mathcal{N}} \succeq 0 \end{array}\right\} \ \Rightarrow\ \sqrt{d_{ij}} \leq \sqrt{d_{ik}} + \sqrt{d_{kj}},\ \ \ i \neq j \neq k \tag{1137}$$

We now show that is indeed the case: Let T be the *leading principal submatrix* in $\mathbb{S}^2$ of $-V_{\mathcal{N}}^{\mathrm{T}} D V_{\mathcal{N}}$ (upper left 2×2 submatrix from (1133));

$$T \triangleq \begin{bmatrix} d_{12} & \frac{1}{2}(d_{12} + d_{13} - d_{23}) \\ \frac{1}{2}(d_{12} + d_{13} - d_{23}) & d_{13} \end{bmatrix} \tag{1138}$$

Submatrix T must be positive (semi)definite whenever $-V_{\mathcal{N}}^{\mathrm{T}} D V_{\mathcal{N}}$ is. (§A.3.1.0.4, §5.8.3) Now we have,

$$\begin{array}{l} -V_{\mathcal{N}}^{\mathrm{T}} D V_{\mathcal{N}} \succeq 0 \ \Rightarrow\ T \succeq 0 \ \Leftrightarrow\ \lambda_1 \geq \lambda_2 \geq 0 \\ -V_{\mathcal{N}}^{\mathrm{T}} D V_{\mathcal{N}} \succ 0 \ \Rightarrow\ T \succ 0 \ \Leftrightarrow\ \lambda_1 \geq \lambda_2 > 0 \end{array} \tag{1139}$$

where λ_1 and λ_2 are the eigenvalues of T, real due only to symmetry of T:

$$\begin{array}{l} \lambda_1 = \frac{1}{2}\left(d_{12} + d_{13} + \sqrt{d_{23}^2 - 2(d_{12} + d_{13})d_{23} + 2(d_{12}^2 + d_{13}^2)}\right) \in \mathbb{R} \\ \lambda_2 = \frac{1}{2}\left(d_{12} + d_{13} - \sqrt{d_{23}^2 - 2(d_{12} + d_{13})d_{23} + 2(d_{12}^2 + d_{13}^2)}\right) \in \mathbb{R} \end{array} \tag{1140}$$

Nonnegativity of eigenvalue λ_1 is guaranteed by only nonnegativity of the d_{ij} which in turn is guaranteed by matrix inequality (1132). Inequality between the eigenvalues in (1139) follows from only realness of the d_{ij}. Since λ_1 always equals or exceeds λ_2, conditions for positive (semi)definiteness of submatrix T can be completely determined by examining λ_2 the smaller of its two eigenvalues. A triangle inequality is made apparent when we express T eigenvalue nonnegativity in terms of D matrix entries; *videlicet,*

$$T \succeq 0 \ \Leftrightarrow\ \det T = \lambda_1 \lambda_2 \geq 0,\ \ \ d_{12}, d_{13} \geq 0 \qquad \text{(c)}$$
$$\Leftrightarrow$$
$$\lambda_2 \geq 0 \qquad \text{(b)} \tag{1141}$$
$$\Leftrightarrow$$
$$|\sqrt{d_{12}} - \sqrt{d_{23}}| \ \leq\ \sqrt{d_{13}} \ \leq\ \sqrt{d_{12}} + \sqrt{d_{23}} \qquad \text{(a)}$$

Triangle inequality (1141a) (*confer*(1036)(1153)), in terms of three rooted entries from D, is equivalent to metric property 4

$$
\begin{aligned}
\sqrt{d_{13}} &\leq \sqrt{d_{12}} + \sqrt{d_{23}} \\
\sqrt{d_{23}} &\leq \sqrt{d_{12}} + \sqrt{d_{13}} \\
\sqrt{d_{12}} &\leq \sqrt{d_{13}} + \sqrt{d_{23}}
\end{aligned}
\qquad (1142)
$$

for the corresponding points x_1, x_2, x_3 from some length-N list.[5.38]

5.8.2.1 Comment

Given D whose dimension N equals or exceeds 3, there are $N!/(3!(N-3)!)$ distinct triangle inequalities in total like (1036) that must be satisfied, of which each d_{ij} is involved in $N-2$, and each point x_i is in $(N-1)!/(2!(N-1-2)!)$. We have so far revealed only one of those triangle inequalities; namely, (1141a) that came from T (1138). Yet we claim if $-V_\mathcal{N}^T D V_\mathcal{N} \succeq 0$ then all triangle inequalities will be satisfied simultaneously;

$$
|\sqrt{d_{ik}} - \sqrt{d_{kj}}| \;\leq\; \sqrt{d_{ij}} \;\leq\; \sqrt{d_{ik}} + \sqrt{d_{kj}} \,, \quad i<k<j
\qquad (1143)
$$

(There are no more.) To verify our claim, we must prove the matrix inequality $-V_\mathcal{N}^T D V_\mathcal{N} \succeq 0$ to be a sufficient test of all the triangle inequalities; more efficient, we mention, for larger N:

5.8.2.1.1 Shore.
The columns of $\Xi_r V_\mathcal{N} \Xi_c$ hold a basis for $\mathcal{N}(\mathbf{1}^T)$ when Ξ_r and Ξ_c are permutation matrices. In other words, any permutation of the rows or columns of $V_\mathcal{N}$ leaves its range and nullspace unchanged; *id est*, $\mathcal{R}(\Xi_r V_\mathcal{N} \Xi_c) = \mathcal{R}(V_\mathcal{N}) = \mathcal{N}(\mathbf{1}^T)$ (975). Hence, two distinct matrix inequalities can be equivalent tests of the positive semidefiniteness of D on $\mathcal{R}(V_\mathcal{N})$; *id est*, $-V_\mathcal{N}^T D V_\mathcal{N} \succeq 0 \;\Leftrightarrow\; -(\Xi_r V_\mathcal{N} \Xi_c)^T D (\Xi_r V_\mathcal{N} \Xi_c) \succeq 0$. By properly choosing permutation matrices,[5.39] the leading principal submatrix $T_\Xi \in \mathbb{S}^2$ of $-(\Xi_r V_\mathcal{N} \Xi_c)^T D (\Xi_r V_\mathcal{N} \Xi_c)$ may be loaded with the entries of D needed to test any particular triangle inequality (similarly to (1133)-(1141)). Because all the triangle inequalities can be individually tested using a test equivalent to the lone matrix inequality $-V_\mathcal{N}^T D V_\mathcal{N} \succeq 0$, it logically follows that the lone matrix inequality tests all those triangle inequalities simultaneously. We conclude that $-V_\mathcal{N}^T D V_\mathcal{N} \succeq 0$ is a sufficient test for the fourth property of the Euclidean metric, triangle inequality. ♦

5.8.2.2 Strict triangle inequality

Without exception, all the inequalities in (1141) and (1142) can be made strict while their corresponding implications remain true. The then strict inequality (1141a) or (1142) may be interpreted as a *strict triangle inequality* under which collinear arrangement of points is

---

[5.38] Accounting for symmetry property 3, the fourth metric property demands three inequalities be satisfied per one of type (1141a). The first of those inequalities in (1142) is self evident from (1141a), while the two remaining follow from the left-hand side of (1141a) and the fact for scalars, $|a| \leq b \Leftrightarrow a \leq b$ and $-a \leq b$.

[5.39] To individually test triangle inequality $|\sqrt{d_{ik}} - \sqrt{d_{kj}}| \leq \sqrt{d_{ij}} \leq \sqrt{d_{ik}} + \sqrt{d_{kj}}$ for particular i, k, j, set $\Xi_r(i,1) = \Xi_r(k,2) = \Xi_r(j,3) = 1$ and $\Xi_c = I$.

not allowed. [229, §24/6, p.322] Hence by similar reasoning, $-V_{\mathcal{N}}^{\mathrm{T}}DV_{\mathcal{N}} \succ 0$ is a sufficient test of all the strict triangle inequalities; *id est*,

$$\left.\begin{array}{r} \delta(D) = \mathbf{0} \\ D^{\mathrm{T}} = D \\ -V_{\mathcal{N}}^{\mathrm{T}}DV_{\mathcal{N}} \succ 0 \end{array}\right\} \Rightarrow \sqrt{d_{ij}} < \sqrt{d_{ik}} + \sqrt{d_{kj}} \ , \quad i \neq j \neq k \tag{1144}$$

5.8.3 $-V_{\mathcal{N}}^{\mathrm{T}}DV_{\mathcal{N}}$ nesting

From (1138) observe that $T = -V_{\mathcal{N}}^{\mathrm{T}}DV_{\mathcal{N}}|_{N\leftarrow 3}$. In fact, for $D \in \mathbb{EDM}^{N}$, the leading principal submatrices of $-V_{\mathcal{N}}^{\mathrm{T}}DV_{\mathcal{N}}$ form a nested sequence (by inclusion) whose members are individually positive semidefinite [164] [208] [338] and have the same form as T; *videlicet*,[5.40]

$$-V_{\mathcal{N}}^{\mathrm{T}}DV_{\mathcal{N}}|_{N\leftarrow 1} = [\emptyset] \tag{o}$$

$$-V_{\mathcal{N}}^{\mathrm{T}}DV_{\mathcal{N}}|_{N\leftarrow 2} = [d_{12}] \in \mathbb{S}_{+} \tag{a}$$

$$-V_{\mathcal{N}}^{\mathrm{T}}DV_{\mathcal{N}}|_{N\leftarrow 3} = \begin{bmatrix} d_{12} & \frac{1}{2}(d_{12}+d_{13}-d_{23}) \\ \frac{1}{2}(d_{12}+d_{13}-d_{23}) & d_{13} \end{bmatrix} = T \in \mathbb{S}_{+}^{2} \tag{b}$$

$$-V_{\mathcal{N}}^{\mathrm{T}}DV_{\mathcal{N}}|_{N\leftarrow 4} = \begin{bmatrix} d_{12} & \frac{1}{2}(d_{12}+d_{13}-d_{23}) & \frac{1}{2}(d_{12}+d_{14}-d_{24}) \\ \frac{1}{2}(d_{12}+d_{13}-d_{23}) & d_{13} & \frac{1}{2}(d_{13}+d_{14}-d_{34}) \\ \frac{1}{2}(d_{12}+d_{14}-d_{24}) & \frac{1}{2}(d_{13}+d_{14}-d_{34}) & d_{14} \end{bmatrix} \tag{c}$$

$$\vdots$$

$$-V_{\mathcal{N}}^{\mathrm{T}}DV_{\mathcal{N}}|_{N\leftarrow i} = \begin{bmatrix} -V_{\mathcal{N}}^{\mathrm{T}}DV_{\mathcal{N}}|_{N\leftarrow i-1} & \nu(i) \\ \nu(i)^{\mathrm{T}} & d_{1i} \end{bmatrix} \in \mathbb{S}_{+}^{i-1} \tag{d}$$

$$\vdots$$

$$-V_{\mathcal{N}}^{\mathrm{T}}DV_{\mathcal{N}} = \begin{bmatrix} -V_{\mathcal{N}}^{\mathrm{T}}DV_{\mathcal{N}}|_{N\leftarrow N-1} & \nu(N) \\ \nu(N)^{\mathrm{T}} & d_{1N} \end{bmatrix} \in \mathbb{S}_{+}^{N-1} \tag{e}$$

$$\tag{1145}$$

where

$$\nu(i) \triangleq \frac{1}{2}\begin{bmatrix} d_{12}+d_{1i}-d_{2i} \\ d_{13}+d_{1i}-d_{3i} \\ \vdots \\ d_{1,i-1}+d_{1i}-d_{i-1,i} \end{bmatrix} \in \mathbb{R}^{i-2}, \quad i > 2 \tag{1146}$$

Hence, the leading principal submatrices of EDM D must also be EDMs.[5.41]

[5.40] $-VDV|_{N\leftarrow 1} = 0 \in \mathbb{S}_{+}^{0}$ (§B.4.1)
[5.41] In fact, each and every principal submatrix of an EDM D is another EDM. [241, §4.1]

Bordered symmetric matrices in the form (1145d) are known to have *intertwined* [338, §6.4] (or *interlaced* [208, §4.3] [335, §IV.4.1]) eigenvalues; *(confer* §5.11.1) that means, for the particular submatrices (1145a) and (1145b),

$$\lambda_2 \leq d_{12} \leq \lambda_1 \tag{1147}$$

where d_{12} is the eigenvalue of submatrix (1145a) and λ_1, λ_2 are the eigenvalues of T (1145b) (1138). Intertwining in (1147) predicts that should d_{12} become 0, then λ_2 must go to 0.[5.42] The eigenvalues are similarly intertwined for submatrices (1145b) and (1145c);

$$\gamma_3 \leq \lambda_2 \leq \gamma_2 \leq \lambda_1 \leq \gamma_1 \tag{1148}$$

where γ_1, γ_2, γ_3 are the eigenvalues of submatrix (1145c). Intertwining likewise predicts that should λ_2 become 0 (a possibility revealed in §5.8.3.1), then γ_3 must go to 0. Combining results so far for $N = 2, 3, 4$: (1147) (1148)

$$\gamma_3 \leq \lambda_2 \leq d_{12} \leq \lambda_1 \leq \gamma_1 \tag{1149}$$

The preceding logic extends by induction through the remaining members of the sequence (1145).

5.8.3.1 Tightening the triangle inequality

Now we apply Schur complement from §A.4 to tighten the triangle inequality from (1137) in case: cardinality $N = 4$. We find that the gains by doing so are modest. From (1145) we identify:

$$\begin{bmatrix} A & B \\ B^{\mathrm{T}} & C \end{bmatrix} \triangleq -V_{\mathcal{N}}^{\mathrm{T}} D V_{\mathcal{N}}|_{N \leftarrow 4} \tag{1150}$$

$$A \triangleq T = -V_{\mathcal{N}}^{\mathrm{T}} D V_{\mathcal{N}}|_{N \leftarrow 3} \tag{1151}$$

both positive semidefinite by assumption, where $B = \nu(4)$ (1146), and $C = d_{14}$. Using nonstrict $CC^{\dagger}$-form (1592), $C \succeq 0$ by assumption (§5.8.1) and $CC^{\dagger} = I$. So by the *positive semidefinite ordering of eigenvalues theorem* (§A.3.1.0.1),

$$-V_{\mathcal{N}}^{\mathrm{T}} D V_{\mathcal{N}}|_{N \leftarrow 4} \succeq 0 \ \Leftrightarrow \ T \succeq d_{14}^{-1} \nu(4) \nu(4)^{\mathrm{T}} \ \Rightarrow \ \begin{cases} \lambda_1 \geq d_{14}^{-1} \|\nu(4)\|^2 \\ \lambda_2 \geq 0 \end{cases} \tag{1152}$$

where $\{d_{14}^{-1}\|\nu(4)\|^2, 0\}$ are the eigenvalues of $d_{14}^{-1}\nu(4)\nu(4)^{\mathrm{T}}$ while λ_1, λ_2 are the eigenvalues of T.

5.8.3.1.1 Example. *Small completion problem,* II.
Applying the inequality for λ_1 in (1152) to the *small completion problem* on page 358 Figure **127**, the lower bound on $\sqrt{d_{14}}$ (1.236 in (961)) is tightened to 1.289. The correct value of $\sqrt{d_{14}}$ to three significant figures is 1.414. □

[5.42]If d_{12} were 0, eigenvalue λ_2 becomes 0 (1140) because d_{13} must then be equal to d_{23}; *id est,* $d_{12} = 0 \ \Leftrightarrow \ x_1 = x_2$. (§5.4)

5.8.4   Affine dimension reduction in two dimensions

(*confer* §5.14.4) The leading principal 2×2 submatrix T of $-V_\mathcal{N}^{\mathrm T}DV_\mathcal{N}$ has largest eigenvalue λ_1 (1140) which is a convex function of D .[5.43] λ_1 can never be 0 unless $d_{12}=d_{13}=d_{23}=0$. Eigenvalue λ_1 can never be negative while the d_{ij} are nonnegative. The remaining eigenvalue λ_2 (1140) is a concave function of D that becomes 0 only at the upper and lower bounds of triangle inequality (1141a) and its equivalent forms: (*confer*(1143))

$$|\sqrt{d_{12}}-\sqrt{d_{23}}| \;\le\; \sqrt{d_{13}} \;\le\; \sqrt{d_{12}}+\sqrt{d_{23}} \qquad \text{(a)}$$
$$\Leftrightarrow$$
$$|\sqrt{d_{12}}-\sqrt{d_{13}}| \;\le\; \sqrt{d_{23}} \;\le\; \sqrt{d_{12}}+\sqrt{d_{13}} \qquad \text{(b)} \qquad (1153)$$
$$\Leftrightarrow$$
$$|\sqrt{d_{13}}-\sqrt{d_{23}}| \;\le\; \sqrt{d_{12}} \;\le\; \sqrt{d_{13}}+\sqrt{d_{23}} \qquad \text{(c)}$$

In between those bounds, λ_2 is strictly positive; otherwise, it would be negative but prevented by the condition $T\succeq0$.

When λ_2 becomes 0, it means triangle $\triangle_{123}$ has collapsed to a line segment; a potential reduction in affine dimension r . The same logic is valid for any particular principal 2×2 submatrix of $-V_\mathcal{N}^{\mathrm T}DV_\mathcal{N}$, hence applicable to other triangles.

5.9   Bridge: Convex polyhedra to EDMs

The criteria for the existence of an EDM include, by definition (968) (1038), the properties imposed upon its entries d_{ij} by the Euclidean metric. From §5.8.1 and §5.8.2, we know there is a relationship of matrix criteria to those properties. Here is a snapshot of what we are sure: for $i,j,k\in\{1\dots N\}$ (*confer* §5.2)

$$\begin{array}{l}\sqrt{d_{ij}}\ge0,\;\; i\ne j\\ \sqrt{d_{ij}}=0,\;\; i=j\\ \sqrt{d_{ij}}=\sqrt{d_{ji}}\\ \sqrt{d_{ij}}\le\sqrt{d_{ik}}+\sqrt{d_{kj}},\;\; i\ne j\ne k\end{array} \quad\Leftarrow\quad \begin{array}{c}-V_\mathcal{N}^{\mathrm T}DV_\mathcal{N}\succeq0\\ \delta(D)=\mathbf0\\ D^{\mathrm T}=D\end{array} \qquad(1154)$$

all implied by $D\in\mathbb{EDM}^N$. In words, these four Euclidean metric properties are necessary conditions for D to be a distance matrix. At the moment, we have no converse. As of concern in §5.3, we have yet to establish metric requirements beyond the four Euclidean metric properties that would allow D to be certified an EDM or might facilitate polyhedron or list reconstruction from an incomplete EDM. We deal with this problem in §5.14. Our present goal is to establish *ab initio* the necessary and sufficient matrix criteria that will

---

[5.43]The largest eigenvalue of any symmetric matrix is always a convex function of its entries, while the smallest eigenvalue is always concave. [62, exmp.3.10] In our particular case, say $\underline{d}\triangleq\begin{bmatrix}d_{12}\\d_{13}\\d_{23}\end{bmatrix}\in\mathbb{R}^3$. Then the Hessian (1852) $\nabla^2\lambda_1(\underline{d})\succeq0$ certifies convexity whereas $\nabla^2\lambda_2(\underline{d})\preceq0$ certifies concavity. Each Hessian has rank equal to 1. The respective gradients $\nabla\lambda_1(\underline{d})$ and $\nabla\lambda_2(\underline{d})$ are nowhere $\mathbf0$ and can be uniquely defined.

subsume all the Euclidean metric properties and any further requirements[5.44] for all $N > 1$ (§5.8.3); *id est,*

$$\left. \begin{array}{r} -V_{\mathcal{N}}^{\mathrm{T}} D V_{\mathcal{N}} \succeq 0 \\ D \in \mathbb{S}_h^N \end{array} \right\} \quad \Leftrightarrow \quad D \in \mathbb{EDM}^N \qquad (987)$$

or for EDM definition (1047),

$$\left. \begin{array}{r} \Omega \succeq 0 \\ \sqrt{\delta(d)} \succeq 0 \end{array} \right\} \quad \Leftrightarrow \quad D = \mathbf{D}(\Omega, d) \in \mathbb{EDM}^N \qquad (1155)$$

5.9.1 Geometric arguments

5.9.1.0.1 Definition. *Elliptope*:   [244] [241, §2.3] [116, §31.5]
a unique bounded immutable convex Euclidean body in $\mathbb{S}^n$; intersection of positive semidefinite cone $\mathbb{S}_+^n$ with that set of n hyperplanes defined by unity main diagonal;

$$\mathcal{E}^n \triangleq \mathbb{S}_+^n \cap \{ \Phi \in \mathbb{S}^n \mid \delta(\Phi) = \mathbf{1} \} \qquad (1156)$$

`a.k.a` the set of all *correlation matrices* of dimension

$$\dim \mathcal{E}^n = n(n-1)/2 \quad \text{in } \mathbb{R}^{n(n+1)/2} \qquad (1157)$$

An elliptope $\mathcal{E}^n$ is not a polyhedron, in general, but has some polyhedral faces and an infinity of vertices.[5.45] Of those, 2^{n-1} vertices (some extreme points of the elliptope) are extreme directions $y y^{\mathrm{T}}$ of the positive semidefinite cone, where the entries of vector $y \in \mathbb{R}^n$ belong to $\{\pm 1\}$ and exercise every combination. Each of the remaining vertices has rank, greater than one, belonging to the set $\{k > 0 \mid k(k+1)/2 \leq n\}$. Each and every face of an elliptope is exposed. $\triangle$

In fact, any positive semidefinite matrix whose entries belong to $\{\pm 1\}$ is a rank-one correlation matrix; and *vice versa*:[5.46]

5.9.1.0.2 Theorem. *Elliptope vertices rank-one.*   [129, §2.1.1]
For $Y \in \mathbb{S}^n$, $y \in \mathbb{R}^n$, and all $i, j \in \{1 \dots n\}$ *(confer* §2.3.1.0.1)

$$Y \succeq 0, \quad Y_{ij} \in \{\pm 1\} \qquad \Leftrightarrow \qquad Y = y y^{\mathrm{T}}, \quad y_i \in \{\pm 1\} \qquad (1158)$$

$\diamond$

[5.44]In 1935, Schoenberg [320, (1)] first extolled matrix product $-V_{\mathcal{N}}^{\mathrm{T}} D V_{\mathcal{N}}$ (1133) (predicated on symmetry and self-distance) specifically incorporating $V_{\mathcal{N}}$, albeit algebraically. He showed: nonnegativity $-y^{\mathrm{T}} V_{\mathcal{N}}^{\mathrm{T}} D V_{\mathcal{N}} y \geq 0$, for all $y \in \mathbb{R}^{N-1}$, is necessary and sufficient for D to be an EDM. Gower [168, §3] remarks how surprising it is that such a fundamental property of Euclidean geometry was obtained so late.
[5.45]Laurent defines vertex distinctly from the sense herein (§2.6.1.0.1); she defines *vertex* as a point with full-dimensional (nonempty interior) normal cone (§E.10.3.2.1). Her definition excludes point C in Figure 33, for example.
[5.46]As there are few equivalent conditions for rank constraints, this device is rather important for relaxing integer, combinatorial, or Boolean problems.

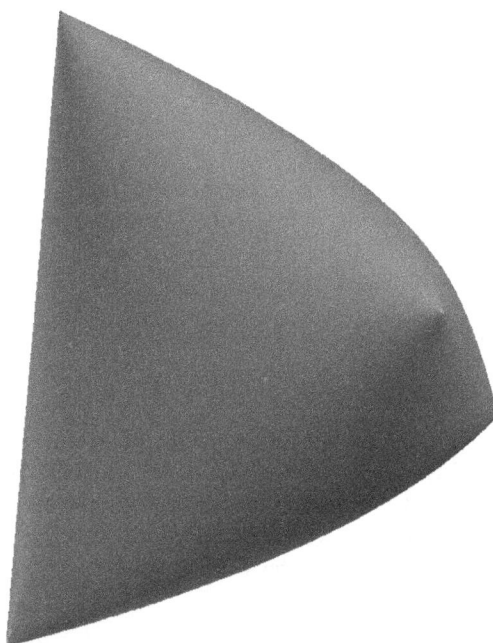

Figure 141: Elliptope $\mathcal{E}^3$ in isometrically isomorphic $\mathbb{R}^6$ (projected on $\mathbb{R}^3$) is a convex body that appears to possess some kind of symmetry in this dimension; it resembles a malformed pillow in the shape of a bulging tetrahedron. Elliptope relative boundary is not *smooth* and comprises all set members (1156) having at least one 0 eigenvalue. [244, §2.1] This elliptope has an infinity of vertices, but there are only four vertices corresponding to a rank-1 matrix. Those yy^{T}, evident in the illustration, have binary vector $y \in \mathbb{R}^3$ with entries in $\{\pm 1\}$.

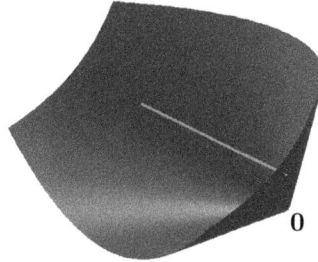

Figure 142: Elliptope $\mathcal{E}^2$ in isometrically isomorphic $\mathbb{R}^3$ is a line segment illustrated interior to positive semidefinite cone $\mathbb{S}_+^2$ (Figure 44).

The elliptope for dimension $n=2$ is a line segment in isometrically isomorphic $\mathbb{R}^{n(n+1)/2}$ (Figure 142). Obviously, $\mathrm{cone}(\mathcal{E}^n) \neq \mathbb{S}_+^n$. The elliptope for dimension $n=3$ is realized in Figure 141.

5.9.1.0.3 Lemma. *Hypersphere.*   (*confer* bullet p.366) [17, §4]
Matrix $\Psi = [\Psi_{ij}] \in \mathbb{S}^N$ belongs to the elliptope in $\mathbb{S}^N$ iff there exist N points p on the boundary of a hypersphere in $\mathbb{R}^{\mathrm{rank}\,\Psi}$ having radius 1 such that

$$\|p_i - p_j\|^2 = 2(1 - \Psi_{ij}) , \quad i,j=1 \ldots N \tag{1159}$$

$\diamond$

There is a similar theorem for Euclidean distance matrices:
We derive matrix criteria for D to be an EDM, validating (987) using simple geometry; distance to the polyhedron formed by the convex hull of a list of points (76) in Euclidean space $\mathbb{R}^n$.

5.9.1.0.4 EDM assertion.
D is a Euclidean distance matrix if and only if $D \in \mathbb{S}_h^N$ and distances-square from the origin

$$\{\|p(y)\|^2 = -y^{\mathrm{T}} V_{\mathcal{N}}^{\mathrm{T}} D V_{\mathcal{N}} \, y \mid y \in \mathcal{S} - \beta\} \tag{1160}$$

correspond to points p in some bounded convex polyhedron

$$\mathcal{P} - \alpha = \{p(y) \mid y \in \mathcal{S} - \beta\} \tag{1161}$$

having N or fewer vertices embedded in an r-dimensional subspace $\mathcal{A} - \alpha$ of $\mathbb{R}^n$, where $\alpha \in \mathcal{A} = \mathrm{aff}\,\mathcal{P}$ and where domain of linear surjection $p(y)$ is the unit simplex $\mathcal{S} \subset \mathbb{R}_+^{N-1}$ shifted such that its vertex at the origin is translated to $-\beta$ in $\mathbb{R}^{N-1}$. When $\beta = 0$, then $\alpha = x_1$. $\diamond$

In terms of $V_\mathcal{N}$, the unit simplex (292) in $\mathbb{R}^{N-1}$ has an equivalent representation:

$$\mathcal{S} = \{s \in \mathbb{R}^{N-1} \mid \sqrt{2}V_\mathcal{N}\, s \succeq -e_1\} \tag{1162}$$

where e_1 is as in (984). Incidental to the *EDM assertion,* shifting the unit-simplex domain in $\mathbb{R}^{N-1}$ translates the polyhedron $\mathcal{P}$ in $\mathbb{R}^n$. Indeed, there is a map from vertices of the unit simplex to members of the list generating $\mathcal{P}$;

$$p \quad : \quad \mathbb{R}^{N-1} \qquad \rightarrow \qquad \mathbb{R}^n$$

$$p\left(\left\{\begin{array}{c} -\beta \\ e_1 - \beta \\ e_2 - \beta \\ \vdots \\ e_{N-1} - \beta \end{array}\right\}\right) = \left\{\begin{array}{c} x_1 - \alpha \\ x_2 - \alpha \\ x_3 - \alpha \\ \vdots \\ x_N - \alpha \end{array}\right\} \tag{1163}$$

5.9.1.0.5 Proof. *EDM assertion.*

($\Rightarrow$) We demonstrate that if D is an EDM, then each distance-square $\|p(y)\|^2$ described by (1160) corresponds to a point p in some embedded polyhedron $\mathcal{P} - \alpha$. Assume D is indeed an EDM; *id est,* D can be made from some list X of N unknown points in Euclidean space $\mathbb{R}^n$; $D = \mathbf{D}(X)$ for $X \in \mathbb{R}^{n \times N}$ as in (968). Since D is translation invariant (§5.5.1), we may shift the affine hull $\mathcal{A}$ of those unknown points to the origin as in (1103). Then take any point p in their convex hull (86);

$$\mathcal{P} - \alpha = \{p = (X - Xb\mathbf{1}^\mathrm{T})a \mid a^\mathrm{T}\mathbf{1} = 1,\ a \succeq 0\} \tag{1164}$$

where $\alpha = Xb \in \mathcal{A} \Leftrightarrow b^\mathrm{T}\mathbf{1} = 1$. Solutions to $a^\mathrm{T}\mathbf{1} = 1$ are:[5.47]

$$a \in \left\{e_1 + \sqrt{2}V_\mathcal{N}\, s \mid s \in \mathbb{R}^{N-1}\right\} \tag{1165}$$

where e_1 is as in (984). Similarly, $b = e_1 + \sqrt{2}V_\mathcal{N}\,\beta$.

$$\begin{aligned} \mathcal{P} - \alpha &= \{p = X(I - (e_1 + \sqrt{2}V_\mathcal{N}\beta)\mathbf{1}^\mathrm{T})(e_1 + \sqrt{2}V_\mathcal{N}\, s) \mid \sqrt{2}V_\mathcal{N}\, s \succeq -e_1\} \\ &= \{p = X\sqrt{2}V_\mathcal{N}(s - \beta) \mid \sqrt{2}V_\mathcal{N}\, s \succeq -e_1\} \end{aligned} \tag{1166}$$

that describes the domain of $p(s)$ as the unit simplex

$$\mathcal{S} = \{s \mid \sqrt{2}V_\mathcal{N}\, s \succeq -e_1\} \subset \mathbb{R}_+^{N-1} \tag{1162}$$

Making the substitution $s - \beta \leftarrow y$

$$\mathcal{P} - \alpha = \{p = X\sqrt{2}V_\mathcal{N}\, y \mid y \in \mathcal{S} - \beta\} \tag{1167}$$

Point p belongs to a convex polyhedron $\mathcal{P} - \alpha$ embedded in an r-dimensional subspace of $\mathbb{R}^n$ because the convex hull of any list forms a polyhedron, and because the translated

---

[5.47]Since $\mathcal{R}(V_\mathcal{N}) = \mathcal{N}(\mathbf{1}^\mathrm{T})$ and $\mathcal{N}(\mathbf{1}^\mathrm{T}) \perp \mathcal{R}(\mathbf{1})$, then over all $s \in \mathbb{R}^{N-1}$, $V_\mathcal{N}\, s$ is a hyperplane through the origin orthogonal to $\mathbf{1}$. Thus the solutions $\{a\}$ constitute a hyperplane orthogonal to the vector $\mathbf{1}$, and offset from the origin in $\mathbb{R}^N$ by any particular solution; in this case, $a = e_1$.

affine hull $\mathcal{A} - \alpha$ contains the translated polyhedron $\mathcal{P} - \alpha$ (1106) and the origin (when $\alpha \in \mathcal{A}$), and because $\mathcal{A}$ has dimension r by definition (1108). Now, any distance-square from the origin to the polyhedron $\mathcal{P} - \alpha$ can be formulated

$$\{p^{\mathrm{T}}p \;=\; \|p\|^2 \;=\; 2y^{\mathrm{T}}V_{\mathcal{N}}^{\mathrm{T}}X^{\mathrm{T}}XV_{\mathcal{N}}\,y \mid y \in \mathcal{S} - \beta\} \tag{1168}$$

Applying (1115) to (1168) we get (1160).

($\Leftarrow$) To validate the *EDM assertion* in the reverse direction, we prove: If each distance-square $\|p(y)\|^2$ (1160) on the shifted unit-simplex $\mathcal{S} - \beta \subset \mathbb{R}^{N-1}$ corresponds to a point $p(y)$ in some embedded polyhedron $\mathcal{P} - \alpha$, then D is an EDM. The r-dimensional subspace $\mathcal{A} - \alpha \subseteq \mathbb{R}^n$ is spanned by

$$p(\mathcal{S} - \beta) = \mathcal{P} - \alpha \tag{1169}$$

because $\mathcal{A} - \alpha = \mathrm{aff}(\mathcal{P} - \alpha) \supseteq \mathcal{P} - \alpha$ (1106). So, outside domain $\mathcal{S} - \beta$ of linear surjection $p(y)$, simplex complement $\backslash\mathcal{S} - \beta \subset \mathbb{R}^{N-1}$ must contain domain of the distance-square $\|p(y)\|^2 = p(y)^{\mathrm{T}}p(y)$ to remaining points in subspace $\mathcal{A} - \alpha$; *id est*, to the polyhedron's relative exterior $\backslash\mathcal{P} - \alpha$. For $\|p(y)\|^2$ to be nonnegative on the entire subspace $\mathcal{A} - \alpha$, $-V_{\mathcal{N}}^{\mathrm{T}}DV_{\mathcal{N}}$ must be positive semidefinite and is assumed symmetric;[5.48]

$$-V_{\mathcal{N}}^{\mathrm{T}}DV_{\mathcal{N}} \;\triangleq\; \Theta_{\mathrm{p}}^{\mathrm{T}}\Theta_{\mathrm{p}} \tag{1170}$$

where[5.49] $\Theta_{\mathrm{p}} \in \mathbb{R}^{m \times N-1}$ for some $m \geq r$. Because $p(\mathcal{S} - \beta)$ is a convex polyhedron, it is necessarily a set of linear combinations of points from some length-N list because every convex polyhedron having N or fewer vertices can be generated that way (§2.12.2). Equivalent to (1160) are

$$\{p^{\mathrm{T}}p \mid p \in \mathcal{P} - \alpha\} \;=\; \{p^{\mathrm{T}}p = y^{\mathrm{T}}\Theta_{\mathrm{p}}^{\mathrm{T}}\Theta_{\mathrm{p}}\,y \mid y \in \mathcal{S} - \beta\} \tag{1171}$$

Because $p \in \mathcal{P} - \alpha$ may be found by factoring (1171), the list Θ_{p} is found by factoring (1170). A unique EDM can be made from that list using inner-product form definition $\mathbf{D}(\Theta)|_{\Theta=\Theta_{\mathrm{p}}}$ (1038). That EDM will be identical to D if $\delta(D) = \mathbf{0}$, by injectivity of $\mathbf{D}$ (1089). $\qquad\blacklozenge$

5.9.2 Necessity and sufficiency

From (1131) we learned that matrix inequality $-V_{\mathcal{N}}^{\mathrm{T}}DV_{\mathcal{N}} \succeq 0$ is a necessary test for D to be an EDM. In §5.9.1, the connection between convex polyhedra and EDMs was pronounced by the *EDM assertion*; the matrix inequality together with $D \in \mathbb{S}_h^N$ became a sufficient test when the *EDM assertion* demanded that every bounded convex polyhedron have a corresponding EDM. For all $N > 1$ (§5.8.3), the matrix criteria for the existence of an EDM in (987), (1155), and (963) are therefore necessary and sufficient and subsume all the Euclidean metric properties and further requirements.

[5.48]The antisymmetric part $\left(-V_{\mathcal{N}}^{\mathrm{T}}DV_{\mathcal{N}} - (-V_{\mathcal{N}}^{\mathrm{T}}DV_{\mathcal{N}})^{\mathrm{T}}\right)/2$ is annihilated by $\|p(y)\|^2$. By the same reasoning, any positive (semi)definite matrix A is generally assumed symmetric because only the symmetric part $(A + A^{\mathrm{T}})/2$ survives the test $y^{\mathrm{T}}Ay \geq 0$. [208, §7.1]
[5.49]$A^{\mathrm{T}} = A \succeq 0 \;\Leftrightarrow\; A = R^{\mathrm{T}}R$ for some real matrix R. [338, §6.3]

5.9.3   Example revisited

Now we apply the necessary and sufficient EDM criteria (987) to an earlier problem.

5.9.3.0.1   Example. *Small completion problem,* III.   (*confer* §5.8.3.1.1)
Continuing Example 5.3.0.0.2 pertaining to Figure **127** where $N=4$, distance-square d_{14}
is ascertainable from the matrix inequality $-V_{\mathcal{N}}^{\mathrm{T}}DV_{\mathcal{N}}\succeq 0$. Because all distances in (960)
are known except $\sqrt{d_{14}}$, we may simply calculate the smallest eigenvalue of $-V_{\mathcal{N}}^{\mathrm{T}}DV_{\mathcal{N}}$
over a range of d_{14} as in Figure **143**. We observe a unique value of d_{14} satisfying (987)
where the abscissa axis is tangent to the hypograph of the smallest eigenvalue. Since the
smallest eigenvalue of a symmetric matrix is known to be a concave function (§5.8.4), we
calculate its second partial derivative with respect to d_{14} evaluated at 2 and find $-1/3$.
We conclude there are no other satisfying values of d_{14}. Further, that value of d_{14} does
not meet an upper or lower bound of a triangle inequality like (1143), so neither does it
cause the collapse of any triangle. Because the smallest eigenvalue is 0, affine dimension
r of any point list corresponding to D cannot exceed $N-2$. (§5.7.1.1)   □

5.10   EDM-entry composition

Laurent [241, §2.3] applies results from Schoenberg (1938) [321] to show certain nonlinear
compositions of individual EDM entries yield EDMs; in particular,

$$\begin{aligned} D \in \mathbb{EDM}^N \quad &\Leftrightarrow \quad [1 - e^{-\alpha d_{ij}}] \in \mathbb{EDM}^N \quad &\forall\, \alpha > 0 \quad &\text{(a)}\\ &\Leftrightarrow \quad [e^{-\alpha d_{ij}}] \in \mathcal{E}^N \quad &\forall\, \alpha > 0 \quad &\text{(b)} \end{aligned} \qquad (1172)$$

where $D=[d_{ij}]$ and $\mathcal{E}^N$ is the elliptope (1156).

5.10.0.0.1   Proof. (Monique Laurent, 2003)   [321] (*confer* [233])

> **Lemma 2.1.** from *A Tour d'Horizon ... on Completion Problems.*
> [241] For $D=[d_{ij}\,,\ i,j=1\ldots N]\in\mathbb{S}_h^N$ and $\mathcal{E}^N$ the elliptope in $\mathbb{S}^N$ (§5.9.1.0.1), the
> following assertions are equivalent:
>
> **(i)** $D \in \mathbb{EDM}^N$
>
> **(ii)** $e^{-\alpha D} \triangleq [e^{-\alpha d_{ij}}] \in \mathcal{E}^N$ for all $\alpha > 0$
>
> **(iii)** $\mathbf{1}\mathbf{1}^{\mathrm{T}} - e^{-\alpha D} \triangleq [1 - e^{-\alpha d_{ij}}] \in \mathbb{EDM}^N$ for all $\alpha > 0$ ◇

1) Equivalence of Lemma 2.1 (i) (ii) is stated in Schoenberg's Theorem 1 [321, p.527].

smallest eigenvalue

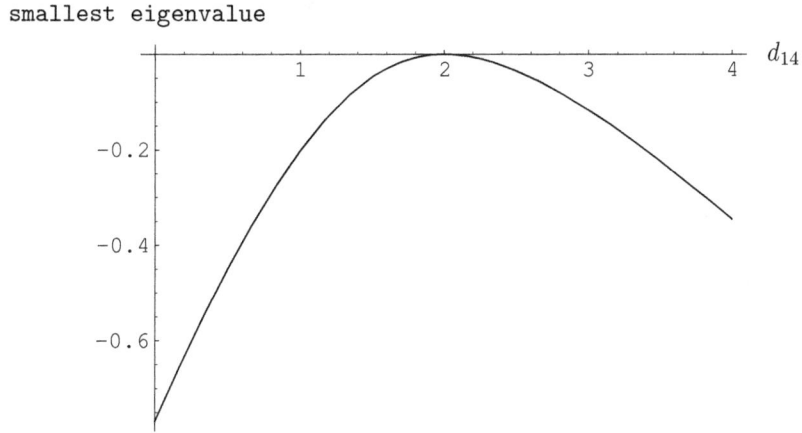

Figure 143: Smallest eigenvalue of $-V_\mathcal{N}^\mathrm{T} D V_\mathcal{N}$ makes it a PSD matrix for only one value of d_{14}: 2.

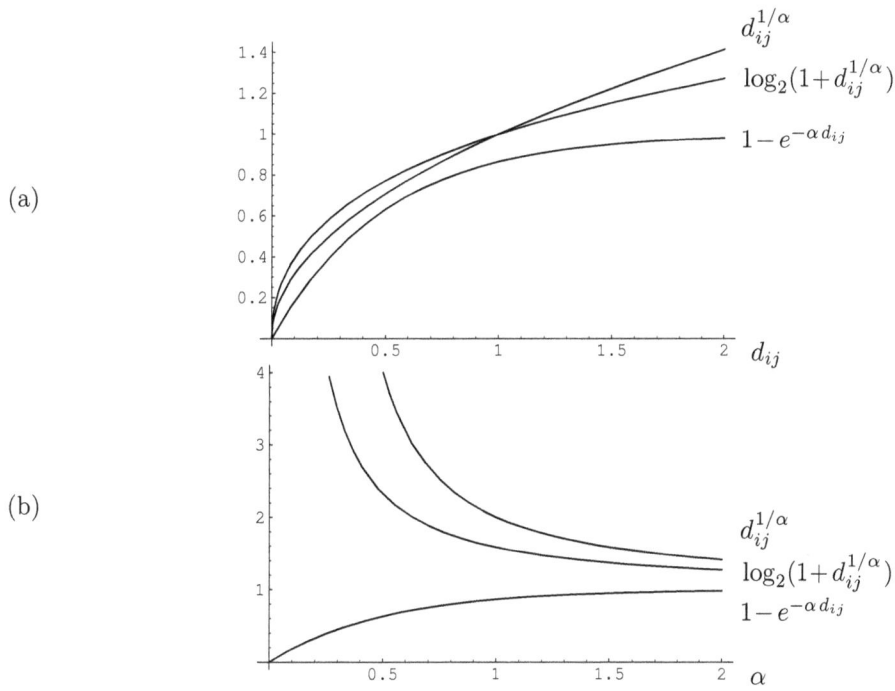

(a)

$$d_{ij}^{1/\alpha}$$
$$\log_2(1+d_{ij}^{1/\alpha})$$
$$1-e^{-\alpha d_{ij}}$$

(b)

$$d_{ij}^{1/\alpha}$$
$$\log_2(1+d_{ij}^{1/\alpha})$$
$$1-e^{-\alpha d_{ij}}$$

Figure 144: Some entrywise EDM compositions: **(a)** $\alpha = 2$. Concave nondecreasing in d_{ij}. **(b)** Trajectory convergence in α for $d_{ij} = 2$.

2) (ii) $\Rightarrow$ (iii) can be seen from the statement in the beginning of section 3, saying that a distance space embeds in L_2 iff some associated matrix is PSD. We reformulate it:

Let $d = (d_{ij})_{i,j=0,1\ldots N}$ be a distance space on $N+1$ points (*i.e.*, symmetric hollow matrix of order $N+1$) and let $p = (p_{ij})_{i,j=1\ldots N}$ be the symmetric matrix of order N related by:

(A) $2p_{ij} = d_{0i} + d_{0j} - d_{ij}$ for $i, j = 1 \ldots N$

or equivalently

(B) $d_{0i} = p_{ii}$, $d_{ij} = p_{ii} + p_{jj} - 2p_{ij}$ for $i, j = 1 \ldots N$

Then d embeds in L_2 iff p is a positive semidefinite matrix iff d is of negative type (second half page 525/top of page 526 in [321]).

For the implication from (ii) to (iii), set: $p = e^{-\alpha d}$ and define d' from p using (B) above. Then d' is a distance space on $N+1$ points that embeds in L_2. Thus its subspace of N points also embeds in L_2 and is precisely $1 - e^{-\alpha d}$.

Note that (iii) $\Rightarrow$ (ii) cannot be read immediately from this argument since (iii) involves the subdistance of d' on N points (and not the full d' on $N+1$ points).

3) Show (iii) $\Rightarrow$ (i) by using the series expansion of the function $1 - e^{-\alpha d}$: the constant term cancels, α factors out; there remains a summation of d plus a multiple of α. Letting α go to 0 gives the result.

This is not explicitly written in Schoenberg, but he also uses such an argument; expansion of the exponential function then $\alpha \to 0$ (first proof on [321, p.526]). $\blacklozenge$

Schoenberg's results [321, §6 thm.5] (*confer* [233, p.108-109]) also suggest certain finite positive roots of EDM entries produce EDMs; specifically,

$$D \in \mathbb{EDM}^N \;\Leftrightarrow\; [d_{ij}^{1/\alpha}] \in \mathbb{EDM}^N \quad \forall \alpha > 1 \tag{1173}$$

The special case $\alpha = 2$ is of interest because it corresponds to absolute distance; *e.g.*,

$$D \in \mathbb{EDM}^N \Rightarrow \sqrt[\circ]{D} \in \mathbb{EDM}^N \tag{1174}$$

Assuming that points constituting a corresponding list X are distinct (1135), then it follows: for $D \in \mathbb{S}_h^N$

$$\lim_{\alpha \to \infty} [d_{ij}^{1/\alpha}] \;=\; \lim_{\alpha \to \infty} [1 - e^{-\alpha d_{ij}}] \;=\; -E \triangleq \mathbf{1}\mathbf{1}^{\mathrm{T}} - I \tag{1175}$$

Negative elementary matrix $-E$ (§B.3) is: relatively interior to the EDM cone (§6.5), on its axis, and terminal to respective trajectories (1172a) and (1173) as functions of α. Both trajectories are confined to the EDM cone; in engineering terms, the EDM cone is an *invariant set* [318] to either trajectory. Further, if D is not an EDM but for some particular α_{p} it becomes an EDM, then for all greater values of α it remains an EDM.

5.10.0.0.2 Exercise. *Concave nondecreasing EDM-entry composition.*
Given EDM $D = [d_{ij}]$, empirical evidence suggests that the composition $[\log_2(1 + d_{ij}^{1/\alpha})]$ is also an EDM for each fixed $\alpha \geq 1$ [*sic*]. Its concavity in d_{ij} is illustrated in Figure 144 together with functions from (1172a) and (1173). Prove whether it holds more generally: Any concave nondecreasing composition of individual EDM entries d_{ij} on $\mathbb{R}_+$ produces another EDM. $\blacktriangledown$

5.10.0.0.3 Exercise. *Taxicab distance matrix as EDM.*
Determine whether taxicab distance matrices ($\mathbf{D}_1(X)$ in Example 3.7.0.0.2) are all numerically equivalent to EDMs. Explain why or why not. ▼

5.10.1 EDM by elliptope

(*confer*(994)) For some $\kappa \in \mathbb{R}_+$ and $C \in \mathbb{S}_+^N$ in elliptope $\mathcal{E}^N$ (§5.9.1.0.1), Alfakih asserts: any given EDM D is expressible [9] [116, §31.5]

$$D = \kappa(\mathbf{11}^\mathrm{T} - C) \in \mathbb{EDM}^N \tag{1176}$$

This expression exhibits nonlinear combination of variables κ and C. We therefore propose a different expression requiring redefinition of the elliptope (1156) by scalar parametrization;

$$\mathcal{E}_t^n \triangleq \mathbb{S}_+^n \cap \{\Phi \in \mathbb{S}^n \mid \delta(\Phi) = t\mathbf{1}\} \tag{1177}$$

where, of course, $\mathcal{E}^n = \mathcal{E}_1^n$. Then any given EDM D is expressible

$$D = t\mathbf{11}^\mathrm{T} - \mathfrak{E} \in \mathbb{EDM}^N \tag{1178}$$

which is linear in variables $t \in \mathbb{R}_+$ and $\mathfrak{E} \in \mathcal{E}_t^N$.

5.11 EDM indefiniteness

By known result (§A.7.2) regarding a 0-valued entry on the main diagonal of a symmetric positive semidefinite matrix, there can be no positive or negative semidefinite EDM except the **0** matrix because $\mathbb{EDM}^N \subseteq \mathbb{S}_h^N$ (967) and

$$\mathbb{S}_h^N \cap \mathbb{S}_+^N = \mathbf{0} \tag{1179}$$

the origin. So when $D \in \mathbb{EDM}^N$, there can be no factorization $D = A^\mathrm{T}A$ or $-D = A^\mathrm{T}A$. [338, §6.3] Hence eigenvalues of an EDM are neither all nonnegative or all nonpositive; an EDM is indefinite and possibly invertible.

5.11.1 EDM eigenvalues, congruence transformation

For any symmetric $-D$, we can characterize its eigenvalues by congruence transformation: [338, §6.3]

$$-W^\mathrm{T}DW = -\begin{bmatrix} V_\mathcal{N}^\mathrm{T} \\ \mathbf{1}^\mathrm{T} \end{bmatrix} D \begin{bmatrix} V_\mathcal{N} & \mathbf{1} \end{bmatrix} = -\begin{bmatrix} V_\mathcal{N}^\mathrm{T}DV_\mathcal{N} & V_\mathcal{N}^\mathrm{T}D\mathbf{1} \\ \mathbf{1}^\mathrm{T}DV_\mathcal{N} & \mathbf{1}^\mathrm{T}D\mathbf{1} \end{bmatrix} \in \mathbb{S}^N \tag{1180}$$

Because

$$W \triangleq \begin{bmatrix} V_\mathcal{N} & \mathbf{1} \end{bmatrix} \in \mathbb{R}^{N \times N} \tag{1181}$$

is full-rank, then (1597)

$$\text{inertia}(-D) = \text{inertia}(-W^\mathrm{T}DW) \tag{1182}$$

the congruence (1180) has the same number of positive, zero, and negative eigenvalues as $-D$. Further, if we denote by $\{\gamma_i\ ,\ i=1\ldots N-1\}$ the eigenvalues of $-V_{\mathcal{N}}^{\mathrm{T}}DV_{\mathcal{N}}$ and denote eigenvalues of the congruence $-W^{\mathrm{T}}DW$ by $\{\zeta_i\ ,\ i=1\ldots N\}$ and if we arrange each respective set of eigenvalues in nonincreasing order, then by theory of *interlacing eigenvalues for bordered symmetric matrices* [208, §4.3] [338, §6.4] [335, §IV.4.1]

$$\zeta_N \le \gamma_{N-1} \le \zeta_{N-1} \le \gamma_{N-2} \le \cdots \le \gamma_2 \le \zeta_2 \le \gamma_1 \le \zeta_1 \tag{1183}$$

When $D\in\mathbb{EDM}^N$, then $\gamma_i\ge 0\ \forall i$ (1531) because $-V_{\mathcal{N}}^{\mathrm{T}}DV_{\mathcal{N}}\succeq 0$ as we know. That means the congruence must have $N-1$ nonnegative eigenvalues; $\zeta_i\ge 0$, $i=1\ldots N-1$. The remaining eigenvalue ζ_N cannot be nonnegative because then $-D$ would be positive semidefinite, an impossibility; so $\zeta_N<0$. By congruence, nontrivial $-D$ must therefore have exactly one negative eigenvalue;[5.50] [116, §2.4.5]

$$D\in\mathbb{EDM}^N \Rightarrow \begin{cases} \lambda(-D)_i \ge 0, & i=1\ldots N-1 \\ \left(\sum_{i=1}^{N}\lambda(-D)_i = 0\right) \\ D\in\mathbb{S}_h^N\cap\mathbb{R}_+^{N\times N} \end{cases} \tag{1184}$$

where the $\lambda(-D)_i$ are nonincreasingly ordered eigenvalues of $-D$ whose sum must be 0 only because $\operatorname{tr}D=0$ [338, §5.1]. The eigenvalue summation condition, therefore, can be considered redundant. Even so, all these conditions are insufficient to determine whether some given $H\in\mathbb{S}_h^N$ is an EDM, as shown by counterexample.[5.51]

5.11.1.0.1   Exercise.   *Spectral inequality.*
Prove whether it holds: for $D=[d_{ij}]\in\mathbb{EDM}^N$

$$\lambda(-D)_1 \ge d_{ij} \ge \lambda(-D)_{N-1} \qquad \forall i\ne j \tag{1185}$$

▼

Terminology: a *spectral cone* is a convex cone containing all *eigenspectra* [233, p.365] [335, p.26] corresponding to some set of matrices. Any positive semidefinite matrix, for example, possesses a vector of nonnegative eigenvalues corresponding to an eigenspectrum contained in a spectral cone that is a nonnegative orthant.

[5.50] All entries of the corresponding eigenvector must have the same sign, with respect to each other, [91, p.116] because that eigenvector is the *Perron vector* corresponding to *spectral radius*; [208, §8.3.1] the predominant characteristic of square nonnegative matrices. Unlike positive semidefinite matrices, nonnegative matrices are guaranteed only to have at least one nonnegative eigenvalue.
[5.51] When $N=3$, for example, the symmetric hollow matrix

$$H = \begin{bmatrix} 0 & 1 & 1 \\ 1 & 0 & 5 \\ 1 & 5 & 0 \end{bmatrix} \in \mathbb{S}_h^N\cap\mathbb{R}_+^{N\times N}$$

is not an EDM, although $\lambda(-H) = [5\ \ 0.3723\ \ -5.3723]^{\mathrm{T}}$ conforms to (1184).

5.11.2 Spectral cones

Denoting the eigenvalues of Cayley-Menger matrix $\begin{bmatrix} 0 & \mathbf{1}^T \\ \mathbf{1} & -D \end{bmatrix} \in \mathbb{S}^{N+1}$ by

$$\lambda\left(\begin{bmatrix} 0 & \mathbf{1}^T \\ \mathbf{1} & -D \end{bmatrix}\right) \in \mathbb{R}^{N+1} \qquad (1186)$$

we have the Cayley-Menger form (§5.7.3.0.1) of necessary and sufficient conditions for $D \in \mathbb{EDM}^N$ from the literature: [190, §3][5.52] [75, §3] [116, §6.2] (*confer* (987) (963))

$$D \in \mathbb{EDM}^N \Leftrightarrow \left\{ \begin{array}{c} \lambda\left(\begin{bmatrix} 0 & \mathbf{1}^T \\ \mathbf{1} & -D \end{bmatrix}\right)_i \geq 0, \quad i=1\dots N \\ D \in \mathbb{S}_h^N \end{array} \right\} \Leftrightarrow \left\{ \begin{array}{c} -V_{\mathcal{N}}^T D V_{\mathcal{N}} \succeq 0 \\ D \in \mathbb{S}_h^N \end{array} \right. \qquad (1187)$$

These conditions say the Cayley-Menger form has one and only one negative eigenvalue. When D is an EDM, eigenvalues $\lambda\left(\begin{bmatrix} 0 & \mathbf{1}^T \\ \mathbf{1} & -D \end{bmatrix}\right)$ belong to that particular orthant in $\mathbb{R}^{N+1}$ having the $N+1^{\text{th}}$ coordinate as sole negative coordinate[5.53]:

$$\begin{bmatrix} \mathbb{R}_+^N \\ \mathbb{R}_- \end{bmatrix} = \text{cone}\{e_1, e_2, \cdots e_N, -e_{N+1}\} \qquad (1188)$$

5.11.2.1 Cayley-Menger *versus* Schoenberg

Connection to the Schoenberg criterion (987) is made when the Cayley-Menger form is further partitioned:

$$\begin{bmatrix} 0 & \mathbf{1}^T \\ \mathbf{1} & -D \end{bmatrix} = \begin{bmatrix} \begin{bmatrix} 0 & 1 \\ 1 & 0 \end{bmatrix} & \begin{bmatrix} \mathbf{1}^T \\ -D_{1,2:N} \end{bmatrix} \\ [\mathbf{1} \quad -D_{2:N,1}] & -D_{2:N,2:N} \end{bmatrix} \qquad (1189)$$

Matrix $D \in \mathbb{S}_h^N$ is an EDM if and only if the Schur complement of $\begin{bmatrix} 0 & 1 \\ 1 & 0 \end{bmatrix}$ (§A.4) in this partition is positive semidefinite; [17, §1] [223, §3] *id est*, (*confer* (1133))

$$D \in \mathbb{EDM}^N$$
$$\Leftrightarrow$$
$$-D_{2:N,2:N} - [\mathbf{1} \quad -D_{2:N,1}]\begin{bmatrix} 0 & 1 \\ 1 & 0 \end{bmatrix}\begin{bmatrix} \mathbf{1}^T \\ -D_{1,2:N} \end{bmatrix} = -2V_{\mathcal{N}}^T D V_{\mathcal{N}} \succeq 0 \qquad (1190)$$
$$\textbf{and}$$
$$D \in \mathbb{S}_h^N$$

Positive semidefiniteness of that Schur complement insures nonnegativity ($D \in \mathbb{R}_+^{N\times N}$, §5.8.1), whereas *complementary inertia* (1599) insures existence of that lone negative eigenvalue of the Cayley-Menger form.

Now we apply results from chapter 2 with regard to polyhedral cones and their duals.

[5.52]Recall: for $D \in \mathbb{S}_h^N$, $-V_{\mathcal{N}}^T D V_{\mathcal{N}} \succeq 0$ subsumes nonnegativity property 1 (§5.8.1).
[5.53]Empirically, all except one entry of the corresponding eigenvector have the same sign with respect to each other.

5.11.2.2   Ordered eigenspectra

Conditions (1187) specify eigenvalue membership to the smallest pointed polyhedral *spectral cone* for $\begin{bmatrix} 0 & \mathbf{1}^{\mathrm{T}} \\ \mathbf{1} & -\mathrm{EDM}^N \end{bmatrix}$:

$$
\begin{aligned}
\mathcal{K}_\lambda &\triangleq \{\zeta \in \mathbb{R}^{N+1} \mid \zeta_1 \geq \zeta_2 \geq \cdots \geq \zeta_N \geq 0 \geq \zeta_{N+1} \, , \ \mathbf{1}^{\mathrm{T}}\zeta = 0 \} \\
&= \mathcal{K}_{\mathcal{M}} \cap \begin{bmatrix} \mathbb{R}_+^N \\ \mathbb{R}_- \end{bmatrix} \cap \partial\mathcal{H} \\
&= \lambda\left(\begin{bmatrix} 0 & \mathbf{1}^{\mathrm{T}} \\ \mathbf{1} & -\mathrm{EDM}^N \end{bmatrix}\right)
\end{aligned}
\tag{1191}
$$

where

$$
\partial\mathcal{H} \triangleq \{\zeta \in \mathbb{R}^{N+1} \mid \mathbf{1}^{\mathrm{T}}\zeta = 0\}
\tag{1192}
$$

is a hyperplane through the origin, and $\mathcal{K}_{\mathcal{M}}$ is the monotone cone (§2.13.9.4.3, implying ordered eigenspectra) which is full-dimensional but is not pointed;

$$
\mathcal{K}_{\mathcal{M}} = \{\zeta \in \mathbb{R}^{N+1} \mid \zeta_1 \geq \zeta_2 \geq \cdots \geq \zeta_{N+1}\} \tag{435}
$$

$$
\mathcal{K}_{\mathcal{M}}^* = \{[\, e_1 - e_2 \quad e_2 - e_3 \quad \cdots \quad e_N - e_{N+1}\,]\, a \mid a \succeq 0\} \subset \mathbb{R}^{N+1} \tag{436}
$$

So because of the hyperplane,

$$
\dim \mathrm{aff}\, \mathcal{K}_\lambda = \dim \partial\mathcal{H} = N \tag{1193}
$$

indicating $\mathcal{K}_\lambda$ is not full-dimensional. Defining

$$
A \triangleq \begin{bmatrix} e_1^{\mathrm{T}} - e_2^{\mathrm{T}} \\ e_2^{\mathrm{T}} - e_3^{\mathrm{T}} \\ \vdots \\ e_N^{\mathrm{T}} - e_{N+1}^{\mathrm{T}} \end{bmatrix} \in \mathbb{R}^{N \times N+1}, \qquad
B \triangleq \begin{bmatrix} e_1^{\mathrm{T}} \\ e_2^{\mathrm{T}} \\ \vdots \\ e_N^{\mathrm{T}} \\ -e_{N+1}^{\mathrm{T}} \end{bmatrix} \in \mathbb{R}^{N+1 \times N+1} \tag{1194}
$$

we have the halfspace-description:

$$
\mathcal{K}_\lambda = \{\zeta \in \mathbb{R}^{N+1} \mid A\zeta \succeq 0, \ B\zeta \succeq 0, \ \mathbf{1}^{\mathrm{T}}\zeta = 0\} \tag{1195}
$$

From this and (443) we get a vertex-description for a pointed spectral cone that is not full-dimensional:

$$
\mathcal{K}_\lambda = \left\{ V_{\mathcal{N}} \left(\begin{bmatrix} \hat{A} \\ \hat{B} \end{bmatrix} V_{\mathcal{N}} \right)^\dagger b \mid b \succeq 0 \right\} \tag{1196}
$$

where $V_{\mathcal{N}} \in \mathbb{R}^{N+1 \times N}$, and where [*sic*]

$$
\hat{B} = e_N^{\mathrm{T}} \in \mathbb{R}^{1 \times N+1} \tag{1197}
$$

and

$$\hat{A} = \begin{bmatrix} e_1^{\mathrm{T}} - e_2^{\mathrm{T}} \\ e_2^{\mathrm{T}} - e_3^{\mathrm{T}} \\ \vdots \\ e_{N-1}^{\mathrm{T}} - e_N^{\mathrm{T}} \end{bmatrix} \in \mathbb{R}^{N-1 \times N+1} \tag{1198}$$

hold those rows of A and B corresponding to conically independent rows (§2.10) in $\begin{bmatrix} A \\ B \end{bmatrix} V_{\mathcal{N}}$.

Conditions (1187) can be equivalently restated in terms of a spectral cone for Euclidean distance matrices:

$$D \in \mathrm{EDM}^N \quad \Leftrightarrow \quad \begin{cases} \lambda\left(\begin{bmatrix} 0 & \mathbf{1}^{\mathrm{T}} \\ \mathbf{1} & -D \end{bmatrix} \right) \in \mathcal{K}_{\mathcal{M}} \cap \begin{bmatrix} \mathbb{R}_+^N \\ \mathbb{R}_- \end{bmatrix} \cap \partial\mathcal{H} \\ D \in \mathbb{S}_h^N \end{cases} \tag{1199}$$

Vertex-description of the dual spectral cone is, (313)

$$\begin{aligned} \mathcal{K}_\lambda^* &= \overline{\mathcal{K}_{\mathcal{M}}^* + \begin{bmatrix} \mathbb{R}_+^N \\ \mathbb{R}_- \end{bmatrix}^* + \partial\mathcal{H}^*} \subseteq \mathbb{R}^{N+1} \\ &= \left\{ \begin{bmatrix} A^{\mathrm{T}} & B^{\mathrm{T}} & \mathbf{1} & -\mathbf{1} \end{bmatrix} b \mid b \succeq 0 \right\} = \left\{ \begin{bmatrix} \hat{A}^{\mathrm{T}} & \hat{B}^{\mathrm{T}} & \mathbf{1} & -\mathbf{1} \end{bmatrix} a \mid a \succeq 0 \right\} \end{aligned} \tag{1200}$$

From (1196) and (444) we get a halfspace-description:

$$\mathcal{K}_\lambda^* = \{ y \in \mathbb{R}^{N+1} \mid (V_{\mathcal{N}}^{\mathrm{T}} [\hat{A}^{\mathrm{T}} \; \hat{B}^{\mathrm{T}}])^\dagger V_{\mathcal{N}}^{\mathrm{T}} y \succeq 0 \} \tag{1201}$$

This polyhedral dual spectral cone $\mathcal{K}_\lambda^*$ is closed, convex, full-dimensional because $\mathcal{K}_\lambda$ is pointed, but is not pointed because $\mathcal{K}_\lambda$ is not full-dimensional.

5.11.2.3 Unordered eigenspectra

Spectral cones are not unique; eigenspectra ordering can be rendered benign within a cone by presorting a vector of eigenvalues into nonincreasing order.[5.54] Then things simplify: Conditions (1187) now specify eigenvalue membership to the spectral cone

$$\begin{aligned} \lambda\left(\begin{bmatrix} 0 & \mathbf{1}^{\mathrm{T}} \\ \mathbf{1} & -\mathrm{EDM}^N \end{bmatrix} \right) &= \begin{bmatrix} \mathbb{R}_+^N \\ \mathbb{R}_- \end{bmatrix} \cap \partial\mathcal{H} \\ &= \{ \zeta \in \mathbb{R}^{N+1} \mid B\zeta \succeq 0, \; \mathbf{1}^{\mathrm{T}}\zeta = 0 \} \end{aligned} \tag{1202}$$

where B is defined in (1194), and $\partial\mathcal{H}$ in (1192). From (443) we get a vertex-description

---

[5.54]Eigenspectra ordering (represented by a cone having monotone description such as (1191)) becomes benign in (1413), for example, where projection of a given presorted vector on the nonnegative orthant in a subspace is equivalent to its projection on the monotone nonnegative cone in that same subspace; equivalence is a consequence of presorting.

for a pointed spectral cone not full-dimensional:

$$\lambda\left(\begin{bmatrix} 0 & \mathbf{1}^{\mathrm{T}} \\ \mathbf{1} & -\mathrm{EDM}^N \end{bmatrix}\right) = \left\{ V_{\mathcal{N}}(\tilde{B}V_{\mathcal{N}})^{\dagger} b \mid b \succeq 0 \right\}$$
$$= \left\{ \begin{bmatrix} I \\ -\mathbf{1}^{\mathrm{T}} \end{bmatrix} b \mid b \succeq 0 \right\} \tag{1203}$$

where $V_{\mathcal{N}} \in \mathbb{R}^{N+1 \times N}$ and

$$\tilde{B} \triangleq \begin{bmatrix} e_1^{\mathrm{T}} \\ e_2^{\mathrm{T}} \\ \vdots \\ e_N^{\mathrm{T}} \end{bmatrix} \in \mathbb{R}^{N \times N+1} \tag{1204}$$

holds only those rows of B corresponding to conically independent rows in $BV_{\mathcal{N}}$.

For presorted eigenvalues, (1187) can be equivalently restated

$$D \in \mathrm{EDM}^N \quad \Leftrightarrow \quad \begin{cases} \lambda\left(\begin{bmatrix} 0 & \mathbf{1}^{\mathrm{T}} \\ \mathbf{1} & -D \end{bmatrix}\right) \in \begin{bmatrix} \mathbb{R}_+^N \\ \mathbb{R}_- \end{bmatrix} \cap \partial\mathcal{H} \\ D \in \mathbb{S}_h^N \end{cases} \tag{1205}$$

Vertex-description of the dual spectral cone is, (313)

$$\lambda\left(\begin{bmatrix} 0 & \mathbf{1}^{\mathrm{T}} \\ \mathbf{1} & -\mathrm{EDM}^N \end{bmatrix}\right)^* = \begin{bmatrix} \mathbb{R}_+^N \\ \mathbb{R}_- \end{bmatrix} + \partial\mathcal{H}^* \subseteq \mathbb{R}^{N+1}$$
$$= \left\{ \begin{bmatrix} B^{\mathrm{T}} & \mathbf{1} & -\mathbf{1} \end{bmatrix} b \mid b \succeq 0 \right\} = \left\{ \begin{bmatrix} \tilde{B}^{\mathrm{T}} & \mathbf{1} & -\mathbf{1} \end{bmatrix} a \mid a \succeq 0 \right\} \tag{1206}$$

From (444) we get a halfspace-description:

$$\lambda\left(\begin{bmatrix} 0 & \mathbf{1}^{\mathrm{T}} \\ \mathbf{1} & -\mathrm{EDM}^N \end{bmatrix}\right)^* = \{ y \in \mathbb{R}^{N+1} \mid (V_{\mathcal{N}}^{\mathrm{T}}\tilde{B}^{\mathrm{T}})^{\dagger} V_{\mathcal{N}}^{\mathrm{T}} y \succeq 0 \}$$
$$= \{ y \in \mathbb{R}^{N+1} \mid [\, I \;\; -\mathbf{1} \,] y \succeq 0 \} \tag{1207}$$

This polyhedral dual spectral cone is closed, convex, full-dimensional but is not pointed. (Notice that any nonincreasingly ordered eigenspectrum belongs to this dual spectral cone.)

5.11.2.4 Dual cone *versus* dual spectral cone

An open question regards the relationship of convex cones and their duals to the corresponding spectral cones and their duals. A positive semidefinite cone, for example, is selfdual. Both the nonnegative orthant and the monotone nonnegative cone are spectral cones for it. When we consider the nonnegative orthant, then that spectral cone for the selfdual positive semidefinite cone is also selfdual.

5.12 List reconstruction

The term *metric multidimensional scaling*[5.55] [263] [108] [362] [106] [266] [91] refers to any reconstruction of a list $X \in \mathbb{R}^{n \times N}$ in Euclidean space from interpoint distance information, possibly incomplete (§6.7), ordinal (§5.13.2), or specified perhaps only by bounding-constraints (§5.4.2.2.12) [360]. Techniques for reconstruction are essentially methods for optimally embedding an unknown list of points, corresponding to given Euclidean distance data, in an affine subset of desired or minimum dimension. The oldest known precursor is called *principal component analysis* [171] which analyzes the correlation matrix (§5.9.1.0.1); [54, §22] a.k.a, *Karhunen-Loéve transform* in the digital signal processing literature.

Isometric reconstruction (§5.5.3) of point list X is best performed by eigenvalue decomposition of a Gram matrix; for then, numerical errors of factorization are easily spotted in the eigenvalues: Now we consider how rotation/reflection and translation invariance factor into a reconstruction.

5.12.1 x_1 at the origin. $V_{\mathcal{N}}$

At the stage of reconstruction, we have $D \in \mathbb{EDM}^N$ and wish to find a generating list (§2.3.2) for polyhedron $\mathcal{P} - \alpha$ by factoring Gram matrix $-V_{\mathcal{N}}^{\mathrm{T}} D V_{\mathcal{N}}$ (985) as in (1170). One way to factor $-V_{\mathcal{N}}^{\mathrm{T}} D V_{\mathcal{N}}$ is via diagonalization of symmetric matrices; [338, §5.6] [208] (§A.5.1, §A.3)

$$-V_{\mathcal{N}}^{\mathrm{T}} D V_{\mathcal{N}} \triangleq Q \Lambda Q^{\mathrm{T}} \qquad (1208)$$

$$Q \Lambda Q^{\mathrm{T}} \succeq 0 \;\Leftrightarrow\; \Lambda \succeq 0 \qquad (1209)$$

where $Q \in \mathbb{R}^{N-1 \times N-1}$ is an orthogonal matrix containing eigenvectors while $\Lambda \in \mathbb{S}^{N-1}$ is a diagonal matrix containing corresponding nonnegative eigenvalues ordered by nonincreasing value. From the diagonalization, identify the list using (1115);

$$-V_{\mathcal{N}}^{\mathrm{T}} D V_{\mathcal{N}} = 2 V_{\mathcal{N}}^{\mathrm{T}} X^{\mathrm{T}} X V_{\mathcal{N}} \triangleq Q \sqrt{\Lambda} \, Q_{\mathrm{p}}^{\mathrm{T}} Q_{\mathrm{p}} \sqrt{\Lambda} \, Q^{\mathrm{T}} \qquad (1210)$$

where $\sqrt{\Lambda} \, Q_{\mathrm{p}}^{\mathrm{T}} Q_{\mathrm{p}} \sqrt{\Lambda} \triangleq \Lambda = \sqrt{\Lambda}\sqrt{\Lambda}$ and where $Q_{\mathrm{p}} \in \mathbb{R}^{n \times N-1}$ is unknown as is its dimension n. Rotation/reflection is accounted for by Q_{p} yet only its first r columns are necessarily orthonormal.[5.56] Assuming membership to the unit simplex $y \in \mathcal{S}$ (1167), then point $p = X \sqrt{2} V_{\mathcal{N}} y = Q_{\mathrm{p}} \sqrt{\Lambda} \, Q^{\mathrm{T}} y$ in $\mathbb{R}^n$ belongs to the translated polyhedron

$$\mathcal{P} - x_1 \qquad (1211)$$

[5.55]Scaling [358] *means making a scale,* i.e., *a numerical representation of qualitative data. If the scale is multidimensional, it's* multidimensional scaling.   −Jan de Leeuw
A goal of multidimensional scaling is to find a low-dimensional representation of list X so that distances between its elements best fit a given set of measured pairwise *dissimilarities*. When data comprises measurable distances, then reconstruction is termed *metric multidimensional scaling*. In one dimension, N coordinates in X define the scale.

[5.56]Recall r signifies affine dimension. Q_{p} is not necessarily an orthogonal matrix. Q_{p} is constrained such that only its first r columns are necessarily orthonormal because there are only r nonzero eigenvalues in Λ when $-V_{\mathcal{N}}^{\mathrm{T}} D V_{\mathcal{N}}$ has rank r (§5.7.1.1). Remaining columns of Q_{p} are arbitrary.

whose generating list constitutes the columns of (1109)

$$
\begin{aligned}
\begin{bmatrix} \mathbf{0} & X\sqrt{2}V_{\mathcal{N}} \end{bmatrix} &= \begin{bmatrix} \mathbf{0} & Q_{\mathrm{p}}\sqrt{\Lambda}\,Q^{\mathrm{T}} \end{bmatrix} \in \mathbb{R}^{n \times N} \\
&= \begin{bmatrix} \mathbf{0} & x_2 - x_1 & x_3 - x_1 & \cdots & x_N - x_1 \end{bmatrix}
\end{aligned}
\tag{1212}
$$

The scaled auxiliary matrix $V_{\mathcal{N}}$ represents that translation. A simple choice for Q_{p} has n set to $N-1$; *id est*, $Q_{\mathrm{p}} = I$. Ideally, each member of the generating list has at most r nonzero entries; r being, affine dimension

$$
\operatorname{rank} V_{\mathcal{N}}^{\mathrm{T}} D V_{\mathcal{N}} = \operatorname{rank} Q_{\mathrm{p}}\sqrt{\Lambda}\,Q^{\mathrm{T}} = \operatorname{rank}\Lambda = r
\tag{1213}
$$

Each member then has at least $N-1-r$ zeros in its higher-dimensional coordinates because $r \leq N-1$. (1121) To truncate those zeros, choose n equal to affine dimension which is the smallest n possible because $XV_{\mathcal{N}}$ has rank $r \leq n$ (1117).[5.57] In that case, the simplest choice for Q_{p} is $\begin{bmatrix} I & \mathbf{0} \end{bmatrix}$ having dimension $r \times N-1$.

We may wish to verify the list (1212) found from the diagonalization of $-V_{\mathcal{N}}^{\mathrm{T}} D V_{\mathcal{N}}$. Because of rotation/reflection and translation invariance (§5.5), EDM D can be uniquely made from that list by calculating: (968)

$$
\mathbf{D}(X) = \mathbf{D}(X\begin{bmatrix} \mathbf{0} & \sqrt{2}V_{\mathcal{N}} \end{bmatrix}) = \mathbf{D}(Q_{\mathrm{p}}\begin{bmatrix} \mathbf{0} & \sqrt{\Lambda}\,Q^{\mathrm{T}} \end{bmatrix}) = \mathbf{D}(\begin{bmatrix} \mathbf{0} & \sqrt{\Lambda}\,Q^{\mathrm{T}} \end{bmatrix})
\tag{1214}
$$

This suggests a way to find EDM D given $-V_{\mathcal{N}}^{\mathrm{T}} D V_{\mathcal{N}}$ (*confer* (1093))

$$
D = \begin{bmatrix} 0 \\ \delta(-V_{\mathcal{N}}^{\mathrm{T}} D V_{\mathcal{N}}) \end{bmatrix}\mathbf{1}^{\mathrm{T}} + \mathbf{1}\begin{bmatrix} 0 & \delta(-V_{\mathcal{N}}^{\mathrm{T}} D V_{\mathcal{N}})^{\mathrm{T}} \end{bmatrix} - 2\begin{bmatrix} 0 & \mathbf{0}^{\mathrm{T}} \\ \mathbf{0} & -V_{\mathcal{N}}^{\mathrm{T}} D V_{\mathcal{N}} \end{bmatrix}
\tag{1089}
$$

5.12.2   0 geometric center. V

Alternatively we may perform reconstruction using auxiliary matrix V (§B.4.1) and Gram matrix $-VDV\frac{1}{2}$ (989) instead; to find a generating list for polyhedron

$$
\mathcal{P} - \alpha_c
\tag{1215}
$$

whose geometric center has been translated to the origin. Redimensioning diagonalization factors $Q, \Lambda \in \mathbb{R}^{N \times N}$ and unknown $Q_{\mathrm{p}} \in \mathbb{R}^{n \times N}$, (1116)

$$
-VDV = 2VX^{\mathrm{T}}XV \triangleq Q\sqrt{\Lambda}\,Q_{\mathrm{p}}^{\mathrm{T}}Q_{\mathrm{p}}\sqrt{\Lambda}\,Q^{\mathrm{T}} \triangleq Q\Lambda Q^{\mathrm{T}}
\tag{1216}
$$

where the geometrically centered generating list constitutes (*confer* (1212))

$$
\begin{aligned}
XV &= \tfrac{1}{\sqrt{2}}Q_{\mathrm{p}}\sqrt{\Lambda}\,Q^{\mathrm{T}} \in \mathbb{R}^{n \times N} \\
&= \begin{bmatrix} x_1 - \tfrac{1}{N}X\mathbf{1} & x_2 - \tfrac{1}{N}X\mathbf{1} & x_3 - \tfrac{1}{N}X\mathbf{1} & \cdots & x_N - \tfrac{1}{N}X\mathbf{1} \end{bmatrix}
\end{aligned}
\tag{1217}
$$

[5.57] If we write $Q^{\mathrm{T}} = \begin{bmatrix} q_1^{\mathrm{T}} \\ \vdots \\ q_{N-1}^{\mathrm{T}} \end{bmatrix}$ as rowwise eigenvectors, $\Lambda = \begin{bmatrix} \lambda_1 & & & & \mathbf{0} \\ & \ddots & & & \\ & & \lambda_r & & \\ & & & 0 & \\ & & & & \ddots \\ \mathbf{0} & & & & 0 \end{bmatrix}$ in terms of

eigenvalues, and $Q_{\mathrm{p}} = \begin{bmatrix} q_{\mathrm{p}1} & \cdots & q_{\mathrm{p}N-1} \end{bmatrix}$ as column vectors, then $Q_{\mathrm{p}}\sqrt{\Lambda}\,Q^{\mathrm{T}} = \sum\limits_{i=1}^{r} \sqrt{\lambda_i}\, q_{\mathrm{p}i}q_i^{\mathrm{T}}$ is a sum of r linearly independent rank-one matrices (§B.1.1). Hence the summation has rank r.

where $\alpha_c = \frac{1}{N} X \mathbf{1}$. (§5.5.1.0.1) The simplest choice for Q_p is $[\,I \quad \mathbf{0}\,] \in \mathbb{R}^{r \times N}$.

Now EDM D can be uniquely made from the list found: (968)

$$\mathbf{D}(X) = \mathbf{D}(XV) = \mathbf{D}(\frac{1}{\sqrt{2}} Q_\mathrm{p} \sqrt{\Lambda} Q^\mathrm{T}) = \mathbf{D}(\sqrt{\Lambda} Q^\mathrm{T}) \frac{1}{2} \qquad (1218)$$

This EDM is, of course, identical to (1214). Similarly to (1089), from $-VDV$ we can find EDM D (*confer* (1080))

$$D = \delta(-VDV\tfrac{1}{2})\mathbf{1}^\mathrm{T} + \mathbf{1}\delta(-VDV\tfrac{1}{2})^\mathrm{T} - 2(-VDV\tfrac{1}{2}) \qquad (1079)$$

5.13 Reconstruction examples

5.13.1 Isometric reconstruction

5.13.1.0.1 Example. *Cartography.*
The most fundamental application of EDMs is to reconstruct relative point position given only interpoint distance information. Drawing a map of the United States is a good illustration of isometric reconstruction (§5.4.2.2.10) from complete distance data. We obtained latitude and longitude information for the coast, border, states, and Great Lakes from the *usalo atlas data file* within MATLAB Mapping Toolbox; conversion to Cartesian coordinates (x, y, z) via:

$$
\begin{aligned}
\phi &\triangleq \pi/2 - \text{latitude} \\
\theta &\triangleq \text{longitude} \\
x &= \sin(\phi)\cos(\theta) \\
y &= \sin(\phi)\sin(\theta) \\
z &= \cos(\phi)
\end{aligned}
\qquad (1219)
$$

We used 64% of the available map data to calculate EDM D from $N = 5020$ points. The original (decimated) data and its isometric reconstruction via (1210) are shown in Figure 145a-d. [389, MATLAB code] The eigenvalues computed for (1208) are

$$\lambda(-V_\mathcal{N}^\mathrm{T} D V_\mathcal{N}) = [199.8 \quad 152.3 \quad 2.465 \quad 0 \quad 0 \quad 0 \quad \cdots \;]^\mathrm{T} \qquad (1220)$$

The 0 eigenvalues have absolute numerical error on the order of 2E-13; meaning, the EDM data indicates three dimensions ($r = 3$) are required for reconstruction to nearly machine precision. $\qquad \square$

5.13.2 Isotonic reconstruction

Sometimes only comparative information about distance is known (Earth is closer to the Moon than it is to the Sun). Suppose, for example, EDM D for three points is unknown:

$$
D = [d_{ij}] = \begin{bmatrix} 0 & d_{12} & d_{13} \\ d_{12} & 0 & d_{23} \\ d_{13} & d_{23} & 0 \end{bmatrix} \in \mathbb{S}_h^3 \qquad (957)
$$

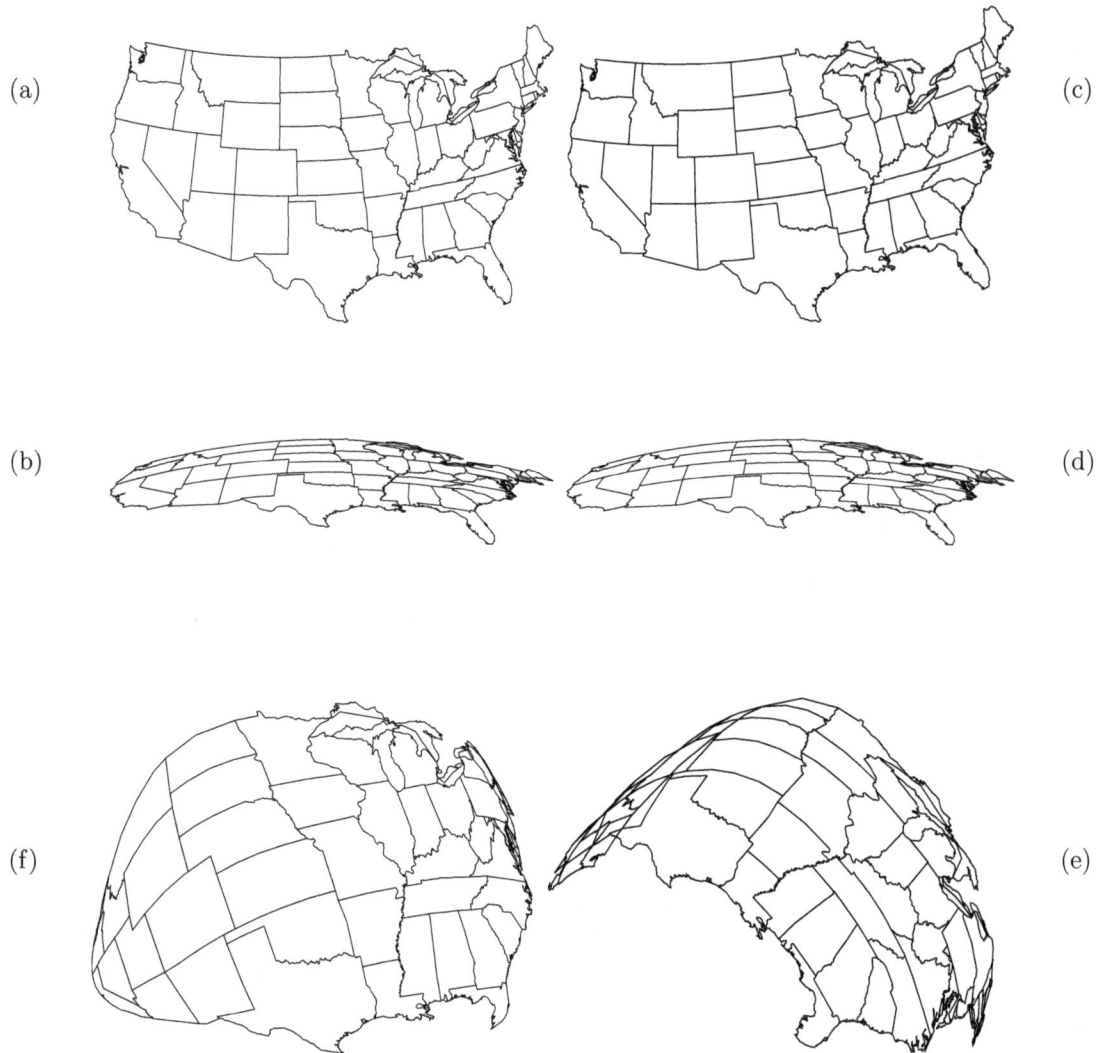

Figure 145: Map of United States of America showing some state boundaries and the Great Lakes. All plots made by connecting 5020 points. Any difference in scale in (a) through (d) is artifact of plotting routine.
(a) Shows original map made from decimated (latitude, longitude) data.
(b) Original map data rotated (freehand) to highlight curvature of Earth.
(c) Map isometrically reconstructed from an EDM (from distance only).
(d) Same reconstructed map illustrating curvature.
(e)(f) Two views of one isotonic reconstruction (from comparative distance); problem (1228) with no sort constraint $\Pi\underline{d}$ (and no hidden line removal).

but comparative distance data is available:

$$d_{13} \geq d_{23} \geq d_{12} \qquad (1221)$$

With vectorization $\underline{d} = [d_{12} \ d_{13} \ d_{23}]^{\mathrm{T}} \in \mathbb{R}^{\mathbf{3}}$, we express the comparative data as the nonincreasing sorting

$$\Pi \underline{d} \;=\; \begin{bmatrix} 0 & 1 & 0 \\ 0 & 0 & 1 \\ 1 & 0 & 0 \end{bmatrix} \begin{bmatrix} d_{12} \\ d_{13} \\ d_{23} \end{bmatrix} \;=\; \begin{bmatrix} d_{13} \\ d_{23} \\ d_{12} \end{bmatrix} \in \mathcal{K}_{\mathcal{M}+} \qquad (1222)$$

where Π is a given permutation matrix expressing known sorting action on the entries of unknown EDM D, and $\mathcal{K}_{\mathcal{M}+}$ is the monotone nonnegative cone (§2.13.9.4.2)

$$\mathcal{K}_{\mathcal{M}+} = \{z \mid z_1 \geq z_2 \geq \cdots \geq z_{N(N-1)/2} \geq 0\} \subseteq \mathbb{R}_+^{N(N-1)/2} \qquad (428)$$

where $N(N-1)/2 = 3$ for the present example. From sorted vectorization (1222) we create the *sort-index matrix*

$$O \;=\; \begin{bmatrix} 0 & 1^2 & 3^2 \\ 1^2 & 0 & 2^2 \\ 3^2 & 2^2 & 0 \end{bmatrix} \in \mathbb{S}_h^{\mathbf{3}} \cap \mathbb{R}_+^{\mathbf{3 \times 3}} \qquad (1223)$$

generally defined

$$O_{ij} \triangleq k^2 \mid d_{ij} = (\Xi \Pi \underline{d})_k \,, \qquad j \neq i \qquad (1224)$$

where Ξ is a permutation matrix (1819) completely reversing order of vector entries.

Replacing EDM data with indices-square of a nonincreasing sorting like this is, of course, a heuristic we invented and may be regarded as a nonlinear introduction of much noise into the Euclidean distance matrix. For large data sets, this heuristic makes an otherwise intense problem computationally tractable; we see an example in relaxed problem (1229).

Any process of reconstruction that leaves comparative distance information intact is called *ordinal multidimensional scaling* or *isotonic reconstruction*. Beyond rotation, reflection, and translation error, (§5.5) list reconstruction by isotonic reconstruction is subject to error in absolute scale (*dilation*) and distance ratio. Yet Borg & Groenen argue: [54, §2.2] reconstruction from complete comparative distance information for a large number of points is as highly constrained as reconstruction from an EDM; the larger the number, the smaller the optimal solution set; whereas,

$$\text{isotonic solution set} \supseteq \text{isometric solution set} \qquad (1225)$$

5.13.2.1 Isotonic cartography

To test Borg & Groenen's conjecture, suppose we make a complete sort-index matrix $O \in \mathbb{S}_h^N \cap \mathbb{R}_+^{N \times N}$ for the map of the USA and then substitute O in place of EDM D in the reconstruction process of §5.12. Whereas EDM D returned only three significant eigenvalues (1220), the sort-index matrix O is generally not an EDM (certainly not an

$\lambda(-V_{\mathcal{N}}^{\mathrm{T}}OV_{\mathcal{N}})_j$

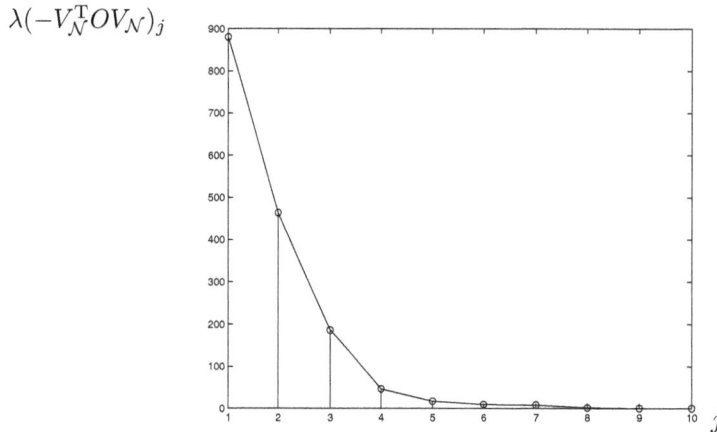

Figure 146: Largest ten eigenvalues, of $-V_{\mathcal{N}}^{\mathrm{T}}OV_{\mathcal{N}}$ for map of USA, sorted by nonincreasing value.

EDM with corresponding affine dimension 3) so returns many more. The eigenvalues, calculated with absolute numerical error approximately **5E-7**, are plotted in Figure **146**:

$$\lambda(-V_{\mathcal{N}}^{\mathrm{T}}OV_{\mathcal{N}}) = [880.1 \quad 463.9 \quad 186.1 \quad 46.20 \quad 17.12 \quad 9.625 \quad 8.257 \quad 1.701 \quad 0.7128 \quad 0.6460 \;\cdots\;]^{\mathrm{T}}$$
$$(1226)$$

The extra eigenvalues indicate that affine dimension corresponding to an EDM near O is likely to exceed 3. To realize the map, we must simultaneously reduce that dimensionality and find an EDM D closest to O in some sense[5.58] while maintaining the known comparative distance relationship. For example: given permutation matrix Π expressing the known sorting action like (1222) on entries

$$\underline{d} \triangleq \frac{1}{\sqrt{2}} \operatorname{dvec} D = \begin{bmatrix} d_{12} \\ d_{13} \\ d_{23} \\ d_{14} \\ d_{24} \\ d_{34} \\ \vdots \\ d_{N-1,N} \end{bmatrix} \in \mathbb{R}^{N(N-1)/2} \qquad (1227)$$

of unknown $D \in \mathbb{S}_h^N$, we can make sort-index matrix O input to the optimization problem

$$\begin{array}{ll} \underset{D}{\text{minimize}} & \|-V_{\mathcal{N}}^{\mathrm{T}}(D-O)V_{\mathcal{N}}\|_{\mathrm{F}} \\ \text{subject to} & \operatorname{rank} V_{\mathcal{N}}^{\mathrm{T}}DV_{\mathcal{N}} \le 3 \\ & \Pi\underline{d} \in \mathcal{K}_{\mathcal{M}+} \\ & D \in \mathbb{EDM}^N \end{array} \qquad (1228)$$

[5.58] a problem explored more in §7.

that finds the EDM D (corresponding to affine dimension not exceeding 3 in isomorphic dvec $\mathbb{EDM}^N \cap \Pi^T \mathcal{K}_{\mathcal{M}+}$) closest to O in the sense of Schoenberg (987).

Analytical solution to this problem, ignoring the sort constraint $\Pi \underline{d} \in \mathcal{K}_{\mathcal{M}+}$, is known [362]: we get the convex optimization [*sic*] (§7.1)

$$
\begin{aligned}
&\underset{D}{\text{minimize}} && \|-V_\mathcal{N}^T(D-O)V_\mathcal{N}\|_F \\
&\text{subject to} && \operatorname{rank} V_\mathcal{N}^T D V_\mathcal{N} \leq 3 \\
&&& D \in \mathbb{EDM}^N
\end{aligned}
\tag{1229}
$$

Only the three largest nonnegative eigenvalues in (1226) need be retained to make list (1212); the rest are discarded. The reconstruction from EDM D found in this manner is plotted in Figure 145e-f. (In the MATLAB code on $\mathcal{W}_{\imath\kappa\imath}$*mization* [382], matrix O is normalized by $(N(N-1)/2)^2$.) From these plots it becomes obvious that inclusion of the sort constraint is necessary for isotonic reconstruction.

That sort constraint demands: any optimal solution $D^\star$ must possess the known comparative distance relationship that produces the original ordinal distance data O (1224). Ignoring the sort constraint, apparently, violates it. Yet even more remarkable is how much the map reconstructed using only ordinal data still resembles the original map of the USA after suffering the many violations produced by solving relaxed problem (1229). This suggests the simple reconstruction techniques of §5.12 are robust to a significant amount of noise.

5.13.2.2 Isotonic solution with sort constraint

Because problems involving rank are generally difficult, we will partition (1228) into two problems we know how to solve and then alternate their solution until convergence:

$$
\begin{aligned}
&\underset{D}{\text{minimize}} && \|-V_\mathcal{N}^T(D-O)V_\mathcal{N}\|_F \\
&\text{subject to} && \operatorname{rank} V_\mathcal{N}^T D V_\mathcal{N} \leq 3 &&& \text{(a)} \\
&&& D \in \mathbb{EDM}^N
\end{aligned}
$$

$$
\begin{aligned}
&\underset{\sigma}{\text{minimize}} && \|\sigma - \Pi \underline{d}\| \\
&\text{subject to} && \sigma \in \mathcal{K}_{\mathcal{M}+} &&& \text{(b)}
\end{aligned}
\tag{1230}
$$

where sort-index matrix O (a given constant in (a)) becomes an implicit vector variable ϱ_i solving the ith instance of (1230b)

$$
\frac{1}{\sqrt{2}} \operatorname{dvec} O_i = \varrho_i \triangleq \Pi^T \sigma^\star \in \mathbb{R}^{N(N-1)/2} , \qquad i \in \{1,2,3\ldots\}
\tag{1231}
$$

As mentioned in discussion of relaxed problem (1229), a closed-form solution to problem (1230a) exists. Only the first iteration of (1230a) sees the original sort-index matrix O whose entries are nonnegative whole numbers; *id est,* $O_0 = O \in \mathbb{S}_h^N \cap \mathbb{R}_+^{N \times N}$ (1224). Subsequent iterations i take the previous solution of (1230b) as input

$$
O_i = \operatorname{dvec}^{-1}(\sqrt{2}\,\varrho_i) \in \mathbb{S}^N
\tag{1232}
$$

real successors, estimating distance-square not order, to the sort-index matrix O.

New convex problem (1230b) finds the unique minimum-distance projection of $\Pi\underline{d}$ on the monotone nonnegative cone $\mathcal{K}_{\mathcal{M}+}$. By defining

$$Y^{\dagger\mathrm{T}} = [e_1 - e_2 \quad e_2 - e_3 \quad e_3 - e_4 \quad \cdots \quad e_m] \in \mathbb{R}^{m \times m} \qquad (429)$$

where $m \triangleq N(N-1)/2$, we may rewrite (1230b) as an equivalent quadratic program; a convex problem in terms of the halfspace-description of $\mathcal{K}_{\mathcal{M}+}$:

$$\begin{aligned}
&\underset{\sigma}{\text{minimize}} && (\sigma - \Pi\underline{d})^{\mathrm{T}}(\sigma - \Pi\underline{d}) \\
&\text{subject to} && Y^{\dagger}\sigma \succeq 0
\end{aligned} \qquad (1233)$$

This quadratic program can be converted to a semidefinite program via Schur-form (§3.5.2); we get the equivalent problem

$$\begin{aligned}
&\underset{t \in \mathbb{R},\, \sigma}{\text{minimize}} && t \\
&\text{subject to} && \begin{bmatrix} tI & \sigma - \Pi\underline{d} \\ (\sigma - \Pi\underline{d})^{\mathrm{T}} & 1 \end{bmatrix} \succeq 0 \\
& && Y^{\dagger}\sigma \succeq 0
\end{aligned} \qquad (1234)$$

5.13.2.3 Convergence

In §E.10 we discuss convergence of alternating projection on intersecting convex sets in a Euclidean vector space; convergence to a point in their intersection. Here the situation is different for two reasons:

Firstly, sets of positive semidefinite matrices having an upper bound on rank are generally not convex. Yet in §7.1.4.0.1 we prove (1230a) is equivalent to a projection of nonincreasingly ordered eigenvalues on a subset of the nonnegative orthant:

$$\begin{aligned}
&\underset{D}{\text{minimize}} && \|-V_{\mathcal{N}}^{\mathrm{T}}(D-O)V_{\mathcal{N}}\|_{\mathrm{F}} & &\underset{\Upsilon}{\text{minimize}} && \|\Upsilon - \Lambda\|_{\mathrm{F}} \\
&\text{subject to} && \text{rank}\, V_{\mathcal{N}}^{\mathrm{T}}DV_{\mathcal{N}} \le 3 & \equiv &\text{subject to} && \delta(\Upsilon) \in \begin{bmatrix} \mathbb{R}_+^3 \\ \mathbf{0} \end{bmatrix} \\
& && D \in \mathbb{EDM}^N
\end{aligned} \qquad (1235)$$

where $-V_{\mathcal{N}}^{\mathrm{T}}DV_{\mathcal{N}} \triangleq U\Upsilon U^{\mathrm{T}} \in \mathbb{S}^{N-1}$ and $-V_{\mathcal{N}}^{\mathrm{T}}OV_{\mathcal{N}} \triangleq Q\Lambda Q^{\mathrm{T}} \in \mathbb{S}^{N-1}$ are ordered diagonalizations (§A.5). It so happens: optimal orthogonal $U^\star$ always equals Q given. Linear operator $T(A) = U^{\star\mathrm{T}}AU^\star$, acting on square matrix A, is an isometry because Frobenius' norm is orthogonally invariant (48). This isometric isomorphism T thus maps a nonconvex problem to a convex one that preserves distance.

Secondly, the second half (1230b) of the *alternation* takes place in a different vector space; $\mathbb{S}_h^N$ (*versus* $\mathbb{S}^{N-1}$). From §5.6 we know these two vector spaces are related by an isomorphism, $\mathbb{S}^{N-1} = \mathbf{V}_{\mathcal{N}}(\mathbb{S}_h^N)$ (1098), but not by an isometry.

We have, therefore, no guarantee from theory of alternating projection that the alternation (1230) converges to a point, in the set of all EDMs corresponding to affine dimension not in excess of 3, belonging to $\text{dvec}\,\mathbb{EDM}^N \cap \Pi^{\mathrm{T}}\mathcal{K}_{\mathcal{M}+}$.

5.13.2.4 Interlude

Map reconstruction from comparative distance data, isotonic reconstruction, would also prove invaluable to stellar cartography where absolute interstellar distance is difficult to acquire. But we have not yet implemented the second half (1233) of alternation (1230) for USA map data because memory-demands exceed capability of our computer.

5.13.2.4.1 Exercise. *Convergence of isotonic solution by alternation.*
Empirically demonstrate convergence, discussed in §5.13.2.3, on a smaller data set.

▼

It would be remiss not to mention another method of solution to this isotonic reconstruction problem: Once again we assume only comparative distance data like (1221) is available. Given known set of indices $\mathcal{I}$

$$
\begin{array}{ll}
\underset{D}{\text{minimize}} & \text{rank}\, V D V \\
\text{subject to} & d_{ij} \leq d_{kl} \leq d_{mn} \qquad \forall (i,j,k,l,m,n) \in \mathcal{I} \\
& D \in \mathbb{EDM}^N
\end{array}
\qquad (1236)
$$

this problem minimizes affine dimension while finding an EDM whose entries satisfy known comparative relationships. Suitable rank heuristics are discussed in §4.4.1 and §7.2.2 that will transform this to a convex optimization problem.

Using contemporary computers, even with a rank heuristic in place of the objective function, this problem formulation is more difficult to compute than the relaxed counterpart problem (1229). That is because there exist efficient algorithms to compute a selected few eigenvalues and eigenvectors from a very large matrix. Regardless, it is important to recognize: the optimal solution set for this problem (1236) is practically always different from the optimal solution set for its counterpart, problem (1228).

5.14 Fifth property of Euclidean metric

We continue now with the question raised in §5.3 regarding necessity for at least one requirement more than the four properties of the Euclidean metric (§5.2) to certify realizability of a bounded convex polyhedron or to reconstruct a generating list for it from incomplete distance information. There we saw that the four Euclidean metric properties are necessary for $D \in \mathbb{EDM}^N$ in the case $N = 3$, but become insufficient when cardinality N exceeds 3 (regardless of affine dimension).

5.14.1 Recapitulate

In the particular case $N = 3$, $-V_{\mathcal{N}}^{\mathrm{T}} D V_{\mathcal{N}} \succeq 0$ (1139) and $D \in \mathbb{S}_h^3$ are necessary and sufficient conditions for D to be an EDM. By (1141), triangle inequality is then the only Euclidean condition bounding the necessarily nonnegative d_{ij} ; and those bounds are tight. That means the first four properties of the Euclidean metric are necessary and sufficient conditions for D to be an EDM in the case $N=3$; for $i, j \in \{1, 2, 3\}$

$$
\left.
\begin{aligned}
\sqrt{d_{ij}} &\geq 0, \quad i \neq j \\
\sqrt{d_{ij}} &= 0, \quad i = j \\
\sqrt{d_{ij}} &= \sqrt{d_{ji}} \\
\sqrt{d_{ij}} &\leq \sqrt{d_{ik}} + \sqrt{d_{kj}}, \quad i \neq j \neq k
\end{aligned}
\right\}
\quad \Leftrightarrow \quad
\begin{aligned}
-V_{\mathcal{N}}^{\mathrm{T}} D V_{\mathcal{N}} &\succeq 0 \\
D &\in \mathbb{S}_h^3
\end{aligned}
\quad \Leftrightarrow \quad D \in \mathbb{EDM}^3 \qquad (1237)
$$

Yet those four properties become insufficient when $N > 3$.

5.14.2   Derivation of the fifth

Correspondence between the triangle inequality and the EDM was developed in §5.8.2 where a triangle inequality (1141a) was revealed within the leading principal 2×2 submatrix of $-V_{\mathcal{N}}^{\mathrm{T}} D V_{\mathcal{N}}$ when positive semidefinite. Our choice of the leading principal submatrix was arbitrary; actually, a unique triangle inequality like (1036) corresponds to any one of the $(N-1)! / (2!(N-1-2)!)$ principal 2×2 submatrices.[5.59] Assuming $D \in \mathbb{S}_h^4$ and $-V_{\mathcal{N}}^{\mathrm{T}} D V_{\mathcal{N}} \in \mathbb{S}^3$, then by the *positive (semi)definite principal submatrices theorem* (§A.3.1.0.4) it is sufficient to prove: all d_{ij} are nonnegative, all triangle inequalities are satisfied, and $\det(-V_{\mathcal{N}}^{\mathrm{T}} D V_{\mathcal{N}})$ is nonnegative. When $N = 4$, in other words, that nonnegative determinant becomes the fifth and last Euclidean metric requirement for $D \in \mathbb{EDM}^N$. We now endeavor to ascribe geometric meaning to it.

5.14.2.1   Nonnegative determinant

By (1042) when $D \in \mathbb{EDM}^4$, $-V_{\mathcal{N}}^{\mathrm{T}} D V_{\mathcal{N}}$ is equal to inner product (1037),

$$
\Theta^{\mathrm{T}} \Theta =
\begin{bmatrix}
d_{12} & \sqrt{d_{12} d_{13}} \cos \theta_{213} & \sqrt{d_{12} d_{14}} \cos \theta_{214} \\
\sqrt{d_{12} d_{13}} \cos \theta_{213} & d_{13} & \sqrt{d_{13} d_{14}} \cos \theta_{314} \\
\sqrt{d_{12} d_{14}} \cos \theta_{214} & \sqrt{d_{13} d_{14}} \cos \theta_{314} & d_{14}
\end{bmatrix}
\qquad (1238)
$$

Because Euclidean space is an inner-product space, the more concise inner-product form of the determinant is admitted;

$$
\det(\Theta^{\mathrm{T}} \Theta) = -d_{12} d_{13} d_{14} \left(\cos(\theta_{213})^2 + \cos(\theta_{214})^2 + \cos(\theta_{314})^2 - 2 \cos \theta_{213} \cos \theta_{214} \cos \theta_{314} - 1 \right) \qquad (1239)
$$

The determinant is nonnegative if and only if

$$
\cos \theta_{214} \cos \theta_{314} - \sqrt{\sin(\theta_{214})^2 \sin(\theta_{314})^2} \leq \cos \theta_{213} \leq \cos \theta_{214} \cos \theta_{314} + \sqrt{\sin(\theta_{214})^2 \sin(\theta_{314})^2}
$$
$$
\Leftrightarrow
$$
$$
\cos \theta_{213} \cos \theta_{314} - \sqrt{\sin(\theta_{213})^2 \sin(\theta_{314})^2} \leq \cos \theta_{214} \leq \cos \theta_{213} \cos \theta_{314} + \sqrt{\sin(\theta_{213})^2 \sin(\theta_{314})^2}
$$
$$
\Leftrightarrow
$$
$$
\cos \theta_{213} \cos \theta_{214} - \sqrt{\sin(\theta_{213})^2 \sin(\theta_{214})^2} \leq \cos \theta_{314} \leq \cos \theta_{213} \cos \theta_{214} + \sqrt{\sin(\theta_{213})^2 \sin(\theta_{214})^2}
$$
$$
(1240)
$$

[5.59]There are fewer principal 2×2 submatrices in $-V_{\mathcal{N}}^{\mathrm{T}} D V_{\mathcal{N}}$ than there are triangles made by four or more points because there are $N! / (3!(N-3)!)$ triangles made by point triples. The triangles corresponding to those submatrices all have vertex x_1. (*confer* §5.8.2.1)

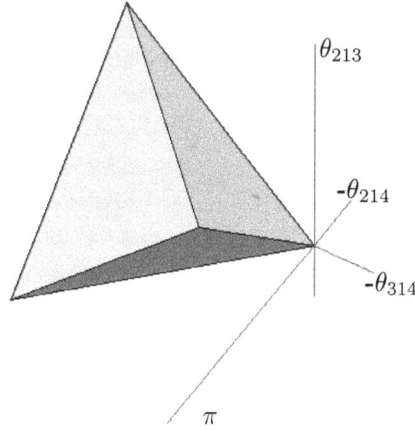

Figure 147: The *relative-angle inequality tetrahedron* (1243) bounding $\mathbb{EDM}^4$ is regular; drawn in entirety. Each angle θ (1034) must belong to this solid to be realizable.

which simplifies, for $0 \le \theta_{i1\ell}, \theta_{\ell 1j}, \theta_{i1j} \le \pi$ and all $i \ne j \ne \ell \in \{2,3,4\}$, to

$$\cos(\theta_{i1\ell} + \theta_{\ell 1j}) \ \le \ \cos\theta_{i1j} \ \le \ \cos(\theta_{i1\ell} - \theta_{\ell 1j}) \tag{1241}$$

Analogously to triangle inequality (1153), the determinant is 0 upon equality on either side of (1241) which is tight. Inequality (1241) can be equivalently written linearly as a triangle inequality between relative angles [421, §1.4];

$$\begin{aligned} |\theta_{i1\ell} - \theta_{\ell 1j}| \ &\le \ \theta_{i1j} \ \le \ \theta_{i1\ell} + \theta_{\ell 1j} \\ \theta_{i1\ell} + \theta_{\ell 1j} &+ \theta_{i1j} \le 2\pi \\ 0 \le \theta_{i1\ell}, &\theta_{\ell 1j}, \theta_{i1j} \le \pi \end{aligned} \tag{1242}$$

Generalizing this:

5.14.2.1.1 Fifth property of Euclidean metric - restatement.
Relative-angle inequality.   (*confer* §5.3.1.0.1) [50] [51, p.17, p.107] [241, §3.1]
Augmenting the four fundamental Euclidean metric properties in $\mathbb{R}^n$, for all $i,j,\ell \ne k \in \{1 \ldots N\}$, $i < j < \ell$, and for $N \ge 4$ distinct points $\{x_k\}$, the inequalities

$$\begin{aligned} |\theta_{ik\ell} - \theta_{\ell kj}| \ &\le \ \theta_{ikj} \ \le \ \theta_{ik\ell} + \theta_{\ell kj} &\text{(a)} \\ \theta_{ik\ell} + \theta_{\ell kj} &+ \theta_{ikj} \le 2\pi &\text{(b)} \\ 0 \le \theta_{ik\ell}, &\theta_{\ell kj}, \theta_{ikj} \le \pi &\text{(c)} \end{aligned} \tag{1243}$$

where $\theta_{ikj} = \theta_{jki}$ is the angle between vectors at vertex x_k (as defined in (1034) and illustrated in Figure 128), must be satisfied at each point x_k regardless of affine dimension. $\diamond$

Because point labelling is arbitrary, this fifth Euclidean metric requirement must apply to each of the N points as though each were in turn labelled x_1; hence the new index k in (1243). Just as the triangle inequality is the ultimate test for realizability of only three points, the relative-angle inequality is the ultimate test for only four. For four distinct points, the triangle inequality remains a necessary although penultimate test; (§5.4.3)

$$\begin{matrix}\text{Four Euclidean metric properties (§5.2).} \\ \text{Angle } \theta \text{ inequality (962) or (1243).}\end{matrix} \quad \Leftrightarrow \quad \begin{matrix}-V_{\mathcal{N}}^{\mathrm{T}} D V_{\mathcal{N}} \succeq 0 \\ D \in \mathbb{S}_h^4\end{matrix} \quad \Leftrightarrow \quad D = \mathbf{D}(\Theta) \in \mathbb{EDM}^4 \quad (1244)$$

The relative-angle inequality, for this case, is illustrated in Figure 147.

5.14.2.2　Beyond the fifth metric property

When cardinality N exceeds 4, the first four properties of the Euclidean metric and the relative-angle inequality together become insufficient conditions for realizability. In other words, the four Euclidean metric properties and relative-angle inequality remain necessary but become a sufficient test only for positive semidefiniteness of all the principal 3×3 submatrices [*sic*] in $-V_{\mathcal{N}}^{\mathrm{T}} D V_{\mathcal{N}}$. Relative-angle inequality can be considered the ultimate test only for realizability at each vertex x_k of each and every purported tetrahedron constituting a hyperdimensional body.

When $N = 5$ in particular, relative-angle inequality becomes the penultimate Euclidean metric requirement while nonnegativity of then unwieldy $\det(\Theta^{\mathrm{T}}\Theta)$ corresponds (by the *positive (semi)definite principal submatrices theorem* in §A.3.1.0.4) to the sixth and last Euclidean metric requirement. Together these six tests become necessary and sufficient, and so on.

Yet for all values of N, only assuming nonnegative d_{ij}, relative-angle matrix inequality in (1155) is necessary and sufficient to certify realizability; (§5.4.3.1)

$$\begin{matrix}\text{Euclidean metric property 1 (§5.2).} \\ \text{Angle matrix inequality } \Omega \succeq 0 \text{ (1043).}\end{matrix} \quad \Leftrightarrow \quad \begin{matrix}-V_{\mathcal{N}}^{\mathrm{T}} D V_{\mathcal{N}} \succeq 0 \\ D \in \mathbb{S}_h^N\end{matrix} \quad \Leftrightarrow \quad D = \mathbf{D}(\Omega, d) \in \mathbb{EDM}^N \quad (1245)$$

Like matrix criteria (963), (987), and (1155), the relative-angle matrix inequality and nonnegativity property subsume all the Euclidean metric properties and further requirements.

5.14.3　Path not followed

As a means to test for realizability of four or more points, an intuitively appealing way to augment the four Euclidean metric properties is to recognize generalizations of the triangle inequality: In the case of cardinality $N = 4$, the three-dimensional analogue to triangle & distance is tetrahedron & facet-area, while in case $N = 5$ the four-dimensional analogue is polychoron & facet-volume, *ad infinitum*. For N points, $N + 1$ metric properties are required.

5.14.3.1 $N = 4$

Each of the four facets of a general tetrahedron is a triangle and its relative interior. Suppose we identify each facet of the tetrahedron by its area-square: c_1, c_2, c_3, c_4. Then analogous to metric property 4, we may write a tight[5.60] area inequality for the facets

$$\sqrt{c_i} \leq \sqrt{c_j} + \sqrt{c_k} + \sqrt{c_\ell}, \quad i \neq j \neq k \neq \ell \in \{1, 2, 3, 4\} \tag{1246}$$

which is a generalized "triangle" inequality [233, §1.1] that follows from

$$\sqrt{c_i} = \sqrt{c_j} \cos\varphi_{ij} + \sqrt{c_k} \cos\varphi_{ik} + \sqrt{c_\ell} \cos\varphi_{i\ell} \tag{1247}$$

[248] [401, *Law of Cosines*] where φ_{ij} is the dihedral angle at the common edge between triangular facets i and j.

If D is the EDM corresponding to the whole tetrahedron, then area-square of the ith triangular facet has a convenient formula in terms of $D_i \in \mathbb{EDM}^{N-1}$ the EDM corresponding to that particular facet: From the *Cayley-Menger determinant*[5.61] for simplices, [401] [135] [168, §4] [90, §3.3] the ith facet area-square for $i \in \{1 \ldots N\}$ is (§A.4.1)

$$c_i = \frac{-1}{2^{N-2}(N-2)!^2} \det\begin{bmatrix} 0 & \mathbf{1}^{\text{T}} \\ \mathbf{1} & -D_i \end{bmatrix} \tag{1248}$$

$$= \frac{(-1)^N}{2^{N-2}(N-2)!^2} \det D_i \left(\mathbf{1}^{\text{T}} D_i^{-1} \mathbf{1}\right) \tag{1249}$$

$$= \frac{(-1)^N}{2^{N-2}(N-2)!^2} \mathbf{1}^{\text{T}} \text{cof}(D_i)^{\text{T}} \mathbf{1} \tag{1250}$$

where D_i is the ith principal $N-1 \times N-1$ submatrix[5.62] of $D \in \mathbb{EDM}^N$, and $\text{cof}(D_i)$ is the $N-1 \times N-1$ matrix of *cofactors* [338, §4] corresponding to D_i. The number of principal 3×3 submatrices in D is, of course, equal to the number of triangular facets in the tetrahedron; four $(N!/(3!(N-3)!))$ when $N = 4$.

5.14.3.1.1 Exercise. *Sufficiency conditions for an EDM of four points.*
Triangle inequality (property 4) and area inequality (1246) are conditions necessary for D to be an EDM. Prove their sufficiency in conjunction with the remaining three Euclidean metric properties. ▼

5.14.3.2 $N = 5$

Moving to the next level, we might encounter a Euclidean body called *polychoron*: a bounded polyhedron in four dimensions.[5.63] Our polychoron has five $(N!/(4!(N-4)!))$

[5.60]The upper bound is met when all angles in (1247) are simultaneously 0; that occurs, for example, if one point is relatively interior to the convex hull of the three remaining.

[5.61] whose foremost characteristic is: the determinant vanishes if and only if affine dimension does not equal penultimate cardinality; *id est*, $\det\begin{bmatrix} 0 & \mathbf{1}^{\text{T}} \\ \mathbf{1} & -D \end{bmatrix} = 0 \Leftrightarrow r < N-1$ where D is any EDM (§5.7.3.0.1). Otherwise, the determinant is negative.

[5.62]Every principal submatrix of an EDM remains an EDM. [241, §4.1]

[5.63]The simplest polychoron is called a *pentatope* [401]; a regular simplex hence convex. (A *pentahedron* is a three-dimensional body having five vertices.)

facets, each of them a general tetrahedron whose volume-square c_i is calculated using the same formula; (1248) where D is the EDM corresponding to the polychoron, and D_i is the EDM corresponding to the ith facet (the principal 4×4 submatrix of $D \in \mathbb{EDM}^N$ corresponding to the ith tetrahedron). The analogue to triangle & distance is now polychoron & facet-volume. We could then write another generalized "triangle" inequality like (1246) but in terms of facet volume; [407, §IV]

$$\sqrt{c_i} \leq \sqrt{c_j} + \sqrt{c_k} + \sqrt{c_\ell} + \sqrt{c_m} \,, \quad i \neq j \neq k \neq \ell \neq m \in \{1 \ldots 5\} \qquad (1251)$$

5.14.3.2.1 Exercise. *Sufficiency for an EDM of five points.*
For $N = 5$, triangle (distance) inequality (§5.2), area inequality (1246), and volume inequality (1251) are conditions necessary for D to be an EDM. Prove their sufficiency.
▼

5.14.3.3 Volume of simplices

There is no known formula for the volume of a bounded general convex polyhedron expressed either by halfspace or vertex-description. [419, §2.1] [290, p.173] [239] [179] [180] Volume is a concept germane to $\mathbb{R}^3$; in higher dimensions it is called *content*. Applying the *EDM assertion* (§5.9.1.0.4) and a result from [62, p.407], a general nonempty simplex (§2.12.3) in $\mathbb{R}^{N-1}$ corresponding to an EDM $D \in \mathbb{S}_h^N$ has content

$$\sqrt{c} = \text{content}(\mathcal{S}) \sqrt{\det(-V_\mathcal{N}^{\mathrm{T}} D V_\mathcal{N})} \qquad (1252)$$

where content-square of the unit simplex $\mathcal{S} \subset \mathbb{R}^{N-1}$ is proportional to its Cayley-Menger determinant;

$$\text{content}(\mathcal{S})^2 = \frac{-1}{2^{N-1}(N-1)!^2} \det \begin{bmatrix} 0 & \mathbf{1}^{\mathrm{T}} \\ \mathbf{1} & -\mathbf{D}([\mathbf{0} \ e_1 \ e_2 \ \cdots \ e_{N-1}]) \end{bmatrix} \qquad (1253)$$

where $e_i \in \mathbb{R}^{N-1}$ and the EDM operator used is $\mathbf{D}(X)$ (968).

5.14.3.3.1 Example. *Pyramid.*
A formula for volume of a pyramid is known;[5.64] it is $\frac{1}{3}$ the product of its base area with its height. [229] The pyramid in Figure 148 has volume $\frac{1}{3}$. To find its volume using EDMs, we must first decompose the pyramid into simplicial parts. Slicing it in half along the plane containing the line segments corresponding to radius R and height h we find the vertices of one simplex,

$$X = \begin{bmatrix} 1/2 & 1/2 & -1/2 & 0 \\ 1/2 & -1/2 & -1/2 & 0 \\ 0 & 0 & 0 & 1 \end{bmatrix} \in \mathbb{R}^{n \times N} \qquad (1254)$$

where $N = n + 1$ for any nonempty simplex in $\mathbb{R}^n$. The volume of this simplex is half that of the entire pyramid; *id est*, $\sqrt{c} = \frac{1}{6}$ found by evaluating (1252). □

With that, we conclude digression of path.

[5.64]Pyramid volume is independent of the paramount vertex position as long as its height remains constant.

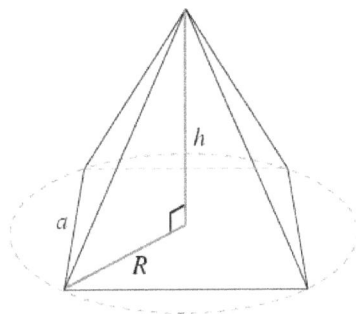

Figure 148: Length of one-dimensional face a equals height $h = a = 1$ of this convex nonsimplicial pyramid in $\mathbb{R}^3$ with square base inscribed in a circle of radius R centered at the origin. [401, *Pyramid*]

5.14.4   Affine dimension reduction in three dimensions

(*confer* §5.8.4) The determinant of any $M \times M$ matrix is equal to the product of its M eigenvalues. [338, §5.1] When $N = 4$ and $\det(\Theta^{\mathrm{T}}\Theta)$ is 0, that means one or more eigenvalues of $\Theta^{\mathrm{T}}\Theta \in \mathbb{R}^{3 \times 3}$ are 0. The determinant will go to 0 whenever equality is attained on either side of (962), (1243a), or (1243b), meaning that a tetrahedron has collapsed to a lower affine dimension; *id est*, $r = \operatorname{rank}\Theta^{\mathrm{T}}\Theta = \operatorname{rank}\Theta$ is reduced below $N - 1$ exactly by the number of 0 eigenvalues (§5.7.1.1).

In solving completion problems of any size N where one or more entries of an EDM are unknown, therefore, dimension r of the affine hull required to contain the unknown points is potentially reduced by selecting distances to attain equality in (962) or (1243a) or (1243b).

5.14.4.1   *Exemplum redux*

We now apply the *fifth Euclidean metric property* to an earlier problem:

5.14.4.1.1   Example. *Small completion problem*, IV.   (*confer* §5.9.3.0.1)
Returning again to Example 5.3.0.0.2 that pertains to Figure **127** where $N = 4$, distance-square d_{14} is ascertainable from the fifth Euclidean metric property. Because all distances in (960) are known except $\sqrt{d_{14}}$, then $\cos\theta_{123} = 0$ and $\theta_{324} = 0$ result from identity (1034). Applying (962),

$$
\begin{aligned}
\cos(\theta_{123} + \theta_{324}) \;\leq\; &\cos\theta_{124} \;\leq\; \cos(\theta_{123} - \theta_{324}) \\
0 \;\leq\; &\cos\theta_{124} \;\leq\; 0
\end{aligned}
\tag{1255}
$$

It follows again from (1034) that d_{14} can only be 2. As explained in this subsection, affine dimension r cannot exceed $N - 2$ because equality is attained in (1255).   □

Chapter 6

Cone of distance matrices

For $N > 3$, the cone of EDMs is no longer a circular cone and the geometry becomes complicated...

— Hayden, Wells, Liu, & Tarazaga (1991) [191, §3]

In the subspace of symmetric matrices $\mathbb{S}^N$, we know that the convex cone of Euclidean distance matrices $\mathbb{EDM}^N$ (the EDM cone) does not intersect the positive semidefinite cone $\mathbb{S}^N_+$ (PSD cone) except at the origin, their only vertex; there can be no positive or negative semidefinite EDM. (1179) [241]

$$\mathbb{EDM}^N \cap \mathbb{S}^N_+ = \mathbf{0} \qquad (1256)$$

Even so, the two convex cones can be related. In §6.8.1 we prove the equality

$$\mathbb{EDM}^N = \mathbb{S}^N_h \cap \left(\mathbb{S}^{N\perp}_c - \mathbb{S}^N_+ \right) \qquad (1349)$$

a resemblance to EDM definition (968) where

$$\mathbb{S}^N_h = \left\{ A \in \mathbb{S}^N \mid \delta(A) = \mathbf{0} \right\} \qquad (66)$$

is the symmetric hollow subspace (§2.2.3) and where

$$\mathbb{S}^{N\perp}_c = \{ u\mathbf{1}^{\mathrm{T}} + \mathbf{1}u^{\mathrm{T}} \mid u \in \mathbb{R}^N \} \qquad (2096)$$

is the orthogonal complement of the geometric center subspace (§E.7.2.0.2)

$$\mathbb{S}^N_c = \{ Y \in \mathbb{S}^N \mid Y\mathbf{1} = \mathbf{0} \} \qquad (2094)$$

CITATION: Dattorro, *Convex Optimization & Euclidean Distance Geometry*, $\mathcal{M}\varepsilon\beta oo$ Publishing USA, 2005, v2015.02.15.

6.0.1 gravity

Equality (1349) is equally important as the known isomorphisms (1087) (1088) (1099) (1100) relating the EDM cone $\mathbb{EDM}^N$ to positive semidefinite cone $\mathbb{S}_+^{N-1}$ (§5.6.2.1) or to an $N(N-1)/2$-dimensional face of $\mathbb{S}_+^N$ (§5.6.1.1).[6.1] But those isomorphisms have never led to this equality relating whole cones $\mathbb{EDM}^N$ and $\mathbb{S}_+^N$.

Equality (1349) is not obvious from the various EDM definitions such as (968) or (1272) because inclusion must be proved algebraically in order to establish equality; $\mathbb{EDM}^N \supseteq \mathbb{S}_h^N \cap (\mathbb{S}_c^{N\perp} - \mathbb{S}_+^N)$. We will instead prove (1349) using purely geometric methods.

6.0.2 highlight

In §6.8.1.7 we show: the Schoenberg criterion for discriminating Euclidean distance matrices

$$D \in \mathbb{EDM}^N \;\Leftrightarrow\; \begin{cases} -V_\mathcal{N}^T D V_\mathcal{N} \in \mathbb{S}_+^{N-1} \\ D \in \mathbb{S}_h^N \end{cases} \qquad (987)$$

is a discretized membership relation (§2.13.4, dual generalized inequalities) between the EDM cone and its ordinary dual.

6.1 Defining EDM cone

We invoke a popular matrix criterion to illustrate correspondence between the EDM and PSD cones belonging to the ambient space of symmetric matrices:

$$D \in \mathbb{EDM}^N \;\Leftrightarrow\; \begin{cases} -VDV \in \mathbb{S}_+^N \\ D \in \mathbb{S}_h^N \end{cases} \qquad (991)$$

where $V \in \mathbb{S}^N$ is the geometric centering matrix (§B.4). The set of all EDMs of dimension $N \times N$ forms a closed convex cone $\mathbb{EDM}^N$ because any pair of EDMs satisfies the definition of a convex cone (175); *videlicet*, for each and every $\zeta_1, \zeta_2 \geq 0$ (§A.3.1.0.2)

$$\begin{matrix} \zeta_1 VD_1V + \zeta_2 VD_2V \succeq 0 \\ \zeta_1 D_1 + \zeta_2 D_2 \in \mathbb{S}_h^N \end{matrix} \quad\Leftarrow\quad \begin{matrix} VD_1V \succeq 0, & VD_2V \succeq 0 \\ D_1 \in \mathbb{S}_h^N, & D_2 \in \mathbb{S}_h^N \end{matrix} \qquad (1257)$$

and convex cones are invariant to inverse linear transformation [315, p.22].

[6.1]Because both positive semidefinite cones are frequently in play, dimension is explicit.

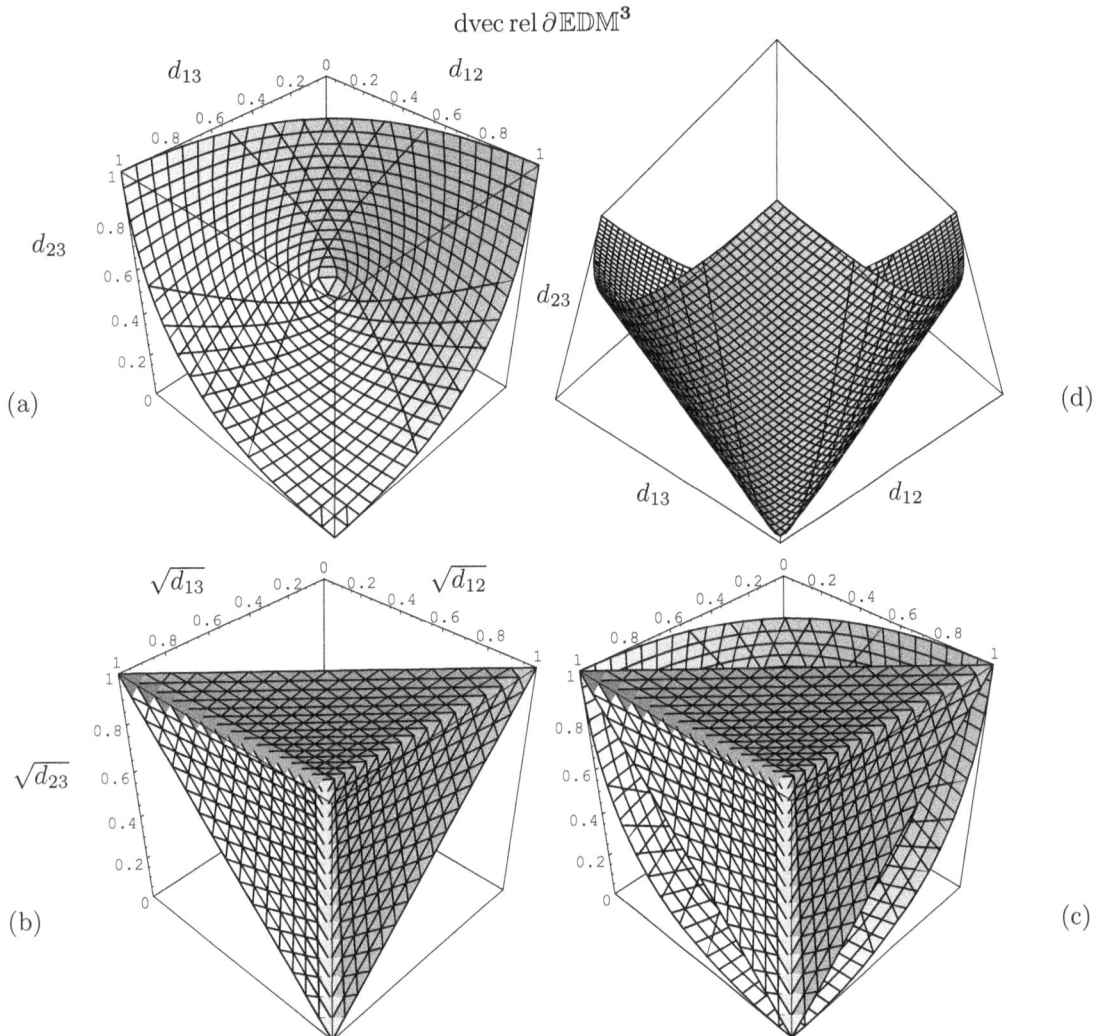

Figure 149: Relative boundary (tiled) of EDM cone $\mathbb{EDM}^3$ drawn truncated in isometrically isomorphic subspace $\mathbb{R}^3$. **(a)** EDM cone drawn in usual distance-square coordinates d_{ij}. View is from interior toward origin. Unlike positive semidefinite cone, EDM cone is not selfdual; neither is it proper in ambient symmetric subspace (dual EDM cone for this example belongs to isomorphic $\mathbb{R}^6$). **(b)** Drawn in its natural coordinates $\sqrt{d_{ij}}$ (absolute distance), cone remains convex (*confer* §5.10); intersection of three halfspaces (1142) whose partial boundaries each contain origin. Cone geometry becomes nonconvex (nonpolyhedral) in higher dimension. (§6.3) **(c)** Two coordinate systems artificially superimposed. Coordinate transformation from d_{ij} to $\sqrt{d_{ij}}$ appears a topological contraction. **(d)** Sitting on its vertex **0**, pointed $\mathbb{EDM}^3$ is a circular cone having axis of revolution $\mathrm{dvec}(-E) = \mathrm{dvec}(\mathbf{1}\mathbf{1}^T - I)$ (1175) (73). (Rounded vertex is plot artifact.)

6.1.0.0.1 Definition. *Cone of Euclidean distance matrices.*
In the subspace of symmetric matrices, the set of all Euclidean distance matrices forms a unique immutable pointed closed convex cone called the *EDM cone*: for $N > 0$

$$
\begin{aligned}
\mathbb{EDM}^N &\triangleq \left\{ D \in \mathbb{S}_h^N \mid -VDV \in \mathbb{S}_+^N \right\} \\
&= \bigcap_{z \in \mathcal{N}(\mathbf{1}^{\mathrm{T}})} \left\{ D \in \mathbb{S}^N \mid \langle zz^{\mathrm{T}}, -D \rangle \geq 0, \ \delta(D) = \mathbf{0} \right\}
\end{aligned}
\tag{1258}
$$

The EDM cone in isomorphic $\mathbb{R}^{N(N+1)/2}$ [*sic*] is the intersection of an infinite number (when $N > 2$) of halfspaces about the origin and a finite number of hyperplanes through the origin in vectorized variable $D = [d_{ij}]$. Hence $\mathbb{EDM}^N$ is not full-dimensional with respect to $\mathbb{S}^N$ because it is confined to the symmetric hollow subspace $\mathbb{S}_h^N$. The EDM cone relative interior comprises

$$
\begin{aligned}
\mathrm{rel\,int}\,\mathbb{EDM}^N &= \bigcap_{z \in \mathcal{N}(\mathbf{1}^{\mathrm{T}})} \left\{ D \in \mathbb{S}^N \mid \langle zz^{\mathrm{T}}, -D \rangle > 0, \ \delta(D) = \mathbf{0} \right\} \\
&= \left\{ D \in \mathbb{EDM}^N \mid \mathrm{rank}(VDV) = N - 1 \right\}
\end{aligned}
\tag{1259}
$$

while its relative boundary comprises

$$
\begin{aligned}
\mathrm{rel\,\partial}\mathbb{EDM}^N &= \left\{ D \in \mathbb{EDM}^N \mid \langle zz^{\mathrm{T}}, -D \rangle = 0 \ \text{for some } z \in \mathcal{N}(\mathbf{1}^{\mathrm{T}}) \right\} \\
&= \left\{ D \in \mathbb{EDM}^N \mid \mathrm{rank}(VDV) < N - 1 \right\}
\end{aligned}
\tag{1260}
$$

$\triangle$

This cone is more easily visualized in the isomorphic vector subspace $\mathbb{R}^{N(N-1)/2}$ corresponding to $\mathbb{S}_h^N$:

In the case $N = 1$ point, the EDM cone is the origin in $\mathbb{R}^0$.

In the case $N = 2$, the EDM cone is the nonnegative real line in $\mathbb{R}$; a halfline in a subspace of the realization in Figure 153.

The EDM cone in the case $N = 3$ is a circular cone in $\mathbb{R}^3$ illustrated in Figure 149(a)(d); rather, the set of all matrices

$$
D = \begin{bmatrix} 0 & d_{12} & d_{13} \\ d_{12} & 0 & d_{23} \\ d_{13} & d_{23} & 0 \end{bmatrix} \in \mathbb{EDM}^3
\tag{1261}
$$

makes a circular cone in this dimension. In this case, the first four Euclidean metric properties are necessary and sufficient tests to certify realizability of triangles; (1237). Thus triangle inequality property 4 describes three halfspaces (1142) whose intersection makes a polyhedral cone in $\mathbb{R}^3$ of realizable $\sqrt{d_{ij}}$ (absolute distance); an isomorphic subspace representation of the set of all EDMs D in the natural coordinates

$$
\sqrt[\circ]{D} \triangleq \begin{bmatrix} 0 & \sqrt{d_{12}} & \sqrt{d_{13}} \\ \sqrt{d_{12}} & 0 & \sqrt{d_{23}} \\ \sqrt{d_{13}} & \sqrt{d_{23}} & 0 \end{bmatrix}
\tag{1262}
$$

illustrated in Figure 149b.

6.2 Polyhedral bounds

The convex cone of EDMs is nonpolyhedral in d_{ij} for $N > 2$; *e.g.*, Figure 149a. Still we found necessary and sufficient bounding polyhedral relations consistent with EDM cones for cardinality $N = 1, 2, 3, 4$:

$N = 3$. Transforming distance-square coordinates d_{ij} by taking their positive square root provides the polyhedral cone in Figure 149b; polyhedral because an intersection of three halfspaces in natural coordinates $\sqrt{d_{ij}}$ is provided by triangle inequalities (1142). This polyhedral cone implicitly encompasses necessary and sufficient metric properties: nonnegativity, self-distance, symmetry, and triangle inequality.

$N = 4$. Relative-angle inequality (1243) together with four Euclidean metric properties are necessary and sufficient tests for realizability of tetrahedra. (1244) Albeit relative angles θ_{ikj} (1034) are nonlinear functions of the d_{ij}, relative-angle inequality provides a regular tetrahedron in $\mathbb{R}^3$ [*sic*] (Figure 147) bounding angles θ_{ikj} at vertex x_k consistently with $\mathbb{EDM}^4$.[6.2]

Yet were we to employ the procedure outlined in §5.14.3 for making generalized triangle inequalities, then we would find all the necessary and sufficient d_{ij}-transformations for generating bounding polyhedra consistent with EDMs of any higher dimension ($N > 3$).

6.3 $\sqrt{\text{EDM}}$ cone is not convex

For some applications, like a molecular conformation problem (Figure 3, Figure 138) or multidimensional scaling [105] [363], absolute distance $\sqrt{d_{ij}}$ is the preferred variable. Taking square root of the entries in all EDMs D of dimension N, we get another cone but not a convex cone when $N > 3$ (Figure 149b): [91, §4.5.2]

$$\sqrt{\mathbb{EDM}^N} \triangleq \{ \sqrt[\circ]{D} \mid D \in \mathbb{EDM}^N \} \tag{1263}$$

where $\sqrt[\circ]{D}$ is defined like (1262).   It is a cone simply because any cone is completely constituted by rays emanating from the origin: (§2.7) Any given ray $\{\zeta \Gamma \in \mathbb{R}^{N(N-1)/2} \mid \zeta \geq 0\}$ remains a ray under entrywise square root: $\{\sqrt[\circ]{\zeta \Gamma} \in \mathbb{R}^{N(N-1)/2} \mid \zeta \geq 0\}$. It is already established that

$$D \in \mathbb{EDM}^N \;\Rightarrow\; \sqrt[\circ]{D} \in \mathbb{EDM}^N \qquad (1174)$$

But because of how $\sqrt{\mathbb{EDM}^N}$ is defined, it is obvious that (*confer* §5.10)

$$D \in \mathbb{EDM}^N \;\Leftrightarrow\; \sqrt[\circ]{D} \in \sqrt{\mathbb{EDM}^N} \tag{1264}$$

Were $\sqrt{\mathbb{EDM}^N}$ convex, then given $\sqrt[\circ]{D_1}, \sqrt[\circ]{D_2} \in \sqrt{\mathbb{EDM}^N}$ we would expect their conic combination $\sqrt[\circ]{D_1} + \sqrt[\circ]{D_2}$ to be a member of $\sqrt{\mathbb{EDM}^N}$. That is easily proven false by

---

[6.2]Still, property-4 triangle inequalities (1142) corresponding to each principal 3×3 submatrix of $-V_N^{\mathrm{T}} D V_N$ demand that the corresponding $\sqrt{d_{ij}}$ belong to a polyhedral cone like that in Figure 149b.

counterexample via (1264), for then $(\sqrt[\circ]{D_1} + \sqrt[\circ]{D_2}) \circ (\sqrt[\circ]{D_1} + \sqrt[\circ]{D_2})$ would need to be a member of $\mathbb{EDM}^N$.

Notwithstanding,

$$\sqrt{\mathbb{EDM}^N} \subseteq \mathbb{EDM}^N \tag{1265}$$

by (1174) (Figure **149**), and we learn how to transform a nonconvex *proximity problem* in the natural coordinates $\sqrt{d_{ij}}$ to a convex optimization in §7.2.1.

6.4   EDM definition in 11^T

Any EDM D corresponding to affine dimension r has representation

$$\mathbf{D}(V_{\mathcal{X}}) \triangleq \delta(V_{\mathcal{X}}V_{\mathcal{X}}^T)1^T + 1\delta(V_{\mathcal{X}}V_{\mathcal{X}}^T)^T - 2V_{\mathcal{X}}V_{\mathcal{X}}^T \in \mathbb{EDM}^N \tag{1266}$$

where $\mathcal{R}(V_{\mathcal{X}} \in \mathbb{R}^{N \times r}) \subseteq \mathcal{N}(1^T) = 1^\perp$

$$V_{\mathcal{X}}^T V_{\mathcal{X}} = \delta^2(V_{\mathcal{X}}^T V_{\mathcal{X}}) \quad \textbf{and} \quad V_{\mathcal{X}} \text{ is full-rank with orthogonal columns.} \tag{1267}$$

Equation (1266) is simply the standard EDM definition (968) with a centered list X as in (1055); Gram matrix $X^T X$ has been replaced with the subcompact singular value decomposition (§A.6.2)[6.3]

$$V_{\mathcal{X}}V_{\mathcal{X}}^T \equiv V^T X^T X V \in \mathbb{S}_c^N \cap \mathbb{S}_+^N \tag{1268}$$

This means: inner product $V_{\mathcal{X}}^T V_{\mathcal{X}}$ is an $r \times r$ diagonal matrix Σ of nonzero singular values.

Vector $\delta(V_{\mathcal{X}}V_{\mathcal{X}}^T)$ may me decomposed into complementary parts by projecting it on orthogonal subspaces $1^\perp$ and $\mathcal{R}(1)$: namely,

$$P_{1^\perp}\big(\delta(V_{\mathcal{X}}V_{\mathcal{X}}^T)\big) = V\delta(V_{\mathcal{X}}V_{\mathcal{X}}^T) \tag{1269}$$

$$P_1\big(\delta(V_{\mathcal{X}}V_{\mathcal{X}}^T)\big) = \frac{1}{N}11^T\delta(V_{\mathcal{X}}V_{\mathcal{X}}^T) \tag{1270}$$

Of course

$$\delta(V_{\mathcal{X}}V_{\mathcal{X}}^T) = V\delta(V_{\mathcal{X}}V_{\mathcal{X}}^T) + \frac{1}{N}11^T\delta(V_{\mathcal{X}}V_{\mathcal{X}}^T) \tag{1271}$$

by (990). Substituting this into EDM definition (1266), we get the Hayden, Wells, Liu, & Tarazaga EDM formula [191, §2]

$$\mathbf{D}(V_{\mathcal{X}}, y) \triangleq y1^T + 1y^T + \frac{\lambda}{N}11^T - 2V_{\mathcal{X}}V_{\mathcal{X}}^T \in \mathbb{EDM}^N \tag{1272}$$

where

$$\lambda \triangleq 2\|V_{\mathcal{X}}\|_F^2 = 1^T\delta(V_{\mathcal{X}}V_{\mathcal{X}}^T)2 \quad \textbf{and} \quad y \triangleq \delta(V_{\mathcal{X}}V_{\mathcal{X}}^T) - \frac{\lambda}{2N}1 = V\delta(V_{\mathcal{X}}V_{\mathcal{X}}^T) \tag{1273}$$

[6.3]Subcompact SVD: $V_{\mathcal{X}}V_{\mathcal{X}}^T \triangleq Q\sqrt{\Sigma}\sqrt{\Sigma}Q^T \equiv V^T X^T X V$. So $V_{\mathcal{X}}^T$ is not necessarily XV (§5.5.1.0.1), although affine dimension $r = \text{rank}(V_{\mathcal{X}}^T) = \text{rank}(XV)$. (1112)

and $y = \mathbf{0}$ if and only if $\mathbf{1}$ is an eigenvector of EDM D. Scalar λ becomes an eigenvalue when corresponding eigenvector $\mathbf{1}$ exists.[6.4]

Then the particular dyad sum from (1272)

$$y\mathbf{1}^{\mathrm{T}} + \mathbf{1}y^{\mathrm{T}} + \frac{\lambda}{N}\mathbf{1}\mathbf{1}^{\mathrm{T}} \in \mathbb{S}_c^{N\perp} \tag{1274}$$

must belong to the orthogonal complement of the geometric center subspace (p.630), whereas $V_{\mathcal{X}}V_{\mathcal{X}}^{\mathrm{T}} \in \mathbb{S}_c^N \cap \mathbb{S}_+^N$ (1268) belongs to the positive semidefinite cone in the geometric center subspace.

Proof. We validate eigenvector $\mathbf{1}$ and eigenvalue λ.
($\Rightarrow$) Suppose $\mathbf{1}$ is an eigenvector of EDM D. Then because

$$V_{\mathcal{X}}^{\mathrm{T}}\mathbf{1} = \mathbf{0} \tag{1275}$$

it follows

$$D\mathbf{1} = \delta(V_{\mathcal{X}}V_{\mathcal{X}}^{\mathrm{T}})\mathbf{1}^{\mathrm{T}}\mathbf{1} + \mathbf{1}\delta(V_{\mathcal{X}}V_{\mathcal{X}}^{\mathrm{T}})^{\mathrm{T}}\mathbf{1} = N\delta(V_{\mathcal{X}}V_{\mathcal{X}}^{\mathrm{T}}) + \|V_{\mathcal{X}}\|_{\mathrm{F}}^2\mathbf{1}$$
$$\Rightarrow \delta(V_{\mathcal{X}}V_{\mathcal{X}}^{\mathrm{T}}) \propto \mathbf{1} \tag{1276}$$

For some $\kappa \in \mathbb{R}_+$

$$\delta(V_{\mathcal{X}}V_{\mathcal{X}}^{\mathrm{T}})^{\mathrm{T}}\mathbf{1} = N\kappa = \mathrm{tr}(V_{\mathcal{X}}^{\mathrm{T}}V_{\mathcal{X}}) = \|V_{\mathcal{X}}\|_{\mathrm{F}}^2 \quad \Rightarrow \quad \delta(V_{\mathcal{X}}V_{\mathcal{X}}^{\mathrm{T}}) = \frac{1}{N}\|V_{\mathcal{X}}\|_{\mathrm{F}}^2\mathbf{1} \tag{1277}$$

so $y = \mathbf{0}$.
($\Leftarrow$) Now suppose $\delta(V_{\mathcal{X}}V_{\mathcal{X}}^{\mathrm{T}}) = \frac{\lambda}{2N}\mathbf{1}$; *id est*, $y = \mathbf{0}$. Then

$$D = \frac{\lambda}{N}\mathbf{1}\mathbf{1}^{\mathrm{T}} - 2V_{\mathcal{X}}V_{\mathcal{X}}^{\mathrm{T}} \in \mathrm{EDM}^N \tag{1278}$$

$\mathbf{1}$ is an eigenvector with corresponding eigenvalue λ. $\qquad\qquad\qquad\blacklozenge$

6.4.1 Range of EDM D

From §B.1.1 pertaining to linear independence of dyad sums: If the transpose halves of all the dyads in the sum (1266)[6.5] make a linearly independent set, then the nontranspose halves constitute a basis for the range of EDM D. Saying this mathematically: For $D \in \mathrm{EDM}^N$

$$\begin{aligned}
\mathcal{R}(D) &= \mathcal{R}([\,\delta(V_{\mathcal{X}}V_{\mathcal{X}}^{\mathrm{T}}) \quad \mathbf{1} \quad V_{\mathcal{X}}\,]) &\Leftarrow \quad \mathrm{rank}([\,\delta(V_{\mathcal{X}}V_{\mathcal{X}}^{\mathrm{T}}) \quad \mathbf{1} \quad V_{\mathcal{X}}\,]) = 2 + r \\
\mathcal{R}(D) &= \mathcal{R}([\,\mathbf{1} \quad V_{\mathcal{X}}\,]) &\Leftarrow \quad \text{otherwise}
\end{aligned} \tag{1279}$$

[6.4] *e.g.*, when $X = I$ in EDM definition (968).
[6.5] Identifying columns $V_{\mathcal{X}} \triangleq [\,v_1 \cdots v_r\,]$, then $V_{\mathcal{X}}V_{\mathcal{X}}^{\mathrm{T}} = \sum_i v_i v_i^{\mathrm{T}}$ is also a sum of dyads.

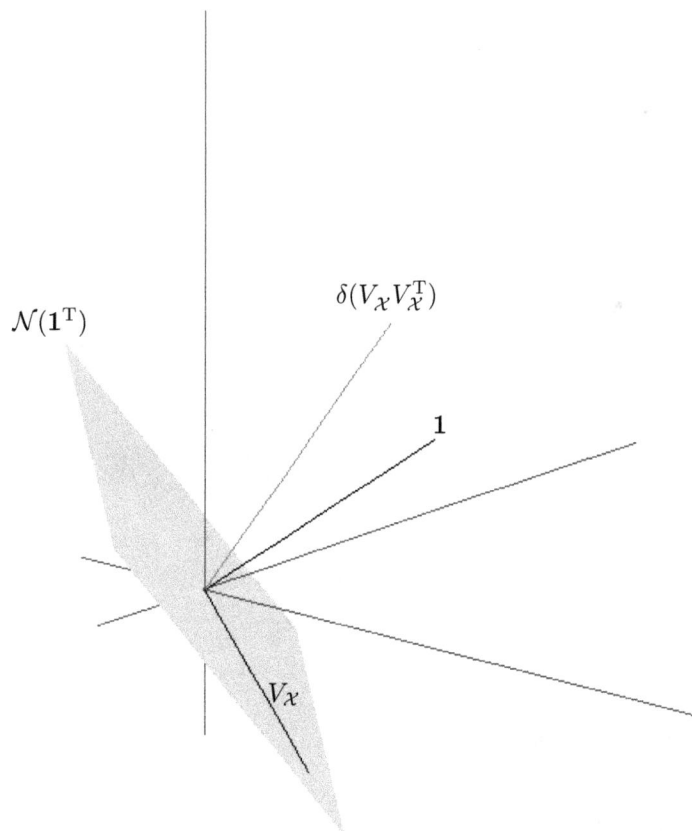

Figure 150: Example of $V_{\mathcal{X}}$ selection to make an EDM corresponding to cardinality $N = 3$ and affine dimension $r = 1$; $V_{\mathcal{X}}$ is a vector in nullspace $\mathcal{N}(\mathbf{1}^{\mathrm{T}}) \subset \mathbb{R}^3$. Nullspace of $\mathbf{1}^{\mathrm{T}}$ is hyperplane in $\mathbb{R}^3$ (drawn truncated) having normal $\mathbf{1}$. Vector $\delta(V_{\mathcal{X}} V_{\mathcal{X}}^{\mathrm{T}})$ may or may not be in plane spanned by $\{\mathbf{1}, V_{\mathcal{X}}\}$, but belongs to nonnegative orthant which is strictly supported by $\mathcal{N}(\mathbf{1}^{\mathrm{T}})$.

6.4.1.0.1 Proof. We need that condition under which the rank equality is satisfied: We know $\mathcal{R}(V_{\mathcal{X}})\perp\mathbf{1}$, but what is the relative geometric orientation of $\delta(V_{\mathcal{X}}V_{\mathcal{X}}^{\mathrm{T}})$? $\delta(V_{\mathcal{X}}V_{\mathcal{X}}^{\mathrm{T}})\succeq0$ because $V_{\mathcal{X}}V_{\mathcal{X}}^{\mathrm{T}}\succeq0$, and $\delta(V_{\mathcal{X}}V_{\mathcal{X}}^{\mathrm{T}})\propto\mathbf{1}$ remains possible (1276); this means $\delta(V_{\mathcal{X}}V_{\mathcal{X}}^{\mathrm{T}})\notin\mathcal{N}(\mathbf{1}^{\mathrm{T}})$ simply because it has no negative entries. (Figure 150) If the projection of $\delta(V_{\mathcal{X}}V_{\mathcal{X}}^{\mathrm{T}})$ on $\mathcal{N}(\mathbf{1}^{\mathrm{T}})$ does not belong to $\mathcal{R}(V_{\mathcal{X}})$, then that is a necessary and sufficient condition for linear independence (l.i.) of $\delta(V_{\mathcal{X}}V_{\mathcal{X}}^{\mathrm{T}})$ with respect to $\mathcal{R}([\,\mathbf{1}\ \ V_{\mathcal{X}}\,])$; *id est,*

$$
\begin{aligned}
V\delta(V_{\mathcal{X}}V_{\mathcal{X}}^{\mathrm{T}}) &\neq V_{\mathcal{X}}\,a \quad\text{for any } a\in\mathbb{R}^r\\
(I-\tfrac{1}{N}\mathbf{1}\mathbf{1}^{\mathrm{T}})\delta(V_{\mathcal{X}}V_{\mathcal{X}}^{\mathrm{T}}) &\neq V_{\mathcal{X}}\,a\\
\delta(V_{\mathcal{X}}V_{\mathcal{X}}^{\mathrm{T}})-\tfrac{1}{N}\|V_{\mathcal{X}}\|_{\mathrm{F}}^2\mathbf{1} &\neq V_{\mathcal{X}}\,a\\
\delta(V_{\mathcal{X}}V_{\mathcal{X}}^{\mathrm{T}})-\tfrac{\lambda}{2N}\mathbf{1}=y &\neq V_{\mathcal{X}}\,a \ \Leftrightarrow\ \{\mathbf{1},\ \delta(V_{\mathcal{X}}V_{\mathcal{X}}^{\mathrm{T}}),\ V_{\mathcal{X}}\}\ \text{is l.i.}
\end{aligned}
\tag{1280}
$$

When this condition is violated (when (1273) $y=V_{\mathcal{X}}\,a_{\mathrm{p}}$ for some particular $a\in\mathbb{R}^r$), on the other hand, then from (1272) we have

$$
\begin{aligned}
\mathcal{R}\big(D=y\mathbf{1}^{\mathrm{T}}+\mathbf{1}y^{\mathrm{T}}+\tfrac{\lambda}{N}\mathbf{1}\mathbf{1}^{\mathrm{T}}-2V_{\mathcal{X}}V_{\mathcal{X}}^{\mathrm{T}}\big) &= \mathcal{R}\big((V_{\mathcal{X}}\,a_{\mathrm{p}}+\tfrac{\lambda}{N}\mathbf{1})\mathbf{1}^{\mathrm{T}}+(\mathbf{1}a_{\mathrm{p}}^{\mathrm{T}}-2V_{\mathcal{X}})V_{\mathcal{X}}^{\mathrm{T}}\big)\\
&= \mathcal{R}([\,V_{\mathcal{X}}\,a_{\mathrm{p}}+\tfrac{\lambda}{N}\mathbf{1}\quad \mathbf{1}a_{\mathrm{p}}^{\mathrm{T}}-2V_{\mathcal{X}}\,])\\
&= \mathcal{R}([\,\mathbf{1}\ \ V_{\mathcal{X}}\,])
\end{aligned}
\tag{1281}
$$

An example of such a violation is (1278) where, in particular, $a_{\mathrm{p}}=\mathbf{0}$.   ♦

Then a statement parallel to (1279) is, for $D\in\mathbb{EDM}^N$ (Theorem 5.7.3.0.1)

$$
\begin{aligned}
\mathrm{rank}(D)=r+2 &\ \Leftrightarrow\ y\notin\mathcal{R}(V_{\mathcal{X}})\ (\Leftrightarrow\ \mathbf{1}^{\mathrm{T}}D^{\dagger}\mathbf{1}=0)\\
\mathrm{rank}(D)=r+1 &\ \Leftrightarrow\ y\in\mathcal{R}(V_{\mathcal{X}})\ (\Leftrightarrow\ \mathbf{1}^{\mathrm{T}}D^{\dagger}\mathbf{1}\neq0)
\end{aligned}
\tag{1282}
$$

6.4.2 Boundary constituents of EDM cone

Expression (1266) has utility in forming the set of all EDMs corresponding to affine dimension r:

$$
\begin{aligned}
&\left\{D\in\mathbb{EDM}^N\ \middle|\ \mathrm{rank}(VDV)=r\right\}\\
&= \left\{\mathbf{D}(V_{\mathcal{X}})\ \middle|\ V_{\mathcal{X}}\in\mathbb{R}^{N\times r},\ \mathrm{rank}\,V_{\mathcal{X}}=r,\ V_{\mathcal{X}}^{\mathrm{T}}V_{\mathcal{X}}=\delta^2(V_{\mathcal{X}}^{\mathrm{T}}V_{\mathcal{X}}),\ \mathcal{R}(V_{\mathcal{X}})\subseteq\mathcal{N}(\mathbf{1}^{\mathrm{T}})\right\}
\end{aligned}
\tag{1283}
$$

whereas $\{D\in\mathbb{EDM}^N\ |\ \mathrm{rank}(VDV)\leq r\}$ is the closure of this same set;

$$
\left\{D\in\mathbb{EDM}^N\ \middle|\ \mathrm{rank}(VDV)\leq r\right\}=\overline{\left\{D\in\mathbb{EDM}^N\ \middle|\ \mathrm{rank}(VDV)=r\right\}}
\tag{1284}
$$

For example,

$$
\begin{aligned}
\mathrm{rel}\,\partial\mathbb{EDM}^N &= \left\{D\in\mathbb{EDM}^N\ \middle|\ \mathrm{rank}(VDV)<N-1\right\}\\
&= \bigcup_{r=0}^{N-2}\left\{D\in\mathbb{EDM}^N\ \middle|\ \mathrm{rank}(VDV)=r\right\}
\end{aligned}
\tag{1285}
$$

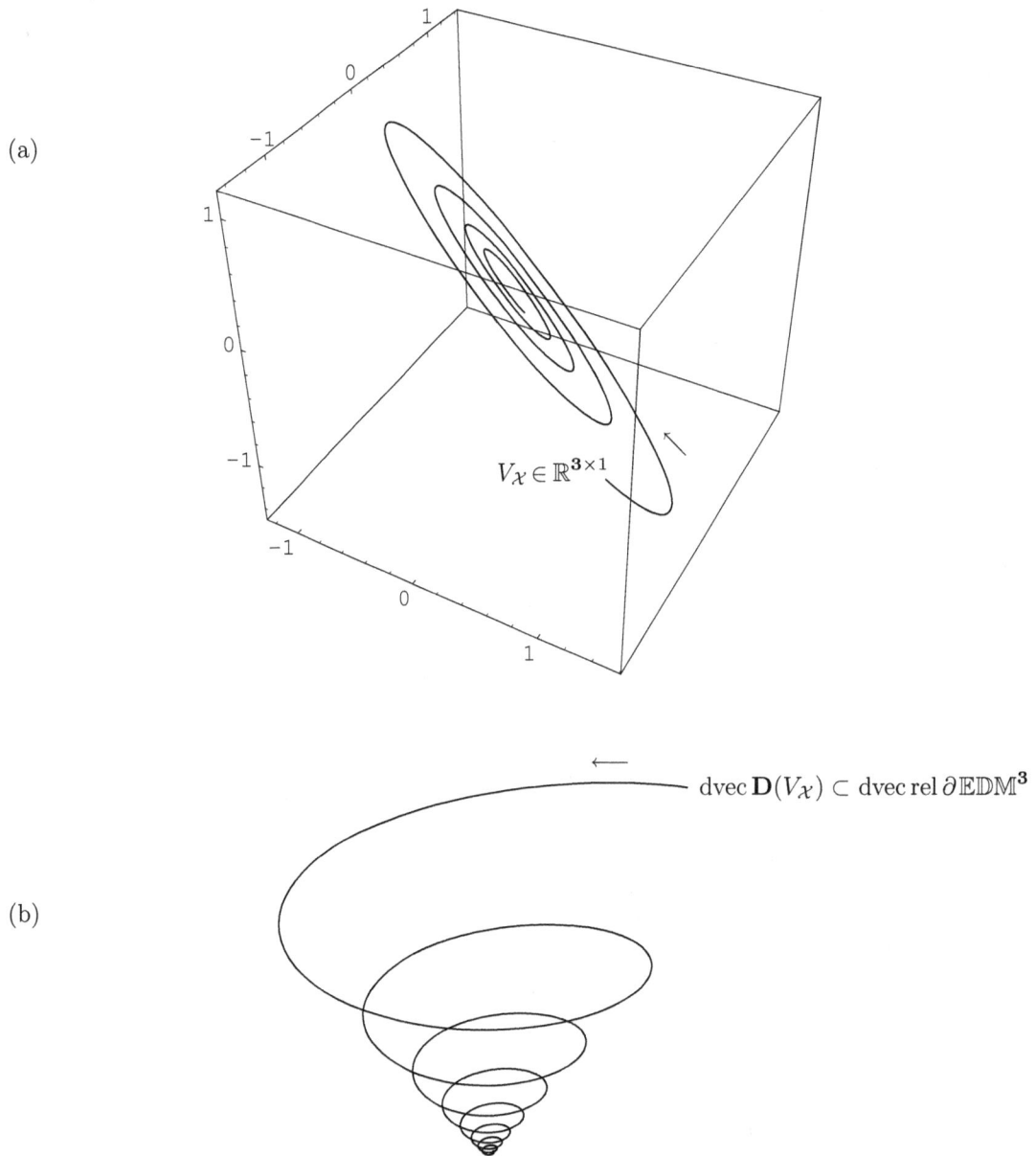

Figure 151: (a) Vector $V_{\mathcal{X}}$ from Figure 150 spirals in $\mathcal{N}(\mathbf{1}^{\mathrm{T}}) \subset \mathbb{R}^3$ decaying toward origin. (Spiral is two-dimensional in vector space $\mathbb{R}^3$.) (b) Corresponding trajectory $\mathbf{D}(V_{\mathcal{X}})$ on EDM cone relative boundary creates a vortex also decaying toward origin. There are two complete orbits on EDM cone boundary about axis of revolution for every single revolution of $V_{\mathcal{X}}$ about origin. (Vortex is three-dimensional in isometrically isomorphic $\mathbb{R}^3$.)

None of these are necessarily convex sets, although

$$
\begin{aligned}
\mathbb{EDM}^N &= \overline{\bigcup_{r=0}^{N-1} \left\{ D \in \mathbb{EDM}^N \mid \mathrm{rank}(VDV) = r \right\}} \\
&= \overline{\left\{ D \in \mathbb{EDM}^N \mid \mathrm{rank}(VDV) = N-1 \right\}} \\
\mathrm{rel\,int}\,\mathbb{EDM}^N &= \left\{ D \in \mathbb{EDM}^N \mid \mathrm{rank}(VDV) = N-1 \right\}
\end{aligned}
\tag{1286}
$$

are pointed convex cones.

When cardinality $N=3$ and affine dimension $r=2$, for example, the relative interior $\mathrm{rel\,int}\,\mathbb{EDM}^3$ is realized via (1283). (§6.5)

When $N=3$ and $r=1$, the relative boundary of the EDM cone $\mathrm{dvec\,rel}\,\partial\mathbb{EDM}^3$ is realized in isomorphic $\mathbb{R}^3$ as in Figure 149d. This figure could be constructed via (1284) by spiraling vector $V_{\mathcal{X}}$ tightly about the origin in $\mathcal{N}(1^{\mathrm{T}})$; as can be imagined with aid of Figure 150. Vectors close to the origin in $\mathcal{N}(1^{\mathrm{T}})$ are correspondingly close to the origin in $\mathbb{EDM}^N$. As vector $V_{\mathcal{X}}$ orbits the origin in $\mathcal{N}(1^{\mathrm{T}})$, the corresponding EDM orbits the axis of revolution while remaining on the boundary of the circular cone $\mathrm{dvec\,rel}\,\partial\mathbb{EDM}^3$. (Figure 151)

6.4.3 Faces of EDM cone

Like the positive semidefinite cone, EDM cone faces are EDM cones.

6.4.3.0.1 Exercise. *Isomorphic faces.*
Prove that in high cardinality N, any set of EDMs made via (1283) or (1284) with particular affine dimension r is isomorphic with any set admitting the same affine dimension but made in lower cardinality. ▼

6.4.3.1 smallest face that contains an EDM

Now suppose we are given a particular EDM $\mathbf{D}(V_{\mathcal{X}_{\mathrm{p}}}) \in \mathbb{EDM}^N$ corresponding to affine dimension r and parametrized by $V_{\mathcal{X}_{\mathrm{p}}}$ in (1266). The EDM cone's smallest face that contains $\mathbf{D}(V_{\mathcal{X}_{\mathrm{p}}})$ is

$$
\begin{aligned}
&\mathcal{F}\!\left(\mathbb{EDM}^N \ni \mathbf{D}(V_{\mathcal{X}_{\mathrm{p}}}) \right) \\
&= \overline{\left\{ \mathbf{D}(V_{\mathcal{X}}) \mid V_{\mathcal{X}} \in \mathbb{R}^{N \times r},\ \mathrm{rank}\,V_{\mathcal{X}} = r,\ V_{\mathcal{X}}^{\mathrm{T}} V_{\mathcal{X}} = \delta^2(V_{\mathcal{X}}^{\mathrm{T}} V_{\mathcal{X}}),\ \mathcal{R}(V_{\mathcal{X}}) \subseteq \mathcal{R}(V_{\mathcal{X}_{\mathrm{p}}}) \right\}} \\
&\simeq \mathbb{EDM}^{r+1}
\end{aligned}
\tag{1287}
$$

which is isomorphic[6.6] with convex cone $\mathbb{EDM}^{r+1}$, hence of dimension

$$
\dim \mathcal{F}\!\left(\mathbb{EDM}^N \ni \mathbf{D}(V_{\mathcal{X}_{\mathrm{p}}}) \right) = (r+1)r/2
\tag{1288}
$$

[6.6]The fact that the smallest face is isomorphic with another EDM cone (perhaps smaller than $\mathbb{EDM}^N$) is implicit in [191, §2].

in isomorphic $\mathbb{R}^{N(N-1)/2}$. Not all dimensions are represented; *e.g.*, the EDM cone has no two-dimensional faces.

When cardinality $N = 4$ and affine dimension $r = 2$ so that $\mathcal{R}(V_{\mathcal{X}_p})$ is any two-dimensional subspace of three-dimensional $\mathcal{N}(\mathbf{1}^{\mathrm{T}})$ in $\mathbb{R}^4$, for example, then the corresponding face of $\mathbb{EDM}^4$ is isometrically isomorphic with: (1284)

$$\mathbb{EDM}^3 = \{D \in \mathbb{EDM}^3 \mid \mathrm{rank}(VDV) \le 2\} \simeq \mathcal{F}(\mathbb{EDM}^4 \ni \mathbf{D}(V_{\mathcal{X}_p})) \qquad (1289)$$

Each two-dimensional subspace of $\mathcal{N}(\mathbf{1}^{\mathrm{T}})$ corresponds to another three-dimensional face.

Because each and every principal submatrix of an EDM in $\mathbb{EDM}^N$ (§5.14.3) is another EDM [241, §4.1], for example, then each principal submatrix belongs to a particular face of $\mathbb{EDM}^N$.

6.4.3.2 extreme directions of EDM cone

In particular, extreme directions (§2.8.1) of $\mathbb{EDM}^N$ correspond to affine dimension $r = 1$ and are simply represented: for any particular cardinality $N \ge 2$ (§2.8.2) and each and every nonzero vector z in $\mathcal{N}(\mathbf{1}^{\mathrm{T}})$

$$\begin{aligned} \Gamma &\triangleq (z \circ z)\mathbf{1}^{\mathrm{T}} + \mathbf{1}(z \circ z)^{\mathrm{T}} - 2zz^{\mathrm{T}} \in \mathbb{EDM}^N \\ &= \delta(zz^{\mathrm{T}})\mathbf{1}^{\mathrm{T}} + \mathbf{1}\delta(zz^{\mathrm{T}})^{\mathrm{T}} - 2zz^{\mathrm{T}} \end{aligned} \qquad (1290)$$

is an extreme direction corresponding to a one-dimensional face of the EDM cone $\mathbb{EDM}^N$ that is a ray in isomorphic subspace $\mathbb{R}^{N(N-1)/2}$.

Proving this would exercise the fundamental definition (186) of extreme direction. Here is a sketch: Any EDM may be represented

$$\mathbf{D}(V_{\mathcal{X}}) = \delta(V_{\mathcal{X}} V_{\mathcal{X}}^{\mathrm{T}})\mathbf{1}^{\mathrm{T}} + \mathbf{1}\delta(V_{\mathcal{X}} V_{\mathcal{X}}^{\mathrm{T}})^{\mathrm{T}} - 2V_{\mathcal{X}} V_{\mathcal{X}}^{\mathrm{T}} \in \mathbb{EDM}^N \qquad (1266)$$

where matrix $V_{\mathcal{X}}$ (1267) has orthogonal columns. For the same reason (1548) that zz^{T} is an extreme direction of the positive semidefinite cone (§2.9.2.7) for any particular nonzero vector z, there is no conic combination of distinct EDMs (each conically independent of Γ (§2.10)) equal to Γ.   ∎

6.4.3.2.1 Example. *Biorthogonal expansion of an EDM.*
(*confer* §2.13.7.1.1) When matrix D belongs to the EDM cone, nonnegative coordinates for biorthogonal expansion are the eigenvalues $\lambda \in \mathbb{R}^N$ of $-VDV\frac{1}{2}$: For any $D \in \mathbb{S}_h^N$ it holds

$$D = \delta\left(-VDV\tfrac{1}{2}\right)\mathbf{1}^{\mathrm{T}} + \mathbf{1}\delta\left(-VDV\tfrac{1}{2}\right)^{\mathrm{T}} - 2\left(-VDV\tfrac{1}{2}\right) \qquad (1079)$$

By diagonalization $-VDV\frac{1}{2} \triangleq Q\Lambda Q^{\mathrm{T}} \in \mathbb{S}_c^N$ (§A.5.1) we may write

$$\begin{aligned} D &= \delta\left(\sum_{i=1}^N \lambda_i\, q_i q_i^{\mathrm{T}}\right)\mathbf{1}^{\mathrm{T}} + \mathbf{1}\delta\left(\sum_{i=1}^N \lambda_i\, q_i q_i^{\mathrm{T}}\right)^{\mathrm{T}} - 2\sum_{i=1}^N \lambda_i\, q_i q_i^{\mathrm{T}} \\ &= \sum_{i=1}^N \lambda_i \left(\delta(q_i q_i^{\mathrm{T}})\mathbf{1}^{\mathrm{T}} + \mathbf{1}\delta(q_i q_i^{\mathrm{T}})^{\mathrm{T}} - 2q_i q_i^{\mathrm{T}}\right) \end{aligned} \qquad (1291)$$

where q_i is the ith eigenvector of $-VDV\frac{1}{2}$ arranged columnar in orthogonal matrix

$$Q = [\,q_1 \quad q_2 \quad \cdots \quad q_N\,] \in \mathbb{R}^{N \times N} \qquad (399)$$

and where $\{\delta(q_iq_i^{\text{T}})\mathbf{1}^{\text{T}} + \mathbf{1}\delta(q_iq_i^{\text{T}})^{\text{T}} - 2q_iq_i^{\text{T}}\,,\ i=1\ldots N\}$ are extreme directions of some pointed polyhedral cone $\mathcal{K} \subset \mathbb{S}_h^N$ and extreme directions of $\mathbb{EDM}^N$. Invertibility of (1291)

$$
\begin{aligned}
-VDV\tfrac{1}{2} &= -V\sum_{i=1}^{N}\lambda_i\left(\delta(q_iq_i^{\text{T}})\mathbf{1}^{\text{T}} + \mathbf{1}\delta(q_iq_i^{\text{T}})^{\text{T}} - 2q_iq_i^{\text{T}}\right)V\tfrac{1}{2} \\
&= \sum_{i=1}^{N}\lambda_i\, q_iq_i^{\text{T}}
\end{aligned}
\qquad (1292)
$$

implies linear independence of those extreme directions. Then biorthogonal expansion is expressed

$$\operatorname{dvec}D = YY^{\dagger}\operatorname{dvec}D = Y\,\lambda\!\left(-VDV\tfrac{1}{2}\right) \qquad (1293)$$

where

$$Y \triangleq \left[\operatorname{dvec}\!\left(\delta(q_iq_i^{\text{T}})\mathbf{1}^{\text{T}} + \mathbf{1}\delta(q_iq_i^{\text{T}})^{\text{T}} - 2q_iq_i^{\text{T}}\right),\ i=1\ldots N\right] \in \mathbb{R}^{N(N-1)/2 \times N} \qquad (1294)$$

When D belongs to the EDM cone in the subspace of symmetric hollow matrices, unique coordinates $Y^{\dagger}\operatorname{dvec}D$ for this biorthogonal expansion must be the nonnegative eigenvalues λ of $-VDV\frac{1}{2}$. This means D simultaneously belongs to the EDM cone and to the pointed polyhedral cone $\operatorname{dvec}\mathcal{K} = \operatorname{cone}(Y)$. $\qquad\square$

6.4.3.3 open question

Result (1288) is analogous to that for the positive semidefinite cone (222), although the question remains open whether all faces of $\mathbb{EDM}^N$ (whose dimension is less than dimension of the cone) are exposed like they are for the positive semidefinite cone.[6.7] (§2.9.2.3) [355]

6.5 Correspondence to PSD cone $\mathbb{S}_+^{N-1}$

Hayden, Wells, Liu, & Tarazaga [191, §2] assert one-to-one correspondence of EDMs with positive semidefinite matrices in the symmetric subspace. Because $\operatorname{rank}(VDV)\leq N-1$ (§5.7.1.1), that PSD cone corresponding to the EDM cone can only be $\mathbb{S}_+^{N-1}$. [8, §18.2.1] To clearly demonstrate this correspondence, we invoke inner-product form EDM definition

$$
\mathbf{D}(\Phi) = \begin{bmatrix} 0 \\ \delta(\Phi) \end{bmatrix}\mathbf{1}^{\text{T}} + \mathbf{1}\begin{bmatrix} 0 & \delta(\Phi)^{\text{T}} \end{bmatrix} - 2\begin{bmatrix} 0 & \mathbf{0}^{\text{T}} \\ \mathbf{0} & \Phi \end{bmatrix} \in \mathbb{EDM}^N
$$
$$\Leftrightarrow$$
$$\Phi \succeq 0$$
$$\qquad (1097)$$

Then the EDM cone may be expressed

$$\mathbb{EDM}^N = \left\{\mathbf{D}(\Phi)\ \middle|\ \Phi \in \mathbb{S}_+^{N-1}\right\} \qquad (1295)$$

[6.7]Elementary example of face not exposed is given by the closed convex set in Figure 33 and in Figure 43.

Hayden & Wells' assertion can therefore be equivalently stated in terms of an inner-product form EDM operator

$$\mathbf{D}(\mathbb{S}_+^{N-1}) = \mathbb{EDM}^N \qquad (1099)$$

$$\mathbf{V}_{\mathcal{N}}(\mathbb{EDM}^N) = \mathbb{S}_+^{N-1} \qquad (1100)$$

identity (1100) holding because $\mathcal{R}(V_{\mathcal{N}}) = \mathcal{N}(\mathbf{1}^{\mathrm{T}})$ (975), linear functions $\mathbf{D}(\Phi)$ and $\mathbf{V}_{\mathcal{N}}(D) = -V_{\mathcal{N}}^{\mathrm{T}} D V_{\mathcal{N}}$ (§5.6.2.1) being mutually inverse.

In terms of affine dimension r, Hayden & Wells claim particular correspondence between PSD and EDM cones:

$r = N-1$: Symmetric hollow matrices $-D$ positive definite on $\mathcal{N}(\mathbf{1}^{\mathrm{T}})$ correspond to points relatively interior to the EDM cone.

$r < N-1$: Symmetric hollow matrices $-D$ positive semidefinite on $\mathcal{N}(\mathbf{1}^{\mathrm{T}})$, where $-V_{\mathcal{N}}^{\mathrm{T}} D V_{\mathcal{N}}$ has at least one 0 eigenvalue, correspond to points on the relative boundary of the EDM cone.

$r = 1$: Symmetric hollow nonnegative matrices rank-one on $\mathcal{N}(\mathbf{1}^{\mathrm{T}})$ correspond to extreme directions (1290) of the EDM cone; *id est*, for some nonzero vector u (§A.3.1.0.7)

$$\left.\begin{array}{r} \operatorname{rank} V_{\mathcal{N}}^{\mathrm{T}} D V_{\mathcal{N}} = 1 \\[4pt] D \in \mathbb{S}_h^N \cap \mathbb{R}_+^{N \times N} \end{array}\right\} \Leftrightarrow \begin{array}{c} D \in \mathbb{EDM}^N \\[4pt] D \text{ is an extreme direction} \end{array} \Leftrightarrow \left\{\begin{array}{c} -V_{\mathcal{N}}^{\mathrm{T}} D V_{\mathcal{N}} \equiv uu^{\mathrm{T}} \\[4pt] D \in \mathbb{S}_h^N \end{array}\right. \qquad (1296)$$

6.5.0.0.1   Proof.   Case $r = 1$ is easily proved: From the nonnegativity development in §5.8.1, extreme direction (1290), and Schoenberg criterion (987), we need show only sufficiency; *id est*, prove

$$\left.\begin{array}{r} \operatorname{rank} V_{\mathcal{N}}^{\mathrm{T}} D V_{\mathcal{N}} = 1 \\[4pt] D \in \mathbb{S}_h^N \cap \mathbb{R}_+^{N \times N} \end{array}\right\} \Rightarrow \begin{array}{c} D \in \mathbb{EDM}^N \\[4pt] D \text{ is an extreme direction} \end{array}$$

Any symmetric matrix D satisfying the rank condition must have the form, for $z, q \in \mathbb{R}^N$ and nonzero $z \in \mathcal{N}(\mathbf{1}^{\mathrm{T}})$,

$$D = \pm(\mathbf{1}q^{\mathrm{T}} + q\mathbf{1}^{\mathrm{T}} - 2zz^{\mathrm{T}}) \qquad (1297)$$

because (§5.6.2.1, *confer* §E.7.2.0.2)

$$\mathcal{N}(\mathbf{V}_{\mathcal{N}}(D)) = \{\mathbf{1}q^{\mathrm{T}} + q\mathbf{1}^{\mathrm{T}} \mid q \in \mathbb{R}^N\} \subseteq \mathbb{S}^N \qquad (1298)$$

Hollowness demands $q = \delta(zz^{\mathrm{T}})$ while nonnegativity demands choice of positive sign in (1297). Matrix D thus takes the form of an extreme direction (1290) of the EDM cone. ♦

The foregoing proof is not extensible in rank: An EDM with corresponding affine dimension r has the general form, for $\{z_i \in \mathcal{N}(\mathbf{1}^{\mathrm{T}}), \ i = 1 \ldots r\}$ an independent set,

$$D = \mathbf{1}\delta\left(\sum_{i=1}^r z_i z_i^{\mathrm{T}}\right)^{\mathrm{T}} + \delta\left(\sum_{i=1}^r z_i z_i^{\mathrm{T}}\right)\mathbf{1}^{\mathrm{T}} - 2\sum_{i=1}^r z_i z_i^{\mathrm{T}} \in \mathbb{EDM}^N \qquad (1299)$$

The EDM so defined relies principally on the sum $\sum z_i z_i^{\mathrm{T}}$ having positive summand coefficients $(\Leftrightarrow -V_{\mathcal{N}}^{\mathrm{T}} D V_{\mathcal{N}} \succeq 0)^{6.8}$. Then it is easy to find a sum incorporating negative coefficients while meeting rank, nonnegativity, and symmetric hollowness conditions but not positive semidefiniteness on subspace $\mathcal{R}(V_{\mathcal{N}})$; *e.g.*, from page 416,

$$-V \begin{bmatrix} 0 & 1 & 1 \\ 1 & 0 & 5 \\ 1 & 5 & 0 \end{bmatrix} V\frac{1}{2} = z_1 z_1^{\mathrm{T}} - z_2 z_2^{\mathrm{T}} \tag{1300}$$

6.5.0.0.2 Example. *Extreme rays* versus *rays on the boundary*.

The EDM $D = \begin{bmatrix} 0 & 1 & 4 \\ 1 & 0 & 1 \\ 4 & 1 & 0 \end{bmatrix}$ is an extreme direction of $\mathbb{EDM}^3$ where $u = \begin{bmatrix} 1 \\ 2 \end{bmatrix}$ in (1296).

Because $-V_{\mathcal{N}}^{\mathrm{T}} D V_{\mathcal{N}}$ has eigenvalues $\{0, 5\}$, the ray whose direction is D also lies on the relative boundary of $\mathbb{EDM}^3$.

In exception, EDM $D = \kappa \begin{bmatrix} 0 & 1 \\ 1 & 0 \end{bmatrix}$, for any particular $\kappa > 0$, is an extreme direction of $\mathbb{EDM}^2$ but $-V_{\mathcal{N}}^{\mathrm{T}} D V_{\mathcal{N}}$ has only one eigenvalue: $\{\kappa\}$. Because $\mathbb{EDM}^2$ is a ray whose relative boundary (§2.6.1.4.1) is the origin, this conventional boundary does not include D which belongs to the relative interior in this dimension. (§2.7.0.0.1)   □

6.5.1 Gram-form correspondence to $\mathbb{S}_+^{N-1}$

With respect to $\mathbf{D}(G) = \delta(G)\mathbf{1}^{\mathrm{T}} + \mathbf{1}\delta(G)^{\mathrm{T}} - 2G$ (980) the linear Gram-form EDM operator, results in §5.6.1 provide [2, §2.6]

$$\mathbb{EDM}^N = \mathbf{D}\Big(\mathbf{V}(\mathbb{EDM}^N)\Big) \equiv \mathbf{D}\Big(V_{\mathcal{N}}\,\mathbb{S}_+^{N-1} V_{\mathcal{N}}^{\mathrm{T}}\Big) \tag{1301}$$

$$V_{\mathcal{N}}\,\mathbb{S}_+^{N-1} V_{\mathcal{N}}^{\mathrm{T}} \equiv \mathbf{V}\Big(\mathbf{D}\Big(V_{\mathcal{N}}\,\mathbb{S}_+^{N-1} V_{\mathcal{N}}^{\mathrm{T}}\Big)\Big) = \mathbf{V}(\mathbb{EDM}^N) \triangleq -V\,\mathbb{EDM}^N V\tfrac{1}{2} = \mathbb{S}_c^N \cap \mathbb{S}_+^N \tag{1302}$$

a one-to-one correspondence between $\mathbb{EDM}^N$ and $\mathbb{S}_+^{N-1}$.

6.5.2 EDM cone by elliptope

Having defined the elliptope parametrized by scalar $t > 0$

$$\mathcal{E}_t^N = \mathbb{S}_+^N \cap \{\Phi \in \mathbb{S}^N \mid \delta(\Phi) = t\mathbf{1}\} \tag{1177}$$

then following Alfakih [9] we have

$$\mathbb{EDM}^N = \overline{\mathrm{cone}\{\mathbf{1}\mathbf{1}^{\mathrm{T}} - \mathcal{E}_1^N\}} = \overline{\{t(\mathbf{1}\mathbf{1}^{\mathrm{T}} - \mathcal{E}_1^N) \mid t \geq 0\}} \tag{1303}$$

Identification $\mathcal{E}^N = \mathcal{E}_1^N$ equates the standard elliptope (§5.9.1.0.1, Figure 141) to our parametrized elliptope.

6.8 $(\Leftarrow)$ For $a_i \in \mathbb{R}^{N-1}$, let $z_i = V_{\mathcal{N}}^{\dagger\mathrm{T}} a_i$.

$$\text{dvec rel}\,\partial\,\mathbb{EDM}^{\mathbf{3}}$$

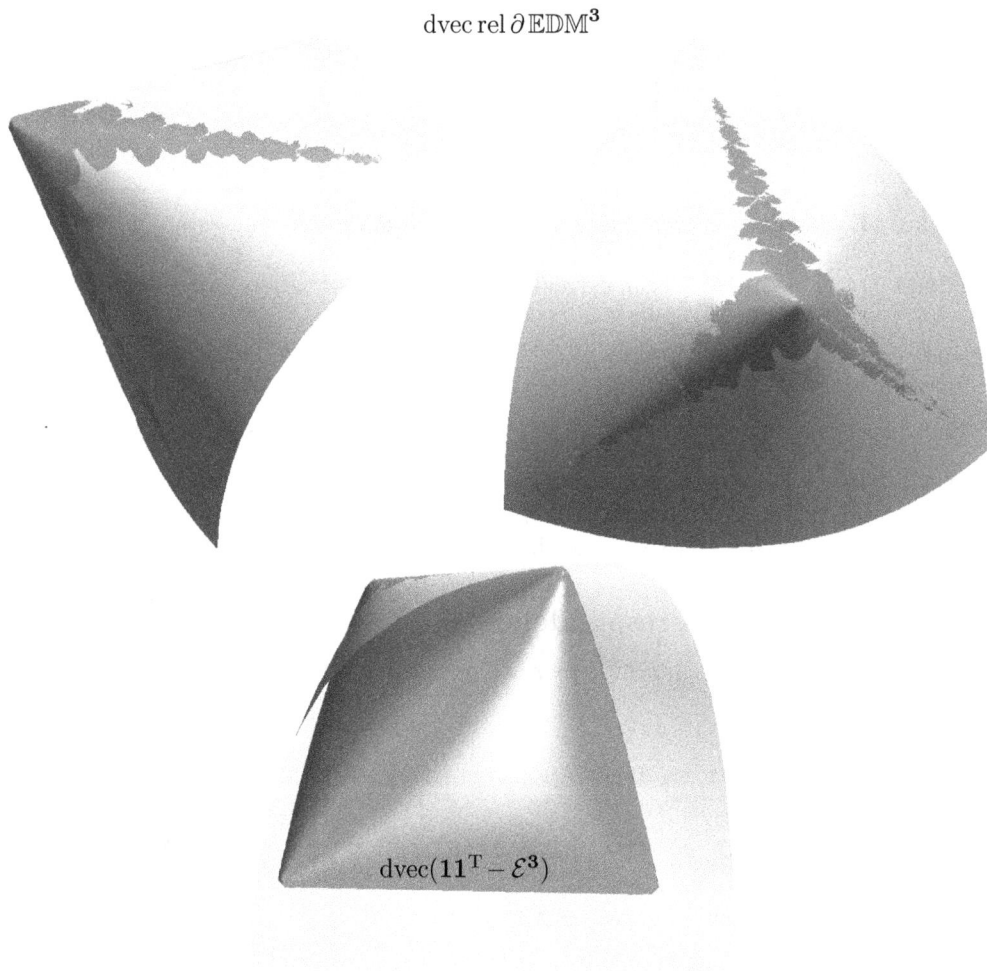

$$\text{dvec}(\mathbf{11}^{\mathrm{T}} - \mathcal{E}^{\mathbf{3}})$$

$$\mathbb{EDM}^N = \overline{\text{cone}\{\mathbf{11}^{\mathrm{T}} - \mathcal{E}^N\}} = \overline{\{t(\mathbf{11}^{\mathrm{T}} - \mathcal{E}^N)\mid t \geq 0\}} \qquad (1303)$$

Figure 152: Three views of translated negated elliptope $\mathbf{11}^{\mathrm{T}} - \mathcal{E}^{\mathbf{3}}_1$ (*confer* Figure 141) shrouded by truncated EDM cone. Fractal on EDM cone relative boundary is numerical artifact belonging to intersection with elliptope relative boundary. The fractal is trying to convey existence of a neighborhood about the origin where the translated elliptope boundary and EDM cone boundary intersect.

6.5.2.0.1   **Expository.**   *Normal cone, tangent cone, elliptope.*
Define $\mathcal{T}_{\mathcal{E}}(\mathbf{11}^{\mathrm{T}})$ to be the *tangent cone* to the elliptope $\mathcal{E}$ at point $\mathbf{11}^{\mathrm{T}}$; *id est*,

$$\mathcal{T}_{\mathcal{E}}(\mathbf{11}^{\mathrm{T}}) \triangleq \overline{\{t(\mathcal{E} - \mathbf{11}^{\mathrm{T}}) \mid t \geq 0\}} \tag{1304}$$

The normal cone $\mathcal{K}_{\mathcal{E}}^{\perp}(\mathbf{11}^{\mathrm{T}})$ to the elliptope at $\mathbf{11}^{\mathrm{T}}$ is a closed convex cone defined (§E.10.3.2.1, Figure 187)

$$\mathcal{K}_{\mathcal{E}}^{\perp}(\mathbf{11}^{\mathrm{T}}) \triangleq \{B \mid \langle B,\, \Phi - \mathbf{11}^{\mathrm{T}}\rangle \leq 0,\ \Phi \in \mathcal{E}\} \tag{1305}$$

The *polar cone* of any set $\mathcal{K}$ is the closed convex cone (*confer* (296))

$$\mathcal{K}^{\circ} \triangleq \{B \mid \langle B,\, A\rangle \leq 0,\ \text{for all } A \in \mathcal{K}\} \tag{1306}$$

The normal cone is well known to be the polar of the tangent cone,

$$\mathcal{K}_{\mathcal{E}}^{\perp}(\mathbf{11}^{\mathrm{T}}) = \mathcal{T}_{\mathcal{E}}(\mathbf{11}^{\mathrm{T}})^{\circ} \tag{1307}$$

and *vice versa*; [205, §A.5.2.4]

$$\mathcal{K}_{\mathcal{E}}^{\perp}(\mathbf{11}^{\mathrm{T}})^{\circ} = \mathcal{T}_{\mathcal{E}}(\mathbf{11}^{\mathrm{T}}) \tag{1308}$$

From Deza & Laurent [116, p.535] we have the EDM cone

$$\mathbb{EDM} = -\mathcal{T}_{\mathcal{E}}(\mathbf{11}^{\mathrm{T}}) \tag{1309}$$

The polar EDM cone is also expressible in terms of the elliptope. From (1307) we have

$$\mathbb{EDM}^{\circ} = -\mathcal{K}_{\mathcal{E}}^{\perp}(\mathbf{11}^{\mathrm{T}}) \tag{1310}$$

$\bigstar$

In §5.10.1 we proposed the expression for EDM D

$$D = t\mathbf{11}^{\mathrm{T}} - \mathfrak{E} \in \mathbb{EDM}^{N} \tag{1178}$$

where $t \in \mathbb{R}_+$ and $\mathfrak{E}$ belongs to the parametrized elliptope $\mathcal{E}_t^N$. We further propose, for any particular $t > 0$

$$\mathbb{EDM}^{N} = \overline{\mathrm{cone}\{t\mathbf{11}^{\mathrm{T}} - \mathcal{E}_t^N\}} \tag{1311}$$

Proof (pending). $\blacksquare$

6.5.2.0.2   **Exercise.**   *EDM cone from elliptope.*
Relationship of the translated negated elliptope with the EDM cone is illustrated in Figure 152. Prove whether it holds that

$$\mathbb{EDM}^{N} = \overline{\lim_{t \to \infty} t\mathbf{11}^{\mathrm{T}} - \mathcal{E}_t^N} \tag{1312}$$

$\blacktriangledown$

6.6   Vectorization & projection interpretation

In §E.7.2.0.2 we learn: $-VDV$ can be interpreted as orthogonal projection [6, §2] of vectorized $-D \in \mathbb{S}_h^N$ on the subspace of geometrically centered symmetric matrices

$$
\begin{aligned}
\mathbb{S}_c^N &= \{G \in \mathbb{S}^N \mid G\mathbf{1} = \mathbf{0}\} & (2094)\\
&= \{G \in \mathbb{S}^N \mid \mathcal{N}(G) \supseteq \mathbf{1}\} = \{G \in \mathbb{S}^N \mid \mathcal{R}(G) \subseteq \mathcal{N}(\mathbf{1}^T)\} & \\
&= \{VYV \mid Y \in \mathbb{S}^N\} \subset \mathbb{S}^N & (2095)\\
&\equiv \{V_\mathcal{N} A V_\mathcal{N}^T \mid A \in \mathbb{S}^{N-1}\} &
\end{aligned}
\qquad (1070)
$$

because elementary auxiliary matrix V is an orthogonal projector (§B.4.1). Yet there is another useful projection interpretation:

Revising the fundamental matrix criterion for membership to the EDM cone (963),[6.9]

$$
\left.\begin{aligned}
\langle zz^T, -D \rangle \geq 0 \quad \forall zz^T \mid \mathbf{1}^T zz^T = \mathbf{0}\\
D \in \mathbb{S}_h^N
\end{aligned}\right\} \Leftrightarrow D \in \mathbb{EDM}^N
\qquad (1313)
$$

this is equivalent, of course, to the Schoenberg criterion

$$
\left.\begin{aligned}
-V_\mathcal{N}^T D V_\mathcal{N} \succeq 0\\
D \in \mathbb{S}_h^N
\end{aligned}\right\} \Leftrightarrow D \in \mathbb{EDM}^N
\qquad (987)
$$

because $\mathcal{N}(\mathbf{1}\mathbf{1}^T) = \mathcal{R}(V_\mathcal{N})$. When $D \in \mathbb{EDM}^N$, correspondence (1313) means $-z^T Dz$ is proportional to a nonnegative coefficient of orthogonal projection (§E.6.4.2, Figure 154) of $-D$ in isometrically isomorphic $\mathbb{R}^{N(N+1)/2}$ on the range of each and every vectorized (§2.2.2.1) symmetric dyad (§B.1) in the nullspace of $\mathbf{1}\mathbf{1}^T$; id est, on each and every member of

$$
\begin{aligned}
\mathcal{T} &\triangleq \left\{\operatorname{svec}(zz^T) \mid z \in \mathcal{N}(\mathbf{1}\mathbf{1}^T) = \mathcal{R}(V_\mathcal{N})\right\} \subset \operatorname{svec} \partial \mathbb{S}_+^N\\
&= \left\{\operatorname{svec}(V_\mathcal{N} vv^T V_\mathcal{N}^T) \mid v \in \mathbb{R}^{N-1}\right\}
\end{aligned}
\qquad (1314)
$$

whose dimension is
$$
\dim \mathcal{T} = N(N-1)/2 \qquad (1315)
$$

The set of all symmetric dyads $\{zz^T \mid z \in \mathbb{R}^N\}$ constitute the extreme directions of the positive semidefinite cone (§2.8.1, §2.9) $\mathbb{S}_+^N$, hence lie on its boundary. Yet only those dyads in $\mathcal{R}(V_\mathcal{N})$ are included in the test (1313), thus only a subset $\mathcal{T}$ of all vectorized extreme directions of $\mathbb{S}_+^N$ is observed.

In the particularly simple case $D \in \mathbb{EDM}^2 = \{D \in \mathbb{S}_h^2 \mid d_{12} \geq 0\}$, for example, only one extreme direction of the PSD cone is involved:

$$
zz^T = \begin{bmatrix} 1 & -1\\ -1 & 1 \end{bmatrix}
\qquad (1316)
$$

Any nonnegative scaling of vectorized zz^T belongs to the set $\mathcal{T}$ illustrated in Figure 153 and Figure 154.

[6.9] $\mathcal{N}(\mathbf{1}\mathbf{1}^T) = \mathcal{N}(\mathbf{1}^T)$ and $\mathcal{R}(zz^T) = \mathcal{R}(z)$

$$\begin{bmatrix} d_{11} & d_{12} \\ d_{12} & d_{22} \end{bmatrix}$$

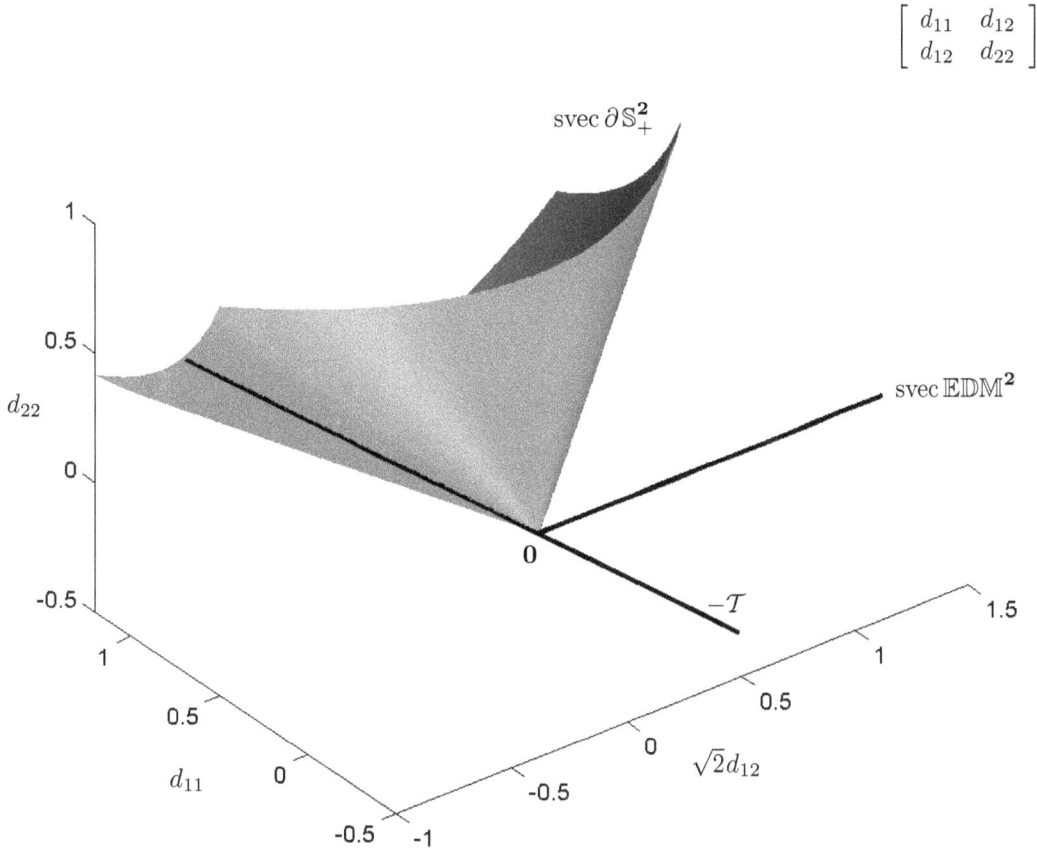

$$\mathcal{T} \triangleq \left\{ \operatorname{svec}(zz^{\mathrm{T}}) \mid z \in \mathcal{N}(\mathbf{11}^{\mathrm{T}}) = \kappa \begin{bmatrix} 1 \\ -1 \end{bmatrix}, \ \kappa \in \mathbb{R} \right\} \subset \operatorname{svec} \partial \mathbb{S}_+^2$$

Figure 153: Truncated boundary of positive semidefinite cone $\mathbb{S}_+^2$ in isometrically isomorphic $\mathbb{R}^3$ (via svec (56)) is, in this dimension, constituted solely by its extreme directions. Truncated cone of Euclidean distance matrices $\mathbb{EDM}^2$ in isometrically isomorphic subspace $\mathbb{R}$. Relative boundary of EDM cone is constituted solely by matrix $\mathbf{0}$. Halfline $\mathcal{T} = \{\kappa^2[1 \ -\sqrt{2} \ 1]^{\mathrm{T}} \mid \kappa \in \mathbb{R}\}$ on PSD cone boundary depicts that lone extreme ray (1316) on which orthogonal projection of $-D$ must be positive semidefinite if D is to belong to $\mathbb{EDM}^2$. aff cone $\mathcal{T} = \operatorname{svec} \mathbb{S}_c^2$. (1321) Dual EDM cone is halfspace in $\mathbb{R}^3$ whose bounding hyperplane has inward-normal svec $\mathbb{EDM}^2$.

$$\begin{bmatrix} d_{11} & d_{12} \\ d_{12} & d_{22} \end{bmatrix}$$

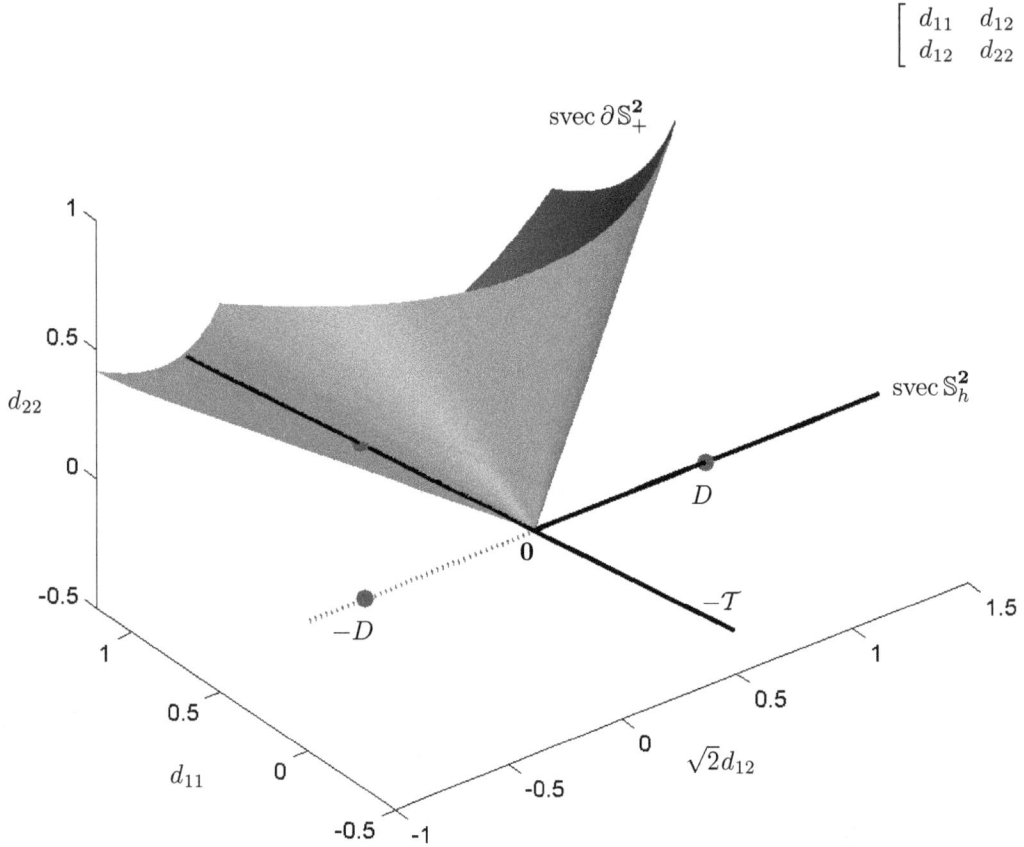

Projection of vectorized $-D$ on range of vectorized zz^{T} :

$$P_{\mathrm{svec}\, zz^{\mathrm{T}}}\bigl(\mathrm{svec}(-D)\bigr) \;=\; \frac{\langle zz^{\mathrm{T}},\, -D\rangle}{\langle zz^{\mathrm{T}},\, zz^{\mathrm{T}}\rangle}\, zz^{\mathrm{T}}$$

$$D \in \mathbb{EDM}^{N} \;\Leftrightarrow\; \begin{cases} \langle zz^{\mathrm{T}}, -D\rangle \geq 0 & \forall zz^{\mathrm{T}} \mid \mathbf{1}\mathbf{1}^{\mathrm{T}} zz^{\mathrm{T}} = \mathbf{0} \\ D \in \mathbb{S}_{h}^{N} \end{cases} \qquad (1313)$$

Figure 154: Given-matrix D is assumed to belong to symmetric hollow subspace $\mathbb{S}_{h}^{2}$; a line in this dimension. Negating D, we find its polar along $\mathbb{S}_{h}^{2}$. Set $\mathcal{T}$ (1314) has only one ray member in this dimension; not orthogonal to $\mathbb{S}_{h}^{2}$. Orthogonal projection of $-D$ on $\mathcal{T}$ (indicated by half dot) has nonnegative projection coefficient. Matrix D must therefore be an EDM.

6.6.1 Face of PSD cone $\mathbb{S}_+^N$ containing V

In any case, set $\mathcal{T}$ (1314) constitutes the vectorized extreme directions of an $N(N-1)/2$-dimensional face of the PSD cone $\mathbb{S}_+^N$ containing auxiliary matrix V; a face isomorphic with $\mathbb{S}_+^{N-1} = \mathbb{S}_+^{\operatorname{rank}V}$ (§2.9.2.3).

To show this, we must first find the smallest face that contains auxiliary matrix V and then determine its extreme directions. From (221),

$$
\begin{aligned}
\mathcal{F}\left(\mathbb{S}_+^N \ni V\right) &= \{W \in \mathbb{S}_+^N \mid \mathcal{N}(W) \supseteq \mathcal{N}(V)\} = \{W \in \mathbb{S}_+^N \mid \mathcal{N}(W) \supseteq \mathbf{1}\} \\
&= \{VYV \succeq 0 \mid Y \in \mathbb{S}^N\} \equiv \{V_{\mathcal{N}}BV_{\mathcal{N}}^{\mathrm{T}} \mid B \in \mathbb{S}_+^{N-1}\} \\
&\simeq \mathbb{S}_+^{\operatorname{rank}V} = -V_{\mathcal{N}}^{\mathrm{T}}\mathrm{EDM}^N V_{\mathcal{N}}
\end{aligned}
\tag{1317}
$$

where the equivalence $\equiv$ is from §5.6.1 while isomorphic equality $\simeq$ with transformed EDM cone is from (1100). Projector V belongs to $\mathcal{F}\left(\mathbb{S}_+^N \ni V\right)$ because $V_{\mathcal{N}}V_{\mathcal{N}}^{\dagger}V_{\mathcal{N}}^{\dagger\mathrm{T}}V_{\mathcal{N}}^{\mathrm{T}} = V$. (§B.4.3) Each and every rank-one matrix belonging to this face is therefore of the form:

$$
V_{\mathcal{N}}\upsilon\upsilon^{\mathrm{T}}V_{\mathcal{N}}^{\mathrm{T}} \mid \upsilon \in \mathbb{R}^{N-1}
\tag{1318}
$$

Because $\mathcal{F}\left(\mathbb{S}_+^N \ni V\right)$ is isomorphic with a positive semidefinite cone $\mathbb{S}_+^{N-1}$, then $\mathcal{T}$ constitutes the vectorized extreme directions of $\mathcal{F}$, the origin constitutes the extreme points of $\mathcal{F}$, and auxiliary matrix V is some convex combination of those extreme points and directions by the *extremes theorem* (§2.8.1.1.1). $\blacklozenge$

In fact the smallest face, that contains auxiliary matrix V, of the PSD cone $\mathbb{S}_+^N$ is the intersection with the geometric center subspace (2094) (2095);

$$
\begin{aligned}
\mathcal{F}\left(\mathbb{S}_+^N \ni V\right) &= \operatorname{cone}\left\{V_{\mathcal{N}}\upsilon\upsilon^{\mathrm{T}}V_{\mathcal{N}}^{\mathrm{T}} \mid \upsilon \in \mathbb{R}^{N-1}\right\} \\
&= \mathbb{S}_c^N \cap \mathbb{S}_+^N \\
&\equiv \{X \succeq 0 \mid \langle X, \mathbf{1}\mathbf{1}^{\mathrm{T}}\rangle = 0\} \quad (1676)
\end{aligned}
\tag{1319}
$$

In isometrically isomorphic $\mathbb{R}^{N(N+1)/2}$

$$
\operatorname{svec}\mathcal{F}\left(\mathbb{S}_+^N \ni V\right) = \operatorname{cone}\mathcal{T}
\tag{1320}
$$

related to $\mathbb{S}_c^N$ by

$$
\operatorname{aff}\operatorname{cone}\mathcal{T} = \operatorname{svec}\mathbb{S}_c^N
\tag{1321}
$$

6.6.2 EDM criteria in $\mathbf{1}\mathbf{1}^{\mathrm{T}}$

(*confer* §6.4, (994)) Laurent specifies an elliptope trajectory condition for EDM cone membership: [241, §2.3]

$$
D \in \mathrm{EDM}^N \Leftrightarrow [1 - e^{-\alpha d_{ij}}] \in \mathrm{EDM}^N \ \forall\alpha > 0
\tag{1172a}
$$

From the parametrized elliptope $\mathcal{E}_t^N$ in §6.5.2 and §5.10.1 we propose

$$D \in \mathbb{EDM}^N \quad \Leftrightarrow \quad \exists \left.\begin{array}{c} t \in \mathbb{R}_+ \\ \mathfrak{E} \in \mathcal{E}_t^N \end{array}\right\} \ni D = t\mathbf{1}\mathbf{1}^\mathrm{T} - \mathfrak{E} \qquad (1322)$$

Chabrillac & Crouzeix [75, §4] prove a different criterion they attribute to Finsler (1937) [148]. We apply it to EDMs: for $D \in \mathbb{S}_h^N$ (1120)

$$-V_\mathcal{N}^\mathrm{T} D V_\mathcal{N} \succ 0 \quad \Leftrightarrow \quad \exists \kappa > 0 \ni -D + \kappa \mathbf{1}\mathbf{1}^\mathrm{T} \succ 0$$
$$\Leftrightarrow \qquad\qquad\qquad\qquad (1323)$$
$$D \in \mathbb{EDM}^N \text{ with corresponding affine dimension } r = N-1$$

This *Finsler criterion* has geometric interpretation in terms of the vectorization & projection already discussed in connection with (1313). With reference to Figure **153**, the offset $\mathbf{1}\mathbf{1}^\mathrm{T}$ is simply a direction orthogonal to $\mathcal{T}$ in isomorphic $\mathbb{R}^\mathbf{3}$. Intuitively, translation of $-D$ in direction $\mathbf{1}\mathbf{1}^\mathrm{T}$ is like orthogonal projection on $\mathcal{T}$ insofar as similar information can be obtained.

When the Finsler criterion (1323) is applied despite lower affine dimension, the constant κ can go to infinity making the test $-D + \kappa\mathbf{1}\mathbf{1}^\mathrm{T} \succeq 0$ impractical for numerical computation. Chabrillac & Crouzeix invent a criterion for the semidefinite case, but is no more practical: for $D \in \mathbb{S}_h^N$

$$D \in \mathbb{EDM}^N$$
$$\Leftrightarrow \qquad\qquad\qquad\qquad\qquad (1324)$$
$$\exists \kappa_\mathrm{p} > 0 \ni \forall \kappa \geq \kappa_\mathrm{p} \, , \ -D - \kappa\mathbf{1}\mathbf{1}^\mathrm{T} \ [sic] \text{ has exactly one negative eigenvalue}$$

6.7 A geometry of completion

> *It is not known how to proceed if one wishes to restrict the dimension of the Euclidean space in which the configuration of points may be constructed.*
>
> −Michael W. Trosset (2000) [361, §1]

Given an incomplete noiseless EDM, intriguing is the question of whether a list in $X \in \mathbb{R}^{n \times N}$ (76) may be reconstructed and under what circumstances reconstruction is unique. [2] [4] [5] [6] [8] [17] [68] [211] [223] [240] [241] [242]

If one or more entries of a particular EDM are fixed, then geometric interpretation of the feasible set of completions is the intersection of the EDM cone $\mathbb{EDM}^N$ in isomorphic subspace $\mathbb{R}^{N(N-1)/2}$ with as many hyperplanes as there are fixed symmetric entries.[6.10] Assuming a nonempty intersection, then the number of completions is generally infinite, and those corresponding to particular affine dimension $r < N-1$ belong to some generally nonconvex subset of that intersection (*confer* §2.9.2.9.2) that can be as small as a point.

[6.10]Depicted in Figure **155a** is an intersection of the EDM cone $\mathbb{EDM}^3$ with a single hyperplane representing the set of all EDMs having one fixed symmetric entry. This representation is practical because it is easily combined with or replaced by other convex constraints; *e.g.*, slab inequalities in (781) that admit bounding of noise processes.

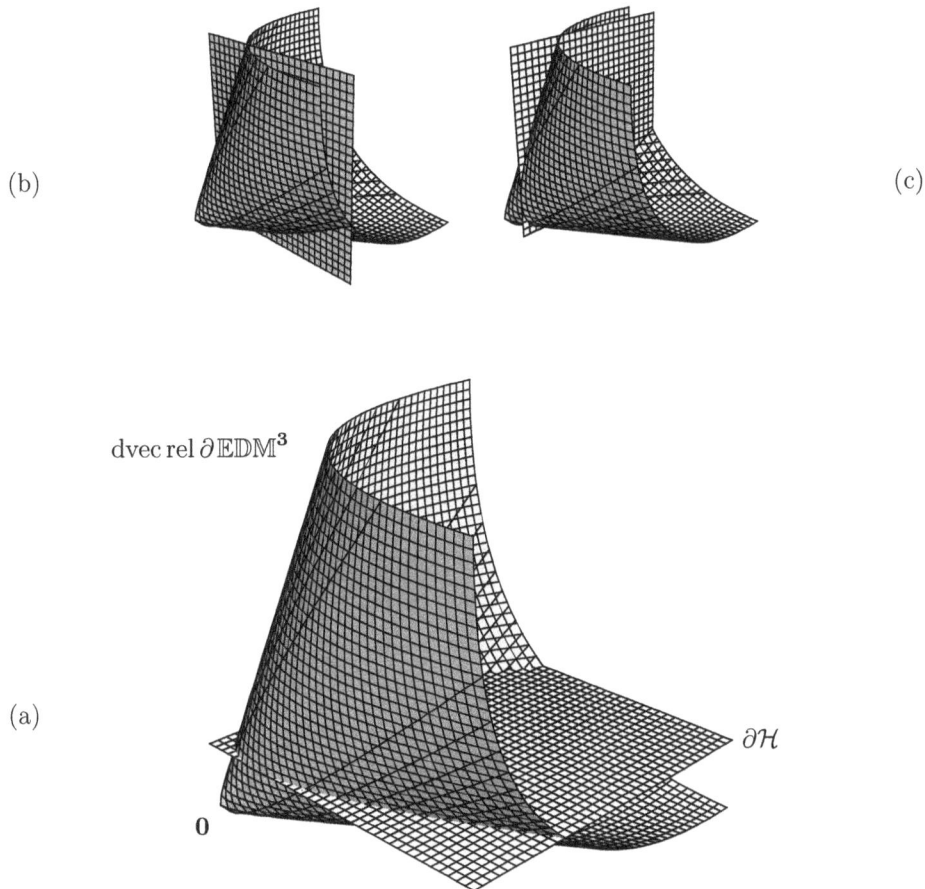

Figure 155:   (a) In isometrically isomorphic subspace $\mathbb{R}^3$, intersection of $\mathbb{EDM}^3$ with hyperplane $\partial\mathcal{H}$ representing one fixed symmetric entry $d_{23} = \kappa$ (both drawn truncated, rounded vertex is artifact of plot).   EDMs in this dimension corresponding to affine dimension 1 comprise relative boundary of EDM cone (§6.5). Since intersection illustrated includes a nontrivial subset of cone's relative boundary, then it is apparent there exist infinitely many EDM completions corresponding to affine dimension 1.   In this dimension it is impossible to represent a unique nonzero completion corresponding to affine dimension 1, for example, using a single hyperplane because any hyperplane supporting relative boundary at a particular point Γ contains an entire ray $\{\zeta\Gamma \mid \zeta \geq 0\}$ belonging to rel$\partial\mathbb{EDM}^3$ by Lemma 2.8.0.0.1.   (b) $d_{13} = \kappa$.   (c) $d_{12} = \kappa$.

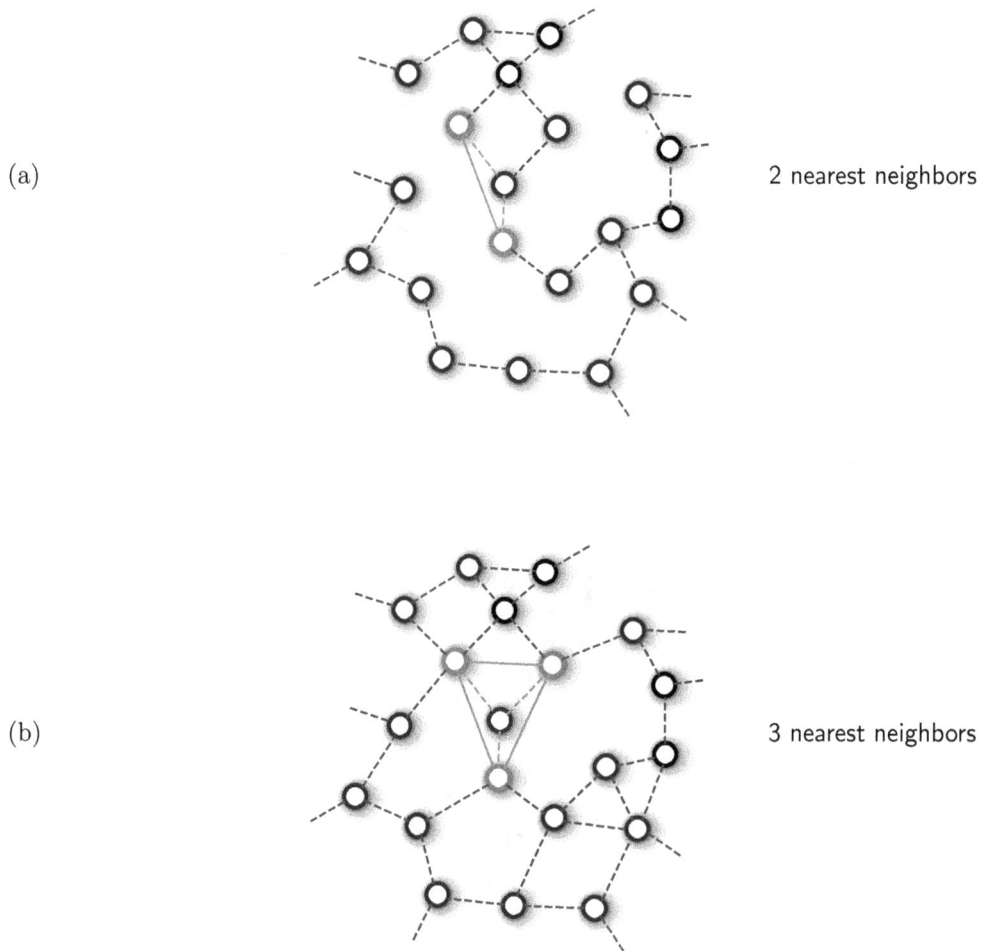

(a)   2 nearest neighbors

(b)   3 nearest neighbors

Figure 156: Neighborhood graph (dashed) with one dimensionless EDM subgraph completion (solid) superimposed (but not obscuring dashed). Local view of a few dense samples ○ from relative interior of some arbitrary Euclidean manifold whose affine dimension appears two-dimensional in this neighborhood. All line segments measure absolute distance. Dashed line segments help visually locate nearest neighbors; suggesting, best number of nearest neighbors can be greater than value of embedding dimension after topological transformation. (*confer* [218, §2]) Solid line segments represent completion of EDM subgraph from available distance data for an arbitrarily chosen sample and its nearest neighbors. Each distance from EDM subgraph becomes distance-square in corresponding EDM submatrix.

6.7.0.0.1   Example.   *Maximum variance unfolding.*   [400]

A process minimizing affine dimension (§2.1.5) of certain kinds of Euclidean *manifold* by topological transformation can be posed as a completion problem (*confer* §E.10.2.1.2). Weinberger & Saul, who originated the technique, specify an applicable manifold in three dimensions by analogy to an ordinary sheet of paper (*confer* §2.1.6); imagine, we find it deformed from flatness in some way introducing neither holes, tears, or self-intersections. [400, §2.2] The physical process is intuitively described as *unfurling, unfolding, diffusing, decompacting,* or *unraveling.* In particular instances, the process is a sort of flattening by stretching until taut (but not by crushing); *e.g.,* unfurling a three-dimensional Euclidean body resembling a billowy national flag reduces that manifold's affine dimension to $r = 2$.

Data input to the proposed process originates from distances between neighboring relatively dense samples of a given manifold. Figure 156 realizes a densely sampled neighborhood; called, *neighborhood graph.* Essentially, the algorithmic process preserves local isometry between *nearest neighbors* allowing distant neighbors to excurse expansively by "maximizing variance" (Figure 5). The common number of nearest neighbors to each sample is a data-dependent algorithmic parameter whose minimum value connects the graph. The dimensionless *EDM subgraph* between each sample and its nearest neighbors is completed from available data and included as input; one such EDM subgraph completion is drawn superimposed upon the neighborhood graph in Figure 156.[6.11] The consequent dimensionless EDM graph comprising all the subgraphs is incomplete, in general, because the neighbor number is relatively small; incomplete even though it is a superset of the neighborhood graph. Remaining distances (those not graphed at all) are squared then made variables within the algorithm; it is this variability that admits unfurling.

To demonstrate, consider untying the *trefoil knot* drawn in Figure 157a. A corresponding EDM $D = [d_{ij} \ , \ i, j = 1 \ldots N]$ employing only 2 nearest neighbors is banded having the incomplete form

$$D = \begin{bmatrix} 0 & \check{d}_{12} & \check{d}_{13} & ? & \cdots & ? & \check{d}_{1,N-1} & \check{d}_{1N} \\ \check{d}_{12} & 0 & \check{d}_{23} & \check{d}_{24} & \ddots & ? & ? & \check{d}_{2N} \\ \check{d}_{13} & \check{d}_{23} & 0 & \check{d}_{34} & \ddots & ? & ? & ? \\ ? & \check{d}_{24} & \check{d}_{34} & 0 & \ddots & \ddots & ? & ? \\ \vdots & \ddots & \ddots & \ddots & \ddots & \ddots & \ddots & ? \\ ? & ? & ? & \ddots & \ddots & 0 & \check{d}_{N-2,N-1} & \check{d}_{N-2,N} \\ \check{d}_{1,N-1} & ? & ? & ? & \ddots & \check{d}_{N-2,N-1} & 0 & \check{d}_{N-1,N} \\ \check{d}_{1N} & \check{d}_{2N} & ? & ? & ? & \check{d}_{N-2,N} & \check{d}_{N-1,N} & 0 \end{bmatrix} \qquad (1325)$$

where $\check{d}_{ij}$ denotes a given fixed distance-square. The unfurling algorithm can be expressed

---

[6.11]Local reconstruction of point position, from the EDM submatrix corresponding to a complete dimensionless EDM subgraph, is unique to within an isometry (§5.6, §5.12).

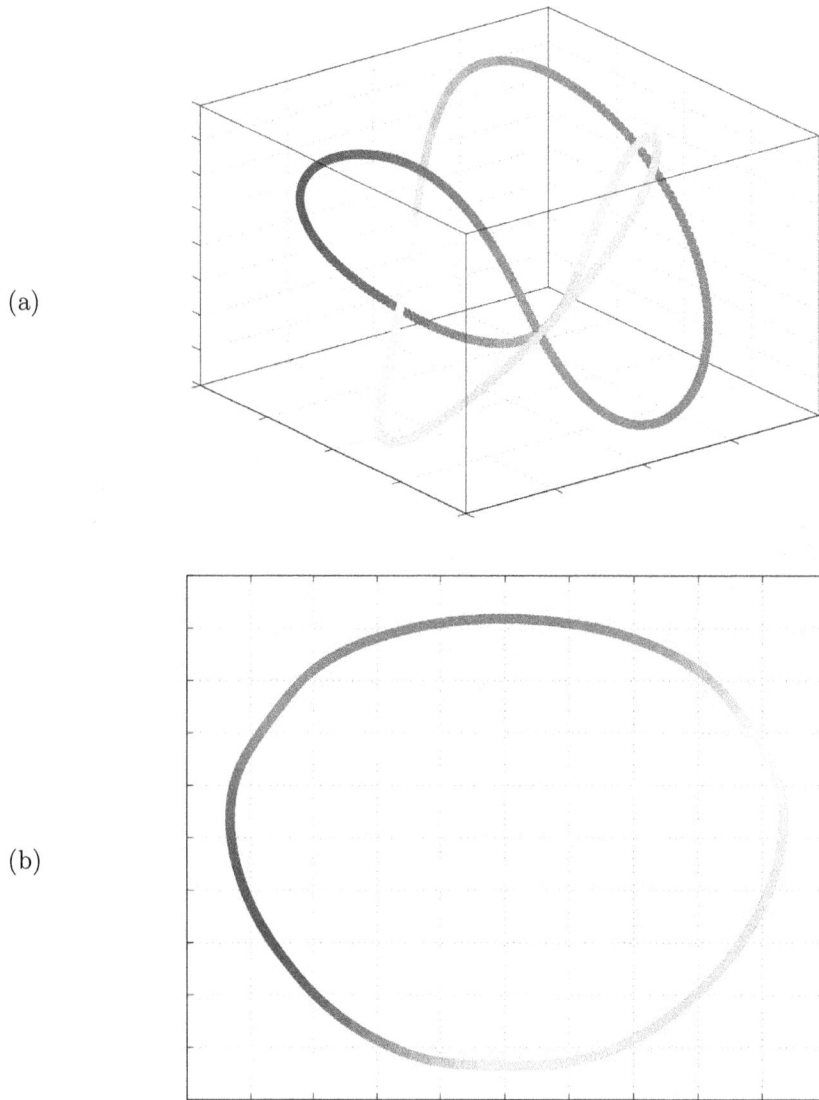

(a)

(b)

Figure 157: (a) *Trefoil knot* in $\mathbb{R}^3$ from Weinberger & Saul [400].   (b) Topological transformation algorithm employing 4 nearest neighbors and $N = 539$ samples reduces affine dimension of knot to $r = 2$. Choosing instead 2 nearest neighbors would make this embedding more circular.

as an optimization problem; constrained total distance-square maximization:

$$
\begin{aligned}
\underset{D}{\text{maximize}} \quad & \langle -V\,,\,D\rangle \\
\text{subject to} \quad & \langle D\,,\,e_i e_j^{\mathrm{T}} + e_j e_i^{\mathrm{T}}\rangle \tfrac{1}{2} = \check{d}_{ij} \qquad \forall (i,j)\in\mathcal{I} \\
& \operatorname{rank}(VDV) = 2 \\
& D \in \mathbb{EDM}^N
\end{aligned}
\tag{1326}
$$

where $e_i \in \mathbb{R}^N$ is the ith member of the standard basis, where set $\mathcal{I}$ indexes the given distance-square data like that in (1325), where $V \in \mathbb{R}^{N\times N}$ is the geometric centering matrix (§B.4.1), and where

$$
\langle -V\,,\,D\rangle \;=\; \operatorname{tr}(-VDV) \;=\; 2\operatorname{tr} G \;=\; \frac{1}{N}\sum_{i,j} d_{ij} \tag{992}
$$

where G is the Gram matrix producing D assuming $G\mathbf{1}=\mathbf{0}$.

If the (rank) constraint on affine dimension is ignored, then problem (1326) becomes convex, a corresponding solution $D^\star$ can be found, and a nearest rank-2 solution is then had by ordered eigenvalue decomposition of $-VD^\star V$ followed by *spectral projection* (§7.1.3) on $\begin{bmatrix}\mathbb{R}^2_+\\ \mathbf{0}\end{bmatrix} \subset \mathbb{R}^N$. This two-step process is necessarily suboptimal. Yet because the decomposition for the trefoil knot reveals only two dominant eigenvalues, the spectral projection is nearly benign. Such a reconstruction of point position (§5.12) utilizing 4 nearest neighbors is drawn in Figure 157b; a low-dimensional embedding of the trefoil knot.

This problem (1326) can, of course, be written equivalently in terms of Gram matrix G, facilitated by (998); *videlicet*, for Φ_{ij} as in (966)

$$
\begin{aligned}
\underset{G\in\mathbb{S}^N_c}{\text{maximize}} \quad & \langle I\,,\,G\rangle \\
\text{subject to} \quad & \langle G\,,\,\Phi_{ij}\rangle = \check{d}_{ij} \qquad \forall (i,j)\in\mathcal{I} \\
& \operatorname{rank} G = 2 \\
& G \succeq 0
\end{aligned}
\tag{1327}
$$

The advantage to converting EDM to Gram is: Gram matrix G is a bridge between point list X and EDM D; constraints on any or all of these three variables may now be introduced. (Example 5.4.2.2.8) Confinement to the geometric center subspace $\mathbb{S}^N_c$ (implicit constraint $G\mathbf{1}=\mathbf{0}$) keeps G independent of its translation-invariant subspace $\mathbb{S}^{N\perp}_c$ (§5.5.1.1, Figure 159) so as not to become numerically unbounded.

A problem dual to *maximum variance unfolding problem* (1327) (less the Gram rank constraint) has been called the *fastest mixing Markov process*. That dual has simple interpretations in graph and circuit theory and in mechanical and thermal systems, explored in [347], and has direct application to quick calculation of *PageRank* by search engines [237, §4]. Optimal Gram rank turns out to be tightly bounded above by minimum multiplicity of the second smallest eigenvalue of a dual optimal variable. □

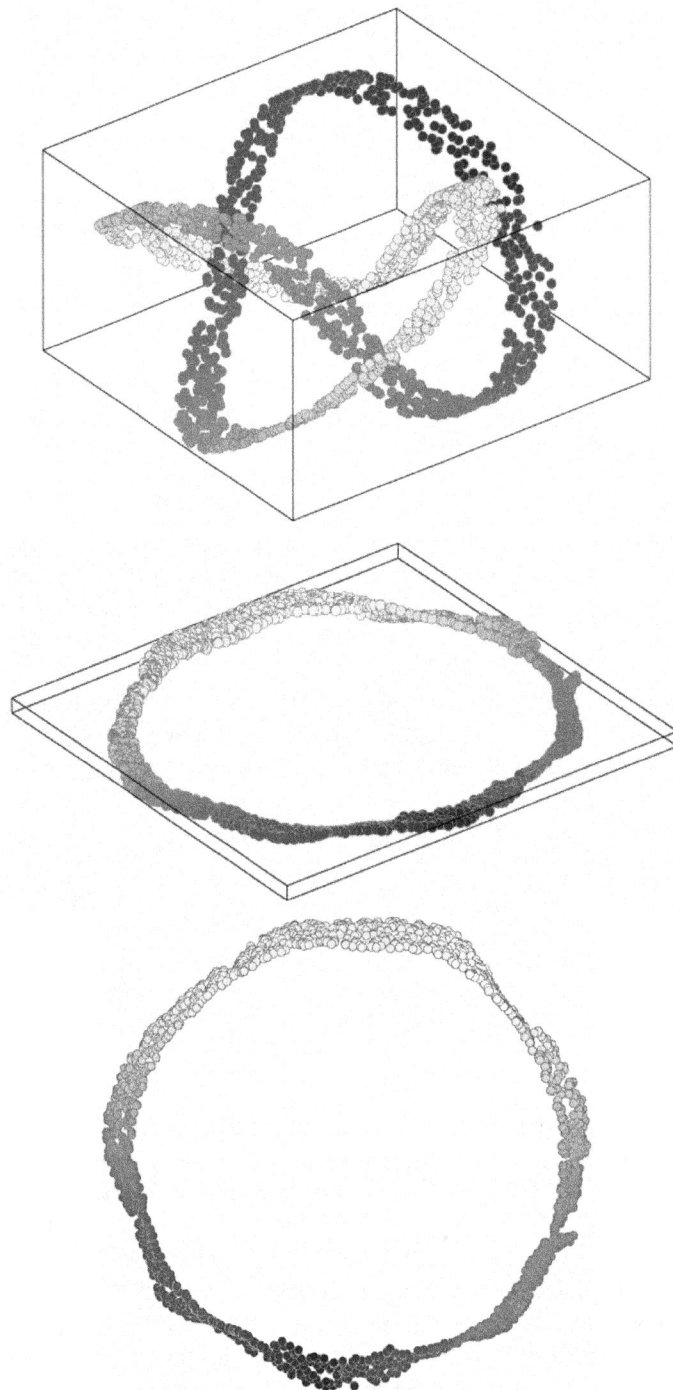

Figure 158: *Trefoil ribbon*; courtesy, Kilian Weinberger. Same topological transformation algorithm with 5 nearest neighbors and $N = 1617$ samples.

6.8 Dual EDM cone

6.8.1 Ambient $\mathbb{S}^N$

We consider finding the ordinary dual EDM cone in ambient space $\mathbb{S}^N$ where $\mathbb{EDM}^N$ is pointed, closed, convex, but not full-dimensional. The set of all EDMs in $\mathbb{S}^N$ is a closed convex cone because it is the intersection of halfspaces about the origin in vectorized variable D (§2.4.1.1.1, §2.7.2):

$$\mathbb{EDM}^N = \bigcap_{\substack{z \in \mathcal{N}(\mathbf{1}^T) \\ i=1...N}} \left\{ D \in \mathbb{S}^N \mid \langle e_i e_i^T, D \rangle \geq 0, \; \langle e_i e_i^T, D \rangle \leq 0, \; \langle zz^T, -D \rangle \geq 0 \right\} \quad (1328)$$

By definition (296), dual cone $\mathcal{K}^*$ comprises each and every vector inward-normal to a hyperplane supporting convex cone $\mathcal{K}$. The dual EDM cone in the ambient space of symmetric matrices is therefore expressible as the aggregate of every conic combination of inward-normals from (1328):

$$
\begin{aligned}
\mathbb{EDM}^{N*} &= \operatorname{cone}\{e_i e_i^T, \; -e_j e_j^T \mid i,j = 1 \ldots N\} \; - \operatorname{cone}\{zz^T \mid \mathbf{1}^T z z^T = \mathbf{0}\} \\
&= \{\sum_{i=1}^N \zeta_i e_i e_i^T - \sum_{j=1}^N \xi_j e_j e_j^T \mid \zeta_i, \xi_j \geq 0\} \; - \operatorname{cone}\{zz^T \mid \mathbf{1}^T z z^T = \mathbf{0}\} \\
&= \{\delta(u) \mid u \in \mathbb{R}^N\} - \operatorname{cone}\left\{ V_{\mathcal{N}} v v^T V_{\mathcal{N}}^T \mid v \in \mathbb{R}^{N-1}, \, (\|v\| = 1) \right\} \subset \mathbb{S}^N \\
&= \{\delta^2(Y) - V_{\mathcal{N}} \Psi V_{\mathcal{N}}^T \mid Y \in \mathbb{S}^N, \; \Psi \in \mathbb{S}_+^{N-1}\}
\end{aligned}
\quad (1329)
$$

The EDM cone is not selfdual in ambient $\mathbb{S}^N$ because its affine hull belongs to a proper subspace

$$\operatorname{aff} \mathbb{EDM}^N = \mathbb{S}_h^N \quad (1330)$$

The ordinary dual EDM cone cannot, therefore, be pointed. (§2.13.1.1)

When $N = 1$, the EDM cone is the point at the origin in $\mathbb{R}$. Auxiliary matrix $V_{\mathcal{N}}$ is empty $[\emptyset]$, and dual cone $\mathbb{EDM}^*$ is the real line.

When $N = 2$, the EDM cone is a nonnegative real line in isometrically isomorphic $\mathbb{R}^3$; there $\mathbb{S}_h^2$ is a real line containing the EDM cone. Dual cone $\mathbb{EDM}^{2*}$ is the particular halfspace in $\mathbb{R}^3$ whose boundary has inward-normal $\mathbb{EDM}^2$. Diagonal matrices $\{\delta(u)\}$ in (1329) are represented by a hyperplane through the origin $\{\underline{d} \mid [0 \; 1 \; 0]\underline{d} = 0\}$ while the term $\operatorname{cone}\{V_{\mathcal{N}} v v^T V_{\mathcal{N}}^T\}$ is represented by the halfline $\mathcal{T}$ in Figure 153 belonging to the positive semidefinite cone boundary. The dual EDM cone is formed by translating the hyperplane along the negative semidefinite halfline $-\mathcal{T}$; the union of each and every translation. (*confer* §2.10.2.0.1)

When cardinality N exceeds 2, the dual EDM cone can no longer be polyhedral simply because the EDM cone cannot. (§2.13.1.1)

6.8.1.1 EDM cone and its dual in ambient $\mathbb{S}^N$

Consider the two convex cones

$$\mathcal{K}_1 \triangleq \mathbb{S}_h^N$$
$$\mathcal{K}_2 \triangleq \bigcap_{y\in\mathcal{N}(\mathbf{1}^{\mathrm{T}})}\left\{A\in\mathbb{S}^N \mid \langle yy^{\mathrm{T}},-A\rangle \geq 0\right\}$$
$$= \left\{A\in\mathbb{S}^N \mid -z^{\mathrm{T}}VAVz \geq 0 \quad \forall zz^{\mathrm{T}}(\succeq 0)\right\}$$
$$= \left\{A\in\mathbb{S}^N \mid -VAV \succeq 0\right\} \tag{1331}$$

so

$$\mathcal{K}_1 \cap \mathcal{K}_2 = \mathbb{EDM}^N \tag{1332}$$

Dual cone $\mathcal{K}_1^* = \mathbb{S}_h^{N\perp} \subseteq \mathbb{S}^N$ (72) is the subspace of diagonal matrices. From (1329) via (313),

$$\mathcal{K}_2^* = -\operatorname{cone}\left\{V_{\mathcal{N}}vv^{\mathrm{T}}V_{\mathcal{N}}^{\mathrm{T}} \mid v\in\mathbb{R}^{N-1}\right\} \subset \mathbb{S}^N \tag{1333}$$

Gaffke & Mathar [152, §5.3] observe that projection on $\mathcal{K}_1$ and $\mathcal{K}_2$ have simple closed forms: Projection on subspace $\mathcal{K}_1$ is easily performed by symmetrization and zeroing the main diagonal or *vice versa*, while projection of $H\in\mathbb{S}^N$ on $\mathcal{K}_2$ is

$$P_{\mathcal{K}_2}H = H - P_{\mathbb{S}_+^N}(VHV) \tag{1334}$$

Proof. First, we observe membership of $H-P_{\mathbb{S}_+^N}(VHV)$ to $\mathcal{K}_2$ because

$$P_{\mathbb{S}_+^N}(VHV) - H = \left(P_{\mathbb{S}_+^N}(VHV) - VHV\right) + (VHV - H) \tag{1335}$$

The term $P_{\mathbb{S}_+^N}(VHV) - VHV$ necessarily belongs to the (dual) positive semidefinite cone by Theorem E.9.2.0.1. $V^2=V$, hence

$$-V\left(H-P_{\mathbb{S}_+^N}(VHV)\right)V \succeq 0 \tag{1336}$$

by Corollary A.3.1.0.5.
   Next, we require

$$\langle P_{\mathcal{K}_2}H - H , P_{\mathcal{K}_2}H\rangle = 0 \tag{1337}$$

Expanding,

$$\langle -P_{\mathbb{S}_+^N}(VHV), H-P_{\mathbb{S}_+^N}(VHV)\rangle = 0 \tag{1338}$$
$$\langle P_{\mathbb{S}_+^N}(VHV),(P_{\mathbb{S}_+^N}(VHV)-VHV)+(VHV-H)\rangle = 0 \tag{1339}$$
$$\langle P_{\mathbb{S}_+^N}(VHV),(VHV-H)\rangle = 0 \tag{1340}$$

Product VHV belongs to the geometric center subspace; (§E.7.2.0.2)

$$VHV\in\mathbb{S}_c^N = \{Y\in\mathbb{S}^N \mid \mathcal{N}(Y)\supseteq\mathbf{1}\} \tag{1341}$$

Diagonalize $VHV \triangleq Q\Lambda Q^{\mathrm{T}}$ (§A.5) whose nullspace is spanned by the eigenvectors corresponding to 0 eigenvalues by Theorem A.7.3.0.1. Projection of VHV on the PSD cone (§7.1) simply zeroes negative eigenvalues in diagonal matrix Λ. Then

$$\mathcal{N}(P_{\mathbb{S}_+^N}(VHV)) \supseteq \mathcal{N}(VHV) \ (\supseteq \mathcal{N}(V)) \tag{1342}$$

from which it follows:

$$P_{\mathbb{S}_+^N}(VHV) \in \mathbb{S}_c^N \tag{1343}$$

so $P_{\mathbb{S}_+^N}(VHV) \perp (VHV - H)$ because $VHV - H \in \mathbb{S}_c^{N\perp}$.

Finally, we must have $P_{\mathcal{K}_2}H - H = -P_{\mathbb{S}_+^N}(VHV) \in \mathcal{K}_2^*$. From §6.6.1 we know dual cone $\mathcal{K}_2^* = -\mathcal{F}\left(\mathbb{S}_+^N \ni V\right)$ is the negative of the positive semidefinite cone's smallest face that contains auxiliary matrix V. Thus $P_{\mathbb{S}_+^N}(VHV) \in \mathcal{F}\left(\mathbb{S}_+^N \ni V\right) \Leftrightarrow \mathcal{N}(P_{\mathbb{S}_+^N}(VHV)) \supseteq \mathcal{N}(V)$ (§2.9.2.3) which was already established in (1342). $\blacklozenge$

From results in §E.7.2.0.2, we know matrix product VHV is the orthogonal projection of $H \in \mathbb{S}^N$ on the geometric center subspace $\mathbb{S}_c^N$. Thus the projection product

$$P_{\mathcal{K}_2}H = H - P_{\mathbb{S}_+^N}P_{\mathbb{S}_c^N}H \tag{1344}$$

6.8.1.1.1 Lemma. *Projection on PSD cone $\cap$ geometric center subspace.*

$$P_{\mathbb{S}_+^N \cap \mathbb{S}_c^N} = P_{\mathbb{S}_+^N}P_{\mathbb{S}_c^N} \tag{1345}$$

$\diamond$

Proof. For each and every $H \in \mathbb{S}^N$, projection of $P_{\mathbb{S}_c^N}H$ on the positive semidefinite cone remains in the geometric center subspace

$$\begin{aligned}
\mathbb{S}_c^N &= \{G \in \mathbb{S}^N \mid G\mathbf{1} = \mathbf{0}\} &&(2094) \\
&= \{G \in \mathbb{S}^N \mid \mathcal{N}(G) \supseteq \mathbf{1}\} = \{G \in \mathbb{S}^N \mid \mathcal{R}(G) \subseteq \mathcal{N}(\mathbf{1}^{\mathrm{T}})\} &&(1070) \\
&= \{VYV \mid Y \in \mathbb{S}^N\} \subset \mathbb{S}^N &&(2095)
\end{aligned}$$

That is because: eigenvectors of $P_{\mathbb{S}_c^N}H$ corresponding to 0 eigenvalues span its nullspace $\mathcal{N}(P_{\mathbb{S}_c^N}H)$. (§A.7.3.0.1) To project $P_{\mathbb{S}_c^N}H$ on the positive semidefinite cone, its negative eigenvalues are zeroed. (§7.1.2) The nullspace is thereby expanded while eigenvectors originally spanning $\mathcal{N}(P_{\mathbb{S}_c^N}H)$ remain intact. Because the geometric center subspace is invariant to projection on the PSD cone, then the rule for projection on a convex set in a subspace governs (§E.9.5, projectors do not commute) and statement (1345) follows directly. $\blacklozenge$

From the lemma it follows

$$\{P_{\mathbb{S}_+^N}P_{\mathbb{S}_c^N}H \mid H \in \mathbb{S}^N\} = \{P_{\mathbb{S}_+^N \cap \mathbb{S}_c^N}H \mid H \in \mathbb{S}^N\} \tag{1346}$$

$$\mathbb{EDM}^2 = \mathbb{S}_h^2 \cap \left(\mathbb{S}_c^{2\perp} - \mathbb{S}_+^2 \right)$$

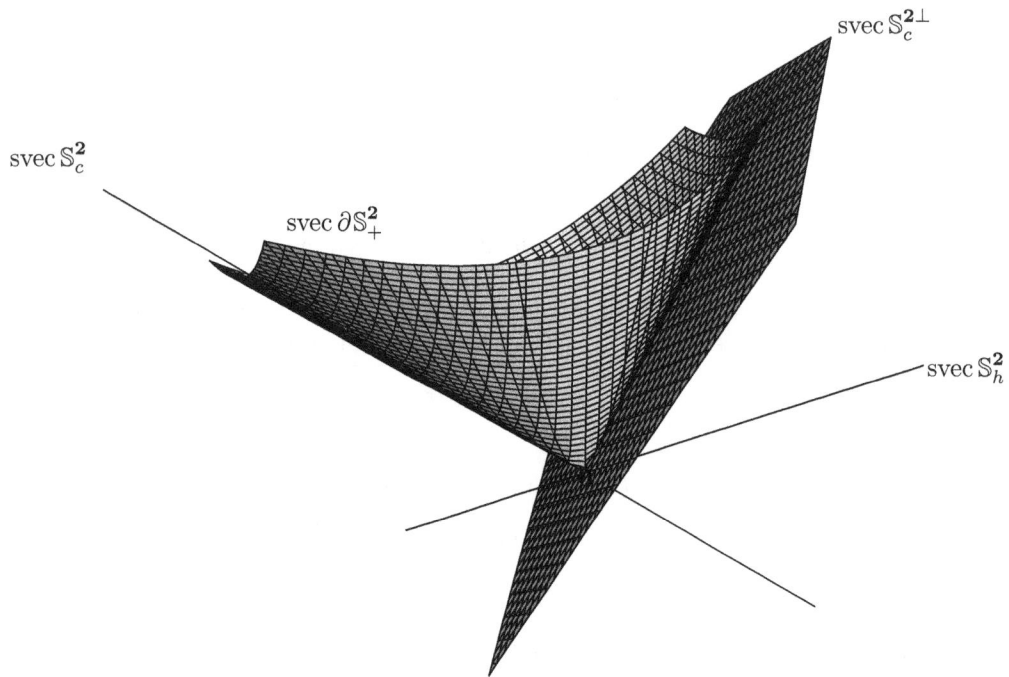

Figure 159: Orthogonal complement $\mathbb{S}_c^{2\perp}$ (2096) (§B.2) of geometric center subspace (a plane in isometrically isomorphic $\mathbb{R}^3$; drawn is a tiled fragment) apparently supporting positive semidefinite cone. (Rounded vertex is artifact of plot.) Line svec $\mathbb{S}_c^2 =$ aff cone $\mathcal{T}$ (1321) intersects svec $\partial\mathbb{S}_+^2$, also drawn in Figure **153**; it runs along PSD cone boundary. (*confer* Figure **140**)

$$\mathbb{EDM}^2 = \mathbb{S}_h^2 \cap \left(\mathbb{S}_c^{2\perp} - \mathbb{S}_+^2 \right)$$

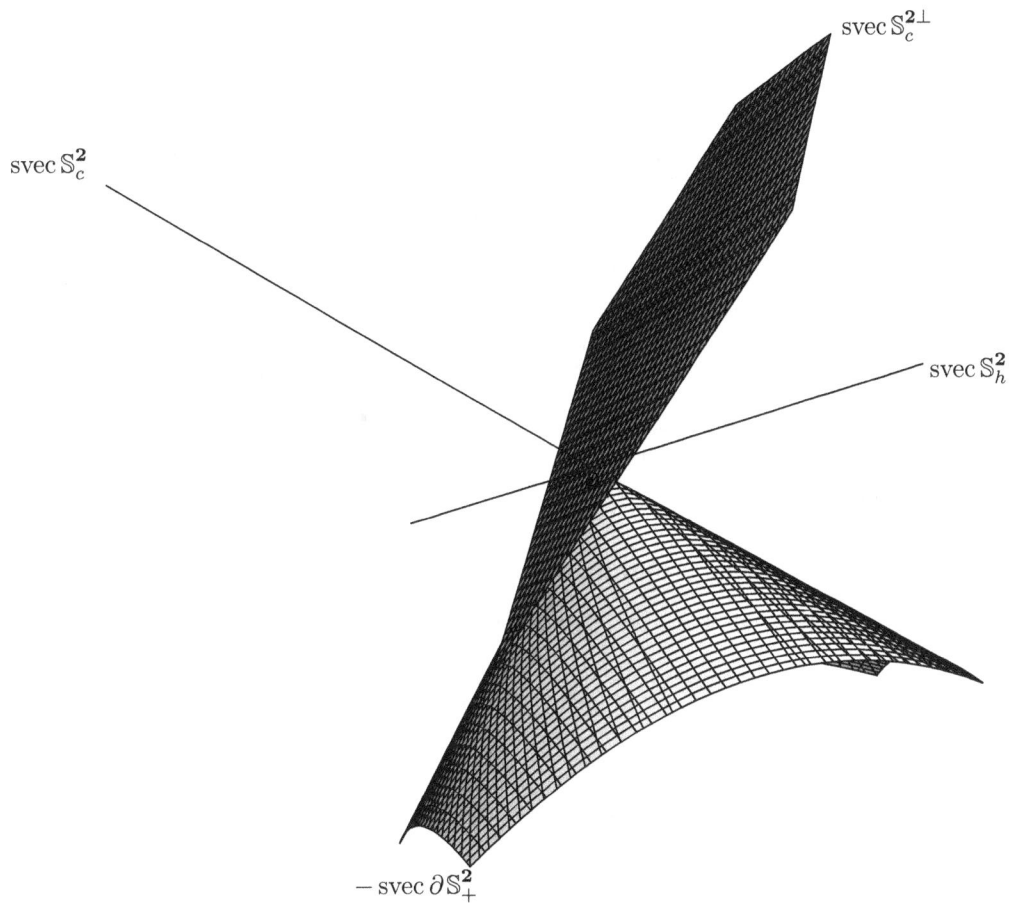

Figure 160: EDM cone construction in isometrically isomorphic $\mathbb{R}^3$ by adding polar PSD cone to $\text{svec}\,\mathbb{S}_c^{2\perp}$. Difference $\text{svec}\left(\mathbb{S}_c^{2\perp} - \mathbb{S}_+^2 \right)$ is halfspace partially bounded by $\text{svec}\,\mathbb{S}_c^{2\perp}$. EDM cone is nonnegative halfline along $\text{svec}\,\mathbb{S}_h^2$ in this dimension.

Then from (2121)

$$-\left(\mathbb{S}_c^N \cap \mathbb{S}_+^N\right)^* = \{H - P_{\mathbb{S}_+^N} P_{\mathbb{S}_c^N} H \mid H \in \mathbb{S}^N\} \tag{1347}$$

From (313) we get closure of a vector sum

$$\mathcal{K}_2 = -\left(\mathbb{S}_c^N \cap \mathbb{S}_+^N\right)^* = \mathbb{S}_c^{N\perp} - \mathbb{S}_+^N \tag{1348}$$

therefore the equality [102]

$$\mathbb{EDM}^N = \mathcal{K}_1 \cap \mathcal{K}_2 = \mathbb{S}_h^N \cap \left(\mathbb{S}_c^{N\perp} - \mathbb{S}_+^N\right) \tag{1349}$$

whose veracity is intuitively evident, in hindsight, [91, p.109] from the most fundamental EDM definition (968). Formula (1349) is not a matrix criterion for membership to the EDM cone, it is not an EDM definition, and it is not an equivalence between EDM operators or an isomorphism. Rather, it is a recipe for constructing the EDM cone whole from large Euclidean bodies: the positive semidefinite cone, orthogonal complement of the geometric center subspace, and symmetric hollow subspace. A realization of this construction in low dimension is illustrated in Figure 159 and Figure 160.

The dual EDM cone follows directly from (1349) by standard properties of cones (§2.13.1.1):

$$\mathbb{EDM}^{N*} = \overline{\mathcal{K}_1^* + \mathcal{K}_2^*} = \mathbb{S}_h^{N\perp} - \mathbb{S}_c^N \cap \mathbb{S}_+^N \tag{1350}$$

which bears strong resemblance to (1329).

6.8.1.2 nonnegative orthant contains $\mathbb{EDM}^N$

That $\mathbb{EDM}^N$ is a proper subset of the nonnegative orthant is not obvious from (1349). We wish to verify

$$\mathbb{EDM}^N = \mathbb{S}_h^N \cap \left(\mathbb{S}_c^{N\perp} - \mathbb{S}_+^N\right) \subset \mathbb{R}_+^{N \times N} \tag{1351}$$

While there are many ways to prove this, it is sufficient to show that all entries of the extreme directions of $\mathbb{EDM}^N$ must be nonnegative; *id est*, for any particular nonzero vector $z = [z_i , \ i=1\ldots N] \in \mathcal{N}(\mathbf{1}^T)$ (§6.4.3.2),

$$\delta(zz^T)\mathbf{1}^T + \mathbf{1}\delta(zz^T)^T - 2zz^T \geq \mathbf{0} \tag{1352}$$

where the inequality denotes entrywise comparison. The inequality holds because the i,j^{th} entry of an extreme direction is squared: $(z_i - z_j)^2$.

We observe that the dyad $2zz^T \in \mathbb{S}_+^N$ belongs to the positive semidefinite cone, the doublet

$$\delta(zz^T)\mathbf{1}^T + \mathbf{1}\delta(zz^T)^T \in \mathbb{S}_c^{N\perp} \tag{1353}$$

to the orthogonal complement (2096) of the geometric center subspace, while their difference is a member of the symmetric hollow subspace $\mathbb{S}_h^N$.   ♦

Here is an algebraic method to prove nonnegativity: Suppose we are given $A \in \mathbb{S}_c^{N\perp}$ and $B = [B_{ij}] \in \mathbb{S}_+^N$ and $A - B \in \mathbb{S}_h^N$. Then we have, for some vector u, $A = u\mathbf{1}^T + \mathbf{1}u^T = [A_{ij}] = [u_i + u_j]$ and $\delta(B) = \delta(A) = 2u$. Positive semidefiniteness of B requires nonnegativity $A - B \geq \mathbf{0}$ because

$$(e_i - e_j)^T B(e_i - e_j) = (B_{ii} - B_{ij}) - (B_{ji} - B_{jj}) = 2(u_i + u_j) - 2B_{ij} \geq 0 \qquad (1354)$$

$\blacklozenge$

6.8.1.3 Dual Euclidean distance matrix criterion

Conditions necessary for membership of a matrix $D^* \in \mathbb{S}^N$ to the dual EDM cone $\mathbb{EDM}^{N*}$ may be derived from (1329): $D^* \in \mathbb{EDM}^{N*} \Rightarrow D^* = \delta(y) - V_{\mathcal{N}} A V_{\mathcal{N}}^T$ for some vector y and positive semidefinite matrix $A \in \mathbb{S}_+^{N-1}$. This in turn implies $\delta(D^*\mathbf{1}) = \delta(y)$. Then, for $D^* \in \mathbb{S}^N$

$$D^* \in \mathbb{EDM}^{N*} \Leftrightarrow \delta(D^*\mathbf{1}) - D^* \succeq 0 \qquad (1355)$$

where, for any symmetric matrix D^*

$$\delta(D^*\mathbf{1}) - D^* \in \mathbb{S}_c^N \qquad (1356)$$

To show sufficiency of the matrix criterion in (1355), recall Gram-form EDM operator

$$\mathbf{D}(G) = \delta(G)\mathbf{1}^T + \mathbf{1}\delta(G)^T - 2G \qquad (980)$$

where Gram matrix G is positive semidefinite by definition, and recall the self-adjointness property of the main-diagonal linear operator δ (§A.1):

$$\langle D, D^* \rangle = \langle \delta(G)\mathbf{1}^T + \mathbf{1}\delta(G)^T - 2G, D^* \rangle = \langle G, \delta(D^*\mathbf{1}) - D^* \rangle 2 \qquad (998)$$

Assuming $\langle G, \delta(D^*\mathbf{1}) - D^* \rangle \geq 0$ (1566), then we have known membership relation (§2.13.2.0.1)

$$D^* \in \mathbb{EDM}^{N*} \Leftrightarrow \langle D, D^* \rangle \geq 0 \quad \forall D \in \mathbb{EDM}^N \qquad (1357)$$

$\blacklozenge$

Elegance of this matrix criterion (1355) for membership to the dual EDM cone is lack of any other assumptions except D^* be symmetric:[6.12] Linear Gram-form EDM operator $\mathbf{D}(Y)$ (980) has adjoint, for $Y \in \mathbb{S}^N$

$$\mathbf{D}^T(Y) \triangleq (\delta(Y\mathbf{1}) - Y)2 \qquad (1358)$$

Then from (1357) and (981) we have: [91, p.111]

$$\begin{aligned}
\mathbb{EDM}^{N*} &= \{D^* \in \mathbb{S}^N \mid \langle D, D^* \rangle \geq 0 \quad \forall D \in \mathbb{EDM}^N\} \\
&= \{D^* \in \mathbb{S}^N \mid \langle \mathbf{D}(G), D^* \rangle \geq 0 \quad \forall G \in \mathbb{S}_+^N\} \\
&= \{D^* \in \mathbb{S}^N \mid \langle G, \mathbf{D}^T(D^*) \rangle \geq 0 \,\forall G \in \mathbb{S}_+^N\} \\
&= \{D^* \in \mathbb{S}^N \mid \delta(D^*\mathbf{1}) - D^* \succeq 0\}
\end{aligned} \qquad (1359)$$

the dual EDM cone expressed in terms of the adjoint operator. A dual EDM cone determined this way is illustrated in Figure 162.

[6.12]Recall: Schoenberg criterion (987) for membership to the EDM cone requires membership to the symmetric hollow subspace.

6.8.1.3.1   Exercise.   *Dual EDM spectral cone.*

Find a spectral cone as in §5.11.2 corresponding to $\mathbb{EDM}^{N*}$.   ▼

6.8.1.4   Nonorthogonal components of dual EDM

Now we tie construct (1350) for the dual EDM cone together with the matrix criterion (1355) for dual EDM cone membership. For any $D^* \in \mathbb{S}^N$ it is obvious:

$$\delta(D^* \mathbf{1}) \in \mathbb{S}_h^{N\perp} \tag{1360}$$

any diagonal matrix belongs to the subspace of diagonal matrices (67). We know when $D^* \in \mathbb{EDM}^{N*}$

$$\delta(D^* \mathbf{1}) - D^* \in \mathbb{S}_c^N \cap \mathbb{S}_+^N \tag{1361}$$

this adjoint expression (1358) belongs to that face (1319) of the positive semidefinite cone $\mathbb{S}_+^N$ in the geometric center subspace. Any nonzero dual EDM

$$D^* = \delta(D^* \mathbf{1}) - (\delta(D^* \mathbf{1}) - D^*) \in \mathbb{S}_h^{N\perp} \ominus \mathbb{S}_c^N \cap \mathbb{S}_+^N = \mathbb{EDM}^{N*} \tag{1362}$$

can therefore be expressed as the difference of two linearly independent (when vectorized) nonorthogonal components (Figure 140, Figure 161).

6.8.1.5   Affine dimension complementarity

From §6.8.1.3 we have, for some $A \in \mathbb{S}_+^{N-1}$ (*confer*(1361))

$$\delta(D^* \mathbf{1}) - D^* = V_{\mathcal{N}} A V_{\mathcal{N}}^{\mathrm{T}} \in \mathbb{S}_c^N \cap \mathbb{S}_+^N \tag{1363}$$

if and only if D^* belongs to the dual EDM cone. Call $\mathrm{rank}(V_{\mathcal{N}} A V_{\mathcal{N}}^{\mathrm{T}})$ *dual affine dimension*. Empirically, we find a complementary relationship in affine dimension between the projection of some arbitrary symmetric matrix H on the polar EDM cone, $\mathbb{EDM}^{N\circ} = -\mathbb{EDM}^{N*}$, and its projection on the EDM cone; *id est*, the optimal solution of [6.13]

$$\begin{array}{ll} \underset{D^\circ \in \mathbb{S}^N}{\text{minimize}} & \|D^\circ - H\|_{\mathrm{F}} \\ \text{subject to} & D^\circ - \delta(D^\circ \mathbf{1}) \succeq 0 \end{array} \tag{1364}$$

has dual affine dimension complementary to affine dimension corresponding to the optimal solution of

$$\begin{array}{ll} \underset{D \in \mathbb{S}_h^N}{\text{minimize}} & \|D - H\|_{\mathrm{F}} \\ \text{subject to} & -V_{\mathcal{N}}^{\mathrm{T}} D V_{\mathcal{N}} \succeq 0 \end{array} \tag{1365}$$

[6.13]This polar projection can be solved quickly (without semidefinite programming) via Lemma 6.8.1.1.1; rewriting,

$$\begin{array}{ll} \underset{D^\circ \in \mathbb{S}^N}{\text{minimize}} & \|(D^\circ - \delta(D^\circ \mathbf{1})) - (H - \delta(D^\circ \mathbf{1}))\|_{\mathrm{F}} \\ \text{subject to} & D^\circ - \delta(D^\circ \mathbf{1}) \succeq 0 \end{array}$$

which is the projection of affinely transformed optimal solution $H - \delta(D^{\circ\star} \mathbf{1})$ on $\mathbb{S}_c^N \cap \mathbb{S}_+^N$;

$$D^{\circ\star} - \delta(D^{\circ\star} \mathbf{1}) = P_{\mathbb{S}_+^N} P_{\mathbb{S}_c^N}(H - \delta(D^{\circ\star} \mathbf{1}))$$

Foreknowledge of an optimal solution $D^{\circ\star}$ as argument to projection suggests recursion.

$$D^\circ = \delta(D^\circ \mathbf{1}) + (D^\circ - \delta(D^\circ \mathbf{1})) \in \mathbb{S}_h^{N\perp} \oplus \mathbb{S}_c^N \cap \mathbb{S}_+^N = \mathbb{EDM}^{N^\circ}$$

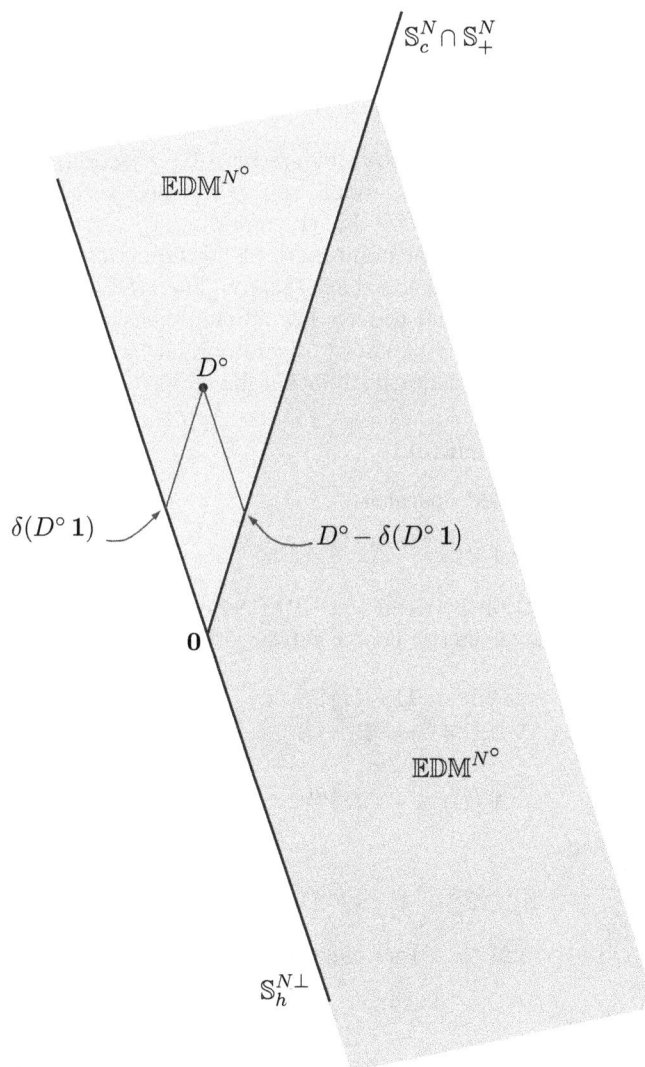

Figure 161: Hand-drawn abstraction of polar EDM cone (drawn truncated). Any member D° of polar EDM cone can be decomposed into two linearly independent nonorthogonal components: $\delta(D^\circ \mathbf{1})$ and $D^\circ - \delta(D^\circ \mathbf{1})$.

Precisely,

$$\text{rank}(D^{\circ\star} - \delta(D^{\circ\star}\mathbf{1})) + \text{rank}(V_{\mathcal{N}}^{\mathrm{T}}D^{\star}V_{\mathcal{N}}) = N-1 \qquad (1366)$$

and $\text{rank}(D^{\circ\star} - \delta(D^{\circ\star}\mathbf{1})) \leq N-1$ because vector $\mathbf{1}$ is always in the nullspace of rank's argument. This is similar to the known result for projection on the selfdual positive semidefinite cone and its polar:

$$\text{rank}\, P_{-\mathbb{S}_+^N}H + \text{rank}\, P_{\mathbb{S}_+^N}H = N \qquad (1367)$$

When low affine dimension is a desirable result of projection on the EDM cone, projection on the polar EDM cone should be performed instead. Convex polar problem (1364) can be solved for $D^{\circ\star}$ by transforming to an equivalent Schur-form semidefinite program (§3.5.2). Interior-point methods for numerically solving semidefinite programs tend to produce high-rank solutions. (§4.1.2) Then $D^\star = H - D^{\circ\star} \in \mathbb{EDM}^N$ by Corollary E.9.2.2.1, and $D^\star$ will tend to have low affine dimension. This approach breaks when attempting projection on a cone subset discriminated by affine dimension or rank, because then we have no complementarity relation like (1366) or (1367) (§7.1.4.1).

6.8.1.6 EDM cone is not selfdual

In §5.6.1.1, via Gram-form EDM operator

$$\mathbf{D}(G) = \delta(G)\mathbf{1}^{\mathrm{T}} + \mathbf{1}\delta(G)^{\mathrm{T}} - 2G \ \in \mathbb{EDM}^N \quad \Leftarrow \quad G \succeq 0 \qquad (980)$$

we established clear connection between the EDM cone and that face (1319) of positive semidefinite cone $\mathbb{S}_+^N$ in the geometric center subspace:

$$\mathbb{EDM}^N = \mathbf{D}(\mathbb{S}_c^N \cap \mathbb{S}_+^N) \qquad (1087)$$
$$\mathbf{V}(\mathbb{EDM}^N) = \mathbb{S}_c^N \cap \mathbb{S}_+^N \qquad (1088)$$

where

$$\mathbf{V}(D) = -VDV\tfrac{1}{2} \qquad (1076)$$

In §5.6.1 we established

$$\mathbb{S}_c^N \cap \mathbb{S}_+^N = V_{\mathcal{N}}\mathbb{S}_+^{N-1}V_{\mathcal{N}}^{\mathrm{T}} \qquad (1074)$$

Then from (1355), (1363), and (1329) we can deduce

$$\delta(\mathbb{EDM}^{N^*}\mathbf{1}) - \mathbb{EDM}^{N^*} = V_{\mathcal{N}}\mathbb{S}_+^{N-1}V_{\mathcal{N}}^{\mathrm{T}} = \mathbb{S}_c^N \cap \mathbb{S}_+^N \qquad (1368)$$

which, by (1087) and (1088), means the EDM cone can be related to the dual EDM cone by an equality:

$$\mathbb{EDM}^N = \mathbf{D}\Big(\delta(\mathbb{EDM}^{N^*}\mathbf{1}) - \mathbb{EDM}^{N^*}\Big) \qquad (1369)$$

$$\mathbf{V}(\mathbb{EDM}^N) = \delta(\mathbb{EDM}^{N^*}\mathbf{1}) - \mathbb{EDM}^{N^*} \qquad (1370)$$

This means projection $-\mathbf{V}(\mathbb{EDM}^N)$ of the EDM cone on the geometric center subspace $\mathbb{S}_c^N$ (§E.7.2.0.2) is a linear transformation of the dual EDM cone: $\mathbb{EDM}^{N^*} - \delta(\mathbb{EDM}^{N^*}\mathbf{1})$. Secondarily, it means the EDM cone is not selfdual in $\mathbb{S}^N$.

6.8.1.7 Schoenberg criterion is discretized membership relation

We show the Schoenberg criterion

$$
\left.\begin{array}{r}
-V_{\mathcal{N}}^{\mathrm{T}} D V_{\mathcal{N}} \in \mathbb{S}_+^{N-1} \\
D \in \mathbb{S}_h^N
\end{array}\right\} \quad \Leftrightarrow \quad D \in \mathrm{EDM}^N \qquad (987)
$$

to be a discretized membership relation (§2.13.4) between a closed convex cone $\mathcal{K}$ and its dual $\mathcal{K}^*$ like

$$
\langle y, x \rangle \geq 0 \ \text{ for all } \ y \in \mathcal{G}(\mathcal{K}^*) \ \Leftrightarrow \ x \in \mathcal{K} \qquad (365)
$$

where $\mathcal{G}(\mathcal{K}^*)$ is any set of generators whose conic hull constructs closed convex dual cone $\mathcal{K}^*$:

The Schoenberg criterion is the same as

$$
\left.\begin{array}{r}
\langle zz^{\mathrm{T}}, -D \rangle \geq 0 \quad \forall zz^{\mathrm{T}} \mid \mathbf{1}^{\mathrm{T}} zz^{\mathrm{T}} = \mathbf{0} \\
D \in \mathbb{S}_h^N
\end{array}\right\} \Leftrightarrow D \in \mathrm{EDM}^N \qquad (1313)
$$

which, by (1314), is the same as

$$
\left.\begin{array}{r}
\langle zz^{\mathrm{T}}, -D \rangle \geq 0 \quad \forall zz^{\mathrm{T}} \in \left\{ V_{\mathcal{N}} v v^{\mathrm{T}} V_{\mathcal{N}}^{\mathrm{T}} \mid v \in \mathbb{R}^{N-1} \right\} \\
D \in \mathbb{S}_h^N
\end{array}\right\} \Leftrightarrow D \in \mathrm{EDM}^N \qquad (1371)
$$

where the zz^{T} constitute a set of generators $\mathcal{G}$ for the positive semidefinite cone's smallest face $\mathcal{F}\!\left(\mathbb{S}_+^N \ni V\right)$ (§6.6.1) that contains auxiliary matrix V. From the aggregate in (1329) we get the ordinary membership relation, assuming only $D \in \mathbb{S}^N$ [205, p.58]

$$
\langle D^*, D \rangle \geq 0 \quad \forall D^* \in \mathrm{EDM}^{N*} \Leftrightarrow D \in \mathrm{EDM}^N \qquad (1372)
$$

$$
\langle D^*, D \rangle \geq 0 \quad \forall D^* \in \{\delta(u) \mid u \in \mathbb{R}^N\} - \mathrm{cone}\left\{ V_{\mathcal{N}} v v^{\mathrm{T}} V_{\mathcal{N}}^{\mathrm{T}} \mid v \in \mathbb{R}^{N-1} \right\} \Leftrightarrow D \in \mathrm{EDM}^N
$$

Discretization (365) yields:

$$
\langle D^*, D \rangle \geq 0 \quad \forall D^* \in \{ e_i e_i^{\mathrm{T}}, -e_j e_j^{\mathrm{T}}, -V_{\mathcal{N}} v v^{\mathrm{T}} V_{\mathcal{N}}^{\mathrm{T}} \mid i, j = 1 \dots N, \ v \in \mathbb{R}^{N-1} \} \Leftrightarrow D \in \mathrm{EDM}^N
$$
$$
(1373)
$$

Because $\left\langle \{\delta(u) \mid u \in \mathbb{R}^N\}, D \right\rangle \geq 0 \Leftrightarrow D \in \mathbb{S}_h^N$, we can restrict observation to the symmetric hollow subspace without loss of generality. Then for $D \in \mathbb{S}_h^N$

$$
\langle D^*, D \rangle \geq 0 \quad \forall D^* \in \left\{ -V_{\mathcal{N}} v v^{\mathrm{T}} V_{\mathcal{N}}^{\mathrm{T}} \mid v \in \mathbb{R}^{N-1} \right\} \Leftrightarrow D \in \mathrm{EDM}^N \qquad (1374)
$$

this discretized membership relation becomes (1371); identical to the Schoenberg criterion.

Hitherto a correspondence between the EDM cone and a face of a PSD cone, the Schoenberg criterion is now accurately interpreted as a discretized membership relation between the EDM cone and its ordinary dual.

6.8.2   Ambient $\mathbb{S}_h^N$

When instead we consider the ambient space of symmetric hollow matrices (1330), then still we find the EDM cone is not selfdual for $N > 2$. The simplest way to prove this is as follows:

Given a set of generators $\mathcal{G} = \{\Gamma\}$ (1290) for the pointed closed convex EDM cone, the *discretized membership theorem* in §2.13.4.2.1 asserts that members of the dual EDM cone in the ambient space of symmetric hollow matrices can be discerned via discretized membership relation:

$$
\begin{aligned}
\text{EDM}^{N^*} \cap \mathbb{S}_h^N &\triangleq \{D^* \in \mathbb{S}_h^N \mid \langle \Gamma, D^* \rangle \geq 0 \quad \forall \Gamma \in \mathcal{G}(\text{EDM}^N)\} \\
&= \{D^* \in \mathbb{S}_h^N \mid \langle \delta(zz^{\mathrm{T}})\mathbf{1}^{\mathrm{T}} + \mathbf{1}\delta(zz^{\mathrm{T}})^{\mathrm{T}} - 2zz^{\mathrm{T}}, D^* \rangle \geq 0 \; \forall z \in \mathcal{N}(\mathbf{1}^{\mathrm{T}})\} \quad (1375) \\
&= \{D^* \in \mathbb{S}_h^N \mid \langle \mathbf{1}\delta(zz^{\mathrm{T}})^{\mathrm{T}} - zz^{\mathrm{T}}, D^* \rangle \geq 0 \; \forall z \in \mathcal{N}(\mathbf{1}^{\mathrm{T}})\}
\end{aligned}
$$

By comparison

$$
\text{EDM}^N = \{D \in \mathbb{S}_h^N \mid \langle -zz^{\mathrm{T}}, D \rangle \geq 0 \; \forall z \in \mathcal{N}(\mathbf{1}^{\mathrm{T}})\} \qquad (1376)
$$

the term $\delta(zz^{\mathrm{T}})^{\mathrm{T}} D^* \mathbf{1}$ foils any hope of selfdualness in ambient $\mathbb{S}_h^N$.   ◆

To find the dual EDM cone in ambient $\mathbb{S}_h^N$ per §2.13.9.4 we prune the aggregate in (1329) describing the ordinary dual EDM cone, removing any member having nonzero main diagonal:

$$
\begin{aligned}
\text{EDM}^{N^*} \cap \mathbb{S}_h^N &= \text{cone}\left\{\delta^2(V_{\mathcal{N}} \upsilon \upsilon^{\mathrm{T}} V_{\mathcal{N}}^{\mathrm{T}}) - V_{\mathcal{N}} \upsilon \upsilon^{\mathrm{T}} V_{\mathcal{N}}^{\mathrm{T}} \mid \upsilon \in \mathbb{R}^{N-1}\right\} \\
&= \{\delta^2(V_{\mathcal{N}} \Psi V_{\mathcal{N}}^{\mathrm{T}}) - V_{\mathcal{N}} \Psi V_{\mathcal{N}}^{\mathrm{T}} \mid \Psi \in \mathbb{S}_+^{N-1}\}
\end{aligned} \qquad (1377)
$$

When $N = 1$, the EDM cone and its dual in ambient $\mathbb{S}_h$ each comprise the origin in isomorphic $\mathbb{R}^0$; thus, selfdual in this dimension. (*confer* (104))

When $N = 2$, the EDM cone is the nonnegative real line in isomorphic $\mathbb{R}$. (Figure **153**) EDM^{2^*} in $\mathbb{S}_h^2$ is identical, thus selfdual in this dimension. This result is in agreement with (1375), verified directly: for all $\kappa \in \mathbb{R}$, $z = \kappa \begin{bmatrix} 1 \\ -1 \end{bmatrix}$ and $\delta(zz^{\mathrm{T}}) = \kappa^2 \begin{bmatrix} 1 \\ 1 \end{bmatrix} \Rightarrow d_{12}^* \geq 0$.

The first case adverse to selfdualness $N = 3$ may be deduced from Figure **149**; the EDM cone is a circular cone in isomorphic $\mathbb{R}^3$ corresponding to no rotation of Lorentz cone (178) (the selfdual circular cone). Figure **162** illustrates the EDM cone and its dual in ambient $\mathbb{S}_h^3$; no longer selfdual.

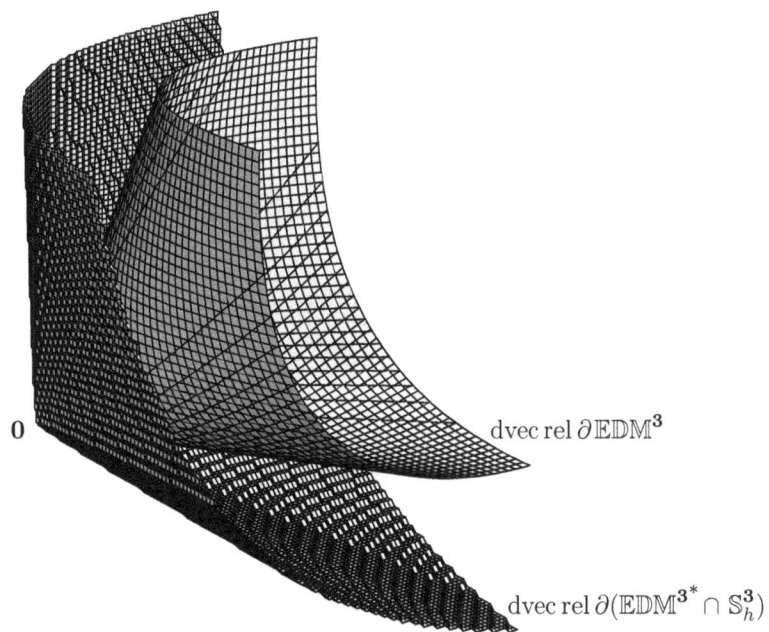

$$D^* \in \mathbb{EDM}^{N^*} \Leftrightarrow \delta(D^*\mathbf{1}) - D^* \succeq 0 \qquad (1355)$$

Figure 162: Ordinary dual EDM cone projected on $\mathbb{S}_h^{\mathbf{3}}$ shrouds $\mathbb{EDM}^{\mathbf{3}}$; drawn tiled in isometrically isomorphic $\mathbb{R}^{\mathbf{3}}$. (It so happens: intersection $\mathbb{EDM}^{N^*} \cap \mathbb{S}_h^N$ (§2.13.9.3) is identical to projection of dual EDM cone on $\mathbb{S}_h^N$.)

6.9   Theorem of the alternative

In §2.13.2.1.1 we showed how alternative systems of generalized inequality can be derived from closed convex cones and their duals. This section is, therefore, a fitting postscript to the discussion of the dual EDM cone.

6.9.0.0.1   Theorem.   *EDM alternative.*   [169, §1]
Given $D \in \mathbb{S}_h^N$

$$D \in \mathbb{EDM}^N$$

$$\text{or in the alternative}$$

$$\exists\, z \ \text{ such that } \left\{ \begin{array}{l} \mathbf{1}^{\mathrm{T}} z = 1 \\ Dz = \mathbf{0} \end{array} \right. \tag{1378}$$

In words, either $\mathcal{N}(D)$ intersects hyperplane $\{z \,|\, \mathbf{1}^{\mathrm{T}}z{=}1\}$ or D is an EDM; the alternatives are incompatible. $\diamond$

When D is an EDM [265, §2]

$$\mathcal{N}(D) \subset \mathcal{N}(\mathbf{1}^{\mathrm{T}}) = \{z \mid \mathbf{1}^{\mathrm{T}}z = 0\} \tag{1379}$$

Because [169, §2] (§E.0.1)

$$\begin{aligned} DD^{\dagger}\mathbf{1} &= \mathbf{1} \\ \mathbf{1}^{\mathrm{T}}D^{\dagger}D &= \mathbf{1}^{\mathrm{T}} \end{aligned} \tag{1380}$$

then

$$\mathcal{R}(\mathbf{1}) \subset \mathcal{R}(D) \tag{1381}$$

6.10   Postscript

We provided an equality (1349) relating the convex cone of Euclidean distance matrices to the convex cone of positive semidefinite matrices. Projection on a positive semidefinite cone, constrained by an upper bound on rank, is easy and well known; [133] simply, a matter of truncating a list of eigenvalues. Projection on a positive semidefinite cone with such a rank constraint is, in fact, a convex optimization problem. (§7.1.4)

In the past, it was difficult to project on the EDM cone under a constraint on rank or affine dimension. A surrogate method was to invoke the Schoenberg criterion (987) and then project on a positive semidefinite cone under a rank constraint bounding affine dimension from above. But a solution acquired that way is necessarily suboptimal.

In §7.3.3 we present a method for projecting directly on the EDM cone under a constraint on rank or affine dimension.

Chapter 7

Proximity problems

In the "extremely large-scale case" (N of order of tens and hundreds of thousands), [iteration cost $\mathrm{O}(N^3)$] *rules out all advanced convex optimization techniques, including all known polynomial time algorithms.*

−Arkadi Nemirovski (2004)

A problem common to various sciences is to find the Euclidean distance matrix (EDM) $D \in \mathbb{EDM}^N$ closest in some sense to a given complete matrix of measurements H under a constraint on affine dimension $0 \leq r \leq N-1$ (§2.3.1, §5.7.1.1); rather, r is bounded above by desired affine dimension ρ.

7.0.1 Measurement matrix H

Ideally, we want a given matrix of measurements $H \in \mathbb{R}^{N \times N}$ to conform with the first three Euclidean metric properties (§5.2); to belong to the intersection of the orthant of nonnegative matrices $\mathbb{R}_+^{N \times N}$ with the symmetric hollow subspace $\mathbb{S}_h^N$ (§2.2.3.0.1). Geometrically, we want H to belong to the polyhedral cone (§2.12.1.0.1)

$$\mathcal{K} \triangleq \mathbb{S}_h^N \cap \mathbb{R}_+^{N \times N} \tag{1382}$$

Yet in practice, H can possess significant measurement uncertainty (noise).

Sometimes realization of an optimization problem demands that its input, the given matrix H, possess some particular characteristics; perhaps symmetry and hollowness or nonnegativity. When that H given does not have the desired properties, then we must impose them upon H prior to optimization:

- When *measurement matrix* H is not symmetric or hollow, taking its symmetric hollow part is equivalent to orthogonal projection on the symmetric hollow subspace $\mathbb{S}_h^N$.

- When measurements of distance in H are negative, zeroing negative entries effects unique minimum-distance projection on the orthant of nonnegative matrices $\mathbb{R}_+^{N \times N}$ in isomorphic $\mathbb{R}^{N^2}$ (§E.9.2.2.3).

CITATION: Dattorro, *Convex Optimization & Euclidean Distance Geometry*, $\mathcal{M}\varepsilon\beta oo$ Publishing USA, 2005, v2015.02.15.

7.0.1.1   Order of imposition

Since convex cone $\mathcal{K}$ (1382) is the intersection of an orthant with a subspace, we want to project on that subset of the orthant belonging to the subspace; on the nonnegative orthant in the symmetric hollow subspace that is, in fact, the intersection. For that reason alone, unique minimum-distance projection of H on $\mathcal{K}$ (that member of $\mathcal{K}$ closest to H in isomorphic $\mathbb{R}^{N^2}$ in the Euclidean sense) can be attained by first taking its symmetric hollow part, and only then clipping negative entries of the result to 0; *id est*, there is only one correct *order of projection,* in general, on an orthant intersecting a subspace:

- project on the subspace, then project the result on the orthant in that subspace. (*confer* §E.9.5)

In contrast, order of projection on an intersection of subspaces is arbitrary.

That order-of-projection rule applies more generally, of course, to the intersection of any convex set $\mathcal{C}$ with any subspace. Consider the *proximity problem*[7.1] over convex feasible set $\mathbb{S}_h^N \cap \mathcal{C}$ given nonsymmetric nonhollow $H \in \mathbb{R}^{N \times N}$:

$$\begin{array}{cl} \underset{B \in \mathbb{S}_h^N}{\text{minimize}} & \|B - H\|_{\mathrm{F}}^2 \\ \text{subject to} & B \in \mathcal{C} \end{array} \tag{1383}$$

a convex optimization problem. Because the symmetric hollow subspace $\mathbb{S}_h^N$ is orthogonal to the antisymmetric antihollow subspace $\mathbb{R}_h^{N \times N \perp}$ (§2.2.3), then for $B \in \mathbb{S}_h^N$

$$\operatorname{tr}\left(B^{\mathrm{T}}\left(\frac{1}{2}(H - H^{\mathrm{T}}) + \delta^2(H)\right)\right) = 0 \tag{1384}$$

so the objective function is equivalent to

$$\|B - H\|_{\mathrm{F}}^2 \equiv \left\|B - \left(\frac{1}{2}(H + H^{\mathrm{T}}) - \delta^2(H)\right)\right\|_{\mathrm{F}}^2 + \left\|\frac{1}{2}(H - H^{\mathrm{T}}) + \delta^2(H)\right\|_{\mathrm{F}}^2 \tag{1385}$$

This means the antisymmetric antihollow part of given matrix H would be ignored by minimization with respect to symmetric hollow variable B under Frobenius' norm; *id est*, minimization proceeds as though given the symmetric hollow part of H.

This action of Frobenius' norm (1385) is effectively a Euclidean projection (minimum-distance projection) of H on the symmetric hollow subspace $\mathbb{S}_h^N$ prior to minimization. Thus minimization proceeds inherently following the correct order for projection on $\mathbb{S}_h^N \cap \mathcal{C}$. Therefore we may either assume $H \in \mathbb{S}_h^N$, or take its symmetric hollow part prior to optimization.

[7.1] There are two equivalent interpretations of projection (§E.9): one finds a set normal, the other, minimum distance between a point and a set. Here we realize the latter view.

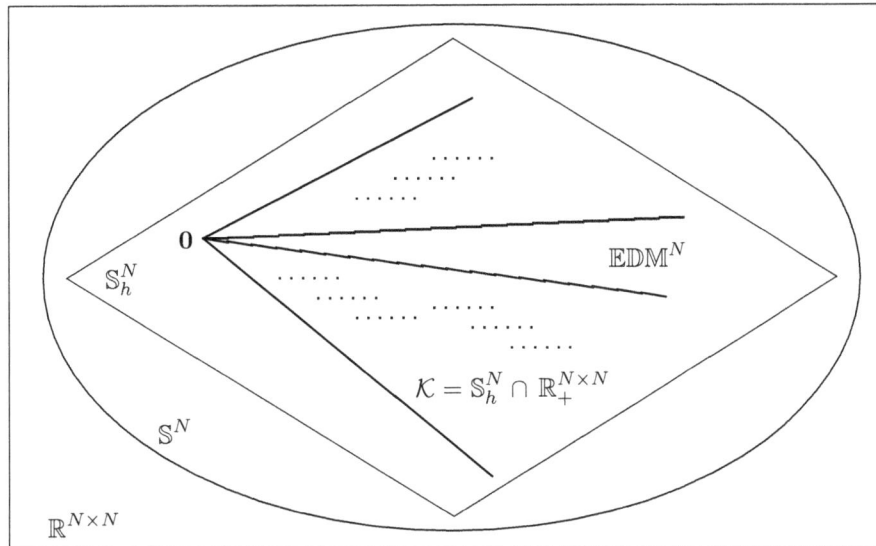

Figure 163: Pseudo-Venn diagram: The EDM cone belongs to the intersection of the symmetric hollow subspace with the nonnegative orthant; $\mathbb{EDM}^N \subseteq \mathcal{K}$ (967). $\mathbb{EDM}^N$ cannot exist outside $\mathbb{S}_h^N$, but $\mathbb{R}_+^{N \times N}$ does.

7.0.1.2 Flagrant input error under nonnegativity demand

More pertinent to the optimization problems presented herein where

$$\mathcal{C} \triangleq \mathbb{EDM}^N \subseteq \mathcal{K} = \mathbb{S}_h^N \cap \mathbb{R}_+^{N \times N} \tag{1386}$$

then should some particular realization of a proximity problem demand input H be nonnegative, and were we only to zero negative entries of a nonsymmetric nonhollow input H prior to optimization, then the ensuing projection on $\mathbb{EDM}^N$ would be guaranteed incorrect (out of order).

Now comes a surprising fact: Even were we to correctly follow the order-of-projection rule and provide $H \in \mathcal{K}$ prior to optimization, then the ensuing projection on $\mathbb{EDM}^N$ will be incorrect whenever input H has negative entries and some proximity problem demands nonnegative input H.

This is best understood referring to Figure 163: Suppose nonnegative input H is demanded, and then the problem realization correctly projects its input first on $\mathbb{S}_h^N$ and then directly on $\mathcal{C} = \mathbb{EDM}^N$. That demand for nonnegativity effectively requires imposition of $\mathcal{K}$ on input H prior to optimization so as to obtain correct order of projection (on $\mathbb{S}_h^N$ first). Yet such an imposition prior to projection on $\mathbb{EDM}^N$ generally introduces an *elbow* into the path of projection (illustrated in Figure 164) caused by the technique itself; that being, a particular proximity problem realization requiring nonnegative input.

Any procedure for imposition of nonnegativity on input H can only be incorrect in this circumstance. There is no resolution unless input H is guaranteed nonnegative with no tinkering. Otherwise, we have no choice but to employ a different problem realization; one not demanding nonnegative input.

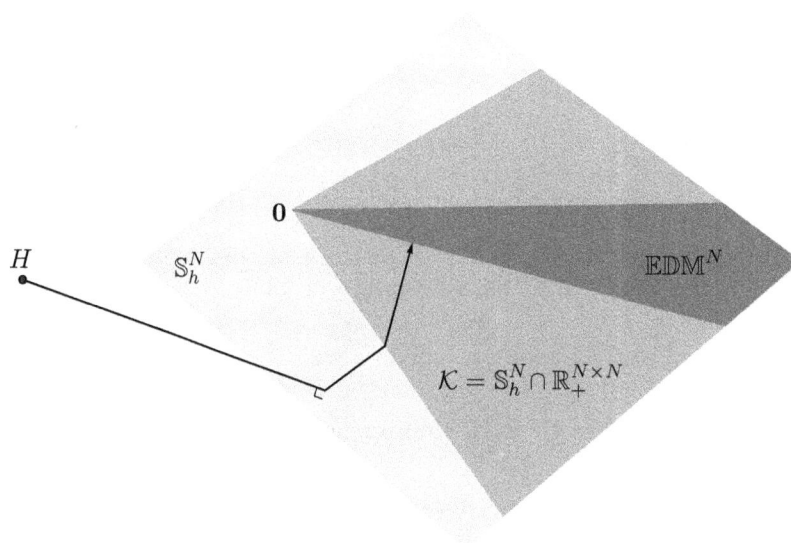

Figure 164: Pseudo-Venn diagram from Figure **163** showing elbow placed in path of projection of H on $\mathbb{EDM}^N \subset \mathbb{S}_h^N$ by an optimization problem demanding nonnegative input matrix H. The first two line segments leading away from H result from correct order-of-projection required to provide nonnegative H prior to optimization. Were H nonnegative, then its projection on $\mathbb{S}_h^N$ would instead belong to $\mathcal{K}$; making the elbow disappear. (*confer* Figure **178**)

7.0.2 Lower bound

Most of the problems we encounter in this chapter have the general form:

$$
\begin{array}{ll}
\underset{B}{\text{minimize}} & \|B - A\|_{\text{F}} \\
\text{subject to} & B \in \mathcal{C}
\end{array}
\tag{1387}
$$

where $A \in \mathbb{R}^{m \times n}$ is given data. This particular objective denotes Euclidean projection (§E) of vectorized matrix A on the set $\mathcal{C}$ which may or may not be convex. When $\mathcal{C}$ is convex, then the projection is unique minimum-distance because Frobenius' norm when squared is a strictly convex function of variable B and because the optimal solution is the same regardless of the square (512). When $\mathcal{C}$ is a subspace, then the direction of projection is orthogonal to $\mathcal{C}$.

Denoting by $A \triangleq U_A \Sigma_A Q_A^{\text{T}}$ and $B \triangleq U_B \Sigma_B Q_B^{\text{T}}$ their full singular value decompositions (whose singular values are always nonincreasingly ordered (§A.6)), there exists a tight lower bound on the objective over the manifold of orthogonal matrices;

$$
\|\Sigma_B - \Sigma_A\|_{\text{F}} \leq \underset{U_A, U_B, Q_A, Q_B}{\inf} \|B - A\|_{\text{F}}
\tag{1388}
$$

This least lower bound holds more generally for any orthogonally invariant norm on $\mathbb{R}^{m \times n}$ (§2.2.1) including the Frobenius and spectral norm [335, §II.3]. [208, §7.4.51]

7.0.3 Problem approach

Problems traditionally posed in terms of point position $\{x_i \in \mathbb{R}^n, \ i = 1 \ldots N\}$ such as

$$
\underset{\{x_i\}}{\text{minimize}} \sum_{i, j \in \mathcal{I}} \left(\|x_i - x_j\| - h_{ij}\right)^2
\tag{1389}
$$

or

$$
\underset{\{x_i\}}{\text{minimize}} \sum_{i, j \in \mathcal{I}} \left(\|x_i - x_j\|^2 - h_{ij}\right)^2
\tag{1390}
$$

(where $\mathcal{I}$ is an abstract set of indices and h_{ij} is given data) are everywhere converted herein to the distance-square variable D or to Gram matrix G; the Gram matrix acting as bridge between position and distance. (That conversion is performed regardless of whether known data is complete.) Then the techniques of chapter 5 or chapter 6 are applied to find relative or absolute position. This approach is taken because we prefer introduction of rank constraints into convex problems rather than searching a googol of local minima in nonconvex problems like (1389) [107] (§3.6.4.0.3, §7.2.2.7.1) or (1390).

7.0.4 Three prevalent proximity problems

There are three statements of the closest-EDM problem prevalent in the literature, the multiplicity due primarily to choice of projection on the EDM *versus* positive semidefinite (PSD) cone and vacillation between the distance-square variable d_{ij} *versus* absolute distance $\sqrt{d_{ij}}$. In their most fundamental form, the three prevalent proximity problems are (1391.1), (1391.2), and (1391.3): [349] for $D \triangleq [d_{ij}]$ and $\sqrt[\circ]{D} \triangleq [\sqrt{d_{ij}}]$

$$
\begin{array}{lll}
(1) & \begin{array}{ll} \underset{D}{\text{minimize}} & \|-V(D-H)V\|_{\text{F}}^2 \\ \text{subject to} & \operatorname{rank} VDV \le \rho \\ & D \in \mathbb{EDM}^N \end{array} & \begin{array}{ll} \underset{\sqrt[\circ]{D}}{\text{minimize}} & \|\sqrt[\circ]{D}-H\|_{\text{F}}^2 \\ \text{subject to} & \operatorname{rank} VDV \le \rho \\ & \sqrt[\circ]{D} \in \sqrt{\mathbb{EDM}^N} \end{array} & (2) \\[4em]
(3) & \begin{array}{ll} \underset{D}{\text{minimize}} & \|D-H\|_{\text{F}}^2 \\ \text{subject to} & \operatorname{rank} VDV \le \rho \\ & D \in \mathbb{EDM}^N \end{array} & \begin{array}{ll} \underset{\sqrt[\circ]{D}}{\text{minimize}} & \|-V(\sqrt[\circ]{D}-H)V\|_{\text{F}}^2 \\ \text{subject to} & \operatorname{rank} VDV \le \rho \\ & \sqrt[\circ]{D} \in \sqrt{\mathbb{EDM}^N} \end{array} & (4)
\end{array}
$$

$$(1391)$$

where we have made explicit an imposed upper bound ρ on affine dimension

$$ r \;=\; \operatorname{rank} V_{\mathcal{N}}^{\text{T}} D V_{\mathcal{N}} \;=\; \operatorname{rank} VDV \qquad (1122) $$

that is benign when $\rho = N-1$ or H were realizable with $r \le \rho$. Problems (1391.2) and (1391.3) are Euclidean projections of a vectorized matrix H on an EDM cone (§6.3), whereas problems (1391.1) and (1391.4) are Euclidean projections of a vectorized matrix $-VHV$ on a PSD cone.[7.2] Problem (1391.4) is not posed in the literature because it has limited theoretical foundation.[7.3]

Analytical solution to (1391.1) is known in closed form for any bound ρ although, as the problem is stated, it is a convex optimization only in the case $\rho = N-1$. We show in §7.1.4 how (1391.1) becomes a convex optimization problem for any ρ when transformed to the spectral domain. When expressed as a function of point list in a matrix X as in (1389), problem (1391.2) becomes a variant of what is known in statistics literature as a *stress problem.* [54, p.34] [105] [363] Problems (1391.2) and (1391.3) are convex optimization problems in D for the case $\rho = N-1$ wherein (1391.3) becomes equivalent to (1390). Even with the rank constraint removed from (1391.2), we will see that the convex problem remaining inherently minimizes affine dimension.

Generally speaking, each problem in (1391) produces a different result because there is no isometry relating them. Of the various auxiliary V-matrices (§B.4), the geometric centering matrix V (990) appears in the literature most often although $V_{\mathcal{N}}$ (974) is the auxiliary matrix naturally consequent to Schoenberg's seminal exposition [320]. Substitution of any auxiliary matrix or its pseudoinverse into these problems produces another valid problem.

Substitution of $V_{\mathcal{N}}^{\text{T}}$ for left-hand V in (1391.1), in particular, produces a different result because

$$
\begin{array}{ll}
\underset{D}{\text{minimize}} & \|-V(D-H)V\|_{\text{F}}^2 \\
\text{subject to} & D \in \mathbb{EDM}^N
\end{array}
\qquad (1392)
$$

[7.2]Because $-VHV$ is orthogonal projection of $-H$ on the geometric center subspace $\mathbb{S}_c^N$ (§E.7.2.0.2), problems (1391.1) and (1391.4) may be interpreted as oblique (nonminimum distance) projections of $-H$ on a positive semidefinite cone.

[7.3] $D \in \mathbb{EDM}^N \Rightarrow \sqrt[\circ]{D} \in \mathbb{EDM}^N$, $-V\sqrt[\circ]{D}V \in \mathbb{S}_+^N$ (§5.10) finds D to attain Euclidean distance of vectorized $-VHV$ to the positive semidefinite cone in ambient isometrically isomorphic $\mathbb{R}^{N(N+1)/2}$, whereas

$$\begin{array}{ll}\underset{D}{\text{minimize}} & \|-V_\mathcal{N}^{\mathrm{T}}(D-H)V_\mathcal{N}\|_{\mathrm{F}}^2 \\ \text{subject to} & D \in \mathbb{EDM}^N\end{array} \tag{1393}$$

attains Euclidean distance of vectorized $-V_\mathcal{N}^{\mathrm{T}}HV_\mathcal{N}$ to the positive semidefinite cone in isometrically isomorphic subspace $\mathbb{R}^{N(N-1)/2}$; quite different projections[7.4] regardless of whether affine dimension is constrained. But substitution of auxiliary matrix $V_\mathcal{W}^{\mathrm{T}}$ (§B.4.3) or $V_\mathcal{N}^{\dagger}$ yields the same result as (1391.1) because $V = V_\mathcal{W}V_\mathcal{W}^{\mathrm{T}} = V_\mathcal{N}V_\mathcal{N}^{\dagger}$; *id est*,

$$\begin{aligned}\|-V(D-H)V\|_{\mathrm{F}}^2 &= \|-V_\mathcal{W}V_\mathcal{W}^{\mathrm{T}}(D-H)V_\mathcal{W}V_\mathcal{W}^{\mathrm{T}}\|_{\mathrm{F}}^2 &= \|-V_\mathcal{W}^{\mathrm{T}}(D-H)V_\mathcal{W}\|_{\mathrm{F}}^2 \\ &= \|-V_\mathcal{N}V_\mathcal{N}^{\dagger}(D-H)V_\mathcal{N}V_\mathcal{N}^{\dagger}\|_{\mathrm{F}}^2 &= \|-V_\mathcal{N}^{\dagger}(D-H)V_\mathcal{N}\|_{\mathrm{F}}^2\end{aligned} \tag{1394}$$

We see no compelling reason to prefer one particular auxiliary V-matrix over another. Each has its own coherent interpretations; *e.g.*, §5.4.2, §6.6, §B.4.5. Neither can we say any particular problem formulation produces generally better results than another.[7.5]

7.1 First prevalent problem: Projection on PSD cone

This first problem

$$\left.\begin{array}{ll}\underset{D}{\text{minimize}} & \|-V_\mathcal{N}^{\mathrm{T}}(D-H)V_\mathcal{N}\|_{\mathrm{F}}^2 \\ \text{subject to} & \text{rank}\, V_\mathcal{N}^{\mathrm{T}}DV_\mathcal{N} \leq \rho \\ & D \in \mathbb{EDM}^N\end{array}\right\} \quad \text{Problem 1} \tag{1395}$$

poses a Euclidean projection of $-V_\mathcal{N}^{\mathrm{T}}HV_\mathcal{N}$ in subspace $\mathbb{S}^{N-1}$ on a generally nonconvex subset (when $\rho < N-1$) of the positive semidefinite cone boundary $\partial\mathbb{S}_+^{N-1}$ whose elemental matrices have rank no greater than desired affine dimension ρ (§5.7.1.1). Problem 1 finds the closest EDM D in the sense of Schoenberg. (987) [320] As it is stated, this optimization problem is convex only when desired affine dimension is largest $\rho = N-1$ although its analytical solution is known [263, thm.14.4.2] for all nonnegative $\rho \leq N-1$.[7.6]

We assume only that the given measurement matrix H is symmetric;[7.7]

$$H \in \mathbb{S}^N \tag{1396}$$

[7.4]The isomorphism $T(Y) = V_\mathcal{N}^{\dagger\mathrm{T}}YV_\mathcal{N}^{\dagger}$ onto $\mathbb{S}_c^N = \{VXV \mid X \in \mathbb{S}^N\}$ relates the map in (1393) to that in (1392), but is not an isometry.

[7.5]All four problem formulations (1391) produce identical results when affine dimension r, implicit to a realizable measurement matrix H, does not exceed desired affine dimension ρ; because, the optimal objective value will vanish ($\|\star\| = 0$).

[7.6]being first pronounced in the context of multidimensional scaling by Mardia [262] in 1978 who attributes the generic result (§7.1.2) to Eckart & Young, 1936 [133].

[7.7]Projection in Problem 1 is on a rank ρ subset of the positive semidefinite cone $\mathbb{S}_+^{N-1}$ (§2.9.2.1) in the subspace of symmetric matrices $\mathbb{S}^{N-1}$. It is wrong here to zero the main diagonal of given H because first projecting H on the symmetric hollow subspace places an elbow in the path of projection in Problem 1. (Figure 164)

Arranging the eigenvalues λ_i of $-V_\mathcal{N}^{\mathrm{T}} H V_\mathcal{N}$ in nonincreasing order for all i, $\lambda_i \geq \lambda_{i+1}$ with v_i the corresponding i^{th} eigenvector, then an optimal solution to Problem 1 is [362, §2]

$$-V_\mathcal{N}^{\mathrm{T}} D^\star V_\mathcal{N} = \sum_{i=1}^{\rho} \max\{0,\, \lambda_i\}\, v_i v_i^{\mathrm{T}} \tag{1397}$$

where

$$-V_\mathcal{N}^{\mathrm{T}} H V_\mathcal{N} \triangleq \sum_{i=1}^{N-1} \lambda_i\, v_i v_i^{\mathrm{T}} \in \mathbb{S}^{N-1} \tag{1398}$$

is an eigenvalue decomposition and

$$D^\star \in \mathbb{EDM}^N \tag{1399}$$

is an optimal Euclidean distance matrix.

In §7.1.4 we show how to transform Problem 1 to a convex optimization problem for any ρ.

7.1.1 Closest-EDM Problem 1, convex case

7.1.1.0.1 Proof. *Solution* (1397), *convex case.*
When desired affine dimension is unconstrained, $\rho = N-1$, the rank function disappears from (1395) leaving a convex optimization problem; a simple unique minimum-distance projection on the positive semidefinite cone $\mathbb{S}_+^{N-1}$: *videlicet*

$$\begin{array}{ll} \underset{D\in\mathbb{S}_h^N}{\text{minimize}} & \|-V_\mathcal{N}^{\mathrm{T}}(D-H)V_\mathcal{N}\|_{\mathrm{F}}^2 \\ \text{subject to} & -V_\mathcal{N}^{\mathrm{T}} D V_\mathcal{N} \succeq 0 \end{array} \tag{1400}$$

by (987). Because

$$\mathbb{S}^{N-1} = -V_\mathcal{N}^{\mathrm{T}} \mathbb{S}_h^N V_\mathcal{N} \tag{1091}$$

then the necessary and sufficient conditions for projection in isometrically isomorphic $\mathbb{R}^{N(N-1)/2}$ on the selfdual (376) positive semidefinite cone $\mathbb{S}_+^{N-1}$ are:[7.8] (§E.9.2.0.1) (1673) (*confer*(2128))

$$\begin{array}{l} -V_\mathcal{N}^{\mathrm{T}} D^\star V_\mathcal{N} \succeq 0 \\ -V_\mathcal{N}^{\mathrm{T}} D^\star V_\mathcal{N} \left(-V_\mathcal{N}^{\mathrm{T}} D^\star V_\mathcal{N} + V_\mathcal{N}^{\mathrm{T}} H V_\mathcal{N}\right) = \mathbf{0} \\ -V_\mathcal{N}^{\mathrm{T}} D^\star V_\mathcal{N} + V_\mathcal{N}^{\mathrm{T}} H V_\mathcal{N} \succeq 0 \end{array} \tag{1401}$$

Symmetric $-V_\mathcal{N}^{\mathrm{T}} H V_\mathcal{N}$ is diagonalizable hence decomposable in terms of its eigenvectors v and eigenvalues λ as in (1398). Therefore (*confer*(1397))

$$-V_\mathcal{N}^{\mathrm{T}} D^\star V_\mathcal{N} = \sum_{i=1}^{N-1} \max\{0,\, \lambda_i\} v_i v_i^{\mathrm{T}} \tag{1402}$$

[7.8]The Karush-Kuhn-Tucker (KKT) optimality conditions [288, p.328] [62, §5.5.3] for problem (1400) are identical to these conditions for projection on a convex cone.

satisfies (1401), optimally solving (1400). To see that, recall: these eigenvectors constitute an orthogonal set and

$$-V_{\mathcal{N}}^{\mathrm{T}}D^{\star}V_{\mathcal{N}} + V_{\mathcal{N}}^{\mathrm{T}}HV_{\mathcal{N}} = -\sum_{i=1}^{N-1}\min\{0,\,\lambda_i\}v_iv_i^{\mathrm{T}} \qquad (1403)$$

$\blacklozenge$

7.1.2 generic problem

Prior to determination of $D^{\star}$, analytical solution (1397) to Problem 1 is equivalent to solution of a generic rank-constrained projection problem: Given desired affine dimension ρ and

$$A \triangleq -V_{\mathcal{N}}^{\mathrm{T}}HV_{\mathcal{N}} = \sum_{i=1}^{N-1}\lambda_i\,v_iv_i^{\mathrm{T}} \in \mathbb{S}^{N-1} \qquad (1398)$$

Euclidean projection on a rank ρ subset of a PSD cone (on a generally nonconvex subset of the PSD cone boundary $\partial\mathbb{S}_+^{N-1}$ when $\rho < N-1$)

$$\left.\begin{array}{ll}\underset{B\in\mathbb{S}^{N-1}}{\text{minimize}} & \|B-A\|_{\mathrm{F}}^2 \\ \text{subject to} & \operatorname{rank}B \leq \rho \\ & B \succeq 0\end{array}\right\} \quad \text{Generic 1} \qquad (1404)$$

has well known optimal solution (Eckart & Young) [133]

$$B^{\star} \triangleq -V_{\mathcal{N}}^{\mathrm{T}}D^{\star}V_{\mathcal{N}} = \sum_{i=1}^{\rho}\max\{0,\,\lambda_i\}\,v_iv_i^{\mathrm{T}} \in \mathbb{S}^{N-1} \qquad (1397)$$

Once optimal $B^{\star}$ is found, the technique of §5.12 can be used to determine a uniquely corresponding optimal Euclidean distance matrix $D^{\star}$; a unique correspondence by injectivity arguments in §5.6.2.

7.1.2.1 Projection on rank ρ subset of PSD cone

Because Problem 1 is the same as

$$\begin{array}{ll}\underset{D\in\mathbb{S}_h^N}{\text{minimize}} & \|-V_{\mathcal{N}}^{\mathrm{T}}(D-H)V_{\mathcal{N}}\|_{\mathrm{F}}^2 \\ \text{subject to} & \operatorname{rank}V_{\mathcal{N}}^{\mathrm{T}}DV_{\mathcal{N}} \leq \rho \\ & -V_{\mathcal{N}}^{\mathrm{T}}DV_{\mathcal{N}} \succeq 0\end{array} \qquad (1405)$$

and because (1091) provides invertible mapping to the generic problem, then Problem 1 is truly a Euclidean projection of vectorized $-V_{\mathcal{N}}^{\mathrm{T}}HV_{\mathcal{N}}$ on that generally nonconvex subset of symmetric matrices (belonging to the positive semidefinite cone $\mathbb{S}_+^{N-1}$) having rank no greater than desired affine dimension ρ;[7.9] called rank ρ subset: (260)

$$\mathbb{S}_+^{N-1}\setminus\mathbb{S}_+^{N-1}(\rho+1) = \{X\in\mathbb{S}_+^{N-1} \mid \operatorname{rank}X \leq \rho\} \qquad (216)$$

[7.9]Recall: affine dimension is a lower bound on embedding (§2.3.1), equal to dimension of the smallest affine set in which points from a list X corresponding to an EDM D can be embedded.

7.1.3   Choice of spectral cone

Spectral projection substitutes projection on a polyhedral cone, containing a complete set of eigenspectra, in place of projection on a convex set of diagonalizable matrices; *e.g.*, (1417). In this section we develop a method of spectral projection for constraining rank of positive semidefinite matrices in a proximity problem like (1404). We will see why an orthant turns out to be the best choice of spectral cone, and why presorting is critical.

Define a nonlinear permutation-operator $\pi(x) : \mathbb{R}^n \to \mathbb{R}^n$ that sorts its vector argument x into nonincreasing order.

7.1.3.0.1   Definition.   *Spectral projection.*

Let R be an orthogonal matrix and Λ a nonincreasingly ordered diagonal matrix of eigenvalues. *Spectral projection* means unique minimum-distance projection of a rotated $(R$, §B.5.5) nonincreasingly ordered (π) vector (δ) of eigenvalues

$$\pi\big(\delta(R^{\mathrm{T}}\Lambda R)\big) \tag{1406}$$

on a polyhedral cone containing all eigenspectra corresponding to a rank ρ subset of a positive semidefinite cone (§2.9.2.1) or the EDM cone (in Cayley-Menger form, §5.11.2.3).

$\triangle$

In the simplest and most common case, projection on a positive semidefinite cone, orthogonal matrix R equals I (§7.1.4.0.1) and diagonal matrix Λ is ordered during diagonalization (§A.5.1). Then spectral projection simply means projection of $\delta(\Lambda)$ on a subset of the nonnegative orthant, as we shall now ascertain:

It is curious how nonconvex Problem 1 has such a simple analytical solution (1397). Although solution to generic problem (1404) is well known since 1936 [133], its equivalence was observed in 1997 [362, §2] to projection of an ordered vector of eigenvalues (in diagonal matrix Λ) on a subset of the monotone nonnegative cone (§2.13.9.4.2)

$$\mathcal{K}_{\mathcal{M}+} = \{v \mid v_1 \geq v_2 \geq \cdots \geq v_{N-1} \geq 0\} \subseteq \mathbb{R}_+^{N-1} \tag{428}$$

Of interest, momentarily, is only the smallest convex subset of the monotone nonnegative cone $\mathcal{K}_{\mathcal{M}+}$ containing every nonincreasingly ordered eigenspectrum corresponding to a rank ρ subset of the positive semidefinite cone $\mathbb{S}_+^{N-1}$; *id est*,

$$\mathcal{K}_{\mathcal{M}+}^{\rho} \triangleq \{v \in \mathbb{R}^{\rho} \mid v_1 \geq v_2 \geq \cdots \geq v_{\rho} \geq 0\} \subseteq \mathbb{R}_+^{\rho} \tag{1407}$$

a pointed polyhedral cone, a ρ-dimensional convex subset of the monotone nonnegative cone $\mathcal{K}_{\mathcal{M}+} \subseteq \mathbb{R}_+^{N-1}$ having property, for λ denoting eigenspectra,

$$\begin{bmatrix} \mathcal{K}_{\mathcal{M}+}^{\rho} \\ \mathbf{0} \end{bmatrix} = \pi(\lambda(\text{rank } \rho \text{ subset})) \subseteq \mathcal{K}_{\mathcal{M}+}^{N-1} \triangleq \mathcal{K}_{\mathcal{M}+} \tag{1408}$$

For each and every elemental eigenspectrum

$$\gamma \in \lambda(\text{rank } \rho \text{ subset}) \subseteq \mathbb{R}_+^{N-1} \tag{1409}$$

of the rank ρ subset (ordered or unordered in λ), there is a nonlinear surjection $\pi(\gamma)$ onto $\mathcal{K}_{\mathcal{M}+}^{\rho}$.

7.1.3.0.2 Exercise. *Smallest spectral cone.*
Prove that there is no convex subset of $\mathcal{K}_{\mathcal{M}+}$ smaller than $\mathcal{K}_{\mathcal{M}+}^{\rho}$ containing every ordered eigenspectrum corresponding to the rank ρ subset of a positive semidefinite cone (§2.9.2.1).

▼

7.1.3.0.3 Proposition. (Hardy-Littlewood-Pólya) *Inequalities.*   [185, §X]
[56, §1.2] Any vectors σ and γ in $\mathbb{R}^{N-1}$ satisfy a tight inequality

$$\pi(\sigma)^{\mathrm{T}}\pi(\gamma) \geq \sigma^{\mathrm{T}}\gamma \geq \pi(\sigma)^{\mathrm{T}}\Xi\,\pi(\gamma) \tag{1410}$$

where Ξ is the order-reversing permutation matrix defined in (1819), and permutator $\pi(\gamma)$ is a nonlinear function that sorts vector γ into nonincreasing order thereby providing the greatest upper bound and least lower bound with respect to every possible sorting. ◇

7.1.3.0.4 Corollary. *Monotone nonnegative sort.*
Any given vectors $\sigma, \gamma \in \mathbb{R}^{N-1}$ satisfy a tight Euclidean distance inequality

$$\|\pi(\sigma) - \pi(\gamma)\| \leq \|\sigma - \gamma\| \tag{1411}$$

where nonlinear function $\pi(\gamma)$ sorts vector γ into nonincreasing order thereby providing the least lower bound with respect to every possible sorting. ◇

Given $\gamma \in \mathbb{R}^{N-1}$

$$\inf_{\sigma \in \mathbb{R}_+^{N-1}} \|\sigma - \gamma\| = \inf_{\sigma \in \mathbb{R}_+^{N-1}} \|\pi(\sigma) - \pi(\gamma)\| = \inf_{\sigma \in \mathbb{R}_+^{N-1}} \|\sigma - \pi(\gamma)\| = \inf_{\sigma \in \mathcal{K}_{\mathcal{M}+}} \|\sigma - \pi(\gamma)\| \tag{1412}$$

Yet for γ representing an arbitrary vector of eigenvalues, because

$$\inf_{\sigma \in \left[\begin{smallmatrix}\mathbb{R}_+^{\rho}\\\mathbf{0}\end{smallmatrix}\right]} \|\sigma - \gamma\|^2 \geq \inf_{\sigma \in \left[\begin{smallmatrix}\mathbb{R}_+^{\rho}\\\mathbf{0}\end{smallmatrix}\right]} \|\sigma - \pi(\gamma)\|^2 = \inf_{\sigma \in \left[\begin{smallmatrix}\mathcal{K}_{\mathcal{M}+}^{\rho}\\\mathbf{0}\end{smallmatrix}\right]} \|\sigma - \pi(\gamma)\|^2 \tag{1413}$$

then projection of γ on the eigenspectra of a rank ρ subset can be tightened simply by presorting γ into nonincreasing order.

Proof. Simply because $\pi(\gamma)_{1:\rho} \succeq \pi(\gamma_{1:\rho})$

$$\begin{aligned} \inf_{\sigma \in \left[\begin{smallmatrix}\mathbb{R}_+^{\rho}\\\mathbf{0}\end{smallmatrix}\right]} \|\sigma - \gamma\|^2 &= \gamma_{\rho+1:N-1}^{\mathrm{T}}\gamma_{\rho+1:N-1} + \inf_{\sigma \in \mathbb{R}_+^{N-1}} \|\sigma_{1:\rho} - \gamma_{1:\rho}\|^2 \\ &= \gamma^{\mathrm{T}}\gamma + \inf_{\sigma \in \mathbb{R}_+^{N-1}} \sigma_{1:\rho}^{\mathrm{T}}\sigma_{1:\rho} - 2\sigma_{1:\rho}^{\mathrm{T}}\gamma_{1:\rho} \\ &\geq \gamma^{\mathrm{T}}\gamma + \inf_{\sigma \in \mathbb{R}_+^{N-1}} \sigma_{1:\rho}^{\mathrm{T}}\sigma_{1:\rho} - 2\sigma_{1:\rho}^{\mathrm{T}}\pi(\gamma)_{1:\rho} \end{aligned} \tag{1414}$$

$$\inf_{\sigma \in \left[\begin{smallmatrix}\mathbb{R}_+^{\rho}\\\mathbf{0}\end{smallmatrix}\right]} \|\sigma - \gamma\|^2 \geq \inf_{\sigma \in \left[\begin{smallmatrix}\mathbb{R}_+^{\rho}\\\mathbf{0}\end{smallmatrix}\right]} \|\sigma - \pi(\gamma)\|^2$$

◆

7.1.3.1 Orthant is best spectral cone for Problem 1

This means unique minimum-distance projection of γ on the nearest spectral member of the rank ρ subset is tantamount to presorting γ into nonincreasing order. Only then does unique spectral projection on a subset $\mathcal{K}^{\rho}_{\mathcal{M}+}$ of the monotone nonnegative cone become equivalent to unique spectral projection on a subset $\mathbb{R}^{\rho}_{+}$ of the nonnegative orthant (which is simpler); in other words, unique minimum-distance projection of sorted γ on the nonnegative orthant in a ρ-dimensional subspace of $\mathbb{R}^{N}$ is indistinguishable from its projection on the subset $\mathcal{K}^{\rho}_{\mathcal{M}+}$ of the monotone nonnegative cone in that same subspace.

7.1.4 Closest-EDM Problem 1, "nonconvex" case

Proof of solution (1397), for projection on a rank ρ subset of the positive semidefinite cone $\mathbb{S}^{N-1}_{+}$, can be algebraic in nature. [362, §2] Here we derive that known result but instead using a more geometric argument via spectral projection on a polyhedral cone (subsuming the proof in §7.1.1). In so doing, we demonstrate how nonconvex Problem 1 is transformed to a convex optimization:

7.1.4.0.1 Proof. *Solution* (1397), *nonconvex case.*
As explained in §7.1.2, we may instead work with the more facile generic problem (1404). With diagonalization of unknown

$$B \triangleq U \Upsilon U^{\mathrm{T}} \in \mathbb{S}^{N-1} \tag{1415}$$

given desired affine dimension $0 \le \rho \le N-1$ and diagonalizable

$$A \triangleq Q \Lambda Q^{\mathrm{T}} \in \mathbb{S}^{N-1} \tag{1416}$$

having eigenvalues in Λ arranged in nonincreasing order, by (48) the generic problem is equivalent to

$$
\begin{array}{ll}
\underset{B \in \mathbb{S}^{N-1}}{\text{minimize}} & \|B - A\|_{\mathrm{F}}^{2} \\
\text{subject to} & \operatorname{rank} B \le \rho \\
& B \succeq 0
\end{array}
\quad \equiv \quad
\begin{array}{ll}
\underset{R,\, \Upsilon}{\text{minimize}} & \|\Upsilon - R^{\mathrm{T}} \Lambda R\|_{\mathrm{F}}^{2} \\
\text{subject to} & \operatorname{rank} \Upsilon \le \rho \\
& \Upsilon \succeq 0 \\
& R^{-1} = R^{\mathrm{T}}
\end{array}
\tag{1417}
$$

where

$$R \triangleq Q^{\mathrm{T}} U \in \mathbb{R}^{N-1 \times N-1} \tag{1418}$$

in U on the set of orthogonal matrices is a bijection. We propose solving (1417) by instead solving the problem sequence:

$$
\begin{array}{ll}
\underset{\Upsilon}{\text{minimize}} & \|\Upsilon - R^{\mathrm{T}} \Lambda R\|_{\mathrm{F}}^{2} \\
\text{subject to} & \operatorname{rank} \Upsilon \le \rho \\
& \Upsilon \succeq 0
\end{array}
\qquad (a)
$$

$$
\begin{array}{ll}
\underset{R}{\text{minimize}} & \|\Upsilon^{\star} - R^{\mathrm{T}} \Lambda R\|_{\mathrm{F}}^{2} \\
\text{subject to} & R^{-1} = R^{\mathrm{T}}
\end{array}
\qquad (b)
$$

$$\tag{1419}$$

Problem (1419a) is equivalent to: (1) orthogonal projection of $R^{\mathrm{T}}\Lambda R$ on an $N-1$-dimensional subspace of isometrically isomorphic $\mathbb{R}^{N(N-1)/2}$ containing $\delta(\Upsilon) \in \mathbb{R}_+^{N-1}$, (2) nonincreasingly ordering the result, (3) unique minimum-distance projection of the ordered result on $\begin{bmatrix} \mathbb{R}_+^\rho \\ \mathbf{0} \end{bmatrix}$. (§E.9.5) Projection on that $N-1$-dimensional subspace amounts to zeroing $R^{\mathrm{T}}\Lambda R$ at all entries off the main diagonal; thus, the equivalent sequence leading with a spectral projection:

$$
\begin{aligned}
& \underset{\Upsilon}{\text{minimize}} \quad \| \, \delta(\Upsilon) - \pi\big(\delta(R^{\mathrm{T}}\Lambda R)\big) \, \|^2 \\
& \text{subject to} \quad \delta(\Upsilon) \in \begin{bmatrix} \mathbb{R}_+^\rho \\ \mathbf{0} \end{bmatrix}
\end{aligned} \qquad \text{(a)}
$$

$$
\begin{aligned}
& \underset{R}{\text{minimize}} \quad \|\Upsilon^\star - R^{\mathrm{T}}\Lambda R\|_{\mathrm{F}}^2 \\
& \text{subject to} \quad R^{-1} = R^{\mathrm{T}}
\end{aligned} \qquad \text{(b)}
$$

$$(1420)$$

Because any permutation matrix is an orthogonal matrix, $\delta(R^{\mathrm{T}}\Lambda R) \in \mathbb{R}^{N-1}$ can always be arranged in nonincreasing order without loss of generality; hence, permutation operator π. Unique minimum-distance projection of vector $\pi\big(\delta(R^{\mathrm{T}}\Lambda R)\big)$ on the ρ-dimensional subset $\begin{bmatrix} \mathbb{R}_+^\rho \\ \mathbf{0} \end{bmatrix}$ of nonnegative orthant $\mathbb{R}_+^{N-1}$ requires: (§E.9.2.0.1)

$$
\begin{aligned}
\delta(\Upsilon^\star)_{\rho+1:N-1} &= \mathbf{0} \\
\delta(\Upsilon^\star) &\succeq 0 \\
\delta(\Upsilon^\star)^{\mathrm{T}}\big(\delta(\Upsilon^\star) - \pi(\delta(R^{\mathrm{T}}\Lambda R))\big) &= 0 \\
\delta(\Upsilon^\star) - \pi(\delta(R^{\mathrm{T}}\Lambda R)) &\succeq 0
\end{aligned}
\qquad (1421)
$$

which are necessary and sufficient conditions. Any value $\Upsilon^\star$ satisfying conditions (1421) is optimal for (1420a). So

$$
\delta(\Upsilon^\star)_i = \begin{cases} \max\Big\{0,\, \pi\big(\delta(R^{\mathrm{T}}\Lambda R)\big)_i\Big\}, & i=1\ldots\rho \\ 0, & i=\rho+1\ldots N-1 \end{cases}
\qquad (1422)
$$

specifies an optimal solution. The lower bound on the objective with respect to R in (1420b) is tight: by (1388)

$$
\| \, |\Upsilon^\star| - |\Lambda| \, \|_{\mathrm{F}} \leq \|\Upsilon^\star - R^{\mathrm{T}}\Lambda R\|_{\mathrm{F}}
\qquad (1423)
$$

where $|\ |$ denotes absolute entry-value. For selection of $\Upsilon^\star$ as in (1422), this lower bound is attained when (*confer* §C.4.2.2)

$$
R^\star = I
\qquad (1424)
$$

which is the known solution. ♦

7.1.4.1 Significance

Importance of this well-known [133] optimal solution (1397) for projection on a rank ρ subset of a positive semidefinite cone should not be dismissed:

- Problem 1, as stated, is generally nonconvex. This analytical solution at once encompasses projection on a rank ρ subset (216) of the positive semidefinite cone (generally, a nonconvex subset of its boundary) from either the exterior or interior of that cone.[7.10] By problem transformation to the spectral domain, projection on a rank ρ subset becomes a convex optimization problem.

- This solution is closed form.

- This solution is equivalent to projection on a polyhedral cone in the spectral domain (spectral projection, projection on a spectral cone §7.1.3.0.1); a necessary and sufficient condition (§A.3.1) for membership of a symmetric matrix to a rank ρ subset of a positive semidefinite cone (§2.9.2.1).

- Because $U^\star = Q$, a minimum-distance projection on a rank ρ subset of the positive semidefinite cone is a positive semidefinite matrix orthogonal (in the Euclidean sense) to direction of projection.[7.11]

- For the convex case problem (1400), this solution is always unique. Otherwise, distinct eigenvalues (multiplicity 1) in Λ guarantees uniqueness of this solution by the reasoning in §A.5.0.1 .[7.12]

7.1.5 Problem 1 in spectral norm, convex case

When instead we pose the matrix 2-norm (spectral norm) in Problem 1 (1395) for the convex case $\rho = N-1$, then the new problem

$$\begin{array}{ll} \underset{D}{\text{minimize}} & \|-V_\mathcal{N}^T(D-H)V_\mathcal{N}\|_2 \\ \text{subject to} & D \in \mathbb{EDM}^N \end{array} \tag{1425}$$

is convex although its solution is not necessarily unique;[7.13] giving rise to nonorthogonal projection (§E.1) on the positive semidefinite cone $\mathbb{S}_+^{N-1}$. Indeed, its solution set includes the Frobenius solution (1397) for the convex case whenever $-V_\mathcal{N}^T H V_\mathcal{N}$ is a normal matrix. [190, §1] [183] [62, §8.1.1] Proximity problem (1425) is equivalent to

$$\begin{array}{ll} \underset{\mu,\, D}{\text{minimize}} & \mu \\ \text{subject to} & -\mu I \preceq -V_\mathcal{N}^T(D-H)V_\mathcal{N} \preceq \mu I \\ & D \in \mathbb{EDM}^N \end{array} \tag{1426}$$

by (1795) where

$$\mu^\star = \max_i \left\{ \left|\lambda\big(-V_\mathcal{N}^T(D^\star-H)V_\mathcal{N}\big)_i\right| , \quad i=1\dots N-1 \right\} \in \mathbb{R}_+ \tag{1427}$$

the minimized largest absolute eigenvalue (due to matrix symmetry).

For lack of unique solution here, we prefer the Frobenius rather than spectral norm.

[7.10]Projection on the boundary from the interior of a convex Euclidean body is generally a nonconvex problem.

[7.11]But Theorem E.9.2.0.1 for unique projection on a closed convex cone does not apply here because the direction of projection is not necessarily a member of the dual PSD cone. This occurs, for example, whenever positive eigenvalues are truncated.

[7.12]Uncertainty of uniqueness prevents the erroneous conclusion that a rank ρ subset (216) were a convex body by the *Bunt-Motzkin theorem* (§E.9.0.0.1).

[7.13]For each and every $|t| \leq 2$, for example, $\begin{bmatrix} 2 & 0 \\ 0 & t \end{bmatrix}$ has the same spectral-norm value.

7.2 Second prevalent problem: Projection on EDM cone in $\sqrt{d_{ij}}$

Let

$$\sqrt[\circ]{D} \triangleq [\sqrt{d_{ij}}] \in \mathcal{K} = \mathbb{S}_h^N \cap \mathbb{R}_+^{N \times N} \qquad (1428)$$

be an unknown matrix of absolute distance; *id est,*

$$D = [d_{ij}] \triangleq \sqrt[\circ]{D} \circ \sqrt[\circ]{D} \in \mathbb{EDM}^N \qquad (1429)$$

where $\circ$ denotes Hadamard product. The second prevalent proximity problem is a Euclidean projection (in the natural coordinates $\sqrt{d_{ij}}$) of matrix H on a nonconvex subset of the boundary of the nonconvex cone of Euclidean absolute-distance matrices rel $\partial\sqrt{\mathbb{EDM}^N}$: (§6.3, *confer* Figure 149b)

$$\left.\begin{array}{ll} \underset{\sqrt[\circ]{D}}{\text{minimize}} & \|\sqrt[\circ]{D} - H\|_F^2 \\ \text{subject to} & \text{rank } V_N^T D V_N \leq \rho \\ & \sqrt[\circ]{D} \in \sqrt{\mathbb{EDM}^N} \end{array}\right\} \quad \text{Problem 2} \qquad (1430)$$

where

$$\sqrt{\mathbb{EDM}^N} = \{\sqrt[\circ]{D} \mid D \in \mathbb{EDM}^N\} \qquad (1263)$$

This statement of the second proximity problem is considered difficult to solve because of the constraint on desired affine dimension ρ (§5.7.2) and because the objective function

$$\|\sqrt[\circ]{D} - H\|_F^2 = \sum_{i,j}(\sqrt{d_{ij}} - h_{ij})^2 \qquad (1431)$$

is expressed in the natural coordinates; projection on a doubly nonconvex set.

Our solution to this second problem prevalent in the literature requires measurement matrix H to be nonnegative;

$$H = [h_{ij}] \in \mathbb{R}_+^{N \times N} \qquad (1432)$$

If the H matrix given has negative entries, then the technique of solution presented here becomes invalid. As explained in §7.0.1, projection of H on $\mathcal{K} = \mathbb{S}_h^N \cap \mathbb{R}_+^{N \times N}$ (1382) prior to application of this proposed solution is incorrect.

7.2.1   Convex case

When $\rho = N - 1$, the rank constraint vanishes and a convex problem that is equivalent to (1389) emerges:[7.14]

$$\begin{array}{ll} \underset{\sqrt[\circ]{D}}{\text{minimize}} & \|\sqrt[\circ]{D} - H\|_{\mathrm{F}}^2 \\ \text{subject to} & \sqrt[\circ]{D} \in \sqrt{\mathbb{EDM}^N} \end{array} \quad \Leftrightarrow \quad \begin{array}{ll} \underset{D}{\text{minimize}} & \displaystyle\sum_{i,j} d_{ij} - 2h_{ij}\sqrt{d_{ij}} + h_{ij}^2 \\ \text{subject to} & D \in \mathbb{EDM}^N \end{array} \qquad (1433)$$

For any fixed i and j, the argument of summation is a convex function of d_{ij} because (for nonnegative constant h_{ij}) the negative square root is convex in nonnegative d_{ij} and because $d_{ij} + h_{ij}^2$ is affine (convex). Because the sum of any number of convex functions in D remains convex [62, §3.2.1] and because the feasible set is convex in D, we have a convex optimization problem:

$$\begin{array}{ll} \underset{D}{\text{minimize}} & \mathbf{1}^{\mathrm{T}}(D - 2H \circ \sqrt[\circ]{D})\mathbf{1} + \|H\|_{\mathrm{F}}^2 \\ \text{subject to} & D \in \mathbb{EDM}^N \end{array} \qquad (1434)$$

The objective function being a sum of strictly convex functions is, moreover, strictly convex in D on the nonnegative orthant. Existence of a unique solution $D^\star$ for this second prevalent problem depends upon nonnegativity of H and a convex feasible set (§3.1.2).[7.15]

7.2.1.1   Equivalent semidefinite program, Problem 2, convex case

Convex problem (1433) is numerically solvable for its global minimum using an interior-point method [419] [297] [284] [410] [11] [149]. We translate (1433) to an equivalent semidefinite program (SDP) for a pedagogical reason made clear in §7.2.2.2 and because there exist readily available computer programs for numerical solution [173] [412] [413] [368] [34] [411] [357] [343].

Substituting a new matrix variable $Y \triangleq [y_{ij}] \in \mathbb{R}_+^{N \times N}$

$$h_{ij}\sqrt{d_{ij}} \leftarrow y_{ij} \qquad (1435)$$

Boyd proposes: problem (1433) is equivalent to the semidefinite program

$$\begin{array}{ll} \underset{D,\,Y}{\text{minimize}} & \displaystyle\sum_{i,j} d_{ij} - 2y_{ij} + h_{ij}^2 \\ \text{subject to} & \begin{bmatrix} d_{ij} & y_{ij} \\ y_{ij} & h_{ij}^2 \end{bmatrix} \succeq 0, \quad i,j = 1 \dots N \\ & D \in \mathbb{EDM}^N \end{array} \qquad (1436)$$

[7.14] still thought to be a nonconvex problem as late as 1997 [363] even though discovered convex by de Leeuw in 1993. [105] [54, §13.6] Yet using methods from §3, it can be easily ascertained: $\|\sqrt[\circ]{D} - H\|_{\mathrm{F}}$ is not convex in D.

[7.15] The transformed problem in variable D no longer describes Euclidean projection on an EDM cone. Otherwise we might erroneously conclude $\sqrt{\mathbb{EDM}^N}$ were a convex body by the *Bunt-Motzkin theorem* (§E.9.0.0.1).

To see that, recall $d_{ij} \geq 0$ is implicit to $D \in \mathbb{EDM}^N$ (§5.8.1, (987)). So when $H \in \mathbb{R}_+^{N \times N}$ is nonnegative as assumed,

$$\begin{bmatrix} d_{ij} & y_{ij} \\ y_{ij} & h_{ij}^2 \end{bmatrix} \succeq 0 \iff h_{ij}\sqrt{d_{ij}} \geq \sqrt{y_{ij}^2} \tag{1437}$$

Minimization of the objective function implies maximization of y_{ij} that is bounded above. Hence nonnegativity of y_{ij} is implicit to (1436) and, as desired, $y_{ij} \to h_{ij}\sqrt{d_{ij}}$ as optimization proceeds. ◆

If the given matrix H is now assumed symmetric and nonnegative,

$$H = [h_{ij}] \in \mathbb{S}^N \cap \mathbb{R}_+^{N \times N} \tag{1438}$$

then $Y = H \circ \sqrt[\circ]{D}$ must belong to $\mathcal{K} = \mathbb{S}_h^N \cap \mathbb{R}_+^{N \times N}$ (1382). Because $Y \in \mathbb{S}_h^N$ (§B.4.2 *no.20*), then

$$\|\sqrt[\circ]{D} - H\|_{\mathrm{F}}^2 = \sum_{i,j} d_{ij} - 2y_{ij} + h_{ij}^2 = -N \operatorname{tr}(V(D - 2Y)V) + \|H\|_{\mathrm{F}}^2 \tag{1439}$$

So convex problem (1436) is equivalent to the semidefinite program

$$\begin{aligned} &\underset{D,\,Y}{\text{minimize}} && -\operatorname{tr}(V(D - 2Y)V) \\ &\text{subject to} && \begin{bmatrix} d_{ij} & y_{ij} \\ y_{ij} & h_{ij}^2 \end{bmatrix} \succeq 0, \quad N \geq j > i = 1 \dots N-1 \\ & && Y \in \mathbb{S}_h^N \\ & && D \in \mathbb{EDM}^N \end{aligned} \tag{1440}$$

where the constants h_{ij}^2 and N have been dropped arbitrarily from the objective.

7.2.1.2 Gram-form semidefinite program, Problem 2, convex case

There is great advantage to expressing problem statement (1440) in Gram-form because Gram matrix G is a bidirectional bridge between point list X and distance matrix D; *e.g.*, §5.4.2.2.8, §6.7.0.0.1. This way, problem convexity can be maintained while simultaneously constraining point list X, Gram matrix G, and distance matrix D at our discretion.

Convex problem (1440) may be equivalently written via linear bijective (§5.6.1) EDM operator $\mathbf{D}(G)$ (980);

$$\begin{aligned} &\underset{G \in \mathbb{S}_c^N,\; Y \in \mathbb{S}_h^N}{\text{minimize}} && -\operatorname{tr}(V(\mathbf{D}(G) - 2Y)V) \\ &\text{subject to} && \begin{bmatrix} \langle \Phi_{ij}, G \rangle & y_{ij} \\ y_{ij} & h_{ij}^2 \end{bmatrix} \succeq 0, \quad N \geq j > i = 1 \dots N-1 \\ & && G \succeq 0 \end{aligned} \tag{1441}$$

where distance-square $D = [d_{ij}] \in \mathbb{S}_h^N$ (964) is related to Gram matrix entries $G = [g_{ij}] \in \mathbb{S}_c^N \cap \mathbb{S}_+^N$ by

$$\begin{aligned} d_{ij} &= g_{ii} + g_{jj} - 2g_{ij} \\ &= \langle \Phi_{ij}, G \rangle \end{aligned} \tag{979}$$

where

$$\Phi_{ij} = (e_i - e_j)(e_i - e_j)^{\mathrm{T}} \in \mathbb{S}_+^N \qquad (966)$$

Confinement of G to the geometric center subspace provides numerical stability and no loss of generality (*confer*(1327)); implicit constraint $G\mathbf{1} = \mathbf{0}$ is otherwise unnecessary.

To include constraints on the list $X \in \mathbb{R}^{n \times N}$, we would first rewrite (1441)

$$\begin{array}{cl}
\underset{G \in \mathbb{S}_c^N,\; Y \in \mathbb{S}_h^N,\; X \in \mathbb{R}^{n \times N}}{\text{minimize}} & -\operatorname{tr}(V(\mathbf{D}(G) - 2Y)V) \\[2ex]
\text{subject to} & \begin{bmatrix} \langle \Phi_{ij}\,,\, G \rangle & y_{ij} \\ y_{ij} & h_{ij}^2 \end{bmatrix} \succeq 0, \quad N \geq j > i = 1 \ldots N-1 \\[3ex]
& \begin{bmatrix} I & X \\ X^{\mathrm{T}} & G \end{bmatrix} \succeq 0 \\[2ex]
& X \in \mathcal{C}
\end{array} \qquad (1442)$$

and then add the constraints, realized here in abstract membership to some convex set $\mathcal{C}$. This problem realization includes a convex relaxation of the nonconvex constraint $G = X^{\mathrm{T}}X$ and, if desired, more constraints on G could be added. This technique is discussed in §5.4.2.2.8.

7.2.2   Minimization of affine dimension in Problem 2

When desired affine dimension ρ is diminished, the rank function becomes reinserted into problem (1436) that is then rendered difficult to solve because feasible set $\{D, Y\}$ loses convexity in $\mathbb{S}_h^N \times \mathbb{R}^{N \times N}$. Indeed, the rank function is quasiconcave (§3.8) on the positive semidefinite cone; (§2.9.2.9.2) *id est*, its sublevel sets are not convex.

7.2.2.1   Rank minimization heuristic

A remedy developed in [268] [139] [140] [138] introduces convex envelope of the quasiconcave rank function: (Figure 165)

7.2.2.1.1   Definition. *Convex envelope.*   [204]
Convex envelope cenv f of a function $f : \mathcal{C} \to \mathbb{R}$ is defined to be the largest convex function g such that $g \leq f$ on convex domain $\mathcal{C} \subseteq \mathbb{R}^n$. [7.16]   △

[7.16]Provided $f \not\equiv +\infty$ and there exists an affine function $h \leq f$ on $\mathbb{R}^n$, then the convex envelope is equal to the convex conjugate (the *Legendre-Fenchel transform*) of the convex conjugate of f; *id est*, the conjugate-conjugate function f^{**}. [205, §E.1]

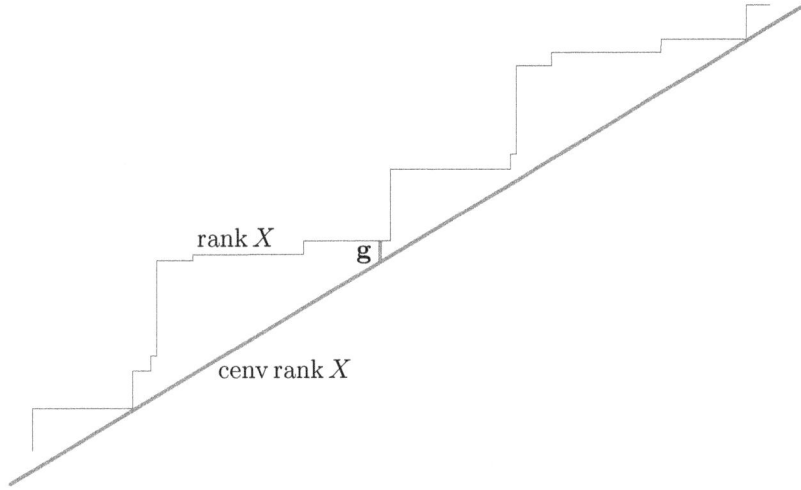

Figure 165: Abstraction of convex envelope of rank function. Rank is a quasiconcave function on positive semidefinite cone, but its convex envelope is the largest convex function whose epigraph contains it. Vertical bar labelled **g** measures a trace/rank gap; *id est*, rank found always exceeds estimate; large decline in trace required here for only small decrease in rank.

- [139] [138] Convex envelope of rank function: for σ_i a singular value, (1647)

$$\text{cenv}(\text{rank}\,A) \text{ on } \{A\in\mathbb{R}^{m\times n} \mid \|A\|_2\leq\kappa\} = \frac{1}{\kappa}\mathbf{1}^\mathrm{T}\sigma(A) = \frac{1}{\kappa}\,\text{tr}\sqrt{A^\mathrm{T}A} \qquad (1443)$$

$$\text{cenv}(\text{rank}\,A) \text{ on } \{A \text{ normal} \mid \|A\|_2\leq\kappa\} = \frac{1}{\kappa}\|\lambda(A)\|_1 = \frac{1}{\kappa}\,\text{tr}\sqrt{A^\mathrm{T}A} \qquad (1444)$$

$$\text{cenv}(\text{rank}\,A) \text{ on } \{A\in\mathbb{S}_+^n \mid \|A\|_2\leq\kappa\} = \frac{1}{\kappa}\mathbf{1}^\mathrm{T}\lambda(A) = \frac{1}{\kappa}\,\text{tr}(A) \qquad (1445)$$

A properly scaled trace thus represents the best convex lower bound on rank for positive semidefinite matrices. The idea, then, is to substitute convex envelope for rank of some variable $A\in\mathbb{S}_+^M$ (§A.6.5)

$$\text{rank}\,A \leftarrow \text{cenv}(\text{rank}\,A) \propto \text{tr}\,A = \sum_i \sigma(A)_i = \sum_i \lambda(A)_i \qquad (1446)$$

which is equivalent to the sum of all eigenvalues or singular values.

- [138] Convex envelope of the cardinality function is proportional to the 1-norm:

$$\text{cenv}(\text{card}\,x) \text{ on } \{x\in\mathbb{R}^n \mid \|x\|_\infty\leq\kappa\} = \frac{1}{\kappa}\|x\|_1 \qquad (1447)$$

$$\text{cenv}(\text{card}\,x) \text{ on } \{x\in\mathbb{R}_+^n \mid \|x\|_\infty\leq\kappa\} = \frac{1}{\kappa}\mathbf{1}^\mathrm{T}x \qquad (1448)$$

7.2.2.2 Applying trace rank-heuristic to Problem 2

Substituting rank envelope for rank function in Problem 2, for $D \in \mathbb{EDM}^N$ (*confer*(1122))

$$\text{cenv rank}(-V_{\mathcal{N}}^{\mathrm{T}}DV_{\mathcal{N}}) = \text{cenv rank}(-VDV) \; \propto \; -\text{tr}(VDV) \qquad (1449)$$

and for desired affine dimension $\rho \leq N-1$ and nonnegative H [*sic*] we get a convex optimization problem

$$\begin{array}{ll} \underset{D}{\text{minimize}} & \|\sqrt[\circ]{D} - H\|_{\mathrm{F}}^2 \\ \text{subject to} & -\text{tr}(VDV) \leq \kappa\rho \\ & D \in \mathbb{EDM}^N \end{array} \qquad (1450)$$

where $\kappa \in \mathbb{R}_+$ is a constant determined by cut-and-try. The equivalent semidefinite program makes κ variable: for nonnegative and symmetric H

$$\begin{array}{ll} \underset{D,\,Y,\,\kappa}{\text{minimize}} & \kappa\rho + 2\,\text{tr}(VYV) \\ \text{subject to} & \begin{bmatrix} d_{ij} & y_{ij} \\ y_{ij} & h_{ij}^2 \end{bmatrix} \succeq 0, \quad N \geq j > i = 1 \ldots N-1 \\ & -\text{tr}(VDV) \leq \kappa\rho \\ & Y \in \mathbb{S}_h^N \\ & D \in \mathbb{EDM}^N \end{array} \qquad (1451)$$

which is the same as (1440), the problem with no explicit constraint on affine dimension. As the present problem is stated, the desired affine dimension ρ yields to the variable scale factor κ; ρ is effectively ignored.

Yet this result is an illuminant for problem (1440) and it equivalents (all the way back to (1433)): When the given measurement matrix H is nonnegative and symmetric, finding the closest EDM D as in problem (1433), (1436), or (1440) implicitly entails minimization of affine dimension (*confer* §5.8.4, §5.14.4). Those non$-$rank-constrained problems are each inherently equivalent to cenv(rank)-minimization problem (1451), in other words, and their optimal solutions are unique because of the strictly convex objective function in (1433).

7.2.2.3 Rank-heuristic insight

Minimization of affine dimension by use of this trace rank-heuristic (1449) tends to find a list configuration of least energy; rather, it tends to optimize compaction of the reconstruction by minimizing total distance. (992) It is best used where some physical equilibrium implies such an energy minimization; *e.g.*, [361, §5].

For this Problem 2, the trace rank-heuristic arose naturally in the objective in terms of V. We observe: V (in contrast to $V_{\mathcal{N}}^{\mathrm{T}}$) spreads energy over all available distances (§B.4.2 *no.*20, contrast *no.*22) although the rank function itself is insensitive to choice of auxiliary matrix.

Trace rank-heuristic (1445) is useless when a main diagonal is constrained to be constant. Such would be the case were optimization over an elliptope (§5.4.2.2.1), or when the diagonal represents a Boolean vector; *e.g.*, §4.2.3.1.1, §4.6.0.0.9.

7.2.2.4 Rank minimization heuristic beyond convex envelope

Fazel, Hindi, & Boyd [140] [415] [141] propose a rank heuristic more potent than trace (1446) for problems of rank minimization;

$$\operatorname{rank} Y \;\leftarrow\; \log \det(Y + \varepsilon I) \tag{1452}$$

the concave surrogate function $\log \det$ in place of quasiconcave $\operatorname{rank} Y$ (§2.9.2.9.2) when $Y \in \mathbb{S}_+^n$ is variable and where ε is a small positive constant. They propose minimization of the surrogate by substituting a sequence comprising infima of a linearized surrogate about the current estimate Y_i ; *id est*, from the first-order Taylor series expansion about Y_i on some open interval of $\|Y\|_2$ (§D.1.7)

$$\log \det(Y + \varepsilon I) \approx \log \det(Y_i + \varepsilon I) + \operatorname{tr}\!\big((Y_i + \varepsilon I)^{-1}(Y - Y_i)\big) \tag{1453}$$

we make the surrogate sequence of infima over bounded convex feasible set $\mathcal{C}$

$$\arg \inf_{Y \in \mathcal{C}} \operatorname{rank} Y \;\leftarrow\; \lim_{i \to \infty} Y_{i+1} \tag{1454}$$

where, for $i = 0 \ldots$

$$Y_{i+1} = \arg \inf_{Y \in \mathcal{C}} \operatorname{tr}\!\big((Y_i + \varepsilon I)^{-1} Y\big) \tag{1455}$$

a matrix analogue to the reweighting scheme disclosed in [213, §4.11.3]. Choosing $Y_0 = I$, the first step becomes equivalent to finding the infimum of $\operatorname{tr} Y$; the trace rank-heuristic (1446). The intuition underlying (1455) is the new term in the argument of trace; specifically, $(Y_i + \varepsilon I)^{-1}$ weights Y so that relatively small eigenvalues of Y found by the infimum are made even smaller.

To see that, substitute the nonincreasingly ordered diagonalizations

$$\begin{aligned} Y_i + \varepsilon I &\triangleq Q(\Lambda + \varepsilon I)Q^{\mathrm{T}} &\quad \text{(a)} \\ Y &\triangleq U \Upsilon U^{\mathrm{T}} &\quad \text{(b)} \end{aligned} \tag{1456}$$

into (1455). Then from (1792) we have,

$$\begin{aligned} \inf_{\Upsilon \in U^{\star\mathrm{T}}\mathcal{C}U^{\star}} \delta\big((\Lambda + \varepsilon I)^{-1}\big)^{\mathrm{T}} \delta(\Upsilon) \;&=\; \inf_{\Upsilon \in U^{\mathrm{T}}\mathcal{C}U} \; \inf_{R^{\mathrm{T}}=R^{-1}} \operatorname{tr}\!\big((\Lambda + \varepsilon I)^{-1} R^{\mathrm{T}} \Upsilon R\big) \\ &\leq\; \inf_{Y \in \mathcal{C}} \operatorname{tr}\!\big((Y_i + \varepsilon I)^{-1} Y\big) \end{aligned} \tag{1457}$$

where $R \triangleq Q^{\mathrm{T}}U$ in U on the set of orthogonal matrices is a bijection. The role of ε is, therefore, to limit maximum weight; the smallest entry on the main diagonal of Υ gets the largest weight. ♦

7.2.2.5 Applying $\log\det$ rank-heuristic to Problem 2

When the $\log\det$ rank-heuristic is inserted into Problem 2, problem (1451) becomes the problem sequence in i

$$
\begin{array}{ll}
\underset{D,\,Y,\,\kappa}{\text{minimize}} & \kappa\,\rho + 2\,\mathrm{tr}(VYV) \\[2mm]
\text{subject to} & \begin{bmatrix} d_{jl} & y_{jl} \\ y_{jl} & h_{jl}^{2} \end{bmatrix} \succeq 0, \quad l > j = 1\ldots N-1 \\[4mm]
& -\mathrm{tr}\big((-VD_iV + \varepsilon I)^{-1}VDV\big) \le \kappa\,\rho \\[2mm]
& Y \in \mathbb{S}_h^N \\[1mm]
& D \in \mathbb{EDM}^N
\end{array}
\tag{1458}
$$

where $D_{i+1} \triangleq D^{\star} \in \mathbb{EDM}^N$ and $D_0 \triangleq \mathbf{1}\mathbf{1}^{\mathrm{T}} - I$.

7.2.2.6 Tightening this $\log\det$ rank-heuristic

Like the trace method, this $\log\det$ technique for constraining rank offers no provision for meeting a predetermined upper bound ρ. Yet since eigenvalues are simply determined, $\lambda(Y_i + \varepsilon I) = \delta(\Lambda + \varepsilon I)$, we may certainly force selected weights to ε^{-1} by manipulating diagonalization (1456a). Empirically we find this sometimes leads to better results, although affine dimension of a solution cannot be guaranteed.

7.2.2.7 Cumulative summary of rank heuristics

We have studied a perturbation method of rank reduction in §4.3 as well as the trace heuristic (convex envelope method §7.2.2.1.1) and $\log\det$ heuristic in §7.2.2.4. There is another good contemporary method called LMIRank [293] based on alternating projection (§E.10).[7.17]

7.2.2.7.1 Example. *Unidimensional scaling.*
We apply the convex iteration method from §4.4.1 to numerically solve an instance of Problem 2; a method empirically superior to the foregoing convex envelope and $\log\det$ heuristics for rank regularization and enforcing affine dimension.

Unidimensional scaling, [107] a historically practical application of multidimensional scaling (§5.12), entails solution of an optimization problem having local minima whose multiplicity varies as the factorial of point-list cardinality; geometrically, it means reconstructing a list constrained to lie in one affine dimension. In terms of point list, the nonconvex problem is: given nonnegative symmetric matrix $H = [h_{ij}] \in \mathbb{S}^N \cap \mathbb{R}_+^{N \times N}$ (1438) whose entries h_{ij} are all known,

$$
\underset{\{x_i \in \mathbb{R}\}}{\text{minimize}} \sum_{i,\,j=1}^{N} \big(|x_i - x_j| - h_{ij}\big)^2 \tag{1389}
$$

called a *raw stress* problem [54, p.34] which has an implicit constraint on dimensional embedding of points $\{x_i \in \mathbb{R}, \ i = 1\ldots N\}$. This problem has proven NP-hard; *e.g.*, [74].

[7.17] that does not solve the *ball packing* problem presented in §5.4.2.2.6.

As always, we first transform variables to distance-square $D \in \mathbb{S}_h^N$; so begin with convex problem (1440) on page 495

$$
\begin{aligned}
&\underset{D,Y}{\text{minimize}} \quad -\operatorname{tr}(V(D-2Y)V) \\
&\text{subject to} \quad \begin{bmatrix} d_{ij} & y_{ij} \\ y_{ij} & h_{ij}^2 \end{bmatrix} \succeq 0, \quad N \geq j > i = 1 \ldots N-1 \\
&\qquad\qquad Y \in \mathbb{S}_h^N \\
&\qquad\qquad D \in \mathbb{EDM}^N \\
&\qquad\qquad \operatorname{rank} V_{\mathcal{N}}^{\mathrm{T}} D V_{\mathcal{N}} = 1
\end{aligned}
\tag{1459}
$$

that becomes equivalent to (1389) by making explicit the constraint on affine dimension via rank. The iteration is formed by moving the dimensional constraint to the objective:

$$
\begin{aligned}
&\underset{D,Y}{\text{minimize}} \quad -\langle V(D-2Y)V , I\rangle - w\langle V_{\mathcal{N}}^{\mathrm{T}} D V_{\mathcal{N}} , W\rangle \\
&\text{subject to} \quad \begin{bmatrix} d_{ij} & y_{ij} \\ y_{ij} & h_{ij}^2 \end{bmatrix} \succeq 0, \quad N \geq j > i = 1 \ldots N-1 \\
&\qquad\qquad Y \in \mathbb{S}_h^N \\
&\qquad\qquad D \in \mathbb{EDM}^N
\end{aligned}
\tag{1460}
$$

where w (≈ 10) is a positive scalar just large enough to make $\langle V_{\mathcal{N}}^{\mathrm{T}} D V_{\mathcal{N}} , W\rangle$ vanish to within some numerical precision, and where direction matrix W is an optimal solution to semidefinite program (1791a)

$$
\begin{aligned}
&\underset{W}{\text{minimize}} \quad -\langle V_{\mathcal{N}}^{\mathrm{T}} D^\star V_{\mathcal{N}} , W\rangle \\
&\text{subject to} \quad 0 \preceq W \preceq I \\
&\qquad\qquad \operatorname{tr} W = N-1
\end{aligned}
\tag{1461}
$$

one of which is known in closed form. Semidefinite programs (1460) and (1461) are iterated until convergence in the sense defined on page 266. This iteration is not a projection method. (§4.4.1.1) Convex problem (1460) is neither a relaxation of unidimensional scaling problem (1459); instead, problem (1460) is a convex equivalent to (1459) at convergence of the iteration.

Jan de Leeuw provided us with some test data

$$
H = \begin{bmatrix}
0.000000 & 5.235301 & 5.499274 & 6.404294 & 6.486829 & 6.263265 \\
5.235301 & 0.000000 & 3.208028 & 5.840931 & 3.559010 & 5.353489 \\
5.499274 & 3.208028 & 0.000000 & 5.679550 & 4.020339 & 5.239842 \\
6.404294 & 5.840931 & 5.679550 & 0.000000 & 4.862884 & 4.543120 \\
6.486829 & 3.559010 & 4.020339 & 4.862884 & 0.000000 & 4.618718 \\
6.263265 & 5.353489 & 5.239842 & 4.543120 & 4.618718 & 0.000000
\end{bmatrix}
\tag{1462}
$$

and a globally optimal solution

$$
\begin{aligned}
X^\star &= \begin{bmatrix} -4.981494 & -2.121026 & -1.038738 & 4.555130 & 0.764096 & 2.822032 \end{bmatrix} \\
&= \begin{bmatrix} x_1^\star & x_2^\star & x_3^\star & x_4^\star & x_5^\star & x_6^\star \end{bmatrix}
\end{aligned}
\tag{1463}
$$

found by searching 6! local minima of (1389) [107]. By iterating convex problems (1460) and (1461) about twenty times (initial $W = \mathbf{0}$) we find a global infimum 98.12812 of stress problem (1389), and by (1212) we find a corresponding one-dimensional point list that is a rigid transformation in $\mathbb{R}$ of $X^\star$.

Here we found the infimum to accuracy of the given data, but that ceases to hold as problem size increases. Because of machine numerical precision and an interior-point method of solution, we speculate, accuracy degrades quickly as problem size increases beyond this.   □

7.3 Third prevalent problem:
Projection on EDM cone in d_{ij}

In summary, we find that the solution to problem [(1391.3) p.484] is difficult and depends on the dimension of the space as the geometry of the cone of EDMs becomes more complex.

−Hayden, Wells, Liu, & Tarazaga (1991) [191, §3]

Reformulating Problem 2 (p.493) in terms of EDM D changes the problem considerably:

$$\left.\begin{array}{ll} \underset{D}{\text{minimize}} & \|D - H\|_{\mathrm{F}}^2 \\ \text{subject to} & \operatorname{rank} V_{\mathcal{N}}^{\mathrm{T}} D V_{\mathcal{N}} \le \rho \\ & D \in \mathbb{EDM}^N \end{array}\right\} \quad \text{Problem 3} \qquad (1464)$$

This third prevalent proximity problem is a Euclidean projection of given matrix H on a generally nonconvex subset ($\rho < N-1$) of $\partial\mathbb{EDM}^N$ the boundary of the convex cone of Euclidean distance matrices relative to subspace $\mathbb{S}_h^N$ (Figure 149d). Because coordinates of projection are distance-square and H now presumably holds distance-square measurements, numerical solution to Problem 3 is generally different than that of Problem 2.

For the moment, we need make no assumptions regarding measurement matrix H.

7.3.1 Convex case

$$\begin{array}{ll} \underset{D}{\text{minimize}} & \|D - H\|_{\mathrm{F}}^2 \\ \text{subject to} & D \in \mathbb{EDM}^N \end{array} \qquad (1465)$$

When the rank constraint disappears (for $\rho = N-1$), this third problem becomes obviously convex because the feasible set is then the entire EDM cone and because the objective function

$$\|D - H\|_{\mathrm{F}}^2 = \sum_{i,j}(d_{ij} - h_{ij})^2 \qquad (1466)$$

is a strictly convex quadratic in D ;[7.18]

$$
\begin{array}{ll}
\underset{D}{\text{minimize}} & \sum_{i,j} d_{ij}^2 - 2h_{ij}\,d_{ij} \,+\, h_{ij}^2 \\
\text{subject to} & D \in \mathbb{EDM}^N
\end{array}
\tag{1467}
$$

Optimal solution $D^\star$ is therefore unique, as expected, for this simple projection on the EDM cone equivalent to (1390).

7.3.1.1 Equivalent semidefinite program, Problem 3, convex case

In the past, this convex problem was solved numerically by means of alternating projection. (Example 7.3.1.1.1) [159] [152] [191, §1] We translate (1467) to an equivalent semidefinite program because we have a good solver:

Assume the given measurement matrix H to be nonnegative and symmetric;[7.19]

$$
H = [h_{ij}] \in \mathbb{S}^N \cap \mathbb{R}_+^{N \times N} \tag{1438}
$$

We then propose: Problem (1467) is equivalent to the semidefinite program, for

$$
\partial \triangleq [d_{ij}^2] = D \circ D \tag{1468}
$$

a matrix of distance-square squared,

$$
\begin{array}{ll}
\underset{\partial\,,\,D}{\text{minimize}} & -\operatorname{tr}(V(\partial - 2\,H \circ D)V) \\[2mm]
\text{subject to} & \begin{bmatrix} \partial_{ij} & d_{ij} \\ d_{ij} & 1 \end{bmatrix} \succeq 0, \quad N \ge j > i = 1 \dots N-1 \\[4mm]
& D \in \mathbb{EDM}^N \\[1mm]
& \partial \in \mathbb{S}_h^N
\end{array}
\tag{1469}
$$

where

$$
\begin{bmatrix} \partial_{ij} & d_{ij} \\ d_{ij} & 1 \end{bmatrix} \succeq 0 \ \Leftrightarrow\ \partial_{ij} \ge d_{ij}^2 \tag{1470}
$$

Symmetry of input H facilitates trace in the objective (§B.4.2 *no.*20), while its nonnegativity causes $\partial_{ij} \to d_{ij}^2$ as optimization proceeds.

[7.18]For nonzero $Y \in \mathbb{S}_h^N$ and some open interval of $t \in \mathbb{R}$ (§3.7.3.0.2, §D.2.3)

$$
\frac{d^2}{dt^2}\|(D + t\,Y) - H\|_{\mathrm{F}}^2 \,=\, 2\operatorname{tr} Y^{\mathrm{T}} Y > 0 \qquad \blacklozenge
$$

[7.19]If that H given has negative entries, then the technique of solution presented here becomes invalid. Projection of H on $\mathcal{K}$ (1382) prior to application of this proposed technique, as explained in §7.0.1, is incorrect.

7.3.1.1.1   Example.   *Alternating projection on nearest EDM.*
By solving (1469) we confirm the result from an example given by Glunt, Hayden, Hong, & Wells [159, §6] who found analytical solution to convex optimization problem (1465) for particular cardinality $N=3$ by using the alternating projection method of von Neumann (§E.10):

$$H = \begin{bmatrix} 0 & 1 & 1 \\ 1 & 0 & 9 \\ 1 & 9 & 0 \end{bmatrix}, \qquad D^\star = \begin{bmatrix} 0 & \frac{19}{9} & \frac{19}{9} \\ \frac{19}{9} & 0 & \frac{76}{9} \\ \frac{19}{9} & \frac{76}{9} & 0 \end{bmatrix} \tag{1471}$$

The original problem (1465) of projecting H on the EDM cone is transformed to an equivalent iterative sequence of projections on the two convex cones (1331) from §6.8.1.1. Using ordinary alternating projection, input H goes to $D^\star$ with an accuracy of four decimal places in about 17 iterations. Affine dimension corresponding to this optimal solution is $r=1$.

Obviation of semidefinite programming's computational expense is the principal advantage of this alternating projection technique.   □

7.3.1.2   Schur-form semidefinite program, Problem 3 convex case

Semidefinite program (1469) can be reformulated by moving the objective function in

$$\begin{array}{ll} \underset{D}{\text{minimize}} & \|D - H\|_{\text{F}}^2 \\ \text{subject to} & D \in \mathbb{EDM}^N \end{array} \tag{1465}$$

to the constraints. This makes an equivalent epigraph form of the problem: for any measurement matrix H

$$\begin{array}{ll} \underset{t\in\mathbb{R},\ D}{\text{minimize}} & t \\ \text{subject to} & \|D - H\|_{\text{F}}^2 \leq t \\ & D \in \mathbb{EDM}^N \end{array} \tag{1472}$$

We can transform this problem to an equivalent Schur-form semidefinite program; (§3.5.2)

$$\begin{array}{ll} \underset{t\in\mathbb{R},\ D}{\text{minimize}} & t \\ \text{subject to} & \begin{bmatrix} tI & \text{vec}(D-H) \\ \text{vec}(D-H)^{\text{T}} & 1 \end{bmatrix} \succeq 0 \\ & D \in \mathbb{EDM}^N \end{array} \tag{1473}$$

characterized by great sparsity and structure. The advantage of this SDP is lack of conditions on input H; *e.g.*, negative entries would invalidate any solution provided by (1469). (§7.0.1.2)

7.3.1.3 Gram-form semidefinite program, Problem 3 convex case

Further, this problem statement may be equivalently written in terms of a Gram matrix via linear bijective (§5.6.1) EDM operator $\mathbf{D}(G)$ (980);

$$
\begin{array}{cl}
\underset{G\in\mathbb{S}_c^N,\ t\in\mathbb{R}}{\text{minimize}} & t \\
\text{subject to} & \begin{bmatrix} tI & \text{vec}(\mathbf{D}(G)-H) \\ \text{vec}(\mathbf{D}(G)-H)^{\mathrm{T}} & 1 \end{bmatrix} \succeq 0 \\
& G\succeq 0
\end{array}
\qquad (1474)
$$

To include constraints on the list $X\in\mathbb{R}^{n\times N}$, we would rewrite this:

$$
\begin{array}{cl}
\underset{G\in\mathbb{S}_c^N,\ t\in\mathbb{R},\ X\in\mathbb{R}^{n\times N}}{\text{minimize}} & t \\
\text{subject to} & \begin{bmatrix} tI & \text{vec}(\mathbf{D}(G)-H) \\ \text{vec}(\mathbf{D}(G)-H)^{\mathrm{T}} & 1 \end{bmatrix} \succeq 0 \\
& \begin{bmatrix} I & X \\ X^{\mathrm{T}} & G \end{bmatrix} \succeq 0 \\
& X\in\mathcal{C}
\end{array}
\qquad (1475)
$$

where $\mathcal{C}$ is some abstract convex set. This technique is discussed in §5.4.2.2.8.

7.3.1.4 Dual interpretation, projection on EDM cone

From §E.9.1.1 we learn that projection on a convex set has a dual form. In the circumstance $\mathcal{K}$ is a convex cone and point x exists exterior to the cone or on its boundary, distance to the nearest point Px in $\mathcal{K}$ is found as the optimal value of the objective

$$
\begin{array}{cl}
\|x-Px\| = \underset{a}{\text{maximize}} & a^{\mathrm{T}}x \\
\text{subject to} & \|a\|\leq 1 \\
& a\in\mathcal{K}^{\circ}
\end{array}
\qquad (2112)
$$

where $\mathcal{K}^{\circ}$ is the polar cone.

Applying this result to (1465), we get a convex optimization for any given symmetric matrix H exterior to or on the EDM cone boundary:

$$
\begin{array}{cl}
\underset{D}{\text{minimize}} & \|D-H\|_{\mathrm{F}}^2 \\
\text{subject to} & D\in\mathbb{EDM}^N
\end{array}
\equiv
\begin{array}{cl}
\underset{A^{\circ}}{\text{maximize}} & \langle A^{\circ},\,H\rangle \\
\text{subject to} & \|A^{\circ}\|_{\mathrm{F}}\leq 1 \\
& A^{\circ}\in\mathbb{EDM}^{N^{\circ}}
\end{array}
\qquad (1476)
$$

Then from (2114) projection of H on cone $\mathbb{EDM}^N$ is

$$
D^{\star} = H - A^{\circ\star}\langle A^{\circ\star},\,H\rangle
\qquad (1477)
$$

Critchley proposed, instead, projection on the polar EDM cone in his 1980 thesis [91, p.113]: In that circumstance, by projection on the algebraic complement (§E.9.2.2.1),

$$
D^{\star} = A^{\star}\langle A^{\star},\,H\rangle
\qquad (1478)
$$

which is equal to (1477) when $A^\star$ solves

$$
\begin{aligned}
\underset{A}{\text{maximize}} \quad & \langle A \, , \, H \rangle \\
\text{subject to} \quad & \|A\|_{\text{F}} = 1 \\
& A \in \mathbb{EDM}^N
\end{aligned}
\tag{1479}
$$

This projection of symmetric H on polar cone $\mathbb{EDM}^{N^\circ}$ can be made a convex problem, of course, by relaxing the equality constraint ($\|A\|_{\text{F}} \le 1$).

7.3.2 Minimization of affine dimension in Problem 3

When desired affine dimension ρ is diminished, Problem 3 (1464) is difficult to solve [191, §3] because the feasible set in $\mathbb{R}^{N(N-1)/2}$ loses convexity. By substituting rank envelope (1449) into Problem 3, then for any given H we get a convex problem

$$
\begin{aligned}
\underset{D}{\text{minimize}} \quad & \|D - H\|_{\text{F}}^2 \\
\text{subject to} \quad & -\operatorname{tr}(VDV) \le \kappa \rho \\
& D \in \mathbb{EDM}^N
\end{aligned}
\tag{1480}
$$

where $\kappa \in \mathbb{R}_+$ is a constant determined by cut-and-try. Given κ, problem (1480) is a convex optimization having unique solution in any desired affine dimension ρ; an approximation to Euclidean projection on that nonconvex subset of the EDM cone containing EDMs with corresponding affine dimension no greater than ρ.

The SDP equivalent to (1480) does not move κ into the variables as on page 498: for nonnegative symmetric input H and distance-square squared variable ∂ as in (1468),

$$
\begin{aligned}
\underset{\partial \, , \, D}{\text{minimize}} \quad & -\operatorname{tr}(V(\partial - 2\,H \circ D)V) \\
\text{subject to} \quad & \begin{bmatrix} \partial_{ij} & d_{ij} \\ d_{ij} & 1 \end{bmatrix} \succeq 0, \quad N \ge j > i = 1 \ldots N-1 \\
& -\operatorname{tr}(VDV) \le \kappa \rho \\
& D \in \mathbb{EDM}^N \\
& \partial \in \mathbb{S}_h^N
\end{aligned}
\tag{1481}
$$

That means we will not see equivalence of this cenv(rank)-minimization problem to the non$-$rank-constrained problems (1467) and (1469) like we saw for its counterpart (1451) in Problem 2.

Another approach to affine dimension minimization is to project instead on the polar EDM cone; discussed in §6.8.1.5.

7.3.3 Constrained affine dimension, Problem 3

When one desires affine dimension diminished further below what can be achieved via cenv(rank)-minimization as in (1481), spectral projection can be considered a natural means in light of its successful application to projection on a rank ρ subset of the positive semidefinite cone in §7.1.4.

Yet it is wrong here to zero eigenvalues of $-VDV$ or $-VGV$ or a variant to reduce affine dimension, because that particular method comes from projection on a positive semidefinite cone (1417); zeroing those eigenvalues here in Problem 3 would place an elbow in the projection path (Figure 164) thereby producing a result that is necessarily suboptimal. Problem 3 is instead a projection on the EDM cone whose associated spectral cone is considerably different. (§5.11.2.3) Proper choice of spectral cone is demanded by diagonalization of that variable argument to the objective:

7.3.3.1 Cayley-Menger form

We use Cayley-Menger composition of the Euclidean distance matrix to solve a problem that is the same as Problem 3 (1464): (§5.7.3.0.1)

$$
\begin{aligned}
\underset{D}{\text{minimize}} \quad & \left\| \begin{bmatrix} 0 & \mathbf{1}^{\mathrm{T}} \\ \mathbf{1} & -D \end{bmatrix} - \begin{bmatrix} 0 & \mathbf{1}^{\mathrm{T}} \\ \mathbf{1} & -H \end{bmatrix} \right\|_{\mathrm{F}}^{2} \\
\text{subject to} \quad & \operatorname{rank} \begin{bmatrix} 0 & \mathbf{1}^{\mathrm{T}} \\ \mathbf{1} & -D \end{bmatrix} \le \rho + 2 \\
& D \in \mathbb{EDM}^{N}
\end{aligned}
\tag{1482}
$$

a projection of H on a generally nonconvex subset (when $\rho < N-1$) of the Euclidean distance matrix cone boundary rel $\partial\mathbb{EDM}^{N}$; *id est*, projection from the EDM cone interior or exterior on a subset of its relative boundary (§6.5, (1260)).

Rank of an optimal solution is intrinsically bounded above and below;

$$
2 \le \operatorname{rank} \begin{bmatrix} 0 & \mathbf{1}^{\mathrm{T}} \\ \mathbf{1} & -D^{\star} \end{bmatrix} \le \rho + 2 \le N + 1
\tag{1483}
$$

Our proposed strategy for low-rank solution is projection on that subset of a spectral cone $\lambda\left(\begin{bmatrix} 0 & \mathbf{1}^{\mathrm{T}} \\ \mathbf{1} & -\mathbb{EDM}^{N} \end{bmatrix} \right)$ (§5.11.2.3) corresponding to affine dimension not in excess of that ρ desired; *id est*, spectral projection on

$$
\begin{bmatrix} \mathbb{R}_{+}^{\rho+1} \\ \mathbf{0} \\ \mathbb{R}_{-} \end{bmatrix} \cap \partial\mathcal{H} \subset \mathbb{R}^{N+1}
\tag{1484}
$$

where

$$
\partial\mathcal{H} = \{ \lambda \in \mathbb{R}^{N+1} \mid \mathbf{1}^{\mathrm{T}}\lambda = 0 \}
\tag{1192}
$$

is a hyperplane through the origin. This pointed polyhedral cone (1484), to which membership subsumes the rank constraint, is not full-dimensional.

Given desired affine dimension $0 \leq \rho \leq N-1$ and diagonalization (§A.5) of unknown EDM D

$$\begin{bmatrix} 0 & \mathbf{1}^{\mathrm{T}} \\ \mathbf{1} & -D \end{bmatrix} \triangleq U \Upsilon U^{\mathrm{T}} \in \mathbb{S}_h^{N+1} \tag{1485}$$

and given symmetric H in diagonalization

$$\begin{bmatrix} 0 & \mathbf{1}^{\mathrm{T}} \\ \mathbf{1} & -H \end{bmatrix} \triangleq Q \Lambda Q^{\mathrm{T}} \in \mathbb{S}^{N+1} \tag{1486}$$

having eigenvalues arranged in nonincreasing order, then by (1205) problem (1482) is equivalent to

$$\begin{aligned} \underset{\Upsilon, R}{\text{minimize}} \quad & \left\| \delta(\Upsilon) - \pi\big(\delta(R^{\mathrm{T}}\Lambda R)\big) \right\|^2 \\ \text{subject to} \quad & \delta(\Upsilon) \in \begin{bmatrix} \mathbb{R}_+^{\rho+1} \\ \mathbf{0} \\ \mathbb{R}_- \end{bmatrix} \cap \partial \mathcal{H} \\ & \delta(QR\Upsilon R^{\mathrm{T}}Q^{\mathrm{T}}) = \mathbf{0} \\ & R^{-1} = R^{\mathrm{T}} \end{aligned} \tag{1487}$$

where π is the permutation operator from §7.1.3 arranging its vector argument in nonincreasing order,[7.20] where

$$R \triangleq Q^{\mathrm{T}} U \in \mathbb{R}^{N+1 \times N+1} \tag{1488}$$

in U on the set of orthogonal matrices is a bijection, and where $\partial \mathcal{H}$ insures one negative eigenvalue. Hollowness constraint $\delta(QR\Upsilon R^{\mathrm{T}}Q^{\mathrm{T}}) = \mathbf{0}$ makes problem (1487) difficult by making the two variables dependent.

Our plan is to instead divide problem (1487) into two and then iterate their solution:

$$\begin{aligned} \underset{\Upsilon}{\text{minimize}} \quad & \left\| \delta(\Upsilon) - \pi\big(\delta(R^{\mathrm{T}}\Lambda R)\big) \right\|^2 \\ \text{subject to} \quad & \delta(\Upsilon) \in \begin{bmatrix} \mathbb{R}_+^{\rho+1} \\ \mathbf{0} \\ \mathbb{R}_- \end{bmatrix} \cap \partial \mathcal{H} \end{aligned} \quad \text{(a)}$$
$$\tag{1489}$$
$$\begin{aligned} \underset{R}{\text{minimize}} \quad & \| R \Upsilon^\star R^{\mathrm{T}} - \Lambda \|_{\mathrm{F}}^2 \\ \text{subject to} \quad & \delta(QR\Upsilon^\star R^{\mathrm{T}}Q^{\mathrm{T}}) = \mathbf{0} \\ & R^{-1} = R^{\mathrm{T}} \end{aligned} \quad \text{(b)}$$

[7.20]Recall, any permutation matrix is an orthogonal matrix.

Proof. We justify disappearance of the hollowness constraint in convex optimization problem (1489a): From the arguments in §7.1.3 with regard to π the permutation operator, cone membership constraint $\delta(\Upsilon) \in \begin{bmatrix} \mathbb{R}_+^{\rho+1} \\ \mathbf{0} \\ \mathbb{R}_- \end{bmatrix} \cap \partial\mathcal{H}$ from (1489a) is equivalent to

$$\delta(\Upsilon) \in \begin{bmatrix} \mathbb{R}_+^{\rho+1} \\ \mathbf{0} \\ \mathbb{R}_- \end{bmatrix} \cap \partial\mathcal{H} \cap \mathcal{K}_\mathcal{M} \tag{1490}$$

where $\mathcal{K}_\mathcal{M}$ is the monotone cone (§2.13.9.4.3). Membership of $\delta(\Upsilon)$ to the *polyhedral cone of majorization* (Theorem A.1.2.0.1)

$$\mathcal{K}_{\lambda\delta}^* = \partial\mathcal{H} \cap \mathcal{K}_{\mathcal{M}+}^* \tag{1507}$$

where $\mathcal{K}_{\mathcal{M}+}^*$ is the dual monotone nonnegative cone (§2.13.9.4.2), is a condition (in absence of a hollowness constraint) that would insure existence of a symmetric hollow matrix $\begin{bmatrix} 0 & \mathbf{1}^\mathrm{T} \\ \mathbf{1} & -D \end{bmatrix}$. Curiously, intersection of this feasible superset $\begin{bmatrix} \mathbb{R}_+^{\rho+1} \\ \mathbf{0} \\ \mathbb{R}_- \end{bmatrix} \cap \partial\mathcal{H} \cap \mathcal{K}_\mathcal{M}$ from (1490) with the cone of majorization $\mathcal{K}_{\lambda\delta}^*$ is a benign operation; *id est,*

$$\partial\mathcal{H} \cap \mathcal{K}_{\mathcal{M}+}^* \cap \mathcal{K}_\mathcal{M} = \partial\mathcal{H} \cap \mathcal{K}_\mathcal{M} \tag{1491}$$

verifiable by observing conic dependencies (§2.10.3) among the aggregate of halfspace-description normals. The cone membership constraint in (1489a) therefore inherently insures existence of a symmetric hollow matrix. $\blacklozenge$

Optimization (1489b) would be a Procrustes problem (§C.4) were it not for the hollowness constraint; it is, instead, a minimization over the intersection of the nonconvex manifold of orthogonal matrices with another nonconvex set in variable R specified by the hollowness constraint.

We solve problem (1489b) by a method introduced in §4.6.0.0.2: Define $R = [\, r_1 \cdots r_{N+1}\,] \in \mathbb{R}^{N+1 \times N+1}$ and make the assignment

$$G = \begin{bmatrix} r_1 \\ \vdots \\ r_{N+1} \\ 1 \end{bmatrix} [\, r_1^\mathrm{T} \cdots r_{N+1}^\mathrm{T} \ 1\,] \qquad \in \mathbb{S}^{(N+1)^2+1}$$

$$= \begin{bmatrix} R_{11} & \cdots & R_{1,N+1} & r_1 \\ \vdots & \ddots & & \vdots \\ R_{1,N+1}^\mathrm{T} & & R_{N+1,N+1} & r_{N+1} \\ r_1^\mathrm{T} & \cdots & r_{N+1}^\mathrm{T} & 1 \end{bmatrix} \triangleq \begin{bmatrix} r_1 r_1^\mathrm{T} & \cdots & r_1 r_{N+1}^\mathrm{T} & r_1 \\ \vdots & \ddots & & \vdots \\ r_{N+1} r_1^\mathrm{T} & & r_{N+1} r_{N+1}^\mathrm{T} & r_{N+1} \\ r_1^\mathrm{T} & \cdots & r_{N+1}^\mathrm{T} & 1 \end{bmatrix} \tag{1492}$$

where $R_{ij} \triangleq r_i r_j^{\mathrm{T}} \in \mathbb{R}^{N+1 \times N+1}$ and $\Upsilon_{ii}^\star \in \mathbb{R}$. Since $R\,\Upsilon^\star R^{\mathrm{T}} = \sum\limits_{i=1}^{N+1} \Upsilon_{ii}^\star R_{ii}$ then problem (1489b) is equivalently expressed:

$$\begin{array}{ll}
\underset{R_{ii} \in \mathbb{S},\, R_{ij},\, r_i}{\text{minimize}} & \left\| \sum\limits_{i=1}^{N+1} \Upsilon_{ii}^\star R_{ii} - \Lambda \right\|_{\mathrm{F}}^2 \\
\text{subject to} & \operatorname{tr} R_{ii} = 1, \qquad\qquad\qquad\qquad\qquad i = 1 \ldots N+1 \\
& \operatorname{tr} R_{ij} = 0, \qquad\qquad\qquad\qquad\qquad i < j = 2 \ldots N+1 \\
& G = \begin{bmatrix} R_{11} & \cdots & R_{1,N+1} & r_1 \\ \vdots & \ddots & & \vdots \\ R_{1,N+1}^{\mathrm{T}} & & R_{N+1,N+1} & r_{N+1} \\ r_1^{\mathrm{T}} & \cdots & r_{N+1}^{\mathrm{T}} & 1 \end{bmatrix} (\succeq 0) \\
& \delta\!\left(Q \sum\limits_{i=1}^{N+1} \Upsilon_{ii}^\star R_{ii}\, Q^{\mathrm{T}} \right) = \mathbf{0} \\
& \operatorname{rank} G = 1
\end{array} \qquad (1493)$$

The rank constraint is regularized by method of convex iteration developed in §4.4. Problem (1493) is partitioned into two convex problems:

$$\begin{array}{ll}
\underset{R_{ij},\, r_i}{\text{minimize}} & \left\| \sum\limits_{i=1}^{N+1} \Upsilon_{ii}^\star R_{ii} - \Lambda \right\|_{\mathrm{F}}^2 + \langle G,\, W \rangle \\
\text{subject to} & \operatorname{tr} R_{ii} = 1, \qquad\qquad\qquad\qquad\qquad i = 1 \ldots N+1 \\
& \operatorname{tr} R_{ij} = 0, \qquad\qquad\qquad\qquad\qquad i < j = 2 \ldots N+1 \\
& G = \begin{bmatrix} R_{11} & \cdots & R_{1,N+1} & r_1 \\ \vdots & \ddots & & \vdots \\ R_{1,N+1}^{\mathrm{T}} & & R_{N+1,N+1} & r_{N+1} \\ r_1^{\mathrm{T}} & \cdots & r_{N+1}^{\mathrm{T}} & 1 \end{bmatrix} \succeq 0 \\
& \delta\!\left(Q \sum\limits_{i=1}^{N+1} \Upsilon_{ii}^\star R_{ii}\, Q^{\mathrm{T}} \right) = \mathbf{0}
\end{array} \qquad (1494)$$

and

$$\begin{array}{ll}
\underset{W \in \mathbb{S}^{(N+1)^2+1}}{\text{minimize}} & \langle G^\star,\, W \rangle \\
\text{subject to} & 0 \preceq W \preceq I \\
& \operatorname{tr} W = (N+1)^2
\end{array} \qquad (1495)$$

then iterated with convex problem (1489a) until a rank-1 G matrix is found and the objective of (1489a) is minimized. An optimal solution to (1495) is known in closed form (p.567). The hollowness constraint in (1494) may cause numerical infeasibility; in that case, it can be moved to the objective within an added weighted norm. Conditions for termination of the iteration would then comprise a vanishing norm of hollowness.

7.4 Conclusion

The importance and application of solving rank- or cardinality-constrained problems are enormous, a conclusion generally accepted *gratis* by the mathematics and engineering communities. Rank-constrained semidefinite programs arise in many vital feedback and control problems [178], optics [76] (Figure **123**), and communications [146] [260] (Figure **105**). *For example, one might be interested in the minimal order dynamic output feedback which stabilizes a given linear time invariant plant* (*this problem is considered to be among the most important open problems in control*). [269] Rank and cardinality constraints also arise naturally in combinatorial optimization (§4.6.0.0.11, Figure **116**), and find application to face recognition (Figure **4**), cartography (Figure **145**), video surveillance, latent semantic indexing [237], sparse or low-rank matrix completion for preference models and collaborative filtering, multidimensional scaling or principal component analysis (§5.12), medical imaging (Figure **111**), digital filter design with time domain constraints [386], molecular conformation (Figure **138**), sensor-network localization and wireless location (Figure **88**), *etcetera*.

There has been little progress in spectral projection since the discovery by Eckart & Young in 1936 [133] leading to a formula for projection on a rank ρ subset of a positive semidefinite cone (§2.9.2.1). [150] The only closed-form spectral method presently available for solving proximity problems, having a constraint on rank, is based on their discovery (Problem 1, §7.1, §5.13).

- One popular recourse is intentional misapplication of Eckart & Young's result by introducing spectral projection on a positive semidefinite cone into Problem 3 via $\mathbf{D}(G)$ (980), for example. [74] Since Problem 3 instead demands spectral projection on the EDM cone, any solution acquired that way is necessarily suboptimal.

- A second recourse is problem redesign: A presupposition to all proximity problems in this chapter is that matrix H is given. We considered H having various properties such as nonnegativity, symmetry, hollowness, or lack thereof. It was assumed that if H did not already belong to the EDM cone, then we wanted an EDM closest to H in some sense; *id est*, input-matrix H was assumed corrupted somehow. For practical problems, it withstands reason that such a proximity problem could instead be reformulated so that some or all entries of H were unknown but bounded above and below by known limits; the norm objective is thereby eliminated as in the development beginning on page 278. That particular redesign (*the art*, p.8), in terms of the Gram-matrix bridge between point-list X and EDM D, at once encompasses proximity and completion problems.

- A third recourse is to apply the method of convex iteration just like we did in §7.2.2.7.1. This technique is applicable to any semidefinite problem requiring a rank constraint; it places a regularization term in the objective that enforces the rank constraint.

Appendix A

Linear algebra

A.1 Main-diagonal δ operator, λ, tr, vec

We introduce notation δ denoting the main-diagonal linear self-adjoint operator. When linear function δ operates on a square matrix $A \in \mathbb{R}^{N \times N}$, $\delta(A)$ returns a vector composed of all the entries from the main diagonal in the natural order;

$$\delta(A) \in \mathbb{R}^N \tag{1496}$$

Operating on a vector $y \in \mathbb{R}^N$, δ naturally returns a diagonal matrix;

$$\delta(y) \in \mathbb{S}^N \tag{1497}$$

Operating recursively on a vector $\Lambda \in \mathbb{R}^N$ or diagonal matrix $\Lambda \in \mathbb{S}^N$, $\delta(\delta(\Lambda))$ returns Λ itself;

$$\delta^2(\Lambda) \equiv \delta(\delta(\Lambda)) \triangleq \Lambda \tag{1498}$$

Defined in this manner, main-diagonal linear operator δ is *self-adjoint* [233, §3.10, §9.5-1];[A.1] *videlicet*, (§2.2)

$$\delta(A)^{\mathrm{T}} y = \langle \delta(A), y \rangle = \langle A, \delta(y) \rangle = \mathrm{tr}\big(A^{\mathrm{T}} \delta(y)\big) \tag{1499}$$

A.1.1 Identities

This δ notation is efficient and unambiguous as illustrated in the following examples where: $A \circ B$ denotes Hadamard product [208] [164, §1.1.4] of matrices of like size, $\otimes$ Kronecker product [172] (§D.1.2.1), y a vector, X a matrix, e_i the ith member of the standard basis for $\mathbb{R}^n$, $\mathbb{S}_h^N$ the symmetric hollow subspace, $\sigma(A)$ a vector of (nonincreasingly) ordered singular values of matrix A, and $\lambda(A)$ a vector of nonincreasingly ordered eigenvalues:

[A.1]Linear operator $T : \mathbb{R}^{m \times n} \to \mathbb{R}^{M \times N}$ is self-adjoint when, $\forall\, X_1, X_2 \in \mathbb{R}^{m \times n}$

$$\langle T(X_1), X_2 \rangle = \langle X_1, T(X_2) \rangle$$

CITATION: Dattorro, *Convex Optimization & Euclidean Distance Geometry*, $\mathcal{M}\varepsilon\beta oo$ Publishing USA, 2005, v2015.02.15.

1. $\delta(A) = \delta(A^{\mathrm{T}})$

2. $\operatorname{tr}(A) = \operatorname{tr}(A^{\mathrm{T}}) = \delta(A)^{\mathrm{T}}\mathbf{1} = \langle I, A \rangle$

3. $\delta(cA) = c\,\delta(A)$ $c \in \mathbb{R}$

4. $\operatorname{tr}(cA) = c\operatorname{tr}(A) = c\,\mathbf{1}^{\mathrm{T}}\lambda(A)$ $c \in \mathbb{R}$

5. $\operatorname{vec}(cA) = c\operatorname{vec}(A)$ $c \in \mathbb{R}$

6. $A \circ cB = cA \circ B$ $c \in \mathbb{R}$

7. $A \otimes cB = cA \otimes B$ $c \in \mathbb{R}$

8. $\delta(A + B) = \delta(A) + \delta(B)$

9. $\operatorname{tr}(A + B) = \operatorname{tr}(A) + \operatorname{tr}(B)$

10. $\operatorname{vec}(A + B) = \operatorname{vec}(A) + \operatorname{vec}(B)$

11. $(A + B) \circ C = A \circ C + B \circ C$
    $A \circ (B + C) = A \circ B + A \circ C$

12. $(A + B) \otimes C = A \otimes C + B \otimes C$
    $A \otimes (B + C) = A \otimes B + A \otimes C$

13. $\operatorname{sgn}(c)\,\lambda(|c|A) = c\,\lambda(A)$ $c \in \mathbb{R}$

14. $\operatorname{sgn}(c)\,\sigma(|c|A) = c\,\sigma(A)$ $c \in \mathbb{R}$

15. $\operatorname{tr}(c\sqrt{A^{\mathrm{T}}A}) = c\operatorname{tr}\sqrt{A^{\mathrm{T}}A} = c\,\mathbf{1}^{\mathrm{T}}\sigma(A)$ $c \in \mathbb{R}$

16. $\pi(\delta(A)) = \lambda(I \circ A)$ where π is the presorting function.

17. $\delta(AB) = (A \circ B^{\mathrm{T}})\mathbf{1} = (B^{\mathrm{T}} \circ A)\mathbf{1}$

18. $\delta(AB)^{\mathrm{T}} = \mathbf{1}^{\mathrm{T}}(A^{\mathrm{T}} \circ B) = \mathbf{1}^{\mathrm{T}}(B \circ A^{\mathrm{T}})$

19. $\delta(uv^{\mathrm{T}}) = \begin{bmatrix} u_1 v_1 \\ \vdots \\ u_N v_N \end{bmatrix} = u \circ v\,, \quad u, v \in \mathbb{R}^N$

20. $\operatorname{tr}(A^{\mathrm{T}}B) = \operatorname{tr}(AB^{\mathrm{T}}) = \operatorname{tr}(BA^{\mathrm{T}}) = \operatorname{tr}(B^{\mathrm{T}}A)$
    $= \mathbf{1}^{\mathrm{T}}(A \circ B)\mathbf{1} = \mathbf{1}^{\mathrm{T}}\delta(AB^{\mathrm{T}}) = \delta(A^{\mathrm{T}}B)^{\mathrm{T}}\mathbf{1} = \delta(BA^{\mathrm{T}})^{\mathrm{T}}\mathbf{1} = \delta(B^{\mathrm{T}}A)^{\mathrm{T}}\mathbf{1}$

21. $D = [d_{ij}] \in \mathbb{S}_h^N\,, \quad H = [h_{ij}] \in \mathbb{S}_h^N\,, \quad V = I - \frac{1}{N}\mathbf{1}\mathbf{1}^{\mathrm{T}} \in \mathbb{S}^N$ *(confer §B.4.2 no.20)*
    $N\operatorname{tr}(-V(D \circ H)V) = \operatorname{tr}(D^{\mathrm{T}}H) = \mathbf{1}^{\mathrm{T}}(D \circ H)\mathbf{1} = \operatorname{tr}(\mathbf{1}\mathbf{1}^{\mathrm{T}}(D \circ H)) = \sum_{i,j} d_{ij}\,h_{ij}$

22. $\operatorname{tr}(\Lambda A) = \delta(\Lambda)^{\mathrm{T}}\delta(A)\,, \quad \delta^2(\Lambda) \triangleq \Lambda \in \mathbb{S}^N$

23. $\quad y^{\mathrm{T}}B\,\delta(A) = \operatorname{tr}(B\,\delta(A)y^{\mathrm{T}}) = \operatorname{tr}(\delta(B^{\mathrm{T}}y)A) = \operatorname{tr}(A\,\delta(B^{\mathrm{T}}y))$
    $\quad = \delta(A)^{\mathrm{T}}B^{\mathrm{T}}y = \operatorname{tr}(y\,\delta(A)^{\mathrm{T}}B^{\mathrm{T}}) = \operatorname{tr}(A^{\mathrm{T}}\delta(B^{\mathrm{T}}y)) = \operatorname{tr}(\delta(B^{\mathrm{T}}y)A^{\mathrm{T}})$

24. $\delta^2(A^{\mathrm{T}}A) = \sum_i e_i e_i^{\mathrm{T}} A^{\mathrm{T}} A e_i e_i^{\mathrm{T}}$

25. $\delta\big(\delta(A)\mathbf{1}^{\mathrm{T}}\big) = \delta\big(\mathbf{1}\,\delta(A)^{\mathrm{T}}\big) = \delta(A)$

26. $\delta(A\mathbf{1})\mathbf{1} = \delta(A\mathbf{1}\mathbf{1}^{\mathrm{T}}) = A\mathbf{1}$, $\delta(y)\mathbf{1} = \delta(y\mathbf{1}^{\mathrm{T}}) = y$

27. $\delta(I\mathbf{1}) = \delta(\mathbf{1}) = I$

28. $\delta(e_i e_j^{\mathrm{T}}\mathbf{1}) = \delta(e_i) = e_i e_i^{\mathrm{T}}$

29. For $\zeta = [\zeta_i] \in \mathbb{R}^k$ and $x = [x_i] \in \mathbb{R}^k$, $\sum_i \zeta_i/x_i = \zeta^{\mathrm{T}}\delta(x)^{-1}\mathbf{1}$

30. $\begin{aligned} \mathrm{vec}(A \circ B) &= \mathrm{vec}(A) \circ \mathrm{vec}(B) = \delta(\mathrm{vec}\,A)\,\mathrm{vec}(B) \\ &= \mathrm{vec}(B) \circ \mathrm{vec}(A) = \delta(\mathrm{vec}\,B)\,\mathrm{vec}(A) \end{aligned}$ (42)(1879)

31. $\mathrm{vec}(AXB) = (B^{\mathrm{T}} \otimes A)\,\mathrm{vec}\,X$ (not ^{H})

32. $\mathrm{vec}(BXA) = (A^{\mathrm{T}} \otimes B)\,\mathrm{vec}\,X$

33. $\begin{aligned} \mathrm{tr}(AXBX^{\mathrm{T}}) &= \mathrm{vec}(X)^{\mathrm{T}}\mathrm{vec}(AXB) = \mathrm{vec}(X)^{\mathrm{T}}(B^{\mathrm{T}} \otimes A)\,\mathrm{vec}\,X \quad [172] \\ &= \delta\big(\mathrm{vec}(X)\,\mathrm{vec}(X)^{\mathrm{T}}(B^{\mathrm{T}} \otimes A)\big)^{\mathrm{T}}\mathbf{1} \end{aligned}$

34. $\begin{aligned} \mathrm{tr}(AX^{\mathrm{T}}BX) &= \mathrm{vec}(X)^{\mathrm{T}}\mathrm{vec}(BXA) = \mathrm{vec}(X)^{\mathrm{T}}(A^{\mathrm{T}} \otimes B)\,\mathrm{vec}\,X \\ &= \delta\big(\mathrm{vec}(X)\,\mathrm{vec}(X)^{\mathrm{T}}(A^{\mathrm{T}} \otimes B)\big)^{\mathrm{T}}\mathbf{1} \end{aligned}$

35. For any permutation matrix Ξ and dimensionally compatible vector y or matrix A

$$\delta(\Xi\,y) = \Xi\,\delta(y)\,\Xi^{\mathrm{T}} \qquad\qquad (1500)$$
$$\delta(\Xi\,A\,\Xi^{\mathrm{T}}) = \Xi\,\delta(A) \qquad\qquad (1501)$$

So given any permutation matrix Ξ and any dimensionally compatible matrix B, for example,

$$\delta^2(B) = \Xi\,\delta^2(\Xi^{\mathrm{T}}B\,\Xi)\,\Xi^{\mathrm{T}} \qquad\qquad (1502)$$

36. $A \otimes \mathbf{1} = \mathbf{1} \otimes A = A$

37. $A \otimes (B \otimes C) = (A \otimes B) \otimes C$

38. $(A \otimes B)(C \otimes D) = AC \otimes BD$

39. For A a vector, $(A \otimes B) = (A \otimes I)B$

40. For B a row vector, $(A \otimes B) = A(I \otimes B)$

41. $(A \otimes B)^{\mathrm{T}} = A^{\mathrm{T}} \otimes B^{\mathrm{T}}$

42. $(A \otimes B)^{-1} = A^{-1} \otimes B^{-1}$

43. $\mathrm{tr}(A \otimes B) = \mathrm{tr}\,A\,\mathrm{tr}\,B$

44. For $A \in \mathbb{R}^{m \times m}$, $B \in \mathbb{R}^{n \times n}$, $\det(A \otimes B) = \det^n(A)\det^m(B)$

45. There exist permutation matrices Ξ_1 and Ξ_2 such that [172, p.28]

$$A \otimes B = \Xi_1 (B \otimes A) \Xi_2 \qquad (1503)$$

46. For eigenvalues $\lambda(A) \in \mathbb{C}^n$ and eigenvectors $v(A) \in \mathbb{C}^{n \times n}$ such that $A = v\delta(\lambda)v^{-1} \in \mathbb{R}^{n \times n}$

$$\lambda(A \otimes B) = \lambda(A) \otimes \lambda(B) , \quad v(A \otimes B) = v(A) \otimes v(B) \qquad (1504)$$

47. Given *analytic function* [84] $f : \mathbb{C}^{n \times n} \to \mathbb{C}^{n \times n}$, then $f(I \otimes A) = I \otimes f(A)$ and $f(A \otimes I) = f(A) \otimes I$ [172, p.28]

A.1.2   Majorization

A.1.2.0.1   Theorem.   (Schur) *Majorization.*   [421, §7.4] [208, §4.3]
[209, §5.5] Let $\lambda \in \mathbb{R}^N$ denote a given vector of eigenvalues and let $\delta \in \mathbb{R}^N$ denote a given vector of main diagonal entries, both arranged in nonincreasing order. Then

$$\exists A \in \mathbb{S}^N \ni \lambda(A) = \lambda \ \textbf{and} \ \delta(A) = \delta \ \Leftarrow \ \lambda - \delta \in \mathcal{K}_{\lambda\delta}^* \qquad (1505)$$

and conversely

$$A \in \mathbb{S}^N \ \Rightarrow \ \lambda(A) - \delta(A) \in \mathcal{K}_{\lambda\delta}^* \qquad (1506)$$

The difference belongs to the pointed polyhedral cone of majorization (not a full-dimensional cone, *confer* (312))

$$\mathcal{K}_{\lambda\delta}^* \triangleq \mathcal{K}_{\mathcal{M}+}^* \cap \{\zeta \mathbf{1} \mid \zeta \in \mathbb{R}\}^* \qquad (1507)$$

where $\mathcal{K}_{\mathcal{M}+}^*$ is the dual monotone nonnegative cone (434), and where the dual of the line is a hyperplane; $\partial \mathcal{H} = \{\zeta \mathbf{1} \mid \zeta \in \mathbb{R}\}^* = \mathbf{1}^\perp$.   ◇

Majorization cone $\mathcal{K}_{\lambda\delta}^*$ is naturally consequent to the definition of majorization; *id est*, vector $y \in \mathbb{R}^N$ *majorizes* vector x if and only if

$$\sum_{i=1}^k x_i \leq \sum_{i=1}^k y_i \quad \forall 1 \leq k \leq N \qquad (1508)$$

and

$$\mathbf{1}^\mathrm{T} x = \mathbf{1}^\mathrm{T} y \qquad (1509)$$

Under these circumstances, rather, vector x is majorized by vector y.

In the particular circumstance $\delta(A) = \mathbf{0}$ we get:

A.1.2.0.2   Corollary.   *Symmetric hollow majorization.*
Let $\lambda \in \mathbb{R}^N$ denote a given vector of eigenvalues arranged in nonincreasing order. Then

$$\exists A \in \mathbb{S}_h^N \ni \lambda(A) = \lambda \ \Leftarrow \ \lambda \in \mathcal{K}_{\lambda\delta}^* \qquad (1510)$$

and conversely

$$A \in \mathbb{S}_h^N \ \Rightarrow \ \lambda(A) \in \mathcal{K}_{\lambda\delta}^* \qquad (1511)$$

where $\mathcal{K}_{\lambda\delta}^*$ is defined in (1507).   ◇

A.2 Semidefiniteness: domain of test

The most fundamental necessary, sufficient, and definitive test for positive semidefiniteness of matrix $A \in \mathbb{R}^{n \times n}$ is: [209, §1]

$$x^{\mathrm{T}} A\, x \geq 0 \quad \text{for each and every } x \in \mathbb{R}^n \text{ such that } \|x\| = 1 \qquad (1512)$$

Traditionally, authors demand evaluation over broader domain; namely, over all $x \in \mathbb{R}^n$ which is sufficient but unnecessary. Indeed, that standard textbook requirement is far over-reaching because if $x^{\mathrm{T}} A\, x$ is nonnegative for particular $x = x_{\mathrm{p}}$, then it is nonnegative for any αx_{p} where $\alpha \in \mathbb{R}$. Thus, only normalized x in $\mathbb{R}^n$ need be evaluated.

Many authors add the further requirement that the domain be complex; the broadest domain. By so doing, only *Hermitian matrices* ($A^{\mathrm{H}} = A$ where superscript ^{H} denotes conjugate transpose)[A.2] are admitted to the set of positive semidefinite matrices (1515); an unnecessary prohibitive condition.

A.2.1 Symmetry *versus* semidefiniteness

We call (1512) *the most fundamental test* of positive semidefiniteness. Yet some authors instead say, for real A and complex domain $\{x \in \mathbb{C}^n\}$, the complex test $x^{\mathrm{H}} A x \geq 0$ is most fundamental. That complex broadening of the domain of test causes nonsymmetric real matrices to be excluded from the set of positive semidefinite matrices. Yet admitting nonsymmetric real matrices or not is a matter of preference[A.3] unless that complex test is adopted, as we shall now explain.

Any real square matrix A has a representation in terms of its symmetric and antisymmetric parts; *id est*,

$$A = \frac{(A + A^{\mathrm{T}})}{2} + \frac{(A - A^{\mathrm{T}})}{2} \qquad (53)$$

Because, for all real A, the antisymmetric part vanishes under real test,

$$x^{\mathrm{T}} \frac{(A - A^{\mathrm{T}})}{2}\, x = 0 \qquad (1513)$$

only the symmetric part of A, $(A + A^{\mathrm{T}})/2$, has a role determining positive semidefiniteness. Hence the oft-made presumption that only symmetric matrices may be positive semidefinite is, of course, erroneous under (1512). Because eigenvalue-signs of a symmetric matrix translate unequivocally to its semidefiniteness, the eigenvalues that determine semidefiniteness are always those of the *symmetrized* matrix. (§A.3) For that reason, and because symmetric (or Hermitian) matrices must have real eigenvalues, the convention adopted in the literature is that semidefinite matrices are synonymous with symmetric semidefinite matrices. Certainly misleading under (1512), that presumption is typically bolstered with compelling examples from the physical sciences where symmetric matrices occur within the mathematical exposition of natural phenomena.[A.4] [144, §52]

[A.2] Hermitian symmetry is the complex analogue; the real part of a Hermitian matrix is symmetric while its imaginary part is antisymmetric. A Hermitian matrix has real eigenvalues and real main diagonal.

[A.3] Golub & Van Loan [164, §4.2.2], for example, admit nonsymmetric real matrices.

[A.4] Symmetric matrices are certainly pervasive in the our chosen subject as well.

Perhaps a better explanation of this pervasive presumption of symmetry comes from
Horn & Johnson [208, §7.1] whose perspective[A.5] is the complex matrix, thus necessitating
the complex domain of test throughout their treatise. They explain, if $A \in \mathbb{C}^{n \times n}$

> *... and if $x^{\mathrm{H}}Ax$ is real for all $x \in \mathbb{C}^n$, then A is Hermitian. Thus, the
> assumption that A is Hermitian is not necessary in the definition of positive
> definiteness. It is customary, however.*

Their comment is best explained by noting, the real part of $x^{\mathrm{H}}Ax$ comes from the
Hermitian part $(A + A^{\mathrm{H}})/2$ of A;

$$\mathrm{re}(x^{\mathrm{H}}Ax) = x^{\mathrm{H}}\frac{A + A^{\mathrm{H}}}{2}x \tag{1514}$$

rather,

$$x^{\mathrm{H}}Ax \in \mathbb{R} \ \Leftrightarrow \ A^{\mathrm{H}} = A \tag{1515}$$

because the imaginary part of $x^{\mathrm{H}}Ax$ comes from the anti-Hermitian part $(A - A^{\mathrm{H}})/2$;

$$\mathrm{im}(x^{\mathrm{H}}Ax) = x^{\mathrm{H}}\frac{A - A^{\mathrm{H}}}{2}x \tag{1516}$$

that vanishes for nonzero x if and only if $A = A^{\mathrm{H}}$. So the Hermitian symmetry assumption
is unnecessary, according to Horn & Johnson, not because nonHermitian matrices could
be regarded positive semidefinite, rather because nonHermitian (includes nonsymmetric
real) matrices are not comparable on the real line under $x^{\mathrm{H}}Ax$. Yet that complex edifice
is dismantled in the test of real matrices (1512) because the domain of test is no longer
necessarily complex; meaning, $x^{\mathrm{T}}Ax$ will certainly always be real, regardless of symmetry,
and so real A will always be comparable.

In summary, if we limit the domain of test to all x in $\mathbb{R}^n$ as in (1512), then nonsymmetric
real matrices are admitted to the realm of semidefinite matrices because they become
comparable on the real line. One important exception occurs for rank-one matrices
$\Psi = uv^{\mathrm{T}}$ where u and v are real vectors: Ψ is positive semidefinite if and only if $\Psi = uu^{\mathrm{T}}$.
(§A.3.1.0.7)

We might choose to expand the domain of test to all x in $\mathbb{C}^n$ so that only symmetric
matrices would be comparable. The alternative to expanding domain of test is to assume
all matrices of interest to be symmetric; that is commonly done, hence the synonymous
relationship with semidefinite matrices.

A.2.1.0.1 Example. *Nonsymmetric positive definite product.*
Horn & Johnson assert and Zhang agrees:

> *If $A, B \in \mathbb{C}^{n \times n}$ are positive definite, then we know that the product AB
> is positive definite if and only if AB is Hermitian.*   [208, §7.6 prob.10]
> [421, §6.2, §3.2]

[A.5]A totally complex perspective is not necessarily more advantageous. The positive semidefinite cone,
for example, is not selfdual (§2.13.5) in the ambient space of Hermitian matrices. [201, §II]

Implicitly in their statement, A and B are assumed individually Hermitian and the domain of test is assumed complex.

We prove that assertion to be false for real matrices under (1512) that adopts a real domain of test.

$$A^{\mathrm{T}} = A = \begin{bmatrix} 3 & 0 & -1 & 0 \\ 0 & 5 & 1 & 0 \\ -1 & 1 & 4 & 1 \\ 0 & 0 & 1 & 4 \end{bmatrix}, \qquad \lambda(A) = \begin{bmatrix} 5.9 \\ 4.5 \\ 3.4 \\ 2.0 \end{bmatrix} \tag{1517}$$

$$B^{\mathrm{T}} = B = \begin{bmatrix} 4 & 4 & -1 & -1 \\ 4 & 5 & 0 & 0 \\ -1 & 0 & 5 & 1 \\ -1 & 0 & 1 & 4 \end{bmatrix}, \qquad \lambda(B) = \begin{bmatrix} 8.8 \\ 5.5 \\ 3.3 \\ 0.24 \end{bmatrix} \tag{1518}$$

$$(AB)^{\mathrm{T}} \neq AB = \begin{bmatrix} 13 & 12 & -8 & -4 \\ 19 & 25 & 5 & 1 \\ -5 & 1 & 22 & 9 \\ -5 & 0 & 9 & 17 \end{bmatrix}, \qquad \lambda(AB) = \begin{bmatrix} 36. \\ 29. \\ 10. \\ 0.72 \end{bmatrix} \tag{1519}$$

$$\tfrac{1}{2}(AB + (AB)^{\mathrm{T}}) = \begin{bmatrix} 13 & 15.5 & -6.5 & -4.5 \\ 15.5 & 25 & 3 & 0.5 \\ -6.5 & 3 & 22 & 9 \\ -4.5 & 0.5 & 9 & 17 \end{bmatrix}, \quad \lambda\left(\tfrac{1}{2}(AB + (AB)^{\mathrm{T}})\right) = \begin{bmatrix} 36. \\ 30. \\ 10. \\ 0.014 \end{bmatrix} \tag{1520}$$

Whenever $A \in \mathbb{S}_+^n$ and $B \in \mathbb{S}_+^n$, then $\lambda(AB) = \lambda(\sqrt{A}\,B\,\sqrt{A})$ will always be a nonnegative vector by (1544) and Corollary A.3.1.0.5. Yet positive definiteness of product AB is certified instead by the nonnegative eigenvalues $\lambda\left(\tfrac{1}{2}(AB + (AB)^{\mathrm{T}})\right)$ in (1520) (§A.3.1.0.1) despite the fact that AB is not symmetric.[A.6] Horn & Johnson and Zhang resolve the anomaly by choosing to exclude nonsymmetric matrices and products; they do so by expanding the domain of test to $\mathbb{C}^n$. □

A.3 Proper statements of positive semidefiniteness

Unlike Horn & Johnson and Zhang, we never adopt a complex domain of test with real matrices. So motivated is our consideration of proper statements of positive semidefiniteness under real domain of test. This restriction, ironically, complicates the facts when compared to corresponding statements for the complex case (found elsewhere [208] [421]).

We state several fundamental facts regarding positive semidefiniteness of real matrix A and the product AB and sum $A + B$ of real matrices under fundamental real test (1512); a few require proof as they depart from the standard texts, while those remaining are well established or obvious.

[A.6]It is a little more difficult to find a counterexample in $\mathbb{R}^{2 \times 2}$ or $\mathbb{R}^{3 \times 3}$; which may have served to advance any confusion.

A.3.0.0.1 Theorem. *Positive (semi)definite matrix.*

$A \in \mathbb{S}^M$ is positive semidefinite if and only if for each and every vector $x \in \mathbb{R}^M$ of unit norm, $\|x\| = 1$,[A.7] we have $x^{\mathrm{T}} A x \geq 0$ (1512);

$$A \succeq 0 \quad \Leftrightarrow \quad \mathrm{tr}(xx^{\mathrm{T}}A) = x^{\mathrm{T}} A x \geq 0 \ \ \forall xx^{\mathrm{T}} \tag{1521}$$

Matrix $A \in \mathbb{S}^M$ is positive definite if and only if for each and every $\|x\| = 1$ we have $x^{\mathrm{T}} A x > 0$;

$$A \succ 0 \quad \Leftrightarrow \quad \mathrm{tr}(xx^{\mathrm{T}}A) = x^{\mathrm{T}} A x > 0 \ \ \forall xx^{\mathrm{T}}, \ xx^{\mathrm{T}} \neq \mathbf{0} \tag{1522}$$

$\diamond$

Proof. Statements (1521) and (1522) are each a particular instance of dual generalized inequalities (§2.13.2) with respect to the positive semidefinite cone; *videlicet*, [369]

$$\begin{aligned} A \succeq 0 \quad &\Leftrightarrow \quad \langle xx^{\mathrm{T}}, A \rangle \geq 0 \ \ \forall xx^{\mathrm{T}}(\succeq 0) \\ A \succ 0 \quad &\Leftrightarrow \quad \langle xx^{\mathrm{T}}, A \rangle > 0 \ \ \forall xx^{\mathrm{T}}(\succeq 0), \ xx^{\mathrm{T}} \neq \mathbf{0} \end{aligned} \tag{1523}$$

This says: positive semidefinite matrix A must belong to the normal side of every hyperplane whose normal is an extreme direction of the positive semidefinite cone. Relations (1521) and (1522) remain true when xx^{T} is replaced with "for each and every" positive semidefinite matrix $X \in \mathbb{S}_+^M$ (§2.13.5) of unit norm, $\|X\| = 1$, as in

$$\begin{aligned} A \succeq 0 \quad &\Leftrightarrow \quad \mathrm{tr}(XA) \geq 0 \ \ \forall X \in \mathbb{S}_+^M \\ A \succ 0 \quad &\Leftrightarrow \quad \mathrm{tr}(XA) > 0 \ \ \forall X \in \mathbb{S}_+^M, \ X \neq \mathbf{0} \end{aligned} \tag{1524}$$

But that condition is more than what is necessary. By the *discretized membership theorem* in §2.13.4.2.1, the extreme directions xx^{T} of the positive semidefinite cone constitute a minimal set of generators necessary and sufficient for discretization of dual generalized inequalities (1524) certifying membership to that cone.   ◆

A.3.1   Semidefiniteness, eigenvalues, nonsymmetric

When $A \in \mathbb{R}^{n \times n}$, let $\lambda\left(\frac{1}{2}(A + A^{\mathrm{T}})\right) \in \mathbb{R}^n$ denote eigenvalues of the symmetrized matrix[A.8] arranged in nonincreasing order.

- By positive semidefiniteness of $A \in \mathbb{R}^{n \times n}$ we mean,[A.9] [278, §1.3.1] (*confer* §A.3.1.0.1)

$$x^{\mathrm{T}} A x \geq 0 \ \ \forall x \in \mathbb{R}^n \quad \Leftrightarrow \quad A + A^{\mathrm{T}} \succeq 0 \quad \Leftrightarrow \quad \lambda(A + A^{\mathrm{T}}) \succeq 0 \tag{1525}$$

[A.7]The traditional condition requiring all $x \in \mathbb{R}^M$ for defining positive (semi)definiteness is actually more than what is necessary. The set of norm-1 vectors is necessary and sufficient to establish positive semidefiniteness; actually, any particular norm and any nonzero norm-constant will work.

[A.8]The symmetrization of A is $(A + A^{\mathrm{T}})/2$. $\lambda\left(\frac{1}{2}(A + A^{\mathrm{T}})\right) = \lambda(A + A^{\mathrm{T}})/2$.

[A.9]Strang agrees [338, p.334] it is not $\lambda(A)$ that requires observation. Yet he is mistaken by proposing the Hermitian part alone $x^{\mathrm{H}}(A + A^{\mathrm{H}})x$ be tested, because the anti-Hermitian part does not vanish under complex test unless A is Hermitian. (1516)

- (§2.9.0.1)

$$A \succeq 0 \ \Rightarrow \ A^{\mathrm{T}} = A \tag{1526}$$

$$A \succeq B \ \Leftrightarrow \ A - B \succeq 0 \ \not\Rightarrow \ A \succeq 0 \ \text{or} \ B \succeq 0 \tag{1527}$$

$$x^{\mathrm{T}} A x \geq 0 \ \forall x \ \not\Rightarrow \ A^{\mathrm{T}} = A \tag{1528}$$

- Matrix symmetry is not intrinsic to positive semidefiniteness;

$$A^{\mathrm{T}} = A, \ \ \lambda(A) \succeq 0 \ \Rightarrow \ x^{\mathrm{T}} A x \geq 0 \ \forall x \tag{1529}$$

$$\lambda(A) \succeq 0 \ \Leftarrow \ A^{\mathrm{T}} = A, \ \ x^{\mathrm{T}} A x \geq 0 \ \forall x \tag{1530}$$

- If $A^{\mathrm{T}} = A$ then

$$\lambda(A) \succeq 0 \ \Leftrightarrow \ A \succeq 0 \tag{1531}$$

meaning, matrix A belongs to the positive semidefinite cone in the subspace of symmetric matrices if and only if its eigenvalues belong to the nonnegative orthant.

$$\langle A \, , \, A \rangle = \langle \lambda(A) \, , \, \lambda(A) \rangle \tag{45}$$

- For $\mu \in \mathbb{R}$, $A \in \mathbb{R}^{n \times n}$, and vector $\lambda(A) \in \mathbb{C}^n$ holding the ordered eigenvalues of A

$$\lambda(\mu I + A) = \mu \mathbf{1} + \lambda(A) \tag{1532}$$

Proof: $A = MJM^{-1}$ and $\mu I + A = M(\mu I + J)M^{-1}$ where J is the *Jordan form* for A; [338, §5.6, App.B] *id est*, $\delta(J) = \lambda(A)$, so $\lambda(\mu I + A) = \delta(\mu I + J)$ because $\mu I + J$ is also a Jordan form. $\quad\blacklozenge$

By similar reasoning,

$$\lambda(I + \mu A) = \mathbf{1} + \lambda(\mu A) \tag{1533}$$

For vector $\sigma(A)$ holding the singular values of any matrix A

$$\sigma(I + \mu A^{\mathrm{T}} A) = \pi(|\mathbf{1} + \mu \sigma(A^{\mathrm{T}} A)|) \tag{1534}$$

$$\sigma(\mu I + A^{\mathrm{T}} A) = \pi(|\mu \mathbf{1} + \sigma(A^{\mathrm{T}} A)|) \tag{1535}$$

where π is the nonlinear permutation-operator sorting its vector argument into nonincreasing order.

- For $A \in \mathbb{S}^M$ and each and every $\|w\| = 1$ [208, §7.7 prob.9]

$$w^{\mathrm{T}} A w \leq \mu \ \Leftrightarrow \ A \preceq \mu I \ \Leftrightarrow \ \lambda(A) \preceq \mu \mathbf{1} \tag{1536}$$

- [208, §2.5.4] (*confer* (44))

$$A \ \text{is normal matrix} \ \Leftrightarrow \ \|A\|_{\mathrm{F}}^2 = \lambda(A)^{\mathrm{T}} \lambda(A) \tag{1537}$$

- For $A \in \mathbb{R}^{m \times n}$

$$A^{\mathrm{T}}A \succeq 0, \qquad AA^{\mathrm{T}} \succeq 0 \tag{1538}$$

because, for dimensionally compatible vector x, $\quad x^{\mathrm{T}}A^{\mathrm{T}}Ax = \|Ax\|_2^2$, $x^{\mathrm{T}}AA^{\mathrm{T}}x = \|A^{\mathrm{T}}x\|_2^2$.

- For $A \in \mathbb{R}^{n \times n}$ and $c \in \mathbb{R}$

$$\mathrm{tr}(cA) \;=\; c\,\mathrm{tr}(A) \;=\; c\mathbf{1}^{\mathrm{T}}\lambda(A) \qquad (\S\mathrm{A.1.1}\ no.4)$$

For m a nonnegative integer, (1960)

$$\det(A^m) = \prod_{i=1}^{n} \lambda(A)_i^m \tag{1539}$$

$$\mathrm{tr}(A^m) = \sum_{i=1}^{n} \lambda(A)_i^m \tag{1540}$$

- For A diagonalizable ($\S$A.5), $A = S\Lambda S^{-1}$, (*confer* [338, p.255])

$$\mathrm{rank}\,A = \mathrm{rank}\,\delta(\lambda(A)) = \mathrm{rank}\,\Lambda \tag{1541}$$

meaning, rank is equal to the number of nonzero eigenvalues in vector

$$\lambda(A) \triangleq \delta(\Lambda) \tag{1542}$$

by the 0 *eigenvalues theorem* ($\S$A.7.3.0.1).

- (Ky Fan) For $A, B \in \mathbb{S}^n$ [56, $\S$1.2] (*confer*(1832))

$$\mathrm{tr}(AB) \leq \lambda(A)^{\mathrm{T}}\lambda(B) \qquad (1818)$$

with equality (Theobald) when A and B are simultaneously diagonalizable [208] with the same ordering of eigenvalues.

- For $A \in \mathbb{R}^{m \times n}$ and $B \in \mathbb{R}^{n \times m}$

$$\mathrm{tr}(AB) = \mathrm{tr}(BA) \tag{1543}$$

and η eigenvalues of the product and commuted product are identical, including their multiplicity; [208, $\S$1.3.20] *id est*,

$$\lambda(AB)_{1:\eta} = \lambda(BA)_{1:\eta}, \qquad \eta \triangleq \min\{m, n\} \tag{1544}$$

Any eigenvalues remaining are zero. By the 0 *eigenvalues theorem* ($\S$A.7.3.0.1),

$$\mathrm{rank}(AB) = \mathrm{rank}(BA), \qquad AB \text{ and } BA \text{ diagonalizable} \tag{1545}$$

- For any compatible matrices A, B [208, $\S$0.4]

$$\min\{\mathrm{rank}\,A, \mathrm{rank}\,B\} \geq \mathrm{rank}(AB) \tag{1546}$$

- For $A,B \in \mathbb{S}_+^n$ (*confer*(257))

$$\operatorname{rank} A + \operatorname{rank} B \geq \operatorname{rank}(A+B) \geq \min\{\operatorname{rank} A,\operatorname{rank} B\} \geq \operatorname{rank}(AB) \quad (1547)$$

- For linearly independent matrices $A,B \in \mathbb{S}_+^n$ (§2.1.2, $\mathcal{R}(A) \cap \mathcal{R}(B) = \mathbf{0}$, $\mathcal{R}(A^{\mathrm{T}}) \cap \mathcal{R}(B^{\mathrm{T}}) = \mathbf{0}$, §B.1.1),

$$\operatorname{rank} A + \operatorname{rank} B = \operatorname{rank}(A+B) > \min\{\operatorname{rank} A,\operatorname{rank} B\} \geq \operatorname{rank}(AB) \quad (1548)$$

- Because $\mathcal{R}(A^{\mathrm{T}}A) = \mathcal{R}(A^{\mathrm{T}})$ and $\mathcal{R}(AA^{\mathrm{T}}) = \mathcal{R}(A)$ (p.548), for any $A \in \mathbb{R}^{m \times n}$

$$\operatorname{rank}(AA^{\mathrm{T}}) = \operatorname{rank}(A^{\mathrm{T}}A) = \operatorname{rank} A = \operatorname{rank} A^{\mathrm{T}} \quad (1549)$$

- For $A \in \mathbb{R}^{m \times n}$ having no nullspace, and for any $B \in \mathbb{R}^{n \times k}$

$$\operatorname{rank}(AB) = \operatorname{rank}(B) \quad (1550)$$

Proof. For any compatible matrix C, $\mathcal{N}(CAB) \supseteq \mathcal{N}(AB) \supseteq \mathcal{N}(B)$ is obvious. By assumption $\exists A^\dagger \ni A^\dagger A = I$. Let $C = A^\dagger$, then $\mathcal{N}(AB) = \mathcal{N}(B)$ and the stated result follows by conservation of dimension (1666). $\blacklozenge$

- For $A \in \mathbb{S}^n$ and any nonsingular matrix Y

$$\operatorname{inertia}(A) = \operatorname{inertia}(YAY^{\mathrm{T}}) \quad (1551)$$

a.k.a *Sylvester's law of inertia.* (1597) [116, §2.4.3]

- For $A,B \in \mathbb{R}^{n \times n}$ square, [208, §0.3.5]

$$\det(AB) = \det(BA) \quad (1552)$$
$$\det(AB) = \det A \det B \quad (1553)$$

Yet for $A \in \mathbb{R}^{m \times n}$ and $B \in \mathbb{R}^{n \times m}$ [78, p.72]

$$\det(I + AB) = \det(I + BA) \quad (1554)$$

- For $A,B \in \mathbb{S}^n$, product AB is symmetric iff AB is commutative;

$$(AB)^{\mathrm{T}} = AB \;\Leftrightarrow\; AB = BA \quad (1555)$$

Proof. ($\Rightarrow$) Suppose $AB = (AB)^{\mathrm{T}}$.
$(AB)^{\mathrm{T}} = B^{\mathrm{T}}A^{\mathrm{T}} = BA$. $AB = (AB)^{\mathrm{T}} \Rightarrow AB = BA$.
($\Leftarrow$) Suppose $AB = BA$.
$BA = B^{\mathrm{T}}A^{\mathrm{T}} = (AB)^{\mathrm{T}}$. $AB = BA \Rightarrow AB = (AB)^{\mathrm{T}}$. $\blacklozenge$

Commutativity alone is insufficient for product symmetry. [338, p.26] Matrix symmetry alone is insufficient for product symmetry.

- Diagonalizable matrices $A, B \in \mathbb{R}^{n \times n}$ commute if and only if they are simultaneously diagonalizable. [208, §1.3.12] A product of diagonal matrices is always commutative.

- For $A, B \in \mathbb{R}^{n \times n}$ and $AB = BA$

$$x^T A x \geq 0,\ x^T B x \geq 0\ \forall x \ \Rightarrow\ \lambda(A+A^T)_i\, \lambda(B+B^T)_i \geq 0\ \forall i\ \not\Leftrightarrow\ x^T A B x \geq 0\ \forall x \quad (1556)$$

the negative result arising because of the schism between the product of eigenvalues $\lambda(A + A^T)_i\, \lambda(B + B^T)_i$ and the eigenvalues of the symmetrized matrix product $\lambda(AB + (AB)^T)_i$. For example, X^2 is generally not positive semidefinite unless matrix X is symmetric; then (1538) applies. Simply substituting symmetric matrices changes the outcome:

- For $A, B \in \mathbb{S}^n$ and $AB = BA$

$$A \succeq 0,\ B \succeq 0 \ \Rightarrow\ \lambda(AB)_i = \lambda(A)_i\, \lambda(B)_i \geq 0\ \forall i\ \Leftrightarrow\ AB \succeq 0 \quad (1557)$$

Positive semidefiniteness of commutative A and B is sufficient but not necessary for positive semidefiniteness of product AB.

Proof. Because all symmetric matrices are diagonalizable, (§A.5.1) [338, §5.6] we have $A = S \Lambda S^T$ and $B = T \Delta T^T$, where Λ and Δ are real diagonal matrices while S and T are orthogonal matrices. Because $(AB)^T = AB$, then T must equal S, [208, §1.3] and the eigenvalues of A are ordered in the same way as those of B; *id est,* $\lambda(A)_i = \delta(\Lambda)_i$ and $\lambda(B)_i = \delta(\Delta)_i$ correspond to the same eigenvector.
($\Rightarrow$) Assume $\lambda(A)_i\, \lambda(B)_i \geq 0$ for $i = 1 \ldots n$. $AB = S \Lambda \Delta S^T$ is symmetric and has nonnegative eigenvalues contained in diagonal matrix $\Lambda \Delta$ by assumption; hence positive semidefinite by (1525). Now assume $A, B \succeq 0$. That, of course, implies $\lambda(A)_i\, \lambda(B)_i \geq 0$ for all i because all the individual eigenvalues are nonnegative.
($\Leftarrow$) Suppose $AB = S \Lambda \Delta S^T \succeq 0$. Then $\Lambda \Delta \succeq 0$ by (1525), and so all products $\lambda(A)_i\, \lambda(B)_i$ must be nonnegative; meaning, $\mathrm{sgn}(\lambda(A)) = \mathrm{sgn}(\lambda(B))$. We may, therefore, conclude nothing about the semidefiniteness of A and B. $\blacklozenge$

- For $A, B \in \mathbb{S}^n$ and $A \succeq 0,\ B \succeq 0$ (Example A.2.1.0.1)

$$AB = BA \ \Rightarrow\ \lambda(AB)_i = \lambda(A)_i\, \lambda(B)_i \geq 0\ \forall i\ \Rightarrow\ AB \succeq 0 \quad (1558)$$

$$AB = BA \ \Rightarrow\ \lambda(AB)_i \geq 0,\ \lambda(A)_i\, \lambda(B)_i \geq 0\ \forall i\ \Leftrightarrow\ AB \succeq 0 \quad (1559)$$

- For $A, B \in \mathbb{S}^n$ [208, §7.7 prob.3] [209, §4.2.13, §5.2.1]

$$A \succeq 0,\ B \succeq 0 \ \Rightarrow\ A \otimes B \succeq 0 \quad (1560)$$
$$A \succeq 0,\ B \succeq 0 \ \Rightarrow\ A \circ B \succeq 0 \quad (1561)$$
$$A \succ 0,\ B \succ 0 \ \Rightarrow\ A \otimes B \succ 0 \quad (1562)$$
$$A \succ 0,\ B \succ 0 \ \Rightarrow\ A \circ B \succ 0 \quad (1563)$$

where Kronecker and Hadamard products are symmetric.

- For $A, B \in \mathbb{S}^n$, (1531) $A \succeq 0 \Leftrightarrow \lambda(A) \succeq 0$ yet

$$A \succeq 0 \Rightarrow \delta(A) \succeq 0 \tag{1564}$$

$$A \succeq 0 \Rightarrow \operatorname{tr} A \geq 0 \tag{1565}$$

$$A \succeq 0, \ B \succeq 0 \Rightarrow \operatorname{tr} A \operatorname{tr} B \geq \operatorname{tr}(AB) \geq 0 \tag{1566}$$

[421, §6.2] Because $A \succeq 0$, $B \succeq 0 \Rightarrow \lambda(AB) = \lambda(\sqrt{A} B \sqrt{A}) \succeq 0$ by (1544) and Corollary A.3.1.0.5, then we have $\operatorname{tr}(AB) \geq 0$.

$$A \succeq 0 \Leftrightarrow \operatorname{tr}(AB) \geq 0 \quad \forall B \succeq 0 \tag{377}$$

- For $A, B, C \in \mathbb{S}^n$ (Löwner)

$$
\begin{array}{ll}
A \preceq B, \ B \preceq C \Rightarrow A \preceq C & \text{(transitivity)} \\
A \preceq B \Leftrightarrow A + C \preceq B + C & \text{(additivity)} \\
A \preceq B, \ A \succeq B \Rightarrow A = B & \text{(antisymmetry)} \\
A \preceq A & \text{(reflexivity)}
\end{array}
\tag{1567}
$$

$$
\begin{array}{ll}
A \preceq B, \ B \prec C \Rightarrow A \prec C & \text{(strict transitivity)} \\
A \prec B \Leftrightarrow A + C \prec B + C & \text{(strict additivity)}
\end{array}
\tag{1568}
$$

- For $A, B \in \mathbb{R}^{n \times n}$

$$x^{\mathrm{T}} A x \geq x^{\mathrm{T}} B x \ \forall x \Rightarrow \operatorname{tr} A \geq \operatorname{tr} B \tag{1569}$$

Proof. $x^{\mathrm{T}} A x \geq x^{\mathrm{T}} B x \ \forall x \Leftrightarrow \lambda((A-B) + (A-B)^{\mathrm{T}})/2 \succeq 0 \Rightarrow$
$\operatorname{tr}(A + A^{\mathrm{T}} - (B + B^{\mathrm{T}}))/2 = \operatorname{tr}(A - B) \geq 0$. There is no converse. $\blacklozenge$

- For $A, B \in \mathbb{S}^n$ [421, §6.2] (Theorem A.3.1.0.4)

$$A \succeq B \Rightarrow \operatorname{tr} A \geq \operatorname{tr} B \tag{1570}$$

$$A \succeq B \Rightarrow \delta(A) \succeq \delta(B) \tag{1571}$$

There is no converse, and restriction to the positive semidefinite cone does not improve the situation. All-strict versions hold.

$$A \succeq B \succeq 0 \Rightarrow \operatorname{rank} A \geq \operatorname{rank} B \tag{1572}$$

$$A \succeq B \succeq 0 \Rightarrow \det A \geq \det B \geq 0 \tag{1573}$$

$$A \succ B \succeq 0 \Rightarrow \det A > \det B \geq 0 \tag{1574}$$

- For $A, B \in \operatorname{int} \mathbb{S}_+^n$ [34, §4.2] [208, §7.7.4]

$$A \succeq B \Leftrightarrow A^{-1} \preceq B^{-1}, \qquad A \succ 0 \Leftrightarrow A^{-1} \succ 0 \tag{1575}$$

- For $A, B \in \mathbb{S}^n$ [421, §6.2]

$$
\begin{array}{c}
A \succeq B \succeq 0 \Rightarrow \sqrt{A} \succeq \sqrt{B} \\
A \succeq 0 \Leftarrow A^{1/2} \succeq 0
\end{array}
\tag{1576}
$$

- For $A, B \in \mathbb{S}^n$ and $AB = BA$ [421, §6.2 prob.3]

$$A \succeq B \succeq 0 \Rightarrow A^k \succeq B^k, \quad k = 1, 2, \ldots \tag{1577}$$

A.3.1.0.1 Theorem. *Positive semidefinite ordering of eigenvalues.*
For $A,B\in\mathbb{R}^{M\times M}$, place the eigenvalues of each symmetrized matrix into the respective vectors $\lambda\big(\tfrac{1}{2}(A+A^{\mathrm{T}})\big),\lambda\big(\tfrac{1}{2}(B+B^{\mathrm{T}})\big)\in\mathbb{R}^M$. Then [338, §6]

$$x^{\mathrm{T}}Ax\geq 0\ \ \forall x\qquad\Leftrightarrow\ \lambda(A+A^{\mathrm{T}})\succeq 0\qquad(1578)$$

$$x^{\mathrm{T}}Ax> 0\ \ \forall x\neq\mathbf{0}\Leftrightarrow\ \lambda(A+A^{\mathrm{T}})\succ 0\qquad(1579)$$

because $x^{\mathrm{T}}(A-A^{\mathrm{T}})x=0$.   (1513) Now arrange the entries of $\lambda\big(\tfrac{1}{2}(A+A^{\mathrm{T}})\big)$ and $\lambda\big(\tfrac{1}{2}(B+B^{\mathrm{T}})\big)$ in nonincreasing order so $\lambda\big(\tfrac{1}{2}(A+A^{\mathrm{T}})\big)_1$ holds the largest eigenvalue of symmetrized A while $\lambda\big(\tfrac{1}{2}(B+B^{\mathrm{T}})\big)_1$ holds the largest eigenvalue of symmetrized B, and so on. Then [208, §7.7 prob.1 prob.9] for $\kappa\in\mathbb{R}$

$$\begin{aligned}x^{\mathrm{T}}Ax\geq x^{\mathrm{T}}Bx\ \forall x&\Rightarrow\ \lambda(A+A^{\mathrm{T}})\succeq\lambda(B+B^{\mathrm{T}})\\ x^{\mathrm{T}}Ax\geq x^{\mathrm{T}}Ix\kappa\ \forall x&\Leftrightarrow\ \lambda\big(\tfrac{1}{2}(A+A^{\mathrm{T}})\big)\succeq\kappa\mathbf{1}\end{aligned}\qquad(1580)$$

Now let $A,B\in\mathbb{S}^M$ have diagonalizations $A=Q\Lambda Q^{\mathrm{T}}$ and $B=U\Upsilon U^{\mathrm{T}}$ with $\lambda(A)=\delta(\Lambda)$ and $\lambda(B)=\delta(\Upsilon)$ arranged in nonincreasing order. Then

$$A\succeq B\Leftrightarrow\lambda(A-B)\succeq 0\qquad(1581)$$

$$A\succeq B\Rightarrow\lambda(A)\succeq\lambda(B)\qquad(1582)$$

$$A\succeq B\not\Leftarrow\lambda(A)\succeq\lambda(B)\qquad(1583)$$

$$S^{\mathrm{T}}AS\succeq B\Leftarrow\lambda(A)\succeq\lambda(B)\qquad(1584)$$

where $S=QU^{\mathrm{T}}$. [421, §7.5]   ⋄

A.3.1.0.2 Theorem. (Weyl) *Eigenvalues of sum.*   [208, §4.3.1]
For $A,B\in\mathbb{R}^{M\times M}$, place the eigenvalues of each symmetrized matrix into the respective vectors $\lambda\big(\tfrac{1}{2}(A+A^{\mathrm{T}})\big),\lambda\big(\tfrac{1}{2}(B+B^{\mathrm{T}})\big)\in\mathbb{R}^M$ in nonincreasing order so $\lambda\big(\tfrac{1}{2}(A+A^{\mathrm{T}})\big)_1$ holds the largest eigenvalue of symmetrized A while $\lambda\big(\tfrac{1}{2}(B+B^{\mathrm{T}})\big)_1$ holds the largest eigenvalue of symmetrized B, and so on. Then, for any $k\in\{1\dots M\}$

$$\lambda(A+A^{\mathrm{T}})_k+\lambda(B+B^{\mathrm{T}})_M\leq\lambda\big((A+A^{\mathrm{T}})+(B+B^{\mathrm{T}})\big)_k\leq\lambda(A+A^{\mathrm{T}})_k+\lambda(B+B^{\mathrm{T}})_1\quad(1585)$$

⋄

Weyl's theorem establishes: concavity of the smallest λ_M and convexity of the largest eigenvalue λ_1 of a symmetric matrix, via (496), and positive semidefiniteness of a sum of positive semidefinite matrices; for $A,B\in\mathbb{S}_+^M$

$$\lambda(A)_k+\lambda(B)_M\leq\lambda(A+B)_k\leq\lambda(A)_k+\lambda(B)_1\qquad(1586)$$

Because $\mathbb{S}_+^M$ is a convex cone (§2.9.0.0.1), then by (175)

$$A,B\succeq 0\Rightarrow\zeta A+\xi B\succeq 0\ \text{ for all }\zeta,\xi\geq 0\qquad(1587)$$

A.3.1.0.3 Corollary. *Eigenvalues of sum and difference.* [208, §4.3]
For $A \in \mathbb{S}^M$ and $B \in \mathbb{S}_+^M$, place the eigenvalues of each matrix into respective vectors $\lambda(A), \lambda(B) \in \mathbb{R}^M$ in nonincreasing order so $\lambda(A)_1$ holds the largest eigenvalue of A while $\lambda(B)_1$ holds the largest eigenvalue of B, and so on. Then, for any $k \in \{1 \ldots M\}$

$$\lambda(A - B)_k \; \leq \; \lambda(A)_k \; \leq \; \lambda(A + B)_k \qquad (1588)$$

$\diamond$

When B is rank-one positive semidefinite, the eigenvalues interlace; *id est*, for $B = qq^{\mathrm{T}}$

$$\lambda(A)_{k-1} \; \leq \; \lambda(A - qq^{\mathrm{T}})_k \; \leq \; \lambda(A)_k \; \leq \; \lambda(A + qq^{\mathrm{T}})_k \; \leq \; \lambda(A)_{k+1} \qquad (1589)$$

A.3.1.0.4 Theorem. *Positive (semi)definite principal submatrices.*[A.10]

- $A \in \mathbb{S}^M$ is positive semidefinite if and only if all M principal submatrices of dimension $M-1$ are positive semidefinite and $\det A$ is nonnegative.

- $A \in \mathbb{S}^M$ is positive definite if and only if any one principal submatrix of dimension $M-1$ is positive definite and $\det A$ is positive. $\diamond$

If any one principal submatrix of dimension $M-1$ is not positive definite, conversely, then A can neither be. Regardless of symmetry, if $A \in \mathbb{R}^{M \times M}$ is positive (semi)definite, then the determinant of each and every principal submatrix is (nonnegative) positive. [278, §1.3.1]

A.3.1.0.5 Corollary. *Positive (semi)definite symmetric products.* [208, p.399]

- If $A \in \mathbb{S}^M$ is positive definite and any particular dimensionally compatible matrix Z has no nullspace, then $Z^{\mathrm{T}}AZ$ is positive definite.

- If matrix $A \in \mathbb{S}^M$ is positive (semi)definite then, for any matrix Z of compatible dimension, $Z^{\mathrm{T}}AZ$ is positive semidefinite.

- $A \in \mathbb{S}^M$ is positive (semi)definite if and only if there exists a nonsingular Z such that $Z^{\mathrm{T}}AZ$ is positive (semi)definite.

- If $A \in \mathbb{S}^M$ is positive semidefinite and singular it remains possible, for some skinny $Z \in \mathbb{R}^{M \times N}$ with $N < M$, that $Z^{\mathrm{T}}AZ$ becomes positive definite.[A.11] $\diamond$

We can deduce from these, given nonsingular matrix Z and any particular dimensionally compatible Y: matrix $A \in \mathbb{S}^M$ is positive semidefinite if and only if $\begin{bmatrix} Z^{\mathrm{T}} \\ Y^{\mathrm{T}} \end{bmatrix} A \begin{bmatrix} Z & Y \end{bmatrix}$ is positive semidefinite. In other words, from the Corollary it follows: for dimensionally compatible Z

[A.10] A recursive condition for positive (semi)definiteness, this theorem is a synthesis of facts from [208, §7.2] [338, §6.3] (*confer* [278, §1.3.1]).
[A.11] Using the interpretation in §E.6.4.3, this means coefficients of orthogonal projection of vectorized A on a subset of extreme directions from $\mathbb{S}_+^M$ determined by Z can be positive.

- $A \succeq 0 \;\Leftrightarrow\; Z^{\mathrm{T}}AZ \succeq 0$ **and** Z^{T} has a left inverse

Products such as $Z^{\dagger}Z$ and $ZZ^{\dagger}$ are symmetric and positive semidefinite although, given $A \succeq 0$, $Z^{\dagger}AZ$ and $ZAZ^{\dagger}$ are neither necessarily symmetric or positive semidefinite.

A.3.1.0.6 Theorem. *Symmetric projector semidefinite.*   [20, §III] [21, §6] [228, p.55] For symmetric idempotent matrices P and R

$$P, R \succeq 0$$

$$P \succeq R \;\Leftrightarrow\; \mathcal{R}(P) \supseteq \mathcal{R}(R) \;\Leftrightarrow\; \mathcal{N}(P) \subseteq \mathcal{N}(R) \tag{1590}$$

Projector P is never positive definite [340, §6.5 prob.20] unless it is the Identity matrix.
                                                                                     $\diamond$

A.3.1.0.7 Theorem. *Symmetric positive semidefinite.*   [208, p.400]
Given real matrix Ψ with rank $\Psi = 1$

$$\Psi \succeq 0 \;\Leftrightarrow\; \Psi = uu^{\mathrm{T}} \tag{1591}$$

where u is some real vector; *id est*, symmetry is necessary and sufficient for positive semidefiniteness of a rank-1 matrix. $\diamond$

    Proof. Any rank-one matrix must have the form $\Psi = uv^{\mathrm{T}}$. (§B.1) Suppose Ψ is symmetric; *id est*, $v = u$. For all $y \in \mathbb{R}^{M}$, $y^{\mathrm{T}}uu^{\mathrm{T}}y \geq 0$. Conversely, suppose uv^{T} is positive semidefinite. We know that can hold if and only if $uv^{\mathrm{T}} + vu^{\mathrm{T}} \succeq 0 \;\Leftrightarrow\;$ for all normalized $y \in \mathbb{R}^{M}$, $2\,y^{\mathrm{T}}uv^{\mathrm{T}}y \geq 0$; but that is possible only if $v = u$. $\blacklozenge$

The same does not hold true for matrices of higher rank, as Example A.2.1.0.1 shows.

A.4 Schur complement

Consider *Schur-form* partitioned matrix G: Given $A^{\mathrm{T}} = A$ and $C^{\mathrm{T}} = C$, then [60]

$$G = \begin{bmatrix} A & B \\ B^{\mathrm{T}} & C \end{bmatrix} \succeq 0$$

$$\begin{aligned} &\Leftrightarrow\; A \succeq 0, \;\; B^{\mathrm{T}}(I - AA^{\dagger}) = \mathbf{0}, \;\; C - B^{\mathrm{T}}A^{\dagger}B \succeq 0 \\ &\Leftrightarrow\; C \succeq 0, \;\; B(I - CC^{\dagger}) = \mathbf{0}, \;\; A - BC^{\dagger}B^{\mathrm{T}} \succeq 0 \end{aligned} \tag{1592}$$

where $A^{\dagger}$ denotes the Moore-Penrose (pseudo)inverse (§E). In the first instance, $I - AA^{\dagger}$ is a symmetric projection matrix orthogonally projecting on $\mathcal{N}(A^{\mathrm{T}})$. (2003) It is apparently required

$$\mathcal{R}(B) \perp \mathcal{N}(A^{\mathrm{T}}) \tag{1593}$$

which precludes $A = \mathbf{0}$ when B is any nonzero matrix. Note that $A \succ 0 \Rightarrow A^{\dagger} = A^{-1}$; thereby, the projection matrix vanishes. Likewise, in the second instance, $I - CC^{\dagger}$ projects orthogonally on $\mathcal{N}(C^{\mathrm{T}})$. It is required

$$\mathcal{R}(B^{\mathrm{T}}) \perp \mathcal{N}(C^{\mathrm{T}}) \tag{1594}$$

which precludes $C = \mathbf{0}$ for B nonzero. Again, $C \succ 0 \;\Rightarrow\; C^\dagger = C^{-1}$. So we get, for A or C nonsingular,

$$G = \begin{bmatrix} A & B \\ B^{\mathrm{T}} & C \end{bmatrix} \succeq 0$$

$$\Leftrightarrow$$

$$A \succ 0, \quad C - B^{\mathrm{T}}A^{-1}B \succeq 0$$
$$\textbf{or}$$
$$C \succ 0, \quad A - BC^{-1}B^{\mathrm{T}} \succeq 0$$

(1595)

When A is full-rank then, for all B of compatible dimension, $\mathcal{R}(B)$ is in $\mathcal{R}(A)$. Likewise, when C is full-rank, $\mathcal{R}(B^{\mathrm{T}})$ is in $\mathcal{R}(C)$. Thus the flavor, for A and C nonsingular,

$$G = \begin{bmatrix} A & B \\ B^{\mathrm{T}} & C \end{bmatrix} \succ 0$$

$$\Leftrightarrow A \succ 0, \quad C - B^{\mathrm{T}}A^{-1}B \succ 0$$
$$\Leftrightarrow C \succ 0, \quad A - BC^{-1}B^{\mathrm{T}} \succ 0$$

(1596)

where $C - B^{\mathrm{T}}A^{-1}B$ is called the *Schur complement of A in G*, while the *Schur complement of C in G* is $A - BC^{-1}B^{\mathrm{T}}$. [157, §4.8]

Origin of the term *Schur complement* is from complementary *inertia*: [116, §2.4.4] Define

$$\mathrm{inertia}\left(G \in \mathbb{S}^M\right) \triangleq \{p, z, n\}$$

(1597)

where p, z, n respectively represent number of positive, zero, and negative eigenvalues of G; *id est*,

$$M = p + z + n$$

(1598)

Then, when A is invertible,

$$\mathrm{inertia}(G) = \mathrm{inertia}(A) + \mathrm{inertia}(C - B^{\mathrm{T}}A^{-1}B)$$

(1599)

and when C is invertible,

$$\mathrm{inertia}(G) = \mathrm{inertia}(C) + \mathrm{inertia}(A - BC^{-1}B^{\mathrm{T}})$$

(1600)

A.4.0.0.1 Example. *Equipartition inertia.*   [56, §1.2 exer.17]
When $A = C = \mathbf{0}$, denoting nonincreasingly ordered singular values of matrix $B \in \mathbb{R}^{m \times m}$ by $\sigma(B) \in \mathbb{R}_+^m$, then we have eigenvalues

$$\lambda(G) = \lambda\left(\begin{bmatrix} \mathbf{0} & B \\ B^{\mathrm{T}} & \mathbf{0} \end{bmatrix}\right) = \begin{bmatrix} \sigma(B) \\ -\Xi\,\sigma(B) \end{bmatrix}$$

(1601)

and

$$\mathrm{inertia}(G) = \mathrm{inertia}(B^{\mathrm{T}}B) + \mathrm{inertia}(-B^{\mathrm{T}}B)$$

(1602)

where Ξ is the order-reversing permutation matrix defined in (1819).   □

A.4.0.0.2   Example.   *Nonnegative polynomial.*   [34, p.163]

Quadratic multivariate polynomial $x^T A\, x + 2b^T x + c$ is a convex function of vector x if and only if $A \succeq 0$, but sublevel set $\{x \mid x^T A\, x + 2b^T x + c \leq 0\}$ is convex if $A \succeq 0$ yet not *vice versa*. Schur-form positive semidefiniteness is sufficient for polynomial convexity but necessary and sufficient for nonnegativity:

$$\begin{bmatrix} A & b \\ b^T & c \end{bmatrix} \succeq 0 \ \Leftrightarrow \ [x^T \ 1] \begin{bmatrix} A & b \\ b^T & c \end{bmatrix} \begin{bmatrix} x \\ 1 \end{bmatrix} \geq 0 \ \Leftrightarrow \ x^T A\, x + 2b^T x + c \geq 0 \tag{1603}$$

All is extensible to univariate polynomials; *e.g.*, $x \triangleq [t^n \ t^{n-1} \ t^{n-2} \cdots t]^T$.   □

A.4.0.0.3   Example.   *Schur-form fractional function trace minimization.*

From (1565),

$$\begin{bmatrix} A & B \\ B^T & C \end{bmatrix} \succeq 0 \qquad \Rightarrow \quad \mathrm{tr}(A + C) \geq 0$$

$$\Updownarrow$$

$$\begin{bmatrix} A & \mathbf{0} \\ \mathbf{0}^T & C - B^T A^{-1} B \end{bmatrix} \succeq 0 \ \Rightarrow \ \mathrm{tr}(C - B^T A^{-1} B) \geq 0 \tag{1604}$$

$$\Updownarrow$$

$$\begin{bmatrix} A - B C^{-1} B^T & \mathbf{0} \\ \mathbf{0}^T & C \end{bmatrix} \succeq 0 \ \Rightarrow \ \mathrm{tr}(A - B C^{-1} B^T) \geq 0$$

Since $\mathrm{tr}(C - B^T A^{-1} B) \geq 0 \ \Leftrightarrow \ \mathrm{tr}\, C \geq \mathrm{tr}(B^T A^{-1} B) \geq 0$ for example, then minimization of $\mathrm{tr}\, C$ is necessary and sufficient for minimization of $\mathrm{tr}(C - B^T A^{-1} B)$ when both are under constraint $\begin{bmatrix} A & B \\ B^T & C \end{bmatrix} \succeq 0$.   □

A.4.0.1   Schur-form nullspace basis

From (1592),

$$G = \begin{bmatrix} A & B \\ B^T & C \end{bmatrix} \succeq 0$$

$$\Updownarrow$$

$$\begin{bmatrix} A & \mathbf{0} \\ \mathbf{0}^T & C - B^T A^\dagger B \end{bmatrix} \succeq 0 \qquad \text{and} \qquad B^T (I - A A^\dagger) = \mathbf{0} \tag{1605}$$

$$\Updownarrow$$

$$\begin{bmatrix} A - B C^\dagger B^T & \mathbf{0} \\ \mathbf{0}^T & C \end{bmatrix} \succeq 0 \qquad \text{and} \qquad B (I - C C^\dagger) = \mathbf{0}$$

These facts plus Moore-Penrose condition (§E.0.1) provide a partial basis:

$$\mathrm{basis}\,\mathcal{N}\left(\begin{bmatrix} A & B \\ B^T & C \end{bmatrix} \right) \supseteq \begin{bmatrix} I - A A^\dagger & \mathbf{0} \\ \mathbf{0}^T & I - C C^\dagger \end{bmatrix} \tag{1606}$$

A.4.0.1.1 Example. *Sparse Schur conditions.*
Setting matrix A to the Identity simplifies the Schur conditions. One consequence relates definiteness of three quantities:

$$\begin{bmatrix} I & B \\ B^{\mathrm{T}} & C \end{bmatrix} \succeq 0 \;\; \Leftrightarrow \;\; C - B^{\mathrm{T}}B \succeq 0 \;\; \Leftrightarrow \;\; \begin{bmatrix} I & \mathbf{0} \\ \mathbf{0}^{\mathrm{T}} & C - B^{\mathrm{T}}B \end{bmatrix} \succeq 0 \qquad (1607)$$

$\square$

A.4.0.1.2 Exercise. *Eigenvalues λ of sparse Schur-form.*
Prove: given $C - B^{\mathrm{T}}B = \mathbf{0}$, for $B \in \mathbb{R}^{m \times n}$ and $C \in \mathbb{S}^n$

$$\lambda\left(\begin{bmatrix} I & B \\ B^{\mathrm{T}} & C \end{bmatrix}\right)_i = \begin{cases} 1 + \lambda(C)_i, & 1 \leq i \leq n \\ 1, & n < i \leq m \\ 0, & \text{otherwise} \end{cases} \qquad (1608)$$

▼

A.4.0.1.3 Theorem. *Rank of partitioned matrices.* [421, §2.2 prob.7]
When symmetric matrix A is invertible and C is symmetric,

$$\mathrm{rank}\begin{bmatrix} A & B \\ B^{\mathrm{T}} & C \end{bmatrix} = \mathrm{rank}\begin{bmatrix} A & \mathbf{0} \\ \mathbf{0}^{\mathrm{T}} & C - B^{\mathrm{T}}A^{-1}B \end{bmatrix} \qquad (1609)$$
$$= \mathrm{rank}\,A + \mathrm{rank}(C - B^{\mathrm{T}}A^{-1}B)$$

equals rank of main diagonal block A plus rank of its Schur complement. Similarly, when symmetric matrix C is invertible and A is symmetric,

$$\mathrm{rank}\begin{bmatrix} A & B \\ B^{\mathrm{T}} & C \end{bmatrix} = \mathrm{rank}\begin{bmatrix} A - B\,C^{-1}B^{\mathrm{T}} & \mathbf{0} \\ \mathbf{0}^{\mathrm{T}} & C \end{bmatrix} \qquad (1610)$$
$$= \mathrm{rank}(A - B\,C^{-1}B^{\mathrm{T}}) + \mathrm{rank}\,C$$

◇

Proof. The first assertion (1609) holds if and only if [208, §0.4.6c]

$$\exists \text{ nonsingular } X, Y \ni X\begin{bmatrix} A & B \\ B^{\mathrm{T}} & C \end{bmatrix}Y = \begin{bmatrix} A & \mathbf{0} \\ \mathbf{0}^{\mathrm{T}} & C - B^{\mathrm{T}}A^{-1}B \end{bmatrix} \qquad (1611)$$

Let [208, §7.7.6]

$$Y = X^{\mathrm{T}} = \begin{bmatrix} I & -A^{-1}B \\ \mathbf{0}^{\mathrm{T}} & I \end{bmatrix} \qquad (1612)$$

♦

From Corollary A.3.1.0.3 eigenvalues are related by

$$0 \preceq \lambda(C - B^{\mathrm{T}}A^{-1}B) \preceq \lambda(C) \qquad (1613)$$
$$0 \preceq \lambda(A - B\,C^{-1}B^{\mathrm{T}}) \preceq \lambda(A) \qquad (1614)$$

which means

$$\text{rank}(C - B^T A^{-1} B) \leq \text{rank}\, C \qquad (1615)$$

$$\text{rank}(A - B C^{-1} B^T) \leq \text{rank}\, A \qquad (1616)$$

Therefore

$$\text{rank} \begin{bmatrix} A & B \\ B^T & C \end{bmatrix} \leq \text{rank}\, A + \text{rank}\, C \qquad (1617)$$

A.4.0.1.4 Lemma. *Rank of Schur-form block.*   [140] [138]
Matrix $B \in \mathbb{R}^{m \times n}$ has $\text{rank}\, B \leq \rho$ if and only if there exist matrices $A \in \mathbb{S}^m$ and $C \in \mathbb{S}^n$ such that

$$\text{rank} \begin{bmatrix} A & 0 \\ 0^T & C \end{bmatrix} \leq 2\rho \quad \textbf{and} \quad G = \begin{bmatrix} A & B \\ B^T & C \end{bmatrix} \succeq 0 \qquad (1618)$$

$\diamond$

Schur-form positive semidefiniteness alone implies $\text{rank}\, A \geq \text{rank}\, B$ and $\text{rank}\, C \geq \text{rank}\, B$. But, even in absence of semidefiniteness, we must always have $\text{rank}\, G \geq \text{rank}\, A$, $\text{rank}\, B$, $\text{rank}\, C$ by fundamental linear algebra.

A.4.1   Determinant

$$G = \begin{bmatrix} A & B \\ B^T & C \end{bmatrix} \qquad (1619)$$

We consider again a matrix G partitioned like (1592), but not necessarily positive (semi)definite, where A and C are symmetric.

- When A is invertible,

$$\det G = \det A \;\det(C - B^T A^{-1} B) \qquad (1620)$$

When C is invertible,

$$\det G = \det C \;\det(A - B C^{-1} B^T) \qquad (1621)$$

- When B is full-rank and skinny, $C = 0$, and $A \succeq 0$, then [62, §10.1.1]

$$\det G \neq 0 \;\Leftrightarrow\; A + BB^T \succ 0 \qquad (1622)$$

When B is a (column) vector, then for all $C \in \mathbb{R}$ and all A of dimension compatible with G

$$\det G = \det(A)C - B^T A_{\text{cof}}^T B \qquad (1623)$$

while for $C \neq 0$

$$\det G = C \det\!\left(A - \frac{1}{C} BB^T\right) \qquad (1624)$$

where A_{cof} is the matrix of cofactors [338, §4] corresponding to A.

- When B is full-rank and fat, $A = \mathbf{0}$, and $C \succeq 0$, then

$$\det G \neq 0 \quad \Leftrightarrow \quad C + B^{\mathrm{T}}B \succ 0 \qquad (1625)$$

When B is a row-vector, then for $A \neq 0$ and all C of dimension compatible with G

$$\det G = A \det(C - \frac{1}{A}B^{\mathrm{T}}B) \qquad (1626)$$

while for all $A \in \mathbb{R}$

$$\det G = \det(C)A - B\,C_{\mathrm{cof}}^{\mathrm{T}}B^{\mathrm{T}} \qquad (1627)$$

where C_{cof} is the matrix of cofactors corresponding to C.

A.5 Eigenvalue decomposition

All square matrices have associated eigenvalues λ and eigenvectors; if not square, $Ax = \lambda_i x$ becomes impossible dimensionally. Eigenvectors must be nonzero. Prefix *eigen* is from the German; in this context meaning, something akin to "characteristic". [335, p.14]

When a square matrix $X \in \mathbb{R}^{m \times m}$ is *diagonalizable*, [338, §5.6] then

$$X = S\Lambda S^{-1} = [\, s_1 \cdots s_m \,] \Lambda \begin{bmatrix} w_1^{\mathrm{T}} \\ \vdots \\ w_m^{\mathrm{T}} \end{bmatrix} = \sum_{i=1}^{m} \lambda_i\, s_i\, w_i^{\mathrm{T}} \qquad (1628)$$

where $\{s_i \in \mathcal{N}(X - \lambda_i I) \subseteq \mathbb{C}^m\}$ are l.i. (right-)eigenvectors constituting the columns of $S \in \mathbb{C}^{m \times m}$ defined by

$$XS = S\Lambda \qquad \textbf{rather} \qquad Xs_i \triangleq \lambda_i s_i\,, \quad i = 1\ldots m \qquad (1629)$$

$\{w_i \in \mathcal{N}(X^{\mathrm{T}} - \lambda_i I) \subseteq \mathbb{C}^m\}$ are linearly independent *left-eigenvectors* of X (eigenvectors of X^{T}) constituting the rows of S^{-1} defined by [208]

$$S^{-1}X = \Lambda S^{-1} \qquad \textbf{rather} \qquad w_i^{\mathrm{T}}X \triangleq \lambda_i w_i^{\mathrm{T}}\,, \quad i = 1\ldots m \qquad (1630)$$

and where $\{\lambda_i \in \mathbb{C}\}$ are eigenvalues (1542)

$$\delta(\lambda(X)) = \Lambda \in \mathbb{C}^{m \times m} \qquad (1631)$$

corresponding to both left and right eigenvectors; *id est,* $\lambda(X) = \lambda(X^{\mathrm{T}})$.

There is no connection between diagonalizability and invertibility of X. [338, §5.2] Diagonalizability is guaranteed by a full set of linearly independent eigenvectors, whereas invertibility is guaranteed by all nonzero eigenvalues.

$$\begin{array}{c} \text{distinct eigenvalues} \;\Rightarrow\; \text{l.i. eigenvectors} \;\Leftrightarrow\; \text{diagonalizable} \\ \text{not diagonalizable} \;\Rightarrow\; \text{repeated eigenvalue} \end{array} \qquad (1632)$$

$\begin{bmatrix} 1 & 0 & -1 \\ -1 & 1 & 0 \\ 3 & -1 & -2 \end{bmatrix}$ is not diagonalizable, for example, having three 0-eigenvalues which are hard to compute with accuracy better than $\mathtt{1E\text{-}6}$. (Yates, D'Errico)

A.5.0.0.1   Theorem.   *Real eigenvector.*
Eigenvectors of a real matrix corresponding to real eigenvalues must be real.   ◇

Proof. $Ax = \lambda x$.   Given $\lambda = \lambda^*$, $x^{\mathrm H}Ax = \lambda x^{\mathrm H}x = \lambda\|x\|^2 = x^{\mathrm T}A\,x^* \;\Rightarrow\; x = x^*$, where $x^{\mathrm H}=x^{*\mathrm T}$. The converse is equally simple.   ♦

A.5.0.1   Uniqueness

From the *fundamental theorem of algebra,* [225] which guarantees existence of zeros for a given polynomial, it follows: eigenvalues, including their multiplicity, for a given square matrix are unique; meaning, there is no other set of eigenvalues for that matrix. (Conversely, many different matrices may share the same unique set of eigenvalues; *e.g.,* for any X, $\lambda(X) = \lambda(X^{\mathrm T})$.)

Uniqueness of eigenvectors, in contrast, disallows multiplicity of the same direction:

A.5.0.1.1   Definition.   *Unique eigenvectors.*
When eigenvectors are *unique,* we mean: unique to within a real nonzero scaling, and their directions are distinct.   △

If S is a matrix of eigenvectors of X as in (1628), for example, then $-S$ is certainly another matrix of eigenvectors decomposing X with the same eigenvalues. Although directions are distinct, eigenvectors $-S$ are equivalent to eigenvectors S by Definition A.5.0.1.1.

For any square matrix, the eigenvector corresponding to a distinct eigenvalue is unique; [335, p.220]

$$\text{distinct eigenvalues} \;\Rightarrow\; \text{eigenvectors unique} \tag{1633}$$

Eigenvectors corresponding to a repeated eigenvalue are not unique for a diagonalizable matrix;

$$\text{repeated eigenvalue} \;\Rightarrow\; \text{eigenvectors not unique} \tag{1634}$$

Proof follows from the observation: any linear combination of distinct eigenvectors of diagonalizable X, corresponding to a particular eigenvalue, produces another eigenvector. For eigenvalue λ whose multiplicity[A.12] $\dim\mathcal{N}(X-\lambda I)$ exceeds 1, in other words, any choice of independent vectors from $\mathcal{N}(X-\lambda I)$ (of the same multiplicity) constitutes eigenvectors corresponding to λ.   ♦

Caveat diagonalizability insures linear independence which implies existence of distinct eigenvectors. We may conclude, for diagonalizable matrices,

$$\text{distinct eigenvalues} \;\Leftrightarrow\; \text{eigenvectors unique} \tag{1635}$$

[A.12]A matrix is diagonalizable iff *algebraic multiplicity* (number of occurrences of same eigenvalue) equals *geometric multiplicity* $\dim\mathcal{N}(X-\lambda I) = m - \mathrm{rank}(X-\lambda I)$ [335, p.15] (number of *Jordan blocks* w.r.t λ or number of corresponding l.i. eigenvectors).

A.5.0.2 Invertible matrix

When diagonalizable matrix $X \in \mathbb{R}^{m \times m}$ is *nonsingular* (no zero eigenvalues), then it has an inverse obtained simply by inverting eigenvalues in (1628):

$$X^{-1} = S\Lambda^{-1}S^{-1} \tag{1636}$$

A.5.0.3 eigenmatrix

The (right-)eigenvectors $\{s_i\}$ (1628) are naturally orthogonal $w_i^{\mathrm{T}}s_j = 0$ to left-eigenvectors $\{w_i\}$ except, for $i = 1 \ldots m$, $w_i^{\mathrm{T}}s_i = 1$; called a *biorthogonality condition* [374, §2.2.4] [208] because neither set of left or right eigenvectors is necessarily an orthogonal set. Consequently, each dyad from a diagonalization is an independent (§B.1.1) nonorthogonal projector because

$$s_i w_i^{\mathrm{T}} s_i w_i^{\mathrm{T}} = s_i w_i^{\mathrm{T}} \tag{1637}$$

(whereas the dyads of singular value decomposition are not inherently projectors (*confer* (1644))).

Dyads of eigenvalue decomposition can be termed *eigenmatrices* because

$$X s_i w_i^{\mathrm{T}} = \lambda_i s_i w_i^{\mathrm{T}} \tag{1638}$$

Sum of the eigenmatrices is the Identity;

$$\sum_{i=1}^{m} s_i w_i^{\mathrm{T}} = I \tag{1639}$$

A.5.1 Symmetric matrix diagonalization

The set of *normal matrices* is, precisely, that set of all real matrices having a complete orthonormal set of eigenvectors; [421, §8.1] [340, prob.10.2.31] *id est*, any matrix X for which $XX^{\mathrm{T}} = X^{\mathrm{T}}X$; [164, §7.1.3] [335, p.3] *e.g.*, symmetric, orthogonal, and circulant matrices [174]. All normal matrices are diagonalizable.

A symmetric matrix is a special normal matrix whose eigenvalues Λ must be real[A.13] and whose eigenvectors S can be chosen to make a real orthonormal set; [340, §6.4] [338, p.315] *id est*, for $X \in \mathbb{S}^m$

$$X = S\Lambda S^{\mathrm{T}} = [s_1 \cdots s_m] \Lambda \begin{bmatrix} s_1^{\mathrm{T}} \\ \vdots \\ s_m^{\mathrm{T}} \end{bmatrix} = \sum_{i=1}^{m} \lambda_i s_i s_i^{\mathrm{T}} \tag{1640}$$

where $\delta^2(\Lambda) = \Lambda \in \mathbb{S}^m$ (§A.1) and $S^{-1} = S^{\mathrm{T}} \in \mathbb{R}^{m \times m}$ (orthogonal matrix, §B.5.2) because of symmetry: $S\Lambda S^{-1} = S^{-\mathrm{T}}\Lambda S^{\mathrm{T}}$. By 0 *eigenvalues theorem* A.7.3.0.1,

$$\begin{aligned} \mathcal{R}\{s_i \,|\, \lambda_i \neq 0\} &= \mathcal{R}(A) = \mathcal{R}(A^{\mathrm{T}}) \\ \mathcal{R}\{s_i \,|\, \lambda_i = 0\} &= \mathcal{N}(A^{\mathrm{T}}) = \mathcal{N}(A) \end{aligned} \tag{1641}$$

[A.13]**Proof.** Suppose λ_i is an eigenvalue corresponding to eigenvector s_i of real $A = A^{\mathrm{T}}$. Then $s_i^{\mathrm{H}}As_i = s_i^{\mathrm{T}}As_i^*$ (by transposition) $\Rightarrow s_i^{*\mathrm{T}}\lambda_i s_i = s_i^{\mathrm{T}}\lambda_i^* s_i^*$ because $(As_i)^* = (\lambda_i s_i)^*$ by assumption. So we have $\lambda_i \|s_i\|^2 = \lambda_i^* \|s_i\|^2$. There is no converse. ◆

A.5.1.1 Diagonal matrix diagonalization

Because arrangement of eigenvectors and their corresponding eigenvalues is arbitrary, we almost always arrange eigenvalues in nonincreasing order as is the convention for singular value decomposition. Then to diagonalize a symmetric matrix that is already a diagonal matrix, orthogonal matrix S becomes a permutation matrix.

A.5.1.2 Invertible symmetric matrix

When symmetric matrix $X \in \mathbb{S}^m$ is nonsingular (invertible), then its inverse (obtained by inverting eigenvalues in (1640)) is also symmetric:

$$X^{-1} = S\Lambda^{-1}S^{\mathrm{T}} \in \mathbb{S}^m \tag{1642}$$

A.5.1.3 Positive semidefinite matrix square root

When $X \in \mathbb{S}^m_+$, its unique positive semidefinite matrix square root is defined

$$\sqrt{X} \triangleq S\sqrt{\Lambda}\,S^{\mathrm{T}} \in \mathbb{S}^m_+ \tag{1643}$$

where the square root of nonnegative diagonal matrix $\sqrt{\Lambda}$ is taken entrywise and positive. Then $X = \sqrt{X}\sqrt{X}$.

A.6 Singular value decomposition, SVD

A.6.1 Compact SVD

[164, §2.5.4] For any $A \in \mathbb{R}^{m \times n}$

$$A = U\Sigma Q^{\mathrm{T}} = [\,u_1 \cdots u_\eta\,]\, \Sigma \begin{bmatrix} q_1^{\mathrm{T}} \\ \vdots \\ q_\eta^{\mathrm{T}} \end{bmatrix} = \sum_{i=1}^{\eta} \sigma_i\, u_i q_i^{\mathrm{T}}$$

$$U \in \mathbb{R}^{m \times \eta}, \quad \Sigma \in \mathbb{R}^{\eta \times \eta}_+, \quad Q \in \mathbb{R}^{n \times \eta}$$
$$U^{\mathrm{T}}U = I, \qquad\qquad Q^{\mathrm{T}}Q = I$$

$$\tag{1644}$$

where U and Q are always skinny-or-square real, each having orthonormal columns, and where

$$\eta \triangleq \min\{m, n\} \tag{1645}$$

Square matrix Σ is diagonal (§A.1.1)

$$\delta^2(\Sigma) = \Sigma \in \mathbb{R}^{\eta \times \eta}_+ \tag{1646}$$

holding the singular values $\{\sigma_i \in \mathbb{R}\}$ of A which are always arranged in nonincreasing order by convention and are related to eigenvalues λ by[A.14]

$$
\sigma(A)_i = \sigma(A^{\mathrm{T}})_i = \begin{cases} \sqrt{\lambda(A^{\mathrm{T}}A)_i} = \sqrt{\lambda(AA^{\mathrm{T}})_i} = \lambda\left(\sqrt{A^{\mathrm{T}}A}\right)_i = \lambda\left(\sqrt{AA^{\mathrm{T}}}\right)_i > 0, & 1 \leq i \leq \rho \\ 0, & \rho < i \leq \eta \end{cases}
$$

(1647)

of which the last $\eta - \rho$ are 0 ,[A.15] where

$$
\rho \triangleq \operatorname{rank} A = \operatorname{rank} \Sigma
$$

(1648)

A point sometimes lost: Any real matrix may be decomposed in terms of its real singular values $\sigma(A) \in \mathbb{R}^\eta$ and real matrices U and Q as in (1644), where [164, §2.5.3]

$$
\begin{aligned}
\mathcal{R}\{u_i \mid \sigma_i \neq 0\} &= \mathcal{R}(A) \\
\mathcal{R}\{u_i \mid \sigma_i = 0\} &\subseteq \mathcal{N}(A^{\mathrm{T}}) \\
\mathcal{R}\{q_i \mid \sigma_i \neq 0\} &= \mathcal{R}(A^{\mathrm{T}}) \\
\mathcal{R}\{q_i \mid \sigma_i = 0\} &\subseteq \mathcal{N}(A)
\end{aligned}
$$

(1649)

A.6.2 Subcompact SVD

Some authors allow only nonzero singular values. In that case the compact decomposition can be made smaller; it can be redimensioned in terms of rank ρ because, for any $A \in \mathbb{R}^{m \times n}$

$$
\rho = \operatorname{rank} A = \operatorname{rank} \Sigma = \max\{i \in \{1 \dots \eta\} \mid \sigma_i \neq 0\} \leq \eta
$$

(1650)

- There are η singular values. For any flavor SVD, rank is equivalent to the number of nonzero singular values on the main diagonal of Σ .

Now

$$
A = U\Sigma Q^{\mathrm{T}} = [\, u_1 \cdots u_\rho \,] \, \Sigma \begin{bmatrix} q_1^{\mathrm{T}} \\ \vdots \\ q_\rho^{\mathrm{T}} \end{bmatrix} = \sum_{i=1}^{\rho} \sigma_i u_i q_i^{\mathrm{T}}
$$

$$
U \in \mathbb{R}^{m \times \rho}, \quad \Sigma \in \mathbb{R}_+^{\rho \times \rho}, \quad Q \in \mathbb{R}^{n \times \rho}
$$

$$
U^{\mathrm{T}}U = I, \qquad\qquad Q^{\mathrm{T}}Q = I
$$

(1651)

where the main diagonal of diagonal matrix Σ has no 0 entries, and

$$
\begin{aligned}
\mathcal{R}\{u_i\} &= \mathcal{R}(A) \\
\mathcal{R}\{q_i\} &= \mathcal{R}(A^{\mathrm{T}})
\end{aligned}
$$

(1652)

[A.14]When matrix A is normal, $\sigma(A) = |\lambda(A)|$. [421, §8.1]

[A.15]For $\eta = n$, $\sigma(A) = \sqrt{\lambda(A^{\mathrm{T}}A)} = \lambda\left(\sqrt{A^{\mathrm{T}}A}\right)$.

For $\eta = m$, $\sigma(A) = \sqrt{\lambda(AA^{\mathrm{T}})} = \lambda\left(\sqrt{AA^{\mathrm{T}}}\right)$.

A.6.3   Full SVD

Another common and useful expression of the SVD makes U and Q square; making the decomposition larger than compact SVD. Completing the nullspace bases in U and Q from (1649) provides what is called the *full singular value decomposition* of $A \in \mathbb{R}^{m \times n}$ [338, App.A]. Orthonormal real matrices U and Q become orthogonal matrices (§B.5):

$$
\begin{aligned}
\mathcal{R}\{u_i \,|\, \sigma_i \neq 0\} &= \mathcal{R}(A) \\
\mathcal{R}\{u_i \,|\, \sigma_i = 0\} &= \mathcal{N}(A^{\mathrm{T}}) \\
\mathcal{R}\{q_i \,|\, \sigma_i \neq 0\} &= \mathcal{R}(A^{\mathrm{T}}) \\
\mathcal{R}\{q_i \,|\, \sigma_i = 0\} &= \mathcal{N}(A)
\end{aligned}
\tag{1653}
$$

For any matrix A having rank $\rho \ (= \operatorname{rank} \Sigma)$

$$
A = U \Sigma Q^{\mathrm{T}} = [\, u_1 \cdots u_m \,] \, \Sigma \begin{bmatrix} q_1^{\mathrm{T}} \\ \vdots \\ q_n^{\mathrm{T}} \end{bmatrix} = \sum_{i=1}^{\eta} \sigma_i \, u_i q_i^{\mathrm{T}}
$$

$$
= \begin{bmatrix} m \times \rho \ \text{basis}\,\mathcal{R}(A) & m \times m - \rho \ \text{basis}\,\mathcal{N}(A^{\mathrm{T}}) \end{bmatrix}
\begin{bmatrix} \sigma_1 & & & \\ & \sigma_2 & & \\ & & \ddots & \end{bmatrix}
\begin{bmatrix} \left(n \times \rho \ \text{basis}\,\mathcal{R}(A^{\mathrm{T}}) \right)^{\mathrm{T}} \\ \left(n \times n - \rho \ \text{basis}\,\mathcal{N}(A) \right)^{\mathrm{T}} \end{bmatrix}
$$

$$
U \in \mathbb{R}^{m \times m}, \quad \Sigma \in \mathbb{R}_{+}^{m \times n}, \quad Q \in \mathbb{R}^{n \times n}
$$

$$
U^{\mathrm{T}} = U^{-1}, \qquad\qquad\qquad Q^{\mathrm{T}} = Q^{-1}
\tag{1654}
$$

where upper limit of summation η is defined in (1645). Matrix Σ is no longer necessarily square, now padded with respect to (1646) by $m - \eta$ zero rows or $n - \eta$ zero columns; the nonincreasingly ordered (possibly 0) singular values appear along its main diagonal as for compact SVD (1647).

An important geometrical interpretation of SVD is given in Figure 166 for $m = n = 2$: The image of the unit sphere under any $m \times n$ matrix multiplication is an ellipse. Considering the three factors of the SVD separately, note that Q^{T} is a pure rotation of the circle. Figure 166 shows how the axes q_1 and q_2 are first rotated by Q^{T} to coincide with the coordinate axes. Second, the circle is stretched by Σ in the directions of the coordinate axes to form an ellipse. The third step rotates the ellipse by U into its final position. Note how q_1 and q_2 are rotated to end up as u_1 and u_2 , the principal axes of the final ellipse. A direct calculation shows that $Aq_j = \sigma_j u_j$. Thus q_j is first rotated to coincide with the jth coordinate axis, stretched by a factor σ_j , and then rotated to point in the direction of u_j . All of this is beautifully illustrated for 2×2 matrices by the MATLAB *code* `eigshow.m` *(see [341]).*

A direct consequence of the geometric interpretation is that the largest singular value σ_1 measures the "magnitude" of A (its 2-norm):

$$
\|A\|_2 = \sup_{\|x\|_2 = 1} \|Ax\|_2 = \sigma_1
\tag{1655}
$$

This means that $\|A\|_2$ is the length of the longest principal semiaxis of the ellipse.

$$A = U\Sigma Q^{\mathrm{T}} = [\, u_1 \cdots u_m \,] \, \Sigma \begin{bmatrix} q_1^{\mathrm{T}} \\ \vdots \\ q_n^{\mathrm{T}} \end{bmatrix} = \sum_{i=1}^{\eta} \sigma_i \, u_i q_i^{\mathrm{T}}$$

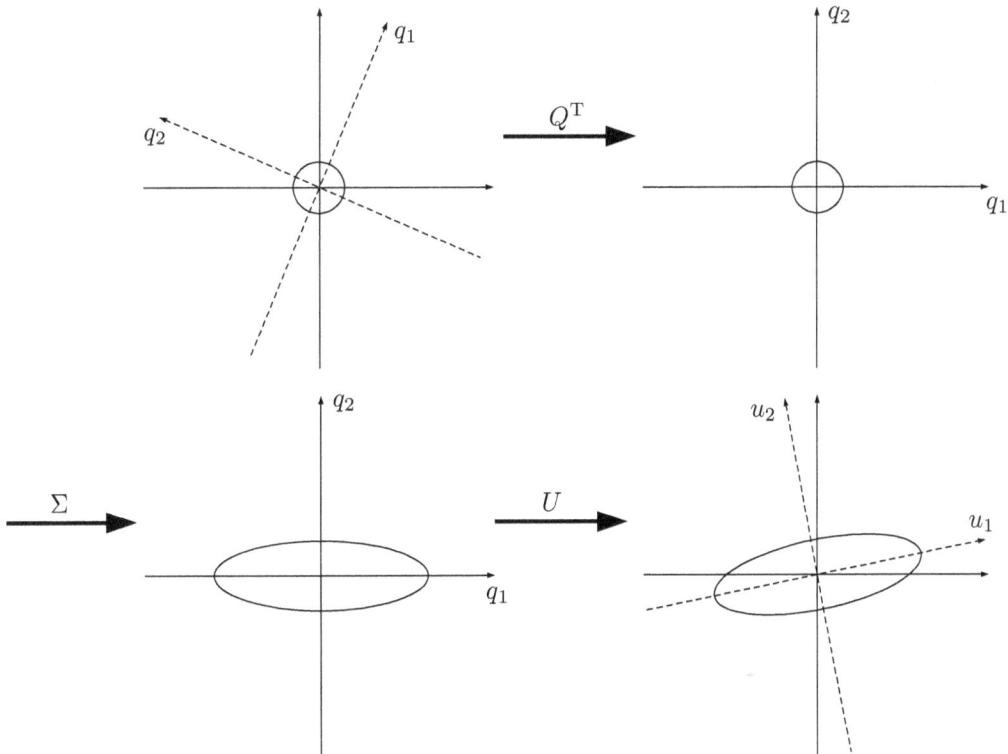

Figure 166: Geometrical interpretation of full SVD [277]: Image of circle $\{x \in \mathbb{R}^2 \mid \|x\|_2 = 1\}$ under matrix multiplication Ax is, in general, an ellipse. For the example illustrated, $U \triangleq [\, u_1 \ u_2 \,] \in \mathbb{R}^{2 \times 2}$, $Q \triangleq [\, q_1 \ q_2 \,] \in \mathbb{R}^{2 \times 2}$.

Expressions for U, Q, and Σ follow readily from (1654),

$$AA^{\mathrm{T}}U = U\Sigma\Sigma^{\mathrm{T}} \quad \text{and} \quad A^{\mathrm{T}}AQ = Q\Sigma^{\mathrm{T}}\Sigma \tag{1656}$$

demonstrating that the columns of U are the eigenvectors of AA^{T} and the columns of Q are the eigenvectors of $A^{\mathrm{T}}A$.   −Muller, Magaia, & Herbst [277]

A.6.4   Pseudoinverse by SVD

Matrix pseudoinverse (§E) is nearly synonymous with singular value decomposition because of the elegant expression, given $A = U\Sigma Q^{\mathrm{T}} \in \mathbb{R}^{m \times n}$

$$A^{\dagger} = Q\Sigma^{\dagger\mathrm{T}}U^{\mathrm{T}} \in \mathbb{R}^{n \times m} \tag{1657}$$

that applies to all three flavors of SVD, where $\Sigma^{\dagger}$ simply inverts nonzero entries of matrix Σ.

Given symmetric matrix $A \in \mathbb{S}^n$ and its diagonalization $A = S\Lambda S^{\mathrm{T}}$ (§A.5.1), its pseudoinverse simply inverts all nonzero eigenvalues:

$$A^{\dagger} = S\Lambda^{\dagger}S^{\mathrm{T}} \in \mathbb{S}^n \tag{1658}$$

A.6.5   SVD of symmetric matrices

From (1647) and (1643) for $A = A^{\mathrm{T}}$

$$\sigma(A)_i = \begin{cases} \sqrt{\lambda(A^2)_i} = \lambda\left(\sqrt{A^2}\right)_i = |\lambda(A)_i| > 0, & 1 \le i \le \rho \\ 0, & \rho < i \le \eta \end{cases} \tag{1659}$$

A.6.5.0.1   Definition. *Step function.*   *(confer §4.3.2.0.1)*
Define the signum-like quasilinear function $\psi : \mathbb{R}^n \to \mathbb{R}^n$ that takes value 1 corresponding to a 0-valued entry in its argument:

$$\psi(a) \triangleq \left[\lim_{x_i \to a_i} \frac{x_i}{|x_i|} = \begin{cases} 1, & a_i \ge 0 \\ -1, & a_i < 0 \end{cases}, \quad i = 1 \ldots n \right] \in \mathbb{R}^n \tag{1660}$$

$\triangle$

Eigenvalue signs of a symmetric matrix having diagonalization $A = S\Lambda S^{\mathrm{T}}$ (1640) can be absorbed either into real U or real Q from the full SVD; [359, p.34] *(confer §C.4.2.1)*

$$A = S\Lambda S^{\mathrm{T}} = S\delta(\psi(\delta(\Lambda)))\,|\Lambda|S^{\mathrm{T}} \triangleq U\,\Sigma Q^{\mathrm{T}} \in \mathbb{S}^n \tag{1661}$$

or

$$A = S\Lambda S^{\mathrm{T}} = S|\Lambda|\,\delta(\psi(\delta(\Lambda)))S^{\mathrm{T}} \triangleq U\Sigma\,Q^{\mathrm{T}} \in \mathbb{S}^n \tag{1662}$$

where matrix of singular values $\Sigma = |\Lambda|$ denotes entrywise absolute value of diagonal eigenvalue matrix Λ.

A.7 Zeros

A.7.1 norm zero

For any given norm, by definition,

$$\|x\|_\ell = 0 \quad \Leftrightarrow \quad x = \mathbf{0} \tag{1663}$$

Consequently, a generally nonconvex constraint in x like $\|Ax - b\| = \kappa$ becomes convex when $\kappa = 0$.

A.7.2 0 entry

If a positive semidefinite matrix $A = [A_{ij}] \in \mathbb{R}^{n \times n}$ has a 0 entry A_{ii} on its main diagonal, then $A_{ij} + A_{ji} = 0 \;\; \forall j$. [278, §1.3.1]

Any symmetric positive semidefinite matrix having a 0 entry on its main diagonal must be $\mathbf{0}$ along the entire row and column to which that 0 entry belongs. [164, §4.2.8] [208, §7.1 prob.2] From which it follows: for $A \in \mathbb{S}_+^n$

$$\delta(A) = \mathbf{0} \;\Leftrightarrow\; A = \mathbf{0} \tag{1664}$$

$$\mathrm{tr}(A) = 0 \;\Leftrightarrow\; A = \mathbf{0} \tag{1665}$$

A.7.3 *0 eigenvalues theorem*

This theorem is simple, powerful, and widely applicable:

A.7.3.0.1 Theorem. *Number of 0 eigenvalues.*
  For any matrix $A \in \mathbb{R}^{m \times n}$

$$\mathrm{rank}(A) + \dim \mathcal{N}(A) = n \tag{1666}$$

by conservation of dimension. [208, §0.4.4]

  For any square matrix $A \in \mathbb{R}^{m \times m}$, number of 0 eigenvalues is at least equal to $\dim \mathcal{N}(A)$

$$\dim \mathcal{N}(A) \leq \text{number of 0 eigenvalues} \leq m \tag{1667}$$

while all eigenvectors corresponding to those 0 eigenvalues belong to $\mathcal{N}(A)$. [338, §5.1][A.16]

[A.16]We take as given the well-known fact that the number of 0 eigenvalues cannot be less than dimension of the nullspace. We offer an example of the converse:

$$A = \begin{bmatrix} 1 & 0 & 1 & 0 \\ 0 & 0 & 1 & 0 \\ 0 & 0 & 0 & 0 \\ 1 & 0 & 0 & 0 \end{bmatrix}$$

$\dim \mathcal{N}(A) = 2$, $\lambda(A) = [0\ 0\ 0\ 1]^{\mathrm{T}}$; three eigenvectors in the nullspace but only two are independent. The right-hand side of (1667) is tight for nonzero matrices; *e.g.*, (§B.1) dyad $uv^{\mathrm{T}} \in \mathbb{R}^{m \times m}$ has m 0-eigenvalues when $u \in v^\perp$.

542   *APPENDIX A. LINEAR ALGEBRA*

For diagonalizable matrix A (§A.5), the number of 0 eigenvalues is precisely $\dim \mathcal{N}(A)$ while the corresponding eigenvectors span $\mathcal{N}(A)$. Real and imaginary parts of the eigenvectors remaining span $\mathcal{R}(A)$.

(TRANSPOSE.)
Likewise, for any matrix $A \in \mathbb{R}^{m \times n}$

$$\operatorname{rank}(A^{\mathrm{T}}) + \dim \mathcal{N}(A^{\mathrm{T}}) = m \tag{1668}$$

For any square $A \in \mathbb{R}^{m \times m}$, number of 0 eigenvalues is at least equal to $\dim \mathcal{N}(A^{\mathrm{T}}) = \dim \mathcal{N}(A)$ while all left-eigenvectors (eigenvectors of A^{T}) corresponding to those 0 eigenvalues belong to $\mathcal{N}(A^{\mathrm{T}})$.

For diagonalizable A, number of 0 eigenvalues is precisely $\dim \mathcal{N}(A^{\mathrm{T}})$ while the corresponding left-eigenvectors span $\mathcal{N}(A^{\mathrm{T}})$. Real and imaginary parts of the left-eigenvectors remaining span $\mathcal{R}(A^{\mathrm{T}})$. $\diamond$

Proof. First we show, for a diagonalizable matrix, the number of 0 eigenvalues is precisely the dimension of its nullspace while the eigenvectors corresponding to those 0 eigenvalues span the nullspace:

Any diagonalizable matrix $A \in \mathbb{R}^{m \times m}$ must possess a complete set of linearly independent eigenvectors. If A is full-rank (invertible), then all $m = \operatorname{rank}(A)$ eigenvalues are nonzero. [338, §5.1]

Suppose $\operatorname{rank}(A) < m$. Then $\dim \mathcal{N}(A) = m - \operatorname{rank}(A)$. Thus there is a set of $m - \operatorname{rank}(A)$ linearly independent vectors spanning $\mathcal{N}(A)$. Each of those can be an eigenvector associated with a 0 eigenvalue because A is diagonalizable $\Leftrightarrow \exists\, m$ linearly independent eigenvectors. [338, §5.2] Eigenvectors of a real matrix corresponding to 0 eigenvalues must be real.[A.17] Thus A has at least $m - \operatorname{rank}(A)$ eigenvalues equal to 0.

Now suppose A has more than $m - \operatorname{rank}(A)$ eigenvalues equal to 0. Then there are more than $m - \operatorname{rank}(A)$ linearly independent eigenvectors associated with 0 eigenvalues, and each of those eigenvectors must be in $\mathcal{N}(A)$. Thus there are more than $m - \operatorname{rank}(A)$ linearly independent vectors in $\mathcal{N}(A)$; a contradiction.

Diagonalizable A therefore has $\operatorname{rank}(A)$ nonzero eigenvalues and exactly $m - \operatorname{rank}(A)$ eigenvalues equal to 0 whose corresponding eigenvectors span $\mathcal{N}(A)$.

By similar argument, the left-eigenvectors corresponding to 0 eigenvalues span $\mathcal{N}(A^{\mathrm{T}})$.

Next we show when A is diagonalizable, the real and imaginary parts of its eigenvectors (corresponding to nonzero eigenvalues) span $\mathcal{R}(A)$:

The (right-)eigenvectors of a diagonalizable matrix $A \in \mathbb{R}^{m \times m}$ are linearly independent if and only if the left-eigenvectors are. So, matrix A has a representation in terms of its right- and left-eigenvectors; from the diagonalization (1628), assuming 0 eigenvalues are ordered last,

$$A = \sum_{i=1}^{m} \lambda_i\, s_i w_i^{\mathrm{T}} = \sum_{\substack{i=1 \\ \lambda_i \neq 0}}^{k \leq m} \lambda_i\, s_i w_i^{\mathrm{T}} \tag{1669}$$

From the *linearly independent dyads theorem* (§B.1.1.0.2), the dyads $\{s_i w_i^{\mathrm{T}}\}$ must be independent because each set of eigenvectors are; hence $\operatorname{rank} A = k$, the number of nonzero

---

[A.17]**Proof.** Let $*$ denote complex conjugation. Suppose $A = A^*$ and $As_i = \mathbf{0}$. Then $s_i = s_i^* \Rightarrow As_i = As_i^* \Rightarrow As_i^* = \mathbf{0}$. Conversely, $As_i^* = \mathbf{0} \Rightarrow As_i = As_i^* \Rightarrow s_i = s_i^*$.   ♦

eigenvalues. Complex eigenvectors and eigenvalues are common for real matrices, and must come in complex conjugate pairs for the summation to remain real. Assume that conjugate pairs of eigenvalues appear in sequence. Given any particular conjugate pair from (1669), we get the partial summation

$$\begin{aligned} \lambda_i\, s_i w_i^{\mathrm{T}} + \lambda_i^*\, s_i^* w_i^{*\mathrm{T}} &= 2\,\mathrm{re}(\lambda_i\, s_i w_i^{\mathrm{T}}) \\ &= 2\big(\mathrm{re}\, s_i\; \mathrm{re}(\lambda_i\, w_i^{\mathrm{T}}) - \mathrm{im}\, s_i\; \mathrm{im}(\lambda_i\, w_i^{\mathrm{T}})\big) \end{aligned} \tag{1670}$$

where[A.18] $\lambda_i^* \triangleq \lambda_{i+1}$, $s_i^* \triangleq s_{i+1}$, and $w_i^* \triangleq w_{i+1}$. Then (1669) is equivalently written

$$A = 2 \sum_{\substack{i \\ \lambda \in \mathbb{C} \\ \lambda_i \neq 0}} \mathrm{re}\, s_{2i}\; \mathrm{re}(\lambda_{2i}\, w_{2i}^{\mathrm{T}}) - \mathrm{im}\, s_{2i}\; \mathrm{im}(\lambda_{2i}\, w_{2i}^{\mathrm{T}}) \; + \sum_{\substack{j \\ \lambda \in \mathbb{R} \\ \lambda_j \neq 0}} \lambda_j\, s_j w_j^{\mathrm{T}} \tag{1671}$$

The summation (1671) shows: A is a linear combination of real and imaginary parts of its (right-)eigenvectors corresponding to nonzero eigenvalues. The k vectors $\{\mathrm{re}\, s_i \in \mathbb{R}^m,\ \mathrm{im}\, s_i \in \mathbb{R}^m \mid \lambda_i \neq 0,\ i \in \{1 \ldots m\}\}$ must therefore span the range of diagonalizable matrix A .

The argument is similar regarding span of the left-eigenvectors. ♦

A.7.4 0 trace and matrix product

For $X, A \in \mathbb{R}_+^{M \times N}$ (39)

$$\mathrm{tr}(X^{\mathrm{T}} A) = 0 \;\Leftrightarrow\; X \circ A = A \circ X = \mathbf{0} \tag{1672}$$

For $X, A \in \mathbb{S}_+^M$ [34, §2.6.1 exer.2.8] [368, §3.1]

$$\mathrm{tr}(XA) = 0 \;\Leftrightarrow\; XA = AX = \mathbf{0} \tag{1673}$$

Proof. ($\Leftarrow$) Suppose $XA = AX = \mathbf{0}$. Then $\mathrm{tr}(XA) = 0$ is obvious.
($\Rightarrow$) Suppose $\mathrm{tr}(XA) = 0$. $\mathrm{tr}(XA) = \mathrm{tr}(\sqrt{A}\, X \sqrt{A})$ whose argument is positive semidefinite by Corollary A.3.1.0.5. Trace of any square matrix is equivalent to the sum of its eigenvalues. Eigenvalues of a positive semidefinite matrix can total 0 if and only if each and every nonnegative eigenvalue is 0. The only positive semidefinite matrix, having all 0 eigenvalues, resides at the origin; (*confer* (1697)) *id est,*

$$\sqrt{A}\, X \sqrt{A} = \left(\sqrt{X}\sqrt{A}\right)^{\mathrm{T}} \sqrt{X}\sqrt{A} = \mathbf{0} \tag{1674}$$

implying $\sqrt{X}\sqrt{A} = \mathbf{0}$ which in turn implies $\sqrt{X}(\sqrt{X}\sqrt{A})\sqrt{A} = XA = \mathbf{0}$. Arguing similarly yields $AX = \mathbf{0}$. ♦

Diagonalizable matrices A and X are *simultaneously diagonalizable* if and only if they are commutative under multiplication; [208, §1.3.12] *id est*, iff they share a complete set of eigenvectors.

[A.18]Complex conjugate of w is denoted w^*. Conjugate transpose is denoted $w^{\mathrm{H}} = w^{*\mathrm{T}}$.

A.7.4.0.1   Example.   *An equivalence in nonisomorphic spaces.*
Identity (1673) leads to an unusual equivalence relating convex geometry to traditional
linear algebra: The convex sets, given $A \succeq 0$

$$\{X \mid \langle X, A \rangle = 0\} \cap \{X \succeq 0\} \equiv \{X \mid \mathcal{N}(X) \supseteq \mathcal{R}(A)\} \cap \{X \succeq 0\} \qquad (1675)$$

(one expressed in terms of a hyperplane, the other in terms of nullspace and range) are
equivalent only when symmetric matrix A is positive semidefinite.

   We might apply this equivalence to the geometric center subspace, for example,

$$\begin{aligned} \mathbb{S}_c^M &= \{Y \in \mathbb{S}^M \mid Y\mathbf{1} = \mathbf{0}\} \\ &= \{Y \in \mathbb{S}^M \mid \mathcal{N}(Y) \supseteq \mathbf{1}\} = \{Y \in \mathbb{S}^M \mid \mathcal{R}(Y) \subseteq \mathcal{N}(\mathbf{1}^T)\} \end{aligned} \qquad (2094)$$

from which we derive (*confer* (1074))

$$\mathbb{S}_c^M \cap \mathbb{S}_+^M \equiv \{X \succeq 0 \mid \langle X, \mathbf{1}\mathbf{1}^T \rangle = 0\} \qquad (1676)$$

$\square$

A.7.5   Zero definite

The domain over which an arbitrary real matrix A is zero definite can exceed its left and
right nullspaces. For any positive semidefinite matrix $A \in \mathbb{R}^{M \times M}$ (for $A + A^T \succeq 0$)

$$\{x \mid x^T A x = 0\} = \mathcal{N}(A + A^T) \qquad (1677)$$

because $\exists R \ni A + A^T = R^T R$, $\|Rx\| = 0 \Leftrightarrow Rx = \mathbf{0}$, and $\mathcal{N}(A + A^T) = \mathcal{N}(R)$.
Then given any particular vector x_p, $x_p^T A x_p = 0 \Leftrightarrow x_p \in \mathcal{N}(A + A^T)$. For any positive
definite matrix A (for $A + A^T \succ 0$)

$$\{x \mid x^T A x = 0\} = \mathbf{0} \qquad (1678)$$

Further, [421, §3.2 prob.5]

$$\{x \mid x^T A x = 0\} = \mathbb{R}^M \Leftrightarrow A^T = -A \qquad (1679)$$

while

$$\{x \mid x^H A x = 0\} = \mathbb{C}^M \Leftrightarrow A = \mathbf{0} \qquad (1680)$$

   The positive semidefinite matrix

$$A = \begin{bmatrix} 1 & 2 \\ 0 & 1 \end{bmatrix} \qquad (1681)$$

for example, has no nullspace. Yet

$$\{x \mid x^T A x = 0\} = \{x \mid \mathbf{1}^T x = 0\} \subset \mathbb{R}^2 \qquad (1682)$$

which is the nullspace of the symmetrized matrix. Symmetric matrices are not spared
from the excess; *videlicet,*

$$B = \begin{bmatrix} 1 & 2 \\ 2 & 1 \end{bmatrix} \qquad (1683)$$

has eigenvalues $\{-1,3\}$, no nullspace, but is zero definite on[A.19]

$$\mathcal{X} \triangleq \{x \in \mathbb{R}^2 \mid x_2 = (-2 \pm \sqrt{3}\,)x_1\} \tag{1684}$$

A.7.5.0.1 Proposition. (Sturm/Zhang) *Dyad-decompositions.* [344, §5.2]
Let positive semidefinite matrix $X \in \mathbb{S}_+^M$ have rank ρ. Then given symmetric matrix $A \in \mathbb{S}^M$, $\langle A, X \rangle = 0$ if and only if there exists a dyad-decomposition

$$X = \sum_{j=1}^{\rho} x_j x_j^{\mathrm{T}} \tag{1685}$$

satisfying
$$\langle A, x_j x_j^{\mathrm{T}} \rangle = 0 \text{ for each and every } j \in \{1 \ldots \rho\} \tag{1686}$$
$\diamond$

The dyad-decomposition of X proposed is generally not that obtained from a standard diagonalization by eigenvalue decomposition, unless $\rho = 1$ or the given matrix A is simultaneously diagonalizable (§A.7.4) with X. That means, elemental dyads in decomposition (1685) constitute a generally nonorthogonal set. Sturm & Zhang give a simple procedure for constructing the dyad-decomposition [396] where matrix A may be regarded as a parameter.

A.7.5.0.2 Example. *Dyad.*
The dyad $uv^{\mathrm{T}} \in \mathbb{R}^{M \times M}$ (§B.1) is zero definite on all x for which either $x^{\mathrm{T}}u = 0$ or $x^{\mathrm{T}}v = 0$;
$$\{x \mid x^{\mathrm{T}}uv^{\mathrm{T}}x = 0\} = \{x \mid x^{\mathrm{T}}u = 0\} \cup \{x \mid v^{\mathrm{T}}x = 0\} \tag{1687}$$
id est, on $u^{\perp} \cup v^{\perp}$. Symmetrizing the dyad does not change the outcome:
$$\{x \mid x^{\mathrm{T}}(uv^{\mathrm{T}} + vu^{\mathrm{T}})x/2 = 0\} = \{x \mid x^{\mathrm{T}}u = 0\} \cup \{x \mid v^{\mathrm{T}}x = 0\} \tag{1688}$$
□

[A.19]These two lines represent the limit in the union of two generally distinct hyperbolae; *id est*, for matrix B and set $\mathcal{X}$ as defined
$$\lim_{\varepsilon \to 0^+} \{x \in \mathbb{R}^2 \mid x^{\mathrm{T}}Bx = \varepsilon\} = \mathcal{X}$$

Appendix B

Simple matrices

Mathematicians also attempted to develop algebra of vectors but there was no natural definition of the product of two vectors that held in arbitrary dimensions. The first vector algebra that involved a noncommutative vector product (that is, v×w need not equal w×v) was proposed by Hermann Grassmann in his book Ausdehnungslehre *(1844). Grassmann's text also introduced the product of a column matrix and a row matrix, which resulted in what is now called a simple or a rank-one matrix. In the late 19th century the American mathematical physicist Willard Gibbs published his famous treatise on vector analysis. In that treatise Gibbs represented general matrices, which he called* dyadics, *as sums of simple matrices, which Gibbs called* dyads. *Later the physicist P. A. M. Dirac introduced the term "bra-ket" for what we now call the scalar product of a "bra" (row) vector times a "ket" (column) vector and the term "ket-bra" for the product of a ket times a bra, resulting in what we now call a simple matrix, as above. Our convention of identifying column matrices and vectors was introduced by physicists in the 20th century.*

−Suddhendu Biswas [48, p.2]

B.1 Rank-one matrix (dyad)

Any matrix formed from the unsigned outer product of two vectors,

$$\Psi = uv^{\mathrm{T}} \in \mathbb{R}^{M \times N} \tag{1689}$$

where $u \in \mathbb{R}^M$ and $v \in \mathbb{R}^N$, is rank-one and called a *dyad*. Conversely, any rank-one matrix must have the form Ψ. [208, prob.1.4.1] Product $-uv^{\mathrm{T}}$ is a *negative dyad*. For matrix products AB^{T}, in general, we have

$$\mathcal{R}(AB^{\mathrm{T}}) \subseteq \mathcal{R}(A) \,, \qquad \mathcal{N}(AB^{\mathrm{T}}) \supseteq \mathcal{N}(B^{\mathrm{T}}) \tag{1690}$$

CITATION: Dattorro, *Convex Optimization & Euclidean Distance Geometry*, $\mathcal{M}\varepsilon\beta oo$ Publishing USA, 2005, v2015.02.15.

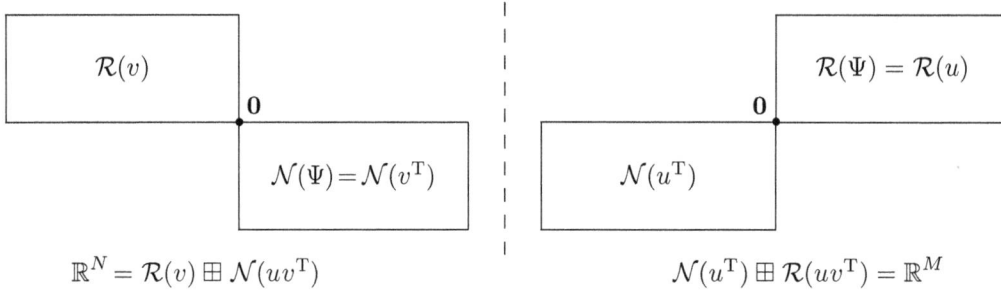

$$\mathbb{R}^N = \mathcal{R}(v) \boxplus \mathcal{N}(uv^{\mathrm{T}}) \qquad\qquad\qquad \mathcal{N}(u^{\mathrm{T}}) \boxplus \mathcal{R}(uv^{\mathrm{T}}) = \mathbb{R}^M$$

Figure 167: The four fundamental subspaces [340, §3.6] of any dyad $\Psi = uv^{\mathrm{T}} \in \mathbb{R}^{M \times N}$: $\mathcal{R}(v) \perp \mathcal{N}(\Psi)$ & $\mathcal{N}(u^{\mathrm{T}}) \perp \mathcal{R}(\Psi)$. $\Psi(x) \triangleq uv^{\mathrm{T}}x$ is a linear mapping from $\mathbb{R}^N$ to $\mathbb{R}^M$. Map from $\mathcal{R}(v)$ to $\mathcal{R}(u)$ is bijective. [338, §3.1]

with equality when $B = A$ [338, §3.3, §3.6][B.1] or respectively when B is invertible and $\mathcal{N}(A) = \mathbf{0}$. Yet for all nonzero dyads we have

$$\mathcal{R}(uv^{\mathrm{T}}) = \mathcal{R}(u) , \qquad \mathcal{N}(uv^{\mathrm{T}}) = \mathcal{N}(v^{\mathrm{T}}) \equiv v^{\perp} \tag{1691}$$

where $\dim v^{\perp} = N - 1$.

It is obvious a dyad can be $\mathbf{0}$ only when u or v is $\mathbf{0}$;

$$\Psi = uv^{\mathrm{T}} = \mathbf{0} \;\Leftrightarrow\; u = \mathbf{0} \;\text{ or }\; v = \mathbf{0} \tag{1692}$$

The matrix 2-norm for Ψ is equivalent to Frobenius' norm;

$$\|\Psi\|_2 = \sigma_1 = \|uv^{\mathrm{T}}\|_{\mathrm{F}} = \|uv^{\mathrm{T}}\|_2 = \|u\| \, \|v\| \tag{1693}$$

When u and v are normalized, the pseudoinverse is the transposed dyad. Otherwise,

$$\Psi^{\dagger} = (uv^{\mathrm{T}})^{\dagger} = \frac{vu^{\mathrm{T}}}{\|u\|^2 \, \|v\|^2} \tag{1694}$$

When dyad $uv^{\mathrm{T}} \in \mathbb{R}^{N \times N}$ is square, uv^{T} has at least $N - 1$ 0-eigenvalues and corresponding eigenvectors spanning $v^{\perp}$. The remaining eigenvector u spans the range of uv^{T} with corresponding eigenvalue

$$\lambda = v^{\mathrm{T}}u = \mathrm{tr}(uv^{\mathrm{T}}) \in \mathbb{R} \tag{1695}$$

Determinant is a product of the eigenvalues; so, it is always true that

$$\det \Psi = \det(uv^{\mathrm{T}}) = 0 \tag{1696}$$

[B.1]**Proof.** $\mathcal{R}(AA^{\mathrm{T}}) \subseteq \mathcal{R}(A)$ is obvious.

$$\begin{aligned} \mathcal{R}(AA^{\mathrm{T}}) &= \{AA^{\mathrm{T}}y \mid y \in \mathbb{R}^m\} \\ &\supseteq \{AA^{\mathrm{T}}y \mid A^{\mathrm{T}}y \in \mathcal{R}(A^{\mathrm{T}})\} = \mathcal{R}(A) \;\text{ by (142)} \end{aligned}$$

$\blacklozenge$

When $\lambda = 1$, the square dyad is a nonorthogonal projector projecting on its range ($\Psi^2 = \Psi$, §E.6); a *projector dyad*. It is quite possible that $u \in v^\perp$ making the remaining eigenvalue instead 0;[B.2] $\lambda = 0$ together with the first $N-1$ 0-eigenvalues; *id est*, it is possible uv^{T} were nonzero while all its eigenvalues are 0. The matrix

$$\begin{bmatrix} 1 \\ -1 \end{bmatrix} \begin{bmatrix} 1 & 1 \end{bmatrix} = \begin{bmatrix} 1 & 1 \\ -1 & -1 \end{bmatrix} \tag{1697}$$

for example, has two 0-eigenvalues. In other words, eigenvector u may simultaneously be a member of the nullspace and range of the dyad. The explanation is, simply, because u and v share the same dimension, $\dim u = M = \dim v = N$:

Proof. Figure 167 shows the four fundamental subspaces for the dyad. Linear operator $\Psi : \mathbb{R}^N \to \mathbb{R}^M$ provides a map between vector spaces that remain distinct when $M = N$;

$$u \in \mathcal{R}(uv^{\mathrm{T}})$$
$$u \in \mathcal{N}(uv^{\mathrm{T}}) \Leftrightarrow v^{\mathrm{T}}u = 0 \tag{1698}$$
$$\mathcal{R}(uv^{\mathrm{T}}) \cap \mathcal{N}(uv^{\mathrm{T}}) = \emptyset$$

$\blacklozenge$

B.1.0.1 rank-one modification

For $A \in \mathbb{R}^{M \times N}$, $x \in \mathbb{R}^N$, $y \in \mathbb{R}^M$, and $y^{\mathrm{T}}Ax \neq 0$ [212, §2.1]

$$\mathrm{rank}\left(A - \frac{Axy^{\mathrm{T}}A}{y^{\mathrm{T}}Ax}\right) = \mathrm{rank}(A) - 1 \tag{1699}$$

Given nonsingular matrix $A \in \mathbb{R}^{N \times N} \ni 1 + v^{\mathrm{T}}A^{-1}u \neq 0$, [157, §4.11.2] [227, App.6] [421, §2.3 prob.16] (Sherman-Morrison-Woodbury)

$$(A + uv^{\mathrm{T}})^{-1} = A^{-1} - \frac{A^{-1}uv^{\mathrm{T}}A^{-1}}{1 + v^{\mathrm{T}}A^{-1}u} \tag{1700}$$

which is unstable numerically, by Saunders' reckoning: *"We know how to update matrix factorizations reliably (even if the input or output matrix is singular) but, in general, there's no stable way to update a matrix inverse."*

B.1.0.2 dyad symmetry

In the specific circumstance that $v = u$, then $uu^{\mathrm{T}} \in \mathbb{R}^{N \times N}$ is symmetric, rank-one, and positive semidefinite having exactly $N-1$ 0-eigenvalues. In fact, (Theorem A.3.1.0.7)

$$uv^{\mathrm{T}} \succeq 0 \Leftrightarrow v = u \tag{1701}$$

and the remaining eigenvalue is almost always positive;

$$\lambda = u^{\mathrm{T}}u = \mathrm{tr}(uu^{\mathrm{T}}) > 0 \quad \text{unless} \quad u = \mathbf{0} \tag{1702}$$

[B.2]A dyad is not always diagonalizable (§A.5) because its eigenvectors are not necessarily independent.

When $\lambda = 1$, the dyad becomes an orthogonal projector.

Matrix

$$\begin{bmatrix} \Psi & u \\ u^{\mathrm{T}} & 1 \end{bmatrix} \tag{1703}$$

for example, is rank-1 positive semidefinite if and only if $\Psi = uu^{\mathrm{T}}$.

B.1.1 Dyad independence

Now we consider a sum of dyads like (1689) as encountered in diagonalization and singular value decomposition:

$$\mathcal{R}\left(\sum_{i=1}^{k} s_i w_i^{\mathrm{T}}\right) = \sum_{i=1}^{k} \mathcal{R}(s_i w_i^{\mathrm{T}}) = \sum_{i=1}^{k} \mathcal{R}(s_i) \; \Leftarrow \; w_i \; \forall i \; \text{are l.i.} \tag{1704}$$

range of summation is the vector sum of ranges.[B.3] (Theorem B.1.1.1.1) Under the assumption the dyads are linearly independent (l.i.), then vector sums are unique (p.666): for $\{w_i\}$ l.i. and $\{s_i\}$ l.i.

$$\mathcal{R}\left(\sum_{i=1}^{k} s_i w_i^{\mathrm{T}}\right) = \mathcal{R}(s_1 w_1^{\mathrm{T}}) \oplus \ldots \oplus \mathcal{R}(s_k w_k^{\mathrm{T}}) = \mathcal{R}(s_1) \oplus \ldots \oplus \mathcal{R}(s_k) \tag{1705}$$

B.1.1.0.1 Definition. *Linearly independent dyads.*   [217, p.29 thm.11]
[346, p.2] The set of k dyads

$$\{s_i w_i^{\mathrm{T}} \mid i = 1 \ldots k\} \tag{1706}$$

where $s_i \in \mathbb{C}^M$ and $w_i \in \mathbb{C}^N$, is said to be linearly independent iff

$$\mathrm{rank}\left(SW^{\mathrm{T}} \triangleq \sum_{i=1}^{k} s_i w_i^{\mathrm{T}}\right) = k \tag{1707}$$

where $S \triangleq [s_1 \cdots s_k] \in \mathbb{C}^{M \times k}$ and $W \triangleq [w_1 \cdots w_k] \in \mathbb{C}^{N \times k}$. $\triangle$

Dyad independence does not preclude existence of a nullspace $\mathcal{N}(SW^{\mathrm{T}})$, as defined, nor does it imply SW^{T} were full-rank. In absence of assumption of independence, generally, $\mathrm{rank}\, SW^{\mathrm{T}} \leq k$. Conversely, any rank-$k$ matrix can be written in the form SW^{T} by singular value decomposition. (§A.6)

B.1.1.0.2 Theorem. *Linearly independent (l.i.) dyads.*
Vectors $\{s_i \in \mathbb{C}^M, \; i = 1 \ldots k\}$ are l.i. and vectors $\{w_i \in \mathbb{C}^N, \; i = 1 \ldots k\}$ are l.i. if and only if dyads $\{s_i w_i^{\mathrm{T}} \in \mathbb{C}^{M \times N}, \; i = 1 \ldots k\}$ are l.i. $\diamond$

Proof. Linear independence of k dyads is identical to definition (1707).

[B.3]Move of range $\mathcal{R}$ to inside summation admitted by linear independence of $\{w_i\}$.

($\Rightarrow$) Suppose $\{s_i\}$ and $\{w_i\}$ are each linearly independent sets. Invoking Sylvester's rank inequality, [208, §0.4] [421, §2.4]

$$\operatorname{rank} S + \operatorname{rank} W - k \ \leq\ \operatorname{rank}(SW^{\mathrm{T}}) \ \leq\ \min\{\operatorname{rank} S,\ \operatorname{rank} W\} \ (\leq k) \tag{1708}$$

Then $k \leq \operatorname{rank}(SW^{\mathrm{T}}) \leq k$ which implies the dyads are independent.
($\Leftarrow$) Conversely, suppose $\operatorname{rank}(SW^{\mathrm{T}}) = k$. Then

$$k \leq \min\{\operatorname{rank} S,\ \operatorname{rank} W\} \leq k \tag{1709}$$

implying the vector sets are each independent. $\blacklozenge$

B.1.1.1 Biorthogonality condition, Range and Nullspace of Sum

Dyads characterized by biorthogonality condition $W^{\mathrm{T}}S = I$ are independent; *id est*, for $S \in \mathbb{C}^{M \times k}$ and $W \in \mathbb{C}^{N \times k}$, if $W^{\mathrm{T}}S = I$ then $\operatorname{rank}(SW^{\mathrm{T}}) = k$ by the *linearly independent dyads theorem* because (*confer* §E.1.1)

$$W^{\mathrm{T}}S = I \ \Rightarrow\ \operatorname{rank} S = \operatorname{rank} W = k \leq M = N \tag{1710}$$

To see that, we need only show: $\mathcal{N}(S) = \mathbf{0} \ \Leftrightarrow\ \exists\, B \ni BS = I$.[B.4]
($\Leftarrow$) Assume $BS = I$. Then $\mathcal{N}(BS) = \mathbf{0} = \{x \,|\, BSx = \mathbf{0}\} \supseteq \mathcal{N}(S)$. (1690)
($\Rightarrow$) If $\mathcal{N}(S) = \mathbf{0}$ then S must be full-rank skinny-or-square.
$\therefore \ \exists\, A, B, C \ni \begin{bmatrix} B \\ C \end{bmatrix} [S\ A] = I$ (*id est*, $[S\ A]$ is invertible) $\ \Rightarrow\ BS = I$. Left inverse B is given as W^{T} here. Because of reciprocity with S, it immediately follows: $\mathcal{N}(W) = \mathbf{0} \ \Leftrightarrow\ \exists\, S \ni S^{\mathrm{T}}W = I$. $\blacklozenge$

Dyads produced by diagonalization, for example, are independent because of their inherent biorthogonality. (§A.5.0.3) The converse is generally false; *id est*, linearly independent dyads are not necessarily biorthogonal.

B.1.1.1.1 Theorem. *Nullspace and range of dyad sum.*
Given a sum of dyads represented by SW^{T} where $S \in \mathbb{C}^{M \times k}$ and $W \in \mathbb{C}^{N \times k}$

$$\begin{aligned} \mathcal{N}(SW^{\mathrm{T}}) = \mathcal{N}(W^{\mathrm{T}}) \ &\Leftarrow\ \exists\, B \ni BS = I \\ \mathcal{R}(SW^{\mathrm{T}}) = \mathcal{R}(S) \ &\Leftarrow\ \exists\, Z \ni W^{\mathrm{T}}Z = I \end{aligned} \tag{1711}$$

$\diamond$

Proof. ($\Rightarrow$) $\mathcal{N}(SW^{\mathrm{T}}) \supseteq \mathcal{N}(W^{\mathrm{T}})$ and $\mathcal{R}(SW^{\mathrm{T}}) \subseteq \mathcal{R}(S)$ are obvious.
($\Leftarrow$) Assume the existence of a left inverse $B \in \mathbb{R}^{k \times N}$ and a right inverse $Z \in \mathbb{R}^{N \times k}$.[B.5]

$$\mathcal{N}(SW^{\mathrm{T}}) = \{x \,|\, SW^{\mathrm{T}}x = \mathbf{0}\} \subseteq \{x \,|\, BSW^{\mathrm{T}}x = \mathbf{0}\} = \mathcal{N}(W^{\mathrm{T}}) \tag{1712}$$
$$\mathcal{R}(SW^{\mathrm{T}}) = \{SW^{\mathrm{T}}x \,|\, x \in \mathbb{R}^N\} \supseteq \{SW^{\mathrm{T}}Zy \,|\, Zy \in \mathbb{R}^N\} = \mathcal{R}(S) \tag{1713}$$

$\blacklozenge$

[B.4]Left inverse is not unique, in general.
[B.5]By counterexample, the theorem's converse cannot be true; *e.g.*, $S = W = [\,\mathbf{1}\ \ \mathbf{0}\,]$.

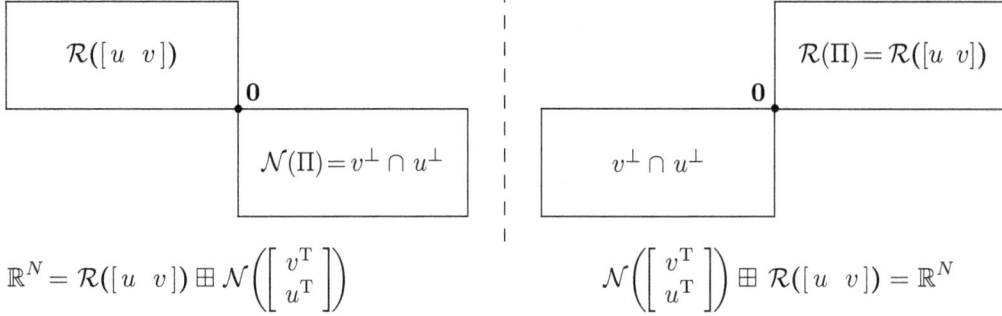

$$\mathbb{R}^N = \mathcal{R}([\,u\ \ v\,]) \boxplus \mathcal{N}\left(\left[\begin{array}{c} v^{\mathrm{T}} \\ u^{\mathrm{T}} \end{array}\right]\right) \qquad\qquad \mathcal{N}\left(\left[\begin{array}{c} v^{\mathrm{T}} \\ u^{\mathrm{T}} \end{array}\right]\right) \boxplus \mathcal{R}([\,u\ \ v\,]) = \mathbb{R}^N$$

Figure 168: The four fundamental subspaces [340, §3.6] of a doublet $\Pi = uv^{\mathrm{T}} + vu^{\mathrm{T}} \in \mathbb{S}^N$. $\Pi(x) = (uv^{\mathrm{T}} + vu^{\mathrm{T}})x$ is a linear bijective mapping from $\mathcal{R}([\,u\ \ v\,])$ to $\mathcal{R}([\,u\ \ v\,])$.

B.2   Doublet

Consider a sum of two linearly independent square dyads, one a transposition of the other:

$$\Pi = uv^{\mathrm{T}} + vu^{\mathrm{T}} = [\,u\ \ v\,] \left[\begin{array}{c} v^{\mathrm{T}} \\ u^{\mathrm{T}} \end{array}\right] = SW^{\mathrm{T}} \in \mathbb{S}^N \tag{1714}$$

where $u, v \in \mathbb{R}^N$. Like the dyad, a doublet can be $\mathbf{0}$ only when u or v is $\mathbf{0}$;

$$\Pi = uv^{\mathrm{T}} + vu^{\mathrm{T}} = \mathbf{0} \ \Leftrightarrow\ u = \mathbf{0} \ \text{ or } \ v = \mathbf{0} \tag{1715}$$

By assumption of independence, a nonzero doublet has two nonzero eigenvalues

$$\lambda_1 \triangleq u^{\mathrm{T}}v + \|uv^{\mathrm{T}}\| \ , \qquad \lambda_2 \triangleq u^{\mathrm{T}}v - \|uv^{\mathrm{T}}\| \tag{1716}$$

where $\lambda_1 > 0 > \lambda_2$, with corresponding eigenvectors

$$x_1 \triangleq \frac{u}{\|u\|} + \frac{v}{\|v\|} \ , \qquad x_2 \triangleq \frac{u}{\|u\|} - \frac{v}{\|v\|} \tag{1717}$$

spanning the doublet range. Eigenvalue λ_1 cannot be 0 unless u and v have opposing directions, but that is antithetical since then the dyads would no longer be independent. Eigenvalue λ_2 is 0 if and only if u and v share the same direction, again antithetical. Generally we have $\lambda_1 > 0$ and $\lambda_2 < 0$, so Π is indefinite.

By the *nullspace and range of dyad sum theorem*, doublet Π has $N-2$ zero-eigenvalues remaining and corresponding eigenvectors spanning $\mathcal{N}\left(\left[\begin{array}{c} v^{\mathrm{T}} \\ u^{\mathrm{T}} \end{array}\right]\right)$. We therefore have

$$\mathcal{R}(\Pi) = \mathcal{R}([\,u\ \ v\,]) \ , \qquad \mathcal{N}(\Pi) = v^{\perp} \cap u^{\perp} \tag{1718}$$

of respective dimension 2 and $N-2$.

B.3   Elementary matrix

A matrix of the form

$$E = I - \zeta uv^{\mathrm{T}} \in \mathbb{R}^{N \times N} \tag{1719}$$

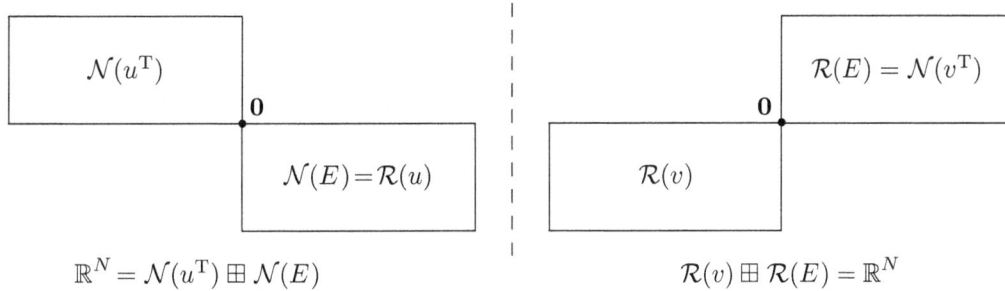

Figure 169: $v^{\mathrm{T}}u = 1/\zeta$. The four fundamental subspaces [340, §3.6] of elementary matrix E as a linear mapping $E(x) = \left(I - \dfrac{uv^{\mathrm{T}}}{v^{\mathrm{T}}u}\right)x$.

where $\zeta \in \mathbb{R}$ is finite and $u,v \in \mathbb{R}^N$, is called an *elementary matrix* or a *rank-one modification of the Identity.* [210] Any elementary matrix in $\mathbb{R}^{N \times N}$ has $N-1$ eigenvalues equal to 1 corresponding to real eigenvectors that span $v^\perp$. The remaining eigenvalue

$$\lambda = 1 - \zeta v^{\mathrm{T}}u \tag{1720}$$

corresponds to eigenvector u .[B.6] From [227, App.7.A.26] the determinant:

$$\det E = 1 - \mathrm{tr}\big(\zeta uv^{\mathrm{T}}\big) = \lambda \tag{1721}$$

If $\lambda \neq 0$ then E is invertible; [157] (*confer* §B.1.0.1)

$$E^{-1} = I + \frac{\zeta}{\lambda}uv^{\mathrm{T}} \tag{1722}$$

Eigenvectors corresponding to 0 eigenvalues belong to $\mathcal{N}(E)$, and the number of 0 eigenvalues must be at least $\dim \mathcal{N}(E)$ which, here, can be at most one. (§A.7.3.0.1) The nullspace exists, therefore, when $\lambda=0$; *id est*, when $v^{\mathrm{T}}u=1/\zeta$; rather, whenever u belongs to hyperplane $\{z \in \mathbb{R}^N \mid v^{\mathrm{T}}z=1/\zeta\}$. Then (when $\lambda=0$) elementary matrix E is a nonorthogonal projector projecting on its range ($E^2=E$, §E.1) and $\mathcal{N}(E)=\mathcal{R}(u)$; eigenvector u spans the nullspace when it exists. By conservation of dimension, $\dim \mathcal{R}(E)=N-\dim \mathcal{N}(E)$. It is apparent from (1719) that $v^\perp \subseteq \mathcal{R}(E)$, but $\dim v^\perp=N-1$. Hence $\mathcal{R}(E) \equiv v^\perp$ when the nullspace exists, and the remaining eigenvectors span it.

In summary, when a nontrivial nullspace of E exists,

$$\mathcal{R}(E) = \mathcal{N}(v^{\mathrm{T}}), \qquad \mathcal{N}(E) = \mathcal{R}(u), \qquad v^{\mathrm{T}}u = 1/\zeta \tag{1723}$$

illustrated in Figure 169, which is opposite to the assignment of subspaces for a dyad (Figure 167). Otherwise, $\mathcal{R}(E) = \mathbb{R}^N$.

When $E = E^{\mathrm{T}}$, the spectral norm is

$$\|E\|_2 = \max\{1 \,, |\lambda|\} \tag{1724}$$

[B.6]Elementary matrix E is not always diagonalizable because eigenvector u need not be independent of the others; *id est,* $u \in v^\perp$ is possible.

B.3.1   Householder matrix

An elementary matrix is called a Householder matrix when it has the defining form, for nonzero vector u [164, §5.1.2] [157, §4.10.1] [338, §7.3] [208, §2.2]

$$H = I - 2\frac{uu^{\mathrm{T}}}{u^{\mathrm{T}}u} \in \mathbb{S}^{N} \qquad (1725)$$

which is a symmetric orthogonal (reflection) matrix $(H^{-1}\!=\!H^{\mathrm{T}}\!=\!H$ (§B.5.3)). Vector u is normal to an $N-1$-dimensional subspace $u^{\perp}$ through which this particular H effects pointwise reflection; *e.g.*, $Hu^{\perp}\!=\!u^{\perp}$ while $Hu\!=\!-u$.

    Matrix H has $N-1$ orthonormal eigenvectors spanning that reflecting subspace $u^{\perp}$ with corresponding eigenvalues equal to 1. The remaining eigenvector u has corresponding eigenvalue -1; so

$$\det H = -1 \qquad (1726)$$

Due to symmetry of H, the matrix 2-norm (the spectral norm) is equal to the largest eigenvalue-magnitude. A Householder matrix is thus characterized,

$$H^{\mathrm{T}} = H\,, \qquad H^{-1} = H^{\mathrm{T}}\,, \qquad \|H\|_{2} = 1\,, \qquad H \not\succeq 0 \qquad (1727)$$

    For example, the permutation matrix

$$\Xi \;=\; \begin{bmatrix} 1 & 0 & 0 \\ 0 & 0 & 1 \\ 0 & 1 & 0 \end{bmatrix} \qquad (1728)$$

is a Householder matrix having $u\!=\![0\ \ 1\ \ {-1}]^{\mathrm{T}}/\sqrt{2}$. Not all permutation matrices are Householder matrices, although all permutation matrices are orthogonal matrices (§B.5.2, $\Xi^{\mathrm{T}}\Xi = I$) [338, §3.4] because they are made by permuting rows and columns of the Identity matrix. Neither are all symmetric permutation matrices Householder matrices; *e.g.*,

$\Xi = \begin{bmatrix} 0 & 0 & 0 & 1 \\ 0 & 0 & 1 & 0 \\ 0 & 1 & 0 & 0 \\ 1 & 0 & 0 & 0 \end{bmatrix}$ (1819) is not a Householder matrix.

B.4   Auxiliary V-matrices

B.4.1   Auxiliary projector matrix V

It is convenient to define a matrix V that arises naturally as a consequence of translating the geometric center α_{c} (§5.5.1.0.1) of some list X to the origin. In place of $X - \alpha_{c}\mathbf{1}^{\mathrm{T}}$ we may write XV as in (1054) where

$$V \;=\; I - \frac{1}{N}\mathbf{1}\mathbf{1}^{\mathrm{T}} \in \mathbb{S}^{N} \qquad (990)$$

is an elementary matrix called the *geometric centering matrix*.

Any elementary matrix in $\mathbb{R}^{N \times N}$ has $N-1$ eigenvalues equal to 1. For the particular elementary matrix V, the Nth eigenvalue equals 0. The number of 0 eigenvalues must equal $\dim \mathcal{N}(V) = 1$, by the 0 *eigenvalues theorem* (§A.7.3.0.1), because $V = V^{\text{T}}$ is diagonalizable. Because

$$V \mathbf{1} = \mathbf{0} \qquad (1729)$$

the nullspace $\mathcal{N}(V) = \mathcal{R}(\mathbf{1})$ is spanned by the eigenvector $\mathbf{1}$. The remaining eigenvectors span $\mathcal{R}(V) \equiv \mathbf{1}^{\perp} = \mathcal{N}(\mathbf{1}^{\text{T}})$ that has dimension $N-1$.

Because

$$V^2 = V \qquad (1730)$$

and $V^{\text{T}} = V$, elementary matrix V is also a projection matrix (§E.3) projecting orthogonally on its range $\mathcal{N}(\mathbf{1}^{\text{T}})$ which is a hyperplane containing the origin in $\mathbb{R}^N$

$$V = I - \mathbf{1}(\mathbf{1}^{\text{T}}\mathbf{1})^{-1}\mathbf{1}^{\text{T}} \qquad (1731)$$

The $\{0,1\}$ eigenvalues also indicate diagonalizable V is a projection matrix. [421, §4.1 thm.4.1] Symmetry of V denotes orthogonal projection; from (2009),

$$V^{\text{T}} = V \,, \qquad V^{\dagger} = V \,, \qquad \|V\|_2 = 1 \,, \qquad V \succeq 0 \qquad (1732)$$

Matrix V is also circulant [174].

B.4.1.0.1 Example. *Relationship of Auxiliary to Householder matrix.*
Let $H \in \mathbb{S}^N$ be a Householder matrix (1725) defined by

$$u = \begin{bmatrix} 1 \\ \vdots \\ 1 \\ 1 + \sqrt{N} \end{bmatrix} \in \mathbb{R}^N \qquad (1733)$$

Then we have [159, §2]

$$V = H \begin{bmatrix} I & \mathbf{0} \\ \mathbf{0}^{\text{T}} & 0 \end{bmatrix} H \qquad (1734)$$

Let $D \in \mathbb{S}_h^N$ and define

$$-HDH \triangleq - \begin{bmatrix} A & b \\ b^{\text{T}} & c \end{bmatrix} \qquad (1735)$$

where b is a vector. Then because H is nonsingular (§A.3.1.0.5) [190, §3]

$$-VDV = -H \begin{bmatrix} A & \mathbf{0} \\ \mathbf{0}^{\text{T}} & 0 \end{bmatrix} H \succeq 0 \;\Leftrightarrow\; -A \succeq 0 \qquad (1736)$$

and affine dimension is $r = \operatorname{rank} A$ when D is a Euclidean distance matrix. $\qquad \square$

B.4.2   Schoenberg auxiliary matrix $V_{\mathcal{N}}$

1. $V_{\mathcal{N}} = \dfrac{1}{\sqrt{2}} \begin{bmatrix} -\mathbf{1}^{\mathrm{T}} \\ I \end{bmatrix} \in \mathbb{R}^{N \times N-1}$

2. $V_{\mathcal{N}}^{\mathrm{T}} \mathbf{1} = \mathbf{0}$

3. $I - e_1 \mathbf{1}^{\mathrm{T}} = \begin{bmatrix} \mathbf{0} & \sqrt{2}V_{\mathcal{N}} \end{bmatrix}$

4. $\begin{bmatrix} \mathbf{0} & \sqrt{2}V_{\mathcal{N}} \end{bmatrix} V_{\mathcal{N}} = V_{\mathcal{N}}$

5. $\begin{bmatrix} \mathbf{0} & \sqrt{2}V_{\mathcal{N}} \end{bmatrix} V = V$

6. $V \begin{bmatrix} \mathbf{0} & \sqrt{2}V_{\mathcal{N}} \end{bmatrix} = \begin{bmatrix} \mathbf{0} & \sqrt{2}V_{\mathcal{N}} \end{bmatrix}$

7. $\begin{bmatrix} \mathbf{0} & \sqrt{2}V_{\mathcal{N}} \end{bmatrix} \begin{bmatrix} \mathbf{0} & \sqrt{2}V_{\mathcal{N}} \end{bmatrix} = \begin{bmatrix} \mathbf{0} & \sqrt{2}V_{\mathcal{N}} \end{bmatrix}$

8. $\begin{bmatrix} \mathbf{0} & \sqrt{2}V_{\mathcal{N}} \end{bmatrix}^{\dagger} = \begin{bmatrix} 0 & \mathbf{0}^{\mathrm{T}} \\ \mathbf{0} & I \end{bmatrix} V$

9. $\begin{bmatrix} \mathbf{0} & \sqrt{2}V_{\mathcal{N}} \end{bmatrix}^{\dagger} V = \begin{bmatrix} \mathbf{0} & \sqrt{2}V_{\mathcal{N}} \end{bmatrix}^{\dagger}$

10. $\begin{bmatrix} \mathbf{0} & \sqrt{2}V_{\mathcal{N}} \end{bmatrix} \begin{bmatrix} \mathbf{0} & \sqrt{2}V_{\mathcal{N}} \end{bmatrix}^{\dagger} = V$

11. $\begin{bmatrix} \mathbf{0} & \sqrt{2}V_{\mathcal{N}} \end{bmatrix}^{\dagger} \begin{bmatrix} \mathbf{0} & \sqrt{2}V_{\mathcal{N}} \end{bmatrix} = \begin{bmatrix} 0 & \mathbf{0}^{\mathrm{T}} \\ \mathbf{0} & I \end{bmatrix}$

12. $\begin{bmatrix} \mathbf{0} & \sqrt{2}V_{\mathcal{N}} \end{bmatrix} \begin{bmatrix} 0 & \mathbf{0}^{\mathrm{T}} \\ \mathbf{0} & I \end{bmatrix} = \begin{bmatrix} \mathbf{0} & \sqrt{2}V_{\mathcal{N}} \end{bmatrix}$

13. $\begin{bmatrix} 0 & \mathbf{0}^{\mathrm{T}} \\ \mathbf{0} & I \end{bmatrix} \begin{bmatrix} \mathbf{0} & \sqrt{2}V_{\mathcal{N}} \end{bmatrix} = \begin{bmatrix} 0 & \mathbf{0}^{\mathrm{T}} \\ \mathbf{0} & I \end{bmatrix}$

14. $\begin{bmatrix} V_{\mathcal{N}} & \frac{1}{\sqrt{2}}\mathbf{1} \end{bmatrix}^{-1} = \begin{bmatrix} V_{\mathcal{N}}^{\dagger} \\ \frac{\sqrt{2}}{N}\mathbf{1}^{\mathrm{T}} \end{bmatrix}$

15. $V_{\mathcal{N}}^{\dagger} = \sqrt{2} \begin{bmatrix} -\frac{1}{N}\mathbf{1} & I-\frac{1}{N}\mathbf{1}\mathbf{1}^{\mathrm{T}} \end{bmatrix} \in \mathbb{R}^{N-1 \times N} ,$ $\left(I-\frac{1}{N}\mathbf{1}\mathbf{1}^{\mathrm{T}} \in \mathbb{S}^{N-1} \right)$

16. $V_{\mathcal{N}}^{\dagger} \mathbf{1} = \mathbf{0}$

17. $V_{\mathcal{N}}^{\dagger} V_{\mathcal{N}} = I$

18. $V^{\mathrm{T}} = V = V_{\mathcal{N}} V_{\mathcal{N}}^{\dagger} = I - \frac{1}{N}\mathbf{1}\mathbf{1}^{\mathrm{T}} \in \mathbb{S}^{N}$

19. $-V_{\mathcal{N}}^{\dagger}(\mathbf{1}\mathbf{1}^{\mathrm{T}} - I)V_{\mathcal{N}} = I ,$ $\left(\mathbf{1}\mathbf{1}^{\mathrm{T}} - I \in \mathbb{EDM}^{N} \right)$

20. $D = [d_{ij}] \in \mathbb{S}_h^N$ (992)

$$\text{tr}(-VDV) = \text{tr}(-VD) = \text{tr}(-V_{\mathcal{N}}^\dagger DV_{\mathcal{N}}) = \tfrac{1}{N}\mathbf{1}^\text{T}D\,\mathbf{1} = \tfrac{1}{N}\text{tr}(\mathbf{11}^\text{T}D) = \tfrac{1}{N}\sum_{i,j} d_{ij}$$

Any elementary matrix $E \in \mathbb{S}^N$ of the particular form

$$E = k_1 I - k_2\,\mathbf{11}^\text{T} \tag{1737}$$

where k_1, $k_2 \in \mathbb{R}$,[B.7] will make $\text{tr}(-ED)$ proportional to $\sum d_{ij}$.

21. $D = [d_{ij}] \in \mathbb{S}^N$

$$\text{tr}(-VDV) = \tfrac{1}{N}\sum_{\substack{i,j\\i\neq j}} d_{ij} - \tfrac{N-1}{N}\sum_i d_{ii} = \tfrac{1}{N}\mathbf{1}^\text{T}D\,\mathbf{1} - \text{tr}\,D$$

22. $D = [d_{ij}] \in \mathbb{S}_h^N$

$$\text{tr}(-V_{\mathcal{N}}^\text{T}DV_{\mathcal{N}}) = \sum_j d_{1j}$$

23. For $Y \in \mathbb{S}^N$

$$V(Y - \delta(Y\mathbf{1}))V = Y - \delta(Y\mathbf{1})$$

B.4.3 Orthonormal auxiliary matrix $V_{\mathcal{W}}$

The skinny matrix

$$V_{\mathcal{W}} \triangleq \begin{bmatrix} \frac{-1}{\sqrt{N}} & \frac{-1}{\sqrt{N}} & \cdots & \frac{-1}{\sqrt{N}} \\ 1+\frac{-1}{N+\sqrt{N}} & \frac{-1}{N+\sqrt{N}} & \cdots & \frac{-1}{N+\sqrt{N}} \\ \frac{-1}{N+\sqrt{N}} & \ddots & \ddots & \frac{-1}{N+\sqrt{N}} \\ \vdots & \ddots & \ddots & \vdots \\ \frac{-1}{N+\sqrt{N}} & \frac{-1}{N+\sqrt{N}} & \cdots & 1+\frac{-1}{N+\sqrt{N}} \end{bmatrix} \in \mathbb{R}^{N\times N-1} \tag{1738}$$

has $\mathcal{R}(V_{\mathcal{W}}) = \mathcal{N}(\mathbf{1}^\text{T})$ and orthonormal columns. [6] We defined three auxiliary V-matrices: V, $V_{\mathcal{N}}$ (974), and $V_{\mathcal{W}}$ sharing some attributes listed in Table B.4.4. For example, V can be expressed

$$V = V_{\mathcal{W}}V_{\mathcal{W}}^\text{T} = V_{\mathcal{N}}V_{\mathcal{N}}^\dagger \tag{1739}$$

but $V_{\mathcal{W}}^\text{T}V_{\mathcal{W}} = I$ means V is an orthogonal projector (2006) and

$$V_{\mathcal{W}}^\dagger = V_{\mathcal{W}}^\text{T}, \qquad \|V_{\mathcal{W}}\|_2 = 1, \qquad V_{\mathcal{W}}^\text{T}\mathbf{1} = \mathbf{0} \tag{1740}$$

[B.7]If k_1 is $1-\rho$ while k_2 equals $-\rho \in \mathbb{R}$, then all eigenvalues of E for $-1/(N-1) < \rho < 1$ are guaranteed positive and therefore E is guaranteed positive definite. [309]

B.4.4   Auxiliary V-matrix Table

	$\dim V$	$\mathrm{rank}\,V$	$\mathcal{R}(V)$	$\mathcal{N}(V^{\mathrm{T}})$	$V^{\mathrm{T}}V$	VV^{T}	$VV^{\dagger}$
V	$N \times N$	$N-1$	$\mathcal{N}(\mathbf{1}^{\mathrm{T}})$	$\mathcal{R}(\mathbf{1})$	V	V	V
$V_{\mathcal{N}}$	$N \times (N-1)$	$N-1$	$\mathcal{N}(\mathbf{1}^{\mathrm{T}})$	$\mathcal{R}(\mathbf{1})$	$\frac{1}{2}(I + \mathbf{1}\mathbf{1}^{\mathrm{T}})$	$\frac{1}{2}\begin{bmatrix} N-1 & -\mathbf{1}^{\mathrm{T}} \\ -\mathbf{1} & I \end{bmatrix}$	V
$V_{\mathcal{W}}$	$N \times (N-1)$	$N-1$	$\mathcal{N}(\mathbf{1}^{\mathrm{T}})$	$\mathcal{R}(\mathbf{1})$	I	V	V

B.4.5   More auxiliary matrices

Mathar shows [265, §2] that any elementary matrix (§B.3) of the form

$$V_{\mathcal{S}} = I - b\,\mathbf{1}^{\mathrm{T}} \in \mathbb{R}^{N \times N} \tag{1741}$$

such that $b^{\mathrm{T}}\mathbf{1} = 1$ (*confer* [168, §2]), is an auxiliary V-matrix having

$$\begin{aligned}
\mathcal{R}(V_{\mathcal{S}}^{\mathrm{T}}) &= \mathcal{N}(b^{\mathrm{T}}), & \mathcal{R}(V_{\mathcal{S}}) &= \mathcal{N}(\mathbf{1}^{\mathrm{T}}) \\
\mathcal{N}(V_{\mathcal{S}}) &= \mathcal{R}(b), & \mathcal{N}(V_{\mathcal{S}}^{\mathrm{T}}) &= \mathcal{R}(\mathbf{1})
\end{aligned} \tag{1742}$$

Given $X \in \mathbb{R}^{n \times N}$, the choice $b = \frac{1}{N}\mathbf{1}$ ($V_{\mathcal{S}} = V$) minimizes $\|X(I - b\,\mathbf{1}^{\mathrm{T}})\|_{\mathrm{F}}$. [170, §3.2.1]

B.5   Orthomatrices

B.5.1   Orthonormal matrix

Property $Q^{\mathrm{T}}Q = I$ completely defines orthonormal matrix $Q \in \mathbb{R}^{n \times k}$ $(k \le n)$; a skinny-or-square full-rank matrix characterized by nonexpansivity (2010)

$$\|Q^{\mathrm{T}}x\|_2 \le \|x\|_2 \quad \forall\, x \in \mathbb{R}^n, \qquad \|Qy\|_2 = \|y\|_2 \quad \forall\, y \in \mathbb{R}^k \tag{1743}$$

and preservation of vector inner-product

$$\langle Qy,\, Qz \rangle = \langle y,\, z \rangle \tag{1744}$$

B.5.2   Orthogonal matrix & vector rotation

An orthogonal matrix is a square orthonormal matrix. Property $Q^{-1} = Q^{\mathrm{T}}$ completely defines orthogonal matrix $Q \in \mathbb{R}^{n \times n}$ employed to effect vector rotation; [338, §2.6, §3.4] [340, §6.5] [208, §2.1] for any $x \in \mathbb{R}^n$

$$\|Qx\|_2 = \|x\|_2 \tag{1745}$$

In other words, the 2-norm is orthogonally invariant. Any antisymmetric matrix constructs an orthogonal matrix; *id est*, for $A = -A^{\mathrm{T}}$

$$Q = (I + A)^{-1}(I - A) \tag{1746}$$

A *unitary matrix* is a complex generalization of orthogonal matrix; conjugate transpose defines it: $U^{-1} = U^{\mathrm{H}}$. An orthogonal matrix is simply a real unitary matrix.[B.8]

Orthogonal matrix Q is a normal matrix further characterized by spectral norm:

$$Q^{-1} = Q^{\mathrm{T}} , \qquad \|Q\|_2 = 1 \tag{1747}$$

Applying characterization (1747) to Q^{T} we see it too is an orthogonal matrix. Hence the rows and columns of Q respectively form an orthonormal set. Normalcy guarantees diagonalization (§A.5.1) so, for $Q \triangleq S\Lambda S^{\mathrm{H}}$

$$S\Lambda^{-1}S^{\mathrm{H}} = S^*\Lambda S^{\mathrm{T}} \left(= S\Lambda^* S^{\mathrm{H}}\right) , \qquad \|\delta(\Lambda)\|_\infty = 1 \tag{1748}$$

characterizes an orthogonal matrix in terms of eigenvalues and eigenvectors.

All permutation matrices Ξ, for example, are nonnegative orthogonal matrices; and *vice versa*. Product or Kronecker product of any permutation matrices remains a permutator. Any product of permutation matrix with orthogonal matrix remains orthogonal. In fact, any product of orthogonal matrices AQ remains orthogonal by definition. Given any other dimensionally compatible orthogonal matrix U, the mapping $g(A) = U^{\mathrm{T}}AQ$ is a bijection on the domain of orthogonal matrices (a nonconvex manifold of dimension $\frac{1}{2}n(n-1)$ [53]). [245, §2.1] [246]

The largest magnitude entry of an orthogonal matrix is 1; for each and every $j \in 1 \ldots n$

$$\begin{aligned} \|Q(j,:)\|_\infty &\leq 1 \\ \|Q(:,j)\|_\infty &\leq 1 \end{aligned} \tag{1749}$$

Each and every eigenvalue of a (real) orthogonal matrix has magnitude 1 $(\Lambda^{-1} = \Lambda^*)$

$$\lambda(Q) \in \mathbb{C}^n , \qquad |\lambda(Q)| = \mathbf{1} \tag{1750}$$

but only the Identity matrix can be simultaneously orthogonal and positive definite. Orthogonal matrices have complex eigenvalues in conjugate pairs: so $\det Q = \pm 1$.

B.5.3 Reflection

A matrix for pointwise reflection is defined by imposing symmetry upon the orthogonal matrix; *id est*, a *reflection matrix* is completely defined by $Q^{-1} = Q^{\mathrm{T}} = Q$. The reflection matrix is a symmetric orthogonal matrix, and *vice versa,* characterized:

$$Q^{\mathrm{T}} = Q , \qquad Q^{-1} = Q^{\mathrm{T}} , \qquad \|Q\|_2 = 1 \tag{1751}$$

The Householder matrix (§B.3.1) is an example of symmetric orthogonal (reflection) matrix.

Reflection matrices have eigenvalues equal to ± 1 and so $\det Q = \pm 1$. It is natural to expect a relationship between reflection and projection matrices because all projection matrices have eigenvalues belonging to $\{0, 1\}$. In fact, any reflection matrix Q is related to some orthogonal projector P by [210, §1 prob.44]

$$Q = I - 2P \tag{1752}$$

[B.8]Orthogonal and unitary matrices are called *unitary linear operators*.

Figure 170: *Gimbal*: a mechanism imparting three degrees of dimensional freedom to a Euclidean body suspended at the device's center. Each ring is free to rotate about one axis. (Courtesy of The MathWorks Inc.)

Yet P is, generally, neither orthogonal or invertible. (§E.3.2)

$$\lambda(Q) \in \mathbb{R}^n, \qquad |\lambda(Q)| = \mathbf{1} \tag{1753}$$

Reflection is with respect to $\mathcal{R}(P)^{\perp}$. Matrix $2P - I$ represents antireflection.

Every orthogonal matrix can be expressed as the product of a rotation and a reflection. The collection of all orthogonal matrices of particular dimension does not form a convex set.

B.5.4 Rotation of range and rowspace

Given orthogonal matrix Q, column vectors of a matrix X are simultaneously rotated about the origin via product QX. In three dimensions ($X \in \mathbb{R}^{\mathbf{3} \times N}$), the precise meaning of rotation is best illustrated in Figure **170** where the gimbal aids visualization of what is achievable; mathematically, (§5.5.2.0.1)

$$Q = \begin{bmatrix} \cos\theta & 0 & -\sin\theta \\ 0 & 1 & 0 \\ \sin\theta & 0 & \cos\theta \end{bmatrix} \begin{bmatrix} 1 & 0 & 0 \\ 0 & \cos\psi & -\sin\psi \\ 0 & \sin\psi & \cos\psi \end{bmatrix} \begin{bmatrix} \cos\phi & -\sin\phi & 0 \\ \sin\phi & \cos\phi & 0 \\ 0 & 0 & 1 \end{bmatrix} \tag{1754}$$

B.5.4.0.1   Example.   *One axis of revolution.*

Partition   an $n+1$-dimensional   Euclidean   space $\mathbb{R}^{n+1} \triangleq \begin{bmatrix} \mathbb{R}^n \\ \mathbb{R} \end{bmatrix}$ and   define   an n-dimensional subspace

$$\mathcal{R} \triangleq \{\lambda \in \mathbb{R}^{n+1} \mid \mathbf{1}^{\mathrm{T}}\lambda = 0\} \tag{1755}$$

(a hyperplane through the origin). We want an orthogonal matrix that rotates a list in the columns of matrix $X \in \mathbb{R}^{n+1 \times N}$ through the dihedral angle between $\mathbb{R}^n$ and $\mathcal{R}$ (§2.4.3)

$$\sphericalangle(\mathbb{R}^n, \mathcal{R}) = \arccos\left(\frac{\langle e_{n+1}, \mathbf{1}\rangle}{\|e_{n+1}\| \|\mathbf{1}\|}\right) = \arccos\left(\frac{1}{\sqrt{n+1}}\right) \text{ radians} \tag{1756}$$

The vertex-description of the nonnegative orthant in $\mathbb{R}^{n+1}$ is

$$\{[\, e_1 \quad e_2 \cdots e_{n+1}\,]\, a \mid a \succeq 0\} \;=\; \{a \succeq 0\} \;=\; \mathbb{R}_+^{n+1} \subset \mathbb{R}^{n+1} \qquad (1757)$$

Consider rotation of these vertices via orthogonal matrix

$$Q \;\triangleq\; [\, \mathbf{1}\tfrac{1}{\sqrt{n+1}} \quad \Xi V_{\mathcal{W}} \,]\Xi \in \mathbb{R}^{n+1 \times n+1} \qquad (1758)$$

where permutation matrix $\Xi \in \mathbb{S}^{n+1}$ is defined in (1819), and $V_{\mathcal{W}} \in \mathbb{R}^{n+1 \times n}$ is the orthonormal auxiliary matrix defined in §B.4.3. This particular orthogonal matrix is selected because it rotates any point in subspace $\mathbb{R}^n$ about one axis of revolution onto $\mathcal{R}$; *e.g.*, rotation $Q e_{n+1}$ aligns the last standard basis vector with subspace normal $\mathcal{R}^\perp = \mathbf{1}$. The rotated standard basis vectors remaining are orthonormal spanning $\mathcal{R}$. $\qquad\square$

Another interpretation of product QX is rotation/reflection of $\mathcal{R}(X)$. Rotation of X as in QXQ^{T} is a simultaneous rotation/reflection of range and rowspace.[B.9]

Proof. Any matrix can be expressed as a singular value decomposition $X = U\Sigma W^{\mathrm{T}}$ (1644) where $\delta^2(\Sigma) = \Sigma$, $\mathcal{R}(U) \supseteq \mathcal{R}(X)$, and $\mathcal{R}(W) \supseteq \mathcal{R}(X^{\mathrm{T}})$. $\qquad\blacklozenge$

B.5.5 Matrix rotation

Orthogonal matrices are also employed to rotate/reflect other matrices like vectors: [164, §12.4.1] Given orthogonal matrix Q, the product $Q^{\mathrm{T}}A$ will rotate $A \in \mathbb{R}^{n \times n}$ in the Euclidean sense in $\mathbb{R}^{n^2}$ because Frobenius' norm is orthogonally invariant (§2.2.1);

$$\|Q^{\mathrm{T}}A\|_{\mathrm{F}} \;=\; \sqrt{\operatorname{tr}(A^{\mathrm{T}}QQ^{\mathrm{T}}A)} \;=\; \|A\|_{\mathrm{F}} \qquad (1759)$$

(likewise for AQ). Were A symmetric, such a rotation would depart from $\mathbb{S}^n$. One remedy is to instead form product $Q^{\mathrm{T}}AQ$ because

$$\|Q^{\mathrm{T}}AQ\|_{\mathrm{F}} \;=\; \sqrt{\operatorname{tr}(Q^{\mathrm{T}}A^{\mathrm{T}}QQ^{\mathrm{T}}AQ)} \;=\; \|A\|_{\mathrm{F}} \qquad (1760)$$

By §A.1.1 *no.*31,

$$\operatorname{vec} Q^{\mathrm{T}}AQ = (Q \otimes Q)^{\mathrm{T}} \operatorname{vec} A \qquad (1761)$$

which is a rotation of the vectorized A matrix because Kronecker product of any orthogonal matrices remains orthogonal; *e.g.*, by §A.1.1 *no.*38,

$$(Q \otimes Q)^{\mathrm{T}}(Q \otimes Q) = I \qquad (1762)$$

Matrix A is *orthogonally equivalent* to B if $B = S^{\mathrm{T}}AS$ for some orthogonal matrix S. Every square matrix, for example, is orthogonally equivalent to a matrix having equal entries along the main diagonal. [208, §2.2, prob.3]

[B.9]The product $Q^{\mathrm{T}}AQ$ can be regarded as a coordinate transformation; *e.g.*, given linear map $y = Ax : \mathbb{R}^n \to \mathbb{R}^n$ and orthogonal Q, the transformation $Qy = AQx$ is a rotation/reflection of range and rowspace (141) of matrix A where $Qy \in \mathcal{R}(A)$ and $Qx \in \mathcal{R}(A^{\mathrm{T}})$ (142).

Appendix C

Some analytical optimal results

People have been working on Optimization since the ancient Greeks [Zenodorus, circa 200BC] learned that a string encloses the most area when it is formed into the shape of a circle.

−ROMAN POLYAK

We speculate that optimization problems possessing analytical solution have convex transformation or constructive global optimality conditions, perhaps yet unknown; *e.g.*, §7.1.4, (1791), §C.3.0.1.

C.1   Properties of infima

-

$$\begin{aligned}
\inf \emptyset &\triangleq \ \ \infty \\
\sup \emptyset &\triangleq -\infty
\end{aligned} \tag{1763}$$

- Given $f(x) : \mathcal{X} \rightarrow \mathbb{R}$ defined on arbitrary set $\mathcal{X}$ [205, §0.1.2]

$$\begin{aligned}
\inf_{x \in \mathcal{X}} f(x) &= - \sup_{x \in \mathcal{X}} -f(x) \\
\sup_{x \in \mathcal{X}} f(x) &= - \inf_{x \in \mathcal{X}} -f(x)
\end{aligned} \tag{1764}$$

$$\begin{aligned}
\arg\inf_{x \in \mathcal{X}} f(x) &= \arg\sup_{x \in \mathcal{X}} -f(x) \\
\arg\sup_{x \in \mathcal{X}} f(x) &= \arg\inf_{x \in \mathcal{X}} -f(x)
\end{aligned} \tag{1765}$$

- Given scalar κ and $f(x) : \mathcal{X} \rightarrow \mathbb{R}$ and $g(x) : \mathcal{X} \rightarrow \mathbb{R}$ defined on arbitrary set $\mathcal{X}$ [205, §0.1.2]

$$\begin{aligned}
\inf_{x \in \mathcal{X}} (\kappa + f(x)) &= \kappa + \inf_{x \in \mathcal{X}} f(x) \\
\arg\inf_{x \in \mathcal{X}} (\kappa + f(x)) &= \arg\inf_{x \in \mathcal{X}} f(x)
\end{aligned} \tag{1766}$$

563
CITATION: Dattorro, *Convex Optimization & Euclidean Distance Geometry*, $\mathcal{M}\varepsilon\beta oo$ Publishing USA, 2005, v2015.02.15.

$$\left.\begin{array}{r} \inf\limits_{x\in\mathcal{X}} \kappa\, f(x) = \kappa \inf\limits_{x\in\mathcal{X}} f(x) \\[2mm] \arg\inf\limits_{x\in\mathcal{X}} \kappa\, f(x) = \arg\inf\limits_{x\in\mathcal{X}} f(x) \end{array}\right\}, \quad \kappa > 0 \tag{1767}$$

$$\inf_{x\in\mathcal{X}} (f(x) + g(x)) \geq \inf_{x\in\mathcal{X}} f(x) + \inf_{x\in\mathcal{X}} g(x) \tag{1768}$$

- Given $f(x) : \mathcal{X} \to \mathbb{R}$ defined on arbitrary set $\mathcal{X}$

$$\arg\inf_{x\in\mathcal{X}} |f(x)| = \arg\inf_{x\in\mathcal{X}} f(x)^2 \tag{1769}$$

- Given $f(x) : \mathcal{X} \cup \mathcal{Y} \to \mathbb{R}$ and arbitrary sets $\mathcal{X}$ and $\mathcal{Y}$ [205, §0.1.2]

$$\mathcal{X} \subset \mathcal{Y} \Rightarrow \inf_{x\in\mathcal{X}} f(x) \geq \inf_{x\in\mathcal{Y}} f(x) \tag{1770}$$

$$\inf_{x\in\mathcal{X}\cup\mathcal{Y}} f(x) = \min\{\inf_{x\in\mathcal{X}} f(x),\ \inf_{x\in\mathcal{Y}} f(x)\} \tag{1771}$$

$$\inf_{x\in\mathcal{X}\cap\mathcal{Y}} f(x) \geq \max\{\inf_{x\in\mathcal{X}} f(x),\ \inf_{x\in\mathcal{Y}} f(x)\} \tag{1772}$$

C.2 Trace, singular and eigen values

- For $A \in \mathbb{R}^{m\times n}$ and $\sigma(A)$ denoting its singular values, the *nuclear norm* (Ky Fan norm) $\|A\|_2^*$ of matrix A (*confer* (44), (1647), [209, p.200]) is

$$\sum_i \sigma(A)_i = \operatorname{tr}\sqrt{A^{\mathrm{T}}A} = \|A\|_2^* = \sup_{\|X\|_2\leq 1} \operatorname{tr}(X^{\mathrm{T}}A) = \begin{array}{c} \text{maximize} \\ X\in\mathbb{R}^{m\times n} \end{array} \operatorname{tr}(X^{\mathrm{T}}A) \tag{1773}$$

$$\text{subject to} \quad \begin{bmatrix} I & X \\ X^{\mathrm{T}} & I \end{bmatrix} \succeq 0$$

$$= \tfrac{1}{2} \begin{array}{c} \text{minimize} \\ X\in\mathbb{S}^m,\ Y\in\mathbb{S}^n \end{array} \operatorname{tr} X + \operatorname{tr} Y$$

$$\text{subject to} \quad \begin{bmatrix} X & A \\ A^{\mathrm{T}} & Y \end{bmatrix} \succeq 0$$

This nuclear norm is convex[C.1] and dual to spectral norm. [209, p.214] [62, §A.1.6] Given singular value decomposition $A = S\Sigma Q^{\mathrm{T}} \in \mathbb{R}^{m\times n}$ (A.6), then $X^\star = SQ^{\mathrm{T}} \in \mathbb{R}^{m\times n}$ is an optimal solution to maximization (*confer* §2.3.2.0.5) while $X^\star = S\Sigma S^{\mathrm{T}} \in \mathbb{S}^m$ and $Y^\star = Q\Sigma Q^{\mathrm{T}} \in \mathbb{S}^n$ is an optimal solution to minimization [139]. Srebro [330] asserts

$$\begin{aligned} \sum_i \sigma(A)_i &= \tfrac{1}{2} \begin{array}{c} \text{minimize} \\ U,V \end{array} \quad \|U\|_{\mathrm{F}}^2 + \|V\|_{\mathrm{F}}^2 \\ &\qquad \text{subject to} \quad A = UV^{\mathrm{T}} \\[2mm] &= \begin{array}{c} \text{minimize} \\ U,V \end{array} \quad \|U\|_{\mathrm{F}}\|V\|_{\mathrm{F}} \\ &\qquad \text{subject to} \quad A = UV^{\mathrm{T}} \end{aligned} \tag{1774}$$

[C.1] discernible as envelope of the rank function (1443) or as supremum of functions linear in A (Figure 75).

- For $A \in \mathbb{R}^{m \times n}$ and $\sigma(A)_1$ connoting spectral norm,

$$
\sigma(A)_1 = \sqrt{\lambda(A^{\mathrm{T}}A)_1} = \|A\|_2 = \sup_{\|x\|=1} \|Ax\|_2 = \begin{array}{c} \underset{t \in \mathbb{R}}{\text{minimize}} \quad t \\ \text{subject to} \quad \begin{bmatrix} tI & A \\ A^{\mathrm{T}} & tI \end{bmatrix} \succeq 0 \end{array} \tag{591}
$$

By confining dyad uv^{T} to the unit nuclear norm ball (94),

$$
\sigma(A)_1 = \|A\|_2 = \sup_{\|u\|=1,\,\|v\|=1} u^{\mathrm{T}}Av = \begin{array}{c} \underset{Z \in \mathbb{R}^{m \times n},\, X \in \mathbb{S}^m,\, Y \in \mathbb{S}^n}{\text{maximize}} \quad \mathrm{tr}(Z^{\mathrm{T}}A) \\ \text{subject to} \quad \mathrm{tr}\,X + \mathrm{tr}\,Y \leq 2 \\ \begin{bmatrix} X & Z \\ Z^{\mathrm{T}} & Y \end{bmatrix} \succeq 0 \end{array} \tag{1775}
$$

with corresponding left and right singular vectors (optimal) $u^{\star}$ and $v^{\star}$. Applying (1782) to a result of Lanczos [163, p.207],

$$
\sigma(A)_1 = \|A\|_2 = \sup_{\left\| \begin{bmatrix} u \\ v \end{bmatrix} \right\| = 1} \begin{bmatrix} u \\ v \end{bmatrix}^{\mathrm{T}} \begin{bmatrix} \mathbf{0} & A \\ A^{\mathrm{T}} & \mathbf{0} \end{bmatrix} \begin{bmatrix} u \\ v \end{bmatrix} = \begin{array}{c} \underset{X \in \mathbb{S}_+^{m+n}}{\text{maximize}} \quad \mathrm{tr}\left(X \begin{bmatrix} \mathbf{0} & A \\ A^{\mathrm{T}} & \mathbf{0} \end{bmatrix} \right) \\ \text{subject to} \quad \mathrm{tr}\,X = 1 \end{array} \tag{1776}
$$

$$
= \begin{array}{c} \underset{t \in \mathbb{R}}{\text{minimize}} \quad t \\ \text{subject to} \quad \begin{bmatrix} \mathbf{0} & A \\ A^{\mathrm{T}} & \mathbf{0} \end{bmatrix} \preceq tI \end{array}
$$

whose corresponding left and right singular vectors are $\sqrt{2}u^{\star}$ and $\sqrt{2}v^{\star}$.

C.2.0.0.1 Exercise. *Optimal matrix factorization.*
Prove (1774).[C.2] ▼

- For $X \in \mathbb{S}^m$, $Y \in \mathbb{S}^n$, $A \in \mathcal{C} \subseteq \mathbb{R}^{m \times n}$ for set $\mathcal{C}$ convex, and $\sigma(A)$ denoting the singular values of A [139, §3],

$$
\begin{array}{c} \underset{A}{\text{minimize}} \quad \sum_i \sigma(A)_i \\ \text{subject to} \quad A \in \mathcal{C} \end{array} \equiv \begin{array}{c} \frac{1}{2} \underset{A,X,Y}{\text{minimize}} \quad \mathrm{tr}\,X + \mathrm{tr}\,Y \\ \text{subject to} \quad \begin{bmatrix} X & A \\ A^{\mathrm{T}} & Y \end{bmatrix} \succeq 0 \\ A \in \mathcal{C} \end{array} \tag{1777}
$$

For feasible set $\mathcal{C}$ equal to the unit nuclear norm ball (94),

$$
\begin{array}{c} \text{find} \quad A \\ \text{subject to} \quad A \in \{Z \in \mathbb{R}^{m \times n} \mid \sum_i \sigma(Z)_i \leq 1\} \end{array} \equiv \begin{array}{c} \underset{A,X,Y}{\text{find}} \quad A \\ \text{subject to} \quad \mathrm{tr}\,X + \mathrm{tr}\,Y \leq 2 \\ \begin{bmatrix} X & A \\ A^{\mathrm{T}} & Y \end{bmatrix} \succeq 0 \end{array} \tag{1778}
$$

[C.2]Hint: Write $A = S\Sigma Q^{\mathrm{T}} \in \mathbb{R}^{m \times n}$ and

$$
\begin{bmatrix} X & A \\ A^{\mathrm{T}} & Y \end{bmatrix} = \begin{bmatrix} U \\ V \end{bmatrix} \begin{bmatrix} U^{\mathrm{T}} & V^{\mathrm{T}} \end{bmatrix} \succeq 0
$$

Show $U^{\star} = S\sqrt{\Sigma} \in \mathbb{R}^{m \times \min\{m,n\}}$ and $V^{\star} = Q\sqrt{\Sigma} \in \mathbb{R}^{n \times \min\{m,n\}}$, hence $\|U^{\star}\|_{\mathrm{F}}^2 = \|V^{\star}\|_{\mathrm{F}}^2$.

- For $A \in \mathbb{S}^N_+$ and $\beta \in \mathbb{R}$

$$
\beta \operatorname{tr} A = \begin{array}{l} \underset{X \in \mathbb{S}^N}{\text{maximize}} \quad \operatorname{tr}(XA) \\ \text{subject to} \quad X \preceq \beta I \end{array} \tag{1779}
$$

But the following statement is numerically stable, preventing an unbounded solution in direction of a 0 eigenvalue:

$$
\begin{array}{l} \underset{X \in \mathbb{S}^N}{\text{maximize}} \quad \operatorname{sgn}(\beta) \operatorname{tr}(XA) \\ \text{subject to} \quad X \preceq |\beta| I \\ \qquad\qquad\;\; X \succeq -|\beta| I \end{array} \tag{1780}
$$

where $\beta \operatorname{tr} A = \operatorname{tr}(X^\star A)$. If $\beta \geq 0$, then $(X \succeq -|\beta| I) \leftarrow (X \succeq 0)$.

- For symmetric $A \in \mathbb{S}^N$, its smallest and largest eigenvalue in $\lambda(A) \in \mathbb{R}^N$ are respectively [11, §4.1] [43, §I.6.15] [208, §4.2] [245, §2.1] [246]

$$
\underset{i}{\min}\{\lambda(A)_i\} = \underset{\|x\|=1}{\inf} x^{\mathrm{T}} A x = \begin{array}{l} \underset{X \in \mathbb{S}^N_+}{\text{minimize}} \quad \operatorname{tr}(XA) \\ \text{subject to} \quad \operatorname{tr} X = 1 \end{array} = \begin{array}{l} \underset{t \in \mathbb{R}}{\text{maximize}} \quad t \\ \text{subject to} \quad A \succeq t I \end{array} \tag{1781}
$$

$$
\underset{i}{\max}\{\lambda(A)_i\} = \underset{\|x\|=1}{\sup} x^{\mathrm{T}} A x = \begin{array}{l} \underset{X \in \mathbb{S}^N_+}{\text{maximize}} \quad \operatorname{tr}(XA) \\ \text{subject to} \quad \operatorname{tr} X = 1 \end{array} = \begin{array}{l} \underset{t \in \mathbb{R}}{\text{minimize}} \quad t \\ \text{subject to} \quad A \preceq t I \end{array} \tag{1782}
$$

whereas

$$
\lambda_N I \preceq A \preceq \lambda_1 I \tag{1783}
$$

The largest eigenvalue λ_1 is always convex in $A \in \mathbb{S}^N$ because, given particular x, $x^{\mathrm{T}} A x$ is linear in matrix A; supremum of a family of linear functions is convex, as illustrated in Figure 75. So for $A, B \in \mathbb{S}^N$, $\lambda_1(A + B) \leq \lambda_1(A) + \lambda_1(B)$. (1585) Similarly, the smallest eigenvalue λ_N of any symmetric matrix is a concave function of its entries; $\lambda_N(A + B) \geq \lambda_N(A) + \lambda_N(B)$. (1585) For v_1 a normalized eigenvector of A corresponding to the largest eigenvalue, and v_N a normalized eigenvector corresponding to the smallest eigenvalue,

$$
v_N = \arg \underset{\|x\|=1}{\inf} x^{\mathrm{T}} A x \tag{1784}
$$

$$
v_1 = \arg \underset{\|x\|=1}{\sup} x^{\mathrm{T}} A x \tag{1785}
$$

- For $A \in \mathbb{S}^N$ having eigenvalues $\lambda(A) \in \mathbb{R}^N$, consider the unconstrained nonconvex optimization that is a projection on the rank-1 subset (§2.9.2.1, §3.6.0.0.1) of the boundary of positive semidefinite cone $\mathbb{S}^N_+$: Defining $\lambda_1 \triangleq \max_i\{\lambda(A)_i\}$ and corresponding eigenvector v_1

$$
\underset{x}{\text{minimize}}\; \|xx^{\mathrm{T}} - A\|_{\mathrm{F}}^2 = \underset{x}{\text{minimize}}\; \operatorname{tr}(xx^{\mathrm{T}}(x^{\mathrm{T}}x) - 2Axx^{\mathrm{T}} + A^{\mathrm{T}}A)
$$

$$
= \begin{cases} \|\lambda(A)\|^2, & \lambda_1 \leq 0 \\ \|\lambda(A)\|^2 - \lambda_1^2, & \lambda_1 > 0 \end{cases} \tag{1786}
$$

$$\arg\underset{x}{\text{minimize}} \ \|xx^{\mathrm{T}} - A\|_{\mathrm{F}}^{2} = \begin{cases} \mathbf{0} \ , & \lambda_1 \leq 0 \\ v_1\sqrt{\lambda_1} \ , & \lambda_1 > 0 \end{cases} \tag{1787}$$

Proof. This is simply the Eckart & Young solution from §7.1.2:

$$x^{\star}x^{\star\mathrm{T}} = \begin{cases} \mathbf{0} \ , & \lambda_1 \leq 0 \\ \lambda_1 \, v_1 v_1^{\mathrm{T}} \ , & \lambda_1 > 0 \end{cases} \tag{1788}$$

Given nonincreasingly ordered diagonalization $A = Q\Lambda Q^{\mathrm{T}}$ where $\Lambda = \delta(\lambda(A))$ (§A.5), then (1786) has minimum value

$$\underset{x}{\text{minimize}} \ \|xx^{\mathrm{T}}{-}A\|_{\mathrm{F}}^{2} = \begin{cases} \|Q\Lambda Q^{\mathrm{T}}\|_{\mathrm{F}}^{2} \ = \ \|\delta(\Lambda)\|^{2} \ , & \lambda_1 \leq 0 \\[3ex] \left\| Q\left(\begin{bmatrix} \lambda_1 & & \\ & 0 & \\ & & \ddots \\ & & & 0 \end{bmatrix} - \Lambda\right)Q^{\mathrm{T}}\right\|_{\mathrm{F}}^{2} = \left\|\begin{bmatrix} \lambda_1 \\ 0 \\ \vdots \\ 0 \end{bmatrix} - \delta(\Lambda)\right\|^{2} \ , & \lambda_1 > 0 \end{cases} \tag{1789}$$

$\blacklozenge$

C.2.0.0.2 Exercise. *Rank-1 approximation.*
Given symmetric matrix $A \in \mathbb{S}^{N}$, prove:

$$\begin{aligned} v_1 \ = \ &\arg\underset{x}{\text{minimize}} && \|xx^{\mathrm{T}} - A\|_{\mathrm{F}}^{2} \\ &\text{subject to} && \|x\| = 1 \end{aligned} \tag{1790}$$

where v_1 is a normalized eigenvector of A corresponding to its largest eigenvalue. What is the objective's optimal value? $\blacktriangledown$

- (Ky Fan, 1949) For $B \in \mathbb{S}^{N}$ whose eigenvalues $\lambda(B) \in \mathbb{R}^{N}$ are arranged in nonincreasing order, and for $1 \leq k \leq N$ [11, §4.1] [221] [208, §4.3.18] [368, §2] [245, §2.1] [246]

$$\begin{aligned} \sum_{i=N-k+1}^{N} \lambda(B)_i \ = \ \underset{\substack{U \in \mathbb{R}^{N \times k} \\ U^{\mathrm{T}}U = I}}{\inf} \ \text{tr}(UU^{\mathrm{T}}B) = \ &\underset{X \in \mathbb{S}_+^N}{\text{minimize}} && \text{tr}(XB) & \text{(a)} \\ &\text{subject to} && X \preceq I \\ & && \text{tr}\,X = k \\ = \ &\underset{\mu \in \mathbb{R}\,,\, Z \in \mathbb{S}_+^N}{\text{maximize}} && (k-N)\mu + \text{tr}(B-Z) & \text{(b)} \\ &\text{subject to} && \mu I + Z \succeq B \end{aligned} \tag{1791}$$

$$\begin{aligned} \sum_{i=1}^{k} \lambda(B)_i \ = \ \underset{\substack{U \in \mathbb{R}^{N \times k} \\ U^{\mathrm{T}}U = I}}{\sup} \ \text{tr}(UU^{\mathrm{T}}B) = \ &\underset{X \in \mathbb{S}_+^N}{\text{maximize}} && \text{tr}(XB) & \text{(c)} \\ &\text{subject to} && X \preceq I \\ & && \text{tr}\,X = k \\ = \ &\underset{\mu \in \mathbb{R}\,,\, Z \in \mathbb{S}_+^N}{\text{minimize}} && k\mu + \text{tr}\,Z & \text{(d)} \\ &\text{subject to} && \mu I + Z \succeq B \end{aligned}$$

Given ordered diagonalization $B = Q\Lambda Q^{\mathrm{T}}$, (§A.5.1) then an optimal U for the infimum is $U^\star = Q(:, N-k+1:N) \in \mathbb{R}^{N\times k}$ whereas $U^\star = Q(:, 1:k) \in \mathbb{R}^{N\times k}$ for the supremum is more reliably computed. In both cases, $X^\star = U^\star U^{\star\mathrm{T}}$. Optimization (a) searches the convex hull of outer product UU^{T} of all $N \times k$ orthonormal matrices. (§2.3.2.0.1)

- For $B \in \mathbb{S}^N$ whose eigenvalues $\lambda(B) \in \mathbb{R}^N$ are arranged in nonincreasing order, and for diagonal matrix $\Upsilon \in \mathbb{S}^k$ whose diagonal entries are arranged in nonincreasing order where $1 \le k \le N$, we utilize the main-diagonal δ operator's self-adjointness property (1499): [12, §4.2]

$$\sum_{i=1}^{k} \Upsilon_{ii} \lambda(B)_{N-i+1} = \inf_{\substack{U \in \mathbb{R}^{N\times k} \\ U^{\mathrm{T}}U = I}} \mathrm{tr}(\Upsilon U^{\mathrm{T}}BU) = \inf_{\substack{U \in \mathbb{R}^{N\times k} \\ U^{\mathrm{T}}U = I}} \delta(\Upsilon)^{\mathrm{T}}\delta(U^{\mathrm{T}}BU) \tag{1792}$$

$$\begin{aligned} = \underset{V_i \in \mathbb{S}^N}{\text{minimize}} \quad & \mathrm{tr}\left(B\sum_{i=1}^{k}(\Upsilon_{ii}-\Upsilon_{i+1,i+1})V_i\right) \\ \text{subject to} \quad & \mathrm{tr}\,V_i = i\,, && i=1\ldots k \\ & I \succeq V_i \succeq 0\,, && i=1\ldots k \end{aligned}$$

where $\Upsilon_{k+1,k+1} \triangleq 0$. We speculate,

$$\sum_{i=1}^{k} \Upsilon_{ii} \lambda(B)_i = \sup_{\substack{U \in \mathbb{R}^{N\times k} \\ U^{\mathrm{T}}U = I}} \mathrm{tr}(\Upsilon U^{\mathrm{T}}BU) = \sup_{\substack{U \in \mathbb{R}^{N\times k} \\ U^{\mathrm{T}}U = I}} \delta(\Upsilon)^{\mathrm{T}}\delta(U^{\mathrm{T}}BU) \tag{1793}$$

Alizadeh shows: [11, §4.2]

$$\begin{aligned} \sum_{i=1}^{k} \Upsilon_{ii} \lambda(B)_i = \underset{\mu \in \mathbb{R}^k,\, Z_i \in \mathbb{S}^N}{\text{minimize}} \quad & \sum_{i=1}^{k} i\mu_i + \mathrm{tr}\,Z_i \\ \text{subject to} \quad & \mu_i I + Z_i - (\Upsilon_{ii}-\Upsilon_{i+1,i+1})B \succeq 0\,, && i=1\ldots k \\ & Z_i \succeq 0\,, && i=1\ldots k \\[4pt] = \underset{V_i \in \mathbb{S}^N}{\text{maximize}} \quad & \mathrm{tr}\left(B\sum_{i=1}^{k}(\Upsilon_{ii}-\Upsilon_{i+1,i+1})V_i\right) \\ \text{subject to} \quad & \mathrm{tr}\,V_i = i\,, && i=1\ldots k \\ & I \succeq V_i \succeq 0\,, && i=1\ldots k \end{aligned} \tag{1794}$$

where $\Upsilon_{k+1,k+1} \triangleq 0$.

- The largest eigenvalue magnitude μ of $A \in \mathbb{S}^N$

$$\begin{aligned} \max_i \{|\lambda(A)_i|\} = \underset{\mu \in \mathbb{R}}{\text{minimize}} \quad & \mu \\ \text{subject to} \quad & -\mu I \preceq A \preceq \mu I \end{aligned} \tag{1795}$$

is minimized over convex set $\mathcal{C}$ by semidefinite program: (*confer* §7.1.5)

$$
\begin{array}{ll}
\underset{A}{\text{minimize}} & \|A\|_2 \\
\text{subject to} & A \in \mathcal{C}
\end{array}
\quad \equiv \quad
\begin{array}{ll}
\underset{\mu,\,A}{\text{minimize}} & \mu \\
\text{subject to} & -\mu I \preceq A \preceq \mu I \\
& A \in \mathcal{C}
\end{array}
\tag{1796}
$$

id est,

$$
\mu^\star \triangleq \max_i \{\, |\lambda(A^\star)_i| \,,\; i = 1 \ldots N \,\} \in \mathbb{R}_+
\tag{1797}
$$

- For $B \in \mathbb{S}^N$ whose eigenvalues $\lambda(B) \in \mathbb{R}^N$ are arranged in nonincreasing order, let $\Pi\,\lambda(B)$ be a permutation of eigenvalues $\lambda(B)$ such that their absolute value becomes arranged in nonincreasing order: $|\Pi\,\lambda(B)|_1 \geq |\Pi\,\lambda(B)|_2 \geq \cdots \geq |\Pi\,\lambda(B)|_N$. Then, for $1 \leq k \leq N$ [11, §4.3][C.3]

$$
\sum_{i=1}^{k} |\Pi\,\lambda(B)|_i \;=\;
\begin{array}{ll}
\underset{\mu \in \mathbb{R},\, Z \in \mathbb{S}^N_+}{\text{minimize}} & k\mu + \operatorname{tr} Z \\
\text{subject to} & \mu I + Z + B \succeq 0 \\
& \mu I + Z - B \succeq 0
\end{array}
\;=\;
\begin{array}{ll}
\underset{V,\,W \in \mathbb{S}^N_+}{\text{maximize}} & \langle B,\, V - W \rangle \\
\text{subject to} & I \succeq V,\, W \\
& \operatorname{tr}(V + W) = k
\end{array}
\tag{1798}
$$

For diagonal matrix $\Upsilon \in \mathbb{S}^k$ whose diagonal entries are arranged in nonincreasing order where $1 \leq k \leq N$

$$
\sum_{i=1}^{k} \Upsilon_{ii} |\Pi\,\lambda(B)|_i \;=\;
\begin{array}{ll}
\underset{\mu \in \mathbb{R}^k,\, Z_i \in \mathbb{S}^N}{\text{minimize}} & \sum_{i=1}^{k} i\mu_i + \operatorname{tr} Z_i \\
\text{subject to} & \mu_i I + Z_i + (\Upsilon_{ii} - \Upsilon_{i+1,i+1})B \succeq 0, \quad i = 1 \ldots k \\
& \mu_i I + Z_i - (\Upsilon_{ii} - \Upsilon_{i+1,i+1})B \succeq 0, \quad i = 1 \ldots k \\
& Z_i \succeq 0, \qquad\qquad\qquad\qquad\qquad\quad\; i = 1 \ldots k
\end{array}
$$

$$
=\;
\begin{array}{ll}
\underset{V_i,\,W_i \in \mathbb{S}^N}{\text{maximize}} & \operatorname{tr}\!\left(B \sum_{i=1}^{k} (\Upsilon_{ii} - \Upsilon_{i+1,i+1})(V_i - W_i) \right) \\
\text{subject to} & \operatorname{tr}(V_i + W_i) = i, \qquad i = 1 \ldots k \\
& I \succeq V_i \succeq 0, \qquad\qquad i = 1 \ldots k \\
& I \succeq W_i \succeq 0, \qquad\qquad i = 1 \ldots k
\end{array}
\tag{1799}
$$

where $\Upsilon_{k+1,k+1} \triangleq 0$.

C.2.0.0.3 Exercise. *Weighted sum of largest eigenvalues.*
Prove (1793). ▼

[C.3]We eliminate a redundant positive semidefinite variable from Alizadeh's minimization. There exist typographical errors in [300, (6.49) (6.55)] for this minimization.

- For $A, B \in \mathbb{S}^N$ whose eigenvalues $\lambda(A), \lambda(B) \in \mathbb{R}^N$ are respectively arranged in nonincreasing order, and for nonincreasingly ordered diagonalizations $A = W_A \Upsilon W_A^T$ and $B = W_B \Lambda W_B^T$ [206] [245, §2.1] [246]

$$\lambda(A)^T \lambda(B) \;=\; \sup_{\substack{U \in \mathbb{R}^{N \times N} \\ U^T U = I}} \mathrm{tr}(A^T U^T B U) \;\geq\; \mathrm{tr}(A^T B) \qquad (1818)$$

($confer\,(1823)$) where optimal U is

$$U^\star \;=\; W_B W_A^T \in \mathbb{R}^{N \times N} \qquad (1815)$$

We can push that upper bound higher using a result in §C.4.2.1:

$$|\lambda(A)|^T |\lambda(B)| \;=\; \sup_{\substack{U \in \mathbb{C}^{N \times N} \\ U^H U = I}} \mathrm{re}\,\mathrm{tr}(A^T U^H B U) \qquad (1800)$$

For step function ψ as defined in (1660), optimal U becomes

$$U^\star = W_B \sqrt{\delta(\psi(\delta(\Lambda)))}^{\,H} \sqrt{\delta(\psi(\delta(\Upsilon)))}\, W_A^T \in \mathbb{C}^{N \times N} \qquad (1801)$$

C.3   Orthogonal Procrustes problem

Given matrices $A, B \in \mathbb{R}^{n \times N}$, their product having full singular value decomposition (§A.6.3)

$$A B^T \triangleq U \Sigma Q^T \in \mathbb{R}^{n \times n} \qquad (1802)$$

then an optimal solution $R^\star$ to the orthogonal Procrustes problem

$$\begin{array}{ll} \underset{R}{\text{minimize}} & \|A - R^T B\|_F \\ \text{subject to} & R^T = R^{-1} \end{array} \qquad (1803)$$

maximizes $\mathrm{tr}(A^T R^T B)$ over the nonconvex manifold of orthogonal matrices: [208, §7.4.8]

$$R^\star = Q U^T \in \mathbb{R}^{n \times n} \qquad (1804)$$

A necessary and sufficient condition for optimality

$$A B^T R^\star \succeq 0 \qquad (1805)$$

holds whenever $R^\star$ is an orthogonal matrix. [170, §4]

Optimal solution $R^\star$ can reveal rotation/reflection (§5.5.2, §B.5) of one list in the columns of matrix A with respect to another list in B. Solution is unique if $\mathrm{rank}\, B V_{\mathcal{N}} = n$. [116, §2.4.1] In the case that A is a vector and permutation of B, solution $R^\star$ is not necessarily a permutation matrix (§4.6.0.0.3) although the optimal objective will be zero. More generally, the optimal value for objective of minimization is

$$\begin{aligned} \mathrm{tr}\big(A^T A + B^T B - 2 A B^T R^\star\big) &= \mathrm{tr}(A^T A) + \mathrm{tr}(B^T B) - 2\,\mathrm{tr}(U \Sigma U^T) \\ &= \|A\|_F^2 + \|B\|_F^2 - 2\delta(\Sigma)^T \mathbf{1} \end{aligned} \qquad (1806)$$

while the optimal value for corresponding trace maximization is

$$\sup_{R^T=R^{-1}} \text{tr}(A^TR^TB) = \text{tr}(A^TR^{\star T}B) = \delta(\Sigma)^T\mathbf{1} \geq \text{tr}(A^TB) \qquad (1807)$$

The same optimal solution $R^\star$ solves

$$\begin{array}{ll} \underset{R}{\text{maximize}} & \|A + R^TB\|_F \\ \text{subject to} & R^T = R^{-1} \end{array} \qquad (1808)$$

C.3.0.1 Procrustes relaxation

By replacing its feasible set with the convex hull of orthogonal matrices (Example 2.3.2.0.5), we relax Procrustes problem (1803) to a convex problem

$$\begin{array}{lll} \underset{R}{\text{minimize}} \quad \|A - R^TB\|_F^2 & = & \text{tr}(A^TA + B^TB) - 2\underset{R}{\text{maximize}} \quad \text{tr}(A^TR^TB) \\ \text{subject to} \quad R^T = R^{-1} & & \text{subject to} \quad \begin{bmatrix} I & R \\ R^T & I \end{bmatrix} \succeq 0 \end{array} \qquad (1809)$$

whose adjusted objective must always equal Procrustes'.[C.4]

C.3.1 Effect of translation

Consider the impact on problem (1803) of DC offset in known lists $A, B \in \mathbb{R}^{n \times N}$. Rotation of B there is with respect to the origin, so better results may be obtained if offset is first accounted. Because the geometric centers of the lists AV and BV are the origin, instead we solve

$$\begin{array}{ll} \underset{R}{\text{minimize}} & \|AV - R^TBV\|_F \\ \text{subject to} & R^T = R^{-1} \end{array} \qquad (1810)$$

where $V \in \mathbb{S}^N$ is the geometric centering matrix (§B.4.1). Now we define the full singular value decomposition

$$AVB^T \triangleq U\Sigma Q^T \in \mathbb{R}^{n \times n} \qquad (1811)$$

and an optimal rotation matrix

$$R^\star = QU^T \in \mathbb{R}^{n \times n} \qquad (1804)$$

The desired result is an optimally rotated offset list

$$R^{\star T}BV + A(I - V) \approx A \qquad (1812)$$

which most closely matches the list in A. Equality is attained when the lists are precisely related by a rotation/reflection and an offset. When $R^{\star T}B = A$ or $B\mathbf{1} = A\mathbf{1} = \mathbf{0}$, this result (1812) reduces to $R^{\star T}B \approx A$.

[C.4] (because orthogonal matrices are the extreme points of this hull) and whose optimal numerical solution (SDPT3 [357]) [173] is reliably observed to be orthogonal for $n \leq N$.

C.3.1.1   Translation of extended list

Suppose an optimal rotation matrix $R^\star \in \mathbb{R}^{n \times n}$ were derived as before from matrix $B \in \mathbb{R}^{n \times N}$, but B is part of a larger list in the columns of $[\,C \;\; B\,] \in \mathbb{R}^{n \times M+N}$ where $C \in \mathbb{R}^{n \times M}$. In that event, we wish to apply the rotation/reflection and translation to the larger list. The expression supplanting the approximation in (1812) makes $\mathbf{1}^{\mathrm{T}}$ of compatible dimension;

$$R^{\star \mathrm{T}}[\,C - B\mathbf{11}^{\mathrm{T}}\tfrac{1}{N} \quad BV\,] \;+\; A\mathbf{11}^{\mathrm{T}}\tfrac{1}{N} \tag{1813}$$

id est, $C - B\mathbf{11}^{\mathrm{T}}\tfrac{1}{N} \in \mathbb{R}^{n \times M}$ and $A\mathbf{11}^{\mathrm{T}}\tfrac{1}{N} \in \mathbb{R}^{n \times M+N}$.

C.4   Two-sided orthogonal Procrustes

C.4.0.1   Minimization

Given symmetric $A, B \in \mathbb{S}^N$, each having diagonalization (§A.5.1)

$$A \triangleq Q_A \Lambda_A Q_A^{\mathrm{T}} , \qquad B \triangleq Q_B \Lambda_B Q_B^{\mathrm{T}} \tag{1814}$$

where eigenvalues are arranged in their respective diagonal matrix Λ in nonincreasing order, then an optimal solution [134]

$$R^\star = Q_B Q_A^{\mathrm{T}} \in \mathbb{R}^{N \times N} \tag{1815}$$

to the two-sided orthogonal Procrustes problem

$$\begin{array}{ll} \underset{R}{\text{minimize}} & \|A - R^{\mathrm{T}}BR\|_{\mathrm{F}} \\ \text{subject to} & R^{\mathrm{T}} = R^{-1} \end{array} \;=\; \begin{array}{ll} \underset{R}{\text{minimize}} & \mathrm{tr}\big(A^{\mathrm{T}}A - 2A^{\mathrm{T}}R^{\mathrm{T}}BR + B^{\mathrm{T}}B\big) \\ \text{subject to} & R^{\mathrm{T}} = R^{-1} \end{array} \tag{1816}$$

maximizes $\mathrm{tr}(A^{\mathrm{T}}R^{\mathrm{T}}BR)$ over the nonconvex manifold of orthogonal matrices. Optimal product $R^{\star \mathrm{T}}BR^\star$ has the eigenvectors of A but the eigenvalues of B. [170, §7.5.1] The optimal value for the objective of minimization is, by (48)

$$\|Q_A \Lambda_A Q_A^{\mathrm{T}} - R^{\star \mathrm{T}} Q_B \Lambda_B Q_B^{\mathrm{T}} R^\star\|_{\mathrm{F}} \;=\; \|Q_A(\Lambda_A - \Lambda_B)Q_A^{\mathrm{T}}\|_{\mathrm{F}} \;=\; \|\Lambda_A - \Lambda_B\|_{\mathrm{F}} \tag{1817}$$

while the corresponding trace maximization has optimal value

$$\sup_{R^{\mathrm{T}}=R^{-1}} \mathrm{tr}(A^{\mathrm{T}}R^{\mathrm{T}}BR) \;=\; \mathrm{tr}(A^{\mathrm{T}}R^{\star \mathrm{T}}BR^\star) \;=\; \mathrm{tr}(\Lambda_A \Lambda_B) \;\geq\; \mathrm{tr}(A^{\mathrm{T}}B) \tag{1818}$$

The lower bound on inner product of eigenvalues is due to Fan (p.522).

C.4.0.2   Maximization

Any permutation matrix is an orthogonal matrix. Defining a row- and column-swapping permutation matrix (a reflection matrix, §B.5.3)

$$\Xi \;=\; \Xi^{\mathrm{T}} \;=\; \begin{bmatrix} \mathbf{0} & & & 1 \\ & & \cdot & \\ & 1 & & \\ 1 & & & \mathbf{0} \end{bmatrix} \tag{1819}$$

then an optimal solution $R^\star$ to the maximization problem [*sic*]

$$
\begin{array}{ll}
\underset{R}{\text{maximize}} & \|A - R^{\mathrm{T}}BR\|_{\mathrm{F}} \\
\text{subject to} & R^{\mathrm{T}} = R^{-1}
\end{array}
\tag{1820}
$$

minimizes $\operatorname{tr}(A^{\mathrm{T}}R^{\mathrm{T}}BR)$: [206] [245, §2.1] [246]

$$
R^\star = Q_B \Xi Q_A^{\mathrm{T}} \in \mathbb{R}^{N \times N}
\tag{1821}
$$

The optimal value for the objective of maximization is

$$
\begin{aligned}
\|Q_A \Lambda_A Q_A^{\mathrm{T}} - R^{\star\mathrm{T}} Q_B \Lambda_B Q_B^{\mathrm{T}} R^\star\|_{\mathrm{F}} &= \|Q_A \Lambda_A Q_A^{\mathrm{T}} - Q_A \Xi^{\mathrm{T}} \Lambda_B \Xi Q_A^{\mathrm{T}}\|_{\mathrm{F}} \\
&= \|\Lambda_A - \Xi \Lambda_B \Xi\|_{\mathrm{F}}
\end{aligned}
\tag{1822}
$$

while the corresponding trace minimization has optimal value

$$
\inf_{R^{\mathrm{T}} = R^{-1}} \operatorname{tr}(A^{\mathrm{T}}R^{\mathrm{T}}BR) = \operatorname{tr}(A^{\mathrm{T}}R^{\star\mathrm{T}}BR^\star) = \operatorname{tr}(\Lambda_A \Xi \Lambda_B \Xi)
\tag{1823}
$$

C.4.1 Procrustes' relation to linear programming

Although these two-sided Procrustes problems are nonconvex, there is a connection with linear programming [12, §3] [245, §2.1] [246]: Given $A, B \in \mathbb{S}^N$, this semidefinite program in S and T

$$
\begin{array}{llll}
\underset{R}{\text{minimize}} & \operatorname{tr}(A^{\mathrm{T}}R^{\mathrm{T}}BR) &=& \underset{S,\,T \in \mathbb{S}^N}{\text{maximize}} \quad \operatorname{tr}(S + T) \\
\text{subject to} & R^{\mathrm{T}} = R^{-1} & & \text{subject to} \quad A^{\mathrm{T}} \otimes B - I \otimes S - T \otimes I \succeq 0
\end{array}
\tag{1824}
$$

(where $\otimes$ signifies Kronecker product (§D.1.2.1)) has optimal objective value (1823). These two problems in (1824) are strong duals (§2.13.1.0.3). Given ordered diagonalizations (1814), make the observation:

$$
\inf_R \operatorname{tr}(A^{\mathrm{T}}R^{\mathrm{T}}BR) = \inf_{\hat{R}} \operatorname{tr}(\Lambda_A \hat{R}^{\mathrm{T}} \Lambda_B \hat{R})
\tag{1825}
$$

because $\hat{R} \triangleq Q_B^{\mathrm{T}} R Q_A$ on the set of orthogonal matrices (which includes the permutation matrices) is a bijection. This means, basically, diagonal matrices of eigenvalues Λ_A and Λ_B may be substituted for A and B, so only the main diagonals of S and T come into play;

$$
\begin{array}{ll}
\underset{S,T \in \mathbb{S}^N}{\text{maximize}} & \mathbf{1}^{\mathrm{T}} \delta(S + T) \\
\text{subject to} & \delta(\Lambda_A \otimes (\Xi \Lambda_B \Xi) - I \otimes S - T \otimes I) \succeq 0
\end{array}
\tag{1826}
$$

a linear program in $\delta(S)$ and $\delta(T)$ having the same optimal objective value as the semidefinite program (1824).

We relate their results to Procrustes problem (1816) by manipulating signs (1764) and permuting eigenvalues:

$$
\begin{array}{llll}
\underset{R}{\text{maximize}} & \operatorname{tr}(A^{\mathrm{T}}R^{\mathrm{T}}BR) &=& \underset{S,\,T \in \mathbb{S}^N}{\text{minimize}} \quad \mathbf{1}^{\mathrm{T}} \delta(S + T) \\
\text{subject to} & R^{\mathrm{T}} = R^{-1} & & \text{subject to} \quad \delta(I \otimes S + T \otimes I - \Lambda_A \otimes \Lambda_B) \succeq 0 \\
\\
&&=& \underset{S,\,T \in \mathbb{S}^N}{\text{minimize}} \quad \operatorname{tr}(S + T) \\
&&& \text{subject to} \quad I \otimes S + T \otimes I - A^{\mathrm{T}} \otimes B \succeq 0
\end{array}
\tag{1827}
$$

This formulation has optimal objective value identical to that in (1818).

C.4.2   Two-sided orthogonal Procrustes via SVD

By making left- and right-side orthogonal matrices independent, we can push the upper bound on trace (1818) a little further: Given real matrices A, B each having full singular value decomposition (§A.6.3)

$$A \triangleq U_A \Sigma_A Q_A^{\mathrm{T}} \in \mathbb{R}^{m \times n} \,, \qquad B \triangleq U_B \Sigma_B Q_B^{\mathrm{T}} \in \mathbb{R}^{m \times n} \tag{1828}$$

then a well-known optimal solution $R^\star$, $S^\star$ to the problem

$$\begin{array}{ll}
\underset{R\,,\,S}{\text{minimize}} & \|A - SBR\|_{\mathrm{F}} \\
\text{subject to} & R^{\mathrm{H}} = R^{-1} \\
& S^{\mathrm{H}} = S^{-1}
\end{array} \tag{1829}$$

maximizes $\mathrm{re\,tr}(A^{\mathrm{T}} SBR)$: [322] [296] [53] [200] optimal orthogonal matrices

$$S^\star = U_A U_B^{\mathrm{H}} \in \mathbb{R}^{m \times m} \,, \qquad R^\star = Q_B Q_A^{\mathrm{H}} \in \mathbb{R}^{n \times n} \tag{1830}$$

[*sic*] are not necessarily unique [208, §7.4.13] because the feasible set is not convex. The optimal value for the objective of minimization is, by (48)

$$\|U_A \Sigma_A Q_A^{\mathrm{H}} - S^\star U_B \Sigma_B Q_B^{\mathrm{H}} R^\star\|_{\mathrm{F}} \;=\; \|U_A(\Sigma_A - \Sigma_B)Q_A^{\mathrm{H}}\|_{\mathrm{F}} \;=\; \|\Sigma_A - \Sigma_B\|_{\mathrm{F}} \tag{1831}$$

while the corresponding trace maximization has optimal value [43, §III.6.12]

$$\sup_{\substack{R^{\mathrm{H}}=R^{-1} \\ S^{\mathrm{H}}=S^{-1}}} |\mathrm{tr}(A^{\mathrm{T}} SBR)| \;=\; \sup_{\substack{R^{\mathrm{H}}=R^{-1} \\ S^{\mathrm{H}}=S^{-1}}} \mathrm{re\,tr}(A^{\mathrm{T}} SBR) \;=\; \mathrm{re\,tr}(A^{\mathrm{T}} S^\star BR^\star) \;=\; \mathrm{tr}(\Sigma_A^{\mathrm{T}} \Sigma_B) \;\geq\; \mathrm{tr}(A^{\mathrm{T}} B) \tag{1832}$$

for which it is necessary

$$A^{\mathrm{T}} S^\star BR^\star \succeq 0 \,, \qquad BR^\star A^{\mathrm{T}} S^\star \succeq 0 \tag{1833}$$

The lower bound on inner product of singular values in (1832) is due to von Neumann. Equality is attained if $U_A^{\mathrm{H}} U_B = I$ and $Q_B^{\mathrm{H}} Q_A = I$.

C.4.2.1   Symmetric matrices

Now optimizing over the complex manifold of unitary matrices (§B.5.2), the upper bound on trace (1818) is thereby raised: Suppose we are given diagonalizations for (real) symmetric A, B (§A.5)

$$A = W_A \Upsilon W_A^{\mathrm{T}} \in \mathbb{S}^n \,, \qquad \delta(\Upsilon) \in \mathcal{K}_{\mathcal{M}} \tag{1834}$$

$$B = W_B \Lambda W_B^{\mathrm{T}} \in \mathbb{S}^n \,, \qquad \delta(\Lambda) \in \mathcal{K}_{\mathcal{M}} \tag{1835}$$

having their respective eigenvalues in diagonal matrices $\Upsilon, \Lambda \in \mathbb{S}^n$ arranged in nonincreasing order (membership to the monotone cone $\mathcal{K}_{\mathcal{M}}$ (435)). Then by splitting eigenvalue signs, we invent a symmetric SVD-like decomposition

$$A \triangleq U_A \Sigma_A Q_A^{\mathrm{H}} \in \mathbb{S}^n \,, \qquad B \triangleq U_B \Sigma_B Q_B^{\mathrm{H}} \in \mathbb{S}^n \tag{1836}$$

where U_A, U_B, Q_A, $Q_B \in \mathbb{C}^{n \times n}$ are unitary matrices defined by (*confer* §A.6.5)

$$U_A \triangleq W_A \sqrt{\delta(\psi(\delta(\Upsilon)))}, \quad Q_A \triangleq W_A \sqrt{\delta(\psi(\delta(\Upsilon)))}^{\mathrm{H}}, \quad \Sigma_A = |\Upsilon| \qquad (1837)$$

$$U_B \triangleq W_B \sqrt{\delta(\psi(\delta(\Lambda)))}, \quad Q_B \triangleq W_B \sqrt{\delta(\psi(\delta(\Lambda)))}^{\mathrm{H}}, \quad \Sigma_B = |\Lambda| \qquad (1838)$$

where step function ψ is defined in (1660). In this circumstance,

$$S^\star = U_A U_B^{\mathrm{H}} = R^{\star\mathrm{T}} \in \mathbb{C}^{n \times n} \qquad (1839)$$

optimal matrices (1830) now unitary are related by transposition. The optimal value of objective (1831) is

$$\|U_A \Sigma_A Q_A^{\mathrm{H}} - S^\star U_B \Sigma_B Q_B^{\mathrm{H}} R^\star\|_{\mathrm{F}} = \| \, |\Upsilon| - |\Lambda| \, \|_{\mathrm{F}} \qquad (1840)$$

while the corresponding optimal value of trace maximization (1832) is

$$\sup_{\substack{R^{\mathrm{H}}=R^{-1} \\ S^{\mathrm{H}}=S^{-1}}} \mathrm{re}\,\mathrm{tr}(A^{\mathrm{T}}SBR) = \mathrm{tr}(|\Upsilon|\,|\Lambda|) \qquad (1841)$$

C.4.2.2   Diagonal matrices

Now suppose A and B are diagonal matrices

$$A = \Upsilon = \delta^2(\Upsilon) \in \mathbb{S}^n, \quad \delta(\Upsilon) \in \mathcal{K}_\mathcal{M} \qquad (1842)$$

$$B = \Lambda = \delta^2(\Lambda) \in \mathbb{S}^n, \quad \delta(\Lambda) \in \mathcal{K}_\mathcal{M} \qquad (1843)$$

both having their respective main diagonal entries arranged in nonincreasing order:

$$\begin{array}{ll} \underset{R\,,\,S}{\text{minimize}} & \|\Upsilon - S\Lambda R\|_{\mathrm{F}} \\ \text{subject to} & R^{\mathrm{H}} = R^{-1} \\ & S^{\mathrm{H}} = S^{-1} \end{array} \qquad (1844)$$

Then we have a symmetric decomposition from unitary matrices as in (1836) where

$$U_A \triangleq \sqrt{\delta(\psi(\delta(\Upsilon)))}, \quad Q_A \triangleq \sqrt{\delta(\psi(\delta(\Upsilon)))}^{\mathrm{H}}, \quad \Sigma_A = |\Upsilon| \qquad (1845)$$

$$U_B \triangleq \sqrt{\delta(\psi(\delta(\Lambda)))}, \quad Q_B \triangleq \sqrt{\delta(\psi(\delta(\Lambda)))}^{\mathrm{H}}, \quad \Sigma_B = |\Lambda| \qquad (1846)$$

Procrustes solution (1830) again sees the transposition relationship

$$S^\star = U_A U_B^{\mathrm{H}} = R^{\star\mathrm{T}} \in \mathbb{C}^{n \times n} \qquad (1839)$$

but both optimal unitary matrices are now themselves diagonal. So,

$$S^\star \Lambda R^\star = \delta(\psi(\delta(\Upsilon)))\Lambda\,\delta(\psi(\delta(\Lambda))) = \delta(\psi(\delta(\Upsilon)))|\Lambda| \qquad (1847)$$

C.5 Nonconvex quadratics

[354, §2] [329, §2] Given $A \in \mathbb{S}^n$, $b \in \mathbb{R}^n$, and $\rho > 0$

$$\begin{array}{ll} \underset{x}{\text{minimize}} & \frac{1}{2}x^{\mathrm{T}}Ax + b^{\mathrm{T}}x \\ \text{subject to} & \|x\| \leq \rho \end{array} \tag{1848}$$

is a nonconvex problem for symmetric A unless $A \succeq 0$. But necessary and sufficient global optimality conditions are known for any symmetric A: vector $x^\star$ solves (1848) iff $\exists$ Lagrange multiplier $\lambda^\star \geq 0$ ∍

$$\begin{array}{rl} \textbf{i)} & (A + \lambda^\star I)x^\star = -b \\ \textbf{ii)} & \lambda^\star(\|x^\star\| - \rho) = 0, \qquad \|x^\star\| \leq \rho \\ \textbf{iii)} & A + \lambda^\star I \succeq 0 \end{array} \tag{1849}$$

$\lambda^\star$ is unique.[C.5] $x^\star$ is unique if $A + \lambda^\star I \succ 0$. Equality constrained problem

$$\begin{array}{ll} \underset{x}{\text{minimize}} & \frac{1}{2}x^{\mathrm{T}}Ax + b^{\mathrm{T}}x \\ \text{subject to} & \|x\| = \rho \end{array} \quad \Leftrightarrow \quad \begin{array}{rl} \textbf{i)} & (A + \lambda^\star I)x^\star = -b \\ \textbf{ii)} & \|x^\star\| = \rho \\ \textbf{iii)} & A + \lambda^\star I \succeq 0 \end{array} \tag{1850}$$

is nonconvex for any symmetric A matrix. $x^\star$ solves (1850) iff $\exists$ $\lambda^\star \in \mathbb{R}$ satisfying the associated conditions. $\lambda^\star$ and $x^\star$ are unique as before.

Hiriart-Urruty disclosed global optimality conditions in 1998 [203] for maximizing a convexly constrained convex quadratic; a nonconvex problem [315, §32].

[C.5]The first two are necessary KKT conditions [62, §5.5.3] while condition **iii** governs passage to nonconvex global optimality.

Appendix D

Matrix calculus

From too much study, and from extreme passion, cometh madnesse.

−Isaac Newton [158, §5]

D.1 Directional derivative, Taylor series

D.1.1 Gradients

Gradient of a differentiable real function $f(x) : \mathbb{R}^K \to \mathbb{R}$ with respect to its vector argument is defined uniquely in terms of partial derivatives

$$\nabla f(x) \triangleq \begin{bmatrix} \frac{\partial f(x)}{\partial x_1} \\ \frac{\partial f(x)}{\partial x_2} \\ \vdots \\ \frac{\partial f(x)}{\partial x_K} \end{bmatrix} \in \mathbb{R}^K \tag{1851}$$

while the second-order gradient of the twice differentiable real function with respect to its vector argument is traditionally called the *Hessian*;

$$\nabla^2 f(x) \triangleq \begin{bmatrix} \frac{\partial^2 f(x)}{\partial x_1^2} & \frac{\partial^2 f(x)}{\partial x_1 \partial x_2} & \cdots & \frac{\partial^2 f(x)}{\partial x_1 \partial x_K} \\ \frac{\partial^2 f(x)}{\partial x_2 \partial x_1} & \frac{\partial^2 f(x)}{\partial x_2^2} & \cdots & \frac{\partial^2 f(x)}{\partial x_2 \partial x_K} \\ \vdots & \vdots & \ddots & \vdots \\ \frac{\partial^2 f(x)}{\partial x_K \partial x_1} & \frac{\partial^2 f(x)}{\partial x_K \partial x_2} & \cdots & \frac{\partial^2 f(x)}{\partial x_K^2} \end{bmatrix} \in \mathbb{S}^K \tag{1852}$$

The gradient of vector-valued function $v(x) : \mathbb{R} \to \mathbb{R}^N$ on real domain is a row-vector

$$\nabla v(x) \triangleq \begin{bmatrix} \frac{\partial v_1(x)}{\partial x} & \frac{\partial v_2(x)}{\partial x} & \cdots & \frac{\partial v_N(x)}{\partial x} \end{bmatrix} \in \mathbb{R}^N \tag{1853}$$

CITATION: Dattorro, *Convex Optimization & Euclidean Distance Geometry*, $\mathcal{M}\varepsilon\beta oo$ Publishing USA, 2005, v2015.02.15.

while the second-order gradient is

$$\nabla^2 v(x) \triangleq \left[\begin{array}{cccc} \frac{\partial^2 v_1(x)}{\partial x^2} & \frac{\partial^2 v_2(x)}{\partial x^2} & \cdots & \frac{\partial^2 v_N(x)}{\partial x^2} \end{array}\right] \in \mathbb{R}^N \qquad (1854)$$

Gradient of vector-valued function $h(x) : \mathbb{R}^K \to \mathbb{R}^N$ on vector domain is

$$\nabla h(x) \triangleq \left[\begin{array}{cccc} \frac{\partial h_1(x)}{\partial x_1} & \frac{\partial h_2(x)}{\partial x_1} & \cdots & \frac{\partial h_N(x)}{\partial x_1} \\ \frac{\partial h_1(x)}{\partial x_2} & \frac{\partial h_2(x)}{\partial x_2} & \cdots & \frac{\partial h_N(x)}{\partial x_2} \\ \vdots & \vdots & & \vdots \\ \frac{\partial h_1(x)}{\partial x_K} & \frac{\partial h_2(x)}{\partial x_K} & \cdots & \frac{\partial h_N(x)}{\partial x_K} \end{array}\right] \qquad (1855)$$

$$= \left[\begin{array}{cccc} \nabla h_1(x) & \nabla h_2(x) & \cdots & \nabla h_N(x) \end{array}\right] \in \mathbb{R}^{K \times N}$$

while the second-order gradient has a three-dimensional written representation dubbed *cubix* ;[D.1]

$$\nabla^2 h(x) \triangleq \left[\begin{array}{cccc} \nabla \frac{\partial h_1(x)}{\partial x_1} & \nabla \frac{\partial h_2(x)}{\partial x_1} & \cdots & \nabla \frac{\partial h_N(x)}{\partial x_1} \\ \nabla \frac{\partial h_1(x)}{\partial x_2} & \nabla \frac{\partial h_2(x)}{\partial x_2} & \cdots & \nabla \frac{\partial h_N(x)}{\partial x_2} \\ \vdots & \vdots & & \vdots \\ \nabla \frac{\partial h_1(x)}{\partial x_K} & \nabla \frac{\partial h_2(x)}{\partial x_K} & \cdots & \nabla \frac{\partial h_N(x)}{\partial x_K} \end{array}\right] \qquad (1856)$$

$$= \left[\begin{array}{cccc} \nabla^2 h_1(x) & \nabla^2 h_2(x) & \cdots & \nabla^2 h_N(x) \end{array}\right] \in \mathbb{R}^{K \times N \times K}$$

where the gradient of each real entry is with respect to vector x as in (1851).

The gradient of real function $g(X) : \mathbb{R}^{K \times L} \to \mathbb{R}$ on matrix domain is

$$\nabla g(X) \triangleq \left[\begin{array}{cccc} \frac{\partial g(X)}{\partial X_{11}} & \frac{\partial g(X)}{\partial X_{12}} & \cdots & \frac{\partial g(X)}{\partial X_{1L}} \\ \frac{\partial g(X)}{\partial X_{21}} & \frac{\partial g(X)}{\partial X_{22}} & \cdots & \frac{\partial g(X)}{\partial X_{2L}} \\ \vdots & \vdots & & \vdots \\ \frac{\partial g(X)}{\partial X_{K1}} & \frac{\partial g(X)}{\partial X_{K2}} & \cdots & \frac{\partial g(X)}{\partial X_{KL}} \end{array}\right] \in \mathbb{R}^{K \times L}$$

$$(1857)$$

$$= \left[\begin{array}{c} \nabla_{X(:,1)}\, g(X) \\ \nabla_{X(:,2)}\, g(X) \\ \ddots \\ \nabla_{X(:,L)}\, g(X) \end{array}\right] \in \mathbb{R}^{K \times 1 \times L}$$

where gradient $\nabla_{X(:,i)}$ is with respect to the ith column of X. The strange appearance of (1857) in $\mathbb{R}^{K \times 1 \times L}$ is meant to suggest a third dimension perpendicular to the page (not

---

[D.1]The word *matrix* comes from the Latin for *womb*; related to the prefix *matri-* derived from *mater* meaning *mother*.

a diagonal matrix). The second-order gradient has representation

$$
\nabla^2 g(X) \triangleq
\begin{bmatrix}
\nabla \frac{\partial g(X)}{\partial X_{11}} & \nabla \frac{\partial g(X)}{\partial X_{12}} & \cdots & \nabla \frac{\partial g(X)}{\partial X_{1L}} \\
\nabla \frac{\partial g(X)}{\partial X_{21}} & \nabla \frac{\partial g(X)}{\partial X_{22}} & \cdots & \nabla \frac{\partial g(X)}{\partial X_{2L}} \\
\vdots & \vdots & & \vdots \\
\nabla \frac{\partial g(X)}{\partial X_{K1}} & \nabla \frac{\partial g(X)}{\partial X_{K2}} & \cdots & \nabla \frac{\partial g(X)}{\partial X_{KL}}
\end{bmatrix}
\in \mathbb{R}^{K \times L \times K \times L}
$$

(1858)

$$
=
\begin{bmatrix}
\nabla \nabla_{X(:,1)} g(X) & & & \\
& \nabla \nabla_{X(:,2)} g(X) & & \\
& & \ddots & \\
& & & \nabla \nabla_{X(:,L)} g(X)
\end{bmatrix}
\in \mathbb{R}^{K \times 1 \times L \times K \times L}
$$

where the gradient ∇ is with respect to matrix X.

Gradient of vector-valued function $g(X) : \mathbb{R}^{K \times L} \to \mathbb{R}^N$ on matrix domain is a *cubix*

$$
\nabla g(X) \triangleq
\begin{bmatrix}
\nabla_{X(:,1)} g_1(X) & \nabla_{X(:,1)} g_2(X) & \cdots & \nabla_{X(:,1)} g_N(X) \\
\nabla_{X(:,2)} g_1(X) & \nabla_{X(:,2)} g_2(X) & \cdots & \nabla_{X(:,2)} g_N(X) \\
\ddots & \ddots & & \ddots \\
\nabla_{X(:,L)} g_1(X) & \nabla_{X(:,L)} g_2(X) & \cdots & \nabla_{X(:,L)} g_N(X)
\end{bmatrix}
$$

(1859)

$$
= [\nabla g_1(X) \ \ \nabla g_2(X) \ \ \cdots \ \ \nabla g_N(X)] \in \mathbb{R}^{K \times N \times L}
$$

while the second-order gradient has a five-dimensional representation;

$$
\nabla^2 g(X) \triangleq
\begin{bmatrix}
\nabla \nabla_{X(:,1)} g_1(X) & \nabla \nabla_{X(:,1)} g_2(X) & \cdots & \nabla \nabla_{X(:,1)} g_N(X) \\
\nabla \nabla_{X(:,2)} g_1(X) & \nabla \nabla_{X(:,2)} g_2(X) & \cdots & \nabla \nabla_{X(:,2)} g_N(X) \\
\ddots & \ddots & & \ddots \\
\nabla \nabla_{X(:,L)} g_1(X) & \nabla \nabla_{X(:,L)} g_2(X) & \cdots & \nabla \nabla_{X(:,L)} g_N(X)
\end{bmatrix}
$$

(1860)

$$
= [\nabla^2 g_1(X) \ \ \nabla^2 g_2(X) \ \ \cdots \ \ \nabla^2 g_N(X)] \in \mathbb{R}^{K \times N \times L \times K \times L}
$$

The gradient of matrix-valued function $g(X) : \mathbb{R}^{K \times L} \to \mathbb{R}^{M \times N}$ on matrix domain has a four-dimensional representation called *quartix (fourth-order tensor)*

$$
\nabla g(X) \triangleq
\begin{bmatrix}
\nabla g_{11}(X) & \nabla g_{12}(X) & \cdots & \nabla g_{1N}(X) \\
\nabla g_{21}(X) & \nabla g_{22}(X) & \cdots & \nabla g_{2N}(X) \\
\vdots & \vdots & & \vdots \\
\nabla g_{M1}(X) & \nabla g_{M2}(X) & \cdots & \nabla g_{MN}(X)
\end{bmatrix}
\in \mathbb{R}^{M \times N \times K \times L}
$$

(1861)

while the second-order gradient has a six-dimensional representation

$$\nabla^2 g(X) \triangleq \begin{bmatrix} \nabla^2 g_{11}(X) & \nabla^2 g_{12}(X) & \cdots & \nabla^2 g_{1N}(X) \\ \nabla^2 g_{21}(X) & \nabla^2 g_{22}(X) & \cdots & \nabla^2 g_{2N}(X) \\ \vdots & \vdots & & \vdots \\ \nabla^2 g_{M1}(X) & \nabla^2 g_{M2}(X) & \cdots & \nabla^2 g_{MN}(X) \end{bmatrix} \in \mathbb{R}^{M \times N \times K \times L \times K \times L} \quad (1862)$$

and so on.

D.1.2   Product rules for matrix-functions

Given dimensionally compatible matrix-valued functions of matrix variable $f(X)$ and $g(X)$

$$\nabla_X \big(f(X)^{\mathrm{T}} g(X)\big) = \nabla_X(f)\, g \,+\, \nabla_X(g)\, f \quad (1863)$$

while [54, §8.3] [323]

$$\nabla_X \operatorname{tr}\big(f(X)^{\mathrm{T}} g(X)\big) = \nabla_X \Big(\operatorname{tr}\big(f(X)^{\mathrm{T}} g(Z)\big) \,+\, \operatorname{tr}\big(g(X)\, f(Z)^{\mathrm{T}}\big) \Big)\Big|_{Z \leftarrow X} \quad (1864)$$

These expressions implicitly apply as well to scalar-, vector-, or matrix-valued functions of scalar, vector, or matrix arguments.

D.1.2.0.1   Example.   *Cubix.*
Suppose $f(X): \mathbb{R}^{2\times2} \to \mathbb{R}^2 = X^{\mathrm{T}} a$ and $g(X): \mathbb{R}^{2\times2} \to \mathbb{R}^2 = Xb$. We wish to find

$$\nabla_X \big(f(X)^{\mathrm{T}} g(X)\big) = \nabla_X\, a^{\mathrm{T}} X^2 b \quad (1865)$$

using the product rule. Formula (1863) calls for

$$\nabla_X\, a^{\mathrm{T}} X^2 b = \nabla_X(X^{\mathrm{T}} a)\, Xb \,+\, \nabla_X(Xb)\, X^{\mathrm{T}} a \quad (1866)$$

Consider the first of the two terms:

$$\begin{aligned} \nabla_X(f)\, g &= \nabla_X(X^{\mathrm{T}} a)\, Xb \\ &= \big[\, \nabla(X^{\mathrm{T}} a)_1 \quad \nabla(X^{\mathrm{T}} a)_2 \,\big]\, Xb \end{aligned} \quad (1867)$$

The gradient of $X^{\mathrm{T}} a$ forms a cubix in $\mathbb{R}^{2\times2\times2}$; a.k.a, *third-order tensor.*

$$\nabla_X(X^{\mathrm{T}} a)\, Xb = \begin{bmatrix} \frac{\partial(X^{\mathrm{T}} a)_1}{\partial X_{11}} \cdots\cdots\cdots\cdots \frac{\partial(X^{\mathrm{T}} a)_2}{\partial X_{11}} \\ \quad \frac{\partial(X^{\mathrm{T}} a)_1}{\partial X_{12}} \cdots\cdots \frac{\partial(X^{\mathrm{T}} a)_2}{\partial X_{12}} \\ \frac{\partial(X^{\mathrm{T}} a)_1}{\partial X_{21}} \cdots\cdots \frac{\partial(X^{\mathrm{T}} a)_2}{\partial X_{21}} \\ \quad \frac{\partial(X^{\mathrm{T}} a)_1}{\partial X_{22}} \cdots\cdots \frac{\partial(X^{\mathrm{T}} a)_2}{\partial X_{22}} \end{bmatrix} \begin{bmatrix} (Xb)_1 \\ \\ (Xb)_2 \end{bmatrix} \in \mathbb{R}^{2\times1\times2} \quad (1868)$$

Because gradient of the product (1865) requires total change with respect to change in each entry of matrix X, the Xb vector must make an inner product with each vector in that second dimension of the cubix indicated by dotted line segments;

$$
\nabla_X (X^{\mathrm{T}} a) \, Xb = \left[\begin{array}{cc} a_1 & 0 \\ & 0 \quad a_1 \\ a_2 & 0 \\ & 0 \quad a_2 \end{array} \right] \left[\begin{array}{c} b_1 X_{11} + b_2 X_{12} \\ b_1 X_{21} + b_2 X_{22} \end{array} \right] \in \mathbb{R}^{\mathbf{2 \times 1 \times 2}}
$$

$$
= \left[\begin{array}{cc} a_1 (b_1 X_{11} + b_2 X_{12}) & a_1 (b_1 X_{21} + b_2 X_{22}) \\ a_2 (b_1 X_{11} + b_2 X_{12}) & a_2 (b_1 X_{21} + b_2 X_{22}) \end{array} \right] \in \mathbb{R}^{\mathbf{2 \times 2}}
$$

$$
= ab^{\mathrm{T}} X^{\mathrm{T}} \tag{1869}
$$

where the cubix appears as a complete $2 \times 2 \times 2$ matrix. In like manner for the second term $\nabla_X (g) \, f$

$$
\nabla_X (Xb) \, X^{\mathrm{T}} a = \left[\begin{array}{cc} b_1 & 0 \\ & b_2 \quad 0 \\ 0 & b_1 \\ & 0 \quad b_2 \end{array} \right] \left[\begin{array}{c} X_{11} a_1 + X_{21} a_2 \\ X_{12} a_1 + X_{22} a_2 \end{array} \right] \in \mathbb{R}^{\mathbf{2 \times 1 \times 2}}
$$

$$
= X^{\mathrm{T}} ab^{\mathrm{T}} \in \mathbb{R}^{\mathbf{2 \times 2}} \tag{1870}
$$

The solution

$$
\nabla_X \, a^{\mathrm{T}} X^2 b = ab^{\mathrm{T}} X^{\mathrm{T}} + X^{\mathrm{T}} ab^{\mathrm{T}} \tag{1871}
$$

can be found from Table **D.2.1** or verified using (1864). $\qquad\qquad\square$

D.1.2.1 Kronecker product

A partial remedy for venturing into *hyperdimensional* matrix representations, such as the cubix or quartix, is to first vectorize matrices as in (37). This device gives rise to the Kronecker product of matrices $\otimes$; a.k.a, *tensor product*. Although it sees reversal in the literature, [334, §2.1] we adopt the definition: for $A \in \mathbb{R}^{m \times n}$ and $B \in \mathbb{R}^{p \times q}$

$$
B \otimes A \triangleq \left[\begin{array}{cccc} B_{11} A & B_{12} A & \cdots & B_{1q} A \\ B_{21} A & B_{22} A & \cdots & B_{2q} A \\ \vdots & \vdots & & \vdots \\ B_{p1} A & B_{p2} A & \cdots & B_{pq} A \end{array} \right] \in \mathbb{R}^{pm \times qn} \tag{1872}
$$

for which $A \otimes 1 = 1 \otimes A = A$ (real unity acts like Identity).

One advantage to vectorization is existence of the traditional two-dimensional matrix representation (*second-order tensor*) for the second-order gradient of a real function with respect to a vectorized matrix. For example, from §A.1.1 *no.*33 (§D.2.1) for square $A, B \in \mathbb{R}^{n \times n}$ [172, §5.2] [12, §3]

$$
\nabla^2_{\mathrm{vec}\,X} \mathrm{tr}(AXBX^{\mathrm{T}}) = \nabla^2_{\mathrm{vec}\,X} \mathrm{vec}(X)^{\mathrm{T}} (B^{\mathrm{T}} \otimes A) \, \mathrm{vec}\,X = B \otimes A^{\mathrm{T}} + B^{\mathrm{T}} \otimes A \in \mathbb{R}^{n^2 \times n^2} \tag{1873}
$$

To disadvantage is a large new but known set of algebraic rules (§A.1.1) and the fact that its mere use does not generally guarantee two-dimensional matrix representation of gradients.

Another application of the Kronecker product is to reverse order of appearance in a matrix product: Suppose we wish to weight the columns of a matrix $S \in \mathbb{R}^{M \times N}$, for example, by respective entries w_i from the main diagonal in

$$W \triangleq \begin{bmatrix} w_1 & & \mathbf{0} \\ & \ddots & \\ \mathbf{0}^T & & w_N \end{bmatrix} \in \mathbb{S}^N \tag{1874}$$

A conventional means for accomplishing column weighting is to multiply S by diagonal matrix W on the right-hand side:

$$SW = S \begin{bmatrix} w_1 & & \mathbf{0} \\ & \ddots & \\ \mathbf{0}^T & & w_N \end{bmatrix} = \begin{bmatrix} S(:,1)w_1 & \cdots & S(:,N)w_N \end{bmatrix} \in \mathbb{R}^{M \times N} \tag{1875}$$

To reverse product order such that diagonal matrix W instead appears to the left of S: for $I \in \mathbb{S}^M$ (Sze Wan)

$$SW = (\delta(W)^T \otimes I) \begin{bmatrix} S(:,1) & \mathbf{0} & & \mathbf{0} \\ \mathbf{0} & S(:,2) & \ddots & \\ & \ddots & \ddots & \mathbf{0} \\ \mathbf{0} & & \mathbf{0} & S(:,N) \end{bmatrix} \in \mathbb{R}^{M \times N} \tag{1876}$$

To instead weight the rows of S via diagonal matrix $W \in \mathbb{S}^M$, for $I \in \mathbb{S}^N$

$$WS = \begin{bmatrix} S(1,:) & \mathbf{0} & & \mathbf{0} \\ \mathbf{0} & S(2,:) & \ddots & \\ & \ddots & \ddots & \mathbf{0} \\ \mathbf{0} & & \mathbf{0} & S(M,:) \end{bmatrix} (\delta(W) \otimes I) \in \mathbb{R}^{M \times N} \tag{1877}$$

For any matrices of like size, $S, Y \in \mathbb{R}^{M \times N}$

$$S \circ Y = \begin{bmatrix} \delta(Y(:,1)) & \cdots & \delta(Y(:,N)) \end{bmatrix} \begin{bmatrix} S(:,1) & \mathbf{0} & & \mathbf{0} \\ \mathbf{0} & S(:,2) & \ddots & \\ & \ddots & \ddots & \mathbf{0} \\ \mathbf{0} & & \mathbf{0} & S(:,N) \end{bmatrix} \in \mathbb{R}^{M \times N} \tag{1878}$$

which converts a Hadamard product into a standard matrix product. In the special case that $S = s$ and $Y = y$ are vectors in $\mathbb{R}^M$

$$s \circ y = \delta(s)y \tag{1879}$$

$$\begin{aligned} s^T \otimes y &= ys^T \\ s \otimes y^T &= sy^T \end{aligned} \tag{1880}$$

D.1.3 Chain rules for composite matrix-functions

Given dimensionally compatible matrix-valued functions of matrix variable $f(X)$ and $g(X)$ [225, §15.7]

$$\nabla_X \, g\big(f(X)^{\mathrm{T}}\big) = \nabla_X f^{\mathrm{T}} \, \nabla_f \, g \qquad (1881)$$

$$\nabla_X^2 \, g\big(f(X)^{\mathrm{T}}\big) \;=\; \nabla_X \big(\nabla_X f^{\mathrm{T}} \, \nabla_f \, g\big) \;=\; \nabla_X^2 f \, \nabla_f \, g \;+\; \nabla_X f^{\mathrm{T}} \, \nabla_f^2 \, g \, \nabla_X f \qquad (1882)$$

D.1.3.1 Two arguments

$$\nabla_X \, g\big(f(X)^{\mathrm{T}}, \, h(X)^{\mathrm{T}}\big) = \nabla_X f^{\mathrm{T}} \, \nabla_f \, g \;+\; \nabla_X h^{\mathrm{T}} \, \nabla_h \, g \qquad (1883)$$

D.1.3.1.1 Example. *Chain rule for two arguments.*   [42, §1.1]

$$g\big(f(x)^{\mathrm{T}}, \, h(x)^{\mathrm{T}}\big) = \big(f(x) + h(x)\big)^{\mathrm{T}} A\big(f(x) + h(x)\big) \qquad (1884)$$

$$f(x) = \begin{bmatrix} x_1 \\ \varepsilon x_2 \end{bmatrix}, \qquad h(x) = \begin{bmatrix} \varepsilon x_1 \\ x_2 \end{bmatrix} \qquad (1885)$$

$$\nabla_x \, g\big(f(x)^{\mathrm{T}}, \, h(x)^{\mathrm{T}}\big) = \begin{bmatrix} 1 & 0 \\ 0 & \varepsilon \end{bmatrix} (A + A^{\mathrm{T}})(f + h) + \begin{bmatrix} \varepsilon & 0 \\ 0 & 1 \end{bmatrix} (A + A^{\mathrm{T}})(f + h) \qquad (1886)$$

$$\nabla_x \, g\big(f(x)^{\mathrm{T}}, \, h(x)^{\mathrm{T}}\big) = \begin{bmatrix} 1 + \varepsilon & 0 \\ 0 & 1 + \varepsilon \end{bmatrix} (A + A^{\mathrm{T}})\left(\begin{bmatrix} x_1 \\ \varepsilon x_2 \end{bmatrix} + \begin{bmatrix} \varepsilon x_1 \\ x_2 \end{bmatrix} \right) \qquad (1887)$$

$$\lim_{\varepsilon \to 0} \nabla_x \, g\big(f(x)^{\mathrm{T}}, \, h(x)^{\mathrm{T}}\big) = (A + A^{\mathrm{T}})x \qquad (1888)$$

from Table D.2.1.   □

These foregoing formulae remain correct when gradient produces hyperdimensional representation:

D.1.4 First directional derivative

Assume that a differentiable function $g(X) : \mathbb{R}^{K \times L} \to \mathbb{R}^{M \times N}$ has continuous first- and second-order gradients ∇g and $\nabla^2 g$ over $\operatorname{dom} g$ which is an open set. We seek simple expressions for the first and second directional derivatives in direction $Y \in \mathbb{R}^{K \times L}$: respectively, $\overset{\to Y}{dg} \in \mathbb{R}^{M \times N}$ and $\overset{\to Y}{dg^2} \in \mathbb{R}^{M \times N}$.

Assuming that the limit exists, we may state the partial derivative of the mn^{th} entry of g with respect to the kl^{th} entry of X;

$$\frac{\partial g_{mn}(X)}{\partial X_{kl}} \;=\; \lim_{\Delta t \to 0} \frac{g_{mn}(X + \Delta t \, e_k e_l^{\mathrm{T}}) - g_{mn}(X)}{\Delta t} \in \mathbb{R} \qquad (1889)$$

where e_k is the kth standard basis vector in $\mathbb{R}^K$ while e_l is the lth standard basis vector in $\mathbb{R}^L$. The total number of partial derivatives equals $KLMN$ while the gradient is defined in their terms; the mn^{th} entry of the gradient is

$$
\nabla g_{mn}(X) = \begin{bmatrix}
\frac{\partial g_{mn}(X)}{\partial X_{11}} & \frac{\partial g_{mn}(X)}{\partial X_{12}} & \cdots & \frac{\partial g_{mn}(X)}{\partial X_{1L}} \\
\frac{\partial g_{mn}(X)}{\partial X_{21}} & \frac{\partial g_{mn}(X)}{\partial X_{22}} & \cdots & \frac{\partial g_{mn}(X)}{\partial X_{2L}} \\
\vdots & \vdots & & \vdots \\
\frac{\partial g_{mn}(X)}{\partial X_{K1}} & \frac{\partial g_{mn}(X)}{\partial X_{K2}} & \cdots & \frac{\partial g_{mn}(X)}{\partial X_{KL}}
\end{bmatrix} \in \mathbb{R}^{K \times L} \qquad (1890)
$$

while the gradient is a quartix

$$
\nabla g(X) = \begin{bmatrix}
\nabla g_{11}(X) & \nabla g_{12}(X) & \cdots & \nabla g_{1N}(X) \\
\nabla g_{21}(X) & \nabla g_{22}(X) & \cdots & \nabla g_{2N}(X) \\
\vdots & \vdots & & \vdots \\
\nabla g_{M1}(X) & \nabla g_{M2}(X) & \cdots & \nabla g_{MN}(X)
\end{bmatrix} \in \mathbb{R}^{M \times N \times K \times L} \qquad (1891)
$$

By simply rotating our perspective of a four-dimensional representation of gradient matrix, we find one of three useful transpositions of this quartix (connoted $^{\text{T}_1}$):

$$
\nabla g(X)^{\text{T}_1} = \begin{bmatrix}
\frac{\partial g(X)}{\partial X_{11}} & \frac{\partial g(X)}{\partial X_{12}} & \cdots & \frac{\partial g(X)}{\partial X_{1L}} \\
\frac{\partial g(X)}{\partial X_{21}} & \frac{\partial g(X)}{\partial X_{22}} & \cdots & \frac{\partial g(X)}{\partial X_{2L}} \\
\vdots & \vdots & & \vdots \\
\frac{\partial g(X)}{\partial X_{K1}} & \frac{\partial g(X)}{\partial X_{K2}} & \cdots & \frac{\partial g(X)}{\partial X_{KL}}
\end{bmatrix} \in \mathbb{R}^{K \times L \times M \times N} \qquad (1892)
$$

When the limit for $\Delta t \in \mathbb{R}$ exists, it is easy to show by substitution of variables in (1889)

$$
\frac{\partial g_{mn}(X)}{\partial X_{kl}} Y_{kl} = \lim_{\Delta t \to 0} \frac{g_{mn}(X + \Delta t\, Y_{kl}\, e_k e_l^{\text{T}}) - g_{mn}(X)}{\Delta t} \in \mathbb{R} \qquad (1893)
$$

which may be interpreted as the change in g_{mn} at X when the change in X_{kl} is equal to Y_{kl}, the kl^{th} entry of any $Y \in \mathbb{R}^{K \times L}$. Because the total change in $g_{mn}(X)$ due to Y is the sum of change with respect to each and every X_{kl}, the mn^{th} entry of the directional derivative is the corresponding total differential [225, §15.8]

$$
dg_{mn}(X)|_{dX \to Y} = \sum_{k,l} \frac{\partial g_{mn}(X)}{\partial X_{kl}} Y_{kl} = \operatorname{tr}\big(\nabla g_{mn}(X)^{\text{T}} Y\big) \qquad (1894)
$$

$$
= \sum_{k,l} \lim_{\Delta t \to 0} \frac{g_{mn}(X + \Delta t\, Y_{kl}\, e_k e_l^{\text{T}}) - g_{mn}(X)}{\Delta t} \qquad (1895)
$$

$$
= \lim_{\Delta t \to 0} \frac{g_{mn}(X + \Delta t\, Y) - g_{mn}(X)}{\Delta t} \qquad (1896)
$$

$$
= \frac{d}{dt}\bigg|_{t=0} g_{mn}(X + t\, Y) \qquad (1897)
$$

where $t \in \mathbb{R}$. Assuming finite Y, equation (1896) is called the *Gâteaux differential* [41, App.A.5] [205, §D.2.1] [367, §5.28] whose existence is implied by existence of the *Fréchet differential* (the sum in (1894)). [256, §7.2] Each may be understood as the change in g_{mn} at X when the change in X is equal in magnitude and direction to Y.[D.2] Hence the directional derivative,

$$
\overset{\rightarrow Y}{dg}(X) \triangleq \left.\begin{bmatrix} dg_{11}(X) & dg_{12}(X) & \cdots & dg_{1N}(X) \\ dg_{21}(X) & dg_{22}(X) & \cdots & dg_{2N}(X) \\ \vdots & \vdots & & \vdots \\ dg_{M1}(X) & dg_{M2}(X) & \cdots & dg_{MN}(X) \end{bmatrix}\right|_{dX \to Y} \in \mathbb{R}^{M \times N}
$$

$$
= \begin{bmatrix} \mathrm{tr}\big(\nabla g_{11}(X)^{\mathrm{T}} Y\big) & \mathrm{tr}\big(\nabla g_{12}(X)^{\mathrm{T}} Y\big) & \cdots & \mathrm{tr}\big(\nabla g_{1N}(X)^{\mathrm{T}} Y\big) \\ \mathrm{tr}\big(\nabla g_{21}(X)^{\mathrm{T}} Y\big) & \mathrm{tr}\big(\nabla g_{22}(X)^{\mathrm{T}} Y\big) & \cdots & \mathrm{tr}\big(\nabla g_{2N}(X)^{\mathrm{T}} Y\big) \\ \vdots & \vdots & & \vdots \\ \mathrm{tr}\big(\nabla g_{M1}(X)^{\mathrm{T}} Y\big) & \mathrm{tr}\big(\nabla g_{M2}(X)^{\mathrm{T}} Y\big) & \cdots & \mathrm{tr}\big(\nabla g_{MN}(X)^{\mathrm{T}} Y\big) \end{bmatrix} \tag{1898}
$$

$$
= \begin{bmatrix} \sum_{k,l} \frac{\partial g_{11}(X)}{\partial X_{kl}} Y_{kl} & \sum_{k,l} \frac{\partial g_{12}(X)}{\partial X_{kl}} Y_{kl} & \cdots & \sum_{k,l} \frac{\partial g_{1N}(X)}{\partial X_{kl}} Y_{kl} \\ \sum_{k,l} \frac{\partial g_{21}(X)}{\partial X_{kl}} Y_{kl} & \sum_{k,l} \frac{\partial g_{22}(X)}{\partial X_{kl}} Y_{kl} & \cdots & \sum_{k,l} \frac{\partial g_{2N}(X)}{\partial X_{kl}} Y_{kl} \\ \vdots & \vdots & & \vdots \\ \sum_{k,l} \frac{\partial g_{M1}(X)}{\partial X_{kl}} Y_{kl} & \sum_{k,l} \frac{\partial g_{M2}(X)}{\partial X_{kl}} Y_{kl} & \cdots & \sum_{k,l} \frac{\partial g_{MN}(X)}{\partial X_{kl}} Y_{kl} \end{bmatrix}
$$

from which it follows

$$
\overset{\rightarrow Y}{dg}(X) = \sum_{k,l} \frac{\partial g(X)}{\partial X_{kl}} Y_{kl} \tag{1899}
$$

Yet for all $X \in \mathrm{dom}\, g$, any $Y \in \mathbb{R}^{K \times L}$, and some open interval of $t \in \mathbb{R}$

$$
g(X + tY) = g(X) + t\,\overset{\rightarrow Y}{dg}(X) + o(t^2) \tag{1900}
$$

which is the first-order Taylor series expansion about X. [225, §18.4] [156, §2.3.4] Differentiation with respect to t and subsequent t-zeroing isolates the second term of expansion. Thus differentiating and zeroing $g(X + tY)$ in t is an operation equivalent to individually differentiating and zeroing every entry $g_{mn}(X + tY)$ as in (1897). So the directional derivative of $g(X) : \mathbb{R}^{K \times L} \to \mathbb{R}^{M \times N}$ in any direction $Y \in \mathbb{R}^{K \times L}$ evaluated at $X \in \mathrm{dom}\, g$ becomes

$$
\overset{\rightarrow Y}{dg}(X) = \left.\frac{d}{dt}\right|_{t=0} g(X + tY) \in \mathbb{R}^{M \times N} \tag{1901}
$$

[284, §2.1, §5.4.5] [34, §6.3.1] which is simplest. In case of a real function $g(X) : \mathbb{R}^{K \times L} \to \mathbb{R}$

$$
\overset{\rightarrow Y}{dg}(X) = \mathrm{tr}\big(\nabla g(X)^{\mathrm{T}} Y\big) \tag{1923}
$$

[D.2]Although Y is a matrix, we may regard it as a vector in $\mathbb{R}^{KL}$.

$$v \triangleq \begin{bmatrix} \nabla_x f(\alpha) \\ \overrightarrow{\nabla_x} f(\alpha) \\ \frac{1}{2} df(\alpha) \end{bmatrix}$$

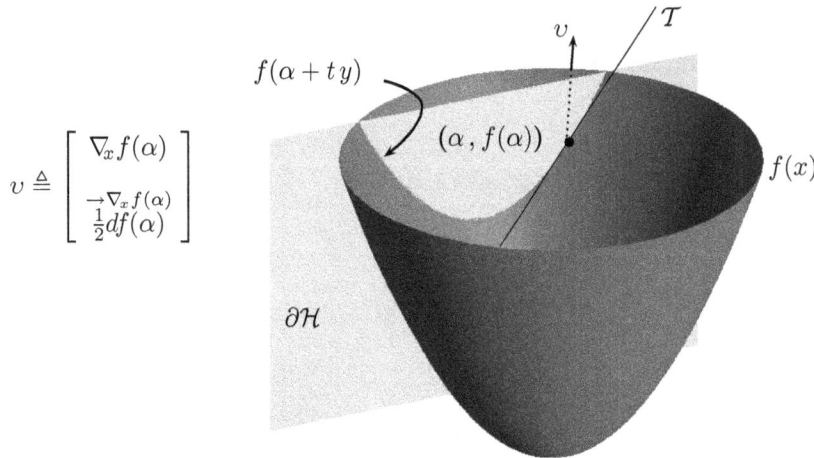

Figure 171: Strictly convex quadratic bowl in $\mathbb{R}^2 \times \mathbb{R}$; $f(x) = x^T x : \mathbb{R}^2 \to \mathbb{R}$ *versus* x on some open disc in $\mathbb{R}^2$. Plane slice $\partial \mathcal{H}$ is perpendicular to function domain. Slice intersection with domain connotes bidirectional vector y. Slope of tangent line $\mathcal{T}$ at point $(\alpha, f(\alpha))$ is value of $\nabla_x f(\alpha)^T y$ directional derivative (1926) at α in slice direction y. Negative gradient $-\nabla_x f(x) \in \mathbb{R}^2$ is direction of *steepest descent*. [403] [225, §15.6] [156] When vector $v \in \mathbb{R}^3$ entry v_3 is half directional derivative in gradient direction at α and when $\begin{bmatrix} v_1 \\ v_2 \end{bmatrix} = \nabla_x f(\alpha)$, then $-v$ points directly toward bowl bottom.

In case $g(X) : \mathbb{R}^K \to \mathbb{R}$

$$\overrightarrow{dg}^Y(X) = \nabla g(X)^T Y \qquad (1926)$$

Unlike gradient, directional derivative does not expand dimension; directional derivative (1901) retains the dimensions of g. The derivative with respect to t makes the directional derivative resemble ordinary calculus (§D.2); *e.g.*, when $g(X)$ is linear, $\overrightarrow{dg}^Y(X) = g(Y)$. [256, §7.2]

D.1.4.1   Interpretation of directional derivative

In the case of any differentiable real function $g(X) : \mathbb{R}^{K \times L} \to \mathbb{R}$, the directional derivative of $g(X)$ at X in any direction Y yields the slope of g along the line $\{X + tY \mid t \in \mathbb{R}\}$ through its domain evaluated at $t = 0$. For higher-dimensional functions, by (1898), this slope interpretation can be applied to each entry of the directional derivative.

Figure 171, for example, shows a plane slice of a real convex bowl-shaped function $f(x)$ along a line $\{\alpha + ty \mid t \in \mathbb{R}\}$ through its domain. The slice reveals a one-dimensional real function of t; $f(\alpha + ty)$. The directional derivative at $x = \alpha$ in direction y is the slope of $f(\alpha + ty)$ with respect to t at $t = 0$. In the case of a real function having vector argument $h(X) : \mathbb{R}^K \to \mathbb{R}$, its directional derivative in the normalized direction of its gradient is the gradient magnitude. (1926) For a real function of real variable, the directional derivative evaluated at any point in the function domain is just the slope of that function there scaled by the real direction. (*confer* §3.6)

Directional derivative generalizes our one-dimensional notion of derivative to a multidimensional domain. When direction Y coincides with a member of the standard Cartesian basis $e_k e_l^{\mathrm{T}}$ (60), then a single partial derivative $\partial g(X)/\partial X_{kl}$ is obtained from directional derivative (1899); such is each entry of gradient $\nabla g(X)$ in equalities (1923) and (1926), for example.

D.1.4.1.1 Theorem. *Directional derivative optimality condition.*
[256, §7.4] Suppose $f(X) : \mathbb{R}^{K \times L} \to \mathbb{R}$ is minimized on convex set $\mathcal{C} \subseteq \mathbb{R}^{K \times L}$ by $X^\star$, and the directional derivative of f exists there. Then for all $X \in \mathcal{C}$

$$\overset{\to X - X^\star}{df}(X) \geq 0 \tag{1902}$$

$\diamond$

D.1.4.1.2 Example. *Simple bowl.*
Bowl function (Figure 171)

$$f(x) : \mathbb{R}^K \to \mathbb{R} \triangleq (x - a)^{\mathrm{T}}(x - a) - b \tag{1903}$$

has function offset $-b \in \mathbb{R}$, axis of revolution at $x = a$, and positive definite Hessian (1852) everywhere in its domain (an open *hyperdisc* in $\mathbb{R}^K$); *id est*, strictly convex quadratic $f(x)$ has unique global minimum equal to $-b$ at $x = a$. A vector $-v$ based anywhere in $\mathrm{dom}\, f \times \mathbb{R}$ pointing toward the unique bowl-bottom is specified:

$$v \propto \begin{bmatrix} x - a \\ f(x) + b \end{bmatrix} \in \mathbb{R}^K \times \mathbb{R} \tag{1904}$$

Such a vector is

$$v = \begin{bmatrix} \nabla_x f(x) \\ \frac{1}{2} \overset{\to \nabla_x f(x)}{df}(x) \end{bmatrix} \tag{1905}$$

since the gradient is

$$\nabla_x f(x) = 2(x - a) \tag{1906}$$

and the directional derivative in direction of the gradient is (1926)

$$\overset{\to \nabla_x f(x)}{df}(x) = \nabla_x f(x)^{\mathrm{T}} \nabla_x f(x) = 4(x - a)^{\mathrm{T}}(x - a) = 4(f(x) + b) \tag{1907}$$

$\square$

D.1.5   Second directional derivative

By similar argument, it so happens: the second directional derivative is equally simple. Given $g(X) : \mathbb{R}^{K \times L} \to \mathbb{R}^{M \times N}$ on open domain,

$$\nabla \frac{\partial g_{mn}(X)}{\partial X_{kl}} = \frac{\partial \nabla g_{mn}(X)}{\partial X_{kl}} = \begin{bmatrix} \frac{\partial^2 g_{mn}(X)}{\partial X_{kl} \partial X_{11}} & \frac{\partial^2 g_{mn}(X)}{\partial X_{kl} \partial X_{12}} & \cdots & \frac{\partial^2 g_{mn}(X)}{\partial X_{kl} \partial X_{1L}} \\ \frac{\partial^2 g_{mn}(X)}{\partial X_{kl} \partial X_{21}} & \frac{\partial^2 g_{mn}(X)}{\partial X_{kl} \partial X_{22}} & \cdots & \frac{\partial^2 g_{mn}(X)}{\partial X_{kl} \partial X_{2L}} \\ \vdots & \vdots & & \vdots \\ \frac{\partial^2 g_{mn}(X)}{\partial X_{kl} \partial X_{K1}} & \frac{\partial^2 g_{mn}(X)}{\partial X_{kl} \partial X_{K2}} & \cdots & \frac{\partial^2 g_{mn}(X)}{\partial X_{kl} \partial X_{KL}} \end{bmatrix} \in \mathbb{R}^{K \times L} \qquad (1908)$$

$$\nabla^2 g_{mn}(X) = \begin{bmatrix} \nabla \frac{\partial g_{mn}(X)}{\partial X_{11}} & \nabla \frac{\partial g_{mn}(X)}{\partial X_{12}} & \cdots & \nabla \frac{\partial g_{mn}(X)}{\partial X_{1L}} \\ \nabla \frac{\partial g_{mn}(X)}{\partial X_{21}} & \nabla \frac{\partial g_{mn}(X)}{\partial X_{22}} & \cdots & \nabla \frac{\partial g_{mn}(X)}{\partial X_{2L}} \\ \vdots & \vdots & & \vdots \\ \nabla \frac{\partial g_{mn}(X)}{\partial X_{K1}} & \nabla \frac{\partial g_{mn}(X)}{\partial X_{K2}} & \cdots & \nabla \frac{\partial g_{mn}(X)}{\partial X_{KL}} \end{bmatrix} \in \mathbb{R}^{K \times L \times K \times L}$$

$$= \begin{bmatrix} \frac{\partial \nabla g_{mn}(X)}{\partial X_{11}} & \frac{\partial \nabla g_{mn}(X)}{\partial X_{12}} & \cdots & \frac{\partial \nabla g_{mn}(X)}{\partial X_{1L}} \\ \frac{\partial \nabla g_{mn}(X)}{\partial X_{21}} & \frac{\partial \nabla g_{mn}(X)}{\partial X_{22}} & \cdots & \frac{\partial \nabla g_{mn}(X)}{\partial X_{2L}} \\ \vdots & \vdots & & \vdots \\ \frac{\partial \nabla g_{mn}(X)}{\partial X_{K1}} & \frac{\partial \nabla g_{mn}(X)}{\partial X_{K2}} & \cdots & \frac{\partial \nabla g_{mn}(X)}{\partial X_{KL}} \end{bmatrix} \qquad (1909)$$

Rotating our perspective, we get several views of the second-order gradient:

$$\nabla^2 g(X) = \begin{bmatrix} \nabla^2 g_{11}(X) & \nabla^2 g_{12}(X) & \cdots & \nabla^2 g_{1N}(X) \\ \nabla^2 g_{21}(X) & \nabla^2 g_{22}(X) & \cdots & \nabla^2 g_{2N}(X) \\ \vdots & \vdots & & \vdots \\ \nabla^2 g_{M1}(X) & \nabla^2 g_{M2}(X) & \cdots & \nabla^2 g_{MN}(X) \end{bmatrix} \in \mathbb{R}^{M \times N \times K \times L \times K \times L} \qquad (1910)$$

$$\nabla^2 g(X)^{\mathrm{T}_1} = \begin{bmatrix} \nabla \frac{\partial g(X)}{\partial X_{11}} & \nabla \frac{\partial g(X)}{\partial X_{12}} & \cdots & \nabla \frac{\partial g(X)}{\partial X_{1L}} \\ \nabla \frac{\partial g(X)}{\partial X_{21}} & \nabla \frac{\partial g(X)}{\partial X_{22}} & \cdots & \nabla \frac{\partial g(X)}{\partial X_{2L}} \\ \vdots & \vdots & & \vdots \\ \nabla \frac{\partial g(X)}{\partial X_{K1}} & \nabla \frac{\partial g(X)}{\partial X_{K2}} & \cdots & \nabla \frac{\partial g(X)}{\partial X_{KL}} \end{bmatrix} \in \mathbb{R}^{K \times L \times M \times N \times K \times L} \qquad (1911)$$

$$\nabla^2 g(X)^{\mathrm{T}_2} = \begin{bmatrix} \frac{\partial \nabla g(X)}{\partial X_{11}} & \frac{\partial \nabla g(X)}{\partial X_{12}} & \cdots & \frac{\partial \nabla g(X)}{\partial X_{1L}} \\ \frac{\partial \nabla g(X)}{\partial X_{21}} & \frac{\partial \nabla g(X)}{\partial X_{22}} & \cdots & \frac{\partial \nabla g(X)}{\partial X_{2L}} \\ \vdots & \vdots & & \vdots \\ \frac{\partial \nabla g(X)}{\partial X_{K1}} & \frac{\partial \nabla g(X)}{\partial X_{K2}} & \cdots & \frac{\partial \nabla g(X)}{\partial X_{KL}} \end{bmatrix} \in \mathbb{R}^{K \times L \times K \times L \times M \times N} \qquad (1912)$$

Assuming the limits exist, we may state the partial derivative of the mn^{th} entry of g with respect to the kl^{th} and ij^{th} entries of X;

$$\frac{\partial^2 g_{mn}(X)}{\partial X_{kl}\,\partial X_{ij}} = \lim_{\Delta\tau,\Delta t\to 0}\frac{g_{mn}(X+\Delta t\,e_k e_l^{\text{T}}+\Delta\tau\,e_i e_j^{\text{T}})-g_{mn}(X+\Delta t\,e_k e_l^{\text{T}})-\left(g_{mn}(X+\Delta\tau\,e_i e_j^{\text{T}})-g_{mn}(X)\right)}{\Delta\tau\,\Delta t} \quad (1913)$$

Differentiating (1893) and then scaling by Y_{ij}

$$\frac{\partial^2 g_{mn}(X)}{\partial X_{kl}\,\partial X_{ij}}Y_{kl}Y_{ij} = \lim_{\Delta t\to 0}\frac{\partial g_{mn}(X+\Delta t\,Y_{kl}\,e_k e_l^{\text{T}})-\partial g_{mn}(X)}{\partial X_{ij}\,\Delta t}Y_{ij} \quad (1914)$$

$$= \lim_{\Delta\tau,\Delta t\to 0}\frac{g_{mn}(X+\Delta t\,Y_{kl}\,e_k e_l^{\text{T}}+\Delta\tau\,Y_{ij}\,e_i e_j^{\text{T}})-g_{mn}(X+\Delta t\,Y_{kl}\,e_k e_l^{\text{T}})-\left(g_{mn}(X+\Delta\tau\,Y_{ij}\,e_i e_j^{\text{T}})-g_{mn}(X)\right)}{\Delta\tau\,\Delta t}$$

which can be proved by substitution of variables in (1913). The mn^{th} second-order total differential due to any $Y\in\mathbb{R}^{K\times L}$ is

$$d^2 g_{mn}(X)|_{dX\to Y} = \sum_{i,j}\sum_{k,l}\frac{\partial^2 g_{mn}(X)}{\partial X_{kl}\,\partial X_{ij}}Y_{kl}Y_{ij} = \text{tr}\left(\nabla_X\,\text{tr}\left(\nabla g_{mn}(X)^{\text{T}}Y\right)^{\text{T}}Y\right) \quad (1915)$$

$$= \sum_{i,j}\lim_{\Delta t\to 0}\frac{\partial g_{mn}(X+\Delta t\,Y)-\partial g_{mn}(X)}{\partial X_{ij}\,\Delta t}Y_{ij} \quad (1916)$$

$$= \lim_{\Delta t\to 0}\frac{g_{mn}(X+2\Delta t\,Y)-2g_{mn}(X+\Delta t\,Y)+g_{mn}(X)}{\Delta t^2} \quad (1917)$$

$$= \frac{d^2}{dt^2}\bigg|_{t=0}g_{mn}(X+t\,Y) \quad (1918)$$

Hence the second directional derivative,

$$\overset{\to Y}{dg^2}(X) \triangleq \begin{bmatrix} d^2 g_{11}(X) & d^2 g_{12}(X) & \cdots & d^2 g_{1N}(X) \\ d^2 g_{21}(X) & d^2 g_{22}(X) & \cdots & d^2 g_{2N}(X) \\ \vdots & \vdots & & \vdots \\ d^2 g_{M1}(X) & d^2 g_{M2}(X) & \cdots & d^2 g_{MN}(X) \end{bmatrix}\bigg|_{dX\to Y} \in\mathbb{R}^{M\times N}$$

$$= \begin{bmatrix} \text{tr}\left(\nabla\text{tr}(\nabla g_{11}(X)^{\text{T}}Y)^{\text{T}}Y\right) & \text{tr}\left(\nabla\text{tr}(\nabla g_{12}(X)^{\text{T}}Y)^{\text{T}}Y\right) & \cdots & \text{tr}\left(\nabla\text{tr}(\nabla g_{1N}(X)^{\text{T}}Y)^{\text{T}}Y\right) \\ \text{tr}\left(\nabla\text{tr}(\nabla g_{21}(X)^{\text{T}}Y)^{\text{T}}Y\right) & \text{tr}\left(\nabla\text{tr}(\nabla g_{22}(X)^{\text{T}}Y)^{\text{T}}Y\right) & \cdots & \text{tr}\left(\nabla\text{tr}(\nabla g_{2N}(X)^{\text{T}}Y)^{\text{T}}Y\right) \\ \vdots & \vdots & & \vdots \\ \text{tr}\left(\nabla\text{tr}(\nabla g_{M1}(X)^{\text{T}}Y)^{\text{T}}Y\right) & \text{tr}\left(\nabla\text{tr}(\nabla g_{M2}(X)^{\text{T}}Y)^{\text{T}}Y\right) & \cdots & \text{tr}\left(\nabla\text{tr}(\nabla g_{MN}(X)^{\text{T}}Y)^{\text{T}}Y\right) \end{bmatrix}$$

$$= \begin{bmatrix} \sum_{i,j}\sum_{k,l}\frac{\partial^2 g_{11}(X)}{\partial X_{kl}\,\partial X_{ij}}Y_{kl}Y_{ij} & \sum_{i,j}\sum_{k,l}\frac{\partial^2 g_{12}(X)}{\partial X_{kl}\,\partial X_{ij}}Y_{kl}Y_{ij} & \cdots & \sum_{i,j}\sum_{k,l}\frac{\partial^2 g_{1N}(X)}{\partial X_{kl}\,\partial X_{ij}}Y_{kl}Y_{ij} \\ \sum_{i,j}\sum_{k,l}\frac{\partial^2 g_{21}(X)}{\partial X_{kl}\,\partial X_{ij}}Y_{kl}Y_{ij} & \sum_{i,j}\sum_{k,l}\frac{\partial^2 g_{22}(X)}{\partial X_{kl}\,\partial X_{ij}}Y_{kl}Y_{ij} & \cdots & \sum_{i,j}\sum_{k,l}\frac{\partial^2 g_{2N}(X)}{\partial X_{kl}\,\partial X_{ij}}Y_{kl}Y_{ij} \\ \vdots & \vdots & & \vdots \\ \sum_{i,j}\sum_{k,l}\frac{\partial^2 g_{M1}(X)}{\partial X_{kl}\,\partial X_{ij}}Y_{kl}Y_{ij} & \sum_{i,j}\sum_{k,l}\frac{\partial^2 g_{M2}(X)}{\partial X_{kl}\,\partial X_{ij}}Y_{kl}Y_{ij} & \cdots & \sum_{i,j}\sum_{k,l}\frac{\partial^2 g_{MN}(X)}{\partial X_{kl}\,\partial X_{ij}}Y_{kl}Y_{ij} \end{bmatrix} \quad (1919)$$

from which it follows

$$\overset{\rightarrow Y}{dg^2}(X) = \sum_{i,j}\sum_{k,l} \frac{\partial^2 g(X)}{\partial X_{kl}\,\partial X_{ij}} Y_{kl} Y_{ij} = \sum_{i,j} \frac{\partial}{\partial X_{ij}}\, \overset{\rightarrow Y}{dg}(X)\, Y_{ij} \qquad (1920)$$

Yet for all $X \in \operatorname{dom} g$, any $Y \in \mathbb{R}^{K \times L}$, and some open interval of $t \in \mathbb{R}$

$$g(X + tY) = g(X) \,+\, t\,\overset{\rightarrow Y}{dg}(X) \,+\, \frac{1}{2!}t^2 \overset{\rightarrow Y}{dg^2}(X) \,+\, o(t^3) \qquad (1921)$$

which is the second-order Taylor series expansion about X. [225, §18.4] [156, §2.3.4] Differentiating twice with respect to t and subsequent t-zeroing isolates the third term of the expansion. Thus differentiating and zeroing $g(X + tY)$ in t is an operation equivalent to individually differentiating and zeroing every entry $g_{mn}(X + tY)$ as in (1918). So the second directional derivative of $g(X) : \mathbb{R}^{K \times L} \to \mathbb{R}^{M \times N}$ becomes [284, §2.1, §5.4.5] [34, §6.3.1]

$$\overset{\rightarrow Y}{dg^2}(X) = \frac{d^2}{dt^2}\bigg|_{t=0} g(X + tY) \in \mathbb{R}^{M \times N} \qquad (1922)$$

which is again simplest. (*confer* (1901)) Directional derivative retains the dimensions of g.

D.1.6 directional derivative expressions

In the case of a real function $g(X) : \mathbb{R}^{K \times L} \to \mathbb{R}$, all its directional derivatives are in $\mathbb{R}$:

$$\overset{\rightarrow Y}{dg}(X) = \operatorname{tr}\!\big(\nabla g(X)^{\mathrm T} Y\big) \qquad (1923)$$

$$\overset{\rightarrow Y}{dg^2}(X) = \operatorname{tr}\!\Big(\nabla_X \operatorname{tr}\!\big(\nabla g(X)^{\mathrm T} Y\big)^{\mathrm T} Y\Big) = \operatorname{tr}\!\Big(\nabla_X \overset{\rightarrow Y}{dg}(X)^{\mathrm T} Y\Big) \qquad (1924)$$

$$\overset{\rightarrow Y}{dg^3}(X) = \operatorname{tr}\!\Big(\nabla_X \operatorname{tr}\!\big(\nabla_X \operatorname{tr}\!\big(\nabla g(X)^{\mathrm T} Y\big)^{\mathrm T} Y\big)^{\mathrm T} Y\Big) = \operatorname{tr}\!\Big(\nabla_X \overset{\rightarrow Y}{dg^2}(X)^{\mathrm T} Y\Big) \qquad (1925)$$

In the case $g(X) : \mathbb{R}^K \to \mathbb{R}$ has vector argument, they further simplify:

$$\overset{\rightarrow Y}{dg}(X) = \nabla g(X)^{\mathrm T} Y \qquad (1926)$$

$$\overset{\rightarrow Y}{dg^2}(X) = Y^{\mathrm T} \nabla^2 g(X) Y \qquad (1927)$$

$$\overset{\rightarrow Y}{dg^3}(X) = \nabla_X\!\big(Y^{\mathrm T} \nabla^2 g(X) Y\big)^{\mathrm T} Y \qquad (1928)$$

and so on.

D.1.7 Taylor series

Series expansions of the differentiable matrix-valued function $g(X)$, of matrix argument, were given earlier in (1900) and (1921). Assuming $g(X)$ has continuous first-, second-, and third-order gradients over the open set $\text{dom}\, g$, then for $X \in \text{dom}\, g$ and any $Y \in \mathbb{R}^{K \times L}$ the complete Taylor series is expressed on some open interval of $\mu \in \mathbb{R}$

$$g(X + \mu Y) = g(X) \;+\; \mu\, \overset{\to Y}{dg}(X) \;+\; \frac{1}{2!}\mu^2\, \overset{\to Y}{dg^2}(X) \;+\; \frac{1}{3!}\mu^3\, \overset{\to Y}{dg^3}(X) \;+\; o(\mu^4) \qquad (1929)$$

or on some open interval of $\|Y\|_2$

$$g(Y) = g(X) \;+\; \overset{\to Y-X}{dg}(X) \;+\; \frac{1}{2!}\overset{\to Y-X}{dg^2}(X) \;+\; \frac{1}{3!}\overset{\to Y-X}{dg^3}(X) \;+\; o(\|Y\|^4) \qquad (1930)$$

which are third-order expansions about X. The *mean value theorem* from calculus is what insures finite order of the series. [225] [42, §1.1] [41, App.A.5] [205, §0.4] These somewhat unbelievable formulae imply that a function can be determined over the whole of its domain by knowing its value and all its directional derivatives at a single point X.

D.1.7.0.1 Example. *Inverse-matrix function.*
Say $g(Y) = Y^{-1}$. From the table on page 596,

$$\overset{\to Y}{dg}(X) = \left.\frac{d}{dt}\right|_{t=0} g(X + tY) = -X^{-1}YX^{-1} \qquad (1931)$$

$$\overset{\to Y}{dg^2}(X) = \left.\frac{d^2}{dt^2}\right|_{t=0} g(X + tY) = 2X^{-1}YX^{-1}YX^{-1} \qquad (1932)$$

$$\overset{\to Y}{dg^3}(X) = \left.\frac{d^3}{dt^3}\right|_{t=0} g(X + tY) = -6X^{-1}YX^{-1}YX^{-1}YX^{-1} \qquad (1933)$$

Let's find the Taylor series expansion of g about $X = I$: Since $g(I) = I$, for $\|Y\|_2 < 1$ ($\mu = 1$ in (1929))

$$g(I + Y) \;=\; (I + Y)^{-1} = I - Y + Y^2 - Y^3 + \dots \qquad (1934)$$

If Y is small, $(I + Y)^{-1} \approx I - Y$.[D.3] Now we find Taylor series expansion about X:

$$g(X + Y) \;=\; (X + Y)^{-1} = X^{-1} - X^{-1}YX^{-1} + 2X^{-1}YX^{-1}YX^{-1} - \dots \qquad (1935)$$

If Y is small, $(X + Y)^{-1} \approx X^{-1} - X^{-1}YX^{-1}$. $\qquad\qquad\qquad\qquad\qquad$ $\square$

D.1.7.0.2 Exercise. *log det*. $\qquad\qquad\qquad\qquad\qquad\qquad$ (*confer* [62, p.644])
Find the first three terms of a Taylor series expansion for $\log \det Y$. Specify an open interval over which the expansion holds in vicinity of X. $\qquad\qquad\qquad$ ▼

[D.3]Had we instead set $g(Y) = (I + Y)^{-1}$, then the equivalent expansion would have been about $X = \mathbf{0}$.

D.1.8   Correspondence of gradient to derivative

From the foregoing expressions for directional derivative, we derive a relationship between gradient with respect to matrix X and derivative with respect to real variable t :

D.1.8.1   first-order

Removing evaluation at $t = 0$ from (1901),[D.4] we find an expression for the directional derivative of $g(X)$ in direction Y evaluated anywhere along a line $\{X + tY \mid t \in \mathbb{R}\}$ intersecting dom g

$$\overset{\rightarrow Y}{dg}(X + tY) = \frac{d}{dt}g(X + tY) \tag{1936}$$

In the general case $g(X) : \mathbb{R}^{K \times L} \to \mathbb{R}^{M \times N}$, from (1894) and (1897) we find

$$\operatorname{tr}\big(\nabla_X\, g_{mn}(X + tY)^{\mathrm{T}}Y\big) = \frac{d}{dt}g_{mn}(X + tY) \tag{1937}$$

which is valid at $t = 0$, of course, when $X \in$ dom g. In the important case of a real function $g(X) : \mathbb{R}^{K \times L} \to \mathbb{R}$, from (1923) we have simply

$$\operatorname{tr}\big(\nabla_X\, g(X + tY)^{\mathrm{T}}Y\big) = \frac{d}{dt}g(X + tY) \tag{1938}$$

When, additionally, $g(X) : \mathbb{R}^K \to \mathbb{R}$ has vector argument,

$$\nabla_X\, g(X + tY)^{\mathrm{T}}Y = \frac{d}{dt}g(X + tY) \tag{1939}$$

D.1.8.1.1   Example.   *Gradient.*
$g(X) = w^{\mathrm{T}}X^{\mathrm{T}}Xw$, $X \in \mathbb{R}^{K \times L}$, $w \in \mathbb{R}^L$. Using the tables in §D.2,

$$\begin{aligned}
\operatorname{tr}\big(\nabla_X\, g(X + tY)^{\mathrm{T}}Y\big) &= \operatorname{tr}\big(2ww^{\mathrm{T}}(X^{\mathrm{T}} + tY^{\mathrm{T}})Y\big) & (1940) \\
&= 2w^{\mathrm{T}}(X^{\mathrm{T}}Y + tY^{\mathrm{T}}Y)w & (1941)
\end{aligned}$$

Applying the equivalence (1938),

$$\begin{aligned}
\frac{d}{dt}g(X + tY) &= \frac{d}{dt}w^{\mathrm{T}}(X + tY)^{\mathrm{T}}(X + tY)w & (1942) \\
&= w^{\mathrm{T}}\big(X^{\mathrm{T}}Y + Y^{\mathrm{T}}X + 2t\,Y^{\mathrm{T}}Y\big)w & (1943) \\
&= 2w^{\mathrm{T}}(X^{\mathrm{T}}Y + tY^{\mathrm{T}}Y)w & (1944)
\end{aligned}$$

which is the same as (1941); hence, equivalence is demonstrated.

[D.4]Justified by replacing X with $X + tY$ in (1894)-(1896); beginning,

$$dg_{mn}(X + tY)\big|_{dX \to Y} = \sum_{k,\,l} \frac{\partial g_{mn}(X + tY)}{\partial X_{kl}}Y_{kl}$$

It is easy to extract $\nabla g(X)$ from (1944) knowing only (1938):

$$
\begin{aligned}
\operatorname{tr}\!\big(\nabla_X\, g(X + t\,Y)^{\mathrm{T}} Y\big) &= 2w^{\mathrm{T}}(X^{\mathrm{T}}Y + t\,Y^{\mathrm{T}}Y)w \\
&= 2\operatorname{tr}\!\big(ww^{\mathrm{T}}(X^{\mathrm{T}} + t\,Y^{\mathrm{T}})Y\big) \\
\operatorname{tr}\!\big(\nabla_X\, g(X)^{\mathrm{T}} Y\big) &= 2\operatorname{tr}\!\big(ww^{\mathrm{T}}X^{\mathrm{T}}Y\big) \\
&\Leftrightarrow \\
\nabla_X\, g(X) &= 2Xww^{\mathrm{T}}
\end{aligned}
\tag{1945}
$$

$\square$

D.1.8.2   second-order

Likewise removing the evaluation at $t = 0$ from (1922),

$$
\overset{\to Y}{dg^2}(X + t\,Y) = \frac{d^2}{dt^2} g(X + t\,Y)
\tag{1946}
$$

we can find a similar relationship between second-order gradient and second derivative: In the general case $g(X) : \mathbb{R}^{K \times L} \to \mathbb{R}^{M \times N}$ from (1915) and (1918),

$$
\operatorname{tr}\!\left(\nabla_X \operatorname{tr}\!\big(\nabla_X\, g_{mn}(X + t\,Y)^{\mathrm{T}} Y\big)^{\mathrm{T}} Y\right) = \frac{d^2}{dt^2} g_{mn}(X + t\,Y)
\tag{1947}
$$

In the case of a real function $g(X) : \mathbb{R}^{K \times L} \to \mathbb{R}$ we have, of course,

$$
\operatorname{tr}\!\left(\nabla_X \operatorname{tr}\!\big(\nabla_X\, g(X + t\,Y)^{\mathrm{T}} Y\big)^{\mathrm{T}} Y\right) = \frac{d^2}{dt^2} g(X + t\,Y)
\tag{1948}
$$

From (1927), the simpler case, where the real function $g(X) : \mathbb{R}^K \to \mathbb{R}$ has vector argument,

$$
Y^{\mathrm{T}} \nabla_X^2 g(X + t\,Y) Y = \frac{d^2}{dt^2} g(X + t\,Y)
\tag{1949}
$$

D.1.8.2.1   Example.   *Second-order gradient.*
Given real function $g(X) = \log \det X$ having domain $\operatorname{int}\mathbb{S}_+^K$, we want to find $\nabla^2 g(X) \in \mathbb{R}^{K \times K \times K \times K}$. From the tables in §D.2,

$$
h(X) \triangleq \nabla g(X) = X^{-1} \in \operatorname{int}\mathbb{S}_+^K
\tag{1950}
$$

so $\nabla^2 g(X) = \nabla h(X)$. By (1937) and (1900), for $Y \in \mathbb{S}^K$

$$
\operatorname{tr}\!\big(\nabla h_{mn}(X)^{\mathrm{T}} Y\big) = \left.\frac{d}{dt}\right|_{t=0} h_{mn}(X + t\,Y)
\tag{1951}
$$

$$
= \left(\left.\frac{d}{dt}\right|_{t=0} h(X + t\,Y)\right)_{mn}
\tag{1952}
$$

$$
= \left(\left.\frac{d}{dt}\right|_{t=0} (X + t\,Y)^{-1}\right)_{mn}
\tag{1953}
$$

$$
= -\big(X^{-1} Y X^{-1}\big)_{mn}
\tag{1954}
$$

Setting Y to a member of $\{e_k e_l^{\mathrm{T}} \in \mathbb{R}^{K \times K} \mid k, l = 1 \ldots K\}$, and employing a property (39) of the trace function we find

$$\nabla^2 g(X)_{mnkl} = \operatorname{tr}\big(\nabla h_{mn}(X)^{\mathrm{T}} e_k e_l^{\mathrm{T}}\big) = \nabla h_{mn}(X)_{kl} = -\big(X^{-1} e_k e_l^{\mathrm{T}} X^{-1}\big)_{mn} \qquad (1955)$$

$$\nabla^2 g(X)_{kl} = \nabla h(X)_{kl} = -\big(X^{-1} e_k e_l^{\mathrm{T}} X^{-1}\big) \in \mathbb{R}^{K \times K} \qquad (1956)$$

$$\square$$

From all these first- and second-order expressions, we may generate new ones by evaluating both sides at arbitrary t (in some open interval) but only after the differentiation.

D.2 Tables of gradients and derivatives

- Results may be numerically proven by Romberg extrapolation. [111] When proving results for symmetric matrices algebraically, it is critical to take gradients ignoring symmetry and to then substitute symmetric entries afterward. [172] [66]

- $a, b \in \mathbb{R}^n$, $\quad x, y \in \mathbb{R}^k$, $\quad A, B \in \mathbb{R}^{m \times n}$, $\quad X, Y \in \mathbb{R}^{K \times L}$, $\quad t, \mu \in \mathbb{R}$, $i, j, k, \ell, K, L, m, n, M, N$ are integers, unless otherwise noted.

- x^μ means $\delta(\delta(x)^\mu)$ for $\mu \in \mathbb{R}$; *id est*, entrywise vector exponentiation. δ is the main-diagonal linear operator (1496). $x^0 \triangleq \mathbf{1}$, $X^0 \triangleq I$ if square.

- $\frac{d}{dx} \triangleq \begin{bmatrix} \frac{d}{dx_1} \\ \vdots \\ \frac{d}{dx_k} \end{bmatrix}$, $\quad \overset{\rightarrow y}{dg}(x)$, $\quad \overset{\rightarrow y}{dg^2}(x)$ (directional derivatives §D.1), $\quad \log x$, $\quad e^x$, $\quad |x|$,

  $\operatorname{sgn} x$, x/y (Hadamard quotient), $\sqrt[\circ]{x}$ (entrywise square root), *etcetera*, are maps $f : \mathbb{R}^k \to \mathbb{R}^k$ that maintain dimension; *e.g.*, (§A.1.1)

$$\frac{d}{dx} x^{-1} \triangleq \nabla_x \mathbf{1}^{\mathrm{T}} \delta(x)^{-1} \mathbf{1} \qquad (1957)$$

- For A a scalar or square matrix, we have the Taylor series [78, §3.6]

$$e^A \triangleq \sum_{k=0}^{\infty} \frac{1}{k!} A^k \qquad (1958)$$

Further, [338, §5.4]

$$e^A \succ 0 \qquad \forall A \in \mathbb{S}^m \qquad (1959)$$

- For all square A and integer k

$$\det{}^k A = \det A^k \qquad (1960)$$

D.2.1 algebraic

$$\nabla_x\, x = \nabla_x\, x^{\mathrm{T}} = I \in \mathbb{R}^{k \times k}$$

$$\nabla_x(Ax - b) = A^{\mathrm{T}}$$

$$\nabla_x\big(x^{\mathrm{T}}A - b^{\mathrm{T}}\big) = A$$

$$\nabla_x(Ax - b)^{\mathrm{T}}(Ax - b) = 2A^{\mathrm{T}}(Ax - b)$$
$$\nabla_x^2(Ax - b)^{\mathrm{T}}(Ax - b) = 2A^{\mathrm{T}}A$$

$$\nabla_x\|Ax - b\| = A^{\mathrm{T}}(Ax - b)/\|Ax - b\|$$

$$\nabla_x z^{\mathrm{T}}|Ax - b| = A^{\mathrm{T}}\delta(z)\,\mathrm{sgn}(Ax - b)\,,\ \ z_i \neq 0 \Rightarrow (Ax - b)_i \neq 0$$
$$\nabla_x \mathbf{1}^{\mathrm{T}} f\big(|Ax - b|\big) = A^{\mathrm{T}}\delta\!\left(\frac{df(y)}{dy}\bigg|_{y=|Ax-b|}\right)\mathrm{sgn}(Ax - b)$$

$$\nabla_x\big(x^{\mathrm{T}}Ax + 2x^{\mathrm{T}}By + y^{\mathrm{T}}Cy\big) = \big(A + A^{\mathrm{T}}\big)x + 2By$$
$$\nabla_x(x + y)^{\mathrm{T}}A(x + y) = (A + A^{\mathrm{T}})(x + y)$$
$$\nabla_x^2\big(x^{\mathrm{T}}Ax + 2x^{\mathrm{T}}By + y^{\mathrm{T}}Cy\big) = A + A^{\mathrm{T}}$$

$$\nabla_x\, a^{\mathrm{T}}x^{\mathrm{T}}xb = 2xa^{\mathrm{T}}b$$

$$\nabla_x\, a^{\mathrm{T}}xx^{\mathrm{T}}b = (ab^{\mathrm{T}} + ba^{\mathrm{T}})x$$

$$\nabla_x\, a^{\mathrm{T}}x^{\mathrm{T}}xa = 2xa^{\mathrm{T}}a$$

$$\nabla_x\, a^{\mathrm{T}}xx^{\mathrm{T}}a = 2aa^{\mathrm{T}}x$$

$$\nabla_x\, a^{\mathrm{T}}yx^{\mathrm{T}}b = ba^{\mathrm{T}}y$$

$$\nabla_x\, a^{\mathrm{T}}y^{\mathrm{T}}xb = yb^{\mathrm{T}}a$$

$$\nabla_x\, a^{\mathrm{T}}xy^{\mathrm{T}}b = ab^{\mathrm{T}}y$$

$$\nabla_x\, a^{\mathrm{T}}x^{\mathrm{T}}yb = ya^{\mathrm{T}}b$$

$$\nabla_X X = \nabla_X X^{\mathrm{T}} \triangleq I \in \mathbb{R}^{K \times L \times K \times L} \qquad \text{(Identity)}$$

$$\nabla_X\, a^{\mathrm{T}}Xb = \nabla_X\, b^{\mathrm{T}}X^{\mathrm{T}}a = ab^{\mathrm{T}}$$

$$\nabla_X\, a^{\mathrm{T}}X^2 b = X^{\mathrm{T}}ab^{\mathrm{T}} + ab^{\mathrm{T}}X^{\mathrm{T}}$$

$$\nabla_X\, a^{\mathrm{T}}X^{-1}b = -X^{-\mathrm{T}}ab^{\mathrm{T}}X^{-\mathrm{T}}$$

$$\nabla_X(X^{-1})_{kl} = \frac{\partial X^{-1}}{\partial X_{kl}} = -X^{-1}e_k e_l^{\mathrm{T}}X^{-1},\quad \begin{array}{l}\textit{confer}\\ (1892)\\ (1956)\end{array}$$

$$\nabla_X\, a^{\mathrm{T}}X^{\mathrm{T}}Xb = X(ab^{\mathrm{T}} + ba^{\mathrm{T}})$$

$$\nabla_X\, a^{\mathrm{T}}XX^{\mathrm{T}}b = (ab^{\mathrm{T}} + ba^{\mathrm{T}})X$$

$$\nabla_X\, a^{\mathrm{T}}X^{\mathrm{T}}Xa = 2Xaa^{\mathrm{T}}$$

$$\nabla_X\, a^{\mathrm{T}}XX^{\mathrm{T}}a = 2aa^{\mathrm{T}}X$$

$$\nabla_X\, a^{\mathrm{T}}YX^{\mathrm{T}}b = ba^{\mathrm{T}}Y$$

$$\nabla_X\, a^{\mathrm{T}}Y^{\mathrm{T}}Xb = Yab^{\mathrm{T}}$$

$$\nabla_X\, a^{\mathrm{T}}XY^{\mathrm{T}}b = ab^{\mathrm{T}}Y$$

$$\nabla_X\, a^{\mathrm{T}}X^{\mathrm{T}}Yb = Yba^{\mathrm{T}}$$

algebraic continued

$$\frac{d}{dt}(X + tY) = Y$$

$$\frac{d}{dt}B^{\mathrm{T}}(X + tY)^{-1}A = -B^{\mathrm{T}}(X + tY)^{-1}Y(X + tY)^{-1}A$$
$$\frac{d}{dt}B^{\mathrm{T}}(X + tY)^{-\mathrm{T}}A = -B^{\mathrm{T}}(X + tY)^{-\mathrm{T}}Y^{\mathrm{T}}(X + tY)^{-\mathrm{T}}A$$
$$\frac{d}{dt}B^{\mathrm{T}}(X + tY)^{\mu}A = \dots, \quad -1 \le \mu \le 1, \quad X, Y \in \mathbb{S}_+^M$$

$$\frac{d^2}{dt^2}B^{\mathrm{T}}(X + tY)^{-1}A = \quad 2B^{\mathrm{T}}(X + tY)^{-1}Y(X + tY)^{-1}Y(X + tY)^{-1}A$$
$$\frac{d^3}{dt^3}B^{\mathrm{T}}(X + tY)^{-1}A = -6B^{\mathrm{T}}(X + tY)^{-1}Y(X + tY)^{-1}Y(X + tY)^{-1}Y(X + tY)^{-1}A$$

$$\frac{d}{dt}\big((X + tY)^{\mathrm{T}}A(X + tY)\big) = Y^{\mathrm{T}}AX + X^{\mathrm{T}}AY + 2\,t\,Y^{\mathrm{T}}AY$$
$$\frac{d^2}{dt^2}\big((X + tY)^{\mathrm{T}}A(X + tY)\big) = 2\,Y^{\mathrm{T}}AY$$
$$\frac{d}{dt}\big((X + tY)^{\mathrm{T}}A(X + tY)\big)^{-1}$$
$$= -\big((X + tY)^{\mathrm{T}}A(X + tY)\big)^{-1}(Y^{\mathrm{T}}AX + X^{\mathrm{T}}AY + 2\,Y^{\mathrm{T}}AY)\big((X + tY)^{\mathrm{T}}A(X + tY)\big)^{-1}$$
$$\frac{d}{dt}\big((X + tY)A(X + tY)\big) = YAX + XAY + 2\,t\,YAY$$
$$\frac{d^2}{dt^2}\big((X + tY)A(X + tY)\big) = 2\,YAY$$

D.2.1.0.1 Exercise. *Expand these tables.*
Provide unfinished table entries indicated by … throughout §D.2.   ▼

D.2.1.0.2 Exercise. *log*.   (§D.1.7)
Find the first four terms of the Taylor series expansion for $\log x$ about $x = 1$. Prove that
$\log x \le x - 1$; alternatively, plot the supporting hyperplane to the hypograph of $\log x$ at
$\begin{bmatrix} x \\ \log x \end{bmatrix} = \begin{bmatrix} 1 \\ 0 \end{bmatrix}$.   ▼

D.2.2   trace Kronecker

$$\nabla_{\mathrm{vec}\,X}\,\mathrm{tr}(AXBX^{\mathrm{T}}) = \nabla_{\mathrm{vec}\,X}\,\mathrm{vec}(X)^{\mathrm{T}}(B^{\mathrm{T}} \otimes A)\,\mathrm{vec}\,X = (B \otimes A^{\mathrm{T}} + B^{\mathrm{T}} \otimes A)\,\mathrm{vec}\,X$$

$$\nabla_{\mathrm{vec}\,X}^2\,\mathrm{tr}(AXBX^{\mathrm{T}}) = \nabla_{\mathrm{vec}\,X}^2\,\mathrm{vec}(X)^{\mathrm{T}}(B^{\mathrm{T}} \otimes A)\,\mathrm{vec}\,X = B \otimes A^{\mathrm{T}} + B^{\mathrm{T}} \otimes A$$

D.2.3 trace

$\nabla_x \, \mu \, x = \mu I$

$\nabla_X \operatorname{tr} \mu X = \nabla_X \mu \operatorname{tr} X = \mu I$

$\nabla_x \mathbf{1}^{\mathrm{T}} \delta(x)^{-1} \mathbf{1} = \frac{d}{dx} x^{-1} = -x^{-2}$
$\nabla_x \mathbf{1}^{\mathrm{T}} \delta(x)^{-1} y = -\delta(x)^{-2} y$

$\nabla_X \operatorname{tr} X^{-1} = -X^{-2\mathrm{T}}$
$\nabla_X \operatorname{tr}(X^{-1}Y) = \nabla_X \operatorname{tr}(YX^{-1}) = -X^{-\mathrm{T}}Y^{\mathrm{T}}X^{-\mathrm{T}}$

$\frac{d}{dx} x^{\mu} = \mu x^{\mu-1}$

$\nabla_X \operatorname{tr} X^{\mu} = \mu X^{\mu-1}, \qquad\qquad\qquad X \in \mathbb{S}^{M}$

$\nabla_X \operatorname{tr} X^{j} = j X^{(j-1)\mathrm{T}}$

$\nabla_x (b - a^{\mathrm{T}}x)^{-1} = (b - a^{\mathrm{T}}x)^{-2} a$

$\nabla_X \operatorname{tr}\big((B - AX)^{-1}\big) = \big((B - AX)^{-2} A\big)^{\mathrm{T}}$

$\nabla_x (b - a^{\mathrm{T}}x)^{\mu} = -\mu(b - a^{\mathrm{T}}x)^{\mu-1} a$

$\nabla_x \, x^{\mathrm{T}}y = \nabla_x \, y^{\mathrm{T}}x = y$

$\nabla_X \operatorname{tr}(X^{\mathrm{T}}Y) = \nabla_X \operatorname{tr}(YX^{\mathrm{T}}) = \nabla_X \operatorname{tr}(Y^{\mathrm{T}}X) = \nabla_X \operatorname{tr}(XY^{\mathrm{T}}) = Y$

$\nabla_X \operatorname{tr}(AXBX^{\mathrm{T}}) = \nabla_X \operatorname{tr}(XBX^{\mathrm{T}}A) = A^{\mathrm{T}}XB^{\mathrm{T}} + AXB$
$\nabla_X \operatorname{tr}(AXBX) = \nabla_X \operatorname{tr}(XBXA) = A^{\mathrm{T}}X^{\mathrm{T}}B^{\mathrm{T}} + B^{\mathrm{T}}X^{\mathrm{T}}A^{\mathrm{T}}$

$\nabla_X \operatorname{tr}(AXAXAX) = \nabla_X \operatorname{tr}(XAXAXA) = 3(AXAXA)^{\mathrm{T}}$

$\nabla_X \operatorname{tr}(YX^{k}) = \nabla_X \operatorname{tr}(X^{k}Y) = \sum_{i=0}^{k-1} \big(X^{i}YX^{k-1-i}\big)^{\mathrm{T}}$

$\nabla_X \operatorname{tr}(Y^{\mathrm{T}}XX^{\mathrm{T}}Y) = \nabla_X \operatorname{tr}(X^{\mathrm{T}}YY^{\mathrm{T}}X) = 2YY^{\mathrm{T}}X$
$\nabla_X \operatorname{tr}(Y^{\mathrm{T}}X^{\mathrm{T}}XY) = \nabla_X \operatorname{tr}(XYY^{\mathrm{T}}X^{\mathrm{T}}) = 2XYY^{\mathrm{T}}$

$\nabla_X \operatorname{tr}\big((X + Y)^{\mathrm{T}}(X + Y)\big) = 2(X + Y) = \nabla_X \|X + Y\|_{\mathrm{F}}^{2}$
$\nabla_X \operatorname{tr}\big((X + Y)(X + Y)\big) = 2(X + Y)^{\mathrm{T}}$

$\nabla_X \operatorname{tr}(A^{\mathrm{T}}XB) = \nabla_X \operatorname{tr}(X^{\mathrm{T}}AB^{\mathrm{T}}) = AB^{\mathrm{T}}$
$\nabla_X \operatorname{tr}(A^{\mathrm{T}}X^{-1}B) = \nabla_X \operatorname{tr}(X^{-\mathrm{T}}AB^{\mathrm{T}}) = -X^{-\mathrm{T}}AB^{\mathrm{T}}X^{-\mathrm{T}}$

$\nabla_X \, a^{\mathrm{T}}Xb = \nabla_X \operatorname{tr}(ba^{\mathrm{T}}X) = \nabla_X \operatorname{tr}(Xba^{\mathrm{T}}) = ab^{\mathrm{T}}$
$\nabla_X \, b^{\mathrm{T}}X^{\mathrm{T}}a = \nabla_X \operatorname{tr}(X^{\mathrm{T}}ab^{\mathrm{T}}) = \nabla_X \operatorname{tr}(ab^{\mathrm{T}}X^{\mathrm{T}}) = ab^{\mathrm{T}}$
$\nabla_X \, a^{\mathrm{T}}X^{-1}b = \nabla_X \operatorname{tr}(X^{-\mathrm{T}}ab^{\mathrm{T}}) = -X^{-\mathrm{T}}ab^{\mathrm{T}}X^{-\mathrm{T}}$
$\nabla_X \, a^{\mathrm{T}}X^{\mu}b = \ldots$

trace continued

$\frac{d}{dt} \operatorname{tr} g(X+tY) = \operatorname{tr} \frac{d}{dt} g(X+tY)$ [209, p.491]

$\frac{d}{dt} \operatorname{tr}(X+tY) = \operatorname{tr} Y$

$\frac{d}{dt} \operatorname{tr}^j(X+tY) = j \operatorname{tr}^{j-1}(X+tY) \operatorname{tr} Y$

$\frac{d}{dt} \operatorname{tr}(X+tY)^j = j \operatorname{tr}\big((X+tY)^{j-1}Y\big)$ $(\forall j)$

$\frac{d}{dt} \operatorname{tr}((X+tY)Y) = \operatorname{tr} Y^2$

$\frac{d}{dt} \operatorname{tr}\big((X+tY)^k Y\big) = \frac{d}{dt} \operatorname{tr}(Y(X+tY)^k) = k \operatorname{tr}\big((X+tY)^{k-1}Y^2\big) , \quad k \in \{0,1,2\}$

$\frac{d}{dt} \operatorname{tr}\big((X+tY)^k Y\big) = \frac{d}{dt} \operatorname{tr}(Y(X+tY)^k) = \operatorname{tr} \sum_{i=0}^{k-1} (X+tY)^i Y(X+tY)^{k-1-i} Y$

$\frac{d}{dt} \operatorname{tr}\big((X+tY)^{-1}Y\big) \quad = -\operatorname{tr}\big((X+tY)^{-1}Y(X+tY)^{-1}Y\big)$
$\frac{d}{dt} \operatorname{tr}\big(B^{\mathrm{T}}(X+tY)^{-1}A\big) = -\operatorname{tr}\big(B^{\mathrm{T}}(X+tY)^{-1}Y(X+tY)^{-1}A\big)$
$\frac{d}{dt} \operatorname{tr}\big(B^{\mathrm{T}}(X+tY)^{-\mathrm{T}}A\big) = -\operatorname{tr}\big(B^{\mathrm{T}}(X+tY)^{-\mathrm{T}}Y^{\mathrm{T}}(X+tY)^{-\mathrm{T}}A\big)$
$\frac{d}{dt} \operatorname{tr}\big(B^{\mathrm{T}}(X+tY)^{-k}A\big) = ... , \quad k>0$
$\frac{d}{dt} \operatorname{tr}\big(B^{\mathrm{T}}(X+tY)^{\mu}A\big) \ = ... , \quad -1 \le \mu \le 1, \ \ X,Y \in \mathbb{S}_+^M$

$\frac{d^2}{dt^2} \operatorname{tr}\big(B^{\mathrm{T}}(X+tY)^{-1}A\big) = 2 \operatorname{tr}\big(B^{\mathrm{T}}(X+tY)^{-1}Y(X+tY)^{-1}Y(X+tY)^{-1}A\big)$

$\frac{d}{dt} \operatorname{tr}\big((X+tY)^{\mathrm{T}}A(X+tY)\big) = \operatorname{tr}\big(Y^{\mathrm{T}}AX + X^{\mathrm{T}}AY + 2t Y^{\mathrm{T}}AY\big)$
$\frac{d^2}{dt^2} \operatorname{tr}\big((X+tY)^{\mathrm{T}}A(X+tY)\big) = 2 \operatorname{tr}\big(Y^{\mathrm{T}}AY\big)$
$\frac{d}{dt} \operatorname{tr}\Big(\big((X+tY)^{\mathrm{T}}A(X+tY)\big)^{-1}\Big)$
$\quad = -\operatorname{tr}\Big(\big((X+tY)^{\mathrm{T}}A(X+tY)\big)^{-1}(Y^{\mathrm{T}}AX + X^{\mathrm{T}}AY + 2t Y^{\mathrm{T}}AY)\big((X+tY)^{\mathrm{T}}A(X+tY)\big)^{-1}\Big)$
$\frac{d}{dt} \operatorname{tr}\big((X+tY)A(X+tY)\big) = \operatorname{tr}(YAX + XAY + 2t YAY)$
$\frac{d^2}{dt^2} \operatorname{tr}\big((X+tY)A(X+tY)\big) = 2 \operatorname{tr}(YAY)$

D.2.4 logarithmic determinant

$x \succ 0$, $\det X > 0$ on some neighborhood of X, and $\det(X + tY) > 0$ on some open interval of t; otherwise, $\log(\ )$ would be discontinuous. [84, p.75]

$\frac{d}{dx} \log x = x^{-1}$	$\nabla_X \log \det X = X^{-\mathrm{T}}$ $\nabla_X^2 \log \det(X)_{kl} = \frac{\partial X^{-\mathrm{T}}}{\partial X_{kl}} = -\left(X^{-1} e_k e_l^\mathrm{T} X^{-1}\right)^\mathrm{T}$, $\;confer\,(1909)(1956)$
$\frac{d}{dx} \log x^{-1} = -x^{-1}$	$\nabla_X \log \det X^{-1} = -X^{-\mathrm{T}}$
$\frac{d}{dx} \log x^\mu = \mu x^{-1}$	$\nabla_X \log \det{}^\mu X = \mu X^{-\mathrm{T}}$
	$\nabla_X \log \det X^\mu = \mu X^{-\mathrm{T}}$
	$\nabla_X \log \det X^k = \nabla_X \log \det{}^k X = k X^{-\mathrm{T}}$
	$\nabla_X \log \det{}^\mu(X + tY) = \mu(X + tY)^{-\mathrm{T}}$
$\nabla_x \log(a^\mathrm{T} x + b) = a \frac{1}{a^\mathrm{T} x + b}$	$\nabla_X \log \det(AX + B) = A^\mathrm{T}(AX + B)^{-\mathrm{T}}$
	$\nabla_X \log \det(I \pm A^\mathrm{T} X A) = \pm A(I \pm A^\mathrm{T} X A)^{-\mathrm{T}} A^\mathrm{T}$
	$\nabla_X \log \det(X + tY)^k = \nabla_X \log \det{}^k(X + tY) = k(X + tY)^{-\mathrm{T}}$
	$\frac{d}{dt} \log \det(X + tY) = \mathrm{tr}\left((X + tY)^{-1} Y\right)$
	$\frac{d^2}{dt^2} \log \det(X + tY) = -\mathrm{tr}\left((X + tY)^{-1} Y (X + tY)^{-1} Y\right)$
	$\frac{d}{dt} \log \det(X + tY)^{-1} = -\mathrm{tr}\left((X + tY)^{-1} Y\right)$
	$\frac{d^2}{dt^2} \log \det(X + tY)^{-1} = \mathrm{tr}\left((X + tY)^{-1} Y (X + tY)^{-1} Y\right)$
	$\frac{d}{dt} \log \det\left(\delta(A(x + ty) + a)^2 + \mu I\right)$ $= \mathrm{tr}\left(\left(\delta(A(x + ty) + a)^2 + \mu I\right)^{-1} 2\delta(A(x + ty) + a)\delta(Ay)\right)$

D.2.5　determinant

$$\nabla_X \det X = \nabla_X \det X^{\mathrm{T}} = \det(X)X^{-\mathrm{T}}$$

$$\nabla_X \det X^{-1} = -\det(X^{-1})X^{-\mathrm{T}} = -\det(X)^{-1}X^{-\mathrm{T}}$$

$$\nabla_X \det{}^{\mu} X = \mu \det{}^{\mu}(X)X^{-\mathrm{T}}$$

$$\nabla_X \det X^{\mu} = \mu \det(X^{\mu})X^{-\mathrm{T}}$$

$$\nabla_X \det X^k = k\det{}^{k-1}(X)\big(\mathrm{tr}(X)I - X^{\mathrm{T}}\big)\ , \qquad\qquad X \in \mathbb{R}^{2\times 2}$$

$$\nabla_X \det X^k = \nabla_X \det{}^k X = k\det(X^k)X^{-\mathrm{T}} = k\det{}^k(X)X^{-\mathrm{T}}$$

$$\nabla_X \det{}^{\mu}(X + tY) = \mu \det{}^{\mu}(X + tY)(X + tY)^{-\mathrm{T}}$$

$$\nabla_X \det(X + tY)^k = \nabla_X \det{}^k(X + tY) = k\det{}^k(X + tY)(X + tY)^{-\mathrm{T}}$$

$$\tfrac{d}{dt}\det(X + tY) = \det(X + tY)\,\mathrm{tr}((X + tY)^{-1}Y)$$

$$\tfrac{d^2}{dt^2}\det(X + tY) = \det(X + tY)\big(\mathrm{tr}^2\big((X + tY)^{-1}Y\big) - \mathrm{tr}((X + tY)^{-1}Y(X + tY)^{-1}Y)\big)$$

$$\tfrac{d}{dt}\det(X + tY)^{-1} = -\det(X + tY)^{-1}\,\mathrm{tr}((X + tY)^{-1}Y)$$

$$\tfrac{d^2}{dt^2}\det(X + tY)^{-1} = \det(X + tY)^{-1}\big(\mathrm{tr}^2((X + tY)^{-1}Y) + \mathrm{tr}((X + tY)^{-1}Y(X + tY)^{-1}Y)\big)$$

$$\tfrac{d}{dt}\det{}^{\mu}(X + tY) = \mu \det{}^{\mu}(X + tY)\,\mathrm{tr}((X + tY)^{-1}Y)$$

D.2.6　logarithmic

Matrix logarithm.

$$\tfrac{d}{dt}\log(X + tY)^{\mu} = \mu Y(X + tY)^{-1} = \mu(X + tY)^{-1}Y\ , \qquad XY = YX$$

$$\tfrac{d}{dt}\log(I - tY)^{\mu} = -\mu Y(I - tY)^{-1} = -\mu(I - tY)^{-1}Y \qquad \text{[209, p.493]}$$

D.2.7 exponential

Matrix exponential. [78, §3.6, §4.5] [338, §5.4]

$$\nabla_X e^{\operatorname{tr}(Y^{\mathrm{T}}X)} = \nabla_X \det e^{Y^{\mathrm{T}}X} = e^{\operatorname{tr}(Y^{\mathrm{T}}X)} Y \qquad\qquad (\forall\, X, Y)$$

$$\nabla_X \operatorname{tr} e^{YX} = e^{Y^{\mathrm{T}}X^{\mathrm{T}}} Y^{\mathrm{T}} = Y^{\mathrm{T}} e^{X^{\mathrm{T}}Y^{\mathrm{T}}}$$

$$\nabla_x \mathbf{1}^{\mathrm{T}} e^{Ax} = A^{\mathrm{T}} e^{Ax}$$

$$\nabla_x \mathbf{1}^{\mathrm{T}} e^{|Ax|} = A^{\mathrm{T}} \delta(\operatorname{sgn}(Ax)) e^{|Ax|} \qquad\qquad (Ax)_i \neq 0$$

$$\nabla_x \log(\mathbf{1}^{\mathrm{T}} e^x) = \frac{1}{\mathbf{1}^{\mathrm{T}} e^x}\, e^x$$

$$\nabla_x^2 \log(\mathbf{1}^{\mathrm{T}} e^x) = \frac{1}{\mathbf{1}^{\mathrm{T}} e^x}\left(\delta(e^x) - \frac{1}{\mathbf{1}^{\mathrm{T}} e^x}\, e^x e^{x\mathrm{T}}\right)$$

$$\nabla_x \prod_{i=1}^{k} x_i^{\frac{1}{k}} = \frac{1}{k}\left(\prod_{i=1}^{k} x_i^{\frac{1}{k}}\right)\mathbf{1}/x$$

$$\nabla_x^2 \prod_{i=1}^{k} x_i^{\frac{1}{k}} = -\frac{1}{k}\left(\prod_{i=1}^{k} x_i^{\frac{1}{k}}\right)\left(\delta(x)^{-2} - \frac{1}{k}(1/x)(1/x)^{\mathrm{T}}\right)$$

$$\frac{d}{dt} e^{tY} = e^{tY} Y = Y e^{tY}$$

$$\frac{d}{dt} e^{X+tY} = e^{X+tY} Y = Y e^{X+tY}, \qquad\qquad XY = YX$$

$$\frac{d^2}{dt^2} e^{X+tY} = e^{X+tY} Y^2 = Y e^{X+tY} Y = Y^2 e^{X+tY}, \quad XY = YX$$

$$\frac{d^j}{dt^j} e^{\operatorname{tr}(X+tY)} = e^{\operatorname{tr}(X+tY)} \operatorname{tr}^j(Y)$$

Appendix E

Projection

Rob Reiner's "A Few Good Men" is one of those movies that tells you what it's going to do, does it, and then tells you what it did. It doesn't think the audience is very bright.

$-$Roger Ebert, 1992

For any $A \in \mathbb{R}^{m \times n}$, the *pseudoinverse* [208, §7.3 prob.9] [256, §6.12 prob.19] [164, §5.5.4] [338, App.A]

$$A^\dagger \triangleq \lim_{t \to 0^+} (A^{\mathrm{T}}A + t\,I)^{-1}A^{\mathrm{T}} = \lim_{t \to 0^+} A^{\mathrm{T}}(AA^{\mathrm{T}} + t\,I)^{-1} \in \mathbb{R}^{n \times m} \qquad (1961)$$

is a unique matrix from the optimal solution set to $\underset{X}{\text{minimize}} \|AX - I\|_{\mathrm{F}}^2$ (§3.6.0.0.2). For any $t > 0$

$$I - A(A^{\mathrm{T}}A + t\,I)^{-1}A^{\mathrm{T}} = t(AA^{\mathrm{T}} + t\,I)^{-1} \qquad (1962)$$

Equivalently, pseudoinverse $A^\dagger$ is that unique matrix satisfying the *Moore-Penrose conditions*: [210, §1.3] [401]

1.	$AA^\dagger A = A$	3.	$(AA^\dagger)^{\mathrm{T}} = AA^\dagger$
2.	$A^\dagger AA^\dagger = A^\dagger$	4.	$(A^\dagger A)^{\mathrm{T}} = A^\dagger A$

which are necessary and sufficient to establish the pseudoinverse whose principal action is to injectively map $\mathcal{R}(A)$ onto $\mathcal{R}(A^{\mathrm{T}})$ (Figure 172). Conditions 1 and 3 are necessary and sufficient for $AA^\dagger$ to be the orthogonal projector on $\mathcal{R}(A)$, while conditions 2 and 4 hold iff $A^\dagger A$ is the orthogonal projector on $\mathcal{R}(A^{\mathrm{T}})$.

Range and nullspace of the pseudoinverse [273] [335, §III.1 exer.1]

$$\mathcal{R}(A^\dagger) = \mathcal{R}(A^{\mathrm{T}}), \qquad \mathcal{R}(A^{\dagger\mathrm{T}}) = \mathcal{R}(A) \qquad (1963)$$
$$\mathcal{N}(A^\dagger) = \mathcal{N}(A^{\mathrm{T}}), \qquad \mathcal{N}(A^{\dagger\mathrm{T}}) = \mathcal{N}(A) \qquad (1964)$$

can be derived by singular value decomposition (§A.6).

CITATION: Dattorro, *Convex Optimization & Euclidean Distance Geometry*, $\mathcal{M}\varepsilon\beta oo$ Publishing USA, 2005, v2015.02.15.

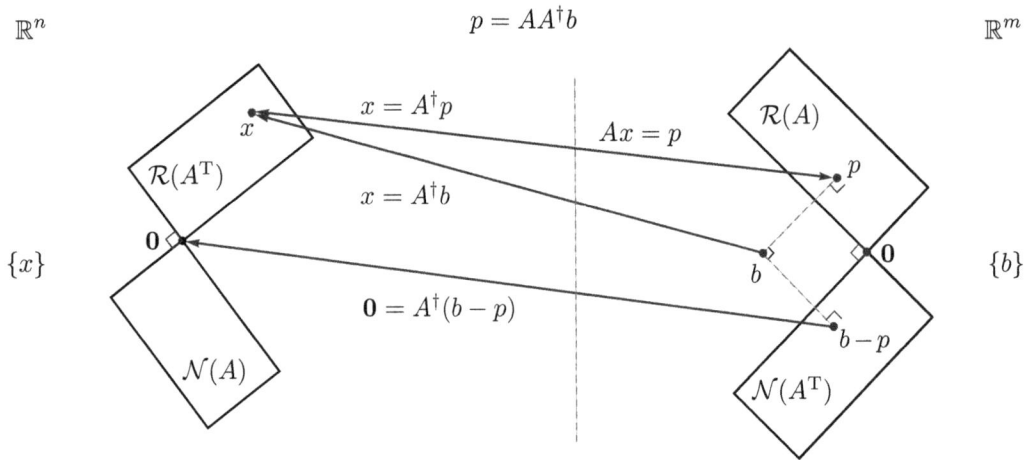

Figure 172: (*confer* Figure 16) Pseudoinverse $A^\dagger \in \mathbb{R}^{n \times m}$ action: [338, p.449] Component of b in $\mathcal{N}(A^T)$ maps to $\mathbf{0}$, while component of b in $\mathcal{R}(A)$ maps to rowspace $\mathcal{R}(A^T)$. For any $A \in \mathbb{R}^{m \times n}$, inversion is bijective $\forall p \in \mathcal{R}(A)$. $x = A^\dagger b \Leftrightarrow x \in \mathcal{R}(A^T)$ & $b - Ax \perp \mathcal{R}(A)$ $\Leftrightarrow x \perp \mathcal{N}(A)$ & $b - Ax \in \mathcal{N}(A^T)$. [49]

The following relations reliably hold without qualification:

a. $A^{T\dagger} = A^{\dagger T}$
b. $A^{\dagger\dagger} = A$
c. $(AA^T)^\dagger = A^{\dagger T}A^\dagger$
d. $(A^T A)^\dagger = A^\dagger A^{\dagger T}$
e. $(AA^\dagger)^\dagger = AA^\dagger$
f. $(A^\dagger A)^\dagger = A^\dagger A$

Yet for arbitrary A, B it is generally true that $(AB)^\dagger \neq B^\dagger A^\dagger$:

E.0.0.0.1 Theorem. *Pseudoinverse of product.*   [175] [58] [247, exer.7.23]
For $A \in \mathbb{R}^{m \times n}$ and $B \in \mathbb{R}^{n \times k}$

$$(AB)^\dagger = B^\dagger A^\dagger \tag{1965}$$

if and only if

$$\mathcal{R}(A^T AB) \subseteq \mathcal{R}(B) \quad \textbf{and} \quad \mathcal{R}(BB^T A^T) \subseteq \mathcal{R}(A^T) \tag{1966}$$
$\diamond$

Pseudoinverse of normalized vector u is the vector transpose. Otherwise,

$$u^\dagger = \frac{u^T}{\|u\|^2} \tag{1967}$$

$U^\dagger = U^T$ for orthonormal (including the orthogonal) matrices U. So, for orthonormal matrices U, Q and arbitrary A

$$(UAQ^T)^\dagger = QA^\dagger U^T \tag{1968}$$

E.0.0.0.2 Exercise. *Kronecker inverse.*
Prove:

$$(A \otimes B)^\dagger = A^\dagger \otimes B^\dagger \tag{1969}$$

▼

E.0.1 Logical deductions

When A is invertible, $A^\dagger = A^{-1}$; so $A^\dagger A = A A^\dagger = I$.
Otherwise, for $A \in \mathbb{R}^{m \times n}$ $\qquad\qquad$ [157, §5.3.3.1] [247, §7] [303]

g. $A^\dagger A = I$,	$A^\dagger = (A^\mathrm{T}A)^{-1}A^\mathrm{T}$,	rank $A = n$
h. $AA^\dagger = I$,	$A^\dagger = A^\mathrm{T}(AA^\mathrm{T})^{-1}$,	rank $A = m$
i. $A^\dagger A \omega = \omega$,		$\omega \in \mathcal{R}(A^\mathrm{T})$
j. $AA^\dagger v = v$,		$v \in \mathcal{R}(A)$
k. $A^\dagger A = AA^\dagger$,		A normal
l. $A^{k\dagger} = A^{\dagger k}$,		k an integer, A normal

Equivalent to the corresponding Moore-Penrose condition:

1. $A^\mathrm{T} = A^\mathrm{T}AA^\dagger$ $\qquad$ **or** $\qquad$ $A^\mathrm{T} = A^\dagger AA^\mathrm{T}$
2. $A^{\dagger\mathrm{T}} = A^{\dagger\mathrm{T}}A^\dagger A$ $\qquad$ **or** $\qquad$ $A^{\dagger\mathrm{T}} = AA^\dagger A^{\dagger\mathrm{T}}$

When A is symmetric, $A^\dagger$ is symmetric and (§A.6)

$$A \succeq 0 \iff A^\dagger \succeq 0 \tag{1970}$$

E.0.1.0.1 Example. *Solution to classical linear equation $Ax = b$.*
In §2.5.1.1, the solution set to matrix equation $Ax = b$ was represented as an intersection of hyperplanes. Regardless of rank of A or its shape (fat or skinny), interpretation as a hyperplane intersection describing a possibly empty affine set generally holds. If matrix A is rank deficient or fat, there is an infinity of solutions x when $b \in \mathcal{R}(A)$. A unique solution occurs when the hyperplanes intersect at a single point.

Given arbitrary matrix A (of any rank and dimension) and vector b not necessarily in $\mathcal{R}(A)$, we wish to find a best solution $x^\star$ to

$$Ax \approx b \tag{1971}$$

in a Euclidean sense by solving an algebraic expression for orthogonal projection of b on $\mathcal{R}(A)$

$$\underset{x}{\text{minimize}} \; \|Ax - b\|^2 \tag{1972}$$

Necessary and sufficient condition for optimal solution to this unconstrained optimization is the so-called *normal equation* that results from zeroing the convex objective's gradient: (§D.2.1)

$$A^\mathrm{T}A\,x = A^\mathrm{T}b \tag{1973}$$

normal because error vector $b - Ax$ is perpendicular to $\mathcal{R}(A)$; *id est*, $A^\mathrm{T}(b - Ax) = \mathbf{0}$. Given any matrix A and any vector b, the normal equation is solvable exactly; always so, because $\mathcal{R}(A^\mathrm{T}A) = \mathcal{R}(A^\mathrm{T})$ and $A^\mathrm{T}b \in \mathcal{R}(A^\mathrm{T})$.

Given particular $\tilde{x} \in \mathcal{R}(A^{\mathrm{T}})$ solving (1973), then (Figure **172**) it is necessarily unique and $\tilde{x} = x^{\star} = A^{\dagger}b$. When A is skinny-or-square full-rank, normal equation (1973) can be solved exactly by inversion:

$$x^{\star} = (A^{\mathrm{T}}A)^{-1}A^{\mathrm{T}}b \equiv A^{\dagger}b \tag{1974}$$

For matrix A of arbitrary rank and shape, on the other hand, $A^{\mathrm{T}}A$ might not be invertible. Yet the normal equation can always be solved exactly by: (1961)

$$x^{\star} = \lim_{t \to 0^{+}}(A^{\mathrm{T}}A + t\,I)^{-1}A^{\mathrm{T}}b = A^{\dagger}b \tag{1975}$$

invertible for any positive value of t by (1532). The exact inversion (1974) and this pseudoinverse solution (1975) each solve a limited regularization

$$\lim_{t \to 0^{+}}\underset{x}{\text{minimize}}\; \|Ax - b\|^{2} + t\,\|x\|^{2} \equiv \lim_{t \to 0^{+}}\underset{x}{\text{minimize}}\; \left\| \begin{bmatrix} A \\ \sqrt{t}\,I \end{bmatrix} x - \begin{bmatrix} b \\ \mathbf{0} \end{bmatrix} \right\|^{2} \tag{1976}$$

simultaneously providing least squares solution to (1972) and the classical *least norm* solution[E.1] [338, App.A.4] [49] (*confer* §3.2.0.0.1)

$$\begin{aligned} \arg\underset{x}{\text{minimize}}\quad & \|x\|^{2} \\ \text{subject to}\quad & Ax = AA^{\dagger}b \end{aligned} \tag{1977}$$

where $AA^{\dagger}b$ is the orthogonal projection of vector b on $\mathcal{R}(A)$. Least norm solution can be interpreted as orthogonal projection of the origin $\mathbf{0}$ on affine subset $\mathcal{A} = \{x \mid Ax = AA^{\dagger}b\}$; (§E.5.0.0.5, §E.5.0.0.6)

$$\begin{aligned} \arg\underset{x}{\text{minimize}}\quad & \|x - \mathbf{0}\|^{2} \\ \text{subject to}\quad & x \in \mathcal{A} \end{aligned} \tag{1978}$$

equivalently, maximization of the Euclidean ball until it kisses $\mathcal{A}$; rather, $\arg\text{dist}(\mathbf{0},\mathcal{A})$.

$$\square$$

E.1 Idempotent matrices

Projection matrices are square and defined by *idempotence*, $P^{2} = P$; [338, §2.6] [210, §1.3] equivalent to the condition, P be diagonalizable [208, §3.3 prob.3] with eigenvalues $\phi_{i} \in \{0, 1\}$. [421, §4.1 thm.4.1] Idempotent matrices are not necessarily symmetric. The transpose of an idempotent matrix remains idempotent; $P^{\mathrm{T}}P^{\mathrm{T}} = P^{\mathrm{T}}$. Solely excepting $P = I$, all projection matrices are neither orthogonal (§B.5) or invertible. [338, §3.4] The collection of all projection matrices of particular dimension does not form a convex set.

Suppose we wish to project nonorthogonally (*obliquely*) on the range of any particular matrix $A \in \mathbb{R}^{m \times n}$. All idempotent matrices projecting nonorthogonally on $\mathcal{R}(A)$ may be expressed:

$$P = A(A^{\dagger} + BZ^{\mathrm{T}}) \in \mathbb{R}^{m \times m} \tag{1979}$$

[E.1]This means: optimal solutions of lesser norm than the so-called *least norm* solution (1977) can be obtained (at expense of approximation $Ax \approx b$ hence, of perpendicularity) by ignoring the limiting operation and introducing finite positive values of t into (1976).

where $\mathcal{R}(P)\!=\!\mathcal{R}(A)$,[E.2] $B\in\mathbb{R}^{n\times k}$ for positive integer k is arbitrary, and $Z\in\mathbb{R}^{m\times k}$ is any matrix whose range is in $\mathcal{N}(A^{\mathrm{T}})$; *id est,*

$$A^{\mathrm{T}}Z \;=\; A^{\dagger}Z \;=\; \mathbf{0} \tag{1980}$$

Evidently, the collection of nonorthogonal projectors projecting on $\mathcal{R}(A)$ is an affine subset

$$\mathcal{P}_k = \left\{ A(A^{\dagger} + BZ^{\mathrm{T}}) \mid B\in\mathbb{R}^{n\times k}\right\} \tag{1981}$$

When matrix A in (1979) is skinny full-rank ($A^{\dagger}A\!=\!I$) or has orthonormal columns ($A^{\mathrm{T}}A\!=\!I$), either property leads to a biorthogonal characterization of nonorthogonal projection:

E.1.1 Biorthogonal characterization of projector

Any nonorthogonal projector $P^2\!=\!P\in\mathbb{R}^{m\times m}$ projecting on nontrivial $\mathcal{R}(U)$ can be defined by a biorthogonality condition $Q^{\mathrm{T}}U\!=\!I$; the *biorthogonal decomposition* of P being (*confer* (1979))[E.3]

$$P = UQ^{\mathrm{T}}, \qquad Q^{\mathrm{T}}U = I \tag{1982}$$

where[E.4] (§B.1.1.1)

$$\mathcal{R}(P)\!=\!\mathcal{R}(U) , \qquad \mathcal{N}(P)\!=\!\mathcal{N}(Q^{\mathrm{T}}) \tag{1983}$$

and where generally (*confer* (2009))[E.5]

$$P^{\mathrm{T}} \neq P, \qquad P^{\dagger} \neq P, \qquad \|P\|_2 \neq 1, \qquad P \nsucceq 0 \tag{1984}$$

and P is not nonexpansive (2010) (2189).

($\Leftarrow$) To verify assertion (1982) we observe: because idempotent matrices are diagonalizable (§A.5), [208, §3.3 prob.3] they must have the form (1628)

$$P \;=\; S\Phi S^{-1} \;=\; \sum_{i=1}^{m} \phi_i\, s_i w_i^{\mathrm{T}} \;=\; \sum_{i=1}^{k\,\leq\, m} s_i w_i^{\mathrm{T}} \tag{1985}$$

[E.2]**Proof.** $\mathcal{R}(P)\subseteq\mathcal{R}(A)$ is obvious [338, §3.6]. By (141) and (142),

$$\begin{aligned}\mathcal{R}(A^{\dagger}+BZ^{\mathrm{T}}) &= \{(A^{\dagger}+BZ^{\mathrm{T}})y \mid y\in\mathbb{R}^m\} \\ &\supseteq \{(A^{\dagger}+BZ^{\mathrm{T}})y \mid y\in\mathcal{R}(A)\} = \mathcal{R}(A^{\mathrm{T}})\end{aligned}$$
$$\begin{aligned}\mathcal{R}(P) &= \{A(A^{\dagger}+BZ^{\mathrm{T}})y \mid y\in\mathbb{R}^m\} \\ &\supseteq \{A(A^{\dagger}+BZ^{\mathrm{T}})y \mid (A^{\dagger}+BZ^{\mathrm{T}})y\in\mathcal{R}(A^{\mathrm{T}})\} = \mathcal{R}(A)\end{aligned} \qquad \blacklozenge$$

[E.3] $A \leftarrow U$, $A^{\dagger}+BZ^{\mathrm{T}} \leftarrow Q^{\mathrm{T}}$
[E.4]**Proof.** Obviously, $\mathcal{R}(P)\subseteq\mathcal{R}(U)$. Because $Q^{\mathrm{T}}U = I$

$$\begin{aligned}\mathcal{R}(P) &= \{UQ^{\mathrm{T}}x \mid x\in\mathbb{R}^m\} \\ &\supseteq \{UQ^{\mathrm{T}}Uy \mid y\in\mathbb{R}^k\} = \mathcal{R}(U)\end{aligned} \qquad \blacklozenge$$

[E.5]Orthonormal decomposition (2006) (*confer* §E.3.4) is a special case of biorthogonal decomposition (1982) characterized by (2009). So, these characteristics (1984) are not necessary conditions for biorthogonality.

that is a sum of $k = \operatorname{rank} P$ independent projector dyads (idempotent dyads, §B.1.1, §E.6.2.1) where $\phi_i \in \{0,1\}$ are the eigenvalues of P [421, §4.1 thm.4.1] in diagonal matrix $\Phi \in \mathbb{R}^{m \times m}$ arranged in nonincreasing order, and where $s_i, w_i \in \mathbb{R}^m$ are the right- and left-eigenvectors of P, respectively, which are independent and real.[E.6] Therefore

$$U \triangleq S(:,1\!:\!k) = [\,s_1 \cdots s_k\,] \in \mathbb{R}^{m \times k} \tag{1986}$$

is the full-rank matrix $S \in \mathbb{R}^{m \times m}$ having $m-k$ columns truncated (corresponding to 0 eigenvalues), while

$$Q^{\mathrm{T}} \triangleq S^{-1}(1\!:\!k,:) = \begin{bmatrix} w_1^{\mathrm{T}} \\ \vdots \\ w_k^{\mathrm{T}} \end{bmatrix} \in \mathbb{R}^{k \times m} \tag{1987}$$

is matrix S^{-1} having the corresponding $m-k$ rows truncated. By the 0 *eigenvalues theorem* (§A.7.3.0.1), $\mathcal{R}(U) = \mathcal{R}(P)$, $\mathcal{R}(Q) = \mathcal{R}(P^{\mathrm{T}})$, and

$$\begin{aligned}
\mathcal{R}(P) &= \operatorname{span}\{s_i \mid \phi_i = 1 \ \forall i\} \\
\mathcal{N}(P) &= \operatorname{span}\{s_i \mid \phi_i = 0 \ \forall i\} \\
\mathcal{R}(P^{\mathrm{T}}) &= \operatorname{span}\{w_i \mid \phi_i = 1 \ \forall i\} \\
\mathcal{N}(P^{\mathrm{T}}) &= \operatorname{span}\{w_i \mid \phi_i = 0 \ \forall i\}
\end{aligned} \tag{1988}$$

Thus biorthogonality $Q^{\mathrm{T}}U = I$ is a necessary condition for idempotence, and so the collection of nonorthogonal projectors projecting on $\mathcal{R}(U)$ is the affine subset $\mathcal{P}_k = U\mathcal{Q}_k^{\mathrm{T}}$ where $\mathcal{Q}_k = \{Q \mid Q^{\mathrm{T}}U = I, \ Q \in \mathbb{R}^{m \times k}\}$.

($\Rightarrow$) Biorthogonality is a sufficient condition for idempotence;

$$P^2 = \sum_{i=1}^{k} s_i w_i^{\mathrm{T}} \sum_{j=1}^{k} s_j w_j^{\mathrm{T}} = P \tag{1989}$$

id est, if the cross-products are annihilated, then $P^2 = P$. $\blacklozenge$

Nonorthogonal projection of x on $\mathcal{R}(P)$ has expression like a biorthogonal expansion,

$$Px = UQ^{\mathrm{T}}x = \sum_{i=1}^{k} w_i^{\mathrm{T}}x\, s_i \tag{1990}$$

When the domain is restricted to range of P, say $x = U\xi$ for $\xi \in \mathbb{R}^k$, then $x = Px = UQ^{\mathrm{T}}U\xi = U\xi$ and expansion is unique because the eigenvectors are linearly independent. Otherwise, any component of x in $\mathcal{N}(P) = \mathcal{N}(Q^{\mathrm{T}})$ will be annihilated. Direction of nonorthogonal projection is orthogonal to $\mathcal{R}(Q) \Leftrightarrow Q^{\mathrm{T}}U = I$; *id est*, for $Px \in \mathcal{R}(U)$ (*confer* (2000))

$$Px - x \ \perp \ \mathcal{R}(Q) \text{ in } \mathbb{R}^m \tag{1991}$$

[E.6]Eigenvectors of a real matrix corresponding to real eigenvalues must be real. (§A.5.0.0.1)

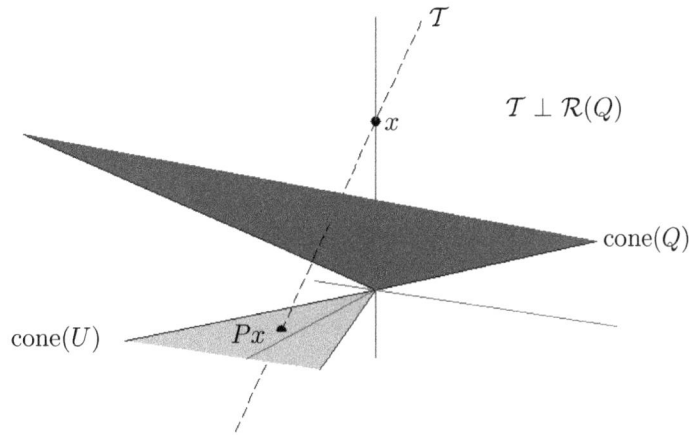

Figure 173: Nonorthogonal projection of $x \in \mathbb{R}^3$ on $\mathcal{R}(U) = \mathbb{R}^2$ under biorthogonality condition; *id est*, $Px = UQ^{\mathrm{T}}x$ such that $Q^{\mathrm{T}}U = I$. Any point along imaginary line $\mathcal{T}$ connecting x to Px will be projected nonorthogonally on Px with respect to horizontal plane constituting $\mathbb{R}^2 = \mathrm{aff}\,\mathrm{cone}(U)$ in this example. Extreme directions of $\mathrm{cone}(U)$ correspond to two columns of U; likewise for $\mathrm{cone}(Q)$. For purpose of illustration, we truncate each conic hull by truncating coefficients of conic combination at unity. Conic hull $\mathrm{cone}(Q)$ is headed upward at an angle, out of plane of page. Nonorthogonal projection would fail were $\mathcal{N}(Q^{\mathrm{T}})$ in $\mathcal{R}(U)$ (were $\mathcal{T}$ parallel to a line in $\mathcal{R}(U)$).

E.1.1.0.1 Example. *Illustration of nonorthogonal projector.*
Figure 173 shows $\operatorname{cone}(U)$, conic hull of the columns of

$$U = \begin{bmatrix} 1 & 1 \\ -0.5 & 0.3 \\ 0 & 0 \end{bmatrix} \tag{1992}$$

from nonorthogonal projector $P = UQ^{\mathrm{T}}$. Matrix U has a limitless number of left inverses because $\mathcal{N}(U^{\mathrm{T}})$ is nontrivial. Similarly depicted is left inverse Q^{T} from (1979)

$$\begin{aligned} Q = U^{\dagger \mathrm{T}} + ZB^{\mathrm{T}} &= \begin{bmatrix} 0.3750 & 0.6250 \\ -1.2500 & 1.2500 \\ 0 & 0 \end{bmatrix} + \begin{bmatrix} 0 \\ 0 \\ 1 \end{bmatrix} \begin{bmatrix} 0.5 & 0.5 \end{bmatrix} \\ &= \begin{bmatrix} 0.3750 & 0.6250 \\ -1.2500 & 1.2500 \\ 0.5000 & 0.5000 \end{bmatrix} \end{aligned} \tag{1993}$$

where $Z \in \mathcal{N}(U^{\mathrm{T}})$ and matrix B is selected arbitrarily; *id est,* $Q^{\mathrm{T}}U = I$ because U is full-rank.

Direction of projection on $\mathcal{R}(U)$ is orthogonal to $\mathcal{R}(Q)$. Any point along line $\mathcal{T}$ in the figure, for example, will have the same projection. Were matrix Z instead equal to $\mathbf{0}$, then $\operatorname{cone}(Q)$ would become the relative dual to $\operatorname{cone}(U)$ (sharing the same affine hull; §2.13.8, *confer* Figure 57a). In that case, projection $Px = UU^{\dagger}x$ of x on $\mathcal{R}(U)$ becomes orthogonal projection (and unique minimum-distance).   □

E.1.2   Idempotence summary

Nonorthogonal subspace-projector P is a (convex) linear operator defined by idempotence or biorthogonal decomposition (1982), but characterized not by symmetry or positive semidefiniteness or nonexpansivity (2010).

E.2 $I - P$, Projection on algebraic complement

It follows from the diagonalizability of idempotent matrices that $I - P$ must also be a projection matrix because it too is idempotent, and because it may be expressed

$$I - P = S(I - \Phi)S^{-1} = \sum_{i=1}^{m} (1 - \phi_i) s_i w_i^{\mathrm{T}} \tag{1994}$$

where $(1 - \phi_i) \in \{1, 0\}$ are the eigenvalues of $I - P$ (1533) whose eigenvectors s_i, w_i are identical to those of P in (1985). A consequence of that complementary relationship of eigenvalues is the fact, [350, §2] [345, §2] for subspace projector $P = P^2 \in \mathbb{R}^{m \times m}$

$$\begin{aligned} \mathcal{R}(P) &= \operatorname{span}\{s_i \mid \phi_i = 1 \;\; \forall i\} = \operatorname{span}\{s_i \mid (1 - \phi_i) = 0 \;\; \forall i\} = \mathcal{N}(I - P) \\ \mathcal{N}(P) &= \operatorname{span}\{s_i \mid \phi_i = 0 \;\; \forall i\} = \operatorname{span}\{s_i \mid (1 - \phi_i) = 1 \;\; \forall i\} = \mathcal{R}(I - P) \\ \mathcal{R}(P^{\mathrm{T}}) &= \operatorname{span}\{w_i \mid \phi_i = 1 \;\; \forall i\} = \operatorname{span}\{w_i \mid (1 - \phi_i) = 0 \;\; \forall i\} = \mathcal{N}(I - P^{\mathrm{T}}) \\ \mathcal{N}(P^{\mathrm{T}}) &= \operatorname{span}\{w_i \mid \phi_i = 0 \;\; \forall i\} = \operatorname{span}\{w_i \mid (1 - \phi_i) = 1 \;\; \forall i\} = \mathcal{R}(I - P^{\mathrm{T}}) \end{aligned} \tag{1995}$$

that is easy to see from (1985) and (1994). Idempotent $I-P$ therefore projects vectors on its range: $\mathcal{N}(P)$. Because all eigenvectors of a real idempotent matrix are real and independent, the algebraic complement of $\mathcal{R}(P)$ [233, §3.3] is equivalent to $\mathcal{N}(P)$;[E.7] *id est,*

$$\mathcal{R}(P) \oplus \mathcal{N}(P) = \mathcal{R}(P^{\mathrm{T}}) \oplus \mathcal{N}(P^{\mathrm{T}}) = \mathcal{R}(P^{\mathrm{T}}) \oplus \mathcal{N}(P) = \mathcal{R}(P) \oplus \mathcal{N}(P^{\mathrm{T}}) = \mathbb{R}^m \quad (1996)$$

because $\mathcal{R}(P) \oplus \mathcal{R}(I-P) = \mathbb{R}^m$. For idempotent $P \in \mathbb{R}^{m \times m}$, consequently,

$$\operatorname{rank} P + \operatorname{rank}(I-P) = m \qquad (1997)$$

E.2.0.0.1 Theorem. *Rank/Trace.* [421, §4.1 prob.9] *(confer* (2014))

$$P^2 = P$$
$$\Leftrightarrow \qquad\qquad (1998)$$
$$\operatorname{rank} P = \operatorname{tr} P \quad \textbf{and} \quad \operatorname{rank}(I-P) = \operatorname{tr}(I-P)$$

$\diamond$

E.2.1 Universal projector characteristic

Although projection is not necessarily orthogonal and $\mathcal{R}(P) \not\perp \mathcal{R}(I-P)$ in general, still for any projector P and any $x \in \mathbb{R}^m$

$$Px + (I-P)x = x \qquad (1999)$$

must hold where $\mathcal{R}(I-P) = \mathcal{N}(P)$ is the algebraic complement of $\mathcal{R}(P)$. Algebraic complement of closed convex cone $\mathcal{K}$, for example, is the negative dual cone $-\mathcal{K}^*$. (2120) (Figure 177)

E.3 Symmetric idempotent matrices

When idempotent matrix P is symmetric, P is an orthogonal projector. In other words, the direction of projection of point $x \in \mathbb{R}^m$ on subspace $\mathcal{R}(P)$ is orthogonal to $\mathcal{R}(P)$; *id est,* for $P^2 = P \in \mathbb{S}^m$ and projection $Px \in \mathcal{R}(P)$

$$Px - x \ \perp \ \mathcal{R}(P) \text{ in } \mathbb{R}^m \qquad (2000)$$

(confer (1991)) Perpendicularity is a necessary and sufficient condition for orthogonal projection on a subspace. [112, §4.9]

A condition equivalent to (2000) is: Norm of direction $x-Px$ is the infimum over all nonorthogonal projections of x on $\mathcal{R}(P)$; [256, §3.3] for $P^2 = P \in \mathbb{S}^m$, $\mathcal{R}(P) = \mathcal{R}(A)$, matrices A, B, Z and positive integer k as defined for (1979), and given $x \in \mathbb{R}^m$

$$\|x - Px\|_2 = \|x - AA^\dagger x\|_2 = \inf_{B \in \mathbb{R}^{n \times k}} \|x - A(A^\dagger + BZ^{\mathrm{T}})x\|_2 = \operatorname{dist}(x, \mathcal{R}(P)) \quad (2001)$$

[E.7]The same phenomenon occurs with symmetric (nonidempotent) matrices, for example. When summands in $A \oplus B = \mathbb{R}^m$ are orthogonal vector spaces, the algebraic complement is the orthogonal complement.

The infimum is attained for $\mathcal{R}(B) \subseteq \mathcal{N}(A)$ over any affine subset of nonorthogonal projectors (1981) indexed by k.

Proof is straightforward: The vector 2-norm is a convex function. Setting gradient of the norm-square to $\mathbf{0}$, applying §D.2,

$$\left(A^{\mathrm{T}}ABZ^{\mathrm{T}} - A^{\mathrm{T}}(I - AA^{\dagger})\right) xx^{\mathrm{T}}A = \mathbf{0}$$
$$\Leftrightarrow \qquad\qquad (2002)$$
$$A^{\mathrm{T}}ABZ^{\mathrm{T}}xx^{\mathrm{T}}A = \mathbf{0}$$

because $A^{\mathrm{T}} = A^{\mathrm{T}}AA^{\dagger}$. Projector $P = AA^{\dagger}$ is therefore unique; the minimum-distance projector is the orthogonal projector, and *vice versa*.   ◆

We get $P = AA^{\dagger}$ so this projection matrix must be symmetric. Then for any matrix $A \in \mathbb{R}^{m \times n}$, symmetric idempotent P projects a given vector x in $\mathbb{R}^m$ orthogonally on $\mathcal{R}(A)$. Under either condition (2000) or (2001), the projection Px is unique minimum-distance; for subspaces, perpendicularity and minimum-distance conditions are equivalent.

E.3.1   Four subspaces

We summarize the orthogonal projectors projecting on the four fundamental subspaces: for $A \in \mathbb{R}^{m \times n}$

$$
\begin{array}{llll}
A^{\dagger}A & : \mathbb{R}^n & \text{on} & \left(\mathcal{R}(A^{\dagger}A) \quad = \mathcal{R}(A^{\mathrm{T}})\right) \\
AA^{\dagger} & : \mathbb{R}^m & \text{on} & \left(\mathcal{R}(AA^{\dagger}) \quad = \mathcal{R}(A)\right) \\
I - A^{\dagger}A & : \mathbb{R}^n & \text{on} & \left(\mathcal{R}(I - A^{\dagger}A) = \mathcal{N}(A)\right) \\
I - AA^{\dagger} & : \mathbb{R}^m & \text{on} & \left(\mathcal{R}(I - AA^{\dagger}) = \mathcal{N}(A^{\mathrm{T}})\right)
\end{array}
\qquad (2003)
$$

Given a known subspace, matrix A is neither unique or necessarily full-rank. Despite that, a basis for each fundamental subspace comprises the linearly independent column vectors from its associated symmetric projection matrix:

$$
\begin{array}{lllll}
\text{basis}\,\mathcal{R}(A^{\mathrm{T}}) & \subseteq & A^{\dagger}A & \subseteq & \mathcal{R}(A^{\mathrm{T}}) \\
\text{basis}\,\mathcal{R}(A) & \subseteq & AA^{\dagger} & \subseteq & \mathcal{R}(A) \\
\text{basis}\,\mathcal{N}(A) & \subseteq & I - A^{\dagger}A & \subseteq & \mathcal{N}(A) \\
\text{basis}\,\mathcal{N}(A^{\mathrm{T}}) & \subseteq & I - AA^{\dagger} & \subseteq & \mathcal{N}(A^{\mathrm{T}})
\end{array}
\qquad (2004)
$$

For completeness:[E.8] (1995)

$$
\begin{array}{ll}
\mathcal{N}(A^{\dagger}A) & = \mathcal{N}(A) \\
\mathcal{N}(AA^{\dagger}) & = \mathcal{N}(A^{\mathrm{T}}) \\
\mathcal{N}(I - A^{\dagger}A) & = \mathcal{R}(A^{\mathrm{T}}) \\
\mathcal{N}(I - AA^{\dagger}) & = \mathcal{R}(A)
\end{array}
\qquad (2005)
$$

[E.8]**Proof** is by singular value decomposition (§A.6.2): $\mathcal{N}(A^{\dagger}A) \subseteq \mathcal{N}(A)$ is obvious. Conversely, suppose $A^{\dagger}Ax = \mathbf{0}$. Then $x^{\mathrm{T}}A^{\dagger}Ax = x^{\mathrm{T}}QQ^{\mathrm{T}}x = \|Q^{\mathrm{T}}x\|^2 = 0$ where $A = U\Sigma Q^{\mathrm{T}}$ is the subcompact singular value decomposition. Because $\mathcal{R}(Q) = \mathcal{R}(A^{\mathrm{T}})$, then $x \in \mathcal{N}(A)$ which implies $\mathcal{N}(A^{\dagger}A) \supseteq \mathcal{N}(A)$. $\therefore \mathcal{N}(A^{\dagger}A) = \mathcal{N}(A)$.   ◆

E.3.2 Orthogonal characterization of projector

Any symmetric projector $P^2 = P \in \mathbb{S}^m$ projecting on nontrivial $\mathcal{R}(Q)$ can be defined by the orthonormality condition $Q^\mathrm{T}Q = I$. When skinny matrix $Q \in \mathbb{R}^{m \times k}$ has orthonormal columns, then $Q^\dagger = Q^\mathrm{T}$ by the Moore-Penrose conditions. Hence, any P having an *orthonormal decomposition* (§E.3.4)

$$P = QQ^\mathrm{T}, \qquad Q^\mathrm{T}Q = I \tag{2006}$$

where [338, §3.3] (1690)

$$\mathcal{R}(P) = \mathcal{R}(Q), \qquad \mathcal{N}(P) = \mathcal{N}(Q^\mathrm{T}) \tag{2007}$$

is an orthogonal projector projecting on $\mathcal{R}(Q)$ having, for $Px \in \mathcal{R}(Q)$ (*confer* (1991))

$$Px - x \perp \mathcal{R}(Q) \text{ in } \mathbb{R}^m \tag{2008}$$

From (2006), orthogonal projector P is obviously positive semidefinite (§A.3.1.0.6); necessarily, (*confer* (1984))

$$P^\mathrm{T} = P, \qquad P^\dagger = P, \qquad \|P\|_2 = 1, \qquad P \succeq 0 \tag{2009}$$

and $\|Px\| = \|QQ^\mathrm{T}x\| = \|Q^\mathrm{T}x\|$ because $\|Qy\| = \|y\| \; \forall y \in \mathbb{R}^k$. All orthogonal projectors are therefore *nonexpansive* because

$$\sqrt{\langle Px, x \rangle} = \|Px\| = \|Q^\mathrm{T}x\| \leq \|x\| \quad \forall x \in \mathbb{R}^m \tag{2010}$$

the Bessel inequality, [112] [233] with equality when $x \in \mathcal{R}(Q)$.

From the diagonalization of idempotent matrices (1985) on page 607

$$P = S\Phi S^\mathrm{T} = \sum_{i=1}^{m} \phi_i \, s_i s_i^\mathrm{T} = \sum_{i=1}^{k \leq m} s_i s_i^\mathrm{T} \tag{2011}$$

orthogonal projection of point x on $\mathcal{R}(P)$ has expression like an orthogonal expansion [112, §4.10]

$$Px = QQ^\mathrm{T}x = \sum_{i=1}^{k} s_i^\mathrm{T} x \, s_i \tag{2012}$$

where

$$Q = S(:, 1:k) = [\, s_1 \cdots s_k \,] \in \mathbb{R}^{m \times k} \tag{2013}$$

and where the s_i [*sic*] are orthonormal eigenvectors of symmetric idempotent P. When the domain is restricted to range of P, say $x = Q\xi$ for $\xi \in \mathbb{R}^k$, then $x = Px = QQ^\mathrm{T}Q\xi = Q\xi$ and expansion is unique. Otherwise, any component of x in $\mathcal{N}(Q^\mathrm{T})$ will be annihilated.

E.3.2.0.1   Theorem.   *Symmetric rank/trace.*   (*confer* (1998) (1537))

$$P^{\mathrm{T}} = P, \quad P^2 = P$$
$$\Leftrightarrow$$
$$\operatorname{rank} P = \operatorname{tr} P = \|P\|_{\mathrm{F}}^2 \quad \text{and} \quad \operatorname{rank}(I - P) = \operatorname{tr}(I - P) = \|I - P\|_{\mathrm{F}}^2 \qquad (2014)$$

$$\diamond$$

Proof.   We take as given Theorem E.2.0.0.1 establishing idempotence. We have left only to show $\operatorname{tr} P = \|P\|_{\mathrm{F}}^2 \Rightarrow P^{\mathrm{T}} = P$, established in [421, §7.1].   ♦

E.3.3   Summary, symmetric idempotent

(*confer* §E.1.2) Orthogonal projector P is a (convex) linear operator defined [205, §A.3.1] by idempotence and symmetry, and characterized by positive semidefiniteness and nonexpansivity. The algebraic complement (§E.2) to $\mathcal{R}(P)$ becomes the orthogonal complement $\mathcal{R}(I - P)$; *id est,* $\mathcal{R}(P) \perp \mathcal{R}(I - P)$.

E.3.4   Orthonormal decomposition

When $Z = \mathbf{0}$ in the general nonorthogonal projector $A(A^{\dagger} + BZ^{\mathrm{T}})$ (1979), an orthogonal projector results (for any matrix A) characterized principally by idempotence and symmetry. Any real orthogonal projector may, in fact, be represented by an orthonormal decomposition such as (2006). [210, §1 prob.42]

To verify that assertion for the four fundamental subspaces (2003), we need only to express A by subcompact singular value decomposition (§A.6.2): From pseudoinverse (1657) of $A = U\Sigma Q^{\mathrm{T}} \in \mathbb{R}^{m \times n}$

$$\begin{aligned}
AA^{\dagger} &= U\Sigma\Sigma^{\dagger}U^{\mathrm{T}} = UU^{\mathrm{T}}, & A^{\dagger}A &= Q\Sigma^{\dagger}\Sigma Q^{\mathrm{T}} = QQ^{\mathrm{T}} \\
I - AA^{\dagger} &= I - UU^{\mathrm{T}} = U^{\perp}U^{\perp\mathrm{T}}, & I - A^{\dagger}A &= I - QQ^{\mathrm{T}} = Q^{\perp}Q^{\perp\mathrm{T}}
\end{aligned} \qquad (2015)$$

where $U^{\perp} \in \mathbb{R}^{m \times m - \operatorname{rank} A}$ holds columnar an orthonormal basis for the orthogonal complement of $\mathcal{R}(U)$, and likewise for $Q^{\perp} \in \mathbb{R}^{n \times n - \operatorname{rank} A}$. Existence of an orthonormal decomposition is sufficient to establish idempotence and symmetry of an orthogonal projector (2006).   ♦

E.3.5   Unifying trait of all projectors: direction

Whereas nonorthogonal projectors possess only a biorthogonal decomposition (§E.1.1), relation (2015) shows: orthogonal projectors simultaneously possess a biorthogonal decomposition ($AA^{\dagger}$ whence $Px = AA^{\dagger}x$) and an orthonormal decomposition (UU^{T} whence $Px = UU^{\mathrm{T}}x$). Orthogonal projection of a point is unique but its expansion is not; *e.g.,* A can have dependent columns.

E.3.5.1   orthogonal projector, orthonormal decomposition

Consider orthogonal expansion of $x \in \mathcal{R}(U)$:

$$x = UU^{\mathrm{T}}x = \sum_{i=1}^{n} u_i u_i^{\mathrm{T}} x \tag{2016}$$

a sum of one-dimensional orthogonal projections (§E.6.3) where

$$U \triangleq [\, u_1 \cdots u_n \,] \qquad \text{and} \qquad U^{\mathrm{T}}U = I \tag{2017}$$

and where the subspace projector has two expressions: (2015)

$$AA^{\dagger} \triangleq UU^{\mathrm{T}} \tag{2018}$$

where $A \in \mathbb{R}^{m \times n}$ has rank n. The direction of projection of x on u_j for some $j \in \{1 \dots n\}$, for example, is orthogonal to u_j but parallel to a vector in the span of all remaining vectors constituting the columns of U ;

$$\begin{aligned} u_j^{\mathrm{T}}(u_j u_j^{\mathrm{T}} x - x) &= 0 \\ u_j u_j^{\mathrm{T}} x - x = u_j u_j^{\mathrm{T}} x - UU^{\mathrm{T}} x &\in \mathcal{R}(\{u_i \,|\, i = 1 \dots n\, , \, i \neq j\}) \end{aligned} \tag{2019}$$

E.3.5.2   orthogonal projector, biorthogonal decomposition

We get a similar result for biorthogonal expansion of $x \in \mathcal{R}(A)$. Define

$$A \triangleq [\, a_1 \;\; a_2 \;\; \cdots \;\; a_n \,] \in \mathbb{R}^{m \times n} \tag{2020}$$

and rows of the pseudoinverse[E.9]

$$A^{\dagger} \triangleq \begin{bmatrix} a_1^{*\mathrm{T}} \\ a_2^{*\mathrm{T}} \\ \vdots \\ a_n^{*\mathrm{T}} \end{bmatrix} \in \mathbb{R}^{n \times m} \tag{2021}$$

under biorthogonality condition $A^{\dagger}A = I$. In biorthogonal expansion (§2.13.8)

$$x = AA^{\dagger}x = \sum_{i=1}^{n} a_i a_i^{*\mathrm{T}} x \tag{2022}$$

the direction of projection of x on a_j for some particular $j \in \{1 \dots n\}$, for example, is orthogonal to a_j^* and parallel to a vector in the span of all the remaining vectors constituting the columns of A ;

$$\begin{aligned} a_j^{*\mathrm{T}}(a_j a_j^{*\mathrm{T}} x - x) &= 0 \\ a_j a_j^{*\mathrm{T}} x - x = a_j a_j^{*\mathrm{T}} x - AA^{\dagger} x &\in \mathcal{R}(\{a_i \,|\, i = 1 \dots n\, , \, i \neq j\}) \end{aligned} \tag{2023}$$

[E.9]Notation * in this context connotes extreme direction of a dual cone; *e.g.*, (406) or Example E.5.0.0.3.

E.3.5.3   nonorthogonal projector, biorthogonal decomposition

Because the result in §E.3.5.2 is independent of matrix symmetry $AA^\dagger = (AA^\dagger)^{\mathrm{T}}$, we must get the same result for any nonorthogonal projector characterized by a biorthogonality condition; namely, for nonorthogonal projector $P = UQ^{\mathrm{T}}$ (1982) under biorthogonality condition $Q^{\mathrm{T}}U = I$, in biorthogonal expansion of $x \in \mathcal{R}(U)$

$$x = UQ^{\mathrm{T}}x = \sum_{i=1}^{k} u_i q_i^{\mathrm{T}} x \tag{2024}$$

where

$$U \triangleq [\, u_1 \cdots u_k \,] \in \mathbb{R}^{m \times k}$$

$$Q^{\mathrm{T}} \triangleq \begin{bmatrix} q_1^{\mathrm{T}} \\ \vdots \\ q_k^{\mathrm{T}} \end{bmatrix} \in \mathbb{R}^{k \times m} \tag{2025}$$

the direction of projection of x on u_j is orthogonal to q_j and parallel to a vector in the span of the remaining u_i :

$$q_j^{\mathrm{T}}(u_j q_j^{\mathrm{T}} x - x) = 0$$
$$u_j q_j^{\mathrm{T}} x - x = u_j q_j^{\mathrm{T}} x - UQ^{\mathrm{T}} x \in \mathcal{R}(\{u_i \,|\, i = 1 \ldots k\,,\, i \neq j\}) \tag{2026}$$

E.4   Algebra of projection on affine subsets

Let $P_{\mathcal{A}}x$ denote orthogonal or nonorthogonal projection of x on affine subset $\mathcal{A} \triangleq \mathcal{R} + \alpha$ where $\mathcal{R}$ is a subspace and $\alpha \in \mathcal{A}$. Let $P_{\mathcal{R}}x$ denote projection of x on $\mathcal{R}$ in the same direction. Then, because $\mathcal{R}$ is parallel to $\mathcal{A}$, it holds:

$$\begin{aligned} P_{\mathcal{A}}x = P_{\mathcal{R}+\alpha}x &= P_{\mathcal{R}}x + (I - P_{\mathcal{R}})\alpha \\ &= P_{\mathcal{R}}(x - \alpha) + \alpha \end{aligned} \tag{2027}$$

Orthogonal or nonorthogonal subspace-projector $P_{\mathcal{R}}$ is a linear operator ($P_{\mathcal{A}}$ is not). $P_{\mathcal{R}}(x + y) = P_{\mathcal{R}}x$ whenever $y \perp \mathcal{R}$ and $P_{\mathcal{R}}$ is an orthogonal projector.

E.4.0.0.1   Theorem.   *Orthogonal projection on affine subset.* [112, §9.26]
Let $\mathcal{A} = \mathcal{R} + \alpha$ be an affine subset where $\alpha \in \mathcal{A}$, and let $\mathcal{R}^{\perp}$ be the orthogonal complement of subspace $\mathcal{R}$. Then $P_{\mathcal{A}}x$ is the orthogonal projection of x on $\mathcal{A}$ if and only if

$$P_{\mathcal{A}}x \in \mathcal{A}\,, \quad \langle P_{\mathcal{A}}x - x\,,\, a - \alpha \rangle = 0 \quad \forall\, a \in \mathcal{A} \tag{2028}$$

or if and only if

$$P_{\mathcal{A}}x \in \mathcal{A}\,, \quad P_{\mathcal{A}}x - x \in \mathcal{R}^{\perp} \tag{2029}$$

$$\diamond$$

E.4.0.0.2 Example. *Intersection of affine subsets.*
When designing an optimization problem, intersection of sets is easy to express when the individual sets themselves are. Suppose $\mathcal{A} = \{x \mid Ax = b\}$ and $\mathcal{C} = \{x \mid Cx = d\}$ denote two affine subsets. Then membership to their intersection

$$\mathcal{A} \cap \mathcal{C} = \left\{ x \mid \begin{bmatrix} A \\ C \end{bmatrix} x = \begin{bmatrix} b \\ d \end{bmatrix} \right\} \tag{2030}$$

is realized as solution to simultaneous equations. In 1937, Kaczmarz instead proposed alternating projection (§E.10) on hyperplanes (whose normals constitute the rows of matrices A and C) as a means to overcome numerical instability when solving very large systems. The intersection may then be described as fixed points of projection on affine subsets, assuming $\mathcal{A} \cap \mathcal{C} \neq \emptyset$

$$\mathcal{A} \cap \mathcal{C} = \{x \mid P_{\mathcal{C}} P_{\mathcal{A}} x = x\} \tag{2031}$$

an affine system of equations, by (2027), extensible to M intersecting hyperplanes (§E.5.0.0.8):

$$\mathcal{A} \cap \mathcal{C} = \{x \mid P_M \cdots P_1 x = x\} \tag{2032}$$

$\square$

E.5 Projection examples

E.5.0.0.1 Example. *Orthogonal projection on orthogonal basis.*
Orthogonal projection on a subspace can instead be accomplished by orthogonally projecting on the individual members of an orthogonal basis for that subspace. Suppose, for example, matrix $A \in \mathbb{R}^{m \times n}$ holds an orthonormal basis for $\mathcal{R}(A)$ in its columns; $A \triangleq [a_1 \; a_2 \; \cdots \; a_n]$. Then orthogonal projection of vector $x \in \mathbb{R}^n$ on $\mathcal{R}(A)$ is a sum of one-dimensional orthogonal projections

$$Px = AA^\dagger x = A(A^{\mathrm{T}}A)^{-1}A^{\mathrm{T}}x = AA^{\mathrm{T}}x = \sum_{i=1}^{n} a_i a_i^{\mathrm{T}} x \tag{2033}$$

where each symmetric dyad $a_i a_i^{\mathrm{T}}$ is an orthogonal projector projecting on $\mathcal{R}(a_i)$. (§E.6.3) Because $\|x - Px\|$ is minimized by orthogonal projection, Px is considered to be the best approximation (in the Euclidean sense) to x from the set $\mathcal{R}(A)$. [112, §4.9] $\square$

E.5.0.0.2 Example. *Orthogonal projection on span of nonorthogonal basis.*
Orthogonal projection on a subspace can also be accomplished by projecting nonorthogonally on the individual members of any nonorthogonal basis for that subspace. This interpretation is in fact the principal application of the pseudoinverse we discussed. Now suppose matrix A holds a nonorthogonal basis for $\mathcal{R}(A)$ in its columns,

$$A = [a_1 \; a_2 \; \cdots \; a_n] \in \mathbb{R}^{m \times n} \tag{2020}$$

and define the rows $a_i^{*\mathrm{T}}$ of its pseudoinverse $A^\dagger$ as in (2021). Then orthogonal projection of vector $x \in \mathbb{R}^n$ on $\mathcal{R}(A)$ is a sum of one-dimensional nonorthogonal projections

$$Px \;=\; AA^\dagger x \;=\; \sum_{i=1}^{n} a_i a_i^{*\mathrm{T}} x \tag{2034}$$

where each nonsymmetric dyad $a_i a_i^{*\mathrm{T}}$ is a nonorthogonal projector projecting on $\mathcal{R}(a_i)$, (§E.6.1) idempotent because of biorthogonality condition $A^\dagger A = I$.

The projection Px is regarded as the best approximation to x from the set $\mathcal{R}(A)$, as it was in Example E.5.0.0.1. $\square$

E.5.0.0.3 Example. *Biorthogonal expansion as nonorthogonal projection.*
Biorthogonal expansion can be viewed as a sum of components, each a nonorthogonal projection on the range of an extreme direction of a pointed polyhedral cone $\mathcal{K}$; *e.g.*, Figure 174.

Suppose matrix $A \in \mathbb{R}^{m \times n}$ holds a nonorthogonal basis for $\mathcal{R}(A)$ in its columns as in (2020), and the rows of pseudoinverse $A^\dagger$ are defined as in (2021). Assuming the most general biorthogonality condition $(A^\dagger + BZ^\mathrm{T})A = I$ with BZ^T defined as for (1979), then biorthogonal expansion of vector x is a sum of one-dimensional nonorthogonal projections; for $x \in \mathcal{R}(A)$

$$x \;=\; A(A^\dagger + BZ^\mathrm{T})x \;=\; AA^\dagger x \;=\; \sum_{i=1}^{n} a_i a_i^{*\mathrm{T}} x \tag{2035}$$

where each dyad $a_i a_i^{*\mathrm{T}}$ is a nonorthogonal projector projecting on $\mathcal{R}(a_i)$. (§E.6.1) The extreme directions of $\mathcal{K} = \mathrm{cone}(A)$ are $\{a_1 \ldots a_n\}$ the linearly independent columns of A, while the extreme directions $\{a_1^* \ldots a_n^*\}$ of relative dual cone $\mathcal{K}^* \cap \mathrm{aff}\,\mathcal{K} = \mathrm{cone}(A^{\dagger\mathrm{T}})$ (§2.13.9.4) correspond to the linearly independent (§B.1.1.1) rows of $A^\dagger$. Directions of nonorthogonal projection are determined by the pseudoinverse; *id est*, direction of projection $a_i a_i^{*\mathrm{T}} x - x$ on $\mathcal{R}(a_i)$ is orthogonal to a_i^*.[E.10]

Because the extreme directions of this cone $\mathcal{K}$ are linearly independent, component projections are unique in the sense:

- There is only one linear combination of extreme directions of $\mathcal{K}$ that yields a particular point $x \in \mathcal{R}(A)$ whenever

$$\mathcal{R}(A) \;=\; \mathrm{aff}\,\mathcal{K} \;=\; \mathcal{R}(a_1) \oplus \mathcal{R}(a_2) \oplus \ldots \oplus \mathcal{R}(a_n) \tag{2036}$$

                                                                                                        $\square$

E.5.0.0.4 Example. *Nonorthogonal projection on elementary matrix.*
Suppose $P\mathcal{y}$ is a linear nonorthogonal projector projecting on subspace $\mathcal{Y}$, and suppose the range of a vector u is linearly independent of $\mathcal{Y}$; *id est*, for some other subspace $\mathcal{M}$ containing $\mathcal{Y}$ suppose

$$\mathcal{M} \;=\; \mathcal{R}(u) \oplus \mathcal{Y} \tag{2037}$$

[E.10]This remains true in high dimension although only a little more difficult to visualize in $\mathbb{R}^3$; *confer*, Figure 65.

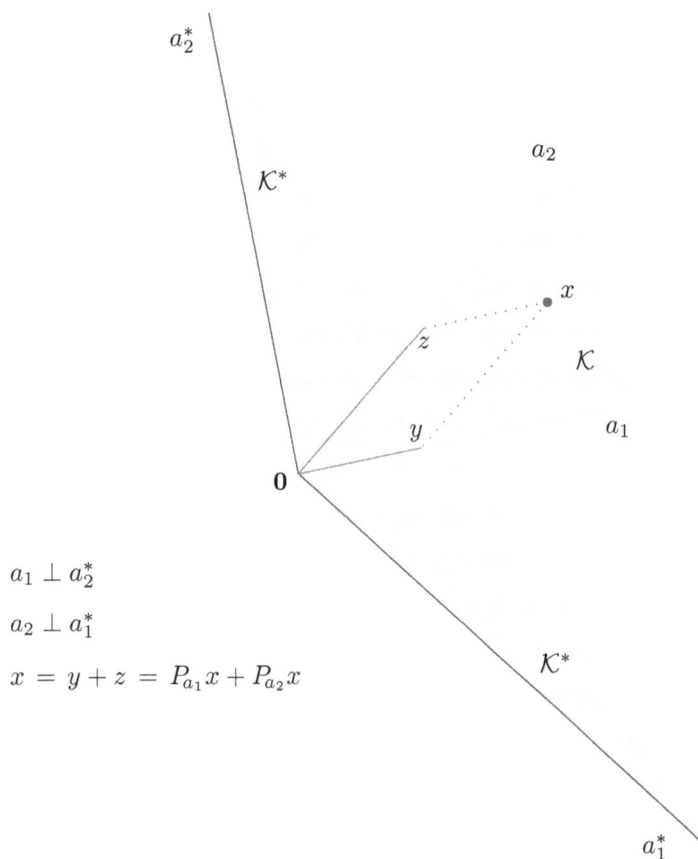

Figure 174: (*confer* Figure 64) Biorthogonal expansion of point $x \in \text{aff} \, \mathcal{K}$ is found by projecting x nonorthogonally on extreme directions of polyhedral cone $\mathcal{K} \subset \mathbb{R}^2$. (Dotted lines of projection bound this translated negated cone.) Direction of projection P_{a_1} on extreme direction a_1 is orthogonal to extreme direction a_1^* of dual cone $\mathcal{K}^*$ and parallel to a_2 (§E.3.5); similarly, direction of projection P_{a_2} on a_2 is orthogonal to a_2^* and parallel to a_1. Point x is sum of nonorthogonal projections: x on $\mathcal{R}(a_1)$ and x on $\mathcal{R}(a_2)$. Expansion is unique because extreme directions of $\mathcal{K}$ are linearly independent. Were a_1 orthogonal to a_2, then $\mathcal{K}$ would be identical to $\mathcal{K}^*$ and nonorthogonal projections would become orthogonal.

Assuming $P_\mathcal{M}x = P_ux + P_yx$ holds, then it follows for vector $x \in \mathcal{M}$

$$P_ux = x - P_yx \,, \qquad P_yx = x - P_ux \tag{2038}$$

nonorthogonal projection of x on $\mathcal{R}(u)$ can be determined from nonorthogonal projection of x on $\mathcal{Y}$, and *vice versa*.

Such a scenario is realizable were there some arbitrary basis for $\mathcal{Y}$ populating a full-rank skinny-or-square matrix A

$$A \triangleq [\, \text{basis}\, \mathcal{Y} \quad u \,] \in \mathbb{R}^{N \times n+1} \tag{2039}$$

Then $P_\mathcal{M} = AA^\dagger$ fulfills the requirements, with $P_u = A(:, n+1)A^\dagger(n+1,:)$ and $P_y = A(:, 1{:}n)A^\dagger(1{:}n,:)$. Observe, $P_\mathcal{M}$ is an orthogonal projector whereas P_y and P_u are nonorthogonal projectors.

Now suppose, for example, P_y is an elementary matrix (§B.3); in particular,

$$P_y = I - e_1\mathbf{1}^\mathrm{T} = \begin{bmatrix} \mathbf{0} & \sqrt{2}V_\mathcal{N} \end{bmatrix} \in \mathbb{R}^{N \times N} \tag{2040}$$

where $\mathcal{Y} = \mathcal{N}(\mathbf{1}^\mathrm{T})$. We have $\mathcal{M} = \mathbb{R}^N$, $A = [\sqrt{2}V_\mathcal{N} \quad e_1]$, and $u = e_1$. Thus $P_u = e_1\mathbf{1}^\mathrm{T}$ is a nonorthogonal projector projecting on $\mathcal{R}(u)$ in a direction parallel to a vector in $\mathcal{Y}$ (§E.3.5), and $P_yx = x - e_1\mathbf{1}^\mathrm{T}x$ is a nonorthogonal projection of x on $\mathcal{Y}$ in a direction parallel to u. $\qquad\square$

E.5.0.0.5 Example. *Projecting origin on a hyperplane.*
(*confer* §2.4.2.0.2) Given the hyperplane representation having $b \in \mathbb{R}$ and nonzero normal $a \in \mathbb{R}^m$

$$\partial\mathcal{H} = \{y \mid a^\mathrm{T}y = b\} \subset \mathbb{R}^m \qquad (114)$$

orthogonal projection of the origin $P\mathbf{0}$ on that hyperplane is the unique optimal solution to a minimization problem: (2001)

$$\begin{aligned} \|P\mathbf{0} - \mathbf{0}\|_2 &= \inf_{y \in \partial\mathcal{H}} \|y - \mathbf{0}\|_2 \\ &= \inf_{\xi \in \mathbb{R}^{m-1}} \|Z\xi + x\|_2 \end{aligned} \tag{2041}$$

where x is any solution to $a^\mathrm{T}y = b$, and where the columns of $Z \in \mathbb{R}^{m \times m-1}$ constitute a basis for $\mathcal{N}(a^\mathrm{T})$ so that $y = Z\xi + x \in \partial\mathcal{H}$ for all $\xi \in \mathbb{R}^{m-1}$.

The infimum can be found by setting the gradient (with respect to ξ) of the strictly convex norm-square to $\mathbf{0}$. We find the minimizing argument

$$\xi^\star = -(Z^\mathrm{T}Z)^{-1}Z^\mathrm{T}x \tag{2042}$$

so

$$y^\star = \big(I - Z(Z^\mathrm{T}Z)^{-1}Z^\mathrm{T}\big)x \tag{2043}$$

and from (2003)

$$P\mathbf{0} = y^\star = a(a^\mathrm{T}a)^{-1}a^\mathrm{T}x = \frac{a}{\|a\|}\frac{a^\mathrm{T}}{\|a\|}x = \frac{a}{\|a\|^2}a^\mathrm{T}x \triangleq A^\dagger Ax = a\frac{b}{\|a\|^2} \tag{2044}$$

In words, any point x in the hyperplane $\partial\mathcal{H}$ projected on its normal a (*confer* (2068)) yields that point $y^\star$ in the hyperplane closest to the origin. $\qquad\square$

E.5.0.0.6   Example.   *Projection on affine subset.*
The technique of Example E.5.0.0.5 is extensible. Given an intersection of hyperplanes

$$\mathcal{A} = \{y \mid Ay = b\} \subset \mathbb{R}^m \tag{2045}$$

where each row of $A \in \mathbb{R}^{m \times n}$ is nonzero and $b \in \mathcal{R}(A)$, then the orthogonal projection Px of any point $x \in \mathbb{R}^n$ on $\mathcal{A}$ is the solution to a minimization problem:

$$
\begin{aligned}
\|Px - x\|_2 &= \inf_{y \in \mathcal{A}} \quad \|y - x\|_2 \\
&= \inf_{\xi \in \mathbb{R}^{n - \text{rank} A}} \|Z\xi + y_{\text{p}} - x\|_2
\end{aligned}
\tag{2046}
$$

where y_{p} is any solution to $Ay = b$, and where the columns of $Z \in \mathbb{R}^{n \times n - \text{rank} A}$ constitute a basis for $\mathcal{N}(A)$ so that $y = Z\xi + y_{\text{p}} \in \mathcal{A}$ for all $\xi \in \mathbb{R}^{n - \text{rank} A}$. When rank of fat full-rank A is $n - 1$, then $\mathcal{A}$ describes a line; when rank $A = 1$, then $\mathcal{A}$ describes a hyperplane.

The infimum is found by setting the gradient of the strictly convex norm-square to $\mathbf{0}$. The minimizing argument is

$$\xi^\star = -(Z^{\text{T}}Z)^{-1}Z^{\text{T}}(y_{\text{p}} - x) \tag{2047}$$

so

$$y^\star = \big(I - Z(Z^{\text{T}}Z)^{-1}Z^{\text{T}}\big)(y_{\text{p}} - x) + x \tag{2048}$$

and from (2003),

$$
\begin{aligned}
Px = y^\star &= x - A^{\dagger}(Ax - b) \\
&= (I - A^{\dagger}A)x + A^{\dagger}A y_{\text{p}}
\end{aligned}
\tag{2049}
$$

which is a projection of x on $\mathcal{N}(A)$ then translated perpendicularly with respect to the nullspace until it meets the affine subset $\mathcal{A}$.   □

E.5.0.0.7   Example.   *Projection on affine subset, vertex-description.*
Suppose now we instead describe the affine subset $\mathcal{A}$ in terms of some given minimal set of generators arranged columnar in $X \in \mathbb{R}^{n \times N}$ (76); *id est*,

$$\mathcal{A} = \text{aff} X = \{Xa \mid a^{\text{T}}\mathbf{1} = 1\} \subseteq \mathbb{R}^n \tag{78}$$

Here *minimal set* means $XV_{\mathcal{N}} = [\,x_2 - x_1 \;\; x_3 - x_1 \;\; \cdots \;\; x_N - x_1\,]/\sqrt{2}$ (1040) is full-rank (§2.4.2.2) where $V_{\mathcal{N}} \in \mathbb{R}^{N \times N - 1}$ is the Schoenberg auxiliary matrix (§B.4.2). For $N = 2$ affinely independent generators, $\mathcal{A}$ represents a line; for $N = n$ affinely independent generators, $\mathcal{A}$ represents a hyperplane. Then the orthogonal projection Px of any point $x \in \mathbb{R}^n$ on $\mathcal{A}$ is the solution to a minimization problem:

$$
\begin{aligned}
\|Px - x\|_2 &= \inf_{a^{\text{T}}\mathbf{1} = 1} \|Xa - x\|_2 \\
&= \inf_{\xi \in \mathbb{R}^{N - 1}} \|X(V_{\mathcal{N}}\xi + a_{\text{p}}) - x\|_2
\end{aligned}
\tag{2050}
$$

where a_{p} is any solution to $a^{\text{T}}\mathbf{1} = 1$. We find the minimizing argument

$$\xi^\star = -(V_{\mathcal{N}}^{\text{T}}X^{\text{T}}XV_{\mathcal{N}})^{-1}V_{\mathcal{N}}^{\text{T}}X^{\text{T}}(Xa_{\text{p}} - x) \tag{2051}$$

and so the orthogonal projection is [211, §3]

$$Px = Xa^\star = (I - XV_\mathcal{N}(XV_\mathcal{N})^\dagger)Xa_\mathrm{p} + XV_\mathcal{N}(XV_\mathcal{N})^\dagger x \qquad (2052)$$

a projection of point x on $\mathcal{R}(XV_\mathcal{N})$ then translated perpendicularly with respect to that range until it meets the affine subset $\mathcal{A}$. □

E.5.0.0.8   Example. *Projecting on hyperplane, halfspace, slab.*
Given hyperplane representation having $b \in \mathbb{R}$ and nonzero normal $a \in \mathbb{R}^m$

$$\partial\mathcal{H} = \{y \mid a^\mathrm{T}y = b\} \subset \mathbb{R}^m \qquad (114)$$

orthogonal projection of any point $x \in \mathbb{R}^m$ on that hyperplane [212, §3.1] is unique:

$$Px = x - a(a^\mathrm{T}a)^{-1}(a^\mathrm{T}x - b) \qquad (2053)$$

Orthogonal projection of x on the halfspace parametrized by $b \in \mathbb{R}$ and nonzero normal $a \in \mathbb{R}^m$

$$\mathcal{H}_- = \{y \mid a^\mathrm{T}y \leq b\} \subset \mathbb{R}^m \qquad (106)$$

is the point

$$Px = x - a(a^\mathrm{T}a)^{-1}\max\{0,\ a^\mathrm{T}x - b\} \qquad (2054)$$

Orthogonal projection of x on the convex slab (Figure 11), for $c < b$

$$\{y \mid c \leq a^\mathrm{T}y \leq b\} \subset \mathbb{R}^m \qquad (2055)$$

is the point [152, §5.1]

$$Px = x - a(a^\mathrm{T}a)^{-1}\big(\max\{0,\ a^\mathrm{T}x - b\} - \max\{0,\ c - a^\mathrm{T}x\}\big) \qquad (2056)$$

□

E.6   Vectorization interpretation, projection on a matrix

E.6.1   Nonorthogonal projection on a vector

Nonorthogonal projection of vector x on the range of vector y is accomplished using a normalized dyad P_0 (§B.1); *videlicet*,

$$\frac{\langle z, x\rangle}{\langle z, y\rangle}y = \frac{z^\mathrm{T}x}{z^\mathrm{T}y}y = \frac{yz^\mathrm{T}}{z^\mathrm{T}y}x \triangleq P_0 x \qquad (2057)$$

where $\langle z,x\rangle/\langle z,y\rangle$ is the coefficient of projection on y. Because $P_0^2 = P_0$ and $\mathcal{R}(P_0) = \mathcal{R}(y)$, rank-one matrix P_0 is a nonorthogonal projector dyad projecting on $\mathcal{R}(y)$. Direction of nonorthogonal projection is orthogonal to z ; *id est*,

$$P_0 x - x \perp \mathcal{R}(P_0^\mathrm{T}) \qquad (2058)$$

E.6.2 Nonorthogonal projection on vectorized matrix

Formula (2057) is extensible. Given $X, Y, Z \in \mathbb{R}^{m \times n}$, we have the one-dimensional nonorthogonal projection of X in isomorphic $\mathbb{R}^{mn}$ on the range of vectorized Y : (§2.2)

$$\frac{\langle Z \, , \, X \rangle}{\langle Z \, , \, Y \rangle} \, Y \, , \qquad \langle Z \, , \, Y \rangle \neq 0 \tag{2059}$$

where $\langle Z \, , \, X \rangle / \langle Z \, , \, Y \rangle$ is the coefficient of projection. The inequality accounts for the fact: projection on $\mathcal{R}(\operatorname{vec} Y)$ is in a direction orthogonal to $\operatorname{vec} Z$.

E.6.2.1 Nonorthogonal projection on dyad

Now suppose we have nonorthogonal projector dyad

$$P_0 = \frac{yz^{\mathrm{T}}}{z^{\mathrm{T}}y} \in \mathbb{R}^{m \times m} \tag{2060}$$

Analogous to (2057), for $X \in \mathbb{R}^{m \times m}$

$$P_0 X P_0 = \frac{yz^{\mathrm{T}}}{z^{\mathrm{T}}y} X \frac{yz^{\mathrm{T}}}{z^{\mathrm{T}}y} = \frac{z^{\mathrm{T}}Xy}{(z^{\mathrm{T}}y)^2} \, yz^{\mathrm{T}} = \frac{\langle zy^{\mathrm{T}}, X \rangle}{\langle zy^{\mathrm{T}}, yz^{\mathrm{T}} \rangle} \, yz^{\mathrm{T}} \tag{2061}$$

is a nonorthogonal projection of matrix X on the range of vectorized dyad P_0 ; from which it follows:

$$P_0 X P_0 = \frac{z^{\mathrm{T}}Xy}{z^{\mathrm{T}}y} \frac{yz^{\mathrm{T}}}{z^{\mathrm{T}}y} = \left\langle \frac{zy^{\mathrm{T}}}{z^{\mathrm{T}}y} \, , \, X \right\rangle \frac{yz^{\mathrm{T}}}{z^{\mathrm{T}}y} = \langle P_0^{\mathrm{T}}, X \rangle \, P_0 = \frac{\langle P_0^{\mathrm{T}}, X \rangle}{\langle P_0^{\mathrm{T}}, P_0 \rangle} \, P_0 \tag{2062}$$

Yet this relationship between matrix product and vector inner-product only holds for a dyad projector. When nonsymmetric projector P_0 is rank-one as in (2060), therefore,

$$\mathcal{R}(\operatorname{vec} P_0 X P_0) = \mathcal{R}(\operatorname{vec} P_0) \text{ in } \mathbb{R}^{m^2} \tag{2063}$$

and

$$P_0 X P_0 - X \perp P_0^{\mathrm{T}} \text{ in } \mathbb{R}^{m^2} \tag{2064}$$

E.6.2.1.1 Example. λ *as coefficients of nonorthogonal projection.*
Any diagonalization (§A.5)

$$X = S\Lambda S^{-1} = \sum_{i=1}^{m} \lambda_i \, s_i w_i^{\mathrm{T}} \in \mathbb{R}^{m \times m} \tag{1628}$$

may be expressed as a sum of one-dimensional nonorthogonal projections of X, each on the range of a vectorized eigenmatrix $P_j \triangleq s_j w_j^{\mathrm{T}}$;

$$
\begin{aligned}
X &= \sum_{i,j=1}^{m} \langle (Se_i e_j^{\mathrm{T}} S^{-1})^{\mathrm{T}}, X \rangle \, Se_i e_j^{\mathrm{T}} S^{-1} \\
&= \sum_{j=1}^{m} \langle (s_j w_j^{\mathrm{T}})^{\mathrm{T}}, X \rangle \, s_j w_j^{\mathrm{T}} \; + \; \sum_{\substack{i,j=1 \\ j \neq i}}^{m} \langle (Se_i e_j^{\mathrm{T}} S^{-1})^{\mathrm{T}}, S\Lambda S^{-1} \rangle \, Se_i e_j^{\mathrm{T}} S^{-1} \\
&= \sum_{j=1}^{m} \langle (s_j w_j^{\mathrm{T}})^{\mathrm{T}}, X \rangle \, s_j w_j^{\mathrm{T}} \\
&\triangleq \sum_{j=1}^{m} \langle P_j^{\mathrm{T}}, X \rangle \, P_j \;=\; \sum_{j=1}^{m} s_j w_j^{\mathrm{T}} X s_j w_j^{\mathrm{T}} \;=\; \sum_{j=1}^{m} P_j X P_j \\
&= \sum_{j=1}^{m} \lambda_j \, s_j w_j^{\mathrm{T}}
\end{aligned}
\tag{2065}
$$

This biorthogonal expansion of matrix X is a sum of nonorthogonal projections because the term outside the projection coefficient $\langle \; \rangle$ is not identical to the inside-term. (§E.6.4) The eigenvalues λ_j are coefficients of nonorthogonal projection of X, while the remaining $M(M-1)/2$ coefficients (for $i \neq j$) are zeroed by projection. When P_j is rank-one as in (2065),

$$
\mathcal{R}(\mathrm{vec}\, P_j X P_j) \;=\; \mathcal{R}(\mathrm{vec}\, s_j w_j^{\mathrm{T}}) \;=\; \mathcal{R}(\mathrm{vec}\, P_j) \text{ in } \mathbb{R}^{m^2}
\tag{2066}
$$

and

$$
P_j X P_j - X \;\perp\; P_j^{\mathrm{T}} \text{ in } \mathbb{R}^{m^2}
\tag{2067}
$$

Were matrix X symmetric, then its eigenmatrices would also be. So the one-dimensional projections would become orthogonal. (§E.6.4.1.1) $\square$

E.6.3   Orthogonal projection on a vector

The formula for orthogonal projection of vector x on the range of vector y (*one-dimensional projection*) is basic *analytic geometry*; [13, §3.3] [338, §3.2] [374, §2.2] [408, §1-8]

$$
\frac{\langle y, x \rangle}{\langle y, y \rangle} y \;=\; \frac{y^{\mathrm{T}} x}{y^{\mathrm{T}} y} y \;=\; \frac{yy^{\mathrm{T}}}{y^{\mathrm{T}} y} x \;\triangleq\; P_1 x
\tag{2068}
$$

where $\langle y, x \rangle / \langle y, y \rangle$ is the coefficient of projection on $\mathcal{R}(y)$. An equivalent description is: Vector $P_1 x$ is the orthogonal projection of vector x on $\mathcal{R}(P_1) = \mathcal{R}(y)$. Rank-one matrix P_1 is a projection matrix because $P_1^2 = P_1$. The direction of projection is orthogonal

$$
P_1 x - x \;\perp\; \mathcal{R}(P_1)
\tag{2069}
$$

because $P_1^{\mathrm{T}} = P_1$.

E.6.4 Orthogonal projection on a vectorized matrix

From (2068), given instead $X, Y \in \mathbb{R}^{m \times n}$, we have the one-dimensional orthogonal projection of matrix X in isomorphic $\mathbb{R}^{mn}$ on the range of vectorized Y : (§2.2)

$$\frac{\langle Y, X \rangle}{\langle Y, Y \rangle} Y \tag{2070}$$

where $\langle Y, X \rangle / \langle Y, Y \rangle$ is the coefficient of projection.

For orthogonal projection, the term outside the vector inner-products $\langle \ \rangle$ must be identical to the terms inside in three places.

E.6.4.1 Orthogonal projection on dyad

There is opportunity for insight when Y is a dyad yz^{T} (§B.1): Instead given $X \in \mathbb{R}^{m \times n}$, $y \in \mathbb{R}^m$, and $z \in \mathbb{R}^n$

$$\frac{\langle yz^{\mathrm{T}}, X \rangle}{\langle yz^{\mathrm{T}}, yz^{\mathrm{T}} \rangle} yz^{\mathrm{T}} = \frac{y^{\mathrm{T}} X z}{y^{\mathrm{T}} y \, z^{\mathrm{T}} z} yz^{\mathrm{T}} \tag{2071}$$

is the one-dimensional orthogonal projection of X in isomorphic $\mathbb{R}^{mn}$ on the range of vectorized yz^{T}. To reveal the obscured symmetric projection matrices P_1 and P_2 we rewrite (2071):

$$\frac{y^{\mathrm{T}} X z}{y^{\mathrm{T}} y \, z^{\mathrm{T}} z} yz^{\mathrm{T}} = \frac{yy^{\mathrm{T}}}{y^{\mathrm{T}} y} X \frac{zz^{\mathrm{T}}}{z^{\mathrm{T}} z} \triangleq P_1 X P_2 \tag{2072}$$

So for projector dyads, projection (2072) is the orthogonal projection in $\mathbb{R}^{mn}$ if and only if projectors P_1 and P_2 are symmetric;[E.11] in other words,

- for orthogonal projection on the range of a vectorized dyad yz^{T}, the term outside the vector inner-products $\langle \ \rangle$ in (2071) must be identical to the terms inside in three places.

When P_1 and P_2 are rank-one symmetric projectors as in (2072), (37)

$$\mathcal{R}(\mathrm{vec}\, P_1 X P_2) = \mathcal{R}(\mathrm{vec}\, yz^{\mathrm{T}}) \text{ in } \mathbb{R}^{mn} \tag{2073}$$

and

$$P_1 X P_2 - X \perp yz^{\mathrm{T}} \text{ in } \mathbb{R}^{mn} \tag{2074}$$

When $y = z$ then $P_1 = P_2 = P_2^{\mathrm{T}}$ and

$$P_1 X P_1 = \langle P_1, X \rangle P_1 = \frac{\langle P_1, X \rangle}{\langle P_1, P_1 \rangle} P_1 \tag{2075}$$

[E.11] For diagonalizable $X \in \mathbb{R}^{m \times m}$ (§A.5), its orthogonal projection in isomorphic $\mathbb{R}^{m^2}$ on the range of vectorized $yz^{\mathrm{T}} \in \mathbb{R}^{m \times m}$ becomes:

$$P_1 X P_2 = \sum_{i=1}^{m} \lambda_i \, P_1 \, s_i \, w_i^{\mathrm{T}} P_2$$

When $\mathcal{R}(P_1) = \mathcal{R}(w_j)$ and $\mathcal{R}(P_2) = \mathcal{R}(s_j)$, the jth dyad term from the diagonalization is isolated but only, in general, to within a scale factor because neither set of left or right eigenvectors is necessarily orthonormal unless X is a normal matrix [421, §3.2]. Yet when $\mathcal{R}(P_2) = \mathcal{R}(s_k)$, $k \neq j \in \{1 \dots m\}$, then $P_1 X P_2 = \mathbf{0}$.

meaning, $P_1 X P_1$ is equivalent to orthogonal projection of matrix X on the range of vectorized projector dyad P_1. Yet this relationship between matrix product and vector inner-product does not hold for general symmetric projector matrices.

E.6.4.1.1 Example. *Eigenvalues λ as coefficients of orthogonal projection.*
Let $\mathcal{C}$ represent any convex subset of subspace $\mathbb{S}^M$, and let $\mathcal{C}_1$ be any element of $\mathcal{C}$. Then $\mathcal{C}_1$ can be expressed as the orthogonal expansion

$$\mathcal{C}_1 = \sum_{i=1}^{M} \sum_{\substack{j=1 \\ j \geq i}}^{M} \langle E_{ij}, \mathcal{C}_1 \rangle \, E_{ij} \ \in \mathcal{C} \subset \mathbb{S}^M \tag{2076}$$

where $E_{ij} \in \mathbb{S}^M$ is a member of the standard orthonormal basis for $\mathbb{S}^M$ (59). This expansion is a sum of one-dimensional orthogonal projections of $\mathcal{C}_1$; each projection on the range of a vectorized standard basis matrix. Vector inner-product $\langle E_{ij}, \mathcal{C}_1 \rangle$ is the coefficient of projection of $\operatorname{svec} \mathcal{C}_1$ on $\mathcal{R}(\operatorname{svec} E_{ij})$.

When $\mathcal{C}_1$ is any member of a convex set $\mathcal{C}$ whose dimension is L, *Carathéodory's theorem* [116] [315] [205] [41] [42] guarantees that no more than $L+1$ affinely independent members from $\mathcal{C}$ are required to faithfully represent $\mathcal{C}_1$ by their linear combination.[E.12]

Dimension of $\mathbb{S}^M$ is $L = M(M+1)/2$ in isometrically isomorphic $\mathbb{R}^{M(M+1)/2}$. Yet because any symmetric matrix can be diagonalized, (§A.5.1) $\mathcal{C}_1 \in \mathbb{S}^M$ is a linear combination of its M eigenmatrices $q_i q_i^{\mathrm{T}}$ (§A.5.0.3) weighted by its eigenvalues λ_i;

$$\mathcal{C}_1 = Q \Lambda Q^{\mathrm{T}} = \sum_{i=1}^{M} \lambda_i \, q_i q_i^{\mathrm{T}} \tag{2077}$$

where $\Lambda \in \mathbb{S}^M$ is a diagonal matrix having $\delta(\Lambda)_i = \lambda_i$, and $Q = [q_1 \cdots q_M]$ is an orthogonal matrix in $\mathbb{R}^{M \times M}$ containing corresponding eigenvectors.

To derive eigenvalue decomposition (2077) from expansion (2076), M standard basis matrices E_{ij} are rotated (§B.5) into alignment with the M eigenmatrices $q_i q_i^{\mathrm{T}}$ of $\mathcal{C}_1$ by applying a *similarity transformation*; [338, §5.6]

$$\{Q E_{ij} Q^{\mathrm{T}}\} = \left\{ \begin{array}{ll} q_i q_i^{\mathrm{T}}, & i = j = 1 \ldots M \\ \frac{1}{\sqrt{2}}\left(q_i q_j^{\mathrm{T}} + q_j q_i^{\mathrm{T}}\right), & 1 \leq i < j \leq M \end{array} \right\} \tag{2078}$$

which remains an orthonormal basis for $\mathbb{S}^M$. Then remarkably

---

[E.12]Carathéodory's theorem guarantees existence of a biorthogonal expansion for any element in $\operatorname{aff} \mathcal{C}$ when $\mathcal{C}$ is any pointed closed convex cone.

$$
\begin{aligned}
\mathcal{C}_1 &= \sum_{\substack{i,j=1 \\ j \geq i}}^{M} \langle Q E_{ij} Q^{\mathrm{T}}, \mathcal{C}_1 \rangle \, Q E_{ij} Q^{\mathrm{T}} \\
&= \sum_{i=1}^{M} \langle q_i q_i^{\mathrm{T}}, \mathcal{C}_1 \rangle \, q_i q_i^{\mathrm{T}} \;+\; \sum_{\substack{i,j=1 \\ j > i}}^{M} \langle Q E_{ij} Q^{\mathrm{T}}, Q \Lambda Q^{\mathrm{T}} \rangle \, Q E_{ij} Q^{\mathrm{T}} \\
&= \sum_{i=1}^{M} \langle q_i q_i^{\mathrm{T}}, \mathcal{C}_1 \rangle \, q_i q_i^{\mathrm{T}} \\
&\triangleq \sum_{i=1}^{M} \langle P_i , \mathcal{C}_1 \rangle \, P_i \;=\; \sum_{i=1}^{M} q_i q_i^{\mathrm{T}} \mathcal{C}_1 \, q_i q_i^{\mathrm{T}} \;=\; \sum_{i=1}^{M} P_i \mathcal{C}_1 P_i \\
&= \sum_{i=1}^{M} \lambda_i \, q_i q_i^{\mathrm{T}}
\end{aligned}
\tag{2079}
$$

this orthogonal expansion becomes the diagonalization; still a sum of one-dimensional orthogonal projections. The eigenvalues

$$
\lambda_i = \langle q_i q_i^{\mathrm{T}}, \mathcal{C}_1 \rangle
\tag{2080}
$$

are clearly coefficients of projection of $\mathcal{C}_1$ on the range of each vectorized eigenmatrix. (*confer* §E.6.2.1.1) The remaining $M(M-1)/2$ coefficients $(i \neq j)$ are zeroed by projection. When P_i is rank-one symmetric as in (2079),

$$
\mathcal{R}(\operatorname{svec} P_i \mathcal{C}_1 P_i) \;=\; \mathcal{R}(\operatorname{svec} q_i q_i^{\mathrm{T}}) \;=\; \mathcal{R}(\operatorname{svec} P_i) \text{ in } \mathbb{R}^{M(M+1)/2}
\tag{2081}
$$

and

$$
P_i \mathcal{C}_1 P_i - \mathcal{C}_1 \perp P_i \text{ in } \mathbb{R}^{M(M+1)/2}
\tag{2082}
$$

$\square$

E.6.4.2 Positive semidefiniteness test as orthogonal projection

For any given $X \in \mathbb{R}^{m \times m}$ the familiar quadratic construct $y^{\mathrm{T}} X y \geq 0$, over broad domain, is a fundamental test for positive semidefiniteness. (§A.2) It is a fact that $y^{\mathrm{T}} X y$ is always proportional to a coefficient of orthogonal projection; letting z in formula (2071) become $y \in \mathbb{R}^m$, then $P_2 = P_1 = y y^{\mathrm{T}} / y^{\mathrm{T}} y = y y^{\mathrm{T}} / \| y y^{\mathrm{T}} \|_2$ (*confer* (1693)) and formula (2072) becomes

$$
\frac{\langle y y^{\mathrm{T}}, X \rangle}{\langle y y^{\mathrm{T}}, y y^{\mathrm{T}} \rangle} \, y y^{\mathrm{T}} = \frac{y^{\mathrm{T}} X y}{y^{\mathrm{T}} y} \frac{y y^{\mathrm{T}}}{y^{\mathrm{T}} y} = \frac{y y^{\mathrm{T}}}{y^{\mathrm{T}} y} X \frac{y y^{\mathrm{T}}}{y^{\mathrm{T}} y} \triangleq P_1 X P_1
\tag{2083}
$$

By (2070), product $P_1 X P_1$ is the one-dimensional orthogonal projection of X in isomorphic $\mathbb{R}^{m^2}$ on the range of vectorized P_1 because, for $\operatorname{rank} P_1 = 1$ and $P_1^2 = P_1 \in \mathbb{S}^m$ (*confer* (2062))

$$
P_1 X P_1 = \frac{y^{\mathrm{T}} X y}{y^{\mathrm{T}} y} \frac{y y^{\mathrm{T}}}{y^{\mathrm{T}} y} = \left\langle \frac{y y^{\mathrm{T}}}{y^{\mathrm{T}} y} , X \right\rangle \frac{y y^{\mathrm{T}}}{y^{\mathrm{T}} y} = \langle P_1 , X \rangle \, P_1 = \frac{\langle P_1 , X \rangle}{\langle P_1 , P_1 \rangle} P_1
\tag{2084}
$$

The coefficient of orthogonal projection $\langle P_1 , X \rangle = y^{\mathrm{T}} X y/(y^{\mathrm{T}} y)$ is also known as *Rayleigh's quotient*.[E.13] When P_1 is rank-one symmetric as in (2083),

$$\mathcal{R}(\operatorname{vec} P_1 X P_1) = \mathcal{R}(\operatorname{vec} P_1) \text{ in } \mathbb{R}^{m^2} \tag{2085}$$

and

$$P_1 X P_1 - X \perp P_1 \text{ in } \mathbb{R}^{m^2} \tag{2086}$$

The test for positive semidefiniteness, then, is a test for nonnegativity of the coefficient of orthogonal projection of X on the range of each and every vectorized extreme direction yy^{T} (§2.8.1) from the positive semidefinite cone in the ambient space of symmetric matrices.

E.6.4.3 $PXP \succeq 0$

In some circumstances, it may be desirable to limit the domain of test $y^{\mathrm{T}} X y \geq 0$ for positive semidefiniteness; *e.g.*, $\{\|y\| = 1\}$. Another example of limiting domain-of-test is central to Euclidean distance geometry: For $\mathcal{R}(V) = \mathcal{N}(\mathbf{1}^{\mathrm{T}})$, the test $-VDV \succeq 0$ determines whether $D \in \mathbb{S}_h^N$ is a Euclidean distance matrix. The same test may be stated: For $D \in \mathbb{S}_h^N$ (and optionally $\|y\| = 1$)

$$D \in \mathbb{EDM}^N \;\Leftrightarrow\; -y^{\mathrm{T}} D y = \langle yy^{\mathrm{T}}, -D \rangle \geq 0 \quad \forall y \in \mathcal{R}(V) \tag{2087}$$

The test $-VDV \succeq 0$ is therefore equivalent to a test for nonnegativity of the coefficient of orthogonal projection of $-D$ on the range of each and every vectorized extreme direction yy^{T} from the positive semidefinite cone $\mathbb{S}_+^N$ such that $\mathcal{R}(yy^{\mathrm{T}}) = \mathcal{R}(y) \subseteq \mathcal{R}(V)$. (Validity of this result is independent of whether V is itself a projection matrix.)

E.7 Projection on matrix subspaces

E.7.1 PXP misinterpretation for higher-rank P

For a projection matrix P of rank greater than 1, PXP is generally not commensurate with $\frac{\langle P, X \rangle}{\langle P, P \rangle} P$ as is the case for projector dyads (2084). Yet for a symmetric idempotent

---

[E.13] When y becomes the jth eigenvector s_j of diagonalizable X, for example, $\langle P_1 , X \rangle$ becomes the jth eigenvalue: [201, §III]

$$\langle P_1 , X \rangle|_{y=s_j} = \frac{s_j^{\mathrm{T}} \left(\sum_{i=1}^{m} \lambda_i s_i w_i^{\mathrm{T}} \right) s_j}{s_j^{\mathrm{T}} s_j} = \lambda_j$$

Similarly for $y = w_j$, the jth left-eigenvector,

$$\langle P_1 , X \rangle|_{y=w_j} = \frac{w_j^{\mathrm{T}} \left(\sum_{i=1}^{m} \lambda_i s_i w_i^{\mathrm{T}} \right) w_j}{w_j^{\mathrm{T}} w_j} = \lambda_j$$

A quandary may arise regarding the potential annihilation of the antisymmetric part of X when $s_j^{\mathrm{T}} X s_j$ is formed. Were annihilation to occur, it would imply the eigenvalue thus found came instead from the symmetric part of X. The quandary is resolved recognizing that diagonalization of real X admits complex eigenvectors; hence, annihilation could only come about by forming $\operatorname{re}(s_j^{\mathrm{H}} X s_j) = s_j^{\mathrm{H}}(X + X^{\mathrm{T}}) s_j/2$ [208, §7.1] where $(X + X^{\mathrm{T}})/2$ is the symmetric part of X , and s_j^{H} denotes conjugate transpose.

matrix P of any rank we are tempted to say " PXP is the orthogonal projection of $X \in \mathbb{S}^m$ on $\mathcal{R}(\text{vec}\,P)$ ". The fallacy is: $\text{vec}\,PXP$ does not necessarily belong to the range of vectorized P ; the most basic requirement for projection on $\mathcal{R}(\text{vec}\,P)$.

E.7.2 Orthogonal projection on matrix subspaces

With $A_1 \in \mathbb{R}^{m \times n}$, $B_1 \in \mathbb{R}^{n \times k}$, $Z_1 \in \mathbb{R}^{m \times k}$, $A_2 \in \mathbb{R}^{p \times n}$, $B_2 \in \mathbb{R}^{n \times k}$, $Z_2 \in \mathbb{R}^{p \times k}$ as defined for nonorthogonal projector (1979), and defining

$$P_1 \triangleq A_1 A_1^\dagger \in \mathbb{S}^m \,, \qquad P_2 \triangleq A_2 A_2^\dagger \in \mathbb{S}^p \qquad (2088)$$

then, given compatible X

$$\|X - P_1 X P_2\|_{\mathrm{F}} = \inf_{B_1,\,B_2 \in \mathbb{R}^{n \times k}} \|X - A_1(A_1^\dagger + B_1 Z_1^{\mathrm{T}})X(A_2^{\dagger\mathrm{T}} + Z_2 B_2^{\mathrm{T}})A_2^{\mathrm{T}}\|_{\mathrm{F}} \qquad (2089)$$

As for all subspace projectors, range of the projector is the subspace on which projection is made: $\{P_1 Y P_2 \mid Y \in \mathbb{R}^{m \times p}\}$. For projectors P_1 and P_2 of any rank, altogether, this means projection $P_1 X P_2$ is unique minimum-distance, orthogonal

$$P_1 X P_2 - X \perp \{P_1 Y P_2 \mid Y \in \mathbb{R}^{m \times p}\} \text{ in } \mathbb{R}^{mp} \qquad (2090)$$

and P_1 and P_2 must each be symmetric (*confer* (2072)) to attain the infimum.

E.7.2.0.1 Proof. *Minimum Frobenius norm* (2089).
Defining $P \triangleq A_1(A_1^\dagger + B_1 Z_1^{\mathrm{T}})$,

$$\begin{aligned}
&\inf_{B_1,B_2} \|X - A_1(A_1^\dagger + B_1 Z_1^{\mathrm{T}})X(A_2^{\dagger\mathrm{T}} + Z_2 B_2^{\mathrm{T}})A_2^{\mathrm{T}}\|_{\mathrm{F}}^2 \\
&= \inf_{B_1,B_2} \|X - PX(A_2^{\dagger\mathrm{T}} + Z_2 B_2^{\mathrm{T}})A_2^{\mathrm{T}}\|_{\mathrm{F}}^2 \\
&= \inf_{B_1,B_2} \mathrm{tr}\Big((X^{\mathrm{T}} - A_2(A_2^\dagger + B_2 Z_2^{\mathrm{T}})X^{\mathrm{T}}P^{\mathrm{T}})(X - PX(A_2^{\dagger\mathrm{T}} + Z_2 B_2^{\mathrm{T}})A_2^{\mathrm{T}})\Big) \\
&= \inf_{B_1,B_2} \mathrm{tr}\Big(X^{\mathrm{T}}X - X^{\mathrm{T}}PX(A_2^{\dagger\mathrm{T}} + Z_2 B_2^{\mathrm{T}})A_2^{\mathrm{T}} - A_2(A_2^\dagger + B_2 Z_2^{\mathrm{T}})X^{\mathrm{T}}P^{\mathrm{T}}X \\
&\qquad\qquad + A_2(A_2^\dagger + B_2 Z_2^{\mathrm{T}})X^{\mathrm{T}}P^{\mathrm{T}}PX(A_2^{\dagger\mathrm{T}} + Z_2 B_2^{\mathrm{T}})A_2^{\mathrm{T}}\Big)
\end{aligned} \qquad (2091)$$

Necessary conditions for a global minimum are $\nabla_{B_1} = \mathbf{0}$ and $\nabla_{B_2} = \mathbf{0}$. Terms not containing B_2 in (2091) will vanish from gradient ∇_{B_2} ; (§D.2.3)

$$\begin{aligned}
&\nabla_{B_2} \mathrm{tr}\Big(-X^{\mathrm{T}}PXZ_2 B_2^{\mathrm{T}}A_2^{\mathrm{T}} - A_2 B_2 Z_2^{\mathrm{T}}X^{\mathrm{T}}P^{\mathrm{T}}X + A_2 A_2^\dagger X^{\mathrm{T}}P^{\mathrm{T}}PXZ_2 B_2^{\mathrm{T}}A_2^{\mathrm{T}} \\
&\qquad\qquad + A_2 B_2 Z_2^{\mathrm{T}}X^{\mathrm{T}}P^{\mathrm{T}}PXA_2^{\dagger\mathrm{T}}A_2^{\mathrm{T}} + A_2 B_2 Z_2^{\mathrm{T}}X^{\mathrm{T}}P^{\mathrm{T}}PXZ_2 B_2^{\mathrm{T}}A_2^{\mathrm{T}}\Big) \\
&= -2A_2^{\mathrm{T}}X^{\mathrm{T}}PXZ_2 + 2A_2^{\mathrm{T}}A_2 A_2^\dagger X^{\mathrm{T}}P^{\mathrm{T}}PXZ_2 + 2A_2^{\mathrm{T}}A_2 B_2 Z_2^{\mathrm{T}}X^{\mathrm{T}}P^{\mathrm{T}}PXZ_2 \\
&= A_2^{\mathrm{T}}\Big(-X^{\mathrm{T}} + A_2 A_2^\dagger X^{\mathrm{T}}P^{\mathrm{T}} + A_2 B_2 Z_2^{\mathrm{T}}X^{\mathrm{T}}P^{\mathrm{T}}\Big)PXZ_2 \\
&= \mathbf{0} \qquad\qquad\qquad\qquad \Leftrightarrow \\
&\qquad\qquad \mathcal{R}(B_1) \subseteq \mathcal{N}(A_1) \quad \textbf{and} \quad \mathcal{R}(B_2) \subseteq \mathcal{N}(A_2)
\end{aligned} \qquad (2092)$$

(or $Z_2 = \mathbf{0}$) because $A^{\mathrm{T}} = A^{\mathrm{T}}AA^\dagger$. Symmetry requirement (2088) is implicit. Were instead $P^{\mathrm{T}} \triangleq (A_2^{\dagger\mathrm{T}} + Z_2 B_2^{\mathrm{T}})A_2^{\mathrm{T}}$ and the gradient with respect to B_1 observed, then similar results are obtained. The projector is unique. Perpendicularity (2090) establishes uniqueness [112, §4.9] of projection $P_1 X P_2$ on a matrix subspace. The minimum-distance projector is the orthogonal projector, and *vice versa.*   ◆

E.7.2.0.2   Example. *PXP redux & $\mathcal{N}(\mathbf{V})$.*

Suppose we define a subspace of $m \times n$ matrices, each elemental matrix having columns constituting a list whose geometric center (§5.5.1.0.1) is the origin in $\mathbb{R}^m$:

$$\begin{aligned}
\mathbb{R}_c^{m \times n} &\triangleq \{Y \in \mathbb{R}^{m \times n} \mid Y\mathbf{1} = \mathbf{0}\} \\
&= \{Y \in \mathbb{R}^{m \times n} \mid \mathcal{N}(Y) \supseteq \mathbf{1}\} = \{Y \in \mathbb{R}^{m \times n} \mid \mathcal{R}(Y^{\mathrm{T}}) \subseteq \mathcal{N}(\mathbf{1}^{\mathrm{T}})\} \\
&= \{XV \mid X \in \mathbb{R}^{m \times n}\} \subset \mathbb{R}^{m \times n}
\end{aligned} \tag{2093}$$

the *nonsymmetric geometric center subspace.* Further suppose $V \in \mathbb{S}^n$ is a projection matrix having $\mathcal{N}(V) = \mathcal{R}(\mathbf{1})$ and $\mathcal{R}(V) = \mathcal{N}(\mathbf{1}^{\mathrm{T}})$. Then linear mapping $T(X) = XV$ is the orthogonal projection of any $X \in \mathbb{R}^{m \times n}$ on $\mathbb{R}_c^{m \times n}$ in the Euclidean (vectorization) sense because V is symmetric, $\mathcal{N}(XV) \supseteq \mathbf{1}$, and $\mathcal{R}(VX^{\mathrm{T}}) \subseteq \mathcal{N}(\mathbf{1}^{\mathrm{T}})$.

Now suppose we define a subspace of symmetric $n \times n$ matrices each of whose columns constitute a list having the origin in $\mathbb{R}^n$ as geometric center,

$$\begin{aligned}
\mathbb{S}_c^n &\triangleq \{Y \in \mathbb{S}^n \mid Y\mathbf{1} = \mathbf{0}\} \\
&= \{Y \in \mathbb{S}^n \mid \mathcal{N}(Y) \supseteq \mathbf{1}\} = \{Y \in \mathbb{S}^n \mid \mathcal{R}(Y) \subseteq \mathcal{N}(\mathbf{1}^{\mathrm{T}})\}
\end{aligned} \tag{2094}$$

the *geometric center subspace.* Further suppose $V \in \mathbb{S}^n$ is a projection matrix, the same as before. Then VXV is the orthogonal projection of any $X \in \mathbb{S}^n$ on $\mathbb{S}_c^n$ in the Euclidean sense (2090) because V is symmetric, $VXV\mathbf{1} = \mathbf{0}$, and $\mathcal{R}(VXV) \subseteq \mathcal{N}(\mathbf{1}^{\mathrm{T}})$. Two-sided projection is necessary only to remain in the ambient symmetric matrix subspace. Then

$$\mathbb{S}_c^n = \{VXV \mid X \in \mathbb{S}^n\} \subset \mathbb{S}^n \tag{2095}$$

has $\dim \mathbb{S}_c^n = n(n-1)/2$ in isomorphic $\mathbb{R}^{n(n+1)/2}$. We find its orthogonal complement as the aggregate of all negative directions of orthogonal projection on $\mathbb{S}_c^n$: the *translation-invariant subspace* (§5.5.1.1)

$$\begin{aligned}
\mathbb{S}_c^{n\perp} &\triangleq \{X - VXV \mid X \in \mathbb{S}^n\} \subset \mathbb{S}^n \\
&= \{u\mathbf{1}^{\mathrm{T}} + \mathbf{1}u^{\mathrm{T}} \mid u \in \mathbb{R}^n\}
\end{aligned} \tag{2096}$$

characterized by doublet $u\mathbf{1}^{\mathrm{T}} + \mathbf{1}u^{\mathrm{T}}$ (§B.2).[E.14] Defining geometric center mapping

---

[E.14]**Proof.**

$$\begin{aligned}
\{X - VXV \mid X \in \mathbb{S}^n\} &= \{X - (I - \tfrac{1}{n}\mathbf{1}\mathbf{1}^{\mathrm{T}})X(I - \mathbf{1}\mathbf{1}^{\mathrm{T}}\tfrac{1}{n}) \mid X \in \mathbb{S}^n\} \\
&= \{\tfrac{1}{n}\mathbf{1}\mathbf{1}^{\mathrm{T}}X + X\mathbf{1}\mathbf{1}^{\mathrm{T}}\tfrac{1}{n} - \tfrac{1}{n}\mathbf{1}\mathbf{1}^{\mathrm{T}}X\mathbf{1}\mathbf{1}^{\mathrm{T}}\tfrac{1}{n} \mid X \in \mathbb{S}^n\}
\end{aligned}$$

Because $\{X\mathbf{1} \mid X \in \mathbb{S}^n\} = \mathbb{R}^n$,

$$\begin{aligned}
\{X - VXV \mid X \in \mathbb{S}^n\} &= \{\mathbf{1}\zeta^{\mathrm{T}} + \zeta\mathbf{1}^{\mathrm{T}} - \mathbf{1}\mathbf{1}^{\mathrm{T}}(\mathbf{1}^{\mathrm{T}}\zeta\tfrac{1}{n}) \mid \zeta \in \mathbb{R}^n\} \\
&= \{\mathbf{1}\zeta^{\mathrm{T}}(I - \mathbf{1}\mathbf{1}^{\mathrm{T}}\tfrac{1}{2n}) + (I - \tfrac{1}{2n}\mathbf{1}\mathbf{1}^{\mathrm{T}})\zeta\mathbf{1}^{\mathrm{T}} \mid \zeta \in \mathbb{R}^n\}
\end{aligned}$$

where $I - \tfrac{1}{2n}\mathbf{1}\mathbf{1}^{\mathrm{T}}$ is invertible.   ◆

$$\mathbf{V}(X) = -VXV\frac{1}{2} \qquad (1076)$$

consistently with (1076), then $\mathcal{N}(\mathbf{V}) = \mathcal{R}(I - \mathbf{V})$ on domain $\mathbb{S}^n$ analogously to vector projectors (§E.2); *id est*,

$$\mathcal{N}(\mathbf{V}) = \mathbb{S}_c^{n\perp} \qquad (2097)$$

a subspace of $\mathbb{S}^n$ whose dimension is $\dim\mathbb{S}_c^{n\perp} = n$ in isomorphic $\mathbb{R}^{n(n+1)/2}$. Intuitively, operator $\mathbf{V}$ is an orthogonal projector; any argument duplicitously in its range is a fixed point. So, this symmetric operator's nullspace must be orthogonal to its range.

Now compare the subspace of symmetric matrices having all zeros in the first row and column

$$
\begin{aligned}
\mathbb{S}_1^n &\triangleq \{Y \in \mathbb{S}^n \mid Ye_1 = \mathbf{0}\} \\
&= \left\{ \begin{bmatrix} 0 & \mathbf{0}^{\mathrm{T}} \\ \mathbf{0} & I \end{bmatrix} X \begin{bmatrix} 0 & \mathbf{0}^{\mathrm{T}} \\ \mathbf{0} & I \end{bmatrix} \mid X \in \mathbb{S}^n \right\} \\
&= \left\{ \begin{bmatrix} \mathbf{0} & \sqrt{2}V_{\mathcal{N}} \end{bmatrix}^{\mathrm{T}} Z \begin{bmatrix} \mathbf{0} & \sqrt{2}V_{\mathcal{N}} \end{bmatrix} \mid Z \in \mathbb{S}^N \right\}
\end{aligned}
\qquad (2098)
$$

where $P = \begin{bmatrix} 0 & \mathbf{0}^{\mathrm{T}} \\ \mathbf{0} & I \end{bmatrix}$ is an orthogonal projector. Then, similarly, PXP is the orthogonal projection of any $X \in \mathbb{S}^n$ on $\mathbb{S}_1^n$ in the Euclidean sense (2090), and

$$
\begin{aligned}
\mathbb{S}_1^{n\perp} &\triangleq \left\{ \begin{bmatrix} 0 & \mathbf{0}^{\mathrm{T}} \\ \mathbf{0} & I \end{bmatrix} X \begin{bmatrix} 0 & \mathbf{0}^{\mathrm{T}} \\ \mathbf{0} & I \end{bmatrix} - X \mid X \in \mathbb{S}^n \right\} \subset \mathbb{S}^n \\
&= \{ue_1^{\mathrm{T}} + e_1u^{\mathrm{T}} \mid u \in \mathbb{R}^n\}
\end{aligned}
\qquad (2099)
$$

Obviously, $\mathbb{S}_1^n \oplus \mathbb{S}_1^{n\perp} = \mathbb{S}^n$. $\qquad\qquad\qquad\qquad\qquad\qquad\qquad\qquad\qquad\qquad\square$

E.8 Range/Rowspace interpretation

For idempotent matrices P_1 and P_2 of any rank, $P_1XP_2^{\mathrm{T}}$ is a projection of $\mathcal{R}(X)$ on $\mathcal{R}(P_1)$ and a projection of $\mathcal{R}(X^{\mathrm{T}})$ on $\mathcal{R}(P_2)$: For any given $X = U\Sigma Q^{\mathrm{T}} = \sum_{i=1}^{\eta} \sigma_i u_i q_i^{\mathrm{T}} \in \mathbb{R}^{m\times p}$, as in compact SVD (1644),

$$P_1XP_2^{\mathrm{T}} = \sum_{i=1}^{\eta} \sigma_i P_1 u_i q_i^{\mathrm{T}} P_2^{\mathrm{T}} = \sum_{i=1}^{\eta} \sigma_i P_1 u_i (P_2 q_i)^{\mathrm{T}} \qquad (2100)$$

where $\eta \triangleq \min\{m, p\}$. Recall $u_i \in \mathcal{R}(X)$ and $q_i \in \mathcal{R}(X^{\mathrm{T}})$ when the corresponding singular value σ_i is nonzero. (§A.6.1) So P_1 projects u_i on $\mathcal{R}(P_1)$ while P_2 projects q_i on $\mathcal{R}(P_2)$; *id est*, the range and rowspace of any X are respectively projected on the ranges of P_1 and P_2 .[E.15]

[E.15] When P_1 and P_2 are symmetric and $\mathcal{R}(P_1) = \mathcal{R}(u_j)$ and $\mathcal{R}(P_2) = \mathcal{R}(q_j)$, then the jth dyad term from the singular value decomposition of X is isolated by the projection. Yet if $\mathcal{R}(P_2) = \mathcal{R}(q_\ell)$, $\ell \neq j \in \{1\ldots\eta\}$, then $P_1XP_2 = \mathbf{0}$.

E.9 Projection on convex set

Thus far we have discussed only projection on subspaces. Now we generalize, considering projection on arbitrary convex sets in Euclidean space; convex because projection is, then, unique minimum-distance and a convex optimization problem:

For projection $P_{\mathcal{C}}x$ of point x on any closed set $\mathcal{C} \subseteq \mathbb{R}^n$ it is obvious:

$$\mathcal{C} \equiv \{P_{\mathcal{C}}x \mid x \in \mathbb{R}^n\} = \{x \in \mathbb{R}^n \mid P_{\mathcal{C}}x = x\} \tag{2101}$$

where $P_{\mathcal{C}}$ is a projection operator that is convex when $\mathcal{C}$ is convex. [62, p.88]

If $\mathcal{C} \subseteq \mathbb{R}^n$ is a closed convex set, then for each and every $x \in \mathbb{R}^n$ there exists a unique point $P_{\mathcal{C}}x$ belonging to $\mathcal{C}$ that is closest to x in the Euclidean sense. Like (2001), unique projection Px (or $P_{\mathcal{C}}x$) of a point x on convex set $\mathcal{C}$ is that point in $\mathcal{C}$ closest to x ; [256, §3.12]

$$\|x - Px\|_2 = \inf_{y \in \mathcal{C}} \|x - y\|_2 = \operatorname{dist}(x, \mathcal{C}) \tag{2102}$$

There exists a converse (in finite-dimensional Euclidean space):

E.9.0.0.1   Theorem. (Bunt-Motzkin) *Convex set if projections unique.* [399, §7.5] [202] If $\mathcal{C} \subseteq \mathbb{R}^n$ is a nonempty closed set and if for each and every x in $\mathbb{R}^n$ there is a unique Euclidean projection Px of x on $\mathcal{C}$ belonging to $\mathcal{C}$, then $\mathcal{C}$ is convex.   ⋄

Borwein & Lewis propose, for closed convex set $\mathcal{C}$ [56, §3.3 exer.12d]

$$\nabla\|x - Px\|_2^2 = 2(x - Px) \tag{2103}$$

for any point x whereas, for $x \notin \mathcal{C}$

$$\nabla\|x - Px\|_2 = (x - Px)\,\|x - Px\|_2^{-1} \tag{2104}$$

E.9.0.0.2   Exercise. *Norm gradient.*
Prove (2103) and (2104). (Not proved in [56].)   ▼

A well-known equivalent characterization of projection on a convex set is a generalization of the perpendicularity condition (2000) for projection on a subspace:

E.9.1   Dual interpretation of projection on convex set

E.9.1.0.1   Definition. *Normal vector.*   [315, p.15]
Vector z is *normal* to convex set $\mathcal{C}$ at point $Px \in \mathcal{C}$ if

$$\langle z, y - Px \rangle \leq 0 \quad \forall y \in \mathcal{C} \tag{2105}$$

△

A convex set has at least one nonzero normal at each of its boundary points. [315, p.100] (Figure **68**) Hence, the *normal* or *dual* interpretation of projection:

E.9.1.0.2   Theorem.   *Unique minimum-distance projection.*   [205, §A.3.1]
[256, §3.12] [112, §4.1] [80] (Figure 181b p.648) Given a closed convex set $\mathcal{C} \subseteq \mathbb{R}^n$, point
Px is the unique projection of a given point $x \in \mathbb{R}^n$ on $\mathcal{C}$ (Px is that point in $\mathcal{C}$ nearest
x) if and only if

$$Px \in \mathcal{C}\ , \qquad \langle x - Px\,,\, y - Px \rangle \leq 0 \quad \forall\, y \in \mathcal{C} \tag{2106}$$

$$\diamond$$

As for subspace projection, convex operator P is idempotent in the sense: for each and
every $x \in \mathbb{R}^n$, $P(Px) = Px$. Yet operator P is nonlinear;

- Projector P is a linear operator if and only if convex set $\mathcal{C}$ (on which projection is
  made) is a subspace. (§E.4)

E.9.1.0.3   Theorem.   *Unique projection via normal cone.*[E.16]   [112, §4.3]
Given closed convex set $\mathcal{C} \subseteq \mathbb{R}^n$, point Px is the unique projection of a given point $x \in \mathbb{R}^n$
on $\mathcal{C}$ if and only if

$$Px \in \mathcal{C}\ , \qquad Px - x \in (\mathcal{C} - Px)^* \tag{2107}$$

In other words, Px is that point in $\mathcal{C}$ nearest x if and only if $Px - x$ belongs to that cone
dual to translate $\mathcal{C} - Px$. $\diamond$

E.9.1.1   Dual interpretation as optimization

Deutsch [114, thm.2.3] [115, §2] and Luenberger [256, p.134] carry forward Nirenberg's
dual interpretation of projection [287] as solution to a maximization problem: Minimum
distance from a point $x \in \mathbb{R}^n$ to a convex set $\mathcal{C} \subset \mathbb{R}^n$ can be found by maximizing distance
from x to hyperplane $\partial\mathcal{H}$ over the set of all hyperplanes separating x from $\mathcal{C}$. Existence
of a separating hyperplane (§2.4.2.7) presumes point x lies on the boundary or exterior
to set $\mathcal{C}$.

The optimal separating hyperplane is characterized by the fact it also supports $\mathcal{C}$. Any
hyperplane supporting $\mathcal{C}$ (Figure 30a) has form

$$\underline{\partial\mathcal{H}}_- = \left\{ y \in \mathbb{R}^n \mid a^{\mathrm{T}} y = \sigma_{\mathcal{C}}(a) \right\} \qquad (129)$$

where the support function is convex, defined

$$\sigma_{\mathcal{C}}(a) = \sup_{z \in \mathcal{C}} a^{\mathrm{T}} z \qquad (554)$$

When point x is finite and set $\mathcal{C}$ contains finite points, under this projection
interpretation, if the supporting hyperplane is a separating hyperplane then the support
function is finite. From Example E.5.0.0.8, projection $P_{\underline{\partial\mathcal{H}}_-} x$ of x on any given
supporting hyperplane $\underline{\partial\mathcal{H}}_-$ is

$$P_{\underline{\partial\mathcal{H}}_-} x = x - a(a^{\mathrm{T}}a)^{-1}\!\left(a^{\mathrm{T}}x - \sigma_{\mathcal{C}}(a)\right) \tag{2108}$$

[E.16] $-(\mathcal{C} - Px)^*$ is the normal cone to set $\mathcal{C}$ at point Px. (§E.10.3.2)

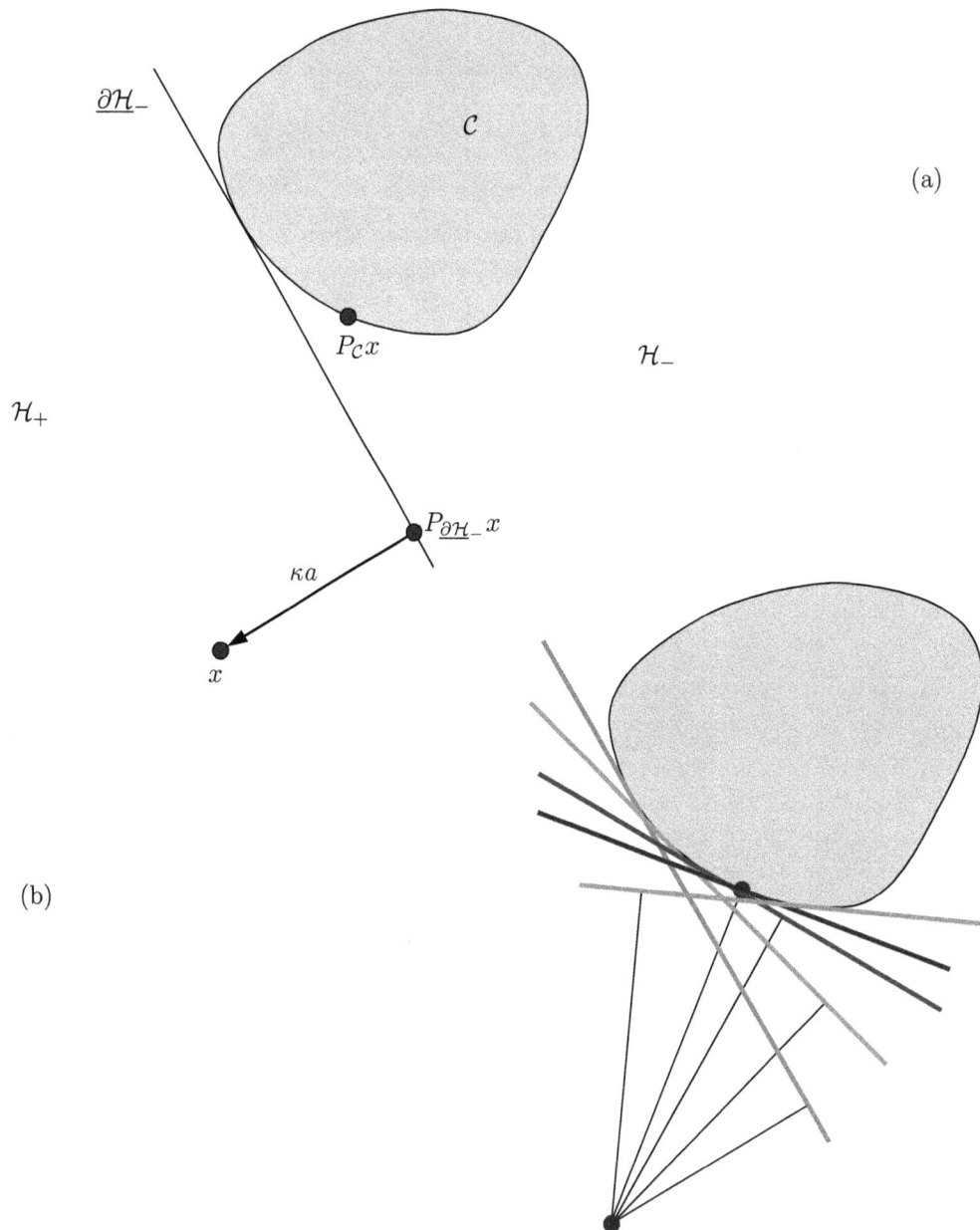

Figure 175: Dual interpretation of projection of point x on convex set $\mathcal{C}$ in $\mathbb{R}^2$. **(a)** $\kappa = (a^{\mathrm{T}}a)^{-1}(a^{\mathrm{T}}x - \sigma_{\mathcal{C}}(a))$ **(b)** Minimum distance from x to $\mathcal{C}$ is found by maximizing distance to all hyperplanes supporting $\mathcal{C}$ and separating it from x. A convex problem for any convex set, distance of maximization is unique.

With reference to Figure 175, identifying

$$\mathcal{H}_+ = \{y \in \mathbb{R}^n \mid a^\mathrm{T}y \geq \sigma_\mathcal{C}(a)\} \qquad (107)$$

then

$$\|x - P_\mathcal{C}x\| = \sup_{\partial\mathcal{H}_- \mid x \in \mathcal{H}_+} \|x - P_{\partial\mathcal{H}_-}x\| = \sup_{a \mid x \in \mathcal{H}_+} \|a(a^\mathrm{T}a)^{-1}(a^\mathrm{T}x - \sigma_\mathcal{C}(a))\|$$
$$= \sup_{a \mid x \in \mathcal{H}_+} \tfrac{1}{\|a\|}|a^\mathrm{T}x - \sigma_\mathcal{C}(a)| \qquad (2109)$$

which can be expressed as a convex optimization, for arbitrary positive constant τ

$$\|x - P_\mathcal{C}x\| = \frac{1}{\tau}\begin{array}{c}\text{maximize} \\ a\end{array} \quad a^\mathrm{T}x - \sigma_\mathcal{C}(a) \\ \text{subject to} \quad \|a\| \leq \tau \qquad (2110)$$

The unique minimum-distance projection on convex set $\mathcal{C}$ is therefore

$$P_\mathcal{C}x = x - a^\star\big(a^{\star\mathrm{T}}x - \sigma_\mathcal{C}(a^\star)\big)\frac{1}{\tau^2} \qquad (2111)$$

where optimally $\|a^\star\| = \tau$.

E.9.1.1.1 Exercise. *Dual projection technique on polyhedron.*
Test that projection paradigm from Figure 175 on any convex polyhedral set. ▼

E.9.1.2 Dual interpretation of projection on cone

In the circumstance set $\mathcal{C}$ is a closed convex cone $\mathcal{K}$ and there exists a hyperplane separating given point x from $\mathcal{K}$, then optimal $\sigma_\mathcal{K}(a^\star)$ takes value 0 [205, §C.2.3.1]. So problem (2110) for projection of x on $\mathcal{K}$ becomes

$$\|x - P_\mathcal{K}x\| = \frac{1}{\tau}\begin{array}{c}\text{maximize} \\ a\end{array} \quad a^\mathrm{T}x \\ \text{subject to} \quad \|a\| \leq \tau \\ a \in \mathcal{K}^\circ \qquad (2112)$$

The norm inequality in (2112) can be handled by Schur complement (§3.5.2). Normals a to all hyperplanes supporting $\mathcal{K}$ belong to the polar cone $\mathcal{K}^\circ = -\mathcal{K}^*$ by definition: (318)

$$a \in \mathcal{K}^\circ \iff \langle a, x \rangle \leq 0 \text{ for all } x \in \mathcal{K} \qquad (2113)$$

Projection on cone $\mathcal{K}$ is

$$P_\mathcal{K}x = (I - \frac{1}{\tau^2}a^\star a^{\star\mathrm{T}})x \qquad (2114)$$

whereas projection on the polar cone $-\mathcal{K}^*$ is (§E.9.2.2.1)

$$P_{\mathcal{K}^\circ}x = x - P_\mathcal{K}x = \frac{1}{\tau^2}a^\star a^{\star\mathrm{T}}x \qquad (2115)$$

Negating vector a , this maximization problem (2112) becomes a minimization (the same problem) and the polar cone becomes the dual cone:

$$\|x - P_{\mathcal{K}}x\| = -\frac{1}{\tau} \begin{array}{l} \text{minimize} \quad a^{\mathrm{T}}x \\ \quad a \\ \text{subject to} \quad \|a\| \leq \tau \\ \qquad\qquad a \in \mathcal{K}^* \end{array} \tag{2116}$$

E.9.2 Projection on cone

When convex set $\mathcal{C}$ is a cone, there is a finer statement of optimality conditions:

E.9.2.0.1 Theorem. *Unique projection on cone.*   [205, §A.3.2]
Let $\mathcal{K} \subseteq \mathbb{R}^n$ be a closed convex cone, and $\mathcal{K}^*$ its dual (§2.13.1). Then Px is the unique minimum-distance projection of $x \in \mathbb{R}^n$ on $\mathcal{K}$ if and only if

$$Px \in \mathcal{K} , \quad \langle Px - x , Px \rangle = 0 , \quad Px - x \in \mathcal{K}^* \tag{2117}$$

$$\diamond$$

In words, Px is the unique minimum-distance projection of x on $\mathcal{K}$ if and only if

1) projection Px lies in $\mathcal{K}$

2) direction $Px-x$ is orthogonal to the projection Px

3) direction $Px-x$ lies in the dual cone $\mathcal{K}^*$.

As the theorem is stated, it admits projection on $\mathcal{K}$ not full-dimensional; *id est*, on closed convex cones in a proper subspace of $\mathbb{R}^n$.

Projection on $\mathcal{K}$ of any point $x \in -\mathcal{K}^*$, belonging to the negative dual cone, is the origin. By (2117): the set of all points reaching the origin, when projecting on $\mathcal{K}$, constitutes the negative dual cone; `a.k.a,` the polar cone

$$\mathcal{K}^{\circ} = -\mathcal{K}^* = \{x \in \mathbb{R}^n \mid Px = \mathbf{0}\} \tag{2118}$$

E.9.2.1 Relationship to subspace projection

Conditions 1 and 2 of the theorem are common with orthogonal projection on a subspace $\mathcal{R}(P)$:

Condition 1 corresponds to the most basic requirement; namely, the projection $Px \in \mathcal{R}(P)$ belongs to the subspace (*confer* (2101))

$$\mathcal{K} = \{Px \mid x \in \mathbb{R}^n\} \triangleq \mathcal{R}(P) \tag{2119}$$

Recall the perpendicularity requirement for projection on a subspace:

$$Px - x \perp \mathcal{R}(P) \quad \textbf{or} \quad Px - x \in \mathcal{R}(P)^{\perp} \tag{2000}$$

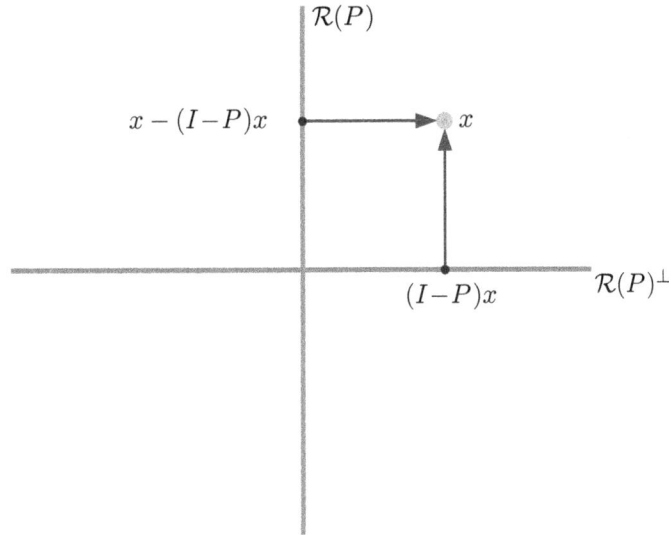

Figure 176: (*confer* Figure 86, Figure 177) Given orthogonal projection $(I-P)x$ of x on orthogonal complement $\mathcal{R}(P)^\perp$, projection on $\mathcal{R}(P)$ is immediate: $x-(I-P)x$.

which corresponds to condition 2.

Yet condition 3 is a generalization of subspace projection; *id est*, for unique minimum-distance projection on a closed convex cone $\mathcal{K}$, polar cone $-\mathcal{K}^*$ (Figure 177) plays the role $\mathcal{R}(P)^\perp$ plays for subspace projection (Figure 176: $P_\mathcal{R}x = x - P_{\mathcal{R}^\perp}\,x$). Indeed, $-\mathcal{K}^*$ is the algebraic complement in the orthogonal vector sum (p.666) [275] [205, §A.3.2.5]

$$\mathcal{K} \boxplus -\mathcal{K}^* = \mathbb{R}^n \ \Leftrightarrow \ \text{cone } \mathcal{K} \text{ is closed and convex} \qquad (2120)$$

Given unique minimum-distance projection Px on $\mathcal{K}$ satisfying the theorem (§E.9.2.0.1), then by projection on the algebraic complement via $I-P$ in §E.2 we have

$$-\mathcal{K}^* = \{x - Px \mid x \in \mathbb{R}^n\} = \{x \in \mathbb{R}^n \mid Px = \mathbf{0}\} = \mathcal{N}(P) \qquad (2121)$$

consequent to Moreau (2124). Converse (2121)(2119) $\Rightarrow$ (2120) holds as well. Recalling that any subspace is a closed convex cone[E.17]

$$\mathcal{K} = \mathcal{R}(P) \ \Leftrightarrow \ -\mathcal{K}^* = \mathcal{R}(P)^\perp \qquad (2122)$$

meaning, when a cone is a subspace $\mathcal{R}(P)$, then the dual cone becomes its orthogonal complement $\mathcal{R}(P)^\perp$. In this circumstance, condition 3 becomes coincident with condition 2.

Properties of projection on cones, following in §E.9.2.2, further generalize to subspaces by: (4)

$$\mathcal{K} = \mathcal{R}(P) \ \Leftrightarrow \ -\mathcal{K} = \mathcal{R}(P) \qquad (2123)$$

[E.17] but a proper subspace is not a proper cone (§2.7.2.2.1).

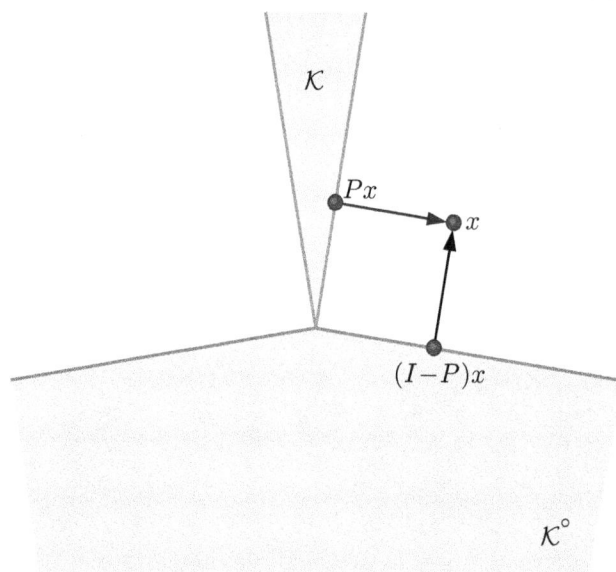

Figure 177: (*confer* Figure 176) Given minimum-distance projection $(I-P)x$ of x on negative dual cone $\mathcal{K}^{\circ}$, projection on $\mathcal{K}$ is immediate: $x - (I-P)x = Px$.

E.9.2.2 Salient properties: Projection Px on closed convex cone $\mathcal{K}$

[205, §A.3.2] [112, §5.6] For $x, x_1, x_2 \in \mathbb{R}^n$

1. $P_{\mathcal{K}}(\alpha x) = \alpha\, P_{\mathcal{K}} x \quad \forall \alpha \ge 0$ (nonnegative homogeneity)

2. $\|P_{\mathcal{K}} x\| \le \|x\|$

3. $P_{\mathcal{K}} x = \mathbf{0} \ \Leftrightarrow\ x \in -\mathcal{K}^*$

4. $P_{\mathcal{K}}(-x) = -P_{-\mathcal{K}} x$

5. (Jean-Jacques Moreau (1962)) [275]

$$x = x_1 + x_2, \quad x_1 \in \mathcal{K}, \quad x_2 \in -\mathcal{K}^*, \quad x_1 \perp x_2$$
$$\Leftrightarrow \tag{2124}$$
$$x_1 = P_{\mathcal{K}} x, \quad x_2 = P_{-\mathcal{K}^*} x$$

6. $\mathcal{K} = \{x - P_{-\mathcal{K}^*} x \mid x \in \mathbb{R}^n\} = \{x \in \mathbb{R}^n \mid P_{-\mathcal{K}^*} x = \mathbf{0}\}$

7. $-\mathcal{K}^* = \{x - P_{\mathcal{K}} x \mid x \in \mathbb{R}^n\} = \{x \in \mathbb{R}^n \mid P_{\mathcal{K}} x = \mathbf{0}\}$ (2121)

E.9.2.2.1 Corollary. *$I-P$ for cones.*   (*confer* §E.2)
Denote by $\mathcal{K} \subseteq \mathbb{R}^n$ a closed convex cone, and call $\mathcal{K}^*$ its dual. Then $x - P_{-\mathcal{K}^*} x$ is the unique minimum-distance projection of $x \in \mathbb{R}^n$ on $\mathcal{K}$ if and only if $P_{-\mathcal{K}^*} x$ is the unique minimum-distance projection of x on $-\mathcal{K}^*$ the polar cone.   ◇

Proof. Assume $x_1 = P_{\mathcal{K}}x$. Then by Theorem E.9.2.0.1 we have

$$x_1 \in \mathcal{K}, \qquad x_1 - x \perp x_1, \qquad x_1 - x \in \mathcal{K}^* \qquad (2125)$$

Now assume $x - x_1 = P_{-\mathcal{K}^*}x$. Then we have

$$x - x_1 \in -\mathcal{K}^*, \qquad -x_1 \perp x - x_1, \qquad -x_1 \in -\mathcal{K} \qquad (2126)$$

But these two assumptions are apparently identical. We must therefore have

$$x - P_{-\mathcal{K}^*}x = x_1 = P_{\mathcal{K}}x \qquad (2127)$$

♦

E.9.2.2.2 Corollary. *Unique projection via dual or normal cone.*
[112, §4.7] (§E.10.3.2, *confer* Theorem E.9.1.0.3) Given point $x \in \mathbb{R}^n$ and closed convex cone $\mathcal{K} \subseteq \mathbb{R}^n$, the following are equivalent statements:

1. point Px is the unique minimum-distance projection of x on $\mathcal{K}$

2. $Px \in \mathcal{K}, \qquad x - Px \in -(\mathcal{K} - Px)^* = -\mathcal{K}^* \cap (Px)^\perp$

3. $Px \in \mathcal{K}, \qquad \langle x - Px, Px \rangle = 0, \qquad \langle x - Px, y \rangle \leq 0 \ \forall y \in \mathcal{K}$ ◇

E.9.2.2.3 Example. *Unique projection on nonnegative orthant.*
(*confer* (1401)) From Theorem E.9.2.0.1, to project matrix $H \in \mathbb{R}^{m \times n}$ on the selfdual orthant (§2.13.5.1) of nonnegative matrices $\mathbb{R}_+^{m \times n}$ in isomorphic $\mathbb{R}^{mn}$, the necessary and sufficient conditions are:

$$\begin{aligned} &H^\star \geq \mathbf{0} \\ &\operatorname{tr}\left((H^\star - H)^{\mathrm{T}} H^\star\right) = 0 \\ &H^\star - H \geq \mathbf{0} \end{aligned} \qquad (2128)$$

where the inequalities denote entrywise comparison. The optimal solution $H^\star$ is simply H having all its negative entries zeroed;

$$H_{ij}^\star = \max\{H_{ij}, 0\}, \quad i,j \in \{1 \ldots m\} \times \{1 \ldots n\} \qquad (2129)$$

Now suppose the nonnegative orthant is translated by $T \in \mathbb{R}^{m \times n}$; *id est*, consider $\mathbb{R}_+^{m \times n} + T$. Then projection on the translated orthant is [112, §4.8]

$$H_{ij}^\star = \max\{H_{ij}, T_{ij}\} \qquad (2130)$$

□

E.9.2.2.4 Example. *Unique projection on truncated convex cone.*
Consider the problem of projecting a point x on a closed convex cone that is artificially bounded; really, a bounded convex polyhedron having a vertex at the origin:

$$\begin{aligned} \underset{y \in \mathbb{R}^N}{\text{minimize}} \quad & \|x - Ay\|_2 \\ \text{subject to} \quad & y \succeq 0 \\ & \|y\|_\infty \leq 1 \end{aligned} \qquad (2131)$$

where the convex cone has vertex-description (§2.12.2.0.1), for $A\in\mathbb{R}^{n\times N}$

$$\mathcal{K} = \{Ay \mid y \succeq 0\} \tag{2132}$$

and where $\|y\|_\infty \le 1$ is the artificial bound. This is a convex optimization problem having no known closed-form solution, in general. It arises, for example, in the fitting of hearing aids designed around a programmable graphic equalizer (a filter bank whose only adjustable parameters are gain per band each bounded above by unity). [98] The problem is equivalent to a Schur-form semidefinite program (§3.5.2)

$$\begin{array}{ll} \underset{y\in\mathbb{R}^N,\ t\in\mathbb{R}}{\text{minimize}} & t \\ \text{subject to} & \begin{bmatrix} tI & x - Ay \\ (x - Ay)^{\mathrm{T}} & t \end{bmatrix} \succeq 0 \\ & 0 \preceq y \preceq \mathbf{1} \end{array} \tag{2133}$$

$\square$

E.9.3   nonexpansivity

E.9.3.0.1   Theorem. *Nonexpansivity.*   [181, §2] [112, §5.3]
When $\mathcal{C}\subset\mathbb{R}^n$ is an arbitrary closed convex set, projector P projecting on $\mathcal{C}$ is nonexpansive in the sense: for any vectors $x,y\in\mathbb{R}^n$

$$\|Px - Py\| \le \|x - y\| \tag{2134}$$

with equality when $x - Px = y - Py$.[E.18] $\diamond$

   Proof. [55]

$$\begin{aligned} \|x - y\|^2 &= \|Px - Py\|^2 + \|(I - P)x - (I - P)y\|^2 \\ &\quad + 2\langle x - Px,\, Px - Py\rangle + 2\langle y - Py,\, Py - Px\rangle \end{aligned} \tag{2135}$$

Nonnegativity of the last two terms follows directly from the *unique minimum-distance projection theorem* (§E.9.1.0.2). $\blacklozenge$

   The foregoing proof reveals another flavor of nonexpansivity; for each and every $x,y\in\mathbb{R}^n$

$$\|Px - Py\|^2 + \|(I - P)x - (I - P)y\|^2 \le \|x - y\|^2 \tag{2136}$$

Deutsch shows yet another: [112, §5.5]

$$\|Px - Py\|^2 \le \langle x - y,\, Px - Py\rangle \tag{2137}$$

[E.18]This condition for equality corrects an error in [80] (where the norm is applied to each side of the condition given here) easily revealed by counterexample.

E.9.4 Easy projections

- To project any matrix $H \in \mathbb{R}^{n \times n}$ orthogonally in Euclidean/Frobenius sense on subspace of symmetric matrices $\mathbb{S}^n$ in isomorphic $\mathbb{R}^{n^2}$, take symmetric part of H; (§2.2.2.0.1) *id est,* $(H + H^T)/2$ is the projection.

- To project any $H \in \mathbb{R}^{n \times n}$ orthogonally on symmetric hollow subspace $\mathbb{S}^n_h$ in isomorphic $\mathbb{R}^{n^2}$ (§2.2.3.0.1, §7.0.1), take symmetric part then zero all entries along main diagonal or *vice versa* (because this is projection on intersection of two subspaces); *id est,* $(H + H^T)/2 - \delta^2(H)$.

- To project a matrix on nonnegative orthant $\mathbb{R}^{m \times n}_+$, simply clip all negative entries to 0. Likewise, projection on nonpositive orthant $\mathbb{R}^{m \times n}_-$ sees all positive entries clipped to 0. Projection on other orthants is equally simple with appropriate clipping.

- Projecting on hyperplane, halfspace, slab: §E.5.0.0.8.

- Projection of $y \in \mathbb{R}^n$ on Euclidean ball $\mathcal{B} = \{x \in \mathbb{R}^n \mid \|x - a\| \leq c\}$: for $y \neq a$, $P_\mathcal{B}y = \frac{c}{\|y - a\|}(y - a) + a$.

- Clipping in excess of $|1|$ each entry of point $x \in \mathbb{R}^n$ is equivalent to unique minimum-distance projection of x on a hypercube centered at the origin. (*confer* §E.10.3.2)

- Projection of $x \in \mathbb{R}^n$ on a (rectangular) *hyperbox*: [62, §8.1.1]

$$\mathcal{C} = \{y \in \mathbb{R}^n \mid l \preceq y \preceq u, \ l \prec u\} \tag{2138}$$

$$P(x)_{k=0 \dots n} = \begin{cases} l_k \ , & x_k \leq l_k \\ x_k \ , & l_k \leq x_k \leq u_k \\ u_k \ , & x_k \geq u_k \end{cases} \tag{2139}$$

- Orthogonal projection of x on a *Cartesian subspace*, whose basis is some given subset of the Cartesian axes, zeroes entries corresponding to the remaining (complementary) axes.

- Projection of x on set of all cardinality-k vectors $\{y \mid \text{card } y \leq k\}$ keeps k entries of greatest magnitude and clips to 0 those remaining.

- Unique minimum-distance projection of $H \in \mathbb{S}^n$ on positive semidefinite cone $\mathbb{S}^n_+$ in Euclidean/Frobenius sense is accomplished by eigenvalue decomposition (diagonalization) followed by clipping all negative eigenvalues to 0.

- Unique minimum-distance projection on generally nonconvex subset of all matrices belonging to $\mathbb{S}^n_+$ having rank not exceeding ρ (§2.9.2.1) is accomplished by clipping all negative eigenvalues to 0 and zeroing smallest nonnegative eigenvalues keeping only ρ largest. (§7.1.2)

- Unique minimum-distance projection, in Euclidean/Frobenius sense, of $H \in \mathbb{R}^{m \times n}$ on generally nonconvex subset of all $m \times n$ matrices of rank no greater than k is singular value decomposition (§A.6) of H having all singular values beyond k^{th} zeroed. This is also a solution to projection in sense of spectral norm. [335, p.79, p.208]

- Projection of a real vector on the monotone nonnegative cone is identical to its projection on the monotone cone followed by clipping all negative entries of the result to 0. [282, §5]

- Projection on monotone nonnegative cone $\mathcal{K}_{\mathcal{M}+} \subset \mathbb{R}^n$ in less than one cycle (in sense of alternating projections §E.10): [384].

- Fast projection on a simplicial cone: [391].

- Projection on closed convex cone $\mathcal{K}$ of any point $x \in -\mathcal{K}^*$, belonging to polar cone, is equivalent to projection on origin. (§E.9.2)

- $P_{\mathbb{S}_+^N \cap \mathbb{S}_c^N} = P_{\mathbb{S}_+^N} P_{\mathbb{S}_c^N}$ \qquad\qquad (1345)

- $P_{\mathbb{R}_+^{N \times N} \cap \mathbb{S}_h^N} = P_{\mathbb{R}_+^{N \times N}} P_{\mathbb{S}_h^N}$ \quad (§7.0.1.1)

- $P_{\mathbb{R}_+^{N \times N} \cap \mathbb{S}^N} = P_{\mathbb{R}_+^{N \times N}} P_{\mathbb{S}^N}$ \quad (§E.9.5)

E.9.4.0.1 Exercise. *Projection on spectral norm ball.*
Find the unique minimum-distance projection on the convex set of all $m \times n$ matrices whose largest singular value does not exceed 1; *id est*, on $\{X \in \mathbb{R}^{m \times n} \mid \|X\|_2 \le 1\}$ the spectral norm ball (§2.3.2.0.5). ▼

E.9.4.1   notes

Projection on Lorentz (second-order) cone: [62, exer.8.3c].
Deutsch [115] provides an algorithm for projection on polyhedral cones.
Youla [420, §2.5] lists eleven "useful projections", of square-integrable uni- and bivariate real functions on various convex sets, in closed form.
Unique minimum-distance projection on an ellipsoid: Example 4.6.0.0.1.
Numerical algorithms for projection on the nonnegative simplex and 1-norm ball are in [130].

E.9.5   Projection on convex set in subspace

Suppose a convex set $\mathcal{C}$ is contained in some subspace $\mathbb{R}^n$. Then unique minimum-distance projection of any point in $\mathbb{R}^n \oplus \mathbb{R}^{n\perp}$ on $\mathcal{C}$ can be accomplished by first projecting orthogonally on that subspace, and then projecting the result on $\mathcal{C}$; [112, §5.14] *id est*, the ordered product of two individual projections that is not commutable.

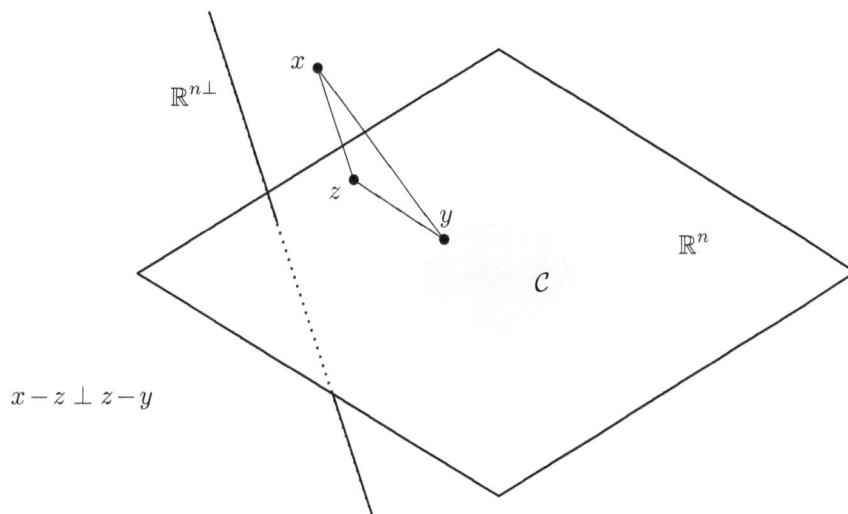

Figure 178: Closed convex set $\mathcal{C}$ belongs to subspace $\mathbb{R}^n$ (shown bounded in sketch and drawn without proper perspective). Point y is unique minimum-distance projection of x on $\mathcal{C}$; equivalent to product of orthogonal projection of x on $\mathbb{R}^n$ and minimum-distance projection of result z on $\mathcal{C}$.

Proof. ($\Leftarrow$) To show that, suppose unique minimum-distance projection $P_{\mathcal{C}}x$ on $\mathcal{C} \subset \mathbb{R}^n$ is y as illustrated in Figure 178;

$$\|x - y\| \leq \|x - q\| \quad \forall q \in \mathcal{C} \tag{2140}$$

Further suppose $P_{\mathbb{R}^n}x$ equals z. By the *Pythagorean theorem*

$$\|x - y\|^2 = \|x - z\|^2 + \|z - y\|^2 \tag{2141}$$

because $x - z \perp z - y$. (2000) [256, §3.3] Then point $y = P_{\mathcal{C}}x$ is the same as $P_{\mathcal{C}}z$ because

$$\|z - y\|^2 = \|x - y\|^2 - \|x - z\|^2 \leq \|z - q\|^2 = \|x - q\|^2 - \|x - z\|^2 \quad \forall q \in \mathcal{C} \tag{2142}$$

which holds by assumption (2140).
($\Rightarrow$) Now suppose $z = P_{\mathbb{R}^n}x$ and

$$\|z - y\| \leq \|z - q\| \quad \forall q \in \mathcal{C} \tag{2143}$$

meaning $y = P_{\mathcal{C}}z$. Then point y is identical to $P_{\mathcal{C}}x$ because

$$\|x - y\|^2 = \|x - z\|^2 + \|z - y\|^2 \leq \|x - q\|^2 = \|x - z\|^2 + \|z - q\|^2 \quad \forall q \in \mathcal{C} \tag{2144}$$

by assumption (2143).   ◆

This proof is extensible via translation argument. (§E.4) Unique minimum-distance projection on a convex set contained in an affine subset is, therefore, similarly accomplished.

Projecting matrix $H \in \mathbb{R}^{n \times n}$ on convex cone $\mathcal{K} = \mathbb{S}^n \cap \mathbb{R}_+^{n \times n}$ in isomorphic $\mathbb{R}^{n^2}$ can be accomplished, for example, by first projecting on $\mathbb{S}^n$ and only then projecting the result on $\mathbb{R}_+^{n \times n}$ (*confer* §7.0.1). This is because that projection product is equivalent to projection on the subset of the nonnegative orthant in the symmetric matrix subspace.

E.10 Alternating projection

Alternating projection is an iterative technique for finding a point in the intersection of a number of arbitrary closed convex sets $\mathcal{C}_k$, or for finding the distance between two nonintersecting closed convex sets. Because it can sometimes be difficult or inefficient to compute the intersection or express it analytically, one naturally asks whether it is possible to instead project (unique minimum-distance) alternately on the individual $\mathcal{C}_k$; often easier and what motivates adoption of this technique. Once a cycle of alternating projections (an *iteration*) is complete, we then *iterate* (repeat the cycle) until convergence. If the intersection of two closed convex sets is empty, then by *convergence* we mean the *iterate* (the result after a cycle of alternating projections) settles to a point of minimum distance separating the sets.

While alternating projection can find the point in the nonempty intersection closest to a given point b, it does not necessarily find the closest point. Finding that closest point is made dependable by an elegantly simple enhancement via correction to the alternating projection technique: this *Dykstra algorithm* (2183) for projection on the intersection is one of the most beautiful projection algorithms ever discovered. It is accurately interpreted as the discovery of what alternating projection originally sought to accomplish: unique minimum-distance projection on the nonempty intersection of a number of arbitrary closed convex sets $\mathcal{C}_k$. Alternating projection is, in fact, a special case of the Dykstra algorithm whose discussion we defer until §E.10.3.

E.10.0.1   commutative projectors

A product of projection operators is generally not another projector. Given two arbitrary convex sets $\mathcal{C}_1$ and $\mathcal{C}_2$ and their respective minimum-distance projection operators P_1 and P_2 : If projectors commute $(P_1 P_2 = P_2 P_1)$ for each and every $x \in \mathbb{R}^n$, then it is easy to show $P_1 P_2 x \in \mathcal{C}_1 \cap \mathcal{C}_2$ and $P_2 P_1 x \in \mathcal{C}_1 \cap \mathcal{C}_2$. When projectors commute, their product is a projector; some point in the intersection can be found in a finite number of steps. While commutativity is a sufficient condition, it is not necessary; *e.g.*, §6.8.1.1.1.

When $\mathcal{C}_1$ and $\mathcal{C}_2$ are subspaces, in particular, projectors P_1 and P_2 commute if and only if $P_1 P_2 = P_{\mathcal{C}_1 \cap \mathcal{C}_2}$ or iff $P_2 P_1 = P_{\mathcal{C}_1 \cap \mathcal{C}_2}$ or iff $P_1 P_2$ is the orthogonal projection on a Euclidean subspace. [112, lem.9.2] Subspace projectors will commute, for example, when $P_1(\mathcal{C}_2) \subset \mathcal{C}_2$ or $P_2(\mathcal{C}_1) \subset \mathcal{C}_1$ or $\mathcal{C}_1 \subset \mathcal{C}_2$ or $\mathcal{C}_2 \subset \mathcal{C}_1$ or $\mathcal{C}_1 \perp \mathcal{C}_2$. When subspace projectors commute, this means we can find a point from the intersection of those subspaces in a finite number of steps; in fact, the closest point. Orthogonal projection on orthogonal subspaces (or intersecting orthogonal affine subsets), in particular, can be performed in any order to find the closest point in their intersection in a number of steps = number of subspaces (or affine subsets).

E.10.0.1.1   Theorem. *Kronecker projector.* [334, §2.7]
Given any projection matrices P_1 and P_2 (subspace projectors), then

$$P_1 \otimes P_2 \quad \textbf{and} \quad P_1 \otimes I \tag{2145}$$

are projection matrices. The product preserves symmetry if present. ◇

E.10.0.2   noncommutative projectors

Typically, one considers the method of alternating projection when projectors do not commute; *id est*, when $P_1 P_2 \neq P_2 P_1$.

The iconic example for noncommutative projectors illustrated in Figure 179 shows the iterates converging to the closest point in the intersection of two arbitrary convex sets. Yet simple examples like Figure 180 reveal that noncommutative alternating projection does not always yield the closest point, although we shall show it always yields some point in the intersection or a point that attains the distance between two convex sets.

Alternating projection is also known as *successive projection* [184] [181] [64], *cyclic projection* [152] [266, §3.2], *successive approximation* [80], or *projection on convex sets* [332] [333, §6.4]. It is traced back to von Neumann (1933) [285] and later Wiener [404] who showed that higher iterates of a product of two orthogonal projections on subspaces converge at each point in the ambient space to the unique minimum-distance projection on the intersection of the two subspaces. More precisely, if $\mathcal{R}_1$ and $\mathcal{R}_2$ are closed subspaces

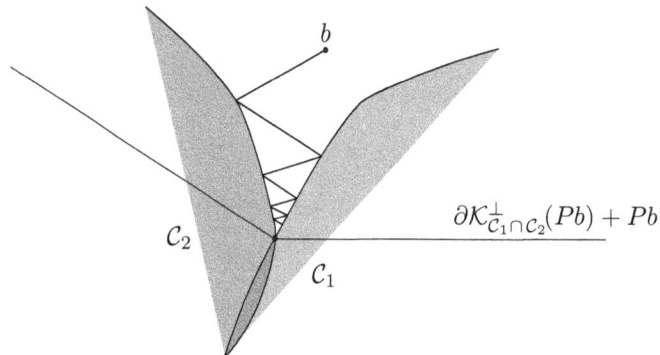

Figure 179: First several alternating projections in von Neumann-style projection (2156), of point b, converging on closest point Pb in intersection of two closed convex sets in $\mathbb{R}^2$: $\mathcal{C}_1$ and $\mathcal{C}_2$ are partially drawn in vicinity of their intersection. Pointed normal cone $\mathcal{K}^\perp$ (448) is translated to Pb, the unique minimum-distance projection of b on intersection. For this particular example, it is possible to start anywhere in a large neighborhood of b and still converge to Pb. Alternating projections are themselves robust with respect to significant noise because they belong to translated normal cone.

of a Euclidean space and P_1 and P_2 respectively denote orthogonal projection on $\mathcal{R}_1$ and $\mathcal{R}_2$, then for each vector b in that space,

$$\lim_{i\to\infty} (P_1 P_2)^i b = P_{\mathcal{R}_1 \cap \mathcal{R}_2} b \qquad (2146)$$

Deutsch [112, thm.9.8, thm.9.35] shows rate of convergence for subspaces to be *geometric* [419, §1.4.4]; bounded above by $\kappa^{2i+1}\|b\|$, $i=0,1,2\ldots$, where $0\leq\kappa<1$:

$$\| (P_1 P_2)^i b - P_{\mathcal{R}_1 \cap \mathcal{R}_2} b \| \leq \kappa^{2i+1}\|b\| \qquad (2147)$$

This means convergence can be slow when κ is close to 1. Rate of convergence on intersecting halfspaces is also geometric. [113] [304]

This von Neumann sense of alternating projection may be applied to convex sets that are not subspaces, although convergence is not necessarily to the unique minimum-distance projection on the intersection. Figure 179 illustrates one application where convergence is reasonably geometric and the result is the unique minimum-distance projection. Figure 180, in contrast, demonstrates convergence in one iteration to a fixed point (of the projection product)[E.19] in the intersection of two halfspaces; a.k.a, feasibility problem. It was Dykstra who in 1983 [131] (§E.10.3) first solved this projection problem.

[E.19] *Fixed point* of a mapping $T: \mathbb{R}^n \to \mathbb{R}^n$ is a point x whose image is identical under the map; *id est*, $Tx = x$.

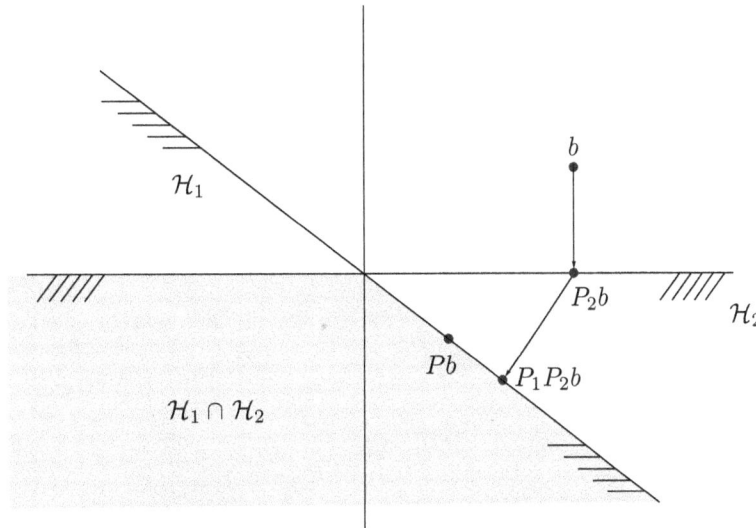

Figure 180: The sets $\{\mathcal{C}_k\}$ in this example comprise two halfspaces $\mathcal{H}_1$ and $\mathcal{H}_2$. The von Neumann-style alternating projection in $\mathbb{R}^2$ quickly converges to $P_1 P_2 b$ (feasibility). Unique minimum-distance projection on the intersection is, of course, Pb.

E.10.0.3   the bullets

Alternating projection has, therefore, various meaning dependent on the application or field of study; it may be interpreted to be: a distance problem, a feasibility problem (von Neumann), or a projection problem (Dykstra):

- **Distance.** Figure 181a-b. Find a unique point of projection $P_1 b \in \mathcal{C}_1$ that attains the distance between any two closed convex sets $\mathcal{C}_1$ and $\mathcal{C}_2$;

$$\|P_1 b - b\| = \operatorname{dist}(\mathcal{C}_1, \mathcal{C}_2) \triangleq \inf_{z \in \mathcal{C}_2} \|P_1 z - z\| \tag{2148}$$

- **Feasibility.** Figure 181c, $\bigcap \mathcal{C}_k \neq \emptyset$. Given a number L of indexed closed convex sets $\mathcal{C}_k \subset \mathbb{R}^n$, find any fixed point in their intersection by iterating (i) a projection product starting from b;

$$\left(\prod_{i=1}^{\infty} \prod_{k=1}^{L} P_k \right) b \in \bigcap_{k=1}^{L} \mathcal{C}_k \tag{2149}$$

- **Optimization.** Figure 181c, $\bigcap \mathcal{C}_k \neq \emptyset$. Given a number of indexed closed convex sets $\mathcal{C}_k \subset \mathbb{R}^n$, uniquely project a given point b on $\bigcap \mathcal{C}_k$;

$$\|Pb - b\| = \inf_{x \in \bigcap \mathcal{C}_k} \|x - b\| \tag{2150}$$

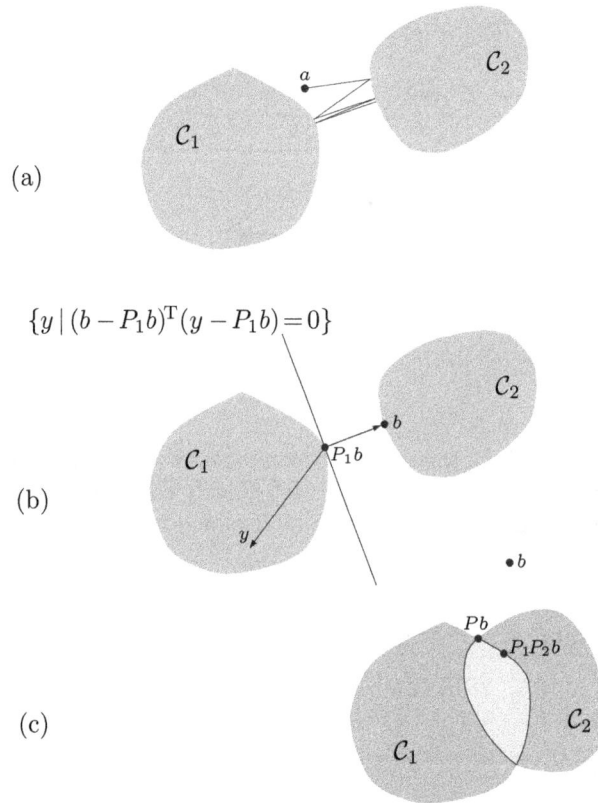

Figure 181:
(a) (**distance**) Intersection of two convex sets in $\mathbb{R}^2$ is empty. Method of alternating projection would be applied to find that point in $\mathcal{C}_1$ nearest $\mathcal{C}_2$.
(b) (**distance**)   Given $b \in \mathcal{C}_2$, then $P_1 b \in \mathcal{C}_1$
is nearest b iff $(y - P_1 b)^{\mathrm{T}}(b - P_1 b) \le 0 \;\forall\, y \in \mathcal{C}_1$ by the *unique minimum-distance projection theorem* (§E.9.1.0.2). When $P_1 b$ attains the distance between the two sets, hyperplane $\{y \mid (b - P_1 b)^{\mathrm{T}}(y - P_1 b) = 0\}$ separates $\mathcal{C}_1$ from $\mathcal{C}_2$. [62, §2.5.1]
(c) (**0 distance**) Intersection is nonempty.
(**optimization**) We may want the point Pb in $\bigcap \mathcal{C}_k$ nearest point b .
(**feasibility**) We may instead be satisfied with a fixed point of the projection product $P_1 P_2 b$ in $\bigcap \mathcal{C}_k$.

E.10.1   Distance and existence

Existence of a fixed point is established:

E.10.1.0.1   Theorem. *Distance.*   [80]
Given any two closed convex sets $\mathcal{C}_1$ and $\mathcal{C}_2$ in $\mathbb{R}^n$, then $P_1 b \in \mathcal{C}_1$ is a fixed point of projection product $P_1 P_2$ if and only if $P_1 b$ is a point of $\mathcal{C}_1$ nearest $\mathcal{C}_2$. $\diamond$

Proof. ($\Rightarrow$) Given fixed point $a = P_1 P_2 a \in \mathcal{C}_1$ with $b \triangleq P_2 a \in \mathcal{C}_2$ in tandem so that $a = P_1 b$, then by the *unique minimum-distance projection theorem* (§E.9.1.0.2)

$$
\begin{aligned}
(b-a)^{\mathrm{T}}(u-a) \leq 0 \quad &\forall u \in \mathcal{C}_1 \\
(a-b)^{\mathrm{T}}(v-b) \leq 0 \quad &\forall v \in \mathcal{C}_2 \\
\Leftrightarrow & \\
\|a-b\| \leq \|u-v\| \quad &\forall u \in \mathcal{C}_1 \text{ and } \forall v \in \mathcal{C}_2
\end{aligned}
\tag{2151}
$$

by Cauchy-Schwarz inequality [315]

$$
|\langle x, y \rangle| \leq \|x\| \, \|y\|
\tag{2152}
$$

(with equality iff $x = \kappa y$ where $\kappa \in \mathbb{R}$ (33) [233, p.137]).
($\Leftarrow$) Suppose $a \in \mathcal{C}_1$ and $\|a - P_2 a\| \leq \|u - P_2 u\| \; \forall u \in \mathcal{C}_1$ and we choose $u = P_1 P_2 a$. Then

$$
\|u - P_2 u\| = \|P_1 P_2 a - P_2 P_1 P_2 a\| \leq \|a - P_2 a\| \; \Leftrightarrow \; a = P_1 P_2 a
\tag{2153}
$$

Thus $a = P_1 b$ (with $b = P_2 a \in \mathcal{C}_2$) is a fixed point in $\mathcal{C}_1$ of the projection product $P_1 P_2$.[E.20]
$\blacklozenge$

E.10.2   Feasibility and convergence

The set of all fixed points of any nonexpansive mapping is a closed convex set. [160, lem.3.4] [28, §1] The projection product $P_1 P_2$ is nonexpansive by Theorem E.9.3.0.1 because, for any vectors $x, a \in \mathbb{R}^n$

$$
\|P_1 P_2 \, x - P_1 P_2 \, a\| \; \leq \; \|P_2 x - P_2 a\| \; \leq \; \|x - a\|
\tag{2154}
$$

If the intersection of two closed convex sets $\mathcal{C}_1 \cap \mathcal{C}_2$ is empty, then the iterates converge to a point of minimum distance, a fixed point of the projection product. Otherwise, convergence is to some fixed point in their intersection (a feasible solution) whose existence is guaranteed by virtue of the fact that each and every point in the convex intersection is in one-to-one correspondence with fixed points of the nonexpansive projection product.

Bauschke & Borwein [28, §2] argue that any sequence monotonic in the sense of Fejér is convergent:[E.21]

[E.20]Point $b = P_2 a$ can be shown, similarly, to be a fixed point of product $P_2 P_1$.
[E.21]Other authors prove convergence by different means; *e.g.*, [181] [64].

E.10.2.0.1   Definition. *Fejér monotonicity.*   [276]
Given closed convex set $\mathcal{C} \neq \emptyset$, then a *sequence* $x_i \in \mathbb{R}^n$, $i = 0, 1, 2 \ldots$, is monotonic in the sense of Fejér with respect to $\mathcal{C}$ iff

$$\|x_{i+1} - c\| \leq \|x_i - c\| \quad \text{for all } i \geq 0 \text{ and } \text{each and every } c \in \mathcal{C} \qquad (2155)$$

$$\triangle$$

Given $x_0 \triangleq b$, if we express each iteration of alternating projection by

$$x_{i+1} = P_1 P_2 \, x_i \,, \quad i = 0, 1, 2 \ldots \qquad (2156)$$

and define any fixed point $a = P_1 P_2 \, a$, then sequence x_i is Fejér monotone with respect to fixed point a because

$$\|P_1 P_2 \, x_i - a\| \leq \|x_i - a\| \quad \forall i \geq 0 \qquad (2157)$$

by nonexpansivity. The nonincreasing sequence $\|P_1 P_2 \, x_i - a\|$ is bounded below hence convergent because any bounded monotonic sequence in $\mathbb{R}$ is convergent; [264, §1.2] [42, §1.1] $P_1 P_2 \, x_{i+1} = P_1 P_2 \, x_i = x_{i+1}$. Sequence x_i therefore converges to some fixed point. If the intersection $\mathcal{C}_1 \cap \mathcal{C}_2$ is nonempty, convergence is to some point there by the *distance theorem*. Otherwise, x_i converges to a point in $\mathcal{C}_1$ of minimum distance to $\mathcal{C}_2$.

E.10.2.0.2   Example. *Hyperplane/orthant intersection.*
Find a feasible solution (2149) belonging to the nonempty intersection of two convex sets: given $A \in \mathbb{R}^{m \times n}$, $\beta \in \mathcal{R}(A)$

$$\mathcal{C}_1 \cap \mathcal{C}_2 = \mathbb{R}^n_+ \cap \mathcal{A} = \{y \mid y \succeq 0\} \cap \{y \mid Ay = \beta\} \subset \mathbb{R}^n \qquad (2158)$$

the nonnegative orthant intersecting affine subset $\mathcal{A}$ (an intersection of hyperplanes). Projection of an iterate $x_i \in \mathbb{R}^n$ on $\mathcal{A}$ is calculated

$$P_2 \, x_i = x_i - A^{\mathrm{T}} (A A^{\mathrm{T}})^{-1} (A x_i - \beta) \qquad (2049)$$

while, thereafter, projection of the result on the orthant is simply

$$x_{i+1} = P_1 P_2 \, x_i = \max\{\mathbf{0}, P_2 \, x_i\} \qquad (2159)$$

where the maximum is entrywise (§E.9.2.2.3).

One realization of this problem in $\mathbb{R}^2$ is illustrated in Figure 182: For $A = [\, 1 \quad 1 \,]$, $\beta = 1$, and $x_0 = b = [\, -3 \quad 1/2 \,]^{\mathrm{T}}$, iterates converge to a feasible solution $Pb = [\, 0 \quad 1 \,]^{\mathrm{T}}$.

To give a more palpable sense of convergence in higher dimension, we do this example again but now we compute an alternating projection for the case $A \in \mathbb{R}^{400 \times 1000}$, $\beta \in \mathbb{R}^{400}$, and $b \in \mathbb{R}^{1000}$, all of whose entries are independently and randomly set to a uniformly distributed real number in the interval $[-1, 1]$. Convergence is illustrated in Figure 183.   □

This application of alternating projection to feasibility is extensible to any finite number of closed convex sets.

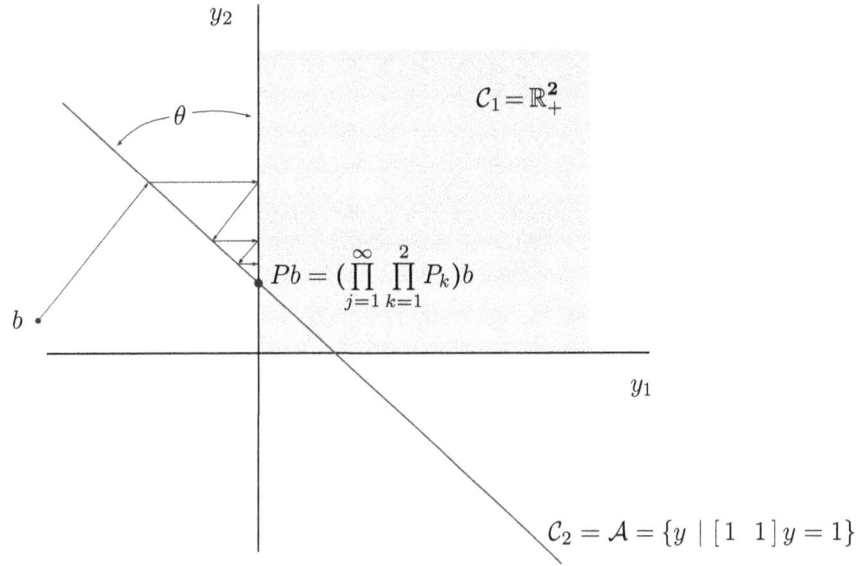

Figure 182: From Example E.10.2.0.2 in $\mathbb{R}^2$, showing von Neumann-style alternating projection to find feasible solution belonging to intersection of nonnegative orthant with hyperplane $\mathcal{A}$. Point Pb lies at intersection of hyperplane with ordinate axis. In this particular example, feasible solution found is coincidentally optimal. Rate of convergence depends upon angle θ; as it becomes more acute, convergence slows. [181, §3]

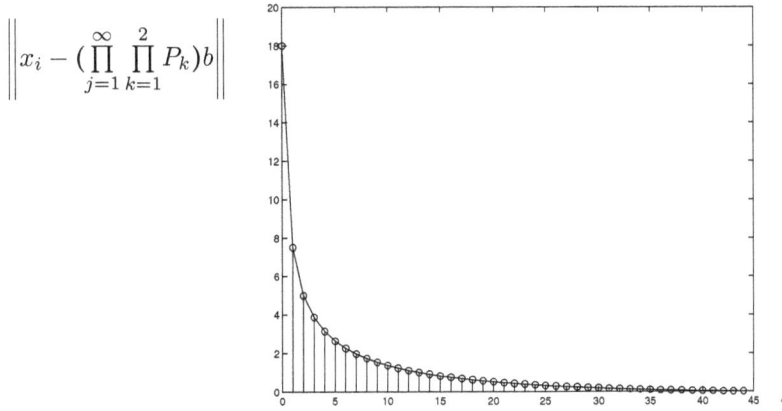

Figure 183: Example E.10.2.0.2 in $\mathbb{R}^{1000}$; geometric convergence of iterates in norm.

E.10.2.0.3   Example. *Under- and over-projection.*   [59, §3]
Consider the following variation of alternating projection: We begin with some point
$x_0 \in \mathbb{R}^n$ then project that point on convex set $\mathcal{C}$ and then project that same point x_0 on
convex set $\mathcal{D}$. To the first iterate we assign $x_1 = (P_{\mathcal{C}}(x_0) + P_{\mathcal{D}}(x_0))\frac{1}{2}$. More generally,

$$x_{i+1} = (P_{\mathcal{C}}(x_i) + P_{\mathcal{D}}(x_i))\frac{1}{2}, \quad i = 0, 1, 2 \ldots \tag{2160}$$

Because the Cartesian product of convex sets remains convex, (§2.1.8) we can reformulate
this problem.
   Consider the convex set

$$\mathcal{S} \triangleq \begin{bmatrix} \mathcal{C} \\ \mathcal{D} \end{bmatrix} \tag{2161}$$

representing Cartesian product $\mathcal{C} \times \mathcal{D}$. Now, those two projections $P_{\mathcal{C}}$ and $P_{\mathcal{D}}$ are
equivalent to one projection on the Cartesian product; *id est*,

$$P_{\mathcal{S}}\left(\begin{bmatrix} x_i \\ x_i \end{bmatrix} \right) = \begin{bmatrix} P_{\mathcal{C}}(x_i) \\ P_{\mathcal{D}}(x_i) \end{bmatrix} \tag{2162}$$

Define the subspace

$$\mathcal{R} \triangleq \left\{ v \in \begin{bmatrix} \mathbb{R}^n \\ \mathbb{R}^n \end{bmatrix} \ \middle| \ [\, I \ \ -I \,]\, v = \mathbf{0} \right\} \tag{2163}$$

By the results in Example E.5.0.0.6

$$P_{\mathcal{RS}}\left(\begin{bmatrix} x_i \\ x_i \end{bmatrix} \right) = P_{\mathcal{R}}\left(\begin{bmatrix} P_{\mathcal{C}}(x_i) \\ P_{\mathcal{D}}(x_i) \end{bmatrix} \right) = \begin{bmatrix} P_{\mathcal{C}}(x_i) + P_{\mathcal{D}}(x_i) \\ P_{\mathcal{C}}(x_i) + P_{\mathcal{D}}(x_i) \end{bmatrix}\frac{1}{2} \tag{2164}$$

This means the proposed variation of alternating projection is equivalent to an alternation
of projection on convex sets $\mathcal{S}$ and $\mathcal{R}$. If $\mathcal{S}$ and $\mathcal{R}$ intersect, these iterations will
converge to a point in their intersection; hence, to a point in the intersection of $\mathcal{C}$ and $\mathcal{D}$.
   We need not apply equal weighting to the projections, as supposed in (2160). In that
case, definition of $\mathcal{R}$ would change accordingly.   □

E.10.2.1   Relative measure of convergence

Inspired by Fejér monotonicity, the alternating projection algorithm from the example of
convergence illustrated by Figure **183** employs a redundant sequence: The first sequence
(indexed by j) estimates point $(\prod_{j=1}^{\infty} \prod_{k=1}^{L} P_k)b$ in the presumably nonempty intersection of
L convex sets, then the quantity

$$\left\| x_i - \left(\prod_{j=1}^{\infty} \prod_{k=1}^{L} P_k \right)b \right\| \tag{2165}$$

in second sequence x_i is observed per iteration i for convergence. *A priori* knowledge of
a feasible solution (2149) is both impractical and antithetical. We need another measure:

Nonexpansivity implies

$$\left\| \left(\prod_{\ell=1}^{L} P_\ell \right) x_{k,i-1} - \left(\prod_{\ell=1}^{L} P_\ell \right) x_{ki} \right\| = \| x_{ki} - x_{k,i+1} \| \leq \| x_{k,i-1} - x_{ki} \| \qquad (2166)$$

where

$$x_{ki} \triangleq P_k x_{k+1,i} \in \mathbb{R}^n , \qquad x_{L+1,i} \triangleq x_{1,i-1} \qquad (2167)$$

x_{ki} represents unique minimum-distance projection of $x_{k+1,i}$ on convex set k at iteration i. So a good convergence measure is total monotonic sequence

$$\varepsilon_i \triangleq \sum_{k=1}^{L} \| x_{ki} - x_{k,i+1} \| , \qquad i=0,1,2\ldots \qquad (2168)$$

where $\lim\limits_{i\to\infty} \varepsilon_i = 0$ whether or not the intersection is nonempty.

E.10.2.1.1   Example. *Affine subset $\cap$ positive semidefinite cone.*
Consider the problem of finding $X \in \mathbb{S}^n$ that satisfies

$$X \succeq 0 , \qquad \langle A_j , X \rangle = b_j , \quad j=1\ldots m \qquad (2169)$$

given nonzero $A_j \in \mathbb{S}^n$ and real b_j. Here we take $\mathcal{C}_1$ to be the positive semidefinite cone $\mathbb{S}^n_+$ while $\mathcal{C}_2$ is the affine subset of $\mathbb{S}^n$

$$
\begin{aligned}
\mathcal{C}_2 = \mathcal{A} &\triangleq \{ X \mid \operatorname{tr}(A_j X) = b_j , \; j=1\ldots m \} \subseteq \mathbb{S}^n \\
&= \{ X \mid \begin{bmatrix} \operatorname{svec}(A_1)^{\mathrm{T}} \\ \vdots \\ \operatorname{svec}(A_m)^{\mathrm{T}} \end{bmatrix} \operatorname{svec} X = b \} \\
&\triangleq \{ X \mid A \operatorname{svec} X = b \}
\end{aligned}
\qquad (2170)
$$

where $b = [b_j] \in \mathbb{R}^m$, $A \in \mathbb{R}^{m \times n(n+1)/2}$, and symmetric vectorization svec is defined by (56). Projection of iterate $X_i \in \mathbb{S}^n$ on $\mathcal{A}$ is: (§E.5.0.0.6)

$$P_2 \operatorname{svec} X_i = \operatorname{svec} X_i - A^\dagger (A \operatorname{svec} X_i - b) \qquad (2171)$$

Euclidean distance from X_i to $\mathcal{A}$ is therefore

$$\operatorname{dist}(X_i , \mathcal{A}) = \| X_i - P_2 X_i \|_{\mathrm{F}} = \| A^\dagger (A \operatorname{svec} X_i - b) \|_2 \qquad (2172)$$

Projection of $P_2 X_i \triangleq \sum_j \lambda_j q_j q_j^{\mathrm{T}}$ on the positive semidefinite cone (§7.1.2) is found from its eigenvalue decomposition (§A.5.1);

$$P_1 P_2 X_i = \sum_{j=1}^{n} \max\{0 , \lambda_j\} q_j q_j^{\mathrm{T}} \qquad (2173)$$

Distance from $P_2 X_i$ to the positive semidefinite cone is therefore

$$\text{dist}(P_2 X_i \,,\, \mathbb{S}_+^n) \;=\; \|P_2 X_i - P_1 P_2 X_i\|_{\text{F}} \;=\; \sqrt{\sum_{j=1}^{n} (\min\{0,\lambda_j\})^2} \qquad (2174)$$

When the intersection is empty $\mathcal{A} \cap \mathbb{S}_+^n = \emptyset$, the iterates converge to that positive semidefinite matrix closest to $\mathcal{A}$ in the Euclidean sense. Otherwise, convergence is to some point in the nonempty intersection.

Barvinok (§2.9.3.0.1) shows that if a solution to (2169) exists, then there exists an $X \in \mathcal{A} \cap \mathbb{S}_+^n$ such that

$$\text{rank} \, X \leq \left\lfloor \frac{\sqrt{8m+1}-1}{2} \right\rfloor \qquad (272)$$

$\square$

E.10.2.1.2 Example. *Semidefinite matrix completion.*
Continuing Example E.10.2.1.1: When $m \leq n(n+1)/2$ and the A_j matrices are distinct members of the standard orthonormal basis $\{E_{\ell q} \in \mathbb{S}^n\}$ (59)

$$\{A_j \in \mathbb{S}^n, \; j=1\ldots m\} \subseteq \{E_{\ell q}\} = \left\{ \begin{array}{ll} e_\ell e_\ell^{\text{T}} \,, & \ell = q = 1 \ldots n \\ \frac{1}{\sqrt{2}}(e_\ell e_q^{\text{T}} + e_q e_\ell^{\text{T}}), & 1 \leq \ell < q \leq n \end{array} \right\} \qquad (2175)$$

and when the constants b_j are set to constrained entries of variable $X \triangleq [X_{\ell q}] \in \mathbb{S}^n$

$$\{b_j \,,\; j=1\ldots m\} \subseteq \left\{ \begin{array}{ll} X_{\ell q} \,, & \ell = q = 1 \ldots n \\ X_{\ell q} \sqrt{2} \,, & 1 \leq \ell < q \leq n \end{array} \right\} = \{\langle X, E_{\ell q} \rangle\} \qquad (2176)$$

then the equality constraints in (2169) fix individual entries of $X \in \mathbb{S}^n$. Thus the feasibility problem becomes a *positive semidefinite matrix completion problem*. Projection of iterate $X_i \in \mathbb{S}^n$ on $\mathcal{A}$ simplifies to *(confer (2171))*

$$P_2 \, \text{svec} \, X_i = \text{svec} \, X_i - A^{\text{T}}(A \, \text{svec} \, X_i - b) \qquad (2177)$$

From this we can see that orthogonal projection is achieved simply by setting corresponding entries of $P_2 X_i$ to the known entries of X, while the entries of $P_2 X_i$ remaining are set to corresponding entries of the current iterate X_i.

Using this technique, we find a positive semidefinite completion for

$$\begin{bmatrix} 4 & 3 & ? & 2 \\ 3 & 4 & 3 & ? \\ ? & 3 & 4 & 3 \\ 2 & ? & 3 & 4 \end{bmatrix} \qquad (2178)$$

Initializing the unknown entries to 0, they all converge geometrically to 1.5858 (rounded) after about 42 iterations.

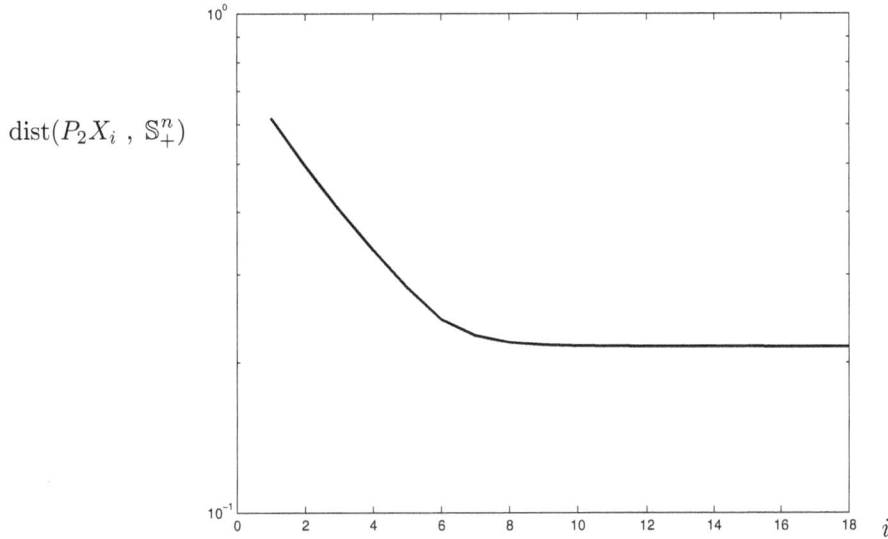

Figure 184: Distance (*confer* (2174)) between positive semidefinite cone and iterate (2177) in affine subset $\mathcal{A}$ (2170) for Laurent's completion problem; initially, decreasing geometrically.

Laurent gives a problem for which no positive semidefinite completion exists: [242]

$$\begin{bmatrix} 1 & 1 & ? & 0 \\ 1 & 1 & 1 & ? \\ ? & 1 & 1 & 1 \\ 0 & ? & 1 & 1 \end{bmatrix} \tag{2179}$$

Initializing unknowns to 0, by alternating projection we find the constrained matrix closest to the positive semidefinite cone,

$$\begin{bmatrix} 1 & 1 & 0.5454 & 0 \\ 1 & 1 & 1 & 0.5454 \\ 0.5454 & 1 & 1 & 1 \\ 0 & 0.5454 & 1 & 1 \end{bmatrix} \tag{2180}$$

and we find the positive semidefinite matrix closest to affine subset $\mathcal{A}$ (2170):

$$\begin{bmatrix} 1.0521 & 0.9409 & 0.5454 & 0.0292 \\ 0.9409 & 1.0980 & 0.9451 & 0.5454 \\ 0.5454 & 0.9451 & 1.0980 & 0.9409 \\ 0.0292 & 0.5454 & 0.9409 & 1.0521 \end{bmatrix} \tag{2181}$$

These matrices (2180) and (2181) attain the Euclidean distance $\mathrm{dist}(\mathcal{A}, \mathbb{S}^n_+)$. Convergence is illustrated in Figure 184. $\qquad\qquad\square$

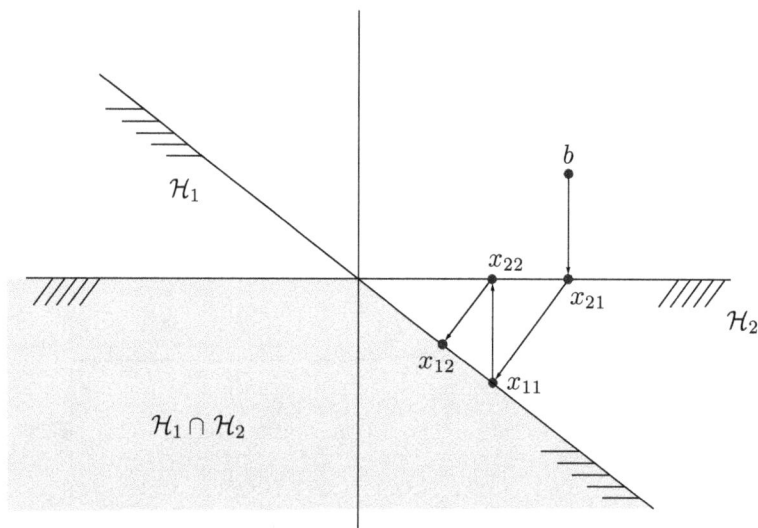

Figure 185: $\mathcal{H}_1$ and $\mathcal{H}_2$ are the same halfspaces as in Figure 180. Dykstra's alternating projection algorithm generates the alternations b, x_{21}, x_{11}, x_{22}, x_{12}, x_{12} ... x_{12}. The path illustrated from b to x_{12} in $\mathbb{R}^2$ terminates at the desired result, Pb in Figure 180. The $\{y_{ki}\}$ correspond to the first two difference vectors drawn (in the first iteration $i=1$), then oscillate between zero and a negative vector thereafter. These alternations are not so robust in presence of noise as for the example in Figure 179.

E.10.3   Optimization and projection

Unique projection on the nonempty intersection of arbitrary convex sets, to find the closest point therein, is a convex optimization problem. The first successful application of alternating projection to this problem is attributed to Dykstra [131] [63] who in 1983 provided an elegant algorithm that prevails today. In 1988, Han [184] rediscovered the algorithm and provided a primal−dual convergence proof. A synopsis of the history of alternating projection[E.22] can be found in [65] where it becomes apparent that Dykstra's work is seminal.

E.10.3.1   Dykstra's algorithm

Assume we are given some point $b \in \mathbb{R}^n$ and closed convex sets $\{\mathcal{C}_k \subset \mathbb{R}^n \mid k=1 \ldots L\}$. Let $x_{ki} \in \mathbb{R}^n$ and $y_{ki} \in \mathbb{R}^n$ respectively denote a *primal* and *dual* vector (whose meaning can be deduced from Figure 185 and Figure 186) associated with set k at iteration i. Initialize

$$ y_{k0} = 0 \;\; \forall k=1 \ldots L \qquad \text{and} \qquad x_{1,0} = b \qquad\qquad (2182) $$

[E.22] For a synopsis of alternating projection applied to distance geometry, see [362, §3.1].

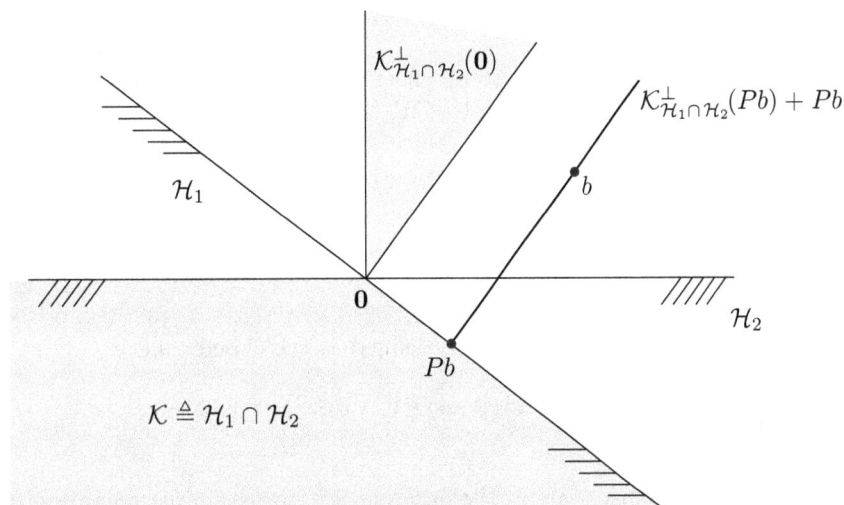

Figure 186: Two examples (truncated): normal cone to $\mathcal{H}_1 \cap \mathcal{H}_2$ at the origin and at point Pb on the boundary. $\mathcal{H}_1$ and $\mathcal{H}_2$ are the same halfspaces from Figure 185. Normal cone at the origin $\mathcal{K}^\perp_{\mathcal{H}_1 \cap \mathcal{H}_2}(\mathbf{0})$ is simply $-\mathcal{K}^*$.

Denoting by $P_k t$ the unique minimum-distance projection of t on $\mathcal{C}_k$, and for convenience $x_{L+1,i} = x_{1,i-1}$ (2167), iterate x_{1i} calculation proceeds:[E.23]

$$
\begin{aligned}
&\texttt{for } i = 1, 2, \ldots \texttt{until convergence } \{ \\
&\quad \texttt{for } k = L \ldots 1 \{ \\
&\quad\quad t = x_{k+1,i} - y_{k,i-1} \\
&\quad\quad x_{ki} = P_k t \\
&\quad\quad y_{ki} = P_k t - t \\
&\quad \} \\
&\}
\end{aligned}
\tag{2183}
$$

Assuming a nonempty intersection, then the iterates converge to the unique minimum-distance projection of point b on that intersection; [112, §9.24]

$$
Pb = \lim_{i \to \infty} x_{1i}
\tag{2184}
$$

In the case that all $\mathcal{C}_k$ are affine, then calculation of y_{ki} is superfluous and the algorithm becomes identical to alternating projection. [112, §9.26] [152, §1] Dykstra's algorithm is so simple, elegant, and represents such a tiny increment in computational intensity over alternating projection, it is nearly always arguably cost effective.

[E.23]We reverse order of projection ($k = L \ldots 1$) in the algorithm for continuity of exposition.

E.10.3.2   Normal cone

Glunt [159, §4] observes that the overall effect of Dykstra's iterative procedure is to drive t toward the translated normal cone to $\bigcap \mathcal{C}_k$ at the solution Pb (translated to Pb). The normal cone gets its name from its graphical construction; which is, loosely speaking, to draw the outward-normals at Pb (Definition E.9.1.0.1) to all the convex sets $\mathcal{C}_k$ touching Pb. Relative interior of the normal cone subtends these normal vectors.

E.10.3.2.1   Definition.   *Normal cone.*   (§2.13.10)
The normal cone [274] [42, p.261] [205, §A.5.2] [56, §2.1] [314, §3] [315, p.15] to any set $\mathcal{S} \subseteq \mathbb{R}^n$ at any particular point $a \in \mathbb{R}^n$ is defined as the closed cone

$$\begin{aligned} \mathcal{K}_{\mathcal{S}}^{\perp}(a) &= \{z \in \mathbb{R}^n \mid z^{\mathrm{T}}(y-a) \leq 0 \ \ \forall y \in \mathcal{S}\} = -(\mathcal{S} - a)^* \\ &= \{z \in \mathbb{R}^n \mid z^{\mathrm{T}}y \leq 0 \ \ \ \forall y \in \mathcal{S} - a\} \end{aligned} \qquad (448)$$

an intersection of halfspaces about the origin in $\mathbb{R}^n$, hence convex regardless of convexity of $\mathcal{S}$; it is the negative dual cone to translate $\mathcal{S} - a$; the set of all vectors normal to $\mathcal{S}$ at a (§E.9.1.0.1). $\triangle$

Examples of normal cone construction are illustrated in Figure 68, Figure 186, Figure 187, and Figure 188. Normal cone at $\mathbf{0}$ in Figure 186 is the vector sum (§2.1.8) of two normal cones; [56, §3.3 exer.10] for $\mathcal{H}_1 \cap \mathrm{int}\, \mathcal{H}_2 \neq \emptyset$

$$\mathcal{K}_{\mathcal{H}_1 \cap \mathcal{H}_2}^{\perp}(\mathbf{0}) = \mathcal{K}_{\mathcal{H}_1}^{\perp}(\mathbf{0}) + \mathcal{K}_{\mathcal{H}_2}^{\perp}(\mathbf{0}) \qquad (2185)$$

This formula applies more generally to other points in the intersection.

The normal cone to any affine set $\mathcal{A}$ at $\alpha \in \mathcal{A}$, for example, is the orthogonal complement of $\mathcal{A} - \alpha$. When $\mathcal{A} = \mathbf{0}$, $\mathcal{K}_{\mathcal{A}}^{\perp}(\mathbf{0}) = \mathcal{A}^{\perp}$ is $\mathbb{R}^n$ the ambient space of $\mathcal{A}$.

Projection of any point in the translated normal cone $\mathcal{K}_{\mathcal{C}}^{\perp}(a \in \mathcal{C}) + a$ on convex set $\mathcal{C}$ is identical to a; in other words, point a is that point in $\mathcal{C}$ closest to any point belonging to the translated normal cone $\mathcal{K}_{\mathcal{C}}^{\perp}(a) + a$; *e.g.*, Theorem E.4.0.0.1. The normal cone to convex cone $\mathcal{K}$ at the origin

$$\mathcal{K}_{\mathcal{K}}^{\perp}(\mathbf{0}) = -\mathcal{K}^* \qquad (2186)$$

is the negative dual cone. Any point belonging to $-\mathcal{K}^*$, projected on $\mathcal{K}$, projects on the origin. More generally, [112, §4.5]

$$\mathcal{K}_{\mathcal{K}}^{\perp}(a) = -(\mathcal{K} - a)^* \qquad (2187)$$

$$\mathcal{K}_{\mathcal{K}}^{\perp}(a \in \mathcal{K}) = -\mathcal{K}^* \cap a^{\perp} \qquad (2188)$$

The normal cone to $\bigcap \mathcal{C}_k$ at Pb in Figure 180 is ray $\{\xi(b - Pb) \mid \xi \geq 0\}$ illustrated in Figure 186. Applying Dykstra's algorithm to that example, convergence to the desired result is achieved in two iterations as illustrated in Figure 185. Yet applying Dykstra's algorithm to the example in Figure 179 does not improve rate of convergence, unfortunately, because the given point b and all the alternating projections already belong to the translated normal cone at the vertex of intersection.

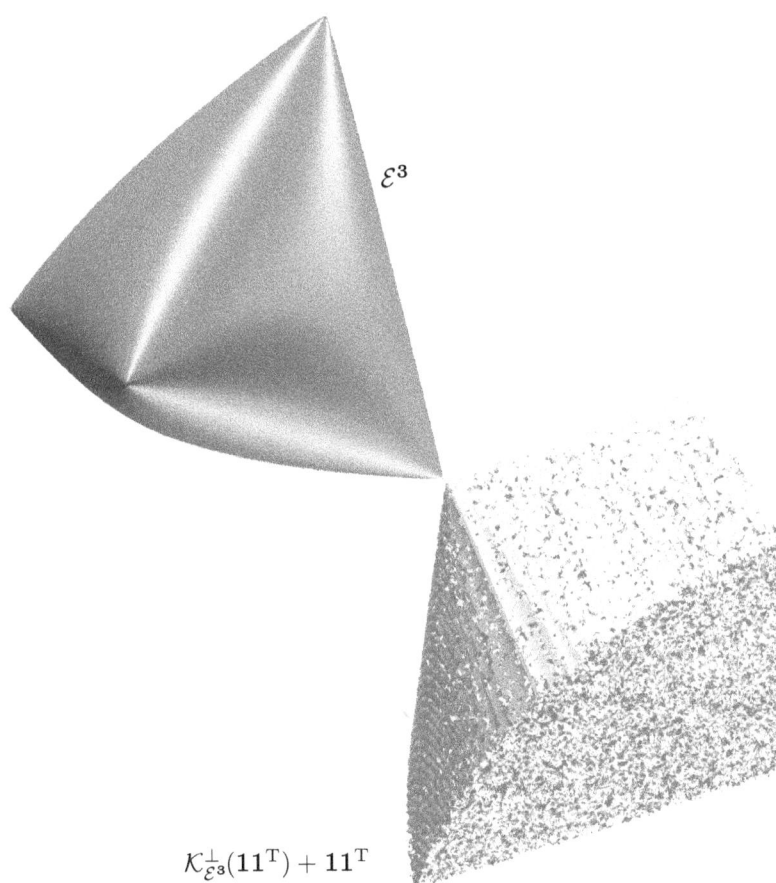

Figure 187: A few renderings (including next page) of normal cone $\mathcal{K}_{\mathcal{E}^3}^{\perp}$ to elliptope $\mathcal{E}^3$ (Figure 141), at point $\mathbf{1}\mathbf{1}^{\mathrm{T}}$, projected on $\mathbb{R}^{\mathbf{3}}$. In [243, fig.2], normal cone is claimed circular in this dimension. (Severe numerical artifacts corrupt boundary and make relative interior corporeal; drawn truncated.)

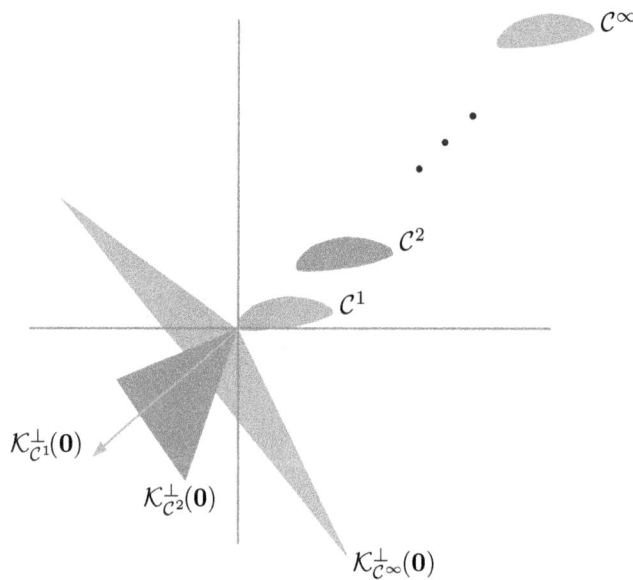

Figure 188: Rough sketch of normal cone to set $\mathcal{C} \subset \mathbb{R}^{\mathbf{2}}$ as $\mathcal{C}$ wanders toward infinity. A point at which a normal cone is determined, here the origin, need not belong to the set. Normal cone to $\mathcal{C}^1$ is a ray. But as $\mathcal{C}$ moves outward, normal cone approaches a halfspace.

E.10.3.3 speculation

Dykstra's algorithm always converges at least as quickly as classical alternating projection, never slower [112], and it succeeds where alternating projection fails. Rate of convergence is wholly dependent on particular geometry of a given problem. From these few examples we surmise, unique minimum-distance projection on *blunt* (not sharp or acute, informally) full-dimensional polyhedral cones may be found by Dykstra's algorithm in few iterations. But total number of alternating projections, constituting those iterations, can never be less than number of convex sets.

APPENDIX E. PROJECTION

Appendix F

Notation, Definitions, Glossary

b (italic $abcdefghijklmnopqrstuvwxyz$)
column vector, scalar, or logical condition

b_i i^{th} *entry of vector* $b = [b_i\,,\ i = 1\dots n]$ or i^{th} *b vector from a set or list* $\{b_j\,,\ j = 1\dots n\}$
or i^{th} *iterate of vector* b

$b_{i:j}$ or $b(i\!:\!j)$, truncated vector comprising i^{th} through j^{th} entry of vector b

$b_k(i\!:\!j)$ or $b_{i:j,\,k}$, truncated vector comprising i^{th} through j^{th} entry of vector b_k

b^{T} vector transpose   or   row vector

b^{H} complex conjugate transpose $b^{*\text{T}}$

A matrix, scalar, or logical condition
(italic $ABCDEFGHIJKLMNOPQRSTUVWXYZ$)

A^{T} Matrix transpose $[A_{ij}] \leftarrow [A_{ji}]$ is a linear operator.
Regarding A as a linear operator, A^{T} is its adjoint.

$A^{-\text{T}}$ matrix transpose of inverse; and *vice versa*, $\left(A^{-1}\right)^{\text{T}} = \left(A^{\text{T}}\right)^{-1}$

A^{T_1} first of various transpositions of a cubix or quartix A (p.584, p.588)

skinny   a skinny matrix; meaning, more rows than columns: $\begin{bmatrix} \\ \\ \end{bmatrix}$.

     When there are more equations than unknowns, we say that the system $Ax = b$ *is* overdetermined. [164, §5.3]

fat   a fat matrix; meaning, more columns than rows: $\begin{bmatrix} & \end{bmatrix}$.
underdetermined

$\mathcal{A}$ some set (calligraphic $\mathcal{ABCDEFGHIJKLMNOPQRSTUVWXYZ}$)

CITATION: Dattorro, *Convex Optimization & Euclidean Distance Geometry*, $\mathcal{M}\varepsilon\beta oo$ Publishing USA, 2005, v2015.02.15.

$\mathcal{F}(\mathcal{C} \ni A)$ smallest face (171) that contains element A of set $\mathcal{C}$

$\mathfrak{F}$ discrete Fourier transform (883)

$\mathcal{G}(\mathcal{K})$ generators (§2.13.4.2.1) of set $\mathcal{K}$; any collection of points and directions whose hull constructs $\mathcal{K}$

$\mathcal{L}^{\nu}_{\nu}$ level set (556)

$\mathcal{L}_{\nu}$ sublevel set (560)

$\mathcal{L}^{\nu}$ superlevel set (646)

A^{-1} inverse of matrix A

$A^{\dagger}$ Moore-Penrose pseudoinverse of matrix A

$\sqrt{}$ positive square root

$\sqrt[\circ]{x}$ entrywise positive square root of vector x

$\sqrt[\ell]{}$ positive ℓ^{th} root

$A^{1/2}$ and $\sqrt{A}$ $A^{1/2}$ is any matrix such that $A^{1/2}A^{1/2} = A$.
For $A \in \mathbb{S}^n_+$, $\sqrt{A} \in \mathbb{S}^n_+$ is unique and $\sqrt{A}\sqrt{A} = A$. [56, §1.2] (§A.5.1.3)

$\sqrt[\circ]{D}$ $= [\sqrt{d_{ij}}]$. (1428) *Hadamard positive square root:* $D = \sqrt[\circ]{D} \circ \sqrt[\circ]{D}$

$\mathfrak{A}$ Euler Fraktur $\mathfrak{ABCDEFGHIJKLMNOPQRSTUVWXYZ}$

$\mathfrak{L}$ Lagrangian (506)

$\mathfrak{E}$ member of elliptope $\mathcal{E}_t$ (1177) parametrized by scalar t

$\mathcal{E}$ elliptope (1156)

E elementary matrix

E_{ij} member of standard orthonormal basis for symmetric (59) or symmetric hollow (75) matrices

A_{ij} or $A(i,j)$, ij^{th} entry of matrix $A = \begin{bmatrix} \cdot & \cdot & \cdot \\ \cdot & \cdot & \cdot \\ \cdot & \cdot & \cdot \end{bmatrix} \begin{smallmatrix} 1 \\ 2 \\ 3 \end{smallmatrix}$

or rank-one matrix $a_i a_j^{\mathrm{T}}$ (§4.8)

$A(i,j)$ A is a function of i and j

A_i i^{th} matrix from a set or i^{th} principal submatrix or i^{th} iterate of A

$A(i,:)$ i^{th} row of matrix A

$A(:,j)$ j^{th} column of matrix A [164, §1.1.8]

$A_{i:j,\,k:\ell}$ or $A(i\!:\!j,\,k\!:\!\ell)$, submatrix taken from i^{th} through j^{th} row and k^{th} through ℓ^{th} column

id est   from the Latin meaning *that is*

e.g.   *exempli gratia,* from the Latin meaning *for sake of example*

videlicet   from the Latin meaning *it is permitted to see*

no.   *number,* from the Latin *numero*

a.i.   affinely independent (§2.4.2.3)

c.i.   conically independent (§2.10)

l.i.   linearly independent

w.r.t   *with respect to*

a.k.a   *also known as*

re   real part

im   imaginary part

$\imath$ or $\jmath$ $\sqrt{-1}$

$\subset\ \supset\ \cap\ \cup$ standard set theory, *subset, superset, intersection, union*

$\in$ membership, *element belongs to,* or *element is a member of*

$\ni$ membership, *contains* as in $\mathcal{C}\ni y$ ($\mathcal{C}$ contains element y)

$\backepsilon$ *such that*

$\exists$ *there exists*

$\therefore$ *therefore*

$\forall$ *for all,* or *over all*

&   (ampersand) *and*

$\mathit{\&}$ (ampersand italic) *and*

$\propto$ *proportional to*

∞ infinity

$\equiv$ *equivalent to*

$\triangleq$ *defined equal to, equal by definition*

$\approx$ *approximately equal to*

$\simeq$ *isomorphic to* or *with*

$\cong$ *congruent to* or *with*

—   Hadamard quotient as in, for $x, y \in \mathbb{R}^n$, $\dfrac{x}{y} \triangleq [\, x_i/y_i \,,\ i=1\ldots n\,] \in \mathbb{R}^n$

$\circ$ Hadamard product of matrices: $x \circ y \triangleq [\, x_i y_i \,,\ i=1\ldots n\,] \in \mathbb{R}^n$

$\otimes$ Kronecker product of matrices (§D.1.2.1)

$\oplus$ vector sum of sets $\mathcal{X} = \mathcal{Y} \oplus \mathcal{Z}$ where every element $x \in \mathcal{X}$ has unique expression $x = y + z$ where $y \in \mathcal{Y}$ and $z \in \mathcal{Z}$; [315, p.19] then summands are *algebraic complements*. $\mathcal{X} = \mathcal{Y} \oplus \mathcal{Z} \Rightarrow \mathcal{X} = \mathcal{Y} + \mathcal{Z}$. Now assume $\mathcal{Y}$ and $\mathcal{Z}$ are nontrivial subspaces. $\mathcal{X} = \mathcal{Y} + \mathcal{Z} \Rightarrow \mathcal{X} = \mathcal{Y} \oplus \mathcal{Z} \Leftrightarrow \mathcal{Y} \cap \mathcal{Z} = \mathbf{0}$ [316, §1.2] [112, §5.8]. Each element from a vector sum (+) of subspaces has unique expression ($\oplus$) when a basis from each subspace is linearly independent of bases from all the other subspaces.

$\ominus$ likewise, the vector difference of sets

$\boxplus$ orthogonal vector sum of sets $\mathcal{X} = \mathcal{Y} \boxplus \mathcal{Z}$ where every element $x \in \mathcal{X}$ has unique orthogonal expression $x = y + z$ where $y \in \mathcal{Y}$, $z \in \mathcal{Z}$, and $y \perp z$. [337, p.51] $\mathcal{X} = \mathcal{Y} \boxplus \mathcal{Z} \Rightarrow \mathcal{X} = \mathcal{Y} + \mathcal{Z}$. If $\mathcal{Z} \subseteq \mathcal{Y}^\perp$ then $\mathcal{X} = \mathcal{Y} \boxplus \mathcal{Z} \Leftrightarrow \mathcal{X} = \mathcal{Y} \oplus \mathcal{Z}$. [112, §5.8] If $\mathcal{Z} = \mathcal{Y}^\perp$ then summands are *orthogonal complements*.

$\pm$ *plus or minus* or *plus and minus*

$\perp$ as in $A \perp B$ meaning A *is orthogonal to* B (and *vice versa*), where A and B are sets, vectors, or matrices. When A and B are vectors (or matrices under Frobenius' norm), $A \perp B \Leftrightarrow \langle A, B \rangle = 0 \Leftrightarrow \|A + B\|^2 = \|A\|^2 + \|B\|^2$

$\backslash$ as in $\backslash \mathcal{A}$ means logical *not* $\mathcal{A}$, or *relative complement of set* $\mathcal{A}$; *id est*, $\backslash \mathcal{A} = \{x \notin \mathcal{A}\}$; *e.g.*, $\mathcal{B} \backslash \mathcal{A} \triangleq \{x \in \mathcal{B} \mid x \notin \mathcal{A}\} \equiv \mathcal{B} \cap \backslash \mathcal{A}$

$\Rightarrow$ or $\Leftarrow$ sufficient or necessary, *implies*, or *is implied by*; *e.g.*,
A is sufficient: $A \Rightarrow B$, A is necessary: $A \Leftarrow B$,
$A \Rightarrow B \Leftrightarrow \backslash A \Leftarrow \backslash B$, $A \Leftarrow B \Leftrightarrow \backslash A \Rightarrow \backslash B$,
if A *then* B,   *if* B *then* A,
A *only if* B. B *only if* A.

$\Leftrightarrow$ *if and only if* (iff) or *corresponds with* or *necessary and sufficient* or *logical equivalence*

is   as in A *is* B means $A \Rightarrow B$; conventional usage of English language imposed by logicians

$\not\Rightarrow$ or $\not\Leftarrow$ insufficient or unnecessary, *does not imply*, or *is not implied by*; *e.g.*,
$A \not\Rightarrow B \Leftrightarrow \backslash A \not\Leftarrow \backslash B$. $A \not\Leftarrow B \Leftrightarrow \backslash A \not\Rightarrow \backslash B$.

$\leftarrow$ *is replaced with*; substitution, assignment

$\rightarrow$ *goes to*, or *approaches*, or *maps to*

$t \to 0^+$ t goes to 0 *from above*; meaning, *from the positive* [205, p.2]

$\vdots \;\; \ddots \;\; \cdots$ as in $1\cdots1$ and $[\,s_1 \cdots s_N\,]$ meaning *continuation*; respectively, ones in a row and a matrix whose columns are s_i for $i=1\ldots N$

$\cdots$ as in $i=1\ldots N$ meaning, i is a *sequence* of successive integers beginning with 1 and ending with N; *id est*, $1\ldots N = 1\!:\!N$

$:$ as in $f: \mathbb{R}^n \to \mathbb{R}^m$ meaning f *is a mapping*,
or sequence of successive integers specified by bounds as in $i\!:\!j = i \ldots j$
(if $j<i$ then sequence is descending)

$f: \mathcal{M} \to \mathcal{R}$ meaning f *is a mapping from ambient space $\mathcal{M}$ to ambient $\mathcal{R}$*, not necessarily denoting either domain or range

$|$ as in $f(x)\,|\,x\in\mathcal{C}$ means *with the condition(s)* or *such that* or *evaluated for*, or as in $\{f(x)\,|\,x\in\mathcal{C}\}$ means *evaluated for each and every x belonging to set $\mathcal{C}$*

$g|_{x_\mathrm{p}}$ *expression g evaluated at x_p*

A, B as in, for example, $A,B \in \mathbb{S}^N$ means $A \in \mathbb{S}^N$ and $B \in \mathbb{S}^N$

(A, B) *open interval* between A and B in $\mathbb{R}$,
or *variable pair* perhaps of disparate dimension

$[A, B]$ *closed interval* or *line segment* between A and B in $\mathbb{R}$

$(\;)$ *hierarchal, parenthetical, optional*

$\{\;\}$ curly braces denote a *set* or *list*, e.g., $\{Xa \mid a\succeq0\}$ *the set comprising Xa evaluated for each and every $a\succeq0$* where membership of a to some space is implicit, a *union*

$\langle\;\rangle$ angle brackets denote *vector inner-product* (33) (38)

$[\;]$ matrix or vector, or quote insertion, or citation

$[d_{ij}]$ matrix whose ij^{th} entry is d_{ij}

$[x_i]$ vector whose i^{th} entry is x_i

x_p particular value of x

x_0 particular instance of x, or initial value of a sequence x_i

x_1 first entry of vector x, or first element of a set or list $\{x_i\}$

x_ε *extreme point*

x_+ vector x whose negative entries are replaced with 0; $x_+ = \frac{1}{2}(x + |x|)$ (534) or *clipped vector x* or *nonnegative part of x*

x_- $x_- \triangleq \frac{1}{2}(x - |x|)$ or *nonpositive part of $x = x_+ + x_-$*

$\check{x}$ known data

$x^\star$ optimal value of variable x. optimal $\Rightarrow$ feasible

x^* *complex conjugate* or *dual variable* or *extreme direction of dual cone*

f^* *convex conjugate function* $f^*(s) = \sup\{\langle s\,,x \rangle - f(x) \mid x \in \operatorname{dom} f\}$

$P_{\mathcal{C}}x$ or Px projection of point x on set $\mathcal{C}$, P is operator or idempotent matrix

$P_k x$ projection of point x on set $\mathcal{C}_k$ or on range of implicit vector

$\delta(A)$ (**a.k.a** diag(A), §A.1) *vector made from main diagonal of A* if A is a matrix; otherwise, *diagonal matrix made from vector A*

$\delta^2(A)$ $\equiv \delta(\delta(A))$. For vector or diagonal matrix Λ, $\delta^2(\Lambda) = \Lambda$

$\delta(A)^2$ $= \delta(A)\delta(A)$ where A is a vector

$\lambda_i(X)$ i^{th} entry of vector λ is function of X

$\lambda(X)_i$ i^{th} entry of vector-valued function of X

$\lambda(A)$ *vector of eigenvalues of matrix A* , (1542) typically arranged in nonincreasing order

$\sigma(A)$ *vector of singular values of matrix A* (always arranged in nonincreasing order), or *support function in direction A*

Σ diagonal matrix of singular values, not necessarily square

$\sum$ sum

$\pi(\gamma)$ nonlinear *permutation operator* (or *presorting function*) arranges vector γ into nonincreasing order (§7.1.3)

Ξ permutation matrix

Π doublet or permutation operator or matrix

$\prod$ product

$\psi(Z)$ signum-like *step function* that returns a scalar for matrix argument (738), it returns a vector for vector argument (1660)

D symmetric hollow matrix of distance-square or *Euclidean distance matrix*

$\mathbf{D}$ Euclidean distance matrix operator

$\mathbf{D}^{\text{T}}(X)$ adjoint operator

$\mathbf{D}(X)^{\text{T}}$ transpose of $\mathbf{D}(X)$

$\mathbf{D}^{-1}(X)$ inverse operator

$\mathbf{D}(X)^{-1}$ inverse of $\mathbf{D}(X)$

$D^\star$ optimal value of variable D

D^* dual to variable D

D° polar variable D

∂ *partial derivative* or *partial differential* or *matrix of distance-square squared* (1468) or *boundary* of set $\mathcal{K}$ as in $\partial\mathcal{K}$ (17) (24)

$\sqrt{d_{ij}}$ (absolute) distance scalar

d_{ij} distance-square scalar, EDM entry

$\mathbf{V}$ geometric centering operator, $\mathbf{V}(D) = -VDV\frac{1}{2}$ (1076)

$\mathbf{V}_{\mathcal{N}}$ $\mathbf{V}_{\mathcal{N}}(D) = -V_{\mathcal{N}}^{\mathrm{T}}DV_{\mathcal{N}}$ (1090)

V $N\times N$ symmetric elementary, auxiliary, projector, geometric centering matrix, $\mathcal{R}(V) = \mathcal{N}(\mathbf{1}^{\mathrm{T}})$, $\mathcal{N}(V) = \mathcal{R}(\mathbf{1})$, $V^2 = V$ (§B.4.1)

$V_{\mathcal{N}}$ $N\times N{-}1$ Schoenberg auxiliary matrix $\mathcal{R}(V_{\mathcal{N}}) = \mathcal{N}(\mathbf{1}^{\mathrm{T}})$, $\mathcal{N}(V_{\mathcal{N}}^{\mathrm{T}}) = \mathcal{R}(\mathbf{1})$ (§B.4.2)

$V_{\mathcal{X}}$ $V_{\mathcal{X}}V_{\mathcal{X}}^{\mathrm{T}} \equiv V^{\mathrm{T}}X^{\mathrm{T}}XV$ (1268)

X point list ((76) having cardinality N) arranged columnar in $\mathbb{R}^{n\times N}$, or set of generators, or extreme directions, or matrix variable

G Gram matrix $X^{\mathrm{T}}X$

r affine dimension

$\mathbf{k}$ number of conically independent generators

$\Bbbk$ raw-data domain of Magnetic Resonance Imaging machine as in $\Bbbk$-*space*

n Euclidean (ambient spatial) dimension of list $X \in \mathbb{R}^{n\times N}$, or integer

N cardinality of list $X \in \mathbb{R}^{n\times N}$, or integer

in *function f in x* means x as argument to f or *x in $\mathcal{C}$* means element x is a member of set $\mathcal{C}$

on *function $f(x)$ on $\mathcal{A}$* means $\mathcal{A}$ is dom f or *relation $\preceq$ on $\mathcal{A}$* means $\mathcal{A}$ is set whose elements are subject to $\preceq$ or *projection of x on $\mathcal{A}$* means $\mathcal{A}$ is body on which projection is made or *operating on vector* identifies argument type to f as "vector"

onto *function $f(x)$ maps onto $\mathcal{M}$* means f over its domain is a surjection with respect to $\mathcal{M}$

over *function $f(x)$ over $\mathcal{C}$* means f evaluated at each and every element of set $\mathcal{C}$

one-to-one	injective map or unique correspondence between sets
epi	function epigraph
dom	function domain
$\mathcal{R}f$	function range
$\mathcal{R}(A)$	the subspace: *range of A*, or span basis $\mathcal{R}(A)$; $\mathcal{R}(A) \perp \mathcal{N}(A^{\mathrm{T}})$
span	as in span $A = \mathcal{R}(A) = \{Ax \mid x \in \mathbb{R}^n\}$ when A is a matrix
basis $\mathcal{R}(A)$	*overcomplete columnar basis for range of A* or *minimal set constituting generators for vertex-description of $\mathcal{R}(A)$* or *linearly independent set of vectors spanning $\mathcal{R}(A)$*
$\mathcal{N}(A)$	the subspace: *nullspace of A*; $\mathcal{N}(A) \perp \mathcal{R}(A^{\mathrm{T}})$
$\mathbb{R}^n$	Euclidean n-dimensional real vector space (nonnegative integer n). $\mathbb{R}^0 = \mathbf{0}$, $\mathbb{R} = \mathbb{R}^1$ or vector space of unspecified dimension. [233] [408]
$\mathbb{R}^{m \times n}$	Euclidean vector space of m by n dimensional real matrices
$\times$	Cartesian product. $\mathbb{R}^{m \times n - m} \triangleq \mathbb{R}^{m \times (n-m)}$. $\mathcal{K}_1 \times \mathcal{K}_2 = \begin{bmatrix} \mathcal{K}_1 \\ \mathcal{K}_2 \end{bmatrix}$
$\begin{bmatrix} \mathbb{R}^m \\ \mathbb{R}^n \end{bmatrix}$	$\mathbb{R}^m \times \mathbb{R}^n = \mathbb{R}^{m+n}$
$\mathbb{C}^n$ or $\mathbb{C}^{n \times n}$	Euclidean complex vector space of respective dimension n and $n \times n$
$\mathbb{R}_+^n$ or $\mathbb{R}_+^{n \times n}$	nonnegative orthant in Euclidean vector space of respective dimension n and $n \times n$
$\mathbb{R}_-^n$ or $\mathbb{R}_-^{n \times n}$	nonpositive orthant in Euclidean vector space of respective dimension n and $n \times n$
$\mathbb{S}^n$	subspace of real symmetric $n \times n$ matrices; the *symmetric matrix subspace*. $\mathbb{S} = \mathbb{S}^1$ or symmetric subspace of unspecified dimension.
$\mathbb{S}^{n\perp}$	orthogonal complement of $\mathbb{S}^n$ in $\mathbb{R}^{n \times n}$, the antisymmetric matrices (51)
$\mathbb{S}_+^n$	convex cone comprising all (real) symmetric positive semidefinite $n \times n$ matrices, the *positive semidefinite cone*
int $\mathbb{S}_+^n$	interior of convex cone comprising all (real) symmetric positive semidefinite $n \times n$ matrices; *id est*, positive definite matrices
$\mathbb{S}_+^n(\rho)$	$= \{X \in \mathbb{S}_+^n \mid \text{rank}\,X \geq \rho\}$ (260) convex set of all positive semidefinite $n \times n$ symmetric matrices whose rank equals or exceeds ρ
EDM^N	cone of $N \times N$ Euclidean distance matrices in the symmetric hollow subspace
$\sqrt{\text{EDM}^N}$	nonconvex cone of $N \times N$ Euclidean absolute distance matrices in the symmetric hollow subspace (§6.3)

CPU   central processing unit

dB   decibel

DC   direct current

EDM   Euclidean distance matrix

PSD   positive semidefinite

SDP   semidefinite program

SVD   singular value decomposition

SNR   signal to noise ratio

$\mathbb{S}_1^n$ subspace comprising all symmetric $n \times n$ matrices having all zeros in first row and column (2098)

$\mathbb{S}_h^n$ subspace comprising all symmetric hollow $n \times n$ matrices (**0** main diagonal), the *symmetric hollow subspace* (66)

$\mathbb{S}_h^{n\perp}$ orthogonal complement of $\mathbb{S}_h^n$ in $\mathbb{S}^n$ (67), the set of all diagonal matrices

$\mathbb{S}_c^n$ subspace comprising all geometrically centered symmetric $n \times n$ matrices; *geometric center subspace* $\mathbb{S}_c^N = \{Y \in \mathbb{S}^N \mid Y\mathbf{1} = \mathbf{0}\}$ (2094)

$\mathbb{S}_c^{n\perp}$ orthogonal complement of $\mathbb{S}_c^n$ in $\mathbb{S}^n$ (2096)

$\mathbb{R}_c^{m \times n}$ subspace comprising all geometrically centered $m \times n$ matrices

$X^\perp$ basis $\mathcal{N}(X^{\mathrm{T}})$ (§2.13.9, §E.3.4)

$x^\perp$ $\mathcal{N}(x^{\mathrm{T}})$; $\{y \in \mathbb{R}^n \mid x^{\mathrm{T}}y = 0\}$ (§2.13.10.1.1)

$\mathcal{R}(P)^\perp$ $\mathcal{N}(P^{\mathrm{T}})$ (fundamental subspace relations (137))

$\mathcal{N}(P)^\perp$ $\mathcal{R}(P^{\mathrm{T}})$

$\mathcal{R}^\perp$ $= \{y \in \mathbb{R}^n \mid \langle x, y \rangle = 0 \ \forall x \in \mathcal{R}\}$ (372). *Orthogonal complement of $\mathcal{R}$ in $\mathbb{R}^n$ when $\mathcal{R}$ is a subspace*

$\mathcal{K}^\perp$ normal cone   (448)

$\mathcal{A}^\perp$ normal cone to affine subset $\mathcal{A}$ (§3.1.2.2.2)

$\mathcal{K}$ *cone*

$\mathcal{K}^*$ *dual cone* $-\mathcal{K}^\circ$

$\mathcal{K}^\circ$ *polar cone* $-\mathcal{K}^*$, or angular *degree* as in 360°

$\mathcal{K}_{\mathcal{M}+}$ monotone nonnegative cone

$\mathcal{K}_{\mathcal{M}}$ monotone cone

$\mathcal{K}_\lambda$	spectral cone
$\mathcal{K}_{\lambda\delta}^*$	cone of majorization
$\mathcal{H}$	halfspace
$\mathcal{H}_-$	halfspace described using an outward-normal (106) to the hyperplane partially bounding it
$\mathcal{H}_+$	halfspace described using an inward-normal (107) to the hyperplane partially bounding it
$\partial\mathcal{H}$	hyperplane; *id est*, partial boundary of halfspace
$\underline{\partial\mathcal{H}}$	supporting hyperplane
$\underline{\partial\mathcal{H}}_-$	a supporting hyperplane having outward-normal with respect to set it supports
$\underline{\partial\mathcal{H}}_+$	a supporting hyperplane having inward-normal with respect to set it supports
$\underline{d}$	vector of distance-square
$\underline{d_{ij}}$	lower bound on distance-square d_{ij}
$\overline{d_{ij}}$	upper bound on distance-square d_{ij}
$\overline{AB}$	closed line segment between points A and B
AB	matrix multiplication of A and B
$\overline{\mathcal{C}}$	*closure of set $\mathcal{C}$*
decomposition	*orthonormal* (2006) page 613, *biorthogonal* (1982) page 607
expansion	*orthogonal* (2016) page 615, *biorthogonal* (402) page 164
vector	*column vector* in $\mathbb{R}^n$
entry	*scalar element or real variable constituting a vector or matrix*
cubix	member of $\mathbb{R}^{M\times N\times L}$
quartix	member of $\mathbb{R}^{M\times N\times L\times K}$
feasible set	most simply, *the set of all variable values satisfying all constraints of an optimization problem*
solution set	most simply, *the set of all optimal solutions to an optimization problem*; a subset of the feasible set and not necessarily a single point
optimal	as in *optimal solution,* means a solution to an optimization problem. An optimal solution is not necessarily unique, but there is no better solution. optimal $\Rightarrow$ feasible

feasible	as in *feasible solution,* means satisfies the ("subject to") constraints of an optimization problem, may or may not be optimal
same	as in *same problem,* means optimal solution set for one problem is identical to optimal solution set of another (without transformation)
equivalent	as in *equivalent problem,* means optimal solution to one problem can be derived from optimal solution to another via suitable transformation
convex	as in *convex problem,* essentially means a convex objective function optimized over a convex set (§4)
objective	The three objectives of Optimization are *minimize, maximize,* and *find*
program	*Semidefinite program* is any convex minimization, maximization, or feasibility problem constraining a variable to a subset of a positive semidefinite cone. *Prototypical semidefinite program* conventionally means: a semidefinite program having linear objective, affine equality constraints, but no inequality constraints except for cone membership. (§4.1.1) *Linear program* is any feasibility problem, or minimization or maximization of a linear objective, constraining the variable to some polyhedron (§2.13.1.0.3)
natural order	with reference to stacking columns in a vectorization means *a vector made from superposing column 1 on top of column 2 then superposing the result on column 3* and so on; as in a vector made from entries of the main diagonal $\delta(A)$ means *taken from left to right and top to bottom*
partial order	relation $\preceq$ is a partial order, on a set, if it possesses reflexivity, antisymmetry, and transitivity (§2.7.2.2)
operator	mapping to a vector space (a multidimensional function)
projector	short for *projection operator*; not necessarily minimum-distance or represented by a matrix
sparsity	ratio of number of nonzero entries to matrix-dimension product
tight	with reference to a bound means *a bound that can be met,* with reference to an inequality means *equality is achievable*
$\ddot{v}$	coefficient vector for two spectral factors (Figure 101, `level 2`)
$\ddot{v}$	coefficient vector corresponding to four spectral factors, `level 3`
$\dddot{v}$	vector containing numerator or denominator coefficients of eight spectral factors; `level 4` in a bifurcation tree
g'	first derivative of possibly multidimensional function with respect to real argument or primed variable closely related to unprimed
g''	second derivative with respect to real argument

$\overset{\rightarrow Y}{dg}$ first directional derivative of possibly multidimensional function g in direction $Y \in \mathbb{R}^{K \times L}$ (maintains dimensions of g)

$\overset{\rightarrow Y}{dg^2}$ second directional derivative of g in direction Y

∇ *gradient* from calculus, ∇f is shorthand for $\nabla_x f(x)$. $\nabla f(y)$ means $\nabla_y f(y)$ or *gradient* $\nabla_x f(y)$ *of* $f(x)$ *with respect to* x *evaluated at* y

∇^2 *second-order gradient*

Δ distance scalar (Figure **26**), or first-order difference matrix (896), or infinitesimal difference operator (§D.1.4)

$\triangle_{ijk}$ triangle made by vertices i , j , and k

I Roman numeral one or capital i

I Identity operator or matrix $I = \delta^2(I)$, $\delta(I) = \mathbf{1}$. Variant: $I_m \triangleq I \in \mathbb{S}^m$

$\mathcal{I}$ *index set,* a discrete set of indices

$\emptyset$ *empty set,* an implicit member of every set

0 real zero

$\mathbf{0}$ *origin* or vector or matrix of zeros

O *sort-index matrix*

O *order of magnitude* information required, or *computational intensity*: $O(N)$ is first order, $O(N^2)$ is second, and so on

1 real one

$\mathbf{1}$ vector of ones. Variant: $\mathbf{1}_m \triangleq \mathbf{1} \in \mathbb{R}^m$

e_i vector whose i^{th} entry is 1 (otherwise 0), i^{th} *member of the standard basis for* $\mathbb{R}^n$ (60)

max *maximum* [205, §0.1.1] or *largest element of a totally ordered set*

$\underset{x}{\text{maximize}}$ *find maximum of objective function w.r.t independent variables* x . Subscript $x \leftarrow x \in \mathcal{C}$ may hold implicit constraints if context clear; *e.g.,* semidefiniteness

arg *argument* of operator or function, or *variable of optimization*

$\sup \mathcal{X}$ *supremum* of totally ordered set $\mathcal{X}$, *least upper bound,* may or may not belong to set [205, §0.1.1]; *e.g.,* range $\mathcal{X}$ of real function

$\arg\sup f(x)$ *argument* x *at supremum of function* f ; not necessarily unique or a member of function domain subject to   specifies constraints of an optimization problem; generally, inequalities and affine equalities. *Subject to* implies: anything not an independent variable is constant, an assignment, or substitution

min   *minimum* [205, §0.1.1] or *smallest element of a totally ordered set*

$\underset{x}{\text{minimize}}$ *find objective function minimum w.r.t independent variables* x.
Subscript $x \leftarrow x \in \mathcal{C}$ may hold implicit constraints if context clear; *e.g.*, semidefiniteness

$\underset{x}{\text{find}}$ *find any feasible solution, specified by the ("subject to") constraints, w.r.t independent variables* x. Subscript $x \leftarrow x \in \mathcal{C}$ may hold implicit constraints if context clear; *e.g.*, semidefiniteness. "Find" denotes a *feasibility problem*; it is the third objective of Optimization

$\inf \mathcal{X}$ *infimum* of totally ordered set $\mathcal{X}$, *greatest lower bound,* may or may not belong to set [205, §0.1.1]; *e.g.*, range $\mathcal{X}$ of real function

$\arg\inf f(x)$ *argument* x *at infimum of function* f; not necessarily unique or a member of function domain

iff   *if and only if*, *necessary and sufficient*; also the meaning indiscriminately attached to appearance of the word "if" in the statement of a mathematical definition, [132, p.106] [264, p.4] an esoteric practice worthy of abolition because of ambiguity thus conferred

rel   relative

int   interior

lim   limit

sgn   signum function or *sign*

round   round to nearest integer

mod   modulus function

tr   matrix trace

rank   as in rank A, *rank of matrix* A; $\dim \mathcal{R}(A)$

dim   dimension, $\dim \mathbb{R}^n = n$, $\dim(x \in \mathbb{R}^n) = n$, $\dim \mathcal{R}(x \in \mathbb{R}^n) = 1$, $\dim \mathcal{R}(A \in \mathbb{R}^{m \times n}) = \text{rank}(A)$

aff   affine hull

dim aff   affine dimension

card   cardinality, *number of nonzero entries* $\text{card} \, x \triangleq \|x\|_0$
or N *is cardinality of list* $X \in \mathbb{R}^{n \times N}$ (p.278)

conv   convex hull (§2.3.2)

cone   conic hull (§2.3.3)

cenv   convex envelope (§7.2.2.1)

content of high-dimensional bounded polyhedron, volume in 3 dimensions, area in 2, and so on

cof   matrix of cofactors corresponding to matrix argument

dist   distance between point or set arguments; *e.g.*, $\mathrm{dist}(x\,,\mathcal{B})$

vec   columnar vectorization of $m \times n$ matrix, Euclidean dimension mn (37)

svec   columnar vectorization of symmetric $n \times n$ matrix, Euclidean dimension $n(n+1)/2$ (56)

dvec   columnar vectorization of symmetric hollow $n \times n$ matrix, Euclidean dimension $n(n-1)/2$ (73)

$\sphericalangle(x\,,y)$ *angle* between vectors x and $y\,$, or *dihedral angle* between affine subsets

$\succeq$ generalized inequality; *e.g.*, $A \succeq 0$ means:

- *vector or matrix A must be expressible in a biorthogonal expansion having nonnegative coordinates with respect to extreme directions of some implicit pointed closed convex cone $\mathcal{K}$ (§2.13.2.0.1, §2.13.7.1.1),*

- **or** *comparison to the origin with respect to some implicit pointed closed convex cone (2.7.2.2),*

- **or** *(when $\mathcal{K} = \mathbb{S}^n_+$) matrix A belongs to the positive semidefinite cone of symmetric matrices (nonnegative eigenvalues, §2.9.0.1),*

- **or** *(when $\mathcal{K} = \mathbb{R}^n_+$) vector A belongs to the nonnegative orthant (each vector entry is nonnegative, §2.3.1.1)*

$\underset{\mathcal{K}}{\succeq}$ as in $x \underset{\mathcal{K}}{\succeq} z$ means $x - z \in \mathcal{K}$ (182)

$\succ$ strict generalized inequality, membership to cone interior; $A \succ 0$ means:

- *vector or matrix A must be expressible in a biorthogonal expansion having positive coordinates with respect to extreme directions of some implicit pointed closed convex cone $\mathcal{K}$ (§2.13.2.0.1, §2.13.7.1.1),*

- **or** *comparison to the origin with respect to the interior of some implicit pointed closed convex cone (2.7.2.2),*

- **or** *(when $\mathcal{K} = \mathbb{S}^n_+$) matrix A belongs to the interior of the positive semidefinite cone of symmetric matrices (positive eigenvalues, §2.9.0.1),*

- **or** (when $\mathcal{K} = \mathbb{R}^n_+$) *vector A belongs to the interior of the nonnegative orthant* (each vector entry is positive, §2.3.1.1)

$\nsucceq$ not positive definite

$\geq$ scalar inequality, *greater than or equal to*; comparison of scalars, or entrywise comparison of vectors or matrices with respect to $\mathbb{R}_+$

nonnegative   for $\alpha \in \mathbb{R}^n$, $\alpha \succeq 0$; *id est*, nonnegative entries when with respect to nonnegative orthant, represents vector on boundary of or interior to pointed closed convex cone $\mathcal{K}$

$>$ *greater than*

positive   for $\alpha \in \mathbb{R}^n$, $\alpha \succ 0$; *id est*, positive (nonzero) entries when with respect to nonnegative orthant, no vector on boundary of pointed closed convex cone $\mathcal{K}$

$\lfloor \ \rfloor$ floor function, $\lfloor x \rfloor$ is greatest integer not exceeding x

$| \ |$ entrywise absolute value of scalars, vectors, and matrices

$\log$ natural (or Napierian) logarithm

$\det$ matrix determinant

$\|x\|$ vector 2-norm   or   *Euclidean norm* $\|x\|_2$

$\|x\|_\ell$ $= \sqrt[\ell]{\sum_{j=1}^{n} |x_j|^\ell}$ vector ℓ-*norm* for $\ell \geq 1$ (convex)

     $\triangleq \sum_{j=1}^{n} |x_j|^\ell$ vector ℓ-*norm* for $0 \leq \ell < 1$ (violates §3.2 *no.*3)

$\|x\|_\infty$ $= \max\{|x_j| \ \forall j\}$ *infinity-norm*

$\|x\|_2^2$ $= x^\mathrm{T} x = \langle x, x \rangle$ *Euclidean norm square*

$\|x\|_1$ $= \mathbf{1}^\mathrm{T} |x|$ 1-norm, *dual infinity-norm*

$\|x\|_0$ $= \mathbf{1}^\mathrm{T} |x|^0$ $(0^0 \triangleq 0)$ 0-norm (§4.5.1), *cardinality of vector x*   (card x)

$\|x\|_{\frac{n}{k}}$ k-*largest norm* (§3.2.2.1)

$\|X\|_2$ $= \sup_{\|a\|=1} \|Xa\|_2 = \sigma_1 = \sqrt{\lambda(X^\mathrm{T} X)_1}$ (591) matrix 2-norm (*spectral norm*), *largest singular value*, [164, p.56]

$$\|Xa\|_2 \leq \|X\|_2 \|a\|_2 \qquad\qquad (2189)$$

For x a vector: $\|\delta(x)\|_2 = \|x\|_\infty$

$\|X\|_2^*$ $= \mathbf{1}^\mathrm{T} \sigma(X)$ *nuclear norm, dual spectral norm*

$\|X\|$ $= \|X\|_\mathrm{F}$ *Frobenius' matrix norm*

- **or** (when $\mathcal{K} = \mathbb{R}_+^n$) *vector A belongs to the interior of the nonnegative orthant* (each vector entry is positive, §2.3.1.1)

$\not\succ$	not positive definite
$\geq$	scalar inequality, *greater than or equal to*; comparison of scalars, or entrywise comparison of vectors or matrices with respect to $\mathbb{R}_+$
nonnegative	for $\alpha \in \mathbb{R}^n$, $\alpha \succeq 0$; *id est*, nonnegative entries when with respect to nonnegative orthant, represents vector on boundary of or interior to pointed closed convex cone $\mathcal{K}$
$>$	*greater than*
positive	for $\alpha \in \mathbb{R}^n$, $\alpha \succ 0$; *id est*, positive (nonzero) entries when with respect to nonnegative orthant, no vector on boundary of pointed closed convex cone $\mathcal{K}$
$\lfloor \ \rfloor$	floor function, $\lfloor x \rfloor$ is greatest integer not exceeding x
$\| \ \|$	entrywise absolute value of scalars, vectors, and matrices
log	natural (or Napierian) logarithm
det	matrix determinant
$\|x\|$	vector 2-norm or *Euclidean norm* $\|x\|_2$

$$\|x\|_\ell \quad = \sqrt[\ell]{\sum_{j=1}^n |x_j|^\ell} \qquad \text{vector } \ell\text{-norm for } \ell \geq 1 \qquad\qquad \text{(convex)}$$

$$\triangleq \sum_{j=1}^n |x_j|^\ell \qquad \text{vector } \ell\text{-norm for } 0 \leq \ell < 1 \qquad \text{(violates §3.2 } no.3)$$

$$\|x\|_\infty \quad = \max\{|x_j| \ \forall j\} \qquad \textit{infinity-norm}$$

$$\|x\|_2^2 \quad = x^\mathrm{T} x = \langle x, x \rangle \qquad \textit{Euclidean norm square}$$

$$\|x\|_1 \quad = \mathbf{1}^\mathrm{T} |x| \qquad\qquad \text{1-norm, } \textit{dual infinity-norm}$$

$$\|x\|_0 \quad = \mathbf{1}^\mathrm{T} |x|^0 \quad (0^0 \triangleq 0) \quad \text{0-norm (§4.5.1), } \textit{cardinality of vector } x \qquad (\text{card } x)$$

$$\|x\|_{\frac{n}{k}} \qquad\qquad\qquad\qquad k\text{-}\textit{largest norm (§3.2.2.1)}$$

$$\|X\|_2 \quad = \sup_{\|a\|=1} \|Xa\|_2 = \sigma_1 = \sqrt{\lambda(X^\mathrm{T} X)_1} \ (591) \text{ matrix 2-norm (}\textit{spectral norm}), \textit{ largest}$$
$$\textit{singular value, } [164, \text{p.56}]$$

$$\|Xa\|_2 \leq \|X\|_2 \|a\|_2 \qquad\qquad\qquad (2189)$$

For x a vector: $\|\delta(x)\|_2 = \|x\|_\infty$

$$\|X\|_2^* \quad = \mathbf{1}^\mathrm{T} \sigma(X) \qquad\qquad \textit{nuclear norm, dual spectral norm}$$

$$\|X\| \quad = \|X\|_\mathrm{F} \qquad\qquad \textit{Frobenius' matrix norm}$$

Bibliography

[1] Edwin A. Abbott. *Flatland: A Romance of Many Dimensions*. Seely & Co., London, sixth edition, 1884.

[2] Suliman Al-Homidan and Henry Wolkowicz. Approximate and exact completion problems for Euclidean distance matrices using semidefinite programming. *Linear Algebra and its Applications*, 406:109–141, September 2005.
http://citeseerx.ist.psu.edu/viewdoc/summary?doi=10.1.1.165.5577

[3] Faiz A. Al-Khayyal and James E. Falk. Jointly constrained biconvex programming. *Mathematics of Operations Research*, 8(2):273–286, May 1983.
http://www.convexoptimization.com/TOOLS/Falk.pdf

[4] Abdo Y. Alfakih. On the uniqueness of Euclidean distance matrix completions. *Linear Algebra and its Applications*, 370:1–14, 2003.

[5] Abdo Y. Alfakih. On the uniqueness of Euclidean distance matrix completions: the case of points in general position. *Linear Algebra and its Applications*, 397:265–277, 2005.

[6] Abdo Y. Alfakih, Amir Khandani, and Henry Wolkowicz. Solving Euclidean distance matrix completion problems via semidefinite programming. *Computational Optimization and Applications*, 12(1):13–30, January 1999.
http://citeseerx.ist.psu.edu/viewdoc/summary?doi=10.1.1.14.8307

[7] Abdo Y. Alfakih and Henry Wolkowicz. On the embeddability of weighted graphs in Euclidean spaces. Research Report CORR 98-12, Department of Combinatorics and Optimization, University of Waterloo, May 1998.
http://citeseerx.ist.psu.edu/viewdoc/summary?doi=10.1.1.50.1160
Erratum: p.394 herein.

[8] Abdo Y. Alfakih and Henry Wolkowicz. Matrix completion problems. In Henry Wolkowicz, Romesh Saigal, and Lieven Vandenberghe, editors, *Handbook of Semidefinite Programming: Theory, Algorithms, and Applications*, chapter 18. Kluwer, 2000.
http://www.convexoptimization.com/TOOLS/Handbook.pdf

[9] Abdo Y. Alfakih and Henry Wolkowicz. Two theorems on Euclidean distance matrices and Gale transform. *Linear Algebra and its Applications*, 340:149–154, 2002.

[10] Farid Alizadeh. *Combinatorial Optimization with Interior Point Methods and Semi-Definite Matrices*. PhD thesis, University of Minnesota, Computer Science Department, Minneapolis Minnesota USA, October 1991.

[11] Farid Alizadeh. Interior point methods in semidefinite programming with applications to combinatorial optimization. *SIAM Journal on Optimization*, 5(1):13–51, February 1995.

[12] Kurt Anstreicher and Henry Wolkowicz. On Lagrangian relaxation of quadratic matrix constraints. *SIAM Journal on Matrix Analysis and Applications*, 22(1):41–55, 2000.

[13] Howard Anton. *Elementary Linear Algebra*. Wiley, second edition, 1977.

[14] James Aspnes, David Goldenberg, and Yang Richard Yang. On the computational complexity of sensor network localization. In *Proceedings of the First International Workshop on Algorithmic*

Aspects of Wireless Sensor Networks (ALGOSENSORS), volume 3121 of *Lecture Notes in Computer Science*, pages 32–44, Turku Finland, July 2004. Springer-Verlag.
cs-www.cs.yale.edu/homes/aspnes/localization-abstract.html

[15] D. Avis and K. Fukuda. A pivoting algorithm for convex hulls and vertex enumeration of arrangements and polyhedra. *Discrete and Computational Geometry*, 8:295–313, 1992.
http://www.convexoptimization.com/TOOLS/AvisFukuda.pdf

[16] Christine Bachoc and Frank Vallentin. New upper bounds for kissing numbers from semidefinite programming. *Journal of the American Mathematical Society*, 21(3):909–924, July 2008.
http://arxiv.org/abs/math/0608426

[17] Mihály Bakonyi and Charles R. Johnson. The Euclidean distance matrix completion problem. *SIAM Journal on Matrix Analysis and Applications*, 16(2):646–654, April 1995.

[18] Keith Ball. An elementary introduction to modern convex geometry. In Silvio Levy, editor, *Flavors of Geometry*, volume 31, chapter 1, pages 1–58. MSRI Publications, 1997.
www.msri.org/publications/books/Book31/files/ball.pdf

[19] Richard G. Baraniuk. Compressive sensing [lecture notes]. *IEEE Signal Processing Magazine*, 24(4):118–121, July 2007.
http://www.convexoptimization.com/TOOLS/GeometryCardinality.pdf

[20] George Phillip Barker. Theory of cones. *Linear Algebra and its Applications*, 39:263–291, 1981.

[21] George Phillip Barker and David Carlson. Cones of diagonally dominant matrices. *Pacific Journal of Mathematics*, 57(1):15–32, 1975.

[22] George Phillip Barker and James Foran. Self-dual cones in Euclidean spaces. *Linear Algebra and its Applications*, 13:147–155, 1976.

[23] Dror Baron, Michael B. Wakin, Marco F. Duarte, Shriram Sarvotham, and Richard G. Baraniuk. Distributed compressed sensing. Technical Report ECE-0612, Rice University, Electrical and Computer Engineering Department, December 2006.
http://citeseerx.ist.psu.edu/viewdoc/summary?doi=10.1.1.85.4789

[24] Alexander I. Barvinok. Problems of distance geometry and convex properties of quadratic maps. *Discrete & Computational Geometry*, 13(2):189–202, 1995.

[25] Alexander I. Barvinok. A remark on the rank of positive semidefinite matrices subject to affine constraints. *Discrete & Computational Geometry*, 25(1):23–31, 2001.
http://citeseerx.ist.psu.edu/viewdoc/summary?doi=10.1.1.34.4627

[26] Alexander I. Barvinok. *A Course in Convexity*. American Mathematical Society, 2002.

[27] Alexander I. Barvinok. Approximating orthogonal matrices by permutation matrices. *Pure and Applied Mathematics Quarterly*, 2:943–961, 2006.

[28] Heinz H. Bauschke and Jonathan M. Borwein. On projection algorithms for solving convex feasibility problems. *SIAM Review*, 38(3):367–426, September 1996.

[29] Steven R. Bell. *The Cauchy Transform, Potential Theory, and Conformal Mapping*. CRC Press, 1992.

[30] Jean Bellissard and Bruno Iochum. Homogeneous and facially homogeneous self-dual cones. *Linear Algebra and its Applications*, 19:1–16, 1978.

[31] Richard Bellman and Ky Fan. On systems of linear inequalities in Hermitian matrix variables. In Victor L. Klee, editor, *Convexity*, volume VII of *Proceedings of Symposia in Pure Mathematics*, pages 1–11. American Mathematical Society, 1963.

[32] Adi Ben-Israel. Linear equations and inequalities on finite dimensional, real or complex, vector spaces: A unified theory. *Journal of Mathematical Analysis and Applications*, 27:367–389, 1969.

[33] Adi Ben-Israel. Motzkin's transposition theorem, and the related theorems of Farkas, Gordan and Stiemke. In Michiel Hazewinkel, editor, *Encyclopaedia of Mathematics*. Springer-Verlag, 2001.
http://www.convexoptimization.com/TOOLS/MOTZKIN.pdf

[34] Aharon Ben-Tal and Arkadi Nemirovski. *Lectures on Modern Convex Optimization: Analysis, Algorithms, and Engineering Applications*. SIAM, 2001.

[35] Aharon Ben-Tal and Arkadi Nemirovski. Non-Euclidean restricted memory level method for large-scale convex optimization. *Mathematical Programming*, 102(3):407–456, January 2005.
http://www2.isye.gatech.edu/~nemirovs/Bundle-Mirror_rev_fin.pdf

[36] John J. Benedetto and Paulo J.S.G. Ferreira editors. *Modern Sampling Theory: Mathematics and Applications*. Birkhäuser, 2001.

[37] Christian R. Berger, Javier Areta, Krishna Pattipati, and Peter Willett. Compressed sensing − A look beyond linear programming. In *Proceedings of the IEEE International Conference on Acoustics, Speech, and Signal Processing*, pages 3857–3860, 2008.

[38] Radu Berinde, Anna C. Gilbert, Piotr Indyk, Howard J. Karloff, and Martin J. Strauss. Combining geometry and combinatorics: A unified approach to sparse signal recovery. In 46th *Annual Allerton Conference on Communication, Control, and Computing*, pages 798–805. IEEE, September 2008.
http://arxiv.org/abs/0804.4666

[39] Abraham Berman. *Cones, Matrices, and Mathematical Programming*, volume 79 of *Lecture Notes in Economics and Mathematical Systems*. Springer-Verlag, 1973.

[40] Abraham Berman and Naomi Shaked-Monderer. *Completely Positive Matrices*. World Scientific, 2003.

[41] Dimitri P. Bertsekas. *Nonlinear Programming*. Athena Scientific, second edition, 1999.

[42] Dimitri P. Bertsekas, Angelia Nedić, and Asuman E. Ozdaglar. *Convex Analysis and Optimization*. Athena Scientific, 2003.

[43] Rajendra Bhatia. *Matrix Analysis*. Springer-Verlag, 1997.

[44] Pratik Biswas, Tzu-Chen Liang, Kim-Chuan Toh, Yinyu Ye, and Ta-Chung Wang. Semidefinite programming approaches for sensor network localization with noisy distance measurements. *IEEE Transactions on Automation Science and Engineering*, 3(4):360–371, October 2006.

[45] Pratik Biswas, Tzu-Chen Liang, Ta-Chung Wang, and Yinyu Ye. Semidefinite programming based algorithms for sensor network localization. *ACM Transactions on Sensor Networks*, 2(2):188–220, May 2006.
http://www.stanford.edu/~yyye/combined_rev3.pdf

[46] Pratik Biswas, Kim-Chuan Toh, and Yinyu Ye. A distributed SDP approach for large-scale noisy anchor-free graph realization with applications to molecular conformation. *SIAM Journal on Scientific Computing*, 30(3):1251–1277, March 2008.
http://www.stanford.edu/~yyye/SISC-molecule-revised-2.pdf

[47] Pratik Biswas and Yinyu Ye. Semidefinite programming for ad hoc wireless sensor network localization. In *Proceedings of the Third International Symposium on Information Processing in Sensor Networks* (IPSN), pages 46–54. IEEE, April 2004.
http://www.stanford.edu/~yyye/adhocn4.pdf

[48] Suddhendu Biswas. *Textbook of Matrix Algebra*. PHI, 3rd edition, 2012.
http://books.google.com/books?id=FxoA6Q2UJKwC
http://www.convexoptimization.com/TOOLS/Vitulli.pdf

[49] Åke Björck and Tommy Elfving. Accelerated projection methods for computing pseudoinverse solutions of systems of linear equations. *BIT Numerical Mathematics*, 19(2):145–163, June 1979.
http://www.convexoptimization.com/TOOLS/BIT.pdf

[50] Leonard M. Blumenthal. A note on the four-point property. *Bulletin of the American Mathematical Society*, 39(6):423–426, 1933.
http://www.convexoptimization.com/TOOLS/euclid.pdf

[51] Leonard M. Blumenthal. *Theory and Applications of Distance Geometry*. Oxford University Press, 1953.

[52] Radu Ioan Boţ, Ernö Robert Csetnek, and Gert Wanka. Regularity conditions via quasi-relative interior in convex programming. *SIAM Journal on Optimization*, 19(1):217–233, 2008.
http://www.convexoptimization.com/TOOLS/Wanka.pdf

[53] A. W. Bojanczyk and A. Lutoborski. The Procrustes problem for orthogonal Stiefel matrices. *SIAM Journal on Scientific Computing*, 21(4):1291–1304, December 1999.
http://citeseerx.ist.psu.edu/viewdoc/summary?doi=10.1.1.15.9063

[54] Ingwer Borg and Patrick Groenen. *Modern Multidimensional Scaling.* Springer-Verlag, 1997.

[55] Jonathan M. Borwein and Heinz Bauschke. Projection algorithms and monotone operators, 1998. http://www.cecm.sfu.ca/~jborwein/projections4.pdf

[56] Jonathan M. Borwein and Adrian S. Lewis. *Convex Analysis and Nonlinear Optimization: Theory and Examples.* Springer-Verlag, 2000. http://www.convexoptimization.com/TOOLS/Borwein.pdf

[57] Jonathan M. Borwein and Warren B. Moors. Stability of closedness of convex cones under linear mappings. *Journal of Convex Analysis*, 16(3):699–705, 2009. http://www.convexoptimization.com/TOOLS/BorweinMoors.pdf

[58] Richard Bouldin. The pseudo-inverse of a product. *SIAM Journal on Applied Mathematics*, 24(4):489–495, June 1973. http://www.convexoptimization.com/TOOLS/Pseudoinverse.pdf

[59] Stephen Boyd and Jon Dattorro. Alternating projections, 2003. http://www.stanford.edu/class/ee392o/alt_proj.pdf

[60] Stephen Boyd, Laurent El Ghaoui, Eric Feron, and Venkataramanan Balakrishnan. *Linear Matrix Inequalities in System and Control Theory.* SIAM, 1994.

[61] Stephen Boyd, Seung-Jean Kim, Lieven Vandenberghe, and Arash Hassibi. A tutorial on geometric programming. *Optimization and Engineering*, 8(1):1389–4420, March 2007. http://www.stanford.edu/~boyd/papers/gp_tutorial.html

[62] Stephen Boyd and Lieven Vandenberghe. *Convex Optimization.* Cambridge University Press, 2004. http://www.stanford.edu/~boyd/cvxbook

[63] James P. Boyle and Richard L. Dykstra. A method for finding projections onto the intersection of convex sets in Hilbert spaces. In R. Dykstra, T. Robertson, and F. T. Wright, editors, *Advances in Order Restricted Statistical Inference*, pages 28–47. Springer-Verlag, 1986.

[64] Lev M. Brègman. The method of successive projection for finding a common point of convex sets. *Soviet Mathematics*, 162(3):487–490, 1965. AMS translation of Doklady Akademii Nauk SSSR, 6:688-692.

[65] Lev M. Brègman, Yair Censor, Simeon Reich, and Yael Zepkowitz-Malachi. Finding the projection of a point onto the intersection of convex sets via projections onto halfspaces. *Journal of Approximation Theory*, 124(2):194–218, October 2003. http://www.optimization-online.org/DB_FILE/2003/06/669.pdf

[66] Mike Brookes. Matrix reference manual: Matrix calculus, 2002. http://www.ee.ic.ac.uk/hp/staff/dmb/matrix/intro.html

[67] Richard A. Brualdi. *Combinatorial Matrix Classes.* Cambridge University Press, 2006.

[68] Jian-Feng Cai, Emmanuel J. Candès, and Zuowei Shen. A singular value thresholding algorithm for matrix completion, October 2008. http://arxiv.org/abs/0810.3286

[69] Emmanuel J. Candès. Compressive sampling. In *Proceedings of the International Congress of Mathematicians*, volume III, pages 1433–1452, Madrid, August 2006. European Mathematical Society. http://www.acm.caltech.edu/~emmanuel/papers/CompressiveSampling.pdf

[70] Emmanuel J. Candès and Justin K. Romberg. ℓ_1-MAGIC : Recovery of sparse signals via convex programming, October 2005. http://users.ece.gatech.edu/~justin/l1magic/downloads/l1magic.pdf

[71] Emmanuel J. Candès, Justin K. Romberg, and Terence Tao. Robust uncertainty principles: Exact signal reconstruction from highly incomplete frequency information. *IEEE Transactions on Information Theory*, 52(2):489–509, February 2006. http://arxiv.org/abs/math.NA/0409186 (2004 DRAFT) http://www.acm.caltech.edu/~emmanuel/papers/ExactRecovery.pdf

[72] Emmanuel J. Candès, Michael B. Wakin, and Stephen P. Boyd. Enhancing sparsity by reweighted ℓ_1 minimization. *Journal of Fourier Analysis and Applications*, 14(5-6):877–905, October 2008.
http://arxiv.org/abs/0711.1612v1

[73] Michael Carter, Holly Hui Jin, Michael Saunders, and Yinyu Ye. SPASELOC: An adaptive subproblem algorithm for scalable wireless sensor network localization. *SIAM Journal on Optimization*, 17(4):1102–1128, December 2006.
http://www.convexoptimization.com/TOOLS/JinSIAM.pdf
Erratum: p.272 herein.

[74] Lawrence Cayton and Sanjoy Dasgupta. Robust Euclidean embedding. In *Proceedings of the $23^{\rm rd}$ International Conference on Machine Learning* (ICML), Pittsburgh Pennsylvania USA, 2006.
http://cseweb.ucsd.edu/~lcayton/robEmb.pdf

[75] Yves Chabrillac and Jean-Pierre Crouzeix. Definiteness and semidefiniteness of quadratic forms revisited. *Linear Algebra and its Applications*, 63:283–292, 1984.

[76] Manmohan K. Chandraker, Sameer Agarwal, Fredrik Kahl, David Nistér, and David J. Kriegman. Autocalibration via rank-constrained estimation of the absolute quadric. In *Proceedings of the IEEE Computer Society Conference on Computer Vision and Pattern Recognition*, June 2007.
http://vision.ucsd.edu/~manu/cvpr07_quadric.pdf

[77] Rick Chartrand. Exact reconstruction of sparse signals via nonconvex minimization. *IEEE Signal Processing Letters*, 14(10):707–710, October 2007.
math.lanl.gov/Research/Publications/Docs/chartrand-2007-exact.pdf

[78] Chi-Tsong Chen. *Linear System Theory and Design.* Oxford University Press, 1999.

[79] Scott ShaoBing Chen, David L. Donoho, and Michael A. Saunders. Atomic decomposition by basis pursuit. *SIAM Journal on Scientific Computing*, 20(1):33–61, 1998.

[80] Ward Cheney and Allen A. Goldstein. Proximity maps for convex sets. *Proceedings of the American Mathematical Society*, 10:448–450, 1959. Erratum: p.640 herein.

[81] Mung Chiang. *Geometric Programming for Communication Systems.* Now, 2005.

[82] Stéphane Chrétien. An alternating l1 approach to the compressed sensing problem, September 2009.
http://arxiv.org/PS_cache/arxiv/pdf/0809/0809.0660v3.pdf

[83] Steven Chu. Autobiography from *Les Prix Nobel*, 1997.
nobelprize.org/nobel_prizes/physics/laureates/1997/chu-autobio.html

[84] Ruel V. Churchill and James Ward Brown. *Complex Variables and Applications.* McGraw-Hill, fifth edition, 1990.

[85] Jon F. Claerbout and Francis Muir. Robust modeling of erratic data. *Geophysics*, 38:826–844, 1973.

[86] J. H. Conway and N. J. A. Sloane. *Sphere Packings, Lattices and Groups.* Springer-Verlag, third edition, 1999.

[87] John B. Conway. *A Course in Functional Analysis.* Springer-Verlag, second edition, 1990.

[88] Jose A. Costa, Neal Patwari, and Alfred O. Hero III. Distributed weighted-multidimensional scaling for node localization in sensor networks. *ACM Transactions on Sensor Networks*, 2(1):39–64, February 2006.
http://www.convexoptimization.com/TOOLS/Costa.pdf

[89] Richard W. Cottle, Jong-Shi Pang, and Richard E. Stone. *The Linear Complementarity Problem.* Academic Press, 1992.

[90] G. M. Crippen and T. F. Havel. *Distance Geometry and Molecular Conformation.* Wiley, 1988.

[91] Frank Critchley. *Multidimensional scaling: a critical examination and some new proposals.* PhD thesis, University of Oxford, Nuffield College, 1980.

[92] Frank Critchley. On certain linear mappings between inner-product and squared-distance matrices. *Linear Algebra and its Applications*, 105:91–107, 1988.
http://www.convexoptimization.com/TOOLS/Critchley.pdf

[93] Ronald E. Crochiere and Lawrence R. Rabiner. *Multirate Digital Signal Processing.* Prentice-Hall, 1983.

[94] Lawrence B. Crowell. *Quantum Fluctuations of Spacetime*. World Scientific, 2005.

[95] Joachim Dahl, Bernard H. Fleury, and Lieven Vandenberghe. Approximate maximum-likelihood estimation using semidefinite programming. In *Proceedings of the IEEE International Conference on Acoustics, Speech, and Signal Processing*, volume VI, pages 721–724, April 2003.
http://www.convexoptimization.com/TOOLS/Dahl.pdf

[96] George B. Dantzig. *Linear Programming and Extensions*.
Princeton University Press, 1963 (Rand Corporation).

[97] Alexandre d'Aspremont, Laurent El Ghaoui, Michael I. Jordan, and Gert R. G. Lanckriet. A direct formulation for sparse PCA using semidefinite programming. In Lawrence K. Saul, Yair Weiss, and Léon Bottou, editors, *Advances in Neural Information Processing Systems 17*, pages 41–48. MIT Press, Cambridge Massachusetts USA, 2005.
http://books.nips.cc/papers/files/nips17/NIPS2004_0645.pdf

[98] Jon Dattorro. Constrained least squares fit of a filter bank to an arbitrary magnitude frequency response, 1991.
http://ccrma.stanford.edu/~dattorro/Hearing.htm
http://ccrma.stanford.edu/~dattorro/PhiLS.pdf

[99] Jon Dattorro. Effect Design, part 1: Reverberator and other filters. *Journal of the Audio Engineering Society*, 45(9):660–684, September 1997.
http://www.convexoptimization.com/TOOLS/EffectDesignPart1.pdf

[100] Jon Dattorro. Effect Design, part 2: Delay-line modulation and chorus. *Journal of the Audio Engineering Society*, 45(10):764–788, October 1997.
http://www.convexoptimization.com/TOOLS/EffectDesignPart2.pdf

[101] Jon Dattorro. Effect Design, part 3: Oscillators: Sinusoidal and pseudonoise. *Journal of the Audio Engineering Society*, 50(3):115–146, March 2002.
http://www.convexoptimization.com/TOOLS/EffectDesignPart3.pdf

[102] Jon Dattorro. Equality relating Euclidean distance cone to positive semidefinite cone. *Linear Algebra and its Applications*, 428(11+12):2597–2600, June 2008.
http://www.convexoptimization.com/TOOLS/DattorroLAA.pdf

[103] Joel Dawson, Stephen Boyd, Mar Hershenson, and Thomas Lee. Optimal allocation of local feedback in multistage amplifiers via geometric programming. *IEEE Transactions on Circuits and Systems I: Fundamental Theory and Applications*, 48(1):1–11, January 2001.
http://www.stanford.edu/~boyd/papers/fdbk_alloc.html

[104] Etienne de Klerk. *Aspects of Semidefinite Programming: Interior Point Algorithms and Selected Applications*. Kluwer Academic Publishers, 2002.

[105] Jan de Leeuw. Fitting distances by least squares. UCLA Statistics Series Technical Report No. 130, Interdivisional Program in Statistics, UCLA, Los Angeles California USA, 1993.
http://citeseerx.ist.psu.edu/viewdoc/summary?doi=10.1.1.50.7472

[106] Jan de Leeuw. Multidimensional scaling. In *International Encyclopedia of the Social & Behavioral Sciences*, pages 13512–13519. Elsevier, 2004.
http://www.convexoptimization.com/TOOLS/leeuw274.pdf

[107] Jan de Leeuw. Unidimensional scaling. In Brian S. Everitt and David C. Howell, editors, *Encyclopedia of Statistics in Behavioral Science*, volume 4, pages 2095–2097. Wiley, 2005.
http://www.convexoptimization.com/TOOLS/deLeeuw2005.pdf

[108] Jan de Leeuw and Willem Heiser. Theory of multidimensional scaling. In P. R. Krishnaiah and L. N. Kanal, editors, *Handbook of Statistics*, volume 2, chapter 13, pages 285–316. North-Holland Publishing, Amsterdam, 1982.

[109] Erik D. Demaine and Martin L. Demaine. Jigsaw puzzles, edge matching, and polyomino packing: Connections and complexity. *Graphs and Combinatorics*, 23(supplement 1):195–208, Springer-Verlag, Tokyo, June 2007.

[110] Erik D. Demaine, Francisco Gomez-Martin, Henk Meijer, David Rappaport, Perouz Taslakian, Godfried T. Toussaint, Terry Winograd, and David R. Wood. The distance geometry of music. *Computational Geometry: Theory and Applications*, 42(5):429–454, July 2009.
http://www.convexoptimization.com/TOOLS/dgm.pdf

[111] John R. D'Errico. DERIVEST, 2007.
http://www.convexoptimization.com/TOOLS/DERIVEST.pdf

[112] Frank Deutsch. *Best Approximation in Inner Product Spaces*. Springer-Verlag, 2001.

[113] Frank Deutsch and Hein Hundal. The rate of convergence of Dykstra's cyclic projections algorithm: The polyhedral case. *Numerical Functional Analysis and Optimization*, 15:537–565, 1994.

[114] Frank Deutsch and Peter H. Maserick. Applications of the Hahn-Banach theorem in approximation theory. *SIAM Review*, 9(3):516–530, July 1967.

[115] Frank Deutsch, John H. McCabe, and George M. Phillips. Some algorithms for computing best approximations from convex cones. *SIAM Journal on Numerical Analysis*, 12(3):390–403, June 1975.

[116] Michel Marie Deza and Monique Laurent. *Geometry of Cuts and Metrics*. Springer-Verlag, 1997.

[117] Sudhakar Waman Dharmadhikari and Kumar Joag-Dev. *Unimodality, Convexity, and Applications*. Academic Press, 1988.

[118] Carolyn Pillers Dobler. A matrix approach to finding a set of generators and finding the polar (dual) of a class of polyhedral cones. *SIAM Journal on Matrix Analysis and Applications*, 15(3):796–803, July 1994. Erratum: p.181 herein.

[119] Elizabeth D. Dolan, Robert Fourer, Jorge J. Moré, and Todd S. Munson. Optimization on the NEOS server. *SIAM News*, 35(6):4,8,9, August 2002.

[120] Bruce Randall Donald. 3-D structure in chemistry and molecular biology, 1998.
http://www.cs.duke.edu/brd/Teaching/Previous/Bio

[121] David L. Donoho. Neighborly polytopes and sparse solution of underdetermined linear equations. Technical Report 2005-04, Stanford University, Department of Statistics, January 2005.
www-stat.stanford.edu/~donoho/Reports/2005/NPaSSULE-01-28-05.pdf

[122] David L. Donoho. Compressed sensing. *IEEE Transactions on Information Theory*, 52(4):1289–1306, April 2006.
www-stat.stanford.edu/~donoho/Reports/2004/CompressedSensing091604.pdf

[123] David L. Donoho. High-dimensional centrally symmetric polytopes with neighborliness proportional to dimension. *Discrete & Computational Geometry*, 35(4):617–652, May 2006.

[124] David L. Donoho, Michael Elad, and Vladimir Temlyakov. Stable recovery of sparse overcomplete representations in the presence of noise. *IEEE Transactions on Information Theory*, 52(1):6–18, January 2006.
www-stat.stanford.edu/~donoho/Reports/2004/StableSparse-Donoho-etal.pdf

[125] David L. Donoho and Philip B. Stark. Uncertainty principles and signal recovery. *SIAM Journal on Applied Mathematics*, 49(3):906–931, June 1989.

[126] David L. Donoho and Jared Tanner. Neighborliness of randomly projected simplices in high dimensions. *Proceedings of the National Academy of Sciences*, 102(27):9452–9457, July 2005.

[127] David L. Donoho and Jared Tanner. Sparse nonnegative solution of underdetermined linear equations by linear programming. *Proceedings of the National Academy of Sciences*, 102(27):9446–9451, July 2005.

[128] David L. Donoho and Jared Tanner. Counting faces of randomly projected polytopes when the projection radically lowers dimension. *Journal of the American Mathematical Society*, 22(1):1–53, January 2009.

[129] Miguel Nuno Ferreira Fialho dos Anjos. *New Convex Relaxations for the Maximum Cut and VLSI Layout Problems*. PhD thesis, University of Waterloo, Ontario Canada, Department of Combinatorics and Optimization, 2001.
http://etd.uwaterloo.ca/etd/manjos2001.pdf

[130] John Duchi, Shai Shalev-Shwartz, Yoram Singer, and Tushar Chandra. Efficient projections onto the ℓ_1-ball for learning in high dimensions. In *Proceedings of the 25th International Conference on Machine Learning* (ICML), pages 272–279, Helsinki Finland, July 2008. Association for Computing Machinery (ACM).
http://icml2008.cs.helsinki.fi/papers/361.pdf

[131] Richard L. Dykstra. An algorithm for restricted least squares regression. *Journal of the American Statistical Association*, 78(384):837–842, 1983.

[132] Peter J. Eccles. *An Introduction to Mathematical Reasoning: numbers, sets and functions.* Cambridge University Press, 1997.

[133] Carl Eckart and Gale Young. The approximation of one matrix by another of lower rank. *Psychometrika*, 1(3):211–218, September 1936.
www.convexoptimization.com/TOOLS/eckart&young.1936.pdf

[134] Alan Edelman, Tomás A. Arias, and Steven T. Smith. The geometry of algorithms with orthogonality constraints. *SIAM Journal on Matrix Analysis and Applications*, 20(2):303–353, 1998.

[135] John J. Edgell. Graphics calculator applications on 4-D constructs, 1996.
http://archives.math.utk.edu/ICTCM/EP-9/C47/pdf/paper.pdf

[136] Ivar Ekeland and Roger Témam. *Convex Analysis and Variational Problems.* SIAM, 1999.

[137] Julius Farkas. Theorie der einfachen Ungleichungen. *Journal für die reine und angewandte Mathematik*, 124:1–27, 1902.
http://gdz.sub.uni-goettingen.de/no_cache/dms/load/img/?IDDOC=261361

[138] Maryam Fazel. *Matrix Rank Minimization with Applications.* PhD thesis, Stanford University, Department of Electrical Engineering, March 2002.
http://faculty.washington.edu/mfazel/thesis-final.pdf

[139] Maryam Fazel, Haitham Hindi, and Stephen P. Boyd. A rank minimization heuristic with application to minimum order system approximation. In *Proceedings of the American Control Conference*, volume 6, pages 4734–4739. American Automatic Control Council (AACC), June 2001.
http://www.stanford.edu/~boyd/papers/nucnorm.html

[140] Maryam Fazel, Haitham Hindi, and Stephen P. Boyd. Log-det heuristic for matrix rank minimization with applications to Hankel and Euclidean distance matrices. In *Proceedings of the American Control Conference*, volume 3, pages 2156–2162. American Automatic Control Council (AACC), June 2003.
http://faculty.washington.edu/mfazel/acc03_final.pdf

[141] Maryam Fazel, Haitham Hindi, and Stephen P. Boyd. Rank minimization and applications in system theory. In *Proceedings of the American Control Conference*, volume 4, pages 3273–3278. American Automatic Control Council (AACC), June 2004.
http://faculty.washington.edu/mfazel/acc04-tutorial.pdf

[142] J. T. Feddema, R. H. Byrne, J. J. Harrington, D. M. Kilman, C. L. Lewis, R. D. Robinett, B. P. Van Leeuwen, and J. G. Young. Advanced mobile networking, sensing, and controls. Sandia Report SAND2005-1661, Sandia National Laboratories, Albuquerque New Mexico USA, March 2005.
www.prod.sandia.gov/cgi-bin/techlib/access-control.pl/2005/051661.pdf

[143] César Fernández, Thomas Szyperski, Thierry Bruyère, Paul Ramage, Egon Mösinger, and Kurt Wüthrich. NMR solution structure of the pathogenesis-related protein P14a. *Journal of Molecular Biology*, 266:576–593, 1997.

[144] Richard Phillips Feynman, Robert B. Leighton, and Matthew L. Sands. *The Feynman Lectures on Physics: Commemorative Issue*, volume I. Addison-Wesley, 1989.

[145] Anthony V. Fiacco and Garth P. McCormick. *Nonlinear Programming: Sequential Unconstrained Minimization Techniques.* SIAM, 1990.

[146] Önder Filiz and Aylin Yener. Rank constrained temporal-spatial filters for CDMA systems with base station antenna arrays. In *Proceedings of the Johns Hopkins University Conference on Information Sciences and Systems*, March 2003.
http://wcan.ee.psu.edu/papers/filiz-yener_ciss03.pdf

[147] P. A. Fillmore and J. P. Williams. Some convexity theorems for matrices. *Glasgow Mathematical Journal*, 12:110–117, 1971.

[148] Paul Finsler. Über das Vorkommen definiter und semidefiniter Formen in Scharen quadratischer Formen. *Commentarii Mathematici Helvetici*, 9:188–192, 1937.

[149] Anders Forsgren, Philip E. Gill, and Margaret H. Wright. Interior methods for nonlinear optimization. *SIAM Review*, 44(4):525–597, 2002.

[150] Shmuel Friedland and Anatoli Torokhti. Generalized rank-constrained matrix approximations. *SIAM Journal on Matrix Analysis and Applications*, 29(2):656–659, March 2007.
http://arxiv.org/pdf/math.OC/0603674.pdf

[151] Ragnar Frisch. The multiplex method for linear programming. *SANKHYĀ: The Indian Journal of Statistics*, 18(3/4):329–362, September 1957.
http://www.convexoptimization.com/TOOLS/Frisch57.pdf

[152] Norbert Gaffke and Rudolf Mathar. A cyclic projection algorithm via duality. *Metrika*, 36:29–54, 1989.

[153] Jérôme Galtier. Semi-definite programming as a simple extension to linear programming: convex optimization with queueing, equity and other telecom functionals. In *3ème Rencontres Francophones sur les Aspects Algorithmiques des Télécommunications* (AlgoTel), pages 21–28. INRIA, Saint Jean de Luz FRANCE, May 2001.
http://citeseerx.ist.psu.edu/viewdoc/summary?doi=10.1.1.8.7378

[154] Laurent El Ghaoui. EE 227A: Convex Optimization and Applications, Lecture 11 − October 3. University of California, Berkeley, Fall 2006. Scribe: Nikhil Shetty.
http://www.convexoptimization.com/TOOLS/Ghaoui.pdf

[155] Laurent El Ghaoui and Silviu-Iulian Niculescu, editors. *Advances in Linear Matrix Inequality Methods in Control*. SIAM, 2000.

[156] Philip E. Gill, Walter Murray, and Margaret H. Wright. *Practical Optimization*. Academic Press, 1981.

[157] Philip E. Gill, Walter Murray, and Margaret H. Wright. *Numerical Linear Algebra and Optimization*, volume 1. Addison-Wesley, 1991.

[158] James Gleik. *Isaac Newton*. Pantheon Books, 2003.

[159] W. Glunt, Tom L. Hayden, S. Hong, and J. Wells. An alternating projection algorithm for computing the nearest Euclidean distance matrix. *SIAM Journal on Matrix Analysis and Applications*, 11(4):589–600, 1990.

[160] K. Goebel and W. A. Kirk. *Topics in Metric Fixed Point Theory*. Cambridge University Press, 1990.

[161] Michel X. Goemans and David P. Williamson. Improved approximation algorithms for maximum cut and satisfiability problems using semidefinite programming. *Journal of the Association for Computing Machinery*, 42(6):1115–1145, November 1995.
http://www.convexoptimization.com/TOOLS/maxcut-jacm.pdf

[162] D. Goldfarb and K. Scheinberg. Interior point trajectories in semidefinite programming. *SIAM Journal on Optimization*, 8(4):871–886, 1998.

[163] Gene Golub and William Kahan. Calculating the singular values and pseudo-inverse of a matrix. *SIAM Journal on Numerical Analysis,* Series **B**, 2(2):205–224, 1965.
http://www.convexoptimization.com/TOOLS/GolubKahan.pdf

[164] Gene H. Golub and Charles F. Van Loan. *Matrix Computations*. Johns Hopkins, third edition, 1996.

[165] P. Gordan. Ueber die auflösung linearer gleichungen mit reellen coefficienten. *Mathematische Annalen*, 6:23–28, 1873.

[166] Irina F. Gorodnitsky and Bhaskar D. Rao. Sparse signal reconstruction from limited data using FOCUSS: A re-weighted minimum norm algorithm. *IEEE Transactions on Signal Processing*, 45(3):600–616, March 1997.
http://www.convexoptimization.com/TOOLS/focuss.pdf

[167] John Clifford Gower. Adding a point to vector diagrams in multivariate analysis. *Biometrika*, 55(3):582–585, 1968.
http://www.convexoptimization.com/TOOLS/GOWER1968.pdf

[168] John Clifford Gower. Euclidean distance geometry. *The Mathematical Scientist*, 7:1–14, 1982.
http://www.convexoptimization.com/TOOLS/Gower2.pdf

[169] John Clifford Gower. Properties of Euclidean and non-Euclidean distance matrices. *Linear Algebra and its Applications*, 67:81–97, 1985.
      http://www.convexoptimization.com/TOOLS/Gower1.pdf

[170] John Clifford Gower and Garmt B. Dijksterhuis. *Procrustes Problems*. Oxford University Press, 2004.

[171] John Clifford Gower and David J. Hand. *Biplots*. Chapman & Hall, 1996.

[172] Alexander Graham. *Kronecker Products and Matrix Calculus with Applications*. Ellis Horwood Limited, 1981.

[173] Michael Grant, Stephen Boyd, and Yinyu Ye.
      cvx: MATLAB software for disciplined convex programming, 2007.
      http://www.stanford.edu/~boyd/cvx

[174] Robert M. Gray. Toeplitz and circulant matrices: A review. *Foundations and Trends in Communications and Information Theory*, 2(3):155–239, 2006.
      http://www-ee.stanford.edu/~gray/toeplitz.pdf

[175] T. N. E. Greville. Note on the generalized inverse of a matrix product. *SIAM Review*, 8:518–521, 1966.

[176] Rémi Gribonval and Morten Nielsen. Sparse representations in unions of bases. *IEEE Transactions on Information Theory*, 49(12):3320–3325, December 2003.
      http://people.math.aau.dk/~mnielsen/reprints/sparse_unions.pdf

[177] Rémi Gribonval and Morten Nielsen. Highly sparse representations from dictionaries are unique and independent of the sparseness measure. *Applied and Computational Harmonic Analysis*, 22(3):335–355, May 2007.
      http://www.convexoptimization.com/TOOLS/R-2003-16.pdf

[178] Karolos M. Grigoriadis and Eric B. Beran. Alternating projection algorithms for linear matrix inequalities problems with rank constraints. In Laurent El Ghaoui and Silviu-Iulian Niculescu, editors, *Advances in Linear Matrix Inequality Methods in Control*, chapter 13, pages 251–267. SIAM, 2000.

[179] Peter Gritzmann and Victor Klee. On the complexity of some basic problems in computational convexity: II. volume and mixed volumes. Technical Report TR:94-31, DIMACS, Rutgers University, 1994.
      http://dimacs.rutgers.edu/TechnicalReports/TechReports/1994/94-31.ps

[180] Peter Gritzmann and Victor Klee. On the complexity of some basic problems in computational convexity: II. volume and mixed volumes. In T. Bisztriczky, P. McMullen, R. Schneider, and A. Ivić Weiss, editors, *Polytopes: Abstract, Convex and Computational*, pages 373–466. Kluwer Academic Publishers, 1994.

[181] L. G. Gubin, B. T. Polyak, and E. V. Raik. The method of projections for finding the common point of convex sets. *U.S.S.R. Computational Mathematics and Mathematical Physics*, 7(6):1–24, 1967.

[182] Osman Güler and Yinyu Ye. Convergence behavior of interior-point algorithms. *Mathematical Programming*, 60(2):215–228, 1993.

[183] P. R. Halmos. Positive approximants of operators. *Indiana University Mathematics Journal*, 21:951–960, 1972.

[184] Shih-Ping Han. A successive projection method. *Mathematical Programming*, 40:1–14, 1988.

[185] Godfrey H. Hardy, John E. Littlewood, and George Pólya. *Inequalities*. Cambridge University Press, second edition, 1952.

[186] Arash Hassibi and Mar Hershenson. Automated optimal design of switched-capacitor filters. In *Proceedings of the Conference on Design, Automation, and Test in Europe*, page 1111, March 2002.

[187] Johan Håstad. Some optimal inapproximability results. In *Proceedings of the Twenty-Ninth Annual ACM Symposium on Theory of Computing* (STOC), pages 1–10, El Paso Texas USA, 1997. Association for Computing Machinery (ACM).
      http://citeseerx.ist.psu.edu/viewdoc/summary?doi=10.1.1.16.183 , 2002.

[188] Johan Håstad. Some optimal inapproximability results. *Journal of the Association for Computing Machinery*, 48(4):798–859, July 2001.

[189] Timothy F. Havel and Kurt Wüthrich. An evaluation of the combined use of nuclear magnetic resonance and distance geometry for the determination of protein conformations in solution. *Journal of Molecular Biology*, 182:281–294, 1985.

[190] Tom L. Hayden and Jim Wells. Approximation by matrices positive semidefinite on a subspace. *Linear Algebra and its Applications*, 109:115–130, 1988.

[191] Tom L. Hayden, Jim Wells, Wei-Min Liu, and Pablo Tarazaga. The cone of distance matrices. *Linear Algebra and its Applications*, 144:153–169, 1991.

[192] Uwe Helmke and John B. Moore. *Optimization and Dynamical Systems*. Springer-Verlag, 1994.

[193] Bruce Hendrickson. Conditions for unique graph realizations. *SIAM Journal on Computing*, 21(1):65–84, February 1992.

[194] T. Herrmann, Peter Güntert, and Kurt Wüthrich. Protein NMR structure determination with automated NOE assignment using the new software CANDID and the torsion angle dynamics algorithm DYANA. *Journal of Molecular Biology*, 319(1):209–227, May 2002.

[195] Mar Hershenson. Design of pipeline analog-to-digital converters via geometric programming. In *Proceedings of the IEEE/ACM International Conference on Computer Aided Design* (ICCAD), pages 317–324, November 2002.

[196] Mar Hershenson. Efficient description of the design space of analog circuits. In *Proceedings of the* 40th *ACM/IEEE Design Automation Conference*, pages 970–973, June 2003.

[197] Mar Hershenson, Stephen Boyd, and Thomas Lee. Optimal design of a CMOS OpAmp via geometric programming. *IEEE Transactions on Computer-Aided Design of Integrated Circuits and Systems*, 20(1):1–21, January 2001.
http://www.stanford.edu/~boyd/papers/opamp.html

[198] Mar Hershenson, Dave Colleran, Arash Hassibi, and Navraj Nandra. Synthesizable full custom mixed-signal IP. *Electronics Design Automation Consortium* (EDA), 2002.
http://www.eda.org/edps/edp02/PAPERS/edp02-s6_2.pdf

[199] Mar Hershenson, Sunderarajan S. Mohan, Stephen Boyd, and Thomas Lee. Optimization of inductor circuits via geometric programming. In *Proceedings of the* 36th *ACM/IEEE Design Automation Conference*, pages 994–998, June 1999.
http://www.stanford.edu/~boyd/papers/inductor_opt.html

[200] Nick Higham. Matrix Procrustes problems, 1995.
http://www.convexoptimization.com/TOOLS/procrust94.ps
Lecture notes.

[201] Richard D. Hill and Steven R. Waters. On the cone of positive semidefinite matrices. *Linear Algebra and its Applications*, 90:81–88, 1987.

[202] Jean-Baptiste Hiriart-Urruty. Ensembles de Tchebychev *vs.* ensembles convexes: l'état de la situation vu via l'analyse convexe non lisse. *Annales des Sciences Mathématiques du Québec*, 22(1):47–62, 1998.

[203] Jean-Baptiste Hiriart-Urruty. Global optimality conditions in maximizing a convex quadratic function under convex quadratic constraints. *Journal of Global Optimization*, 21(4):445–455, December 2001.
http://www.convexoptimization.com/TOOLS/Jean_Baptiste.pdf

[204] Jean-Baptiste Hiriart-Urruty and Claude Lemaréchal. *Convex Analysis and Minimization Algorithms II: Advanced Theory and Bundle Methods*. Springer-Verlag, second edition, 1996.

[205] Jean-Baptiste Hiriart-Urruty and Claude Lemaréchal. *Fundamentals of Convex Analysis*. Springer-Verlag, 2001.

[206] Alan J. Hoffman and Helmut W. Wielandt. The variation of the spectrum of a normal matrix. *Duke Mathematical Journal*, 20:37–40, 1953.

[207] Alfred Horn. Doubly stochastic matrices and the diagonal of a rotation matrix. *American Journal of Mathematics*, 76(3):620–630, July 1954.
http://www.convexoptimization.com/TOOLS/AHorn.pdf

[208] Roger A. Horn and Charles R. Johnson. *Matrix Analysis*. Cambridge University Press, 1987.

[209] Roger A. Horn and Charles R. Johnson. *Topics in Matrix Analysis*. Cambridge University Press, 1994.

[210] Alston S. Householder. *The Theory of Matrices in Numerical Analysis*. Dover, 1975.

[211] Hong-Xuan Huang, Zhi-An Liang, and Panos M. Pardalos. Some properties for the Euclidean distance matrix and positive semidefinite matrix completion problems. *Journal of Global Optimization*, 25(1):3–21, January 2003.
http://www.convexoptimization.com/TOOLS/pardalos.pdf

[212] Lawrence Hubert, Jacqueline Meulman, and Willem Heiser. Two purposes for matrix factorization: A historical appraisal. *SIAM Review*, 42(1):68–82, 2000.
http://www.convexoptimization.com/TOOLS/hubert.pdf

[213] Xiaoming Huo. *Sparse Image Representation via Combined Transforms*. PhD thesis, Stanford University, Department of Statistics, August 1999.
www.convexoptimization.com/TOOLS/ReweightingFrom1999XiaomingHuo.pdf

[214] Robert J. Marks II, editor. *Advanced Topics in Shannon Sampling and Interpolation Theory*. Springer-Verlag, 1993.

[215] 5W Infographic. Wireless 911. *Technology Review*, 107(5):78–79, June 2004.
http://www.technologyreview.com/communications/13643

[216] George Isac. *Complementarity Problems*. Springer-Verlag, 1992.

[217] Nathan Jacobson. *Lectures in Abstract Algebra, vol. II - Linear Algebra*. Van Nostrand, 1953.

[218] Viren Jain and Lawrence K. Saul. Exploratory analysis and visualization of speech and music by locally linear embedding. In *Proceedings of the IEEE International Conference on Acoustics, Speech, and Signal Processing*, volume 3, pages 984–987, May 2004.
http://www.cs.ucsd.edu/~saul/papers/lle_icassp04.pdf

[219] Joakim Jaldén. Bi-criterion ℓ_1/ℓ_2-norm optimization. Master's thesis, Royal Institute of Technology (KTH), Department of Signals Sensors and Systems, Stockholm Sweden, September 2002.
http://www.convexoptimization.com/TOOLS/JaldenMSThesis.pdf

[220] Joakim Jaldén, Cristoff Martin, and Björn Ottersten. Semidefinite programming for detection in linear systems − Optimality conditions and space-time decoding. In *Proceedings of the IEEE International Conference on Acoustics, Speech, and Signal Processing*, volume IV, pages 9–12, April 2003.
http://www.convexoptimization.com/TOOLS/Jalden.pdf

[221] Florian Jarre. Convex analysis on symmetric matrices. In Henry Wolkowicz, Romesh Saigal, and Lieven Vandenberghe, editors, *Handbook of Semidefinite Programming: Theory, Algorithms, and Applications*, chapter 2. Kluwer, 2000.
http://www.convexoptimization.com/TOOLS/Handbook.pdf

[222] Holly Hui Jin. *Scalable Sensor Localization Algorithms for Wireless Sensor Networks*. PhD thesis, University of Toronto, Graduate Department of Mechanical and Industrial Engineering, 2005.
www.stanford.edu/group/SOL/dissertations/holly-thesis.pdf

[223] Charles R. Johnson and Pablo Tarazaga. Connections between the real positive semidefinite and distance matrix completion problems. *Linear Algebra and its Applications*, 223/224:375–391, 1995.

[224] Charles R. Johnson and Pablo Tarazaga. Binary representation of normalized symmetric and correlation matrices. *Linear and Multilinear Algebra*, 52(5):359–366, 2004.

[225] George B. Thomas, Jr. *Calculus and Analytic Geometry*. Addison-Wesley, fourth edition, 1972.

[226] Mark Kahrs and Karlheinz Brandenburg, editors. *Applications of Digital Signal Processing to Audio and Acoustics*. Kluwer Academic Publishers, 1998.

[227] Thomas Kailath. *Linear Systems*. Prentice-Hall, 1980.

[228] Tosio Kato. *Perturbation Theory for Linear Operators*. Springer-Verlag, 1966.

[229] Paul J. Kelly and Norman E. Ladd. *Geometry*. Scott, Foresman and Company, 1965.

[230] Sunyoung Kim, Masakazu Kojima, Hayato Waki, and Makoto Yamashita. SFSDP: a **S**parse version of **F**ull **S**emi**D**efinite **P**rogramming relaxation for sensor network localization problems. Research Report B-457, Department of Mathematical and Computing Sciences, Tokyo Institute of Technology, Japan, July 2009.
http://math.ewha.ac.kr/~skim/Research/B-457.pdf

[231] Yoonsoo Kim and Mehran Mesbahi. On the rank minimization problem. In *Proceedings of the American Control Conference*, volume 3, pages 2015–2020. American Automatic Control Council (AACC), June 2004.

[232] Ron Kimmel. *Numerical Geometry of Images: Theory, Algorithms, and Applications*. Springer-Verlag, 2003.

[233] Erwin Kreyszig. *Introductory Functional Analysis with Applications*. Wiley, 1989.

[234] Anthony Kuh, Chaopin Zhu, and Danilo Mandic. Sensor network localization using least squares kernel regression. In 10th *International Conference of Knowledge-Based Intelligent Information and Engineering Systems* (KES), volume 4253(III) of *Lecture Notes in Computer Science*, pages 1280–1287, Bournemouth UK, October 2006. Springer-Verlag.
http://www.convexoptimization.com/TOOLS/Kuh.pdf

[235] Harold W. Kuhn. Nonlinear programming: a historical view. In Richard W. Cottle and Carlton E. Lemke, editors, *Nonlinear Programming*, pages 1–26. American Mathematical Society, 1976.

[236] Takahito Kuno, Yasutoshi Yajima, and Hiroshi Konno. An outer approximation method for minimizing the product of several convex functions on a convex set. *Journal of Global Optimization*, 3(3):325–335, September 1993.

[237] Amy N. Langville and Carl D. Meyer. *Google's PageRank and Beyond: The Science of Search Engine Rankings*. Princeton University Press, 2006.

[238] Jean B. Lasserre. A new Farkas lemma for positive semidefinite matrices. *IEEE Transactions on Automatic Control*, 40(6):1131–1133, June 1995.

[239] Jean B. Lasserre and Eduardo S. Zeron. A Laplace transform algorithm for the volume of a convex polytope. *Journal of the Association for Computing Machinery*, 48(6):1126–1140, November 2001.
http://arxiv.org/abs/math/0106168 (title revision).

[240] Monique Laurent. A connection between positive semidefinite and Euclidean distance matrix completion problems. *Linear Algebra and its Applications*, 273:9–22, 1998.

[241] Monique Laurent. A tour d'horizon on positive semidefinite and Euclidean distance matrix completion problems. In Panos M. Pardalos and Henry Wolkowicz, editors, *Topics in Semidefinite and Interior-Point Methods*, pages 51–76. American Mathematical Society, 1998.

[242] Monique Laurent. Matrix completion problems. In Christodoulos A. Floudas and Panos M. Pardalos, editors, *Encyclopedia of Optimization*, volume III (Interior-M), pages 221–229. Kluwer, 2001.
http://homepages.cwi.nl/~monique/files/opt.ps

[243] Monique Laurent and Svatopluk Poljak. On a positive semidefinite relaxation of the cut polytope. *Linear Algebra and its Applications*, 223/224:439–461, 1995.

[244] Monique Laurent and Svatopluk Poljak. On the facial structure of the set of correlation matrices. *SIAM Journal on Matrix Analysis and Applications*, 17(3):530–547, July 1996.

[245] Monique Laurent and Franz Rendl. Semidefinite programming and integer programming. *Optimization Online*, 2002.
http://www.optimization-online.org/DB_HTML/2002/12/585.html

[246] Monique Laurent and Franz Rendl. Semidefinite programming and integer programming. In K. Aardal, George L. Nemhauser, and R. Weismantel, editors, *Discrete Optimization*, volume 12 of *Handbooks in Operations Research and Management Science*, chapter 8, pages 393–514. Elsevier, 2005.

[247] Charles L. Lawson and Richard J. Hanson. *Solving Least Squares Problems*. SIAM, 1995.

[248] Jung Rye Lee. The law of cosines in a tetrahedron. *Journal of the Korea Society of Mathematical Education,* Series **B** (The Pure and Applied Mathematics), 4(1):1–6, 1997.

[249] Claude Lemaréchal. Note on an extension of "Davidon" methods to nondifferentiable functions. *Mathematical Programming*, 7(1):384–387, December 1974.
   http://www.convexoptimization.com/TOOLS/Lemarechal.pdf

[250] Vladimir L. Levin. Quasi-convex functions and quasi-monotone operators. *Journal of Convex Analysis*, 2(1/2):167–172, 1995.

[251] Scott Nathan Levine. *Audio Representations for Data Compression and Compressed Domain Processing*. PhD thesis, Stanford University, Department of Electrical Engineering, 1999.
   http://www-ccrma.stanford.edu/~scottl/thesis/thesis.pdf

[252] Adrian S. Lewis. Eigenvalue-constrained faces. *Linear Algebra and its Applications*, 269:159–181, 1998.

[253] Anhua Lin. *Projection algorithms in nonlinear programming*. PhD thesis, Johns Hopkins University, 2003.

[254] Miguel Sousa Lobo, Lieven Vandenberghe, Stephen Boyd, and Hervé Lebret. Applications of second-order cone programming. *Linear Algebra and its Applications*, 284:193–228, November 1998. Special Issue on Linear Algebra in Control, Signals and Image Processing.
   http://www.stanford.edu/~boyd/socp.html

[255] Lee Lorch and Donald J. Newman. On the composition of completely monotonic functions and completely monotonic sequences and related questions. *Journal of the London Mathematical Society,* second series, 28:31–45, 1983.
   http://www.convexoptimization.com/TOOLS/Lorch.pdf

[256] David G. Luenberger. *Optimization by Vector Space Methods*. Wiley, 1969.

[257] David G. Luenberger. *Introduction to Dynamic Systems: Theory, Models, & Applications*. Wiley, 1979.

[258] David G. Luenberger. *Linear and Nonlinear Programming*. Addison-Wesley, second edition, 1989.

[259] Zhi-Quan Luo, Jos F. Sturm, and Shuzhong Zhang. Superlinear convergence of a symmetric primal-dual path following algorithm for semidefinite programming. *SIAM Journal on Optimization*, 8(1):59–81, 1998.

[260] Zhi-Quan Luo and Wei Yu. An introduction to convex optimization for communications and signal processing. *IEEE Journal On Selected Areas In Communications*, 24(8):1426–1438, August 2006.

[261] Morris Marden. *Geometry of Polynomials*. American Mathematical Society, second edition, 1985.

[262] K. V. Mardia. Some properties of classical multi-dimensional scaling. *Communications in Statistics: Theory and Methods*, A7(13):1233–1241, 1978.

[263] K. V. Mardia, J. T. Kent, and J. M. Bibby. *Multivariate Analysis*. Academic Press, 1979.

[264] Jerrold E. Marsden and Michael J. Hoffman. *Elementary Classical Analysis*. Freeman, second edition, 1995.

[265] Rudolf Mathar. The best Euclidean fit to a given distance matrix in prescribed dimensions. *Linear Algebra and its Applications*, 67:1–6, 1985.

[266] Rudolf Mathar. *Multidimensionale Skalierung*. B. G. Teubner Stuttgart, 1997.

[267] Nathan S. Mendelsohn and A. Lloyd Dulmage. The convex hull of sub-permutation matrices. *Proceedings of the American Mathematical Society*, 9(2):253–254, April 1958.
   http://www.convexoptimization.com/TOOLS/permu.pdf

[268] Mehran Mesbahi and G. P. Papavassilopoulos. On the rank minimization problem over a positive semi-definite linear matrix inequality. *IEEE Transactions on Automatic Control*, 42(2):239–243, February 1997.

[269] Mehran Mesbahi and G. P. Papavassilopoulos. Solving a class of rank minimization problems via semi-definite programs, with applications to the fixed order output feedback synthesis. In *Proceedings of the American Control Conference*, volume 1, pages 77–80. American Automatic Control Council (AACC), June 1997.
http://www.convexoptimization.com/TOOLS/Mesbahi.pdf

[270] Sunderarajan S. Mohan, Mar Hershenson, Stephen Boyd, and Thomas Lee. Simple accurate expressions for planar spiral inductances. *IEEE Journal of Solid-State Circuits*, 34(10):1419–1424, October 1999.
http://www.stanford.edu/~boyd/papers/inductance_expressions.html

[271] Sunderarajan S. Mohan, Mar Hershenson, Stephen Boyd, and Thomas Lee. Bandwidth extension in CMOS with optimized on-chip inductors. *IEEE Journal of Solid-State Circuits*, 35(3):346–355, March 2000.
http://www.stanford.edu/~boyd/papers/bandwidth_ext.html

[272] David Moore, John Leonard, Daniela Rus, and Seth Teller. Robust distributed network localization with noisy range measurements. In *Proceedings of the Second International Conference on Embedded Networked Sensor Systems* (SenSys'04), pages 50–61, Baltimore Maryland USA, November 2004. Association for Computing Machinery (ACM, Winner of the Best Paper Award).
http://rvsn.csail.mit.edu/netloc/sensys04.pdf

[273] E. H. Moore. On the reciprocal of the general algebraic matrix. *Bulletin of the American Mathematical Society*, 26:394–395, 1920. Abstract.

[274] B. S. Mordukhovich. Maximum principle in the problem of time optimal response with nonsmooth constraints. *Journal of Applied Mathematics and Mechanics*, 40:960–969, 1976.

[275] Jean-Jacques Moreau. Décomposition orthogonale d'un espace Hilbertien selon deux cônes mutuellement polaires. *Comptes Rendus de l'Académie des Sciences, Paris*, 255:238–240, 1962.

[276] T. S. Motzkin and I. J. Schoenberg. The relaxation method for linear inequalities. *Canadian Journal of Mathematics*, 6:393–404, 1954.

[277] Neil Muller, Lourenço Magaia, and B. M. Herbst. Singular value decomposition, eigenfaces, and 3D reconstructions. *SIAM Review*, 46(3):518–545, September 2004.

[278] Katta G. Murty and Feng-Tien Yu. *Linear Complementarity, Linear and Nonlinear Programming*. Heldermann Verlag, Internet edition, 1988.
ioe.engin.umich.edu/people/fac/books/murty/linear_complementarity_webbook

[279] Oleg R. Musin. An extension of Delsarte's method. The kissing problem in three and four dimensions. In *Proceedings of the 2004 COE Workshop on Sphere Packings*, Kyushu University Japan, 2005.
http://arxiv.org/abs/math.MG/0512649

[280] Stephen G. Nash and Ariela Sofer. *Linear and Nonlinear Programming*.
McGraw-Hill, 1996.

[281] John Lawrence Nazareth. *Differentiable Optimization and Equation Solving: A Treatise on Algorithmic Science and the Karmarkar Revolution*. Springer-Verlag, 2003.

[282] A. B. Németh and S. Z. Németh. How to project onto the monotone nonnegative cone using pool adjacent violators type algorithms, January 2012.
http://arxiv.org/abs/1201.2343

[283] Arkadi Nemirovski. *Lectures on Modern Convex Optimization*.
http://www.convexoptimization.com/TOOLS/LectModConvOpt.pdf , 2005.

[284] Yurii Nesterov and Arkadii Nemirovskii. *Interior-Point Polynomial Algorithms in Convex Programming*. SIAM, 1994.

[285] John von Neumann. *Functional Operators, Volume II: The Geometry of Orthogonal Spaces*. Princeton University Press, 1950. Reprinted from mimeographed lecture notes first distributed in 1933.

[286] Ewa Niewiadomska-Szynkiewicz and Michał Marks. Optimization schemes for wireless sensor network localization. *International Journal of Applied Mathematics and Computer Science*, 19(2):291–302, 2009.
http://www.convexoptimization.com/TOOLS/Marks.pdf

[287] L. Nirenberg. *Functional Analysis*. New York University, New York, 1961. Lectures given in 1960-1961, notes by Lesley Sibner.

[288] Jorge Nocedal and Stephen J. Wright. *Numerical Optimization*. Springer-Verlag, 1999.

[289] Royal Swedish Academy of Sciences. Nobel prize in chemistry, 2002.
www.nobelprize.org/nobel_prizes/chemistry/laureates/2002/wuthrich.html

[290] C. S. Ogilvy. *Excursions in Geometry*. Dover, 1990. Citation: *Proceedings of the CUPM Geometry Conference*, Mathematical Association of America, No.16 (1967), p.21.

[291] Ricardo Oliveira, João Costeira, and João Xavier. Optimal point correspondence through the use of rank constraints. In *Proceedings of the IEEE Computer Society Conference on Computer Vision and Pattern Recognition*, volume 2, pages 1016–1021, June 2005.

[292] Alan V. Oppenheim and Ronald W. Schafer. *Discrete-Time Signal Processing*. Prentice-Hall, 1989.

[293] Robert Orsi, Uwe Helmke, and John B. Moore. A Newton-like method for solving rank constrained linear matrix inequalities. *Automatica*, 42(11):1875–1882, 2006.
users.rsise.anu.edu.au/~robert/publications/OHM06_extended_version.pdf

[294] Brad Osgood. Notes on the Ahlfors mapping of a multiply connected domain, 2000.
www-ee.stanford.edu/~osgood/Ahlfors-Bergman-Szego.pdf

[295] M. L. Overton and R. S. Womersley. On the sum of the largest eigenvalues of a symmetric matrix. *SIAM Journal on Matrix Analysis and Applications*, 13:41–45, 1992.

[296] Pythagoras Papadimitriou. *Parallel Solution of SVD-Related Problems, With Applications*. PhD thesis, Department of Mathematics, University of Manchester, October 1993.

[297] Panos M. Pardalos and Henry Wolkowicz, editors. *Topics in Semidefinite and Interior-Point Methods*. American Mathematical Society, 1998.

[298] Bradford W. Parkinson, James J. Spilker Jr., Penina Axelrad, and Per Enge, editors. *Global Positioning System: Theory and Applications, Volume I*. American Institute of Aeronautics and Astronautics, 1996.

[299] Gábor Pataki. Cone-LP's and semidefinite programs: Geometry and a simplex-type method. In William H. Cunningham, S. Thomas McCormick, and Maurice Queyranne, editors, *Proceedings of the 5th International Conference on Integer Programming and Combinatorial Optimization (IPCO)*, volume 1084 of *Lecture Notes in Computer Science*, pages 162–174, Vancouver, British Columbia, Canada, June 1996. Springer-Verlag.

[300] Gábor Pataki. On the rank of extreme matrices in semidefinite programs and the multiplicity of optimal eigenvalues. *Mathematics of Operations Research*, 23(2):339–358, 1998.
http://citeseerx.ist.psu.edu/viewdoc/summary?doi=10.1.1.49.3831
Erratum: p.569 herein.

[301] Gábor Pataki. The geometry of semidefinite programming. In Henry Wolkowicz, Romesh Saigal, and Lieven Vandenberghe, editors, *Handbook of Semidefinite Programming: Theory, Algorithms, and Applications*, chapter 3. Kluwer, 2000.
http://www.convexoptimization.com/TOOLS/Handbook.pdf

[302] Teemu Pennanen and Jonathan Eckstein. Generalized Jacobians of vector-valued convex functions. Technical Report RRR 6-97, RUTCOR, Rutgers University, May 1997.
http://rutcor.rutgers.edu/pub/rrr/reports97/06.ps

[303] Roger Penrose. A generalized inverse for matrices. In *Proceedings of the Cambridge Philosophical Society*, volume 51, pages 406–413, 1955.

[304] Chris Perkins. A convergence analysis of Dykstra's algorithm for polyhedral sets. *SIAM Journal on Numerical Analysis*, 40(2):792–804, 2002.

[305] Sam Perlis. *Theory of Matrices*. Addison-Wesley, 1958.

[306] Florian Pfender and Günter M. Ziegler. Kissing numbers, sphere packings, and some unexpected proofs. *Notices of the American Mathematical Society*, 51(8):873–883, September 2004.
http://www.convexoptimization.com/TOOLS/Pfender.pdf

[307] Benjamin Recht. *Convex Modeling with Priors*. PhD thesis, Massachusetts Institute of Technology, Media Arts and Sciences Department, 2006.
http://www.convexoptimization.com/TOOLS/06.06.Recht.pdf

[308] Benjamin Recht, Maryam Fazel, and Pablo A. Parrilo. Guaranteed minimum-rank solutions of linear matrix equations via nuclear norm minimization. *Optimization Online*, June 2007.
http://faculty.washington.edu/mfazel/low-rank-v2.pdf

[309] Alex Reznik. Problem 3.41, from Sergio Verdú. *Multiuser Detection*. Cambridge University Press, 1998.
www.ee.princeton.edu/~verdu/mud/solutions/3/3.41.areznik.pdf , 2001.

[310] Arthur Wayne Roberts and Dale E. Varberg. *Convex Functions*. Academic Press, 1973.

[311] Sara Robinson. Hooked on meshing, researcher creates award-winning triangulation program. *SIAM News*, 36(9), November 2003.
http://www.siam.org/news/news.php?id=370

[312] R. Tyrrell Rockafellar. *Conjugate Duality and Optimization*. SIAM, 1974.

[313] R. Tyrrell Rockafellar. Lagrange multipliers in optimization. *SIAM-American Mathematical Society Proceedings*, 9:145–168, 1976.

[314] R. Tyrrell Rockafellar. Lagrange multipliers and optimality. *SIAM Review*, 35(2):183–238, June 1993.

[315] R. Tyrrell Rockafellar. *Convex Analysis*. Princeton University Press, 1997. First published in 1970.

[316] C. K. Rushforth. Signal restoration, functional analysis, and Fredholm integral equations of the first kind. In Henry Stark, editor, *Image Recovery: Theory and Application*, chapter 1, pages 1–27. Academic Press, 1987.

[317] Peter Santos. Application platforms and synthesizable analog IP. In *Proceedings of the Embedded Systems Conference*, San Francisco California USA, 2003.
http://www.convexoptimization.com/TOOLS/Santos.pdf

[318] Shankar Sastry. *Nonlinear Systems: Analysis, Stability, and Control*.
Springer-Verlag, 1999.

[319] Uwe Schäfer. A linear complementarity problem with a P-matrix. *SIAM Review*, 46(2):189–201, June 2004.

[320] Isaac J. Schoenberg. Remarks to Maurice Fréchet's article "Sur la définition axiomatique d'une classe d'espace distanciés vectoriellement applicable sur l'espace de Hilbert". *Annals of Mathematics*, 36(3):724–732, July 1935.
http://www.convexoptimization.com/TOOLS/Schoenberg2.pdf

[321] Isaac J. Schoenberg. Metric spaces and positive definite functions. *Transactions of the American Mathematical Society*, 44:522–536, 1938.
http://www.convexoptimization.com/TOOLS/Schoenberg3.pdf

[322] Peter H. Schönemann. A generalized solution of the orthogonal Procrustes problem. *Psychometrika*, 31(1):1–10, 1966.

[323] Peter H. Schönemann, Tim Dorcey, and K. Kienapple. Subadditive concatenation in dissimilarity judgements. *Perception and Psychophysics*, 38:1–17, 1985.

[324] Alexander Schrijver. On the history of combinatorial optimization (till 1960). In K. Aardal, G. L. Nemhauser, and R. Weismantel, editors, *Handbook of Discrete Optimization*, pages 1–68. Elsevier, 2005.
http://citeseerx.ist.psu.edu/viewdoc/summary?doi=10.1.1.22.3362

[325] Seymour Sherman. Doubly stochastic matrices and complex vector spaces. *American Journal of Mathematics*, 77(2):245–246, April 1955.
http://www.convexoptimization.com/TOOLS/Sherman.pdf

[326] Joshua A. Singer. *Log-Penalized Linear Regression*. PhD thesis, Stanford University, Department of Electrical Engineering, June 2004.
http://www.convexoptimization.com/TOOLS/Josh.pdf

[327] Anthony Man-Cho So and Yinyu Ye. A semidefinite programming approach to tensegrity theory and realizability of graphs. In *Proceedings of the Seventeenth Annual ACM-SIAM Symposium on Discrete Algorithms* (SODA), pages 766–775, Miami Florida USA, January 2006. Association for Computing Machinery (ACM).
http://www.convexoptimization.com/TOOLS/YeSo.pdf

[328] Anthony Man-Cho So and Yinyu Ye. Theory of semidefinite programming for sensor network localization. *Mathematical Programming,* Series **A** and **B**, 109(2):367–384, January 2007.
http://www.stanford.edu/~yyye/local-theory.pdf

[329] D. C. Sorensen. Newton's method with a model trust region modification. *SIAM Journal on Numerical Analysis*, 19(2):409–426, April 1982.
http://www.convexoptimization.com/TOOLS/soren.pdf

[330] Nathan Srebro, Jason D. M. Rennie, and Tommi S. Jaakkola. Maximum-margin matrix factorization. In Lawrence K. Saul, Yair Weiss, and Léon Bottou, editors, *Advances in Neural Information Processing Systems 17* (NIPS 2004), pages 1329–1336. MIT Press, 2005.

[331] Wolfram Stadler, editor. *Multicriteria Optimization in Engineering and in the Sciences.* Springer-Verlag, 1988.

[332] Henry Stark, editor. *Image Recovery: Theory and Application.* Academic Press, 1987.

[333] Henry Stark. Polar, spiral, and generalized sampling and interpolation. In Robert J. Marks II, editor, *Advanced Topics in Shannon Sampling and Interpolation Theory*, chapter 6, pages 185–218. Springer-Verlag, 1993.

[334] Willi-Hans Steeb. *Matrix Calculus and Kronecker Product with Applications and C++ Programs.* World Scientific, 1997.

[335] Gilbert W. Stewart and Ji-guang Sun. *Matrix Perturbation Theory.* Academic Press, 1990.

[336] Josef Stoer. On the characterization of least upper bound norms in matrix space. *Numerische Mathematik*, 6(1):302–314, December 1964.
http://www.convexoptimization.com/TOOLS/Stoer.pdf

[337] Josef Stoer and Christoph Witzgall. *Convexity and Optimization in Finite Dimensions I.* Springer-Verlag, 1970.

[338] Gilbert Strang. *Linear Algebra and its Applications.* Harcourt Brace, third edition, 1988.

[339] Gilbert Strang. *Calculus.* Wellesley-Cambridge Press, 1992.

[340] Gilbert Strang. *Introduction to Linear Algebra.* Wellesley-Cambridge Press, second edition, 1998.

[341] Gilbert Strang. Course 18.06: Linear algebra, 2004.
http://web.mit.edu/18.06/www

[342] Stefan Straszewicz. Über exponierte Punkte abgeschlossener Punktmengen. *Fundamenta Mathematicae*, 24:139–143, 1935.
http://www.convexoptimization.com/TOOLS/Straszewicz.pdf

[343] Jos F. Sturm. SeDuMi (self-dual minimization).
Software for optimization over symmetric cones, 2003.
http://sedumi.ie.lehigh.edu

[344] Jos F. Sturm and Shuzhong Zhang. On cones of nonnegative quadratic functions. *Mathematics of Operations Research*, 28(2):246–267, May 2003.
www.optimization-online.org/DB_HTML/2001/05/324.html

[345] George P. H. Styan. A review and some extensions of Takemura's generalizations of Cochran's theorem. Technical Report 56, Stanford University, Department of Statistics, September 1982.

[346] George P. H. Styan and Akimichi Takemura. Rank additivity and matrix polynomials. Technical Report 57, Stanford University, Department of Statistics, September 1982.

[347] Jun Sun, Stephen Boyd, Lin Xiao, and Persi Diaconis. The fastest mixing Markov process on a graph and a connection to a maximum variance unfolding problem. *SIAM Review*, 48(4):681–699, December 2006.
http://stanford.edu/~boyd/papers/fmmp.html

[348] Chen Han Sung and Bit-Shun Tam. A study of projectionally exposed cones. *Linear Algebra and its Applications*, 139:225–252, 1990.

[349] Yoshio Takane. On the relations among four methods of multidimensional scaling. *Behaviormetrika*, 4:29–43, 1977.
http://takane.brinkster.net/Yoshio/p008.pdf

[350] Akimichi Takemura. On generalizations of Cochran's theorem and projection matrices. Technical Report 44, Stanford University, Department of Statistics, August 1980.

[351] Yasuhiko Takenaga and Toby Walsh. Tetravex is NP-complete. *Information Processing Letters*, 99(5):171–174, September 2006.
http://arxiv.org/pdf/0903.1147v1.pdf

[352] Dharmpal Takhar, Jason N. Laska, Michael B. Wakin, Marco F. Duarte, Dror Baron, Shriram Sarvotham, Kevin F. Kelly, and Richard G. Baraniuk. A new compressive imaging camera architecture using optical-domain compression. In *Proceedings of the SPIE Conference on Computational Imaging IV*, volume 6065, pages 43–52, February 2006.
http://www.convexoptimization.com/TOOLS/csCamera-SPIE-dec05.pdf

[353] Peng Hui Tan and Lars K. Rasmussen. The application of semidefinite programming for detection in CDMA. *IEEE Journal on Selected Areas in Communications*, 19(8), August 2001.

[354] Pham Dinh Tao and Le Thi Hoai An. A D.C. optimization algorithm for solving the trust-region subproblem. *SIAM Journal on Optimization*, 8(2):476–505, 1998.
http://www.convexoptimization.com/TOOLS/pham.pdf

[355] Pablo Tarazaga. Faces of the cone of Euclidean distance matrices: Characterizations, structure and induced geometry. *Linear Algebra and its Applications*, 408:1–13, 2005.

[356] Robert Tibshirani. Regression shrinkage and selection via the lasso. *Journal of the Royal Statistical Society, Series **B** (Methodological)*, 58(1):267–288, 1996.

[357] Kim-Chuan Toh, Michael J. Todd, and Reha H. Tütüncü. SDPT3 – a MATLAB software for semidefinite-quadratic-linear programming, February 2009.
http://www.math.nus.edu.sg/~mattohkc/sdpt3.html

[358] Warren S. Torgerson. *Theory and Methods of Scaling*. Wiley, 1958.

[359] Lloyd N. Trefethen and David Bau, III. *Numerical Linear Algebra*. SIAM, 1997.

[360] Michael W. Trosset. Applications of multidimensional scaling to molecular conformation. *Computing Science and Statistics*, 29:148–152, 1998.

[361] Michael W. Trosset. Distance matrix completion by numerical optimization. *Computational Optimization and Applications*, 17(1):11–22, October 2000.

[362] Michael W. Trosset. Extensions of classical multidimensional scaling: Computational theory. convexoptimization.com/TOOLS/TrossetPITA.pdf , 2001. Revision of technical report entitled "Computing distances between convex sets and subsets of the positive semidefinite matrices" first published in 1997.

[363] Michael W. Trosset and Rudolf Mathar. On the existence of nonglobal minimizers of the STRESS criterion for metric multidimensional scaling. In *Proceedings of the Statistical Computing Section*, pages 158–162. American Statistical Association, 1997.

[364] Joshua Trzasko and Armando Manduca. Highly undersampled magnetic resonance image reconstruction via homotopic ℓ_0-minimization. *IEEE Transactions on Medical Imaging*, 28(1):106–121, January 2009.
http://www.convexoptimization.com/TOOLS/SparseRecon.pdf

[365] Joshua Trzasko, Armando Manduca, and Eric Borisch. Sparse MRI reconstruction via multiscale L_0-continuation. In *Proceedings of the IEEE 14th Workshop on Statistical Signal Processing*, pages 176–180, August 2007.
http://www.convexoptimization.com/TOOLS/LOMRI.pdf

[366] P. P. Vaidyanathan. *Multirate Systems and Filter Banks*. Prentice-Hall, 1993.

[367] Jan van Tiel. *Convex Analysis, an Introductory Text*. Wiley, 1984.

[368] Lieven Vandenberghe and Stephen Boyd. Semidefinite programming. *SIAM Review*, 38(1):49–95, March 1996.

[369] Lieven Vandenberghe and Stephen Boyd. Connections between semi-infinite and semidefinite programming. In R. Reemtsen and J.-J. Rückmann, editors, *Semi-Infinite Programming*, chapter 8, pages 277–294. Kluwer Academic Publishers, 1998.

[370] Lieven Vandenberghe and Stephen Boyd. Applications of semidefinite programming. *Applied Numerical Mathematics*, 29(3):283–299, March 1999.

[371] Lieven Vandenberghe, Stephen Boyd, and Shao-Po Wu. Determinant maximization with linear matrix inequality constraints. *SIAM Journal on Matrix Analysis and Applications*, 19(2):499–533, April 1998.

[372] Robert J. Vanderbei. Convex optimization: Interior-point methods and applications. Lecture notes, 1999.
http://orfe.princeton.edu/~rvdb/pdf/talks/pumath/talk.pdf

[373] Richard S. Varga. *Geršgorin and His Circles*. Springer-Verlag, 2004.

[374] Martin Vetterli and Jelena Kovačević. *Wavelets and Subband Coding*. Prentice-Hall, 1995.

[375] Martin Vetterli, Pina Marziliano, and Thierry Blu. Sampling signals with finite rate of innovation. *IEEE Transactions on Signal Processing*, 50(6):1417–1428, June 2002.
http://bigwww.epfl.ch/publications/vetterli0201.pdf
http://citeseerx.ist.psu.edu/viewdoc/summary?doi=10.1.1.19.5630

[376] È. B. Vinberg. The theory of convex homogeneous cones. *Transactions of the Moscow Mathematical Society*, 12:340–403, 1963. American Mathematical Society and London Mathematical Society joint translation, 1965.

[377] *Wikimization*. An alternative proof without Moreau's theorem.
convexoptimization.com/wikimization/index.php/Complementarity_problem

[378] *Wikimization*. Boolean feasibility.
convexoptimization.com/wikimization/index.php/Matlab_for_Convex_Optimization

[379] *Wikimization*. Compressive sampling of images by convex iteration.
convexoptimization.com/wikimization/index.php/Matlab_for_Convex_Optimization

[380] *Wikimization*. Conic independence.
convexoptimization.com/wikimization/index.php/Matlab_for_Convex_Optimization

[381] *Wikimization*. Convex iteration rank-1, 2013.
convexoptimization.com/wikimization/index.php/Matlab_for_Convex_Optimization

[382] *Wikimization*. EDM using ordinal data.
convexoptimization.com/wikimization/index.php/Matlab_for_Convex_Optimization

[383] *Wikimization*. Fast max cut.
convexoptimization.com/wikimization/index.php/Matlab_for_Convex_Optimization

[384] *Wikimization*. Fast projection on monotone nonnegative cone.
convexoptimization.com/wikimization/index.php/Projection_on_Polyhedral_Cone

[385] *Wikimization*. Fermat point.
http://www.convexoptimization.com/wikimization/index.php/Fermat_point

[386] *Wikimization*. Filter design by convex iteration.
convexoptimization.com/wikimization/index.php/Filter_design_by_convex_iteration

[387] *Wikimization*. Fixed point problems.
convexoptimization.com/wikimization/index.php/Complementarity_problem

[388] *Wikimization*. High-order polynomials.
convexoptimization.com/wikimization/index.php/Matlab_for_Convex_Optimization

[389] *Wıкımization*. Map of the USA.
convexoptimization.com/wikimization/index.php/Matlab_for_Convex_Optimization

[390] *Wıкımization*. MATLAB for convex optimization & Euclidean distance geometry.
convexoptimization.com/wikimization/index.php/Matlab_for_Convex_Optimization

[391] *Wıкımization*. Projection on simplicial cones.
convexoptimization.com/wikimization/index.php/Projection_on_Polyhedral_Cone

[392] *Wıкımization*. Rank reduction subroutine.
convexoptimization.com/wikimization/index.php/Matlab_for_Convex_Optimization

[393] *Wıкımization*. Sampling sparsity.
convexoptimization.com/wikimization/index.php/Sampling_Sparsity

[394] *Wıкımization*. Signal dropout problem.
convexoptimization.com/wikimization/index.php/Matlab_for_Convex_Optimization

[395] *Wıкımization*. Singular value decomposition (SVD) by rank-1 transformation.
convexoptimization.com/wikimization/index.php/Matlab_for_Convex_Optimization

[396] *Wıкımization*. Sturm & Zhang's procedure for constructing dyad-decomposition.
convexoptimization.com/wikimization/index.php/Matlab_for_Convex_Optimization

[397] *Wıкımization*. isedm().
convexoptimization.com/wikimization/index.php/Matlab_for_Convex_Optimization

[398] Michael B. Wakin, Jason N. Laska, Marco F. Duarte, Dror Baron, Shriram Sarvotham, Dharmpal Takhar, Kevin F. Kelly, and Richard G. Baraniuk. An architecture for compressive imaging. In *Proceedings of the IEEE International Conference on Image Processing* (ICIP), pages 1273–1276, October 2006.
http://www.convexoptimization.com/TOOLS/CSCam-ICIP06.pdf

[399] Roger Webster. *Convexity*. Oxford University Press, 1994.

[400] Kilian Q. Weinberger and Lawrence K. Saul. Unsupervised learning of image manifolds by semidefinite programming. In *Proceedings of the IEEE Computer Society Conference on Computer Vision and Pattern Recognition*, volume 2, pages 988–995, 2004.
http://www.cs.ucsd.edu/~saul/papers/sde_cvpr04.pdf

[401] Eric W. Weisstein. Mathworld – *A Wolfram Web Resource*, 2007.
http://mathworld.wolfram.com

[402] D. J. White. An analogue derivation of the dual of the general Fermat problem. *Management Science*, 23(1):92–94, September 1976.
http://www.convexoptimization.com/TOOLS/White.pdf

[403] Bernard Widrow and Samuel D. Stearns. *Adaptive Signal Processing*. Prentice-Hall, 1985.

[404] Norbert Wiener. On factorization of matrices. *Commentarii Mathematici Helvetici*, 29:97–111, 1955.

[405] Ami Wiesel, Yonina C. Eldar, and Shlomo Shamai (Shitz). Semidefinite relaxation for detection of 16-QAM signaling in MIMO channels. *IEEE Signal Processing Letters*, 12(9):653–656, September 2005.

[406] Michael P. Williamson, Timothy F. Havel, and Kurt Wüthrich. Solution conformation of proteinase inhibitor IIA from bull seminal plasma by ^{1}H nuclear magnetic resonance and distance geometry. *Journal of Molecular Biology*, 182:295–315, 1985.

[407] Willie W. Wong. Cayley-Menger determinant and generalized N-dimensional Pythagorean theorem, November 2003. Application of Linear Algebra: Notes on Talk given to Princeton University Math Club.
http://www.convexoptimization.com/TOOLS/gp-r.pdf

[408] William Wooton, Edwin F. Beckenbach, and Frank J. Fleming. *Modern Analytic Geometry*. Houghton Mifflin, 1975.

[409] Margaret H. Wright. The interior-point revolution in optimization: History, recent developments, and lasting consequences. *Bulletin of the American Mathematical Society*, 42(1):39–56, January 2005.

[410] Stephen J. Wright. *Primal-Dual Interior-Point Methods*. SIAM, 1997.

[411] Shao-Po Wu. *max-det Programming with Applications in Magnitude Filter Design*. A dissertation submitted to the department of Electrical Engineering, Stanford University, December 1997.

[412] Shao-Po Wu and Stephen Boyd. `sdpsol`: A parser/solver for semidefinite programming and determinant maximization problems with matrix structure, May 1996.
http://www.stanford.edu/~boyd/old_software/SDPSOL.html

[413] Shao-Po Wu and Stephen Boyd. `sdpsol`: A parser/solver for semidefinite programs with matrix structure. In Laurent El Ghaoui and Silviu-Iulian Niculescu, editors, *Advances in Linear Matrix Inequality Methods in Control*, chapter 4, pages 79–91. SIAM, 2000.
http://www.stanford.edu/~boyd/sdpsol.html

[414] Shao-Po Wu, Stephen Boyd, and Lieven Vandenberghe. FIR filter design via spectral factorization and convex optimization. In Biswa Nath Datta, editor, *Applied and Computational Control, Signals, and Circuits*, volume 1, chapter 5, pages 215–246. Birkhäuser, 1998.
http://www.stanford.edu/~boyd/papers/magdes.html

[415] Naoki Yamamoto and Maryam Fazel. Computational approach to quantum encoder design for purity optimization. *Physical Review A*, 76(1), July 2007.
http://arxiv.org/abs/quant-ph/0606106

[416] David D. Yao, Shuzhong Zhang, and Xun Yu Zhou. Stochastic linear-quadratic control via primal-dual semidefinite programming. *SIAM Review*, 46(1):87–111, March 2004. Erratum: p.203 herein.

[417] Yinyu Ye. Semidefinite programming for Euclidean distance geometric optimization. Lecture notes, 2003.
http://www.stanford.edu/class/ee392o/EE392o-yinyu-ye.pdf

[418] Yinyu Ye. Convergence behavior of central paths for convex homogeneous self-dual cones, 1996.
http://www.stanford.edu/~yyye/yyye/ye.ps

[419] Yinyu Ye. *Interior Point Algorithms: Theory and Analysis*. Wiley, 1997.

[420] D. C. Youla. Mathematical theory of image restoration by the method of convex projection. In Henry Stark, editor, *Image Recovery: Theory and Application*, chapter 2, pages 29–77. Academic Press, 1987.

[421] Fuzhen Zhang. *Matrix Theory: Basic Results and Techniques*. Springer-Verlag, 1999.

[422] Günter M. Ziegler. Kissing numbers: Surprises in dimension four. *Emissary*, pages 4–5, Spring 2004.
http://www.msri.org/communications/emissary

Index

A searchable color pdfBook

Dattorro, *Convex Optimization & Euclidean Distance Geometry,*
$\mathcal{M}\varepsilon\beta oo$ Publishing USA, 2005, v2015.02.15.

is available online from Meboo Publishing.

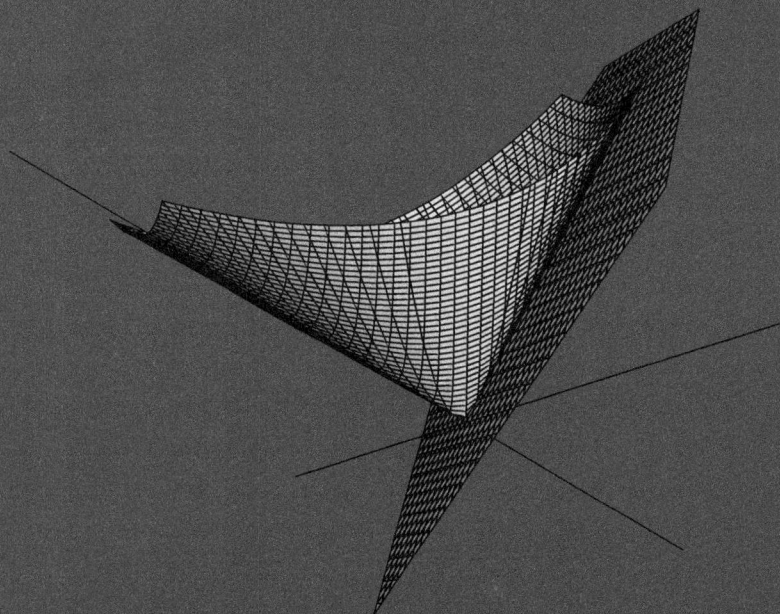

Convex Optimization & Euclidean Distance Geometry, Dattorro

ISBN 0-9764013-0-4

59999

9 780976 401308

Meβoo Publishing USA